Applied Underwater Acoustics

Applied Underwater Acoustics

Leif Bjørnø

UltraTech Holding, Taastrup, Denmark

Edited by

Thomas H. Neighbors III

David Bradley

ELSEVIER elsevier.com

Elsevier
Radarweg 29, PO Box 211, 1000 AE Amsterdam, Netherlands
The Boulevard, Langford Lane, Kidlington, Oxford OX5 1GB, United Kingdom
50 Hampshire Street, 5th Floor, Cambridge, MA 02139, United States

Cover image: HISAS 1030—produced image of a sunken WWII oil tanker copyright ©Kongsberg Maritime AS and Norwegian Defence Research Establishment (FFI). Reprinted with permission.

Notices

Knowledge and best practice in this field are constantly changing. As new research and experience broaden our understanding, changes in research methods, professional practices, or medical treatment may become necessary.

Practitioners and researchers must always rely on their own experience and knowledge in evaluating and using any information, methods, compounds, or experiments described herein. In using such information or methods they should be mindful of their own safety and the safety of others, including parties for whom they have a professional responsibility.

To the fullest extent of the law, neither the Publisher nor the authors, contributors, or editors, assume any liability for any injury and/or damage to persons or property as a matter of products liability, negligence or otherwise, or from any use or operation of any methods, products, instructions, or ideas contained in the material herein.

British Library Cataloguing-in-Publication Data
A catalogue record for this book is available from the British Library

Library of Congress Cataloging-in-Publication Data
A catalog record for this book is available from the Library of Congress

ISBN: 978-0-12-811240-3

For information on all Elsevier publications
visit our website at https://www.elsevier.com/

  Working together
to grow libraries in
developing countries

www.elsevier.com • www.bookaid.org

Publisher: John Fedor
Acquisition Editor: Anita Koch
Editorial Project Manager: Amy Clark
Production Project Manager: Paul Prasad Chandramohan
Designer: Maria Inês Cruz

Typeset by TNQ Books and Journals

This book is dedicated to the memory of

Professor Walter G. Mayer
Department of Physics
Georgetown University
Washington, D.C.

Contents

viii   Contents

List of Contributors

D.A. Abraham
Ellicott City, MD, United States

L. Bjørnø
UltraTech Holding, Taastrup, Denmark

Ph. Blondel
Department of Physics, University of Bath, Bath, United Kingdom

M.J. Buckingham
Scripps Institution of Oceanography, University of California, San Diego, La Jolla, CA, United States

A. Caiti
University of Pisa, Pisa, Italy

N.R. Chapman
University of Victoria, Victoria, BC, Canada

B. Dushaw
University of Washington, Seattle, WA, United States

D. Fattaccioli
LMA-CNRS, Marseille and DGA Naval Systems, Toulon, France

P. Gambogi
University of Pisa, Pisa, Italy

A. Gavrilov
Curtin University, Perth, WA, Australia

G. Grelowska
Gdansk University of Technology, Gdansk, Poland

P. Grenard
CTBTO, Vienna International Centre, Vienna, Austria

G. Haralabus
CTBTO, Vienna International Centre, Vienna, Austria

R.A. Hazelwood
R&V Hazelwood Associates LLP, Guildford, United Kingdom

S. Ivansson
Swedish Defence Research Agency, Stockholm, Sweden

D.R. Jackson
Applied Physics Laboratory, University of Washington, Seattle, WA, United States

E. Kozaczka
Gdansk University of Technology, Gdansk, Poland

P.A. Lepper
Loughborough University, Loughborough, Leicestershire, United Kingdom

J.F. Lynch
Woods Hole Oceanographic Institution, Woods Hole, MA, United States

P. Mikhalevsky
Leidos Inc., Arlington, VA, United States

J.L. Miksis-Olds
University of New Hampshire, Durham, NH, United States

T.H. Neighbors III
Leidos Corporation (Retired), Bellevue, WA, United States

A.E. Newhall
Woods Hole Oceanographic Institution, Woods Hole, MA, United States

L. Pautet
CTBTO, Vienna International Centre, Vienna, Austria

M. Prior
CTBTO, Vienna International Centre, Vienna, Austria

M.D. Richardson
Marine Geosciences Division, Naval Research Laboratory, Stennis Space Center, MS, United States

S.P. Robinson
National Physical Laboratory, Teddington, United Kingdom

D. Scaradozzi
University of Pisa, Pisa, Italy

J.-P. Sessarego
LMA-CNRS, Marseille and DGA Naval Systems, Toulon, France

C.C. Tsimenidis
Newcastle University, Newcastle upon Tyne, United Kingdom

M. Zampolli
CTBTO, Vienna International Centre, Vienna, Austria

Preface

The preface is a personal introduction to the book, with some comments about the contents as seen by the author, the intent of the work, and recognition of those who have contributed. The last is easy: the team of authors who contributed chapters or sections of chapters are, without qualification, an outstanding group of experts who have given their time and knowledge to make this book a valuable and useful text for those working in the world of undersea science and technology. They have been a joy to work with and are complemented and thanked for their efforts. From conversations with Leif, we know his intent was to provide the undersea community with a science-based text, yet an easily understood and practical reference to the details of Underwater Acoustics. From these same conversations, it was clear his concept was to draw on multiple expertises, vice singular authorship, as he felt strongly that the readership would benefit from the depth that a group of experts would provide.

The introduction is "short and sweet": Consider the following to be a practical compendium of the knowledge of Underwater Acoustics; it is meant to be a working document that readers can draw on to accomplish their specific task and a reference base for further study, if required.

Leif wished to dedicate this book to Walter G. Mayer, late Professor of Physics at Georgetown University, Washington, D.C., USA, a close personal friend and colleague for many years.

I thank Irina, my wife, for her support, encouragement, and patience; her presence at my side **is** critical in endeavors like this.

For Leif Bjørnø,

Tom Neighbors and Dave Bradley

General Characteristics of the Underwater Environment

L. Bjørnø[1,†], **M.J. Buckingham**[2]

UltraTech Holding, Taastrup, Denmark[1]; Scripps Institution of Oceanography, University of California, San Diego, La Jolla, CA, United States[2]

1.1 INTRODUCTION

Over the past about 100 years the exploitation of the seas and their resources has continuously increased. Acoustic waves have turned out to be a very useful tool for detecting resources and objects in the water column and on the seafloor. Other methods have been used with varying degrees of success depending on the objects to be detected or investigated. These methods include magnetics, magnetic anomaly detection, where minor changes in the earth's magnetic field due to presence of an object can be measured; optical methods; electric field changes; hydrodynamics such as pressure changes; thermal methods; and electromagnetic waves. While *radar* is very useful for detection of objects above water, electromagnetic radar waves are strongly absorbed in seawater. While electromagnetic waves in the visible frequency band from 4 to $8 \cdot 10^{14}$ Hz are much less absorbed, with a minimum absorption coefficient of $3 \cdot 10^{-3}$ cm^{-1} in the green-blue light near 455 nm wavelength (i.e., $6.59 \cdot 10^{14}$ Hz), electromagnetic wave absorption in the normally used radar bands is several orders of magnitude higher than in the visible band. Seawater salt contains magnesium that makes the water conduct electricity since the Mg_2^+ cation constitutes 3.7% of seawater salt. A 1 GHz radar wave in the ultra-high frequency (UHF) band with a 0.3 m wavelength has a 1400 dB/m absorption coefficient while the same wavelength in the 5 kHz sound wave has a $3 \cdot 10^{-4}$ dB/m absorption coefficient. Therefore, radar systems are not useful for detecting objects under water.

Underwater sound is used in many applications, such as *hydrography, off-shore activities, dredging, defense and security, marine research,* and *fishery.* Hydrography includes harbor and river surveys, bathymetric surveys, flood damage assessment, engineering inspection, pipeline and cable route surveys, exclusive economic zone (EEZ) mapping, breakwater mapping, and so on. Off-shore activities include pipeline and cable installation and inspection, leakage detection, route and site surveys, subsea structure installation support, renewables, remotely operated vehicle (ROV) intervention guidance, decommissioning, reconnaissance surveys,

---

†30 March 1937—24 October 2015.

Applied Underwater Acoustics. http://dx.doi.org/10.1016/B978-0-12-811240-3.00001-1

search and recovery, oil and gas prospecting, and prospecting for minerals and resources on and in the seafloor. Dredging includes sonars used by rock and stone dump vessels, excavator and trailing suction hopper dredgers, cutter suction and bucket dredgers, clamshell grab cranes and underwater plow vessels, and placement support. Defense and security includes mine counter measures, submarine and torpedo detection, obstacle avoidance, search and recovery, underwater communication, vessel and fleet protection, waterside security, diver detection, and so on. Marine research includes environmental monitoring, ambient noise measurements, marine archeology, marine mammal research, and fishery research. Fishery includes fishery operations, fish school detection, catch monitoring and control, trawl position control, phytoplankton and zooplankton investigations, communication between monitoring sensors on fishing gear and the fishing vessel, seabed mapping, bottom discrimination, and so on.

The counterpart to radar above water is *sonar* under water. SONAR is the acronym for sound navigation and ranging. It was originally used during World War II as an analog to the name "radar" and as a replacement for the name "asdics" for underwater detection systems using sound, which were used by the British Royal Navy during World War I. The two most common sonar types are *passive* and *active*. In a passive sonar system, the acoustic signal originates at a *target* and propagates to a *receiver*, where the acoustic signal is converted to an electrical signal for processing. In an active sonar system, an electrical signal is converted to an acoustical signal by a *transmitter* and the sound waves propagate from the transmitter to a target and back to a receiver, where conversion from acoustical to electrical signal takes place followed by electronic signal processing. Signal processing is aimed at enhancing the return signal from the target or reducing the noise in which the return signal may be embedded, as discussed in Chapter 11. The transmitter is normally called *the projector* and the receiver is called *the hydrophone*, as discussed in Chapter 10. If the return signal—the echo—from a target is detected, the position and the potential target movement are determined by the time delay of the echo from the target and the direction of the echo, respectively. The speed of a moving target can be estimated from the frequency shift—*the Doppler shift*—in the echo from the target, as discussed in Chapter 2.

When a sound wave is produced in water it propagates from the site where it is produced. Sound sources can be natural, such as breaking waves, rain falling on the water surface, seismic activities in the seafloor, and so on, or man-made such as sonar signals, underwater explosions, ship noise, and so on, as discussed in Chapter 6. During propagation the sound signal is exposed to a number of processes which may change the sound signal and its propagation, such as sound signal amplitude attenuation due to absorption, divergence, and scattering, as discussed in Chapter 4. Scattering takes place during the sound wave's interaction with the sea surface, seafloor, and inhomogeneities in the water column, as discussed in Chapter 5. These inhomogeneities can be natural, such as plankton, fish and sea mammals, and variations in the sea temperature and salinity. Scattering and reflection of sound signals may cause sound waves to follow different paths, producing multi-path sound propagation, which can make detection of objects in the water column and on the seafloor difficult. The scattering of underwater sound may lead to reverberation which limits detection. Use of advanced signal processing on the transmitted and received signal opens up the possibility to avoid or reduce the degradation of the propagated sound signal, as discussed in Chapter 11. Ambient noise in the sea can also become a limiting factor for signal detection. The sound signal received by a hydrophone carries information about the signal source and what the signal has encountered while propagating from the source to the hydrophone. The signal received by the hydrophone is processed to extract information of value to the user. This complicated "underwater world," where sound propagation is influenced by many individual sources with effect on the sound signal's amplitude, phase, and spectral composition, is the basis for this book, "Applied Underwater Acoustics."

Each chapter is introduced with a section giving the necessary definitions and describing the physical background for the subsequent sections of the chapter. The man-made sources of sound from sonar systems of various types are described in Chapter 10. This chapter also describes the different transducer types, their charge forming elements, and their geometries. Chapter 10 illuminates the sonar types available today, characteristic features, as well as their design, calculation, and calibration. Hydrophones, including array types, and their characteristics are also a part of Chapter 10.

The sound wave propagation through the water and the different factors which influence the propagation path are discussed in several chapters. The oceanographic features with influence on sound propagation are illuminated in Chapter 2. Chapter 2 also includes definitions and describes important acoustic wave concepts, such as wave geometries, divergence, convergence, reflection and transmission at interfaces, refraction and diffraction, and propagation through inhomogeneous media.

Chapter 3 discusses the capability to calculate sound propagation in the sea using available models.

Absorption of sound in fresh and in seawater is caused by the several mechanisms described in Chapter 4. The interplay of these mechanisms and their dependence on frequency are discussed in detail, and the best formulations for calculating sound absorption are provided.

When a sound wave in the sea hits a boundary, such as the seafloor, sea surface, or an object in the water column, the sound wave is reflected and scattered. Chapter 5 describes the scattering dependence on the geometry of the scattering object and its surface qualities. Useful expressions for scattering calculations including perturbation approximations and the Helmholtz−Kirchhoff method are provided. Also scattering from one and two scales of surface roughness are presented. Chapter 5 provides an in-depth discussion of scattering which can lead to reverberation, which in turn can limit sound signal reception in the sea.

Chapter 6 discusses ambient noise in the sea produced by natural sources, such as seismic activities, breaking waves, bubbles formed near the sea surface, precipitation, biological activities, ice, and man-made sources, such as shipping, prospecting for oil and gas, and so on. The spectra, directivity, and ambient noise coherence are presented, and self-noise produced by the ship making, the noise measurements, and procedures for noise reduction are an integral part of Chapter 6. The temporal and spatial variability of noise and statistical methods for characterizing noise are emphasized.

Sound propagation in shallow water is strongly influenced by the physical properties and geometries of the sea surface and seafloor. These boundaries form a sound channel through which the underwater sound is guided. It is possible by using information about the boundaries to produce models for calculating sound propagation through the channel. Many experiments have given valuable information about sound propagation in shallow water, the continental shelf, and ice-covered water. Chapter 7 provides up-to-date results and procedures for measuring and calculating sound propagation in shallow water.

The seafloor has frequently the strongest influence on sound propagation in seawater, in particular in shallow waters. This influence is produced by the nature of the seafloor sediments, their elastic qualities and porosity, and the seafloor surface geometry. Also, rocks and boulders on and in the seafloor influence reflection and scattering from the seafloor. Practical models for calculating scattering from the seafloor at high and at low sound frequencies are provided in Chapter 8. The chapter also includes an in-depth discussion of the physical properties important for seafloor sound propagation, reflection, and scattering. Methods for measuring sediment geoacoustic properties, seafloor roughness spectra, and statistics for seafloor heterogeneity including methods for seafloor identification and characterization by using sonar are presented in Chapter 8.

Underwater sound is used to investigate oceanographic and environmental sea qualities. Sound velocity profiles are created by variations with water depth in temperature, salinity, and pressure, which form sound ducts in the sea that can be used for sound propagation over great distances. This sound propagation is used to detect and describe oceanographic phenomena such as gyres and eddies, fronts, influx of warmer into colder water, and water flow with different salinity, by measuring acoustic signal arrival time to known positions around an acoustic source. This process, which is named *tomography*, is used for studies at basin scale down to shorter distances in shallow water. Acoustic tomography is described in Chapter 9. The *acoustic thermometry* of the ocean, where long-time variations in the ocean temperature are detected by measurements of the arrival time of coded acoustic signals propagated over thousands of kilometers, is also an aspect of tomography. In general acoustic signals gather information about the qualities of the materials in which they have propagated. This information can be unveiled through *inversion* procedures, where return signal processing can inform us about seafloor qualities and characteristics of the water column and help perform rapid environmental assessment. Inversion procedures are discussed in Chapter 9.

When underwater signals have been picked up by a hydrophone or a hydrophone array these signals are processed. Frequently the desired signal is embedded in noise such as ambient noise or reverberation. To detect the signal it is necessary to filter the received signal from noise and to amplify the desired signal before the detection and estimation process is performed. As of 2016, several signal processing "tools" are available to the underwater acoustician. These tools and their applications are described in-depth in Chapter 11.

Underwater acoustic methods are used extensively to detect the type and magnitude of biomass in the sea. Studies range from very small−scale phyto- and zooplankton, over various species of fish to sea mammals. Systems for catch monitoring and control and habitat mapping are described in Chapter 12, which also includes target strengths of single fish and fish shoals and the acoustic models used for studies. Sound produced by certain fish types, and the sensitivity of marine fish and mammals to underwater sound are also discussed in Chapter 12.

In general, sound propagation is considered a linear process. However, higher sound signal amplitudes may produce nonlinear processes such as harmonic distortion and acoustic saturation. Nonlinear processes are also found in focused sound fields and in cavitation, a local bubble formation process formed by pressures below the hydrostatic pressure. Nonlinear underwater acoustics includes the use of parametric acoustic arrays for sound generation and reception and underwater explosions used for prospecting for oil, gas, and minerals. Finite-amplitude underwater sound is discussed in Chapter 13.

Chapter 14 describes a series of underwater sound applications for marine renewables, underwater surveillance networks, investigations of soundscapes, characterization of noise from ships and production platforms, nuclear-test-ban treaty monitoring, underwater communication and networks, unmanned vehicles for surveillance and monitoring, underwater archeology, investigations in polar environments, warning against seismic activities and against tsunamis, model experiments in water tanks, and seafloor application of ocean observatories.

Section 1.2 of this chapter provides a brief history of underwater acoustics. Section 1.3 presents the international system of units used in the book, followed in Section 1.4 with a discussion on the use of the decibel scale. Section 1.5 covers the features of oceanography including sound speed profiles, thermoclines, arctic regions, deep isothermal layers, expressions for the speed of sound, surface waves, internal waves, bubbles from wave breaking, ocean acidification, deep-ocean hydrothermal flows, eddies, fronts and large-scale turbulence, and diurnal and seasonal changes. Section 1.6 discusses the sonar equation which is fundamental to underwater acoustics. Section 1.7 contains a list of the acronyms. The chapter concludes with the list of references.

1.2 A BRIEF EXPOSITION OF THE HISTORY OF UNDERWATER ACOUSTICS

Underwater acoustics is one of the fastest growing fields of research and development in acoustics. This is reflected by the increasing number of publications each year in international journals and conference proceedings. The relations between underwater acoustics and other fields of importance to the international community such as oceanography, meteorology, seismology, fishery, oil and gas industry, communication, shipping, defense, and security are becoming closer. The comprehensive activity in underwater acoustics is based on research and development over more than two millennia, spawned by human curiosity about the sea and its ability to support sound wave propagation.

1.2.1 UNDERWATER ACOUSTICS BEFORE 1912

As far as we know today [1] the work on underwater acoustics was started by the Greek philosopher *Aristotle* (384−322 BC) who was the first to note that sound could be heard in water as well as in air. In 1490 the Italian scientist and artist *Leonardo da Vinci* (1452−1519) wrote in his notebook, "If you cause your ship to stop and place the head of a long tube in the water and place the other extremity to your ear, you will hear ships at great distances." Of course, the ambient noise level in lakes and seas was much lower during his days than today, when several kinds of ships and offshore activities pollute the seas with noise. About 100 years later, the English philosopher Francis Bacon (1551−1626) in his work *Historia Naturalis et Experimentalis* supported the idea, that water is the principal medium by which sounds originating therein reach a human observer standing nearby. In the 17th, 18th, and early 19th centuries, several scientists became interested in light, as well as sound, transmitted in air and water. The Dutch astronomer Willebrord Snellius (1580−1626) worked on the refraction of light. Snell's Law follows from Fermat's principle of least time, which in turn follows from propagation of light as waves. This concept was contradictory to Sir Issac Newton's (1643−1727) assumption that light propagated as particles. Christiaan Huygens (1629−1695) formulated the principle named after him that each point on a wave front is an origin for spherical elementary waves and the wave front propagates as the envelope surface of the elementary waves. This principle is important for understanding the sound propagation in water. The interference between waves was studied by Joseph von Fraunhofer (1787−1826) and Augustin-Jean Fresnel (1788−1827). The mathematical tools to describe sound propagation in water were formed during the 17th through 19th centuries. G.W. Leibnitz (1646−1716) formulated the notation for differentiation and the rules for integration. Other contributions to the mathematical foundation for underwater acoustics today were formulated by various scientists, such as Daniel Bernoulli (1700−1782), Leonhard Euler (1707−1783), J.R. d'Alembert (1717−1783), J.-L. Lagrange (1736−1813), P.-S. Laplace (1749−1827), A.-M. Legendre (1752−1833), J.B.J. Fourier (1768−1830), S.D. Poisson (1781−1840), and Hermann von Helmholtz (1821−1894). Important contributions to the instruments used in underwater acoustics arise from H.C. Ørsted (1777−1851), who discovered the electromagnetism, and J.P. Joule (1818−1889), who contributed to the discovery of the magnetostrictive effect. The discovery of the piezoelectric effect in 1880 was based on works by Henri Becquerel (1852−1908) and the brothers, Paul-Jacques Curie (1856−1941) and Pierre Curie (1859−1906).

Direct sound speed measurements in fresh and saltwater, and comparing these measurements with the speed of sound in air were also performed by several scientists in the 18th and 19th centuries. Sound sources included bells, gunpowder, hunting horns, and human voices. The scientists' ears usually served as receivers. In 1743, J.A. Nollet (1700−1770) conducted a series of experiments to prove that water is compressible. With his head underwater, he heard a pistol shot, bell, whistle, and loud shouts. He noted that the intensity of the sound decreased only a little with depth, thus indicating that the loss mostly occurred at the water surface. In 1780, Alexander Monro (1733−1817) tested his ability to hear sounds underwater. He used a large and a small bell, which he sounded both in air and in water. The bells could be heard in water. However, he found that the pitch sounded lower in water than in air. He also attempted to compare the speed of sound in air and in water, and he concluded that the two sound speeds seemed to be the same.

The breakthrough in sound speed measurement came in September 1826, when the Swiss physicist J.D. *Colladon* (1802−1893) and the French mathematician J.K.F. Sturm (1803−1855) made the first widely known measurement of the *sound speed in water* on Lake Geneva at a water temperature of 8°C. A bell hanging down from a boat was used as transmitter, and when striking the bell a flash of light was made by igniting gunpowder. This flash could be seen by Colladon in a boat about 10 miles from the transmitter. He started his watch when he saw the flash and stopped it, when he heard the sound signal in the water about 10 s later. His receiver was a trumpet designed with one end in the water and the other in his ear. By using this rather primitive setup they were able to measure the sound speed in water at 8°C as 1435 m/s, which is only about 3 m/s less than today's accepted value [2]. From the sound speed and water density they could determine the bulk modulus of the water.

During the years 1830−1860 scientists started thinking about applications of underwater sound. Questions such as "Can the echo of a sound pulse in water be used for determination of the water depth or the distance between ships?" or "Can the communication between ships be improved by underwater transmission of sound?" were posed. The frustration in relation to the use of underwater sound for depth measurements is obvious from M.F. Maury's (1806−1873) words in Chapter 12 of his book *Physical Geography of the Sea*, 6th ed. 1859, where he says, "Attempts to fathom the ocean, by both sound and pressure, had been made, but out in blue water every trial was only a failure repeated. The most ingenious and beautiful contrivances for deep-sea sounding were resorted to. By exploding petards, or ringing bells in the deep sea, when the winds were hushed and all was still, the echo or reverberation from the bottom might, it be held, be heard, and the depth determined from the rate at which sound travels through water. But though the concussion took place many feet below the surface, echo was silent, and no answer was received from the bottom."

During the latter half of the 19th century, when the maritime world changed from sail to engine driven ships and wood was replaced by steel in ship construction, concern was expressed about safe navigation in fog and the danger of collision with other ships or icebergs. John Tyndall (1820−1893) in England and Joseph Henry (1797−1878) in the USA in separate investigations found sound propagation in air to be unreliable and in 1876 recommended to the lighthouse authorities in both countries that they adopt high-power siren warning installations at all major lighthouses. From 1873 joint experiments took place and a large-scale steam-driven siren was built at the South Foreland lighthouse in England driven by a steam pressure of $5 \cdot 10^5$ Pa and 100 to 400 Hz frequencies were investigated [3]. Sound transmission conditions, however, caused problems. Wind speed and temperature gradients over the sound propagation path caused strong variations in the sound detection distance. The possible advantages of signaling by sound in water were taken up again in the late 1880s by Lucien Blake and Thomas Alva Edison (1847−1931) in the USA. Edison invented an underwater device for communication between ships; however, for some unknown reason the US government lost interest in his invention.

Submerged bells on lightships were introduced to a large extent during the last years of the 19th century. The sound from these bells could be detected at a great distance through a stethoscope or by using simple microphones mounted on a ship's hull. When the ship was outfitted with two detecting devices, one on each side of the hull, it became possible to determine the possible bearing of the lightship by transmitting the sounds separately to the right and the left ears of the observer. Elisha Gray, who was working with Edison on improving the telephone, recognized that the carbon button microphone in a suitable waterproof container could be used as a hydrophone to receive underwater bell signals. In 1899, Gray and A.J. Mundy were granted a patent on an electrically operated bell for underwater signaling.

1.2.2 THE YEARS 1912 THROUGH 1918

In 1912, the Submarine Signal Company hired the Canadian R.A. Fessenden (1866−1932), to develop a sound source more efficient than pneumatically or electrically operated bells. Fessenden designed and built a moving coil transducer to emit underwater sound. The Fessenden oscillator which was designed somewhat like an electrodynamic loudspeaker, allowed ships to communicate with each other by using Morse code or to detect echoes from underwater objects. The acoustic power transmitted into the water was about 2 kW at a resonance frequency of 540 Hz, and the electroacoustic efficiency was 40−50%. In 1914, the echo location process known as echo ranging was developed to a level where it could locate an iceberg at a distance of 3.2 km. Unfortunately this development came too late to avoid the Titanic disaster.

The outbreak of World War I and the later introduction of unrestricted submarine warfare were the impetus for developing a number of military applications of underwater sound. In France the Russian electrical engineer Constantin Chilowsky collaborated with the physicist Paul Langevin (1872−1946) on a project involving a condenser (electrostatic) projector and a carbon button microphone situated at the focus of a concave acoustic mirror. The first successful underwater acoustic signals were sent across the river Seine in Paris below the Pont National by the end of 1915 [4]. In 1916 Langevin and Chilowsky filed a joint application for a patent based on their method and equipment. In April 1916 they were able to transmit an underwater signal over 2 km and detect at 200 m echoes reflected by an iron plate. Since

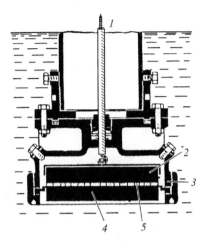

FIGURE 1.1

Langevin's piezoelectric quartz-based transmitter/receiver. (1) connected to a.c. oscillator and amplifier and to the receiver. (2) The steel inner electrode. (3) The watertight container. (4) The steel outer electrode. (5) The layer of 0.004 m thick slices of quartz.
Reproduced from Lasky, M., Review of undersea acoustics to 1950. *J. Acoust. Soc. Amer., **61**, (2), pp. 283–297 (1977), with the permission of the Acoustical Society of America.*

Chilowsky left the project after filing the patent, Paul Langevin, who had moved to Toulon, in 1917, turned his interest to the piezoelectric effect—originally discovered by the Curie brothers in 1880—to develop transmitters and receivers for underwater use. The newly developed vacuum tube amplifier, the Audion valve, was used by Langevin for his quartz receiver, and in 1918 he completed the development of his sandwich-type, steel–quartz–steel transmitter, shown in Fig. 1.1. This transmitter had a resonance frequency of 40 kHz produced by the sandwich consisting of a layer of quartz in the form of a square mosaic 0.004 m thick and 0.2 m in square between two square steel plates each of thickness 0.03 m. This transmitter increased the range for one-way transmission to more than 8 km, and clear submarine echoes were heard in February 1918.

In England Lord Rutherford had assembled a strong group of physicists at University of Manchester. In particular, two persons, who joined this group in 1915 and in 1916, respectively, Albert Beaumont Wood (1890–1964) and the Canadian physicist Robert William Boyle (1883–1955) contributed substantially to sonar development and to underwater acoustics. In 1917 several members of Lord Rutherford's group were moved to Parkeston Quay in the England, where they, under leadership of professor W.H. Bragg, carried out research and development related to underwater echolocation and passive listening under the top secret "ASDIC" project. ASDIC is an acronym for "*A*llied *S*ubmarine *D*etection *I*nvestigation *C*ommittee" [4].

Boyle started out using the Fessenden oscillator, but found early on that its low frequency, around 1 kHz, would not produce the necessary resolution for detecting submarines. After contact was established between Boyle and Langevan in 1917,

Boyle visited France and the scientists shared information. A slab of quartz was sent to Boyle in England, and was tested by Boyle, who in March 1918 achieved an echo from a submarine at nearly 500 m. The first practically working active sonar, or AS-DICS as the British preferred to name it, was built by Boyle in November 1918. It was successfully tested out fitted to a trawler a few days after the armistice on November 11, 1918.

In the USA, Dr. Harvey C. Hayes (1880−1969) had gathered a group of specialists at Naval Experimental Station, New London, with the terms of reference "to devise as quickly as possible the best of available technology to defeat a U-boat." Hayes and his group developed the towed hydrophone assembly called "the Eel," and a passive sonar installation using 48 hydrophones—hull mounted and towed—was tested on a US destroyer. This installation was the most advanced passive sonar system produced during World War I [3].

In 1911 in Kiel, Germany, Karl Heinrich Hecht (1880−1961) developed a hydrodynamic siren source for producing underwater sound. Also, he developed an electromagnetic membrane transmitter, which during World War I was built into several hundred surface ships and submarines.

The German scientist Alexander Behm (1880−1953) successfully tested the first echosounder on the seafloor of the Fjord of Kiel in February 1916. Also, the German engineer Hugo Lichte (1891−1963) performed extensive underwater acoustic studies in which he correctly deduced the effects of temperature, salinity, and pressure on the speed of sound. He predicted in 1919 that upward refraction produced by pressure in deep water should produce extraordinarily long sound listening ranges. This fact was verified many years later.

1.2.3 THE YEARS 1919 THROUGH 1939

During 1918−1940, three uses of underwater acoustics based on wartime experiences were slowly developed extensively. They were echo sounding, sound ranging in the ocean, and seismic prospecting. A significant practical impetus was received from advances in electronics, which made available new methods and devices for amplification, processing, and displaying received underwater signals. M. Marti in 1919 patented a recorder to be used for echo sounding. This recorder, which turned out to be of extreme importance to ocean studies using sound, consisted of a sheet of paper constrained to move slowly beneath a writing pen which traversed the paper from one side to the other perpendicular to the motion of the paper. The pen was driven laterally to the paper motion by an electric signal, whose amplitude was proportional to the output from the underwater sound receiver. By viewing the successive echoes side by side as a function of time, a profile of the seabed could be produced. In 1922 the first long echosounding depth profiles were made while exploring a cable route between France and Algeria.

The need for improved and more robust high-power underwater sound sources instead of the Langevin-type transducers based on quartz or Rochelle salt crystals, led G.W. Pierce in the USA, in 1925, to develop a magnetostrictive oscillator operating at 25 kHz with an emitted sound power of few kilowatts, without the risk of fracturing the oscillating element, frequently found in crystal-based transducers.

During the same period, the US Coast and Geodetic Survey in their attempt to establish geodetic control by horizontal sound ranging was experiencing a strong variability in sound intensity and speed in the sea. Also the Naval Research Laboratory, established in 1923 on a suggestion from Edison, when seeking to improve submarine hunting, working at 20–30 kHz, found the same variability. Some of this variability appeared to show a diurnal cycle, where the equipment in the morning was working according to the specifications while in the afternoon, it did not produce any echoes from submarines, except at very short ranges. The same "afternoon effect" was found in several regions of the ocean. Dr. Harvey Hayes and scientists from the newly established oceanographic institution at Woods Hole, including the institutions head, Columbus Iselin, decided to study these phenomena in more depth.

It soon became clear that the upper parts of the ocean were heated during the day by the sun, thus leaving a layer 4.5–9 m thick with a temperature 1–2°C warmer than the more uniform water layer beneath and with a gradual decrease in temperature with distance from the surface of the sea. Since the appearance of the temperature layer coincided with the signal reception deterioration the scientists concluded that the warm layer caused sound entering the water to bend downward toward the low temperature region thus producing an acoustic shadow zone in which a submarine could hide. This discovery in 1937, which explained the "afternoon effect," achieved through cooperation between acousticians and oceanographers, L. Batchelder of the US Submarine Signal Company and Columbus Iselin, led to the start of a new field of research called acoustical oceanography. The same year Athelstan Spilhaus from MIT invented and build the first bathythermograph, a small torpedo-shaped device that held a temperature sensor and an element to detect changes in static water pressure. By the beginning of World War II, all US naval vessels engaged in antisubmarine work were equipped with the Spilhaus device.

The SONAR developments in the years before World War II were based on exploiting the crystals quartz, Rochelle salt, and tourmaline, along with magnetostrictive Ni-Fe alloys. Single hydrophones, as well as linear and planar arrays, were developed and tested and found their way into the major navies. To provide water tight protection of the charge-forming elements such as quartz and in particular Rochelle salt, which could be dissolved in water, a rubber material with nearly the same acoustical impedance as water, "rho-c" rubber, was developed around 1930. A Rochelle salt-based hydrophone was developed in Germany in 1935 for use onboard warships. For example, the German battle cruiser, *Prinz Eugen*, during World War II was equipped with hydrophones based on Rochelle salt covering the frequency range from 500 Hz to 10 kHz. The hydrophones were mounted on each side of the ship in groups with six hydrophones in each group. Several groups worked together and formed the so-called Gruppen Horch Gerät (GHG), or group listening apparatus.

1.2.4 **THE YEARS 1940 THROUGH 1946**

The outbreak of World War II launched great activity in underwater acoustics research in Europe, USA, and the Far East. The hunt for submarines received high priority. The combination of convoys, aircraft patrols, and ASDIC gear effectively held off most conventional daylight attacks by the initial small number of German submarines (less than 30 when the war started in 1939). This number increased substantially during the war and a total of over 1100 submarines of various sizes were built in Germany during World War II. In addition, the Germans soon learned to launch night attacks on convoys using "wolfpack" techniques. The development of airborne radar, in particular the Allied's monopoly on the 10 cm radar, and the use of improved depth charges and the invention of "hedgehog" techniques, became a great help in hunting down and destroying German submarines, of which Germany lost 781 during the war.

Apart from the development of underwater arms such as the acoustic homing torpedo, acoustic mine, and scanning sonar, a much better understanding of underwater factors influencing sound propagation was established. Concepts, such as target strength, ship self-noise, reverberation of the underwater environment, sound scattering, and sound absorption in seawater, were established and studied. Sound propagation under the influence of vertical variations in sound speed was investigated and modeled using the "*ray theory*" borrowed from the theory of light. Sound propagation in layered media such as the water column and seafloor was calculated by using the *normal mode theory*, developed by C.L. Pekeris in 1941 [5]. Most of the achievements in underwater acoustics during World War II were published in the USA just after the war in 23 reports called "National Defense Research Committee Division 6, Summary Technical Reports." One of these reports entitled "Physics of Sound in the Sea" comprises chapters on deep- and shallow-water acoustic transmission, on intensity fluctuations, and on the explosion as a source of underwater sound.

1.2.5 **THE YEARS AFTER 1946**

The developments in underwater acoustics during the years immediately after World War II were strongly influenced by discoveries and developments during the war. The wartime discoveries and developments in the USA were reported in NDRC's summary reports, of which the summary technical report from Division 6 is of particular interest to underwater acousticians. Books published later on, such as Urick's book titled "Principles of Underwater Sound for Engineers" from 1967, are based substantially on results from the research carried out during World War II.

Maurice Ewing, a professor of geology at Columbia University, had during the war studied the characteristics of low frequency sound propagation in the sea, and he was convinced that it would be possible to propagate sound over hundreds—possibly thousands—of kilometers through the ocean if both source and receiver were appropriately placed. In 1945 he propagated sound from a small explosion over a distance of more than 3000 km from Eleuthera in the Bahamas to Dakar in

West Africa. The sound propagation took place in a ubiquitous permanent sound channel of the deep ocean. Ewing called the channel the SOFAR (SOund Fixing And Ranging) channel. The first application of this discovery was aimed at providing a rescue system for downed-at-sea airmen. From his inflated rubber boat, the airman should drop small cartridges over the side set to explode on the axis of the SOFAR channel situated at about 1200 m depth in the North Atlantic, as shown in Fig. 1.2. Sound from the explosion would be refracted back to the channel axis and the propagation would only be influenced by cylindrical spreading. The signals would then be picked up by hydrophones positioned on the channel axis at various positions off the continental shelves, making it possible by comparing signal arrival times to find the position of the source. Maurice Ewing in the USA and academician L.M. Brekhovskikh in the USSR were studying the undersea sound channel. In the years just after the war Brekhovskikh discovered the existence of the sound channel in the Pacific Ocean by analyzing signals received from underwater explosions in the Sea of Japan.

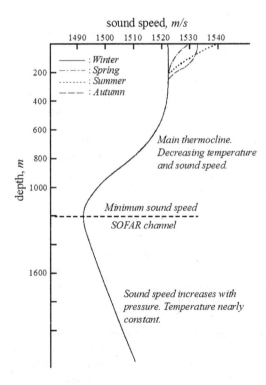

FIGURE 1.2

A characteristic sound velocity profile in the North Atlantic Ocean. The minimum sound speed forming the SOFAR channel is at a depth of about 1200 m. Below the minimum sound velocity, the sound speed increase is caused by increasing pressure. The temperature is nearly constant at about 2°C.

Ewing, together with J.L. Worzel and several other colleagues at Woods Hole Oceanographic Institute studied long-range sound propagation in shallow and deep water by using underwater explosions [6,7]. Based on their data, Chaim Pekeris improved his normal mode propagation theory [5,8] and was able to interpret the shape of the mean dispersion curves measured by Ewing and Worzel. Geometrical dispersion was also described by Tolstoy [9] in a fluid−solid layer. The concept of elastic wave propagation has allowed underwater acousticians to model and understand complex shallow water acoustics. Ewing and Worzel's research also formed the basis for a series of seabed geological structure studies performed mostly in shallow water off the East Coast of the USA. The cooperation established between Ewing's group at Columbia University and the scientists at Woods Hole Oceanographic Institute turned out to be most fruitful for underwater seismology investigations. The "refraction method" and the "continuous seismic profiler" were results of this cooperation.

A group around C.F. Eyring in San Diego, USA, had observed that diffuse echoes were received from the volume of a water column. These echoes were arranged roughly in horizontal layers whose depths were of the order of 400 m at noon, but they migrated to the surface during the twilight and the early evening. At dawn, they migrated downward to complete a daily cycle. Based on help from marine biologists it was possible to show that the responsible scatterers were small planktonic fish that have a swim bladder and live in the deep ocean water regions. The research into the "deep scattering layers" peaked during the 1949−1957 period. Important contributions to marine bioacoustics were made during the subsequent years.

These years also brought important research results related to sound absorption mechanisms in the sea and sources and spectra of ambient noise. The anomalous absorption in seawater below 1 MHz, where the viscous absorption predicted by Stokes in 1845 [10] gave values too low compared with measurements was explained. Liebermann [11] explained the deviation from the Stokes-based viscous absorption dependence on the square of the frequency by the presence of a dilatational viscosity and the influence of relaxational effects, as discussed in Chapter 4. In the 1950s it was shown that the relaxational effects were caused by the presence of magnesium sulfate ($MgSO_4$) [12,13] with a relaxation time below 1 μs, and even lower frequency, <10 kHz, relaxation-caused absorption was discovered in the 1970s to be due to boric acid ($B(OH)_3$) [14,15].

The comprehensive research and development efforts in underwater acoustics after World War II and the development in computer technology after 1960 formed the basis for the nearly explosive development in underwater acoustics from 1960. Among the main trends in underwater acoustics research and development are underwater sound propagation modeling involving mode theory, parabolic equations [16], and finite element methods to include realistic range dependence and surface and bottom effects, reverberation studies, and ambient noise source and directivity studies [17], as also underwater acoustical tomography including the international acoustic thermometry of the ocean climate (ATOC) studies [18], and the coupling between acoustics, oceanography, and meteorology to lead to long term reliable weather forecasts, acoustical studies of biomass in the sea including detection of sea mammals [19], and extensive bottom and subbottom studies for exploitation of minerals, oil, and gas, as well as for cable and tube laying. Also the "Cold War" gave inspiration to a substantial research effort in underwater acoustics and to a considerable development effort in relation to equipment for acoustic studies of "the underwater world." Advanced fixed-positioned warning and surveillance networks such as SOSUS (SOund SUrveillance System) in the Atlantic and Pacific Oceans and ship-towed passive arrays such as SURTASS (Surveillance Towed Array Sensor System), as discussed in Chapter 10, were results of Cold War developments.

The developments in underwater acoustics over the last 20 years are, for instance, reflected in the increasing number of R&D results published in the proceedings of international conferences like the European Conference on Underwater Acoustics (ECUA) from 1991 through 2012 and the Underwater Acoustic Measurements: Technologies and Results (UAM) from 2005 through 2011. From 2013, these conferences are merged into the International Conference and Exhibition on Underwater Acoustics (UA) [20,21].

1.3 INTERNATIONAL STANDARD UNITS

The units used throughout this book are based on the International System of Units, (*Systéme Internationale*, *SI*). The SI includes seven base units and several derived units. Table 1.1 gives the SI base units and Table 1.2 gives the SI derived units and other units used in this book. More units, their magnitude, and relations used in underwater acoustics are given in Table 1.3.

Table 1.1 SI Base Units

Quantity	Name	Symbol
Length	Meter	m
Mass	Kilogram	kg
Time	Second	s
Electric current	Ampere	A
Thermodynamic temperature	Kelvin	K
Amount of substance	Mole	mol
Luminous intensity	Candela	cd

Table 1.2 SI Derived Units and Other Units

Name	Unit	Symbol	Expressed in Terms of Other Units	Expressed in Terms of SI Base Units
Frequency	Hertz	Hz		s^{-1}
Force	Newton	N		$m \cdot kg/s^2$
Pressure	Pascal	Pa	N/m^2	$m^{-1} \cdot kg/s^2$
Energy	Joule	J	$N \cdot m$	$m^2 \cdot kg/s^2$
Power	Watt	W	J/s	$m^2 \cdot kg/s^3$
Electric charge	Coulomb	C		$A \cdot s$
Electric potential	Volt	V	W/A	$m^2 \cdot kg/A \cdot s^3$
Capacitance	Farad	F	C/V	$m^{-2} \cdot kg^{-1} \cdot s^4 \cdot A^2$
Electric resistance	Ohm	Ω	V/A	$m^2 \cdot kg/s^3 \cdot A^2$
Electric conductance	Siemens	S	A/V	$m^{-2} \cdot kg^{-1} \cdot s^3 \cdot A^2$
Magnetic flux	Weber	Wb	$V \cdot s$	$m^2 \cdot kg/s^2 \cdot A$
Magnetic flux density	Tesla	T	Wb/m^2	$kg \cdot s^{-2}/A$
Inductance	Henry	H	Wb/A	$m^2 \cdot kg/(A \cdot s)^2$
Sound intensity		I	$Pa^2/\rho c$	Kg/s^3
Plane angle	Radian	rad		
Solid angle	Steradian	sr		
Area				m^2
Volume				m^3
Sound velocity		c		m/s
Density		ρ		kg/m^3

1.4 THE DECIBEL SCALES

In the analysis of sound propagation, it is customary to use the *decibel* (*dB*) notation to represent the *level* of various quantities relative to a chosen reference value of the quantity. This is a convenient way to handle the wide dynamic range involved in acoustic problems. It also simplifies many system calculations by replacing multiplications with additions of decibel quantities. The decibel is 1/10 of a *bel*, which is a logarithmic unit of a power or an energy ratio. By definition, the decibel (dB) corresponds to 10 times the base-10 logarithm of the ratio of two power or energies. If the two powers are W_1 and W_2 {unit: W}, the power level, *WL*, in the dB scale may be written as:

$$WL = 10 \log_{10}\left(\frac{W_1}{W_2}\right) \tag{1.1}$$

Table 1.3 Other Units Used in Underwater Acoustics

Name	Magnitude	Other Relations
Inch	0.0254 m	
Foot	0.3048 m	
Yard	0.9144 m	
Nautical mile	1852 m	
Knot	1 nautical mile/h	1825 m/h = 0.5144 m/s
Bar	10^5 Pa	10^5 N/m^2
dyn/cm^2	10^{-6} bar	0.1 Pa
Atmosphere	1.01325 bar	$1.01325 \cdot 10^5$ Pa
Pound per square inch (psi)	$6.8948 \cdot 10^{-2}$ bar	$6.8948 \cdot 10^3$ Pa
Rayl		Kg/m$^2 \cdot$s
Kelvin		K
Degree celsius		$^\circ$C = K $-$ 273.15
Degree fahrenheit		$^\circ$F = 1.8°C + 32
Neper		Np: 10$\cdot$log$_e$ to a power ratio
Decibel	Np = 8.7 dB	dB: 10$\cdot$log$_{10}$ to a power ratio

If WL in Eq. (1.1) is 10 dB, W_1 is 10 times higher than W_2. If W_1 is two times larger than W_2, $WL = 3$ dB. The intensity level IL in dB in an acoustic field may be expressed by:

$$IL = 10 \log_{10} \left(\frac{I_1}{I_2} \right) \tag{1.2}$$

where I_1 and I_2 {unit: J/(s$\cdot$m^2)} denote the acoustic intensity (i.e., the acoustic energy flux through a unit area per second), and as the acoustic intensity is proportional with the square of the effective acoustic pressure {unit: Pa}, the acoustic sound pressure level (SPL) in dB may be expressed by:

$$SPL = 20 \log_{10} \left(\frac{p_1}{p_2} \right) \tag{1.3}$$

All quantities with subscript 2 in Eqs. (1.1)−(1.3) are considered as *reference quantities*. There are several units used to specify reference pressure in acoustics. In air acoustics the reference quantity is 20 µPa (micro pascal), which is nearly equivalent to the reference intensity of 10^{-12} W/m^2. In underwater acoustics the most frequently used reference quantity is 1 µPa, which is nearly equivalent to the reference intensity of $6.67 \cdot 10^{-19}$ W/m^2. Sometimes the reference quantity may be 1 µbar ($=10^5$ µPa), which is equivalent to the reference intensity of $6.67 \cdot 10^{-9}$ W/m^2. To avoid confusion, it is important to specify the reference quantity which is being used. A sound pressure level may be expressed as: $SPL = 210$ dB re 1 µPa, which means that acoustic pressure $p_1 = 0.316 \cdot 10^5$ Pa.

The effective acoustic pressure p_e (i.e., the root-mean-square or rms pressure) in a propagating harmonic wave is related to the peak pressure amplitude in the wave p_o through: $p_e = p_o/\sqrt{2}$. For the same effective acoustic pressure the pressure level in water will be 26 dB ($=20\log_{10} 20$) higher than in air.

When using the dB scale the product of several variables, $A \cdot B \cdot C$ in dB, translates into the sum of the dB value of each of the values, $(A \cdot B \cdot C)_{dB} = A_{dB} + B_{dB} + C_{dB}$. However, the level of the sum of the physical quantities in dB is different from the sum of each individual quantity level in dB, $(A + B + C)_{dB} \neq A_{dB} + B_{dB} + C_{dB}$. Therefore, for three SPL values, 170 dB, 180 dB, and 190 dB all re. 1 μPa, the level of their sum SPL value will not be 540 dB, but 190.45 dB, where their individual *acoustic energies* are calculated and their values added, before the sum energy's level in dB is calculated.

1.5 FEATURES OF OCEANOGRAPHY
1.5.1 SOUND SPEED PROFILES

The main factor affecting the propagation of acoustic waves in the ocean is the speed of sound, which has a nominal value of 1500 m/s in temperate and equatorial oceans. Since the ocean is not a homogenous medium, small departures from the nominal value, of the order of 1%, are exhibited by the sound speed. Although these variations in the speed of sound are small, they have a profound effect on acoustic propagation in the ocean. Focusing effects, for example, occur in long-range transmission in the deep ocean due to the depth profile of the sound speed.

The speed of sound at any point in the ocean depends on the local temperature, salinity, and hydrostatic pressure (or depth). In a horizontally stratified ocean, the temperature and salinity, and hence the sound speed, vary with depth but are independent of horizontal range, a situation that is approximated in the deep oceans, away from features such as eddies [22,23] and fronts. Fig. 1.3 shows examples of temperature, salinity, and sound speed profiles [24], taken in September 2012 in the Tonga Trench, South Pacific Ocean, at a point where the maximum depth is 8515 m. Over most of the ocean depth the salinity is essentially uniform, close to 34.85‰ (parts per thousand), and the variations in the sound speed profile are governed primarily by the temperature and the pressure. Immediately beneath the sea surface, the sharply reducing temperature is the controlling factor, which gives rise to a reduction in the sound speed as the depth increases. With increasing depth, however, the temperature stabilizes, becoming more or less independent of depth below 2000 m, while the hydrostatic pressure rises, eventually becoming dominant, causing the gradient of the sound speed profile to change sign from negative to positive. In this cold, pressure-controlled region, the sound speed increases linearly with increasing depth.

The net effect of temperature and pressure is to introduce a minimum into the sound speed profile at a depth somewhere between 500 and 1500 m. Since sound rays are bent toward regions of lower sound speed, the minimum in the profile

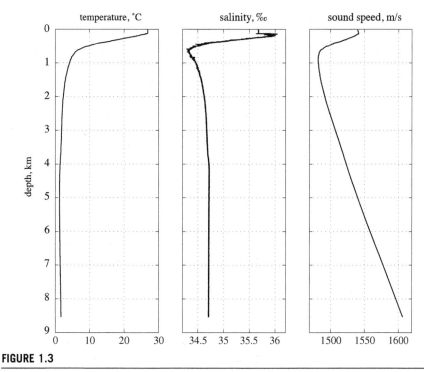

FIGURE 1.3

Temperature, salinity, and sound speed profiles, Tonga Trench, South Pacific Ocean, September 2012.

acts as an acoustic waveguide, known as the deep sound channel, in which energy from a sound source near the minimum may be trapped, allowing it to propagate with little loss to very large ranges [7] (several thousand kilometers). The minimum in the profile is termed the sound channel axis or sometimes the SOFAR channel axis, and at a given latitude it is shallower in the Pacific than the Atlantic by, typically, several hundred meters.

Rather than being measured directly, the sound speed profile is commonly computed from temperature and salinity data recovered from a conductivity—temperature—depth (CTD) probe or from temperature data acquired with a bathy-thermograph (BT) [25]. In the case of the latter, the salinity, which is not returned by the BT, is often estimated for the purpose of computing the sound speed profile from archival salinity data [26]. This practice is satisfactory for many purposes, since the sound speed is only weakly dependent on the salinity. Some CTD probes are capable of descending to the deepest parts of the ocean, although they require a titanium cable for support. Such devices have been used successfully to profile the ocean in the Challenger Deep, Marianna Trench, where the depth is 11 km [27].

A direct measure of the sound speed profile may be obtained using a sound velocity sensor (SVX), an instrument that was developed by Greenspan and Tschiegg [28] and is based on the "sing-around" principle. An ultrasonic pulse is transmitted over a fixed path between a source and a receiver. When the pulse arrives at the receiver, it triggers the transmission of another pulse from the projector. The repetition frequency of the pulses is mainly determined by the travel time over the path length, which is controlled by the speed of sound in the fluid (seawater in the case of ocean profiling) between the source and receiver. A Valeport, custom-designed SVX is capable of operating under the most extreme pressures encountered in the ocean, up to 1200 bar corresponding to a depth of 12 km. (The Challenger Deep in the Mariana Trench is the deepest part of the world's oceans at 11 km.) A sound speed profile taken in September 2012 by the Valeport SVX, extending over the full depth of 8515 m in the Tonga Trench, is illustrated in Fig. 1.4.

Although the scaling differs, it is easy to see that the sound speed profile returned by the SVX in Fig. 1.4 and that computed from the CTD temperature and salinity data and shown in the right hand panel in Fig. 1.3 are essentially identical. Both the CTD and the SVX were mounted on the free-falling (untethered) instrument platform known as Deep Sound [29] as it descended to the bottom of the Tonga Trench. Deep Sound descends under gravity to a pre-assigned depth, at which point

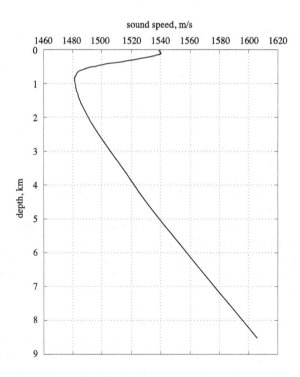

FIGURE 1.4

Sound speed profile returned by an SVX, Tonga Trench, South Pacific Ocean, September 2012.

a burn wire is activated, releasing a drop weight, allowing buoyancy to return the system to the surface. At a speed through the water column of 0.5 m/s, the return travel time is about 9.5 h. All the data acquired during the descent and ascent are stored in onboard solid-state memory and are downloaded after the system returns to the surface. One advantage of deep sound is that, unlike a traditional CTD cast, there is no cable connecting the system to the surface. This is particularly significant when it comes to probing depths in the region of 10 km or greater.

A canonical model of the sound speed profile in the deep ocean has been developed by Munk [30] from an argument based on stability considerations. He assumed an exponential stratification, which is known to be an intrinsic property of the deep oceans, and which is expressed through the Brunt−Väisälä buoyancy frequency [31]:

$$N(z) = N_0 exp\left(-\frac{z}{B}\right), \tag{1.4}$$

where z is depth in kilometers, $B = 1.3$ km is the e-folding depth representing the stratification scale, and $N_0 = 2.8$ cycles per hour is the surface-extrapolated value of $N(z)$. Munk's final expression for the sound speed profile is:

$$c(z) = c_1\left[1 + \varepsilon\left(\eta + e^{-\eta} - 1\right)\right], \quad \varepsilon = 0.0074, \quad \eta = \frac{2(z - z_1)}{B}, \tag{1.5}$$

where z_1 is the depth of the sound channel axis, c_1 is the speed of sound at the axis, and ε is a perturbation parameter that scales with the fractional sound speed gradient in an adiabatic ocean. (A slightly modified version of Eq. (1.5), designated the "deep six profile," has been examined by Miller [32].)

An example of the canonical profile in Eq. (1.5), with the sound channel axis at $z_1 = 1.2$ km, representative of the Atlantic Ocean, is shown in Fig. 1.5. The profile is asymmetrical about the axis, consistent with the form of the measured sound speed profile in Fig. 1.4. At the conjugate depth, sometimes referred to as the critical depth, the sound speed is equal to that at the sea surface, and in the region labeled "depth excess," it is greater than the speed of sound at the sea surface. Any profile showing a depth excess supports convergence zone propagation, whereby steep rays follow deep, upward refracted paths and regions of high intensity known as convergence zones, or caustics [33] are formed near the surface at regular intervals in range, as illustrated in Fig. 1.6. Typically, the range interval between convergence zones in the Atlantic Ocean is of the order of 60 km.

The Munk profile may be expressed in normalized form as follows:

$$\widetilde{c}(\widetilde{z}) = 1 + \varepsilon\left(\eta + e^{-\eta} - 1\right), \quad \eta = 2R(\widetilde{z} - 1), \tag{1.6}$$

where $\widetilde{z} = z/z_1$, $\widetilde{c} = c/c_1$, and the dimensionless parameter $R = z_1/B$ is the ratio of the channel axis depth to the buoyancy scale. Since ε is the same for all abyssal oceans, the only variable parameter in Eq. (1.6) is R, which may take values between 0.6 and 1.2, depending on geographical location. A family of curves, as computed from Eq. (1.6), for R within this range is shown in Fig. 1.7.

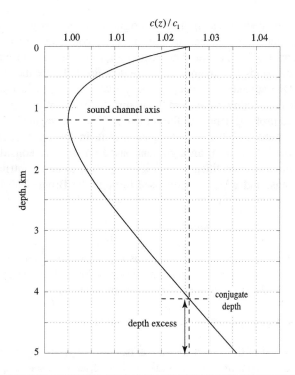

FIGURE 1.5

Munk sound speed profile for Atlantic conditions.

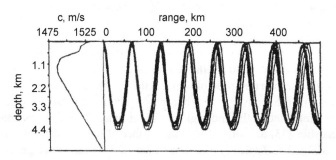

FIGURE 1.6

Convergence zone propagation in an abyssal ocean with a sound speed profile showing a depth excess.

1.5.2 THERMOCLINES

A thermocline is a layer in the ocean in which the temperature changes with depth. In the deep oceans, the temperature decreases sharply with depth for 1000 m or so beneath the surface, as illustrated in Fig. 1.3. This region is known as the main

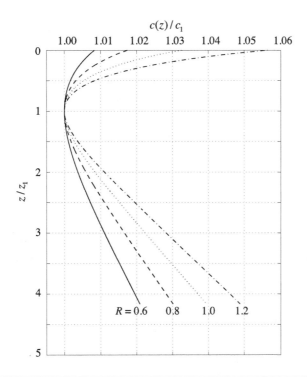

FIGURE 1.7

Normalized Munk profile for four values of the parameter R.

thermocline. Beneath the main thermocline, the axis of the deep sound channel is located at a depth that may be anywhere between 500 and 1500 m, depending on various factors including latitude. The region below the deep sound channel is the deep isothermal layer, where the temperature is essentially independent of depth and the sound speed increases linearly with increasing hydrostatic pressure.

The deep sound channel remains unaffected by surface conditions, particularly solar heating and wind-induced mixing. Immediately beneath the surface, however, down to depths of several hundred meters, the sound speed profile is sensitive to surface inputs, showing diurnal and seasonal variations. In the North Atlantic during winter, for example, the effect of strong winds is to create a well-mixed surface layer of isothermal water. The sound speed in this layer increases linearly with depth, thus forming a surface duct, which acts as an acoustic waveguide (Fig. 1.8A). In the summer months, on the other hand, the effect of solar heating is to create a surface layer of relatively warm water, sometimes known as the secondary or seasonal thermocline (Fig. 1.8B). This is a downward refracting layer that does not support long-range acoustic transmission but instead gives rise to a deep acoustic shadow zone.

Occasionally, surface conditions are such as to create a sound speed profile with two minima (Fig. 1.9). This might occur, for example, after the passage of a storm, when solar heating creates a seasonal thermocline above a well-mixed, deep isothermal layer.

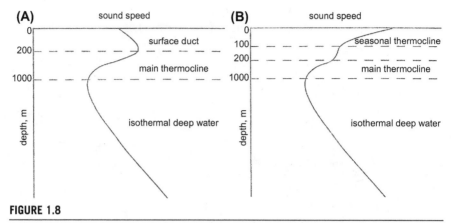

FIGURE 1.8

Schematic of sound speed profile in (A) winter, with a mixed layer to a depth of 200 m and (B) summer, when solar heating creates a seasonal thermocline (depth axes not to scale).

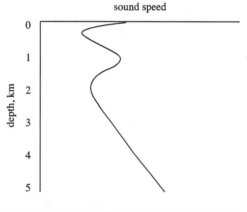

FIGURE 1.9

Example of a sound speed profile showing two minima.

A sound speed profile with permanent double minima, due to the intrusion of warm saline Mediterranean water through the Strait of Gibraltar and into the north Atlantic, is found in the Bay of Biscay. In this case, the shallow and deep minima are typically at depths in the region of 500 and 1500 m, respectively; and the sound speed is lowest at the deeper minimum, implying that sound rays are refracted into it.

1.5.3 ARCTIC REGIONS

The SOFAR channel axis becomes progressively shallower at higher latitudes, eventually reaching the surface in the glaciated polar oceans [34], as illustrated in Fig. 1.10. Thus, there is no deep sound channel in the ice-covered Arctic Ocean

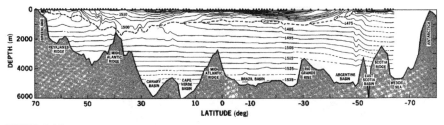

FIGURE 1.10

Sound channel structure in the north and south Atlantic along the 30.5°W meridian. The heavy dashed line depicts the axis of the sound channel and the contours of sound speed are in meters per second.

This figure is from Northrup, J. and Colborn, J. G., Sofar channel axial sound speed and depth in the Atlantic Ocean, J. Geophys. Res., **79**, pp. 5633–5641 (1974).

in the winter, since the minimum temperature is at the surface, where the water is close to freezing ($-2°$C) and the sound speed is about 1440 m/s. However, such a profile acts as a subsurface waveguide, supporting the transmission of sound to long distances with little attenuation [35]. In the absence of ice cover, summer heating may introduce a steep negative gradient into the profile immediately beneath the sea surface, giving rise to a minimum in the profile at a depth of the order of 100 m.

Although receding rapidly due to climate change, the ice cover currently extends from the Arctic Ocean, through the Fram Strait and along the east coast of Greenland, where it forms the Marginal Ice Zone (MIZ). Ice floes in the MIZ, which may be as large as several kilometers across, jostle together under the influence of wind and currents, with leads of open water between (Fig. 1.11). The region south of the Fram Strait, comprising the Greenland, Norwegian, and western Barents Seas, has been designated the Nordic Seas [36], with the term Nordic Basin representing the deep waters of the Nordic Seas. Throughout the Nordic Basin, and regardless of season, all the deep-water sound speed profiles are essentially the same at depths between 1000 and 1500 m; they increase linearly with depth, showing a gradient of $0.016\ \text{s}^{-1}$, and taking a value of about 1470 m/s at a depth of 1400 m. At depths shallower than 1000 m, the profiles are influenced predominantly by either the warm Atlantic water flowing northward through the Faroe-Shetland Channel or the cold Arctic inflow through the Fram Strait, depending on geographical location.

The transition from North Atlantic to Arctic characteristics occurs north of the Faroe-Shetland Channel, where the sound speed profile is typical of the warm North Atlantic water entering the Nordic Basin. At higher latitudes, the physical oceanography becomes quite complicated, with five distinct fronts occurring [37], where cold, southward moving Arctic water meets the northward inflow of warm North Atlantic water. The fronts are stable interfaces between the cold and warm water, the former being less saline but having more or less the same

FIGURE 1.11

Recently fractured ice floes and melt-water pools in the MIZ off the east coast of Greenland. The large central floe is several tens of meters across.

Aerial photograph taken by Buckingham, M. J.

density as the latter. With these reduced density gradients, the fronts in the Nordic Seas, are strong, permanent boundaries in the temperature and salinity fields. A common feature across all the fronts is a steep sound speed gradient in the horizontal, typically of the order of 10^{-4} s^{-1}. In Fig. 1.12, showing five frontal systems in the Nordic Seas, the Svalbard archipelago is top center, Norway is the large landmass to the right, Iceland is the island at bottom left, and Greenland is the large landmass to the left.

Within the Nordic Basin, the sound speed profiles fall broadly into three groups [37]. There are those that are primarily influenced by the warm North Atlantic water on the eastern side of the basin, between the Iceland Gap and Norwegian Coastal fronts. Then there are the profiles that are predominantly characterized by the influx of cold Arctic water from the north, between Svalbard and northern Greenland. The profiles in the third group are found in the transition region between the warm waters flowing to the north and cold, Arctic waters flowing to the south. A selection of summer and winter sound speed profiles for over a dozen locations throughout the Nordic Seas has been presented by Hurdle [38] in his Fig. 4.

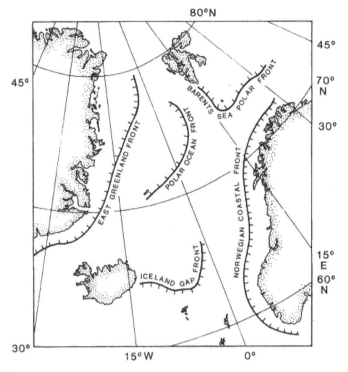

FIGURE 1.12

Five frontal systems in the Nordic Seas. The Svalbard archipelago is top center, Norway is the large landmass to the right, Iceland is the island at bottom left, and Greenland is the large landmass to the left.

This figure is from Johannessen, O. M., Brief Overview of the Physical Oceanography *in The Nordic Seas, Hurdle, B. G. Editor, Springer-Verlag, New York, pp. 103–127 (1986), with permission of Springer.*

Under-ice, winter sound speed profiles tend to increase monotonically with depth beneath the surface. Such a profile may be represented analytically by the inverse-square expression [35]:

$$\frac{1}{c^2(z)} = \frac{1}{c_\infty^2}\left[1 + \frac{z_1^2}{(z+z_s)^2}\right],\tag{1.7}$$

where c is the sound speed at depth z, c_∞ is the sound speed at great depth, z_1 is a measure of the depth of the duct, and z_s is the height of the virtual origin of the sound speed profile above the sea surface. This type of profile is upward refracting, acting as a subsurface acoustic waveguide supporting modal propagation. Fig. 1.13 shows a comparison of the inverse-square profile ($z_1 = 22.94$ m, $z_s = 100$ m, $c_\infty = 1462$ m/s) with sound speed data collected by Meredith et al. [39] from a station (latitude $78°44.0'$ N, longitude $2°56.0'$ W) in the MIZ, Greenland Sea. This was close to the ice edge between the east coast of Greenland and Svalbard (see Fig. 1.12).

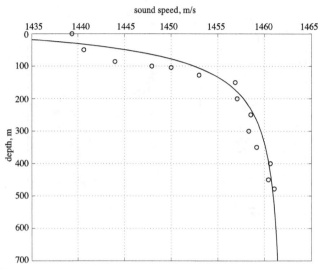

FIGURE 1.13

Comparison of the inverse-square profile, the *solid line*, with CTD data [38] (Cast 19B) from the MIZ, represented by the *circles*.

Two special cases of Eq. (1.7) are the linear profile, $c(z) \propto (z + z_s)$, obtained by setting z_1 proportional to c_∞ and letting c_∞ go to infinity, and the constant profile, which is returned when $z_1 = 0$. More generally, the inverse-square profile provides a useful model for investigating the modal structure of the sound field in the Polar Oceans, because this is one of the few profiles for which an exact, analytical solution of the wave equation exists [35]. This solution consists of a discrete component in the form of an infinite sum of normal modes plus a wavenumber integral representing the continuous component of the field.

The mathematical form of the normal modes in the inverse-square profile is rather unusual, given by Hankel functions of imaginary order and imaginary argument. The mth mode consists of an oscillatory region containing m turning points, and the amplitude of the oscillations shows a small increase just before extinction occurs. Below the extinction depth, the mode decays exponentially with depth. As the mode number, m, increases, the oscillatory part of the mode penetrates deeper into the ocean, with the extinction depth increasing exponentially with m.

Physically, the wavenumber integral for the continuous component of the field represents the direct and surface-reflected arrivals at the receiver. In the limit as $z_1 \to \infty$, representative of a homogenous medium with uniform sound speed, the normal modes all reduce to zero, leaving the wavenumber integral as the only nonzero component of the acoustic field. In this limiting situation, the wavenumber integral evaluates identically to the Lloyd's mirror solution for the field from a point source in a semi-infinite, iso-speed ocean with a pressure release boundary at the surface. In other words, the solution in this limiting case reduces to the field from a point source plus its negative image above the sea surface.

1.5.4 DEEP ISOTHERMAL LAYERS

Much of the water at great depths in the world's oceans originated at the surface in the polar regions of the Atlantic Ocean. Through such processes as evaporation and heat exchange with the atmosphere, the polar-ocean surface water acquires its temperature and salinity characteristics, becoming significantly denser. It is then transported along isopycnals, or contours of constant density, in a process known as thermohaline circulation, to the ocean depths at lower latitudes. Neglecting the effects of compressibility, the temperature and salinity are conservative properties in that their values can only change by mixing with water masses of different temperature and salinity.

Seawater, however, is slightly compressible, otherwise acoustic waves could not propagate through the ocean. If a parcel of seawater at depth were raised to the surface adiabatically, the resultant reduction in pressure would give rise to an increase in the volume and an associated decrease in the temperature of the parcel. The reduced temperature of the parcel at the sea surface is known as the potential temperature. Thus, at any given depth, the potential temperature, θ, is always less than the in situ temperature, T. Strictly, because of compressibility, the in situ temperature is not a conserved quantity, whereas the potential temperature is a conservative property of the ocean.

One effect of the compressibility of seawater on deep and bottom waters is to introduce a positive gradient in the in situ temperature profile at depths below about 5000 m. This is illustrated in Fig. 1.14, which shows an expanded view of the Tonga Trench temperature data in Fig. 1.3. The rate of increase in the in situ temperature with depth is extremely small and for practical purposes is often specified over a vertical distance of 1000 m, in which case it is called the adiabatic temperature gradient. It depends on the in situ temperature, the salinity, and the depth, although in the deep oceans the salinity does not differ much from 34.85‰ and anyway the effect of salinity on the adiabatic temperature increase is very small. For the Tonga Trench in situ temperature profile in Fig. 1.14, the adiabatic temperature gradient at a depth of 8000 m is 0.16°C/km.

Also shown in Fig. 1.14 is the potential temperature θ {unit: °C}, which is essentially constant below 5000 m, indicating that the deep ocean is isothermal with respect to potential temperature. No physics-based analytical expression for the potential temperature exists but an empirical, twelve-term polynomial, based on an integration of the adiabatic temperature gradient and a least-squares fit to experimental data, has been derived by Bryden [40]. The polynomial is of the form:

$$\theta = T - \sum_i \sum_j \sum_k A_{ijk} P^i (S - 35)^j T^k, \qquad (1.8)$$

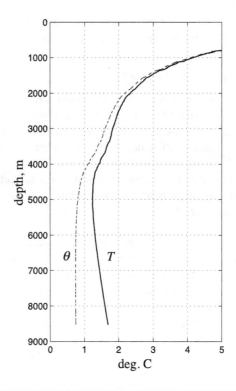

FIGURE 1.14

Temperature (T) and potential temperature (θ) profiles, Tonga Trench, South Pacific Ocean, September 2012.

where the in situ temperature is T {unit: $^{\circ}$C}, the salinity is S {unit: ‰}, the hydrostatic pressure is P {unit: dbar}, and the coefficients take the following values:

$$A_{100} = 0.36504 \times 10^{-4} \qquad A_{200} = 0.89308 \times 10^{-8}$$
$$A_{101} = 0.83198 \times 10^{-5} \qquad A_{201} = -0.31628 \times 10^{-9}$$
$$A_{102} = -0.54065 \times 10^{-7} \qquad A_{202} = 0.21987 \times 10^{-11}$$
$$A_{103} = 0.40274 \times 10^{-9} \qquad A_{210} = -0.41057 \times 10^{-10}$$
$$A_{110} = 0.17439 \times 10^{-5} \qquad A_{300} = -0.16056 \times 10^{-12}$$
$$A_{111} = -0.29778 \times 10^{-7} \qquad A_{301} = 0.50484 \times 10^{-14}.$$

The expression in Eq. (1.8), which was used to compute the potential temperature profile in Fig. 1.14, is valid for salinity in the range 30–40‰, in situ temperature between −2 and 30°C, and pressure from 0 to 10,000 dbar, with the maximum uncertainty in the returned value of θ estimated to be ±0.0256°C.

The potential temperature, rather than the in situ temperature, plays an important role in determining the stability of the water column. In a stratified ocean, the stability is a measure of the resistance to vertical displacement of a parcel of water at any point in the water column. The stability is controlled by the density gradient, and the density depends on the temperature, salinity, and pressure, as expressed by the equation of state. As with the potential temperature, there is no physics-based analytical expression for the equation of state but an empirical polynomial containing 15 terms has been derived by Millero and Poisson [41] from a least-squares fit to experimental data. For the present discussion, however, a simpler, approximate expression for the equation of state, in which all the non-linear terms are neglected, is adequate:

$$\rho - \rho_0 \approx -\alpha(T - T_0) + \beta(S - S_0) + \gamma P, \tag{1.9}$$

where the density, ρ, is in kg/m^3. Taking the values for the constants as $\rho_0 = 1027$ kg/m^3, $T_0 = 10°$C, $S_0 = 35‰$, $\alpha = 0.15$ kg/m^3 per $°$C, $\beta = 0.78$ kg/m^3 per $‰$ salinity, and $\gamma = 0.0045$ kg/m^3 per dbar, the expression in Eq. (1.9) returns the density to within ± 0.5 kg/m^3 or equivalently about 1 part in 2000.

Following a line of thought similar to that for potential temperature, the potential density is defined as the density of a parcel of water after it has been transported adiabatically from its position at depth in the water column to the surface. According to this definition, the potential density, ρ_θ, may be obtained from Eq. (1.9) by setting the pressure equal to zero and replacing the in situ temperature T with the potential temperature θ:

$$\rho_\theta - \rho_0 = -\alpha(\theta - T_0) + \beta(S - S_0), \tag{1.10}$$

It is evident from this expression that the potential density is the in situ density but with the effects of compressibility removed.

Since the effects of compressibility associated with the hydrostatic pressure do not affect the stability, the potential density rather than the in situ density may be used to estimate the stability of the water column. This may be understood by imagining a column of water with uniform temperature and salinity profiles and a height equal to the depth of the ocean. Clearly, the in situ density increases in proportion to depth in the column. If a parcel of water is moved hypothetically from one depth to another, the in situ density of the parcel at its new location is the same as the in situ density of its neighbors. No buoyancy forces act on the parcel tending to make it rise or fall, indicating that changes in the in situ density associated with pressure do not affect the stability of the water column.

Assuming that the compressibility of seawater is independent of temperature (which strictly is not the case), a useful approximate measure of the stability, E, is the normalized gradient of the potential density with respect to depth, z:

$$E \approx \frac{1}{\rho_\theta} \frac{\partial \rho_\theta}{\partial z} \approx \frac{1}{\rho_\theta} \left\{ -\alpha \frac{\partial \theta}{\partial z} + \beta \frac{\partial S}{\partial z} \right\}, \tag{1.11}$$

where the term in parenthesis follows from Eq. (1.10). For the water column to be stable, E should be positive. In the deep water of the Tonga Trench, below about 5000 m, the salinity, shown in Fig. 1.3, is constant in depth and so too is the potential temperature, shown in Fig. 1.14, signifying that both differential terms in Eq. (1.11) are zero.

In this case, Eq. (1.11) returns $E = 0$, leading to the conclusion that at these depths, to the level of approximation in Eq. (1.11), the water column is neutrally stable.

This result, however, bears further examination, since the water column at great depth is expected to be stable. It should be borne in mind that, due to the very low compressibility of seawater, density changes in the deep ocean are extremely small, as a consequence of which the linear approximation for the equation of state in Eq. (1.9) is not sufficiently accurate for the stability calculation. Moreover, the approximation for the stability in Eq. (1.11), in which the temperature dependence of the compressibility is neglected, is also insufficiently precise and needs to be replaced with the exact expression:

$$ E = \frac{1}{\rho} \frac{\partial \rho}{\partial z} - \frac{g}{c^2}, \tag{1.12} $$

where ρ is the in situ density, c is the speed of sound in seawater, and $g = 9.81 \text{ m/s}^2$ is the acceleration due to gravity. Eq. (1.12) is derived by considering the vertical acceleration of a parcel of water that has been displaced from its equilibrium position. For the Tonga Trench, using the Millero and Poisson [40] empirical polynomial for the equation of state, with the SVX data shown in Fig. 1.4 for the sound speed, Eq. (1.12) returns the stability profile shown in Fig. 1.15A. Although these computed

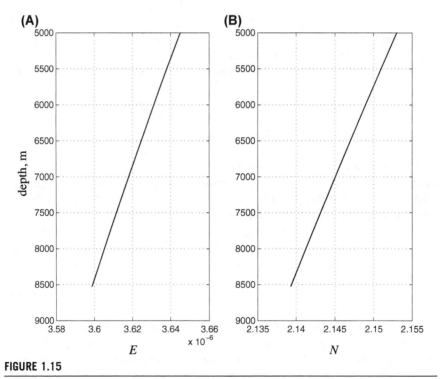

FIGURE 1.15

Tonga Trench stability profiles. (A) Stability, E, from Eq. (1.12) and (B) Brunt–Väisälä frequency, N, from Eq. (1.13).

values of E are small, all are positive, signifying a stable stratification in which vertical movements of seawater are suppressed.

The stability of the ocean is often expressed in terms of the stability frequency, N, also known as the buoyancy frequency and the Brunt–Väisälä frequency:

$$N = \sqrt{gE}. \qquad (1.13)$$

In effect, N is the frequency of vertical oscillation of a parcel of water about its equilibrium position in the water column, and as such, it represents the maximum frequency that can be exhibited by internal waves. Typical values of N in the abyssal oceans and ocean trenches are of the order of a few cycles per hour, as illustrated in Fig. 1.15B, which shows N as computed for the Tonga Trench using the CTD and SVX data shown in Figs. 1.9 and 1.10. Over the limited depth range shown in Fig. 1.15B the Brunt–Väisälä frequency scales almost linearly with the depth, which is consistent with a Taylor expansion of the exponential stratification, expressed in Eq. (1.4), of the canonical Munk sound speed profile [30].

1.5.5 EXPRESSIONS FOR THE SPEED OF SOUND

An SVX based on the sing-around design of Greenspan and Tschiegg [28], provides a direct measure of the speed of sound in a fluid (seawater in the case of the ocean). On rare occasions, sound speed profiles in the deep ocean have been obtained using an SVX [24], but more commonly the sound speed profile is computed from temperature and salinity data or from temperature data in conjunction with an estimated salinity profile. Temperature and salinity profiles may be recovered with a CTD probe or they may be obtained from the vast repository of historical records covering most regions of the world's oceans. Temperature profiles alone are often measured using a ship-deployed expendable bathy–thermograph (XBT) or an aircraft-deployed expendable bathythermograph (AXBT). Once the temperature and salinity profiles have been determined, by whatever means, they can be used to compute the sound speed profile.

The speed of sound in a liquid is given by the expression:

$$c = \sqrt{\frac{1}{\rho K}}, \qquad (1.14)$$

where ρ is the density and K is the adiabatic compressibility of the medium. For seawater, both these quantities show a complicated dependence on temperature and pressure (or depth) and, because seawater is an electrolyte, the compressibility depends on the salinity in a complicated fashion. Ideally, theoretical expressions based on fundamental physical mechanisms would be used to compute ρ and K, and hence the sound speed from Eq. (1.14), but, since no such expressions exist, an alternative approach has been developed.

Instead of physics-based equations, the sound speed is expressed in terms of an empirical equation, a polynomial in the temperature, the salinity, and the hydrostatic pressure or depth. Several such empirical relationships have been formulated [42–52], some dating back to the early 1960s. They were derived from accurate measurements, made under controlled laboratory conditions, of sound speed as a function of temperature, salinity, and pressure. Mackenzie [53] compared these equations with data from many sources and concluded that the empirical equation derived by Del Grosso and Mader [43] is the preferred choice. A similar conclusion was reached by Millero and Kubinski [54] from their own measurements of the speed of sound in standard seawater.

Del Grosso and Mader [43] made careful interferometer measurements (627 points) of the sound speed in standard seawater as a function of temperature over the range 0–40°C, salinity from 30‰ to 41‰, and pressure from 0 to 103.42 MPa. They made a stepwise regression fit of the 627 measurements to a polynomial of 23 variables. Their resultant expression for the sound speed shows an estimated standard error of 0.044 m/s and a mean difference of −0.00001 m/s with respect to the data set. This equation can be written as follows:

$$c = c_0 + a_1 T + a_2 T^2 + a_3 T^3 + a_4 T^4 + a_5 (S - 35) + a_6 P^2 + a_7 P^3$$
$$+ a_8 T(S - 35) + a_9 T(S - 35)^3 + a_{10} T^2 (S - 35) + a_{11} TP$$
$$+ a_{12} TP^2 + a_{13} TP^3 + a_{14} T^2 P^2 + a_{15} T^3 P + a_{16} (S - 35)P$$
$$+ a_{17} (S - 35)P^2 + a_{18} (S - 35)P^3 + a_{19} (S - 35)^2 P + a_{20} (S - 35)^2 P^2$$
$$+ a_{21} (S - 35)^3 + a_{22} T(S - 35)P + a_{23} T^2 (S - 35)P,$$

$$(1.15)$$

where c and c_0 are in m/s, T is temperature in degrees Celsius, S is salinity in parts per thousand, and P is pressure converted to kg/cm^2 using the acceleration due to gravity, $g = 9.80665$ m/s^2. The coefficients in Eq. (1.15) are

$c_0 = 0.140194964197 \times 10^4$	$a_{12} = -0.131201344149 \times 10^{-5}$
$a_1 = 0.500642920686 \times 10^1$	$a_{13} = 0.402773046625 \times 10^{-9}$
$a_2 = -0.583540720391 \times 10^{-1}$	$a_{14} = 0.144431845812 \times 10^{-7}$
$a_3 = 0.349631461597 \times 10^{-3}$	$a_{15} = -0.101232803918 \times 10^{-5}$
$a_4 = -0.164875703289 \times 10^{-5}$	$a_{16} = 0.157450572397 \times 10^{-1}$
$a_5 = 0.134606695473 \times 10^{-1}$	$a_{17} = 0.533284097890 \times 10^{-5}$
$a_6 = -0.593403057954 \times 10^{-4}$	$a_{18} = 0.353545616177 \times 10^{-9}$
$a_7 = -0.209170685764 \times 10^{-7}$	$a_{19} = -0.522655412446 \times 10^{-3}$
$a_8 = -0.112689485296 \times 10^{-1}$	$a_{20} = -0.858535445194 \times 10^{-7}$
$a_9 = -0.583732276240 \times 10^{-6}$	$a_{21} = 0.572274443678 \times 10^{-5}$
$a_{10} = 0.103658662691 \times 10^{-3}$	$a_{22} = -0.117388164634 \times 10^{-4}$
$a_{11} = 0.595900933419 \times 10^{-3}$	$a_{23} = 0.597262459578 \times 10^{-6}$

Del Grosso and Mader [43] point out that less precision than is provided by Eq. (1.15) may be adequate for certain applications. To this end, in addition to Eq. (1.15), they fitted the same data set to polynomials containing 18, 16, and

11 variables (powers of temperature, salinity, and pressure or combinations thereof). These reduced equations return a slightly larger standard error but are sufficiently accurate for many practical purposes.

Mackenzie [47] used the Del Grosso and Mader [43] 18-term equation as "ground truth" to help establish a simpler nine-term equation for the sound speed. This nine-term equation, which has an estimated standard error of 0.070 m/s, has received wide acceptance, taking the form:

$$c = 1448.96 + 4.591T - 5.304 \times 10^{-2}T^2 + 2.374 \times 10^{-4}T^3$$
$$+ 1.340(S - 35) + 1.630 \times 10^{-2}D + 1.675 \times 10^{-7}D^2, \tag{1.16}$$

where c is sound speed in m/s, T {unit: °C} is temperature, S is salinity in parts per thousand, and D {unit: m} is depth. From a practical point of view, the appearance of depth rather than pressure in Eq. (1.16) is convenient, since most historical data recovered from the ocean are in terms of depth.

Eq. (1.16) is valid to depths of 8000 m, for temperatures ranging from −2 to 30°C and salinities from 25 to 40‰. Another simplified equation for the sound speed, containing just seven terms, has been developed by Medwin [55], but it is valid only to depths of less than 1000 m. Mackenzie [47] suggests that Eq. (1.16) rather than longer polynomials is satisfactory for many applications, since the input data from instruments such as AXBTs exhibit errors which exceed those of the equation itself. This is especially true of archival, deep profiles, which often display significant uncertainties. Unless extreme accuracy for research is required, Mackenzie [47] recommends the use of Eq. (1.16).

If, however, an empirical equation is chosen, such as the Del Grosso and Mader [43] Eq. (1.15) that involves pressure rather than depth, then, with data expressed in terms of depth, a conversion from depth to pressure is required. Mackenzie [47] describes a conversion procedure in which the pressure is expressed as an integral of the product of the *in situ* density and the acceleration due to gravity, both of which depend on depth and geographical location. He approximates the acceleration due to gravity by a simple function of latitude and depth, noting that a more complicated equation changed the computed sound speed by only 0.1 m/s at a depth of 10,000 m. The density is represented by an empirical expression involving a ratio of summations of temperature-dependent terms involving about 30 numerical coefficients. Although Mackenzie's formulation [47] for the pressure neglects time-varying factors such as currents, tides, and surface pressures due to storms, as well as ignoring localized anomalies in the gravitational field, it nevertheless provides an accuracy in the computed sound speed to within 0.1 m/s; hence it could be satisfactory for many practical applications where the utmost precision is not required.

Since the early 1980s, there has been a concerted effort to construct a consistent set of algorithms for computing various derived properties of seawater, including the sound speed, from the basic observations of conductivity, temperature, and pressure. In 1983, Fofonov and Millard published a report [56], endorsed by UNESCO, in which the practical salinity scale (PSS-78) and the international equation of state for seawater [57] (EOS-80) are used as the basis for computing the speed of sound along with a number of other properties of seawater. The SeaWater library of EOS-80 is now obsolete, having been superseded by the Gibbs SeaWater (GSW) Oceanographic Toolbox of the International Thermodynamic Equation Of Seawater 2010 [58] (TEOS-10). This new standard has been adopted by the Intergovernmental Oceanographic Commission (IOC), the International Association for the Physical Sciences of the Oceans (IAPSO), and the Scientific Committee on Oceanic Research (SCOR), and has been endorsed by the International Union of Geodesy and Geophysics (IUGG).

The main difference between the old EOS-80 and TEOS-10 is in the treatment of salinity. The dimensionless "Practical Salinity" of EOS-80 has been replaced in TEOS-10 with "Absolute Salinity," which is a mass fraction with units of g/kg. Although the adoption of Absolute Salinity leads to enhanced precision in the computation of certain properties of seawater, the mathematical equations of the new TEOS-10 standard are complicated, containing many coefficients that are specified to 16 significant digits. It is not recommended that these equations be programmed by users; instead, the appropriate codes in MATLAB, C, FORTRAN, and Visual Basic can be found at www.teos-10.org.

With regard to computing sound speed from measurements of conductivity, temperature, and pressure, the deep water profiles from TEOS-10 and from Del Grosso and Mader [43] are the same, within 0.1 m/s, to a depth of 8000 m. Over the same depth range, Mackenzie's 9-term equation [47] differs form TEOS-10 by less than 0.2 m/s.

1.5.6 SURFACE WAVES

When wind blows over the sea surface, waves are created which gain in amplitude as the wind strengthens and the fetch increases. Eventually, with a sufficiently strong wind and long fetch, a state of equilibrium is reached whereby waves break and energy is dissipated as fast as it is generated. This condition is said to represent a *fully developed sea.*

Immediately beneath the rough sea surface, a bubble layer, formed by wave breaking, is of variable thickness but can reach 10 m or more, depending on the wind speed [59]. Since bubbles act as efficient scattering centers, this sub-surface bubble layer may increase acoustic propagation loss significantly [60]. The bubbles also produce sound. Immediately after formation, a bubble is not in equilibrium with its surroundings but attains equilibrium by exhibiting radial oscillations, or "ringing," for the first few tens of milliseconds of its existence [61]. In effect, the bubble acts briefly as an acoustic monopole, which is a highly efficient radiator of sound. The superposition of all such acoustic bubble pulses from wave breaking constitutes the major component of wind-generated ambient noise in the ocean [62–64].

The concentration of bubbles beneath the surface correlates with the displacement of the wind-driven sea surface. The surface elevation is a stochastic process, varying randomly in space and time, with a Gaussian or "normal" probability distribution. At a fixed point on the surface, taking the time-dependent displacement about the mean as $z(t)$, the mean-square value, or variance, is $\overline{z^2(t)}$, where the overbar denotes either a long time average or an ensemble average, the two being the same since the system is ergodic. In general, the variance, which is proportional to the average energy per unit area of sea surface, depends on the position of the fixed point on the surface. With a fully developed sea, however, the surface wave field is spatially homogenous, that is to say, the second-order statistical properties of the displacement, including the variance, are independent of position on the sea surface.

On casually observing the sea surface, it quickly becomes evident that waves of different frequencies have different amplitudes. The spectral characteristics of the sea surface are expressed through the (unilateral) surface wave spectrum, $S(\omega)$, which is defined in terms of the Fourier transform, $Z(\omega)$, of the surface displacement as:

$$S(\omega) = 2\frac{\overline{|Z(\omega)|^2}}{T} \tag{1.17}$$

In this expression, the overbar denotes an ensemble average, T is the observation time of the Fourier transforms contributing to the average, and $\omega \geq 0$ is angular frequency. The factor of 2 in Eq. (1.17) appears because $S(\omega)$ is a single-sided spectrum in which the negative frequency components have been folded into the positive frequencies. In the SI system, the units of $S(\omega)$ are m^2/Hz. The variance, $\overline{z^2(t)}$, is related to $S(\omega)$, through the Wiener−Khintchine theorem [65]:

$$\overline{z^2(t)} = \frac{1}{2\pi} \int_0^\infty S(\omega)d\omega, \tag{1.18}$$

which is consistent with the idea that the variance represents the total energy of the wind-driven wave system. In general, the moments m_n of the power spectrum are defined as:

$$m_n = \int_0^\infty \omega^n S(\omega)d\omega \tag{1.19}$$

and thus the variance of the surface displacement is:

$$\overline{z^2(t)} = \frac{m_0}{2\pi}. \tag{1.20}$$

The condition of the sea surface is commonly expressed as the "sea state," which is a measure of the average wave height, that is, the vertical distance between a crest and a trough. In practice, the sea state is often estimated from a visual observation of the sea surface, but it seems that even experienced observers unwittingly neglect the smaller, higher-frequency waves when arriving at their estimates of the average wave height, since their subjective estimates are less than the values returned by direct measurements. A parameter that corresponds closely to the observers' estimates of the sea state is the *significant wave height*, which is defined as:

$$H_s = 4(m_0)^{1/2}. \tag{1.21}$$

In modern usage, the significant wave height, H_S, has replaced an older measure of the sea state, $H_{1/3}$, the average crest-to-trough height of the highest one-third of the waves, because $H_{1/3}$ is not easily related to any of the standard statistical measures, such as the variance, of the sea surface displacement. For typical wind-driven seas, $H_{1/3}$ is a little less than H_S but by no more than 10%.

The wind-driven wave spectrum exhibits a peak in the frequency band between 0.04 and 0.15 Hz, with higher wind speed corresponding to a lower peak frequency. On the high-frequency side of the peak, the small-scale frequency components are in statistical equilibrium, a requirement for the crests to remain attached to the waves. From a dimensional analysis for a fully developed sea, Phillips [66] has argued that, for the equilibrium range, the wave spectrum decays as ω^{-5}. This high-frequency tail has been incorporated as a feature of the wave-height spectrum of a fully developed sea derived by Pierson and Moskowitz [67] on the basis of the similarity hypothesis of Kitaigorodskii [68]. The essential idea underlying their technique is that the power spectra of fully developed seas, when plotted in a certain dimensionless form, should all be the same. By curve fitting to dimensionless spectral data, rather than from theoretical analysis, they were able to develop an analytical expression giving the wind-speed dependence of the surface-wave spectrum.

The Pierson—Moskowitz wave spectrum can be written in terms of the angular frequency, ω, of the surface-wave Fourier components as:

$$S_{PM}(\omega) = \alpha \frac{g^2}{\omega^5} exp\left[-\beta\left(\frac{g}{U\omega}\right)^4\right], \tag{1.22}$$

where $g = 9.81$ m/s^2 is the acceleration due to gravity, U is the mean wind speed at a height of 19.5 m (the elevation of the anemometers on the weather ships reporting the wind speed data), and $\alpha = 0.0081$ and $\beta = 0.74$ are dimensionless constants. Since $S_{PM}(\omega)$ represents a fully developed sea, it involves only one parameter, the wind speed, but not the fetch. The Pierson—Moskowitz spectrum exhibits a single maximum at a frequency $f_{max} = \omega_{max}/2\pi$, which, by differentiating Eq. (1.22) with respect to ω and equating the result to zero, is found to scale inversely with the wind speed:

$$\omega_{max} = \frac{g}{U}(0.8\beta)^{1/4}. \tag{1.23}$$

According to Eq. (1.23), the quantity $\omega_{max}U/g$ is a constant, which is one of the properties that must hold as a consequence of the Kitaigorodskii similarity argument [68]. When the expression for the peak frequency is substituted back into Eq. (1.22), the exponential function becomes independent of U, and the peak in the wave spectrum scales as the wind speed to the power of five, $S_{PM}(\omega_{max}) \propto \omega_{max}^{-5} \propto U^5$. Since the full expression for the peak in the wave spectrum is:

$$S_{PM}(\omega_{max}) = \alpha \frac{g^2}{\omega_{max}^5} exp\left[-\beta\left(\frac{g}{U\omega_{max}}\right)^4\right], \tag{1.24}$$

a normalized version of the Pierson–Moskowitz wave spectrum can be formulated as:

$$\tilde{S}_{PM}(\tilde{\omega}) = \frac{S_{PM}(\omega)}{S_{PM}(\omega_{max})} = \frac{exp\left[-1.25\left(\tilde{\omega}^{-4} - 1\right)\right]}{\tilde{\omega}^5}, \tag{1.25}$$

where $\tilde{\omega} = \omega/\omega_{max}$ is normalized frequency. An advantage of the normalization in Eq. (1.25) is that U does not appear as a parameter in the expression for the dimensionless wave spectrum, the universal, asymmetrical shape of which is shown in Fig. 1.16.

An integration of the spectrum in Eq. (1.22) over all positive frequencies can be performed with the aid of a simple substitution, yielding the moment m_0 as:

$$m_0 = \int_0^\infty S(\omega)d\omega = \frac{\alpha U^4}{4\beta g^2}. \tag{1.26}$$

The corresponding significant wave height is:

$$H_{sPM} = 4m_0^{1/2} = 2\frac{U^2}{g}\sqrt{\frac{\alpha}{\beta}}, \tag{1.27}$$

which is quadratic in the wind speed, as it must be for consistency with Kitaigorodskii's similarity analysis. Fig. 1.17 illustrates the wind-speed dependence of the peak frequency in Eq. (1.23) and the significant wave height in Eq. (1.27).

In reality, it is rare for intense storms to last long enough and to extend over a sufficiently large area to raise a fully developed sea. To achieve such a condition,

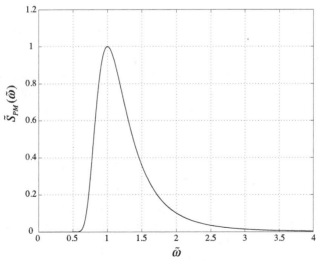

FIGURE 1.16

The normalized Pierson–Moskowitz spectrum.

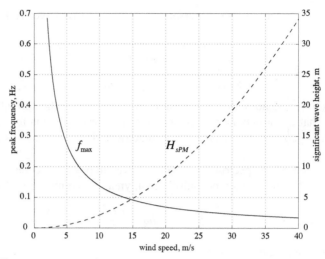

FIGURE 1.17

The peak frequency and significant wave height as functions of wind speed.

a strong wind requires a longer time and a longer fetch than a light wind, and even under light winds, most observed seas are not fully developed. With a wind speed of 25 m/s (about 50 knots), for the sea to be fully developed, the wind must blow for some 3 days over a fetch of 2400 km, circumstances that would be very unusual. With a less than fully developed sea, there is a low-frequency cut-off and the area under the wave spectrum is reduced, corresponding to a smaller average wave height than that under the fully developed condition.

To investigate the effect of a limited fetch on the wave spectrum, a series of wave height measurements was conducted at various distances from the shore in the Joint North Sea Project (JONSWAP) experiment [69,70]. Offshore wind conditions prevailed on a number of occasions during the six-week period of the experiment, allowing a fetch-limited spectrum to be constructed. The JONSWAP spectrum can be formulated in terms of a modified version of the Pierson—Moskowitz expression in Eq. (1.22):

$$S_J(\omega) = \left\{ \hat{\alpha} \frac{g^2}{\omega^5} exp\left[-1.25 \left(\frac{\omega_{Jmax}}{\omega} \right)^4 \right] \right\} \gamma^s, \qquad (1.28)$$

where ω_{Jmax} is the angular frequency of the spectral peak and $\gamma = 3.3$ is a constant, known as the *peak enhancement factor*, with the frequency-dependent exponent:

$$s = exp\left[-\frac{(\omega - \omega_{Jmax})^2}{2\sigma^2 \omega_{Jmax}^2} \right]. \qquad (1.29)$$

Two of the parameters in these expressions, ω_{Jmax} and $\widehat{\alpha}$, depend on both the wind speed at an elevation of 10 m above the sea surface, U_{10}, and the fetch, X. The following empirical expressions have been presented by Hasselmann et al. [71]:

$$\omega_{Jmax} = 2\pi \left\{ 3.5 \frac{g}{U_{10}} \left(\frac{gX}{U_{10}^2} \right)^{-0.33} \right\} \tag{1.30}$$

and

$$\widehat{\alpha} = 2\pi \left\{ 0.076 \left(\frac{gX}{U_{10}^2} \right)^{-0.22} \right\}. \tag{1.31}$$

The remaining parameter, σ, in the expression for the JONSWAP spectrum takes different values in upper and lower frequency ranges:

$$\sigma = \begin{cases} 0.07 & \text{for } \omega \leq \omega_{Jmax} \\ 0.09 & \text{for } \omega > \omega_{Jmax} \end{cases} \tag{1.32}$$

At the peak frequency, the value of the JONSWAP spectrum is:

$$S_J(\omega_{Jmax}) = \widehat{\alpha} \frac{g^2}{\omega_{Jmax}^5} \gamma e^{-5/4} \tag{1.33}$$

which allows the normalized spectrum to be written as:

$$\widetilde{S}_J(\widetilde{\omega}) = \frac{S_J(\omega)}{S_J(\omega_{Jmax})} = \frac{\gamma^{(\widetilde{s}-1)}}{\widetilde{\omega}^5} exp\left[-1.25 \left(\widetilde{\omega}^{-4} - 1 \right) \right], \tag{1.34}$$

where

$$\widetilde{\omega} = \frac{\omega}{\omega_{Jmax}} \tag{1.35}$$

and

$$\widetilde{s} = exp\left[-\frac{(\widetilde{\omega} - 1)^2}{2\sigma^2} \right]. \tag{1.36}$$

A comparison of the normalized JONSWAP and Pierson−Moskowitz spectra is shown in Fig. 1.18. At high frequencies, both spectra decay asymptotically as ω^{-5} but the JONSWAP peak is considerably narrower than that of the Pierson−Moskowitz spectrum. It should be noted that the peak frequencies, ω_{max} and ω_{Jmax}, are not the same, nor are the peak heights, $S_J(\omega_{Jmax})$ and $S_{PM}(\omega_{max})$; the peaks

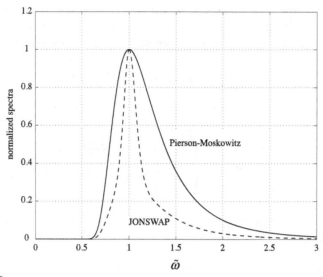

FIGURE 1.18

The normalized JONSWAP and Pierson—Moskowitz spectra.

appear as coincident in Fig. 1.18 solely because of the normalization. The ratio of the peak frequencies is:

$$F = \frac{\omega_{Jmax}}{\omega_{max}} = 25.1 \left(\frac{gX}{U_{10}^2}\right)^{-0.33} \tag{1.37}$$

and the ratio of the peak heights is:

$$R = \frac{S_J(\omega_{Jmax})}{S_{PM}(\omega_{max})} = \frac{\widehat{\alpha}}{\alpha}\gamma\left(\frac{\omega_{max}}{\omega_{Jmax}}\right)^5 = 5.91 \times 10^{-6}\gamma\left(\frac{gX}{U_{10}^2}\right)^{1.43} \tag{1.38}$$

where, for the purpose of the comparison, the wind speed U_{10} in Eqs. (1.30) and (1.31) has been taken as equal to the wind speed U in the Pierson—Moskowitz spectrum. Fig. 1.19 shows F and R plotted as functions of the fetch for a wind speed of 10 m/s.

As the fetch increases, it can be seen in Fig. 1.19 that the peak frequency of the JONSWAP spectrum approaches asymptotically the Pierson—Moskowitz value for a fully developed sea. The magnitude of the JONSWAP peak, however, does not converge to that of the Pierson—Moskowitz spectrum in the limit of a long fetch. For such convergence to occur, the peak enhancement factor, γ, which has been treated as a constant, would have to become fetch dependent, reducing to unity as the fetch increased without limit. This issue has been discussed by Hasselmann et al. [71], who consider that the transition to a fully developed spectrum occurs

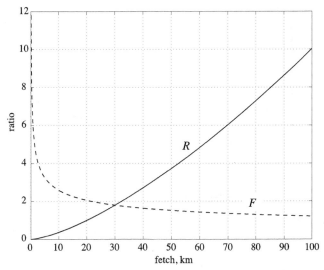

FIGURE 1.19

Ratios of the peak frequencies and peak heights in the JONSWAP and Pierson–Moskowitz wave spectra as functions of the fetch for $U_{10} = 10$ m/s.

in the very final stages of development. For practical purposes, since most seas rarely become fully developed, especially under very high winds, the JONSWAP model, either the version presented above or a slightly refined variant [72–75], provides a reasonable representation of the sea surface wave spectrum.

1.5.7 INTERNAL WAVES

When two fluids of different densities are in contact, a wave can propagate along the interface between them. Familiar examples of this phenomenon include sea surface waves, where the two fluids in question are seawater and air, and oil-and-water toys, which, on being tilted, display slow-motion waves flowing along the interface. At a very basic level, the denser fluid pushes the lighter fluid around when the system is disturbed.

For many purposes, the density of the ocean may be satisfactorily approximated as uniform throughout. The ocean, however, is not homogenous but exhibits a temperature and salinity structure, associated with which are density variations in the vertical and the horizontal, as expressed by the equation of state in Eq. (1.9). Although the range of densities observed in the ocean is small, amounting to just few parts in a thousand, the stratification of the density can have a profound effect on ocean processes, as well as supporting internal waves [76].

A simple two-layer model of the ocean is shown Fig. 1.20A, where the upper and lower layers have densities ρ_1 and ρ_2, respectively. The abrupt change in density at the boundary between the layers could be associated with a thermocline. If a parcel

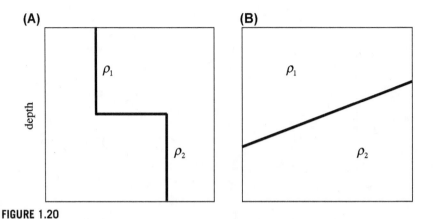

FIGURE 1.20

(A) Two-layer ocean with an abrupt density discontinuity. (B) Disturbed two-layer system with inclined interface.

of water from the lower layer is moved up into the upper layer, then the restoring force, F, is the difference between the weight and the buoyancy, which, neglecting the effect of compressibility, is:

$$F = (\rho_1 - \rho_2)g = \rho_2 g^*,\tag{1.39}$$

where

$$g^* = \frac{(\rho_2 - \rho_1)}{\rho_2} g \tag{1.40}$$

is the *reduced gravity*.

When the two-layer system is disturbed, a wave, known as an *interfacial internal wave*, propagates along the density discontinuity. In Fig. 1.20B, where the interface is shown inclined to the horizontal, the pressure gradient force due to the slope is balanced by the restoring force, but both are scaled down, as are the associated accelerations, by the reduced gravity. For small amplitude displacements the wave is linear, and assuming that both layers are infinitely deep, the wave speed, C, is given by the expression [77]:

$$C^2 = \frac{\rho_2 - \rho_1}{\rho_2 + \rho_1}\left(\frac{g}{\kappa}\right),\tag{1.41}$$

where κ is the wavenumber. When the lower layer is very much denser than the upper layer, as is the case at the sea surface, where the fluid media either side of the interface are seawater and air, then $g^* \approx g$ and the wave speed in Eq. (1.41) reduces correctly to:

$$C^2_{surface} \approx \frac{g}{\kappa}.\tag{1.42}$$

At depth in the ocean, where the density difference between adjacent layers is typically in the region of $1-3$ kg/m^3, the reduced gravity, g^*, is of the order of $g/1000$ and the wave speed is given by:

$$C^2_{internal} \approx \frac{g^*}{2\kappa} \ll C^2_{surface}. \tag{1.43}$$

The inequality in Eq. (1.43) accounts for the slow-motion behavior of internal waves, not only in oil-and-water toys, but also in the ocean, where internal-wave speeds are typically in the region of 0.1 m/s, as opposed to 10 m/s or so for surface gravity waves.

The idealized, discontinuous density profile in Fig. 1.20A may approximate the density structure through a thermocline, but more generally, the density in the ocean changes continuously with depth. It can also vary in the horizontal, due, for example, to flow over bathymetry and outflow of fresh river water into the ocean. In a horizontally stratified ocean in which the density, $\rho(z)$, varies continuously with depth, z, internal waves may propagate not only horizontally but at any angle to the horizontal, including vertically. All that is required is a density gradient across the propagation path. The maximum frequency of such internal waves may be obtained by considering the small vertical displacement of a parcel of fluid, Δz, in a water column in hydrostatic equilibrium. With z representing depth, the restoring force produces an acceleration:

$$\frac{d^2 \Delta z}{dt^2} = -g^* = -g \frac{\rho(z + \Delta z) - \rho(z)}{\rho(z)}, \tag{1.44}$$

which can be written as:

$$\frac{d^2 \Delta z}{dt^2} \approx -\frac{g}{\rho(z)} \left(\frac{d\rho}{dz} \Delta z \right) = -N^2 \Delta z, \tag{1.45}$$

whose solutions for the excursions in displacement are of the form:

$$\Delta z \propto e^{\pm iNt} \quad . \tag{1.46}$$

In these expressions, N, the Brunt−Väisälä or buoyancy frequency, introduced in Eq. (1.13), is given by:

$$N = \left(\frac{g}{\rho} \frac{d\rho}{dz} \right)^{1/2}. \tag{1.47}$$

For internal waves that are not propagating horizontally but whose lines of constant phase are inclined at an angle θ to the vertical, the buoyancy frequency becomes $N\cos\theta$, indicating that N is indeed the maximum possible frequency of oscillation. Typical values of N for the deep ocean are around 2 cph, as shown in Fig. 1.15B, corresponding to wave periods measured in tens of minutes to hours, compared with seconds for surface waves.

Small amplitude internal waves obeying the linear wave equation (Eq. 1.45) for the displacement of surfaces of constant density, or *pycnoclines*, are commonplace in the ocean. These linear internal waves may be produced, for instance, by tidal flow over a sloping seabed; and they propagate randomly in all directions, acting as acoustic scattering agents that redistribute acoustic energy more or less isotropically in the ocean. They also affect the performance of acoustic arrays, since they impose an upper limit on the coherence length of the incident sound field [78].

Large amplitude, highly non-linear internal waves with shorter wavelengths, known as *solitons*, may also be present in the ocean, notably in shallow water. Among other effects [79], solitons act as discrete sound-scattering centers, often introducing a strong azimuthal dependence into the scattered acoustic field. Solitons may also give rise to acoustic mode coupling when the propagation directions of the acoustic field and the internal wave are parallel [80]. On occasion in shallow water, a sequence of solitons, known as a solibore, is observed, taking the form of parallel bands, or wave fronts, of alternating high and low sound speed, with the whole system moving through the ocean along a track normal to the wave fronts. Such a soliton wave train was observed in the Shallow Water Acoustics in Random Media experiment 1995, SWARM-95 [81], conducted in the Mid-Atlantic Bight on the continental shelf off the coast of New Jersey. An interesting observation that emerged from the experiment was that sound propagating from a source to a receiver oriented on a line parallel to the soliton wave fronts can become trapped between the ridges of high sound speed, which act as barriers forming an acoustic duct.

1.5.8 BUBBLES FROM WAVE BREAKING

When a wave breaks on the sea surface, it engulfs a parcel of air, which is fragmented and advected downwards by turbulence, Langmuir circulation [82], and other mechanisms to form an aerated layer immediately beneath the surface. The thickness of this bubble layer is highly irregular but can extend to several tens of meters [83,84]; and the radii of the bubbles within the layer range from about 10 to 500 μm. In the active part of the breaking wave, the steady rate of air entrainment leads to a cascade in which turbulent velocity gradients split larger bubbles into smaller offspring. Scaling arguments [85] lead to a bubble size distribution in the form of a power law:

$$N(a) = N_0 a^{-10/3}, \tag{1.48}$$

where a {unit: μm} is bubble radius, N is the number of bubbles per unit volume with radii between a and $a + da$, and N_0 is the numerical value of N for bubbles of unit radius.

Eqs. (1.48) represents the initial bubble size distribution, but once the bubbles have emerged from the active breaker, the size distribution evolves under the action of buoyancy, dissolution, and residual turbulence in the layer. Buoyancy drives the larger bubbles to the surface. The smaller bubbles quickly acquire an organic coating to become "dirty" bubbles [86]. The downward advection from residual turbulence and the drag associated with the surfactant film on the bubble surface roughly balance the buoyancy, with the result that these smaller bubbles tend to remain in suspension in the water column, eventually disappearing through dissolution. Thus, the bubble size distribution is eroded at both ends of the radius scale.

Garrett et al. [85] developed an expression for the modified bubble size distribution at depth z in the bubble layer. Starting with a number density of the form $N(a) = N_0 a^{-n}$ for $a > a_c$, and zero for smaller bubbles, they assume that immediately after injection from a wave-breaking event, the bubble density is uniform in depth to a distance h below the surface. On taking account of the effects of dissolution on the smaller bubbles and buoyancy on the larger bubbles, they arrive at the expression:

$$N(a,z) = \frac{N_0}{(1-n)DT} \left\{ \left[a^3 + \frac{3D[h-z]}{A} \right]^{(1-n)/3} - a^{1-n} \right\}, \qquad (1.49)$$

where T is the bubble injection periodicity, D is the dissolution rate, which is taken to be independent of bubble radius, and A is a scaling constant representing the bubble rise rate due to buoyancy. Nominal values for the parameters in Eq. (1.49) are $h = 5.5$ m, $T = 60$ s, $A = 1.7 \times 10^6$ m^{-1}s^{-1}, $D = 10^{-6}$ m/s, and $n = 3$. With these assigned values, a volume-scaled version of Eq. (1.49), that is, $a^3 N(a)$, is plotted in Fig. 1.21 for three observation depths, z {unit: m}. Also shown, for comparison of its shape, is a volume-scaled version of the function in Eq. (1.48). The vertical scale in Fig. 1.21 could be converted to absolute units by finding the area under each of the curves, multiplying by $4\pi/3$ and equating the result to the void fraction, that is, the volume of air per unit volume of the two-phase bubbly medium. The area under the curves is found by integrating over the bubble radius between the upper and lower limits set by buoyancy and dissolution, respectively. In effect, this procedure allows the parameter N_0 in Eqs. (1.48) and (1.49) to be determined, and once known, the units of the volume-scaled bubble density can be specified uniquely.

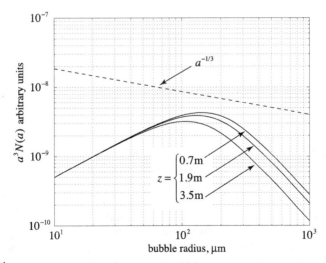

FIGURE 1.21

Volume-scaled bubble size distributions. The *dashed line* is from Eq. (1.48) and the *solid lines* are from Eq. (1.49).

Although there will always be more small bubbles than large bubbles, it can be seen in Fig. 1.21 that most of the volume of the entrained air is in bubbles with radii in the region of 100 μm. This corresponds to a frequency range around 30 kHz where acoustic scattering and attenuation in the near surface bubble layer are most pronounced. Broadband ambient noise, on the other hand, is sensitive to the number density of the bubbles [64], since, on formation, each bubble creates a pulse of sound whose frequency is inversely proportional to the bubble radius.

The presence of air reduces the speed of sound in the bubbly layer immediately beneath the sea surface. If the bubbles are small compared with an acoustic wavelength, the two-phase medium may be treated as a continuum and the granularity associated with the bubbles ignored. Taking the density and bulk modulus of the gas/fluid mixture as ρ and κ, respectively, the sound speed is given by

$$c = \sqrt{\frac{\kappa}{\rho}}. \tag{1.50}$$

Using subscripts 1 and 2 to identify the two constituents, gas and fluid, respectively, the density and bulk modulus can each be expressed as a weighted mean of the corresponding quantities for the gas and the fluid. The weighting coefficients are the volume fractions of the respective materials, that is, β for the gas and $(1 - \beta)$ for the fluid. Thus, the density is:

$$\rho = \beta \rho_1 + (1 - \beta)\rho_2 \tag{1.51}$$

and the bulk modulus is given by:

$$\frac{1}{\kappa} = \beta \frac{1}{\kappa_1} + (1 - \beta) \frac{1}{\kappa_2}. \tag{1.52}$$

When these two equations are substituted into Eq. (1.50), the expression for the sound speed becomes:

$$c = \left\{ \frac{\kappa_1 \kappa_2}{[\beta \kappa_2 + (1 - \beta)\kappa_1][\beta \rho_1 + (1 - \beta)\rho_2]} \right\}^{1/2}, \tag{1.53}$$

which is the Mallock−Wood equation [87,88] for the speed of sound in a two-phase medium in which the two constituents do not react chemically.

When the void fraction, β, takes its limiting values, 1 and 0, respectively, Eq. (1.53) returns sound speeds $c = c_1 = \sqrt{\kappa_1/\rho_1}$, consistent with 100% gas, and $c = c_2 = \sqrt{\kappa_2/\rho_2}$ for 100% fluid. With intermediate values of β, the sound speed in the mixture exhibits a remarkably deep minimum. This is exemplified in Fig. 1.22 representing air in seawater, in which case the densities {unit: kg/m^3} are $\rho_1 = 1.2$ and $\rho_2 = 1024$ and the bulk moduli {unit: Pa} are $\kappa_1 = 1.43 \times 10^5$ and $\kappa_2 = 2.30 \times 10^9$. The minimum in the sound speed, occurring at a void fraction $\beta = 0.496$, is 23.5 m/s, which is significantly less than the sound speed in either air ($c_1 \approx 340$ m/s) or seawater ($c_2 \approx 1500$ m/s). Around the minimum, the two-phase medium is a frothy mixture with the bubbles acting as springs loaded by the inertia of the seawater, which leads to the reduction in the sound speed.

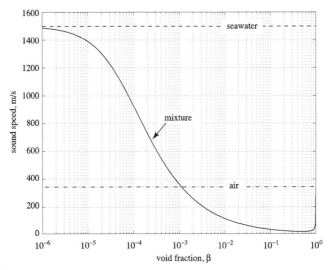

FIGURE 1.22

Sound speed in a bubbly mixture of air and seawater.

The presence of bubbles in the ocean can have a significant effect on acoustic propagation. For instance, a sound wave incident on the sea-surface bubble layer or a mass of air bubbles, such as a ship's wake, will be partially reflected. The reflection coefficient for normal incidence on a thick layer of bubbles is:

$$R = \frac{\rho c - \rho_2 c_2}{\rho c + \rho_2 c_2}, \tag{1.54}$$

which depends on the void fraction through the density ρ and the sound speed c of the mixture. The fraction of energy reflected, R^2, is shown in Fig. 1.23 as a function of the void fraction. As the void fraction increases above 1 part in 1000, the reflected energy rises rapidly, approaching total reflection when the void fraction is unity.

The void fraction in the near-surface bubble layer created by wave breaking is much less than unity, typically taking values around $\beta \approx 10^{-5}$ or less [89]. The concentration of bubbles decreases rapidly with depth in the layer [90], following an exponential or faster decay law, with an e-folding depth of the order of 1 m, thus most of the bubbles lie within 2–3 m of the surface [91]. This entrained air gives rise to a sound speed profile within the layer with a minimum at the surface and increasing, on average, monotonically with depth [59]. Such a profile is upward refracting, since sound rays are bent toward regions of lower sound speed, and thus the quiescent bubble layer acts as a surface waveguide capable of trapping sound that is produced in the wave-breaking process by other, acoustically active bubbles. The trapped sound propagates along the surface duct in the form of normal modes, the shape of which depends on the details of the sound speed profile in the bubbly layer [35,59].

Immediately after they are created, bubbles produce a brief pulse of sound lasting of the order of tens of milliseconds. In fact, the creation of bubbles by wave breaking

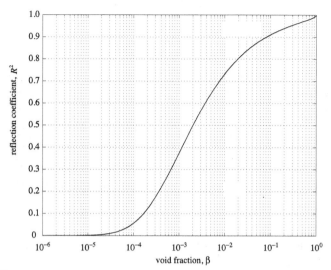

FIGURE 1.23

Reflection coefficient as a function of void fraction.

is a major source of wind-generated noise in the ocean, and as such it formed a central topic in a series of international conferences on sea surface sound [92–95] held in the decade between 1987 and 1997. The primary purpose of these meetings was to identify and understand the physics of natural sources of sound in the ocean.

A bubble is formed when a fluid surface distorts, forming an indentation, which then closes on itself to create a void [96,97]. At the instant of closure, the air trapped within the void is initially at atmospheric pressure, but a pressure differential exists across the bubble wall due to surface tension, otherwise known as the Laplace pressure, and to hydrostatic pressure, which increases as the bubble sinks into the fluid. In response to the excess pressure, the bubble contracts but, because of inertia, overshoots the equilibrium radius, to become overpressurized. It then expands, again passing through the equilibrium radius, and the process continues as a series of radial oscillations, which eventually decay away due to radiation losses and other attenuating mechanisms including thermal conduction and viscosity.

For small excursions about the equilibrium radius, the oscillating bubble is a linear, resonant system. The resonance frequencies of bubbles were investigated in 1933 by Minnaert [98] in a series of classic experiments involving a tank of water, a U-shaped pipette to create individual bubbles, an inverted glass bell and capillary tube to capture and measure the volume of the bubbles, and a tuning fork to help identify the frequencies.

Fig. 1.24, taken from Minnaert's paper, illustrates his basic technique. Water placed in the funnel at A slowly streams through the tube BC to the gas reservoir and flow regulator, CD, which produce a weak flow of gas through the tube DE, leading to the formation of a bubble at the tip of the pipette above E. The bubbles

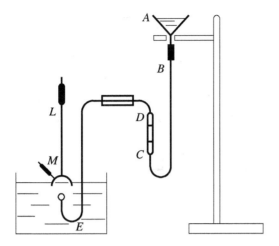

FIGURE 1.24

Minnaert's apparatus for measuring the resonance frequency of bubbles.

were captured by the inverted bell jar and capillary tube, *LM*, allowing their volume to be determined. The resonance frequency of each bubble as released from the pipette was determined by ear, taking for comparison the sound of a tuning fork and estimating fractions of a semi-tone. As Minnaert himself pointed out, "This determination is not very accurate, owing to the considerable difference between the two sounds, the short duration of the bubble sounds, their variation, and their small intensity." He continued: "Moreover the determination of the octave to which the sound belongs is very difficult for me, so that there remains an uncertitude of a factor of two in the absolute values of the vibration numbers." By the end of his paper, however, Minnaert had removed the uncertainty concerning the exact octave of the resonance frequencies, which he did by examining the frequency of the sound from bubbles containing various gases besides air, including hydrogen and butane, and comparing the results with the theoretical expression for the resonance frequency.

The resonance frequency of a bubble exhibiting small-amplitude, linear oscillations about the equilibrium radius, while maintaining a spherical shape, is derived by considering the potential and kinetic energies of the bubble-water system, illustrated in Fig. 1.25.

The displacement of the bubble radius can be represented by:

$$r = r_0 + a \sin \omega_0 t \tag{1.55}$$

and thus the radial velocity is:

$$\frac{dr}{dt} = \dot{r} = a\omega_0 \cos \omega_0 t, \tag{1.56}$$

where r_0 is the equilibrium radius, a is the amplitude of the radial excursions, and ω_0 is the resonance angular frequency, which is to be determined. Assuming that the

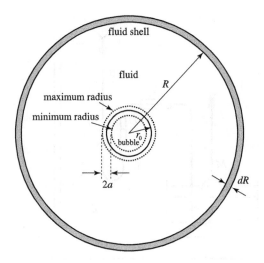

FIGURE 1.25

Bubble-fluid system for deriving the resonance frequency of the bubble.

compression of the gas in the bubble is adiabatic, the ratio of the excess pressure, p, at bubble radius $(r_0 - x)$ to the equilibrium pressure, p_0, at radius r_0 is:

$$\frac{p}{p_0} = \left(\frac{V_0}{V}\right)^{\gamma} = \left(\frac{r_0}{r_0 - x}\right)^{3\gamma}, \tag{1.57}$$

where V and V_0 are the respective volumes, and γ is the ratio of the specific heats at constant pressure and constant volume. Since the amplitude of the oscillations is much less than the bubble radius, a binomial expansion of the term on the right in Eq. (1.57) yields, to first order in the displacement x:

$$p - p_0 = \frac{3\gamma p_0}{r_0} x. \tag{1.58}$$

When the bubble is at minimum volume, the potential energy stored in the gas is:

$$E_{pot} = -\int_{V_0}^{V} (p - p_0) dV, \tag{1.59}$$

which, after conversion to an integral over the radial displacement, x, becomes:

$$E_{pot} = \int_0^a \frac{3\gamma p_0 x}{r_0} 4\pi r_0^2 \, dx = 6\pi \gamma p_0 r_0 a^2. \tag{1.60}$$

This must equal the maximum kinetic energy in the fluid, which is found by considering the kinetic energy in a shell of radius R and thickness dR and integrating over the volume of the fluid:

$$E_{kin} = \frac{\rho}{2} \int_{r_0}^{\infty} 4\pi R^2 \dot{R}^2 \, dR, \tag{1.61}$$

where ρ is the fluid density. To relate the velocity of the shell to that of the bubble surface, consider that the mass of fluid flowing in time Δt through any spherical shell centered on the bubble is:

$$\Delta M(R) = 4\pi R^2 \dot{R}^2 \rho \Delta t. \tag{1.62}$$

Treating the fluid as incompressible, conservation of mass requires that $\Delta M(R)$ be the same for all R, including the surface of the bubble, and therefore at maximum bubble velocity it follows from Eq. (1.56) that:

$$\dot{R} = \frac{r_0^2}{R^2} \dot{r} = \frac{r_0^2}{R^2} a\omega_0, \tag{1.63}$$

which leads to:

$$E_{kin} = 2\pi \rho r_0^3 a^2 \omega_0^2. \tag{1.64}$$

On equating the potential and kinetic energies in Eqs. (1.60) and (1.64), respectively, the resonance frequency of the bubble is found to be:

$$\omega_0 = \frac{1}{r_0} \sqrt{\frac{3p_0 \gamma}{\rho}}, \tag{1.65}$$

which is Minnaert's expression, showing that the resonance frequency scales inversely with the bubble radius. For air bubbles immediately beneath the sea surface, $p_0 = 10^5$ Pa, $\gamma = 1.4$, and $\rho = 1024$ kg/m^3, in which case:

$$f_0 = \frac{\omega_0}{2\pi} = \frac{3.2}{r_0} \text{ Hz}. \tag{1.66}$$

Although involving several approximations and assumptions (adiabatic conditions, negligible surface tension, spherical bubbles, and an incompressible fluid within which all the kinetic energy resides), Eqs. (1.65) and (1.66) accurately represent freely oscillating bubbles with radii of 2 mm or less. Hydrostatic effects tend to distort larger bubbles, which result in shape oscillations, but these are far less efficient in radiating sound than the spherical-volume pulsations of the monopole mode [99–101].

According to Eq (1.66), the frequency of sound radiated by a spherical bubble with the maximum radius of 2 mm is 1.6 kHz. As shown in Fig. 1.21, a sizable fraction of the bubbles produced by wave breaking have radii smaller than the 2 mm upper limit, and these bubbles, oscillating freely, are largely responsible for the Knudsen spectrum [102] of wind-generated ambient noise in the ocean at frequencies above 1.6 kHz. At lower frequencies, a broad, wind-driven, local maximum around 500 Hz is observed in the spectrum of ambient noise [17,103,104]. It seems unlikely that individual bubble oscillations are responsible for this lower-frequency region of the spectrum, since the bubble diameter would have to be of the order of 1 cm. Such large bubbles are not commonly observed beneath breaking waves, because of fragmentation and buoyancy, and even if they were present, their number is not sufficient to account for the observed wind-driven noise.

Several mechanisms have been proposed to account for low-frequency (below 1.6 kHz) wind-driven noise. Between 0.1 and 5 Hz, microseisms produced by nonlinear wave—wave interactions, as postulated by Longuet-Higgins [105], were found by Kibblewhite and Evans [106] to be the dominant source of noise. A theoretical investigation by Isakovitch and Kur'yanov [107] led to the conclusion that random pressure fluctuations on the sea surface produced by turbulent winds contribute to the noise spectrum around 20 Hz. Goncharov [108] suggested that the interaction of surface waves and turbulence generated noise between 10 and 100 Hz.

None of these mechanisms, however, radiates enough sound around 500 Hz to explain the observed wind-driven peak in the ambient noise spectrum. A mechanism that could be responsible for this low-frequency noise is the collective oscillations of a bubble plume [109—115] created by breaking waves at higher sea states. Through the mutual interactions of acoustic and hydrodynamic forces, the micro-bubbles in the plume act as a system of coupled oscillators that can exhibit normal modes at frequencies much lower than the resonance frequencies of the individual bubbles constituting the plume. On a macroscopic level, the two-phase medium, consisting of air bubbles and seawater, may be thought of as a continuum in which, at the low frequencies of interest, where the wavelengths are very much greater than the bubble sizes, the sound speed is governed by the Mallock—Wood equation (Eq. 1.53).

To estimate its lowest resonance frequency, ω_m, the effective medium of the plume is assumed to occupy a spherical volume of radius r_m, as illustrated in Fig. 1.26. The subscript m is used here to identify quantities associated with

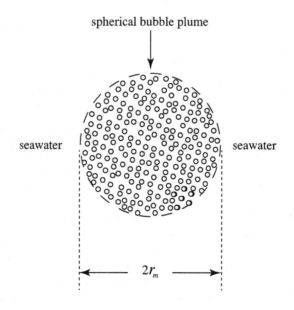

FIGURE 1.26

Spherical bubble plume immersed in seawater.

the air—seawater mixture in the plume. Although the boundary of the sphere is not well defined, the resonance frequency, as with a single bubble, still depends on the dimensions of the oscillatory region. In its lowest mode of oscillation, the plume exhibits small, volumetric expansions and contractions, thereby acting as a monopole radiating sound into the surrounding seawater. In this sense, the spherical bubble plume resembles the single-bubble source, for which the resonance frequency is given by Minnaert's expression in Eq. (1.65).

In fact, the resonance frequency of the plume may be formulated as a modified version of the Minnaert frequency. For typical values of the void fraction in the plume, around $\beta = 10^{-3}$, the density and bulk modulus of the bubbly mixture are, from Eqs. (1.51) and (1.52):

$$\rho_m \approx \rho_f \text{ and } \kappa_m \approx \frac{\kappa_g}{\beta}, \tag{1.67}$$

where the subscripts g and f denote gas and fluid, respectively. Therefore, the sound speed in the mixture is:

$$c_m \approx \sqrt{\frac{\kappa_g}{\beta \rho_f}} . \tag{1.68}$$

Now, by assuming isothermal behavior of the gas, van Wijngaarden [116] has shown, that the sound speed in the mixture is:

$$c_m \approx \sqrt{\frac{P_0}{\beta \rho_f}} , \tag{1.69}$$

where P_0 is the hydrostatic pressure. The isothermal assumption is justified on the basis of the conclusions reached by Hsieh and Plesset [117] from their investigation of the equations for conservation of momentum and energy in the mixture. A comparison of the two expressions for the sound speed in Eqs. (1.68) and (1.69) shows that the hydrostatic pressure approximately equals the bulk modulus of the gas, $P_0 \approx \kappa_g$, a condition that is known to be true. It follows that Minnaert's expression for the resonance frequency of a single bubble, Eq. (1.65), can be rewritten as:

$$\omega_0 = \frac{1}{r_0} \sqrt{\frac{3 \gamma \kappa_g}{\rho_f}} = \frac{1}{r_0} \sqrt{\frac{3 \gamma \rho_g}{\rho_f} c_g^2} . \tag{1.70}$$

The analogous expression for the resonance frequency of the bubble plume, ω_m, is obtained by replacing the single-bubble parameters in Eq. (1.70) with those of the mixture (i.e., $r_0 \rightarrow r_m$, $\rho_g \rightarrow \rho_m$, and $c_g \rightarrow c_m$):

$$\omega_m = \frac{1}{r_m} \sqrt{\frac{3 \rho_m}{\rho_f} c_m^2} = \frac{1}{r_m} \sqrt{\frac{3 P_0}{\beta \rho_f}}, \tag{1.71}$$

where γ has been set to unity, consistent with the isothermal conditions prevailing in the two-phase medium. Eq. (1.71) is the same as the expression developed by Carey and Fitzgerald [113] from an argument involving plane-wave scattering from a spherical bubble plume.

From the ratio of the resonance frequencies:

$$\frac{f_m}{f_0} = \frac{r_0}{r_m}(\beta\gamma)^{-1/2},\tag{1.72}$$

it is evident that the frequency of the plume is much reduced relative to that of a single bubble. For instance, with representative values of the void fraction, $\beta = 10^{-3}$, the radius of the micro-bubbles within the plume, $r_0 = 100$ μm (all assumed to be the same size), and the plume radius, $r_m = 0.5$ m, and with $\gamma = 1.4$, then:

$$f_m \approx 0.0053 f_0.\tag{1.73}$$

Since bubbles of 100 μm radius resonate at $f_0 = 32$ kHz, the resonance frequency of the plume, from Eq. (1.73), is 170 Hz. Smaller plumes or lower void fractions will return higher resonance frequencies and vice versa.

Theoretically, then, bubble plumes created by breaking waves are capable of generating sound at frequencies in the region of several hundred Hz, consistent with the observations of low-frequency wind-driven ambient noise in the ocean. Experimental evidence to support the bubble-plume theory has been provided by Hollett [118], who used the upward-directed end fire beam of a vertical array to monitor a patch of the sea surface and then correlated the acoustic recordings with simultaneous video observations of the same surface patch. Over his experimental bandwidth, from 187.5 to 1000 Hz, he found that breaking waves radiated sound with a more or less flat spectrum. A similar conclusion was reached by Farmer and Vagle [59], who also correlated acoustic and video recordings of bubble plumes created by breaking waves. They observed that wave breaking events radiated sound at frequencies as low as 50 Hz.

A "tipping trough" simulation of a breaking wave, performed by Carey et al. [119], with freshwater and saltwater, under controlled conditions in Seneca Lake and Dodge Pond, returned results similar to those from the open-ocean experiments. A spectral peak in acoustic recordings of the tipping trough events was observed at 100 Hz, which could not be explained by single-bubble oscillations, since the bubble sizes were all far too small to qualify. Carey and colleagues concluded that collective oscillations of the bubbles in the cloud created by each tipping event were responsible for the observed low-frequency radiated sound.

Acoustic resonances of the bubble plume formed by a freshwater jet plunging into a tank of freshwater were observed by Hahn et al. [120] in a laboratory experiment, performed under tightly controlled conditions, in which the fluxes of air and water into the tank were carefully monitored. The jet created a conical bubble plume, which radiated sound at frequencies between 100 Hz and 1 kHz. Up to five, nonuniformly spaced, spectral peaks were observed in the spectrum, associated with the coherent collective oscillations of the bubbles within the plume: the two-phase bubbly medium behaved as a continuum, acting as a resonant conical cavity beneath the jet. The nonuniformly distributed resonances, or eigen frequencies, correspond to the longitudinal modes of oscillation of the conical bubble plume. Thus, it was concluded, as in the open-ocean and tipping-trough experiments, that collective oscillations were responsible for the low-frequency sound observed in the plunging water jet experiments.

Bubble acoustics is an extensive topic with wide ranging ramifications in underwater acoustics applications. For a comprehensive exposition of many aspects of the acoustic bubble, the reader is referred to the text by Leighton [121].

1.5.9 OCEAN ACIDIFICATION

Since the beginning of the industrial revolution the level of carbon dioxide in the atmosphere has been steadily rising, a phenomenon which is attributed mainly to the combustion of fossil fuels, cement production, and deforestation. The pre-industrial level of carbon dioxide (CO_2) in the atmosphere was about 280 parts per million (ppm) [122] but by 1960, when Keeling [123] started making measurements at the top of Mauna Loa volcano in Hawaii, it had risen to 315 ppm. In 2015, the level exceeded 400 ppm [124] for the first time in recorded history. Indeed, this is higher than at any time in the past million, perhaps several million, years [125] and is rising at an ever increasing rate. Currently, the rate of increase of CO_2 in the atmosphere is as much as 30 times faster than natural rates in the geological past.

As early as 1957, before the Keeling record began, Revelle and Suess [126] raised the issue of the absorption of atmospheric carbon dioxide by the ocean. About one-quarter of the annual anthropogenic production of CO_2 is absorbed by the ocean, producing carbonic acid, thus changing the chemistry of seawater in a process called *ocean acidification*. With rising levels of atmospheric carbon dioxide, the partial pressure of CO_2 on either side of the air–sea interface is no longer in equilibrium, resulting in the diffusion of the gas across the interface until equilibrium is restored. At low wind speeds, this relatively slow mechanism of molecular diffusion accounts for most of the gas transfer across the air–sea interface. As the wind picks up, however, to 10 m/s or higher, an appreciable increase in the gas transfer velocity occurs [127], facilitated by surface-wave breaking [128,129], which entrains air that is then fragmented into bubbles and advected by turbulence to depths as great as 10 m.

A significant consequence of the carbon dioxide transfer across the air–sea interface is that the ocean is becoming more acidic as CO_2 levels in the atmosphere rise. The pH scale is a logarithmic measure of acidity, with decreasing numbers representing increased acidity. A solution with pH greater than 7 is basic and one with pH less than 7 is acidic. The pH scale is often said to range from 0 to 14, but extremely strong acids and bases can have a pH less than zero and greater than 14, respectively. It is estimated that surface ocean pH has decreased by 0.1, from 8.25, since the beginning of the industrial revolution, which corresponds to an increase of about 30% in acidity [130]. This enhanced acidity changes the relative abundance of the two forms of boron that exist in the ocean, borate ions $B(OH)_4^-$ and nonionized boric acid $B(OH)_3$.

The ratio of the two boron species is controlled by the pH of the seawater, with the fraction of the ionized molecules decreasing as acidity rises. Because of its charge, the ionized form, which is the larger of the two, carries with it a loose assemblage of water molecules. During the passage of an acoustic wave, this complex of molecules is compressed into a reduced volume, a process which removes energy from the wave, with the molecules relaxing back to their original volume after the wave has passed [15]. Thus, the reduction in the numbers of ionized $B(OH)_4^-$, due to rising acidity, leads to a decrease in sound attenuation, implying an acoustically more transparent ocean [131,132].

The total acoustic attenuation in the ocean is comprised of three additive components: the viscosity of pure water, the relaxation of magnesium sulfate ($MgSO_4$), and the relaxation of borate ions, as described in the preceding paragraph. Francois and Garrison [133] have developed expressions for the frequency dependence of each of these contributions to the attenuation, with the boron component involving the pH of seawater. Their expression for the total attenuation, α, in dB/km, as shown in Fig. 1.27, is discussed in detail in Chapter 4. It is clear from Fig. 1.27 that the

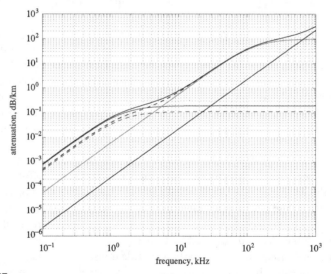

FIGURE 1.27

Frequency dependence of the attenuation in seawater. The *blue line* (dark gray in print versions) is the viscous component, the *green line* (light gray in print versions) is from the relaxation of $MgSO_4$, the *red lines* (gray in print versions) represent the relaxation of the borate ions, and the *black lines* show the total attenuation. With the red and black, the *solid lines* are for pH $= 8.1$ and the *dashed lines* illustrate the reduction that occurs when pH $= 7.8$.

Based on Francois, R. E., and Garrison, G. R., Sound absorption based on ocean measurements: Part II: Boric acid contribution and equation for total absorption, *J. Acoust. Soc. Am.,* **72**, *1879–1890 (1982).*

main effect of the increased acidity associated with a reduction in pH from 8.1 to 7.8 occurs at frequencies below 10 kHz. At 1 kHz and below, the increment of -0.3 in pH reduces the attenuation by approximately 40%.

It has been projected that a pH change of -0.3 in surface waters will occur by the middle of the 21st century, giving rise to a 70% increase in the travel distance of low frequency sound [131]. This increased transparency of the ocean to low-frequency acoustic waves could, it has been suggested [134], have a significant effect on scientific, commercial, and naval operations involving underwater acoustics. On a cautionary note, however, it should be borne in mind that at low frequencies the attenuation in seawater is extremely low (about 2 dB per 100 km at 500 Hz) and, as has been pointed out by Reeder and Chiu [135], 40% of a very small number is itself very small.

In most underwater acoustics applications, seawater attenuation is an almost negligible contributor to the overall loss experienced by a sound wave as it propagates through the ocean. Geometrical spreading (spherical in deep water and cylindrical in shallow), scattering from the sea surface, and reflection from the sea floor are the dominant sources of acoustic loss in the ocean, usually overwhelming seawater attenuation. Reeder and Chiu [135] used transmission loss models to examine the effect of reduced pH, from 8.1 to 7.4, on low frequency acoustic propagation in the ocean. Although this reduction in pH is far greater than ocean chemistry models predict for the remainder of this century [136], their results show that in shallow water there is no observable change in propagation loss and in deep water the difference is less than 0.5 dB. Such a minor change is statistically insignificant compared with the inherent variability of the background noise levels produced by breaking waves and distant shipping. Their central conclusion is that ocean ambient noise levels, as currently represented by the Wenz curves [18], would still be valid after 250 years at the expected rate of increase in ocean acidification.

The effects of ocean acidification on marine biology, as compared with the acoustic environment, are considerably more profound. When CO_2 is absorbed into the ocean, it reduces not only the pH, but also the carbonate ion concentration and the saturation states of calcium carbonate minerals. These minerals are essential building blocks of the skeletons and shells of many marine organisms. Continuing reductions in seawater pH cause many parts of the ocean to become undersaturated in calcium carbonate minerals, one effect of which is to reduce the ability of reef-building corals to construct their skeletons [137,138]. A similar problem exists with shellfish, where the increased solubility of calcium carbonate minerals slows calcification, that is, shell building, which could have serious negative consequences in terms of reduced abundance of commercial species such as clams, oysters, and sea urchins [139].

Increased ocean acidity, especially in conjunction with global warming trends, could also affect other forms of marine organisms [132]. Photosynthesis rates of some cyanobacteria, for example, could be affected by elevated levels of CO_2, and phytoplankton growth could be influenced by changes in acid−base chemistry.

It is possible that toxic algal blooms could become more virulent and abundant; or conversely, the organisms might simply adapt to the changing conditions without an increase in toxicity and possibly even a reduction in numbers. At present, predictions of future trends are speculative, since the interactions among growth rates, nutrient availability, acidification, and rising ocean temperatures are still poorly understood.

1.5.10 DEEP-OCEAN HYDROTHERMAL FLOWS

In May 1976, scientists from Scripps Institution of Oceanography conducted a near-bottom hydrographic and geochemical survey of the Galápagos Rift (1°N, 86°W) using the Deep Tow vehicle of the Scripps Marine Physical Laboratory. They discovered hydrothermal plumes issuing from the axial fissure in the center of the inner rift [140]. Although there had been indirect evidence previously [141], this was the first conclusive identification of hydrothermally circulating seawater in a deep-ocean spreading center. In February and March of 1977, the deep submersible *Alvin* made 24 dives along the axis of the Galápagos Rift, during which visual observations of the hydrothermal plumes [142] were made. Since then, hydrothermal vents have been found at spreading centers and subduction zones in the Atlantic and Pacific Oceans, usually at depths between 2000 and 5000 m, although some shallow-water vents are also known to exist.

The vents are created as near-freezing (2°C) ocean water percolates down through fissures in the lithosphere, where it is heated by molten magma and mixed with hot magmatic water, thus becoming less dense, which allows it to rise and re-emerge into the ocean as a column of superheated water. While mixing with the hydrothermal fluids in the ocean crust, this water, in addition to being heated, becomes highly corrosive, dissolving the surrounding rock and leaching out metals and other elements. The water thereby accumulates a heavy load of mineral particles, which, as they cool after discharge into the oxygen-rich, cold seawater, combine with sulfur to form iron, copper, and zinc sulfides that precipitate out, creating cylindrical, chimney-like structures on the seabed. These chimneys are known as "black smokers" since they vent a billowing black plume of extremely hot seawater containing the tiny, black metal sulfide particles.

"White smokers" formed from precipitates of barium, calcium, and silicon also exist, which are white. The temperature of black smoker plumes is typically around 350°C, whereas white smokers tend to have cooler plumes at 250–300°C. The chimneys, which in extreme cases can grow to a height of 60 m [143], are examples of *focused vents* in which almost all the vent fluid flows out of one small pipe. As long as fluid flows through them, the chimneys continue to grow taller at a rate as high as 30 cm per day. *Diffuse vents* also occur, created when hot fluids mixed with cold seawater spread out over a region of the seafloor without building sulfide chimneys. Such vents are relatively cool, at just a few tens of degrees Celsius above the near-freezing temperature of the surrounding seawater, but they still contain high levels of sulfides and other metal compounds.

The abundance of hydrogen sulfide, CO_2, and methane around a hydrothermal vent supports a vast ecosystem comprised of giant tubeworms, shrimps, mussels, clams, limpets, crabs, snails, fish, and octopi. These relatively complex life forms feed off, or have a symbiotic relationship with, microscopic organisms, autotrophic bacteria at the bottom of the food chain [144], which convert hydrogen sulfide into sulfur to create energy in a process called chemosynthesis. This contrasts with most land-based life forms, which rely for survival on energy from the sun and the conversion of CO_2 into oxygen through the process of photosynthesis. Naturally, the marine organisms that become established around hydrothermal vents rely on the continued existence of the vents for their own survival. Since the chimneys are fragile, they tend to collapse under their own weight if they grow too tall.

A black smoker, being a conduit for the vigorous and turbulent flow of hydrothermal water, with speeds often exceeding 1 m/s, is also a channel for heat and chemical exchange between the ocean and lithosphere. An estimate of these local heat and chemical fluxes can be obtained from the fluid flow rates in the black smoker plumes. Intrusive flow-rate measurements, however, are severely hampered by the extremely high temperatures in and corrosive (acidic) nature of the plume environment. An alternative approach is passive acoustic remote sensing of flow rates using a hydrophone or hydrophone array placed external to the plume. Passive sensing of the energetic flow of black smokers, in conjunction with active methods applied to less energetic diffuse hydrothermal vents [145], could provide flow-rate information on both focused and diffuse vents.

A problem with the passive acoustics approach was that, for some years after their discovery, black smokers were thought to be silent, possibly because the extreme hydrostatic pressure at depths of several thousand meters prevents the formation of bubbles through boiling, even at the elevated temperatures of black smoker plumes. In 1980, however, Riedesel [146] and colleagues deployed an array of five ocean-bottom seismographs, some including hydrophones, in the vicinity of a cluster of black smoker vents that were situated in a narrow band, only 200−500 m wide, along the spreading axis of the East Pacific Rise at 21°N. They observed that "Some of these vents exhibited exit velocities on the order of several meters per second, not unlike the velocity at which water leaves the nozzle of a fire hose." One of their findings was that low-frequency acoustic noise in the water column was 16−64 times higher when the sensor was within 300 m of the black smoker vents than it was 2 km away, implying that the vents were generating a significant acoustic signature as the superheated water streamed into the cold ocean. Whatever the source mechanism, most of the acoustic energy was in the form of infrasound, with frequencies below about 20 Hz.

Acoustic noise in the infrasonic frequency band 2−30 Hz was also observed, by Bibee and Jacobson [147] in July 1985, in a survey of the caldera of Axial Seamount, which is located on the Juan de Fuca Ridge at 46°N. From the acoustic data and the configuration of the sensors, it was concluded that the source of the noise was a low-temperature (35°C) diffuse thermal vent field, rather than a high-temperature (293°C) black smoker that was also in the caldera. A microseismicity study [148] of the area showed earthquake swarm activity in the vicinity of the diffuse field but not near the black smoker, suggesting that the latter may have been inactive during the acoustic recording period.

In a subsequent acoustic survey of Axial Seamount, conducted by Little et al. [149] in September 1987 to determine the feasibility of monitoring hydrothermal vent activity from flow noise generation, no continuous infrasound signals from black smokers were detected but an intermittent signal near 40 Hz, consistent with jet noise, was observed and thought to have originated in a black smoker named "Inferno." Few vigorous black smokers were encountered during the 200+ hours of the survey, but five hydrothermally active areas were crossed by the sensor sled. Consistently elevated sound levels were observed near these sites at frequencies between 15 and 30 Hz.

By the end of the 20th century, the available body of evidence that black smokers produce infrasound associated with the energetic flow of hydrothermal fluids through the chimney was inconclusive. This changed in 2004 and 2005, when experiments were performed around the active black smokers "Sully" and "Puffer" at a depth of about 2200 m in the Main Endeavour vent field on the Juan de Fuca Ridge [150]. At both vents, a hydrophone placed just 0.3 m away from the vent orifice recorded broadband (10–500 Hz) signals that were 10–30 dB above the background ambient noise level, as measured by reference hydrophones located some distance away from the vents themselves. Most of the acoustic energy was found to be below about 100 Hz, consistent with the earlier observations of Riedesel et al. [146] along the East Pacific Rise. The sources of the low frequency sound are uncertain but could include pulsating flow through the orifice, volume changes associated with cooling of the hydrothermal fluid in the mixing region of the emergent jet, fluid heterogeneity in the form of density variations of the flow, and fluid–structure interactions along the inner walls of the chimney. Future measurements using a directional hydrophone array could help to identify the dominant sound-source mechanisms.

The fact that black smokers generate significant levels of sound opens the way to the development of passive acoustic inversion techniques for recovering information about the geological and physical processes occurring within these systems. For example, it may be possible to estimate flow velocity along with the heat and chemical fluxes through the conduit, assuming that a connection can be established between the fluid flow rate and the broadband acoustic signature of the emergent plume. Changes in temperature or chemical composition could affect the density and compressibility of the hydrothermal fluid, and hence could be monitored from the associated tell-tale shifts in the amplitude and frequency of the radiated acoustic signature. Looking ahead, new discoveries of black smokers could perhaps be facilitated by simply listening for the sounds that they produce.

Acoustic images of black smokers and also diffuse hydrothermal vents have been obtained using high-frequency (carrier frequency 333 kHz) sonars [151] mounted on either a deep-diving submersible or a remotely operated vehicle (ROV). The basic idea is that the back-scattered intensity from the active plume-imaging sonar can be used to infer the concentration of particles in the plume, while the Doppler shift returns estimates of the radial component of velocity of the particles. In July 2000, the first long-term acoustic imaging of black smokers, located in the Main Endeavour Field on the Juan de Fuca Ridge, was achieved when ROV JASON, operated by the Deep Submergence Laboratory of the Woods Hole Oceanographic Institution, sat on the ocean floor and continuously imaged the plume complex for 24 h [152].

1.5.11 EDDIES, FRONTS, AND LARGE-SCALE TURBULENCE

Ocean-scale surface currents in both the Atlantic Ocean and the Pacific Ocean, driven mainly by strong winds, follow a circulation pattern in the form of two large anticyclonic gyres that are separated by an equatorial countercurrent. In both oceans, there are strong western boundary currents, the Gulf Stream in the North Atlantic, the Kuroshio Current in the North Pacific, the Brazil Current in the South Atlantic, and the East Australian Current in the South Pacific. At the surface, these currents can be flowing as fast as 2.5 m/s, which is comparable with the speed of many sailing boats. The edges of the anticyclonic gyres are usually marked by steep gradients in temperature and salinity, known as fronts. Elsewhere in the ocean, fronts also occur, notably at shelf breaks and the edge of the outflow from a restricted channel or strait.

Within the basin-scale anticyclonic gyres, the ocean is not motionless. Current speeds can be as high as those in the major current systems themselves, although a current meter anchored within a gyre usually records fluctuations in the flow on a time scale of several days. Such observations indicate that the flow within the gyres is fully turbulent, with its large kinetic energy arising from the presence of turbulent eddies rather than a steady unidirectional flow of the sort that characterizes the major currents such as the Gulf Stream. Observational evidence suggests that the most energetic eddies have length scales from 20 to 200 km, time scales from 7 to 70 days, and characteristics speeds of 0.04−0.4 m/s. In temperate and equatorial oceans, most eddies do not have an obvious surface manifestation, making them difficult to visualize, but in the polar oceans ice may act as a tracer, delineating the surface motion. An example of an ice-covered eddy field in the entrance to Scoresby Sund, the largest fjord on the east coast of Greenland, is shown in Fig. 1.28. These eddies were formed by current flows in the fjord, inward along the northern edge and outward along the southern edge. All of the eddies are cyclonic, with low-pressure centers, as can be seen from the convergence of ice toward the center of each eddy.

One source of the eddy fields, located within the ocean-scale anticyclonic gyres and elsewhere in the oceans, is the major wind-driven currents, of which the Gulf Stream is the best understood. As the Gulf Stream tracks its way across the North Atlantic, it forms large meanders, one of which occasionally closes off to create a large eddy, sometimes referred to as a ring. North of the Gulf Stream, the rings are anticyclonic with warm cores, whereas to the south, they are cyclonic with cold cores. Both the warm core and cold core rings drift southwest against the stream

FIGURE 1.28

Ice floe tracers revealing a field of cyclonic eddies in the Marginal Ice Zone at the entrance to Scoresby Sund, East Greenland.

Aerial photograph taken by Buckingham, M. J.

at a rate of a few centimeters per day. Some of the rings are eventually reabsorbed into the Gulf Stream, but others drift away and break up into smaller eddies, in a process that continues down to ever-smaller eddy scales. Kinetic energy is thereby transferred from the anticyclonic gyres, to the large Gulf Stream rings (in the North Atlantic), down to eddies on a size scale of a few centimeters, at which point molecular viscosity takes over as the main means of dissipating energy. The energy that cascades down through eddies in the *inertial subrange*, that is, smaller than Gulf Stream rings but large enough for viscosity to be negligible, follows the celebrated Kolmogorov "-5/3 law" for the energy spectrum of turbulence:

$$E = C\varepsilon^{2/3}k^{-5/3}, \tag{1.74}$$

where E is a measure of the turbulent kinetic energy, k is the wavenumber associated with the eddy, ε is the rate of dissipation of energy, and the coefficient C is dimensionless, with a value found from experiments to be close to unity. The dissipation rate, ε, is the factor that varies with location.

Another source of mesoscale eddies, comparable in size with the Gulf Stream rings, is the outflow of highly saline Mediterranean water through the Strait of

Gibraltar into the Canary Basin in the northeast Atlantic. The anticyclonic, lens-shaped eddies [153,154] that are spawned by the outflow, known as Meddies, were discovered as anomalies in vertical profiles of salinity ($\sim 0.8\%_0$) and temperature ($\sim 2.5°C$). The Meddies are about 100 km in diameter, 800 m thick and centered on a depth of 1100 m. They may last for two years or longer [155,156], although they are steadily eroded from the edges, top and bottom, losing salt and heat with an *e*-folding time of about one year. A fine structure, present at the outer edges of the Meddies, has a vertical length scale of about 20 m, but is absent in the interior. Radial profiles reveal a core that is in almost solid-body rotation with a period of several days and a velocity that, at the perimeter, can be as high as 34 cm/s. The absence of vorticity indicates that the core of a Meddy is one of those rarely encountered regions of the ocean where the flow is not fully turbulent.

Turbulence, however, is ubiquitous in most ocean environments, generally acting to degrade the performance of underwater acoustic arrays. An expression of the turbulence is a randomly (spatially and temporally) fluctuating sound speed, which gives rise to volume scattering as well as introducing random phase changes into a plane wave as it propagates through the turbulent domain. In the absence of such phase fluctuations, the plane wave is fully coherent along its entire length, or in other words, its *coherence length* is infinite (assuming no boundaries). The signal gain of a line array is then equal to N^2, where N is the number of sensors in the array, which in principle could be indefinitely large. Turbulence, however, reduces the coherence length of an incident sound wave, one effect of which is to impose an upper limit on the useful aperture of a hydrophone receiving array [157].

The coherence length can be determined from numerical modeling, from calculations of wave propagation in random media, or from direct measurement. Carey [157] adopted the measurement approach, finding that the horizontal coherence length in deep water, to source ranges of 500 km, is approximately 100 wavelengths at a frequency of 400 Hz. For a line array with half-wavelength spacing between adjacent elements, this corresponds to a maximum useful value of $N = 200$, giving a signal gain of 40,000 (46 dB). In shallow water with a downward refracting sound speed profile and a sand-silt bottom, an environment in which turbulence is only one of several factors degrading array performance, the horizontal coherence length was found to be of the order of 30 wavelengths for source ranges out to 45 km. In a commemorative paper [158] to Bill Carey, this upper limit of 30 wavelengths is referred to as the "Carey number."

1.5.12 DIURNAL AND SEASONAL CHANGES

Many oceanographic processes exhibit variability, which may be on daily, seasonal, or year-to-year time scales. The tides are an obvious example; and tidal forcing can impose cyclic behavior on other processes, for instance, the acoustic source level of black smokers, which has been observed to exhibit a semidiurnal periodicity of 1.93 cycles per day, corresponding to the 12.42-h tidal component [151]. This amplitude modulation of the emitted sound is related to the discharge rate from the black smoker, which in turn depends on the mechanics of the tidal loading. An example of seasonal variability is the transport of water in the major current systems, for instance, in the Florida Current, that segment of Gulf Stream flowing through the Florida Strait between West Palm Beach and Grand Bahama Island, where the volumetric flow rate is 15–20% greater in summer than winter [159]. Considerable short-term fluctuations are superimposed upon this seasonal cycle.

Seasonal weather conditions affect the sound speed profile, and hence acoustic propagation, within several hundred meters of the sea surface. As illustrated in Fig. 1.8, solar heating in the summer creates a layer of relatively warm water, known as the secondary or seasonal thermocline, which is downward refracting and hence does not support long-range sound propagation but instead gives rise to a deep acoustic shadow zone. During winter months, strong winds create a well-mixed surface layer of isothermal water in which the sound speed increases linearly with depth. This surface duct acts as an acoustic waveguide, supporting propagation out to relatively long ranges.

In addition to seasonal changes, the layer immediately beneath the sea surface, just a few meters deep, is also subject to diurnal variations, warming up during sunny daylight hours and cooling off at night. During the day, solar warming creates a downward refracting sound speed profile in this surface layer, which degrades the performance of surface-ship echo-ranging sonars. This phenomenon, which was first encountered in the early days of active sonar operations, is known as the "afternoon effect" [25] since it is most pronounced later in the day.

Volume reverberation from deep scattering layers in the ocean is another example of diurnal variability, which was reported soon after the end of World War II by Eyring, Christensen, and Raitt [160]. The scattering originates in a diffuse layer, called at the time the ECR layer after its discoverers, although the contemporary name is the *deep scattering layer*. During the day, a high concentration of marine organisms, notably small mesopelagic fish, mostly lantern fish, with swim bladders that act as efficient scatterers of sound, populates the deep scattering layer at a depth of 300–500 m, where the water is cold, dark, and deficient in oxygen. At dusk, a vertical migration toward the surface takes place as the organisms move upward to feed in the nutrient-rich waters of the epipelagic (sunlight illuminated) zone. With the rising of the sun, the organisms return to the darkness of deeper depths, where they are relatively safe from predators.

1.6 SONAR EQUATIONS

The sonar equations combine parameters from the sonar equipment, environment, and target to provide the basis for preliminary sonar performance evaluation. Sonar equations are particularly used for:

1. predicting an existing sonar system's performance under selected conditions, or
2. designing a sonar system to operate under desired conditions.

In the past sonar equations have been used extensively in underwater acoustics; however, they have also been abused since the sonar equations have several limitations, which must be understood to obtain reliable and useful results. Sonar equations express the conservation of acoustic energy and individual terms within the equations provide the magnitude of each term's influence on the overall sonar detection performance. All sonar equation terms are in dimensionless form and expressed using a logarithmic decibel scale where the reference intensity is an assumed plane acoustic wave with an rms pressure of 1 μPa.

1.6.1 DEFINITIONS OF THE SONAR EQUATION TERMS

The *source level* (*SL*) of a projector is based on the intensity measured 1 m from the source's acoustic center. For an *omnidirectional source*, *SL* is given by:

$$SL = 10 \log \left[\frac{source\ intensity\ at\ standard\ distance}{reference\ intensity} \right] \{unit:\ dB\} \quad (1.75)$$

where the reference is the acoustic intensity in a plane wave with a rms pressure of 1 μPa. In Chapter 2, the intensity $I = p^2/\rho c$, where p {unit: Pa} is the rms-pressure and ρc {unit: kg/(m²s)} is the acoustic impedance of the water around the source. In seawater $\rho c \sim 1.5 \cdot 10^6$ kg/(m²s). Thus the reference intensity $I_r = 0.67 \cdot 10^{-18}$ W/m². An omnidirectional source emitting an acoustic power, P {unit: W}, at 1 m distance produces an intensity $I = P/4\pi$ {unit: W/m²}, which makes the source level:

$$SL = 10 \log \left[\frac{I}{I_r} \right] = 10 \log P + 170.8 \ \{unit:\ dB\} \quad (1.76)$$

For a *directional source*, *SL* is influenced by the acoustic power flow in a characteristic direction. Thus the directivity index *DI* {unit: dB} from Eq. (10.55) in Chapter 10 has to be added to Eq. (1.76). The emitted acoustic power P is determined by the projector qualities, including input electrical power, projector's transfer function, and acoustical impedance matching between the projector and its environment.

The *noise level* (*NL*) receives contributions from *self-noise*, that is, noise produced by the measurement platform during the acoustic measurements, *ambient noise*, see Chapter 6, Section 6.6, and *thermal noise*, discussed in Section 6.2.11. *NL* is given by:

$$NL = 10 \log \left[\frac{noise\ intensity\ measured\ by\ omnidirectional\ hydrophone}{reference\ intensity} \right] \{unit:\ dB\}$$

$$(1.77)$$

where the reference is the acoustic intensity of a plane wave in a 1 Hz bandwidth with a rms pressure 1 μPa. Each noise source produces its own contribution to the overall *NL* value. When the noise sources are mutually independent, which frequently is the case, their intensities may be added to produce the total noise level at the receiver terminals.

The *reverberation level (RL)* in dB receives contributions from *surface reverberations*, that is, contributions from the sea surface, seafloor, layers in the water column, and *volume reverberations*, as discussed in Section 5.6. RL_v {unit: dB re 1 µPa} measured at the receiver location due to volume reverberation is given by Eq. (1.78) and RL_s {unit: dB re 1 µPa} due to surface reverberation is given by Eq. (1.79).

$$RL_v = SL - 20 \log r - 2\alpha r + 10 \log \psi + 10 \log\left(\frac{c\tau}{2}\right) + S_{bv} \qquad (1.78)$$

$$RL_s = SL - 30 \log r - 2\alpha r + 10 \log \Phi + 10 \log\left(\frac{c\tau}{2 \cos \theta}\right) + S_{bs} \qquad (1.79)$$

where SL is the source level of the transmitted pulse measured at 1 m source distance {unit: dB rel. 1 µPa}, and r is the distance {unit: m} between the source and the scatterers contributing to the reverberation. α {unit: dB/m} is the absorption coefficient for sound propagation in seawater, τ {unit: s} is the transmitted tone burst duration, and c is the sound velocity in seawater {unit: m/s}. S_{bv} {unit: dB} is the volume scatterer's backscattering strength per unit volume (i.e., the target strength of 1 m^3 of the scattering volume). ψ {unit: steradians} is an ideal solid angle beam width of the source−receiver combination including azimuth and elevation, and is the aperture angle for an ideal beam pattern with a flat response over the angle ψ and no response outside ψ. Φ {unit: radians} is the ideal plane angle beam width of the source−receiver system and θ is the grazing angle. S_{bs} is the backscattering strength {unit: dB} of 1 m^2 of the scattering surface.

The *transmission loss (TL)* is the reduction in acoustic intensity over the signal propagation path. If the acoustic intensity is I_o {unit: W/m^2} at the reference point 1 m from the source and it is I_r at the distance r {unit: m} from the source, the intensity transmission loss is given by:

$$TL = 10 \log\left[\frac{I_o}{I_r}\right] \{\text{unit: dB}\} \qquad (1.80)$$

The acoustic intensity loss is due to geometrical spreading, that is, cylindrical or spherical spreading, and attenuation. The attenuation receives contributions from scattering by inhomogeneities in the propagation path and frequency dependent absorption caused by viscosity, heat conductivity, and relaxation effects, as discussed in Chapter 4. The *TL* value summarizes all transmission loss contributions in a single number.

The *target strength (TS)* represents the backscattered intensity from a target and is given by:

$$TS = 10 \log\left(\frac{I_{bs}}{I_i}\right) \{\text{unit: dB}\} \qquad (1.81)$$

where I_{bs} is the intensity of the backscattered (reflected) acoustic signal measured 1 m from the target's acoustic center, and I_i is the intensity of the acoustic signal incident on the target. For large targets, such as a submarine, the 1 m reference point from the acoustic center may be inside the submarine's hull. Eq. (1.81) may produce positive *TS* values which could lead to the erroneous conclusion, that more acoustic energy is backscattered from the target than is incident. The target strength of various geometrical objects is provided in Chapter 5.

The *directivity index* (*DI*) for a directional projector based on Eq. (10.55), Chapter 10 is:

$$DI = 10 \log D = 10 \log \left(\frac{I_D}{I_o} \right) \{\text{unit: dB}\} \qquad (1.82)$$

where I_D is the acoustic intensity on the acoustic axis of the projector measured at a distance r from the projector. I_o is the intensity at the distance r produced by an omnidirectional source emitting the same acoustic power as the directional projector. D denotes the projector's directivity. For a transducer used as a projector and a hydrophone the transmission and receiver directivity indices are the same.

The *detection threshold* (*DT*) is the ratio of mean signal power to mean noise power in a 1-Hz band measured at the receiver terminals. This results in pre-selected values for *detection probability*, P_d, and *false-alarm probability*, P_{fa}. If S {unit:W} is the mean acoustic signal power at the receiver terminals and N {unit:W} is the noise power in a 1-Hz band at the receiver terminals, *DT* is given by:

$$DT = 10 \log \left(\frac{S}{N} \right) \{\text{unit: dB}\} \qquad (1.83)$$

For *noise-limited reception*, that is, when ambient and self-noise limit detection, the noise and detection thresholds are defined as spectral quantities based on a 1-Hz bandwidth. However, for *reverberation-limited reception* the total noise power over the receiver bandwidth defines the signal-to-noise ratio (SNR) and *DT* according to Eq. (1.83). The SNR is the main parameter which affects receiver performance in target detection and it expresses the relative importance of acoustic power contributions from signal and noise.

When the signal and noise both have a Gaussian distribution with the same standard deviation, σ, the *detection index* (*d*) may be defined based on the detection threshold as:

$$d = \left[\frac{(mean(signal + noise) \; amplitude - mean \; noise \; amplitude)}{\sigma} \right]^2 \qquad (1.84)$$

According to Urick [161], the detection index d is equivalent to the signal-plus-noise to noise ratio of the envelope of the receiver output at the terminals where the amplitude threshold T_{th} is established, as shown in Fig. 1.29. This figure shows the Gaussian distributed probability densities for noise and signal-plus-noise amplitudes together with the detection, P_d, and false-alarm, P_{fa}, probabilities. P_d is the probability a true signal exceeds the detection threshold at the receiver input. P_{fa} is the probability a noise peak in absence of a true signal exceeds the detection threshold at the receiver input. The hatched area to the right of the preselected amplitude

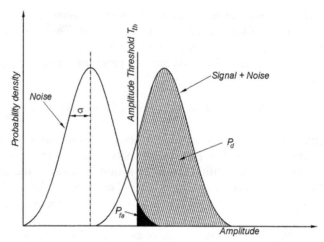

FIGURE 1.29

When the Gaussian distributed probability densities for noise and signal-plus-noise amplitudes have the same standard deviation, σ, the amplitude threshold T_{th} determines detection and false-alarm probabilities and the detection index d. P_d is determined by the hatched area and P_{fa} is determined by the dark area.

Adapted from Figure 12.8, Urick, R. J., Principles of Underwater Sound, 3rd Ed., McGraw-Hill Book Company, 1983, Peninsula Press, Inc., with permission Peninsula Press, Inc.

threshold T_{th} is the probability an amplitude in excess of T_{th} is due to signal plus noise. The hatched area, therefore, represents P_d. The dark area under the noise curve to the right of the T_{th} represents the probability that an amplitude in excess of T_{th} is due to noise alone, and represents P_{fa}. By varying T_{th}, detection and false-alarm probabilities will vary, as will the detection index, d.

The relation between P_d and P_{fa} is given by the *receiver operating characteristic (ROC)* curves shown in Fig. 1.30. For selected values of P_d and P_{fa} the detection index, d, can be read directly from the figure, which determines the likelihood detector threshold for a single detection event. The total number of detection events depends on the sonar type. According to Urban [162], for *passive* sonar operation, the detection event number is the product of the number of receiver channels, the number of frequency resolution cells and, if the detection is performed in the frequency domain, the number of spectra processed per time unit. For *active* sonar operation the detection event number is approximately the product of the number of receiver channels, number of range cells, and number of transmission cycles per unit time.

The *processing gain (PG)* {unit: dB} for a receiver is expressed as the difference between the signal-to-noise ratio in dB at the receiver input to the signal-to-noise ratio in dB at the receiver output, when both ratios are measured over the receiver bandwidth.

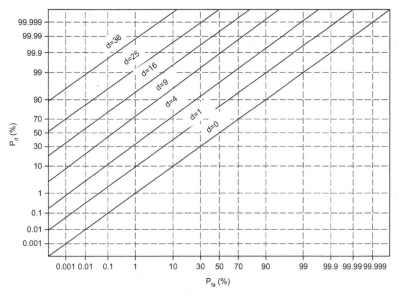

FIGURE 1.30

Receiver operating characteristic (ROC) curves represent the relation between P_d and P_{fa} when both are expressed as percentages. From selected values of P_d and P_{fa} the detection index d can be read from the *ROC curves*.

Adapted from Figure 12.7, Urick, R. J., Principles of Underwater Sound, 3rd Ed., McGraw-Hill Book Company (1983), Peninsula Press, Inc., with permission Peninsula Press, Inc.

1.6.2 SONAR EQUATIONS

When the detection threshold *DT* has been determined from selected detection and false-alarm probabilities, the sonar equation for passive sonar, which is also called the *passive sonar equation*, is given by:

$$SL - TL - NL + DI \geq DT \quad \{\text{unit: dB}\} \tag{1.85}$$

In the passive sonar equation the source level, *SL*, of the distant source, which is the signal level produced at 1 m from the distant source is reduced by the one-way transmission loss, *TL*, between the point 1 m from the distant source and passive receiver, reduced by the local environmental noise level *NL*, and increased by the passive receiver's directivity index, *DI*. If the resulting value exceeds or at least is equal to the detection threshold *DT*, detection occurs. Eq. (1.85) can be used to evaluate a passive sonar's performance for the detection of noise emitted by a distant source or an active transmission from another sonar platform. In the latter case, the passive sonar equation is termed the *intercept sonar equation*, when the passive sonar is optimized for detection of active sonar signals emitted from another sonar platform, and the source level, SL, refers to the active sonar source level determined at a distance 1 m from the active sonar.

Expressions for the detection threshold DT are frequently based on a square law detector acting on a signal imbedded in Gaussian noise. In this case according to Waite [163], DT for a passive sonar is given by Eq. (1.86):

$$DT = 5 \log d - 5 \log BT \quad \{\text{unit: dB}\} \tag{1.86}$$

In the product BT, B is the signal bandwidth and T is the integration time. A broadband sonar is represented by a large BT value. Eq. (1.86) assumes that the detection decision is based on a single time sample or event. However, in practice, detection decisions by operators or by automatic detection systems are based on a number n of successive time samples necessary to make the detection decision. Therefore, the expression for DT in Eq. (1.86) should be modified by including a term $5 \log n$. According to Waite [163], the passive sonar equation for a *broadband square law detector* can be written as:

$$SL - TL - NL + DI - 5 \log d + 5 \log BT + 5 \log n \geq 0 \,\{\text{unit: dB}\} \tag{1.87}$$

If a linear detector is used, DT in Eq. (1.87) becomes $DT = 10 \log d - 10 \log BT$.

The sonar equation for an active sonar, also called *the active sonar equation*, can be written for a *monostatic* sonar, when the signal reception is *noise limited*, as:

$$SL - 2TL + TS - NL + DI \geq DT \,\{\text{unit: dB}\} \tag{1.88}$$

SL, the source level 1 m from the projector, is reduced by the round trip transmission loss, $2TL$, due to the sonar signal's propagation to the target and back to the hydrophone collocated with the projector, increased by the target strength TS related to 1 m from the acoustic center of the target, reduced by the noise level NL measured at the receiver terminals and increased by the hydrophone directivity index DI. If the result is greater than or equal to the detection threshold DT determined by preselected probabilities for detection P_d and false-alarm P_{fa}, the target is detected.

The noise level NL receives contributions from ambient noise, see Chapter 6, and self-noise, see Section 6.6. Self-noise is produced by the platform which carries the sonar, for example, electrical noise, machinery noise, propeller noise, and flow noise. These can be reduced by appropriate platform design and construction procedures. Self-noise is frequently directive, depending on the source. The amplitude very often increases with the platform speed. The self-noise intensity level measured at the hydrophone—water interface is expressed in dB relative to the omnidirectional intensity of a plane wave with pressure amplitude equal to 1 μPa in a 1 Hz band.

For a bistatic active sonar, when the projector position deviates from the hydrophone position, the two transmission paths are different and the term $2\,TL$ in Eq. (1.88) is replaced by $TL_1 + TL_2$ where TL_1 is the transmission loss over the path from the projector to target and TL_2 is the transmission loss over the path from the target to the hydrophone.

For an active sonar the projector's emitted signal is known to the sonar operator and the receiver is designed to match the detected signal by using *matched filter processing*. Two main classes of signals are frequently used by an active sonar, a *continuous wave* (CW) *pulse* of constant frequency and time duration T {unit: s}, or a *frequency modulation (FM) pulse*, where the frequency is changed during the pulse's time duration T {unit: s}. According to Waite [163], since the noise is assumed to be totally uncorrelated with the emitted signal pulse, the detection threshold DT for matched filter processing of an FM signal can be written as:

$$DT = 5 \log d - 10 \log BT \ \{\text{unit: dB}\} \tag{1.89}$$

Eq. (1.89) is based on the coherent replica correlation technique used to process FM signals, where the output of the beam former is correlated with a replica of the emitted signal. After n observations with a matched filter, the detection threshold becomes:

$$DT = 5 \log d - 10 \log BT - 5 \log n \ \{\text{unit: dB}\} \tag{1.90}$$

The broadband processing technique behind Eq. (1.86) for a passive sonar is incoherent, since nothing is known a priori about the unknown source's noise signal.

When perfectly matched filter processing is used in an active sonar for detecting noise-limited return signals, whether based on CW or FM pulses emitted by the projector, after n observations the active sonar equation is given by Eq. (1.91) since the noise-limited case is independent of the receiver bandwidth B:

$$SL - 2TL + TS - NL + DI - 5 \log d + 10 \log T + 5 \log n \geq 0 \ \{\text{unit: dB}\} \tag{1.91}$$

When the active sonar operation is *reverberation limited*, the reverberation level measured by the hydrophone depends on the projector source level, SL, and increases with SL. Since reverberation directivity depends on the environment which scatters the signals from the projector back to the hydrophone, *the reverberation is not isotropic*. As mentioned in Chapter 5, the reverberation may receive contributions from backscattering by surfaces, that is, the sea surface and seafloor, and by water column inhomogeneity. Since reverberation depends on local environmental conditions, it is difficult a priori to predict the reverberation level at a site, even if the site specific conditions are known. In most cases reverberation has to be measured. Moreover, the target will normally be positioned on or in the backscattering surfaces or volumes.

The dominant reverberation source is usually the sea surface, seafloor, or a horizontal layer in the water column. The sonar equation for *surface reverberation–limited* detection is written as:

$$SL - 2TL + TS - RL_S - DT \geq 0 \ \{\text{unit: dB}\} \tag{1.92}$$

where RL_S {unit: dB re 1 μPa} is the surface reverberation level observed at the hydrophone terminals. For monostatic operation this level can be found from

Eq. (1.79). The detection threshold may be expressed by: $DT = 5 \log d - 10 \log BT - 5 \log n$. When ambient noise and/or self-noise also influence the detection together with the reverberation, Eq. (1.92) can be replaced by:

$$SL - 2TL + TS - NL - RL_S - DT \geq 0 \; \{\text{unit: dB}\} \qquad (1.93)$$

When *volume reverberation* due to backscattering from inhomogeneities in the water column is the dominant source limiting detection, the sonar equation is given by:

$$SL - 2TL + TS - RL_V - DT \geq 0 \; \{\text{unit: dB}\} \qquad (1.94)$$

where RL_v {unit: dB re 1 μPa} is the volume reverberation. RL_v for a CW pulse is given by Eq. (1.78).

Surface and volume reverberation levels may be reduced by reducing the effective projector and hydrophone beam widths. This reduction can be achieved by increasing the transmitting and receiving transducer dimensions to many times the transmitted acoustic signal wavelength. An alternative approach is to reduce the sonar side lobe levels by using an appropriate window shading functions discussed in Chapter 11, such as Hamming or Dolph–Chebyshev. The reverberation level produced by the sonar's main and side lobes is modified by platform movements, in particular the yaw of larger surface vessels. Also, the platform movement, sea surface movement, and volume creating the reverberation will produce a Doppler spread in the reverberation.

Signals transmitted by active sonars are normally broadband CW pulses, FM pulses, or pseudorandom noise (PRN) pulses. The sonar system's detection efficiency also depends on the emitted sound pulse's bandwidth, B. The bandwidth of a CW pulse can be increased by reducing the pulse length since $B \sim 1/T$, where T {unit: s} is the pulse's time duration. When T is reduced, the acoustic power in each pulse is also reduced, which reduces the reverberation level. For a CW pulse reduced acoustic power affects the noise influence on detection and could make the detection noise limited. When T is increased the reverberation spectrum is reduced due to a reduction in the emitted signal spectrum. A potential Doppler signal from a target will have its spectrum in the low-power part of the reverberation spectrum, where the influence of ambient noise or self noise may limit detection. The use of matched filter processing, where a replica of the emitted pulse is correlated with the received signal in the matched filter, can also reduce the influence of reverberation on detection. A wideband FM pulse with a bandwidth B will produce a reverberation with its acoustic power spread over the same bandwidth B as the FM pulse. Reducing the pulse time duration increases B and reduces the area of surfaces contributing to reverberation and improves the sonar performance against reverberation. The frequency–time combination in PRN pulses may be changed in a known random manner from emitted sound pulse to emitted sound pulse—often called the ping. Since the random changes are known, matched filter processing can be utilized. In general it is best, if possible, to attempt to ensure that the sonar operation is always noise limited.

Multipath propagation also influences detection. Multipath propagation occurs when signals are propagated along paths other than the straight path between the sonar and its target, such as reflection and propagation at the seafloor or reflection at the sea surface. The reflected and propagated multipath signals are received by the active sonar hydrophones along with the direct return signal from the target. The impact is received signal distortion which extends the apparent range to the target. Multipath propagation increases the total acoustic energy returned to the hydrophone from the target, but without increasing the received reverberation level, and the possibility of detection is therefore enhanced. However, the signal distortion caused by multipath propagation has a negative influence on target classification since it confuses the arrival time and time history of the amplitude structure of the target return signals. Some reduction in multipath effects can be obtained by using narrow beams. The environmental shallow or deep water sound velocity profile has a strong influence on multipath effects.

1.7 ABBREVIATIONS

Table 1.4 is a list of abbreviations used in this chapter along with the page where the abbreviation is defined and first used.

Table 1.4 List of Abbreviations

Abbreviation	Description	Page 1st Usage
ASDIC	Allied Submarine Detection Investigation Committee	9
ASDICS	First practically working active sonar	10
ATOC	Acoustic thermometry of the ocean climate	15
AXBT	Aircraft-deployed expendable bathythermograph	33
BT	Bathythermograph	19
$B(OH)_3$	Boric acid	14
CO_2	Carbon dioxide	57
CTD	Conductivity—temperature—depth	19
dB	Decibel	16
DI	Directivity index	67
DT	Detection threshold	69
ECUA	European Conference on Underwater Acoustics	15
EEZ	Exclusive economic zone	1
GHG	Gruppen Horch Gerät	11
JONSWAP	Joint North Sea Project	40
$MgSO_4$	Magnesium sulfate	14
MIZ	Marginal ice zone	25
NL	Noise level	67
PG	Processing gain	70
RL	Reverberation level	68
ROC	Receiver operating characteristic	70

Continued

Table 1.4 List of Abbreviations—cont'd

Abbreviation	Description	Page 1st Usage
ROV	Remotely operated vehicle	1
SI	Systéme Internationale	15
SL	Source level	67
SOFAR	Sound fixing and ranging	13
SONAR	Sound navigation and ranging	2
SOSUS	Sound surveillance system	15
SURTASS	Surveillance towed array sensor system	15
SVX	Sound velocity sensor	19
SWARM-95	Shallow water acoustics in random media experiment 1995	46
TL	Transmission loss	68
TS	Target strength	68
UA	Underwater acoustics	17
UHF	Ultra high frequency	1
XBT	Expendable bathythermograph	33

ACKNOWLEDGMENT

Chapter 1 was written with support for Michael J. Buckingham from the Office of Naval Research, Code 322OA, under Award Number N00014-14-1-0247.

REFERENCES

[1] Bjørnø, L., *Features of Underwater Acoustics from Aristotle to Our Time*. Acoust. Phys., **49**, (1), pp. 24—30 (2003).

[2] Lindsay, R. B., *Acoustics — Historical and Philisophical Development*. Dowden, Hutchenson and Ross, Stroutsburg, PA, USA (1973).

[3] Lasky, M., *Review of undersea acoustics to 1950*. J. Acoust. Soc. Amer., **61**, (2), pp. 283—297 (1977).

[4] Hackmann, W., *Seek & Strike. Sonar, anti-submarine warfare and the Royal Navy 1914—54*. London: Her Majesty's Stationery Office (1984).

[5] Pekeris, C. L., *Theory of Propagation of Explosive Sound in Shallow Water*. The Geological Society of America, Memoir 27, October 15, 1948.

[6] Worzel, J. L. and Ewing, M., *Explosion Sounds in Shallow Water*. The Geological Society of America, Memoir 27, October 15, 1948.

[7] Ewing, M. and Worzel, J. L., *Long—Range Sound Transmission*. In the Geological Society of America Memoir **27**, *Propagation of Sound in the Ocean*. New York, pp. 1—35 (1948).

[8]   Pekeris, C. L., *Theory of Propagation of Sound in a Half-Space of Variable Sound Velocity under Conditions of Formation of a Shadow Zone*. J. Acoust. Soc. Amer., **18**, (2), pp. 295−315 (1946).

[9]   Tolstoy, I., *Dispersive properties of a fluid layer overlying a semi-infinite elastic solid.* Bull. Seismol. Soc. Amer., **44**, (3), pp. 493−512 (1954).

[10] Craik, A. D. D., *George Gabriel Stokes on water wave theory.* Ann. Rev. Fluid Mech., **37**, pp. 23−42 (2005).

[11] Liebermann, L. N., *The origin of sound absorption in water and sea water.* J. Acoust. Soc. Amer., **20**, pp. 868−873 (1948).

[12] Wilson, O. B. and Leonard, R. W., *Measurements of sound absorption in aqueous salt solutions by a resonator method.* J. Acoust. Soc. Amer., **26**, pp. 223−226 (1954).

[13] Schulkin, M. and Marsh, H. W., *Sound absorption in sea water.* J. Acoust. Soc. Amer., **34**, pp. 864−865 (1962).

[14] Yeager, E., Fisher, F. H., Miceli, J., and Bressel, R., *Origin of the low-frequency sound absorption in sea water.* J. Acoust. Soc. Amer., **53**, pp. 1705−1707 (1973).

[15] Fisher, F. H. and Simmons, V. P., *Sound absorption in sea water.* J. Acoust. Soc. Amer., **62**, pp. 558−564 (1977).

[16] Tappert, F. D., *The parabolic approximation method*, in Keller, J. and Papadakis, J. S. Editors, Wave Propagation and Underwater Acoustics, Springer-Verlag, Berlin, pp. 224−287 (1977).

[17] Wenz, G. M., *Acoustic ambient noise in the ocean: Spectra and sources.* J. Acoust. Soc. Amer., **34**, pp. 1936−1956 (1962).

[18] Munk, W. H., Spindel, R. C., Baggeroer, A. and Birdsall, T. G., *The Heard Island Feasibility Test.* J. Acoust. Soc. Amer., **96**, pp. 2330−2342 (1994).

[19] Stafford, K. M., Fox, C. G. and Clark, D. S., *Long-range detection and localization of blue whale calls in the northern Pacific Ocean.* J. Acoust. Soc. Amer., **104**, pp. 3616−3625 (1998).

[20] Papadakis, J. S. and Bjørnø, L. Editors, "Underwater Acoustics". Proceedings of the 1st International Conference and Exhibition on Underwater Acoustics after the merger of the ECUA and UAM Conferences, Foundation for Research and Technology − Hellas, Corfu, Greece, pp. 1710 (2013).

[21] Papadakis, J. S. and Bjørnø, L. Editors, "Underwater Acoustics". Proceedings of the 2nd International Conference and Exhibition on Underwater Acoustics, Foundation for Research and Technology − Hellas, Rhodes, Greece, pp. 1644 (2014).

[22] Parker, C. E., *Gulf stream rings in the Sargasso Sea*, Deep-Sea Res., **18**, pp. 981−991 (1971).

[23] Saunders, P. M., *Anticyclonic eddies formed from shoreward meanders of the Gulf Stream*, Deep-Sea Res., **18**, pp. 1207−1219 (1971).

[24] Barclay, D. R. and Buckingham, M. J., *On the spatial properties of ambient noise in the Tonga Trench, including the effects of bathymetric shadowing*, J. Acoust. Soc. Am., **136**, pp. 2497−2511 (2014).

[25] Urick, R. J., Principles of Underwater Sound, McGraw-Hill, New York, (1983), Peninsula Publishing.

[26] Sun, C., in OceansObs.09: *Sustained Ocean Observations and Information for Society*, Hall, J. A., Harrison, D. E., and Stammer, D. Editors, ESA Publication WPP-306, Venice, Italy, **2**, (2010).

[27] Taira, K., Yanagimoto, D. and Kitagawa, S., *Deep CTD casts in the Challenger Deep, Mariana Trench*, J. Phys. Ocean., **61**, pp. 447−454 (2005).

[28] Greenspan, M. and Tschiegg, C. E., *Sing-around ultrasonic velocimeter for liquids*, Rev. Sci. Instr., **28**, pp. 897–901 (1957).

[29] Barclay, D. R., Simonet, F. and Buckingham, M. J., *Deep Sound: a free-falling sensor platform for depth-profiling ambient noise in the deep ocean*, Mar. Tech. Soc. J., **43**, pp. 144–150 (2009).

[30] Munk, W. H., *Sound channel in an exponentially stratified ocean, with application to SOFAR*, J. Acoust. Soc. Am., **55**, pp. 220–226 (1974).

[31] Eckart, C., Hydrodynamics of Oceans and Atmospheres, Pergamon, Oxford, (1960).

[32] Miller, J. C., *Oceanic acoustic rays in the deep six sound channel*, J. Acoust. Soc. Am., **71**, pp. 859–862 (1982).

[33] Officer, C. B., Introduction to the Theory of Sound Transmission. McGraw-Hill, New York (1958).

[34] Northrup, J. and Colborn, J. G., *Sofar channel axial sound speed and depth in the Atlantic Ocean*, J. Geophys. Res., **79**, pp. 5633–5641 (1974).

[35] Buckingham, M. J., *On acoustic transmission in ocean-surface waveguides*, Phil. Trans. Roy. Soc. London A, **335**, pp. 513–555 (1991).

[36] Hurdle, B. G., Preface in *The Nordic Seas*, Hurdle, B. G. Editor, Springer-Verlag, New York, pp. v–vi, (1986).

[37] Johannessen, O. M., Brief Overview of the Physical Oceanography in *The Nordic Seas*, Hurdle, B. G. Editor, Springer-Verlag, New York, pp. 103–127, (1986).

[38] Hurdle, B. G., The Sound Speed Structure in *The Nordic Seas*, Hurdle, B. G. Editor, Springer-Verlag, New York, pp. 155–181, (1986).

[39] Meredith, R. W., Bucca, P. J. and McCoy, K., *Environmental Measurements and Analysis: Arctic Acoustics Experiments in the Marginal Ice Zone*, Naval Ocean Research and Development Activity, Report No. 210, pp. 1–70, (August 1989).

[40] Bryden, H. L., *New polynomials for thermal expansion, adiabatic temperature gradient and potential tempaerature of sea water*, Deep-Sea Res., **20**, pp. 401–408 (1973).

[41] Millero, F. J. and Poisson, A., *International one-atmosphere equation of state of seawater*, Deep-Sea Res., **28A**, pp. 625–629 (1981).

[42] Del Grosso, V. A., *Tables of the speed of sound in seawater (with Mediterranean Sea and Red Sea applicability)*, J. Acoust. Soc. Am., **53**, pp. 1384–1401 (1973).

[43] Del Grosso, V. A. and Mader, C. W., *Speed of sound in sea water samples*, J. Acoust. Soc. Am., **52**, pp. 961–974 (1972).

[44] Frye, H. W. and Pugh, J. D., *A new equation for the speed of sound in sea water*, J. Acoust. Soc. Am., **50**, pp. 384–386(L) (1971).

[45] Leroy, C. C., Robinson, S. P. and Goldsmith, M. J., *A new equation for the accurate calculation of sound speed in all oceans*, J. Acoust. Soc. Am., **124**, pp. 2774–2782 (2008).

[46] Leroy, C. C., *Development of simple equations for accurate and more realistic calculation of the speed of sound in seawater*, J. Acoust. Soc. Am., **46**, pp. 216–226 (1969).

[47] Mackenzie, K. V., *Nine term equation for sound speed in the oceans*, J. Acoust. Soc. Am., **70**, pp. 807–812 (1981).

[48] Mackenzie, K. V., *Formulas for the computation of sound speed in sea water*, J. Acoust. Soc. Am., **32**, pp. 100–104 (1960).

[49] Wilson, W. D., *Speed of sound in sea water as a function of temperature, pressure, and salinity*, J. Acoust. Soc. Am., **32**, pp. 641–644 (1960).

[50] Wilson, W. D., *Equation for the speed of sound in seawater*, J. Acoust. Soc. Am., **32**, pp. 1357L (1960).

[51] Wilson, W. D., *Extrapolation of the equation for the speed of sound in sea water*, J. Acoust. Soc. Am., **34**, pp. 866L (1962).

[52] Chen, C.-T. and Millero, F. J., *Speed of sound in seawater at high pressures*, J. Acoust. Soc. Am., **62**, pp. 1129−1135 (1977).

[53] Mackenzie, K. V., *Discussion of sea water sound speed determinations*, J. Acoust. Soc. Am., **70**, pp. 801−806 (1981).

[54] Millero, F. J. and Kubinski, T., *Speed of sound in seawater as a function of temperature and salinity at 1 atm*, J. Acoust. Soc. Am., **57**, pp. 312−319 (1975).

[55] Medwin, H., *Speed of sound in seawater: A simple equation for realistic parameters*, J. Acoust. Soc. Am., **58**, pp. 1318−1319 (1975).

[56] Fofonov, N. P. and Millard Jr., R. C., *Algorithms for Computation of Fundamental Properties of Seawater*, UNESCO Technical Papers in Marine Science, Report No. 44, Division of Marine Sciences, UNESCO, Place du Fontenoy, Paris, pp. 1−54 (1983).

[57] UNESCO, *Background Papers and Supporting Data on the International Equation of State of Seawater 1980*, UNESCO Technical Papers in Marine Science, Report No. 38, pp. 1−192 (1981).

[58] IOC, SCOR and IAPSO, *The International Thermodynamic Equation of Seawater - 2010: Calculation and use of Thermodynamic Properties*, Intergovernmental Oceanographic Commission, Manuals and Guides No. 56, pp. 1−196 (2010).

[59] Farmer, D. M. and Vagle, S., *Waveguide propagation of ambient sound in the ocean-surface bubble layer*, J. Acoust. Soc. Am., **86**, pp. 1897−1908 (1989).

[60] Gilbert, K. E., *A stochastic model for scattering from the near-surface oceanic bubble layer*, J. Acoust. Soc. Am., **94**, pp. 3325−3334 (1993).

[61] Longuet-Higgins, M. S., *Bubble noise spectra*, J. Acoust. Soc. Am., **87**, pp. 652−661 (1990).

[62] Ding, L. and Farmer, D. M., *On the dipole acoustic source level of breaking waves*, J. Acoust. Soc. Am., **96**, pp. 3036−3044 (1994).

[63] Farmer, D. M. and Lemon, D. D., *The influence of bubbles on ambient noise in the ocean at high wind speeds*, J. Phys. Ocean., **14**, pp. 1762−1778 (1984).

[64] Medwin, H. and Beaky, M. M., *Bubble sources of the Knudsen sea noise spectra*, J. Acoust. Soc. Am., **86**, pp. 1124−1130 (1989).

[65] Buckingham, M. J., Noise in Electronic Devices and Systems. Ellis Horwood, Chichester, (1983).

[66] Phillips, O. M., *The equilibrium range in the spectrum of wind-generated waves*, J. Fluid Mech., **4**, pp. 426−434 (1958).

[67] Pierson, Jr., W. J. and Moskowitz, L., *A proposed spectral form for fully developed wind seas based on the similarity theory of S. A. Kitaigorodskii*, J. Geophys. Res., **69**, pp. 5181−5190 (1964).

[68] Kitaigorodskii, S. A., *Applications of the theory of similarity to the analysis of wind-generated wave motion as a stochastic process*, Izvest. Acad. Nauk, SSR, Geophys. Series No. **1**, pp. 105−117 (1961).

[69] Ewing, J. A., *Some results from the Joint North Sea Wave Prtoject of interest to engineers*, in The Dynamics of Marine Vehicles and Structures in Waves, Bishop, R. E. D. and Price, W. G. Editors, Mechanical Engineering Publications Ltd., London, (1975).

[70] Hasselmann, K., Barnett, T. P., Bouws, E., Carlson, H., Cartwright, D. E., Enke, K., Ewing, J. A., Gienapp, H., Hasselmann, D. E., Kruseman, P., Meerburgh, A., Muller, P., Olbers, D. J., Richter, K., Sell, W. and Walden, H., *Measurements of wind wave growth and swell decay during the Joint North Sea Wave Project (JONSWAP)*, Herausgegeben vom Deutsch Hydrographisches Zeitschrift Reihe A (8), Report No. 12, (1973).

[71] Hasselmann, K., Ross, D. B., Müller, P. and Sell, W., *A parametric wave prediction model*, J. Phys. Ocean., **6**, pp. 200−228 (1976).

[72] Aranuvachapun, S., *Parameters of JONSWAP spectral model for surface gravity waves-II. Predictability from real data*, Ocean Eng., **14**, pp. 101−115 (1987).

[73] Battjes, J. A., Zitman, T. J. and Holthuusen, L. H., *A reanalysis of the spectra observed in JONSWAP*, J. Phys. Ocean., **17**, pp. 1288−1295 (1987).

[74] Kumar, V. S. and Kumar, K. A., *Spectral characteristics of high shallow water waves*, Ocean Eng., **35**, pp. 900−911 (2008).

[75] Lewis, A. W. and Allos, R. N., *JONSWAP's parameters: sorting out the inconsistencies*, Ocean Eng., **17**, pp. 409−415 (1990).

[76] Eckart, C., *Internal waves in the ocean*, The Physics of Fluids, **4**, pp. 791−799 (1961).

[77] Knauss, J. A., Introduction to Physical Oceanography, Waveland Press, Long Grove, Illinois, (1997).

[78] Rouseff, D. and Lunkov, A. A., *Modeling the effects of linear shallow-water internal waves on horizontal array coherence*, J. Acoust. Soc. Am., **138**, pp. 2256−2265 (2015).

[79] Apel, J. R., Ostrovsky, L. A., Stepanyanys, Y. A. and Lynch, J. F., *Internal solitons in the ocean and their effect on underwater sound*, J. Acoust. Soc. Am., **121**, pp. 695−722 (2007).

[80] Zhou, J. X., Zhang, X. S. and Rogers, P. J., *Resonant interaction of sound wave with internal solitons in the coastal zone*, J. Acoust. Soc. Am., **90**, pp. 2042−2054 (1991).

[81] Apel, J. R., Badiey, M., Chiu, C. S., Finette, S., Headrick, R. H., Kemp, J., Lynch, J. F., Newhall, A., Orr, M. H., Pasewark, B. H., Tielberger, D., Turgut, A., von der Heydt, K. and Wolf, S., *An overview of the 1995 SWARM shallow-water internal wave acoustic scattering experiment*, IEEE J. Ocean Eng., **22**, pp. 465−500 (1997).

[82] Thorpe, S. A., *The effect of Langmuir circulation on the distribution of submerged bubbles caused by breaking wind waves*, J. Fluid Mech., **142**, pp. 151−170 (1984).

[83] Crawford G. B. and Farmer, D. M. *On the spatial distribution of ocean bubbles*, J. Geophys. Res., **92**, pp. 8231−8243 (1987).

[84] Thorpe, S. A. *On the clouds of bubbles formed by breaking wind-waves in deep water, and their role in air-sea gas transfer*, Phil. Trans. Roy. Soc. A., **304**, pp. 155−210 (1982).

[85] Garrett, C., Li, M. and Farmer, D. M. *The connection between bubble size spectra and energy dissipation rates in the upper ocean*, J. Phys. Ocean., **30**, pp. 2163−2171 (2000).

[86] Eller, A. J. *Damping constants of pulsating bubbles*, J. Acoust. Soc. Am., **47**, pp. 1469−1470 (1970).

[87] Mallock, A., *The damping of sound by frothy liquids*, Proc. Roy. Soc. Lond., Series A **84**, pp. 391−395 (1910).

[88] Wood, A. B., A Textbook of Sound. G. Bell and Sons Ltd., London, (1964).

[89] Thorpe, S. A., *A model of the turbulent diffusion of bubbles below the sea surface*, J. Phys. Ocean., **14**, pp. 841−854 (1984).

[90]   Thorpe, S. A., *On the determination of K_v in the near-surface ocean from acoustic measurements of bubbles*, J. Phys. Ocean., **14**, pp. 855–863 (1984).

[91]   Medwin, H. and Breitz, N. D., *Ambient and transient bubble spectral densities in quiescent seas and under spilling breakers*, J. Geophys. Res., **94**, (C9), 12,751–12,759 (1989).

[92]   Buckingham, M. J. and Potter, J. R. Editors, Sea Surface Sound '94: Proceedings of the III International Meeting on Natural Physical Sources of Underwater Sound. World Scientific, Singapore, pp. 1–494 (1995).

[93]   Kerman, B. R. Editor, Natural Physical Sources of Underwater Sound: Sea Surface Sound (2). Kluwer, Dordrecht, pp. 1–750 (1993).

[94]   Kerman, B. R. Editor, Sea Surface Sound: Natural Mechanisms of Surface Generated Noise in the Ocean. Kluwer, Dordrecht, pp. 1–639 (1988).

[95]   Leighton, T. G. Editor, Natural Physical Processes Associated with Sea Surface Sound. University of Southampton, Southampton, pp. 1–278 (1997).

[96]   Pumphrey, H. C. and Elmore, P. A., *The entrainment of bubbles by drop impacts*, J. Fluid Mech., **220**, pp. 539–567 (1991).

[97]   Pumphrey, H. C. and Ffowcs Williams, J., *Bubbles as sources of ambient noise*, IEEE J. Ocean. Eng., **15**, pp. 268–274 (1990).

[98]   M. Minnaert, J., *On musical air-bubbles and the sounds of running water*, Phil. Mag., **16**, pp. 235–248 (1933).

[99]   Longuet-Higgins, M. S., *Monopole emission of sound by asymmetric bubble oscillations. Part 1. Normal modes*, J. Fluid Mech., **201**, pp. 525–541 (1989).

[100]   Longuet-Higgins, M. S., *Monopole emission of sound by asymmetric bubble oscillations. Part 2. An initial-value problem*, J. Fluid Mech., **201**, pp. 543–565 (1989).

[101]   Strasberg, M., *Gas bubbles as sources of sound in liquids*, J. Acoust. Soc. Am., **28**, pp. 20–26 (1956).

[102]   Knudsen, V. O., Alford, R. S. and Emling, J. W., *Underwater ambient noise*, J. Mar. Res., **7**, pp. 410–429 (1948).

[103]   Gaul, R. D., Knobles, D. P., Shooter, J. A. and Wittenborn, A. F., *Ambient noise analysis of deep-ocean measurements in the Northeast Pacific*, IEEE J. Ocean. Eng., **32**, pp. 497–512 (2007).

[104]   Kerman, B. R., *Underwater sound generation by breaking wind waves*, J. Acoust. Soc. Am., **75**, pp. 149–165 (1984).

[105]   Longuet-Higgins, M. S., *A theory of the origin of microseisms*, Phil. Trans. Roy. Soc., **243**, pp. 1–35 (1950).

[106]   Kibblewhite, A. C. and Ewans, K. C., *Wave-wave interactions, microseisms, and infrasonic ambient noise in the ocean*, J. Acoust. Soc. Am., **78**, pp. 981–994 (1985).

[107]   Isakovich, M. A. and Kur'yanov, B. F., *Theory of low-frequency noise in the ocean*, Sov. Phys. Acoust., **16**, pp. 49–58 (1970).

[108]   Goncharov, V. V., *Sound generation in the ocean by the interaction of surface waves and turbulence*, Izvest., Atmos. Ocean Phys., **6**, pp. 1189–1196 (1970).

[109]   Lu, N. Q., Prosperetti, A. and Yoon, S. W., *Underwater noise emissions from bubble clouds*, IEEE J. Ocean. Eng., **15**, pp. 275–281 (1990).

[110]   Prosperetti, A. *Bubble-related ambient noise in the ocean*, J. Acoust. Soc. Am., **84**, pp. 1042–1054 (1988).

[111]   Prosperetti, A., *Bubble dynamics in oceanic ambient noise* in Sea Surface Sound: Natural Mechanisms of Surface Generated Noise in the Ocean, Kerman, B. R. Editor, Kluwer, Dordrecht, pp. 151–171 (1988).

[112] Prosperetti, A., *Bubble related ambient noise in the ocean*, J. Acoust. Soc. Am., **78**, (Supplement 1), pp. S2 (1985).

[113] Carey, W. M., Fitzgerald, J. W. and Browning, D. G., *Low frequency noise from breaking waves*, in Natural Physical Sources of Ambient Sound: Sea Surface Sound (2), Kerman, B. R. Editor, Kluwer, Cambridge, U. K., pp. 277–304 (1993).

[114] Carey, W. M. and Browning, D., *Low frequency ocean ambient noise: measurements and theory* in Sea Surface Sound: Natural Mechanisms of Surface Generated Noise in the Ocean, Kerman, B. R. Editor, Kluwer, Lerici, Italy, pp. 361–376 (1988).

[115] Carey, W. M. and Fitzgerald, J. W., *Low-frequency noise and bubble plume oscillations*, J. Acoust. Soc. Am., **82**, (Supplement 1), pp. S62 (1987).

[116] van Wijngaarden, L., *One-dimensional flow of liquids containing small gas bubbles*, Ann. Rev. Fluid Mech., **4**, pp. 369–396 (1972).

[117] Hsieh, D.-Y. and Plesset, M. S., *On the propagation of sound in a liquid containing gas bubbles*, Phys. Fluids, **4**, pp. 970–975 (1961).

[118] Hollett, R. *Underwater sound from whitecaps at sea*, J. Acoust. Soc. Am., **85**, (Supplement 1), pp. S145 (1989).

[119] Carey, W. M., Fitzgerald, J. W., Monahan, E. C. and Wang, Q., *Measurement of the sound produced by a tipping trough with fresh and salt water*, J. Acoust. Soc. Am., **93**, pp. 3178–3192 (1993).

[120] Hahn, T. R., Berger, T. K. and Buckingham, M. J., *Acoustic resonances in the bubble plume formed by a plunging water jet*, Proc. Roy. Soc. Lond., Series A, **459**, pp. 1751–1782 (2003).

[121] Leighton, T. G., The Acoustic Bubble. Academic Press, London, pp. 1–613 (1994).

[122] Tans, P., *An accounting of the observed increase in oceanic and atmospheric CO_2 and an outlook for the future*, Oceanography, **22**, pp. 26–35 (2009).

[123] Keeling, C. D., *The concentration and isotropic abundance of carbon dioxide in the atmosphere*, Tellus **XII**, pp. 200–203 (1960).

[124] Dlugokencky, E. and Tans, P., www.esrl.noaa.gov/gmd/ccgg/trends/ (2016).

[125] Kump, L. R., Bralower, T. J. and Ridgwell, A., *Ocean acidification in deep time*, Oceanography, **22**, pp. 94–107 (2009).

[126] Revelle, R. and Suess, H. E., *Carbon dioxide exchange between atmosphere and ocean and the question of an increase of atmospheric CO_2 during the past decades*, Tellus, **9**, pp. 18–27 (1957).

[127] Merlivat, L. and Memery, L., *Gas exchange across an air-water interface: experimental results and modeling of bubble contribution to transfer*, J. Geophys. Res., **88**, pp. 707–724 (1983).

[128] Melville, W. K., *The role of surface-wave breaking in air-sea interaction*, Ann. Rev. Fluid Mech., **28**, pp. 279–321 (1996).

[129] Goddijn-Murphy, L., Woolf, D. K., Callaghan, A. H., Nightingale, P. D. and Shutler, J. D., *A reconciliation of empirical and mechanistic models of the air-sea gas transfer velocity*, J. Geophys. Res., **121**, pp. 1–18, http://dx.doi.org/10.1002/2015JC011096 (2015).

[130] Jacobson, M. Z., *Studying ocean acidification with conservative, stable numerical schemes for non-equilibrium air-ocean exchange and ocean equilibrium chemistry*, J. Geophys. Res., **110**, (D07302), pp. 1–17 (2005).

[131] Brewer, P. G. and Hester, K., *Ocean acidification and the increasing transparency of the ocean to low frequency sound*, Oceanography, **22**, pp. 86–93 (2009).

[132] Doney, S. C., Balch, W. M., Fabry, V. J. and Feely, R. A. *Ocean acidification: a critical emerging problem for the ocean sciences*, Oceanography, **22**, pp. 16−25 (2009).

[133] Francois, R. E. and Garrison, G. R., *Sound absorption based on ocean measurements: Part II: Boric acid contribution and equation for total absorption*, J. Acoust. Soc. Am., **72**, pp. 1879−1890 (1982).

[134] Ilyina, T., Zeebe, R. E. and Brewer, P. G. *Future ocean increasingly transparent to low-frequency sound owing to carbon dioxide emissions*, Nat. Geosci., **3**, pp. 18−22 (2010).

[135] Reeder, D. B. and Chiu, C.-S., *Ocean acidification and its impact on ocean noise: Phenomenology and analysis*, J. Acoust. Soc. Am., **128**, (3), pp. EL137-EL143 (2010).

[136] Caldeira, K. and Wickett, M. E., *Ocean model predictions of chemistry changes from carbon dioxide emissions to the atmosphere and ocean*, J. Geophys. Res., **110**, pp. C09S04, 1−12 (2005).

[137] Cohen, A. L. and Holcomb, M., *Why corals care about ocean acidification: Uncovering the mechanism*, Oceanography, **22**, pp. 118−127 (2009).

[138] Kleypas, J. A. and Yates, K. K., *Coral reefs and ocean acidification*, Oceanography, **22**, pp. 108−117 (2009).

[139] Cooley, A. L., Kite-Powell, H. I. and Doney, S. C., *Ocean acidification potential to alter global marine ecosystem services*, Oceanography, **22**, pp. 172−181 (2009).

[140] Weiss, R. F., Lonsdale, P. F., Lupton, J. E., Bainbridge, A. E. and Craig, H., *Hydrothermal plumes in the Galapagos Rift*, Nature, **267**, pp. 600−603 (1977).

[141] Williams, D. L., von Herzen, R. P., Sclater, J. G. and Andersen, R. N., The Galapagos spreading centre: lithospheric cooling and hydrothermal circulation, Geophys. J. Roy. Astron. Soc., **38**, pp. 587 (1974).

[142] Corliss, J. B., Dymond, J., Gordon, L. I., Edmond, J. M., Von Herzen, R. P., Ballard, R. D., Green, K., Williams, D., Bainbridge, A. E., Crane, K. and van Andel, T. H., *Submarine thermal springs on the Galápagos Rift*, Science, **203**, pp. 1073−1083 (1979).

[143] Perkins, S., *New type of hydrothermal vent looms large*, Sci. News, **160**, pp. 21 (2001).

[144] Lonsdale, P. F., *Clustering of suspension-feeding macrobenthos near abyssal hydrothermal vents at oceanic spreading centers*, Deep Sea Res., **24**, pp. 857−863 (1977).

[145] Duda, T. F. and Trivett, D. A., *Predicted scattering of sound by diffuse hydrothermal vent plumes at mid-ocean ridges*, J. Acoust. Soc. Am., **103**, pp. 330−335 (1998).

[146] Riedesel, M., Orcutt, J. A., Macdonald, K. C. and McClain, J. S., *Microearthquakes in the black smoker hydrothermal field, East Pacific Rise at 21'N*, J. Geophys. Res., **87**, pp. 10,613−10,623 (1982).

[147] Bibee, L. D. and Jacobson, R. S., *Acoustic noise measurements on Axial Seamount, Juan de Fuca Ridge*, Geophys. Res. Lett., **13**, pp. 957−960 (1986).

[148] Jacobson, R. S., Bibee, L. D., Embley, R. W., and Hammond, S. R., *A micro-seismicity survey of Axial Seamount, Juan de Fuca Ridge*, Bull. Seismol. Soc. Am., **77**, pp. 160−172 (1987).

[149] Little, S. H., Stolzenbach, K. D. and Purdy, G. M., *The sound field near hydrothermal vents on Axial Seamount, Juan de Fuca Ridge*, J. Geophys. Res., **95** (B8), pp. 12,917−12,945 (1990).

[150] Crone, T. J., Wilcock, W. S. D., Barclay, A. H. and Parsons, J. D., *The sound generated by mid-ocean ridge black smoker hydrothermal vents*, PLoS One, **1**, pp. 1−11 (2006).

[151] Palmer, D. R. and Rona, P. A., *Acoustical imaging of deep hydrothermal flows* in Sounds in the Sea: From Ocean Acoustics to Acoustical Oceanography, Medwin, H. Editor, Cambridge University Press, Cambridge, pp. 551−563 (2005).

[152] Rona, P. A., Jackson, D. R., Bemis, K. G., Jones, C. D., Milsuzawa, K., Palmer, D. R., and Silver, D., *Acoustics advances study of sea floor hydrothermal flow*, EOS, Trans. Am. Geophys. Union, **83**, pp. 497−502 (2002).

[153] Armi, L. and Zenk, W., *Large lenses of highly saline Mediterranean water*, J. Phys. Ocean., **14**, pp. 1560−1576 (1984).

[154] Armi, L. and Stommel, H., *Four views of a portion of the North Atlantic subtropical gyre*, J. Phys. Ocean., **13**, pp. 828−857 (1983).

[155] Armi, L., Herbert, D., Oakley, N., Price, J., Richardson, P., Rossby, H. J. and Ruddick, B., *Two years in the life of a Mediterranean salt lens*, J. Phys. Ocean., **19**, pp. 354−370 (1989).

[156] Richardson, P. L., Walsh, D., Laurence, A., Schröder, M. and Price, J. F. *Tracking three meddies with SOFAR floats*, J. Phys. Ocean., **19**, pp. 371−383 (1989).

[157] Carey, W. M., *The determination of signal coherence length based on signal coherence and gain measurements in deep and shallow water*, J. Acoust. Soc. Am., **104**, pp. 831−837 (1998).

[158] Lynch, J. F., Duda, T. F. and Colosi, J. A., *Acoustical horizontal array coherence lengths and the "Carey Number"*, Acoust. Today, **10**, pp. 10−19 (2014).

[159] Schott, F. A., Lee, T. N. and Zantopp, R., *Variability of structure and transport of the Florida Current in the periodrange of days to seasonal*, J. Phys. Ocean., **18**, pp. 1209−1230 (1988).

[160] Eyring, C. F., Christensen, R. J. and Raitt, R. W., *Reverberation in the sea*, J. Acoust. Soc. Am., **20**, pp. 462−475 (1948).

[161] Urick, R. J., *Principles of Underwater Sound*, 3rd Ed., McGraw-Hill Book Company (1983), Peninsula Press, Inc.

[162] Urban, H. G., *Handbook of Underwater Acoustic Engineering*. STN ATLAS Elektronik GmbH (2002).

[163] Waite, A. D., *Sonar for Practising Engineers*. 3rd Ed., John Wiley & Sons Ltd. (2002).

Sound Propagation

M.J. Buckingham

Scripps Institution of Oceanography, University of California,
San Diego, La Jolla, CA, United States

2.1 THE CONCEPT OF WAVES

Sound is a pressure wave that propagates through a supporting medium, which may be a fluid, a solid, or a two-phase material, for instance, a bubbly mixture of air and water or marine sediment saturated with seawater. Since it consists of a sequence of compressions and rarefactions traveling in the direction of propagation, a sound wave is variously known as a *longitudinal*, *compressional*, or *dilatational* wave. Longitudinal waves will be familiar to anyone who has played with a slinky toy, where compressions and rarefactions can be seen propagating along the spring. In general, the propagation velocity of sound waves is not necessarily related to the velocity of particle motion in any part of the supporting medium.

When the atoms or molecules of a fluid or solid are displaced, an elastic restoring force tends to return them to their equilibrium positions. This restoring force, combined with the inertia of the material, gives rise to oscillatory vibrations in the form of a sound wave propagating through the medium. An acoustic disturbance in air is audible to most young adults if it contains frequencies between 20 Hz and 20 kHz. Although not audible, *infrasonic* waves, below 20 Hz, and *ultrasonic* waves, above 20 kHz, are still considered to fall within the realm of acoustics.

Infrasound from natural sources is encountered in underwater acoustics when highly energetic seismic events, such as submarine volcanoes, radiate sound into the ocean [1]. Natural sources of ultrasonic radiation are also found in the ocean, notably snapping shrimps, which produce an intense, extremely brief ($\sim 10 \, \mu s$) acoustic pulse with a bandwidth of 200 kHz or higher [2,3]. Around pier pilings and rocky outcrops in warm shallow seas, the sound from colonies of snapping shrimps is often the dominant component of the local ambient noise field. Another source of ultrasound in the ocean is anthropogenic in origin: ultrasonic technology has practical application in the form of high-resolution acoustical imaging systems, which typically operate at frequencies around 1 MHz or above, where wavelengths are of the order of 1 mm. Since turbid waters are acoustically transparent but optically opaque, underwater ultrasonic imaging systems have an advantage over their optical counterparts.

The properties of an acoustic wave field depend on the mechanical properties of the supporting medium, including the density, ρ, and the bulk modulus, κ

Applied Underwater Acoustics. http://dx.doi.org/10.1016/B978-0-12-811240-3.00002-3

(or alternatively its reciprocal, the compressibility $\chi = 1/\kappa$). With increasing density, the propagation speed of the wave decreases, whilst with increasing bulk modulus, or stiffness of the material, the speed increases. The field satisfies a partial differential wave equation, whose solutions relate the wave properties to the physical properties of the supporting medium. In the simplest case, the supporting medium is treated as an inviscid, adiabatic (zero thermal conductivity) fluid in which the only forces governing the acoustic field are mechanical, associated with the elasticity and inertia of the medium.

2.1.1 THE WAVE EQUATION FOR AN INVISCID FLUID

In deriving the wave equation, it is tacitly assumed that the displacements and velocities of particles of the supporting medium are small and that the medium as a whole is at rest, that is to say, it is not undergoing translations or rotations. Three relationships are required in order to establish the wave equation: an equation of state, relating the density and pressure fluctuations; an equation of continuity, ensuring conservation of mass and relating the density fluctuation to the particle velocity; and a force equation, balancing the inertial force against the pressure gradient.

For a perfect gas, the equation of state may be derived from a straightforward thermodynamics argument, but for a fluid, because of the more complicated theoretical procedure, it is more usual to adopt an empirical approach. If ρ_0 and ρ are the densities, respectively, in equilibrium and in the presence of a sound wave, and P_0 and P are the corresponding pressures, then a Taylor expansion for the pressure in terms of the density excursion is

$$P = P_0 + \left(\frac{\partial P}{\partial \rho}\right)_{\rho_0} (\rho - \rho_0) + \frac{1}{2} \left(\frac{\partial^2 P}{\partial \rho^2}\right)_{\rho_0} (\rho - \rho_0)^2 + \cdots \quad . \tag{2.1}$$

For small fluctuations, only the linear term need be retained, allowing the pressure fluctuation, $p = p(x,y,z;t) = P - P_0$, to be written as

$$p \approx \kappa \frac{(\rho - \rho_0)}{\rho_0} \, , \tag{2.2}$$

where

$$\kappa = \rho_0 \left(\frac{\partial P}{\partial \rho}\right)_{\rho_0} \tag{2.3}$$

is the *adiabatic bulk modulus* of the medium. Eq. (2.2) is the *linearized equation of state*, which holds provided the relative density fluctuation, or condensation, is small, that is, $|\rho - \rho_0|/\rho_0 \ll 1$.

Conservation of mass requires that the net flow of mass into an element of volume $dV = dxdydz$ must be zero. Consider the mass fluxes into and out of the fixed

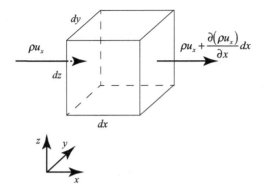

FIGURE 2.1

Mass flow through a fixed volume, $dV = dxdydz$.

volume element shown in Fig. 2.1. The mass flow through the element in the x direction is

$$\left\{ \rho u_x - \left[\rho u_x + \frac{\partial(\rho u_x)}{\partial x} dx \right] \right\} dydz = -\frac{\partial(\rho u_x)}{\partial x} dV \, , \qquad (2.4)$$

where u_x is the particle velocity in the x direction. Eq. (2.4) may be generalized to the net influx in all directions as

$$-\left[\frac{\partial(\rho u_x)}{\partial x} + \frac{\partial(\rho u_y)}{\partial y} + \frac{\partial(\rho u_z)}{\partial z} \right] dV = -\nabla \cdot (\rho \hat{u}) dV \, , \qquad (2.5)$$

where $\nabla \cdot$ is the *divergence operator* and $\hat{u}$ is the velocity vector. All the vectors introduced in the following are similarly identified by a caret. The net rate at which mass increases within the volume is

$$\left(\frac{\partial \rho}{\partial t} \right) dV + \rho_0 Q dV \, ,$$

where $Q = Q(x,y,z;t)$ has dimensions of inverse time and is a measure of the rate at which mass is introduced into unit volume by a source such as a monopole. Since the mass within the volume is conserved, we must have

$$\left(\frac{\partial \rho}{\partial t} \right) + \nabla \cdot (\rho \hat{u}) = -\rho_0 Q \, , \qquad (2.6)$$

which is the *equation of continuity*. The second term on the left in Eq. (2.6) is nonlinear, involving the product of particle velocity and instantaneous density. Bearing in mind that these acoustic variables are small, it is readily shown that, to first order, Eq. (2.6) reduces to

$$\left(\frac{\partial \rho}{\partial t} \right) + \rho_0 \nabla \cdot \hat{u} \approx -\rho_0 Q \, , \qquad (2.7)$$

which is the *linearized equation of continuity*.

To relate the particle velocity, $\widehat{u}$, to the excursions in pressure, p, consider the forces applied to an element of volume, $dV = dx\,dy\,dz$, that *moves with the fluid*. The net force on the element in the x direction due to the difference in pressure on the opposite faces of area $dy\,dz$ is

$$df_x = \left\{ P - \left[P + \left(\frac{\partial P}{\partial x} dx \right) \right] \right\} dy\,dz \,, \tag{2.8}$$

which, when generalized to three dimensions, becomes

$$\begin{aligned} d\widehat{f} &= \widehat{i}\,df_x + \widehat{j}\,df_y + \widehat{k}\,df_z \\ &= -\nabla P \, dV \,, \end{aligned} \tag{2.9}$$

where $\widehat{i}, \widehat{j}, \widehat{k}$ are unit vectors in the x, y, and z directions, respectively. Now, the inertial force on the moving element is the product of its mass and the total derivative of the velocity vector:

$$\rho \frac{d\widehat{u}}{dt} dV = \rho \left\{ \frac{\partial \widehat{u}}{\partial t} + \widehat{u} \cdot \nabla \widehat{u} \right\} dV \,, \tag{2.10}$$

which must be balanced by the pressure-gradient force in Eq. (2.9). On equating the two and retaining terms up to first order, the *linearized force equation* is obtained:

$$\rho_0 \frac{\partial \widehat{u}}{\partial t} = -\nabla p \,. \tag{2.11}$$

By combining Eqs. (2.2), (2.7), and (2.11), a partial differential equation in a single field variable, the pressure p, may be derived:

$$\nabla^2 p - \frac{1}{c^2} \frac{\partial^2 p}{\partial t^2} = \rho_0 \frac{\partial Q}{\partial t} \,, \tag{2.12}$$

where ∇^2 is the Laplacian and

$$c = \sqrt{\frac{\kappa}{\rho_0}} \tag{2.13}$$

is the phase speed of the acoustic waves in the fluid medium. Eq. (2.12) is the linearized, inhomogeneous wave equation for acoustic propagation in a lossless fluid. It is a second-order partial differential equation, and the homogeneous form is obtained when the source term on the right-hand side is set to zero. Since the acoustical pressure is proportional to the density fluctuations, the latter also satisfy a wave equation analogous to that in Eq. (2.12) but with the appropriate source term on the right-hand side.

At this point, suppose that we take the curl ($\nabla \times$) of the linearized force equation in Eq. (2.11), yielding

$$\rho_0 \frac{\partial}{\partial t} (\nabla \times \widehat{u}) = -\nabla \times \nabla p \,. \tag{2.14}$$

Since the curl of a gradient is identically zero, two conditions follow. The first is that

$$\nabla \times \widehat{u} = 0 , \tag{2.15}$$

which leads to the second, namely that $\widehat{u}$ can be expressed as the gradient of a scalar function, ϕ:

$$\widehat{u} = -\nabla \phi , \tag{2.16}$$

where $\phi = \phi(x,y,z;t)$ is defined as the *velocity potential* with dimensions of (length^2/time). Like the pressure and particle velocity, the velocity potential satisfies an inhomogeneous wave equation, in this case:

$$\nabla^2 \phi - \frac{1}{c^2} \frac{\partial^2 \phi}{\partial t^2} = Q . \tag{2.17}$$

From Eq. (2.11), the pressure can be expressed in terms of the velocity potential as

$$p = \rho_0 \frac{\partial \phi}{\partial t} . \tag{2.18}$$

Once a solution for ϕ has been found from the wave equation in Eq. (2.17), the particle velocity and pressure may be determined from Eqs. (2.16) and (2.18), respectively. The physical significance of the result in Eq. (2.16) is that, in an inviscid fluid, an acoustical disturbance exhibits no vorticity, or rotational flow. Thus, no phenomena such as turbulence, boundary layers, or shear waves are encountered in an inviscid fluid.

Suppose that the source transmits a signal with a time signature $f(t)$, then the source function may be expressed as

$$Q = Sf(t) . \tag{2.19}$$

In this case, the wave equation in Eq. (2.17) becomes

$$\nabla^2 \phi - \frac{1}{c^2} \frac{\partial^2 \phi}{\partial t^2} = Sf(t) , \tag{2.20}$$

where $S = S(x,y,z)$ is a measure of the *source strength*. If $f(t)$ has dimensions of inverse time, then the source strength is dimensionless.

2.1.2 THE HELMHOLTZ EQUATION

In order to solve the wave equation for the field, the first step is usually to transform Eq. (2.20) from the time domain to the frequency domain by the application of a Fourier transform with respect to time to both sides. The Fourier transform of the velocity potential is

$$\Phi(\omega) = \int_{-\infty}^{\infty} \phi(t) e^{-i\omega t} dt , \tag{2.21a}$$

where $i = \sqrt{-1}$, ω is angular frequency, $\Phi(\omega) = \Phi(x,y,z;\omega)$, and the inverse transform is

$$\phi(t) = \frac{1}{2\pi} \int_{-\infty}^{\infty} \Phi(\omega)e^{i\omega t}d\omega. \tag{2.21b}$$

The second derivative with respect to time in Eq. (2.20) is Fourier transformed according to

$$\int_{-\infty}^{\infty} \frac{\partial^2 \phi}{\partial t^2} e^{-i\omega t} dt = -\omega^2 \Phi(\omega), \tag{2.22}$$

as may be proved by a double integration by parts. With the aid of these expressions, the Fourier transform of the wave equation in Eq. (2.20) is found to be

$$\nabla^2 \Phi + k^2 \Phi = SF, \tag{2.23}$$

where $F = F(\omega)$ is the Fourier transform of the source time signature $f(t)$ and

$$k = \frac{\omega}{c} \tag{2.24}$$

is the *acoustic wave number*. For an inviscid fluid, the wave number is real but as we shall see later, an effective complex wave number, corresponding to a complex sound speed, provides a convenient way of representing acoustic attenuation in a lossy medium. Eq. (2.23) is the inhomogeneous *Helmholtz equation*.

The effect of the Fourier transformation on the wave equation is to reduce the second derivative with respect to time to an algebraic term. The resultant field, Φ, is a function of frequency, from which the time dependence may be recovered by the application of the inverse transform in Eq. (2.21b).

In the special case of an impulsive source that activates at time $t = 0$, the time signature of the source may be represented as a Dirac delta function, $\delta(t)$, which is perhaps the best known example of a generalized function [4]. The source function then becomes

$$Q = S\delta(t) , \tag{2.25}$$

and the inhomogeneous wave equation in Eq. (2.20) may be written as

$$\nabla^2 g - \frac{1}{c^2} \frac{\partial^2 g}{\partial t^2} = S\delta(t) , \tag{2.26}$$

where the velocity potential $g = g(x,y,z;t)$ is the *impulse response*, otherwise known as *Green's function*.

From the sampling property of the delta function, its Fourier transform is

$$\int_{-\infty}^{\infty} \delta(t)e^{-i\omega t} dt = 1 , \tag{2.27}$$

which is an important result, since it shows that all frequencies are represented equally in the spectrum of a delta function. Thus, for the impulsive source, $F = 1$ and the Helmholtz equation is

$$\nabla^2 G + k^2 G = S, \qquad (2.28)$$

where $G = G(x,y,z;\omega)$ is the Fourier transform of Green's function $g(x,y,z;t)$.

By comparing the two forms of the Helmholtz equation in Eqs. (2.23) and (2.28), bearing in mind that F is independent of position in the medium, it is evident that the field due to a source with an arbitrary time signature is given by

$$\Phi(x, y, z; \omega) = G(x, y, z; \omega)F(\omega) \quad . \qquad (2.29)$$

Thus, in the frequency domain, the field is simply the product of the Fourier transforms of Green's function and the time signature of the source. To return to the time domain, the inverse Fourier transform is applied, yielding

$$\phi(x, y, z; t) = \frac{1}{2\pi} \int_{-\infty}^{\infty} G(x, y, z; \omega)F(\omega)e^{i\omega t}d\omega . \qquad (2.30a)$$

By substituting the Fourier inversion integrals for G and F in the integrand, this expression transforms into a convolution integral [5]:

$$\phi(x, y, z; t) = \frac{1}{2\pi} \int_{-\infty}^{\infty} g(t')f(t - t')dt'$$

$$= \frac{1}{2\pi} \int_{-\infty}^{\infty} g(t - t')f(t')dt' \qquad (2.30b)$$

$$= \frac{1}{2\pi} g(x, y, z; t) \otimes f(t),$$

where $\otimes$ is the convolution operator.

The two forms for the time-dependent field in Eq. (2.30) are equivalent in that they return exactly the same result. However, if the time dependence of the field is required, it is often preferable to work with the inverse Fourier transform of the product, the form in Eq. (2.30a), rather than evaluate the convolution in Eq. (2.30b). If the integral cannot be evaluated analytically, a numerical solution using the fast Fourier transform (FFT) algorithm [6] provides an efficient means of performing the inverse transform.

2.1.3 HARMONIC WAVES

The time signature of a harmonic source of angular frequency ω_0 may be represented as

$$f(t) = \omega_0 \cos(\omega_0 t + \varphi) , \qquad (2.31)$$

where φ is an arbitrary phase. Setting the amplitude to ω_0 ensures that $f(t)$ has dimensions of inverse time. The question of interest now concerns the time dependence of the field at an arbitrary point, (x,y,z), in the supporting medium.

According to Eq. (2.29), in the frequency domain, the field is given by the product of Green's function, G, with the Fourier transform of the source, F. The latter is simply

$$F(\omega) = \frac{\omega_0}{2} \int_{-\infty}^{\infty} \left\{ e^{i\varphi} e^{i(\omega_0 - \omega)t} + e^{-i\varphi} e^{-i(\omega_0 + \omega)t} \right\} dt$$

$$= \pi\omega_0 \left\{ e^{i\varphi} \delta(\omega_0 - \omega) + e^{-i\varphi} \delta(\omega_0 + \omega) \right\},$$

$$(2.32)$$

where the delta function identity

$$2\pi\delta(a) = \int_{-\infty}^{\infty} e^{iax} dx \qquad (2.33)$$

has been used.

From Eq. (2.30a), the field in the time domain is given by the inverse Fourier transform

$$\phi(t) = \frac{\omega_0}{2} \int_{-\infty}^{\infty} \left\{ e^{i\varphi} \delta(\omega_0 - \omega) + e^{-i\varphi} \delta(\omega_0 + \omega) \right\} G(\omega) e^{i\omega t} d\omega$$

$$= \frac{\omega_0}{2} \left\{ G(\omega_0) e^{i(\omega_0 t + \varphi)} + G(-\omega_0) e^{-i(\omega_0 t + \varphi)} \right\}.$$

$$(2.34)$$

Since Green's function, $g(t)$, must be real, its Fourier transform satisfies

$$G(-\omega_0) = G^*(\omega_0), \qquad (2.35)$$

where the asterisk denotes complex conjugation. It follows that the time-dependent field in the supporting medium is given by

$$\phi(x, y, z; t) = \omega_0 Re\left\{ G(x, y, z; \omega_0) e^{i(\omega_0 t + \varphi)} \right\}. \qquad (2.36)$$

By writing G in terms of its amplitude and phase, $\eta = \eta(x,y,z;\omega_0)$,

$$G = |G| e^{i\eta}, \qquad (2.37)$$

the expression for the field in Eq. (2.36) reduces to

$$\phi(x, y, z; t) = \omega_0 |G(x, y, z; \omega_0)| \cos(\omega_0 t + \varphi + \eta). \qquad (2.38)$$

Thus, the field at any point in the medium is given by the harmonic source function but with a modified amplitude and an additional phase delay, both of which depend on the position of the receiver in relation to the source, and both are available from the Fourier transform of Green's function evaluated at the source frequency. Of course, to the level of approximation presented here, the wave equation is linear and hence, everywhere in the medium, the field possesses the same frequency as the harmonic source itself; no nonlinear processes are present that might introduce frequency shifts.

2.1.4 PLANE WAVES

A wave front is a surface of constant phase. If the phase is constant along a plane perpendicular to the direction of propagation, the wave is said to be a *plane wave*. In the ocean, plane waves, or at least very good approximations to plane waves, are commonly encountered far from the source, where the curvature in the wave fronts is negligible. Of course, if the source itself is planar, the waves it generates have plane wavefronts everywhere, assuming a homogeneous medium, including close to the source position. Such a *planar source* may be modeled mathematically as an infinitesimally thin plate of infinite area and uniform thickness, exhibiting thickness vibrations.

Working in Cartesian coordinates, the direction of wave propagation is taken to be along the x-axis, with an impulsive planar source perpendicular to the x-axis at $x = 0$. Symmetry requires that the field be independent of the y- and z-coordinates. The wave equation for Green's function in Eq. (2.26) then reduces to the simpler form

$$\frac{\partial^2 g}{\partial x^2} - \frac{1}{c^2}\frac{\partial^2 g}{\partial t^2} = S_1 \delta(x)\delta(t) \, , \tag{2.39}$$

where $g = g(x;t)$ and the source term on the right represents an instantaneous change in the thickness of the source at time $t = 0$. The *planar source strength*, S_1, is a constant with dimensions of (volume per unit area) = (length). Eq. (2.39) represents plane waves propagating away from the planar source in the positive and negative x directions, and the corresponding Helmholtz equation, obtained by taking the Fourier transform (with respect to time) of both sides of Eq. (2.39), is

$$\frac{\partial^2 G}{\partial x^2} + k^2 G = S_1 \delta(x) \, , \tag{2.40}$$

where $G = G(x;\omega)$ and $k = \omega/c$ is the acoustic wave number. Note that a volume integral of the source term, per unit area of the planar source, returns the source strength, S_1.

A solution for G may be found by taking a second Fourier transform, this time with respect to the spatial coordinate, x. The transform in question takes exactly the same form as the Fourier transform with respect to time in Eq. (2.21a) but with t replaced by x and the angular frequency ω replaced by the wave number s:

$$G_s = \int_{-\infty}^{\infty} G e^{-isx} dx \, , \tag{2.41}$$

whose inverse transform is

$$G = \frac{1}{2\pi} \int_{-\infty}^{\infty} G_s e^{isx} ds \, . \tag{2.42}$$

In these expressions, the transform is identified by using the transform variable, s, in the present case, as a subscript. This convention is particularly convenient when multiple spatial transforms have to be applied, a situation that is encountered when the field is a function of more than one spatial coordinate. The Fourier transform of the second derivative with respect to x in Eq. (2.40), analogous to the expression in Eq. (2.22), is

$$\int_{-\infty}^{\infty} \frac{\partial^2 g}{\partial x^2} e^{-isx} dx = -s^2 G_s , \qquad (2.43)$$

and hence the transformed version of Eq. (2.40) is the algebraic equation

$$G_s = S_1 / \left(k^2 - s^2 \right) . \qquad (2.44)$$

To recover the field, G, in the frequency domain, the inverse transform in Eq. (2.42) is applied to Eq. (2.44), yielding

$$G = -\frac{S_1}{2\pi} \int_{-\infty}^{\infty} \frac{e^{isx}}{s^2 - k^2} ds$$

$$= iS_1 \frac{e^{-ik|x|}}{2k} , \qquad (2.45)$$

where the imaginary part of k is taken to be negative but infinitesimally small. It is readily shown that the pressure $P = P(x;\omega)$ and particle velocity $U = U(x;\omega)$ derived from G are in phase, and the waves show no reduction in amplitude as they propagate away from the source.

In the time domain, the Green's function, g, given by the inverse (temporal) Fourier transform of G, is

$$g = \frac{iS_1}{4\pi} \int_{-\infty}^{\infty} \frac{e^{i(\omega t - k|x|)}}{k} d\omega . \qquad (2.46)$$

Thus, the pressure pulse propagating away from the impulsive planar source in either direction is given by

$$p = \rho_0 \frac{\partial g}{\partial t}$$

$$= -\frac{\rho_0 S_1 c}{4\pi} \int_{-\infty}^{\infty} e^{i(\omega t - k|x|)} d\omega \qquad (2.47)$$

$$= -\frac{\rho_0 S_1 c}{2} \delta\left(t - \frac{|x|}{c} \right) ,$$

where the integral has been evaluated using the theory of generalized functions. Similarly, the particle velocity in the x direction is

$$
\begin{aligned}
u_x &= -\frac{\partial g}{\partial x} \\
&= -\text{sgn}(x)\frac{S_1}{4\pi}\int_{-\infty}^{\infty} e^{i(\omega t - k|x|)}\,d\omega \\
&= -\text{sgn}(x)\frac{S_1}{2}\delta\left(t - \frac{|x|}{c}\right),
\end{aligned}
\tag{2.48}
$$

where the odd function $\text{sgn}(x)$ is a generalized function [4] with a value of 1 for $x > 0$ and -1 for $x < 0$.

A wave of the form in Eq. (2.47), which depends only on the *retarded time* $(t - |x|/c)$, rather than t and x independently, is known as a *retarded potential*. The retarded time is the time when the signal began to propagate from the source, which is equal to the difference between the actual time and the *propagation delay*, $|x|/c$. A one-dimensional pressure pulse, as exemplified by Eq. (2.47), maintains its shape as it propagates through the lossless supporting medium. As we shall see later, such behavior is not maintained in the presence of losses, which introduce pulse distortion that grows with increasing range.

Having developed expressions for Green's function and its temporal Fourier transform, we are now in a position to complete the solution in Eq. (2.38) for the pressure due to a harmonic source of frequency ω_0. From Eq. (2.45), the phase is $\eta = -k_0|x| + \pi/2$, where the wave number $k_0 = \omega_0/c$, and therefore the velocity potential is

$$
\begin{aligned}
\phi(x; t) &= \omega_0|G(x; \omega_0|\cos(\omega_0 t + \eta + \varphi) \\
&= \frac{S_1 c}{2}\sin\left[\omega_0\left(t - \frac{|x|}{c}\right) + \varphi\right],
\end{aligned}
\tag{2.49a}
$$

which is just a phase-shifted version of the harmonic source function. The associated pressure wave, from Eq. (2.18), is

$$
p(x; t) = -\frac{\rho_0 S_1 \omega_0 c}{2}\cos\left[\omega_0\left(t - \frac{|x|}{c}\right) + \varphi\right].
\tag{2.49b}
$$

As with the impulsive source, the pressure has the same shape as the source function but is time-shifted by the propagation delay.

Although the harmonic source function in Eq. (2.31) provides some insight into acoustic propagation in a linear medium, it suffers from an obvious defect: it is not physically realistic because it does not satisfy *causality*. No source could have been running for an infinite time and continue running indefinitely into the future. The field generated by a causal source must itself be causal, that is to say, it must be zero for all times before the source is activated, since a response cannot precede the input.

Consider a causal, harmonic planar source, activated at time $t = 0$, and represented by the function

$$f(t) = u(t)\omega_0 \cos(\omega_0 t + \varphi) \,, \tag{2.50}$$

where $u(t)$, the Heaviside unit step function, is another generalized function, which is zero for $t < 0$ and unity for $t > 0$. As before, the first step toward finding the time dependence of the acoustic field in the medium is to take the Fourier transform of $f(t)$:

$$\begin{aligned}
F(\omega) &= \omega_0 \int_0^\infty \cos(\omega_0 t + \varphi) e^{-i\omega t} dt \\
&= \frac{\omega_0 e^{i\varphi}}{2} \int_0^\infty e^{i(\omega_0 - \omega)t} dt + \frac{\omega_0 e^{-i\varphi}}{2} \int_0^\infty e^{-i(\omega_0 + \omega)t} dt
\end{aligned} \tag{2.51}$$

where the lower limit of zero is consistent with the causal nature of the source. The integrals in this expression may be evaluated using the theory of generalized functions [4], which returns

$$F(\omega) = \frac{\omega_0 e^{i\varphi}}{2} \left\{ \pi\delta(\omega_0 - \omega) + \frac{i}{\omega_0 - \omega} \right\} + \frac{\omega_0 e^{-i\varphi}}{2} \left\{ \pi\delta(\omega_0 + \omega) - \frac{i}{\omega_0 + \omega} \right\}. \tag{2.52}$$

Again, taking the product of F with the Fourier transform of Green's function, G, and performing the inverse Fourier transform yields the time-dependent field. The contribution from the two delta functions is clearly half the noncausal field in Eq. (2.38) and, since the remaining two integrals are conjugates, the total velocity potential may be expressed as

$$\begin{aligned}
\phi(x; t) &= \frac{\omega_0}{2} |G(x; \omega_0)| \cos(\omega_0 t + \varphi + \eta) \\
&\quad - \frac{\omega_0}{2\pi} Re \left\{ i e^{i\varphi} \int_{-\infty}^\infty \frac{G(x; \omega)}{\omega - \omega_0} e^{i\omega t} d\omega \right\}.
\end{aligned} \tag{2.53}$$

With the expression for G in Eq. (2.45), and recalling Eq. (2.35), the velocity potential becomes

$$\phi(x; t) = -\frac{S_1 c}{4} \sin(\omega_0 t - k_0|x| + \varphi) + \frac{\omega_0 S_1 c}{4\pi} Re \left\{ e^{i\varphi} \int_{-\infty}^\infty \frac{e^{i\omega\tau}}{\omega(\omega - \omega_0)} d\omega \right\}, \tag{2.54a}$$

and the corresponding pressure is given by

$$p(x;t) = -\frac{\rho_0\omega_0 S_1 c}{4}\cos(\omega_0 t - k_0|x| + \varphi) + \frac{\rho_0\omega_0 S_1 c}{4\pi} Re\left\{ie^{i\varphi}\int_{-\infty}^{\infty}\frac{e^{i\omega\tau}}{(\omega - \omega_0)}d\omega\right\},$$

(2.54b)

where $\tau = t - |x|/c$ is the retarded time. The integral here is a known form that is available from tables:

$$\int_{-\infty}^{\infty}\frac{e^{i\omega\tau}}{(\omega - \omega_0)}d\omega = i\pi\text{sgn}(\tau)e^{i\omega_0\tau}$$

(2.55)

and hence

$$p(x;t) = -\frac{\rho_0\omega_0 S_1 c}{2}u(t - |x|/c)\cos(\omega_0 t - k_0|x| + \varphi),$$

(2.56)

where we have recognized that the Heaviside function and the sgn function are related as follows:

$$u(\tau) = [1 + \text{sgn}(\tau)]/2 \quad .$$

(2.57)

The acoustic pressure in Eq. (2.56) is strictly causal, with a zero response at the receiver at all times prior to the propagation delay time. Such behavior is characteristic of plane waves in an inviscid fluid. As we shall discuss later, when attenuation is present in the medium, the propagation is still causal, with no arrivals before $t = 0$, but, because the speed of sound is no longer constant but increases as the frequency rises, it is possible for a nonzero response to occur sooner than in the inviscid case.

For plane waves in an inviscid fluid, the pressure from an impulsive source takes a particularly simple form, a delta function in which the argument is the retarded time, as shown in Eq. (2.47). As it happens, this is an example of a situation in which the convolution integral in Eq. (2.30b) provides an easier means of deriving the time dependence of the propagating pressure wave than the Fourier transform approach used above. By taking the derivative with respect to time of the velocity potential in Eq. (2.30b), the pressure for a planar source with arbitrary time signature, $f(t)$, can be expressed as the convolution

$$p(x;t) = \int_{-\infty}^{\infty}p(t - t')f(t')dt'.$$

(2.58)

With the aid of the expression for the pressure in Eq. (2.47), bearing in mind the sampling property of the delta function, the integral in Eq. (2.58) evaluates to

$$p(x;t) = -\frac{\rho_0 S_1 c}{2}f(t - |x|/c).$$

(2.59)

Clearly, when $f(t)$ is given by Eq. (2.50), this result is identical to the expression for the pressure in Eq. (2.56). More generally, Eq. (2.59) states that any plane-wave pressure pulse in an inviscid fluid shows no distortion, retaining exactly the same shape as the time signature of the source that generated it but retarded by the propagation delay.

2.1.5 CYLINDRICAL WAVES

Waves with cylindrical wave fronts are often encountered in underwater acoustics, notably in the far field of a point source in a shallow water channel. A *cylindrical source* may be modeled mathematically as an infinitesimally thin line, infinitely long, exhibiting radial vibrations. A natural choice in the analysis of waves from such a source is the cylindrical coordinate system, with the z-axis coincident with the line source itself and the radial coordinate, $r > 0$, perpendicular to the line of the source. Symmetry dictates that the radiated field must be uniform in azimuth and independent of z.

When expressed in cylindrical coordinates, the wave equation for Green's function in Eq. (2.26) becomes

$$\frac{1}{r}\frac{\partial}{\partial r}\left(r\frac{\partial g}{\partial r}\right) - \frac{1}{c^2}\frac{\partial^2 g}{\partial t^2} = S_2 \frac{\delta(r)}{\pi r}\delta(t) , \qquad (2.60)$$

where $g = g(r;t)$ and the source term on the right represents an instantaneous change in the radius of the line source at time $t = 0$. The *cylindrical source strength*, S_2, is a constant with dimensions of (volume per unit length) = (length)^2. A Fourier transformation with respect to time of both sides of Eq. (2.60) returns the inhomogeneous Helmholtz equation,

$$\frac{1}{r}\frac{\partial}{\partial r}\left(r\frac{\partial G}{\partial r}\right) + k^2 G^2 = S_2 \frac{\delta(r)}{\pi r} , \qquad (2.61)$$

where $G = G(r;\omega)$ is the Fourier transform of $g = g(r;t)$. In Eq. (2.61), the second derivative with respect to time has been converted to an algebraic term, leaving an ordinary inhomogeneous differential equation to be solved for the field, G. Note that a volume integral of the source term for unit length of the line source returns the source strength, S_2.

As with the plane wave problem, the next step is to apply an integral transform to both sides of Eq. (2.61), the appropriate choice in this case being the Hankel transform [7]

$$G_q = \int_0^\infty rG(r;\omega)J_0(qr)dr , \qquad (2.62a)$$

where $J_0(\cdot)$ is the Bessel function of the first kind of order zero and the transform variable, q, is the horizontal wave number. The inverse Hankel transform is

$$G(r;\omega) = \int_0^\infty qG_qJ_0(qr)dq , \qquad (2.62b)$$

and, as may be proved by a double integration by parts, the transform of the second spatial derivative in Eq. (2.61) is

$$\int_0^\infty r\left\{\frac{1}{r}\frac{\partial}{\partial r}\left(r\frac{\partial G}{\partial r}\right)\right\}J_0(qr)dr = -q^2 G_q . \qquad (2.63)$$

This identity, of course, is why the Hankel transform is appropriate for the analysis of cylindrical waves.

After applying the Hankel transform to both sides of Eq. (2.61), it reduces to a simple algebraic equation, taking the form

$$(k^2 - q^2)G_q = \frac{S_2}{2\pi}J_0(0) .$$

(2.64)

where $J_0(0) = 1$. By performing the inverse Hankel transform on G_q, the frequency-dependent field, G, is recovered:

$$G = \frac{S_2}{2\pi} \int_0^\infty \frac{qJ_0(qr)}{(k^2 - q^2)}dq .$$

(2.65)

This integral is a special case of Hankel's discontinuous integral [8], which is a known form, leading to the result

$$G = G(r; \omega) = \frac{iS_2}{4}H_0^{(2)}(kr) ,$$

(2.66a)

where $H_0^{(2)}(\cdot)$ is the Hankel function of the second kind of order zero [9]. The corresponding expressions for the pressure, $P = P(r; \omega)$, and particle velocity, $U = U(r; \omega)$, are

$$P = -\frac{\rho_0\omega S_2}{4}H_0^{(2)}(kr)$$

(2.66b)

and

$$U = \frac{ikS_2}{4}H_1^{(2)}(kr) ,$$

(2.66c)

where, in Eq. (2.66c), a well-known identity [9] has been used to relate the derivative of the Hankel function of order zero to the Hankel function of order unity.

At large ranges, such that $kr \gg 1$, the Hankel functions in these expressions may be approximated by their asymptotic expansions,

$$H_\nu^{(2)}(kr) \approx \sqrt{\frac{2}{\pi kr}}e^{-i[kr-(\nu+0.5)\pi/2]} ,$$

(2.67)

where, in the present case, $\nu = 1$ or 2. Under this far-field condition, the pressure and particle velocity are in quadrature, and all three fields, P, U, and G, decay as $(kr)^{-1/2}$, which is a characteristic of *cylindrical spreading*. At the other extreme, in the near field, where $kr \ll 1$, the Hankel function with $\nu = 0$ diverges logarithmically,

$$H_0^{(2)}(kr) \approx -\frac{2i}{\pi}\ln(kr) ,$$

(2.68a)

whereas with $\nu = 1$ the divergence is faster,

$$H_1^{(2)}(kr) \approx \frac{2i}{\pi kr} .$$

(2.68b)

In both cases, the singularity at $r = 0$ corresponds physically to the receiver being coincident with the line source.

The Green's function, g, is obtained by taking the inverse Fourier transform of Eq. (2.66a), which returns

$$g = g(r; t) = \frac{iS_2}{8\pi} \int_{-\infty}^{\infty} H_0^{(2)}(kr) e^{i\omega t} d\omega .$$

(2.69)

The integral here is not a standard form, but it may be evaluated with the aid of the integral expression for the Hankel function [10],

$$H_0^{(2)}(kr) = -\frac{2}{i\pi} \int_0^{\infty} e^{-ikr \cosh u} du ,$$

(2.70)

which converges provided the imaginary part of the acoustic wave number k, however small, is negative. Under this condition, the losses in the supporting medium may be infinitesimal but not identically zero. By substituting the integral expression for the Hankel function into Eq. (2.69), the Green's function is expressed by the double integral

$$g = -\frac{S_2}{4\pi^2} \int_0^{\infty} \int_{-\infty}^{\infty} e^{i\omega[t - (r/c)\cosh u]} d\omega du .$$

(2.71)

From the theory of generalized functions, the inner integral is just a delta function, that is,

$$g = -\frac{S_2}{2\pi} \int_0^{\infty} \delta[t - (r/c)\cosh u] du ,$$

(2.72)

which, on making the substitution $y = \cosh u$, becomes

$$g = -\frac{S_2 c}{2\pi r} \int_0^{\infty} \delta[y - (ct/r)](y^2 - 1)^{-1/2} dy$$

$$= \begin{cases} -\dfrac{S_2}{2\pi\sqrt{t^2 - (r/c)^2}} & \text{for} \quad t > r/c \\ \\ 0 & \text{for} \quad t < r/c \end{cases} .$$

(2.73)

The corresponding pressure and radial particle velocity are, respectively,

$$p(r; t) = \rho_0 \frac{\partial g}{\partial t}$$

$$= -\frac{\rho_0 S_2}{2\pi} \left(t^2 - \frac{r^2}{c^2} \right)^{-1/2} \left\{ \delta\left(t - \frac{r}{c} \right) - u[t - (r/c)] t \left(t^2 - \frac{r^2}{c^2} \right)^{-1} \right\} ,$$

(2.74)

and

$$u_r(t) = -\frac{\partial g}{\partial r}$$

$$= -\frac{S_2}{2\pi c}\left(t^2 - \frac{r^2}{c^2}\right)^{-1/2}\left\{\delta\left(t - \frac{r}{c}\right) - u\left(t - \frac{r}{c}\right)\frac{r}{c}\left(t^2 - \frac{r^2}{c^2}\right)^{-1}\right\} \qquad (2.75)$$

where $u(\cdot)$ is the Heaviside unit step function. In these expressions for the pressure and particle velocity, a delta function, of opposite polarity to the remainder of the pulse, appears at the propagation delay time.

The pressure pulse in Eq. (2.74) is interesting, not least because it is quite different in character from the analogous plane-wave pressure pulse in an inviscid fluid in Eq. (2.47). The latter depends only on the retarded time and retains the shape of the source signature as it propagates through the medium. By way of contrast, the cylindrically spreading pressure pulse depends on time t and the propagation delay, r/c, individually, which means that it does not retain the shape of the source signature as it propagates. Instead, after the propagation delay at time $t = r/c$, the pulse exhibits an extended tail, or wake, which decays to zero asymptotically as $1/t^2$. The pulse shape (without the leading delta function) is illustrated in Fig. 2.2. If, for comparison, the plane-wave pressure pulse in Eq. (2.47) were

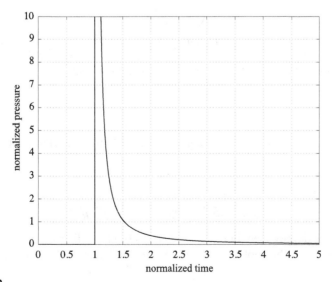

FIGURE 2.2

Cylindrical pressure pulse from Eq. (2.75). The time and pressure are normalized, respectively, to r/c and $\rho_0 S_2 c^2/(2\pi r^2)$.

superimposed on the plot in Fig. 2.2, it would be a delta function located at the normalized time of unity.

2.1.6 SPHERICAL WAVES

Any solid body vibrating in a fluid generates sound waves. If the dimensions of the solid body are very much smaller than the wavelength, it is said to be a *compact source* and the sound radiated into the (homogeneous) surrounding medium is spherically symmetric, that is to say, the wave fronts are spherical surfaces, centered on the source. A compact source may be modeled mathematically as an infinitesimally small point, exhibiting radial vibrations. It is natural to work with spherical polar coordinates when analyzing the sound field from such a source, in which case, with the source at the origin, the field is independent of the polar and azimuthal angles, depending only on the radial coordinate, R.

If the source is impulsive, exhibiting an instantaneous change in radius at time $t = 0$, then the wave equation for the spherically symmetric Green's function, $g = g(R;t)$, may be Fourier transformed into the Helmholtz equation, Eq. (2.28):

$$\frac{1}{R^2}\frac{\partial}{\partial R}\left(R^2\frac{\partial G}{\partial R}\right) + k^2 G = S_3\frac{\delta(R)}{2\pi R^2} \, , \tag{2.76}$$

where $G = G(R;\omega)$ and the *point source strength*, S_3, is a constant with dimensions of $(\text{volume}) = (\text{length})^3$. Note that a volume integral of the source term returns the source strength, S_3.

Although the solution for the spherically symmetric field is well known, it is instructive to derive it from Eq. (2.76) using integral transform techniques. To begin, a new field variable, $W = W(R;\omega)$, is introduced, defined by

$$G = \frac{W}{\sqrt{R}} \, . \tag{2.77}$$

When this expression is substituted into Eq. (2.76), the Helmholtz equation becomes

$$\frac{1}{R}\frac{\partial}{\partial R}\left(R\frac{\partial W}{\partial R}\right) - \frac{W}{4R^2} + k^2 W = \frac{S_3}{2\pi}\frac{\delta(R)}{R^{3/2}} \, . \tag{2.78}$$

Following the development in Eqs. (2.62) and (2.63) for cylindrical waves, a Hankel transform can now be applied to Eq. (2.78), but in this case of order 1/2:

$$W_q = \int_0^\infty RWJ_{1/2}(qR)dR \tag{2.79a}$$

whose inverse is

$$W = \int_0^\infty q W_q J_{1/2}(qR)dq , \qquad (2.79b)$$

where, as previously, the wave number q is the transform variable. The advantage of the formulation in Eq. (2.78) becomes apparent on examining the first two terms on the left:

$$\Delta(R) = \frac{1}{R}\frac{\partial}{\partial R}\left(R\frac{\partial W}{\partial R}\right) - \frac{W}{4R^2} . \qquad (2.80)$$

When the Hankel transform in Eq. (2.79a) is applied to this expression, it takes the familiar form for the transform of a second derivative:

$$\int_0^\infty R\Delta(R)J_{1/2}(qR)dR = -q^2 W_q . \qquad (2.81)$$

Thus, the Hankel transformed version of Eq. (2.78) is the algebraic equation

$$W_q(k^2 - q^2) = \frac{S_3}{4\pi}\sqrt{\frac{2q}{\pi}} , \qquad (2.82)$$

where the right-hand side follows from the condition

$$\lim_{R\to 0}\left[R^{-1/2}J_{1/2}(qR)\right] = \sqrt{2/(q\pi)}\lim_{R\to 0}\left[\frac{\sin(qR)}{R}\right] = \sqrt{2q/\pi} . \qquad (2.83)$$

Now, with the solution for the transformed field available immediately from Eq. (2.82), it remains only to perform the inverse transform to recover W, and hence G:

$$W = \frac{S_3}{4\pi}\sqrt{\frac{2}{\pi}}\int_0^\infty \frac{q^{3/2}}{k^2 - q^2}J_{1/2}(qR)dR . \qquad (2.84)$$

The integral here is a known form, allowing W to be evaluated as

$$W = -\frac{S_3}{4\pi}\frac{e^{-ikR}}{\sqrt{R}} , \qquad (2.85)$$

and thus

$$G = -S_3\frac{e^{-ikR}}{4\pi R} . \qquad (2.86)$$

This is just the classic expression for *spherical spreading*. The associated pressure, $P = P(R;\omega)$, and particle velocity, $U = U(R;\omega)$, are given by the expressions

$$P = -i\omega\rho_0 S_3 \frac{e^{-ikR}}{4\pi R} \tag{2.87a}$$

and

$$U = -S_3(ikR + 1)\frac{e^{-ikR}}{4\pi R^2} . \tag{2.87b}$$

In the far field, where $kR \gg 1$, the pressure and particle velocity are in phase with each other and both decay as $1/R$. Closer to the source, where $kR \ll 1$, they are in quadrature, but the particle velocity now increases as $1/R^2$ with decreasing range, whereas the pressure maintains its $1/R$ dependence.

In the time domain, the Green's function, $g = g(R;t)$, is found by taking the inverse Fourier transform (with respect to frequency) of Eq. (2.86), which yields

$$g = -\frac{S_3}{8\pi^2 R} \int_{-\infty}^{\infty} e^{i\omega\tau} d\omega$$
$$= -S_3 \frac{\delta(\tau)}{4\pi R} , \tag{2.88}$$

where

$$\tau = t - \frac{R}{c} \tag{2.89}$$

is the retarded time. Unlike the cylindrical spreading case, as represented in Eq. (2.73), the pulse in Eq. (2.88) retains its shape as it propagates away from the source. The associated pressure, $p = p(R;t)$, and particle velocity, $u = u(R;t)$, are, respectively,

$$p = \rho_0 \frac{\partial g}{\partial t}$$
$$= -\frac{\rho_0 S_3}{4\pi R} \frac{\partial}{\partial t}\left\{\delta\left(t - \frac{R}{c}\right)\right\} \tag{2.90}$$
$$= \frac{\rho_0 S_3}{4\pi R} \frac{\delta(\tau)}{\tau} ,$$

and

$$u = -\frac{\partial g}{\partial R}$$
$$= \frac{S_3}{4\pi} \frac{\partial}{\partial R}\left\{R^{-1}\delta\left(t - \frac{R}{c}\right)\right\} \tag{2.91}$$
$$= \frac{S_3 \delta(\tau)}{4\pi R c \tau}\left\{1 - \frac{c\tau}{R}\right\} .$$

In deriving these expressions, the delta function identity

$$\frac{\partial}{\partial \tau}\{\delta(\tau)\} = -\frac{\delta(\tau)}{\tau} \tag{2.92}$$

has been used. It should be noted that the pressure pulse in Eq. (2.90) is an odd function of the retarded time, τ, and hence the area under it, known as the *impulse*, is zero [11].

2.1.7 PLANE WAVE DECOMPOSITION OF A SPHERICAL WAVE

Although spherical polar coordinates are the natural choice, the field from a point source may also be derived using the Cartesian coordinate system. The Helmholtz equation may then be written as

$$\frac{\partial G}{\partial x^2} + \frac{\partial G}{\partial y^2} + \frac{\partial G}{\partial z^2} + k^2 G = S_3 \delta(x)\delta(y)\delta(z) , \tag{2.93}$$

where the source strength, S_3, is the same as in Eq. (2.76). Three spatial Fourier transforms are now applied to Eq. (2.93), with respect to x, y, and z, the corresponding transform variables (wave numbers) being u, v, and w, respectively. This procedure returns an algebraic equation for the triply transformed field:

$$-\left(u^2 + v^2 + w^2\right)G_{uvw} + k^2 G_{uvw} = S_3 , \tag{2.94}$$

where a transform variable appearing as a subscript denotes the associated integral transformation, which is a convenient convention when working with multiple transforms [12]. The solution of Eq. (2.94) is

$$G_{uvw} = -\frac{S_3}{u^2 + v^2 + w^2 - k^2} . \tag{2.95}$$

To recover the field, G, the three inverse Fourier transforms must be applied to Eq. (2.95). Beginning with the inverse transform with respect to u, we find that

$$G_{vw} = -\frac{S_3}{2\pi}\int_{-\infty}^{\infty}\frac{e^{iux}}{u^2 - \eta^2}\,du$$

$$= iS_3\,\frac{e^{-i\eta|x|}}{2\eta} , \tag{2.96a}$$

where

$$\eta = \sqrt{k^2 - v^2 - w^2} \tag{2.96b}$$

and the integral has been evaluated using contour integration under the condition $Im(\eta) < 0$. Next, the inverse transform with respect to v is applied, yielding

$$G_w = i\frac{S_3}{4\pi}\int_{-\infty}^{\infty}\left(\xi^2 - v^2\right)^{-1/2}e^{i\left(vy - \sqrt{\xi^2 - v^2}\,|x|\right)}\,dv . \tag{2.97a}$$

where

$$\xi = \sqrt{k^2 - w^2} .$$
(2.97b)

The integral in Eq. (2.97a) can be expressed as a cosine Fourier transform of known form [13], which returns the expression

$$G_w = i\frac{S_3}{4}H_0^{(2)}\left(\xi\sqrt{x^2 + y^2}\right) .$$
(2.98)

The field itself follows from the third Fourier inversion, with respect to w, which may be written as a cosine inverse Fourier transform since G_w is even in w:

$$G = i\frac{S_3}{4\pi}\int_0^\infty H_0^{(2)}\left(\xi\sqrt{x^2 + y^2}\right)\cos wz \, dw .$$
(2.99)

The solution for the field, G, in Eq. (2.99) represents spherical spreading from the point source and must be equal to the solution for the same field expressed in spherical polar coordinates. It therefore follows from Eq. (2.86) that

$$\frac{e^{-ikR}}{R} = -i\int_0^\infty H_0^{(2)}\left(\sqrt{k^2 - w^2}\sqrt{x^2 + y^2}\right)\cos wz \, dw ,$$
(2.100)

where

$$R = \sqrt{x^2 + y^2 + z^2} .$$
(2.101)

In the integrand of Eq. (2.100), the kernel, $\cos wz$, represents a plane wave traveling in the z direction with amplitude given by the Hankel function of the second kind. The integral itself is a superposition of such plane waves. Thus, the integral represents an expansion of the spherical wave on the left in terms of plane waves, in this case propagating in the z direction. Obviously, if the ordering of the three inverse transforms were rearranged such that the third inversion was with respect to either u or v, then the expansion of the spherical wave would have been in terms of plane waves propagating along the x-axis or y-axis, respectively.

The plane-wave expansion of a spherical wave in Eq. (2.100) exemplifies a general technique for evaluating integrals that are not amenable to perhaps more familiar methods, such as contour integration. As illustrated above, an explicit expression for certain integrals can be obtained by solving the wave equation, or another partial differential equation, using two different coordinate systems. The technique works when one of the coordinate systems returns an explicit expression for the field variable, whilst the other returns the integral to be evaluated.

2.2 SOUND PROPAGATION IN A VISCOUS FLUID

In the foregoing discussion, dissipation in the medium supporting the propagation of a sound wave was neglected, apart from allowing the imaginary part of the acoustic wave number to be nonzero but infinitesimally small in order to ensure the convergence of certain integrals. Over shorter ranges and at lower frequencies, the effects of losses in the medium may be negligible but, as frequency increases and ranges become greater, the attenuation experienced by a propagating sound wave may grow to a significant level. In the ocean, the predominant attenuation mechanisms are the molecular relaxation of magnesium sulfate [14] and boric acid [15], as discussed in Section 1.5.9. In pure water, viscosity is the main dissipation mechanism, giving rise to an acoustic attenuation that scales as the square of frequency below a transition frequency, switching to a square root dependence on frequency above the transition frequency.

The classical wave equation for sound propagation in a viscous fluid has long been known, the original derivation dating back to 1845 when it was published by Stokes [16]. This wave equation, in one-dimensional, inhomogeneous form for the velocity-potential Green's function $g = g(x;t)$, is

$$\frac{\partial^2 g}{\partial x^2} - \frac{1}{c_0^2}\frac{\partial^2 g}{\partial t^2} + \gamma\frac{\partial^3 g}{\partial t \partial x^2} = S_1\delta(x)\delta(t) , \tag{2.102}$$

where c_0 is the speed of sound in the fluid in the absence of viscosity, and S_1 is the planar source strength. Eq. (2.102) is a partial differential equation, in which the coefficient of the third-order term, representing viscous dissipation, is given by

$$\gamma = \frac{4\mu}{3\rho_0 c_0^2} , \tag{2.103}$$

where the two parameters μ and ρ_0, representing the fluid, are the dynamic viscosity and density, respectively. Proceeding as in our earlier analyses of the wave equation, the Helmholtz equation is obtained by taking the temporal Fourier transform of Eq. (2.102), which yields

$$\frac{\partial^2 G}{\partial x^2} + k_0^2 G + i\omega\gamma\frac{\partial^2 G}{\partial x^2} = S_1\delta(x) , \tag{2.104}$$

where $G = G(x;\omega)$ is the transform of g and $k_0 = \omega/c_0$ is the (real) acoustic wave number. Solutions of the homogeneous form of Eq. (2.104), that is to say, for harmonic waves, have been discussed by Stokes himself, Stefan [17], Lord Rayleigh [18], and many others.

The presence of the viscous term in Stokes' equation introduces two new effects: the speed of sound is no longer a constant but depends on frequency; and attenuation, which is also frequency dependent, now occurs as sound waves propagate through the fluid. The expressions for the frequency dependence of the sound speed and attenuation in any medium are known as *dispersion formulas*. In the case of a viscous fluid, the dispersion formulas, especially at high frequencies, exhibit properties, which are easily misinterpreted and, indeed, have led to a number of misconceptions appearing in the literature [19]. The issues revolve around the question as to whether the solutions of Stokes' equation satisfy causality. In fact, Stokes' equation is well behaved in the sense that its solutions are strictly causal [20,21], as becomes clearer from an examination of the associated dispersion formulas.

2.2.1 DISPERSION FORMULAS

On combining the two second-order spatial derivatives in Eq. (2.104), the Helmholtz equation can be written as

$$\frac{\partial^2 G}{\partial x^2} + \frac{k_0^2}{(1 + i\omega\gamma)} G = \frac{S_1}{(1 + i\omega\gamma)} \delta(x) . \tag{2.105}$$

By introducing an effective wave number,

$$k = k(\omega) = \frac{k_0}{\sqrt{1 + i\omega\gamma}} , \tag{2.106}$$

the left-hand side of Eq. (2.105) takes on the same form as the Helmholtz equation for an inviscid fluid, Eq. (2.40), the only difference being that k is now a complex function of frequency and the source strength S_1 has been replaced with $S_1/(1 + i\omega\gamma)$. In terms of its real and imaginary parts, the wave number can be expressed in the form

$$k = \frac{\omega}{c(\omega)} - i\alpha(\omega) , \tag{2.107}$$

where $c(\omega)$ and $\alpha(\omega)$ are the sound speed and attenuation coefficient, respectively, at angular frequency ω. Although k is complex, the solution for the field in Eq. (2.45) still holds, but now the x dependence is given by the exponentially decaying function

$$e^{-ik|x|} = e^{-i\omega|x|/c(\omega)} e^{-\alpha(\omega)|x|} . \tag{2.108}$$

Clearly, in the limit as the viscosity goes to zero, Eq. (2.108) reduces to the inviscid solution, since γ is then zero, the wave number in Eq. (2.106) is $k = k_0$, the sound speed, c, becomes c_0, and the attenuation vanishes.

The dispersion formulas for $c(\omega)$ and $\alpha(\omega)$ are obtained by equating the expressions for the complex wave number, k, in Eqs. (2.106) and (2.107). After some standard algebraic manipulations of the radical, the dispersion formulas for sound waves in a viscous fluid are found to be

$$c(\omega) = \frac{c_0}{Re\left[(1 + i\omega\gamma)^{-1/2}\right]}$$

$$= \frac{\sqrt{2}c_0\sqrt{1 + \omega^2\gamma^2}}{\left[\sqrt{1 + \omega^2\gamma^2} + 1\right]^{1/2}} \rightarrow \begin{cases} c_0 & \text{for} \quad |\omega| \ll \gamma^{-1} \\ c_0\sqrt{2|\omega|\gamma} & \text{for} \quad |\omega| \gg \gamma^{-1} \end{cases} \qquad (2.109a)$$

and

$$\alpha(\omega) = -\frac{\omega}{c_0} Im\left[(1 + i\omega\gamma)^{-1/2}\right]$$

$$= \frac{|\omega|}{\sqrt{2}c_0\sqrt{1 + \omega^2\gamma^2}}\left[\sqrt{1 + \omega^2\gamma^2} - 1\right]^{1/2} \rightarrow \begin{cases} \dfrac{\omega^2\gamma}{2c_0} & \text{for} \quad |\omega| \ll \gamma^{-1} \\[2ex] \dfrac{1}{c_0}\sqrt{\dfrac{|\omega|}{2\gamma}} & \text{for} \quad |\omega| \gg \gamma^{-1} \end{cases}$$

$$(2.109b)$$

Note that $c(\omega)$ and $\alpha(\omega)$ are both even functions of frequency. These expressions are plotted in normalized form in Fig. 2.3, illustrating the transition from the low-to high-frequency scaling laws.

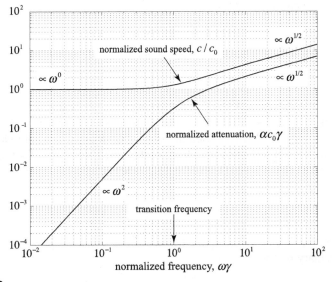

FIGURE 2.3

Dispersion curves for acoustic propagation in a viscous fluid.

Below the transition frequency, the attenuation scales as the square of frequency whilst the sound speed remains constant; above the transition, both the sound speed and the attenuation scale as the square root of frequency. According to Eq. (2.103), the transition frequency itself, γ^{-1}, scales inversely with the viscosity. For water, with viscosity $\mu \sim 10^{-3}$ Pa s, density $\rho_0 \sim 1000$ kg/m^3, and $c_0 \sim 1484$ m/s, the transition frequency is well into the ultrasonic range, taking a value around 260 GHz. A fluid with higher viscosity, for example, glycerol with $\mu \sim 1.41$ Pa s, $\rho_0 \sim 1260$ kg/m^3, and $c_0 \sim 1920$ m/s, exhibits a transition frequency that is considerably lower but still in the ultrasonic band at about 390 MHz.

On closer inspection of the dispersion curves in Fig. 2.3, it is evident that, in the limit of high frequency, the sound speed in a viscous fluid increases without limit. Since waves traveling with infinite speed would arrive at a distant receiver at exactly the same time as they left the source, a condition that could never be encountered in practice, this would seem to suggest that Stokes' equation is not physical. As can be seen in Fig. 2.3, however, those same waves are infinitely attenuated, so never actually depart from the source. Since, in effect, infinitely fast waves are never launched into the fluid, instantaneous arrivals are not a feature of sound propagation in a viscous fluid, and Stokes' equation, in fact, predicts acoustic behavior that is perfectly physical in character. In particular, the solutions of Stokes' equation satisfy causality, which may be demonstrated with the aid of the Kramers–Kronig *dispersion relations* [22–25].

2.2.2 KRAMERS–KRONIG DISPERSION RELATIONS

Consider a real, causal function of time,

$$q(t) = 0 \quad \text{for} \quad t < 0, \tag{2.110}$$

whose Fourier transform is a complex function of frequency,

$$Q(\omega) = R(\omega) + iX(\omega), \tag{2.111}$$

where R and X are the real and imaginary parts of Q. A Fourier transform of a causal function of time, such as $Q(\omega)$, is referred to as a *causal transform*. As a direct consequence of the causality condition in Eq. (2.110), R and X are not independent of one another but are related through two integrals, Hilbert transforms, usually known as the Kramers–Kronig relations but sometimes as the Plemelj formulas [26]. Thus, the real part of Q is

$$R(\omega) = -\frac{1}{\pi} P \int_{-\infty}^{\infty} \frac{X(\omega')}{(\omega' - \omega)} d\omega' \tag{2.112a}$$

and the imaginary part is

$$X(\omega) = \frac{1}{\pi} P \int_{-\infty}^{\infty} \frac{R(\omega')}{(\omega' - \omega)} d\omega', \tag{2.112b}$$

where the P before the integrals denotes the Cauchy principle value. According to these expressions, if the imaginary part of Q is known at all frequencies, then the real part can be found by taking its Hilbert transform, and vice versa. Looked at another way, if the Kramers–Kronig relations are satisfied, then the function $q(t)$ is causal.

The Kramers–Kronig relations are usually proved by contour integration in the complex frequency plane, but this requires that $Q(\omega)$ be both an analytic function and square-integrable. Often, however, neither of these conditions is satisfied by the complex functions encountered in underwater acoustics, for instance, the complex wave number associated with wave propagation in a marine sediment, whose imaginary part, representing the attenuation, follows a frequency power law. Such power laws are neither analytic nor square-integrable, but are in fact causal and do satisfy the Kramers–Kronig relations [27], suggesting that an alternative to the proof by contour integration must exist that does not rely on analyticity and square-integrability.

The Kramers–Kronig relations may be proved by integration along the real frequency axis, without resorting to analytic continuation into the complex plane, the only requirement being that the relevant Fourier transforms exist [27]. To begin, the causal function of time, $q(t)$ in Eq. (2.110), is expressed in terms of its inverse Fourier transform,

$$q(t) = \frac{1}{2\pi} \int_0^\infty Q(\omega') e^{i\omega' t} d\omega' , \tag{2.113}$$

where the lower limit of zero reflects the causality condition in Eq. (2.110). The corresponding inversion integral is

$$Q(\omega) = \int_{-\infty}^\infty q(t) e^{-i\omega t} dt . \tag{2.114}$$

On substituting Eq. (2.113) for $q(t)$ into Eq. (2.114), the Fourier transform becomes the double integral

$$Q(\omega) = \frac{1}{2\pi} \int_0^\infty e^{i(\omega'-\omega)t} dt \int_{-\infty}^\infty Q(\omega') d\omega' , \tag{2.115}$$

which clarifies why the primes were used on the integration variable in Eq. (2.113). From the theory of generalized functions, the integral over time in Eq. (2.115) is

$$\int_0^\infty e^{i(\omega'-\omega)t} dt = \pi\delta(\omega' - \omega) + \frac{i}{(\omega' - \omega)} , \tag{2.116}$$

from which it follows that Eq. (2.115) reduces to

$$Q(\omega) = \frac{i}{\pi} P \int_{-\infty}^\infty \frac{Q(\omega')}{(\omega' - \omega)} d\omega' . \tag{2.117}$$

In effect, this states that the causal transform $Q(\omega)$ is the Hilbert transform of itself. On expressing Q in terms of its real and imaginary parts, as in Eq. (2.111), then equating real part to real part and similarly with the imaginary parts, Eq. (2.117) splits into two equations, the very same Kramers–Kronig relations expressed in Eq. (2.112).

Taking $k = k(\omega)$ as the complex wave number of a propagating wave in a causal, dissipative medium, it may be shown, by considering the passage of the wave through a thin slice of the material, that $-ik(\omega)$ is a causal transform [27], whose real and imaginary parts satisfy the Kramers–Kronig dispersion relations. Conversely, if $-ik(\omega)$ satisfies the Kramers–Kronig relations, the associated wave field must be causal, an idea which is applied in the following to the dispersion formulas derived from Stokes' equation for a sound wave in a viscous fluid.

2.2.3 CAUSALITY AND STOKES' EQUATION

For the case of a viscous fluid, we may set

$$Q(\omega) = -ik(\omega) = -i\frac{\omega}{c_0}(1 + i\omega\gamma)^{-1/2}, \tag{2.118}$$

where the expression on the right derives from Eq. (2.106). The inverse Fourier transform of $Q(\omega)$ can be written in the form

$$q(t) = -\frac{\gamma^{-1/2}}{2\pi c_0}\frac{\partial}{\partial t}\int_{-\infty}^{\infty} \left(\gamma^{-1} + i\omega\right)^{-1/2} e^{i\omega t}\, d\omega, \tag{2.119}$$

where the integral is a standard form [13], allowing $q(t)$ to be evaluated as

$$q(t) = \begin{cases} \dfrac{\gamma^{-1/2}}{c_0\sqrt{\pi}}\dfrac{\partial}{\partial t}\left\{t^{-1/2}e^{-t/\gamma}\right\} & \text{for} \quad t > 0 \\ 0 & \text{for} \quad t < 0 \end{cases}. \tag{2.120}$$

Thus, $q(t)$ is a causal function of time and its transform, $Q(\omega)$, is a causal transform. It follows that the solutions of Stokes' equation for the acoustic field in a viscous fluid must satisfy causality, with no arrivals predicted prior to the onset of the source.

The same conclusion could have been reached, at least in principle, by substituting the real and imaginary parts of $-ik(\omega)$, as represented by the expressions for the sound speed and attenuation in Eq. (2.109), into the Kramers–Kronig dispersion relations in Eq. (2.112). In this particular case, however, there is a difficulty with the Kramers–Kronig relations, in that the Hilbert transforms in Eq. (2.112) are not tractable when $R(\omega) = -\alpha(\omega)$ and $X(\omega) = -\omega/c(\omega)$ take the form for propagation in a viscous fluid, as given by the expressions in Eq. (2.109). This is an example of a situation where causality is satisfied, but the Kramers–Kronig relations fail because the integrals do not converge.

2.2.4 **PULSE PROPAGATION IN A VISCOUS FLUID**

Although the foregoing analysis of the complex wave number confirms that the solutions of Stokes' equation satisfy causality, it provides no information on the shape of transient signals in a viscous fluid. Of particular interest is the nature of the received pulse around the origin of time, $t = 0$ and, in particular, at time $t = 0+$ immediately after the source is activated. Some authors [19,28] have claimed that transient solutions of Stokes' equation are felt instantly throughout the entire fluid domain, implying an infinite speed of nonzero amplitude waves, which would be unphysical. In fact, the transient solutions of Stokes' equation show no instantaneous arrivals but are perfectly well behaved everywhere in the viscous fluid medium [20,21].

To examine the nature of transient acoustic signals in a viscous fluid, we return to the frequency-domain version of Stokes' equation in Eq. (2.105):

$$\frac{\partial^2 G}{\partial x^2} + \frac{k_0^2}{(1 + i\omega\gamma)} G = \frac{S_1}{(1 + i\omega\gamma)} \delta(x) \tag{2.121}$$

where $G = G(x;\omega)$ is the Fourier transform (with respect to time) of the Green's function, $g = g(x;t)$. Eq. (2.121) may be solved for G by taking the Fourier transform with respect to x (as defined in Eq. 2.41) of both sides, which returns the algebraic equation

$$G_s = -\frac{S_1}{\left[s^2(1 + i\omega\gamma) - k_0^2\right]} . \tag{2.122}$$

Two inverse Fourier transforms must be applied to Eq. (2.122), one with respect to wave number, s, and the other with respect to frequency, ω, in order to recover the pulse shape. Apparently, the ordering of the inverse transforms does not matter, or at least, it would not matter if both integrals could be evaluated explicitly, but such is not the case. Whichever order is chosen, only the first inversion integral can be evaluated explicitly but the second cannot. In both cases, the second integral can be approximated, and it is here that some of the confusion in the literature originates. By treating the integral over wave number exactly and approximating the integral over frequency, the approximate solution for the transient signal is noncausal, showing a tail extending into negative times. On the other hand, if the integral over frequency is treated exactly, and the integral over wave number is approximated, the resultant pulse everywhere in the fluid medium is strictly causal, showing a zero response at all negative times with no instantaneous arrivals. The difference between the causal and noncausal results is illustrated below.

It is perhaps natural to perform the inversion over wave number first, as has been the case in the literature [29]. Then, from Eq. (2.122), the Fourier transform of Green's function with respect to time can be written as

$$G = -\frac{S_1}{2\pi(1 + i\omega\gamma)} \int_{-\infty}^{\infty} \frac{e^{isx}}{(s - s_+)(s - s_-)}\, ds \,, \tag{2.123}$$

where the simple poles

$$s_\pm = \pm\frac{k_0}{\sqrt{1 + i\omega\gamma}} \tag{2.124}$$

are the roots of the quadratic denominator in Eq. (2.122). For real ω, s_+ and s_-, respectively, lie in the bottom half and top half of the complex s-plane. Integrating around a D-shaped contour in the top (bottom) half plane for $x > 0$ ($x < 0$) leads to the exact expression

$$G = -\frac{S_1}{2ik_0\sqrt{1 + i\omega\gamma}}\exp\left(-\frac{ik_0|x|}{\sqrt{1 + i\omega\gamma}}\right). \tag{2.125}$$

Notice that, in the limit of zero viscosity, this expression reduces identically to Eq. (2.45) for a plane wave in an inviscid fluid.

Now, converting the velocity potential to pressure, the Fourier inversion over frequency takes the form

$$p(x; t) = \rho_0 \frac{\partial g(x; t)}{\partial x}$$
$$= -\frac{\rho_0 c_0 S_1}{4\pi} \int_{-\infty}^{\infty} (1 + i\omega\gamma)^{-1/2}\exp\left(-\frac{ik_0|x|}{\sqrt{1 + i\omega\gamma}}\right)e^{i\omega t}\, d\omega. \tag{2.126}$$

The integral here cannot be expressed explicitly but may be approximated by expanding the radical in a Taylor series to first order in the frequency:

$$(1 + i\omega\gamma)^{-1/2} = 1 - \frac{i\omega\gamma}{2} + \cdots \tag{2.127}$$

in which case

$$p(x; t) \approx -\frac{\rho_0 c_0 S_1}{4\pi} \int_{-\infty}^{\infty} \exp\left[i\omega\left(t - \frac{|x|}{c_0}\right)\right]\exp\left(-\frac{\omega^2\gamma|x|}{2c_0}\right)d\omega, \tag{2.128}$$

where the radical, in the argument of the exponential, has been approximated by its first-order expansion but, outside the exponential, has been set to unity. The integral in Eq. (2.128) is a known form [30], allowing the approximate solution for the pressure transient to be expressed as

$$p(x; t) \approx -\frac{\rho_0 c_0 S_1}{2}\sqrt{\frac{c_0}{2\pi\gamma|x|}}\exp\left[-\frac{(c_0 t - [x])^2}{2c_0\gamma|x|}\right]. \tag{2.129}$$

This is a symmetrical, Gaussian-shaped pulse, which is obviously noncausal since it exhibits a tail extending into negative times, as well as predicting instantaneous arrivals everywhere in the viscous fluid.

Such behavior is not physical, nor is it consistent with the fact established earlier that the complex wave number in Eq. (2.106) is a causal transform. The problems with the pressure pulse in Eq. (2.129) originate with the Taylor expansion for the radical in Eq. (2.127). This expansion is a low-frequency approximation, in which high frequencies are poorly represented, thus accounting for the noncausal nature of the impulse response in Eq. (2.129).

These difficulties with noncausal behavior are resolved by reversing the order of the Fourier inversions. By choosing the inverse Fourier transform with respect to *frequency* as the first of the two transforms to be applied to Eq. (2.122), all frequencies are then treated correctly. By adopting this approach, the wave number spectrum of Green's function may be written as

$$g_s(t) = \frac{S_1}{2\pi} \int_{-\infty}^{\infty} \frac{e^{i\omega t}}{(\omega - \omega_+)(\omega - \omega_-)} d\omega \,, \tag{2.130}$$

where the simple poles $\omega_\pm$ are the roots of the quadratic denominator in Eq. (2.122):

$$\omega_\pm = \frac{i\gamma c_0^2 \pm \sqrt{4c_0^2 s^2 - \gamma^2 c_0^4 s^4}}{2} \,. \tag{2.131}$$

With s real, both poles lie above the real axis in the top half of the complex ω-plane. From Jordan's lemma and Cauchy's theorem, the D-shaped integration contour used to evaluate Eq. (2.130) for $t < 0$ must be taken around the lower half-plane, which contains no poles and hence the integral is zero. For $t > 0$, the integration is taken around the D-shaped contour in the top half-plane, which encloses the two poles in Eq. (2.131). By adding the residues of these two poles, the wave number spectrum of the field is expressed exactly as

$$g_s(t) = -u(t) \frac{2S_1 c_0^2}{\sqrt{4c_0^2 s^2 - \gamma^2 c_0^4 s^4}} \exp\left(-\frac{\gamma c_0^2 s^2}{2} t\right) \sin\left(\frac{\sqrt{4c_0^2 s^2 - \gamma^2 c_0^4 s^4}}{2} t\right) \,, \tag{2.132}$$

where $u(t)$ is the Heaviside unit step function.

Since, according to Eq. (2.132), every wave number component of the field is identically zero for all negative times, the field itself must satisfy causality. To obtain the Green's function, an inverse Fourier transform with respect to wave number s is applied to Eq. (2.132). The exact solution for the pressure pulse is then found to be

$$p = \rho_0 \frac{\partial g}{\partial t}$$

$$= -u(t)\frac{\rho_0 c_0^2 S_1}{2\pi} \int_{-\infty}^{\infty} e^{isx}\left[\left(1 + \frac{i\gamma c_0 s}{2\chi}\right)\exp\left(ic_0 s\chi - \frac{\gamma c_0^2 s^2}{2}\right)\right. \tag{2.133}$$

$$\left. + \left(1 - \frac{i\gamma c_0 s}{2\chi}\right)\exp\left(-ic_0 s\chi - \frac{\gamma c_0^2 s^2}{2}\right)\right] ds ,$$

where

$$\chi = \sqrt{1 - \frac{\gamma^2 c_0^2 s^2}{4}} . \tag{2.134}$$

Although the integral in Eq. (2.133) cannot be expressed explicitly, if the radical χ is approximated as unity, it reduces to a known form, leading to the approximate, causal, expression for the pressure impulse response

$$p(x;t) \approx -u(t)\frac{\rho_0 c_0 S_1}{4\sqrt{2\pi\gamma t}}\{F(x;t) + F(-x;t)\} , \tag{2.135}$$

where

$$F(x;t) = \left(1 + \frac{x}{c_0 t}\right)\exp\left[-\frac{(c_0 t - x)^2}{2\gamma c_0^2 t}\right] . \tag{2.136}$$

Three normalized pressure pulses are shown in Fig. 2.4 for comparison: the exact, asymmetrical expression in Eq. (2.133), evaluated by numerical integration; the asymmetrical causal approximation in Eq. (2.135); and the noncausal, symmetrical pulse in Eq. (2.129). Time is normalized to the retarded time, $t_0 = |x|/c_0$, where the peak in the noncausal approximation occurs; and pressure is normalized to the noncausal peak pressure, given by $\rho_0 c_0 S_1/(2\sqrt{2\pi\gamma t_0})$. In the example of Fig. 2.4, the value of γ normalized to t_0 is $\gamma/t_0 = 0.3$, which is unrealistically high for most viscous fluids, but it serves to illustrate the differences between the causal and noncausal approximations, and the exact solution.

Immediately after the source is activated, at $t = 0+$, the two asymmetrical, causal pressure pulses in Fig. 2.4 are *maximally flat*, that is, the waveform itself and all its time derivatives are identically zero [20]. Under this condition, the pulse is well behaved physically but does not possess a Taylor expansion about $t = 0$ or, looked at another way, all the coefficients of the Taylor expansion are zero. Thus, the pulse satisfies causality in the strong sense that, not only are there no arrivals before $t = 0$ but also, there are no instantaneous arrivals anywhere in the fluid. Instead, the received pulse rises perfectly smoothly from zero with a steep leading edge, then peaks ahead of the retarded time because it contains high frequency Fourier

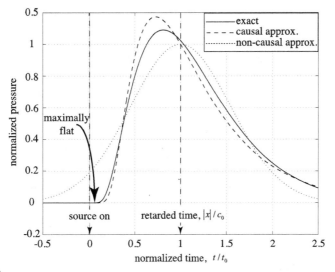

FIGURE 2.4

Pressure pulse in a viscous fluid. Noncausal approximation, Eq. (2.129); causal approximation, Eq. (2.135); and exact integral expression, Eq. (2.133).

components traveling faster than c_0. After the peak, a relatively slow decay occurs as the pulse propagates past the receiver, approaching zero asymptotically as $t \rightarrow \infty$.

The analysis of Stokes' equation for acoustic propagation in a viscous fluid contains a number of subtleties that, if misinterpreted, could lead to the false conclusion that the impulse response is nonphysical. Looking at the dispersion curves in Fig. 2.3, the increase in the sound speed without limit as the frequency rises could be misconstrued as implying that the leading edge of the received pulse will arrive at the same instant as the source is activated. Such behavior would indeed be unphysical but it does not occur because, as illustrated in Fig. 2.3, the infinitely fast Fourier components in the sound speed are also infinitely attenuated and hence do not propagate away from the source. The overall effect of this trade-off between infinite speed and infinite attenuation is an impulse response that, immediately after the source is activated, is maximally flat everywhere in the viscous fluid, thus ensuring that causality is strictly satisfied. Although Stokes himself did not address the question, the solution of his equation [16] for the impulse response of a viscous fluid is physically well behaved, as pictorially illustrated in Fig. 2.4.

2.3 SOUND WAVES AND SHEAR WAVES IN MARINE SEDIMENTS

Most marine sediments are composed of mineral particles, pore fluid, and organic matter, although near-shore wave breaking, or methane seeps in organic-rich muddy sediments [31], may also introduce free gas into the pore spaces. Many marine sediments, however, can be treated as fully saturated, two-phase, unconsolidated granular media, consisting of mineral grains and seawater, in which the grains are in contact but not bonded together. Sediments are broadly classified as sands, silts, and clays, in descending order of particle size, with the mean grain sizes ranging from 2.0 mm for the coarsest sands to less than 1 μm for the finest clays.

Sand grains are noncohesive, having little or no surface attraction to one another, and are maintained in contact, in sandy sediments, by gravity. Clay minerals, on the other hand, occur as cohesive platelets that aggregate into larger particles known as floccules, which are held together by intermolecular van der Waals forces. The porosity, that is, the ratio of the volume of the pore spaces to the total volume, is highest in the very-fine-grained materials, taking values as high as 0.95 in newly deposited clays, before being reduced by consolidation, and is least in the coarser sands, at around 0.37. The latter value is also the porosity of a random "close" packing of uniform spheres [32,33], suggesting that surface roughness has little effect on the packing structure of the larger sand grains.

Sand, silt, and clay sediments all support sound (longitudinal) waves and shear (transverse) waves, with the shear wave speed at least a factor of 10 less than that of the sound wave. Over an extended bandwidth, the attenuation exhibited by both the sound wave [34,35] and the shear wave [36] follows a frequency power law with an exponent close to unity, and the wave speeds both tend to rise with increasing grain size, corresponding to decreasing porosity [37].

2.3.1 THE BIOT THEORY

In 1956, building on his earlier work [38,39], Maurice Biot published a phenomenological theory of wave propagation in a porous elastic solid containing a compressible viscous fluid [40,41]. The theory predicts the existence of three types of wave in the porous medium: a "fast" wave, which is essentially the same as a normal acoustic wave; a second compressional wave, often colloquially designated the "slow" wave, since its phase speed is significantly less than that of the fast wave; and, as in elasticity theory, a rotational or shear wave. The fast and slow waves correspond, respectively, to the pore fluid moving almost in phase and largely out of phase with the solid elastic frame. The shear wave is a natural consequence of the assumed elasticity of the frame.

Since the only dissipation mechanism in the Biot model is viscosity of the interstitial fluid, the fast wave exhibits an attenuation that is characteristic of a viscous fluid: at low frequencies, below the transition frequency, the attenuation scales as the square of frequency, whilst above it is proportional to the square root of frequency, as illustrated in Fig. 2.3. Such behavior is not consistent with experimental observations of the attenuation of sound waves in saturated marine sediments, which, according to Hamilton [35], scales with the first power of frequency. In several reports [42–44], Hamilton asserts that the attenuation, α_p, in dB/m is given by

$$\alpha_p = k_p f , \qquad\qquad (2.137)$$

where f is frequency in kHz. The constant of proportionality, k_p, varies with sediment type, depending on porosity and grain size, taking an approximate value for near-surface sands of $k_p \approx 0.5$. Although the linear frequency scaling in Eq. (2.137) might be matched by the Biot theory over a very limited bandwidth centered on the transition frequency, the theory cannot be made to fit the attenuation data over an extended frequency range.

Biot, in the original development of his theory, had in mind water-saturated rock [40], where the mineral matrix is quite rigid and can be clearly identified as a solid elastic frame. Stoll [45–47], who has adapted Biot's theory in an attempt to reproduce the observed dispersion properties of sound waves and shear waves in unconsolidated sediments, does not say what constitutes the elastic frame in such a granular material. Since the grains are not bonded together but have a degree of mobility relative to one another, the identity of the frame is not entirely clear. It could be postulated, perhaps, that the mineral grains themselves constitute the skeletal frame, with the frictional forces at the grain-to-grain contacts providing some degree of rigidity. Although there is no a priori reason to believe that such a loose granular structure would act as an elastic solid, yet that is exactly the assumption that Stoll makes.

In the context of unconsolidated sediments, a further difficulty with the Biot–Stoll model is that the predicted slow wave, associated with the out-of-phase relative motions of the pore fluid and the elastic frame, does not appear to exist. Admittedly, the slow wave, which is diffusive in character, is expected to be heavily attenuated, making it difficult to detect. Nevertheless, the slow wave has been observed in water-saturated, sintered (consolidated) glass beads [48], but not in unconsolidated (loose) glass beads [49], representative of an unconsolidated sediment. In fact, the slow wave has not been detected in any saturated, unconsolidated granular medium, suggesting that the concept of pore-fluid flow relative to a solid elastic frame, which is the mainstay of the Biot–Stoll model, may not be an appropriate representation of wave propagation in an unconsolidated sediment.

Although Stoll does not specify the nature of the elastic frame in an unconsolidated sediment, he nevertheless assigns to it a shear, or frame, modulus. This, along with the other elastic moduli in the Biot–Stoll model, should strictly be real and independent of frequency. Stoll, however, proceeds by allowing the dynamic moduli in the model to be complex and frequency dependent, with the various frequency dependencies determined by matching the expressions for wave speed and attenuation to experimental data. Such an empirical approach, besides raising questions concerning causality, almost guarantees that the Biot–Stoll model will fit any experimentally measured dispersion curves for waves in unconsolidated sediments, whatever form those curves may happen to take. The model, however, provides little information on the physical processes underlying the observed dispersion behavior.

2.3.2 THE GRAIN-SHEARING THEORY

An alternative theory of wave propagation in unconsolidated marine sediments, known as the grain-shearing (GS) theory, was introduced in its original form in 1997 [50]. A series of refinements [34,37,51−53] developed since then have culminated in the latest version [36] published in 2014. In essence, the GS theory treats the two-phase sediment as a homogeneous continuum in which there is no relative motion between the pore fluid and the mineral grains. There is no assumption that the mineral grains form an elastic frame, only that stresses occur at the grain contacts during the passage of a wave. These stresses are associated with the molecularly thin film of fluid separating asperities (microroughness) on the faces of grains that are in contact and sliding against one another. It is postulated that as the sliding progresses, the phenomenon of strain hardening [54] occurs, whereby there is a continually increasing resistance to the sliding motion as the intergranular shearing proceeds.

Two types of wave are supported by the homogeneous, porous medium, a sound wave and a shear wave, but no slow wave of the type predicted by the Biot theory. The shear wave emerges as a natural consequence of the strain hardening occurring at the intergranular contacts. Both types of wave exhibit frequency dispersion in the wave speed and an attenuation that scales as (nearly) the first power of frequency over an extended bandwidth. In a recent version, designated the VGS theory, viscosity of the pore fluid is included as an additional dissipation mechanism, which modifies the low-frequency (<20 kHz) behavior of the sound speed and attenuation, whilst leaving the shear wave unaffected.

The development of the GS theory [52] is based on the full stress tensor, $\boldsymbol{\sigma}$, in which the terms relate the stress to the rate of strain at a point in the granular medium. The argument follows much the same course as that for an isotropic, viscous fluid [55], except that the stress, instead of being set proportional to the velocity gradient, is given by convolutions representing the mechanism of strain hardening. This leads to the Navier−Stokes equation for the medium, which is then linearized, consistent with the small amplitudes of the particle velocities of the sound and shear waves. From the equation of state, along with continuity of mass, a partial differential equation is obtained for the vector flow field, $\widehat{v}$.

According to Helmholtz's theorem [56], any vector field, $\widehat{v}$, can be expressed as the sum of the gradient of a scalar potential, g, and the curl of a zero-divergence vector potential, $\widehat{A}$:

$$\widehat{v} = \operatorname{grad} g + \operatorname{curl} \widehat{A}; \ \operatorname{div} \widehat{A} = 0 . \tag{2.138}$$

These conditions allow the partial differential equation for the field $\widehat{v}$ to be split into two separate equations, one representing a compressional wave and the other a rotational or shear wave. Both equations may be Fourier transformed with respect to time, to obtain the associated Helmholtz equations, from which the dispersion equations for the compressional and shear waves may be determined.

For the compressional wave, the expressions for the phase speed, $c_p(\omega)$, and attenuation, $\alpha_p(\omega)$, are

$$c_p(\omega) = \frac{c_0}{Re\left[1 + \chi_p(i|\omega|T)^n g(\omega)\right]^{-1/2}} \tag{2.139a}$$

and

$$\alpha_p(\omega) = -\frac{|\omega|}{c_0} Im\left[1 + \chi_p(i|\omega|T)^n g(\omega)\right]^{-1/2} . \tag{2.139b}$$

Similarly for the shear wave, the phase speed, $c_s(\omega)$, and attenuation, $\alpha_s(\omega)$, are given by

$$c_s(\omega) = \frac{\chi_s c_0}{Re[(i|\omega|T)^n]^{-1/2}} = \chi_s c_0 \frac{(|\omega|T)^{n/2}}{\cos(n\pi/4)} \tag{2.140a}$$

and

$$\alpha_s(\omega) = -\frac{|\omega|}{\chi_s c_0} Im[(i|\omega|T)^n]^{-1/2} = \frac{|\omega|}{\chi_s c_0} \frac{\sin(n\pi/4)}{(|\omega|T)^{n/2}} . \tag{2.140b}$$

In these expressions, $0 < n \ll 1$ is the strain hardening index, sometimes known as the material exponent, χ_p and χ_s are dimensionless GS coefficients, somewhat analogous to the Coulomb friction coefficient of classical mechanics, and $T = 1$ s has been included in order to avoid awkward units that would otherwise arise when frequency is raised to a fractional power. In the VGS version of the GS theory, the function $g(\omega)$ in Eq. (2.139), representing the effect of the viscosity of the pore fluid, takes the form

$$g(\omega) = \left(1 + \frac{1}{i\omega\tau}\right)^{-1+n} , \tag{2.141}$$

where $\tau \approx 0.12$ ms is a viscoelastic time constant. In the formulation of the original GS theory, Eq. (2.139) still holds but with $g(\omega)$ set equal to unity; and even in the VGS version, $g(\omega)$ differs from unity only at low frequencies, below about 20 kHz. The dispersion equations for the shear wave in Eq. (2.140) remain the same in both the GS and VGS variants of the GS theory. Note that the units of the attenuation in Eqs. (2.139b) and (2.140b) are Np/m.

Physically, c_0 is the compressional wave speed that would be observed in the absence of grain-to-grain stress relaxation, that is, if the granular medium were a suspension in which there were no contact forces between the grains. The value of c_0 cannot be determined by direct measurement but can be evaluated from the Mallock–Wood equation [57,58] for the equivalent suspension,

$$c_0 = \sqrt{\frac{\kappa_0}{\rho_0}} , \tag{2.142}$$

where the bulk modulus, κ_0, and density, ρ_0, of the two-phase medium are given by the weighted means

$$\rho_0 = N\rho_f + (1 - N)\rho_g \tag{2.143}$$

and

$$k_0^{-1} = N\kappa_f^{-1} + (1 - N)\kappa_g^{-1}, \tag{2.144}$$

in which N is the porosity of the granular medium, ρ_f and ρ_g are, respectively, the densities of the pore fluid and mineral grains, whilst κ_f and κ_g are the corresponding bulk moduli.

The parameters relating to the properties of the pore fluid and the mineral grains can be assigned handbook values, but the three unknown parameters representing the intergranular interactions, the strain hardening index, n, along with the two GS coefficients, χ_p and χ_s, are a little more difficult to specify. Rather like the more familiar Coulomb friction coefficient, it does not seem possible, from theoretical considerations, to express these parameters in terms of known properties of the medium. An alternative approach is to evaluate them by matching the dispersion equations to data at spot frequencies, ω_p and ω_s, respectively, for the compressional and shear wave.

Assuming that experimental values of $c_s(\omega_s)$ and $\alpha_s(\omega_s)$ are available for the shear wave, then n can be found simply by multiplying together the two expressions in Eq. (2.140), to obtain

$$\tan\left(\frac{n\pi}{4}\right) = \frac{c_s(\omega_s)\alpha_s(\omega_s)}{\omega_s}. \tag{2.145}$$

With n known, the shear coefficient, χ_s, can be determined from either expression in Eq. (2.140) for the shear speed, or the shear attenuation. The compressional coefficient, χ_p, can then be found from either of the expressions in Eq. (2.139), assuming that an experimental value of $c_p(\omega_p)$ or $\alpha_p(\omega_p)$ is available.

Although Eq. (2.145) provides a simple basis for evaluating the three unknown GS parameters, it relies on the availability of both speed and attenuation data for the shear wave. In practice, shear wave data may be available for the speed or the attenuation, but usually not both; it is more likely that compressional wave data, $c_p(\omega_p)$ and $\alpha_p(\omega_p)$, will both be known. Assuming this to be the case, and taking ω_p to be a sufficiently high frequency such that $g(\omega_p) \approx 1$, then n can be derived from the compressional wave [34] with a little straightforward algebra. First, a manipulation of the compressional dispersion equations in Eq. (2.139) leads to the expressions

$$\chi_p(\omega_p T)^n \cos\left(\frac{n\pi}{2}\right) = \left[\frac{c_p(\omega_p)}{c_0}\right]^2 \frac{(1 - X^2)}{(1 + X^2)^2} - 1 \tag{2.146a}$$

and

$$\chi_p(\omega_p T)^n \sin\left(\frac{n\pi}{2}\right) = \left[\frac{c_p(\omega_p)}{c_0}\right]^2 \frac{2X}{(1+X^2)^2} , \qquad (2.146b)$$

where

$$X = \frac{c_p(\omega_p)\alpha_p(\omega_p)}{\omega_p} . \qquad (2.147)$$

Eliminating χ_p by division of the expressions in Eq. (2.146) returns

$$\tan\left(\frac{n\pi}{2}\right) = \frac{2X}{(1-X^2) - \left[\frac{c_0}{c_p(\omega_p)}\right]^2 (1+X^2)^2} , \qquad (2.148)$$

which allows n to be evaluated. With n known, χ_p can be found from either of the expressions in Eq. (2.146) and likewise, the value of χ_s may be derived from either of the shear dispersion equations in Eq. (2.140), assuming that one or other of $c_s(\omega_s)$ or $\alpha_s(\omega_s)$ is available as an experimental datum. Once all the parameters in the GS theory have been specified, the dispersion equations, Eqs. (2.139) and (2.140), may be plotted as functions of frequency and compared with data.

During September, October, and November 1999, the Sediment Acoustics Experiment 1999 (SAX99) [59] was conducted off Fort Walton Beach in the northeastern Gulf of Mexico as part of an Office of Naval Research (ONR) research initiative on sound interaction with marine sediments. Broadband, compressional wave data were collected from the medium sand sediment at the experiment site using a variety of measurement techniques; and shear wave data were obtained at a single frequency, 1 kHz. The compressional wave speed and attenuation data acquired during SAX99 are shown in Fig. 2.5, as represented by the symbols [34], along with the GS and VGS dispersion equations from Eq. (2.139) for comparison. In Fig. 2.5A, the ratio of the sound speed in the sediment to the sound speed, c_w, in the water column immediately above the seabed has been plotted, rather than the sediment sound speed itself, since the sound speed ratio is less influenced by fluctuations in temperature at the different times at which the various data sets were collected. At low frequencies, below about 20 kHz, the difference between the GS and VGS formulations is quite evident, with the VGS theory providing a better fit to the data, suggesting that viscosity of the pore fluid is a significant loss mechanism in this lower frequency range. Above 20 kHz, the attenuation data and both versions of the GS theory scale essentially as the first power of frequency. Such a linear scaling with frequency is in accord with Hamilton's reports on the frequency dependence of the compressional wave attenuation, a position that he consistently maintained over a period of several decades [35,42].

Since the SAX99 shear wave data were taken at only a single frequency, it is necessary to look elsewhere for broadband measurements of the shear speed and attenuation in saturated, unconsolidated granular materials. Several authors have published broadband data on the shear attenuation in wet and/or dry sands and glass

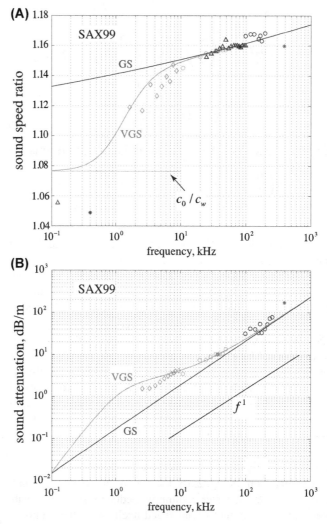

FIGURE 2.5

Compressional-wave dispersion curves for the SAX99 medium sand sediment. Data (*symbols*) are compared with the GS and VGS forms of the GS theory. (A) Sound speed ratio, including c_0/c_w from the Mallock—Wood equation. (B) Sound attenuation, with the line labeled f^1 illustrating a linear scaling with frequency.

beads [60—63] and all their data sets show good agreement [36] with the near-linear frequency scaling predicted by the GS theory, Eq. (2.140b).

Kimura [63] measured not only the attenuation but also the phase speed of the shear wave in water-saturated silica sand at three different temperatures, 5°C, 20°C, and 35°C. He contained the saturated granular material in a long acrylic cylinder, which was regularly vibrated in order to achieve a reproducible porosity.

A vertical stress of 17.6 Pa was applied to the sediment column to ensure good coupling between the granular material and the shear transducers. Comparisons of his data sets with the GS dispersion equations in Eq. (2.140) are shown in Fig. 2.6. The attenuation data, which depend only weakly on temperature, exhibit a near-linear scaling with frequency over the measurement frequency band from 5 to 20 kHz, and this behavior is accurately matched by the GS theory. The shear

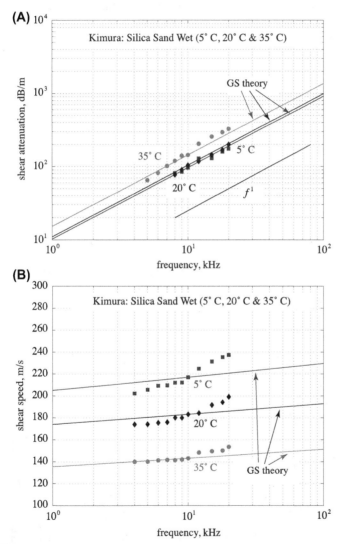

FIGURE 2.6

Shear-wave dispersion curves for water-saturated silica sand at three different temperatures. Kimura's data [63] (*symbols*) are compared with the GS theory (*solid curves*). (A) Shear wave speed. (B) Shear wave attenuation, with the line labeled f^1 illustrating a linear scaling with frequency.

speed data, which show a stronger response to temperature, are in reasonable agreement with the GS theory, although at the two lower temperatures, 20°C and 5°C, and at the higher frequencies, between 10 and 20 kHz, the theory slightly underestimates the data, but in the worst case, 5°C and 20 kHz, the discrepancy is less than 6.7%.

Besides frequency, the VGS dispersion equations in Eqs. (2.139) and (2.140) are functions of the material properties of the porous medium, notably the porosity, the mean grain size, and the overburden pressure, or depth, in the sediment. As detailed

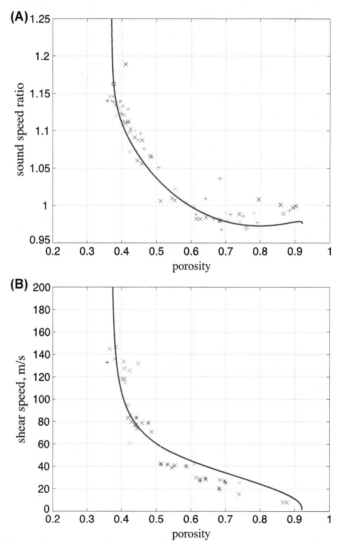

FIGURE 2.7

Sound speed ratio and shear wave speed versus porosity. The *symbols* represent data from a variety of sources [37] and the *solid lines* are from the GS theory, Eqs. (2.139a) and (2.140a).

elsewhere [37], these dependencies appear in the GS theory through the Mallock–Wood [57,58] sound speed, c_0, and the two GS coefficients, χ_p and χ_s. The Hertz theory [64] of elastic bodies in contact provides the means of formulating χ_p and χ_s in terms of the geoacoustic properties of the sediment.

The variation of the compressional and shear wave speeds with porosity is shown in Fig. 2.7 for a variety of sediments ranging between coarse sand and very fine clay. Although they show a little scatter, the data points follow a well-defined trend, which in both cases is mirrored by the theory. The situation is somewhat different in Fig. 2.8, which shows plots similar to those in Fig. 2.7 but for the compressional

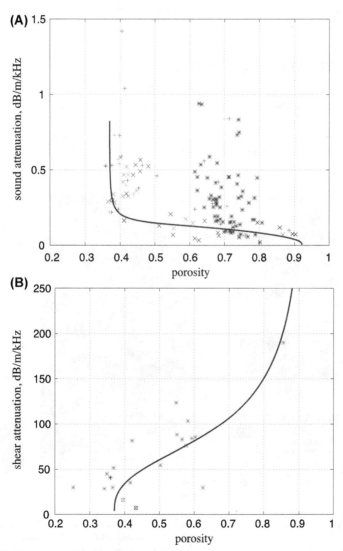

FIGURE 2.8

Sound wave and shear wave attenuation versus porosity. The *symbols* represent data and the *solid lines* the GS theory.

and shear wave attenuations versus porosity. In both the attenuation plots, the data points are quite widely scattered, with most lying above the theoretical curves, which, in effect, form lower bounds on the values that the attenuation can take (if allowance is made for the relatively large measurement errors on the lowest attenuations). Such behavior is consistent with the fact that the theory predicts the *intrinsic attenuation*, due to the irreversible conversion of wave energy into heat, whereas most measurements return the *effective attenuation*, comprised of the intrinsic attenuation plus any additional losses that may be present due to scattering from inhomogeneities such as shell fragments and animal burrows in the sediment. Since the concentration of such inhomogeneities may differ appreciably from one sediment to another, the effective attenuation shows a high degree of variability, even between sediments that are ostensibly similar.

2.4 SOURCE OR RECEIVER IN MOTION

When a sound source and an observer are in relative motion in a medium at rest, the received frequencies are shifted upward on approach and downward on departure. Such upshifts and downshifts in frequency are a commonplace occurrence, exemplified by the sound of the whistle on a passing train, and are named after Johann Christian Doppler who was the first to propose the principle [65]. Fig. 2.9 shows a visual

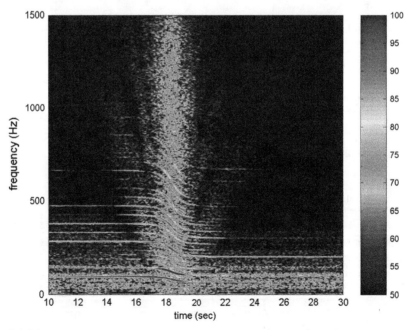

FIGURE 2.9

Sequence of harmonics from the engine and three-bladed propeller of a single-engine light aircraft, a Diamond Star DA40. The color bar represents the power spectrum of the received sound in arbitrary dB.

example of Doppler shifts, in this case, in the harmonics from the engine and propeller of a light aircraft as it flies over a microphone moored 10 m above the sea surface. As the aircraft passes the microphone, at time approximately equal to 18.5 s corresponding to the closest point of approach (CPA), the transition from an upshift to a downshift in frequency is clearly visible. It can also be seen that the magnitude of the Doppler shift is proportional to the intrinsic frequency of the harmonic. Although not evident from the aircraft harmonics in Fig. 2.9, the rate at which the transition occurs, that is, the gradient of the harmonics at the CPA, depends on the speed of the source and the distance between the source and the receiver at the CPA, as discussed below.

2.4.1 DOPPLER FREQUENCY SHIFTS (SOURCE STATIONARY, OBSERVER IN MOTION)

Consider a medium at rest in which the speed of sound is c and through which a receiver is moving in rectilinear motion with speed V, as illustrated in Fig. 2.10. Let a stationary harmonic source emitting an intrinsic frequency f_0 be located a distance z' from the receiver when the latter is at the CPA, the point labeled O in Fig. 2.10. If the angle between the direction of motion and the line connecting the source and observer is θ, then the apparent speed of the waves arriving at the receiver is $c + V \cos \theta$ and the wavelength is

$$\lambda = \frac{c}{f_0} = \frac{c + V \cos \theta}{f} \ .$$

(2.149)

Hence the *apparent* frequency at the receiver is

$$f = f_0 \left(1 + \frac{V}{c} \cos \theta \right) ,$$

(2.150)

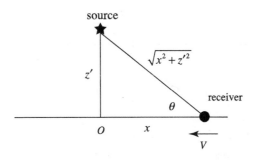

FIGURE 2.10

Stationary source, moving receiver.

which corresponds to a fractional shift in frequency

$$\frac{\Delta f}{f_0} = \frac{f - f_0}{f_0} = \frac{V}{c} \cos \theta .$$ (2.151)

The maximum shift in frequency, occurring when x is large compared with z', is symmetrical about the CPA, being the same on approach ($\theta = 0$) as on departure ($\theta = \pi$):

$$\Delta f_{max} = \pm \frac{f_0 V}{c} .$$ (2.152)

This symmetry is a consequence of the fact that the medium is at rest.

Around the CPA, the rate at which the transition from upshifted to downshifted frequencies occurs can be found by replacing $\cos \theta$ by $x / \sqrt{x^2 + z'^2}$ in Eq. (2.150) to obtain

$$f = f_0 \left(1 + \frac{V}{c} \frac{x}{\sqrt{x^2 + z'^2}} \right).$$ (2.153)

This expression is sketched in Fig. 2.11 for several values of the source offset, z', where it can be seen that there is a reduction in the gradient of f as z' increases. Differentiating with respect to x returns

$$\frac{\partial f}{\partial x} = \frac{f_0 V}{c} \frac{z'^2}{(x^2 + z'^2)^{3/2}} ,$$ (2.154a)

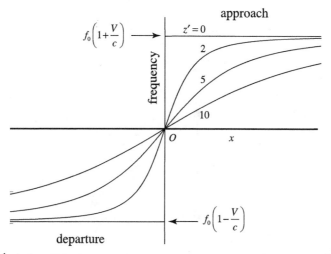

FIGURE 2.11

Moving receiver, stationary source: antisymmetric frequency shifts around the CPA for several values of the source offset z'.

which is zero when $|x| \gg z'$ and takes its maximum value of $f_0 V/cz'$ at $x = 0$. Thus, the gradient of f scales inversely with the offset, z'. In terms of time, t, bearing in mind that $x = -Vt$ (negative t corresponds to positive x and vice versa), the rate at which the transition occurs is

$$\frac{\partial f}{\partial t} = -\frac{f_0 V^2}{c} \frac{z'^2}{(V^2 t^2 + z'^2)^{3/2}} , \qquad (2.154b)$$

where the negative sign is consistent with the reduction in frequency as time progresses.

As the observer passes the origin at $x = 0$, corresponding to the CPA, the value of Δf changes sign. In fact, as can be seen from Eq. (2.153), Δf is an odd function of x, that is to say, the Doppler shifts are antisymmetric about the CPA. In the limit as the source offset z' approaches zero (which actually is not physically possible since the receiver would have to pass through the source), the received frequency can be seen to change instantaneously from $f = f_0\left(1 + \frac{V}{c}\right)$ to $f = f_0\left(1 - \frac{V}{c}\right)$.

2.4.2 DOPPLER FREQUENCY SHIFTS (OBSERVER STATIONARY, SOURCE IN MOTION)

When the source is moving and the observer is stationary, it is convenient to change the frame of reference such that the source is stationary and the receiver is in motion. The problem then becomes similar to that represented in Fig. 2.10, except that now the medium is also in motion, giving rise to an apparent sound speed of $c - V\cos\theta$. When this is used instead of c in Eq. (2.150), the expression for the received frequency with a moving source and a stationary observer is found to be

$$f = \frac{f_0}{\left(1 - \frac{V}{c}\cos\theta\right)} . \qquad (2.155)$$

Similar arguments may now be applied as previously; for instance, the maximum value of $\partial f/\partial x$ is again equal to $f_0 V/cz'$, occurring at the CPA where $x = 0$ and time $t = 0$. The fractional shift in frequency is

$$\frac{\Delta f}{f_0} = \frac{(V/c)\cos\theta}{1 - (V/c)\cos\theta} , \qquad (2.156)$$

but in this case the symmetry between approach and departure has been broken, due to the apparent motion of the medium. On approach ($\theta = 0$), the maximum fractional frequency shift is $V/(c - V)$, whereas on departure ($\theta = \pi$) it takes the smaller

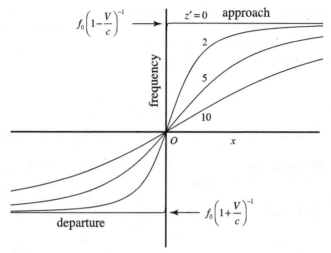

FIGURE 2.12

Moving source, stationary receiver: asymmetrical Doppler shifts around the closest point of approach for several values of the source offset z'.

numerical value $-V/(c + V)$, as illustrated in Fig. 2.12, where Eq. (2.155) is sketched for several values of the source offset, z'.

2.4.3 ACOUSTIC FIELD FROM A MOVING SOURCE

The acoustic field generated by a moving source as it passes a fixed receiver [66,67] may be derived using the same integral transform technique that was introduced earlier in connection with stationary sources and receivers. When the source is moving, its motion must be incorporated into the source term on the right-hand side of the wave equation. As an example of the methodology, consider a line source moving at constant speed V in the x direction past a fixed receiver at $x = 0$, $z = 0$, as illustrated in Fig. 2.13. The source is perpendicular to the plane of the paper (the y direction) and traveling along the track $z = z'$, passing the CPA at time $t = 0$. Thus, negative and positive times, respectively, correspond to the approach and departure phases of the source motion. Although a point source rather than a line source would be more realistic, the addition of the third spatial dimension complicates the mathematical analysis without introducing any fundamentally new physics.

In Cartesian coordinates, the wave equation for the velocity potential, $g = g(x,y,z;t)$, is

$$\frac{\partial^2 g}{\partial x^2} + \frac{\partial^2 g}{\partial z^2} - \frac{1}{c^2}\frac{\partial^2 g}{\partial t^2} = S_2\delta(z - z')\delta(x - Vt)f(t) , \qquad (2.157)$$

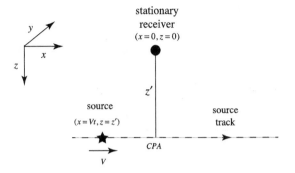

FIGURE 2.13

Coordinate system for a moving line source at a stationary receiver.

where c is the speed of sound in the medium, which is assumed to be homogeneous and isotropic. On the right-hand side, the second delta function accounts for the motion of the source in the x direction, with the function $f(t)$ representing the time dependence of the transmitted waveform. Two Fourier transforms are now applied to both sides of Eq. (2.157), with respect to $x \leftrightarrow s$ and $z \leftrightarrow q$, where the double-headed arrow links each spatial coordinate with its transform variable. These operations convert the two spatial derivatives to algebraic terms:

$$-\left(s^2 + q^2\right)g_{sq} - \frac{1}{c^2}\frac{\partial^2 g_{sq}}{\partial t^2} = S_2 e^{-iqz'}e^{-isVt}f(t) \tag{2.158}$$

where, as before, transform variables used as subscripts identify the corresponding spatial Fourier transforms. A Fourier transform with respect to time returns

$$\left(k^2 - s^2 - q^2\right)G_{sq} = S_2 e^{-iqz'}F(\omega + sV), \tag{2.159}$$

where $F(\omega)$ is the Fourier transform of $f(t)$, ω is angular frequency, and $k = \omega/c$ is the acoustic wave number.

Suppose now that the source is transmitting a harmonic wave of angular frequency ω_0, then

$$f(t) = \omega_0 \cos \omega_0 t , \tag{2.160}$$

where the premultiplier ω_0 gives $f(t)$ the dimensions of $(\text{time})^{-1}$. The Fourier transform of $f(t)$ is

$$F(\omega) = \omega_0 \int_{-\infty}^{\infty} \cos \omega_0 t \, e^{-i\omega t} dt = \pi \omega_0 \{\delta(\omega - \omega_0) + \delta(\omega + \omega_0)\} , \tag{2.161}$$

and hence from Eq. (2.159),

$$G_{sq} = S_2 \pi \omega_0 \frac{e^{-iqz'}}{(k^2 - s^2 - q^2)} \{\delta(\omega - \omega_0 + sV) + \delta(\omega + \omega_0 + sV)\} . \quad (2.162)$$

An inverse Fourier transform with respect to ω returns

$$g_{sq} = \frac{S_2 \omega_0}{2} e^{-iqz'} \left\{ \frac{e^{i(\omega_0 - V)t}}{D_-} + \frac{e^{-i(\omega_0 + V)t}}{D_+} \right\} , \quad (2.163)$$

where

$$D_\pm = \frac{(\omega_0 \pm V)^2}{c^2} - s^2 - q^2 . \quad (2.164)$$

If the imaginary part of the wave number $k_0 = \omega_0/c$, however small, is negative, then the imaginary parts of D_- and D_+, respectively are negative and positive. This distinction is important in the evaluation of the spatial inverse transforms.

To take the inverse transform with respect to $s \leftrightarrow x$, the first step is to complete the square in $D_\pm$, from which it is evident that for each of the integrals there are two simple conjugate poles in the complex s-plane. Integrating around a D-shaped contour in the upper and lower plane, respectively, for negative and positive times, eventually returns the solution

$$g_q = \frac{iS_2 \omega_0}{4\gamma} e^{-iqz'} \left\{ \eta_-^{-1} e^{i\omega_0 t/\gamma^2} e^{-i\eta_- V|t|/\gamma} - \eta_+^{-1} e^{-i\omega_0 t/\gamma^2} e^{i\eta_+ V|t|/\gamma} \right\} \quad (2.165)$$

where x has been set equal to zero, corresponding to the receiver position,

$$\gamma = \sqrt{1 - \frac{V^2}{c^2}} \quad (2.166)$$

and

$$\eta_\pm = \pm\sqrt{q^2 - \frac{\omega_0^2}{c^2 \gamma^2}} . \quad (2.167)$$

The one remaining inverse transform to be taken is with respect to $q \leftrightarrow z$, which, after some algebraic manipulation, can be written as

$$g = \frac{iS_2 \omega_0}{4\pi\gamma} \left\{ e^{i\omega_0 t/\gamma^2} L_- - e^{-i\omega_0 t/\gamma^2} L_+ \right\} , \quad (2.168)$$

where

$$L_\pm = \int_0^\infty \eta_\pm^{-1} \cos qz' \; e^{\pm i\eta_\pm V|t|/\gamma} dq . \quad (2.169)$$

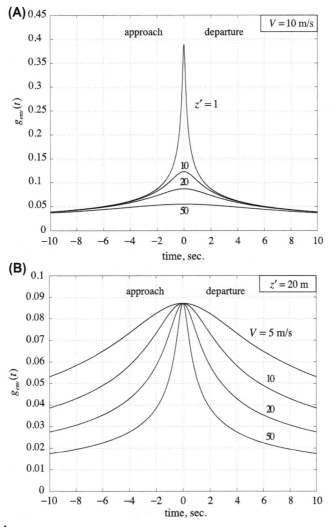

FIGURE 2.14

Envelope of the velocity potential for several values of the source speed and the closest source-to-receiver distance. (A) $V = 10$ m/s, $z' = 1$, 10, 20, 50 m. (B) $z' = 20$ m, $V = 5$, 10, 20, 50 m/s.

and z has been set to zero, corresponding to the receiver position. These integrals are standard forms that may be found in tables, leading to the final expression for the velocity potential at the receiver,

$$g(0,0;t) = i\frac{S_2\omega_0}{8\gamma}\left\{e^{i(\omega_0 t/\gamma^2)}H_0^{(2)}(R) - e^{-i(\omega_0 t/\gamma^2)}H_0^{(1)}(R)\right\}$$

$$= \frac{S_2\omega_0}{4\gamma}\left\{\cos(\omega_0 t/\gamma^2)Y_0(R) - \sin(\omega_0 t/\gamma^2)J_0(R)\right\}$$

(2.170)

where

$$R = \frac{\omega_0}{c\gamma^2}\sqrt{V^2 t^2 + z'^2\gamma^2}\,,$$

(2.171)

$J_0(\cdot)$ and $Y_0(\cdot)$ are, respectively, Bessel functions of the first and second kind of order zero, and

$$H_0^{(1,2)}(\cdot) = J_0(\cdot) \pm iY(\cdot)$$

(2.172)

are the corresponding Hankel functions.

The solution for the velocity potential in Eq. (2.170) is in the form of an amplitude-modulated, rapidly oscillating carrier wave. The relatively slow modulation, or *envelope*, is always positive, as given by the expression

$$g_{env}(t) = \sqrt{J_0^2(R) + Y_0^2(R)}\,,$$

(2.173)

which is plotted in Fig. 2.14 for various values of the source speed, V, and the distance, z', between the source and receiver at the CPA. Since R in Eq. (2.171) is an even function of time, the envelope of the velocity potential is symmetric about $t = 0$, taking the same shape on the approach and departure.

2.5 SOUND REFLECTION AND TRANSMISSION AT A FLUID–FLUID BOUNDARY

The integral-transform technique used above in solving the wave equation for the acoustic field in a homogeneous, unbounded fluid medium has application also when one or more boundaries are present. A case in point is a single fluid–fluid boundary, as illustrated in Fig. 2.15 showing an upper and lower (semi-infinite) fluid with density and sound speed ρ_1, c_1 and ρ_2, c_2, respectively. Stability requires that $\rho_2 > \rho_1$. This simple, two-layer model approximates the seawater–sediment interface in situations where the shear wave in the sediment is weak and the sound speed profile on either side of the boundary is uniform.

To derive the field in both regions from a point source located in the upper medium, a cylindrical coordinate system is chosen, with the z-axis vertical and passing through the source position at $z = z'$, $r = 0$. Setting the interface at $z = 0$, and with z increasing downward, the z coordinate is negative and positive, respectively, in the

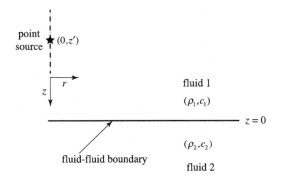

FIGURE 2.15

Fluid–fluid boundary.

upper and lower fluids. Thus, the z-coordinate of the source in the upper fluid, z', is negative. From the symmetry, the field in both domains is axially uniform, that is to say, independent of azimuth. Assuming an impulsive source, the problem now is to solve the wave equations for the Fourier transforms of the velocity potential, $G_1 = G_1(r,z;\omega)$ and $G_2 = G_2(r,z;\omega)$, respectively, in the upper and lower domains subject to the boundary conditions at the fluid–fluid interface.

2.5.1 STRUCTURE OF THE SOLUTION

Since the source is an impulse, a Fourier transform with respect to time, t, converts the wave equations for the acoustic field in each of the two fluids into the following Helmholtz equations:

$$\frac{1}{r}\frac{\partial}{\partial r}\left(\frac{\partial G_1}{\partial r}\right) + \frac{\partial^2 G_1}{\partial z^2} + k_1^2 G_1 = S_3\delta(z - z')\frac{\delta(r)}{\pi r}, \quad \text{for } z \leq 0 , \qquad (2.174a)$$

and

$$\frac{1}{r}\frac{\partial}{\partial r}\left(\frac{\partial G_2}{\partial r}\right) + \frac{\partial^2 G_2}{\partial z^2} + k_2^2 G_2 = 0, \quad \text{for } z > 0 , \qquad (2.174b)$$

where $k_i = \omega/c_i$ is the acoustic wave number in the upper ($i = 1$) and lower ($l = 2$) fluid. In the following analysis, the imaginary part of k_i, however small, is taken to be nonzero.

A Hankel transform of order zero, with properties as defined in Eqs. (2.62) and (2.63), when applied to Eq. (2.174) returns

$$\frac{\partial^2 G_{1q}}{\partial z^2} + \left(k_1^2 - q^2\right)G_{1q} = \frac{S_3}{2\pi}\delta(z - z') \qquad (2.175a)$$

and

$$\frac{\partial^2 G_{2q}}{\partial z^2} + \left(k_2^2 - q^2\right)G_{2q} = 0 \, , \tag{2.175b}$$

where p, the transform variable, is the horizontal wave number. Of the two possible solutions of Eq. (2.175b), the one that decays to zero as z goes to infinity is

$$G_{2q} = G_{2q}(0)e^{-i\eta_2 z} \, , \tag{2.176}$$

where

$$\eta_2 = \sqrt{k_2^2 - p^2}, \quad Im(\eta_2) < 0 \, , \tag{2.177}$$

and $G_{2p}(0)$ is the value of G_{2p} on the boundary, $z = 0$. Taking the derivative of the field with respect to z and setting $z = 0$ returns the condition

$$G'_{2p}(0) = -i\eta_2 G_{2p}(0) \, , \tag{2.178}$$

where the prime denotes $\partial/\partial z$.

The solution of Eq. (2.175a) is found by applying a Fourier transform with respect to z, but over the restricted interval $0 \leq z \leq \infty$ representing the upper fluid. This means that two integrated terms appear in the transformed equation:

$$\left(k_1^2 - p^2 - q^2\right)G_{1pq} + G'_{1p}(0) + iqG_{1p}(0) = \frac{S_3}{2\pi}e^{-iqz'} \, , \tag{2.179}$$

where the transform variable q is the vertical wave number, $G_{1p}(0)$ and $G'_{1p}(0)$ are, respectively, the values of the field and its normal derivative on the boundary, $z = 0$. Eq. (2.179) may be rearranged to take the form

$$G_{1pq} = \frac{(S_3/2\pi)e^{-iqz'} - G'_{1p}(0) - iqG_{1p}(0)}{\left(k_1^2 - p^2 - q^2\right)} \, , \tag{2.180}$$

to which an inverse Fourier transform with respect to $q \leftrightarrow z$ may be applied to obtain

$$G_{1p}(z) = -i\frac{G'_{1p}(0)}{2\eta_1}e^{-i\eta_1|z|} + \frac{G_{1p}(0)}{2}e^{-i\eta_1|z|} + i\frac{S_3}{4\pi\eta_1}e^{-i\eta_1|z-z'|} \, , \tag{2.181}$$

where

$$\eta_1 = \sqrt{k_1^2 - p^2}, \quad Im(\eta_1) < 0. \tag{2.182}$$

The integrals leading to Eq. (2.181) are all standard forms that may be found in tables. When $z = 0$ is substituted into Eq. (2.181), we obtain the following condition on the upper-fluid field at the boundary:

$$G_{1p}(0) = -i\frac{G'_{1p}(0)}{\eta_1} + i\frac{S_3}{2\pi\eta_1}e^{-i\eta_1|z'|} . \tag{2.183}$$

At this point in the discussion, the boundary conditions must be introduced. To prevent fluid particles from accelerating away, there must be continuity of pressure across the boundary, which may be expressed as

$$\rho_1 G_{1p}(0) = \rho_2 G_{2p}(0) \quad ; \tag{2.184}$$

and, to prevent the fluid from ripping apart, the normal component of the particle velocity must also be continuous across the boundary, as expressed by the condition

$$G'_{1p}(0) = G'_{2p}(0) . \tag{2.185}$$

With these two boundary conditions plus the expressions in Eqs. (2.178) and (2.183), the four unknowns, $G_{1p}(0)$, $G_{2p}(0)$, $G'_{1p}(0)$, and $G'_{2p}(0)$, which in effect are constants of integration, may be determined uniquely:

$$G_{1p}(0) = i\frac{S_3}{2\pi}\frac{e^{-i\eta_1|z'|}}{[\eta_1 + b\eta_2]} , \tag{2.186}$$

$$G_{2p}(0) = i\frac{bS_3}{2\pi}\frac{e^{-i\eta_1|z'|}}{[\eta_1 + b\eta_2]} , \tag{2.187}$$

and

$$G'_{1p}(0) = G'_{2p}(0) = \frac{bS_3}{2\pi}\frac{\eta_2 e^{-i\eta_1|z'|}}{[\eta_1 + b\eta_2]} , \tag{2.188}$$

where

$$b = \frac{\rho_1}{\rho_2} \tag{2.189}$$

is the density ratio between the upper and lower fluid.

On substituting for the constants in Eq. (2.181), the expression for the field in the upper fluid may be written as

$$G_{1p}(z) = i\frac{S_3}{4\pi}\left\{\frac{e^{-i\eta_1|z-z'|}}{\eta_1} + V(p)\frac{e^{-i\eta_1(|z|+|z'|)}}{\eta_1}\right\} \tag{2.190}$$

where

$$V(p) = \frac{[\eta_1 - b\eta_2]}{[\eta_1 + b\eta_2]} \tag{2.191}$$

is the reflection coefficient. Similarly, from Eq. (2.176) the field transmitted into the lower fluid is

$$G_{2p}(z) = i\frac{bS_3}{4\pi}T(p)\frac{e^{-i(\eta_1|z'|+\eta_2 z)}}{\eta_1} \,, \tag{2.192}$$

where

$$T(p) = \frac{2\eta_1}{[\eta_1 + b\eta_2]} \tag{2.193}$$

is the transmission coefficient. In Eq. (2.190), the first term in parenthesis represents the incident field and the second term is the field reflected from the boundary. The superposition of these two field components gives rise to an interference structure in the upper fluid, which is the acoustics analogue of a phenomenon in optics known as Lloyd's mirror. From inspection of Eqs. (2.191) and (2.193) it is evident that the reflection and transmission coefficients satisfy the simple relationship

$$1 + V(p) = T(p) \,. \tag{2.194}$$

To recover the field in the frequency domain, an inverse Hankel transform with respect to $p \leftrightarrow r$ must be applied, to obtain from Eq. (2.190),

$$G_1(r, z; \omega) = i\frac{S_3}{4\pi}\{M_{inc} + M_{refl}\} \tag{2.195}$$

and from Eq. (2.192),

$$G_2(r, z; \omega) = i\frac{bS_3}{4\pi}M_{trans} \,. \tag{2.196}$$

In these expressions, the functions M are the inverse Hankel transforms for the incident, reflected and transmitted field. In the case of the incident field, the inversion integral can be evaluated explicitly, taking the form of a spherical wave

$$\begin{aligned} M_{inc} &= \int_0^\infty p\frac{e^{-i\eta_1|z-z'|}}{\eta_1}J_0(pr) \; dp \\ &= i\frac{\exp(-ik_1 R)}{R} \end{aligned} \tag{2.197}$$

where

$$R = \sqrt{r^2 + |z - z'|^2} \,. \tag{2.198}$$

The remaining two inversion integrals are

$$M_{refl} = \int_0^\infty pV(p)\frac{e^{-i\eta_1(|z|+|z'|)}}{\eta_1}J_0(pr)dp \tag{2.199}$$

and

$$M_{trans} = \int_0^\infty pT(p)\frac{e^{-i(\eta_1|z'|+\eta_2 z)}}{\eta_1}J_0(pr)dp \,, \qquad (2.200)$$

neither of which can be expressed explicitly. Of course, the integrals may be evaluated numerically, providing a complete, precise description of the field wherever the source may be positioned, even adjacent to the boundary or in the immediate vicinity of the receiver.

A particular case of interest, however, is with the source far from the boundary, represented by the condition $|k_1 z'| \to \infty$. In this situation, the incident, reflected, and transmitted fields are plane waves and it is possible to develop approximate asymptotic expressions for the integrals M_{refl} and M_{trans}.

2.5.2 THE STATIONARY PHASE APPROXIMATION

Integrals with rapidly oscillating integrands are commonly encountered when integral-transform techniques are applied to wave propagation problems. In 1887, Lord Kelvin (William Thomson), in his theoretical investigation of the angular structure of ship wakes [68], developed an approximation technique for evaluating such integrals that goes by the name of the method of stationary phase. The essential idea behind the technique is that, throughout the range of integration, the contributions to the integral from the oscillations cancel everywhere, except locally around turning points, where the phase is either a maximum or minimum.

The integrals in question can be written in the form

$$I = \int F(\theta)e^{iKs(\theta)}d\theta \qquad (2.201)$$

where $s(\theta)$ is a real phase function and the coefficient K is large compared with unity such that the phase $Ks(\theta)$ varies rapidly, giving rise to an integrand that is highly oscillatory. The relatively slowly varying function $F(\theta)$ may be real or complex. Assuming that at least one turning point exists in the phase function, $s(\theta)$, at say $\theta = \theta_0$, the main contribution to the integral comes from around θ_0.

To derive the stationary phase approximation, the phase function is expanded in a Taylor series about $\theta = \theta_0$:

$$s(\theta) = s(\theta_0) + \frac{(\theta - \theta_0)}{1!}s'(\theta_0) + \frac{(\theta - \theta_0)^2}{2!}s''(\theta_0) + \cdots , \qquad (2.202)$$

where the primes denote differentiation with respect to θ. Since the first derivative is zero at the turning point, the integral in Eq. (2.201) may be expressed as

$$I = \int_{\theta_0-\varepsilon}^{\theta_0+\varepsilon} F(\theta)\exp iK\left[s(\theta_0) + \frac{(\theta - \theta_0)^2}{2!}s''(\theta_0) + \cdots\right]d\theta, \qquad (2.203)$$

where ε is a small quantity whose value depends of the variation of F and s but not on K. The integral in Eq. (2.203) may be simplified by making the change of variable

$$\chi = (\theta - \theta_o)\sqrt{|Ks''(\theta_o)|/2}\,, \tag{2.204}$$

which yields

$$I = \frac{F(\theta_0)\exp[iKs(\theta_0)]}{\sqrt{|Ks''(\theta_0)/2|}} \int_{-\varepsilon\sqrt{|Ks''(\theta_0)/2|}}^{\varepsilon\sqrt{|Ks''(\theta_0)/2|}} \exp \pm i\chi^2 d\chi\,, \tag{2.205}$$

where the sign in the argument of the exponential is chosen to be the same as the sign of the second derivative, $s''(\theta_0)$; and because it is slowly varying, $F(\theta)$ has been set equal to $F(\theta_0)$ and taken outside the integral. In the limit as $K \to \infty$, the integral in Eq. (2.205) becomes a known form,

$$\int_{-\infty}^{\infty} \exp \pm i\chi^2 d\chi = \sqrt{\frac{\pi}{2}}(1 \pm i) = \sqrt{\pi}\exp \pm i\frac{\pi}{4}\,, \tag{2.206}$$

from which it follows that

$$I \approx F(\theta_0)\sqrt{\frac{2\pi}{|Ks''(\theta_0)|}}\exp i\left[Ks(\theta_0) \pm \frac{\pi}{4}\right]. \tag{2.207}$$

This expression is the stationary phase approximation for the integral in Eq. (2.201). If more than one stationary point is present in the phase function, Eq. (2.207) may be used to evaluate the contribution from each and the results simply summed. Should two stationary points coalesce so that the second derivative $s''(\theta_0) = 0$, then a higher-order approximation is required in the Taylor series representation of the phase, leading to an asymptotic solution for the integral involving an Airy function. This type of higher-order approximation has application in connection with caustics occurring at the boundary between an ensonified region and a shadow zone [69].

2.5.3 PLANE-WAVE REFLECTION

The integral for the reflected wave in Eq. (2.199) may be cast into a form suitable for evaluation using the stationary phase approximation by replacing the Bessel function with its asymptotic expansion,

$$J_0(pr) \approx \sqrt{\frac{2}{\pi pr}}\cos\left(pr - \frac{\pi}{4}\right) = \sqrt{\frac{1}{2\pi pr}}\left(e^{-i\pi/4}e^{ipr} + e^{i\pi/4}e^{-ipr}\right). \tag{2.208}$$

It will turn out that, within the limits of integration, a stationary point is present in the phase only with the second of the exponential terms in Eq. (2.208), and therefore Eq. (2.199) may be written in the approximate form

$$M_{refl} \approx \int_0^\infty F_{refl}(p) e^{-iZs(p)} dp ,$$

(2.209)

where

$$F_{refl} = \frac{p}{\eta_1} V(p) \frac{e^{i\pi/4}}{\sqrt{2\pi pr}} ,$$

(2.210)

$$s(p) = \sqrt{k_1^2 - p^2} + p \frac{r}{Z} ,$$

(2.211)

and

$$Z = |z'| + |z| .$$

(2.212)

When the source is far from the boundary, Z is large and the integral in Eq. (2.208) may be evaluated using the stationary phase technique.

The first derivative of the phase function is

$$s'(p) = -\frac{p}{\sqrt{k_1^2 - p^2}} + \frac{r}{Z} ,$$

(2.213)

which, when set equal to zero, has the single positive root

$$p = p_1 = \frac{k_1 r}{\sqrt{r^2 + Z^2}} .$$

(2.214)

At this stationary point, the second derivative of the phase is negative,

$$s''(p_1) = -\frac{(r^2 + Z^2)^{3/2}}{k_1 Z^3} ,$$

(2.215)

the phase itself is

$$s(p_1) = \frac{k_1 \sqrt{r^2 + Z^2}}{Z} ,$$

(2.216)

and the slowly varying function is

$$F_{refl}(p_1) = e^{i\pi/4} \frac{(r^2 + Z^2)^{1/4}}{Z\sqrt{2\pi k_1}} V(p_1) .$$

(2.217)

It follows from the stationary phase expression in Eq. (2.207) that the integral for the reflected wave evaluates to

$$M_{refl} \approx F_{refl}(p_1) \left(\frac{2\pi}{|Zs''(p_1)|} \right)^{1/2} \exp - i \left[k_1 \sqrt{r^2 + Z^2} - \frac{\pi}{4} \right]$$

(2.218)

$$= i \frac{1}{\sqrt{r^2 + Z^2}} V(p_1) \exp - ik_1 \sqrt{r^2 + Z^2} .$$

The function $V(p_1)$ is the Rayleigh *plane-wave reflection coefficient* [18], given by

$$V(p_1) = \frac{\eta_1(p_1) - b\eta_2(p_1)}{\eta_1(p_1) + b\eta_2(p_1)} . \qquad (2.219)$$

If α_{inc} and α_{refl} are the grazing angles of the incident and reflected rays, then

$$\cos \alpha_{inc} = \frac{p_1}{k_1} = \frac{r}{\sqrt{r^2 + Z^2}} = \cos \alpha_{refl} , \qquad (2.220)$$

which states that the angles of incidence and reflection are equal. Setting $\alpha = \alpha_{inc} = \alpha_{refl}$, the reflection coefficient may be expressed as

$$V = \frac{k_1 \sin \alpha - b\sqrt{k_2^2 - k_1^2 \cos^2 \alpha}}{k_1 \sin \alpha + b\sqrt{k_2^2 - k_1^2 \cos^2 \alpha}} . \qquad (2.221)$$

When $c_1 > c_2$, as is commonly the case at the seawater–sediment boundary, then the radical in the expression for the reflection coefficient is zero when

$$k_2 = k_1 \cos \alpha_c , \qquad (2.222)$$

where α_c is the *critical grazing angle* of the fluid–fluid interface. In terms of α_c, the expression in Eq. (2.221) becomes

$$V = \frac{\rho_2 \sin \alpha - \rho_1 \sqrt{\sin^2 \alpha - \sin^2 \alpha_c}}{\rho_2 \sin \alpha + \rho_1 \sqrt{\sin^2 \alpha - \sin^2 \alpha_c}} , \qquad (2.223)$$

where the wave numbers have canceled and the densities have been expressed explicitly. For normal incidence ($\alpha = \pi/2$), the reflection coefficient simplifies to

$$V_{\text{normal}} = \frac{\rho_2 - \rho_1 \cos \alpha_c}{\rho_2 + \rho_1 \cos \alpha_c} . \qquad (2.224)$$

The reflection coefficient may be expressed in terms of its magnitude, $|V|$, and phase, φ:

$$V = |V|e^{i\varphi} . \qquad (2.225)$$

At grazing angles steeper than the critical, the radical in Eq. (2.223) is real and there is no phase change on reflection. For grazing angles less than the critical, the radical is imaginary, and thus the modulus of the reflection coefficient is unity, corresponding to *total reflection* at the boundary. There is, however, a phase change on reflection, given by

$$\tan \varphi = -\frac{2\rho_1\rho_2 \sin \alpha \sqrt{\sin^2 \alpha_c - \sin^2 \alpha}}{\rho_1^2 \sin^2 \alpha_c - (\rho_1^2 + \rho_2^2)\sin^2 \alpha} . \qquad (2.226)$$

To recover φ, the branch of the tangent function must be selected such that when α goes to zero the phase equals $-\pi$, a condition that follows from the expression for the reflection coefficient in Eq. (2.223). A switch to another branch of the arctangent must be made when the denominator in Eq. (2.226) changes sign, which occurs at a grazing angle

$$\alpha = \alpha_0 = \sin^{-1}\left\{\frac{\rho_1 \sin(\alpha_c)}{\sqrt{\rho_1^2 + \rho_2^2}}\right\}, \tag{2.227}$$

at which point the phase change on reflection is $-\pi/2$. Throughout the full range of grazing angles $(0, \pi/2)$, the phase change on reflection

$$\varphi = \begin{cases} -\pi + \tan^{-1}\left[\dfrac{2\rho_1\rho_2 \sin\alpha\sqrt{\sin^2\alpha_c - \sin^2\alpha}}{\rho_1^2 \sin^2\alpha_c - (\rho_1^2 + \rho_2^2)\sin^2\alpha}\right], & 0 \le \alpha < \alpha_0 \\[4mm] -\tan^{-1}\left[\dfrac{2\rho_1\rho_2 \sin\alpha\sqrt{\sin^2\alpha_c - \sin^2\alpha}}{(\rho_1^2 + \rho_2^2)\sin^2\alpha - \rho_1^2 \sin^2\alpha_c}\right], & \alpha_0 \le \alpha < \alpha_c \\[4mm] 0, & \alpha_c < \alpha \le \pi/2 \end{cases} \tag{2.228}$$

where the arctangent functions in these expressions return values in the interval $(0, \pi/2)$.

The magnitude and phase of the reflection coefficient are sketched in Fig. 2.16, where the solid lines represent the lossless case and the dashed line illustrates the

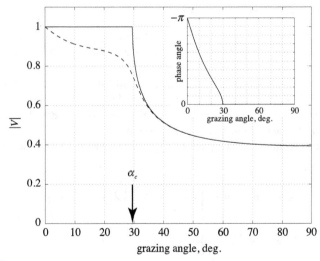

FIGURE 2.16

Magnitude and the phase (inset) of the plane-wave reflection coefficient. The *solid lines* represent lossless fluids and the *dashed line* illustrates the effect of attenuation in the lower fluid.

effect of attenuation in the lower fluid. If absorption is present in the lower fluid, the reflection at grazing angles shallower than the critical cannot be total. Attenuation may be included in Eq. (2.215) by making the imaginary part of the wave number k_2 less than zero. If the lower fluid were representative of a marine sediment, a guide to the level of the attenuation is provided by Hamilton's linear-frequency scaling law in Eq. (2.137). Relatively speaking, the attenuation in the seawater above the sediment usually has a negligible effect on the bottom reflection coefficient.

From examination of Eq. (2.221), it is evidently possible for the reflection coefficient to vanish. This occurs when the numerator in Eq. (2.221) is zero, which occurs at the grazing *angle of intromission*, α_{int}, given by the expression

$$\sin \alpha_{int} = b\sqrt{\frac{n^2 - 1}{1 - b^2}}, \tag{2.229}$$

where $n = c_1/c_2$ is the *index of refraction*. For the case where the density ratio is $b = \rho_1/\rho_2 < 1$, representative of an ocean bottom, the sine function in Eq. (2.229) is real and less than unity only when $1 < n < 1/b$. The only marine sediments to satisfy this condition are the very fine-grained materials, the clays, and silts, which may exhibit a sound speed less than that in the water column, so it is possible that such materials could exhibit an angle of intromission. Otherwise, in the sands and other coarser sedimentary materials, the speed of sound is greater than that in the water column and no angle of intromission can exist.

2.5.4 WESTON'S EFFECTIVE DEPTH

According to the Rayleigh reflection coefficient in Eq. (2.223), for grazing angles less than the critical, total internal reflection is accompanied by a phase change, as given by the expressions in Eq. (2.227). Weston [70] developed a useful, approximate representation of the reflection process, valid for small grazing angles satisfying the condition $\alpha \ll \alpha_c$, in which the phase change on reflection is accommodated geometrically by placing a fictitious pressure-release boundary a distance d beneath the actual interface, as illustrated in Fig. 2.17.

From the first of the expressions in Eq. (2.227), the phase change on reflection is

$$\varphi = -\pi + \tan^{-1}\left\{\frac{2b \sin \alpha \sqrt{\sin^2\alpha_c - \sin^2 \alpha}}{b^2 \sin^2\alpha_c - (1 + b^2)\sin^2 \alpha}\right\} \tag{2.230}$$

$$\approx -\pi + \frac{2 \sin \alpha}{b \sin \alpha_c},$$

where the approximation holds under the condition $\alpha \ll \alpha_c$. Now consider the geometry of the ray reflections shown in Fig. 2.17. The difference in the path length

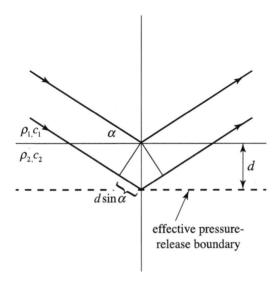

FIGURE 2.17

Ray reflection from a fictitious pressure-release boundary at a distance d beneath the actual interface.

of the two rays is $2d \sin \alpha$, and the phase change experienced by the ray reflected from the displaced pressure release boundary is

$$\varphi_d = -\pi + 2k_1 d \sin \alpha . \tag{2.231}$$

By matching the expressions for φ and φ_d, the reflection from the fictitious interface will have the same phase as the actual reflected ray. This condition is satisfied when the displacement of the effective pressure-release boundary is

$$d = \frac{1}{k_1 b \sin \alpha_c} = \frac{c_1}{\omega b \sin \alpha_c} , \tag{2.232}$$

which is Weston's "effective depth" approximation.

The displacement d is inversely proportional to the frequency but, importantly, is independent of the grazing angle, α. Thus, the same effective depth can be used to represent the reflection of rays incident on the boundary over a range of grazing angles. This particular feature makes the effective depth a useful tool in the development of solutions for the acoustic field in depth-varying shallow-water channels [71], where the grazing angles of the eigen rays associated with the normal modes increase with increasing mode number.

2.5.5 PLANE-WAVE REFRACTION

As with the reflected wave, the integral for the transmitted wave in Eq. (2.200) may also be evaluated using the stationary phase approximation and, as before, only the second of the exponential terms in the expression for the Bessel function in Eq. (2.208) returns a stationary point within the limits of integration. Thus, the integral for the transmitted wave may be written as

$$M_{trans} = \int_0^\infty F_{trans} e^{-iu(p)} \, dp \tag{2.233}$$

where

$$F_{trans} = \frac{p}{\eta_1} T(p) \frac{e^{i\pi/4}}{\sqrt{2\pi r}} \tag{2.234}$$

and the phase function is

$$
\begin{aligned}
u(p) &= \eta_1 |z'| + \eta_2 z + pr \\
&= \sqrt{k_1^2 - p^2} |z'| + \sqrt{k_2^2 - p^2} z + pr \ .
\end{aligned}
\tag{2.235}
$$

This expression exhibits a single turning point within the range of integration, obtained by setting the derivative of the phase function to zero:

$$u'(p) = -\frac{p|z'|}{\sqrt{k_1^2 - p^2}} - \frac{pz}{\sqrt{k_2^2 - p^2}} + r = 0 \ . \tag{2.236}$$

If β is the grazing angle of the refracted ray and α is still the grazing angle of the incident ray, Eq. (2.236) for the turning point is satisfied when

$$p = p_2 = k_1 \cos \alpha = k_2 \cos \beta \ , \tag{2.237}$$

a condition which is familiar as Snell's law. For the case of a fast lower medium, $c_2 > c_1$, a critical grazing angle exists, as defined in Eq. (2.222), and the corresponding grazing angle of refraction, from Eq. (2.237), is equal to zero.

The nature of the transmitted wave differs according to whether the grazing angle of incidence is greater than or less than the critical. Consider first the transmitted wave under the condition of total reflection, which occurs when $\alpha < \alpha_c$. Since the radical in the second term on the right of Eq. (2.236) is then imaginary, the contribution to M_{trans} from the z component in the phase function must be moved to the slowly varying function F_{trans}, which becomes

$$F_{trans} = \frac{p}{\eta_1} T(p) \frac{e^{i\pi/4}}{\sqrt{2\pi r}} e^{-\eta_2 z} \ . \tag{2.238}$$

Correspondingly, the appropriate form of the phase function is

$$u(p) = \eta_1 |z'| + pr \ , \tag{2.239}$$

whose derivative is

$$u'(p) = -\frac{p|z'|}{\sqrt{k_1^2 - p^2}} + r = 0 \ , \tag{2.240}$$

which has the solution for the turning point

$$p = p_2 = \frac{k_1 r}{\sqrt{r^2 + |z'|^2}} .$$

(2.241)

The second derivative is

$$u''(p_2) = -\frac{|z'|}{k_1 \sin^3 \alpha} ,$$

(2.242)

and the phase function itself at the turning point is

$$u(p_2) = k_1 \sqrt{r^2 + |z'|^2} .$$

(2.243)

With these expressions inserted into the stationary phase approximation in Eq. (2.207), the expression for the integral M_{trans} evaluates to

$$M_{trans} = T(p_2) e^{i\pi/2} \sqrt{\frac{\sin \alpha \cos \alpha}{|z'| r}} e^{-k_1 z \sqrt{\cos^2 \alpha - \cos^2 \alpha_c}} e^{-ik_1 \sqrt{r^2 + |z'|^2}}, \quad 0 \leq \alpha < \alpha_c ,$$

(2.244)

in which the transmission coefficient, $T(p_2)$, is complex:

$$T(p_2) = \frac{2 \sin \alpha}{\sin \alpha - ib\sqrt{\cos^2 \alpha - \cos^2 \alpha_c}} .$$

(2.245)

The expression for M_{trans} in Eq. (2.244) represents, not a propagating wave, but rather an evanescent wave that decays exponentially with distance below the interface. The e-folding depth depends on the grazing angle of incidence, taking the form

$$L(\alpha) = \left[k_1 \sqrt{\cos^2 \alpha - \cos^2 \alpha_c} \right]^{-1} .$$

(2.246)

As α approaches the critical grazing angle, $L(\alpha)$ tends to zero; and at grazing incidence,

$$L(0) = \frac{1}{k_1 \sin \alpha_c} .$$

(2.247)

With $c_1 = 1500$ m/s and $\alpha_c = 30$ degrees, at a frequency of 1 kHz, $L(0) \approx 0.47$ m.

When the grazing angle of incidence, α, is greater than the critical, the term in z must be retained in the phase function in Eq. (2.235), since its coefficient, η_2, is real in this case. The stationary phase analysis then proceeds as in the foregoing, leading to the rather cumbersome expression for the transmitted wave,

$$
M_{trans} = T(p_2)e^{i\pi/2} \left\{ \frac{\cos\alpha \sin\alpha \left[\cos^2\alpha_c - \cos^2\alpha\right]^{3/2}}{r(|z'|[\cos^2\alpha_c - \cos^2\alpha] + z\cos^2\alpha_c \sin^3\alpha)} \right\}^{1/2}
$$

$$
\times \exp - ik_1 \left\{ |z'|\sin\alpha + z\sqrt{\cos^2\alpha_c - \cos^2\alpha} + r\sin\alpha \right\}, \alpha_c < \alpha \le \pi/2 ,
$$

(2.248)

where the transmission coefficient is real:

$$
T(p_2) = \frac{2\sin\alpha}{\sin\alpha + b\sqrt{\cos^2\alpha_c - \cos^2\alpha}} .
$$

(2.249)

The expression for the refracted wave in Eq. (2.248) represents a propagating wave that radiates away from the interface into the lower medium.

2.5.6 THE LATERAL WAVE

In deriving the expression for the reflected wave in Eq. (2.218), the fact that the integrand in Eq. (2.199) contains terms that are double valued was neglected. The presence of the radical η_2 in the reflection coefficient, V, gives rise to a second type of wave that must be added to the reflected wave to obtain a complete description of the wave field in the upper fluid. This additional field component is the *lateral wave*, also known as the *head wave*, and is represented here as M_{lat}. Thus, the integral in Eq. (2.218) actually represents, not just the reflected wave, but also the lateral wave, so instead of Eq. (2.218) we have

$$
M_{refl} + M_{lat} = \int_0^\infty pV(p)\frac{e^{-i\eta_1 Z}}{\eta_1} J_0(pr)\, dp ,
$$

(2.250)

where Z is as defined in Eq. (2.212).

Physically, in terms of ray paths, the origin of the lateral wave is as illustrated in Fig. 2.18. A ray that is incident at the interface between the two fluids at the critical grazing angle is refracted, to propagate horizontally at the speed of sound in the lower fluid, where, as it progresses, it continually radiates sound at the critical grazing angle back into the upper fluid. This upward-traveling acoustic field is the lateral wave. In Fig. 2.18, the track of the ray path from source to receiver consists of the three segments, $L_1 = |z'|/\sin\alpha_c$, L_2 and $L_3 = |z|/sin\alpha_c$; and the horizontal range between the source and receiver is

$$
r = L_2 + Z \cot\alpha_c .
$$

(2.251)

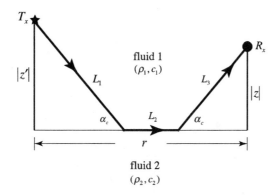

FIGURE 2.18

Ray path of a lateral wave propagating from source, T_x, to receiver, R_x, along segments L_1, L_2, and L_3.

To derive an expression for the lateral wave field, the double-valued nature of the radical η_2 must be taken into account by rearranging the integral in Eq. (2.250) in the complex p-plane. The first step is to express the Bessel function in the integrand as the sum of Hankel functions of the first and second kind:

$$J_0(pr) = \left[H_0^{(1)}(pr) + H_0^{(2)}(pr) \right] \Big/ 2 . \tag{2.252}$$

The integral in Eq. (2.250) can then be expressed as the sum of the integrals

$$I_1 = \int_0^\infty pF(\eta_1, \eta_2)H_0^{(1)}(pr)e^{-i\eta_1 Z}dp \tag{2.253}$$

and

$$I_2 = \int_0^\infty pF(\eta_1, \eta_2)H_0^{(2)}(pr)e^{-i\eta_1 Z}dp , \tag{2.254}$$

where

$$F(\eta_1, \eta_2) = \eta_1^{-1}V(p) = \eta_1^{-1}\frac{\eta_1 - b\eta_2}{\eta_1 + b\eta_2} . \tag{2.255}$$

The radicals η_1 and η_2 in this expression have branch points at $\pm k_1$ and $\pm k_2$, respectively, appearing in the second and fourth quadrants of the p-plane, as shown in Fig. 2.19. Since the integrals in Eqs. (2.253) and (2.254) are taken over the positive range of the horizontal wave number, p, only the branch points in the fourth quadrant are relevant to the derivation of the lateral wave. The two branch cuts sketched in Fig. 2.19 are such that η_1 and η_2 each has an imaginary part that is less than zero everywhere on the top Riemann sheet. On the respective cuts themselves, η_1 and η_2 are real, each taking opposite signs on either side of its cut. This type of branch

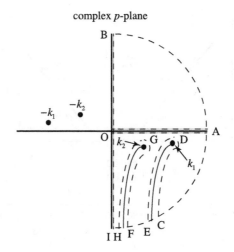

complex p-plane

FIGURE 2.19

Complex p-plane showing branch points, the associated EJP cut lines, and integration contours (dashed lines).

cut is often referred to as an EJP cut, so named after Ewing, Jardetzky, and Press [72], who used it in connection with elastic wave propagation in layered media.

For the integral I_1, a closed contour around the first quadrant is constructed consisting of the positive real axis, OA, an arc of infinite radius, AB, and the negative imaginary axis, BO. Since there are no singularities in the first quadrant and the integral around the arc AB is zero, the integral along the real axis converts to an integral along the imaginary axis:

$$I_1 = \int_0^{i\infty} pF(\eta_1, \eta_2) H_0^{(1)}(pr) e^{-i\eta_1 Z} dp , \qquad (2.256)$$

which, on making the substitution $p = iu$, becomes

$$I_1 = -\int_0^{\infty} uF(\eta_1, \eta_2) H_0^{(1)}(iur) e^{-i\eta_1 Z} du . \qquad (2.257)$$

Now, the Hankel function of imaginary argument can be expressed in terms of the modified Bessel function or real argument, $K_0(ur)$:

$$H_0^{(1)}(iur) = \frac{2}{i\pi} K_0(ur) , \qquad (2.258)$$

and hence

$$I_1 = -\frac{2}{i\pi} \int_0^{\infty} uF(\eta_1, \eta_2) K_0(ur) e^{-i\eta_1 Z} du . \qquad (2.259)$$

Turning now to the integral I_2, a closed contour is constructed around the fourth quadrant consisting of the real axis, OA, the infinite-radius arc, AC, the contour CDE around the k_1 branch cut, the arc EF, the contour FGH around the k_2 branch cut, the arc HI, and the negative imaginary axis, IO. Since the contributions from the infinite-arc segments are zero, the integral along the positive real axis can be written as the sum of the integral along the negative imaginary axis plus the integrals around the two branch cuts:

$$I_2 = -\int_{-i\infty}^{0} pF(\eta_1, \eta_2)H_0^{(2)}(pr)e^{-i\eta_1 Z}dp - I_{CDE} - I_{FGH} . \tag{2.260}$$

From an argument analogous to that for I_1, the integral over the imaginary axis can be converted to a form such that

$$I_2 = \frac{2}{i\pi}\int_0^{\infty} uF(\eta_1, \eta_2)K_0(ur)e^{-i\eta_1 Z}du - I_{CDE} - I_{FGH} \tag{2.261}$$

where the Bessel function identity

$$H_0^{(2)}(-iur) = -\frac{2}{i\pi}K_0(ur) \tag{2.262}$$

has been used.

When I_1 and I_2 are added, the first term on the right of Eq. (2.261) cancels with I_1, leaving

$$I_1 + I_2 = -I_{CDE} - I_{FGH} , \tag{2.263}$$

From a standard stationary phase analysis, the integral I_{CDE} around the k_1 branch cut can be shown to return the expression for the reflected wave in Eq. (2.218). The remaining integral, I_{CDE}, around the k_2 branch cut represents the lateral wave. To derive the expression for the latter, a modified stationary phase argument is used.

To begin, a change of variable, from p to η_2, is made in the integral I_{CDE}. Around the k_2 branch cut, η_2 is real, running from $-\infty$ to $+\infty$, and so, since

$$\eta_2 = \sqrt{k_2^2 - p^2}, \quad -pdp = \eta_2 d\eta_2 , \tag{2.264}$$

$$M_{lat} = -\frac{1}{2}I_{CDE} = -\frac{1}{2}\int_{-\infty}^{\infty} \eta_2 F(\eta_1, \eta_2)e^{-i\eta_1 Z}H_0^{(2)}\left(\sqrt{k_2^2 - \eta_2^2}\, r\right)d\eta_2 . \tag{2.265}$$

The integral here may be expressed over just the positive range of η_2, in which case the integrand contains the difference function

$$F(\eta_1, \eta_2) - F(\eta_1, -\eta_2) = -\frac{4b\eta_2}{\eta_1^2 - b^2\eta_2^2} , \tag{2.266}$$

which follows from Eq. (2.255). For large argument, the Hankel function may be replaced by its asymptotic expansion,

$$H_0^{(2)}\left(\sqrt{k_2^2 - \eta_2^2}\, r\right) \approx \sqrt{\frac{2}{\pi r}}\left(k_2^2 - \eta_2^2\right)^{-1/4} e^{i\pi/4} e^{-i\sqrt{k_2^2 - \eta_2^2}\, r}, \qquad (2.267)$$

and thus the expression for the lateral wave may be approximated as

$$M_{lat} \approx 2b\sqrt{\frac{2}{\pi r}} e^{i\pi/4} \int_0^\infty \left(k_2^2 - \eta_2^2\right)^{-1/4} \frac{\eta_2^2}{\eta_1^2 - b^2\eta_2^2} e^{-i\left(\eta_1 Z + \sqrt{k_2^2 - \eta_2^2}\, r\right)} d\eta_2. \qquad (2.268)$$

The integral in Eq. (2.268) is of a form that is suitable for evaluation using the method of stationary phase, since the integrand consists of a very rapidly oscillating function multiplied by a slowly varying function. A modification to the standard stationary phase technique is required, however, because, at the stationary point, the slowly varying function is zero.

The phase of the exponential function is

$$w(\eta_2) = \eta_1 Z + \sqrt{k_2^2 - \eta_2^2}\, r = \sqrt{\eta_2^2 + k_1^2 \sin^2\alpha_c}\, Z + \sqrt{k_2^2 - \eta_2^2}\, r, \qquad (2.269)$$

and its derivative with respect to η_2 is

$$w'(\eta_2) = \frac{\eta_2 Z}{\sqrt{\eta_2^2 + k_1^2 \sin^2\alpha_c}} - \frac{\eta_2 r}{\sqrt{k_2^2 - \eta_2^2}}, \qquad (2.270)$$

which, when set equal to zero, returns a stationary point at

$$\eta_2 = \eta_{2s} = 0. \qquad (2.271)$$

The second derivative is

$$w''(\eta_2) = \frac{k_1^2 Z \sin^2\alpha_c}{\left(\eta_2^2 + k_1^2 \sin^2\alpha_c\right)^{3/2}} - \frac{k_2^2 r}{\left(k_2^2 - \eta_2^2\right)^{3/2}}, \qquad (2.272)$$

which at the stationary point reduces to

$$w''(0) = \frac{Z}{k_1 \sin\alpha_c} - \frac{r}{k_1 \cos\alpha_c} = -\frac{L_2}{k_1 \cos\alpha_c}; \qquad (2.273)$$

and the phase function itself at the stationary point is

$$w(0) = k_1 Z \sin\alpha_c + k_2 r = k_1(L_1 + L_3) + k_2 L_2, \qquad (2.274)$$

where the last expression follows from Eq. (2.251). On expanding $w(\eta_2)$ in a Taylor series to second order in η_2, and taking all the slowly varying terms, evaluated at the stationary point, outside the integral except the factor η_2^2, which would otherwise return a result of zero, the expression for the lateral wave in Eq. (2.268) becomes

$$M_{lat} \approx \sqrt{\frac{2}{\pi k_1 r \cos \alpha_c}} e^{i\pi/4} \frac{b}{k_1^2 \sin^2 \alpha_c} e^{-i(k_1 Z \sin \alpha_c + k_2 r)} \int_{-\infty}^{\infty} \eta_2^2 e^{-i\eta_2^2 w''(0)/2} d\eta_2 \,,$$

(2.275)

where a factor of 1/2 has been included to account for the fact that the turning point coincides with the lower end point of the integral.

The integral in Eq. (2.275) differs from that in the standard stationary phase approximation through the presence of the factor η_2^2 in the integrand. As with the standard stationary phase technique, the integral can be re-arranged by making the substitution

$$y = w''(0)\eta_2^2/2, \quad dy = w''(0)\eta_2 d\eta_2,$$

(2.276)

which leads to the expression

$$I = \int_{-\infty}^{\infty} \eta_2^2 e^{-i\eta_2^2 w''(0)/2} d\eta_2 = \frac{2^{3/2}}{|w''(0)|^{3/2}} \int_0^{\infty} y^{1/2} e^{\pm iy} dy \,.$$

(2.277)

The second integral here, which may be evaluated using the theory of generalized functions, can be found in Table 1 of Lighthill [4]:

$$I = \frac{\sqrt{2\pi}}{|w''(0)|^{3/2}} e^{\pm i 3\pi/4} \,,$$

(2.278)

where the plus (minus) sign is chosen when $w''(0)$ is negative (positive).

Bearing in mind the result in Eq. (2.273) for the second derivative of the phase function at the stationary point, the final expression for the lateral wave, from Eq. (2.275), can now be written as

$$M_{lat} = -\frac{2b}{\sqrt{r}} \frac{\cos \alpha_c}{k_1 L_2^{3/2} \sin^2 \alpha_c} e^{-i[k_1(L_1+L_3)+k_2 L_2]} \,,$$

(2.279)

where the argument of the exponential function follows from the geometry in Fig. 2.18. Note that at long ranges, such that $r \gg Z$, the horizontal segment $L_2 \approx r$ and thus the amplitude of the lateral wave decays approximately as r^{-2}. This inverse-square law for the amplitude means that the decay of the lateral wave from geometrical spreading is much faster than that of a spherical or cylindrical wave. Nevertheless, lateral waves are widely used for surveying the seabed, since over most of their propagation path they travel at the speed of sound in the lower medium.

2.6 THE "IDEAL" WAVEGUIDE

In shallow water, the acoustic field is trapped between the sea surface and the sea floor, although partial penetration into the seabed may occur, depending on the nature of the bottom boundary conditions. As a first approximation, the two bounding surfaces may be treated as plane, horizontal boundaries, with the sea surface acting as a pressure-release boundary for sound incident from below. This pressure-release condition is a consequence of the 1000:1 density ratio between seawater and air; and the planar representation of the sea surface is reasonable, at least for frequencies up to several hundred hertz, where acoustic scattering from surface roughness is negligible. The bottom boundary is generally more complicated than the sea surface, since the seabed typically supports compressional and shear waves, but in the simplest approximation, it too may be represented as a pressure-release boundary. This idealized waveguide, with two planar, pressure-release boundaries, serves to illustrate some of the properties of acoustic propagation in ocean waveguides.

2.6.1 PLANE WAVES AND NORMAL MODES

The sound field in the waveguide must satisfy the boundary conditions, a constraint that gives rise to propagation in the form of discrete *normal modes*. A normal mode is a pressure distribution in depth, which, in the case of the ideal pressure-release waveguide, takes a value of zero on both boundaries. To understand the origin of the normal modes, consider the channel of depth h shown in Fig. 2.20, where a set of upward-traveling, plane wave fronts is shown propagating along the waveguide from left to right. On encountering the upper boundary, these upward-traveling wave fronts undergo specular reflection, which converts them into downward traveling wave fronts. This process is repeated multiple times, with the total acoustic field being the superposition of the two sets of wave fronts as they propagate along the channel.

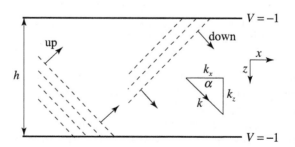

FIGURE 2.20

Ideal waveguide with upward- and downward-traveling, plane wave fronts, denoted by the dashed lines.

The harmonic pressure field due to the upward- and downward-traveling wave trains may be written as

$$p(x,z) = p_0[\sin(k_x x + k_z z + \phi) - \sin(k_x x - k_z z + \phi)], \qquad (2.280)$$

where k_x and k_z are, respectively, the horizontal and vertical wave numbers, p_0 is the amplitude of each wave train, ϕ is an arbitrary phase, and the change in sign of the z-component reflects the fact that the two sets of wave fronts are propagating in opposite vertical directions. A straightforward rearrangement of Eq. (2.280) produces

$$p(x,z) = 2p_0 \cos(k_x x + \phi)\sin k_z z, \qquad (2.281)$$

which clearly satisfies the pressure-release condition on the upper boundary ($z = 0$). On the lower boundary ($z = h$), where the pressure is also zero, we must have

$$\sin k_z h = 0, \qquad (2.282)$$

which is satisfied when

$$k_z = \frac{m\pi}{h}, \quad m = 1, 2, 3\cdots. \qquad (2.283)$$

Thus, the vertical wave number is not continuous but is constrained by the presence of the two boundaries to take discrete values, each of which is identified with a normal mode, with m being the mode number. It follows that the mode shapes are sinusoids in depth, taking the form

$$p(x,z) \propto \sin\left(\frac{m\pi z}{h}\right), \qquad (2.284)$$

which is illustrated in Fig. 2.21 for several of the lowest-order modes. The first mode has one turning point, the second has two, and so on. Note that the mode of order

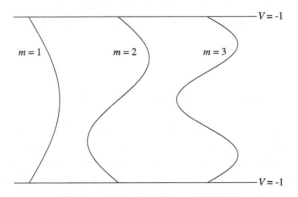

FIGURE 2.21

Mode shapes in a waveguide with pressure-release boundaries.

$m = 0$ does not exist in a channel with two pressure-release boundaries, which is equivalent to saying that its amplitude is zero.

If the acoustic rays, that is, the normals to the wave fronts, propagate at a grazing angle α, then the vertical wave number, k_z, may be expressed in terms of the acoustic wave number, k, as

$$k_z = k \sin \alpha = \frac{2\pi}{\lambda} \sin \alpha \,, \tag{2.285}$$

where λ is the acoustic wavelength. Thus, like k_z, the grazing angle is also constrained to take discrete values:

$$\alpha_m = \sin^{-1}\left(\frac{m\pi}{kh}\right) . \tag{2.286}$$

Since the grazing angle cannot be greater than $\pi/2$, the maximum number of propagating modes is

$$M = \frac{kh}{\pi} = \frac{2h}{\lambda} \,, \tag{2.287}$$

where it is understood that M is to be rounded down to the nearest integer. Evidently, when the channel depth is less than one-half wavelength, there are no propagating modes. Physically, the absence of propagating modes is consistent with the fact that, when $h < \lambda/2$, the pressure-release condition on both boundaries cannot be satisfied.

In terms of frequency f, it is evident from Eq. (2.287) that the mth mode will propagate along the channel provided that

$$f \geq \frac{mc}{2h} \,, \tag{2.288}$$

where c is the speed of sound in the medium. Otherwise, at lower frequencies, the mode does not propagate and is said to be cut off. All the modes are cut off when

$$f < \frac{c}{2h} \,, \tag{2.289}$$

which clearly is just an alternative statement of the half-wavelength cutoff condition in Eq. (2.287).

Although the above discussion relates to a channel with two pressure release boundaries, it is an elementary exercise to extend the arguments to a waveguide with either two rigid boundaries or one pressure-release and one rigid boundary. At a rigid boundary, the normal derivative of the field is zero. Thus, when both boundaries are rigid, the mode shape functions take the form $\cos(m\pi z/h)$, indicating that a mode of order zero can propagate at all frequencies, while the remaining modal cut-off frequencies are the same as with two pressure-release boundaries.

In the case of mixed boundaries, the mode shape functions are $\sin[(m+0.5)\pi z/h]$ and the mth mode will propagate along the channel provided that

$$f \geq (m+0.5)\frac{c}{2h}, \quad m = 0, 1, 2, \cdots . \tag{2.290}$$

2.6.2 THE ACOUSTIC FIELD IN THE IDEAL WAVEGUIDE

The plane-wave analysis of the previous section offers an intuitive description of normal-mode propagation in a channel with pressure-release boundaries, but it does not provide an expression for the full range, depth and frequency dependence of the acoustic field in the channel. For this, it is necessary to solve the Helmholtz equation, which, in cylindrical coordinates, with the z-axis vertical, is

$$\frac{1}{r}\frac{\partial}{\partial r}\left(r\frac{\partial G}{\partial r}\right) + \frac{\partial^2 G}{\partial z^2} + k^2 G = \frac{S_3}{\pi r}\delta(r)\delta(z-z') , \tag{2.291}$$

where $G = G(r,z;\omega)$ is the Fourier transform with respect to time of the velocity potential, k is the acoustic wave number, which is assumed to be uniform throughout the channel, and the term on the right represents a simple point source, with strength S_3, located at range $r = 0$ and depth $z = z'$, as shown in Fig. 2.22. Symmetry dictates that the field be independent of azimuth.

To solve for G, two integral transforms are applied to Eq. (2.291), the first being a Hankel transform, order zero, taken over range, r, as defined in Eq. (2.62), which returns

$$\frac{\partial^2 G_q}{\partial z^2} + \left(k^2 - q^2\right)G_q = \frac{S_3}{2\pi}\delta(z-z') , \tag{2.292}$$

where the horizontal wave number q is the transform variable and we have used the fact that $J_0(0) = 1$. Secondly, because the kernel can be chosen to satisfy the

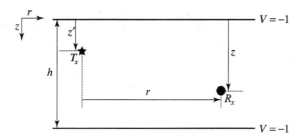

FIGURE 2.22

Source, T_x, and receiver, R_x, in a waveguide with pressure-release boundaries.

pressure-release condition on both boundaries, a Fourier finite sine transform is taken over depth z. The finite sine transform is defined as

$$G_{qv} = \int_0^h G_q \sin(vz)dz , \qquad (2.293)$$

the inverse of which is not an integral but an infinite summation over what will turn out to be the mode number, m:

$$G_q = \frac{2}{h} \sum_{m=1}^{\infty} G_{qv} \sin(vz) , \qquad (2.294)$$

where h is the depth of the channel and the transform variable, v, is

$$v = \frac{m\pi}{h} . \qquad (2.295)$$

From a straightforward double integration by parts, the finite sine transform of the second derivative with respect to z is found to be

$$\int_0^h \frac{\partial^2 G_q}{\partial z^2} \sin(vz)dz = -v^2 G_{qv} . \qquad (2.296)$$

On applying the finite sine transform to Eq. (2.292), an algebraic equation is obtained for the doubly transformed field, the solution of which is

$$G_{qv} = \frac{S_3}{2\pi} \frac{\sin(vz')}{(k^2 - v^2 - q^2)} . \qquad (2.297)$$

The solution for the field itself is recovered by applying the finite-sine and Hankel inverse transforms to Eq. (2.297), which return

$$G = \frac{S_3}{\pi h} \sum_{m=1}^{\infty} \sin(vz)\sin(vz') \int_0^{\infty} \frac{q}{k^2 - v^2 - q^2} J_0(qr)dq$$

$$= i\frac{S_3}{2h} \sum_{m=1}^{\infty} \sin(vz)\sin(vz')H_0^{(2)}\left(\sqrt{k^2 - v^2}\, r\right) , \qquad (2.298)$$

where the integral is available from Tables, and $H_0^{(2)}(\cdot)$ is the Hankel function of the second kind of order zero. The trigonometric terms under the summation can be identified as the mode functions, as sketched in Fig. 2.21, m is the mode number, and the *eigenvalues* are

$$k_m = \sqrt{k^2 - \frac{m^2\pi^2}{h^2}} , \qquad (2.299)$$

which are imaginary for mode numbers such that

$$m > M = \frac{kh}{\pi} = \frac{\omega h}{\pi c} . \qquad (2.300)$$

This condition is identical to that in Eq. (2.287), as derived from the simple plane-wave argument.

For large ranges, where $|k_m r| \gg 1$, the Hankel function is well represented by its asymptotic form,

$$H_0^{(2)}(k_m r) \approx \sqrt{\frac{2}{\pi k_m r}} e^{-i(k_m r - \pi/4)} , \tag{2.301}$$

in which case

$$G \approx i \frac{S_3}{\sqrt{2\pi r h}} e^{i\pi/4} \sum_{m=1}^{\infty} k_m^{-1/2} \sin\left(\frac{m\pi z}{h}\right) \sin\left(\frac{m\pi z'}{h}\right) e^{-i\sqrt{k^2 - \frac{m^2 \pi^2}{h^2}} r} . \tag{2.302}$$

From this expression, it is evident that the propagating modes are those with real eigenvalues, corresponding to mode numbers $m \le M$. Otherwise, with $m > M$, the eigenvalues are (negative) imaginary and the modes are evanescent, decaying exponentially with increasing horizontal range. Moreover, the term preceding the summation indicates that the long-range field exhibits cylindrical spreading, with the amplitudes of the propagating modes decaying as $r^{-1/2}$.

2.6.3 INTERMODAL INTERFERENCE

When two or more modes propagate in the waveguide, intermodal interference occurs, giving rise to a pattern of high and low intensities which has been likened to the Moiré fringes [70] that are observed when two transparent grids of uniformly spaced parallel lines are held together at a slight angle to one another. If many modes are present in the channel, the associated interference pattern may be extremely complicated. The interference phenomenon can be understood, however, by considering the situation where just two propagating modes, m and n, have been excited.

From Eq. (2.302), the intensity of the two-mode acoustic field can be written as the square of the modulus of the pressure,

$$\omega^2 \rho^2 |G|^2 = \frac{1}{r} \left| P_m e^{-ik_m r} + P_n e^{-ik_n r} \right|^2 , \tag{2.303}$$

where ρ is the density of the medium in the channel and

$$P_m = \frac{\omega \rho S_3}{\sqrt{2\pi k_m h}} \sin(\nu z) \sin(\nu z') \tag{2.304}$$

is the amplitude of mode m (excluding the cylindrical spreading term). On expressing the term on the right of Eq. (2.303) explicitly, we find that

$$\omega^2 \rho^2 |G|^2 = \frac{1}{r} \left\{ P_m^2 + P_n^2 + 2 P_m P_n \cos\left(\frac{2\pi r}{L_{mn}}\right) \right\} , \tag{2.305}$$

where

$$L_{mn} = 2\pi |k_m - k_n|^{-1} \tag{2.306}$$

is the *interference length*, that is, the horizontal distance between the modal interference peaks. For lower-order modes such that $k \gg m\pi/h$, the radical in Eq. (2.299) may be approximated by its Taylor expansion to second order in the mode number m, in which case the interference length becomes

$$L_{mn} \approx \frac{4kh^2}{\pi|n^2 - m^2|} \ .$$

(2.307)

As an example, for consecutive mode numbers $m = 1$ and $n = 2$, at a frequency of 100 Hz, in a water column of sound speed $c = 1500$ m/s and depth $h = 20$ m, the interference length is $L_{mn} \approx 71$ m.

2.7 THE PEKERIS CHANNEL

In 1948, Pekeris published a normal-mode analysis [73] of the sound field from a point source in a shallow ocean channel overlying a semi-infinite fluid basement. He took both regions to be homogeneous, with uniform sound speed and density, (ρ_1, c_1) and (ρ_2, c_2), in the upper and lower fluid, respectively. The solution for the field in both regions can be obtained using the integral transform techniques developed above, provided that the boundary conditions—continuity of pressure and continuity of normal component of particle velocity—are handled appropriately. The geometry for the problem is illustrated in Fig. 2.23 showing a channel of depth h in which the source and receiver are at depth z' and z, respectively, and separated by horizontal range r. The sea surface is taken to be a pressure-release boundary for sound incident from below. Cylindrical coordinates with the z-axis vertical

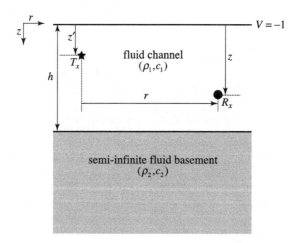

FIGURE 2.23

Source, T_x, and receiver, R_x, in a Pekeris shallow-water channel.

and passing through the source are appropriate for the problem. Symmetry dictates that the field in both domains must be uniform in azimuth; and the radiation conditions require that the field infinitely far from the transmitter in depth and range must go to zero.

2.7.1 THE INTEGRAL-TRANSFORM SOLUTION FOR THE FIELD

The wave equations to be solved for the field in the channel and the basement are

$$\frac{1}{r}\frac{\partial}{\partial r}\left\{r\frac{\partial G_1}{\partial r}\right\} + \frac{\partial^2 G_1}{\partial z^2} + k_1^2 G_1 = \frac{S_3}{\pi r}\delta(r)\delta(z - z') \tag{2.308}$$

and

$$\frac{1}{r}\frac{\partial}{\partial r}\left\{r\frac{\partial G_2}{\partial r}\right\} + \frac{\partial^2 G_2}{\partial z^2} + k_2^2 G_2 = 0 \tag{2.309}$$

where $G_1 = G_1(r,z;\omega)$ and $G_2 = G_2(r,z;\omega)$ are the Fourier transforms with respect to time of the velocity potential in the upper and lower medium, respectively, and S_3 is the source strength. At the bottom boundary, $z = h$, continuity of pressure requires that

$$\rho_1 G_1(h) = \rho_2 G_2(h) , \tag{2.310}$$

and the condition of continuity of the normal component of particle velocity is

$$G_1'(h) = G_2'(h) , \tag{2.311}$$

where the prime denotes differentiation with respect to depth, z, and for brevity only the z dependence has been included in the arguments of the field variables.

The first step in developing the solution for the field in both domains is to apply a Hankel transform with respect to range, as defined in Eq. (2.62), to all four of the above equations. The reduced wave equations then become

$$\frac{\partial^2 G_{1p}}{\partial z^2} + \left(k_1^2 - q^2\right)G_{1p} = \frac{S_3}{2\pi}\delta(z - z') \tag{2.312}$$

and

$$\frac{\partial^2 G_{2p}}{\partial z^2} + \left(k_2^2 - q^2\right)G_{2p} = 0 , \tag{2.313}$$

where k_1 and k_2 are, respectively, the acoustic wave numbers in the channel and the basement, and the transform variable q is the horizontal wave number. Eq. (2.313) can be solved immediately for the transformed field in the lower medium:

$$G_{2q}(z) = G_{2q}(h)e^{-i\eta_2(z-h)} , \tag{2.314}$$

where $G_{2q}(h)$ is the unknown value of the field on the bottom boundary, which is to be determined from the boundary conditions, and

$$\eta_2 = \sqrt{k_2^2 - q^2}, Im(\eta_2) < 0 \ . \tag{2.315}$$

Next, a finite Fourier transform taken over the depth interval $(0,h)$ is applied to Eq. (2.312). The finite Fourier transform of the second derivative returns nonzero integrated terms, which contain additional unknowns that must be determined from the boundary conditions. The analysis involves some quite lengthy algebra, which is not difficult in principle but requires careful treatment of the details, eventually leading to the result

$$G_{1q}(z) = -\frac{S_3}{2\pi} \frac{\sin(\eta_1 z_<)}{\eta_1} \left\{ \frac{\eta_1 \cos \eta_1(h - z_>) + ib\eta_2 \sin \eta_1(h - z_>)}{\eta_1 \cos(\eta_1 h) + ib\eta_2 \sin(\eta_1 h)} \right\} \tag{2.316}$$

where

$$\eta_1 = \sqrt{k_1^2 - q^2}, Im(\eta_1) < 0 \tag{2.317}$$

and $z_>$, $z_<$ are, respectively, the larger and smaller of z and z'. From Eq. (2.314), the field in the bottom is

$$G_{2q}(z) = -\frac{bS_3}{2\pi} \left\{ \frac{\sin(\eta_1 z')e^{-i\eta_2(z-h)}}{\eta_1 \cos(\eta_1 h) + ib\eta_2 \sin(\eta_1 h)} \right\} . \tag{2.318}$$

The parameter b in these expressions is the ratio of the densities in the upper and lower fluids:

$$b = \frac{\rho_1}{\rho_2} < 1 . \tag{2.319}$$

It is readily shown that these solutions for $G_{1q}(z)$ and $G_{2q}(z)$ satisfy the boundary conditions in Eqs. (2.310) and (2.311); and, since $Im(\eta_2) < 0$, the field in the basement decays to zero exponentially in the limit of infinite depth, as required by the radiation condition.

On applying the inverse Hankel transforms to $G_{1q}(z)$ and $G_{2q}(z)$, the following integrals for the frequency-dependent field in the channel and the basement are obtained:

$$G_1(r, z; \omega) = -\frac{S_3}{2\pi} \int_0^\infty q J_0(qr) F_1(\eta_1, \eta_2) dq \tag{2.320}$$

and

$$G_2(r, z; \omega) = -\frac{S_3}{2\pi} \int_0^\infty q J_0(qr) F_2(\eta_1, \eta_2) dq , \tag{2.321}$$

where

$$F_1(\eta_1, \eta_2) = \frac{\sin(\eta_1 z_<)}{\eta_1} \left\{ \frac{\eta_1 \cos \eta_1 (h - z_>) + ib\eta_2 \sin \eta_1 (h - z_>)}{\eta_1 \cos(\eta_1 h) + ib\eta_2 \sin(\eta_1 h)} \right\} \qquad (2.322)$$

and

$$F_2(\eta_1, \eta_2) = \left\{ \frac{\sin(\eta_1 z')e^{-i\eta_2(z-h)}}{\eta_1 \cos(\eta_1 h) + ib\eta_2 \sin(\eta_1 h)} \right\}. \qquad (2.323)$$

Note that both these functions are *even* in η_1 and *mixed* in η_2. The integrals for G_1 and G_2 can be evaluated using a technique similar to that employed in the analysis of the lateral wave in Section 2.5.6. As in Eq. (2.252), the Bessel function is represented as the sum of two Hankel functions, and, as shown in Fig. 2.19, EJP cuts are made in the second and fourth quadrants of the complex wave number plane, which in the present discussion is designated the q-plane.

Proceeding as before, the integral along the positive imaginary axis cancels with the corresponding integral along the negative imaginary axis. Now, however, in the fourth quadrant of the top Riemann sheet, where $Im(\eta_2) < 0$, in addition to the branch line integrals, a finite number, M, of poles exists, corresponding to propagating normal modes. A similar number of poles also exists on the lower Riemann sheet, where $Im(\eta_2) > 0$, but these poles, corresponding to what has been termed "improper" modes [74], are not included in the contour integration around the top sheet.

Since F_1 and F_2 are even in η_1, and they are multiplied in the integrands by η_1, which is odd, each of the integrals around the k_1 cut is identically zero. Around the k_2 cut, however, the integrals are nonzero, since F_1 and F_2 are mixed in η_2. The EJP branch line integrals represent the lateral wave, associated with acoustic rays propagating in the channel at the critical grazing angle. Also included in the EJP integrals is the so-called continuous field, corresponding to relatively steep rays, with grazing angles greater than the critical, which undergo partial reflection and penetration on encountering the bottom. The form of the EJP integrals for the field in the channel and the basement is quite complicated, as discussed by Pekeris [73], who developed approximations for their asymptotic behavior at large r under various environmental conditions.

2.7.2 THE NORMAL MODE SOLUTION

The denominator in the expressions for G_1 and G_2 is

$$D = \eta_1 \cos(\eta_1 h) + ib\eta_2 \sin(\eta_1 h). \qquad (2.324a)$$

This expression has M zeros, corresponding to the eigenvalues, which are the solutions of the *characteristic equation*

$$\eta_1 \cos(\eta_1 h) + ib\eta_2 \sin(\eta_1 h) = 0. \qquad (2.324b)$$

Once the zeros are known the residues of the integrands in Eqs. (2.320) and (2.321) can be determined, allowing the modal component of the field to be fully specified. Before proceeding with the solution for the characteristic equation, however, a general treatment is outlined which leads to expressions for the modal field in terms of the eigenvalues.

If the mth zero is the eigenvalue $q = q_m$, then, to first order, a Taylor expansion of D about q_m is

$$D = (q - q_m)\frac{\partial D}{\partial q}\bigg|_{q=q_m} + \cdots$$

$$= (q - q_m)\frac{q_m}{\eta_{1m}\sin(\eta_{1m}h)}\{\eta_{1m}h - \sin(\eta_{1m}h)\cos(\eta_{1m}h) \tag{2.325}$$

$$-b^2\sin^2(\eta_{1m}h)\tan(\eta_{1m}h)\} + \cdots$$

where

$$\eta_{1m} = \sqrt{k_1^2 - q_m^2} \tag{2.326}$$

and $\eta_2 = \eta_{2m} = \sqrt{k_2^2 - q_m^2}$ has been eliminated by using the condition, obtained from Eq. (2.324b), that

$$\eta_{2m} = \frac{i\eta_{1m}}{b\tan(\eta_{1m}h)}. \tag{2.327}$$

Now, when $q = q_m$, the numerator of the function F_1, in Eq. (2.322) becomes

$$N = \eta_{1m}\cos[\eta_{1m}(h - z_>)] + ib\eta_{2m}\sin[\eta_{1m}(h - z_>)]$$

$$= \sin\eta_{1m}z_>\{\eta_{1m}\sin(\eta_{1m}h) - ib\eta_{2m}\cos(\eta_{1m}h)\}, \tag{2.328}$$

With the aid of the expressions in Eqs. (2.325) and (2.328), the residues of the poles in each integrand can be evaluated, from which the normal mode components of the field in the channel and the basement are found to be

$$G_1(r, z; \omega) = \frac{iS_3}{2}\sum_{m=1}^{M}\frac{\eta_{1m}\sin(\eta_{1m}z)\sin(\eta_{1m}z')H_0^{(2)}\left(\sqrt{k_1^2 - \eta_{1m}^2}\,r\right)}{\eta_{1m}h - \sin(\eta_{1m}h)\cos(\eta_{1m}h) - b^2\sin^2(\eta_{1m}h)\tan(\eta_{1m}h)} \tag{2.329}$$

and

$$G_2(r, z; \omega) = \frac{ibS_3}{2}\sum_{m=1}^{M}\frac{\eta_{1m}\sin(\eta_{1m}h)\sin(\eta_{1m}z')e^{-i\eta_{2m}(z-h)}H_0^{(2)}\left(\sqrt{k_1^2 - \eta_{1m}^2}\,r\right)}{\eta_{1m}h - \sin(\eta_{1m}h)\cos(\eta_{1m}h) - b^2\sin^2(\eta_{1m}h)\tan(\eta_{1m}h)}. \tag{2.330}$$

In Eq. (2.329), because of the symmetry of the two sine functions in the numerator, it is no longer necessary to distinguish between the larger and smaller of z and z'.

The expressions in Eqs. (2.329) and (2.330) are the same as those that were derived originally by Pekeris [73], although he left the limits on the summation signs unspecified. It is clear from the asymptotic expansion of the Hankel function, exemplified in Eq. (2.301), that at long range such that $q_m r \gg 1$, each mode experiences cylindrical spreading, decaying as $r^{-1/2}$. To complete the solution for $G_1(r,z;\omega)$ and $G_2(r,z;\omega)$, it is necessary to solve the characteristic equation in Eq. (2.324b) for its roots, $\eta_1 = \eta_{1m}$, which govern the mode shapes and the eigenvalues. The maximum number, M, of propagating modes also emerges from the characteristic equation.

2.7.3 THE CHARACTERISTIC EQUATION

The characteristic equation, Eq. (2.324b), can be written in the form

$$\tan(\eta_1 h) = \frac{i\eta_1}{b\eta_2} = -\frac{\eta_1}{\sqrt{k_1^2 - k_2^2 - \eta_1^2}} = -\frac{\eta_1}{b\sqrt{k_1^2 \sin^2 \alpha_c - \eta_1^2}} , \qquad (2.331)$$

where α_c is the critical grazing angle, given by $\cos \alpha_c = k_2/k_1$. Eq. (2.331) is a transcendental equation, which can be rearranged to take the form

$$\eta_{1m} h = m\pi - \tan^{-1}\left\{\frac{\eta_{1m}}{b\sqrt{k_1^2 \sin^2 \alpha_c - \eta_{1m}^2}}\right\}, \quad m = 1, 2\cdots, \qquad (2.332)$$

where we have set $\eta_1 = \eta_{1m}$. No general, exact analytical solution of this equation exists.

However, for the radical to be real, we must have

$$\eta_{1m} \leq k_1 \sin \alpha_c . \qquad (2.333)$$

When the equality holds in this expression, the arctangent function in Eq. (2.332) is equal to $\pi/2$ and the mode number m takes its highest possible value, $m = M$. Thus, from Eq. (2.332),

$$M = \frac{k_1 h}{\pi} \sin \alpha_c + \frac{1}{2} , \qquad (2.334)$$

where the expression on the right is understood to be rounded down to the nearest integer, and

$$\eta_{1M} h = \left(M - \frac{1}{2}\right)\pi . \qquad (2.335)$$

According to Eq. (2.335), the shape of the highest-order mode in the water column is the same as if the bottom of the channel were a rigid boundary.

For mode numbers higher than M, solutions of the characteristic equation do exist, albeit determined numerically, but they correspond to nonphysical modes whose amplitudes increase exponentially with depth. As with the "ideal" waveguide, a cut-off frequency exists for each mode in the Pekeris channel, below which the mode is not supported. All modes are cut off when $M < 1$, corresponding to a cut-off frequency

$$f_{cutoff} = \frac{c}{4h \sin \alpha_c}.$$ (2.336)

As an example, at a frequency of 100 Hz, in a channel of depth 50 m, with a sound speed of 1500 m/s and a critical grazing angle $\alpha_c = 25$ degrees, the total number of supported modes, from Eq. (2.334), is $M = 3$; and, from Eq. (2.336), the cut-off frequency below which no modes propagate is 17.7 Hz.

Although the characteristic equation does not possess a general, exact analytical solution for η_{1m}, an approximate solution can be developed, valid for lower-order modes such that η_{1m} is small compared with $\sin\alpha_c$. Under this condition, the arctangent function in Eq. (2.332) may be approximated by the leading order term in its series expansion, and η_{1m}^2 under the radical may be neglected. This leads to the expression

$$\eta_{1m}h \approx m\pi - \frac{\eta_{1m}}{bk_1 \sin \alpha_c},$$ (2.337)

from which it follows that

$$\eta_{1m} \approx \frac{m\pi}{h_e},$$ (2.338)

where

$$h_e = h\left[1 + \frac{1}{k_1 hb \sin \alpha_c}\right].$$ (2.339)

Physically, this result means that, within the limits of the approximation, the mode shapes in the Pekeris channel of depth h are the same as those in an ideal channel with pressure-release boundaries and depth h_e. Clearly, from Eq. (2.339), the depth of the fictitious pressure-release boundary beneath the actual bottom is

$$h_e - h = \frac{1}{k_1 b \sin \alpha_c},$$ (2.340)

which, on comparison with Eq. (2.232), is seen to be precisely Weston's effective depth. It is interesting to note that, since the effective depth is independent of grazing angle, it is also independent of the mode number. Thus, the same effective depth applies to all the modes for which the approximation in Eq. (2.337) holds.

A numerical approach to solving the characteristic equation for its complex roots [75] is based on the Newton–Raphson root-finding algorithm. By a minor modification of Eq. (2.332), the equation to be solved can be written as

$$f(X) = X - \left(m - \frac{1}{2}\right)\pi - \tan^{-1}[g(X)] = 0$$ (2.341)

where

$$X = \eta_{1m}h ,$$ (2.342)

$$g(X) = b\frac{\sqrt{A^2 - X^2}}{X}$$ (2.343)

and

$$A = \sqrt{k_1^2 - k_2^2}h .$$ (2.344)

In this formulation, bottom attenuation is accommodated by retaining the acoustic wave number k_2 explicitly, rather than expressing it in terms of the critical grazing angle, and allowing it to be complex with a negative imaginary component. The derivative of the function in Eq. (2.341) with respect to the complex variable X is

$$f'(X) = \frac{df}{dX} = 1 + \frac{1}{X(1 + g^2)}\left\{g + \frac{b^2}{g}\right\} .$$ (2.345)

If the nth approximation for the required root is X_n, then an improved estimate is

$$X_{n+1} = X_n - \frac{f(X_n)}{f'(X_n)} ,$$ (2.346)

which, with a starting value of $X_0 = (m - 1/2)\pi$, converges to the required complex solution for the vertical wave number, η_{1m}, of mode m after just a few iterations. Once the vertical wave numbers of the modes in the water column have been determined, the eigenvalues may be obtained from the expression

$$q_m = \sqrt{k_1^2 - \eta_{1m}^2} ,$$ (2.347)

The Newton–Raphson algorithm in Eq. (2.346) returns η_{1m} for the propagating modes ($1 \leq m \leq M$), all of which have $Im(\eta_2) < 0$, thus satisfying the radiation condition in the basement. When $m > M$, the Newton–Raphson algorithm still returns roots of the characteristic equation, but all are nonphysical, having $Im(\eta_2) > 0$, corresponding to improper modes, whose amplitudes diverge with increasing depth in the lower medium.

Fig. 2.24 shows the shape of three propagating modes in the upper and lower domains of a Pekeris waveguide, as computed from the Newton–Raphson root-finding algorithm in Eq. (2.346) and from the effective depth approximation in Eq. (2.338). The channel parameters are from Zhang and Tindle [76]: $h = 54$ m, $c_1 = 1500$ m/s, $c_2 = 1600$ m/s, $b = 0.8$, and a bottom attenuation of 0.3125 dB/m/kHz (equivalent to 0.036 Np/m/kHz). At a frequency of 100 Hz, as used in the calculations, only the three modes shown in the figure are supported in the channel; and the effective depth is $h_e = 1.159h = 62.58$ m. The first two modes, as computed from the effective depth approximation are visually indistinguishable from their numerically computed counterparts from the Newton–Raphson algorithm. For mode 3, the

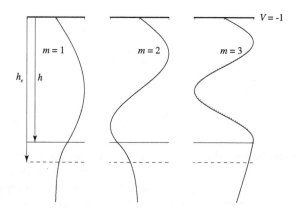

FIGURE 2.24

Mode shapes in a Pekeris two-layer waveguide. The solid modal lines are from the Newton–Raphson algorithm and the (almost indistinguishable) dotted modes are from the effective depth approximation.

difference between the two calculations, although just perceptible, is still extremely small, which is remarkable since the third mode is very close to cutoff. The comparison of the approximated and numerically computed modes in Fig. 2.24 illustrates the fidelity of the effective depth approximation for calculating the modes shapes in the water column. Of course, the effective depth approximation does not provide information on the exponentially decaying modal tails in the basement.

2.8 THREE-DIMENSIONAL PROPAGATION

In a horizontally stratified ocean the sound field from a point source is azimuthally uniform and hence can be described in terms of just two spatial coordinates, horizontal range and depth. Such a field is said to be two-dimensional, even though it occupies a three-dimensional space. When variable stratification is present, associated with nonuniform bathymetry or a laterally varying sound speed, the acoustic field is three-dimensional in that it depends on three spatial coordinates, usually though not always taken to be horizontal range, depth, and azimuth about the source position.

In shallow water with variable bathymetry, the slope of the bottom is often the main cause of horizontal deflections: acoustic rays with wave fronts that are not parallel to the depth contours are, on reflection from the bottom, twisted out of the vertical plane of incidence, thus giving rise to a three-dimensional sound field in the channel. The horizontal projection of a ray that undergoes multiple reflections from the surface and bottom of a depth-varying channel follows a path which, actually, is a sequence of linear or near-linear segments connecting bottom bounces but which, if the bottom slope is small, approximates a continuous curve. The twisting that occurs on reflection from a sloping bottom has been termed *horizontal refraction* by Weston [77], who was the first to analyze the phenomenon.

2.8.1 HORIZONTAL REFRACTION

Ray propagation through a wedge-shaped channel is essentially a problem in three-dimensional geometry. Although the twisting of a ray, or ray bundle, as it reflects from the sloping bottom is difficult to visualize, Weston [77] was able to show, from an argument based on direction cosines, that, in iso-speed water, the horizontal projection of the ray follows a hyperbolic trajectory, as illustrated in Fig. 2.25. As the ray approaches the shoreline, undergoing multiple reflections from the surface and bottom, it turns around and follows a symmetrical path back out toward deeper water.

This curvature in the ray path introduces a bearing error, which is actually double-valued, since a ray group may reach any point in the channel by either of two routes: it may set off upslope and turn around or travel along a direct (slightly curved) path to reach the receiver. Either way, bearing errors arise from horizontal refraction. Confirmation of the bearing error phenomenon was observed in an experiment conducted off the East Australian Continental Slope [78] in which two ships, one towing a source and the other a line array, and initially separated by 34 km, proceeded from shallow water, of depth about 500 m, to deeper water on a divergent but constant line of bearing course. Over a period of half an hour, bearing shifts of tens of degrees were observed in the beam-formed data, which were interpreted as being due to horizontal refraction associated with the sloping bottom.

The total loss of acoustic intensity along a ray path depends strongly on whether the critical angle of the seabed is exceeded at any of the bottom reflections, and also

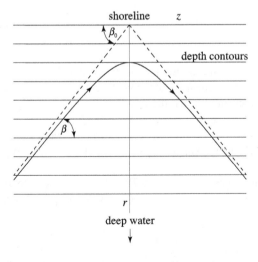

FIGURE 2.25

Plan view showing hyperbolic ray path in an iso-speed, wedge-shaped channel.

on the bottom slope, since the steeper the gradient the fewer reflections occur during the turn-around process. In the East Australian Continental Slope experiment [78], the critical-angle effect was observed to be significant, serving to limit the azimuthal extent of the energy received at certain source–receiver positions.

From his direction-cosine analysis, Weston [77] derived a simple connection between the horizontal and vertical angles associated with the hyperbolic ray trajectory illustrated in Fig. 2.25. He showed that two angles are equal: the horizontal angle, β_0, that the asymptote of the hyperbola makes with the shoreline, and the vertical grazing angle, α_t, of the ray at the turn-around point, where the hyperbolic ray path is closest to the shoreline. This means that all ray bundles approaching the shoreline from deep water from a direction β_0 have the same vertical angle at the point of closest approach to the apex of the wedge, and all experience the same number of bottom reflections. This equality between β_0 and α_t, besides being interesting in its own right, serves to help interpret the modal structure of the wave field in a wedge-shaped channel.

2.8.2 THE "IDEAL" WEDGE

In most channels with range-varying bathymetry, the boundary conditions are such that the acoustic field in the channel is not separable and the wave equation representing the field does not possess a closed-form analytical solution. An exception is an ideal wedge, that is to say, a wedge-like domain in which both boundary planes are perfect reflectors (pressure release, rigid, or mixed). The acoustic field from a point source in such a channel is separable and can be expressed as a sum of uncoupled normal modes whose amplitudes depend strongly on range from the apex and cross-range from the source [69,79]. To obtain the modal solution, a sequence of integral transforms is applied to the wave equation, following a course similar to that in Section 2.6.2 for the field in an ideal waveguide with parallel boundaries.

As with the parallel-sided waveguide, a cylindrical coordinate system is chosen for the ideal wedge problem. Rather than being vertical, however, the z-axis is tilted, running horizontally along the apex of the wedge, range, r, is normal to the apex, and θ is angular depth measured about the apex, as illustrated in Fig. 2.26. The source, T_x, is located at $(r', 0, \theta')$ and the receiver, R_x, at (r, z, θ). In this coordinate system, the coordinate surfaces $\theta = 0$ and $\theta = \theta_0$, where θ_0 is the wedge angle, coincide with the boundaries of the wedge-shaped domain, which is why, with perfect reflectors, the field is separable.

Assuming an impulsive source, the Helmholtz equation to be solved for the Fourier transform (with respect to time) of the velocity-potential Green's function, $G = G(r, z; \omega)$, is

$$\frac{1}{r}\frac{\partial}{\partial r}\left(r\frac{\partial G}{\partial r}\right) + \frac{1}{r^2}\frac{\partial^2 G}{\partial \theta^2} + \frac{\partial^2 G}{\partial z^2} + k^2 G = S_3 \frac{\delta(r - r')}{\pi r}\delta(\theta - \theta')\delta(z), \qquad (2.348)$$

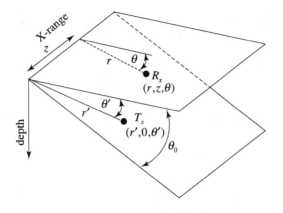

FIGURE 2.26

Cylindrical coordinate system for the ideal wedge problem.

where S_3 is the source strength. For a wedge with Dirichlet (pressure-release) boundaries, the first spatial transform to be applied to Eq. (2.348) is a finite Fourier sine transform taken over angular depth, θ, in which the kernel is $\sin(m\pi\theta/\theta_0)$, chosen to satisfy the boundary conditions. It will turn out that the integer $m = 1, 2, \ldots$ is the mode number. A Hankel transform of order equal to the mode index

$$v = \frac{m\pi}{\theta_0}, \quad m = 1, 2, \cdots, \tag{2.349}$$

is then applied, followed by a cosine Fourier transform over cross-range, z. These transforms yield an algebraic expression for the triply transformed field, which, on applying the three inverse transforms, returns an expression for the velocity potential itself in the form of a sum of uncoupled normal modes [79–81]:

$$G = \frac{S_3 k}{\pi\theta_0} \sum_{m=1}^{\infty} I_v \sin(v\theta)\sin(v\theta'), \tag{2.350}$$

where k is the acoustic wave number, m is the mode number, and the mode amplitude function is

$$I_v = I_v(r, r', z; k) = \frac{1}{k} \int_0^{\infty} p \frac{\exp(-\eta|z|)}{\eta} J_v(pr) J_v(pr') dp. \tag{2.351}$$

In this expression, $J_v(\cdot)$ is a Bessel function of the first kind of order v, the integration variable p is the radial wave number, and

$$\eta = \sqrt{p^2 - k^2}, \quad Im(\eta) < 0 \tag{2.352}$$

is the cross-range wave number. With the multiplier k^{-1} before the integral in Eq. (2.351), the mode amplitude function is dimensionless.

The solution in Eq. (2.350) still holds for a wedge with rigid boundaries provided an additional term, $I_0/2$, representing a zero-order mode, is included and the sine functions are replaced by cosines. With mixed boundaries, one pressure-release and one rigid, the formulation in Eq. (2.350) is again valid but in this case with mode index $\nu = (m + 1/2)\pi/\theta_0$.

The integral in Eq. (2.351) may be rearranged by expressing the Bessel function product in terms of Bessel functions of zero order [82,83]:

$$J_\nu(pr)J_\nu(pr') = \frac{1}{\pi} \int_0^\pi J_0\left(p\sqrt{r^2 + r'^2 - 2rr'\cos A}\right)\cos(\nu A)dA - \frac{1}{\pi}\sin(\nu\pi)$$
$$\times \int_0^\infty J_0\left(p\sqrt{r^2 + r'^2 + 2rr'\cosh B}\right)e^{-\nu B}dB.$$

$$(2.353)$$

On substituting this expression into Eq. (2.351) and evaluating the integral over p, which is a known form, the modal amplitude function reduces to

$$I_\nu = \frac{1}{\pi k}\int_0^\infty \frac{\exp(-ikR)}{R}\cos(\nu A)dA - \frac{\sin(\nu\pi)}{\pi k}\int_0^\infty \frac{\exp(-ikP)}{P}e^{-\nu B}dB , \quad (2.354)$$

where

$$R = R_0\sqrt{1 - 2a\cos A} , \tag{2.355}$$

$$P = R_0\sqrt{1 + 2a\cosh B} , \tag{2.356}$$

$$R_0 = \sqrt{r^2 + r'^2 + z^2} \tag{2.357}$$

and

$$a = \frac{rr'}{R_0^2} . \tag{2.358}$$

The variable a in Eq. (2.358) always lies in the interval $[0,\frac{1}{2}]$, taking its maximum value of $\frac{1}{2}$ when the source and receiver are in the same vertical plane and at the same range from the apex (i.e., when $z = 0$ and $r = r'$). When either r or $r\prime$ is zero, a is also zero, and a approaches zero asymptotically in regions remote from the source.

In Eq. (2.354), the first integral represents the contribution to a particular mode from images of the source in the bounding planes of the wedge, whereas the second integral represents the contribution to the mode from diffraction at the apex of the wedge. When the mode index, ν, is an integer, the diffracted component is identically zero, due to the presence of the factor $\sin\nu\pi$; and even when ν is not an integer, the diffracted component is negligible for the small wedge angles that are characteristic of near-shore ocean channels. For larger wedge angles, greater than π, the wedge is more aptly described as a ridge and the diffracted component is significant.

Only the image mode integral, appropriate to acoustic propagation in narrow ocean wedges is examined here. Writing this integral in its bilateral form, we have

$$I_\nu = \frac{1}{2\pi k R_0}\int_{-\pi}^\pi \frac{\exp[-ikR_0w(A)]}{\sqrt{1 - 2a\cos A}}dA , \tag{2.359}$$

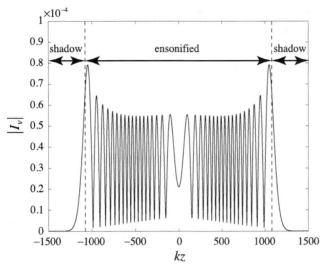

FIGURE 2.27

The image-mode integral in Eq. (2.359) evaluated with $\nu = 200$, $kr' = 1000$, and $kr = 300$. The *vertical dashed lines* depict the caustic at the edges of the ensonified region.

where the phase function is

$$w(A) = -\frac{\nu A}{kR_0} + \sqrt{1 - 2a \cos A} \,. \tag{2.360}$$

An example of the cross-range behavior of $|I_\nu|$, obtained from a numerical evaluation of the integral in Eq. (2.359), is shown in Fig. 2.27. This highly oscillatory curve, which is symmetrical about the source position at $z = 0$, represents the modulus of the complex amplitude of a mode having $\nu = 200$. The central segment of the curve shows a series of *intramodal* interference peaks, which terminate on either side in a shadow zone. A caustic, depicted by the vertical dotted lines, marks the transition between the ensonified region and the shadow.

2.8.3 THE SHADOW EDGE

Since the integral in Eq. (2.359) has a rapidly oscillating integrand, it may be evaluated using the stationary phase approximation [69] for large values of kR_0. The derivative of the phase function with respect to the integration variable, A, is

$$w'(A) = -\frac{\nu}{kR_0} + \frac{a \sin A}{\sqrt{1 - 2a \cos A}} \,, \tag{2.361}$$

which has two roots, corresponding to two stationary points in $w(A)$ within the integration interval $[-\pi,\pi]$:

$$A_{\pm} = \cos^{-1}\left(a\xi^2 \pm \sqrt{a^2\xi^4 - \xi^2 + 1}\right), \tag{2.362}$$

where

$$\xi = \frac{\nu}{kR_0a} . \tag{2.363}$$

The second derivative of the phase function at the two turning points is

$$w''(A_{\pm}) = \frac{a\left(\cos A_{\pm} - a\cos^2 A_{\pm} - a\right)}{(1 - 2a\cos A_{\pm})^{3/2}} , \tag{2.364}$$

which, from the stationary phase expression, allows the I_ν to be approximated as

$$I_\nu \approx \frac{1}{kR_0\sqrt{2\pi kR_0a}}\{F(A_+) + F(A_-)\} , \tag{2.365}$$

where

$$F(A_{\pm}) = \frac{(1 - 2a\cos A_{\pm})^{1/4}}{|\cos A_{\pm} - a\cos^2 A_{\pm} - a|^{1/2}}\exp\left\{-i\left[kR_0w(A_{\pm}) - \nu A_{\pm}\right.\right.$$
$$\left.\left. + \frac{\pi}{4}\text{sgn}w''(A_{\pm})\right]\right\} . \tag{2.366}$$

Within the ensonified region, the expression for the modal amplitude in Eq. (2.365) is an excellent approximation for the intramodal interference structure. It breaks down, however, at the shadow edge, where the two roots, A_+ and A_-, coalesce. This condition occurs when the radical in Eq. (2.362) goes to zero. The two possible solutions for ξ are then

$$\xi_c^2 = \frac{1}{2a_c^2}\left[1 \pm \sqrt{1 - 4a_c^2}\right]$$
$$\approx \frac{1}{2a_c^2}\left[1 \pm (1 - 2a_c^2)\right] \tag{2.367}$$

where the subscript c indicates that the associated variable lies on the caustic, and the approximation implies that $a_c \ll 1$, which will be the case when the source is far from the apex. Where there is a choice of sign in Eq. (2.367), the plus sign must be rejected because it leads to complex values of the root. The locus of the shadow edge is therefore given by the condition

$$\xi_c = \frac{\nu}{kR_{0c}a_c} \approx 1 . \tag{2.368}$$

This expression, representing a hyperbola in the (r,z) plane whose focus is at the source position $(r',0)$, when written in terms of the coordinates (r_c, r', z_c), takes the more familiar form

$$\frac{r_c^2}{r^2}\left[\left(\frac{kr'}{\nu}\right)^2 - 1\right] - \frac{z_c^2}{r'^2} \approx 1 \,. \tag{2.369}$$

The region within this hyperbolic envelope, or *caustic*, is ensonified and beyond is in shadow. For a given mode, the extent of the ensonified region may be expressed in terms of the half-angle β_ν between an asymptote of the hyperbola and the apex of the wedge (the modal equivalent of β_0 in Fig. 2.25), which from Eq. (2.369) is given by

$$\tan \beta_\nu = \sqrt{(kr'/\nu)^2 - 1} \,. \tag{2.370}$$

Thus, the angular width of the ensonified region, as represented by β_ν, *increases* with increasing frequency, source range, and wedge angle, but *decreases* with increasing mode number, m.

The nested structure of the hyperbolic caustics of several modes is illustrated in Fig. 2.28. Since a mode is cut off outside its hyperbolic envelope, the number of propagating modes at any point in the channel depends on the relative positions

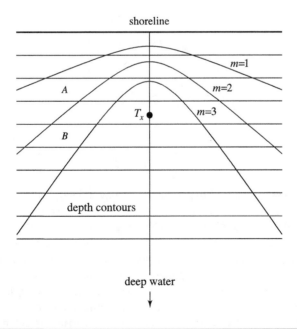

FIGURE 2.28

Hyperbolic envelope of three modes in a wedgelike ocean.

of the source and receiver. For example, a receiver at point A in Fig. 2.28 sees only one mode, whereas at point B there are two.

It is perhaps worth mentioning that, at the caustic, where $a_c = 1/2$ and the two roots in Eq. (2.367) coalesce, the first-order stationary phase analysis breaks down. Under this condition, it is possible to develop a second-order stationary phase approximation, which leads to an accurate, Airy function representation of the modal field through the caustic and into the shadow zone [69].

2.8.4 INTRAMODAL INTERFERENCE

The interference structure in the ensonified region of a mode can be interpreted in terms of modal eigen rays, that is, those rays launched from the source in all azimuthal directions with the same vertical angle. As shown by Weston [77], the horizontal projection of such eigen rays follows a hyperbolic trajectory, the equation of which may be written as

$$\frac{r_\beta^2}{h^2} - \frac{(z_\beta - s)^2}{l^2} = 1 \tag{2.371}$$

where the subscript β indicates that the associated spatial coordinate lies on the locus of an eigen ray trajectory with horizontal launch angle β, as illustrated in Fig. 2.25. From Weston's condition, that the angle between the asymptote of the hyperbolic ray path and the apex of the wedge is equal to the grazing angle of the ray at the turn-around point, the parameters h, l, and s are found to be

$$h = r' \left[\cos^2 \beta + \left(\frac{kr'}{\nu} \right)^2 \sin^2 \beta \right]^{-1/2}, \tag{2.372}$$

$$l = h\sqrt{k^2 h^2 / \nu^2 - 1}, \tag{2.373}$$

and

$$s = \left(\frac{l}{h} \right)^2 r' \tan \beta. \tag{2.374}$$

Fig. 2.29 shows a family of hyperbolic eigen ray trajectories, as computed from Eq. (2.363) for a given mode, represented by the mode index, ν. Also shown in the figure is the caustic delineating the shadow edge, which is another hyperbola, as given by the stationary phase result in Eq. (2.368). It is evident that the locus of the shadow edge, derived from the wave-theoretic analysis, is the bounding envelope of the eigen ray trajectories. This precise correspondence between the eigen ray description of the modal field and the wave-theoretic representation leads naturally to the interpretation of the shadow as a geometrical effect that arises from the three-dimensional phenomenon of horizontal refraction, occurring as eigen rays are reflected from the inclined boundaries of the wedge.

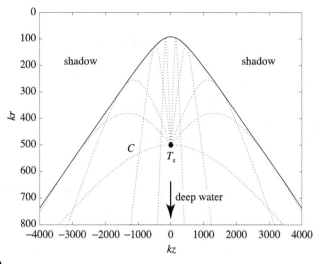

FIGURE 2.29

Modal family of hyperbolic eigen rays (dotted curves), bounded by the hyperbolic caustic (solid curve). The mode index is $\nu = 90$ and the source, T_x, is at $kr' = 500$. Only the eigen rays launched upslope are shown for clarity.

It is apparent from Fig. 2.29 that, at any point within the ensonified region of the modal field, two (and only two) eigen rays intersect, for example, at point C in the figure. Coherent addition of these pairs of eigen rays gives rise to the intrinsically three-dimensional phenomenon of intramodal interference, which creates a highly structured modal field, exactly as predicted from the wave-theoretic analysis and illustrated in Fig. 2.27.

2.8.5 THE PENETRABLE WEDGE

In a wedge-like ocean with a fluid basement, as in the ideal wedge, an acoustic ray launched upslope from a point source experiences multiple reflections from the surface and bottom, becoming progressively steeper as it approaches the turn-around point. In the case of a fast fluid basement, with sound speed greater than that in the water column, the bottom interface possesses a critical grazing angle. If the grazing angle of the ray exceeds the critical before turn-around, the ray will penetrate the bottom and most of the associated energy will be lost to the water column.

The sound field in a penetrable wedge is not separable, even in the cylindrical coordinate system used with the ideal wedge, because the boundary conditions along the fluid–fluid interface are functions of distance from the apex. In this situation, since integral transform techniques breakdown, there is no exact analytical solution for the 3-D acoustic field. An alternative approach is to turn to numerical propagation modeling, although this tends to be computationally intensive. One of the few numerical ocean-acoustic propagation models capable of handling full 3-D acoustic propagation is Porter's BELLHOP3D Gaussian beam-tracing code, which includes out-of-plane effects arising from horizontal refraction [84].

Approximate analytical techniques have also been applied to the penetrable wedge problem [71], based on the concept of Weston's effective depth [70]. In essence, the actual bottom is replaced with a fictitious pressure-release interface beneath and parallel to the fluid—fluid interface. The effective bottom takes account of the phase change that occurs when shallow eigen rays, with grazing angles less than the critical, undergo total internal reflection. With the effective bottom in place, an "effective ideal wedge" is constructed in which the apex is offset from, but the wedge angle is the same as that of, the penetrable wedge. The known modal solution for the ideal wedge is then adapted, by excluding rays that intersect the bottom at grazing angles greater than the critical, to give the spatial and frequency dependence of the mode coefficients in the penetrable wedge.

The modes in the effective ideal wedge are independent of range from the effective apex, but these same modes in the penetrable wedge progressively sink down into the bottom with decreasing range from the actual apex. As with the ideal wedge, the coefficients of the modes in the penetrable wedge exhibit an ensonified region bounded by a hyperbolic envelope, outside of which is a shadow zone. Within the ensonified region, the acoustic field is highly oscillatory, except where eigen rays have been removed in order to account for penetration of the fluid—fluid boundary.

ACKNOWLEDGMENT

Chapter 2 was written with the support of the Office of Naval Research, Code 322OA, under Award Number N00014-14-1-0247.

REFERENCES

[1] Wenz, G. M., *Acoustic ambient noise in the ocean: Spectra and sources*, J. Acoust. Soc. Am., **34**, 1936—1956, (1962).

[2] Au, W. W. L. and Banks, K., *The acoustics of the snapping shrimp Synalpheus parneomeris in Kaneohe Bay*, J. Acoust. Soc. Am., **103**, 41—47, (1998).

[3] Cato, D. H. and Bell, M. J., *Ultrasonic ambient noise in Australian shallow waters at frequencies up to 200 kHz*, DSTO Tech. Rep. No. MRL-TR-91-23, (1992).

[4] Lighthill, M. J. S., *Introduction to Fourier Analysis and Generalised Functions*. Camb. Univ. Press, Cambridge, (1958).

[5] Faulkner, E. A., *Introduction to the Theory of Linear Systems*. Chapman and Hall, London, (1969).

[6] Cooley, J. W. and Tukey, J. W., *An algorithm for machine calculation of complex Fourier series*, Math. Comp., **19**, 297—301, (1965).

[7] Papoulis, A., *Systems and Transforms with Applications in Optics*. McGraw-Hill, New York, (1968).

[8] Watson, G. N., *A Treatise on the Theory of Bessel Functions*. 2nd ed. Camb. Univ. Press, London, (1958).

[9] Abramowitz, M. and Stegun, I. A., *Handbook of Mathematical Functions*. Dover, New York, (1965).

[10] Lebedev, N. N., *Special Functions and their Applications*. Prentice-Hall, Englewood Cliffs, (1965).

[11] Cole, R. H., *Underwater Explosions*. Princeton Univ. Press, Princeton, NJ, (1965).

[12] Carslaw, H. S. and Jaeger, J. C., *Conduction of Heat in Solids*, 2nd ed. Clarendon Press, Oxford, (1959).

[13] Erdélyi, A., *Tables on Integral transforms, Volume 1*. McGraw-Hill, New York, (1954).

[14] Francois, R. E. and Garrison, G. R., *Sound absorption based on ocean measurements: Part I: Pure water and magnesium sulphate contributions*, J. Acoust. Soc. Am., **72**, 896−907, (1982).

[15] Francois, R. E. and Garrison, G. R., *Sound absorption based on ocean measurements: Part II: Boric acid contribution and equation for total absorption*, J. Acoust. Soc. Am., **72**, 1879−1890, (1982).

[16] Stokes, G. G., *On the theories of the internal friction of fluids in motion and of the equilibrium and motion of elastic solids*, Trans. Camb. Phil. Soc., **8**, 287−319, (1845).

[17] Stefan, J., *Über den einflub der inneren reibung in der luft auf die schallbewegung (On the effect of inner friction on the propagation of sound in air)*, Sitzungber Akad. Wien. Math.-Naturwiss. Kl. **53**, 529−527, (1866).

[18] Rayleigh, J. W. S., *The Theory of Sound, vol. II*. Dover, New York, (1945).

[19] Jordan, P. M., Meyer, M. R. and Puri, A., *Causal implications of viscous damping in compressible fluid flows*, Phys. Rev. E **62**, 7918−7926, (2000).

[20] Buckingham, M. J., *Causality, Stokes' wave equation and acoustic pulse propagation in a viscous fluid*, Phys. Rev. E **72**, 026610, (2005).

[21] Buckingham, M. J., *On the transient solutions of three acoustic wave equations: van Wijngaarden's equation, Stokes' equation and the time-dependent diffusion equation*, J. Acoust. Soc. Am., **124**, 1909−1920, (2008).

[22] Kramers, H. A., *Some remarks on the theory of absorption and refraction of X-rays*, Nature **117**, 775, (1926).

[23] Kronig, R. d. L., *On the theory of dispersion of X-rays*, J. Opt. Soc. Am., **12**, 547−557, (1926).

[24] Kramers, H. A., *La diffusion de la lumiere par les atomes (The diffusion of light by atoms)*, Atti del Congr. Internazionale dei Fisíci **2**, 545−557, (1927).

[25] Kronig, R. d. L. and Kramers, H. A., *On the theory of absorption and dispersion of x-rays*, Zeits. für Phys. **48**, 174−179, (1928).

[26] Nussenzveig, H. M., *Causality and Dispersion Relations*. Academic Press, New York, (1972).

[27] Buckingham, M. J., *Wave speed dispersion associated with an attenuation obeying a frequency power law*, J. Acoust. Soc. Am., **138**, 2871−2884, (2015).

[28] Blackstock, D. T., *Transient solution for sound radiated into a viscous fluid*, J. Acoust. Soc. Am., **41**, 1312−1319, (1967).

[29] White, J. E., *Underground Sound: Application of Seismic Waves*. Elsevier, Amsterdam, (1983).

[30] Gradshteyn, I. S. and Ryzhik, I. M., *Tables of Integrals, Series and Products*. Academic Press, New York, (1980).

[31] Richardson, M. D. and Davis, A. M., *Modeling methane-rich sediments of Eckernförde Bay*, Cont. Shelf Res. **18**, 1671−1688, (1998).

[32] Rice, O. K., *On the statistical mechanics of liquids, and the gas of hard elastic spheres*, J. Chem. Phys. **12**, 1−18, (1944).

[33] Wyllie, M. R., Gregory, A. R. and Gardner, L. W., *Elastic wave velocities in heterogeneous and porous media*, Geophysics **21**, 41−70, (1956).

[34] Buckingham, M. J., *On pore-fluid viscosity and the wave properties of saturated granular materials including marine sediments*, J. Acoust. Soc. Am., **122**, 1486−1501, (2007).

[35] Hamilton, E. L., *Acoustic Properties of Sediments*. In *Acoustics and the Ocean Bottom*, edited by Lara-Saenz, A., Cuierra, C. R. and Carbo-Fité, C., Consejo Superior de Investigacions Cientificas, Madrid, pp. 3−58, (1987).

[36] Buckingham, M. J., *Analysis of shear-wave attenuation in unconsolidated sands and glass beads*, J. Acoust. Soc. Am., **136**, 2478−2488, (2014).

[37] Buckingham, M. J., *Compressional and shear wave properties of marine sediments: Comparisons between theory and data*, J. Acoust. Soc. Am., **117**, 137−152, (2005).

[38] Biot, M. A., *General theory of three-dimensional consolidation*, J. Appl. Phys. **12**, 155−164, (1941).

[39] Biot, M. A., *Consolidation settlement under a rectangular load distribution*, J. Appl. Phys. **12**, 426−430, (1941).

[40] Biot, M. A., *Theory of propagation of elastic waves in a fluid-saturated porous solid: I. Low-frequency range*, J. Acoust. Soc. Am., **28**, 168−178, (1956).

[41] Biot, M. A., *Theory of propagation of elastic waves in a fluid-saturated porous solid: II. Higher frequency range*, J. Acoust. Soc. Am., **28**, 179−191, (1956).

[42] Hamilton, E. L., *Compressional-wave attenuation in marine sediments*, Geophysics **37**, 620−646, (1972).

[43] Hamilton, E. L., *Sound attenuation as a function of depth in the sea floor*, J. Acoust. Soc. Am., **59**, 528−535, (1976).

[44] Hamilton, E. L., *Geoacoustic modeling of the sea floor*, J. Acoust. Soc. Am., **68**, 1313−1340, (1980).

[45] Stoll, R. D., *Sediment Acoustics*. Springer-Verlag, Berlin, (1989).

[46] Stoll, R. D. and Bryan, G. M., *Wave attenuation in saturated sediments*, J. Acoust. Soc. Am., **47**, 1440−1447, (1970).

[47] Stoll, R. D., *Acoustic waves in saturated sediments*. In *Physics of Sound in Marine Sediments*, ed. Hampton, L. D., Plenum, New York, pp. 19−39, (1974).

[48] Plona, T. J., *Observation of a second bulk compressional wave in a porous medium at ultrasonic frequencies*, Appl. Phys. Lett., **36**, 259−261, (1980).

[49] Johnson, D. L. and Plona, T. J. *Acoustic slow waves and the consolidation transition*, J. Acoust. Soc. Am., **72**, 556−565, (1982).

[50] Buckingham, M. J., *Theory of acoustic attenuation, dispersion, and pulse propagation in unconsolidated granular materials including marine sediments*, J. Acoust. Soc. Am., **102**, 2579−2596, (1997).

[51] Buckingham, M. J., *Theory of compressional and shear waves in fluid-like marine sediments*, J. Acoust. Soc. Am., **103**, 288−299, (1998).

[52] Buckingham, M. J., *Wave propagation, stress relaxation, and grain-to-grain shearing in saturated, unconsolidated marine sediments*, J. Acoust. Soc. Am., **108**, 2796−2815, (2000).

[53] Buckingham, M. J., *Response to 'Comments on "Pore fluid viscosity and the wave properties of saturated granular materials including marine sediments" [J. Acoust. Soc. Am. 122, 1486−1501, (2007)]'*, J. Acoust. Soc. Am., **127**, 2099−2102, (2010).

[54] Gittus, J., *Creep, Viscoelasticity and Creep Fracture in Solids*. John Wiley, New York, (1975).

[55] Morse, P. M., and Ingard, K. U., *Theoretical Acoustics*. McGraw-Hill, New York, (1968).

[56] Morse, P. M., and Feshbach, H., *Methods of Theoretical Physics: Part 1*. McGraw-Hill, New York, (1953).

[57] Mallock, A., *The damping of sound by frothy liquids*, Proc. Roy. Soc. Lond., Series A **84**, 391−395, (1910).

[58] Wood, A. B., *A Textbook of Sound*, Third ed., G. Bell and Sons Ltd., London, (1964).

[59] Richardson, M. D., Briggs, K. B., Bibee, D. L., Jumars, P. A., Sawyer, W. B., Albert, D. B., Berger, T. K., Buckingham, M. J., Chotiros, N. P., Dahl, P. H., Dewitt, N. T., Fleischer, P., Flood, R., Greenlaw, C. F., Holliday, D. V., Hulbert, M. H., Hutnak, M. P., Jackson, P. D., Jaffe, J. S., Johnson, H. P., Lavoie, D. L., Lyons, A. P., Martens, C. S., McGehee, D. E., Moore, K. D., Orsi, T. H., Piper, J. N, Ray, R. I., Reed, A. H., Self, R. F. L., Schmidt, J. L., Schock, S. G., Simonet, F., Stoll, R. D., Tang, D. J., Thistle, D. E., Thorsos, E. I., Walter, D. J., and Wheatcroft, R. A., *Overview of SAX99: environmental considerations*, IEEE J. Ocean. Eng. **26**, 26−53, (2001).

[60] Bell, D. W., Shear Wave Propagation in Unconsolidated Fluid Saturated Porous Media, Applied Physics Laboratories Report No. ARL-TR-79-31, (1979).

[61] Brunson, B. A., *Shear Wave Attenuation in Unconsolidated Laboratory Sediments*, Ph.D. thesis, Oregon State University, (1983).

[62] Brunson, B. A., and Johnson, R. K., *Laboratory measurements of shear wave attenuation in saturated sand*, J. Acoust. Soc. Am., **68**, 1371−1375, (1980).

[63] Kimura, M., *Shear wave speed dispersion and attenuation in granular marine sediments*, J. Acoust. Soc. Am., **134**, 144−155, (2013).

[64] Timoshenko, S. P., and Goodier, J. N., *Theory of Elasticity*. 3rd ed., McGraw-Hill, New York, (1970).

[65] Doppler, J. C., *Remarks on my theory of the colored light from double stars, with regard to the objections raised by Dr. Ballot of Utrecht*, Annal. der Phys. and Chem. **68**, 1−35, (1846).

[66] Buckingham, M. J., *On the sound field from a moving source in a viscous medium*, J. Acoust. Soc. Am., **114**, 3112−3118, (2003).

[67] Buckingham, M. J., and Giddens, E. M., *Theory of sound propagation from a moving source in a three-layer Pekeris waveguide*, J. Acoust. Soc. Am., **120**, 1825−1841, (2006).

[68] Kelvin, L., *On the waves produced by a single impulse in water of any depth*, Proc. Roy. Soc. Lond. Series A **42**, 80−83, (1887).

[69] Buckingham, M. J., *Theory of acoustic radiation in corners with homogeneous and mixed perfectly reflecting boundaries*, J. Acoust. Soc. Am., **86**, 2273−2291, (1989).

[70] Weston, D. E., *A Moiré fringe analog of sound propagation in shallow water*, J. Acoust. Soc. Am., **32**, 647−654, (1960).

[71] Buckingham, M. J., *Theory of three-dimensional acoustic propagation in a wedge-like ocean with a penetrable bottom*, J. Acoust. Soc. Am., **82**, 198−210, (1987).

[72] Ewing, W. M., Jardetzky, W. S., and Press, F., *Elastic Waves in Layered Media*. McGraw-Hill, New York, (1957).

[73] Pekeris, C. L., Theory of Propagation of Explosive Sound in Shallow Water. In *The Geological Society of America, Memoir 27, Propagation of Sound in the Ocean.* The Geological Society of America, Vol. **27**, pp. 1−117, New York, (1948).

[74] Stickler, D. C., *Normal-mode program with both the discrete and branch line contributions*, J. Acoust. Soc. Am., **57**, 856−861, (1975).

[75] Buckingham, M. J., and Giddens, E. M., *On the acoustic field in a Pekeris waveguide with attenuation in the bottom half-space*, J. Acoust. Soc. Am., **119**, 123−142, (2006).

[76] Zhang, Z. Y., and Tindle, C. T., *Complex effective depth of the ocean bottom*, J. Acoust. Soc. Am., **93**, 205−213, (1993).

[77] Weston, D. E., *Horizontal refraction in a three-dimensional medium of variable stratification*, Proc. Phys. Soc. Lond., **78**, 46−52, (1961).

[78] Doolittle, R., Tolstoy, A., and Buckingham, M. J., *Experimental confirmation of horizontal refraction of cw acoustic radiation from a point source in a wedge-shaped ocean environment*, J. Acoust. Soc. Am., **83**, 2117−2125, (1988).

[79] Bradley, D. L., and Hudimac, A. A., *The Propagation of Sound in a Wedge Shaped Shallow Water Duct*, Naval Ordnance Laboratory, Technical Report No. NOLTR 70−235, (1970).

[80] Buckingham, M. J., *Acoustic Propagation in a Wedge-Shaped Ocean with Perfectly Reflecting Boundaries.* In *Hybrid Formulation of Wave Propagation and Scattering*, ed. Felsen, L. B., Martinus Nijhoff, Dordrecht, pp. 77−105, (1983),.

[81] Buckingham, M. J., *Acoustic Propagation in a Wedge-Shaped Ocean with Perfectly Reflecting Boundaries.* Naval Research Laboratory Report No. NRL Report 8793, (1984).

[82] Dixon, A. L., and Ferrar, W. L., *Integrals for the product of two Bessel functions*, Quart. J. Math. **4**, 193−208, (1933).

[83] Dixon, A. L., and Ferrar, W. L., *Integrals for the product of two Bessel functions (II)*, Quart. J. Math. **4**, 297−304, (1933).

[84] Porter, M. B., *Out-of-plane effects in three-dimensional oceans*, J. Acoust. Soc. Am., **137**, 2419, (2015).

Sound Propagation Modeling

S. Ivansson

Swedish Defence Research Agency, Stockholm, Sweden

Acoustic wave propagation is considered in a medium with water between an upper traction-free (pressure-release) surface and a bottom with sediment layers and a lower homogeneous half-space. In reality, there is an upper air half-space, but the approximation with a free upper boundary is in general adequate. The bottom may be solid, meaning that it supports shear waves. When its shear strength is low, it is preferably treated as fluid.

A left-hand Cartesian coordinate system $(x, y, z) = (x_1, x_2, x_3)$ is introduced in 3-D Euclidean space with basis vectors denoted e_x, e_y, e_z or e_1, e_2, e_3; $e_z = e_3$ points downward. The displacement vector is denoted $u = (u_1, u_2, u_3)$, and $\tau = (\tau_{ij})$ is the symmetric stress tensor. Applying Newton's second law to a small control volume, the equations of motion

$$\rho \frac{\partial^2 u_i}{\partial t^2} = f_i + \sum_j \frac{\partial \tau_{ij}}{\partial x_j} \tag{3.1}$$

appear [1]. The density is here denoted ρ and $f = (f_1, f_2, f_3)$ is the body force per unit volume. All quantities depend on position $x = (x_1, x_2, x_3)$, and u, f, τ depend on time t. Combining Eq. (3.1) with the constitutive relations

$$\tau_{ij} = \lambda \ \text{div} \ u \ \delta_{ij} + \mu \left(\frac{\partial u_i}{\partial x_j} + \frac{\partial u_j}{\partial x_i} \right) \tag{3.2}$$

for isotropic materials, where δ_{ij} is the Kronecker delta while λ and μ are the position-dependent Lamé parameters {unit: N/m^2}, the equations of motion take the form

$$\rho \frac{\partial^2 u_i}{\partial t^2} = f_i + \frac{\partial(\lambda \ \text{div} \ u)}{\partial x_i} + \sum_j \frac{\partial[\mu(\partial u_i/\partial x_j + \partial u_j/\partial x_i)]}{\partial x_j}. \tag{3.3}$$

The divergence (div) operator, as well as gradient (grad) and Laplace (Δ) operators, which follow, act on the spatial variables. Across a medium discontinuity interface, the traction vector $(\tau_{11}n_1 + \tau_{12}n_2 + \tau_{13}n_3, \ \tau_{21}n_1 + \tau_{22}n_2 + \tau_{23}n_3, \ \tau_{31}n_1 + \tau_{32}n_2 + \tau_{33}n_3)$, where $n = (n_1, n_2, n_3)$ is the unit normal vector, is continuous. Across a solid–solid interface, the displacement vector u is also continuous.

Applied Underwater Acoustics. http://dx.doi.org/10.1016/B978-0-12-811240-3.00003-5

Across a fluid–solid or fluid–fluid interface, however, slip is allowed such that only the normal component of $\boldsymbol{u}$ is continuous.

In a fluid region, μ vanishes identically and $\tau_{ij} = -p\,\delta_{ij}$, where

$$p = -\lambda \text{ div } \boldsymbol{u} \tag{3.4}$$

is the acoustic pressure. Eq. (3.3) then readily provides the equation

$$\frac{1}{\lambda}\frac{\partial^2 p}{\partial t^2} = \text{div}\left(\frac{\text{grad } p}{\rho}\right) - \text{div}\left(\frac{\boldsymbol{f}}{\rho}\right) \tag{3.5}$$

for the pressure, resulting in the wave equation

$$\frac{1}{c^2}\frac{\partial^2 p}{\partial t^2} = \Delta p + \rho \text{ grad}\left(\frac{1}{\rho}\right)\cdot\text{grad}(p) - \left[\text{div}(\boldsymbol{f}) + \rho \text{ grad}\left(\frac{1}{\rho}\right)\cdot\boldsymbol{f}\right] \tag{3.6}$$

with sound speed $c = (\lambda/\rho)^{1/2}$. Eq. (3.3) also shows that the displacements in a fluid region can be obtained from the pressure according to

$$\rho\frac{\partial^2 \boldsymbol{u}}{\partial t^2} = \boldsymbol{f} - \text{grad}(p). \tag{3.7}$$

It follows that p and $\rho^{-1}\,\partial p/\partial n$, where $\partial/\partial n$ denotes the normal derivative, are continuous at a fluid–fluid interface across which the medium parameters are discontinuous.

For a harmonic time dependence according to the factor $\exp(-i\omega t)$, where $\omega > 0$ is the angular frequency, a time derivative is simply replaced by multiplication with $-i\omega$. Eq. (3.6) takes the shape of the Helmholtz equation

$$\Delta p + \rho \text{ grad}\left(\frac{1}{\rho}\right)\cdot\text{grad}(p) + \left(\frac{\omega^2}{c^2}\right)p = \text{div}(\boldsymbol{f}) + \rho \text{ grad}\left(\frac{1}{\rho}\right)\cdot\boldsymbol{f}. \tag{3.8}$$

For brevity, the factor $\exp(-i\omega t)$ is typically omitted in the equations. For the symmetric point source at $\boldsymbol{x}_s$ given by $\boldsymbol{f}(\boldsymbol{x}) = -M \text{ grad}(\delta(\boldsymbol{x} - \boldsymbol{x}_s))$, where M {unit: Nm} is the complex moment-tensor amplitude [1] and $\delta(\boldsymbol{x})$ is the Dirac delta function, Eq. (3.8) can be written

$$\Delta\tilde{p} + \rho \text{ grad}\left(\frac{1}{\rho}\right)\cdot\text{grad}(\tilde{p}) + \left(\frac{\omega^2}{c^2}\right)\tilde{p} = \left(\frac{\omega^2}{c^2}\right)M \,\delta_s, \tag{3.9}$$

where $\delta_s(\boldsymbol{x}) = \delta(\boldsymbol{x} - \boldsymbol{x}_s)$, for $\tilde{p}(\boldsymbol{x}) = p(\boldsymbol{x}) + M \,\delta(\boldsymbol{x} - \boldsymbol{x}_s)$ which differs from $p(\boldsymbol{x})$ only at the source. The source term in Eq. (3.9) can be related physically to a point mass source, see Section 4-3 in Ref. [2], at which mass is added and subtracted according to

$$\lim_{V\to\{\boldsymbol{x}_s\}} \rho(\boldsymbol{x}_s)\iint_S \boldsymbol{u}(\boldsymbol{x})\cdot\boldsymbol{n}(\boldsymbol{x}) \,dS(\boldsymbol{x}) = M/c^2(\boldsymbol{x}_s), \tag{3.10}$$

where V is a small volume with surface S, surrounding $\boldsymbol{x}_s$, and $\boldsymbol{n}(\boldsymbol{x})$ is its outward unit normal at $\boldsymbol{x}$. In the particular case when $\boldsymbol{x}_s = (0, 0, z_s)$ and there is symmetry around the z axis, Eq. (3.10) takes the form

$$\lim_{r\to 0} \rho(\boldsymbol{x}_s)2\pi \,r \,u_r(r, z) = M \,\delta(z - z_s)/c^2(\boldsymbol{x}_s), \tag{3.11}$$

where $u_r(r, z)$ is the displacement in the radial horizontal direction at depth z and distance r from the z axis. Eqs. (3.10) and (3.11) can be useful for representing a source term by a boundary condition [3]. In a homogeneous medium, Eq. (3.9) is solved by [2]

$$\tilde{p}(x) = -\left(\frac{\omega^2}{c^2}\right) M \frac{\exp(i(\omega/c)|x - x_s|)}{4\pi|x - x_s|}. \tag{3.12}$$

To model viscoelastic absorption losses, the constitutive relations (3.2) can be amended with time-integral terms involving relaxation functions and the strain history [4]. In the time-harmonic case, it follows that Eq. (3.2) can be used as it is, but with complex Lamé parameters λ and μ, resulting in complex wave speeds. With $\omega > 0$, as assumed here, the bulk modulus $\lambda + 2\mu/3$ and the shear modulus μ must both appear in the fourth quadrant of the complex plane [4]. Relaxation models involving fractional derivatives are useful to account for the frequency dependence of the losses in solids [5].

The physical concepts of energy flux and reciprocity are now briefly introduced in the time-harmonic case for points in a fluid region. Treatments that are more general appear in Refs. [1,4], for example. The time averaged energy flux or intensity vector $\boldsymbol{\Phi}(x)$ at x is [2]:

$$\boldsymbol{\Phi}(x) = \frac{\omega}{2} \text{Im}(p^* u), \tag{3.13}$$

where the asterisk denotes the complex conjugate. For a fluid region without absorption, the surface integral of the normal component of $\boldsymbol{\Phi}(x)$, through a closed surface S that does not contain a source, vanishes due to energy conservation. Together with a Green's identity, Eqs. (3.7) and (3.9) provide a formal verification. Range-averaged properties of acoustic wave propagation can be studied by energy-flux methods (e.g., Ref. [6]).

For the physically relevant boundary conditions, including radiation conditions specifying outgoing waves at infinity (e.g., Refs. [4,7]), the reciprocity principle states that

$$\lambda(y)p(x|y) = \lambda(x)p(y|x), \tag{3.14}$$

where $p(x|y)$ is the pressure at x for a symmetric point source at y and $p(y|x)$ is defined analogously, assuming a common complex moment-tensor amplitude M. This follows from Betti's theorem [1] or, in the fluid case, from Eq. (3.9) and a Green's identity [7].

The following presentation of some common sound propagation modeling methods focuses on frequency-domain methods for the time-harmonic case. However, specific time-domain methods are occasionally mentioned. In general, the time signal resulting from a transient source pulse can be computed from frequency-domain results by Fourier synthesis. An integration over angular frequency ω is performed, including the spectrum of the source pulse, the pressure field $p(x, \omega)$ as computed for many frequencies, and the time-harmonic factors $\exp(-i\omega t)$ with

$\omega > 0$. However, applying a discrete fast Fourier transform necessitates some care. For example, aliasing effects may cause late arrivals to be mixed into the selected computational time window. By utilizing analyticity in ω aliasing can be mitigated by performing the computations for complex ω with a small imaginary part [8]. The fluid–solid case, where (part of) the bottom is treated as solid, is discussed for a few methods while the presentation is restricted to the purely fluid case for others. Some recurring abbreviations and symbols are listed at the end of the chapter.

3.1 RAY MODELS

Approximate solutions to the equations of motion (3.3) can be derived by the ray method, described in most textbooks on acoustic wave propagation, including [9–12]. For transients, the trial solution, involving a travel time function $T(x)$ for a wave front, is

$$u(x,t) = \psi(t - T(x)) \, U(x), \tag{3.15}$$

where the dimensionless pulse function $\psi(t)$ has a sharp onset with a possibly discontinuous second derivative at $t = 0$, for the wave front arrival, and the vector function $U(x)$ is slowly varying. For time-harmonic waves, omitting $\exp(-i\omega t)$, the solution form

$$u(x) = e^{i\omega T(x)} \, U(x) \tag{3.16}$$

is tried, assuming that the angular frequency $\omega > 0$ is large.

In either case [4], the travel time function $T(x)$ satisfies the eikonal equation

$$|\text{grad}(T(x))|^2 = \frac{1}{\alpha^2(x)} \quad \text{or} \quad \frac{1}{\beta^2(x)} \tag{3.17}$$

for $\alpha(x) = [(\lambda(x) + 2\mu(x))/\rho(x)]^{1/2}$ and $\beta(x) = [\mu(x)/\rho(x)]^{1/2}$. The first alternative corresponds to longitudinally polarized compressional waves (P), $U(x) \sim \text{grad}(T(x))$, with velocity $\alpha(x)$, while the second alternative corresponds to transversely polarized shear waves (S), $U(x) \cdot \text{grad}(T(x)) = 0$, with velocity $\beta(x)$. Rays are introduced as curves with local direction $\text{grad}(T(x))$, normal to the wave front. The differential equation of a ray $x(s)$ becomes

$$\frac{d}{ds}\left(\frac{1}{\alpha(x)}\frac{dx}{ds}\right) = \text{grad}\left(\frac{1}{\alpha(x)}\right) \quad \text{and} \quad \frac{d}{ds}\left(\frac{1}{\beta(x)}\frac{dx}{ds}\right) = \text{grad}\left(\frac{1}{\beta(x)}\right) \tag{3.18}$$

in the P- and S-case, respectively, where s is the arc length along the ray. In either case, energy-flux expressions such as Eq. (3.13), in conjunction with the trial solution forms, show that the energy propagates in ray tubes formed by adjacent rays [1,4]. The energy-flux expressions also show that, apart from viscoelastic absorption losses, the displacement amplitude along a ray varies as $(\rho \, \alpha \, dS)^{-1/2}$ and $(\rho \, \beta \, dS)^{-1/2}$ in the P- and S-case, respectively, where dS is local ray-tube cross-section area.

When a wave front reaches an interface with a discontinuity in material properties, P- as well as S-type reflected and refracted wave fronts are created in such a way that the continuous displacement and traction boundary conditions are fulfilled. In particular, Snell's law follows, expressing equality of phase velocities along the interface.

For underwater acoustic applications, the bottom depth should preferably be at least some 10 or 20 wavelengths, since the ray method is a high-frequency approximation. The last example presented in Section 3.5.4 seems to confirm applicability at such conditions, and ray results can often be surprisingly good even at still lower frequencies.

Cerveny [13], with a more comprehensive treatment in Ref. [14], gives a useful account of various ways to trace rays numerically. Of course, standard methods for ordinary differential equations can be applied to the ray equation. In Section 3.1.1, which follows, computational details are given for a particular type of fast analytic ray tracing.

3.1.1 A PARTICULAR TYPE OF ANALYTIC 2-D RAY TRACING

It is now assumed that a point source is located on the z-axis in the water column and that the medium parameters and the bottom depths are rotationally invariant around this axis. All ray tracing may then be done in the xz plane, where the sound speed c in the water is represented by range-invariant profiles within horizontal x segments. In each segment, the variation of $1/c^2$ with depth z is assumed to be piecewise linear. Hence, each ray appears as a sequence of parabolic arcs. Bottom depths are given explicitly at the grid points for the segments, and linear interpolation is used in between. It follows that the intersections of a ray with the bottom can be calculated by solving second-degree algebraic equations.

Complex plane-wave reflection coefficients are computed for the interaction with the bottom. These reflection coefficients are computed for a bottom structure with plane fluid or solid layers locally following the bottom slope at the particular reflection point. Computational methods described in Section 3.2.1 can be used for this purpose. Hence, the reflection coefficients become functions of frequency as well as incidence angle. No ray tracing is performed through the bottom.

The direction vector of a ray is written e_r. In the introduced x, y, z coordinate system,

$$e_r = (\cos\ \varphi, 0, -\sin\ \varphi). \tag{3.19}$$

In particular, φ is the inclination angle, positive upward. The initial values for $x = (x, 0, z)$, and φ at the source point are denoted $x_s = (0, 0, z_s)$ and φ_s, respectively. With $|\varphi_s| < \pi/2$, ray tracing proceeds for $x > 0$. At reflection from a sloping bottom, $\cos\varphi$ may change sign from positive to negative, or vice versa.

Kinematic ray tracing, to determine ray trajectories, is considered in Section 3.1.1.1. The dynamic part, for determining ray-tube cross-section areas, follows in Section 3.1.1.2.

3.1.1.1 Kinematic Ray Tracing

A 2-D rectangle is considered, formed by a horizontal x segment in which the sound speed c depends only on the depth z and a depth interval with a constant gradient for $1/c^2(z)$. Denoting horizontal x change by s, the equations of a ray segment starting at $x_0 = (x_0, 0, z_0)$ with $x_0 \geq 0$ and direction angle φ_0, according to Eq. (3.19), become

$$x(s) = x_0 + s \quad , \quad y(s) = 0 \quad , \quad z(s) = z_0 - s \tan \varphi_0 + \gamma s^2. \tag{3.20}$$

A verification is conveniently obtained by using Snell's law to show that $(dz/dx)^2$ must be a linear function of z. The parameter γ can be expressed as

$$\gamma = \frac{1}{4S^2} \frac{d(1/c^2)}{dz}, \tag{3.21}$$

where $S = (\cos \varphi)/c$, the Snell parameter of the ray, is constant along the ray segment. The variation of the inclination angle φ is given by

$$\tan \varphi(s) = \tan \varphi_0 - 2\gamma s. \tag{3.22}$$

The travel time increment $t(s)$ from $x_0 = (x_0, z_0)$, in the ray segment, becomes

$$t(s) = (s/S) \left[c^{-2}(z_0) - 2S^2 \gamma s (\tan \varphi_0 - 2\gamma s/3) \right]. \tag{3.23}$$

A ray segment proceeds until the next interaction with the sea surface, the sea bottom, a horizontal interface where the gradient of $1/c^2(z)$ changes, or a vertical boundary for the x bottom data grid. Within each bottom grid segment, the bottom depth z is a linear function of x, and the ray Eq. (3.20) provides a second-degree equation in s for the bottom intersection.

Upon reflection or refraction at an interface, the direction vector e_r of the ray changes according to Snell's law to

$$\bar{e}_r = C_q \, e_r + \Gamma e_n = C_q [e_r - (e_r \cdot e_n) e_n] + H e_n, \tag{3.24}$$

where e_n is a unit normal of the interface at the interaction point, C_q is the quotient of the pertinent sound velocities after and before the interaction, and overlining is used to denote the situation directly after the interaction. The constants H and $\Gamma = H - C_q (e_r \cdot e_n)$ are obtained from $|\bar{e}_r| = 1$ and the condition that H and $e_r \cdot e_n$ have the same (opposite) sign upon refraction (reflection). It follows readily that

$$H = \pm \left[1 - C_q^2 \left(1 - (e_r \cdot e_n)^2 \right) \right]^{1/2}. \tag{3.25}$$

In the reflection case, $C_q = 1$, $H = -e_r \cdot e_n$, and Eq. (3.24) simplifies to $\bar{e}_r = e_r - 2(e_r \cdot e_n) e_n$. The postinteraction inclination angle $\bar{\varphi}$ is obtained from $\bar{e}_r = (\cos \bar{\varphi}, 0, -\sin \bar{\varphi})$, the "overline version" of Eq. (3.19).

3.1.1.2 Dynamic Ray Tracing

Along a given ray, a ray-centered 2-D orthonormal coordinate system with coordinates r, u and unit vectors e_r, e_u is introduced. These unit vectors are defined by Eq. (3.19) and

$$\boldsymbol{e}_u = (-\sin \, \varphi, 0, -\cos \, \varphi). \qquad (3.26)$$

The source at $\boldsymbol{x}_s = (0, 0, z_s)$ is assumed to be a three-dimensional point source. Hence, geometrical spreading takes place in the azimuthal as well as vertical directions, and the basic problem is to compute the variation of the quantity

$$A = x\frac{du}{d\varphi_s} \qquad (3.27)$$

along the ray. The geometrical spreading is obtained as a quotient of ray-tube areas by the absolute value of $A/\cos \varphi_s$. Points where A vanishes are called caustics.

The ray is built up by a sequence of parabolic arcs according to Eq. (3.20). For a fixed $\boldsymbol{x}_s = (0, 0, z_s)$, the quantities x_0, z_0, and φ_0 can be viewed as functions of φ_s. Considering a certain point $(x, 0, z)$ on the ray, along with its ray-centered r, u coordinate system, a crucial point is then to determine how the du for the displaced points $(x + dx, 0, z + dz)$ are changed when x_0, z_0, and φ_0 are changed. Differentiation of Eq. (3.20) gives

$$du = -\sin \varphi \, dx_0 - \cos \varphi \left[1 - (2\gamma s \cos \varphi_0)^2\right] dz_0 + \cos \, \varphi(1$$
$$+ \tan \, \varphi_0 \, \tan \, \varphi) \, s \, d\varphi_0. \qquad (3.28)$$

The ray direction vector $\boldsymbol{e}_r$ is changed to $\boldsymbol{e}_r + d\boldsymbol{e}_r$ at $(x + dx, 0, z + dz)$ with

$$d\boldsymbol{e}_r = d\varphi \, \boldsymbol{e}_u, \qquad (3.29)$$

and the corresponding change of the inclination angle φ is obtained from

$$\left(1 + \tan^2 \varphi\right)d\varphi = [1 + \tan \varphi_0(2 \tan \varphi - \tan \varphi_0)]d\varphi_0 - 2\gamma(dx - dx_0)$$
$$+ 8\gamma^2 s \, \cos^2\varphi_0 \, dz_0, \qquad (3.30)$$

as follows by differentiation of Eq. (3.22).

In order to follow the ray segments sequentially, the displacement $(dx, 0, dz)$ of a point $(x, 0, z)$ where the ray is reflected or refracted must be determined. Apparently,

$$(dx, 0, dz) = \kappa \, \boldsymbol{e}_r + du \, \boldsymbol{e}_u, \qquad (3.31)$$

where κ is determined from the condition that $(dx, 0, dz)$ is orthogonal to $\boldsymbol{e}_n$, the unit normal of the interface at $(x, 0, z)$. Writing $\boldsymbol{e}_n = (x_n, 0, z_n)$, Eqs. (3.19) and (3.26) show that

$$(dx, 0, dz) = \frac{du}{x_n \cos \varphi - z_n \sin \varphi}(z_n, 0, -x_n). \qquad (3.32)$$

The differential change $d\overline{\varphi}$ of the postinteraction inclination angle $\overline{\varphi}$ is also needed. By the relation $d\overline{\boldsymbol{e}}_r = d\overline{\varphi} \, \overline{\boldsymbol{e}}_u$, the post-interaction version of Eq. (3.29),

$$d\overline{\varphi} = d\overline{\boldsymbol{e}}_r \cdot \overline{\boldsymbol{e}}_u. \qquad (3.33)$$

Differentiation of Eq. (3.24) gives

$$d\overline{\boldsymbol{e}}_r = C_q \, d\boldsymbol{e}_r + \Gamma \, d\boldsymbol{e}_n + dC_q \, \boldsymbol{e}_r + d\Gamma \, \boldsymbol{e}_n. \qquad (3.34)$$

Here, $de_r = d\varphi\, e_u$ by Eq. (3.29) and de_n vanishes for a linear interface. Since the sound velocities in each grid segment depend on depth z only, $dC_q = (dC_q/dz)\, dz$. Finally, $d\Gamma$ together with dH from Eq. (3.25) additionally involve $d(e_r \cdot e_n) = de_r \cdot e_n + e_r \cdot de_n$. In the reflection case, Eq. (3.34) with $de_n = 0$ simplifies to $d\bar{e}_r = de_r - 2(de_r \cdot e_n)e_n$.

The resulting expression for $d\bar{\varphi}$ shows how a ray tube from the source is changed by interface interaction. Amplitude changes by geometrical spreading can thus be computed.

3.1.1.3 Caustics

The number of caustics along a ray is obtained by summing the contributions from its ray segments, each segment given by equations of type (3.20). Inserting Eq. (3.28) into Eq. (3.27) provides, after some algebra, an expression for $A = A(s)$ as

$$\frac{A(s)}{(x_0 + s)\cos\varphi(s)} = [2\gamma s - \tan\ \varphi_0]\frac{dx_0}{d\varphi_s} - \left[1 - (2\gamma s \cos\ \varphi_0)^2\right]\frac{dz_0}{d\varphi_s}$$
$$+ \left[(1 + \tan^2\varphi_0) - 2\gamma s\ \tan\ \varphi_0\right]s\frac{d\varphi_0}{d\varphi_s}. \tag{3.35}$$

From Eq. (3.22), $\cos\varphi(s) = \pm[1 + (\tan\varphi_0 - 2\gamma s)^2]^{-1/2} \neq 0$. The number of zeroes of $A(s)$ along the ray segment thus agrees with the number of zeroes of the function $[(x_0 + s)\cos\varphi(s)]^{-1}A(s)$, which is a simple second-degree polynomial in s according to Eq. (3.35).

Adding the contributions from each ray segment in turn, the number of zeroes of A and thus the number of caustics along the ray is safely determined. Numerical sign checks for $A(s)$, with small s steps, are not needed. If desired, the caustic locations can of course also be determined, by solving the appearing second-degree equations in s.

3.1.1.4 Coherent Computation of Propagation Loss and Propagation Time Series

The methods described here do not need explicit eigenray computations, which is useful since such computations can be time-consuming. Sorting of rays into families according to number of bottom bounces, etc., compare to (cf.) [15] where such sorting is used, is not needed either.

Consider a receiver point $x_r = (x_r, 0, z_r)$, with $x_r > 0$, at which the response is desired. Rays with inclination-angle separation $\Delta\varphi_s$, differential correspondence $d\varphi_s$, are traced from the source. Each ray is considered to occupy a corresponding ray-centered initial ray-tube area of lobe width $\Delta\varphi_s$ on the unit sphere centered at the source. As the ray proceeds, its energy is assumed to be concentrated uniformly within the cross section of its evolving tube, which determines the change of intensity due to geometrical spreading.

The response at x_r is built up by summing all ray contributions as the ray tracing proceeds. Disregarding some complications for x_r close to the surface or the bottom, a particular ray contributes precisely when x_r is included within its evolving ray tube. An approximation with a locally plane wave front is employed, as supported by Eq. (4.43) in Ref. [1], for example. For a segment of the ray according to Eq. (3.20) and with r, u coordinate frames according to Eqs. (3.19) and (3.26), the conditions become

$$e_r(s) \cdot [(x(s), 0, z(s)) - (x_r, 0, z_r)] = 0 \quad \text{and} \quad |\eta| < \Delta\varphi_s/2, \tag{3.36}$$

where, assuming the stated orthogonality relation, η is an expansion coefficient defined by

$$(x(s), 0, z(s)) - (x_r, 0, z_r) = \eta \ (du/d\varphi_s)e_u(s). \tag{3.37}$$

However, the orthogonality condition of (3.36) leads to a third-degree equation in s. A reasonable approximation is obtained by using a linear approximation of the ray, with direction vector $e_r(x_r - x_0)$, in the vicinity of $(x_r, 0, z(x_r - x_0))$. For high accuracy, needed at high frequencies, Newton's method could be applied subsequently to refine the solution.

At large distances, gaps and overlap may appear between the ray tubes for adjacent rays, as defined by the second condition of (3.36). A way to mitigate such effects is to apply a weaker criterion for contributing rays, for example $|\eta| < F \Delta\varphi_s/2$ for some number $F > 1$, and reduce each contribution, in this case by multiplication with $1/F$.

As the tracing of a particular ray proceeds, all complex reflection and transmission coefficients are multiplied for the desired frequencies to form the frequency-dependent function $\Phi(\omega)$. Additionally (e.g., Section 9-4 in Ref. [2]), a phase shift $-\pi/2$ is incorporated in $\Phi(\omega)$ for each passed caustic. Together with the ray travel time, geometrical spreading, and absorption in the water, these data determine the complex amplitude of the ray contribution to the response at x_r. Summing all ray contributions, the pressure $p(x_r, \omega)$ at x_r for the harmonic time dependence $\exp(-i\omega t)$ with $\omega > 0$, corresponding to a source such that $|x - x_s| p(x, \omega)$ tends to $p_{\text{ref}} r_{\text{ref}}$ as x tends to x_s, cf. Eq. (3.12), fulfills

$$\frac{p(x_r, \omega)}{[\rho(x_r)c(x_r)]^{1/2}} = \sum_m \frac{p_{\text{ref}} \ r_{\text{ref}}}{R_m(x_r|x_s)} \frac{\Phi_m(\omega) \ \exp(i\omega T_m)}{[\rho(x_s)c(x_s)]^{1/2}}. \tag{3.38}$$

Here, p_{ref} is a reference pressure, r_{ref} is a reference range, and the sum on m is for the rays with x_r within their ray tubes. The corresponding travel times and complex coefficient functions are denoted T_m and $\Phi_m(\omega)$, respectively. For a ray m starting in a certain direction from x_s, the corresponding ray-tube area at a point y along the ray is proportional to $[R_m(y|x_s)]^2$, where the positive function $R_m(y|x_s)$, determined by dynamic ray-tracing as described in Section 3.1.1.2, is defined such that $|y - x_s|^{-1} R_m(y|x_s)$ tends to 1 as y tends to x_s initially. The density and the sound speed at position $x = (x, 0, z)$ are denoted $\rho(x)$ and $c(x)$, respectively.

The reciprocity principle from Eq. (3.14) demands, noting Eq. (3.12), that $c(y) R(x|y) = c(x) R(y|x)$ for corresponding rays, as can also be verified directly [16].

Moving from $\exp(-i\omega t)$ to a narrow-band source pulse $\psi(t)$, centered at the positive angular frequency ω_c, it follows by Fourier transformation of Eq. (3.38) with the approximation $\Phi_m(\omega) = \Phi_m(\omega_c) = |\Phi_m(\omega_c)| \, \exp[i\phi_m(\omega_c)]$ for relevant positive ω, that the pressure $p(x_r, t)$ at point x_r and time t can be approximated as

$$\frac{p(x_r, t)}{[\rho(x_r)c(x_r)]^{1/2}} = \sum_m \frac{p_{\text{ref}} \, r_{\text{ref}}}{R_m(x_r|x_s)} \frac{|\Phi_m(\omega_c)|}{[\rho(x_s)c(x_s)]^{1/2}} \left[\cos \phi_m(\omega_c) \, \psi(t - T_m) \right.$$

$$\left. - \sin \phi_m(\omega_c) \, \widetilde{\psi}(t - T_m) \right], \tag{3.39}$$

where $\widetilde{\psi}$ denotes the Hilbert transform of ψ, defined as a principal-value convolution with the function $-1/\pi t$. As x tends to x_s, $|x - x_s| \, p(x, t)$ tends to $p_{\text{ref}} \, r_{\text{ref}} \, \psi(t)$.

Nonvanishing phase shifts $\phi_m(\omega_c)$ may appear because of complex reflection coefficients (in connection with total reflection at an incidence angle larger than the critical one) or passage of caustics. At first sight, the Hilbert-transform contribution in Eq. (3.39) seems to violate causality. A precursor can be understood physically, however, to arise from diffracted arrivals traveling through higher-velocity portions of the medium [17]. Formally, a phase shift $\pm\pi/2$ at passage of an internal caustic should appear since the ray-tube area $A = A(s)$ from Eq. (3.35) changes sign. For a positive pulse, the choice $-\pi/2$ gives a precursor that is also positive, but a negative tail appears.

3.1.2 EXAMPLE

Fig. 3.1 shows the sound-speed profile and some rays for a shallow-water example. The bottom depth is constant (100 m) and there is a source at depth 10 m. Some 40 rays, with start direction within a horizontally centered lobe of width about 7 degrees, are shown by the solid-line curves in the right panel. Steep rays do not contribute very much to the pressure field at long ranges, because they lose much energy by many bottom interactions.

For a range-invariant medium, as this one, each ray is periodic in range. A ray may start upward or downward from the source, and it may hit a receiver from above or from below. This gives four possible up/down patterns, and the cycle number ν indicates how many range periods that have been completed after the receiver was hit the first time with a given up/down pattern. For a receiver depth at 90 m in the present case, rays with $\nu = 0$, 1, and 2 reach maximum ranges of about 1.3, 3.0, and 4.8 km, respectively, as may be realized by considering the dashed-line ray in Fig. 3.1.

Fig. 3.2 shows some corresponding propagation loss (PL) curves, computed coherently as described in Section 3.1.1.4. Bottom reflection coefficients are computed assuming two sediment layers, each 10 m thick. The upper one is fluid with sound speed 1500 m/s, density 1500 kg/m^3, and absorption 1.0 dB/wavelength. The lower one is solid with compressional- and shear-wave speed 1600 and 500 m/s, respectively, density 1900 kg/m^3, and compressional- and shear-wave absorption 0.8 and 2.5 dB/wavelength, respectively. Below the sediment layers, there is a

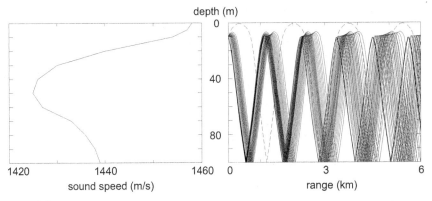

FIGURE 3.1

Left: a shallow-water sound-speed profile. Right: some corresponding rays starting close to the horizontal direction from a source at depth 10 m.

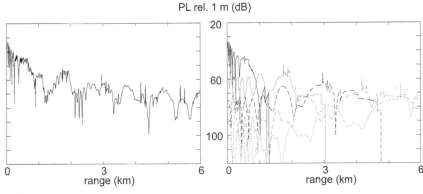

FIGURE 3.2

Coherently computed PL in dB, rel. 1 m, for the 500 Hz example with source and receiver depths 10 and 90 m, respectively. Left: total wave field. Right: partial wave fields for rays with cycle numbers $\nu = 0$ (*solid black*), 1 (*solid gray*), 2 (*dashed black*), 3 (*dashed gray*), and 4 (*dotted black*).

homogeneous solid half-space with density 2400 kg/m^3. Its wave speed and absorption data are 3000 and 1500 m/s, and 0.1 and 0.2 dB/wavelength, for compressional and shear waves, respectively.

The pressure peaks at ranges about 1.8, 3.0, 4.2, and 5.5 km can be tied to rays with cycle numbers $\nu = 1$, 2, 3, and 4, respectively. In general, ray cycling gives rise to zones with increased pressure levels. Such zones, called convergence zones, are particularly important in connection with long-range sound propagation in the ocean (e.g., Ref. [7]).

Fig. 3.2 also indicates some deficiencies with the ray modeling. The ray cycle fields for $v = 0, 1,$ and 2 in the right panel have rather abrupt cutoff ranges, before which too high levels appear. In reality, there is a smoother decay, and sound propagates by diffraction beyond the corresponding geometrical shadow-zone boundaries. Diffraction corrections to the ray modeling have been proposed (e.g., Ref. [7] with additional references given therein). More smooth pressure fields can also be obtained by Gaussian beams; however, the treatment of boundary reflections, for example, causes problems (e.g., Ref. [18]).

The present example is revisited in Sections 3.2.3 and 3.3.5.1, using wave number integration and normal-mode methods, respectively.

3.2 WAVE NUMBER INTEGRATION OR SPECTRAL METHODS

The solution of the basic equations of motion (3.3) is greatly simplified when the medium parameters are independent of the horizontal spatial coordinates x and y. Such a laterally homogeneous, or range-invariant, medium is now assumed. The symmetric point source with $f(x, t) = -M \, \mathrm{grad}(\delta(x - x_s)) \exp(-i\omega t)$ and $\omega > 0$ is stipulated, with $x_s = (0, 0, z_s)$ in the water. Introducing a cylindrical coordinate system (r, ϕ, z) with $r = (x^2 + y^2)^{1/2}$ (for radius) and ϕ (for azimuth) in the horizontal plane, and corresponding basis vectors e_r, e_ϕ, e_z, the wave field is independent of ϕ. Omitting the time-harmonic factor $\exp(-i\omega t)$, as usual, the displacement vector u and the traction vector at horizontal planes T are written

$$u = u(r, z) = u(r, z)e_r + w(r, z)e_z$$
$$T = T(r, z) = \tau^{rz}(r, z)e_r + \tau^{zz}(r, z)e_z, \tag{3.40}$$

respectively. Hankel transforms with respect to r, with horizontal wave number k as transform variable, are now introduced by

$$r_1(k, z) = \int_0^\infty u(r, z) \, J_1(kr)r \, dr$$

$$r_2(k, z) = \int_0^\infty w(r, z) \, J_0(kr)r \, dr$$

$$r_3(k, z) = \int_0^\infty \tau^{rz}(r, z) \, J_1(kr)r \, dr \tag{3.41}$$

$$r_4(k, z) = \int_0^\infty \tau^{zz}(r, z) \, J_0(kr)r \, dr.$$

Written in cylindrical coordinates, the Hankel transformed equations of motion (3.3) result in the four-dimensional first-order ordinary differential equation (ODE) system [1]

$$\mu(z)\frac{\partial r_1(k,z)}{\partial z} = \mu(z)\ k\ r_2(k,z) + r_3(k,z)$$

$$\frac{\partial r_2(k,z)}{\partial z} = -\frac{\lambda(z)}{\lambda(z)+2\mu(z)}k\ r_1(k,z) + \frac{1}{\lambda(z)+2\mu(z)}r_4(k,z)$$

$$\frac{\partial r_3(k,z)}{\partial z} = \left(\frac{4\mu(z)[\lambda(z)+\mu(z)]}{\lambda(z)+2\mu(z)}k^2 - \rho(z)\omega^2\right)r_1(k,z)$$

$$+\frac{\lambda(z)}{\lambda(z)+2\mu(z)}k\ r_4(k,z) - k\ M\ \delta(z-z_s)/2\pi$$

$$\frac{\partial r_4(k,z)}{\partial z} = -\rho(z)\omega^2\ r_2(k,z) - k\ r_3(k,z) + M\ \delta'(z-z_s)/2\pi.$$

$$(3.42)$$

The column vector $r(k, z) = (r_1(k, z), r_2(k, z), r_3(k, z), r_4(k, z))^{\mathrm{T}}$, where ^{T} is the transpose operator, is called the displacement-stress vector {unit: $m^2(m, m, N/m^2, N/m^2)^{\mathrm{T}}$}.

In a fluid region, where $\mu(z)$ and $\tau^{rz}(r, z)$ along with $r_3(k, z)$ vanish identically,

$$r_1(k,z) = \frac{k}{\rho(z)\omega^2}(r_4(k,z) - M\ \delta(z-z_s)/2\pi)$$

$$(3.43)$$

and the remaining Eq. (3.42) reduce to the two-dimensional first-order ODE system

$$\frac{\partial r_2(k,z)}{\partial z} = \left(\frac{1}{\lambda(z)} - \frac{k^2}{\rho(z)\omega^2}\right)(r_4(k,z) - M\ \delta(z-z_s)/2\pi) + \frac{M\ \delta(z-z_s)}{2\pi\lambda(z)}$$

$$\frac{\partial(r_4(k,z) - M\delta(z-z_s)/2\pi)}{\partial z} = -\rho(z)\omega^2\ r_2(k,z).$$

$$(3.44)$$

At the source depth in the water column, $r_4(k, z) - M\ \delta(z - z_s)/2\pi$ is thus continuous while $r_2(k, z)$ has a step discontinuity according to

$$r_2(k,z_s+) - r_2(k,z_s-) = \frac{M}{2\pi\lambda(z_s)}.$$

$$(3.45)$$

Elimination of r_2 shows that $Z(k, z) = r_4(k, z) - M\ \delta(z - z_s)/2\pi$ {unit: $m^2 \cdot N/m^2 = N$} fulfills the depth-separated Helmholtz equation

$$\rho(z)\frac{\partial}{\partial z}\left(\frac{1}{\rho(z)}\frac{\partial Z}{\partial z}\right) + \left(\frac{\omega^2}{c^2(z)} - k^2\right)Z = -\frac{\omega^2}{c^2(z)}\frac{M\ \delta(z-z_s)}{2\pi}$$

$$(3.46)$$

in the fluid, where $c(z) = (\lambda(z)/\rho(z))^{1/2}$ is the sound speed.

Since the pressure $p(r, z)$ in the fluid equals $-\tau^{zz}(r, z)$, it can be computed from the inverse Hankel transform, or wave number integral,

$$p(r,z) = -\int_0^\infty r_4(k,z)J_0(kr)\ k\ dk.$$

$$(3.47)$$

At a fluid—solid interface, $r_1(k, z)$ has a step discontinuity, while the remaining three components of the displacement-stress vector $r(k, z)$ are continuous.

The numerical evaluation of the wave number integral in Eq. (3.47) has two essential ingredients: solution of the ODE systems according to Eqs. (3.42)—(3.45) to get $r_4(k, z)$, and numerical quadrature. Some useful numerical techniques are presented in Sections 3.2.1 and 3.2.2, respectively.

3.2.1 SOLUTION OF THE DEPTH-DEPENDENT ODE SYSTEMS

For definiteness, the fluid region, composed of the water column and possible fluid sediment layers, is assumed to extend from the traction-free surface at $z = 0$ to the depth z_b. Below z_b, there are solid sediments terminated by a homogeneous solid half-space below $z = z_h$. Fig. 3.3 shows the fluid—solid medium schematically.

The propagator-matrix method according to Thomson and Haskell [19,20] is a classical method to solve the depth-dependent ODE systems in Eqs. (3.42) and (3.44). In the solid case, however, a straightforward implementation leads to certain numerical problems. Several methods have been devised to cure this problem. Some of the best known, see reviews in Refs. [7,21], for example, are the global matrix approach [22], the reflection-coefficient matrix method [23—25], the exact finite-element (FE) and recursive stiffness matrix methods, [26,27] and the compound-matrix method [28—31]. The reflection-coefficient method, also called the invariant embedding or Riccati method, is now described in some detail.

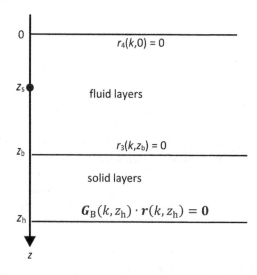

FIGURE 3.3

Range-invariant fluid—solid medium with a depth (z) axis. There is a point source on the z axis at depth z_s. Boundary and interface conditions for the displacement-stress vector $r(k, z)$ are indicated.

3.2.1.1 Recursive Computation of Reflection-Coefficient Matrices for the Solid Bottom

For z, ζ between z_b and z_h, $r(k, z) = P(k; z, \zeta) \cdot r(k, \zeta)$, where $P(k; z, \zeta)$ is the 4×4 propagator matrix corresponding to the ODE system (3.42). By definition [1], for each ζ, the columns of $P(k; z, \zeta)$ satisfy the ODE system with independent variable z and $P(k; \zeta, \zeta) = I$, the identity matrix. The propagator matrix can be used to transport boundary conditions [29], i.e., reformulate them at other depths. In the following, the dependence on the horizontal wave number k is often omitted in the notation. A boundary condition from below,

$$G_B(z_h) \cdot r(z_h) = 0 \tag{3.48}$$

at $z = z_h$, where $G_B(z_h)$ is a 2×4 matrix, can apparently be transported upward to depths z, $z_b \leq z < z_h$, as $G_B(z) \cdot r(z) = 0$ with $G_B(z) = G_B(z_h) \cdot P(z_h, z)$. In particular,

$$G_B(z_b+) \cdot r(z_b+) = 0 \tag{3.49}$$

with $G_B(z_b+) = G_B(z_h) \cdot P(z_h, z_b)$, where z_b+ denotes the lower, solid, side of z_b.

For the common case of a bottom with homogeneous solid sediment layers, the propagator matrix can be given explicitly. Specifically [1], with $\zeta \leq z$, for a homogeneous solid layer between the depths ζ and z, that has thickness $d = z - \zeta$, density ρ, compressional-wave sound speed α, shear-wave speed β, and shear modulus $\mu = \rho\beta^2$,

$$P(z, \zeta) = L \cdot \begin{pmatrix} E & 0 \\ 0 & E^{-1} \end{pmatrix} \cdot L^{-1} \tag{3.50}$$

where $E = \text{diag}\,(e^{-\gamma d}, e^{-\nu d})$, a 2×2 matrix with $\gamma = (k^2 - \omega^2/\alpha^2)^{1/2}$, $\nu = (k^2 - \omega^2/\beta^2)^{1/2}$, and

$$L = \frac{\beta}{\gamma\nu\omega^2} \begin{pmatrix} \alpha k & \beta\nu & \alpha k & \beta\nu \\ \alpha\gamma & \beta k & -\alpha\gamma & -\beta k \\ -2\alpha\mu k\gamma & -\beta\mu(k^2 + \nu^2) & 2\alpha\mu k\gamma & \beta\mu(k^2 + \nu^2) \\ -\alpha\mu(k^2 + \nu^2) & -2\beta\mu k\nu & -\alpha\mu(k^2 + \nu^2) & -2\beta\mu k\nu \end{pmatrix} \tag{3.51}$$

$$L^{-1} = \frac{1}{2\alpha\mu} \begin{pmatrix} 2\beta\mu k\gamma\nu & -\beta\mu\nu(k^2 + \nu^2) & -\beta k\nu & \beta\gamma\nu \\ -\alpha\mu\gamma(k^2 + \nu^2) & 2\alpha\mu k\gamma\nu & \alpha\gamma\nu & -\alpha k\gamma \\ 2\beta\mu k\gamma\nu & \beta\mu\nu(k^2 + \nu^2) & \beta k\nu & \beta\gamma\nu \\ -\alpha\mu\gamma(k^2 + \nu^2) & -2\alpha\mu k\gamma\nu & -\alpha\gamma\nu & -\alpha k\gamma \end{pmatrix}. \tag{3.52}$$

The square roots in the definitions of γ and ν are chosen such that $\text{Re}(\gamma) \geq 0$, with $\text{Im}(\gamma) \leq 0$ if $\text{Re}(\gamma) = 0$, and likewise for ν. Apparently, $i\gamma$ and $i\nu$ can be identified as vertical wave numbers for down-going compressional and shear waves, respectively. At each depth between ζ and z, L^{-1} projects the local displacement-stress vector onto a vector with coefficients for local down- and up-going waves.

The displacement-stress vectors of the local down- and up-going waves are defined by the columns of the layer matrix L.

With N homogeneous solid sediment layers between z_b and z_h, it follows that

$$P(z_h, z_b) = L_N \cdot \begin{pmatrix} E_N & 0 \\ 0 & E_N^{-1} \end{pmatrix} \cdot L_N^{-1} \cdot \ldots \cdot L_1 \cdot \begin{pmatrix} E_1 & 0 \\ 0 & E_1^{-1} \end{pmatrix} \cdot L_1^{-1}, \qquad (3.53)$$

where indices $1, 2, \ldots, N$ are used to identify quantities for the different layers from z_b to z_h. At the interface between layers n and $n+1$, $n = 1, 2, \ldots, N$ with index $n = N+1$ used to denote the half-space below z_h, the displacement-stress vector r can be written

$$r = L_n \cdot \begin{pmatrix} a_n^+ \\ b_n^+ \end{pmatrix} = L_{n+1} \cdot \begin{pmatrix} a_{n+1}^- \\ b_{n+1}^- \end{pmatrix}. \qquad (3.54)$$

The two-dimensional coefficient column vectors $a_1^+, a_2^-, a_2^+, \ldots, a_N^-, a_N^+, a_{N+1}^-$ and $b_1^+, b_2^-, b_2^+, \ldots, b_N^-, b_N^+, b_{N+1}^-$ for local down- and up-going waves, respectively, are defined by these relations. In addition, a_1^- and b_1^- are defined by $r(z_b+) = L_1 \cdot \begin{pmatrix} a_1^- \\ b_1^- \end{pmatrix}$. A 2×4 matrix $G_B(z_h)$ specifying, in the sense of Eq. (3.48), that there are no up-going waves in the homogeneous solid half-space, can now be expressed as

$$G_B(z_h) = (0, I) \cdot L_{N+1}^{-1}. \qquad (3.55)$$

If $G_B(z_b+) = G_B(z_h) \cdot P(z_h, z_b)$ is formed by direct multiplication of $G_B(z_h)$ from the right, by the factor matrices from Eq. (3.53), numerical problems may appear in the evanescent high-frequency regime when some $\gamma_n d_n$ is very large. The corresponding $e^{\gamma_n d_n}$ may dominate, causing $\begin{pmatrix} E_n & 0 \\ 0 & E_n^{-1} \end{pmatrix}$ to act effectively as the rank-deficient matrix $\mathrm{diag}(0, 0, e^{\gamma_n d_n}, 0)$ and the two rows of $G_B(z_b+)$ to become linearly dependent numerically.

Physically, the boundary conditions from below are more naturally expressed in terms of 2×2 reflection matrices R_n^+ and R_n^- for down-going waves: $b_n^+ = R_n^+ \cdot a_n^+$ at the lower interface and $b_n^- = R_n^- \cdot a_n^-$ at the upper interface of layer n, $n = N, \ldots, 1$. This is illustrated in Fig. 3.4, and the response (the reflected up-going waves) is now directly expressed in terms of the excitation (the incident down-going waves). Reflection-matrix propagation upward is performed in a stable way, without exponential growth, according to $R_n^- = E_n \cdot R_n^+ \cdot E_n$, and 2×4 boundary-condition matrices $G_B(z)$ reexpressed as $\left(-R_n^+, I \right)$ and $\left(-R_n^-, I \right)$ maintain full rank.

Formally, $G_B(z_b+)$ is rewritten as

$$G_B(z_b +) = F \cdot \left(-R_1^-, I \right) \cdot L_1^{-1}, \qquad (3.56)$$

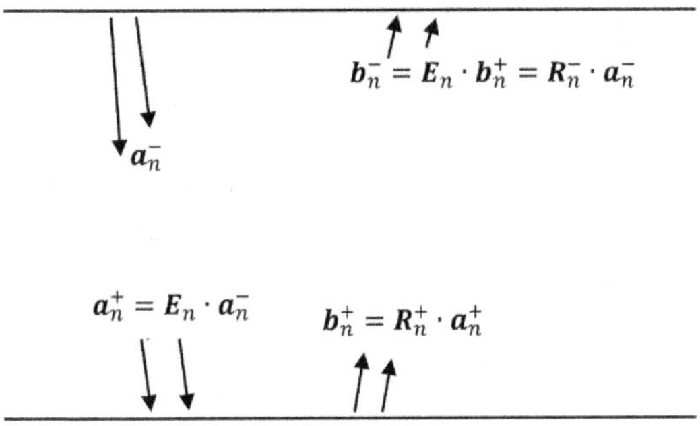

FIGURE 3.4

Solid layer n, $n = N,\ldots,$ 1, with down- and up-going waves and reflection matrices for down-going waves. In the evanescent regime, the two down-going wave components increase upward, while the two up-going ones decrease upward, as indicated by the lengths of the arrows.

where the 2×2 matrix $\boldsymbol{F}$ is defined by

$$\boldsymbol{F} = \boldsymbol{D}_N \cdot \boldsymbol{E}_N^{-1} \cdot \left(\boldsymbol{D}_{N-1} - \boldsymbol{R}_N^- \cdot \boldsymbol{B}_{N-1}\right) \cdot \boldsymbol{E}_{N-1}^{-1} \cdot \ldots \cdot \left(\boldsymbol{D}_1 - \boldsymbol{R}_2^- \cdot \boldsymbol{B}_1\right) \cdot \boldsymbol{E}_1^{-1} \quad (3.57)$$

with 2×2 matrices $\boldsymbol{A}_n$, $\boldsymbol{B}_n$, $\boldsymbol{C}_n$, and $\boldsymbol{D}_n$ defined for $n = 1, 2,\ldots, N$ by

$$\boldsymbol{L}_{n+1}^{-1} \cdot \boldsymbol{L}_n = \begin{pmatrix} \boldsymbol{A}_n & \boldsymbol{B}_n \\ \boldsymbol{C}_n & \boldsymbol{D}_n \end{pmatrix}. \quad (3.58)$$

The 2×2 reflection matrices $\boldsymbol{R}_{N+1}^- = \boldsymbol{0}$, $\boldsymbol{R}_N^+$, $\boldsymbol{R}_N^-,\ldots,$ $\boldsymbol{R}_1^+$, $\boldsymbol{R}_1^-$ for down-going waves are computed recursively by, cf. [32,33],

$$\boldsymbol{R}_n^+ = -\left(\boldsymbol{D}_n - \boldsymbol{R}_{n+1}^- \cdot \boldsymbol{B}_n\right)^{-1} \cdot \left(\boldsymbol{C}_n - \boldsymbol{R}_{n+1}^- \cdot \boldsymbol{A}_n\right), \quad \boldsymbol{R}_n^- = \boldsymbol{E}_n \cdot \boldsymbol{R}_n^+ \cdot \boldsymbol{E}_n \quad (3.59)$$

for $n = N,\ldots,$ 1. The possibly problematic matrices $\boldsymbol{E}_n^{-1}$ are factored out, and the exponentials in a matrix $\boldsymbol{E}_n$ in the evanescent regime typically cause a decrease in magnitude when $\boldsymbol{R}_n^-$ is obtained from $\boldsymbol{R}_n^+$ by Eq. (3.59). As a result, the 2×4 matrices $\left(-\boldsymbol{R}_n^-, \boldsymbol{I}\right)$ forming transported boundary conditions of the type $\boldsymbol{b}_n^- = \boldsymbol{R}_n^- \cdot \boldsymbol{a}_n^-$, $n = N,\ldots,$ 1, indeed have full rank numerically.

Incorporating the fluid–solid constraint $r_3(z_b) = 0$, it follows by Cramer's rule that $\boldsymbol{r}(z_b+)$ is proportional to the vector

$$\left(\begin{vmatrix} h_{12} & h_{14} \\ h_{22} & h_{24} \end{vmatrix}, \; -\begin{vmatrix} h_{11} & h_{14} \\ h_{21} & h_{24} \end{vmatrix}, \; 0, \; \begin{vmatrix} h_{11} & h_{12} \\ h_{21} & h_{22} \end{vmatrix} \right)^{\mathrm{T}}, \quad (3.60)$$

where (h_{ij}), $i = 1, 2, j = 1, 2, 3, 4$ are the elements of the matrix $\left(-\boldsymbol{R}_1^-, \boldsymbol{I}\right) \cdot \boldsymbol{L}_1^{-1}$ from Eq. (3.56). The determinant $\det(\boldsymbol{F})$ can be computed as a product using factor matrices according to Eq. (3.57). It may often be omitted, but it must be kept when an analytic dispersion function, as a function of k, is needed (cf. Section 3.3.1).

A corresponding determinant factor appears in connection with the exact finite-element method (FEM) [26]. Media with inhomogeneous layers can be handled by augmenting the traditional Riccati equation system with a single differential equation for a certain compound-matrix element [32].

A displacement-stress vector proportional to expression (3.60) could be back-propagated downward to compute the field at desired depths in the solid region. In so doing (cf. [29]), it is essential in the evanescent regime to stabilize the procedure by computing the up-going intermediate vectors from the down-going ones according to $b_n^+ = R_n^+ \cdot a_n^+$, $n = 1, ..., N$. Hence, the propagation of the displacement-stress vector downward through layer n, $n = 1, ..., N$, should be done by multiplication from the left with the factor matrices in

$$L_n \cdot \begin{pmatrix} I \\ R_n^+ \end{pmatrix} \cdot (E_n, 0) \cdot L_n^{-1}. \tag{3.61}$$

In this way, the boundary conditions from z_h are properly taken into account, and spurious exponential growth caused by numerical inaccuracies is avoided.

3.2.1.2 Propagator Matrices for the Fluid Region

In the fluid region, with the ODE system (3.44) and its 2×2 propagator matrix $P_F(k; z, \zeta)$, it is sufficient to deal with the two components $r_2(z)$ and $r_4(z)$ of $r(z)$. From below, the two-dimensional column vector $r_B(z_b)$, formed from the second and fourth components of the displacement-stress vector defined by (3.60), can now be propagated from z_b to z_s according to $r_B(z) = (r_{B2}(z), r_{B4}(z))^T = P_F(z, z_b) \cdot r_B(z_b)$ for $z_s \leq z < z_b$. From above, the two-dimensional column vector $r_A(0) = r_{ref}^3 (1, 0)^T$ can be propagated from the traction-free surface at $z = 0$ to z_s according to $r_A(z) = (r_{A2}(z), r_{A4}(z))^T = P_F(z, 0) \cdot r_A(0)$ for $0 < z \leq z_s$.

The propagator matrix can be given explicitly for z and ζ in a homogeneous fluid layer with density ρ and sound speed c as

$$P_F(z, \zeta) = \begin{pmatrix} \cosh(\chi(z - \zeta)) & -\dfrac{\chi}{\rho\omega^2} \sinh(\chi(z - \zeta)) \\ -\dfrac{\rho\omega^2}{\chi} \sinh(\chi(z - \zeta)) & \cosh(\chi(z - \zeta)) \end{pmatrix}, \tag{3.62}$$

where $\chi = (k^2 - \omega^2/c^2)^{1/2}$. Using Airy functions, $P_F(z, \zeta)$ can also be given explicitly for a layer where c^{-2} varies linearly with depth [7,34,35]. (With the high-frequency approximation that compressional and shear waves decouple in a solid layer, Airy functions have been applied for solid gradient layers as well [25].)

The desired displacement-stress vector is proportional to $r_A(z)$ above and $r_B(z)$ below the source depth z_s, respectively. The proportionality coefficients are determined to fulfill continuity of $r_4(z)$ and Eq. (3.45) for $r_2(z)$ at z_s, whereby the determinant of the appearing 2×2 equation system is given by the characteristic function

$$D(k) = \det(r_A(k, z_s), r_B(k, z_s)). \tag{3.63}$$

$D(k)$ can be viewed as a dispersion function, and the wave number integral from Eq. (3.47) takes the form

$$p(r,z) = \int_0^\infty \frac{M}{2\pi\lambda(z_s)} \frac{r_{A4}(k,z_<)r_{B4}(k,z_>)}{D(k)} J_0(kr) \, k \, dk \qquad (3.64)$$

for $0 \leq z_s, z \leq z_b$, where $z_< = \min(z_s, z)$ and $z_> = \max(z_s, z)$.

3.2.1.3 Alternative Treatment of the Fluid Region

Alternatively, the fluid region can be treated with boundary-condition transportation all the way up to the traction-free surface. An advantage is that a vertical source array, with several point sources or a continuous source distribution, can readily be incorporated.

A boundary condition from below can be expressed as $\boldsymbol{g}_B(z_b) \cdot \boldsymbol{r}_F(z_b) = 0$, where $\boldsymbol{r}_F(z) = (r_{F2}(z), r_{F4}(z))^T$ for $0 \leq z \leq z_b$ is the desired displacement-stress vector, and the row vector $\boldsymbol{g}_B(z_b)$ is obtained from the vector components in (3.60) as

$$\boldsymbol{g}_B(z_b) = r_{\text{ref}}^3 \, \det(\boldsymbol{F}) \left(\begin{vmatrix} h_{11} & h_{12} \\ h_{21} & h_{22} \end{vmatrix}, \begin{vmatrix} h_{11} & h_{14} \\ h_{21} & h_{24} \end{vmatrix} \right). \qquad (3.65)$$

This boundary condition can be transported upward to the surface and expressed as

$$\boldsymbol{g}_B(z) \cdot \boldsymbol{r}_F(z) = q(z) \qquad (3.66)$$

at depth z, $0 \leq z \leq z_b$, where $\boldsymbol{g}_B(z) = (g_{B2}(z), g_{B4}(z)) = \boldsymbol{g}_B(z_b) \cdot \boldsymbol{P}_F(z_b, z)$, while $q(z) = 0$ for $z_s < z \leq z_b$ and $q(z) = -\boldsymbol{g}_B(z_s) \cdot (M/2\pi\lambda(z_s), 0)^T$ for $0 \leq z < z_s$. The source contribution is picked up in $q(z)$ at $z = z_s$ according to Eq. (3.45). Contributions from additional point sources can be picked up in the same way, and a continuous source distribution can be handled by integration along the z axis [29]. It follows directly that

$$\boldsymbol{r}_F(0) = (q(0)/g_{B2}(0), 0)^T, \qquad (3.67)$$

where $p_{\text{ref}} \, r_{\text{ref}}^8 \, g_{B2}(0) = p_{\text{ref}} \, r_{\text{ref}}^8 \, g_{B2}(k, 0)$ actually agrees with the dispersion function $D(k)$ from Eq. (3.63) [30].

The displacement-stress vector $\boldsymbol{r}_F(0)$ can subsequently be back-propagated downward to compute the field at desired depths in the fluid region. In so doing, it is essential in the evanescent regime to avoid spurious exponential growth by enforcing the boundary condition (3.66) coming from below. The details (cf. [29]) are now provided for the case of a fluid region with homogeneous layers.

With $\zeta \leq z$, for a homogeneous fluid layer between the depths ζ and z, that has density ρ and sound speed c, the propagator matrix in Eq. (3.62) can be factorized as

$$\boldsymbol{P}_F(z, \zeta) = \boldsymbol{L}_F \cdot \begin{pmatrix} E & 0 \\ 0 & E^{-1} \end{pmatrix} \cdot \boldsymbol{L}_F^{-1}, \qquad (3.68)$$

where $E = e^{-\chi d}$, with $d = z - \zeta$, and

$$L_F = \chi^{-4} \begin{pmatrix} \chi & -\chi \\ \rho\omega^2 & \rho\omega^2 \end{pmatrix}, \quad L_F^{-1} = \frac{\chi^3}{2\rho\omega^2} \begin{pmatrix} \rho\omega^2 & \chi \\ -\rho\omega^2 & \chi \end{pmatrix}. \tag{3.69}$$

The square root in the definition of χ is now chosen such that $\mathrm{Re}(\chi) \geq 0$, with $\mathrm{Im}(\chi) \leq 0$ if $\mathrm{Re}(\chi) = 0$, and $i\chi$ can be identified as a vertical wave number. At each depth between ζ and z, L_F^{-1} projects the local displacement-stress vector onto a vector with local coefficients for down- and up-going waves. The displacement-stress vectors of the down- and up-going waves are defined by the columns of L_F. To accurately fulfill Eq. (3.66) at $z-$, the propagation downward of $r_F(\zeta+)$ to $r_F(z-)$, assuming $z_s \leq \zeta$ or $z \leq z_s$, should be done with matrix multiplications from the left and vector additions according to

$$r_F(z-) = L_F \cdot \begin{pmatrix} 1 \\ R^+ \end{pmatrix} \cdot (E, 0) \cdot L_F^{-1} \cdot r_F(\zeta+) + L_F \cdot \begin{pmatrix} 0 \\ q(z-)/F^+ \end{pmatrix}, \tag{3.70}$$

where F^+ and the reflection coefficient R^+ are defined by $g_B(z) \cdot L_F = F^+(-R^+, 1)$. Eq. (3.45) shows how $r_F(z_s+)$ is obtained from $r_F(z_s-)$ at the source depth.

It can be noted that $g_B(\zeta) = g_B(z) \cdot P_F(z, \zeta) = F^-(-R^-, 1) \cdot L_F^{-1}$, where $F^- = F^+E^{-1}$ and $R^- = ER^+E$ (cf. Eq. 3.59). With L homogeneous fluid layers between z_b and the surface, the boundary-condition vectors $g_B(z) = g_B(z_b) \cdot P_F(z_b, z)$ could be obtained by recursive computation of factors $F_L^+, F_L^-, \ldots, F_1^+, F_1^-$ and reflection coefficients $R_L^+, R_L^-, \ldots, R_1^+, R_1^-$ (cf. Eqs. 3.55–3.59). The computed quantities could be stored for use at the back-propagation step.

3.2.1.4 Final Remarks

The propagator matrices $P(k; z, \zeta)$ and $P_F(k; z, \zeta)$ can be defined for complex horizontal wave numbers k as analytic functions. The factorizations in Eqs. (3.50) and (3.68) break down for k close to the branch points, where $\gamma\nu$ and χ, respectively, vanishes. Other factorizations could be applied [36,37] for improved efficiency; however, in practice the problem is typically not serious. The sediment layers are typically viscoelastic with complex wave speeds, and the corresponding γ, ν, or χ are bounded away from zero for the wave number integration path with real k according to Eq. (3.47). For a fluid layer and k with χ close to zero, there is no problem with exponential growth in the back-propagation of Section 3.2.1.3, and the first term in the right-hand side of Eq. (3.70) can safely be replaced by $P_F(z, \zeta) \cdot r_F(\zeta+)$ with $P_F(z, \zeta)$ according to Eq. (3.62).

In order to avoid overflow and underflow, scaling should preferably be made at propagation, with stored scaling exponents. There is typically no need to worry about spurious exponential growth in the boundary-condition transportation with $g_B(z)$ according to Eq. (3.66). If a strongly dominant exponential base solution is missing, an example would be a thick layer without down-going waves, the problem is typically ill-posed in the sense that a small change in the boundary condition would cause dramatic changes in the solution.

So far, the medium has been terminated at depth by a homogeneous solid half-space. Modifications for other cases are easily devised. For a traction-free or rigid boundary at $z = z_h$, the reflection-matrix recursion is started by $\boldsymbol{R}_N^+ = -\boldsymbol{L}_D^{-1} \cdot \boldsymbol{L}_C$ or $\boldsymbol{R}_N^+ = -\boldsymbol{L}_B^{-1} \cdot \boldsymbol{L}_A$, respectively, where $\boldsymbol{L}_N = \begin{pmatrix} \boldsymbol{L}_A & \boldsymbol{L}_B \\ \boldsymbol{L}_C & \boldsymbol{L}_D \end{pmatrix}$. For a homogeneous fluid half-space below z_b, with density ρ and sound speed c, $\boldsymbol{r}_B(z_b) = r_{\text{ref}}^4 (\chi, \rho\omega^2)^{\text{T}}$ is the appropriate choice, where $\chi = (k^2 - \omega^2/c^2)^{1/2}$ with $\text{Re}(\chi) \geq 0$ and $\text{Im}(\chi) \leq 0$ if $\text{Re}(\chi) = 0$. A traction-free or rigid boundary at $z = z_b$ is incorporated by $\boldsymbol{r}_B(z_b) = r_{\text{ref}}^3 (1, 0)^{\text{T}}$ or $\boldsymbol{r}_B(z_b) = p_{\text{ref}} \, r_{\text{ref}}^2 (0, 1)^{\text{T}}$, respectively.

3.2.2 ADAPTIVE INTEGRATION

The integrand factor $r_4(k, z)$ in Eq. (3.47) is typically highly peaked, there is an example in Section 3.2.3, suggesting an adaptive choice of step lengths for the numerical integration. The factor $J_0(kr)$, on the other hand, is regular but rapidly oscillating. Filon [38] has published a useful quadrature scheme for such oscillating integrands.

Incorporating the Filon scheme, a high-order adaptive algorithm for computing wave number integrals with error control is proposed in Ref. [39]. For the truncated part (K_1, K_2) of the infinite integration interval, with $0 < K_1 < K_2$, the integrand is written as

$$-r_4(k, z)J_0(kr)k = f(k) = f_+(k) + f_-(k) = g_+(k)e^{ikr} + g_-(k)e^{-ikr}, \qquad (3.71)$$

with $f_+(k) = g_+(k)e^{ikr}$, $f_-(k) = g_-(k)e^{-ikr}$, $g_+(k) = -k \, r_4(k,z)H_0^{(1)}(kr)e^{-ikr}/2$, and $g_-(k) = -k \, r_4(k,z)H_0^{(2)}(kr)e^{ikr}/2$. The essential steps of the algorithm are:

1. Initialize a stack of intervals and push (K_1, K_2) onto the stack. Set the current value of the integral estimate $\widehat{V}$ to zero.
2. For the top interval (k_1, k_2) in the stack, set $h_0 = k_2 - k_1$ and $i = 0$. Set the error bound to be used for this interval to $\varepsilon_0 = \varepsilon \, h_0/(K_2 - K_1)$, where ε is a preset error tolerance, and prepare a fresh extrapolation table.
3. Define $h_i = h_0/n$, where $n = 2^i$ (step-size halving), and evaluate

$$S_{i,0} = T\frac{1 - \cos(h_i r)}{(h_i r)^2/2} + \frac{f_+(k_1) - f_+(k_2) + f_-(k_2) - f_-(k_1)}{2} \, i \, h_i \frac{h_i r - \sin(h_i r)}{(h_i r)^2/2},$$

$$(3.72)$$

the trapezoidal Filon estimate, where T is the ordinary trapezoidal estimate

$$T = h_i \left(\frac{f(k_1) + f(k_2)}{2} + \sum_{l=1}^{n-1} f(k_1 + lh_i) \right). \qquad (3.73)$$

4. Add a row to the extrapolation table, starting with $S_{i,0}$ and proceeding for $j = 1$, $2,\dots,$ $\min(i, J)$ with $S_{i,j} = S_{i,j-1} + \left(S_{i,j-1} - S_{i-1,j-1}\right)/\left(2^{2j} - 1\right)$. Provided $h_i < H$, as soon as $\left|S_{i,j-1} - S_{i-1,j-1}\right| < \varepsilon_0$ for an added value with positive i and j, increase $\widehat{V}$ by $S_{i,j}$ and proceed to step 7.
5. If $i < I$, refine the step size by increasing i by one and go to step 3.

6. Split the current interval at its midpoint by changing the top interval in the stack to $(k_1,(k_1 + k_2)/2)$ and pushing the interval $((k_1 + k_2)/2, k_2)$ onto the stack. Then go to step 2.

7. Pop the subinterval stack. Proceed to step 2 if the stack is nonempty. Otherwise, accept the current value of $\widehat{V}$ as the estimate of the integral, with error estimate ε, and terminate.

I and J are preset limits for the extrapolation table, $I = 9$ and $J = 7$ can be useful values. H is a safety parameter for maximum step size, which is set guided by a priori knowledge of the variability of the integrand. The presented scheme uses Richardson extrapolation [40] with step-size halving (Romberg's method). However, in Ref. [39] an extrapolation scheme suggested by Bulirsch−Stoer is applied, with superior computational economy.

For the remaining integration interval $(0, K_1)$, expression (3.71) is not appropriate, since the Hankel functions are singular at the origin. The adaptive scheme can still be applied; however, with Eq. (3.72) replaced by $S_{i,0} = T$. For near-field computations with small r, including $r = 0$, the whole interval $(0, K_2)$ can be treated in this way.

When the source and receiver depths (z_s and z) are close, the integrand decays slowly at infinity. Unless r vanishes or is very small, the remaining integration interval (K_2, ∞) is then best treated by deviation of the integration path into the complex k plane [39]. The $f_+(k)$ integrand term from Eq. (3.71) should be shifted upward into the first quadrant where $H_0^{(1)}(kr)$ decays rapidly, while the $f_-(k)$ term should be shifted downward into the fourth quadrant where $H_0^{(2)}(kr)$ decays rapidly. It is hereby essential that K_2 is large enough to prevent poles of $r_4(k, z)$ from being enclosed by the path shifts.

3.2.3 EXAMPLE

The example from Section 3.1.2 is now revisited. Fig. 3.5 shows the magnitude of the corresponding integrand factor $k\, r_4(k, z)$ appearing in Eq. (3.47). Indeed, $r_4(k, z)$ is highly peaked, and adaptive integration, as suggested in Section 3.2.2, is useful to automatically concentrate the computational work to the significant parts of the integration path. Fig. 5 from Ref. [39] shows an example of the work distribution.

Fig. 3.6 shows some corresponding PL curves, correctly computed by wave number integration. The ray result for the total wave field is repeated from Fig. 3.2 as the gray curve in the left panel. It follows the correct result, the black curve, reasonably well during the first half. There are clear differences at longer ranges, however, where shadow-zone boundaries have been passed, for which the ray method has problems.

The field decomposition in the right panel of Fig. 3.2 is readily obtained with the ray method. It is less apparent how to obtain a corresponding decomposition of the pressure field given by Eq. (3.47). The integrand factor $r_4(k, z)$ can be expanded in a series leading to a certain traveling-wave expansion [41,42]; however, care is needed

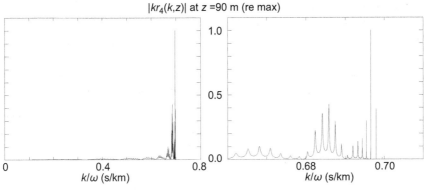

FIGURE 3.5

Relative magnitude of kr_4 (k, z) as a function of real horizontal slowness k/ω for the 500 Hz shallow-water example from Section 3.1.2, with source and receiver depths 10 and 90 m, respectively. The right panel is a magnification of part of the left one.

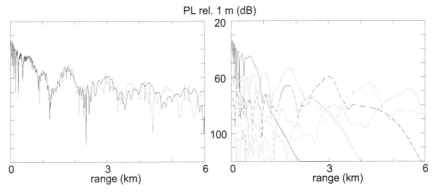

FIGURE 3.6

PL in dB, rel. 1 m, for the 500 Hz example. Left: total wave field by wavenumber integration (black) and rays (gray, repeated from Fig. 3.2). Right: partial wave fields computed with wave number integration, for cycle numbers $\nu = 0$ (*solid black*), 1 (*solid gray*), 2 (*dashed black*), 3 (*dashed gray*), and 4 (*dotted black*).

to obtain a physically useful decomposition with well-defined wave number integrals for its components. The right panel of Fig. 3.6 shows partial wave fields computed by a well-defined recombinant expansion proposed in Ref. [43]. Reflection coefficients, needed for the expansion, are introduced in terms of solutions of the homogeneous version of the depth-separated Helmholtz equation (3.46) that are up- and down-going at certain specification depths. For the example here, these specification depths are chosen at $z_{min} = 50$ m, where the sound-speed profile has its minimum (Fig. 3.1), and local $1/c^2$ linearity is assumed on each side of z_{min} when defining the up- and down-going waves.

Compared with the field decomposition from Fig. 3.2, the one in Fig. 3.6 is much more well-behaved. The pressure peaks at ranges about 1.8, 3.0, 4.2, and 5.5 km are clear, and the field components are smooth without abrupt cutoff ranges.

3.3 NORMAL MODE PROPAGATION MODELS

The wave number integral in Eq. (3.47), for the pressure in the fluid with a symmetric point source according to $f = f(x) = -M \, \mathrm{grad}(\delta(x - x_s))$ where $x_s = (0, 0, z_s)$ is in the water and the time factor $\exp(-i\omega t)$ with $\omega > 0$ is omitted, may be rewritten as

$$p(r, z) = -\frac{1}{2} \int_{-\infty}^{\infty} r_4(k, z) H_0^{(1)}(kr) k \, \mathrm{d}k, \qquad (3.74)$$

since $r_4(k, z)$ is even in k. The integrand is analytic in k, except at the branch cuts induced by the vertical wave numbers $i\gamma$ and $i\nu$ for the lower solid half-space. These branch cuts are typically chosen as the hyperbolic curves where $\mathrm{Re}(\gamma)$ or $\mathrm{Re}(\nu)$ vanishes. According to the residue theorem, Eq. (3.74) can be replaced by residue contributions from poles or modal wave numbers k_m, $m = 1, 2, \ldots$, in the upper half-plane, corresponding to zeroes of the dispersion function $D(k)$ from Eq. (3.63), and branch-cut integrals. The latter are typically negligible at long ranges r, and one obtains from Eq. (3.64)

$$p(r, z) = \frac{iM}{2\lambda(z_s)} \sum_m \frac{r_{A4}(k_m, z_<) r_{B4}(k_m, z_>)}{D'(k_m)} H_0^{(1)}(k_m r) k_m \qquad (3.75)$$

for $0 \le z_s$, $z \le z_b$. The vectors $r_A(k_m, z_s)$ and $r_B(k_m, z_s)$, $m = 1, 2, \ldots$, are linearly dependent since $D(k_m)$ vanishes. Hence, continuous mode functions $r_F(k_m, z)$ {unit: m^2 (m, N/m^2)$^\mathrm{T}$ = (m^3, N)$^\mathrm{T}$} can be defined for $0 \le z \le z_b$, that are proportional to $r_A(k_m, z)$ and $r_B(k_m, z)$ above and below z_s, respectively. The expansion of the pressure field in terms of normal modes alternatively can be derived using the technique of separation of variables [7, 10].

Eq. (3.75) apparently requires that each k_m is a *simple* zero, with a one-dimensional eigenmode space, of the dispersion function $D(k)$. This is true in general, although multiple zeroes may appear for isolated frequencies [44,45].

The modal wave numbers k_m, $m = 1, 2, \ldots$, can be studied as continuous functions of angular frequency ω. As ω tends to zero, each $k_m(\omega)$ tends to a limit value. For most modes, the limit value is finite and nonvanishing, and at low frequency the mode is connected to a particular fluid or solid region in the medium [46,47]. The modes with $\lim_{\omega \to 0} k_m(\omega) = 0$ are finite in number for each fluid–solid medium, and they can be listed explicitly [48]. For a medium including homogeneous solid layers, low-frequency computations of slow modes with $\lim_{\omega \to 0} k_m(\omega) = 0$ and $\lim_{\omega \to 0} k_m(\omega)/\omega = \infty$ necessitate some care, since loss of numerical precision by cancellation may appear, causing the columns of a layer matrix L according to Eq. (3.51) to become linearly dependent numerically. For computations with the compound-matrix method, cancellation-free expressions of the required quantities are derived in Ref. [49].

3.3.1 MODAL WAVE NUMBERS

Several methods have been devised to find the real roots of a dispersion function $D(k)$ in the purely elastic (no absorption) case. For a fluid medium without absorption, elementary Sturm–Liouville theory [50], as applied to Eq. (3.46), shows that all modal wave numbers k_m are real or imaginary, and that the real ones are bounded in magnitude by $\omega/\min(c(z))$. Perturbation techniques have subsequently been applied to determine the shifts of the roots into the complex wave number plane when absorption is included. Particularly for a solid bottom, however, it is difficult to be certain that *all* roots have been found. Furthermore, the accuracy of the perturbation techniques may be insufficient when the absorption is significant. Expanding the modes in terms of the ones for the purely elastic case, Odom [51] has derived a linear generalized eigenvalue equation for the shifted wave numbers. This alternative approach has not yet (2016) been tested numerically, however.

Noting that the dispersion function $D(k)$ is analytic as a function of complex wave number k, winding-number integrals have been used to overcome the mentioned problems [52]. An algorithm with adaptive and automatic splitting of search rectangles has been published in Ref. [53] and discussed in Ref. [54]. It has the following essential steps:

1. Initialize a stack of rectangles in the complex k plane. Push a rectangle R, such that $D(k)$ is analytic in R and the zeroes of $D(k)$ in R are to be computed, onto the stack.
2. Evaluate the argument variation

$$\Delta\Phi = \int_{\partial R_{\text{top}}} d \, \arg(D(k)) \tag{3.76}$$

along the positively oriented boundary ∂R_{top} of the top rectangle R_{top} in the stack. The number of zeroes of $D(k)$ within R_{top} is $N_{\text{top}} = \Delta\Phi/2\pi$. If $N_{\text{top}} = 0$, pop the rectangle stack and proceed to step 6.
3. If the diagonal of R_{top} exceeds 2ε, where ε is a preset error tolerance, proceed to step 4 or 5, depending on whether $N_{\text{top}} = 1$ or $N_{\text{top}} > 1$. Otherwise, accept the midpoint of R_{top} as a zero with multiplicity N_{top} and proceed to step 6.
4. Find the simple zero of $D(k)$ within R_{top}, with accuracy ε, by secant iterations starting from two points in R_{top}. Proceed to step 5 if an estimate falls outside R_{top} or if the successive $D(k)$ values do not decay fast enough.
5. Pop R_{top} from the stack, split this rectangle into two at the midpoints of its longest sides, and push the two subrectangles onto the stack. Proceed to step 2.
6. Terminate if the stack is empty. Otherwise, proceed to step 2.

The argument variation in step 2 is preferably computed using adaptively selected step sizes, aiming at successively selected points k_1, k_2,... counterclockwise around ∂R_{top} such that $|(D(k_{n+1}) - D(k_n))/D(k_n)| \approx q$, where $q < 1$ is a preselected tolerance. Computed values of $D(k)$ should be kept in storage if they are potentially needed in the evaluation of the argument variation along the boundary of a subsequently appearing rectangle.

The rectangle sides are typically parallel to the coordinate axes in the k plane, and the hyperbolic branch cuts for the vertical wave numbers $i\gamma$ and $i\nu$ for the lower solid half-space, indicated in connection with Eqs. (3.74) and (3.75), are not very convenient. $D(k)$ is apparently analytic in rectangles which do not cross the branch cuts, and a useful idea is to choose vertical branch cuts for the computations, parallel to the imaginary axis in the k plane. For a rectangle that crosses a hyperbolic branch cut, computations are made for more than one branch (Riemann sheet), and the zeroes on the "physical sheet" corresponding to the hyperbolic cuts are selected afterward. An alternative approach is to use a "total" dispersion function, obtained as the product of all the $D(k)$s for the individual Riemann sheets, which is an entire function of k (cf. [48,55]).

Section 3.3.5.1 includes an example. The algorithm is actually useful for finding the zeroes of any analytic function.

3.3.2 MODE FUNCTIONS

When the mode functions $r_F(k_m, z)$ are computed, as described above, the source depth z_s only appears as a matching depth z_M for $r_A(k_m, z)$ and $r_B(k_m, z)$. The latter two vector functions may naturally be propagated and defined for all depths z in $0 \leq z \leq z_b$, and there is much latitude in choosing z_M. In fact, different matching depths may be needed for different modes [34]. For example, when $r_A(k_m, z)$ or $r_B(k_m, z)$ is propagated away from a low-velocity zone where the mode is trapped, contamination by spurious exponential growth may be introduced. Indicator functions $I(k_m, z)$, defined for $0 \leq z \leq z_b$ by

$$I(k_m, z) = \frac{|\det(r_A(k_m, z), r_B(k_m, z))|}{|r_A(k_m, z)||r_B(k_m, z)|}, \qquad (3.77)$$

can be used to choose an appropriate z_M for each mode [30]. Mathematically, the numerator in Eq. (3.77) vanishes, at each k_m, for all z. Computationally, however, spurious exponential growth at depths z outside a trapping zone may result in a poor match between $r_A(k_m, z)$ and $r_B(k_m, z)$. For each mode, z_M should preferably be chosen at the depth z for which the computed $I(k_m, z)$ is minimal. With a source at this depth, typically within a possible trapping zone, the mode is well excited. Section 3.3.5.1 includes few examples.

The indicator functions $I(k_m, z)$ have been defined in the fluid region. This is usually sufficient, also for interface waves of Scholte type, for which typically a z_M close to the bottom is obtained. However, consideration of depths z within the solid bottom may occasionally be needed and the indicator functions can be extended as follows. Often for brevity, the dependence on the modal wave number k_m is omitted in the notation.

At depths $z \geq z_b+$, the boundary conditions from above can be expressed as $G_A(z) \cdot r(z) = 0$, where $G_A(z)$ is a 2×4 matrix function with $G_A(z) = G_A(z_b+) \cdot P(z_b, z)$ and

$$G_A(z_b+) = \frac{1}{p_{ref} r_{ref}^2} \begin{pmatrix} 0 & -r_{A4}(z_b)/r_{ref}^3 & 0 & r_{A2}(z_b)/r_{ref}^3 \\ 0 & 0 & 1 & 0 \end{pmatrix}. \qquad (3.78)$$

With N homogeneous solid sediment layers between z_b and z_h, the 2×2 matrices $E_1,..., E_N$, 4×4 matrices $L_1,..., L_{N+1}$, and 2×2 matrices A_1, B_1, C_1, $D_1,..., A_N$, B_N, C_N, D_N are as in Section 3.2.1.1. Starting from $_1R^- = -A_0^{-1} \cdot B_0$, where $G_A(z_b+) \cdot L_1 = (A_0, B_0)$, 2×2 reflection matrices $_1R^-$, $_1R^+$, $_2R^-,..., _NR^+$, $_{N+1}R^-$ for up-going waves are obtained recursively by

$$_nR^+ = E_n \cdot _nR^- \cdot E_n, \quad _{n+1}R^- = (A_n \cdot _nR^+ + B_n) \cdot (C_n \cdot _nR^+ + D_n)^{-1} \quad (3.79)$$

for $n = 1,..., N$. Fig. 3.7 is similar to Fig. 3.4, but for reflection matrices for up- rather than down-going waves in layer n: $a_n^- = _nR^- \cdot b_n^-$ at the upper interface and $a_n^+ = _nR^+ \cdot b_n^+$ at the lower one. The boundary conditions from above are used to express the response (the reflected down-going waves) in terms of the excitation (the incident up-going waves).

At the top of solid layer n, $n = 1,..., N + 1$ with $N + 1$ indicating the half-space, the boundary conditions from above and below on the displacement-stress vector r become

$$\begin{pmatrix} I & -_nR^- \\ -R_n^- & I \end{pmatrix} \cdot L_n^{-1} \cdot r = 0. \quad (3.80)$$

A natural value for the indicator function $I(z)$ at that depth is [32] $I_n = I(z_b-) J_n / J_1$ with

$$J_n = \frac{\left| \det \begin{pmatrix} I & -_nR^- \\ -R_n^- & I \end{pmatrix} \right|}{\| (I, -_nR^-) \| \| (-R_n^-, I) \|} = \frac{\left| 1 + \det(_nR^-) \det(R_n^-) - (_nR^-)^T \circ R_n^- \right|}{\| (I, -_nR^-) \| \| (-R_n^-, I) \|}, \quad (3.81)$$

where $I(z_b-)$ is defined by Eq. (3.77), $\| \cdot \|$ is some conveniently computed matrix norm, and the operator $\circ$ denotes the sum of the products of corresponding elements of two matrices. Mathematically, all I_n vanish, since the boundary conditions are

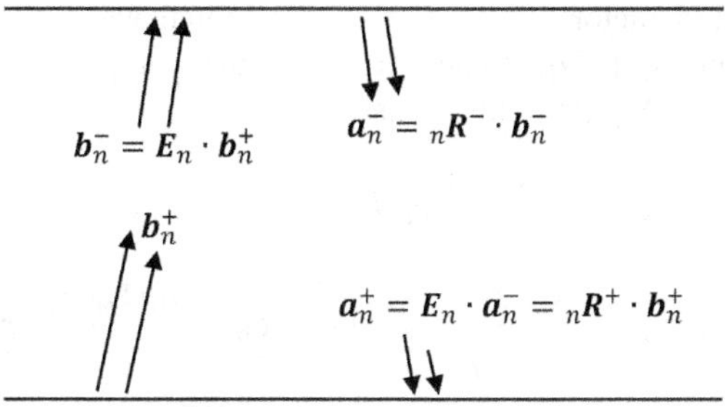

FIGURE 3.7

Solid layer n, $n = 1, 2,..., N$, with up- and down-going waves and reflection matrices for up-going waves. In the evanescent regime, the two up-going wave components increase downward, while the two down-going ones decrease downward.

linearly dependent for a modal wave number k_m. However, computationally spurious exponential growth at depths z outside a trapping zone may give rise to linearly independent boundary-condition row vectors of $G_A(z)$ and $G_B(z)$. To avoid undesired locations outside trapping zones of the matching-depth z_M, it is safe to include the I_n values from the solid bottom when the computed $I(z)$ is minimized, for each mode. Examples appear in Section 3.3.5.1.

If, for a certain mode, a $z_M > z_b$ is selected, $r(z_M)$ is first computed as a nontrivial solution to the linearly dependent conditions $G_A(z_M) \cdot r(z_M) = 0$ and $G_B(z_M) \cdot r(z_M) = 0$. This $r(z_M)$ can subsequently be safely propagated downward according to Eq. (3.61). For safe propagation upward through a solid layer n, the factor matrices in

$$L_n \cdot \begin{pmatrix} {_n}R^- \\ I \end{pmatrix} \cdot (0, E_n) \cdot L_n^{-1} \qquad (3.82)$$

are applied from the left, providing numerical control by the boundary conditions from above. Above z_b, the mode function $r_F(z)$ is proportional to $r_A(z)$, with proportionality factor determined by a match to $r(z_b+)$ as obtained by the propagation upward from z_M.

3.3.3 EXCITATION COEFFICIENTS

Eq. (3.75) includes expressions for excitation coefficients of the modes. However, drawbacks are that the numerical evaluation may become unstable in the evanescent regime and these expressions are not very efficient when results for many source depths are needed. As shown in Refs. [26,30], Eq. (3.75) can be reformulated as

$$p(r,z) = \frac{-iM}{2\lambda(z_s)} \sum_m \frac{r_{F4}(k_m, z_s) r_{F4}(k_m, z)}{N_m} k_m H_0^{(1)}(k_m r) \qquad (3.83)$$

for $0 \leq z_s, z \leq z_b$, where $r(k_m, z) = (r_1(k_m, z), r_2(k_m, z), r_3(k_m, z), r_4(k_m, z))^T$ is the mode function with $r_F(k_m, z) = (r_{F2}(k_m, z), r_{F4}(k_m, z))^T = (r_2(k_m, z), r_4(k_m, z))^T$ and N_m is a normalization factor for the mode with modal wave number k_m, $m = 1$, $2,...$ It is apparent that Eq. (3.83) respects the reciprocity principle from Eq. (3.14). The factors N_m are obtained as

$$\begin{aligned} N_m = &- \int_0^{z_b} w_F(k_m, z) \cdot \frac{\partial A_F(k_m, z)}{\partial k} \cdot r_F(k_m, z) dz \\ &- \int_{z_b}^{z_h} w(k_m, z) \cdot \frac{\partial A(k_m, z)}{\partial k} \cdot r(k_m, z) dz + c_B \cdot \frac{\partial G_B(k_m, z_h)}{\partial k} \cdot r(k_m, z_h), \end{aligned} \qquad (3.84)$$

where A and A_F are the system matrices of the ODE systems (3.42) and (3.44), respectively, $w = (r_3, r_4, -r_1, -r_2)$, $w_F = (r_4, -r_2)$, and c_B is the 1×2 row vector that solves $c_B \cdot G_B(k_m, z_h) = -w(k_m, z_h)$. The equation system for c_B is uniquely solvable [30], since $G_{BL} \cdot (G_{BR})^T$ is symmetric where $G_B = (G_{BL}, G_{BR})$ with $G_B(k_m, z_h)$ given by Eq. (3.55) has full rank.

The integrals in Eq. (3.84) can be expressed using propagator matrices [30]:

$$\int_{\zeta}^{z} w_F(k_m, s) \cdot \frac{\partial A_F(k_m, s)}{\partial k} \cdot r_F(k_m, s) ds = w_F(k_m, z) \cdot \frac{\partial P_F(k_m; z, \zeta)}{\partial k} \cdot r_F(k_m, \zeta) \quad (3.85)$$

for $0 \leq \zeta \leq z \leq z_b$, and

$$\int_{\zeta}^{z} w(k_m, s) \cdot \frac{\partial A(k_m, s)}{\partial k} \cdot r(k_m, s) ds = w(k_m, z) \cdot \frac{\partial P(k_m; z, \zeta)}{\partial k} \cdot r(k_m, \zeta) \quad (3.86)$$

for $z_b \leq \zeta \leq z \leq z_h$. These relations are very useful when, for example, the fluid and solid regions are formed by homogeneous layers for which the propagator matrices and their derivatives can be given in closed form (see Eqs. 3.50, 3.62, and 3.68).

Locally at depths $z \approx \zeta$ in a fluid region for which the vertical wave number $\xi_m(\zeta) = i\chi_m(\zeta) = \left(\omega^2/c^2(\zeta) - k_m^2\right)^{1/2}$ is positive real and at long ranges r, the contribution by mode m can be viewed as a superposition of plane waves of type $\exp[i(k_m r \pm \xi_m(\zeta)z)]$. This follows from Eq. (3.83) together with the asymptotic expression for the Hankel function and the depth-separated Helmholtz equation (3.46). Formally, the WKBJ approximation (e.g., Ref. [1]) can be applied. Thus, it is natural to consider $\varphi_m = \arctan(\xi_m(\zeta)/k_m)$ as the local propagation angle, relative to the horizontal, of mode m at depth ζ. Modes with high propagation angles tend to suffer significant losses by interaction with the typically lossy bottom. A mode stripping effect thus appears at propagation over long distances.

It should be recalled that a symmetric point source according to $f = f(x) = -M$ grad($\delta(x - x_s)$) has been assumed, where $x_s = (0, 0, z_s)$ is in the water and the time factor $\exp(-i\omega t)$ with $\omega > 0$ is omitted. With a vertical source array on the z axis, it is possible to excite one particular mode, while all others are absent [56].

3.3.4 RANGE-DEPENDENT MEDIA

Lateral variation is now allowed among a number of segments in range. Specifically, in the cylindrical coordinate system r, ϕ, z, the medium properties depend on depth z but not on range r or azimuth ϕ within each of the $N + 1$ range segments defined by $0 \leq r \leq r_1, r_1 < r \leq r_2, ..., r_{N-1} < r \leq r_N, r_N < r$. Here, see Fig. 3.8, $r_1, r_2, ..., r_N$

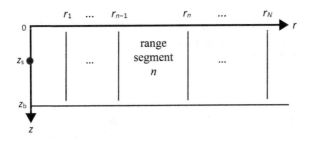

FIGURE 3.8

Cylindrically symmetric range-dependent fluid medium with depth (z) and range (r) axes. There is a point source on the z axis at depth z_s.

with $N \geq 1$ form an increasing sequence of positive range-segment limits. For simplicity, there is a traction-free or rigid boundary at $z = z_b$, common for all segments, and there is no solid region. As before, there is a symmetric time-harmonic point source with angular frequency ω in the water, on the axis ($r = 0$) at depth z_s. Omitting the time factor $\exp(-i\omega t)$ with $\omega > 0$, the body force, per unit volume, is given by $\boldsymbol{f} = -M \, \mathrm{grad}(\delta(r)\delta(z - z_s)/2\pi r)$.

Within each range segment, the pressure field $p(r, z)$ can now be expressed in terms of normal modes. Following Evans [57],

$$p(r,z) = \sum_{m=1}^{\infty} \left[a_{m,n} \widehat{H}_{m,n}^{(1)}(r) + b_{m,n} \widehat{H}_{m,n}^{(2)}(r) \right] Z_{m,n}(z), \qquad (3.87)$$

where $n = n(r)$ is the range-segment index, i.e., $n = 1$ for $0 \leq r \leq r_1$, $n = 2$ for $r_1 < r \leq r_2, \ldots$, $n = N$ for $r_{N-1} < r \leq r_N$, $n = N + 1$ for $r_N < r$. Furthermore, $Z_{m,n}(z)$ is the mode function {unit: N}, i.e., $r_{F4}(k_{m,n}, z)$ for segment n, for the mth mode in the nth range segment, and $k_{m,n}$ is its horizontal wave number. The $\widehat{H}_{m,n}^{(1)}(r)$ and $\widehat{H}_{m,n}^{(2)}(r)$ are Hankel functions, normalized to avoid computational overflow problems. With $r_{N+1} = r_N$, $\widehat{H}_{m,1}^{(1)}(r) = H_0^{(1)}(k_{m,1}r)$ and $\widehat{H}_{m,n}^{(1)}(r) = H_0^{(1)}(k_{m,n}r)/H_0^{(1)}(k_{m,n}r_{n-1})$ for $n = 2, \ldots, N + 1$, and $\widehat{H}_{m,n}^{(2)}(r) = H_0^{(2)}(k_{m,n}r)/H_0^{(2)}(k_{m,n}r_n)$ for $n = 1, 2, \ldots, N + 1$. For each $n = 1, 2, \ldots, N + 1$, with $I_{m,n} = \int \rho_n^{-1}(z) Z_{m,n}^2(z) dz$, the mode functions fulfill the Sturm–Liouville orthogonality relations [50]

$$\int_0^{z_b} \frac{1}{\rho_n(z)} Z_{m1,n}(z) Z_{m2,n}(z) dz = I_{m1,n} \delta_{m1m2}. \qquad (3.88)$$

Here, $\rho_n(z)$ denotes the density function for range segment n. The corresponding Lamé-parameter function $\lambda(z)$, needed below, is denoted $\lambda_n(z)$.

The problem is then to determine the modal expansion coefficients {unit: m^{-2}} $a_{m,n}$ and $b_{m,n}$, $m = 1, 2, \ldots$, $n = 1, 2, \ldots, N + 1$.

3.3.4.1 Equations Relating the Modal Expansion Coefficients

With the horizontal wave numbers $k_{m,n}$ of the modes in the upper half-plane, but not on the negative half-axis, the $a_{m,n}$ and $b_{m,n}$ terms in Eq. (3.87) represent out-going and in-coming components of the pressure field, respectively. It is convenient to define column vectors according to $\boldsymbol{a}_n = (a_{1,n}, a_{2,n}, \ldots)^{\mathrm{T}}$ and $\boldsymbol{b}_n = (b_{1,n}, b_{2,n}, \ldots)^{\mathrm{T}}$, $n = 1, 2, \ldots, N + 1$. There cannot be any in-coming components in the outermost range segment, hence

$$\boldsymbol{b}_{N+1} = \boldsymbol{0}. \qquad (3.89)$$

Recalling that $H_0^{(1)}(k_{m,1}r) + H_0^{(2)}(k_{m,1}r) = 2 J_0(k_{m,1}r)$ is regular at $r = 0$, the excitation-coefficient result of Eq. (3.83) with $N_m = 2k_m I_m/\omega^2$, or Eq. (3.11) together with Eq. (3.88), provide an equation for $\boldsymbol{a}_1$ and $\boldsymbol{b}_1$ according to

$$a_{m,1} - b_{m,1} \Big/ H_0^{(2)}(k_{m,1}r_1) = -\frac{iM\omega^2}{4\lambda_1(z_s)} \frac{Z_{m,1}(z_s)}{I_{m,1}}, \quad m = 1, 2, \dots. \qquad (3.90)$$

For $n = 1, 2, \dots, N$, it is now convenient to introduce the diagonal matrices $\widehat{\boldsymbol{H}}_n^{(1)} = \mathrm{diag}\left(\widehat{H}_{m,n}^{(1)}(r_n), m = 1, 2, ..\right)$ and $\widehat{\boldsymbol{H}}_{n+1}^{(2)} = \mathrm{diag}\left(\widehat{H}_{m,n+1}^{(2)}(r_n), \quad m = 1, 2, \dots\right)$, where $\widehat{\boldsymbol{H}}_{N+1}^{(2)} = \mathbf{I}$ since $r_{N+1} = r_N$. Using continuity of pressure and radial displacement at each range segment interface r_n, $n = 1, 2, \dots, N$, in a Galerkin approach, Eq. (3.87) together with Eq. (3.7) and the orthogonality relations (3.88) show that

$$\begin{pmatrix} \boldsymbol{a}_{n+1} \\ \widehat{\boldsymbol{H}}_{n+1}^{(2)} \cdot \boldsymbol{b}_{n+1} \end{pmatrix} = \begin{pmatrix} \boldsymbol{A}_n & \boldsymbol{B}_n \\ \boldsymbol{C}_n & \boldsymbol{D}_n \end{pmatrix} \cdot \begin{pmatrix} \widehat{\boldsymbol{H}}_n^{(1)} \cdot \boldsymbol{a}_n \\ \boldsymbol{b}_n \end{pmatrix} \qquad (3.91)$$

for $n = 1, 2, \dots, N$. With $\boldsymbol{H}_{n+1} = \left(\boldsymbol{E}_{n+1}^{(1)} + \boldsymbol{E}_{n+1}^{(2)}\right)^{-1}$, the coupling matrices $\boldsymbol{A}_n$, $\boldsymbol{B}_n$, $\boldsymbol{C}_n$, and $\boldsymbol{D}_n$ are

$$\boldsymbol{A}_n = \boldsymbol{H}_{n+1} \cdot \left(\boldsymbol{E}_{n+1}^{(2)} \cdot \boldsymbol{F}_n + \boldsymbol{G}_n \cdot \boldsymbol{\Lambda}_n^{(1)}\right), \quad \boldsymbol{B}_n = \boldsymbol{H}_{n+1} \cdot \left(\boldsymbol{E}_{n+1}^{(2)} \cdot \boldsymbol{F}_n - \boldsymbol{G}_n \cdot \boldsymbol{\Lambda}_n^{(2)}\right),$$

$$\boldsymbol{C}_n = \boldsymbol{H}_{n+1} \cdot \left(\boldsymbol{E}_{n+1}^{(1)} \cdot \boldsymbol{F}_n - \boldsymbol{G}_n \cdot \boldsymbol{\Lambda}_n^{(1)}\right), \quad \boldsymbol{D}_n = \boldsymbol{H}_{n+1} \cdot \left(\boldsymbol{E}_{n+1}^{(1)} \cdot \boldsymbol{F}_n + \boldsymbol{G}_n \cdot \boldsymbol{\Lambda}_n^{(2)}\right),$$

where $\boldsymbol{E}_{n+1}^{(1)} = \mathrm{diag}\left(iH_1^{(1)}(k_{m,n+1}r_n) \big/ H_0^{(1)}(k_{m,n+1}r_n), m = 1, 2, \dots\right)$,
$\boldsymbol{E}_{n+1}^{(2)} = \mathrm{diag}\left(-iH_1^{(2)}(k_{m,n+1}r_n) / H_0^{(2)}(k_{m,n+1}r_n), m = 1, 2, \dots\right)$,
$\boldsymbol{\Lambda}_n^{(1)} = \mathrm{diag}\left(iH_1^{(1)}(k_{m,n}r_n) / H_0^{(1)}(k_{m,n}r_n), m = 1, 2, \dots\right)$, and
$\boldsymbol{\Lambda}_n^{(2)} = \mathrm{diag}\left(-iH_1^{(2)}(k_{m,n}r_n) / H_0^{(2)}(k_{m,n}r_n), m = 1, 2, \dots\right)$

are diagonal matrices that are close to identity matrices for large arguments, and the elements of the mode coupling matrices $\boldsymbol{F}_n$ and $\boldsymbol{G}_n$ are given by

$$(\boldsymbol{F}_n)_{m1,m2} = \frac{1}{I_{m1,n+1}} \int_0^{z_b} \frac{Z_{m1,n+1}(z)Z_{m2,n}(z)}{\rho_{n+1}(z)} dz \qquad (3.92)$$

$$(\boldsymbol{G}_n)_{m1,m2} = \frac{k_{m2,n}}{k_{m1,n+1}} \frac{1}{I_{m1,n+1}} \int_0^{z_b} \frac{Z_{m1,n+1}(z)Z_{m2,n}(z)}{\rho_n(z)} dz. \qquad (3.93)$$

Taken together, Eqs. (3.89)–(3.91) provide a system of $2(N+1)$ linear matrix equations for the $2(N+1)$ vectors $\boldsymbol{a}_n$ and $\boldsymbol{b}_n$, $n = 1, 2, \dots, N+1$. In practice, a truncation is of course made to a finite number of modes, say N_M, in each range segment. A reasonable number of leaky or evanescent modes, having horizontal wave numbers with positive imaginary parts, must typically be included in addition to the propagating modes with positive real horizontal wave numbers.

3.3.4.2 Solution in Terms of Reflection-Coefficient Matrices

The direct propagator-matrix multiplication approach utilized in Ref. [57] is not always reliable for solving the equation system provided by Eqs. (3.89)–(3.91). (It would work for $N_M = 1$, however, when the boundary condition at the source end, Eq. (3.90), is only needed to set a scaling factor, cf. Section 3.2.1.2.) The problem is that the diagonal matrices $\widehat{\boldsymbol{H}}_n^{(1)}$ and $\widehat{\boldsymbol{H}}_{n+1}^{(2)}$ may contain some very small elements leading to an exponential dichotomy with exponentially growing as well as decaying mode amplitudes with range. A decoupling algorithm by Mattheij for solving two-point boundary-value problems is applied in Ref. [58], while the direct global matrix method is applied in Refs. [59,60], and also in Ref. [61], where the global matrix equation system is solved by a particular algorithm for inverting block pentadiagonal matrices. Here, however, the reflection-coefficient method used in Section 3.2.1.1 is adapted for a convenient and numerically stable two-way marching solution. Basically, cf. Section 3 in Ref. [62], reflection matrices are propagated for decreasing r without exponential growth and with maintained full rank. The development in Ref. [62] concerns media with continuous range dependence; reflection-matrix differential equations of Riccati type are derived for expansions using modes for a reference structure.

Eq. (3.89) can be expressed as $\widehat{\boldsymbol{H}}_{N+1}^{(2)} \cdot \boldsymbol{b}_{N+1} = \boldsymbol{R}_{N+1}^- \cdot \boldsymbol{a}_{N+1}$, where $\boldsymbol{R}_{N+1}^- = \boldsymbol{0}$ is interpreted as a reflection-coefficient matrix for out-going waves at range r_N+. Corresponding reflection-coefficient matrices, denoted $\boldsymbol{R}_N^+, \boldsymbol{R}_N^-, \ldots, \boldsymbol{R}_1^+$, are now introduced at ranges r_N-, $r_{N-1}+, \ldots$, r_1-. Specifically, $\boldsymbol{b}_n = \boldsymbol{R}_n^+ \cdot \widehat{\boldsymbol{H}}_n^{(1)} \cdot \boldsymbol{a}_n$ and $\widehat{\boldsymbol{H}}_n^{(2)} \cdot \boldsymbol{b}_n = \boldsymbol{R}_n^- \cdot \boldsymbol{a}_n$ for $n = N, \ldots, 2$, and $\boldsymbol{b}_1 = \boldsymbol{R}_1^+ \cdot \widehat{\boldsymbol{H}}_1^{(1)} \cdot \boldsymbol{a}_1$. Fig. 3.9 is analogous to Fig. 3.4, but layer n is replaced by range segment n, $n = N, \ldots, 2$. Instead of two down-going and two up-going cylindrical waves, there are now lots of out-going and in-coming modes with three of each type indicated in the figure. With the

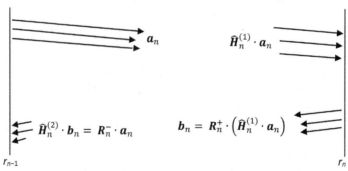

FIGURE 3.9

Range segment n, $n = N, \ldots, 2$, is shown with relevant coefficients for out-going and in-coming waves. In the evanescent regime, the out-going wave components increase inward, while the in-coming ones decrease inward.

reflection matrices, the outer boundary conditions $b_{N+1} = 0$ are naturally used to express the response (the reflected in-coming waves) in terms of the excitation (the incident out-going waves).

Applying Eq. (3.91), the reflection matrices can be computed recursively by

$$R_n^+ = -\left(D_n - R_{n+1}^- \cdot B_n\right)^{-1} \cdot \left(C_n - R_{n+1}^- \cdot A_n\right) \quad , \quad R_n^- = \widehat{H}_n^{(2)} \cdot R_n^+ \cdot \widehat{H}_n^{(1)} \quad (3.94)$$

for $n = N,..., 2$, and

$$R_1^+ = -\left(D_1 - R_2^- \cdot B_1\right)^{-1} \cdot \left(C_1 - R_2^- \cdot A_1\right). \quad (3.95)$$

The recursion by Eqs. (3.94) and (3.95) is quite analogous to the one by Eq. (3.59). No exponential growth is involved, and the exponentials in the asymptotic large-argument expressions for the Hankel functions typically cause a decrease in magnitude when R_n^- is obtained from R_n^+. For a pressure-release or rigid boundary at $r = r_N$, the recursion starts with $R_N^+ = -I$ or $R_N^+ = \left(A_N^{(2)}\right)^{-1} \cdot A_N^{(1)}$, respectively.

When R_1^+ has been determined, $b_1 = R_1^+ \cdot \widehat{H}_1^{(1)} \cdot a_1$ is substituted in Eq. (3.90), which can then be solved for a_1, and b_1 is readily obtained. In analogy to Eqs. (3.61), (3.70), and (3.82), the remaining coefficient vectors are not computed by a direct application of Eq. (3.91) but by stabilized marching, controlled by the outer boundary condition $b_{N+1} = 0$, with matrix multiplications from the right according to

$$\begin{pmatrix} a_{n+1} \\ b_{n+1} \end{pmatrix} = \begin{pmatrix} I \\ R_{n+1}^+ \cdot \widehat{H}_{n+1}^{(1)} \end{pmatrix} \cdot (A_n, B_n) \cdot \begin{pmatrix} \widehat{H}_n^{(1)} \cdot a_n \\ b_n \end{pmatrix} \quad (3.96)$$

for $n = 1, 2,..., N$, where $R_{N+1}^+ = R_{N+1}^- = 0$ (and $\widehat{H}_{N+1}^{(1)} = I$).

3.3.4.3 Final Remarks

The reflection-coefficient method proposed in Section 3.3.4.2 is economical, since matrix inversion is only needed for the rather small matrices of dimension $N_M \times N_M$, where N_M is the number of modes, after truncation, in each range segment. Compared with the two-way (outward followed by inward) marching single-scattering solution proposed in Ref. [63], the reduction to small dimensions is achieved without additional approximations. In the notation here, the outward marching step would start with $b_1 = 0$ and a_1 according to Eq. (3.90), after which Eq. (3.91) would be solved for increasing n setting $b_{n+1} = 0$ all the time, giving $a_{n+1} = \left(A_n - B_n \cdot D_n^{-1} \cdot C_n\right) \cdot \widehat{H}_n^{(1)} \cdot a_n$ and $b_n = -D_n^{-1} \cdot C_n \cdot \widehat{H}_n^{(1)} \cdot a_n$. At a subsequent inward marching step, Eq. (3.91) would be solved for decreasing n temporarily setting $a_n = 0$ all the time but adding the b_n from the outward marching step giving $b_n = D_n^{-1} \cdot \left(\widehat{H}_{n+1}^{(2)} \cdot b_{n+1} - C_n \cdot \widehat{H}_n^{(1)} \cdot a_n\right)$. The approximation could obviously be improved by further outward and inward marching steps.

With appropriate ordering of the modes in the different range segments, all off-diagonal elements in the coupling matrices are ignored in the well-known adiabatic approximation (e.g., Ref. [7]). The resulting diagonal matrices are, of course, easy to invert.

The assumption of a traction-free or rigid lower boundary at $z = z_b$ is useful, since no branch-cut integral is needed for the representation of the field. Undesired reflections from the lower boundary can be reduced by artificially introducing layers at depth with successively increasing absorption. A gradient half-space at depth is also possible [64]. The task of handling media with a solid bottom appears to remain as a research problem.

With this approach, a sloping bottom must be replaced by a number of stair steps. To compute the backward field accurately, the $b_{m,n}$ part of Eq. (3.87), the stair steps should preferably be shorter than a quarter of a wavelength, at least in the up-sloping case [65]. For the forward field, however, the $a_{m,n}$ part of Eq. (3.87), horizontal steps that are several wavelengths long are often sufficient when the bottom slope is gentle [65]. If N_M is proportional to ω, the computational work appears to increase as ω^4 when the frequency is increased, but the power four here may be reduced by utilizing diagonal dominance (cf. the adiabatic approximation) of the $N_M \times N_M$ matrices to be inverted.

As an alternative, the field can be expanded in terms of local modes, defined for each range r by considering the medium locally range-invariant, without using stair steps [66–69]. For a sloping bottom, the corresponding series typically converges very slowly, however, since the local modes do not satisfy the appropriate boundary condition at depth. Modifications of the local modes are desirable and have been attempted [70]. An initial step in Ref. [71], where transformations of the dependent variables are derived to reduce mode couplings, is to make an orthogonal mapping of layers onto rectangles, effectively removing bottom slope in the new coordinates.

3.3.5 EXAMPLES

3.3.5.1 Range-Invariant Media

The perhaps simplest mode-expansion example is a homogeneous water waveguide with sound speed c and density ρ, extending from a traction-free surface at $z = 0$ to a rigid bottom at $z = H$. From Eq. (3.62), the dispersion function can be expressed as $D(k) = \cos(H\xi) = 0$, where $\xi = (\omega^2/c^2 - k^2)^{1/2}$ is the vertical wave number. Thus, the modal wave numbers k_m appear as $k_m = \pm\left[(\omega/c)^2 - ((m - 1/2)\pi/H)^2\right]^{1/2}$, $m = 1, 2,...,$ with corresponding mode functions $Z_m(z) = r_{F4}(k_m, z) = p_{ref} r_{ref}^2 \sin((m - \frac{1}{2})\pi z/H)$.

For a real c, the modal wave number k_m is real and mode m is called propagating when $\omega > \omega_m$, where $\omega_m = (m - \frac{1}{2})\pi c/H$ is the angular cutoff frequency. For smaller ω, k_m is apparently imaginary and the mode is evanescent. Including absorption, $1/c$ moves up into the first quadrant of the complex plane. The derivation of Eq. (3.75) shows that the k_m for the modal expansion should be selected in the upper half-plane, and a consideration of vanishingly small absorption indicates that the real k_m on the positive axis should be selected rather than those on the negative axis. (For more complex media, however, with a solid bottom, there are exceptions to this rule [45].) In the present case, the modal expansion according to Eq. (3.83) takes the simple form

$$p(r,z) = -\frac{iM\omega^2}{2Hc^2} \sum_{m=1}^{\infty} \sin\left(\frac{(m-1/2)\pi z_s}{H}\right) \sin\left(\frac{(m-1/2)\pi z}{H}\right) H_0^{(1)}(k_m r). \quad (3.97)$$

An evanescent mode is apparently exponentially damped with increasing range r, while the amplitude of a propagating mode is reduced by cylindrical spreading only. The number of propagating modes is close to $\omega H/\pi c$. Computations for an up-sloping bottom reveal how the modes are successively lost into bottom, cf. Fig. 3.19.

For more complicated examples, the modes must in general be computed numerically. The 500 Hz shallow-water example from Sections 3.1.2 and 3.2.3 is now revisited, using the techniques of Sections 3.3.1–3.3.3. There are some 650 modal wave numbers in the rectangle defined by $|\mathrm{Re}(k/\omega)| \leq 5.0$ s/km and $0.0 \leq \mathrm{Im}(k/\omega) \leq 5.0$ s/km, and they are all shown in Fig. 3.10. Hyperbolic branch cuts are used, as described at the beginning of Section 3.3. As is the case with the previous more simple example, most of the modal wave numbers gather close to an interval on the positive real axis and close to the imaginary axis. There are 20 modes with $\mathrm{Re}(k/\omega) > 0.75$ s/km. Their mode functions are concentrated in the bottom sediments and, according to Eq. (3.83), they are hardly excited by a shallow source in the water. The mode farthest to the right, with $\mathrm{Re}(k/\omega) \approx 2.36$ s/km, is a Scholte mode (e.g., Ref. [12]), that is concentrated close to the fluid–solid interface in the bottom at depth 110 m. Its mode shape is shown by the gray curve in the right panel of Fig. 3.11.

The modes close to the real axis with $\mathrm{Re}(k/\omega)$ less than 0.75 s/km are typically concentrated in the water column. Farthest to the right among them,

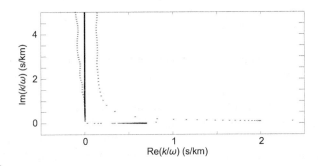

FIGURE 3.10

Modal wave numbers, denoted by *crosses*, for the 500 Hz shallow-water example from Sections 3.1.2 and 3.2.3. (The mirror modes in the lower half-plane, with $\mathrm{Im}(k/\omega) < 0$, are not included.)

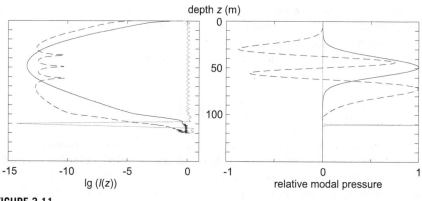

FIGURE 3.11

Right: mode shapes, as real part of pressure, divided by pressure with maximal magnitude, for three modes from Fig. 3.10: Scholte mode (gray), fundamental water mode (*solid black*), and another water mode (*dashed black*). Left: corresponding indicator functions used to select appropriate matching depths (Section 3.3.2).

with $k/\omega \approx 0.7009$ s/km, the black solid curve in the right panel of Fig. 3.11, is the fundamental water mode that is concentrated in the sound channel with low sound speed at depths between 20 and 70 m, cf. the left panel of Fig. 3.1. When excited, it propagates very efficiently, hardly affected by bottom losses.

The peaks in Fig. 3.5 can be related to modes that contribute significantly to the field for the particular source and receiver depths considered there (10 and 90 m, respectively). The highest peak is at $k/\omega \approx 0.6965$ s/km. Indeed, there is a corresponding mode, and the dashed black curve in the right panel of Fig. 3.11 shows its shape.

The left panel shows the indicator functions $I(z)$, cf. Section 3.3.2, used to select appropriate matching depths z_M: 110.0 m for the Scholte mode, 48.4 m for the fundamental water mode, and 71.4 m for the additional water mode. Indeed, minimization of the pertinent $I(z)$ provides a matching depth within the trapping zone of each mode.

According to the asymptotic expression for the Hankel function, the imaginary part $\text{Im}(k)$ of a modal wave number k gives rise to a far-field modal damping by $20 \, \text{Im}(k) \, \lg(e)$ dB/km. Only the 12 modes from Fig. 3.10 with 0.685 s/km $< \text{Re}(k/\omega) < 0.705$ s/km have a damping less than 1 dB/km.

The black curve in the left panel of Fig. 3.12 shows the PL as computed by mode excitation. Only the near field is included, where differences can be seen compared with the wave number integration and ray results shown in the right panel. Perfect agreement with the wave number integration result is achieved by adding the contribution from integration around the hyperbolic branch cuts. This branch-cut contribution, also computed by adaptive wave number integration (with variable transformations to obtain regularity at the branch points), is shown by the gray curve

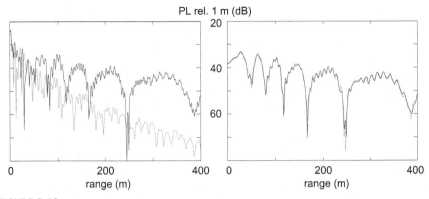

FIGURE 3.12

PL in dB, rel. 1 m, for the 500 Hz example with source and receiver depths 10 and 90 m, respectively. Left: modal part of the wave field (*black*) and branch-cut integral part (*gray*). Right: wave number integration (*black*) and ray (*gray*) results from Figs. 3.2 and 3.6, respectively.

in the left panel. As expected, it decays steadily with range, and it is insignificant in this example beyond a kilometer or so. It can be noted that the near-field ray results in the right panel are very good.

Fig. 3.13 shows the pressure field as a function of depth z for two different source depths (10 and 50 m) and two different source−receiver ranges (1 and 10 km). The

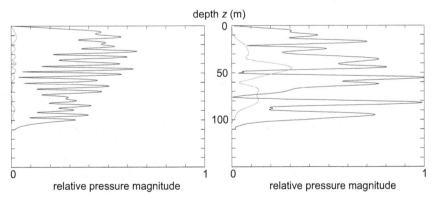

FIGURE 3.13

Pressure field, computed by wave number integration, as function of depth in the water column, and in the fluid sediment layer, for the 500 Hz example. In each panel, the *black curve* is for the range 1 km and the *gray curve* is for the range 10 km. Left: source depth 10 m. Right: source depth 50 m.

mode stripping effect is apparent, with much reduced mode interference at the 10 km range. With source depth 50 m, the fundamental water mode from Fig. 3.11 clearly dominates at 10 km. The water modes are, of course, best excited with a source in the middle of the water column in this case, and the pressure levels are typically higher in the right panel. Observation of the Scholte mode would require source as well as receiver close to the bottom (cf. Eq. 3.83). Lower frequencies and shorter ranges would also be preferable, to reduce the influence of the sediment damping.

The field decomposition in the right panel of Fig. 3.6 is not very convenient to compute by modes. For cycle numbers $\nu \geq 1$, the reflection coefficients needed for the field expansion cause higher-order poles, and the corresponding residues involve higher-order derivatives that are not readily computed. Nevertheless, the modal approach can be useful for a physical understanding of shadow-zone fields [2,9]. A few examples, using the recombinant expansion of Ref. [43] are found in Ref. [72]. Leaky modes, with positive imaginary parts of their horizontal wave numbers k, are needed to achieve the correct field decay at a shadow boundary. Individual leaky-mode contributions can be large at short ranges, however, where the underlying residues series may even diverge [2,73].

3.3.5.2 Range-Dependent Media

With a symmetric point source at $r = 0$, the Helmholtz equation (3.8) becomes

$$\frac{1}{r}\frac{\partial}{\partial r}\left(r\frac{\partial p}{\partial r}\right) + \rho\frac{\partial}{\partial r}\left(\frac{1}{\rho}\right)\frac{\partial p}{\partial r} + \rho\frac{\partial}{\partial z}\left(\frac{1}{\rho}\frac{\partial p}{\partial z}\right) + (\omega^2/c^2)p = 0 \qquad (3.98)$$

for $r > 0$, where $c = c(r, z)$ and $\rho = \rho(r, z)$ are functions of range as well as depth, but not azimuth. In terms of the dependent variable $\tilde{p} = (\rho_0/\rho)^{1/2}p$, where $\rho_0 = \rho_0(z)$ is a range-invariant reference density profile, Eq. (3.98) becomes

$$\frac{1}{r}\frac{\partial}{\partial r}\left(r\frac{\partial\tilde{p}}{\partial r}\right) + \rho_0\frac{\partial}{\partial z}\left(\frac{1}{\rho_0}\frac{\partial\tilde{p}}{\partial z}\right) + (\omega^2/\tilde{c}^2)\tilde{p} = 0, \qquad (3.99)$$

where $\tilde{c}^2$ is defined by $\omega^2/\tilde{c}^2 = \omega^2/c^2 + \Delta\rho/2\rho - 3(\mathrm{grad}(\rho))^2/4\rho^2 - (\partial^2\rho_0/\partial z^2)/2\rho_0 + 3(\partial\rho_0/\partial z)^2/4\rho_0^2$. As is well known [74], p can thus be determined by first solving the Helmholtz equation (3.99), with a range-invariant density.

Restriction is now made to a very simple kind of range-dependence without mode coupling, the one briefly considered in Ref. [75] (Section 7.1.2) with a fluid medium for which $1/c^2(r,z) = 1/c_0^2(z) + S(r)$ and $\rho(r, z) = \rho_0(z)R(r)$, where $c_0(z)$ and $\rho_0(z)$ are range-invariant sound-speed and density profiles, respectively. For simplicity, it is assumed that $S(r) = 0$ and $R(r) = 1$ for large r. At the selected angular frequency $\omega > 0$, the modal wave numbers and mode functions, corresponding to the profiles $c_0(z)$ and $\rho_0(z)$, are denoted $k_{m,0}$ {unit: m^{-1}} and $Z_{m,0}(z)$ {unit: N}, respectively, where $m = 1, 2,...$ The integrals I_m (cf. Eq. 3.88) are defined by $\int \rho_0^{-1}(z)Z_{m1,0}(z)Z_{m2,0}(z)dz = I_{m1}\delta_{m1m2}$.

With $\tilde{p}(r,z) = R^{-1/2}(r)p(r,z)$, the Helmholtz equation(3.99) is now fulfilled with $\tilde{c}(r,z)$ defined by $1/\tilde{c}^2(r,z) = 1/c_0^2(z) + \tilde{S}(r)$, where $\tilde{S}(r) = S(r) + \omega^{-2}\left[R''(r)/2R(r) - 3(R'(r)/2R(r))^2 + R'(r)/2rR(r)\right]$. At each range $r > 0$, $Z_{m,0}(z)$ fulfills the depth-separated Helmholtz equation

$$\rho_0(z)\frac{\partial}{\partial z}\left(\frac{1}{\rho_0(z)}\frac{\partial Z_{m,0}}{\partial z}\right) + \left(\frac{\omega^2}{\tilde{c}^2(r,z)} - \tilde{k}_m^2(r)\right)Z_{m,0} = 0, \qquad (3.100)$$

cf. Eq. (3.46), with horizontal wave number $\tilde{k}_m(r) = \left(k_{m,0}^2 + \omega^2\tilde{S}(r)\right)^{1/2}$, $m = 1$, 2,..., where $\tilde{k}_m(r)$ is in the upper half-plane but not on the negative real axis. It follows, by solving the Helmholtz equation (3.99) with separation of variables, that

$$p(r,z) = R^{1/2}(r)\tilde{p}(r,z) = R^{1/2}(r)\sum_{m=1}^{\infty}H_m(r)Z_{m,0}(z), \qquad (3.101)$$

where $H_m(r)$ {unit: m^{-2}} is proportional to $H_0^{(1)}(k_{m,0}r)$ for large r and satisfies the ODE

$$\frac{d^2H_m(r)}{dr^2} + \frac{1}{r}\frac{dH_m(r)}{dr} + \tilde{k}_m^2(r)H_m(r) = 0 \qquad (3.102)$$

for $r > 0$. The source condition at $r = 0$ determines the proportionality constant, for example by Eq. (3.11). Eq. (3.102) may be written as a first-order ODE system for $H_m(r)$ and $dH_m(r)/dr$. The resulting two-dimensional boundary-value problem can be solved by propagator-matrix techniques involving safe transportation of the boundary-condition at large r [29], possibly in reflection-coefficient form Ref. [32], toward $r = 0$.

For the particular case with $N + 1$ range-invariant segments according to Fig. 3.8, $\tilde{S}(r) = S(r)$ and $R(r)$ in range segment n are denoted S_n and R_n, respectively, for $n = 1, 2,..., N + 1$, and the reflection-coefficient method from Section 3.3.4 can be applied. The modal wave numbers and mode functions in segment n can be written $\tilde{k}_{m,n} = k_{m,n} = \left(k_{m,0}^2 + \omega^2 S_n\right)^{1/2}$ and $Z_{m,n}(z) = R_n^{1/2}Z_{m,0}(z)$, respectively, for $m = 1, 2,..., n = 1, 2,..., N + 1$. In this way, the mode coupling matrices F_n and G_n become simple diagonal matrices: $F_n = \text{diag}\left((R_n/R_{n+1})^{1/2}, m = 1, 2, ...\right)$ and $G_n = \text{diag}\left((k_{m,n}/k_{m,n+1})(R_{n+1}/R_n)^{1/2}, m = 1, 2, ...\right)$.

Some interesting effects of range-dependence can now be illustrated, albeit without mode coupling. Varying S_n among the range segments, a particular mode m may change between being propagating ($k_{m,n}$ close to the positive real axis) or evanescent ($k_{m,n}$ close to the positive imaginary axis). The number of propagating modes may thus vary with range, as they also do for cases with a sloping bottom (cf. the initial example in Section 3.3.5.1, where the number of propagating modes is proportional to the bottom depth H).

The 500 Hz shallow-water example from Sections 3.1.2 and 3.2.3, revisited in the foregoing, is now modified to include range dependence in this particular way. Initially, however, shear-waves are excluded from the bottom to get a purely fluid example, and the lower half-space is replaced by a 60 m (10 wavelengths at 500 Hz) thick gradient layer above a rigid bottom. The absorption in the gradient layer increases to 10 dB/wavelength at its bottom to damp out the down-going waves before they are reflected by the rigid bottom. Propagation loss curves of the type shown in Fig. 3.6 are not visibly affected by these changes. A mode plot such as that in Fig. 3.10 is changed, however. The Scholte mode and the sediment modes with $\mathrm{Re}(k/\omega) > 0.71$ s/km disappear, but they are not noticeably excited by a 500 Hz source in the water column anyway, and a sequence of modes appear close to the previous compressional-wave hyperbolic branch cut (cf. examples in Ref. [53]). In fact, a strong motivation for the modification is to replace the branch-cut integral, which is inconvenient when range-dependence is introduced, by mode contributions.

Fig. 3.14 shows PL curves for two island examples, selected with more regard to simplicity of illustration than realism. For the left panel, a segment with coarse clay is introduced for 2.3 km $< r <$ 2.7 km. In this segment, the medium densities are multiplied with 1.4, i.e., $R(r) = 1.4$ there, and the function $S(r)$ is selected to introduce an absorption of 0.08 dB/wavelength at the source depth (10 m) there. Compared with the black curve in the left panel of Fig. 3.6, the pressure levels are reduced by about 10 dB beyond the island, because of absorption losses in the clay and backward reflections. As shown by the gray curve, however, the backward reflected field (the $b_{m,n}$ part of Eq. 3.87) is rather small in comparison with the forward field (the $a_{m,n}$ part). Before the island, the total field is in fact very similar to the curve from Fig. 3.6.

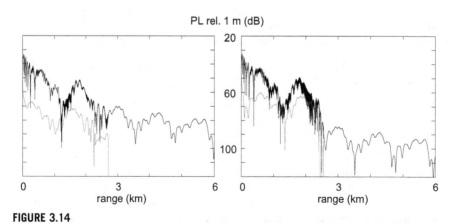

PL rel. 1 m (dB)

FIGURE 3.14

PL in dB, rel. 1 m, for two range-dependence modifications of the 500 Hz example. The *gray curve* in each panel concerns the backward field, i.e., the $b_{m,n}$ part of Eq. (3.87). Left: an island with coarse clay appears for 2.3 km $< r <$ 2.7 km. Right: an island with coarse sand appears for 2.45 km $< r <$ 2.55 km.

The right panel involves a more significant contrast, with coarse sand introduced for 2.45 km $< r <$ 2.55 km. In this segment, the medium densities are multiplied with 2.2, i.e., $R(r) = 2.2$ there, and $S(r)$ is selected to provide a sound speed increase from 1454 to 1800 m/s and an absorption of 0.87 dB/wavelength at the source depth (10 m) there. The backscattered field is now more significant and it interferes with the forward field. The absorption in the sand causes a drop by about 25 dB when the sound wave passes through.

It should be noted that the 2-D modeling in these examples, with azimuthal symmetry, implies that the islands are actually ring-shaped around the source. Hence, enhanced backscattering appears close to the source because of focusing effects. Indeed, the Hankel-function factor for the $b_{m,n}$ component of Eq. (3.87) is singular at $r = 0$. Islands of disc shape give rise to 3-D effects, and Section 3.6.3 provides examples.

3.4 PARABOLIC EQUATION METHODS

The Helmholtz equation (3.98) accommodates out-going as well as in-coming (reflected, backscattered) waves, since a second derivative $\partial^2 p/\partial r^2$ with respect to range is involved. In a medium where the range dependence is weak and smooth, an analysis of mode-coupling equations shows that the energy of the backscattered waves is very small [75] (Section 7.1.3). The clay island example in Section 3.3.5.2 supports this statement. An approximation where the backscattered waves are neglected is thus of interest, with a corresponding parabolic or, with a more correct term, one-way wave equation that only includes first derivatives $\partial p/\partial r$ with respect to range. Its solution could be conveniently marched outward in range, starting from a given field at an initial range. Compared with the coupled-mode approach, even with restriction to the outward marching step in the single-scattering approximation indicated in Section 3.3.4.3, significant gains in computer time are anticipated when the mode structure in all the range segments need not be resolved. The condition of weak and smooth range dependence is often fulfilled in the ocean, and the parabolic equation (PE) method is often a good choice. Recent applications to soundscape research (2014) appear in Ref. [76], for example.

As in Section 3.3.4, a fluid medium with range-invariant segments with common depth z_b is now assumed. It is traction-free or rigid at $z = z_b$, and its lower part is typically occupied by a thick gradient layer, 10 wavelengths or more, where the absorption increases to about 10 dB/wavelength. In this way (cf. the examples in Section 3.3.5.2), the down-going waves are effectively damped out before being reflected at $z = z_b$. Transparent boundary conditions [77] could alternatively be applied, allowing the computational domain to be truncated at a much smaller depth z_b.

Considering a particular range segment, with sound-speed and density profiles $c(z)$ and $\rho(z)$, respectively, Eq. (3.87) suggests, in the weakly range-dependent case,

$$p(r, z) = \sum_m a_m \, Z_m(z) H_0^{(1)}(k_m r) \tag{3.103}$$

as an approximate solution. For this range segment, the k_m are modal wave numbers, in the upper half-plane and on the positive rather than negative real axis, with corresponding mode functions Z_m {unit: N}, and the a_m {unit: m^{-2}} are coefficients.

It is useful to factor out $H_0^{(1)}(k_0 r)$, where k_0 is a positive reference wave number to be commented in the following, and work with the less oscillating function ψ defined by

$$p(r, z) = \psi(r, z) H_0^{(1)}(k_0 r). \tag{3.104}$$

The mode functions Z_m satisfy the homogeneous version of the depth-separated Helmholtz equation (3.46). Hence, in terms of the depth differential operator Γ defined by $\Gamma = k_0^{-2} \left(\rho(z) \partial \left(\rho^{-1}(z) \partial / \partial z \right) / \partial z + \left(\omega^2 / c^2(z) - k_0^2 \right) \right)$,

$$k_0^2 \, \Gamma Z_m = \rho(z) \frac{\partial}{\partial z} \left(\frac{1}{\rho(z)} \frac{\partial Z_m}{\partial z} \right) + \left(\frac{\omega^2}{c^2(z)} - k_0^2 \right) Z_m = \left(k_m^2 - k_0^2 \right) Z_m. \tag{3.105}$$

By Sturm–Liouville theory [50], the mode functions Z_m are complete, and orthogonal to each other, in the space of square-integrable continuous functions on $0 \le z \le z_b$ with weight function $\rho^{-1}(z)$ and with the relevant boundary conditions at $z = 0$ and $z = z_b$ (pressure-release or rigid). A linear operator on this space can thus be defined by its action on each Z_m. For any analytic function f, the operator $f(1 + \Gamma)$ is defined by

$$f(1 + \Gamma) Z_m = f\left((k_m/k_0)^2 \right) Z_m, \tag{3.106}$$

and the image function fulfills the relevant boundary conditions at $z = 0$ and $z = z_b$, since the Z_m do so. In particular, $(1 + \Gamma)^\gamma$ is defined for any real γ by $(1 + \Gamma)^\gamma (Z_m) = (k_m/k_0)^{2\gamma} Z_m$, where the power is defined according to the principal branch of the logarithm.

Aiming at a PE, $\partial p/\partial r$ and $\partial \psi/\partial r$ are now formed. With a restriction to far-field ranges r and modes such that the approximations $d\left(H_0^{(1)}(k_0 r) \right)/dr = i k_0 H_0^{(1)}(k_0 r)$ and $d\left(H_0^{(1)}(k_m r) \right)/dr = i k_m H_0^{(1)}(k_m r)$ hold with good accuracy,

$$\frac{\partial p(r, z)}{\partial r} = i \sum_m a_m k_m Z_m(z) H_0^{(1)}(k_m r)$$

$$= i \, k_0 (1 + \Gamma)^{1/2} p(r, z) \tag{3.107}$$

$$\frac{\partial \psi(r, z)}{\partial r} = i \sum_m a_m (k_m - k_0) Z_m(z) H_0^{(1)}(k_m r) \Big/ H_0^{(1)}(k_0 r)$$

$$= i \, k_0 \left((1 + \Gamma)^{1/2} - 1 \right) \psi(r, z). \tag{3.108}$$

Since the operator $(1 + \Gamma)^{1/2}$ is defined using the mode decomposition, little may seem to have been gained by Eq. (3.108). However, $(1 + \Gamma)^{1/2}$ can be approximated using ordinary depth differential operators. Picking polynomials P and Q, such that $Q(0) = 1$ and $(1 + s)^{1/2} = 1 + P(s)/Q(s)$ holds with good accuracy for small s, corresponding to the quantities $(k_m/k_0)^2 - 1$ where k_0 is chosen close to the k_m of the dominating modes, the PE

$$\frac{\partial(Q(\Gamma)\psi(r,z))}{\partial r} = i k_0 P(\Gamma)\psi(r,z) \tag{3.109}$$

approximates Eq. (3.108) well. It is of order $2J$ with respect to depth derivatives, where J is the maximal degree of the polynomials P and Q, and it is to be solved together with boundary conditions for $\Gamma\psi$, $\Gamma^2\psi,\ldots$, and $\Gamma^{J-1}\psi$, pressure-release or rigid, at the surface $(z = 0)$ and at the bottom $(z = z_b)$, that agree with those for ψ (equivalently p).

The approximation $(1+s)^{1/2} = 1 + s/2$ leads to the standard narrow-angle PE $\partial\psi/\partial r = (i/2k_0)\,(\partial^2\psi/\partial z^2 + \rho\,\partial(1/\rho)/\partial z\,\partial\psi/\partial z + (\omega^2/c^2 - k_0^2)\psi)$. Another common choice is $(1+s)^{1/2} = 1 + (s/2)/(1 + s/4)$, leading to the Claerbout wide-angle PE. It is the first in a series of wide-angle PEs based on rational-function approximations of Padé type of $(1+s)^{1/2}$ [78,79].

Within the considered range segment, Eq. (3.105) shows that

$$\psi(r,z) = \sum_{m=1}^{\infty} \eta_m(r)Z_m(z) \tag{3.110}$$

solves the PE (3.109), provided that each $\eta_m(r)$ is marched according to the ODE

$$\partial\eta_m(r)/\partial r = ik_0 \frac{P(s_m)}{Q(s_m)}\,\eta_m(r), \tag{3.111}$$

with $s_m = \left(k_m^2 - k_0^2\right)/k_0^2$, from a start solution at some range r_0. The phase error of each $\eta_m(r)$ introduced by the approximation $(1+s)^{1/2} = 1 + P(s)/Q(s)$ can be analyzed [80], and it increases with range. For some typical applications, the standard and Claerbout PEs are considered to be accurate for mode propagation angles φ up to about 20 degrees and 35 degrees, respectively, if $k = k_0$ represents horizontal propagation and $s = (k^2 - k_0^2)/k_0^2 = -\sin^2\varphi$.

A formula for selecting the reference wave number k_0 is proposed in Ref. [81]. However, with a very wide-angle PE, perhaps the second from the Padé series, the choice is not critical. Only the far field is of interest, since asymptotic forms of Hankel functions have been used. Relevant mode propagation angles are thus typically close to horizontal, and $k_0 = \omega/\text{Re}(c_s)$ or $k_0 = \omega/\min(\text{Re}(c(r, z)))$ is often a good choice, where c_s is the sound speed at the source and the minimum is taken over r and z in the water column.

3.4.1 INTERFACE CONDITIONS AT THE VERTICAL RANGE-SEGMENT INTERFACES

When $\psi(r, z)$ is given at the left vertical interface of a range-invariant segment, it can be marched to the right interface using the PE (3.109). It is not apparent, however, how $\psi(r, z)$ should be continued across the vertical interface into the next range segment. With the Helmholtz equation, which is second-order in r, the physical interface conditions of continuous pressure and radial displacement can both be enforced. With the PE, which is first-order in r, only one interface condition can be fulfilled.

Some guidance for a reasonable choice may be obtained by considering the time-averaged energy flux vector $\boldsymbol{\Phi}(x)$ from Eq. (3.13) in the absorption-free case with positive real or imaginary k_m and real Z_m [82]. For weak range dependence, the back-scattered field is small, and the integral over depth of the radial energy flux, multiplied with the cylinder circumference $2\pi r$, should be (approximately) maintained when the range r is increased. Specifically,

$$F(r) = \frac{\pi r}{\omega} \int \rho^{-1} \ \mathrm{Im}\left(p^*\frac{\partial p}{\partial r}\right) dz \qquad (3.112)$$

should be independent of r. With p according to Eq. (3.103), for a particular range segment in the far field, where the evanescent modes do not contribute,

$$F(r) = \frac{2}{\omega} \sum_{m:k_m>0} \left|a_m\right|^2 I_m = \frac{\pi k_0 r}{\omega} \int \left|\rho^{-1/2}(1+\Gamma)^{1/4}p\right|^2 dz, \qquad (3.113)$$

by the definitions, $\int \rho^{-1}(z)Z_{m1}(z)Z_{m2}(z)dz = I_{m1}\delta_{m1m2}$ (cf. Eq. 3.88), and the asymptotic form of the Hankel function. In order for $F(r)$ to be maintained across a vertical range-segment interface, for all incident fields, the quantity

$$\rho^{-1/2}(z) \ (1+\Gamma)^{1/4}p(r,z), \qquad (3.114)$$

with different $\rho(z)$ and Γ on the two sides of the interface, should thus be matched at all depths z [82].

When horizontal propagation dominates, as is often the case, the operator $1 + \Gamma$ is well approximated by multiplication with $k_0^{-2}\omega^2/c^2$. In this case, the quantity to be matched across a vertical interface is simplified to

$$(\rho(z)c(z))^{-1/2} \ p(r,z). \qquad (3.115)$$

In practice, matching according to the quantity (3.115) is almost always sufficient [82]. The even simpler alternative with matching of $\rho^{-1/2}(z) \ p(r, z)$ has also proved useful [83].

However, enforcing continuity of $p(r, z)$ across the vertical interfaces is not a good idea [83]. At propagation upslope, the depth-averaged density and sound speed typically increase, and the expression (3.115) indicates that energy would be lost. In the same way, pressure matching would lead to energy gains at propagation downslope.

3.4.2 NUMERICAL SOLUTION METHODS

3.4.2.1 Start Solution

Eqs. (3.107) and (3.108) are derived under a far-field assumption, and the marching in range must start from a given start solution $\psi(r_0, z)$, or $p(r_0, z)$, at a certain far-field range r_0. This start solution could be obtained by finite-difference method (FDM) or FEM (Section 3.5). For a medium that is range-invariant for $r < r_0$ with sound speed profile $c(z)$ and density profile $\rho(z)$, modal wave numbers k_m and mode functions $Z_m(z)$ such that (cf. Eq. 3.88) $\int \rho^{-1}(z)Z_{m1}(z)Z_{m2}(z)dz = I_{m1}\delta_{m1m2}$, the normal-mode solution

$$p(r_0, z) = -\frac{iM\omega^2}{4\rho(z_s)c^2(z_s)} \sum_m \frac{Z_m(z_s)Z_m(z)}{I_m}H_0^{(1)}(k_m r_0) \qquad (3.116)$$

for a point source at range $r = 0$ and depth z_s (cf. Eqs. 3.87 and 3.90) can be used. It should be truncated, however, to only include modes with wave numbers k_m that are well handled by the particular PE to be used (cf. Eq. (3.110) and the comments on mode propagation angles given in that connection). A restriction to modes with k_m not too far from the reference wave number k_0 is typically useful.

Such a modal start solution appears to require a resolution of the mode structure in the initial range segment, containing $r < r_0$. However, as shown by Collins, PE techniques can be used to obtain a "self-starter" that avoids explicitly solving for the modes [84]. In terms of the operator $H_0^{(1)}\left(k_0 r_0(1 + \Gamma)^{1/2}\right)$, defined according to Eq. (3.106), Eq. (3.116) may be rewritten as

$$p(r_0, z) = -\frac{iM}{4}\left(\frac{\omega}{c(z_s)}\right)^2 H_0^{(1)}\left(k_0 r_0(1 + \Gamma)^{1/2}\right)\delta(z - z_s), \qquad (3.117)$$

where the Dirac delta function expansion $\delta(z - z_s) = \rho^{-1}(z_s)\sum_m Z_m(z_s)Z_m(z)/I_m$ has been applied. For large $k_0 r_0$, the asymptotic approximation of the Hankel function gives

$$p(r_0, z) = -\frac{M}{4\pi}\left(\frac{\omega}{c(z_s)}\right)^2\left(\frac{2\pi i}{k_0 r_0}\right)^{1/2}(1 + \Gamma)^{7/4}\exp\left(ik_0 r_0(1 + \Gamma)^{1/2}\right)\sigma(0, z),$$

$$(3.118)$$

where the smooth function $\sigma(0, z)$ is obtained as the solution of the fourth-order ODE

$$(1 + \Gamma)^2\sigma(0, z) = \delta(z - z_s) \qquad (3.119)$$

together with the relevant boundary conditions for $\sigma(0, z)$ and $(1 + \Gamma)\sigma(0, z)$ at $z = 0$ and $z = z_b$ (pressure-release or rigid). At first, $(1 + \Gamma)\sigma(0, z)$ can be determined using the propagator-matrix techniques of Section 3.2.1 for solving the related ODE system (3.44). Subsequently, $\sigma(0, z)$ can be obtained from $(1 + \Gamma)\sigma(0, z)$ using the boundary-condition transportation method of Section 3.2.1.3. For the case with homogeneous layers, elementary ODE theory readily provides analytical solutions.

The function $\sigma(r, z) = \exp(ik_0 r(1 + \Gamma)^{1/2})\sigma(0, z)$ fulfills the PE (3.107), with p replaced by σ, and $\sigma(r_0, z)$ can thus be determined from $\sigma(0, z)$ by marching a

Padé approximation of this PE from $r = 0$ to $r = r_0$. Application of a Padé approximation of the operator $(1 + \Gamma)^{7/4}$ finally provides $p(r_0, z)$.

Thus, the start solution of Eq. (3.118) can be obtained by numerical PE marching methods, which are briefly described in the following, without resolving the mode structure. A smooth $\sigma(0, z)$ is desired to initialize the marching. The suggested operator $(1 + \Gamma)^2$ for regularization of $\delta(z - z_s)$, see Eq. (3.119), is an appropriate choice, although alternatives are occasionally needed [85].

A number of other start solutions for the PE have also been proposed. Several of them are reviewed in Ref. [7].

3.4.2.2 Rational-Function Approximations for the Relevant Operators

A basic Padé approximation of order J, valid for small s, is [78].

$$(1 + s)^{1/2} = 1 + \sum_{j=1}^{J} \frac{a_j s}{1 + b_j s} + O(s^{2J+1}), \tag{3.120}$$

where $a_j = 2(2J + 1)^{-1} \sin^2(j\pi/(2J + 1))$ and $b_j = \cos^2(j\pi/(2J + 1))$, as determined by requiring equality at $s = 0$ of derivatives of order $1, 2,..., 2J$ of the functions involved. Application to Eq. (3.109) provides a PE whose solution can be reduced, by operator splitting techniques [86,87], to combined solutions of the low-order PEs

$$\frac{\partial((1 + b_j\Gamma)\psi(r, z))}{\partial r} = i \, k_0 a_j \, \Gamma\psi(r, z), \tag{3.121}$$

$j = 1, 2,..., J$, together with the relevant boundary conditions for $\psi(0, z)$ at $z = 0$ and $z = z_b$ (pressure-release or rigid). Computations in parallel, with J processors, are possible.

Modes below cutoff, with imaginary wave numbers k_m, cf. the initial example in Section 3.3.5.1, are not handled well by the Padé approximation of Eq. (3.120). The desired exponential decrease of the mode amplitude, with increasing r, does not appear since, with s corresponding to $(k_m/k_0)^2 - 1$, the right-hand side of Eq. (3.120) is still real. A flexible method for deriving rational-function approximations of a general function $g(s)$, with $g(0) = 1$, is proposed in Ref. [88]. Coefficients A_j and B_j, $j = 1, 2,..., J$, are first determined to provide the best least-squares fit of the equations

$$1 + \sum_{j=1}^{J} A_j s_i^j = g(s_i)\left(1 + \sum_{j=1}^{J} B_j s_i^j\right), \tag{3.122}$$

where the s_i are a large number of matching points. An approximation of the type $g(s) = 1 + \sum_{j=1}^{J} a_j s/(1 + b_j s)$ is subsequently obtained by factorizing the two polynomials in Eq. (3.122) and expanding their quotient as partial fractions. For application to Eq. (3.109), with $g(s) = (1 + s)^{1/2}$, most of the s_i are preferably located in an interval around $s = 0$ to provide good phase accuracy for the propagating modes with wave numbers k_m close to k_0, while the remaining ones fulfill

$s_i < -1$ to provide amplitude decay for the evanescent modes with imaginary k_m. For long-range propagation, where good accuracy for small s can be particularly important, derivative conditions at $s = 0$ can be incorporated, cf. the choice of coefficients given in connection with Eq. (3.120).

The approximation of Eq. (3.122) must not only be accurate. There is also a stability issue: when applied to Eq. (3.109), together with some numerical solution method, the solutions $\psi(r, z)$ must remain bounded when r is increased. For a diagonalizable system matrix of the discretized equation, stability is linked to the eigenvalue location in the complex plane [40]. In practice, numerical tests are often useful to check stability.

Recalling the operator definition by Eq. (3.106), the PE (3.108) can be solved analytically, for r and $r + \Delta r$ within the same range segment, by

$$\psi(r + \Delta r, z) = \exp\left(ik_0 \Delta r \left((1 + \Gamma)^{1/2} - 1\right)\right) \psi(r, z). \tag{3.123}$$

In an effort to speed up the computations by allowing larger range steps, Collins [89] applied rational-function approximations to the exponential operator in Eq. (3.123) rather than to the coefficient operator appearing in the PE. With $g(s) = \exp\left(ik_0 \Delta r \left((1 + s)^{1/2} - 1\right)\right)$, the least-squares approach with Eq. (3.122) is directly applicable. The resulting "split-step Padé" solution can be expressed as $\psi(r + \Delta r, z) = \psi(r, z) + \sum_{j=1}^{J} \psi_j(r + \Delta r, z)$, where each function of depth $\psi_j(r + \Delta r, z)$ is obtained by solving the ODE boundary-value problem

$$(1 + b_j \Gamma)\psi_j(r + \Delta r, z) = a_j \Gamma \psi(r, z) \tag{3.124}$$

with the relevant boundary conditions (pressure-release or rigid) at $z = 0$ and $z = z_b$. These computations are also well suited to parallel processing.

As an alternative to the approach leading to the low-order PEs of Eq. (3.121), Eq. (3.120) can be applied to the operator $(1 + \Gamma)^{1/2}$ in the exponent of Eq. (3.123), possibly with modified a_js and b_js as determined by the least-squares approach. Utilizing the Padé approximations $\exp(ik_0 \Delta r a_j s/(1 + b_j s)) = \left(1 + c_j^+ s\right)/\left(1 + c_j^- s\right)$, where $c_j^\pm = b_j \pm ik_0 \Delta r a_j/2$, it follows readily [90] that $\psi(r + \Delta r, z) = \psi_J(r, \Delta r, z)$, where the ψ_j, $j = 0, 1, \ldots, J$, are now defined recursively by $\psi_0(r, \Delta r, z) = \psi(r, z)$ and the ODEs

$$\left(1 + c_j^- \Gamma\right)\psi_j(r, \Delta r, z) = \left(1 + c_j^+ \Gamma\right)\psi_{j-1}(r, \Delta r, z), \tag{3.125}$$

$j = 1, 2, \ldots, J$, with the boundary conditions (pressure-release or rigid) at $z = 0$ and $z = z_b$.

Rational-function approximations are also needed for the interface-condition operator $(1 + \Gamma)^{1/4}$ in expression (3.114) and for the operator $(1 + \Gamma)^{7/4}$ in connection with the self-starter. For the self-starter, the step from $\sigma(0, z)$ to $p(r_0, z)$ in Eq. (3.118) may alternatively be taken by deriving a suitable rational-function approximation of the combined operator $(1 + \Gamma)^{7/4}\exp(ik_0 r_0 (1 + \Gamma)^{1/2})$. The least-squares approach is suitable for all these tasks (cf. [88]).

3.4.2.3 Depth Discretization and Range Integration

The PEs (3.121), as well as the ODEs (3.124) and (3.125), are second-order with respect to depth derivatives. In order to apply FD techniques, a particular range segment is considered, with range-invariant medium properties and density profile $\rho(z)$, and a depth grid is introduced with points $z_j = jh$, $j = 1, 2,..., L$, such that $(L + 1)h = z_b$. For definiteness, the lower boundary at z_b is now also assumed to be traction free.

For any function $\theta(z)$ of depth z, which fulfills the boundary conditions $\theta(0) = \theta(z_b) = 0$ and that $\theta(z)$ as well as $\rho^{-1}(z) \, d\theta(z)/dz$ are continuous also at medium discontinuity depths, the bold-face notation $\boldsymbol{\theta}$ is introduced for the L-dimensional column vector $(\theta(z_1), \theta(z_2),..., \theta(z_L))^{\text{T}}$. The pressure $p(z)$ is an example. As discussed in Section 3.5.1, a convenient discrete approximation of $\Gamma\theta(z)$ appears as $\boldsymbol{T\theta}$, where $\boldsymbol{T}$ is a tridiagonal matrix.

Using the introduced notation for depth discretization in the obvious way, Eqs. (3.124) and (3.125) can directly be rewritten as

$$(\mathbf{I} + b_j\boldsymbol{T})\boldsymbol{\psi}_j(r + \Delta r) = a_j\boldsymbol{T\psi}(r) \tag{3.126}$$

and

$$\left(\mathbf{I} + c_j^-\boldsymbol{T}\right)\boldsymbol{\psi}_j(r, \Delta r) = \left(\mathbf{I} + c_j^+\boldsymbol{T}\right)\boldsymbol{\psi}_{j-1}(r, \Delta r), \tag{3.127}$$

respectively. The resulting linear equation systems for the range marching, with tri-diagonal system matrices, are readily solved by standard numerical techniques [40]. Tests with refined depth grids are often useful to check that convergence is reached.

With Eq. (3.126), the range steps Δr can often be many wavelengths long [89], and refined sampling within a range step can be made surprisingly efficiently [91]. With strong range dependence, the limits for Δr are typically set by the required fineness of the stair-step representation of the medium, but several wavelengths long Δr are often feasible [65]. Eq. (3.127), involving a less accurate expansion of the range propagator of Eq. (3.123), requires much smaller Δr, however [90].

After depth discretization, the PEs (3.121) appear as

$$\frac{\partial((\mathbf{I} + b_j\boldsymbol{T})\boldsymbol{\psi}(r))}{\partial r} = i\,k_0 a_j\,\boldsymbol{T\psi}(r). \tag{3.128}$$

The Crank–Nicolson method [40] is a popular way to do the range marching. It leads to

$$\frac{(\mathbf{I} + b_j\boldsymbol{T})(\boldsymbol{\psi}(r + \Delta r) - \boldsymbol{\psi}(r))}{\Delta r} = i\,k_0 a_j\frac{\boldsymbol{T\psi}(r + \Delta r) + \boldsymbol{T\psi}(r)}{2}, \tag{3.129}$$

and $\boldsymbol{\psi}(r + \Delta r)$ can be determined easily by solving a tridiagonal equation system. The method is second-order accurate in Δr, and it is numerically stable (i.e., the errors are not magnified during the range marching) for arbitrary depth and range steps [40]. Higher-order methods, allowing larger range steps for a prescribed accuracy, can alternatively be used. A fourth-order method by Jeltsch [92] is applied in

Ref. [93] for solving related PEs with local error control and adaptive choice of Δr as the range marching proceeds.

3.4.3 EXTENDED AND ALTERNATIVE PE APPROACHES

3.4.3.1 Extension to Media That Vary Regionwise Smoothly With Range and Depth

With $\Gamma_r = k_0^{-2}\left(\rho(r,z)\partial\left(\rho^{-1}(r,z)\partial/\partial z\right)/\partial z + \left(\omega^2/c^2(r,z) - k_0^2\right)\right)$ replacing Γ, where the index r indicates extension to a possible smooth dependence on range, Eqs. (3.107)–(3.109) are meaningful also for density and sound-speed functions that vary smoothly with range r and depth z. For an analytic function f, $f(1 + \Gamma_r)$ is in this case defined (cf. Eq. 3.106) in terms of local modal wave numbers $k_m(r)$ and mode functions $Z_m(z; r)$ at range r (cf. Section 3.3.4.3). This extension of the equations is not consistent with the matching conditions (3.114) or (3.115) at a vertical interface of medium discontinuity, however. Considering the interface as a limiting case of smooth medium variations in a vertical transition region, range integration over the tiny transition region shows (cf. [94]) that Eqs. (3.107)–(3.109), with bounded right-hand sides, imply continuity across the vertical interface of p, ψ, and $Q(\Gamma_r)\psi$, respectively.

Apparently, the range derivative should be taken of the quantity to be matched across the vertical interfaces. By using the local mode functions $Z_m(z; r)$, the PE

$$\frac{\partial\left(\rho^{-1/2}(r,z)(1 + \Gamma_r)^{1/4}\psi(r,z)\right)}{\partial r} \tag{3.130}$$
$$= ik_0\rho^{-1/2}(r,z)\left((1 + \Gamma_r)^{1/2} - 1\right)(1 + \Gamma_r)^{1/4}\psi(r,z)$$

is readily seen to agree with the PE (3.108) in range-invariant regions and provide the desired match of the quantity (3.114) across a vertical interface.

The PE (3.130) has been proposed and discussed by Godin [94], focusing its reciprocity and energy conservation properties, and implemented by Mikhin [95,96]. In principle, it could be solved by a two-stage procedure: first determining $\Psi(r, z)$ by marching an appropriate starting field in range with the PE $\partial\left(\rho^{-1/2}\Psi\right)/\partial r = ik_0\rho^{-1/2}\left((1 + \Gamma_r)^{1/2} - 1\right)\Psi$, and subsequently determining $\psi(r, z)$ at each receiver range r by solving the depth operator problem $(1 + \Gamma_r)^{1/4}\psi(r,z) = \Psi(r, z)$ there.

Alternatively, rational-function approximations of arbitrarily high accuracy could be applied directly in Eq. (3.130) for the operators $(1 + \Gamma_r)^{1/2} - 1$ and $(1 + \Gamma_r)^{1/4}$. Two simple examples are considered in Ref. [94]: $(1 + s)^{1/2} = 1 + s/2$ together with $(1 + s)^{1/4} = 1$ whereby Eq. (3.130) appears as the narrow-angle PE

$$\partial\left(\rho^{-1/2}(r,z)\psi(r,z)\right)/\partial r = (ik_0/2)\rho^{-1/2}(r,z)\Gamma_r\psi(r,z), \tag{3.131}$$

and $(1+s)^{1/2} = 1 + (s/2)/(1+s/4)$, i.e., Eq. (3.120) with $J = 1$, together with $(1+s)^{1/4} = 1 + s/4$ whereby Eq. (3.130) appears as the generalized Claerbout wide-angle PE

$$\partial\left(\rho^{-1/2}(r,z)(1+\Gamma_r/4)\psi(r,z)\right)\Big/\partial r = (ik_0/2)\rho^{-1/2}(r,z)\Gamma_r\psi(r,z). \qquad (3.132)$$

Considering limiting cases of smooth medium variations in transition regions, it is instructive to note that different rational-function approximations may give rise to different matching conditions at a discontinuity interface between regions within which the medium properties vary continuously. Across a vertical interface, the PEs (3.131) and (3.132) apparently stipulate matching of $\rho^{-1/2}\psi$ and $\rho^{-1/2}(1+\Gamma_r/4)\psi$, respectively. Across a non-vertical interface with slope $dz/dr = h(r)$, Godin [94] shows that ψ is continuous for both PEs (3.131) and (3.132). In addition, $\rho^{-1}(\partial\psi/\partial z - ik_0 h\psi)$ and $\rho^{-1}\partial\psi/\partial z$ are continuous for PEs (3.131) and (3.132), respectively. In the latter case, which involves third-order derivatives normal to the interface when $h \neq 0$, $\rho^{-1/2}(1+\Gamma_r/4)\psi$ turns out to be continuous across any non-horizontal interface.

For the PE (3.131), matching of $\rho^{-1}(\partial\psi/\partial z - ik_0 h\psi)$ across a sloping interface reduces, for a sloping rigid boundary, to a boundary condition proposed in Ref. [97]. For high-order PEs, derivation of appropriate interface conditions between regions appears to be rather tedious. Of course, an alternative is to replace each medium discontinuity interface by a wavelength-wide transition region with smooth parameter variations and apply the appropriate PE, (3.130) or some approximation of it, throughout.

3.4.3.2 Coordinate Transformation Techniques

When a number of stair steps model a sloping bottom, the PE fails simultaneously to accommodate the two physical boundary conditions of continuous pressure and radial displacement on the rises. As an alternative to the single boundary conditions discussed in Section 3.4.1, the coordinates can be rotated to make the PE marching take place parallel to the bottom and allow both physical bottom boundary conditions to be fulfilled [98]. The boundary-condition problem is thereby transferred to the ocean surface, which is now modeled with stair steps. With small stair steps, however, a pressure-release boundary condition is handled rather easily. A bottom with variable slope can be treated by approximation with a sequence of constant-slope regions [99].

The water column, as well as an individual sediment layer, can alternatively be mapped onto a rectangle using curvilinear coordinates. With an orthogonal coordinate transformation, the Helmholtz equation is transformed into an equation without mixed derivatives, which can be approximated by PEs in various ways [93,100]. The physical boundary conditions of continuous pressure and normal displacement can be treated as in the range-invariant case.

Simpler non-orthogonal mappings can also be applied [101,102]. Mixed derivatives appear in the transformed Helmholtz equation; however, they are neglected in order to derive PE approximations. Corrections for the neglected terms are needed in some cases.

3.4.3.3 Two-Way PE Approaches
Although the parabolic approximation is based on neglecting the backscattered field, this field can be included by running the PE inward, toward the source, after the outward PE run [103]. Assuming a medium composed of range-invariant segments, the two-way marching single-scattering approximation solution of the coupled-mode problem, indicated in Section 3.3.4.3, is readily adapted. At each vertical interface, the physical boundary conditions of continuous pressure as well as radial displacement are now taken into account for the field marching. This can be done, since a reflected field component is included in addition to the transmitted one. At the outward PE run, the reflected components are stored, to be added during the following inward run. The range derivatives, needed to compute the radial displacements at both sides of a vertical interface, can be evaluated from depth derivatives, using the PE.

This two-way PE approach entails more computational work, and the reflected field components are typically included only from those parts of the medium where the range dependence is strong, while an interface condition from Section 3.4.1 is used for the remaining parts. The outward and inward PE runs can be iterated to improve the solution. Problems where multiple scattering is important can thereby be treated [104].

3.4.3.4 Extension to Fluid-Solid Media
For the harmonic time dependence factor $\exp(-i\omega t)$ with $\omega > 0$, omitted in the notation, the left-hand side of the equations of motion (3.3) for a fluid-solid medium is replaced by $-\rho\omega^2 u_i$. With a symmetric point source at $x = y = 0$, and cylindrical coordinates, the radial and vertical displacements are denoted u and w, respectively, as in Eq. (3.40). The right-hand side of Eq. (3.3) contains first- and second-order spatial derivatives of u and w. In the range-invariant case, wave number integration and normal-mode solution methods are presented in Sections 3.2 and 3.3. The range-dependent case is also of great interest, for studying sound propagation over hard sloping bottoms, for example.

In the weakly range-dependent case, for a solid region, it is natural to seek a parabolic approximation of the equations of motion, for which the solution can be obtained by range marching. The dependent variables must be chosen with care, partly to allow a convenient parabolic approximation, and partly to allow a convenient handling of boundary conditions, indicated in connection with Eq. (3.3), at interfaces. The choice of $\partial u/\partial r$ and w as dependent variables, proposed in Ref. [105], seems to be preferred today (2016). For a range-invariant range segment, Eq. (3.3) shows that, except at the source,

$$L\frac{\partial^2}{\partial r^2}\begin{pmatrix} \partial u/\partial r \\ w \end{pmatrix} + M\begin{pmatrix} \partial u/\partial r \\ w \end{pmatrix} = \begin{pmatrix} 0 \\ 0 \end{pmatrix},\qquad(3.133)$$

where the 2×2 matrix operators $\boldsymbol{L}$ and $\boldsymbol{M}$ contain $\partial/\partial z$ and $\partial^2/\partial z^2$, but not range derivatives. Proceeding somewhat formally, Eq. (3.133) suggests the PE

$$\frac{\partial}{\partial r} \begin{pmatrix} \partial u/\partial r \\ w \end{pmatrix} = ik_0 (\mathbf{I} + \boldsymbol{\Gamma})^{1/2} \begin{pmatrix} \partial u/\partial r \\ w \end{pmatrix} \tag{3.134}$$

for out-going waves (cf. Eq. 3.107), where k_0 is a reference wave number and $\boldsymbol{\Gamma} = k_0^{-2} (\boldsymbol{L}^{-1}\boldsymbol{M} - k_0^2\mathbf{I})$. Assuming $\boldsymbol{\Gamma}$ to be small in some sense, the operator $(\mathbf{I} + \boldsymbol{\Gamma})^{1/2}$ can be defined, for smooth functions of z fulfilling the relevant boundary conditions, according to the Taylor expansion of $(1 + s)^{1/2}$, and it can be approximated by some rational function. As before, range-invariant segments can approximate a range-dependent medium.

Appropriately modified and extended, the techniques for solving the fluid-medium PE carry over to the fluid–solid case (see, e.g., Ref. [106] and references therein). It is much easier to handle a sloping pressure-release fluid boundary than a sloping fluid–solid interface, and coordinate rotation techniques seem to have attracted increased interest for the fluid–solid case (e.g., Ref. [107]).

3.4.4 EXAMPLES

The simple range-dependent fluid medium from Section 3.3.5.2, with $1/c^2(r, z) = 1/c_0^2(z) + S(r)$ and $\rho(r, z) = \rho_0(z)R(r)$, is now revisited. At the selected angular frequency $\omega > 0$, the basic modal wave numbers and mode functions, corresponding to the profiles $c_0(z)$ and $\rho_0(z)$, are still denoted $k_{m,0}$ and $Z_{m,0}(z)$, respectively. Introducing the wave numbers $k_m(r) = \left(k_{m,0}^2 + \omega^2 S(r)\right)^{1/2}$, $m = 1, 2,...$, where $k_m(r)$ is in the upper half-plane but not on the negative real axis, the PE (3.130) provides the solution

$$p(r, z) = H_0^{(1)}(k_0 r)R^{1/2}(r) \sum_{m=1}^{\infty} \left(\frac{k_0}{k_m(r)}\right)^{1/2} \eta_m(r)Z_{m,0}(z) \tag{3.135}$$

for the pressure field, where $\eta_m(r)$ is marched according to the ODE

$$\partial\eta_m(r)/\partial r = i(k_m(r) - k_0)\eta_m(r) \tag{3.136}$$

from a start solution at some $r_0 > 0$. Hence, $\eta_m(r) = \exp(i\int_{r_0}^{r}(k_m(u) - k_0)du)\eta_m(r_0)$.

When $k_0 r$ is large and $k_m(r)$ varies in the vicinity of k_0, $H_m(r) = H_0^{(1)}(k_0 r)(k_0/k_m(r))^{1/2}\eta_m(r)$ gives an approximate solution of the ODE (3.102) provided that the functions $(dk_m(r)/dr)/k_m^2(r)$, $(d^2k_m(r)/dr^2)/k_m^3(r)$, and $\tilde{k}_m^2(r)/k_m^2(r) - 1$ are small in magnitude. This follows by applying the asymptotic expression for the Hankel function. The smallness conditions, demanding smooth variation with range of $c(r, z)$ as well as $R(r)$, are natural to limit the backscattering, making a PE solution reasonable.

With the narrow-angle PE (3.131), Eqs. (3.135) and (3.136) are modified to

$$p(r,z) = H_0^{(1)}(k_0 r) R^{1/2}(r) \sum_{m=1}^{\infty} \eta_m(r) Z_{m,0}(z) \tag{3.137}$$

$$\partial \eta_m(r)/\partial r = (i/2k_0)\left(k_m^2(r) - k_0^2\right)\eta_m(r), \tag{3.138}$$

while the generalized Claerbout wide-angle PE (3.132) provides

$$p(r,z) = H_0^{(1)}(k_0 r) R^{1/2}(r) \sum_{m=1}^{\infty} \frac{4k_0^2}{3k_0^2 + k_m^2(r)} \eta_m(r) Z_{m,0}(z) \tag{3.139}$$

$$\partial \eta_m(r)/\partial r = 2ik_0 \frac{k_m^2(r) - k_0^2}{3k_0^2 + k_m^2(r)} \eta_m(r). \tag{3.140}$$

With the more general rational-function approximation $1 + P(s)/Q(s)$ of $(1+s)^{1/2}$, where P and Q are polynomials with $Q(0) = 1$, Eq. (3.140) is changed to

$$\partial \eta_m(r)/\partial r = ik_0 \frac{P(s_m(r))}{Q(s_m(r))} \eta_m(r), \tag{3.141}$$

where $s_m(r) = \left(k_m^2(r) - k_0^2\right)/k_0^2 = \left(k_{m,0}^2 - k_0^2 + \omega^2 S(r)\right)/k_0^2$ (cf. Eq. 3.111).

The two 500 Hz island examples from Section 3.3.5.2 are now considered again, but in modified form. As stated, they involve abrupt changes of medium parameters, leading to rather significant backscattering, cf. the right panel of Fig. 3.14, for which the one-way PE method is not suitable. (However, the two-way PE approach indicated in Section 3.4.3.3 could be applied.) The medium parameter changes are now smoothened: the clay island is fully developed for 2.4 km $< r <$ 2.6 km with smooth (linear) adaptations to the surroundings for 2.3 km $< r <$ 2.4 km and 2.6 km $< r <$ 2.7 km, while the sand island is fully developed for 2.45 km $< r <$ 2.55 km (as before) with smooth (linear) adaptations to the surroundings for 2.4 km $< r <$ 2.45 km and 2.55 km $< r <$ 2.6 km.

For these modified examples, the PE method is indeed suitable for calculating the propagation loss. In the adaptation regions, the medium is discretized with 1 m wide range-invariant segments; however, a coarser discretization would suffice (the wavelength is about 3 m). At first, a restriction is made to the contribution by one of the most significant modes, the one with horizontal slowness $k/\omega \approx 0.68581 + 0.00007i$ s/km in the water surroundings. With focus on the range interval between 2 and 3 km, Fig. 3.15 shows corresponding PL results (in dB relative to the total field 1 m from the source), computed by different methods. The reference wave number k_0 for the PE solutions is taken as ω/c_s, where $c_s = 1454$ m/s is the sound speed at the source. A modal starter is used at the range $r_0 = 2$ km. Truncation of the modes that are poorly treated by the PE approximations is not critical in these examples, since those modes are not significant.

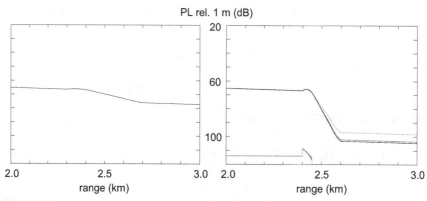

PL rel. 1 m (dB)

FIGURE 3.15

PL in dB, rel. 1 m, with restriction to a particular mode, for the 500 Hz smoothened versions of the clay (left panel) and sand (right panel) island examples from Fig. 3.14, with the island centered at range 2.5 km. The source and receiver depths are 10 and 90 m, respectively. In each panel, the *solid black curve* shows three visually coinciding solutions: full-wave mode solution, and PE results by Eq. (3.130) and "improved equation" (3.132) using Padé order $J = 2$. The *gray dashed* and *solid curves* come from the PEs (3.131) and (3.132), respectively. In the right panel, the backward part of the full-wave solution is shown as an additional solid *gray curve*, at low levels.

The mode considered in Fig. 3.15 propagates almost horizontally at the source depth ($z = 10$ m) in the water. In the sand island, its phase velocity is thus close to 1800 m/s, corresponding to $s = (k^2 - k_0^2)/k_0^2 = -\sin^2\varphi$ with $\varphi \approx 36$ degrees. Hence, it is not surprising that the narrow-angle PE (3.131) fails to model the PL for the sand island case in the right panel of Fig. 3.15 accurately. The generalized Claerbout wide-angle PE (3.132) comes close to the full-wave mode solution, computed according to Section 3.3.4. With the Padé approximation of Eq. (3.120) with $J = 2$ instead of $J = 1$, the result is visually indistinguishable from the mode result. For the less challenging clay island case, all PE curves agree with the full-wave mode result.

The efficacy of matching the quantity (3.114) at the vertical interfaces is rather apparent from these examples. Its density-factor part raises the PE results at the island center ($r = 2.5$ km) with 1.5 and 3.4 dB for the clay and sand cases, respectively. Its $(1 + \Gamma)^{1/4}$ operator part raises the PE (3.130) result at $r = 2.5$ km in the right panel by another 0.5 dB, providing an almost perfect agreement with the full-wave mode result. Indeed, cf. the last paragraph of Section 3.4.1, pressure matching at the vertical interfaces would imply energy losses for these examples.

The pressure levels in Fig. 3.15 decrease significantly at propagation through the fully developed parts of the islands: by about 6 dB between $r = 2.4$ and 2.6 km in the clay case, and by about 26 dB between $r = 2.45$ and 2.55 km in the sand case. With a wavelength of about 3 m, this is reasonably consistent with the assumed absorption

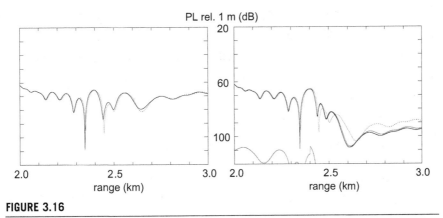

PL rel. 1 m (dB)

FIGURE 3.16

As Fig. 3.15 but for the total fields.

values of 0.08 and 0.87 dB/wavelength for clay and sand, respectively, as expected for this mode.

Fig. 3.16 shows corresponding results including the contributions from all the modes. Interference between the modes gives rise to more complicated PL curves, but similar observations as before can be made concerning the behavior of the different PE approximations. The narrow-angle PE (3.131) clay-island result shows some slight misfit now, which is natural when additional modes, with wave numbers that deviate more from k_0, are incorporated. As in Fig. 3.15, the backscattered fields are much smaller in comparison with the total fields than in Fig. 3.14. For the clay case, they fall just outside the scale, at PL values of about 130 dB.

The mode solution with the single-scattering approximation, indicated in Section 3.3.4.3, has also been tested for these examples. In all cases, the corresponding PL results are visually indistinguishable from those obtained with the full-wave mode solution.

3.5 FINITE-DIFFERENCE AND FINITE-ELEMENT METHODS

The PE approximation, ignoring backscattered waves, is not appropriate for a medium with strong range dependence. As an alternative to coupled modes, general numerical discretization methods of FD or FE type can be applied to solve the Helmholtz equation. With the FDM, derivatives are approximated point-wise by difference quotients involving a discrete computational grid. With the FEM, the physical space is divided into non-overlapping subsets called elements and the sought function is approximated on each element by a polynomial. There are several text books where FDM and FEM are described in detail, and this section merely illustrates the basic ideas for a few wave propagation applications.

3.5.1 ONE-DIMENSIONAL FEM AND FDM FOR PARABOLIC AND NORMAL-MODE EQUATIONS

As an initial illustration of FE and FD methods, tridiagonal matrices T introduced in Section 3.4.2.3 to approximate $\Gamma\theta = k_0^{-2}\left(\rho(\rho^{-1}\theta')' + \left(\omega^2/c^2 - k_0^2\right)\theta\right)$ in connection with PE are now specified. Here, ρ, c, and θ are functions of depth z, the prime notation is used for derivatives with respect to z, and θ as well as $\rho^{-1}\theta'$ are continuous also at medium discontinuity depths. A traction-free bottom is initially assumed. Hence, $\theta(z_b) = \theta(0) = 0$. A depth grid is introduced by $z_j = jh$, $j = 1$, $2,...,L$, such that $(L+1)h = z_b$.

Starting with FEM with linear element functions, θ is approximated by $\theta(z) = \sum_{j=1}^{L} \theta(z_j)S_j(z)$, where the shape functions S_j are defined for $j = 1,..., L$ by $S_j(z) = 1 - |z - z_j|/h$ for $|z - z_j| < h$ and $S_j(z) = 0$ otherwise. An equation $\Gamma\theta = g$, where $g = g(z)$ is piece-wise continuous, is now interpreted in a weak Galerkin sense, with integrals from 0 to z_b, as

$$\frac{\int \Gamma\theta(z)S_i(z)\rho^{-1}(z)dz}{\int S_i(z)\rho^{-1}(z)dz} = \frac{\int g(z)S_i(z)\rho^{-1}(z)dz}{\int S_i(z)\rho^{-1}(z)dz}, \quad i = 1, 2, ..., L. \tag{3.142}$$

With $\boldsymbol{\theta} = (\theta_1, \theta_2,..., \theta_L)^{\mathrm{T}}$ and $\boldsymbol{g} = (g_1, g_2,..., g_L)^{\mathrm{T}}$, where $\theta_i = \theta(z_i)$ and the g_i are defined by the right-hand side of Eq. (3.142), partial integration provides the equation $T\boldsymbol{\theta} = \boldsymbol{g}$, where T is the tridiagonal $L \times L$ matrix with elements (T_{ij}), $i, j = 1, 2,..., L$, given by

$$T_{ij} = \frac{-\int \rho^{-1}(z)S_i'(z)S_j'(z)dz + \int\left(\omega^2/c^2(z) - k_0^2\right)S_i(z)S_j(z)\rho^{-1}(z)dz}{k_0^2 \int S_i(z)\rho^{-1}(z)dz}. \tag{3.143}$$

With $z_{i\pm1/2} = z_i \pm h/2$, and $\theta_0 = \theta_{L+1} = 0$, Taylor expansion of $\rho^{-1}(z)$, $c^{-2}(z)$, and $g(z)$ shows that the ith row of $T\boldsymbol{\theta} = \boldsymbol{g}$, for $i = 1, 2,..., L$, can be expressed as

$$\frac{\rho^{-1}\left(z_{i+1/2}\right)(\theta_{i+1} - \theta_i) - \rho^{-1}\left(z_{i-1/2}\right)(\theta_i - \theta_{i-1})}{k_0^2 h^2 \rho^{-1}(z_i)} + \left(\left(\frac{\omega}{k_0 c(z_i)}\right)^2 - 1\right)\theta_i$$

$$= g(z_i) + O(h^2), \tag{3.144}$$

provided that the functions $\rho(z)$, $c(z)$, and $g(z)$ are twice continuously differentiable in the interval (z_{i-1}, z_{i+1}). Piece-wise spline representations of the density and sound-speed functions are often convenient to achieve this regularity.

Eq. (3.144) is actually a second-order accurate FD approximation of $\Gamma\theta(z_i) = g(z_i)$, which could alternatively have been derived by FDM. With the FEM formulation, the condition that $\rho^{-1}(z)\,d\theta(z)/dz$ is continuous also at a depth z_i where the density is discontinuous appears among the rows of $T\boldsymbol{\theta} = \boldsymbol{g}$ [108] as a natural consequence of the weak formulation. This is readily realized from Taylor expansions downward and upward from z_i. The correspondence to Eq. (3.144) becomes

$$\frac{\rho^{-1}(z_i+)\theta'(z_i+) - \rho^{-1}(z_i-)\theta'(z_i-)}{k_0^2 h(\rho^{-1}(z_i+) + \rho^{-1}(z_i-))\big/2} = O(1), \tag{3.145}$$

which can be viewed as an appropriate approximation of the continuity condition $\rho^{-1}(z_i+)\theta'(z_i+) = \rho^{-1}(z_i-)\theta'(z_i-)$. With FDM, an approximate condition like $\rho^{-1}(z_i+)(\theta_{i+1} - \theta_i) = \rho^{-1}(z_i-)(\theta_i - \theta_{i-1})$, which is of first order, has to be introduced explicitly. The second-order approximation $\rho^{-1}(z_i+)(-\theta_{i+2}/2+2\theta_{i+1} - 3\theta_i/2) = \rho^{-1}(z_i-)(3\theta_i/2 - 2\theta_{i-1}+\theta_{i-2}/2)$, with one-sided differences, could alternatively be applied, but the bandwidth of the corresponding matrix T would increase from three to five.

So far, the lower boundary at z_b has been traction free. With FEM, a rigid-boundary condition $\theta'(z_b) = 0$ appears naturally by inclusion of an additional shape function S_{L+1} defined by $S_{L+1}(z) = 1 - (z_b - z)/h$ for $z_b - h < z < z_b$ and $S_{L+1}(z) = 0$ otherwise. With FDM, an appropriate approximation of the condition $\theta'(z_b) = 0$ is conveniently incorporated by a modified grid such that $(L + 1/2) h = z_b$, and replacement of θ_{L+1} by θ_L.

Higher-order FE and FD approximations can also be used, to allow an increased grid spacing h, for a given accuracy requirement. A fourth-order central FD approximation of $\theta''(z_i)$, relevant for a locally constant density, appears as $(-(\theta_{i+2} + \theta_{i-2})/12 + 4(\theta_{i+1} + \theta_{i-1})/3 - 5\theta_i/2)/h^2$, but the bandwidth of the corresponding matrix T is larger than three. Using the ODE itself, an ODE $\theta'' + G\theta = k_0^2 g$, where the function G corresponds to $\omega^2/c^2 - k_0^2$, can actually be approximated with fourth-order accuracy by a three-term Numerov stencil. The scheme results directly from substitution of $\theta''''(z_i) = k_0^2 g''(z_i) - G(z_i)\theta''(z_i) - 2G'(z_i)\theta'(z_i) - G''(z_i)\theta(z_i)$, expressed as $\theta''''(z_i) = k_0^2(g_{i+1} - 2g_i + g_{i-1})\big/h^2 - G_i(k_0^2 g_i - G_i\theta_i) - (G_{i+1} - G_{i-1})(\theta_{i+1} - \theta_{i-1})\big/2h^2 - (G_{i+1} - 2G_i + G_{i-1})\theta_i\big/h^2 + O(h^2)$, where $g_i = g(z_i)$ and $G_i = G(z_i)$, in $\theta''(z_i) + G(z_i)\theta(z_i) = (\theta_{i+1} - 2\theta_i + \theta_{i-1})\big/h^2 - \theta''''(z_i)h^2\big/12 + G_i\theta_i + O(h^4) = k_0^2 g_i$. The gains with higher-order methods are more significant in higher dimensions, however.

3.5.1.1 Application to Normal Modes

The depth-separated Helmholtz equation for a normal mode can be written $\rho(\rho^{-1}Z')' + (\omega^2/c^2)Z = k^2 Z$ (cf. Eqs. 3.46, 3.100, and 3.105). The operator in the left side can be discretized, using a tridiagonal matrix T, in exactly the same way as the operator Γ, even with fourth-order accuracy using a Numerov scheme.

Together with the boundary conditions at the surface and at the bottom, $Z(0) = 0$, and $Z(z_b) = 0$ in the simple traction-free bottom case, a matrix eigenvalue problem appears for finding the modal wave numbers k. Some efficient solution methods are described in Ref. [109]. At least for fluid media without absorption, they provide alternatives to the winding-number integral and reflection-coefficient methods of Sections 3.3.1 and 3.3.2.

3.5.2 TWO-DIMENSIONAL FEM AND FDM FOR THE HELMHOLTZ EQUATION

FE and FD methods for solving the 2-D Helmholtz equation have been developed. The publications include [110−113] for FEM, and [114−117] for FDM. The present section illustrates how these methods are used to solve the range-dependent propagation problem treated with coupled modes in Section 3.3.4.

A cylindrically symmetric fluid medium is assumed, as shown in Fig. 3.17. It is similar to the medium in Fig. 3.8, with $N = 2$, but lateral variation of a rather general type is now allowed in range segment 2, now denoted Ω, between r_1 and r_2. Specifically, in the cylindrical coordinate system r, ϕ, z, the medium properties depend on depth z but not on azimuth ϕ. There are two range-invariant regions: range segment 1 ($r < r_1$), and range segment 3 ($r_2 < r$). For simplicity, there is a traction-free or rigid boundary at $z = z_b$. As in Section 3.3.4, there is a symmetric time-harmonic point source with angular frequency $\omega > 0$ in the water, on the axis ($r = 0$) at depth z_s. Omitting the time factor $\exp(-i\omega t)$ as usual, the body force, per unit volume, is given by $f = -M \, \mathrm{grad}(\delta(r)\delta(z - z_s)/2\pi r)$.

In Ω, the density and sound-speed functions $\rho(r, z)$ and $c(r, z)$, respectively, vary smoothly within each of two subdomains separated by a smooth interior interface $\partial\Omega_{ws}$: an upper water region Ω_w and a lower sediment region Ω_s. It would be easy to make an extension to several sediment regions.

Within each of range segments 1 and 3, the pressure field $p(r, z)$ can now be expressed in terms of normal modes according to Eq. (3.87). For $n = 1$ and $n = 3$, the mode functions $Z_{m,n}(z)$ together with their horizontal wave numbers $k_{m,n}$, mode integrals $I_{m,n}$, and modal expansion coefficients $a_{m,n}$ and $b_{m,n}$, $m = 1, 2,...$, are defined exactly as in Section 3.3.4. The density functions for the first and third range segments are denoted $\rho_1(z)$ and $\rho_3(z)$, respectively, and $\lambda_1(z)$ is the Lamé-parameter function $\lambda(z)$ for $r < r_1$.

Using Eqs. (3.87) and (3.88), Eq. (3.89) is now reformulated as

$$\frac{\partial p}{\partial r}(r_2-, z) = \frac{\rho(r_2-, z)}{\rho_3(z)}\Gamma_2 p(r_2, \cdot)(z), \tag{3.146}$$

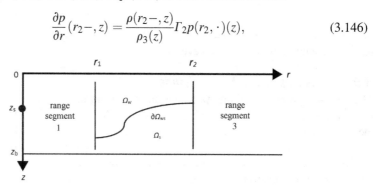

FIGURE 3.17

Range-dependent fluid medium with depth (z) and range (r) axes. There is a point source on the z-axis at depth z_s. In range segment 2, between r_1 and r_2, there is a smooth interior interface $\partial\Omega_{ws}$, which separates an upper water region Ω_w from a lower sediment region Ω_s.

where the linear operator $\Gamma_2 g$, for a general depth function g, is defined by

$$\Gamma_2 g(z) = \sum_{m=1}^{\infty} \frac{Z_{m,3}(z)}{I_{m,3}} \left(\widehat{H}_{m,3}^{(1)}(r_2) \right)^{-1} \frac{\partial \widehat{H}_{m,3}^{(1)}(r_2)}{\partial r} \int_0^{z_b} \frac{g(\zeta) Z_{m,3}(\zeta)}{\rho_3(\zeta)} d\zeta. \tag{3.147}$$

In the same way, Eq. (3.90) is reformulated as

$$\frac{\partial p}{\partial r}(r_1+, z) = \frac{\rho(r_1+, z)}{\rho_1(z)} (\Gamma_1 p(r_1, \cdot)(z) + S_s(z)), \tag{3.148}$$

where, with $\gamma_m = \left(\widehat{H}_{m,1}^{(1)}(r_1) \Big/ H_0^{(2)}(k_{m,1} r_1) + \widehat{H}_{m,1}^{(2)}(r_1) \right)^{-1}$, $\Gamma_1 g$ and S_s are defined by

$$\Gamma_1 g(z) = \sum_{m=1}^{\infty} \gamma_m \frac{Z_{m,1}(z)}{I_{m,1}} \left(\frac{1}{H_0^{(2)}(k_{m,1} r_1)} \frac{\partial \widehat{H}_{m,1}^{(1)}(r_1)}{\partial r} + \frac{\partial \widehat{H}_{m,1}^{(2)}(r_1)}{\partial r} \right)$$

$$\times \int_0^{z_b} \frac{g(\zeta) Z_{m,1}(\zeta)}{\rho_1(\zeta)} d\zeta \tag{3.149}$$

$$S_s(z) = \sum_{m=1}^{\infty} \frac{Z_{m,1}(z)}{I_{m,1}} \left(\gamma_m \frac{\partial \widehat{H}_{m,1}^{(2)}(r_1)}{\partial r} - \left(1 - \frac{\gamma_m}{H_0^{(2)}(k_{m,1} r_1)} \right) \frac{\partial \widehat{H}_{m,1}^{(1)}(r_1)}{\partial r} \right)$$

$$\times \frac{iM\omega^2}{4\lambda_1(z_s)} Z_{m,1}(z_s). \tag{3.150}$$

Eqs. (3.146) and (3.148) provide non-local radiation boundary conditions of Dirichlet-to-Neumann type [118,119] at the vertical boundaries of Ω. By Eqs. (3.7) and (3.8), the problem is then to determine the pressure field $p(r, z)$ in Ω from the Helmholtz equation

$$\text{div} \left(\frac{\text{grad}(p)}{\rho} \right) + \frac{\omega^2}{\rho c^2} p = \frac{1}{r} \frac{\partial}{\partial r} \left(\frac{r}{\rho} \frac{\partial p}{\partial r} \right) + \frac{\partial}{\partial z} \left(\frac{1}{\rho} \frac{\partial p}{\partial z} \right) + \frac{\omega^2}{\rho c^2} p = 0 \tag{3.151}$$

in each of Ω_w and Ω_s, together with the boundary conditions and the interface conditions that p and $\rho^{-1} \partial p / \partial n$ are continuous across $\partial \Omega_{ws}$.

3.5.2.1 FEM Discretization

Nice features of FEM are that an irregular interface $\partial \Omega_{ws}$ is easy to handle, and that the interface condition, as well as the possible rigid-boundary condition at $z = z_b$, appear naturally without having to be included explicitly.

Following [112], a triangulation of Ω is introduced, with triangles of similar side length with maximum h. The nodes are on the boundaries $z = 0$, $z = z_b$, $r = r_1$, $r = r_2$, the interface $\partial \Omega_{ws}$, and in the interiors of Ω_w and Ω_s, such that each triangle is considered to be either in Ω_w or in Ω_s, and such that $\partial \Omega_{ws}$ is well approximated by the triangle sides joining the nodes on $\partial \Omega_{ws}$. Excluding those nodes on $z = 0$ and $z = z_b$ where the essential boundary condition $p(r, z) = 0$ applies, there are L nodes, numbered $j = 1,..., L$. Introducing L triangle-wise linear shape functions S_j, one for each numbered node, p is approximated in Ω by $p(r, z) = \sum_{j=1}^{L} p_j S_j(r, z)$. For each $j = 1,..., L$, S_j equals 1 at its corresponding node, at which the pressure p is denoted p_j, and S_j vanishes at all remaining nodes.

Applying a Green's identity in the standard way, for a torus with cross section Ω in the rz half-plane with $\phi = 0$, a weak, variational, formulation of Eq. (3.151) appears:

$$\iint_{\Omega} \left(\frac{\omega^2}{c^2} S_i p - \frac{\partial S_i}{\partial r} \frac{\partial p}{\partial r} - \frac{\partial S_i}{\partial z} \frac{\partial p}{\partial z} \right) \frac{r}{\rho} dr dz + \int_0^{z_b} r_2 \frac{S_i(r_2, z)}{\rho_3(z)} \Gamma_2 p(r_2, \cdot)(z) dz$$
$$- \int_0^{z_b} r_1 \frac{S_i(r_1, z)}{\rho_1(z)} \Gamma_1 p(r_1, \cdot)(z) dz = \int_0^{z_b} r_1 \frac{S_i(r_1, z)}{\rho_1(z)} S_s(z) dz, \quad i = 1, 2, ..., L.$$

$$(3.152)$$

Inserting $p = \sum_{j=1}^{L} p_j S_j$ in Eq. (3.152), a linear L-dimensional equation system appears, with a sparse and symmetric system matrix, for determination of the nodal pressure values p_j. Practical guidelines for setting up the equation system are given in Ref. [112], for example. The triangle nodes should be enumerated carefully to minimize the bandwidth of the equation system. Spline representations of the density and sound-speed functions, in each of Ω_w and Ω_s, are often convenient when the integrals are computed.

3.5.2.2 FDM Discretization

Application of FDM is facilitated by first mapping each of the regions Ω_w and Ω_s onto rectangles by orthogonal transformations. A method for doing this, which puts some mild requirements on the regularity and curvature on the interface $\partial \Omega_{ws}$, appears in Ref. [120]. In the new coordinates, still denoted r with $r_1 \le r \le r_2$ and z with $0 \le z \le z_b$ for simplicity, $\partial \Omega_{ws}$ is a horizontal line segment at $z = z_{ws}$ across which $\rho^{-1} \partial p / \partial z$ is continuous, and the Helmholtz equation (3.151) appears as [121]

$$\frac{\partial}{\partial r} \left(a \frac{\partial p}{\partial r} \right) + \frac{\partial}{\partial z} \left(\frac{1}{a} \frac{\partial p}{\partial z} \right) + b_r \frac{\partial p}{\partial r} + b_z \frac{\partial p}{\partial z} + \frac{\omega^2}{c^2} ep = 0 \qquad (3.153)$$

with different coefficient functions a, e, b_r, b_z of r, z in each of Ω_w and Ω_s, The functions a, e are determined by the orthogonal transformation, while b_r, b_z involve ρ and its first-order derivatives as well.

A range grid is now introduced by $r_{2,i} = r_1 + (i - 3/2)h_r$, $i = 1, 2,..., L_r$, such that $(L_r - 2)h_r = r_2 - r_1$, and a depth grid by $z_j = jh_z$, $j = 1, 2,..., L_z$, such that $(L_z + 1)h_z = z_b$. This depth grid is adapted to the case with a traction-free boundary at z_b. A grid with $(L_z + 1/2)h_z = z_b$ would be more convenient for a rigid lower boundary. For simplicity, z_{ws} is assumed to coincide with one of the z_j. Otherwise, different h_z should preferably be used for Ω_w and Ω_s. The grid steps h_r and h_z are typically of similar size, and h is defined as their maximum.

Using second-order FD approximations, an L-dimensional linear equation system is subsequently set up, with a sparse but non-symmetric system matrix, for the $L = L_r \, L_z$ grid values $p_{i,j} = p(r_{2,i}, \, z_j)$. Eq. (3.153) does not contain mixed second-order derivatives (a consequence of the orthogonality of the variable transformations), and the one-dimensional approximations from Section 3.5.1 are directly applicable. The integrals involved in the radiation boundary conditions of Eqs. (3.146) and (3.148) can be approximated using the trapezoidal rule. For details, see Ref. [115]. Column-wise enumeration of the grid points is recommended. However, to prepare for preconditioning (see below), the grid points along $\partial\Omega_{ws}$ should preferably be included separately [121]. The equation system gets a nice block-banded structure.

3.5.2.3 Methods to Solve the Linear Equation System and Possibilities to Reduce Its Size

The linear L-dimensional equation system that results from either FEM or FDM discretization can be solved by direct methods, such as Gaussian elimination, only for very small low-frequency problems. Storage is an issue, and iterative methods, which avoid fill-in and take full advantage of the sparsity of the system, are typically needed. Moreover, the large equation systems are typically ill-conditioned, necessitating sophisticated preconditioning methods. Details are not given here; however, some useful solution techniques can be found in the references, for example, Refs. [100,110,122].

With a second-order FE or FD method, the number of grid points needed for a specified accuracy grows as the cube of the frequency [115]. This explains why the equation systems are very large, except for low-frequency problems. The inaccuracy for a coarse grid is caused by numerical dispersion, which can be explained as follows. In the range-invariant case with a homogeneous water waveguide, Eqs. (3.100) and (3.102) show that propagation in the depth as well as range (for large r) direction is governed by the one-dimensional Helmholtz equation $d^2\theta/ds^2 + \kappa^2\theta(s) = 0$. Here κ is the vertical or horizontal wave number, which is considered to be proportional to the frequency, and s is depth z or range r. Basic solutions $\theta(s) = \exp(\pm i\kappa s)$ appear. Approximating the ODE with a central difference scheme $(\theta_{j+1} - 2\theta_j + \theta_{j-1})/h^2 + \kappa^2\theta_j = 0$, however, where $\theta_j = \theta(jh)$ (cf. Section 3.5.1), the basic solutions appear as $\theta_j = (1 - \kappa^2 h^2/2 \pm i\kappa h(1 - \kappa^2 h^2/4)^{1/2})^j$, and the error $\theta_j - \exp(\pm i\kappa jh)$ can be analyzed quantitatively [115].

The required number of grid points can be reduced significantly by applying higher-order methods. With a fourth-order method, as developed in Ref. [122], it grows as the frequency raised to the power 2.5 [123].

An artificial traction-free or rigid boundary has been assumed at depth z_b. To reduce reflections from such a boundary, a 10-wavelengths thick gradient layer with gradually increasing absorption was introduced for the island examples in Section 3.3.5.2. For applications of FEM or FDM, a lot of grid points are needed for such a layer. With perfectly matched layers (PML), first proposed in Ref. [124], artificial echo-reducing absorption layers with much fewer grid points are possible [125].

3.5.3 TIME-DOMAIN MODELING

As pointed out at the beginning of this chapter, the time signal resulting from a transient source pulse in general can be computed from frequency-domain results by Fourier synthesis. A few FE and FD methods for direct time-domain computations, which can often be more convenient, are briefly described in this section.

The cylindrically symmetric fluid medium in Fig. 3.17 is still assumed, but with $r_1 = 0$ and, initially, no absorption and a pressure-release or rigid boundary at $r = r_2$. Furthermore, r_2 as well as z_b are initially assumed large enough to make all reflections from the corresponding boundaries arrive at the receiver points after all arrivals of interest. There is still a symmetric point source in the water, on the axis ($r = 0$) at depth z_s, but its body force, per unit volume, is now given by $f(x, t) = -M(t) \, \mathrm{grad}(\delta(r) \delta(z - z_s)/2\pi r)$, where $M(t)$ is a moment-tensor strength pulse {unit: N m} that vanishes for $t < 0$. By Eqs. (3.6) and (3.7) (cf. Eq. 3.9), the problem is then to determine the pressure field $p(r, z, t)$ in Ω from the wave equation

$$\mathrm{div}\left(\frac{\mathrm{grad}(p)}{\rho}\right) = \frac{1}{r}\frac{\partial}{\partial r}\left(\frac{r}{\rho}\frac{\partial p}{\partial r}\right) + \frac{\partial}{\partial z}\left(\frac{1}{\rho}\frac{\partial p}{\partial z}\right) = \frac{1}{\rho c^2}\frac{\partial^2 p}{\partial t^2} - \frac{1}{\rho c^2}\frac{\delta(r)\delta(z - z_s)}{2\pi r}\frac{d^2 M}{dt^2}$$

(3.154)

together with the boundary and interface conditions.

3.5.3.1 FEM Discretization

The development in Section 3.5.2.1 is modified by introducing time dependence explicitly by $p(r, z, t) = \sum_{j=1}^{L} p_j(t) \, S_j(r, z)$. For a rigid boundary at $r = r_2$, the correspondence to Eq. (3.152) is

$$\iint_\Omega \left(\frac{1}{c^2}S_i\frac{\partial^2 p}{\partial t^2} + \frac{\partial S_i}{\partial r}\frac{\partial p}{\partial r} + \frac{\partial S_i}{\partial z}\frac{\partial p}{\partial z}\right)\frac{r}{\rho}\,drdz = \frac{S_i(0, z_s)}{2\pi\rho(0, z_s)c^2(0, z_s)}\frac{d^2 M}{dt^2},$$

$$i = 1, 2, ..., L.$$

(3.155)

Substituting $p = \sum_{j=1}^{L} p_j \, S_j$ in Eq. (3.152) gives the ODE system

$$A \cdot \frac{d^2 p(t)}{dt^2} + B \cdot p(t) = \frac{d^2 M(t)}{dt^2}b$$

(3.156)

with $p(t) = (p_1(t), p_1(t),..., p_L(t))^\mathrm{T}$ and obvious definitions of the elements of the sparse, positive definite, and symmetric $L \times L$ matrices A (the mass matrix) and B (the stiffness matrix) and the L-dimensional column vector b.

3.5.3.2 FDM Discretization

It is first assumed that $\rho(r, z)$ and $c(r, z)$ vary smoothly throughout Ω, without the discontinuity interface $\partial\Omega_{ws}$. For traction-free boundaries at $r = r_2$ and $z = z_b$, a range grid is introduced by $r_{2,i} = (i - 1)h_r$, $i = 1, 2,..., L_r$, such that $(L_r + 1) h_r = r_2$, and a depth grid by $z_j = jh_z$, $j = 1, 2,..., L_z$, such that $(L_z + 1)h_z = z_b$. The maximum of h_r and h_z is denoted h.

Using FD second-order approximations from Section 3.5.1 for the spatial derivatives, Eq. (3.154) provides the ODE system

$$\frac{d^2\boldsymbol{p}(t)}{dt^2} + \boldsymbol{B} \cdot \boldsymbol{p}(t) = \frac{d^2 M(t)}{dt^2} \boldsymbol{b},\tag{3.157}$$

where $\boldsymbol{p}(t)$ is a column vector containing the $L = L_r L_z$ functions $p_{i,j}(t)$ in some suitable order, $\boldsymbol{B}$ is a sparse $L \times L$ matrix, and $\boldsymbol{b}$ is an L-dimensional source column vector. Two remarks are needed here. For the nodes on the z axis, with $r = 0$, the term $(\rho r)^{-1}\partial p/\partial r$ arising from Eq. (3.154) equals $\rho^{-1}\partial^2 p/\partial r^2$, for which centered FD approximations can be obtained using $p(-r, z, t) = p(r, z, t)$. The second remark concerns the source excitation during the pulse time. One approach [126,127] is to insert some source pressure approximation at, and close to, the boundary nodes of a small, temporarily removed, region (symmetric around the z-axis) surrounding the source. The approximation may be obtained from the analytical result $p(r, z, t) = (d^2 M/dt^2)(t - \tau(r, z))/4\pi c^2(0, z_s)R(r, z)$, where $R(r, z) = (r^2 + |z - z_s|^2)^{1/2}$ and $\tau(r, z) = R(r,z)/c(0, z_s)$, for a homogeneous medium with density $\rho(0, z_s)$ and sound speed $c(0, z_s)$ [1]. However, the source must have become silent and the source region restored when reflected or backscattered energy arrives.

An irregular discontinuity interface $\partial\Omega_{ws}$ is more easily handled with FEM. With FDM, coordinate transformations, such as in Section 3.5.2.2, or irregular grids could be used, but the easiest approach (cf. Section 3.4.3.1) is to replace each medium discontinuity interface by a wavelength-wide transition region with smooth parameter variations.

3.5.3.3 Numerical Dispersion, Time Integration, and Stability

To assess phase errors arising by numerical dispersion, the maximum significant angular frequency ω_{max} of the source pulse spectrum can be studied as indicated in Section 3.5.2.3. With c_{min} as the least $c(r, z)$ in Ω and r_{max} as the maximum propagation range, a condition of type $h^2 < K_d (c_{min}/\omega_{max})^3/r_{max}$ appears, where K_d depends on the error tolerance.

Eqs. (3.156) and (3.157) are discretized in time by $t_n = n\Delta t$ and $\boldsymbol{p}_n = \boldsymbol{p}(t_n)$, $n = -1, 0, 1, \ldots$, with initial conditions $\boldsymbol{p}_{-1} = \boldsymbol{p}_0 = \boldsymbol{0}$. Time marching is subsequently performed according to a central second-order scheme, for example, by approximating $d^2\boldsymbol{p}(t_n)/dt^2$ with $(\boldsymbol{p}_{n+1} - 2\boldsymbol{p}_n + \boldsymbol{p}_{n-1})/(\Delta t)^2$. With Eq. (3.156), an effectively explicit scheme is obtained by lumping the typically diagonally dominant mass matrix $\boldsymbol{A}$ into a diagonal matrix by row (or column) summation. The accuracy of such lumping is studied in Ref. [128].

The stability properties of the resulting scheme can be determined by von Neumann analysis [129], whereby the time evolution of each spatial Fourier component of an introduced error is studied. The magnitude of the corresponding growth factor, determined by the difference equation in time, must not exceed unity. With c_{max} as the largest $c(r, z)$ in Ω, a Courant−Friedrichs−Lewy (CFL) condition of type $c_{max} \Delta t < K_s h$ appears [127], where K_s depends on the spatial scheme. As usual, higher-order FD schemes allow coarser grids.

3.5.3.4 Including Absorption

Viscoelastic absorption in the medium, of interest mainly for the sediment, is easy to include in the frequency domain, by complex Lamé parameters, see the beginning of the chapter. In the time domain, the convolution integral defining the viscoelastic stress–strain relation [1] is more difficult to incorporate. However, an elasticoviscous (Maxwell) behavior [4] is easy to handle. In this special case, Eq. (3.4) is replaced by the constitutive relation

$$\frac{\partial p}{\partial t} + \gamma p = -\lambda \operatorname{div}\left(\frac{\partial \boldsymbol{u}}{\partial t}\right), \tag{3.158}$$

where $\gamma > 0$ is a spatial function characterizing the relation between the elastic response and the creep rate of the medium. Eq. (3.154) is changed by replacing $\partial^2 p/\partial t^2$ in the right-hand side by $\partial^2 p/\partial t^2 + \gamma \, \partial p/\partial t$. An additional term $\boldsymbol{D} \cdot d\boldsymbol{p}(t)/dt$, where $\boldsymbol{D}$ is a sparse, positive definite, and symmetric $L \times L$ matrix, appears in each of Eqs. (3.156) and (3.157). For Eq. (3.157), the damping matrix $\boldsymbol{D}$ is even diagonal. For Eq. (3.156), $\boldsymbol{D}$ can be lumped into a diagonal matrix. Explicit, second-order accurate, conditionally stable time marching is achieved by approximating $d\boldsymbol{p}(t_n)/dt$ with $(\boldsymbol{p}_{n+1} - \boldsymbol{p}_{n-1})/2\Delta t$.

The viscoelastic constitutive relation (3.158) provides a high-frequency absorption, in dB per wavelength, which is inversely proportional to the frequency. More realistic stress–strain relations can be handled by incorporation of ODEs, with respect to time, for memory variables characterizing the strain history [130–132].

PML has been used in the time-domain case as well, to reduce reflections from artificial boundaries [133,134]. The need is not as pressing as in the time-harmonic case, since it may be possible to extract the information of interest before the undesired boundary reflections have reached the receiver. Furthermore, the earliest among the reflections from the artificial boundaries at $r = r_2$ and $z = z_b$ can be eliminated by repeated computations, summing results for traction-free and rigid boundary conditions [135].

3.5.3.5 Some Recent Developments

Eq. (3.155) involves integrals over Ω. By approximating $\iint_\Omega g(r,z)drdz$, for a generic function g, by $\sum_{j=1}^{L} w_j \, g_j(r_j, z_j)$, where the (r_j, z_j) are the nodes of the shape functions S_j and the w_j are certain weights, the resulting mass matrix $\boldsymbol{A}$ in Eq. (3.156) is apparently diagonal directly, without lumping. This idea is further developed and utilized in the spectral FEM, which has become increasingly popular in geophysics and seismology during the last decades [136]. Some applications to ocean acoustics appear in Ref. [137]. In contrast to traditional FEM, spatial functions are approximated by products of high-degree (typically between 5 and 10) Lagrange interpolation polynomials, with control points coinciding with those needed for Gauss–Lobatto–Legendre quadrature.

At times t when the source has become silent, Eq. (3.154) appears as $\operatorname{div}(\operatorname{grad}(p)/\rho) = (\rho c^2)^{-1} \partial^2 p/\partial t^2$, and the divergence theorem shows that $dE(t)/dt = 0$, where

$$E(t) = 2\pi \iint_\Omega \left(\frac{1}{\rho c^2}(\partial p/\partial t)^2 + \frac{1}{\rho}(\operatorname{grad}(p) \cdot \operatorname{grad}(p))\right) r \; drdz. \tag{3.159}$$

The "energy" $E(t)$ is thus constant, controlling the growth of p with time. For certain FD operators, having a summation-by-parts (SBP) symmetry property, stability at time marching can be proved by a corresponding discrete energy method. Boundary conditions can be incorporated by the simultaneous approximation term (SAT) method involving penalty functions. Along with an application to underwater acoustics, [138] gives details of the SBP–SAT technique for studying acoustic wave propagation numerically.

3.5.4 EXAMPLES

This section contains a few examples of FDM solutions of the Helmholtz equation. A difference to Section 3.5.2 is that the 2-D problem is solved in Cartesian xz coordinates rather than cylindrical rz coordinates. Specifically, a line source parallel to the y axis, with body force per unit volume given by $f = -M \text{ grad}(\delta(x)\delta(z - z_s))/r_{ref}$, is considered in a medium that is invariant with respect to y. The solution method for the usual case with a point source in a cylindrically symmetric medium can be used with minor changes [122].

The first example, with a 25 Hz source at depth $z_s = 100$ m in the water, is taken from Ref. [100]. A flat bottom at depth $z = 200$ m is followed by a ridge with apex at $x = 3000$ m, $z = 22$ m, see Fig. 3.18. The water sound speed is 1500 m/s. Two cases are considered for the underlying sediment: (I) sound speed 1700 m/s and density 1200 kg/m^3 and (II) sound speed 4000 m/s and density 2700 kg/m^3. In both cases, the sediment absorption is 0.8 dB/wavelength. At the initial range $x = 0$ m, there are three propagating modes in case I and six in case II. In both cases, the first three of these modes have horizontal slowness $k/\omega \approx 0.66$, 0.64, and 0.61 s/km, respectively. The source excitation is restricted to these three modes, an excitation that can be realized physically by a vertical source array.

Fig. 3.19 shows FDM solutions for the two medium cases of Fig. 3.18. In case I, practically all energy is lost into the bottom during the up-slope propagation. The three propagating modes are lost successively, as narrow beams, as their cutoff

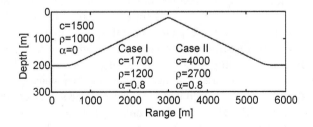

FIGURE 3.18

Geometry, in the xz plane, and medium parameters for a certain waveguide that is invariant with respect to the y coordinate. The sound speed c is in m/s, the density ρ is in kg/m^3, and the absorption, here denoted α, is in dB/wavelength.

*Reproduced with permission from Larsson, E. and Abrahamsson, L., Helmholtz and parabolic equation solutions to a benchmark problem in ocean acoustics, J. Acoust. Soc. Am., **113**, (5), pp. 2446–2454, 2003. Copyright 2003; Acoustical Society of America.*

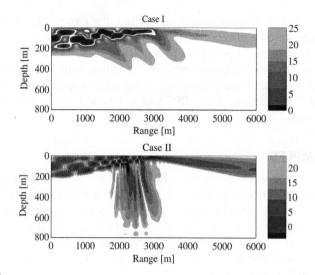

FIGURE 3.19

FDM propagation loss in dB, rel. 1 m, for the medium cases of Fig. 3.18: case I (left) and case II (right). The 25 Hz horizontal line source is at depth 100 m.

Reproduced with permission from Larsson, E. and Abrahamsson, L., Helmholtz and parabolic equation solutions to a benchmark problem in ocean acoustics, *J. Acoust. Soc. Am.,* **113**, *(5), pp. 2446–2454, 2003. Copyright 2003; Acoustical Society of America.*

depths are reached. The harder bottom in case II results in larger bottom reflection coefficients, and a noticeable amount of energy is backscattered into the water column, where it interferes with the out-going waves, instead of being transmitted into the bottom.

Fig. 3.20 shows corresponding results using the Claerbout wide-angle PE. The PE and FDM results for case I are almost identical visually, whereas there are significant differences for case II. In particular, the (one-way) PE fails to account for the nonnegligible amount of energy that is backscattered. According to Ref. [100], the relative error for the PE complex pressure solution, as measured along vertical lines at ranges x less than 1500 m, is about 23% for case II but at most 3% for case I.

The next example, with a 500 Hz source at depth $z_s = 42$ m in the water, is taken from Ref. [122]. The bottom is now very hard, in fact rigid. As shown by Fig. 3.21, its depth is decreasing smoothly from 70 m at range $x = 0$ m to 35 m at $x = 500$ m. The sound speed in the water varies with depth according to a typical summer profile in the Baltic Sea, with a sound-speed minimum close to the source depth creating a sound channel (cf. Fig. 3.1). This time, the source excitation is restricted to the first five propagating modes. Close to the source, these five modes have an almost horizontal propagation direction (cf. Eq. 3.97). Ray theory appears feasible, since the wavelength in the water, about 3 m, is small in comparison with the water depths (cf. Section 3.1). Indeed, the PL result in Fig. 3.21 indicates a field dominated by an initially horizontal beam that is multiply reflected at the bottom and at the surface.

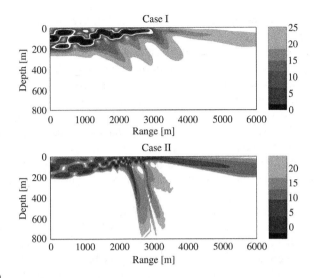

FIGURE 3.20

PE results corresponding to those in Fig. 3.19.

Reproduced with permission from Larsson, E. and Abrahamsson, L., Helmholtz and parabolic equation solutions to a benchmark problem in ocean acoustics, *J. Acoust. Soc. Am., **113**, (5), pp. 2446–2454, 2003. Copyright 2003; Acoustical Society of America.*

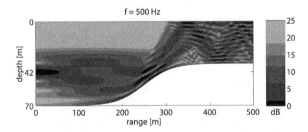

FIGURE 3.21

FDM propagation loss in dB, rel. 1 m, above a smoothly up-sloping rigid bottom shown white. The 500 Hz horizontal line source is at depth 42 m in a sound channel, where it triggers the first five propagating modes.

From Otto, K. and Larsson, E., A flexible solver of the Helmholtz equation for layered media, *In: Proc. European Conference on Computational Fluid Dynamics, Delft, 2006.*

3.6 3-D SOUND PROPAGATION MODELS

In underwater acoustics, it is surprisingly often sufficient to do the modeling in 2-D, neglecting environmental variations in the azimuthal ϕ direction, for a cylindrical r, ϕ, z coordinate system centered at the source. Different environment parameters can of course be set when computations are made for different receiver azimuths from the source, which is called N × 2-D modeling.

Two important cases where 3-D effects may show up are cross-slope propagation over a sloping bottom and propagation around seamounts. In the former case, the azimuthal direction of a ray is changed when it is reflected at the bottom. As illustrated in the left panel of Fig. 3.22, ray trajectories as projected on the horizontal plane appear as if the rays were refracted by a horizontal sound-speed gradient, and the effect is called horizontal refraction. Shadow zones without ray arrivals may appear, as is of course also the case when a cylindrically symmetric seamount or other anomaly, illustrated in the right panel of Fig. 3.22, blocks the sound propagation. Although not captured by classical ray modeling, energy penetrates shadow zones by diffraction.

The basic ray equations at the beginning of Section 3.1 are already given for the 3-D case. The analytic 2-D ray tracing of Section 3.1.1 is readily extended to three-dimensional, although this is not pursued here. It is also possible to extend the PE approaches of Section 3.4 to 3-D. Several papers have appeared on this topic, including Refs. [139−144] with experimental verifications in Ref. [145,146]. Generalizations of the FDM and FEM of Section 3.5 to 3-D are straightforward in principle, but the computational burden increases dramatically.

A trial solution form with normal modes in the vertical direction together with an adiabatic approximation has been used with success in several cases (e.g., Refs. [147−149]). Starting from the Helmholtz equation (3.9), a solution $\tilde{p}(x, y, z)$ is sought in the form

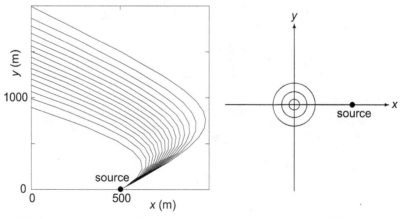

FIGURE 3.22

Horizontal xy planes. Left: apparent horizontal refraction of rays for a homogeneous water column with depth decreasing linearly from 150 m at $x = 0$ m to 50 m at $x = 1000$ m; all rays start at depth 50 m in the same direction horizontally but with different inclination angles. Right: circular depth contours for a seamount that is symmetric around the z axis.

$$\widetilde{p}(x, y, z) = \sum_m \gamma_m(x, y) Z_m(x, y, z), \tag{3.160}$$

where Z_m, $m = 1, 2, \ldots$, are local modes fulfilling the depth-separated Helmholtz equations

$$\rho(x, y, z) \frac{\partial}{\partial z} \left(\frac{1}{\rho(x, y, z)} \frac{\partial Z_m(x, y, z)}{\partial z} \right) + \left(\frac{\omega^2}{c^2(x, y, z)} - k_m^2(x, y) \right) Z_m(x, y, z) = 0 \tag{3.161}$$

together with appropriate boundary conditions and the orthogonality relations

$$\int_0^\infty \frac{1}{\rho(x, y, z)} Z_m(x, y, z) Z_n(x, y, z) dz = I_m(x, y) \delta_{mn}, \tag{3.162}$$

cf. Eq. (3.88), for each x, y. The modal wave numbers are here denoted $k_m(x, y)$. Substituting Eq. (3.160) into Eq. (3.9), integrating over depth after multiplication with Z_n/ρ, and neglecting terms involving horizontal derivatives of ρ or Z_m (assuming appropriate ordering of the modes for different x, y), the "horizontal refraction" equations

$$\frac{\partial^2 \gamma_n}{\partial x^2} + \frac{\partial^2 \gamma_n}{\partial y^2} + k_n^2(x, y) \gamma_n = \left(\frac{\omega^2}{c^2(x_s)} \right) M \delta(x - x_s) \delta(y - y_s) \frac{Z_n(x_s)}{I_n \rho(x_s)} \tag{3.163}$$

appear, $n = 1, 2, \ldots$, one separate 2-D Helmholtz equation for each local vertical mode. Each Eq. (3.163) can be solved by any of the 2-D methods, for example by rays or PE.

For certain environments of great interest with respect to 3-D effects, transform methods can be used to replace the 3-D problem by an ensemble of 2-D problems, which can be solved in parallel. Since there are still restrictions on the environmental variations, these problems are sometimes referred to as 2.5-D rather than 3-D. With the coupled-mode method selected to solve the 2-D problems, details are provided in Sections 3.6.1 and 3.6.2, for a sloping bottom and a cylindrically symmetric anomaly, respectively. Numerically stable and efficient solutions using global matrix as well as marching single-scattering methods have been presented in recent years. Here, however, the method with recursive computation of reflection matrices from Sections 3.2.1 and 3.3.4 is adapted to obtain marching solutions without further approximations.

3.6.1 MODELING HORIZONTAL REFRACTION BY A SLOPING BOTTOM OR CHANGING SOUND-SPEED PROFILE

A medium is now assumed (cf. the particular case in the left panel of Fig. 3.22), for which the medium parameters do not depend on the y coordinate, but lateral variation is allowed among a number of segments along the x axis. Specifically, in the Cartesian coordinate system x, y, z, the medium properties depend on depth z but not on x or y within each of the $N + 1$ segments defined by $x \leq x_1$,

$x_1 < x \le x_2,..., x_{N-1} < x \le x_N, x_N < x$. Here, $x_1, x_2,..., x_N$ with $N \ge 1$ form an increasing sequence of positive segment limits. For simplicity, there is a traction-free or rigid boundary at $z = z_b$, common for all segments, and there is no solid region. As before, there is a time-harmonic point source with angular frequency ω in the water, at $x = y = 0$ and depth z_s. Omitting the time factor $\exp(-i\omega t)$ with $\omega > 0$, the body force per unit volume is $f = -M \, \text{grad}(\delta(x)\delta(y)\delta(z - z_s))$. With the range axis changed to a Cartesian x axis, and the r_n changed to x_n, $n = 1, 2,..., N$, Fig. 3.8 works for this case as well.

3.6.1.1 Fourier Transformation With Respect to the y-Coordinate

In this case, Fourier transformation of Eq. (3.9) provides the field representation

$$p(x, y, z) = \int_{-\infty}^{+\infty} \widehat{p}(x, z; \kappa) e^{i\kappa y} d\kappa, \tag{3.164}$$

where, for each κ and in each x segment, $\widehat{p}(x, z; \kappa) = (2\pi)^{-1} \int_{-\infty}^{+\infty} p(x, y, z) \, e^{-i\kappa y} dy$ {unit: N/m} fulfills the Helmholtz equation

$$\frac{\partial^2 \widetilde{p}}{\partial x^2} + \rho_n(z) \frac{\partial}{\partial z}\left(\frac{1}{\rho_n(z)} \frac{\partial \widetilde{p}}{\partial z}\right) + \left(\frac{\omega^2}{c_n^2(z)} - \kappa^2\right)\widetilde{p} = \left(\frac{\omega^2}{c_n^2(z)}\right) \frac{M\delta(x)\delta(z - z_s)}{2\pi} \tag{3.165}$$

with $\widetilde{p}(x, z; \kappa) = \widehat{p}(x, z; \kappa) + M\delta(x)\delta(z - z_s)/2\pi$. Here, $n = n(x)$ is the segment index, i.e., $n = 1$ for $x \le x_1$, $n = 2$ for $x_1 < x \le x_2,..., n = N$ for $x_{N-1} < x \le x_N$, $n = N + 1$ for $x_N < x$, while $\rho_n(z)$ and $c_n(z)$ are the density and sound-speed functions, respectively, for the nth segment. The pressure function $\widehat{p}(x, z; \kappa) \, e^{i\kappa y}/r_{\text{ref}}$ solves the physical line source problem with body force per unit volume given by $-(M/2\pi r_{\text{ref}}) \, \text{grad}(\delta(x) \, e^{i\kappa y} \, \delta(z - z_s))$.

Within each segment on the x axis, the field $\widehat{p}(x, z; \kappa)$ can now be expressed in terms of normal modes. Following [150], for $x > 0$,

$$\widehat{p}(x, z; \kappa) = \sum_{m=1}^{\infty}\left[a_{m,n}(\kappa)\widehat{E}_{m,n}^{(1)}(x; \kappa) + b_{m,n}(\kappa)\widehat{E}_{m,n}^{(2)}(x; \kappa)\right]Z_{m,n}(z), \tag{3.166}$$

where $Z_{m,n}(z)$ is the mode function {unit: N} for the mth mode in the nth x segment, and $k_{m,n}$ is its horizontal wave number. These modal wave numbers and mode functions, with dependence on x given by the factors $\exp\left(\pm ix\sqrt{k_{m,n}^2 - \kappa^2}\right)$ according to Eq. (3.165), can be computed as described in Sections 3.3.1 and 3.3.2, respectively. The $\widehat{E}_{m,n}^{(1)}(x; \kappa)$ and $\widehat{E}_{m,n}^{(2)}(x; \kappa)$ are normalized exponential functions. With $x_0 = 0$ and $x_{N+1} = x_N$, they are defined by $\widehat{E}_{m,n}^{(1)}(x; \kappa) = \exp\left(i(x - x_{n-1})\right.$ $\left.\sqrt{k_{m,n}^2 - \kappa^2}\right)$ and $\widehat{E}_{m,n}^{(2)}(x; \kappa) = \exp\left(-i(x - x_n)\sqrt{k_{m,n}^2 - \kappa^2}\right)$ for $n = 1, 2,...,$ $N + 1$. For each $n = 1, 2,..., N + 1$, the mode functions fulfill the Sturm–Liouville orthogonality relations given by Eq. (3.88).

For $x < 0$, the field $\widehat{p}(x, z; \kappa)$ can be expressed as

$$\widehat{p}(x, z; \kappa) = \sum_{m=1}^{\infty} B_m(\kappa) \, \exp\left(-ix\sqrt{k_{m,1}^2 - \kappa^2}\right) Z_{m,1}(z). \qquad (3.167)$$

The problem is then to determine the modal expansion coefficients {unit: m^{-1}} $a_{m,n}(\kappa)$ and $b_{m,n}(\kappa)$, $m = 1, 2,..., n = 1, 2,..., N + 1$, and $B_m(\kappa)$, $m = 1, 2,...$, for each κ. Different κ values can apparently be treated independently. In the following description of a solution method, the dependence on κ is in general omitted in the notation.

3.6.1.2 Equations Relating the Modal Expansion Coefficients

With the horizontal wave numbers $\sqrt{k_{m,n}^2 - \kappa^2}$ of the modes in the upper half-plane, but not on the negative half-axis, the $a_{m,n}$ and $b_{m,n}$ terms in Eq. (3.166) represent outgoing and in-coming components of the field, respectively. Column vectors are defined according to $\boldsymbol{a}_n = (a_{1,n}, a_{2,n},...)^{\mathrm{T}}$ and $\boldsymbol{b}_n = (b_{1,n}, b_{2,n},...)^{\mathrm{T}}$, $n = 1, 2,..., N + 1$, and $\boldsymbol{B} = (B_1, B_2,...)^{\mathrm{T}}$. There cannot be any in-coming components in the outermost x segment, hence

$$\boldsymbol{b}_{N+1} = \boldsymbol{0}. \qquad (3.168)$$

Excitation coefficients at the source at $x = 0$ in the first x segment can be derived by considering the range-invariant case, for which Fourier transformation of Eq. (3.165) with respect to x provides a depth-separated Helmholtz equation similar to Eq. (3.46). A slight adaptation of the excitation-coefficient result of Eq. (3.83) (cf. Section 5.2.2 in Ref. [7]) provides expressions for $\boldsymbol{a}_1$ and $\boldsymbol{B}$ according to

$$a_{m,1} = -\frac{iM\omega^2}{4\pi\rho_1(z_s)c_1^2(z_s)\sqrt{k_{m,1}^2 - \kappa^2}} \frac{Z_{m,1}(z_s)}{I_{m,1}}, \quad m = 1, 2, ... \qquad (3.169)$$

$$B_m = a_{m,1} + b_{m,1}e^{ix_1\sqrt{k_{m,1}^2 - \kappa^2}}, \quad m = 1, 2, ... \qquad (3.170)$$

For the range-invariant case, a verification that Eq. (3.169) together with Eqs. (3.164) and (3.166) agree with Eqs. (3.90) and (3.87) can be obtained by expanding the Hankel function in plane waves according to Eqs. (4.13) and (A.3) in Ref. [151].

For $n = 1, 2,..., N + 1$, it is now convenient to introduce the diagonal matrices

$$\widehat{\boldsymbol{E}}_n = \text{diag}\left(\widehat{E}_{m,n}^{(1)}(x_n), \ m = 1, 2, ..\right) = \text{diag}\left(\widehat{E}_{m,n}^{(2)}(x_{n-1}), \ m = 1, 2, ...\right), \quad \text{where}$$

$\widehat{\boldsymbol{E}}_{N+1} = \boldsymbol{I}$ since $x_{N+1} = x_N$. Using continuity of pressure and x-direction displacement at each segment interface x_n, $n = 1, 2,..., N$, Eq. (3.166) together with Eq. (3.7) and the orthogonality relations (3.88) show that

$$\begin{pmatrix} \boldsymbol{a}_{n+1} \\ \widehat{\boldsymbol{E}}_{n+1}\cdot\boldsymbol{b}_{n+1} \end{pmatrix} = \begin{pmatrix} \boldsymbol{A}_n & \boldsymbol{B}_n \\ \boldsymbol{C}_n & \boldsymbol{D}_n \end{pmatrix} \cdot \begin{pmatrix} \widehat{\boldsymbol{E}}_n\cdot\boldsymbol{a}_n \\ \boldsymbol{b}_n \end{pmatrix} \qquad (3.171)$$

for $n = 1, 2,..., N$. The coupling matrices A_n, B_n, C_n, and D_n, can be expressed as $A_n = D_n = (F_n + G_n)/2$ and $B_n = C_n = (F_n - G_n)/2$, where the elements of the mode coupling matrices F_n and G_n appear as in Eqs. (3.92) and (3.93) but with $k_{m2,n}/k_{m1,n+1}$ changed to $\left(k_{m2,n}^2 - \kappa^2\right)^{1/2} / \left(k_{m1,n+1}^2 - \kappa^2\right)^{1/2}$.

Taken together, Eqs. (3.168), (3.169), and (3.171) provide a system of $2(N + 1)$ linear matrix equations for the $2(N + 1)$ vectors a_n and b_n, $n = 1, 2,..., N + 1$. In practice, truncation is of course made to a finite number of modes, say N_M, in each x segment.

3.6.1.3 Solution in Terms of Reflection-Coefficient Matrices

A direct propagator-matrix multiplication approach is not always reliable for solving the obtained equation system (cf. Section 3.3.4.2). The direct global matrix method is applied in Ref. [150]. Here, however, the reflection-coefficient method used in Sections 3.2.1.1 and 3.3.4.2 is adapted for a convenient and numerically stable two-way marching solution.

Eq. (3.168) can be expressed as $\widehat{E}_{N+1} \cdot b_{N+1} = R_{N+1}^- \cdot a_{N+1}$, where $R_{N+1}^- = 0$ is interpreted as a reflection-coefficient matrix for out-going waves at x_N+. Corresponding reflection-coefficient matrices, denoted R_N^+, R_N^-,..., R_1^+, are now introduced at x_N-, $x_{N-1}+$,..., x_1-. Specifically, $b_n = R_n^+ \cdot \widehat{E}_n \cdot a_n$ and $\widehat{E}_n \cdot b_n = R_n^- \cdot a_n$ for $n = N, N - 1,..., 2$, and $b_1 = R_1^+ \cdot \widehat{E}_1 \cdot a_1$. As verified by applying Eq. (3.171), these reflection matrices can be computed recursively by

$$R_n^+ = -\left(D_n - R_{n+1}^- \cdot B_n\right)^{-1} \cdot \left(C_n - R_{n+1}^- \cdot A_n\right), \quad R_n^- = \widehat{E}_n \cdot R_n^+ \cdot \widehat{E}_n \quad (3.172)$$

for $n = N,..., 2$, and

$$R_1^+ = -\left(D_1 - R_2^- \cdot B_1\right)^{-1} \cdot \left(C_1 - R_2^- \cdot A_1\right). \quad (3.173)$$

The recursion by Eqs. (3.172) and (3.173), without exponential growth, is analogous to the ones by Eqs. (3.59), (3.94), and (3.95). For a pressure-release or rigid boundary at $x = x_N$, the recursion starts with $R_N^+ = -I$ or $R_N^+ = I$, respectively.

When R_1^+ has been determined, $b_1 = R_1^+ \cdot \widehat{E}_1 \cdot a_1$ is obtained from a_1 as given by Eq. (3.169), and B is obtained from Eq. (3.170). In analogy to Eqs. (3.61), (3.70), (3.82), and (3.96), the remaining coefficient vectors are not computed by a direct application of Eq. (3.171) but by stabilized marching, controlled by the outer boundary condition $b_{N+1} = 0$, with matrix multiplications from the right according to

$$\begin{pmatrix} a_{n+1} \\ b_{n+1} \end{pmatrix} = \begin{pmatrix} I \\ R_{n+1}^+ \cdot \widehat{E}_{n+1} \end{pmatrix} \cdot (A_n, B_n) \cdot \begin{pmatrix} \widehat{E}_n \cdot a_n \\ b_n \end{pmatrix} \quad (3.174)$$

for $n = 1, 2,..., N$, where $R_{N+1}^+ = R_{N+1}^- = 0$ (and $\widehat{E}_{N+1} = I$).

3.6.1.4 Final Remarks

The incident field in $x < x_1$ is preferably computed according to Eq. (3.83) without decomposition into $\exp(i\kappa y)$ components. It is thus obtained from the range-invariant case, and it corresponds to the terms involving $a_{m,1}$ in Eqs. (3.166) and (3.170).

The remarks in Section 3.3.4.3 are applicable in the present case as well, with the addition that the numerical integration over κ in Eq. (3.164), preferably performed by adaptive integration, causes a more significant increase of the computational work when the frequency is increased. The integration path over real κ should preferably be displaced into the complex plane, to move away from some singularities [152,153].

It is not difficult to make an extension to handle the case when the source is situated in an interior x segment (cf. Section 3.2.1.3 or Section 3.3.2).

3.6.2 MODELING DIFFRACTION AROUND A CYLINDRICALLY SYMMETRIC ANOMALY

A cylindrically symmetric medium is now assumed for which lateral variation is allowed among a number of segments in range. The right panel of Fig. 3.22 provides an illustration. Specifically, in the cylindrical coordinate system r, ϕ, z, the medium properties depend on depth z but not on range r or azimuth ϕ within each of the $N+1$ range segments defined by $r \geq r_1$, $r_1 > r \geq r_2,\ldots$, $r_{N-1} > r \geq r_N$, $r_N > r \geq 0$. Here, see Fig. 3.23, $r_1, r_2,\ldots, r_N$ with $N \geq 1$ form a decreasing sequence of positive range-segment limits. For simplicity, there is a traction-free or rigid boundary at $z = z_b$, common for all segments, and there is no solid region. There is a time-harmonic point source with angular frequency ω in the water, at $r = r_s > r_1$, $\phi = 0$, and depth z_s. Omitting the time factor $\exp(-i\omega t)$ with $\omega > 0$, as usual, the body force per unit volume is $f=-M \,\mathrm{grad}(\delta(r - r_s)\delta(\phi)\delta(z - z_s)/r)$.

3.6.2.1 Fourier Series With Respect to the ϕ Coordinate
In this case, Fourier transformation of Eq. (3.9) provides the field representation

$$p(r,\phi,z) = \sum_{\nu=0}^{\infty} \widehat{p}(r,z;\nu)(2 - \delta_{\nu 0})\cos(\nu\phi), \qquad (3.175)$$

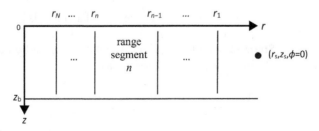

FIGURE 3.23

Range-dependent fluid medium with depth (z) and range (r) axes. There is a point source at range $r_s > r_1$, depth z_s, and azimuth $\phi = 0$.

where, for each ν and in each range segment, $\widehat{p}(r,z;\nu) = \frac{1}{2\pi}\int_0^{2\pi} p(r,\phi,z)\cos(\nu\phi)d\phi = \frac{1}{\pi}\int_0^{\pi} p(r,\phi,z)\cos(\nu\phi)d\phi$ {unit: N/m²} fulfills the Helmholtz equation

$$\frac{1}{r}\frac{\partial}{\partial r}\left(r\frac{\partial\widetilde{p}}{\partial r}\right) + \rho_n(z)\frac{\partial}{\partial z}\left(\frac{1}{\rho_n(z)}\frac{\partial\widetilde{p}}{\partial z}\right) + \left(\frac{\omega^2}{c_n^2(z)} - \frac{\nu^2}{r^2}\right)\widetilde{p}$$
$$= \left(\frac{\omega^2}{c_n^2(z)}\right)\frac{M\ \delta(r-r_s)\delta(z-z_s)}{2\pi r} \tag{3.176}$$

with $\widetilde{p}(r,z;\nu) = \widehat{p}(r,z;\nu) + M\delta(r-r_s)\delta(z-z_s)/2\pi r$. Here, $n=n(r)$ is the range-segment index according to $n=1$ for $r\geq r_1$, $n=2$ for $r_1>r\geq r_2,\ldots,$ $n=N$ for $r_{N-1}>r\geq r_N$, $n=N+1$ for $r_N>r\geq 0$, while $\rho_n(z)$ and $c_n(z)$ are the density and sound-speed functions, respectively, for the nth range segment. Each ν-term in Eq. (3.175) represents the solution to a physical ring-source problem, the one with body force per unit volume given by $-(2-\delta_{\nu 0})M\ \mathrm{grad}(\delta(r-r_s)\cos(\nu\varphi)\delta(z-z_s)/2\pi r)$. The sum including $\nu=0$, $1,\ldots,V$ apparently corresponds to $-M\ \mathrm{grad}\left(\delta(r-r_s)\frac{\sin((V+1/2)\phi)}{\sin(\phi/2)}\delta(z-z_s)/2\pi r\right)$.

Within each range segment, the field $\widehat{p}(r,z;\nu)$ can now be expressed in terms of normal modes. Following [154,155], for $r<r_s$,

$$\widehat{p}(r,z;\nu) = \sum_{m=1}^{\infty}\left[a_{m,n}(\nu)\,\widehat{J}_{m,n}(r;\nu) + b_{m,n}(\nu)\widehat{H}_{m,n}(r;\nu)\right]Z_{m,n}(z), \tag{3.177}$$

where $Z_{m,n}(z)$ is the mode function {unit: N} for the mth mode in the nth range segment, and $k_{m,n}$ is its horizontal wave number. These modal wave numbers and mode functions, with $\widetilde{p}$ in Eq. (3.176) depending on r according to Bessel-function factors $J_\nu(k_{m,n}r)$ or $H_\nu^{(1)}(k_{m,n}r)$, can be computed as described in Sections 3.3.1 and 3.3.2, respectively. The $\widehat{J}_{m,n}(r;\nu)$ and $\widehat{H}_{m,n}(r;\nu)$ are normalized Bessel and Hankel functions, respectively. With $r_0 = r_s$ and $r_{N+1} = r_N$, they are defined by $\widehat{J}_{m,n}(r;\nu) = J_\nu(k_{m,n}r)H_\nu^{(1)}(k_{m,n}r_{n-1})$ and $\widehat{H}_{m,n}(r;\nu) = H_\nu^{(1)}(k_{m,n}r)/H_\nu^{(1)}(k_{m,n}r_n)$ for $n=1, 2,\ldots, N+1$. The function pair J_ν and $H_\nu^{(1)}$ is here preferred to $H_\nu^{(1)}$ and $H_\nu^{(2)}$, because the linear independence of the latter pair is lost numerically for small arguments when ν is large [156]. For each $n=1, 2,\ldots, N+1$, the mode functions fulfill the Sturm–Liouville orthogonality relations given by Eq. (3.88).

For $r>r_s$, the field $\widehat{p}(r,z;\nu)$ can be expressed as

$$\widehat{p}(r,z;\nu) = \sum_{m=1}^{\infty} B_m(\nu)\frac{H_\nu^{(1)}(k_{m,1}r)}{H_\nu^{(1)}(k_{m,1}r_s)}Z_{m,1}(z). \tag{3.178}$$

The problem is then to determine the modal expansion coefficients {unit: m^{-2}} $a_{m,n}(\nu)$ and $b_{m,n}(\nu)$, $m=1, 2,\ldots, n=1, 2,\ldots, N+1$, along with the $B_m(\nu)$, $m=1,$

2,..., for each ν. The different ν can apparently be treated independently. In the following description of a solution method, the dependence on ν is in general omitted in the notation.

3.6.2.2 Equations Relating the Modal Expansion Coefficients

With the horizontal wave numbers $k_{m,n}$ of the modes in the upper half-plane, but not on the negative half-axis, the $a_{m,n}$ and $b_{m,n}$ terms in Eq. (3.177) represent incident and scattered components of the field, respectively. Column vectors are defined according to $\boldsymbol{a}_n = (a_{1,n},\, a_{2,n},...)^{\mathrm{T}}$ and $\boldsymbol{b}_n = (b_{1,n},\, b_{2,n},...)^{\mathrm{T}}$, $n = 1,\, 2,..., N+1$, and $\boldsymbol{B} = (B_1,\, B_2,...)^{\mathrm{T}}$. Regularity of the field at the origin implies that

$$b_{N+1} = 0. \tag{3.179}$$

Excitation coefficients at the source in the first range segment can by derived from the range-invariant case, for which Hankel transformation of Eq. (3.176) provides a depth-separated Helmholtz equation similar to Eq. (3.46). Together with Graf's addition theorem for translation of Hankel functions, the excitation-coefficient result of Eq. (3.83) can be used to provide expressions for $\boldsymbol{a}_1$ and $\boldsymbol{B}$ according to

$$a_{m,1} = -\frac{iM\omega^2}{4\rho_1(z_s)c_1^2(z_s)}\frac{Z_{m,1}(z_s)}{I_{m,1}}, \quad m = 1, 2, ... \tag{3.180}$$

$$B_m = a_{m,1}\,\widehat{J}_{m,1}(r_s) + b_{m,1}\,\widehat{H}_{m,1}(r_s), \quad m = 1, 2, ... \tag{3.181}$$

For $n = 1, 2,..., N$, it is now convenient to introduce the diagonal matrices

$$\widehat{\boldsymbol{J}}_{n+1}^{-} = \mathrm{diag}\left(J_\nu(k_{m,n+1}r_n)H_\nu^{(1)}(k_{m,n+1}r_n),\ m = 1, 2, ...\right),$$

$$\widehat{\boldsymbol{J}}_n^{+} = \mathrm{diag}\left(J_\nu(k_{m,n}r_n)H_\nu^{(1)}(k_{m,n}r_{n-1}),\ m = 1, 2, ...\right),$$

$$\widetilde{\boldsymbol{J}}_{n+1}^{-} = \mathrm{diag}\left(J_\nu'(k_{m,n+1}r_n)iH_\nu^{(1)}(k_{m,n+1}r_n),\ m = 1, 2, ...\right),$$

$$\widetilde{\boldsymbol{J}}_n^{+} = \mathrm{diag}\left(J_\nu'(k_{m,n}r_n)iH_\nu^{(1)}(k_{m,n}r_{n-1}),\ m = 1, 2, ...\right),$$

and

$$\widehat{\boldsymbol{H}}_{n+1} = \mathrm{diag}\left(\widehat{H}_{m,n+1}(r_n),\ m = 1, 2, ...\right),$$

where $\widehat{\boldsymbol{H}}_{N+1} = \boldsymbol{I}$ since $r_{N+1} = r_N$.

Using continuity of pressure and radial displacement at each range segment interface r_n, $n = 1, 2,..., N$, Eq. (3.177) together with Eqs. (3.7) and (3.88) show that

$$\begin{pmatrix} \boldsymbol{a}_{n+1} \\ \widehat{\boldsymbol{H}}_{n+1} \cdot \boldsymbol{b}_{n+1} \end{pmatrix} = \begin{pmatrix} \boldsymbol{A}_n & \boldsymbol{B}_n \\ \boldsymbol{C}_n & \boldsymbol{D}_n \end{pmatrix} \cdot \begin{pmatrix} \boldsymbol{a}_n \\ \boldsymbol{b}_n \end{pmatrix} \tag{3.182}$$

for $n = 1, 2,..., N$. The coupling matrices $\boldsymbol{A}_n$, $\boldsymbol{B}_n$, $\boldsymbol{C}_n$, and $\boldsymbol{D}_n$, can be expressed as

$$A_n = \Lambda_{n+1} \cdot \left(E_{n+1}^- \cdot F_n \cdot \widehat{J}_n^+ + G_n \cdot \widetilde{J}_n^+ \right), \quad B_n = \Lambda_{n+1} \cdot \left(E_{n+1}^- \cdot F_n - G_n \cdot E_n^+ \right),$$

$$C_n = \Lambda_{n+1} \cdot \left(\widetilde{J}_{n+1}^- \cdot F_n \cdot \widehat{J}_n^+ - \widehat{J}_{n+1}^- \cdot G_n \cdot \widetilde{J}_n^+ \right),$$

$$D_n = \Lambda_{n+1} \cdot \left(\widetilde{J}_{n+1}^- \cdot F_n + \widehat{J}_{n+1}^- \cdot G_n \cdot E_n^+ \right),$$

where $E_n^+ = \mathrm{diag}\left(-i\left(H_\nu^{(1)} \right)' (k_{m,n} r_n) / H_\nu^{(1)} (k_{m,n} r_n), \ m = 1, 2, \ldots \right)$ and $E_{n+1}^- = \mathrm{diag}\left(-i\left(H_\nu^{(1)} \right)' (k_{m,n+1} r_n) / H_\nu^{(1)} (k_{m,n+1} r_n), \ m = 1, 2, \ldots \right)$ are diagonal matrices that are close to identity matrices for large arguments, $\Lambda_{n+1} = \mathrm{diag}\left((\pi/2) k_{m,n+1} r_n, m = 1, 2, \ldots \right)$, and the mode coupling matrices F_n and G_n appear as in Eqs. (3.92) and (3.93).

Taken together, Eqs. (3.179), (3.180), and (3.182) provide a system of $2(N + 1)$ linear matrix equations for the $2(N + 1)$ vectors a_n and b_n, $n = 1, 2, \ldots, N + 1$. In practice, truncation is, of course, made to a finite number of modes, say N_M, in each range segment.

3.6.2.3 Solution in Terms of Reflection-Coefficient Matrices

The solution of the obtained equation system is nontrivial (cf. Section 3.3.4.2). Following a previous treatment in Ref. [157], a two-way marching single-scattering global-matrix method is applied in Refs. [154,155], and alternative methods are presented in Refs. [158,159]. Here, however, the reflection-coefficient method used in Sections 3.2.1.1, 3.3.4.2, and 3.6.1.3 is adapted for a convenient and numerically stable two-way marching solution. Basically, reflection matrices are propagated for increasing r without exponential growth and with maintained full rank.

Eq. (3.179) can be expressed as $\widehat{H}_{N+1} \cdot b_{N+1} = R_{N+1}^- \cdot a_{N+1}$, where $R_{N+1}^- = 0$ is interpreted as a reflection-coefficient matrix for incident waves at range r_N-. Corresponding reflection-coefficient matrices, denoted $R_N^+, R_N^-, \ldots, R_1^+$, are now introduced at ranges r_N+, $r_{N-1}-, \ldots$, r_1+. Specifically, $b_n = R_n^+ \cdot a_n$ and $\widehat{H}_n \cdot b_n = R_n^- \cdot a_n$ for $n = N, N - 1, \ldots, 2$, and $b_1 = R_1^+ \cdot a_1$. As verified by applying Eq. (3.182), these reflection matrices can be computed recursively by

$$R_n^+ = -\left(D_n - R_{n+1}^- \cdot B_n \right)^{-1} \cdot \left(C_n - R_{n+1}^- \cdot A_n \right), \quad R_n^- = \widehat{H}_n \cdot R_n^+ \qquad (3.183)$$

for $n = N, \ldots, 2$, and

$$R_1^+ = -\left(D_1 - R_2^- \cdot B_1 \right)^{-1} \cdot \left(C_1 - R_2^- \cdot A_1 \right). \qquad (3.184)$$

The recursion according to Eqs. (3.183) and (3.184), without exponential growth, is quite analogous to the ones according to Eqs. (3.59), (3.94), (3.95), (3.172), and (3.173). For a pressure-release or rigid boundary at $r = r_N$, the recursion starts with $R_N^+ = -\widehat{J}_N^+$ or $R_N^+ = \left(E_N^+ \right)^{-1} \cdot \widetilde{J}_N^+$, respectively.

When R_1^+ has been determined, $b_1 = R_1^+ \cdot a_1$ is obtained from a_1 as given by Eq. (3.180), and B is obtained from Eq. (3.181). In analogy to Eqs. (3.61), (3.70), (3.82), (3.96), and (3.174), the remaining coefficient vectors are not computed by Eq. (3.182) directly but by stabilized marching, controlled by the boundary condition $b_{N+1} = 0$, with matrix multiplications from the right according to

$$\begin{pmatrix} a_{n+1} \\ b_{n+1} \end{pmatrix} = \begin{pmatrix} I \\ R_{n+1}^+ \end{pmatrix} \cdot (A_n, B_n) \cdot \begin{pmatrix} a_n \\ b_n \end{pmatrix} \tag{3.185}$$

for $n = 1, 2, \ldots, N$, where $R_{N+1}^+ = R_{N+1}^- = 0$.

3.6.2.4 Final Remarks

The remarks in Section 3.3.4.3 are applicable in the present case as well, with the modification that the computational work will increase more significantly when the frequency is increased, because more terms are needed for an accurate approximation of the infinite Fourier series in Eq. (3.175). The required number of azimuthal terms can be estimated by $2\pi r_1/\lambda$, where λ is the minimum wavelength in the medium, provided that the incident field (the field without the anomaly) in $r > r_1$ is computed according to Eq. (3.83) without decomposition into azimuthal orders [154].

The recommendation from Ref. [65] of stair steps shorter than a quarter of a wavelength can be alleviated by selecting stair steps of random lengths [155]. Variable stair-step lengths are of interest for the coupled-mode treatments in Sections 3.3.4 and 3.6.1 as well.

It is also possible to handle the case when the source is situated in an interior range segment (cf. Section 3.2.1.3 or Section 3.3.2).

3.6.3 EXAMPLES

The two range-dependent examples from Section 3.3.5.2 are now modified with the ring-shaped islands surrounding the source replaced by more realistic disc shapes with radii 200 and 50 m, respectively (cf. the right panel of Fig. 3.22). Field computations are made as described in Section 3.6.2. However, to facilitate comparison with Figs. 3.6 and 3.14, the results are presented in terms of transformed x, y coordinates, for which the source is at the origin and the island centers are at $(x, y) = (2.5 \text{ km}, 0 \text{ km})$. Some additional examples, also for media without dependence on y as in Section 3.6.1, appear in Ref. [160]. Apart from their intrinsic interest, accurate solutions of 2.5-D problems are useful for benchmarking approximate 3-D propagation models (cf. [161]).

For the left panel of Fig. 3.24, with the clay island, the black curve for PL along the positive x axis, passing through the island, is similar to the one in the left panel of Fig. 3.14, indicating that 2-D modeling may often be sufficient. However, for the sand island case in the right panel, the black curve there, the influence of the island anomaly is now much reduced. Diffraction effects appear, and the pressure field recovers rather well after the drop within and slightly beyond the island. At the longer ranges, the level is now less than 10 dB lower than in the left panel of Fig. 3.6.

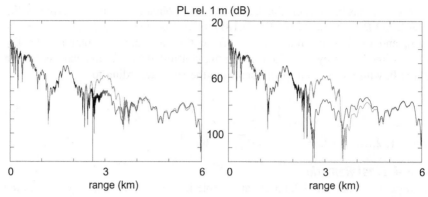

FIGURE 3.24

PL in dB, rel. 1 m, for two modifications of the 500 Hz example involving disc-shaped islands. Left: for the *black curve*, an island with coarse clay appears for | (x, y) − (2.5 km, 0 km) | < 200 m. Right: for the *black curve*, an island with coarse sand appears for | (x, y) − (2.5 km, 0 km) | < 50 m. Propagation loss is considered along the positive x axis. In each panel, a *gray curve* is included for a reciprocity check, with the island center moved to (x, y) = (3.5 km, 0 km).

There is also a gray curve in each of the panels of Fig. 3.24, for which the center of the island is moved from $x = 2.5$ km to $x = 3.5$ km. The similarity to the range-invariant case, the black curve in the left panel of Fig. 3.6, now extends well beyond the range 3 km. To fulfill reciprocity according to Eq. (3.14), the black and gray curves in each of the panels of Fig. 3.24 should meet at the range 6 km. Indeed, this is what they do.

Fig. 3.25 provides near-field examples of the same kind. The island centers are now at $x = 0.3$ km for the black curves and $x = 1.0$ km for the gray curves. This

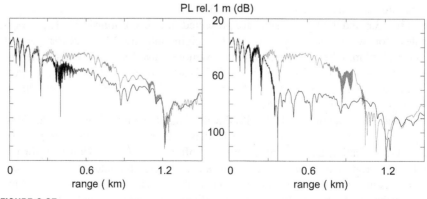

FIGURE 3.25

As Fig. 3.24, but with the island centers moved to (x, y) = (0.3 km, 0 km) for the *black curves* and (x, y) = (1.0 km, 0 km) for the *gray curves*. This time, the black and gray PL curves meet, by reciprocity, at the range 1.3 km.

time, the black and gray curves in each panel meet at the range 1.3 km, as anticipated by reciprocity.

The clay- and sand-island examples have been run with Eq. (3.175) truncated to 451 and 201 azimuthal terms, respectively. Checks with more terms indicate that convergence has been reached. It is expected that the clay-island case requires more ν−terms since the range-segment radius r_1 is larger (200 m as compared with 50 m).

LIST OF ABBREVIATIONS AND SYMBOLS

c	Sound velocity in fluid {unit: m/s}
$\det(\cdot)$, $|\cdot|$	Determinant of a matrix
div	Divergence operator
$\boldsymbol{e}_x$, $\boldsymbol{e}_y$, $\boldsymbol{e}_z$ and $\boldsymbol{e}_1$, $\boldsymbol{e}_2$, $\boldsymbol{e}_3$	Cartesian unit vectors in 3-D Euclidean space
$\boldsymbol{e}_r$, $\boldsymbol{e}_\phi$, $\boldsymbol{e}_z$	Cylindrical unit vectors in 3-D Euclidean space
$\boldsymbol{e}_r$, $\boldsymbol{e}_u$	Unit vectors for ray-centered 2-D coordinates
$\boldsymbol{f} = (f_1, f_2, f_3)$	Body force per unit volume {unit: N/m^3}
FD(M)	Finite difference (method)
FE(M)	Finite element (method)
grad	Gradient operator
$\mathbf{I}$	Identity matrix
$I = \int \rho^{-1} Z^2 dz$	Mode integral for fluid medium {unit: N^{2}m^4/kg}
k	Horizontal wave number {unit: m^{-1}}
M	Complex moment-tensor amplitude of symmetric point source {unit: N m}
$O(\cdot)$	Ordo, for example, $g(h) = O(h^2)$ means that $g(h)/h^2$ is bounded as $h \to 0$
ODE	Ordinary differential equation
p	Acoustic pressure {unit: N/m^2}
p_{ref}	Reference pressure {unit: N/m^2}
PE	Parabolic equation
PL	Propagation loss
PML	Perfectly matched layer
r_{ref}	Reference range {unit: m}
r, ϕ, z	Cylindrical coordinates (radius, azimuth, depth) {units: m, rad, m}
r, u	Ray-centered 2-D coordinates {unit: m}
$\boldsymbol{r}(k, z) = (r_1, r_2, r_3, r_4)^{\mathsf{T}}$	Displacement-stress vector {unit: m^2 (m, m, N/m^2, N/m^2)^{T}}
$\boldsymbol{r}_F(k, z) = (r_2, r_4)^{\mathsf{T}}$	Displacement-stress vector used in fluid {unit: m^2 (m, N/m^2)^{T}}
t	Time {unit: s}
T	Transpose (of a matrix)
$\boldsymbol{u} = (u, v, w) = (u_1, u_2, u_3)$	Displacement vector {unit: m}
$\boldsymbol{x} = (x, y, z) = (x_1, x_2, x_3)$	Position vector {unit: m}
z	Depth coordinate {unit: m}
$Z(z)$	Solution of depth-separated Helmholtz equation {unit: m^{2}N/m^2 = N}
α	Compressional-wave velocity in solid {unit: m/s}

Continued

β	Shear-wave velocity in solid {unit: m/s}
Δ	Laplace operator, Δ can also denote a difference
$\delta(\mathbf{x})$	Dirac delta function in 3-D {unit: m^{-3}}
$\delta(x)$, $\delta(y)$, $\delta(z)$, $\delta(r)$	Dirac delta functions in 1-D {unit: m^{-1}}
$\delta(\phi)$	Dirac delta function in 1-D {unit: rad^{-1}}
δ_{ij}	Kronecker delta ($\delta_{ij} = 1$ when $i = j$, 0 otherwise)
λ, μ	Lamé parameters {unit: N/m^2}, λ can also denote wavelength {unit: m}
$\xi = i\chi$	Vertical wave number in fluid {unit: m^{-1}}
ρ	Density {unit: kg/m^3}
$\tau = (\tau_{ij})$	Stress tensor {unit: N/m^2}
$\boldsymbol{\Phi}$	Time-averaged energy flux or intensity vector {unit: W/m^2}
ω	Angular frequency {unit: rad/s}

ACKNOWLEDGMENTS

My colleagues Ilkka Karasalo (who was to coauthor this chapter originally), Leif Abrahamsson, Brodd-Leif Andersson, Kurt Otto, Jörgen Pihl, and Martin Östberg have helped with valuable comments and suggestions. The comments and suggestions by Leif Bjørnø and his coeditors are also much appreciated.

REFERENCES

[1] Aki, K. and Richards, P. G., *Quantitative Seismology*, University Science Books, Sausalito, 2002.

[2] Pierce, A. D., *Acoustics: An Introduction to Its Physical Principles and Applications*, Acoustical Society of America, New York, 1989.

[3] Westerling, L., *Hydroacoustic wave propagation models* (in Swedish), Swedish Defence Research Agency, Stockholm, 1990.

[4] Hudson, J. A., *The Excitation and Propagation of Elastic Waves*, Cambridge University Press, Cambridge, 1980.

[5] Bagley, L. R. and Torvik, P. J., *A theoretical basis for the application of fractional calculus to viscoelasticity*, J. Rheol., **27**, (3), pp. 201–210, 1983.

[6] Weston, D. E., *Acoustic flux methods for oceanic guided waves*, J. Acoust. Soc. Am. **68**, (1), pp. 287–296, 1980.

[7] Jensen, F. B., Kuperman, W. A., Porter, M. B., and Schmidt, H., *Computational Ocean Acoustics*, Springer, New York, 2011.

[8] Mallick, S. and Frazer, L. N., *Practical aspects of reflectivity modeling*, Geophysics, **52**, (10), pp. 1355–1364, 1987.

[9] Brekhovskikh, L. M., *Waves in Layered Media*, Academic Press, New York, 1980.

[10] Frisk, G. V., *Ocean and Seabed Acoustics, a Theory of Wave Propagation*, Prentice-Hall, New Jersey, 1994.

[11] Lurton, X., *An Introduction to Underwater Acoustics, Principles and Applications*, Springer, London, 2002.

[12] Hovem, M., *Marine Acoustics*, Peninsula Publishing, California, 2012.

[13] Cerveny, V., *Ray tracing algorithms in three-dimensional laterally varying layered structures*. In: *Seismic Tomography with Applications in Global Seismology and Exploration Geophysics*, Nolet, G. (Ed.), D. Reidel Publishing, Dordrecht, pp. 99–133, 1987.

[14] Cerveny, V., *Seismic Ray Theory*, Cambridge University Press, Cambridge, 2001.

[15] Dozier, L. D. and Lallement, P., *Parallel implementation of a 3-D range-dependent ray model for replica field generation*, In: *Full Field Inversion Methods in Ocean and Seismo-Acoustics*, Diachok, O., Caiti, A., Gerstoft, P., and Schmidt, H. (Eds.) Kluwer Academic Publishers, Dordrecht, pp. 45–50, 1995.

[16] Snieder, R. and Chapman, C., *The reciprocity properties of geometrical spreading*, Geophys. J. Int., **132**, (1), pp. 89–95, 1998.

[17] Tolstoy, I., *Phase changes and pulse deformation in acoustics*, J. Acoust. Soc. Am., **44**, (3), pp. 675–683, 1968.

[18] Svensson, E., *Gaussian beam summation in shallow waveguides*, Wave Motion, **45**, (4), pp. 445–456, 2008.

[19] Thomson, W. T., *Transmission of elastic waves through a stratified solid medium*, J. Appl. Phys., **21**, (2), pp. 89–93, 1950.

[20] Haskell, N. A., *The dispersion of surface waves on multilayered media*, Bull. Seism. Soc. Am., **43**, (1), pp. 17–34, 1953.

[21] Buchen, P. W. and Ben-Hador, R., *Free-mode surface-wave computations*, Geophys. J. Int., **124**, (3), pp. 869–887, 1996.

[22] Schmidt, H. and Jensen, F. B., *A full wave solution for propagation in multilayered viscoelastic media with application to Gaussian beam reflection at fluid-solid interfaces*, J. Acoust. Soc. Am., **77**, (3), pp. 813–825, 1985.

[23] Kennett, B. L. N., *Reflections, rays and reverberations*, Bull. Seism. Soc. Am., **65**, (6), pp. 1685–1696, 1974.

[24] Kennett, B. L. N., *Seismic Wave Propagation in Stratified Media*, Cambridge University Press, Cambridge, 1983.

[25] Westwood, E. K., Tindle, C. T. and Chapman, N. R., *A normal mode model for acousto-elastic ocean environments*, J. Acoust. Soc. Am., **100**, (6), pp. 3631–3645, 1996.

[26] Karasalo, I., *Exact finite elements for wave propagation in range-independent fluid-solid media*, J. Sound Vib., **172**, (5), pp. 671–688, 1994.

[27] Wang, L., and Rokhlin, S. I., *Recursive stiffness matrix method for wave propagation in stratified media*, Bull. Seism. Soc. Am., **92**, (3), pp. 1129–1135, 2002.

[28] Dunkin, J. W., *Computation of modal solutions in layered, elastic media at high frequencies*, Bull. Seism. Soc. Am., **55**, (2), pp. 335–358, 1965.

[29] Ivansson, S., *The compound matrix method for multi-point boundary-value problems*, Z. angew. Math. Mech., **77**, (10), pp. 767–776, 1997.

[30] Ivansson, S., *The compound matrix method for multi-point boundary-value problems depending on a parameter*, Z. angew. Math. Mech., **78**, (4), pp. 231–242, 1998.

[31] Chapman, C. H., *Yet another elastic plane-wave, layer-matrix algorithm*, Geophys. J. Int., **154**, (1), pp. 212–223, 2003.

[32] Ivansson, S., *Compound-matrix Riccati method for solving boundary-value problems*, Z. angew. Math. Mech., **83**, (8), pp. 535–548, 2003.

[33] Ivansson, S., *Solving ODE boundary-value problems: safe compound-matrix dimension reduction with the Riccati method*, Int. J. Diff. Eq. Applic., **5**, (4), pp. 339–352, 2002.

[34] Levinson, S. J., Westwood, E. K., Koch, R. A., Mitchell, S. K. and Sheppard, C. V., *An efficient and robust method for underwater acoustic normal-mode computations*, J. Acoust. Soc. Am., **97**, (3), pp. 1576–1585, 1995.

[35] Karasalo, I., and de Winter, J., *Airy function elements for inhomogeneous fluid layers*, In: *Proc. 8th European Conference on Underwater Acoustics*, Jesus, S. M. and Rodrıguez, O. C. (Eds.) Carvoeiro, University of Algarve, pp. 33–38, 2006.

[36] Ivansson, S., *Delta-matrix factorization for fast propagation through solid layers in a fluid-solid medium*, J. Comput. Phys., **108**, (2), pp. 357–367, 1993.

[37] Ivansson, S., *Comment on 'Free-mode surface-wave computations' by P. Buchen and R- Ben-Hador*, Geophys. J. Int., **132**, (3), pp. 725–727, 1998.

[38] Filon, L. N. G., *On a quadrature formula for trigonometric integrals*, Proc. Royal Soc. Edinburgh, **49**, (1), pp. 38–47, 1929.

[39] Ivansson, S., and Karasalo, I., *A high-order adaptive integration method for wave propagation in range-independent fluid-solid media*, J. Acoust. Soc. Am., **92**, (3), pp. 1569–1577, 1992.

[40] Dahlquist, G., Björck, Å. and Anderson, N., *Numerical Methods*, Prentice-Hall, New Jersey, 1974.

[41] Batorsky, D. V., and Felsen, L. B., *Ray-optical calculation of modes excited by sources and scatterers in a weakly inhomogeneous duct*, Radio Science, **6**, (10), pp. 911–973, 1971.

[42] Weinberg, H., *Application of ray theory to acoustic propagation in horizontally stratified oceans*, J. Acoust. Soc. Am., **58**, (1), pp. 97–109, 1975.

[43] Ivansson, S., and Bishop, J., *Travelling-wave representations of diffraction using leaky-mode Green function expansions*, J. Sound Vib., **262**, pp. 1223–1234, 2003.

[44] Evans, R. B., *The existence of generalized eigenfunctions and multiple eigenvalues in underwater acoustics*, J. Acoust. Soc. Am., **92**, (4), pp. 2024–2029, 1992.

[45] Ivansson, S. and Karasalo, I., *Double-root resonances and complex modal slownesses in a fluid-solid medium*, In: *Proc. 4th European Conference on Underwater Acoustics*, Alippi, A. and Cannelli, G. B. (Eds.), CNR-IDAC, Rome, pp. 685–690, 1998.

[46] Ivansson, S., *Low-frequency dispersion-function factorization and classification of P-SV modes by wavenumber limits*, Z. Angew. Math. Mech., **82**, (2), pp. 89–99, 2002.

[47] Ivansson, S., *Mode structure for fluid-solid media as derived by low-frequency asymptotics*, J. Sound Vib., **230**, (2), pp. 411–446, 2000.

[48] Ivansson, S., *A class of low-frequency modes in laterally homogeneous fluid-solid media*, SIAM J. Appl. Math., **58**, (5), pp. 1462–1508, 1998.

[49] Ivansson, S., *Low-frequency slow-wave dispersion computations by compound-matrix propagation*, J. Acoust. Soc. Am., **106**, (1), pp. 61–72, 1999.

[50] Birkhoff G. and Rota, G.-C., *Ordinary Differential Equations*, Xerox, Lexington, 1969.

[51] Odom, R. I., *Travelling wave modes of a plane layered anelastic earth*, Geophys. J. Int., **206**, (2), pp. 993–998, 2016.

[52] Brazier-Smith, P. R. and Scott, J. F. M., *On the determination of the roots of dispersion equations by use of winding number integrals*, J. Sound Vib., **145**, (3), pp. 503–510, 1991.

[53] Ivansson, S. and Karasalo, I., *Computation of modal wavenumbers using an adaptive winding-number integral method with error control*, J. Sound Vib., **161**, (1), pp. 173–180, 1993.

[54] McCollom, B. A. and Collis, J. M., *Root finding in the complex plane for seismo-acoustic propagation scenarios with Green's function solutions*, J. Acoust. Soc. Am., **136**, (3), pp. 1036−1045, 2014.

[55] Cristini, P., *Implementation of a new root finder for KRAKEN*, In: *Proc. 4th European Conference on Underwater Acoustics*, Alippi, A. and Cannelli, G. B. (Eds.), CNR-IDAC, Rome, pp. 685−690, 1998.

[56] Ivansson, S., *Source function to generate an individual mode in a fluid-solid medium*, J. Sound Vib., **186**, (3), pp. 527−534, 1995.

[57] Evans, R. B., *A coupled mode solution for acoustic propagation in a waveguide with stepwise depth variations of a penetrable bottom*, J. Acoust. Soc. Am., **74**, (1), pp. 188−195, 1983.

[58] Evans, R. B., *The decoupling of stepwise coupled modes*, J. Acoust. Soc. Am., **80**, (5), pp. 1414−1418, 1986.

[59] Luo, W. Y., Yang, C., Qin, J. and Zhang, R., *A numerically stable coupled-mode formulation for acoustic propagation in range-dependent waveguides*, Sci. Chin. Phys. Mech. Astron., **55**, (4), pp. 572−588, 2012.

[60] Luo, W.-Y., Yang, C.-M., Qin J.-X., and Zhang, R.-H., *Benchmark solutions for sound propagation in an ideal wedge*, Chin. Phys. B, **22**, (5), 054301, 2013.

[61] Qin, J., Luo, W., Zhang, R., and Yang, C., *Numerical solution of range-dependent acoustic propagation*, Chin. Phys. Lett., **30**, (7), 074301, 2013.

[62] Kennett, B. L. N., *Guided wave propagation in laterally varying media - I. Theoretical development*, Geophys. J. Int., **79**, (1), pp. 235−255, 1984.

[63] Luo, W., Zhang, R. and Schmidt, H., *An efficient and numerically stable coupled-mode solution for range-dependent propagation*, J. Comput. Acoust., **20**, (3), 1250008, 2012.

[64] Westwood, E. K. and Koch, R. A., *Elimination of branch cuts from the normal-mode solution using gradient half spaces*, J. Acoust. Soc. Am., **106**, (5), pp. 2513−2523, 1999.

[65] Jensen, F. B., *On the use of stair steps to approximate bathymetry changes in ocean acoustic models*, J. Acoust. Soc. Am., **104**, (3), pp. 1310−1315, 1998.

[66] Odom, R. I., *A coupled mode examination of irregular waveguides including the continuum spectrum*, Geophys. J. R. astr. Soc., **86**, (2), pp. 425−453, 1986.

[67] Maupin, V., *Surface waves across 2-D structures: a method based on coupled local modes*, Geophys. J. Int., **93**, (1), pp. 173−185, 1988.

[68] Fawcett, J. A., *A derivation of the differential equations of coupled-mode propagation*, J. Acoust. Soc. Am., **92**, (1), pp. 290−295, 1992.

[69] Odom, R. I., Park, M., Mercer, J. A., Crosson, R. S. and Paik, P., *Effects of transverse isotropy on modes and mode coupling in shallow water*, J. Acoust. Soc. Am., **100**, (4), pp. 2079−2092, 1996.

[70] Belibassakis, K. A., Athanassoulis, G. A., Papathanasiou, T. K., Filopoulos, S. P. and Markolefas, S., *Acoustic wave propagation in inhomogeneous, layered waveguides based on modal expansions and hp-FEM*, Wave Motion, **51**, (6), pp. 1021−1043, 2014.

[71] Abrahamsson, L. and Kreiss, H.-O., *Numerical solution of the coupled mode equations in duct acoustics*, J. Comput. Phys., **111**, (1), pp. 1−14, 1994.

[72] Ivansson, S. and Bishop, J., *Single and multiple grazing-ray diffraction as derived by Green's function expansion*, In: *Proc. 5th European Conference on Underwater Acoustics*, Zakharia, M. E., (Ed.), Lyon, European Commision, pp. 21−26, 2000.

[73] Pedersen, M. A. and Gordon, D. F., *Normal-mode and ray theory applied to underwater acoustic conditions of extreme downward refraction*, J. Acoust. Soc. Am., **51**, (1B), pp. 323–368, 1972.

[74] Bergmann, P. G., *The wave equation in a medium with a variable index of refraction*, J. Acoust. Soc. Am., **17**, (4), pp. 329–333, 1946.

[75] Brekhovskikh, L. M. and Godin, O. A., *Acoustics of Layered Media II*, Springer, Berlin, 1992.

[76] Miksis-Olds, J. L., Vernon, J. A. and Heaney, K., *Applying the dynamic soundscape to estimates of signal detection*, In: *Proc. 2nd International Conference and Exhibition on Underwater Acoustics*, Papadakis, J. S., and Bjørnø, L., (Eds.), Rhodes, Greece, pp. 863–870, 2014.

[77] Arnold, A. and Ehrhardt, M., *Discrete transparent boundary conditions for wide angle parabolic equations in underwater acoustics*, J. Comput. Phys., **145**, (2), pp. 611–638, 1998.

[78] Bamberger, A., Engquist, B., Halpern, L. and Joly, P., *Higher order paraxial wave equation approximations in heterogeneous media*, SIAM J. Appl. Math., **48**, (1), pp. 129–154, 1988.

[79] Collins, M. D., *Applications and time-domain solution of higher-order parabolic equations in underwater acoustics*, J. Acoust. Soc. Am., **86**, (3), pp. 1097–1102, 1989.

[80] McDaniel, S. T., *Propagation of normal mode in the parabolic approximation*, J. Acoust. Soc. Am., **57**, (2), pp. 307–311, 1975.

[81] Pierce, A. D., *The natural reference wavenumber for parabolic approximations in ocean acoustics*, Comp. Maths. Appls., **11**, (7/8), pp. 831–841, 1985.

[82] Collins, M. D., *An energy-conserving parabolic equation for elastic media*, J. Acoust. Soc. Am., **94**, (2), pp. 975–982, 1993.

[83] Porter, M. B., Jensen, F. B. and Ferla, C. M., *The problem of energy conservation in one-way models*, J. Acoust. Soc. Am., **89**, (3), pp. 1058–1067, 1991.

[84] Collins, M. D., *A self-starter for the parabolic equation method*, J. Acoust. Soc. Am., **92**, (4), pp. 2069–2074, 1992.

[85] Collins, M. D., *The stabilized self-starter*, J. Acoust. Soc. Am., **106**, (4), pp. 1724–1726, 1999.

[86] Press, W. H., Flannery, B. P., Teukolsky, S. A. and Vetterling, W. T., *Numerical Recipes*, Cambridge University. Press, Cambridge, 1986.

[87] Claerbout, J. F., *Imaging the Earth's Interior*, Blackwell, Palo Alto, 1985.

[88] Cederberg, R. J. and Collins, M. D., *Application of an improved self-starter to geoacoustic inversion*, IEEE J. Ocean. Eng., **22**, (1), pp. 102–109, 1997.

[89] Collins, M. D., *A split-step Padé solution for the parabolic equation method*, J. Acoust. Soc. Am., **93**, (4), pp. 1736–1742, 1993.

[90] Brooke, G. H., Thomson, D. J. and Ebbeson, G. R., *PECan: a Canadian parabolic equation model for underwater sound propagation*, J. Comp. Acoust., **9**, (1), pp. 69–100, 2001.

[91] Collins, M. D., *Generalization of the split-step Padé solution*, J. Acoust. Soc. Am., **96**, (1), pp. 382–385, 1994.

[92] Jeltsch, R., *Multistep methods using higher derivatives and damping at infinity*, Math. Comp., **31**, (137), pp. 124–138, 1977.

[93] Karasalo, I. and Sundström, A., *JEPE - A high-order PE-model for range-dependent fluid media*, In: *Proc. 3rd European Conference on Underwater Acoustics*, Papadakis, J. S., (Ed.), FORTH-IACM, Heraklion, pp. 189–194, 1996.

[94] Godin, O. A., *Reciprocity and energy conservation within the parabolic approxima-tion*, Wave Motion, **29**, (2), pp. 175−194, 1999.

[95] Mikhin, D., *Energy-conserving and reciprocal solutions for higher-order parabolic equations*, J. Comput. Acoust., **9**, (1), pp. 183−203, 2001.

[96] Mikhin, D., *Generalizations of the energy-flux parabolic equation*, J. Comput. Acoust., **13**, (4), pp. 641−665, 2005.

[97] Abrahamsson, L. and Kreiss, H.-O., *Boundary conditions for the parabolic equation in a range-dependent duct*, J. Acoust. Soc. Am., **87**, (6), pp. 2438−2441, 1990.

[98] Collins, M. D., *The rotated parabolic equation and sloping ocean bottoms*, J. Acoust. Soc. Am., **87**, (3), pp. 1035−1037, 1990.

[99] Outing, D. A., Siegmann, W. L., Collins, M. D. and Westwood, E. K., *Generalization of the rotated parabolic equation to variable slopes*, J. Acoust. Soc. Am., **120**, (6), pp. 3534−3538, 2006.

[100] Larsson, E. and Abrahamsson, L., *Helmholtz and parabolic equation solutions to a bench-mark problem in ocean acoustics*, J. Acoust. Soc. Am., **113**, (5), pp. 2446−2454, 2003.

[101] Collins, M. D. and Dacol, D. K., *A mapping approach for handling sloping interfaces*, J. Acoust. Soc. Am., **107**, (4), pp. 1937−1942, 2000.

[102] Metzler, A. M., Moran, D., Collis, J. M., Martin, P. A. and Siegmann, W. L., *A scaled mapping parabolic equation for sloping range-dependent environments*, J. Acoust. Soc. Am., **135**, (3), pp. EL172-EL178, 2014.

[103] Collins, M. D. and Evans, R. B., *A two-way parabolic equation for acoustic backscat-tering in the ocean*, J. Acoust. Soc. Am., **91**, (3), pp. 1357−1368, 1992.

[104] Lingevitch, J. F., Collins, M. D. and Mills, M. J., *A two-way parabolic equation that accounts for multiple scattering*, J. Acoust. Soc. Am., **112**, (2), pp. 476−480, 2002.

[105] Jerzak, W., Siegmann, W. L. and Collins, M. D., *Modeling Rayleigh and Stoneley waves and other interface and boundary effects with the parabolic equation*, J. Acoust. Soc. Am., **117**, (6), pp. 3497−3503, 2005.

[106] Frank, S. D., Collis, J. M. and Odom, R. I., *Elastic parabolic equation solutions for oceanic T-wave generation and propagation from deep seismic sources*, J. Acoust. Soc. Am., **137**, (6), pp. 3534−3543, 2015.

[107] Collis, J. M., Siegmann, W. L., Zampolli, M. and Collins, M. D., *Extension of the rotated elastic parabolic equation to beach and island propagation*, IEEE J. Ocean. Eng., **34**, (4), pp. 617−623, 2009.

[108] Huang, D., *Finite element solution to the parabolic wave equation*, J. Acoust. Soc. Am., **84**, (4), pp. 1405−1413, 1988.

[109] Porter, M. B. and Reiss, E. L., *A numerical method for ocean acoustic normal modes*, J. Acoust. Soc. Am., **76**, (1), pp. 244−252, 1984.

[110] Bayliss, A., Goldstein, C. I. and Turkel, E., *The numerical solution of the Helmholtz equation for wave propagation problems in underwater acoustics*, Comp. Maths. Appls., **11**, (7−8), pp. 655−665, 1985.

[111] Murphy, J. E. and Chin-Bing, S. A., *A finite-element model for ocean acoustic propa-gation and scattering*, J. Acoust. Soc. Am., **86**, (4), pp. 1478−1483, 1989.

[112] Kampanis, N. A. and Dougalis, V. A., *A finite element code for the numerical solution of the Helmholtz equation in axially symmetric waveguides with interfaces*, J. Comput. Acoust.,**7**, (2), pp. 83−110, 1999.

[113] Vendhan, C. P., Diwan, G. C. and Bhattacharya, S. K., *Finite-element modeling of depth and range dependent acoustic propagation in oceanic waveguides*, J. Acoust. Soc. Am., **127**, (6), pp. 3319−3326, 2010.

[114] Stekl, I. and Pratt, R. G., *Accurate viscoelastic modeling by frequency-domain finite differences using rotated operators*, Geophysics, **63**, (5), pp. 1779–1794, 1998.

[115] Otto, K. and Larsson, E., *Iterative solution of the Helmholtz equation by a second-order method*, SIAM J. Matrix Anal. Appl., **21**, (1) pp. 209–229, 1999.

[116] Hustedt, B., Operto, S. and Virieux, J., *Mixed-grid and staggered-grid finite-difference methods for frequency-domain acoustic wave modelling*, Geophys. J. Int., **157**, (3), pp. 1269–1296, 2004.

[117] de Groot-Hedlin, C., *A finite difference solution to the Helmholtz equation in a radially symmetric waveguide: application to near-source scattering in ocean acoustics*, J. Comput. Acoust., **16**, (3), pp. 447–464, 2008.

[118] Fix, G. J. and Marin, S. P., *Variational methods for underwater acoustic problems*, J. Comput. Phys., **28**, (2), pp. 253–270, 1978.

[119] Keller, J. B. and Givoli, D., *Exact non-reflecting boundary conditions*, J. Comput. Phys., **82**, (1), pp. 172–192, 1989.

[120] Abrahamsson, L., *Orthogonal grid generation for two-dimensional ducts*, J. Comput. Appl. Math., **34**, (3), pp. 305–314, 1991.

[121] Larsson, E., *A domain decomposition method for the Helmholtz equation in a multi-layer domain*, SIAM J. Sci. Comput., **20**, (5), pp. 1713–1731, 1999.

[122] Otto, K. and Larsson, E., *A flexible solver of the Helmholtz equation for layered media*, In: *Proc. European Conference on Computational Fluid Dynamics*, Delft, 2006.

[123] Bayliss, A., Goldstein, C. I. and Turkel, E., *On accuracy conditions for the numerical computation of waves*, J. Comput. Phys., **59**, (3), pp. 396–404, 1985.

[124] Bérenger, J.-P., *A perfectly matched layer for the absorption of electromagnetic waves*, J. Comput. Phys., **114**, (2), pp. 185–200, 1994.

[125] Thompson, L. L., *A review of finite-element methods for time-harmonic acoustics*, J. Acoust. Soc. Am., **119**, (3), pp. 1315–1330, 2006.

[126] Kelly, K. R., Ward, R. W., Treitel, S. and Alford, R. M., *Synthetic seismograms: a finite-difference approach*, Geophysics, **41**, (1), pp. 2–27, 1976.

[127] Robertsson, J. O. A. and Blanch, J. O., *Numerical methods, finite difference*, In: *Encyclopedia of Solid Earth Geophysics*, Gupta, H. (Ed.) Springer, pp. 883–892, 2011.

[128] Wu, S. R., *Lumped mass matrix in explicit finite element method for transient dynamics of elasticity*, Comput. Methods Appl. Mech. Engrg., **195**, (44–47), pp. 5983–5994, 2006.

[129] Strikwerda, J. C., *Finite Difference Schemes and Partial Differential Equations*, Chapman & Hall, New York, 1989.

[130] Robertsson, J. O. A., Blanch, J. O. and Symes, W. W., *Viscoelastic finite-difference modelling*, Geophysics, **59**, (9), pp. 1444–1456, 1994.

[131] Blanch, J. O., Robertsson, J. O. A. and Symes, W. W., *Modeling of a constant Q: methodology and algorithm for an efficient and optimally inexpensive viscoelastic technique*, Geophysics, **60**, (1), pp. 176–184, 1995.

[132] Kay, I. and Krebes, E. S., *Applying finite element analysis to the memory variable formulation of wave propagation in anelastic media*, Geophysics, **64**, (1), pp. 300–307, 1999.

[133] Liu, Q.-H. and Tao, J., *The perfectly matched layer for acoustic waves in absorptive media*, J. Acoust. Soc. Am., **102**, (4), pp. 2072–2082, 1997.

[134] Zhao, J.-G. and Shi, R.-Q., *Perfectly matched layer-absorbing boundary condition for finite-element time-domain modeling of elastic wave equations*, Appl. Geophys., **10**, (3), pp. 323–336, 2013.

[135] Smith, W. D., *A nonreflecting plane boundary for wave propagation problems*, J. Comput. Phys., **15**, (4), pp. 492−503, 1974.

[136] Komatitsch, D. and Tromp, J., *Introduction to the spectral element method for three-dimensional seismic wave propagation*, Geophys. J. Int., **139**, (3), pp. 806−822, 1999.

[137] Cristini, P. and Komatitsch, D., *Some illustrative examples of the use of a spectral-element method in ocean acoustics*, J. Acoust. Soc. Am., **131**, (3), pp. EL229-EL235, 2012.

[138] Virta, K. and Mattsson, K., *Acoustic wave propagation in complicated geometries and heterogeneous media*, J. Sci. Comput., **61**, (1), pp. 90−118, 2014.

[139] Lee, D., Botseas, G. and Siegmann, W. L., *Examination of three-dimensional effects using a propagation model with azimuth-coupling capability (FOR3D)*, J. Acoust. Soc. Am., **91**, (6), pp. 3192−3202, 1992.

[140] Fawcett, J. A., *Modeling three-dimensional propagation in an oceanic wedge using parabolic equation methods*, J. Acoust. Soc. Am., **93**, (5), pp. 2627−2632, 1993.

[141] Zhu, D. and Bjørnø, L., *A three-dimensional, two-way, parabolic equation model for acoustic backscattering in a cylindrical coordinate system*, J. Acoust. Soc. Am., **108**, (3), pp. 889−898, 2000.

[142] Sturm, F., *Numerical study of broadband sound pulse propagation in three-dimensional oceanic waveguides*, J. Acoust. Soc. Am., **117**, (3), pp. 1058−1079, 2005.

[143] Lin, Y.-T., Collis, J. M. and Duda, T. F., *A three-dimensional parabolic equation model of sound propagation using higher-order operator splitting and Padé approximants*, J. Acoust. Soc. Am., **132**, (5), pp. EL364-EL370, 2012.

[144] Sturm, F., *Leading-order cross term correction of three-dimensional parabolic equation models*, J. Acoust. Soc. Am., **139**, (1), pp. 263−270, 2016.

[145] Sturm, F., Ivansson, S., Jiang, Y.-M. and Chapman, N. R., *Numerical investigation of out-of-plane sound propagation in a shallow water experiment*, J. Acoust. Soc. Am., **124**, (6), pp. EL341-EL346, 2008.

[146] Sturm, F. and Korakas, A., *Comparisons of laboratory scale measurements of three-dimensional acoustic propagation with solutions by a parabolic equation model*, J. Acoust. Soc. Am., **133**, (1), pp. 108−118, 2013.

[147] Weinberg, H. and Burridge, R., *Horizontal ray theory for ocean acoustics*, J. Acoust. Soc. Am., **55**, (1), pp. 63−79, 1974.

[148] Heaney, K. D., Campbell, R. L. and Murray, J. J., *Comparison of hybrid three-dimensional modeling with measurements on the continental shelf*, J. Acoust. Soc. Am., **131**, (2), pp. 1680−1688, 2012.

[149] Ballard, M. S., *Modeling three-dimensional propagation in a continental shelf environment*, J. Acoust. Soc. Am., **131**, (3), pp. 1969−1977, 2012.

[150] Qin, J.-X., Luo, W.-Y., Zhang, R.-H. and Yang, C.-M., *Three-dimensional sound propagation and scattering in two-dimensional waveguides*, Chin. Phys. Lett., **30**, (11), 114301, 2013.

[151] Boström, A., Kristensson, G. and Ström, S., *Transformation properties of plane, spherical and cylindrical scalar and vector wave functions*, In: *Field Representations and Introduction to Scattering*, Elsevier, Amsterdam, pp. 165−210, 1991.

[152] Fawcett, J. A. and Dawson, T. W., *Fourier synthesis of three-dimensional scattering in a two-dimensional oceanic waveguide using boundary integral equation methods*, J. Acoust. Soc. Am., **88**, (4), pp. 1913−1920, 1990.

[153] Orris, G. J. and Collins, M. D., *The spectral parabolic equation and three-dimensional backscattering*, J. Acoust. Soc. Am., **96**, (3), pp. 1725−1731, 1994.

[154] Luo, W. Y. and Schmidt, H., *Three-dimensional propagation and scattering around a conical seamount*, J. Acoust. Soc. Am., **125**, (1), pp. 52−65, 2009.

[155] Luo, W. Y., Zhang, R. H. and Schmidt, H., *Three-dimensional mode coupling around a seamount*, Sci. China Phys. Mech. Astron., **54**, (9), pp. 1561−1569, 2011.

[156] Ricks, D. C. and Schmidt, H., *A numerically stable global matrix method for cylindrically layered shells excited by ring forces*, J. Acoust. Soc. Am., **95**, (6), pp. 3339−3349, 1994.

[157] Taroudakis, M. I., *A coupled-mode formulation for the solution of the Helmholtz equation in water in the presence of a conical sea-mount*, J. Comput. Acoust., **4**, (1), pp. 101−121, 1996.

[158] Evans, R. B., *Stepwise coupled mode scattering of ambient noise by a cylindrically symmetric seamount*, J. Acoust. Soc. Am., **119**, (1), pp. 161−167, 2006.

[159] Prospathopoulos, A. M., Athanassoulis, G. A. and Belibassakis, K. A., *Underwater acoustic scattering from a radially layered cylindrical obstacle in a 3D ocean waveguide*, J. Sound Vib., **319**, (3−5), pp. 1285−1300, 2009.

[160] Ivansson, S., *Simple illustrations of range-dependence and 3-D effects by normal-mode sound propagation modelling*, In: *Proc. 39th Scandinavian Symposium on Physical Acoustics*, Geilo, Norway, 2016.

[161] Petrov, P. S. and Sturm, F., *An explicit analytical solution for sound propagation in a three-dimensional penetrable wedge with small apex angle*, J. Acoust. Soc. Am., **139**, (3), pp. 1343−1352, 2016.

Absorption of Sound in Seawater

T.H. Neighbors III

Leidos Corporation (Retired), Bellevue, WA, United States

Definitions

The *absorption of sound* in seawater is the frequency-dependent reduction in sound intensity due to the energy loss (conversion of acoustic energy into heat) through viscous and structural relaxation effects and through molecular relaxation processes associated with electrolytes in-solution in the seawater. Relaxation is the process where the medium, here the water, returns to its former state after having been exposed to the pressure variation in a sound wave. Absorption is measured in terms of decibel loss in sound pressure amplitude at a specific frequency over a characteristic length or the number of times the sound pressure amplitude has decreased by 1/e [1 Neper (Np)] of its original value over a characteristic length (i.e., typically, dB/km or Np/km). Neper is a dimensionless unit named after the inventor of the logarithms, John Napier. The *attenuation of sound* in seawater includes absorption and other mechanisms that affect the amplitude of the observed sound wave, such as scattering from particles in suspension, fish, other biological entities, and other fluid inhomogeneities in the water column, as well as changes due to surface and bottom scattering. Also spherical or cylindrical spreading of sound waves leads to attenuation of acoustic signals.

4.1 PHYSICS AND PHENOMENA

The absorption of sound in seawater is dependent upon factors, such as salinity, temperature, pH, and pressure. Water is an associated polar liquid and in *freshwater*, i.e., water free of contaminates, the absorption of sound is due to dissipation of acoustic energy, i.e., the transfer of acoustic energy to thermal energy through the action of shear and bulk viscosity, where the bulk viscosity effect includes the effect of structural relaxation. As suggested by Hall [1] the structural relaxation is due to a structural rearrangement of the water molecules which can exist in two energy states, the higher one being that of closest packing. On the passage of a sound wave through water some molecules break their structural bonds and move from the normal to the close-pack arrangement. This process will involve a time lag and hence acoustic absorption takes place. In seawater other loss mechanisms tend to dominate. Fig. 4.1, which is an updated version of Thorp's plot [2], shows the absorption

Applied Underwater Acoustics. http://dx.doi.org/10.1016/B978-0-12-811240-3.00004-7

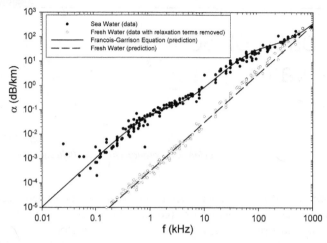

FIGURE 4.1

Absorption of sound in seawater.

*Adapted from W. H. Thorp, Deep-Ocean Attenuation in the Sub- and Low-Kilocycle-per-Second Region, J. Acoust. Soc. Am. **38**, 940, pp. 648–654 (1965) with the permission of the Acoustical Society of America.*

measured in sea trials, conducted over the past several decades, plotted versus the predicted absorption that would occur in freshwater. The solid curve is the absorption of sound in seawater based on the Francois and Garrison equations derived in [3,4] at depth: 50 m, temperature: 10°C, salinity: 32 parts per thousand (‰), and pH: 8. The dashed line is the predicted absorption in freshwater at depth: 50 m and temperature: 10°C. The measured absorption in seawater is significantly higher than that observed in freshwater with the difference increasing at lower frequencies. The road to understanding this anomaly started in 1948 when Lieberman [5] proposed that the excess absorption measured below 1 MHz in seawater at the mouth of the San Diego Bay, USA, relative to freshwater absorption in the El Capitan reservoir was possibly due to a *molecular chemical relaxation process* associated with the pressure-dependent dissociation of the sodium chloride (NaCl) molecules in seawater. In the past several decades, the research on the absorption of sound in seawater has focused on understanding the molecular chemical relaxation processes associated with electrolytes in seawater where ionic dissociation is alternately activated and deactivated by the sound wave's condensation and rarefaction phases, and the influence of acoustic energy on electrolyte association and dissociation rates. Below 1 MHz two distinct molecular chemical relaxation processes appear to dominate. From a few tens of kilohertz to hundreds of kilohertz the excess absorption is due to molecular relaxation processes associated with magnesium sulfate ($MgSO_4$) according to Wilson and Leonard [6] and Murphy et al. [7]. In the range of a few kilohertz to a few tens of kilohertz absorption is due to a combination of boric acid ($B(OH)_3$) and $MgSO_4$ molecular chemical relaxation processes, with the $MgSO_4$ processes dominating. According to Yeager et al. [8] and Fisher and Simmons

[9], from a few hundred hertz to a few kilohertz the $B(OH)_3$ molecular chemical relaxation appears to be the dominant source of observed absorption. In the same range magnesium carbonate also contributes to the absorption to a lesser degree as discussed by Browning and Mellen [10]. Below 100 Hz volume scattering has a dominant effect on the observed signal attenuation, according to Mellen et al. [11].

Section 4.2 contains selected experimental results from data collected over the past few decades in field experiments using explosive and nonexplosive sources and laboratory experiments using resonators; for instance, standing waves in spherical glass spheres, to measure the absorption of sound in artificial and natural seawater. Section 4.3 summarizes the mechanisms associated with sound absorption in freshwater and seawater including the sensitivity of these mechanisms to temperature, pressure, pH, and salinity. Section 4.4 provides the formula developed by Francois and Garrison [3,4] for the absorption of sound as a function of salinity, pH, pressure (depth), and temperature over the range from about 100 Hz to 1 MHz along with the Ainslie and McColm [12] fast running approximation for the Francois and Garrison formulation. Section 4.5 provides a list of symbols and abbreviations used in this chapter including the page of first usage. This is followed by a list of key references.

4.2 EXPERIMENTAL DATA

Field measurements of the absorption of sound using explosive and continuous wave (CW) sources above 10 kHz by Lieberman [5], Murphy et al. [7], Bedzek [13,14], and Garrison et al. [15–17] and below 10 kHz by Thorp [2], Sheehy and Halley [18], Urick [19,20], Kibblewhite and Denham [21,22], Skettering and Leroy [23], Lovett [24], Mellen and Browning [25], Bannister et al. [26], Kibblewhite and Hampton [27], Chow and Turner [28], and Schneider et al. [29] since the late 1940s until 2016 have shown significantly higher absorption levels than could be explained by "classical absorption theory" that only includes the effects of shear and bulk viscosities. The sea trial data were collected globally, spanning locations from the Pacific and Atlantic Oceans to the Arctic. The only correction made in the data plotted in Fig. 4.1 is the respective experimenter's correction for propagation divergence. Some data were collected at relatively shallow depths over a short range and other data were collected in the deep sound channel.

The reduction in pressure, dp, due to absorption in the pressure amplitude of a sound wave propagating in water is proportional to the local sound pressure amplitude, p {unit: Pa} and the distance, dx, the wave has traveled. This relationship is given by:

$$dp = -\alpha_n \, p \, dx \tag{4.1}$$

where the coefficient of proportionality, α_n, is the *amplitude absorption coefficient* which is expressed in Nepers per meter (Np/m), and the sound pressure amplitude $p(x)$ of a plane sound wave after traveling over a distance x, when the original amplitude is P_o, obtained by integrating Eq. (4.1) from $x = 0$ to x is:

$$p(x) = P_o \, e^{-\alpha_n x} \tag{4.2}$$

The acoustic intensity, I {unit: $Pa^2 \cdot m^2 \cdot s/kg$} of a plane sound wave is given by the pressure squared divided by the acoustic impedance, ρc, where ρ is the water density {unit: kg/m^3} and c is the speed of sound in water {unit: m/s}, as discussed in Chapter 2. The acoustic intensity of a plane sound wave propagating over the distance x is reduced from its original intensity I_o to $I(x)$, by the relationship:

$$I(x) = \frac{(P_o \, e^{-\alpha_n x})^2}{2\rho c} = I_o \, e^{-2\alpha_n x} \tag{4.3}$$

When a plane acoustic wave has propagated over a distance $x = 1/\alpha_n$, its pressure amplitude has dropped to $1/e = 0.368$ of the initial pressure amplitude P_o, and the acoustic intensity has dropped to $1/e^2 = 0.135$ of the initial intensity I_o.

Another absorption coefficient, α, for sound wave propagation, is based on the logarithm to the base 10 of the dimensionless ratio of *acoustic intensities*, I_1 and I_2, at two points separated by the distance x, and is expressed in dB[1]/(unit of x):

$$\alpha = \frac{1}{x} 10 \log_{10} \left(\frac{I_1}{I_2} \right) = \frac{1}{x} 20 \log_{10} \left(\frac{P_1}{P_2} \right) \tag{4.4}$$

where the proportionality between the intensity and the square of the pressure amplitude in a plane sound wave has been used. α in Eq. (4.4) has the dimension dB/(unit of x). When x is in meters the unit is dB/m. Likewise when x is in km, the unit is in dB/km, where dB is dimensionless. The relation between the absorption coefficients α_n and α within any consistent unit system based on Eqs. (4.2) and (4.4) is:

$$\alpha = \alpha_n \, 20 \log_{10} e = 8.686 \, \alpha_n \tag{4.5}$$

For sound propagation in a compressible medium like water two types of viscosity, the *dynamic shear viscosity* represented by the shear viscosity coefficient μ {unit: $Pa \cdot s$} and the *dynamic bulk viscosity* represented by the bulk viscosity coefficient μ' {unit: $Pa \cdot s$}, contribute to the sound absorption, see Section 4.3. The dynamic shear viscosity of a fluid represents the ratio of shear stress to strain rate, and is a measure of the diffusion of momentum produced by collisions between molecules from regions of the fluid with different net velocities. Since temperature is a measure of molecular motion, μ depends only on the temperature of the water, and dynamic shear viscosity is active in producing absorption also in a pure longitudinal motion without influence of compression. The dynamic bulk viscosity represents the influence of compression on water molecules, where structural rearrangements of the water molecules between different potential energy states take place with some time delay after the compression phase of the sound wave passes when propagating through water. Bulk viscosity occurs when a fluid is only exposed to compression and expansion (pure dilatation), which leads to a change in volume, and this viscosity is frequently called the *volume viscosity*. The

---

[1]*Decibel*, named after Alexander Graham Bell, see Chapter 1.

absorption caused by the influence of the dynamical shear and bulk viscosities is normally called *classical absorption* and its absorption coefficient is denoted by α_c.

While thermal conduction in most gases contributes substantially to the absorption process during sound wave propagation, the contribution to absorption arising from thermal conduction in water may be neglected. The ratio of the absorption coefficients for thermal conduction losses α_{th} and for viscous losses α_v in water is: $\alpha_{th}/\alpha_v \sim 10^{-3}$, and $\alpha_v \sim \alpha_c$, the classical coefficient of absorption.

The relaxation processes in freshwater connected with the influence of viscosities is characterized by a specific relaxation time τ_r, {unit: s} which represents the time it takes for the water molecules to rearrange (i.e., relax) after the influence of a passing sound wave. In freshwater the viscous relaxation process is very fast, and the relaxation time is very short. For freshwater the period of the sound wave is only comparable to the relaxation time when the frequency is very high. When the acoustic frequency f {unit: Hz} is comparable with the relaxation frequency $f_r = 1/2\pi\tau_r$ {unit: Hz} (i.e., $f \sim f_r$) the influence of relaxation processes on absorption and dispersion (i.e., phase velocity dependence on frequency) appears as discussed in the earlier part of this chapter. However, measurements of dispersion in seawater carried out over several years have shown that the deviation between the phase velocity and the sound velocity is about 1%. Therefore, sound propagation in seawater is considered to be nearly free of dispersion.

If the predicted viscous losses due to the influence of the dynamical shear and bulk viscosities in seawater are subtracted from the measured sound wave propagation absorption losses, an *excess absorption* due to other loss mechanisms is found. Fig. 4.2 plots the excess absorption coefficient, α_e {unit: dB/km} normalized by the sound frequency, f (i.e., α_e/f {unit: dB·s/km}) versus f {unit: Hz}. This spectrum

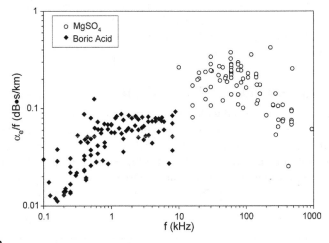

FIGURE 4.2

α_e/f spectrum for data collected by sea trials.

shows two clearly distinct regions where other mechanisms than viscosities are active in the absorption process.

The fairly broad peak in the range of about 30 to 100 kHz represents a chemical relaxation process and is attributed to the change in the association and dissociation rate of the in-solution electrolyte $MgSO_4$, and component species. The peak for $MgSO_4$ may be ascribed to the first step of dissociation that leads from a direct contact of the ions from the $MgSO_4$ molecules to a complex with one water molecule inserted between the ions as:

$$MgO\overset{\diagup H}{\underset{\diagdown H}{}} + SO_4 \rightleftharpoons MgSO_4 + H_2O \qquad (4.6)$$

The second spectral peak that occurs in the few hundred hertz to a few kilohertz range represents another chemical relaxation process and is due to the change in the association and disassociation rate of in-solution H_3BO_3 or $B(OH)_3$,[2] and other component species. Fisher and Simmons [41] showed the relaxation process to be consistent with a two-step equilibrium given by:

$$B(OH)_3 + OH^- \rightleftharpoons B(OH)_3 \cdot OH^- \rightleftharpoons B(OH)_4^- \qquad (4.7)$$

in which the formation of a borate ion involves an intermediate state and the slow second step causes the sound absorption. The absorption is expected to increase with the borate ion concentration.

The factors that influence the peaks in the excess absorption spectra include depth (pressure), temperature, salinity, and pH. Since 1948, the data from sea trials and laboratory experiments such as Wilson and Leonard [6], Yeager et al. [8], Fisher and Simmons [9], Fisher [30], Mellen and Simmons [31], and Mellen et al. [32−35] have been analyzed to determine the factors that affect absorption and the relative importance of these factors in various spectral regions. Two quantities are derived from analyzing the data sets, the frequency at which the excess absorption peak occurs, i.e., the relaxation frequency, f_r {unit: Hz} and the magnitude of α_e/f_r or the excess absorption per wavelength, $\alpha_e \lambda_r$.

4.2.1 ABSORPTION PRESSURE DEPENDENCE

Since the fluctuations associated with the sound pressure wave in seawater affect the equilibrium association and dissociation rates of the in-solution electrolytes, the increase in hydrostatic pressure with depth, which reduces the chemical compressibility of the solution according to Fisher [30], should reduce the absorption coefficient. The absorption of sound measured in seawater and in aqueous solutions of electrolytes in resonators by Fisher [30] shows a consistent decrease with

---

[2]Boric acid (H_3BO_3) is commonly written as $B(OH)_3$ in the literature. The common notation is used in the equations in this chapter to be consistent with the referenced papers.

increasing pressure. Fig. 4.3 shows the 1971 Pacific Ocean sound absorption measurements by Bezdek [13] at 75.8 kHz for two experimental conditions. In Fig. 4.3 the absorption coefficient includes influences from classical absorptions as well as from excess absorption. The absorption coefficient is, therefore, the total absorption coefficient $\alpha_t = \alpha_c + \alpha_e$ {unit: dB/km}.

In the first experimental condition the absorption of sound was measured at 100, 600, 700, 800, and 900 m vertical distances centered at horizontal distances 750, 1350, 1950, 2550, and 3150 m. In each instance the source and receiver were attached to the same instrumentation cable. For the second experimental configuration horizontal measurements were made at depths ranging from 910 to 3350 m with the transmitter and receiver separation varying from about 200 to 1300 m for each data collection. The solid line, where d is the depth in meters, is a simple second-order regression curve that fits the absorption of sound pressure dependence to this data set.

Resonator experiments conducted by Fisher [30] at 100 kHz and 500 kHz on a 0.5 M solution of $MgSO_2$ and a 1.017 M solution of NaCl in the pressure range from about 14.5 pounds per square inch (psi) to about 20,000 psi showed about a factor of 4 decrease in the $MgSO_4$ absorption at resonance due to the increased pressure, and the relaxation frequency appeared to be independent of pressure. Subsequent experiments by Hsu and Fisher [36] with smaller molar concentrations, 0.02 M of $MgSO_4$ and a 0.02 M of $MgSO_4$ in solution with 0.6 M of NaCl, also showed a decrease in the reaction rate (absorption per wavelength) without any change in the relaxation frequency, within the range of experimental error. The influence of $MgSO_4$ on absorption in seawater is remarkable as the amount of $MgSO_4$ is much smaller than the amount of common salt, NaCl. In parts per million (ppm)

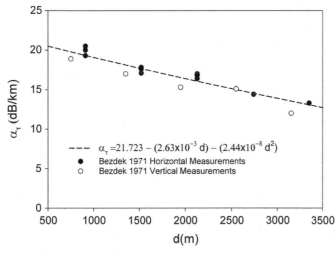

FIGURE 4.3

Total absorption (α_t) pressure dependence based on data extracted from Bezdek's [13] 1971 measurements.

seawater comprises: chlorine (19,000 ppm), sodium (10,500 ppm), magnesium (1350 ppm), and sulfate (885 ppm), as discussed in Chapter 1.

In the frequency region where the excess absorption is dominated by H_3BO_3 the data from sea trials and laboratory experiments had not shown any significant dependence on pressure. This is consistent with the correlation analysis performed by Schulkin and Marsh [37] between low-frequency experimental data and oceanographic variables pH, salinity, temperature, and depth that demonstrated a minimal correlation between excess absorption per wavelength and depth or relaxation frequency and depth.

4.2.2 ABSORPTION TEMPERATURE DEPENDENCE

As the temperature changes, the equilibrium association and dissociation rates of the in-solution electrolytes in seawater change. In the frequency range where $MgSO_4$ dominates the excess absorption contribution, a decrease in temperature affects seawater micro viscosity and reduces the effective mobility of the ions involved in the dissociation of $MgSO_4$ as discussed by Francois and Garrison [3], Schulkin and Marsh [37], and Glotov [38]. Fig. 4.4 illustrates the difference in the measured sound absorption, represented by α_t, in seawater for two data sets, where the primary difference is the temperature. The first data set contains the 1965 and 1966 measurements made by Greene in Arctic waters with the temperature range from about -1.4 to $-1.3°C$, salinity about $32.6‰$, and depth 122 m, as summarized by Francois and Garrison [3]. The second data set was collected in 1971 by Bezdek [13] in the Pacific Ocean with the temperature of about $7°C$, salinity about $34.6‰$, and depth 200 m, also summarized by Francois and Garrison [3]. The difference between the two

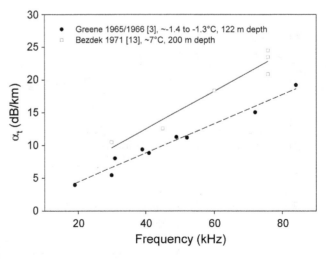

FIGURE 4.4

Temperature dependence of absorption of sound in seawater, Francois and Garrison [3].

regression curves illustrates the change in absorption due to the change in seawater temperature when depth and salinity are approximately the same. Laboratory measurements of the excess absorption in natural seawater and synthesized seawater by Wilson and Leonard [6] show the same type of trend, i.e., the excess absorption and relaxation frequency increase with temperature.

In the low-frequency range where the excess absorption of sound in seawater is dominated by the in-solution boric acid the excess absorption per wavelength does not show any significant temperature dependence as shown by Schulkin and Marsh [37], however, the relaxation frequency is dependent upon temperature. Fig. 4.5, which is derived from Francois and Garrison [4], shows an increase in relaxation frequency with temperature for the five data sets used by Francois and Garrison in developing the boric acid dependency for the sound absorption in seawater equations summarized in Section 4.4. Multiplying the relaxation frequency by $(35/S)^{0.5}$, where S is the salinity {unit: ‰}, removes the boric acid salinity dependence shown in Eq. (4.22) in Section 4.4. The increase in relaxation frequency is consistent with an increase in the absorption of sound in seawater with increased temperature. These data sets shown in Fig. 4.5 include: the Northeast Pacific data collected in September 1973 by Chow and Turner [28], Thorp's [2] measurements in the Atlantic Ocean in 1962, Skretting and Leroy's [23] measurements in the Mediterranean in 1966, Browning's measurements in the Red Sea reported in Mellen et al. [32], and measurements by Mellen et al. [11] in the Gulf of Aden reported in 1974.

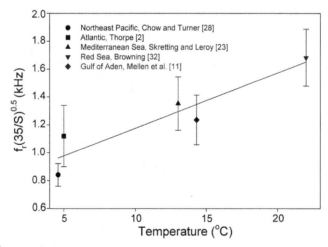

FIGURE 4.5

Relaxation frequency dependence on temperature for absorption measurements are carried out in the frequency range: 0.16–6.0 kHz.

Adapted from Figure 2 in R. E. Francois and G. R. Garrison, Sound absorption based on ocean measurements: Part II: Boric acid contribution and equation for total absorption, J. Acoust. Soc. Am. **72**, *(6), pp. 1879–1890 (1982), with the permission of Acoustical Society of America.*

4.2.3 PH DEPENDENCE OF ABSORPTION

The absorption of sound does not appear to depend upon seawater pH in the frequency range where the excess absorption is dominated by in-solution $MgSO_4$. However, in the low-frequency range where excess absorption is dominated by the in-solution boric acid, sea trials indicate a first-order dependence on pH for the excess absorption per wavelength and relaxation frequency according to Schulkin and Marsh [37] and Mellen and Browning [39]. Until recently laboratory experiments with resonators at low frequencies used electrolyte concentrations in excess of the concentrations found in natural seawater to measure the relaxation frequency and excess absorption. Qui [40] developed a cylindrical resonator technique that allows the measurement of maximum absorption per wavelength and relaxation frequency using normal seawater concentrations. Fig. 4.6, replotted from data presented by Qui [40], illustrates the dependence of the total absorption of sound in seawater on pH over the pH range from 6.4 to 8.6 at 2.91, 6.82, and 10.76 kHz. The measurements were performed with natural seawater with salinity of about 33.91‰ and temperature of about 12.6°C. These data are consistent with the Schulkin and Marsh [37] analysis of the data which shows that the pH is correlated with the concentration of borate ions in the solution.

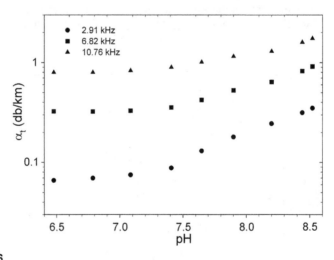

FIGURE 4.6

Sound absorption sensitivity to pH changes at low frequencies.

*Adapted from Figure 9 in X. Qui, A cylindrical resonator method for the investigation of low-frequency sound absorption in sea water, J. Acoust. Soc. Am. **90**, (6), pp. 3263–3270 (1991), with the permission of Acoustical Society of America.*

4.2.4 SALINITY DEPENDENCE

From the analyses of experimental data conducted over 1960s to 1980s by Francois and Garrison [3,4], Schulkin and Marsh [37], and Fisher and Simmons [41], the excess absorption of sound in the frequency region dominated by in-solution $MgSO_4$ appears to have a linear dependence upon salinity. This is, however, not the case in the low-frequency range dominated by the in-solution H_3BO_3. In this regime, the absorption magnitude appears to have almost no dependence on salinity and the relaxation frequency appears to be inversely proportional to the square root of the salinity as discussed in Section 4.4.1.

4.3 SOUND ABSORPTION MECHANISMS
4.3.1 SOUND ABSORPTION IN FRESHWATER

Absorption of sound propagated in freshwater is due to shear and bulk viscous mechanisms as discussed by Lieberman [5] and Urick [42] where the "classical" absorption coefficient, α_c {unit: Np/m}, due to shear viscosity alone is given by the expression developed by Rayleigh [43]:

$$\alpha_c = \frac{8\pi^2 f^2 \eta}{3\rho c^3}. \qquad (4.8)$$

where f is the frequency {unit: Hz}, ρ is the density {unit: kg/m^3}, c is the local sound velocity {unit: m/s}, and η is the dynamical shear viscosity coefficient {unit: Pa·s}.

In freshwater the measured absorption due to viscosity is about three times the absorption due to the dynamical shear viscosity alone. According to Beyer [44], if the *bulk viscosity* influence, represented by the bulk viscosity coefficient η' {unit: Pa·s}, is included together with the shear viscosity, the classical absorption coefficient, α_c {unit: Np/m}, becomes:

$$\alpha_c = \frac{8\pi^2 f^2 \left(\eta + \frac{3}{4}\eta'\right)}{3\rho c^3}, \qquad (4.9)$$

which is consistent with the measured values for freshwater. In pure water at 15°C, $\eta = 0.114$ Pa·s and $\eta'/\eta = 2.81$ according to Litovitz and Davis [45]. Table 4.1 provides a summary of the temperature and pressure dependence of the ratio η'/η.

By combining expressions for absorption and dispersion due to influence of a relaxation process in freshwater from Kinsler et al. [46], the relaxation time τ_r may be expressed by $\tau_r = 2\alpha_c/2\pi f k$, where $k = 2\pi/\lambda$ {unit: m^{-1}} is the wave number and λ {unit: m} is the wavelength. If Eq. (4.9) is inserted into τ_r, the relaxation time {unit: s} in freshwater is given by:

$$\tau_r = \frac{\frac{4\mu}{3} + \mu'}{\rho c^2} \qquad (4.10)$$

For freshwater at 10°C with dynamic shear viscosity $\mu = 1.3 \; 10^{-3}$ {unit: Pa·s} the dynamic bulk viscosity is nearly three times the shear viscosity, the water density

Table 4.1 Dynamical Viscosity Coefficient Ratio's Temperature Dependence at 10^5 Pa and Pressure Dependence at 40°C.

T (°C)	η'/η	Pressure (atm)	η'/η
0	3.11	1	2.68
20	2.80	1000	2.33
40	2.68	2000	2.33
60	2.72		

Extracted with permission from T. A. Litovitz and C. M. Davis, Structural and Shear Relaxation in Liquids, In Physical Acoustics, Vol. II Part A, (W. P. Mason, ed.), Academic Press, New York, 1965.

$\rho = 10^3$ {unit: kg/m^3} and the sound velocity $c = 1476$ {unit: m/s}, the relaxation time $\tau_r = 2.5 \cdot 10^{-12}$ {unit: s}. The very short relaxation time in freshwater means, that absorption in freshwater will obey the frequency-squared relation given by Eq. (4.9) *up to frequencies close to 10^{12} Hz*, or in general, at most frequencies used for measurements.

4.3.2 MOLECULAR CHEMICAL RELAXATION PROCESSES

As shown in Fig. 4.1 the electrolytes in seawater significantly increase propagated sound absorption relative to freshwater. This excess absorption is due to the effect of molecular chemical relaxation processes. Typically, the chemical processes may involve two or more stages of the form:

$$n_1X_1 + n_2X_2 \rightleftarrows n_3X_3 + n_4X_4 \rightleftarrows n_5X_5 + n_6X_6 ... \rightleftarrows n_{i-1}X_{i-1} + n_iX,$$

where the n_i is the molar concentration of chemical species, X_i. According to Lieberman [5] and Urick [47], when a *single chemical process* is involved the *absorption spectrum* can be described in terms of a *single relaxation frequency*, f_r, as

$$\alpha_t = A\frac{f^2 f_r}{f^2 + f_r^2} + Bf^2, \tag{4.11}$$

where the first term represents the excess absorption and the second term is the classical absorption due to dynamic shear and bulk viscosities, that is, $B = \alpha_c/f^2$. According to Wilson and Leonard [6] and Bedzek [13,14], A is a factor characterizing the specific chemical relaxation process that depends on temperature and hydrostatic pressure. In Eq. (4.11), the excess absorption, represented by α_e, due to the chemical relaxation process, is obtained by subtracting viscous losses, represented by α_c from the measured total absorption, represented by α_t.

Absorption data collected under the same in situ temperature, observation depth, salinity, and pH can be used to experimentally determine the relaxation frequency, f_r, and chemical relaxation term, A. For example, Fig. 4.7 is a plot of the excess absorption coefficient associated with the sound absorption data collected by the Applied Physics Laboratory—University of Washington (APL-UW) [7] at Dabob Bay

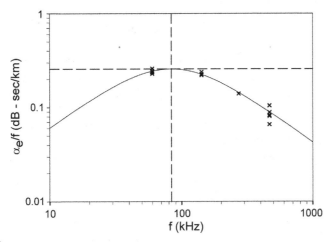

FIGURE 4.7

Approximate MgSO$_4$ induced excess absorption values as a function of frequency (f), A, and f_r based on Dabob Bay data.

Adapted from Figure 5 in S. R. Murphy, G. R. Garrison, and D. S. Potter, Sound Absorption at 50 to 500 kc from Transmission Measurements in the Sea, J. Acoust. Soc. Am. **30**, *(9), pp. 871–875 (1958) with the permission of the Acoustical Society of America.*

between 1953 and 1956 for the frequency range from 60 to 467 kHz, at depths from about 50 m through about 150 m, in seawater with a temperature in the range approximately 8 to 10°C, and salinity of about 30‰. Based on the curve fit, to first order the relaxation frequency is about 85 kHz and the chemical relaxation term A is about 0.5 (dB·s/km).

As shown in Fig. 4.7, α_e/f is a maximum at f_r. When $f \gg f_r$, the pressure change in the acoustic wave is too rapid to affect the chemical reaction rate since the relaxing molecules are not able to respond fast enough to be activated, and a sort of chemical equilibrium is preserved. As a result, the excess absorption coefficient in Eq. (4.11) approaches a constant value Af_r. When $f \ll f_r$, the pressure change in the acoustic wave is so slow that the relaxation process is finished before a new change in pressure takes place, and no particular contributions from the chemical relaxation process influence the absorption coefficient. As the wave frequency decreases the excess absorption coefficient approaches Af^2/f_r. In the case of a solution, such as seawater, which has multiple electrolytes in-solution, multiple relaxation frequencies can affect absorption. If each of the relaxation processes is mutually independent, then the coefficient of total absorption of sound in seawater can be described by the sum of the individual excess absorption coefficient terms and the classical absorption coefficient. Then Eq. (4.11) becomes:

$$\alpha_t = A_1 \frac{f^2 f_1}{f^2 + f_1^2} + A_2 \frac{f^2 f_2}{f^2 + f_2^2} + \cdots + Bf^2, \tag{4.12}$$

where f_1 and f_2 are the relaxation frequencies of the dominant relaxation processes and the dotted terms indicate contributions from additional relaxation processes which produce second-order effects.

The *factors that affect the relaxation frequency* f_r and the coefficient A include temperature, pressure (depth), pH, and salinity. These dependencies have been examined in a series of laboratory experiments using resonators and in situ through various sea trials since the experiments conducted by Liebermann [5].

If Eq. (4.11), without the classical absorption term, is multiplied by the wavelength, λ, the result given in Eq. (4.13) is the excess absorption per wavelength, $\alpha_e \lambda$ {unit: dB or Nepers} by use of Eq. (4.5), where c_o is the local speed of sound:

$$\alpha_e \lambda = A c_o \frac{f f_r}{f^2 + f_r^2}. \tag{4.13}$$

Since the maximum excess absorption per wavelength at the relaxation frequency, $(\alpha_e \lambda)_r$, is given by $A c_o/2$, Eq. (4.13) can be written as:

$$\alpha_e \lambda = 2(\alpha_e \lambda)_r \frac{f f_r}{f^2 + f_r^2}. \tag{4.14}$$

In the high frequency limit the absorption per wavelength $\alpha_e \lambda$ becomes $2(\alpha_e \lambda)_r f_r/f$ which in this limit approaches 0, and in the low frequency limit $\alpha_e \lambda$ becomes $2(\alpha_e \lambda)_r f/f_r$.

4.3.2.1 Temperature Dependence

Fig. 4.8 plots the excess absorption coefficient based on data measured in 1974 and 1975 by APL-UW by Garrison et al. [15,16] at locations off Pt. Barrow. For these data sets the salinity is about 32‰, water temperature is about $-1.6°C$, and CW sources were used to collect the data at a 45 m depth.

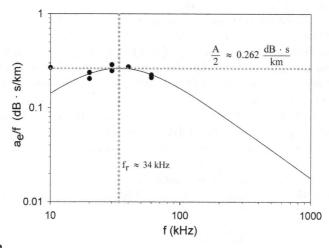

FIGURE 4.8

Excess absorption coefficient measured off Pt. Barrow by Garrison et al. [15,16].

If the slight differences in measurement depth and salinity are ignored, the comparison of Figs. 4.8 and 4.7 shows that the $MgSO_4$ relaxation frequency is highly dependent upon temperature. This dependence is explained by the Glotov's theory [38] which is included by Francois and Garrison [3] and Schulkin and Marsh [37] in their empirical formulas. Glotov [38] relates the $MgSO_4$ ion dissociation to the macroscopic viscosity, μ_m {unit: Pa·s}, of seawater's dependence on temperature and salinity. According to Francois and Garrison [3] for a salinity $S = 30-40\%$ at the temperature T {unit: °C} the dependence of the relaxation frequency f_r {unit: Hz} according to Glotov [38] is given by:

$$f_r(T) = \frac{7.55 \cdot 10^3}{\mu_m}(273 + T)10^{-\left(\frac{934}{273+T}\right)} \tag{4.15}$$

Fig. 4.9, derived from Francois and Garrison [3], plots the $MgSO_4$ relaxation frequency derived from data of several experimenters as a function of temperature and the equation for the predicted $MgSO_4$ relaxation frequency, f_r, is given by:

$$f_r = \frac{8.17 \cdot 10^{(8-1900/\theta)}}{1 + 0.0018(S - 35)}, \tag{4.16}$$

where S is salinity (‰) and θ {unit: K} is $(T + 273)$, where T {unit: °C} is the temperature.

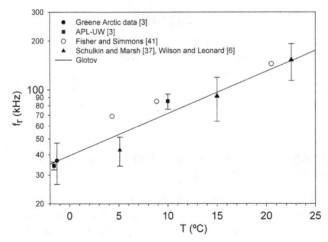

FIGURE 4.9

$MgSO_4$ relaxation frequency dependence on temperature.

Adapted from Figure 8 in R. E. Francois and G. R. Garrison, Sound absorption based on ocean measurements: Part I: Pure water and magnesium sulfate contributions, *J. Acoust. Soc. Am.* **72**, *(3), pp. 896–907 (1982), with the permission of Acoustical Society of America.*

Table 4.2 Pressure Correction Factors for $MgSO_4$ Relaxation Processes

Authors	β_1 [1/(100 kPa)]	β_2 [1/(100 kPa)^2]	Data Source
Hsu–Fisher [36]	-6.39×10^{-4}	0	Laboratory
Francois–Garrison [3]	-1.37×10^{-3}	6.2×10^{-7}	Sea trials/laboratory
Schulkin–Marsh [37]	-6.54×10^{-3}	0	Laboratory
Bezdek [13]	-1.36×10^{-3}	4.47×10^{-7}	Sea trials

4.3.2.2 Pressure Effects

In the frequency range where relaxation processes are dominated by in-solution $MgSO_4$ the absorption of sound is pressure dependent. The pressure dependence is given by the dimensionless parabolic correction factor:

$$K = 1 + \beta_1 P + \beta_2 P^2,$$ (4.17)

where P is pressure {unit: Pa}, and the coefficients β_1 and β_2 are in Pa^{-1} and Pa^{-2}, respectively. Table 4.2 provides examples of coefficients developed over the years including the data source. Correction factors based primarily on laboratory resonator experiments using aqueous electrolyte solutions are significantly higher than those derived from field data. The Francois–Garrison correction factor takes the regression equation based on Bezdek's data and modifies the equation to assure that it follows the trend of the laboratory data at very high pressures.

4.4 FORMULAS AND EXPRESSIONS

Although there have been several formulas [2,37,41,48–52] developed over the past few decades for the absorption of sound in seawater, the formula which appears to provide the best overall agreement with the experimental data is the one developed by Francois and Garrison [3,4]. A simpler approximation for the Francois and Garrison equation by Ainslie and McColm [12] was published in 1998. These approximations are outlined in the following sections.

4.4.1 FRANCOIS AND GARRISON EQUATION FOR SOUND ABSORPTION IN SEAWATER

Based on field and laboratory measurements conducted from the 1940s through the early 1980s Francois and Garrison in 1982 developed a recommended formula for the absorption of sound in seawater as a function of depth (pressure) {unit: m}, temperature {unit: °C}, salinity in parts per thousand {unit: ‰}, frequency {unit: Hz}, and pH for the range from about 100 Hz to over 1 MHz. This formula is of the form:

$$\alpha_t = \frac{A_1 P_1 f_1 f^2}{f^2 + f_1^2} + \frac{A_2 P_2 f_2 f^2}{f^2 + f_2^2} + A_3 P_3 f^2,$$ (4.18)

where α_t is the total absorption {unit: dB/km}, f is the frequency {unit: kHz}, and P_1, P_2, and P_3 are nondimensional correction factors for the pressure. The first term on the right side of the equation is the absorption induced by the loss in acoustic energy due to a propagating sound pressure wave's disruption of the chemical equilibrium of the in-solution $B(OH)_3$, which has the relaxation frequency f_1. The second term on the right side of the equation is the absorption due to a propagating sound pressure wave's disruption of the chemical equilibrium of in-solution $MgSO_4$, which has the relaxation frequency f_2, and the third term on the right side of the equation is the absorption that occurs in pure water, which is only due to shear and bulk viscosity. The part of Eq. (4.18) which covers the range of 10–500 kHz ($MgSO_4$) is based on field measurements in the range of −2 to 22°C, salinity in the range of 30‰ to 35‰, and depths in the range of 0 to 3500 m. Below 10 kHz the term for the boric acid contribution is based on measurements in the range of −2 to 22°C, salinity in the range of 34‰ to 41‰, depths ranging down to 1500 m, and pH in the range of 7.72 to 8.18. Below 200 Hz the equation is likely to deviate from measured data due to the presence of scattering by temperature inhomogeneities and due to diffraction effects leading to losses that increase exponentially with the wavelength, as found by Mellen [52]. Below 1 kHz Kibblewhite and Hampton [27] based on a review of all available data on low-frequency absorption in the sound fixing and ranging (SOFAR) channel, showed that the frequency dependence for α_t {unit: dB/km} at frequencies f {unit: Hz} below 1 kHz could be expressed by the following empirical formula:

$$a_t(f) = a_s + 0.11\, L\frac{f^2}{1+f^2} + 0.011f^2 \qquad (4.19)$$

where α_s represents additional attenuation of the acoustic waves due to scattering and diffraction and L is a coefficient which accounts for the regional variation of boron-related effects in the oceans where absorption measurements took place.

4.4.1.1 Boric Acid Coefficients

The coefficients for the first term in Eq. (4.18) are given by:

$$A_1 = \frac{8.86}{c} \times 10^{(0.78\ \text{pH}-5)}, \left\{\text{unit: } \frac{\text{dB}}{\text{km}\cdot\text{kHz}}\right\} \qquad (4.20)$$

$$P_1 = 1.0, \qquad (4.21)$$

$$f_1 = 2.8(S/35)^{0.5}10^{(4-1245/\theta)}, \{\text{unit: kHz}\} \qquad (4.22)$$

where the speed of sound c is given by:

$$c = 1412 + 3.21 \times T + 1.19 \times S + 0.0167 \times D, \{\text{unit: m/s}\} \qquad (4.23)$$

with the temperature T {unit: °C}, $\theta = 273 + T$ {unit: K}, the salinity S {unit: ‰}, and the depth D {unit: m}.

4.4.1.2 Magnesium Sulfate Coefficients

The coefficients for the second term in Eq. (4.18) are given by:

$$A_2 = 21.44 \frac{S}{c}(1 + 0.025\ T), \left\{ \text{unit:}\ \frac{\text{dB}}{\text{km} \cdot \text{kHz}} \right\} \tag{4.24}$$

$$P_2 = 1 - 1.37 \times 10^{-4}D + 6.2 \times 10^{-9}D^2, \tag{4.25}$$

$$f_2 = \frac{8.17 \times 10^{(8-1990/\theta)}}{1 + 0.0018\ (S - 35)}, \{\text{unit:}\ \text{kHz}\} \tag{4.26}$$

4.4.1.3 Pure Water Contribution

The pure water contribution to the absorption of sound in Francois and Garrison's work, Eq. (4.18), includes two terms—a term when the temperature T is at and below $20°C$ and a term when the temperature is above $20°C$. The coefficient A_3 is:

For $T < 20°C$,

$$A_3 = 4.937 \times 10^{-4} - 2.59 \times 10^{-5}\ T + 9.11 \times 10^{-7}\ T^2 - 1.50 \times 10^{-8}\ T^3,$$

$$\left\{ \text{unit:}\ \frac{\text{dB}}{\text{km} \cdot \text{kHz}^2} \right\}$$

For $T \geq 20°C$,

$$A_3 = 3.964 \times 10^{-4} - 1.146 \times 10^{-5}\ T + 1.45 \times 10^{-7}\ T^2 - 6.5 \times 10^{-10}\ T^3,$$

$$\left\{ \text{unit:}\ \frac{\text{dB}}{\text{km} \cdot \text{kHz}^2} \right\}$$

$$\tag{4.27}$$

The pressure correction factor P_3 is given by:

$$P_3 = 1 - 3.83 \times 10^{-5}D + 4.9 \times 10^{-10}D^2 \tag{4.28}$$

4.4.2 AINSLIE AND MCCOLM SIMPLIFIED EQUATION FOR SOUND ABSORPTION IN SEAWATER

A simplified equation for the absorption of sound in seawater was developed by Ainslie and McColm [12]. This equation is reported to provide predictions of the absorption of sound within 10% of the values predicted by the Francois–Garrison equation as long as water temperature, pH, salinity, and prediction depth D fall into the following ranges:

$$
\begin{aligned}
-6 < T < 35°C \quad &(S = 35\ \text{ppt, pH} = 8,\ z = 0), \\
7.7 < \text{pH} < 8.3 \quad &(T = 10°C,\ S = 35\ \text{ppt},\ z = 0), \\
5 < S < 50\ \text{ppt} \quad &(T = 10°C,\ \text{pH} = 8,\ z = 0), \\
0 < D < 7\ \text{km} \quad &(T = 10°C,\ S = 35\ \text{ppt, pH} = 8).
\end{aligned}
$$

In this approximation, Eq. (4.12) becomes:

$$\alpha_t = \frac{A_1 f_1 f^2}{f^2 + f_1^2} + \frac{A_2 f_2 f^2}{f^2 + f_2^2} + A_3 f^2, \{\text{unit: dB/km}\}. \tag{4.29}$$

The relaxation frequencies for boric acid and $MgSO_4$ are given by

$$f_1 = 0.78 \left(\frac{S}{35}\right)^{1/2} e^{T/26} \ (\text{Boric Acid})\{\text{unit: kHz}\},$$

$$f_2 = 42 \ e^{T/17} \ (MgSO_4)\{\text{unit: kHz}\}, \tag{4.30}$$

where the salinity is S {unit: ‰} and the temperature is T {unit: °C}. The coefficients A_1, A_2, and A_3 then become:

$$A_1 = 0.106 \times e^{(\text{pH}-8)/0.56}, \left\{\text{unit: } \frac{\text{dB}}{\text{km} \cdot \text{kHz}}\right\} \tag{4.31}$$

$$A_2 = 0.52 \left(1 + \frac{T}{43}\right)\left(\frac{S}{35}\right) e^{-z/6}, \left\{\text{unit: } \frac{\text{dB}}{\text{km} \cdot \text{kHz}}\right\} \tag{4.32}$$

$$A_3 = 0.00049 \times e^{-(T/27+z/17)} \left\{\text{unit: } \frac{\text{dB}}{\text{km} \cdot \text{kHz}^2}\right\}. \tag{4.33}$$

4.5 SYMBOLS AND ABBREVIATIONS

Table 4.3 provides a list of symbols and abbreviations, which includes a description of the symbol or abbreviation and the page of first usage.

Table 4.3 List of Symbols and Abbreviations

Symbol or Abbreviation	Description	Page of First Usage
Np	Number of times to decrease by 1/e (1 Neper) of original value	273
NaCl	Sodium chloride	274
$MgSO_4$	Magnesium sulfate	274
$B(OH)_3$, H_3BO_3	Boric acid	274
p	Sound pressure amplitude {unit: Pa}	275
α_n	Amplitude absorption coefficient {unit: Nepers per meter (Np/m)}	275
I	Acoustic intensity {unit: $Pa^2 \cdot m^2 \cdot s/kg$}	276
ρ	Water density {unit: kg/m^3}	276
c	Sound speed in water {unit: m/s}	276

Continued

Table 4.3 List of Symbols and Abbreviations—cont'd

Symbol or Abbreviation	Description	Page of First Usage
dB	Decibel	276
μ	Dynamic shear viscosity {unit: Pa·s}	276
μ'	Dynamic bulk viscosity {unit: Pa·s}	276
α_c	Classical absorption coefficient	277
α_{th}	Thermal conduction absorption coefficient	277
τ_r	Relaxation time {unit: s}	277
f	Frequency {unit: Hz}	277
α_e	Excess absorption coefficient {unit: dB/km}	277
f_r	Relaxation frequency {unit: Hz}	277
α_t	Total absorption coefficient {unit: dB/km}	279
M	mole	279
ppm	Parts per million	279
S	Salinity {unit: %}	281
η	Dynamical shear viscosity coefficient {unit: Pa·s}	283
η'	Bulk viscosity coefficient {unit: Pa·s}	283
k	Wavenumber {unit: m^{-1}}	283
λ	Wavelength {unit: m}	283
μ_m	Macroscopic viscosity {unit: Pa·s}	287
T	Temperature {unit: °C}	287
θ	$T + 273$ {unit: °K}	287
K	Parabolic correction factor {dimensionless}	288
SOFAR	Sound fixing and ranging	289
L	Coefficient for regional boron effects	289
D	Depth {unit: m}	289

REFERENCES

[1] L. Hall, *The Origin of Ultrasonic Absorption in Water*, Phys. Rev. **73**, pp. 775–1781 (1948).

[2] W. H. Thorp, *Deep-Ocean Attenuation in the Sub- and Low-Kilocycle-per-Second Region*, J. Acoust. Soc. Am. **38**, (4), pp. 648–654 (1965).

[3] R. E. Francois and G. R. Garrison, *Sound absorption based on ocean measurements: Part I: Pure water and magnesium sulfate contributions*, J. Acoust. Soc. Am. **72**, (3), pp. 896–907 (1982).

[4] R. E. Francois and G. R. Garrison, *Sound absorption based on ocean measurements: Part II: Boric acid contribution and equation for total absorption*, J. Acoust. Soc. Am. **72**, (6), pp. 1879–1890 (1982).

[5] L. N. Lieberman, *The origin of sound absorption in water and sea water*, J. Acoust. Soc. Am. **20**, (6), pp. 868–878 (1948).

[6] O. B. Wilson, Jr. and R. W. Leonard, *Measurements of Sound Absorption in Aqueous Salt Solutions by a Resonator Method*, J. Acoust. Soc. Am. **26**, (2), pp. 223–226 (1954).

[7] S. R. Murphy, G. R. Garrison, and D. S. Potter, *Sound Absorption at 50 to 500 kc from Transmission Measurements in the Sea*, J. Acoust. Soc. Am. **30**, (9), pp. 871–875 (1958).

[8] E. Yeager, F. H. Fisher, J. Miceli, and R. Bressel, *Origin of the low-frequency sound absorption in sea water*, J. Acoust. Soc. Am. **53**, (6), pp. 1705–1707 (1973).

[9] F. H. Fisher and V. P. Simmons, *Discovery of Boric Acid as Cause of Low Frequency Sound Absorption in the Ocean*, IEEE Ocean **75**, pp. 21–24 (1975).

[10] D. G. Browning and R. H. Mellen, *Attenuation of low-frequency sound in the sea: recent results*, In Progress in Underwater Acoustics, (H. M. Merklinger), Plenum Press, New York, 1987.

[11] R. H. Mellen, D. G. Browning, and J. M. Ross, *Attenuation in randomly inhomogeneous sound channels*, J. Acoust. Soc. Am. **56**, (1), pp. 80–82 (1974).

[12] M. A. Ainslie and James G. McColm, *A simplified formula for viscous and chemical absorption in sea water*, J. Acoust. Soc. Am. **103**, (3), pp. 1671–1672 (1998).

[13] H. F. Bedzek, *Pressure dependence of sound attenuation in the Pacific Ocean*, J. Acoust. Soc. Am. **53**, (3), pp. 782–788 (1973).

[14] H. F. Bedzek, *Pressure dependence of the acoustic relaxation frequency associated with $MgSO_4$ in the ocean*, J. Acoust. Soc. Am. **54**, (4), pp. 1062–1065 (1973).

[15] G. R. Garrison, R. E. Francois, and E. A. Pence, *Sound absorption measurements at 10–60 kHz in near-freezing sea water*, J. Acoust. Soc. Am. **58**, (3), pp. 608–619 (1975).

[16] G. R. Garrison, E. W. Early, and T. Wen, *Additional sound absorption measurements in near-freezing sea water*, J. Acoust. Soc. Am. **59**, (6), pp. 1278–1283 (1976).

[17] G. R. Garrison, R. E. Francois, E. W. Early, and T. Wen, *Sound absorption measurements at 10–650 kHz in Arctic water*, J. Acoust. Soc. Am. **73**, (2), pp. 497–501 (1983).

[18] M. J. Sheehy and R. Halley, *Measurement of the Attenuation of Low-Frequency Underwater Sound*, J. Acoust. Soc. Am. **29**, (4), pp. 464–469 (1957).

[19] R. J. Urick, *Low-Frequency Sound Attenuation in the Deep Ocean*, J. Acoust. Soc. Am. **35**, (9), pp. 1413–1422 (1963).

[20] R. J. Urick, *Long-Range Deep-Sea Attenuation Measurement*, J. Acoust. Soc. Am. **39**, (5), pp. 904–906 (1966).

[21] A. C. Kibblewhite and R. N. Denham, *Long-Range Sound Propagation in the South Tasman Sea*, J. Acoust. Soc. Am. **41**, (2), pp. 401–411 (1967).

[22] A. C. Kibblewhite and R. N. Denham, *Low-frequency Acoustic Attenuation in the South Pacific Ocean*, J. Acoust. Soc. Am. **49**, (3), pp. 810–815 (1971).

[23] A. Skretting and C. C. Leroy, *Sound Attenuation between 200 Hz and 10 kHz*, J. Acoust. Soc. Am. **49**, (1), pp. 278–282 (1970).

[24] J. R. Lovett, *Northeast Pacific sound attenuation using low-frequency CW Sources*, J. Acoust. Soc. Am. **58**, (3), pp. 620–625 (1975).

[25] R. H. Mellen and D. G. Browning, *Low-frequency attenuation in the Pacific Ocean*, J. Acoust. Soc. Am. **59**, (3), pp. 700–702 (1976).

[26] R. W. Bannister, R. N. Denham, K. M. Guthrie, and D. G. Browning, *Project TASMAN TWO: Low-frequency propagation measurements in the South Tasman Sea*, J. Acoust. Soc. Am. **62**, (4), pp. 847−859 (1977).

[27] A. C. Kibblewhite and L. D. Hampton, A review of deep ocean sound attenuation data at very low frequencies, J. Acoust. Soc. Am. **67**, (1), 147−157 (1980).

[28] R. K. Chow and R. G. Turner, *Attenuation of low-frequency sound in the Northeast Pacific Ocean*, J. Acoust. Soc. Am. **72**, (3), pp. 888−891 (1982).

[29] H. G. Schneider, R. Thiele, and P. C. Wille, *Measurement of sound absorption in low salinity water of the Baltic Sea*, J. Acoust. Soc. Am. **77**, (4), pp. 1409−1412 (1985).

[30] F. H. Fisher, *Effect of High Pressure on Sound Absorption and Chemical Equilibrium*, J. Acoust. Soc. Am. **30**, (5), pp. 442−448 (1958).

[31] R. H. Mellen and V. P. Simmons, *Sound absorption in sea water: A third chemical relaxation*, J. Acoust. Soc. Am. **65**, (4), pp. 923−925 (1979).

[32] R. H. Mellen, D. G. Browning, and V. P. Simmons, *Investigation of chemical sound absorption in seawater by the resonator method: Part I*, J. Acoust. Soc. Am. **68**, (1), pp. 248−257 (1980).

[33] R. H. Mellen, D. G. Browning, and V. P. Simmons, *Investigation of chemical sound absorption in sea water by the resonator method: Part II*, J. Acoust. Soc. Am. **69**, (6), pp. 1660−1662 (1981).

[34] R. H. Mellen, D. G. Browning, and V. P. Simmons, *Investigation of chemical sound absorption in sea water by the resonator method: Part III*, J. Acoust. Soc. Am. **70**, (1), pp. 143−148 (1981).

[35] R. H. Mellen, D. G. Browning, and V. P. Simmons, *Investigation of chemical sound absorption in seawater by the resonator method: Part IV*, J. Acoust. Soc. Am. **74**, (3), pp. 987−993 (1983).

[36] C. C. Hsu and F. H. Fisher, *Effect of pressure on sound absorption in synthetic seawater and aqueous solutions of $MgSO_4$*, J. Acoust. Soc. Am. **74**, (2), pp. 564−569 (1983).

[37] M. Schulkin and H. W. Marsh, *Sound absorption in Sea Water*, J. Acoust. Soc. Am. **34**, (6), pp. 864−865 (1962).

[38] V, P. Glotov, *Calculation of the relaxation time for the degree of dissociation of magnesium sulfate in fresh and sea water as a function of temperature*, Sov. Phys. Acoust. **10**, pp. 33−38 (1964).

[39] R. H. Mellen and D. G. Browning, *Variability of low-frequency sound attenuation in the ocean: pH dependence*, J. Acoust. Soc. Am. **61**, (3), pp. 704−706 (1977).

[40] X. Qui, *A cylindrical resonator method for the investigation of low-frequency sound absorption in sea water*, J. Acoust. Soc. Am. **90**, (6), pp. 3263−3270 (1991).

[41] F. H. Fisher and V. P. Simmons, *Sound absorption in sea water*, J. Acoust. Soc. Am. **62**, (3), pp. 558−564 (1977).

[42] R. J. Urick, *Principles of Underwater Sound*, pp. 98−99, McGraw-Hill Book Company, New York, 1975.

[43] J. W. S. Rayleigh, *The Theory of Sound*, Dover Publications, Vol. II, pp. 316, 1945.

[44] R. T. Beyer, *Nonlinear Acoustics*, pg 37, Acoustical Society of America, New York (1977).

[45] T. A. Litovitz and C. M. Davis, *Structural and Shear Relaxation in Liquids*, in Physical Acoustics, Vol. II Part A, (W. P. Mason, ed.), Academic Press, New York, 1965.

[46] L. E. Kinsler, A. R. Frey, A. B. Coppens and J. V. Sanders, Fundamentals of Acoustics. John Wiley & Sons, Inc., New York, 2000.

[47] R. J. Urick, *Principles of Underwater Sound*, pp. 105–106, McGraw-Hill Book Company, New York, 1975.

[48] M. Schulkin and H. W. Marsh, *Low-frequency sound absorption in the ocean*, J. Acoust. Soc. Am. **63**, (1), pp. 43–48 (1978).

[49] H. W. Marsh, *Attenuation of Explosive Sounds in Sea Water*, J. Acoust. Soc. Am. **35**, (11), pp. 1837–1838 (1963).

[50] W. H. Thorp, *Analytic Description of the Low-Frequency Attenuation Coefficient*, J. Acoust. Soc. Am. **42**, (1), pp. 270 (1967).

[51] J. R. Lovett, *Geographic variation of low-frequency sound absorption in the Pacific Ocean*, J. Acoust. Soc. Am. **65**, (1), pp. 253–254 (1979).

[52] R. H. Mellen, Chemical sound absorption in the sea, In L. Bjørnø (Ed.), Underwater Acoustics and Signal Processing, D. Reidel Publishing Comp., Holland, pp. 71–80, 1981.

Scattering of Sound

L. Bjørnø[†]

UltraTech Holding, Taastrup, Denmark

Definitions

Inhomogeneities of different kinds are found everywhere in the sea. They are in the water column and along the boundaries, sea surface, and seafloor. They range in size from tiny dust particles and plankton through bubbles and fish schools to rough seabeds and oceanic ridges. Inhomogeneities deviate in density and sound velocity from their environments and influence sound propagation by intercepting and reradiating incident sound energy. When reflected from a plane and smooth surface reradiated sound energy propagates in a preferred direction, the specular direction. Reradiated sound from inhomogeneities of various sizes, materials, and surface qualities is said to be *scattered*, i.e., it propagates in all directions. The total scattered acoustic energy from all inhomogeneities—apart from the expected echo—arriving at a measurement site is the *reverberation*. The reverberation is normally heard as a long, slowly decaying sound in the sea following the emission of a sonar's ping or by the transient signal from an underwater explosion. When reverberation forms a limit on the use of sonar signals, sonar signal reception is *reverberation limited*, where the received echo-to-noise ratio is determined by reverberation. This chapter deals with the concepts of *scattering* and *reverberation*.

5.1 PHYSICS AND PHENOMENA

When a sound wave propagating in water strikes an interface between regions of different characteristic acoustical impedance, ρc, where ρ and c are the media density {unit: kg/m^3} and sound velocity {unit: m/s}, respectively, it may be *reflected*, *refracted*, and *transmitted*, as discussed in Chapter 2. A plane—or locally plane— wave incident on a plane interface is *reflected* coherently in a direction symmetric with its direction of arrival (*specular reflection*), similar to light reflection from a mirror. Reflected sound is one contributor to the formation of an underwater acoustic *echo*. Another contributor is sound *scattered* from objects in the water column, such as bubbles, fish schools, and submarines; from objects on and in the seabed, such as rocks, boulders, mines, pipelines, and gas pockets; and from objects at or near the

[†]30 March 1937—24 October 2015.

Applied Underwater Acoustics. http://dx.doi.org/10.1016/B978-0-12-811240-3.00005-9

sea surface, such as waves, ice floes, and bubble clouds. Due to their shape, these objects scatter sound energy in all directions and therefore strongly influence echo detection. The echo from a submarine received by sonar may be "buried" in noise formed by the scattered acoustic signals from a broad variety of scatterers producing a high reverberation level. Acoustic energy scattered back toward the sonar is called *backscattering*, while scattering in the direction away from the sonar is called *forward scattering*. Most sonar systems are *monostatic*, i.e., where the transmitter and receiver are located in the same position and detect backscattered echoes. For a *bistatic* sonar, where the transmitter and the receiver are located in different positions, scattering in any direction, including forward scattering may be exploited for echo detection.

In addition to an object's characteristic acoustical impedance, the surface quality of the object has a substantial influence on the amount of energy scattered. For a smooth surface, most energy is *coherently* reflected, while a rough surface scatters a substantial amount of energy *incoherently*. The *correlation distance*, which is the distance between two points on the rough surface showing similar scattering characteristics, influences the incoherent energy's amplitude and propagation direction. The surface roughness influence on the *surface scattering* magnitude is described by the dimensionless *Rayleigh parameter*, R_a, which is the ratio between the root mean square (rms) deviation of the roughness heights, h {unit: m}, from the average height, called the *rms-roughness height* {unit: m}, and the acoustic wavelength λ {unit: m} through the expression discussed by Brekhovskikh and Lysanov [1]:

$$R_a = 2hk \cos \theta, \tag{5.1}$$

where $k = 2\pi/\lambda$ {unit: 1/m} is the acoustic wave number and θ is the angle of incidence, i.e., the angle between the direction of the incoming sound wave and perpendicular to the surface, as shown in Fig. 5.1.

For $R_a \ll 1$, the influence of the surface roughness is small with only slight scattering of sound from the surface taking place and most of the sound energy is coherently reflected in the specular direction. If $R_a \gg 1$, the influence of the roughness is strong, and most sound energy is scattered incoherently in all directions of the upper half space including back toward the source (backscattering).

Eq. (5.1) shows that the relative importance of the specular (coherent) reflected and scattered (incoherent) components of the acoustic field depends on the ratio between a parameter characterizing surface roughness and the acoustic signal's

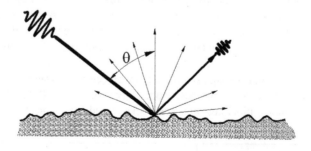

FIGURE 5.1

Plane sound wave incident on and scattered by a rough surface.

wavelength. The degree of roughness on a surface can cover a wide scale of amplitudes with several roughness scales found on the same surface. For example, on the sea surface centimeter scale wind produced capillary waves are superimposed on meter scale swells. On a sandy seabed, flow produced sand ripples with a roughness height of centimeters may be superimposed on a meter scale rocky sub-bottom. Frequently, the roughness height distribution is assumed *isotropic*, and the spatial height distribution spectrum is direction independent; however, in the previous example flow-generated sand ripples possess directivity.

In addition, the water column contains scatterers, such as free gas bubbles of various magnitudes, biomass ranging in size from micron scale like phytoplankton and zooplankton through centimeter scale represented by a broad variety of fish types to meter scale marine mammals. Several biomass representatives in the sea have gas bubbles trapped on or in them, such as tiny bubbles trapped on the surface of plankton and shrimp and the gas in the fish's swim bladder, as discussed by Urick [2] and Medwin and Clay [3]. These gas bubbles produce a strong echo when hit by a sound wave, in particular when fish gather in schools and the school contributes to *volume scattering*, as shown in Fig. 5.2.

The fundamental parameter characterizing surface and volume scattering is the *scattering strength*. The scattering strength, S, is the ratio in dB between the intensity of the sound scattered by a unit surface (1 m^2) or by a unit volume (1 m^3), measured at a distance, 1 m, from the acoustic center of the scattering surface/volume, and the incident acoustic wave intensity. The scattering strength S {in dB for 1 m^2 or 1 m^3} is written as:

$$S = 10 \ \log\left(\frac{I_{sca}}{I_{inc}}\right) \tag{5.2}$$

where I_{sca} {unit: W/m^2} denotes the sound intensity scattered from a unit surface or a unit volume measured at a distance of 1 m from the acoustic center of the

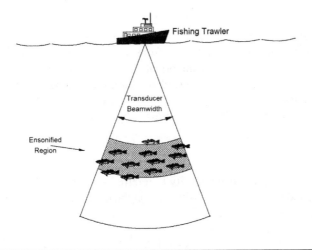

FIGURE 5.2

Ensonified fish school contributing to volume scattering.

surface/volume, and where I_{inc} {unit: W/m^2} is the incident plane acoustic wave intensity at the scattering surface/volume. Depending on the source of scattered acoustic energy, *surface reverberation* or *volume reverberation* is defined in terms of the scattering strength.

The *acoustical scattering cross section*, σ_s, is also a basic parameter in surface/volume scattering. The scattering cross section {unit: m^2} is the ratio of the acoustic power scattered by the surface/volume, measured 1 m from the acoustic center and the incident acoustic intensity. By assuming that all incident acoustic power is scattered isotropically into the space around the scattering surface/volume, an *effective acoustical scattering cross section* {unit: m^2} is defined as:

$$\sigma_{se} = 4\pi R_l^2 \left(\frac{I_{sca}}{I_{inc}}\right),$$ (5.3)

where $R_l = 1$ m.

The *total acoustic scattering cross section*, σ_{st}, refers to the total acoustic power scattered over all angles. This scattering can be either isotropic or anisotropic. Depending on scatterer characteristics and the incident acoustic wave's frequency, the scattering cross section may be larger or smaller than the scatterer's geometrical cross section. For example, the total acoustic scattering cross section from a resonant gas bubble in water may be 10^3 times the geometrical cross section.

The scattering surface or volume also absorbs acoustic power through dissipation processes. The *total acoustic absorption cross section*, σ_{at}, is the ratio between the total acoustic power absorbed and the incident acoustic wave intensity. The incident sound wave loses power through scattering and absorption during the scattering process with the power extracted from the available acoustic power in the incident wave. The *total acoustic extinction cross section*, σ_{et}, represents the ratio between the power lost from the incident acoustic wave by scattering and absorption processes and the incident wave's intensity. The total acoustic extinction cross section is:

$$\sigma_{et} = \sigma_{st} + \sigma_{at}.$$ (5.4)

For a sonar system operating in monostatic mode the *echo intensity*, which is the total acoustic signal intensity returned from a target, consisting of reflected and backscattered signal intensity, is important. Ten times the log of the ratio of the echo intensity, I_e, {unit: W/m^2} received at a monostatic sonar system (or a hydrophone in a bistatic mode) and a reference intensity, I_r, where $I_r = 1$ W/m^2, is the *echo level EL* {in dB rel. 1 W/m^2}, expressed by:

$$EL = 10 \ \log\left(\frac{I_e}{I_r}\right).$$ (5.5)

The ratio between the echo intensity I_e, referred to a distance 1 m from the center of the target, and the incident intensity, I_{inc}, on the target forms the *target's intensity reflection coefficient*, $\boldsymbol{R}_t$, written as:

$$\boldsymbol{R}_t = \left(\frac{I_e}{I_{inc}}\right).$$ (5.6)

Targets can be of different nature and have strongly varying structural qualities. Typical targets include:

- the seafloor ensonified in vertical or oblique directions;
- mines and wrecks on or in the seabed;
- cables and tubes in the seabed;
- mineral nodules on the seafloor;
- single fish or fish schools;
- autonomous underwater vehicles (AUVs) and submarines in the water column;
- surface vessels and bubble clouds on or near the sea surface.

Depending upon the acoustic signal's frequency content, some targets are small compared with the acoustic wavelength and the target will behave like "a point target," where the nature of the target is of minor importance compared with the acoustic field characteristics, the beam width, the signal's time history, and the distance to the target. Other targets have dimensions bigger than the transverse dimension of the acoustic beam, and only a part of the target is ensonified by the beam, for example, the fish school in Fig. 5.2. The magnitude of the ensonified part of the target determines the target reflection coefficient.

The concept of *target strength*, *TS* {unit: dB} is defined through R_t as:

$$TS = 10 \log\left(\frac{I_e}{I_{inc}}\right), \tag{5.7}$$

which depends on the characteristics of the target—materials, geometry, size, and surface structure—and on the features of the acoustic signal, its frequency, angle of incidence on the target, and beam width. The angle at which the target is seen from the sonar position, the aspect angle, may even lead to positive dB values for *TS*, while in most cases negative dB values are obtained. For bistatic sonar operation the orientation of the target relative to the receiver influences the *TS* values.

Target strength measurements are mostly done in terms of the peak pressure or average intensity of the incident and backscattered signals reduced to their values at 1 m distance from the acoustic center of the target. Because in most cases it is impossible to carry out measurements 1 m from the target, the normal practice is to measure the backscattered signal at long range and then use *the sonar equation* (discussed in Chapter 1) to reduce the signal intensity to the value be expected at 1 m from the acoustic center of the target using the expression:

$$EL = SL - 2TL + TS. \tag{5.8}$$

where *EL* is the echo level {unit: dB}, *SL* is the source level {unit: dB}, 2 *TL* is the two-way transmission loss {unit: dB}. From Eq. (5.8), *TS* can be determined when *EL* and *SL* are measured, and *TL* is calculated assuming spherical or cylindrical spreading plus absorption, or is measured by a calibrated transponder. The accurate determination of *TL* is critical for target strength measurements since it is one of the main reasons for variations in measured target strengths. The target strength of smaller objects at shorter ranges may be determined by comparing the echo level from the target with the echo level from a reference target; for instance, a solid sphere of known diameter and of known material. From the measured acoustic signal peak pressure the target strength is:

$$TS_{\text{peak}} = 10 \ \log\left(\frac{p_{bs}^2}{p_{inc}^2}\right), \tag{5.9}$$

where p_{bs} {unit: Pa} and p_{inc} {unit: Pa} denote, respectively, the backscattered signal peak pressure amplitude and the peak pressure in the signal incident on the target. The TS_{peak} values determined by Eq. (5.9) depend strongly on the acoustic signal duration and type. From lower values at shorter signal lengths, it reaches a nearly constant value when the signal length becomes comparable to the target's characteristic dimensions.

The *monostatic mode* is the most frequently used sonar configuration. Echo sounders used for fish detection, bathymetric sounding, and anti-collision measures operate in the monostatic mode, using backscattered signals from ensonified objects. An effective acoustic *backscattering cross section*, σ_{bs}, may be defined similar to Eq. (5.3), where I_{sca} then is the backscattered acoustic intensity, I_{bs}, referred to a distance of 1 m from the center of the ensonified object. If the scattering is omnidirectional, the backscattering cross section, σ_{bs}, is related to the total acoustic scattering cross section σ_{st} by $\sigma_{bs} = \sigma_{st}/4\pi$. Similar to the scattering strength S in expression (5.2), a *backscattering strength S_b* {in dB for 1 m^2 or 1 m^3} may be defined as:

$$S_b = 10 \ \log\left(\frac{I_{bs}}{I_{inc}}\right) \tag{5.10}$$

Depending on the sonar beam width, signal's time duration and distance between the sonar and the target, some targets will be so small, that they are totally engulfed by the acoustic beam. These targets are said to form "point targets," where their scattering qualities are only related to the target's physical and geometrical qualities. Other targets may have such a size, that only parts of the target are ensonified. Single fish and fish schools may be said to be point targets, while depending on the acoustic field, single fish and in particular fish schools may have such an extent, that they are not fully ensonified and behave like extended targets, as shown in Fig. 5.2. The target strength from fish schools receives contributions from multiple scatterers in the ensonified volume. Objects that are on, or in, the seabed may act as point targets, while the seabed and sea surface form extended scattering targets. Section 5.2 discusses scattering from single point-like objects and from multiple objects. Section 5.3 presents scattering from extended, nearly plane, rough surfaces. Section 5.4 presents the theoretical basis for calculating scattering from rough surfaces. Section 5.5 discusses scattering from curved, rough surfaces. Section 5.6 discusses reverberation. Section 5.7 provides a table that lists the symbols and abbreviations used in this chapter and the page that the symbol or abbreviation is defined and first appears. This section is followed by the list of references.

5.2 SCATTERING FROM POINT-LIKE OBJECTS

An object may be said to be perfectly engulfed by an acoustic beam, when all object dimensions are smaller than the first Fresnel zone of the beam. The first Fresnel zone forms the central part, with the same sign of the phase, around the acoustic axis of the diffraction patterns of a beam transmitted by a plane, circular transducer, for instance. Single objects, such as, solid or elastic spheres and other canonically shaped objects, bubbles, single fish, underwater vehicles, and mines may be smaller than the first Fresnel zone and form point-like objects.

5.2.1 SINGLE OBJECTS

5.2.1.1 Rigid and Elastic Spheres

When a *rigid sphere* with radius, a {unit: m}, is ensonified by a high-frequency plane acoustic wave, the apparent cross section, seen from the acoustic source, is independent of the wave's angle of incidence, and has the value πa^2. The acoustic power intercepted by the rigid sphere is $\pi a^2 I_{inc}$. If this power is scattered isotropically, the backscattered intensity, I_{bs}, 1 m from the center of the sphere is $I_{bs} = \pi a^2 I_{inc}/4\pi r^2$, where $r = 1$ m. The rigid sphere's backscattering cross section, σ_{bs} {unit: m^2}, is:

$$\sigma_{bs} = \frac{a^2}{4},$$

(5.11)

where σ_{bs} is independent of the frequency. The simple expression (5.11) is valid for high ka values ($ka \gg 1$), which are found in the *geometrical scattering region*. The target strength, *TS*, of the rigid sphere is:

$$TS = 10 \log\left(\frac{\sigma_{bs}}{A}\right),$$

(5.12)

where $A = 1$ m^2.

A 0.05 m radius rigid sphere has a backscattering cross section $\sigma_{bs} = 6.25 \cdot 10^{-4}$ m^2, and a target strength $TS = -32$ dB. A 2 m radius rigid sphere has a 0 dB target strength.

Solid metal spheres are frequently used for practical sonar calibrations due to their well-defined and frequency-independent backscattering cross section and target strength. However, even minor deviations from a spherical shape can influence the target strength. Metal spheres are made of materials, such as those shown in Table 5.1, whose elastic response can influence the calibration process through three effects:

1. penetration of longitudinal waves into the sphere materials and the reflection of these waves from the back of the sphere;
2. excitation of circumferential interface waves on the sphere's surface, which are similar to Rayleigh waves on the free surface of a solid, and which propagate around the sphere and retransmit signals into the water;

Table 5.1 Density and Elastic Qualities of Materials Used in Spheres for Measuring Backscattering Cross Sections

Material	Density (kg/m^3)	Poisson's Ratio	Compressional Velocity (m/s)	Shear Velocity (m/s)
Beryllium	1870	0.05	12890	8880
Tungsten	19300	0.37	5410	2640
Glass (Pyrex)	2320	0.18	5640	3280
Iron (Armco)	7850	0.29	5960	3240
Steel (stainless)	7900	0.28	5790	3100
Copper (anneal)	8930	0.35	4760	2325
Titanium	4500	0.36	6070	3125
Brass (yellow)	8600	0.37	4700	2110
Lead (rolled)	11400	0.44	2160	700
Aluminum	2700	0.33	6420	3040
Polystyrene	1060	0.34	2350	1120

Data from CRC Handbook of Chemistry and Physics, 56th Ed. 1975/76.

3. formation of creeping waves, which are diffracted waves formed at the edge of the sphere's geometrical shadow zone and which propagate around the sphere, as discussed by Neubauer [4].

The three wave types reradiate acoustic energy into the water around the sphere and to the sonar under calibration. The second wave is dispersive and may lead to scattering resonances, as discussed by Überall et al. [5]. The creeping wave type in particular contributes to the oscillations in the Mie scattering region shown in Fig. 5.3.

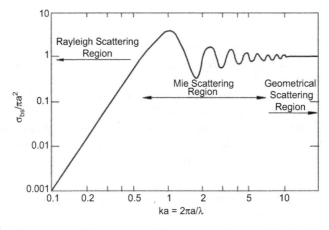

FIGURE 5.3

Ratio of backscattering cross section, σ_{bs}, to the geometric cross section of a rigid, fixed sphere as a function of *ka*.

For lower frequencies, when the wavelength is comparable to or larger than the radius of the sphere (i.e., $ka \ll 1$), scattering is similar to radiation from a source that behaves like the superposition between a monopole and a dipole [3]. For this case Rayleigh [6] has shown that for a fixed, rigid sphere of radius a, the backscattering cross section is:

$$\sigma_{bs} = \frac{25}{36}k^4 a^6,$$

(5.13)

which shows that in the *Rayleigh scattering region, $ka \ll 1$,* the backscattering cross section is dependent on frequency to the fourth power. The scattering cross section is much smaller than the sphere's geometrical cross section, πa^2, since the sound waves that bend around the rigid sphere are nearly unaffected. The limit of validity for Eq. (5.13) is roughly $ka = 1$, when the wavelength is equal to the circumference of the sphere. The transition region between Rayleigh scattering and geometrical scattering, $1 < ka < 10$, is dominated by the interference between the directly reflected wave and a refracted wave "creeping" around the sphere, as discussed by Bjørnø and Bjørnø [7]. This leads to the oscillations in the *TS* values shown in Fig. 5.3. Targets with a more complex geometry like sand grains are often assumed to follow rigid sphere scattering, as discussed by Bjørnø and Bjørnø [7] and by Bjørnø [8].

For an *elastic sphere*, Rayleigh [6] has shown that the influence of the elasticity and density of the sphere, relative to the values for water, on the backscattering cross section σ_{bs} in the *Rayleigh scattering region ($ka \ll 1$)* could be expressed through:

$$\sigma_{bs} = (ka)^4 \left[\frac{\rho_s c_s^2 - \rho_w c_w^2}{3\rho_s c_s^2} + \frac{\rho_s - \rho_w}{2\rho_s + \rho_w} \right]^2 a^2,$$

(5.14)

where ρ and c are density and sound velocity, respectively, and the subscripts s and w denote the sphere and water, respectively. When the sphere's characteristic impedance, $\rho_s c_s$, is large compared to the characteristic impedance of water, Eq. (5.14) reduces to Eq. (5.13). In the case of a gas bubble in water, where $\rho_s c_s \ll \rho_w c_w$, the backscattering cross section σ_{bs} is:

$$\sigma_{bs} = k^4 a^6 \left[\frac{\rho_w c_w^2}{3\rho_s c_s^2} \right]^2,$$

(5.15)

When the air and seawater material parameters are inserted into Eq. (5.15), the target strength of an air bubble in seawater is found to be 74 dB higher than that of a rigid sphere with the same radius at the same acoustic frequency.

5.2.1.2 Gas Bubbles

Gas bubbles present in water have a profound effect on sound scattering. Gas bubbles are formed by rain, hail, and snow falling on the sea surface, as well as by the breakdown of organic materials in the seabed. They are present in large numbers in the water column when released by the biomass, or produced by a gas leak—frequently carrying methane—from the seabed. Waves breaking on the sea surface pump significant amounts of air into the water column to depths up to 15 m and form bubble clouds and plumes in the space near the sea surface.

Near resonance, the gas bubble influence on scattering is substantially enhanced. The scattering cross section σ_s of a *single* gas bubble of radius a for $ka \ll 1$ is given by:

$$\sigma_s = \frac{4\pi a^2}{\left(\frac{f_r^2}{f^2} - 1\right)^2 + \delta^2}, \tag{5.16}$$

where f_r denotes the bubble resonance frequency and δ describes the attenuation effect, which is a function of the frequency f and contributions received from the reradiation of acoustic energy, heat conduction due to the polytropic compression and expansion of the enclosed gas, and viscous dissipation due to viscous forces acting at the gas—water interface, as discussed by Devin [9]. According to Medwin and Clay [3] a simple expression for δ at bubble resonance is $\delta_r = 0.025 f_r^{0.33}$ {f_r in kHz}, which is valid for bubble resonance frequencies f_r in the range 1—100 kHz, as shown in Fig. 5.4.

At resonance, when $f = f_r$, the total acoustical resonance scattering cross section of a single bubble is:

$$\sigma_{sr} = \frac{4\pi a^2}{\delta_r^2}. \tag{5.17}$$

The total scattering cross section for a single bubble of diameter 50 μm and resonance frequency $f_r = 65$ kHz, is nearly 400 times larger than the bubble's geometrical cross section.

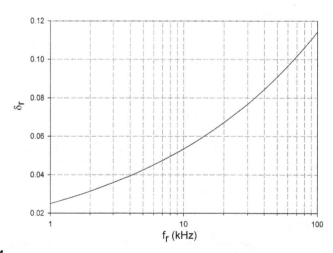

FIGURE 5.4

The attenuation constant δ_r as a function of the resonance frequency f_r for a single bubble.

The resonance frequency of a single bubble is a function of radius, thermodynamic qualities of the gas in the bubble and water depth, where the bubble is pulsating. A simplified expression for the resonance frequency, f_r {unit: Hz} of a single bubble is:

$$f_r = \frac{\sqrt{\frac{3\gamma p_0}{\rho}(1 + 0.1z)}}{2\pi a},$$ (5.18)

where p_0 {unit: Pa} and ρ {unit: kg/m^3} are the ambient pressure acting on the gas bubble at the sea surface and the gas density, respectively, and z {unit: m} is the water depth at which the bubble is pulsating. γ denotes the ratio of specific heats at constant pressure, c_p, and constant volume, c_v. For a diatomic gas, such as air, $\gamma = 1.4$.

The target strength of a single bubble is:

$$TS = 10 \; \log \left[\frac{a^2}{\left(\frac{f_r^2}{f^2} - 1\right)^2 + \delta^2} \right],$$ (5.19)

since the backscattering cross section, σ_{bs}, of the single bubble is $\sigma_{bs} = \sigma_s/4\pi$.

5.2.1.3 Single Fish

When examining scattering from a single fish totally engulfed by the acoustic beam, the gas-filled swim bladder is a significant contributor to the fish's acoustic cross section at and near resonance. Scattering at resonance by swim bladder bearing fish is a major cause of volume reverberation in the ocean at frequencies between 1 and 25 kHz. The swim bladder is used by the fish for adjusting its buoyancy at depth, and there are fish types with closed as well as open (to the stomach) swim bladders.

The swim bladder volume is normally 4–5% of the total fish volume, but the strong impedance mismatch between the gas-filled volume in the fish tissue and seawater contributes substantially to the scattering quality of the swim bladder. Depending on their size and form, which varies by species, the resonance frequency can range from below 500 Hz to above 2500 Hz. The water depth at which the fish is swimming influences the resonance frequency. Increasing water depth leads to an increased resonance frequency. The fish tissue, which increases the stiffness of the swim bladder over that of a free bubble also increases the resonance frequency.

Several models have been developed in order to describe the scattering qualities of fish, most assume the scattering takes place due to bubbles surrounded by spherical shells of various elastic or viscoelastic qualities, as discussed by Love [10,11]. Since the frequency range normally used to study individual fish is broad, studies have included broadband technology, such as the broadband parametric acoustic arrays as reported by Kjærgaard et al. [12]. Reeder et al. [13] performed studies on the influence by anatomical components of the fish, skull, vertebrae, etc. on fish scattering qualities. X-ray and CT scan (swim bladder) images were digitized and incorporated into two scattering models to verify experimental data in the 40–95 kHz frequency range.

The fish size, represented by its length L, and to some degree the acoustic frequency or wavelength are the most important quantities for determining the order of magnitude of target strength TS of individual fish, as discussed by Medwin and Clay [3] and Love [10,11]. A useful relation between the dorsal aspect target strength, $TS_{fish,d}$ {unit: dB relative to a 2 m diameter rigid sphere}, L {unit: cm}, and λ {unit: cm} is:

$$TS_{fish,d} = 10 \log \sigma_{bs} = 19.1 \log L + 0.9 \log \lambda - 69.1. \qquad (5.20)$$

Eq. (5.20) is valid for $0.7 < L/\lambda < 90$. Fish without a swim bladder, such as the mackerel, may have a target strength 10–15 dB lower than a fish with a swim bladder.

According to Love [10] an expression for maximum side aspect target strength, $TS_{fish,s}$ {unit: dB relative to a 2 m diameter rigid sphere} as a function of L {unit: cm} and λ {unit: cm} is:

$$TS_{fish,s} = 22.8 \log L - 2.8 \log \lambda - 72.9, \qquad (5.21)$$

which is valid for $1 < L/\lambda < 130$. Fig. 5.5 shows the dorsal- and side-aspect target strengths as a function of fish length calculated using Eqs. (5.20) and (5.21) at 38 kHz.

5.2.1.4 Canonically Shaped Objects

Plane acoustic wave backscattering from canonical and arbitrarily shaped objects has been studied over the past several decades and due to the great number of shapes available for objects in water, only the simplest shapes have been studied in depth as discussed by Urick [2] and reported by Bjørnø and Bjørnø [7] and Ramirez and Bjørnø [14]. Through Eq. (5.12) the backscattering cross section can be used to determine the target strength for a number of simple shapes, when the wavelength

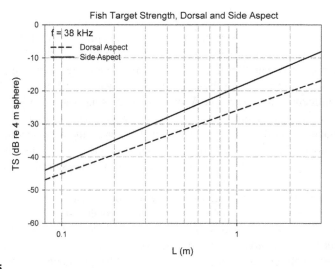

FIGURE 5.5

Target strengths of individual fish as a function of fish length for $f = 38$ kHz.

λ is either much smaller than any characteristic dimension a of the object (i.e., for $ka \gg 1$) or much larger than a (i.e., for $ka \ll 1$) as reported by Urick [2]. In practice $ka \gg 1$ means that $a > 5\lambda$, and $ka \ll 1$ means $a < \lambda/20$. Based on data provided by Urick [2], Table 5.2 provides backscattering cross sections for an incidence angle normal to the target and Tables 5.3 and 5.4 provide the backscattering cross sections for an oblique incidence angle and an incidence angle parallel to the target's symmetry axis.

For more involved targets and for longer sonar pulses, the target strength may be found by dividing the target into components with geometrical shapes like the canonical shapes above and by adding the contributions from all components.

Table 5.2 Backscattering Cross Section for a Normal Incidence Angle

Surface Type	σ_{bs}	Comments
Convex	$\dfrac{a_1 a_2}{4}$	a_1 and a_2 are the surface's principal radii of curvature
Infinitely long, thick cylinder	$\dfrac{ar}{2}$	a and r denote the cylinder radius and range, respectively. $r > a$ and $ka \gg 1$
Infinitely long, thin cylinder	$\dfrac{9\pi^4 a^4 r}{\lambda^2}$	a and r denote the cylinder radius and range, respectively. $r > a$ and $ka \ll 1$
Finite length, thick cylinder	$\dfrac{aL^2}{2\lambda}$	L is the cylinder length, range $r > L^2/\lambda$, and $ka \gg 1$
Circular plate	$\left(\dfrac{4}{3\pi}\right)^2 k^4 a^6$	Radius a, for $ka \ll 1$, where k is the wave number

Table 5.3 Backscattering Cross Section for an Oblique Incidence Angle

Surface Type	σ_{bs}	Comments
Finite cylinder	$\dfrac{aL^2}{2\lambda} \left[\dfrac{\sin(kL \sin\theta)}{kL \sin\theta} \right]^2 \cos^2\theta$	θ measured from the normal to a cylinder with radius, a, and length, L
Rectangular plate	$\left(\dfrac{bd}{\lambda}\right)^2 \left[\dfrac{\sin(kb \sin\theta)}{kb \sin\theta} \right]^2 \cos^2\theta$	Side lengths b and d, where the range $r > b^2/\lambda$, $kb \gg 1$, $b > d$, and θ is measured from the normal in the plane containing b
Circular plate	$\left(\dfrac{\pi a^2}{\lambda}\right)^2 \left[\dfrac{2J_1(2ka \sin\theta)}{2ka \sin\theta} \right]^2 \cos^2\theta$	θ measured from the normal to circular plate of radius a, for $r > a^2/\lambda$, $ka \gg 1$, and $J_1()$ is the Bessel function of order 1
Conical tip	$\left(\dfrac{\lambda}{8\pi}\right)^2 \dfrac{\tan^4\varphi}{\left(1 - \dfrac{\sin^2\theta}{\cos^2\varphi}\right)^3}$	θ measured from the axis of the conical tip and $\theta < \varphi$ where φ is the half angle of the cone
Triangular corner	$\left[\dfrac{L^2}{\lambda(3)^{1/2}} (1 - 7.6 \cdot 10^{-4}\theta^2) \right]^2$	θ measured from the axis of symmetry of a triangular corner with the side lengths L

Table 5.4 Backscattering Cross Section for an Incidence Angle Parallel to the Target's Symmetry Axis

Surface Type	σ_{bs}	Comments
Ellipsoid	$\left(\dfrac{bc}{2a}\right)^2$	Wave incident parallel to axis a of an ellipsoid with semi-major axis a, b, and c for ka, kb, kc $\gg$ 1, and for the range $r \gg a, b, c$
Hemispherical end cap	$\dfrac{a^2}{4}\left[\begin{array}{l}4(1-n)\sin^2(nka)+ \\ n^2\left\{1-\dfrac{\sin(2nka)}{nka}+\left(\dfrac{\sin(nka)}{nka}\right)^2\right\}\end{array}\right]$	a is radius of curvature and na is the axial height of the end cap
Truncated cone	$\dfrac{2}{9\lambda}(H^{1.5}-h^{1.5})^2\dfrac{\sin\theta}{\cos^4\theta}$	Original cone height H {unit: m}, truncation height h_t {unit: m} and vertex angle θ

For shapes like cubes, tetrahedrons, and octahedrons it has been shown by Bjørnø and Bjørnø [7] and Ramirez and Bjørnø [14] that the backscattering cross section may be obtained by multiplying σ_{bs} for a sphere with a form function f_∞, which has to be determined experimentally, as discussed in Section 5.5. Experiments conducted by Bjørnø and Bjørnø [7] have in particular shown a strong dependence of f_∞ on the angle of incidence.

5.2.1.5 Submarines

Two World Wars and the Cold War have focused the interest on backscattering, and thus the target strength, of submarines. While more recent measurements are classified, data from World War 1 and in particular World War 2 are publicly available, see Urick [2]. Backscattering from submarines receives contributions from the water-backed casings at frequencies above 20 kHz; the air-backed pressure hull and conning tower; the fins, stabilizers, rudder, and propeller; and objects inside the external hull. Resonance modes in parts of the submarine excited by an active sonar may contribute to backscattering. Attempts to minimize backscattering have included optimizing the submarine shape or lining the hull with anechoic materials. The *anechoic lining materials* that are used to reduce the target strength are frequently rubber based, such as butyl rubber, and loaded with metal and cork powder, as discussed by Bjørnø and Kjeldgaard [15]. The effect of lining materials is frequency limited and most effective at elevated frequencies, where the material absorption is high. At low frequencies, i.e., below 5 kHz, lining materials are of limited value due to weight and drag problems since to obtain an effective target strength reduction the lining should be at least $\lambda/2$ thick.

A submarine's backscattering is very aspect angle dependent, and this is also true for torpedoes and cylindrically shaped mines. A first approximation for the submarine's beam aspect target strength is obtained by assuming normal sound incidence on a finite length cylinder and using the backscattering cross section given in Table 5.2. An 8 kHz sonar signal normally incident on a 60 m long cylinder with a 4 m radius has a target strength, $TS \sim +46$ dB. For a submarine with a real hull shape this should be considered a maximum value, which is reduced by aspect angle, hull taper, and other deviations from the cylindrical shape. Characteristic dimensions that influence submarine target strengths are given in Table 5.5.

Table 5.5 Characteristic Dimensions and Displacements of Submarines From Various Countries, Extracted From Jane's Fighting Ships 2004−2005 [16]

Submarine Class	Submarine Dimensions (m)	Displacement (Surfaced) (tons)	Country
Kronborg	57.6 × 5.7 × 5.5	1015	Denmark
Saelen	47.2 × 4.7 × 3.8	370	Denmark
Rubis Amethyste	73.6 × 7.6 × 6.4	2410	France
Le Triomphant	138 × 12.5/17 × 12.5	2640	France
Type 206 A	48.6 × 4.6 × 4.5	450	Germany
Type 212 A	55.9 × 7 × 6	1450	Germany
Papanikolis (Typ. 214)	65 × 6.3 × 6.6	1700	Greece
Sanso	64.4 × 6.8 × 5.6	1476	Italy
Oyashio	81.7 × 8.9 × 7.4	2750	Japan
Harushio	77.0 × 10.0 × 7.7	2450	Japan
Typhoon	171.5 × 24.6 × 13	18500	Russia
Delta III	160 × 12 × 8.7	10550	Russia
Oscar II	154 × 18.2 × 9	13900	Russia
Akula	110 × 14 × 10.4	7500	Russia
Sierra	107 × 12.5 × 8.8	7200	Russia
Kilo	72.6 × 9.9 × 6.6	2325	Russia
Vanguard	149.9 × 12.8 × 12	15900 (dived in)	UK
Trafalgar	85.4 × 9.8 × 9.5	4740	UK
Swiftsure	82.9 × 9.8 × 8.5	4400	UK
Astute	97.0 × 11.27 × 10	6500	UK
Ohio	170.7 × 12.8 × 11.1	16600	USA
Virginia	114.9 × 10.4 × 9.3	7800 (dived in)	USA
Seawolf	107.6 × 12.9 × 10.9	8060	USA
Los Angeles	110.3 × 10.1 × 9.9	6082	USA

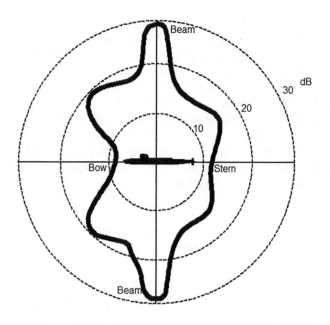

FIGURE 5.6

Submarine target strength "butterfly"-like aspect dependence.

A "butterfly"-like pattern is frequently ascribed to the aspect dependence of the target strength of a submarine, as discussed by Urick [2] and shown in Fig. 5.6, with peak values in the beam direction. Bow and stern directions show dips in target strength due to the influence of hull shadowing effects and the submarine's wake. However, this "clean" butterfly pattern is rarely found. Instead, the target strength aspect dependence shows substantial variations that for the same submarine are frequently time dependent, due to strong influence of phase variations in backscattered signals from various parts of the submarine. The submarine target strength is also influenced by the range to the target, also discussed by Urick [2], where the submarine target strength is frequently found to be smaller at shorter than at longer ranges, by the duration and bandwidth of the sonar signal, and by the potential anechoic lining of the submarine hull.

Backscattering from *torpedoes* and *mines* is strongly influenced by size, shape, structure, and construction materials. Mine shapes may be approximated by spheres or cylinders with hemispherical ends, while torpedoes in general are cylindrical with rounded or flat noses. Therefore, backscattering cross sections of mines and torpedoes to a large extent can be approximated by backscattering expressions for simple geometrical shapes, such as spheres, finite length cylinders, and circular plates. For a 20 kHz continuous wave (CW) signal a torpedo with a 0.5 m diameter, 6 m length, and a flat nose part has an approximate beam aspect target strength of +17.8 dB, and nose aspect target strength of +8.4 dB. If the torpedo nose is rounded (hemispherical) with the same radius of curvature as the cylindrical body, the nose aspect target strength is -18 dB. If the radius of curvature is 0.3 m and the axial height of the end cap is 0.15 m (i.e., $na = 0.5$), the nose aspect target strength would be -21 dB.

5.2.2 MULTIPLE OBJECTS

When the target dimensions are much larger than the first Fresnel zone, the target is no longer be engulfed by the sound beam. This type of target can consist of a large number of single objects, which are only partly insonified by the beam. Examples include schools of fish, bubble clouds, ship wakes, deep scattering layers (DSLs), and suspended sediment particles near the seabed. These all contribute to *volume scattering* due to their presence and spatial extend in the water column. The backscattering cross section of a target consisting of multiple objects not fully insonified by the acoustic beam is governed by the size of the target part insonified by the beam and by the target's backscattering strength per unit volume. The backscattering from a unit volume in a volume distribution of individual scatterers can be considered to be the incoherent sum of contributions from each scatterer present in the target's unit volume $\{1 \text{ m}^3\}$. For a given frequency this backscattering depends on the dimensions, shape of the scatterers, and on the density, sound velocity, structure, and other material qualities of the scatterer. If the individual scatterers are smaller than the wavelength of the acoustic signal, the shape dependence starts to vanish. For a specific frequency and if scatterers are of the same type, i.e., the same structure and consist of the same materials, but are of variable size, the average volume backscattering cross section σ_{bs} for the ensemble of scatterers in a unit volume will depend on a function of their dimensions $\psi(a)$, where a is a characteristic dimension, the average number N_s of scatterers within a the target's unit volume and on the probability density function $f_p(a)$ for the occurrence of individual scatterers with dimension a. The backscattering cross section is the weighted sum of the individual contributions from all scatterers in the unit volume:

$$\sigma_{bs} = N_s \int_0^\infty \psi(a) f_p(a) \, da. \tag{5.22}$$

Eq. (5.22) assumes that individual acoustic intensity contributions from scatterers are mutually independent and can be summed directly. This assumption is true if.

1. multiple scattering, in which signals scattered by one scatterer are rescattered by one or more other scatterers, is negligible;
2. shadow effects where one scatterer is forming a shadow for the acoustic intensity scattered by another scatterer is also negligible.

Studies by Bjørnø and Bjørnø [17] have shown that if the average distance between the scatterers is larger than $10a$ and larger than 10λ, where λ is the wavelength of the acoustic signal, the two conditions may be considered satisfied.

Most volume backscattering is caused by biological organisms and a quantity frequently used to describe volume backscattering is the *column strength*. When the height of the water column from the seabed to sea surface is H, and the backscattering strength of an average unit volume of dimension $1 \times 1 \times 1$ m, is denoted by S_{bv}, the *column backscattering strength* S_{bc} is:

$$S_{bc} = S_{bv} + 10 \log H. \tag{5.23}$$

5.2.2.1 Fish Schools

Fish schools in general have a broad variety of shapes and sizes depending on the fish species. The vertical dimension may range from one to several meters and the horizontal dimensions can exceed tens of meters. Since a fish school normally comprises fish of the same type and size, i.e., the same age, the backscattering cross section of a unit volume of the fish school is based on a summation of the number of individual backscattering contributions of the same magnitude. If the number of fish, of nearly the same size, per m^3 is N and Eq. (5.20) is used for the dorsal aspect target strength of each fish, the target strength per unit volume for a fish school, TS_{sch}, is:

$$TS_{sch} = 10 \log N\sigma_{bs} = 19.1 \log L + 0.9 \log \lambda - 69.1 + 10 \log N, \tag{5.24}$$

where σ_{bs} is the *individual fish backscattering cross section*. In Eq. (5.24) L {unit: cm} is the fish length and λ {unit: cm} is the echo-sounder frequency. Eq. (5.24) is based on the assumption that shadowing effects and multiple scattering are negligible. Depending on the fish species these assumptions may not always be true. The interplay between the fish length L and number N per m^3 may lead to decreasing target strength for an increasing fish size, as the number of fish per unit volume decreases. A frequently used echo-integration technique for assessing the number and size of fish in a fish school is based on Eq. (5.24). An in-depth exposition of fish target strength is provided by Foote [18].

5.2.2.2 Bubble Clouds

Clouds of bubbles are formed in the sea by waves breaking on the sea surface, and the clouds, which include microbubbles, are entrained into a depth of up to 15 m or more (maximum 25 m as presented in Trevorrow [19]) below the sea surface due to turbulence and Langmuir circulations, as discussed by Thorpe [20]. Microbubbles are the small bubbles that remain after the large bubbles formed by breaking waves have risen to the sea surface. Thorpe [20] has found that the bubble density at the sea surface is proportional to the wind speed cubed. Also, wakes from ships form bubble clouds comprising bubbles over a very broad range of sizes. In particular, bubbles of hydraulic diameters between 10 and 400 μm are important in sound propagation and attenuation in seawater. The movement of the bubbles in clouds, caused by propeller created turbulence that have scale sizes ranging from the cross and depth dimensions of the wake down to 100 μm scale sizes, contribute to the difficulties found by attempts to develop an acoustic model for the wakes. A strong turbulent diffusion takes place in a wake and attempts to use two-phase

Navier–Stokes equations including Reynolds stress terms, related to the mean flow gradient, and the turbulent kinetic energy and its rate of dissipation to describe the propeller flow have not yet been brought to a level that will support a reliable flow model. Many factors influence the flow and, therefore, the acoustic conditions near a ship and in a near wake. These factors include the shape, size, type, ship displacement, and number of propellers as discussed by King [21], propeller geometry, number of rotations per minute, degree of cavitation, cavitation threshold, and ocean conditions, such as surface waves, wind speed, currents, temperature, salinity, degree of sea contamination and time from after the wake is formed. The lack of well-documented data and the uncertainty regarding the individual and mutual importance of input data to models for flow near a ship make realistic modeling of flows and the acoustic qualities of wakes extremely difficult, as outlined by Bradley et al. [22]. Several measurements of the backscattering from wakes have been performed; however, a reliable expression for the backscattering cross section of a wake has yet to be established.

Increased low-frequency sound backscattering by the sea surface and the elevated low-frequency ambient noise levels frequently measured in the sea have been attributed to *bubble cloud resonances*. Prosperetti [23] has suggested that due to the reduced stiffness and density in a bubble cloud compared to seawater without bubbles and the substantial dimensions of bubble clouds, frequently several meters, the whole cloud could vibrate as "big bubble" at low frequencies. Laboratory tests by Kozhevnikova and Bjørnø [24] have verified this assumption. Therefore, resonance scattering by bubble clouds cannot be excluded.

The effects on sound propagation by bubbles near the sea surface are refraction of low-frequency sound waves due to reduced sound velocity caused by the bubbles and resonance absorption at higher frequencies. Model calculations by Deane et al. [25] show that resonance absorption caused by the largest bubbles is strongly dependent on frequency and wind speed. The frequency dependence is explained by the concept of the *bubble escape radius*, which is the bubble radius for which the turbulent fluid velocity fluctuation around the bubble and the bubble's terminal velocity in the upper ocean boundary layer are in balance. Bubbles smaller than the escape radius will be trapped in the turbulent layers, while larger bubbles migrate and degas at the surface.

A bubble cloud includes bubbles in a nearly continuous size distribution. If the number of bubbles having radii between a and da are $n(a)da$ in a unit volume of a bubble cloud, the low-frequency backscattering cross section ($ka \ll 1$) per unit volume (1 m^3) based on Eq. (5.16) is:

$$\sigma_{bs} = \int_a \frac{n(a)a^2}{\left(\frac{f_r^2}{f^2} - 1\right)^2 + \delta^2} \, da. \tag{5.25}$$

When $n(a)da$ is known, the backscattering cross section per m^3 of a bubble cloud is determined from Eq. (5.25). According to Medwin and Clay [3] since the major scattering from bubbles in bubble clouds may be ascribed to resonance of bubbles, the bubble cloud's *backscattering cross section* per m^3 is approximately:

$$\sigma_{bs} = \pi a_r^3 \frac{n(a_r)}{2\delta_r},$$ (5.26)

where a_r and δ_r denote the radius of the resonant bubbles and attenuation function at resonance, respectively.

A simple expression for the *volume backscattering strength* S_b near the sea surface at 248 kHz, the frequency used for the measured data reported by Thorpe [20], as a function of the wind speed w {unit: m/s}, may be written as: $S_b = -48 + 2.9(10 \log w)$ {unit: dB}. This expression by Hall [26] is in agreement with Wu's [27] interpretation of Thorpe's data [20].

Apart from a substantial influence on sound scattering and therefore on reverberation, even low volume fractions of bubbles influence compressibility, and to a smaller extent density of water/bubble mixtures and through it, the sound velocity in the water, as discussed by Kozhevnikova and Bjørnø [28]. Seawater containing gas bubbles in general is *dispersive*, i.e., the velocity of sound depends on the acoustic frequency of the propagating signals. Given the number of bubbles, $n(a)$, of radii between a and da, some simple expressions at frequencies much higher and lower than bubble resonance frequencies in the water can used to calculate the velocity of sound in bubbly water.

In the high-frequency limit, the sound velocity in bubbly water is independent of the frequency and bubble size distribution and approaches the velocity of sound in bubble-free seawater. This limit is used in sound velocimeters, where the sound speed is determined by measuring the time for a MHz signal to propagate over a fixed distance between two transducers, a transmitter, and receiver, or between a transducer and reflector, where the transmitter acts as the receiver. For acoustic signal frequencies much lower than the bubble resonance frequencies, the velocity of sound depends on the *void fraction*, β, i.e., the ratio of bubble volume to the total volume, and the compressibility and density of the gas in the bubbles and of the water without bubbles. The low-frequency limit for the velocity of sound, c_{lfreq}, in bubbly water according to Wood [29] is:

$$c_{lfreq} = \sqrt{\frac{E_g E_w}{[\beta E_w + (1-\beta)E_g][\beta \rho_g + (1-\beta)\rho_w]}},$$ (5.27)

where E and ρ denote the bulk modulus of elasticity and the density, respectively, and the indices g and w indicate gas in the bubbles and seawater, respectively. E_w is determined by the relation: $c_w^2 = E_w/\rho_w$, where c_w is given by the expressions for the velocity of sound in seawater without bubbles in Chapter 1. E_g for isentropic wave propagation is found by using the definition of the velocity of sound in a gas, i.e.:

$$c_g^2 = \left(\frac{\partial p}{\partial \rho}\right)_s = \frac{\gamma p_g}{\rho_g} = \frac{E_g}{\rho_g},$$

where p_g is the ambient pressure {unit: pascal, Pa} acting on the gas bubbles and γ is the ratio of specific heats (c_p/c_v). The ambient pressure is dependent on the depth of the bubbles and is given by:

$$p_g = p_0 + \rho_w g z, \qquad (5.28)$$

where p_0 and g symbolize the atmospheric pressure at the sea surface and gravitational constant ($g = 9.81$ m/s^2), respectively, and z {unit: m} is the depth below the sea surface. An estimate of the void fraction β can be made by measuring the velocity of sound in bubbly water and inserting the data into Eq. (5.27), which is valid for all void fractions as long as the signal frequency is much lower than the bubble resonance frequencies. A more comprehensive discussion on bubbles influence on sound wave dispersion and attenuation in seawater is provided in books by Medwin and Clay [3] and Leighton [30].

5.2.2.3 Deep Scattering Layer

Many organisms in the sea are light sensitive since light influences their nutrients. The change in light penetration into the sea from day to night causes these organisms to move up and down in the sea in response to the light intensity. The organisms are mostly of biological origin, such as phytoplankton, zooplankton, and small pelagic fish, and can be found in all seas. During daytime they form layers at depths ranging from about 300 to 600 m. During sunset these biological layers rise to shallow depths about 100 to 150 m and at sunrise they again move back to daytime depths. This diurnal migration of unknown scatterers was observed during World War 2 and explained after the War, as discussed by Urick [2]. The layers of biological organisms, called *the DSL*, have a latitude-dependent effect on sound absorption and scattering with the influence maximum near equator and minimum at the polar regions. There is also an intermediate minimum in the subtropical regions near 30 degrees latitude.

The major backscattering effects from the DSLs are caused by resonances in small bubbles trapped on or in the biological organisms, such as the swim bladder in fish, with resonance frequencies in the range of 1−20 kHz. Due to the vertical migration of the DSL, the ambient pressure acting on the trapped bubbles changes with depth and the DSL backscattered frequency spectrum shows a diurnal change with higher average frequencies during days than during nights. At frequencies above 25 kHz most backscattering is nearly independent of frequency and the scattering is assumed to be of nonresonant. The DSL volume scattering strength, S_{vl}, for 1 m^3 at frequencies near 24 kHz has been reported by Urick [2] to be on the order of −70 to −80 dB. The DSL thickness, H, adds to the DSL scattering strength and the scattering strength, S_H, of a water column of a cross section of 1 m^2 and thickness H is:

$$S_H = S_{vl} + 10 \log H. \qquad (5.29)$$

Fish schools feeding at particular depths in shallow water can also lead to layers of scatterers without the characteristic features of the DSL.

5.2.2.4 Suspended Sediments

During recent years the increasing use of acoustics for underwater monitoring and control has started the exploitation of backscattering from high-frequency signals, 0.5–5 MHz, as a tool used by sedimentologists and coastal engineers to study of sediment entrainment and transport in rivers, harbors, and shallow coastal regions such as in the EU-MAST Project: *TRIDISMA* [31]. In a 2004 work by Thorne and Buckingham [32] the backscattering from suspended glass spheres and seven different sands collected from estuarine, beach, and quarried locations was measured and its characteristics were used to develop a form function and the attenuation through the total scattering cross section. These comprehensive measurements, which broadly covered the Rayleigh, intermediate, and geometrical scattering regimes, were aiming at determining sediment parameters, such as particle size and concentrations by using an inversion algorithm. The data showed that within the Rayleigh scattering region ($ka \ll 1$) scattering from the sand grains is comparable with scattering from spheres of comparable size measured using a standard sieve procedure. For ka values above the Rayleigh region, up to $ka = 3$, divergence was observed between the scattering characteristics of spheres and sand grains. In the geometrical scattering region a nearly constant deviation was observed with enhanced scattering from sand relative to spheres, and a scattering enhancement factor, γ_e, was defined as:

$$\gamma_e = \frac{\beta \chi^3 + 0.5\chi + 3.5}{\chi^3 + 3.5},$$ (5.30)

where $\chi = ka_s$, with a_s the mean particle radius for particles in suspension and k the acoustic wave number. β, a free parameter, independent of χ that represents the upper limit for γ_e, was found to be $\beta = 1.7 \pm 0.3$. The pure spherical shape is represented by $\gamma_e = 1$. The enhancement factor γ_e in the geometrical scattering region provided a simple relationship between the sphere model, which is discussed in Section 5.5, and the measured scattering characteristics of sand suspensions.

5.3 SCATTERING FROM EXTENDED, NEARLY PLANE, ROUGH SURFACES

The scattering, and in particular the backscattering, from the sea surface and the seabed is critical in underwater acoustics. Real sea surfaces and seabeds have a roughness distribution that can be described statistically. A general assumption is that the roughness spectrum (i.e., wave displacements and surface slopes) to first order are approximately described by a normal (Gaussian) distribution. However, a statistical description of a rough surface requires information about the probability density function for the rough surface displacements from a mean surface position, the probability density function for the surface slopes, the rms value of the displacements, and the temporal and spatial spectra of the rough surfaces, and their correlation functions. The higher-order statistical moments are important in describing surface scattering. The spatial spectrum of the topography of the rough surface can be found by a Fourier transform of the surface relief and its features in the same way a time signal's Fourier transform provides its frequency composition. The spatial spectrum is characterized by a spatial wave number $\kappa = 2\pi/\Lambda$, where Λ is the spatial wavelength. If the roughness distribution is totally random the spatial spectrum is continuous and is frequently given an exponential form.

Frequently scattering is considered an additional effect associated with sound reflection from a rough surface since the incident acoustic energy's interaction with a rough surface is divided between transmitted energy and energy returned to the water column, where the latter is divided between coherent (specularly reflected) and incoherent (scattered) energy.

If the rough surface (seabed or sea surface) is randomly rough and the *rms-roughness height, h,* is small compared to the acoustic wavelength, λ, the *coherent (specular reflection) loss* at the rough surface has been shown by Brekhovskikh and Lysanov [1] to be modified by the Rayleigh parameter R_a given in Eq. (5.1):

$$R^*(\theta) = R(\theta)e^{-\frac{R_a^2}{2}} = R(\theta)e^{-2h^2k^2\cos^2\theta}, \tag{5.31}$$

where $R(\theta)$ is the amplitude reflection coefficient for a smooth surface, discussed in Chapter 2, and where $R^*(\theta)$ is the amplitude reflection coefficient modified by the influence of scattering. Since $R(\theta) \sim -1$ at a smooth sea surface, Eq. (5.31) gives for a moderately rough sea surface $R^*(\theta) \simeq -e^{R_a^2/2}$. The degree of coherence in the scattered field given by Eq. (5.31) falls off rapidly with increasing R_a, as shown in Fig. 5.7, because the phases of waves scattered from different parts of a rough surface are random for larger values of R_a and the field is, therefore, incoherent.

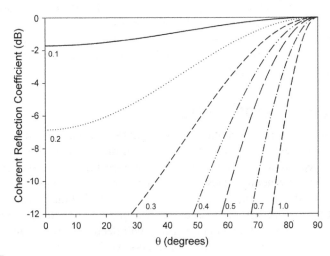

FIGURE 5.7

The reduction in the coherent reflection coefficient $20 \log(R^*(\theta)/R(\theta))$ {unit: dB}, as a function of grazing angle for values of $2h/\lambda$.

The limit for using Eq. (5.31) is in general set to $R_a = \pi/2$, and the coherent reflection coefficient, shown in Fig. 5.7, for that value has been reduced by 10.7 dB. A coherent field is obtained for $R_a \to 0$.

Several successful attempts based on different assumptions have been made during the past to describe the forward- and backscattering from a rough surface. They include *Bragg scattering*, *reflection from facets*, and *Lambert's law*.

5.3.1 BRAGG SCATTERING

Braggs law for X-ray diffraction in crystals, formulated by W.L. Bragg in 1912, is used to describe scattering from rough surfaces where the roughness is very small compared with the acoustic wavelength. Then the scattered signal can be assumed to be a continuum of contributions from points along the rough surface and, in the Bragg diffraction approximation, under a certain grazing angle, θ, the major contributions to the scattered signal will arise from points with a constant distance, d, and these contributions will all be in phase and form a constructive interference pattern. Bragg's relation among θ, d, and the wavelength λ for the acoustic scattering is:

$$2d \sin \theta = \lambda, \tag{5.32}$$

which leads to the spatial roughness wave number $\kappa = 2\pi/d = 2k \sin \theta$, where k is the acoustic wave number. Brekhovskikh and Lysanov [1] provide the following expression for the backscattering from a surface satisfying Eq. (5.32):

$$\sigma_{bs} = 4k^4 \cos^4\theta G(\kappa), \tag{5.33}$$

where $G(\kappa)$ is the spatial roughness spectrum of the rough surface. The fourth power dependence on the acoustic wave number $k = 2\pi/\lambda$ shows the relation to Rayleigh scattering given by Eq. (5.13). The spatial roughness spectrum $G(\kappa)$ can be obtained from Eq. (5.33) by measuring the backscattering cross section under various grazing angles θ and for various frequencies.

5.3.2 REFLECTION FROM FACETS

For the relatively smooth horizontal sea surface at low sea states (1−4) or a seabed consisting of sediments a nearly vertical surface insonification of the surface scattering will appear to come from a mosaic of surface facets (i.e., sections of the surface) all tilted at random angles φ_i with the horizontal. For most facets the tilt angle is very small and the incident sound field is reflected at the facets in specular directions. Under these assumptions the backscattering strength S_b will have a maximum near the vertical direction, while S_b will decrease strongly in directions away from the vertical when the number of facets contributing to the returned sound energy is reduced due to their small slope angles. For the case where the slope angles are assumed to have a normal (Gaussian) distribution, Brekhovskikh and Lysanov

[1] have shown that the backscattering cross section for high-frequency acoustic waves is:

$$\sigma_{bs} = R(\theta) \frac{e^{-\frac{\tan^2\theta}{2\delta^2}}}{8\pi\delta^2 \cos^4\theta}, \tag{5.34}$$

where δ^2 is the variance of the facet slopes, θ is the angle of incidence and $R(\theta)$ is the amplitude reflection coefficient which was discussed in Chapter 2. Eq. (5.34) has no dependence on the incident wave's frequency and can be used to determine the backscattering cross section for near vertical angles of incidence when the local curvature of the facets is much smaller than the acoustic wavelength (i.e., nearly plane facets), and when the frequency is not too low.

5.3.3 LAMBERT'S LAW

While sound penetrates through a rough surface into the medium behind the surface, for instance at a seabed, subsequent acoustic energy reradiation to the water column may take place due to backscattering from inhomogeneities in the seabed. This backscattering is primarily significant at low frequencies and at high grazing angles, while backscattering at higher frequencies can be considered to be primarily due to the rough surface. When sound with intensity I_i is incident under the grazing angle θ on a small surface of area dA of the rough surface, the acoustic power received by dA will be $I_i \sin \theta \, dA$. This power is assumed by *Lambert's law* (sometimes also called Lambert's rule) to be scattered in directions that are a function of $\sin \varphi$, where the grazing angle φ is the scattering angle. The intensity of the scattered sound I_s in the direction of φ according to Urick [2] is:

$$I_s = \mu I_i \sin \theta \sin \varphi \, dA, \tag{5.35}$$

where μ is a constant. For backscattering from a unit area of the rough surface, the backscattering strength S_b may be written as:

$$S_b = 10 \log \mu + 10 \log \sin^2\theta. \tag{5.36}$$

Eq. (5.35) shows that the backscattering strength, which obeys Lambert's law, varies as the square of the sine of the grazing angle. If no acoustic energy is transmitted into the medium behind the rough surface and all energy is scattered and redistributed in the half space on the water side of the rough surface, then $10 \log \mu = -5 \, dB$, which is the backscattering strength for sound normally incidence on the rough surface. Practical data obtained from measurements on various seabed types have shown values of $10 \log \mu$ in the range -10 to -35 dB rel. 1 m^2, which is probably caused partly by energy transmission across the rough surface. In spite of this deviation and the assumptions related to the distribution of the scattered acoustic energy in space, Lambert's law has been used extensively in underwater acoustics for scattering from very rough surfaces, in particular when the direction of sound incidence is different from the scattering direction. Its application to slightly rough surfaces is limited to oblique incidence. However, for slightly rough surfaces, like low sea state sea surfaces or sediment on the seabed, the backscattering expressions under Bragg scattering or reflection by facets may be applied.

Lambert's law is mathematically simple and provides a reasonable expression for rough surfaces' scattering qualities without accounting for any frequency dependence. However, its *limitation* is that it does not provide a physical understanding of the scattering process and does not take into account any fine structure in the grazing angle dependence.

5.3.4 SCATTERING FROM THE SEA SURFACE

The turbulent winds blowing over the sea surface cause extremely irregular wave motions. The displacement of the surface from a mean plane by wind-generated waves is a random function of time and position on the surface. The statistical surface displacement at a fixed position is close to a normal (Gaussian) distribution; however, it has a broader peak than a normal distribution and is slightly skewed, as discussed by Phillips [33]. The two-dimensional wave slope distribution is also close to a normal distribution, which constitutes the first approximation to sea surface statistics. The *frequency spectrum*, $S(\Omega)$, for waves on a fully developed rough sea surface depends on the wind speed w {unit: m/s}, and an expression for energy content as a function of frequency Ω {unit: Hz} and w, as developed by *Pierson and Moskowitz* [34] is:

$$S(\Omega) = \frac{8.1 \times 10^{-3} g^2}{\Omega^5} e^{-0.75\left(\frac{g}{\Omega w}\right)^4}, \tag{5.37}$$

where g {unit: m/s^2} is the acceleration due to gravity.

Forward scattering and backscattering from a rough sea surface have been the subjects of many investigations in the 1950s by Eckart [35], Urick and Hoover [36], Twersky [37], and Garrison et al. [38]. Although these investigations were primarily at high frequencies, the authors noticed a disturbing influence by bubble layers just beneath the sea surface. In the 1960s, Chapman and Harris [39] and Chapman and Scott [40] performed a series of comprehensive studies of backscattering at lower frequencies as a function of wind speed, grazing angle, and acoustic frequency. Omnidirectional hydrophones measured the backscattering signal from explosive sound sources. Based on data covering the frequency range $0.4 \text{ kHz} < f < 6.4 \text{ kHz}$ and grazing angles θ {unit: degrees} less than 40 degrees, the following empirical expression for the *backscattering strength*, S_b, was derived:

$$S_b = 3.3 \beta \log(\theta/30) - 42.4 \log \beta + 2.6, \tag{5.38}$$

where β {unit: dB per grazing angle doubled} is the slope of the scattering strength curve versus wind speed, w {unit: m/s}, and frequency, f {unit: Hz}, is given by the expression:

$$\beta = 107\left(wf^{0.33}\right)^{-0.58}. \tag{5.39}$$

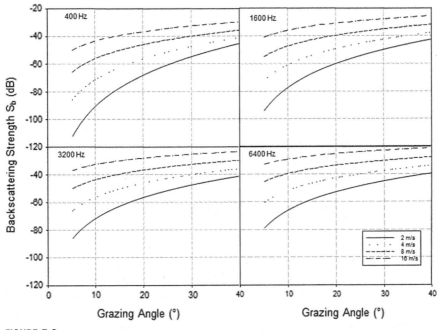

FIGURE 5.8

Sea surface backscattering strength {unit: dB} as a function of grazing angle {unit: degrees} for various frequencies and for wind speeds ranging from 2 to 16 m/s. The curves are based on Chapman and Harris' expressions (5.38) and (5.39) [39].

Eq. (5.38) provides predictions of the backscattering strength at a rough sea surface and, as verified in the recent *Critical Sea Test* by Ogden and Erskine [41], for wind speeds above 4 m/s and frequencies above 300 Hz this agreement is good. At lower wind speeds characteristic of a calmer sea surface and frequencies below 300 Hz perturbation theory provides adequate correlation with current measured data. Eq. (5.38) has found broad application for sea surface backscattering when low-frequency active sonar is used. Fig. 5.8 shows Chapman and Harris' *backscattering strength* [39] as a function of grazing angle for four frequencies, 400, 1600, 3200, and 6400 Hz with the wind speed as parameter at the values 2, 4, 8, and 16 m/s. Fig. 5.8 is based on Eqs. (5.38) and (5.39).

As discussed by Ainslie et al. [42] and Ellis and Crowe [43], for all grazing angles, θ, including angles close to normal incidence, the sea surface backscattering strength, S_b, can be determined by adding a facet scattering term to the Chapman and Harris backscattering strength. This term is obtained by developing an equivalent backscattering cross section, σ_{bs}, which is defined by $\sigma_{bs} = 10^{S_b/10}$:

$$\sigma_{bs}(\theta) = \sigma_{CH}(\theta) + \frac{e^{-\dfrac{\cot^2 \theta}{2\xi^2}}}{8\pi\xi^2 \sin^4 \theta}, \tag{5.40}$$

where the index *CH* refers to Chapman and Harris backscattering strength derived by inverting Eq. (5.38) (i.e., $\sigma_{bs} = 10^{Sb/10}$) where S_b is given by Eq. (5.38). The factor ξ, the facet slope relation from Chapman and Scott [40], is given by $\xi^2 = 0.003 + 0.00512w$, where w is the wind speed {unit: m/s}.

Since the early 1990s, several theories have been proposed to explain *anomalous low-frequency reverberation* including:

- scattering from late time bubble plumes formed by breaking waves with gas void fractions $V_f \sim 10^{-6}$ using cylindrical plumes with ellipsoidal cross sections, McDonald [44], and circular cross sections, Henyey [45];
- scattering from intermediate time bubble clouds with $V_f \sim 10^{-2}$ to 10^{-4} with cloud shapes taken as spheres and ellipsoids, Gragg and Wurmser [46];
- scattering from early to late time bubble structures with $V_f \sim 10^{-2}$ to 10^{-6} with symmetrical and asymmetrical bubble plume and cloud shapes, Sarkar and Prosperetti [47].

In each case, the cloud/plume surface interactions assume a planar pressure release surface. The complexity of the approaches ranges from the use of weak scattering theory to describe the interaction within the scattering volume by McDonald [44] and Henyey [45] to the use of a T-matrix scattering approach by Sarkar and Prosperetti [47]. For each approach, there are regions where the agreement with Chapman and Harris' empirical formulation is reasonable. However, each approach has regions where the agreement deviates from the Chapman–Harris scattering strength formulation. In each approach, the focus is on a scattering volume, or scatterer that is located in a planar ocean. The impact of a nonplanar ocean surface on the scattering process can be demonstrated by calculating the effective cross section of a bubble plume in the presence of a *large-scale surface fluctuation* using the weak scattering theory approach. Fig. 5.9 illustrates the change in the scattering cross section {unit: dB rel. 1 m^2} when the bubble plume is at different locations on a rough sea surface with a wavelength, λ, of about 77 m and wave height about 1.75 m. The calculations were performed by Neighbors and Bjørnø [48] for a plume with a circular base at the ocean surface and with an average plume void fraction e-folding depth of 2.5 m, radius 1 m, length 8 m, and a cosine-shaped scattering volume. In Fig. 5.9, the plume location is moved across a large-scale fluctuation in increments of $\lambda/4$ at the locations shown in the inset. There is about a 30 dB change in the effective scattering cross section at shallow angles when the plume moves from the trough to the crest, the shadowing by the preceding trough reduces the illuminated scattering volume at the plume base ($z = 0$). This effect has been demonstrated by Neighbors et al. [49] through a controlled, two-dimensional scale modeling approach for bubble plume and rough surface scattering that included multiple scattering and shadowing effects. These experiments showed that bubble plume shadowing at low grazing angles has a first-order impact on the scattering strength observed from bubble plumes and the rough sea surface at moderate to high sea states.

As the frequency increases from a few hundred hertz to the low kilohertz regime additional phenomena may contribute and possibly dominate sea surface scattering.

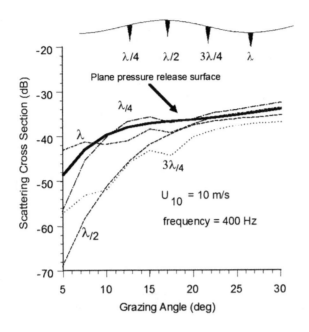

FIGURE 5.9

Impact of bubble plume location on scattering cross section. Inset, location on surface wave.

*Adapted from Figure 6, Neighbors, T.H. and Bjørnø, L., Anomalous Low Frequency Sea Surface Reverberations. Hydroacoustics, **4**, pp. 181–192, 2001, with permission of Journal Hydroacoustics.*

Based on surface reverberation fluctuations at and above 3 kHz it was concluded by McDaniel [50] that resonant scattering in a thin layer of near-resonant microbubbles just below the sea surface, generated by wave breaking, could explain the anomalous higher scattering strength by rough sea surfaces, better than scattering from bubble plumes when the number of wave-breaking events producing bubble plumes were taken into consideration. For frequencies ≥20 kHz scattering from *resonant subsurface microbubbles* has been shown by McDaniel and Gorman [51] to contribute to the anomalous scattering.

In the polar regions, the effect of the *ice cover* on the backscattering of sound under the ice is substantial. Large ridge-like structures extending to depths of some tens of meters interrupt relatively smooth stretches of ice. An under-ice roughness spectrum measured by Mellen [52] over a distance of 10 km using a high-frequency, narrow beam, upward-looking sonar on a submarine, showed an average and an rms depth of the ice keels below the sea level of 5.5 m and 3.2 m, respectively. Some parts of the ice ridges melt during the summer; however, the winter with low temperatures and strong winds refreezes the ice, breaks it, and adds to the magnitude and number of ice ridges. Moreover, the positive gradient on sound velocity in polar regions bends propagating sound waves up toward the underside of the ice layer and contributes to increased scattering. The acoustic backscatter

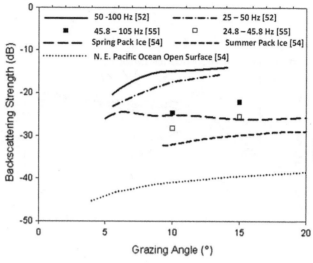

FIGURE 5.10

Under-ice average backscattering strengths as a function of grazing angle and frequency.

levels from Arctic Sea ice are much stronger than observed levels in the open ocean, as discussed by Milne [53], Brown [54], and Brown and Milne [55]. In general, the backscattering level from well-defined under-ice ridges is high compared to the backscattering level by the underside of the ice between the ridges, in particular at low grazing angles.

Backscattering measurements performed in the Norwegian−Greenland Sea by Hayward and Yang [56] using a large aperture vertical hydrophone array with 0.8 kg SUS charges as sound sources, showed that the backscattering strengths decrease with decreasing grazing angle and with decreasing frequency in the frequency range below 1 kHz, as shown in Fig. 5.10.

Fig. 5.10 also shows an about 10 dB difference in level between data from Milne [53] and from Hayward and Yang [56] which can be explained by differences in rms roughness heights for the ice in the two cases. The data by Milne [53] originate from a typical rms roughness of 4 m, while the data by Hayward and Yang [56] are obtained at an rms roughness of 2 m. Moreover, Fig. 5.10 shows the substantial difference in backscattering strength between spring and summer in the frequency range of 1.28 to 2.56 kHz, and the difference of nearly 20 dB between spring-ice backscattering and backscattering from an open ocean surface at wind speeds between 10 and 12.5 m/s, as measured by Brown and Milne [55].

5.3.5 SCATTERING FROM THE SEABED

Scattering from the seabed is frequently complicated to interpret and describe since factors, such as bottom materials, surface roughness, and volume inhomogeneities control the seabed's scattering strength. The upper few meters of seabed sediment layers around the world have been investigated extensively by taking samples for laboratory studies and photographs of the sediment surfaces and by acoustic reflection and scattering studies. The main purpose of this underwater acoustics research has been to relate the geophysical properties of the sediments to their acoustic qualities. Comprehensive studies of the seabed have been performed by Hamilton [57,58] and Hamilton and Bachman [59]. Table 5.6, data extracted from Ref. [59], provides examples of some of their geophysical and acoustical data for various types of sediments. It should be noted that the sediment qualities depend strongly on the site where the measurements have been carried out, and that the values in Table 5.6 may be more indicative than absolute.

It has been, and it still is, customary to divide the backscattering from a seabed into *shallow and deep water measurements* for two main reasons:

1. backscattering measurements from a seabed in shallow water in particular at low grazing angles may receive "contaminating" acoustic power from sea surface backscattering due to the lower water depth;
2. large portions of the deep water abyssal plains are covered with clay along with carbonate and silicon mud (clay and silt mixtures), while a seabed in shallow water is characterized by the presence of a broad variety of sediments ranging from fine clays and silts to coarse sands, pebbles, shell materials, and rocks, see Chapter 8, The Seafloor.

Table 5.6 Characteristic Values of Geophysical Qualities of Sediments From the Continental Shelf and Slope Environment, Including Mean Grain Size {unit: mm}, Porosity {unit: %}, Density {unit: kg/m^3}, Speed of Sound {unit: m/s}, and the Velocity Ratio of the Sediment Speed of Sound to the Speed of Sound 1528 m/s (in Seawater at 23°C With Salinity the Same as in the Sediment Pore Water) as the Reference to Indicate Fast or Slow Bottoms

Sediment Type	Mean Grain Size (mm)	Porosity (%)	Density (kg/m^3)	Sound Speed (m/s)	Velocity Ratio
Coarse sand	0.5285	38.6	2034	1836	1.201
Fine sand	0.1638	44.5	1962	1759	1.152
Very fine sand	0.0988	48.5	1878	1709	1.120
Silty sand	0.0529	54.2	1783	1658	1.086
Sandy silt	0.0340	54.7	1769	1644	1.076
Silt	0.0237	56.2	1740	1615	1.057
Sand–silt–clay	0.0177	66.3	1575	1582	1.036
Clayey silt	0.0071	71.6	1489	1546	1.012
Silty clay	0.0022	73.0	1480	1517	0.990

The dependence of backscattering from a seabed on the materials forming the upper parts of the seabed, with the scattering strength decreasing from rock over sand, silt to clay, has been known from early measurements carried out by McKinney and Anderson [60] and Wong and Chesterman [61]. They found that the backscattering strength was constant for *grazing angles* from nearly 60 to 20 degrees and for grazing angles below 20 degrees the backscattering strength decreased as the grazing angle decreased.

However, later data collected by Jackson and Briggs [62] showed other alternatives for backscattering as a function of grazing angle. At other measurement sites the data for backscattering as a function of grazing angle showed a continuous decrease in the backscattering strength from grazing angles of 90 degrees (normal incidence) to about 60 degrees, then a nearly grazing angle independent backscattering strength down to a grazing angle of nearly 20 degrees, and below 20 degrees a decrease in backscattering strength as the grazing angle decreased. There are two reasons for the discrepancy between earlier and more recent data:

1. the seabed acts as a rough interface, scattering sound both in and out of the plane containing the incident wave;
2. sound is transmitted across the interface and into the seabed, depending upon seabed materials, grazing angle, and frequency. The transmitted sound is absorbed and scattered—also back into the water column—by inhomogeneities and continuous variations in sound velocity and density of the seabed materials.

Inhomogeneities in the seabed include shells, stones, living organisms, minerals, and, in particular, gas bubbles formed by organic materials deteriorating or by methane from deeper parts of the ocean bottom. The inhomogeneities also include discontinuities in acoustical impedance between geological layers of different origin, which frequently form characteristic features of a seabed. The sediment layers may have sound velocity and density profiles on a par with what is found in the water column. Positive gradients in sound velocity between 0.5 and 2 {unit: s^{-1}} are frequently found in the seabed. If the sound velocity gradient in a seabed is assumed to be 1 m/s/m, and the sound velocity in sediments at the interface between water column and seabed is assumed to be 1500 m/s, the sound velocity in the sediments at a depth 100 m below the interface will be 1600 m/s. Only small variations in sediment density are found with depth below the interface, as long as the type of sediment is not changed. Thus, the sound velocity change with depth is the main contributor to the acoustical impedance change. The seabed characteristics found in various parts of the world's oceans cause a substantial variation in backscattering strength as a function of grazing angle, frequency, and bottom conditions. They are the main reasons for the difficulty in establishing general and widely applicable empirical expressions for seabed backscattering.

While it is generally accepted and widely verified that a strong correlation exists between geophysical and acoustical qualities of seabed materials, as illustrated in Table 5.6, it is rare to find a direct correlation between material qualities and the seabed roughness, the seabed slope, and their variances. A sandy seabed may show ripples produced by flow near the seabed, while a seabed consisting of silt, and in particular clay, very rarely shows any surface roughness. Based on backscattering measurements from clay, sand, and gravel bottoms over the complete range of grazing angles Gensane [63] found three backscattering regimes. For grazing angles below 15 degrees the backscattering strength frequently exceeded values predicted by Lambert's law, between 15 and 70 degrees, the curves for backscattering as a function of grazing angle became flatter, and above 70 degrees there was a substantial increase in the backscattering strength. This is consistent with other data reported in the literature by Jackson and Briggs [62] and may be described by three theoretical mechanisms. At grazing angles below the critical angle of the sediment very little acoustic energy penetrates into the bottom, and the scattering is mainly caused by interface roughness. This backscattering is described by the Rayleigh–Rice small roughness perturbation approximation which is discussed in Section 5.4. At intermediate grazing angles the backscattering receives contributions from *roughness scattering* at the interface and from volume scattering in the seabed. For grazing angles near normal incidence, and depending upon the frequency, the surface roughness scattering is the main contributor to backscattering. This backscattering is described by the Helmholtz integral using the Kirchhoff approximation which is discussed in Section 5.4. The frequency dependence at lower frequencies causes a deeper sound penetration into the seabed, due to lower absorption and, therefore, an increased contribution to backscattering from volume inhomogeneities in the seabed. At higher frequencies increased absorption reduces the volume scattering contribution relative to interface scattering. The three scattering regions can be seen in Fig. 5.11, from Jackson and Briggs [62], where backscattering data and model predictions are compared for two seabeds with different geophysical qualities.

Ellis and Crowe [43] have added a facet scattering term to Lambert's law to describe backscattering from the seabed using an approach similar to the one used to add the facet scattering term in Eq. (5.40) to Chapman and Harris' backscattering strength given by Eq. (5.38). For $\sigma_{bs} = 10^{S_b/10}$ the *backscattering strength S_b* by scattering from a seabed may be found with a reasonable accuracy from:

$$\sigma_{bs}(\theta) = \mu \sin^2\theta + \frac{\nu e^{-\frac{\cot^2\theta}{2\xi^2}}}{\sin^4\theta}, \tag{5.41}$$

where ν is the facet strength expressed as $\nu = R(8\pi\xi^2)^{-1}$ by Ainslie et al. [42]. R is the sound intensity reflection coefficient for normal incidence on the seabed, discussed in Chapter 2, ξ is the facet slope and μ is the Lambert constant.

In Fig. 5.11A the seabed consisted of sand with well-organized nearly 0.12 m wavelength ripples. Samples of the seabed down to a depth of 23 cm showed fairly homogeneous materials. The seabed in Fig. 5.11B consisted of sands and gravels embedded in a silty clay matrix. Several coarser components found in the seabed

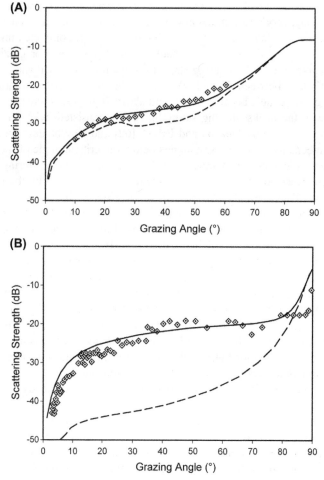

FIGURE 5.11

A comparison between model predictions and backscattering data [62]. (A) Fine sand with ripples measured at 25 kHz. (B) Fine sediments mixed with coarser components measured at 20 kHz.

Reproduced from Figures 10 and 12, Jackson, D.R. and Briggs, K.B., High-frequency bottom backscattering: Roughness versus sediment volume scattering. J. Acoust. Soc. Am., 92, (2), pp. 962–977, 1992, with permission of the Acoustical Society of America.

materials consisted of sand- and gravel-sized mollusk shells, shell fragments, and carbonate rocks. The dashed curve shows model predictions for scattering from a rough surface and the solid curve shows the model predictions for the total backscattering strength from the rough interface and sediment volume scattering. The two contributions are assumed to be uncorrelated.

In Fig. 5.11A roughness scattering dominates at all grazing angles, while in Fig. 5.11B there is a substantial contribution to the total scattering strength from volume scattering. On the test site associated with Fig. 5.11B the scattering strength showed no frequency dependence over the range of 15−45 kHz at a 20 degrees grazing angle.

Fig. 5.11 seems to indicate that roughness scattering from a "harder" bottom is the primary contribution to the scattering strength except at intermediate grazing angles where there are some contributions from volume scattering and that roughness scattering dominates at all grazing angles if the bottom consists of coarse sand or gravel. However, sediment volume scattering is dominant when the seabed consists of relatively fine sediments with main grain sizes less than 0.03 mm. At low frequencies (i.e., 100 Hz to 10 kHz), in the North Atlantic deep waters at the Sohm Abyssal Plain measurements by Mourad and Jackson [64] indicate that the main backscattering contribution comes from volume scattering.

For backscattering from *slow sediments*, i.e., sediments with a sound velocity less than seawater, as found on the Continental and on the Bermuda Rises, volume scattering dominates at grazing angles from about 5 to 75 degrees. A decrease in up to 10 dB in the overall backscattering level occurs when the frequency increases from 100 Hz to 1 kHz, while at higher frequencies, up to 10 kHz, backscattering is largely independent of frequency.

For *fast sediments*, i.e., sediments with sound velocity higher than in seawater, as observed by Mourad and Jackson [64], volume scattering dominates from near the critical grazing angle up to about 70 degrees for the frequency range 100 Hz to 10 kHz. Below the critical angle, volume scattering dominates between 100 Hz and 1 kHz. At higher frequencies, up to 10 kHz, the backscattering strength increases with increasing frequency, which indicates the influence of interface roughness scattering. The results published by Mourad and Jackson [64] provide evidence that in the deep oceans, seabed features that produce volume backscattering, such as strong layering in sediments and inclusion of a basement interface within sediments, control the total backscatter from the ocean bottom. Low-frequency bottom backscattering in deep oceans is particularly sensitive to acoustic frequency; to sound wave attenuation within sediments; to sound speed structure within sediments including upward sound refraction; and to factors that control the distribution, scale, and acoustic response of sediment volume inhomogeneities.

The influence of inhomogeneities in the seafloor like broken shell fragments in the sediments, which are common in many shallow water areas, on the scattering—single as well multiple scattering—has been reported by Lyons [65], who studied the acoustic power backscattered from the shell fragments as a function of the volume of scatterers and the frequencies, 10−100 kHz, in the incident waves. Williams et al. [66] found that rough surface scattering is the dominant scattering mechanism for backscattering from sand sediments in the frequency range 20−50 kHz, and that the Biot poroelastic sediment model could accurately predict the measured backscattering strength.

5.4 THEORETICAL BASIS FOR SCATTERING CALCULATIONS

Most mathematical models for scattering from a randomly rough interface fall into two categories: (1) the small roughness perturbation approximation, frequently referred to as the *Rayleigh–Rice method* and (2) the Helmholtz integral with the Kirchhoff approximation, often called the tangent plane method. The following sections discuss the *perturbation approximation* and the *Helmholtz–Kirchhoff method*.

5.4.1 THE PERTURBATION APPROXIMATION

The perturbation approximation is widely used to calculate low-frequency acoustic scattering from rough surfaces. *Lord Rayleigh* [6,67] made the first contribution to this approximation by calculating scattering for a wave normally incident on a periodic corrugated surface. Later, *Rice* [68] extended the method to include randomly rough surfaces. The perturbation approximation writes the unknown scattered acoustic field as a sum of outgoing plane waves and determines the unknown wave coefficients by satisfying well-defined boundary conditions on the scattering surface. The series converges if the surface is only slightly rough. The method does not take multiple scattering into account since only outgoing waves are considered.

Normally the perturbation approximation assumes a plane mean scattering surface; however, this assumption is not valid when multiple roughness scales are present, as is discussed in Section 5.4.3. When the mean scattering surface is a plane at $z = 0$, the boundary conditions on the scattering surface are known, the surface roughness height over the mean level at $z = 0$ is $\zeta(x,y)$, the rms-roughness height defined in Eq. (5.1) is $h = (<\zeta^2>)^{1/2}$, and the mean value $<\zeta> = 0$, the Taylor series expansion of the wave field function, $f(x,y,\zeta)$, on the scattering surface is:

$$F(x, y, \zeta) = f(x, y, 0) + \zeta \frac{\partial f}{\partial z_{z=0}} + \zeta^2 \frac{\partial^2 f}{\partial z^2_{z=0}} + \zeta^3 \frac{\partial^3 f}{\partial z^3_{z=0}} + \cdots. \tag{5.42}$$

Eq. (5.42) may be used to derive an approximate boundary condition at $z = 0$, which leads to an expression for the unknown scattered field on the mean scattering surface, from which the scattered field at some distance from the scattering surface is given by an integral over $z = 0$. To use the perturbation approximation the boundary conditions on the scattering surface must be satisfied.

Ogilvy [69] has developed a first-order perturbation solution for a rough *pressure-release* surface (Dirichlet boundary conditions) and rough *rigid* surface (Neumann boundary conditions) with the average intensity, $<I_{DI}>$, of a bistatic scattered field from a pressure release rough surface given by

$$< I_{D1} >= 4k^4 \sin^2\theta_1 \quad \sin^2\theta_2 2r^{-2}A_M P(kA, kB), \tag{5.43}$$

where θ_1 and θ_2 are grazing angles for the incident and the scattered waves, respectively, k is the acoustic wave number, and r is the distance between the scattering surface and the field point. A_M is the area of the mean plane of the scattering surface and $P(kA, kB)$ is the scattering surface's spatial power spectrum (or power spectral density function). kA and kB are the x and y components of the change in the wave vector, respectively. For a rigid rough surface the average intensity, $<I_{N1}>$, of a bistatic scattered field is:

$$< I_{N1} >= \frac{4k^4(1 - \cos\theta_1 \, \cos\theta_2 \, \sin\theta_3)^2}{r^2} A_M P(kA, kB), \qquad (5.44)$$

where θ_3 denotes the azimuth angle between the x-axis and the component of the scattered wave vector in the $z = 0$ plane. While Eqs. (5.43) and (5.44) describe the average intensity of the incoherently scattered field, the first-order perturbation solution does not indicate any change in the coherently scattered field due to surface roughness. This change only shows up in the second-order perturbation solution.

Thorsos and Jackson's paper [70] provides a comprehensive treatment of the second- and higher-order perturbation solution. The authors explore the validity of the perturbation approximation by comparing exact results from an integral equation solution with low-order perturbation predictions with higher-order predictions. The authors calculate the bistatic scattering cross section for a rough pressure release with a Gaussian roughness spectrum and with surface height variations only in one direction as shown by the scattering geometry is Fig. 5.12. The Gaussian roughness spectrum is used because at high spatial wave numbers this spectrum falls off more rapidly than a power-law spectrum, which is more typical of rough sea surfaces. A power-law spectrum may have the form $W(K) = \psi K^{-\gamma}$, where $3 < \gamma < 3.5$ for most bottom data, and where K is the spatial wave number. While power-law spectra, in general, are multiscale spectra, the Gaussian spectrum restricts the roughness to a single horizontal spatial scale which is a useful simplification for several purposes.

The spatial Gaussian roughness spectrum, $W(K)$, of the rough surface is given by:

$$W(K) = \frac{\ell h^2}{2\pi^{1/2}} e^{-\frac{K^2 l^2}{4}}, \qquad (5.45)$$

where ℓ the surface correlation length and K is the spatial wave number. h is the surface's rms-roughness height and is related to the roughness spectrum $W(K)$ by:

$$\int_{-\infty}^{+\infty} W(K)dK = h^2. \qquad (5.46)$$

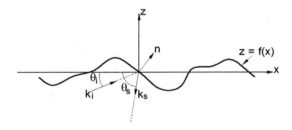

FIGURE 5.12

The scattering geometry for bistatic scattering from a one-dimensional rough surface. The rough surface wave profile is independent of the y-direction. $f(x)$ is the surface height and θ_i and θ_s are the grazing angles for the incident and scattered waves, respectively.

The total acoustic pressure field $p(r)$ including the incident, reflected, and scattered waves for scattering from a rough pressure-release surface can be expanded into a power series in the small parameter $hk \ll 1$ as:

$$p(r) = p_0(r) + (hk)p_1(r) + \frac{(hk)^2}{2!}p_2(r) + \frac{(hk)^3}{3!}p(r) + \cdots, \qquad (5.47)$$

where the pressure-release boundary conditions on the rough surface, $z = f(x)$, require that $p(r) = 0$ on the surface. r denotes the position vector to the field point and k is the acoustic field wave number. $p_0(r)$ is the "flat surface" field including the incident and reflected acoustic waves. The terms $p_1(r)$, $p_2(r)$, and $p_3(r)$ are the first-, second-, and third-order perturbation contributions to the scattered field. If only the first-order perturbation term is included, the scattering cross section σ_{11} according to Thorsos and Jackson [70] is:

$$\sigma_{11} = \frac{4k_{iz}^2 k_{sz}^2}{k} W(k_{ix} - k_{sx}), \qquad (5.48)$$

where k_{iz} and k_{sz} are the wave number z-components for the incident and scattered acoustic wave vectors, respectively, and k_{ix} and k_{sx} are the wave number x-components, respectively. The dependence of σ_{11} on the roughness spectrum $W(K)$ through the factor $W(k_{ix} - k_{sx})$ is the Bragg scattering resonance condition. As shown in Fig. 5.12, the z-direction is vertical to the plane mean surface. The scattering cross section is given by Eq. (5.48), when terms on the order of $(hk)^2$ are included. When second-order perturbation contributions to the scattered acoustic field are included the scattering cross section, σ_{22}, is:

$$\sigma_{22} = \frac{4k_{iz}^2 k_{sz}^2}{k} \int_{-\infty}^{+\infty} \left[W(k_{sx} - k_x)dk_x \cdot \right.$$
$$\left. \left\{ \left| k^2 - k_x^2 \right| + (k^2 - k_x^2)^{1/2} \left[k^2 - (k_{ix} + k_{sx} - k_x)^2 \right]^{1/2 *} \right\} \right\}, \qquad (5.49)$$

where the asterisk * denotes the complex conjugate. The roots in Eq. (5.49) are positive, when their arguments are positive, and positive imaginary when their arguments are negative. The second-order perturbation contribution to the scattered intensity and, therefore, to the scattering cross section is of a magnitude of $(hk)^4$, while $hk \ll 1$.

According to Thorsos and Jackson [70], for very short correlation lengths ℓ, i.e., $k\ell \ll 1$, the rms-roughness slope $s = 2^{1/2}h/\ell$, when the Gaussian roughness spectrum is of the order 1 or less, and the expressions for the scattering cross sections σ_{11} and σ_{22} reduce to:

$$\sigma_{11} \rightarrow \frac{4k_{iz}^2 k_{sz}^2}{2\pi^{1/2}k} \ell h^2 \quad \text{and} \quad \sigma_{22} \rightarrow \frac{s^2 \sigma_{11}}{2^{1/2}}. \qquad (5.50)$$

Thorsos and Jackson [70] emphasize the importance of the correlation length ℓ since the condition $hk \ll 1$ is not sufficient to guarantee the accuracy of the first- and higher-order perturbation contributions to the scattered field. When $k\ell$ becomes larger than 1, the second- and third-order perturbation terms give a larger contribution to the scattered field than the first-order term, due to the influence of $W(K)$ on the scattering cross sections.

5.4.2 THE HELMHOLTZ–KIRCHHOFF METHOD

While the perturbation approximation only applies to scattering from rough surfaces with small rms roughness heights and with small surface slopes, scattering from the rougher surfaces frequently found on the sea surface and seabed can be calculated by the Helmholtz–Kirchhoff method, when the rough interface radii of curvature are large compared with the acoustic wavelength. The derivation of the Helmholtz–Kirchhoff integral can be found in many textbooks such as Medwin and Clay [3] and Morse and Feshback [71]. The Helmholtz–Kirchhoff integral for the field, p_s, when a harmonic wave is scattered by a rough surface of area dS is:

$$p_s(r_o) = \frac{\ell}{4\pi} \int_S \left\{ p_s(r) \frac{\partial}{\partial n} \frac{e^{ikr_1}}{r_1} - \frac{e^{ikr_1}}{r_1} \frac{\partial}{\partial n} p_s(r) \right\} dS, \tag{5.51}$$

where r_o is the position vector of the field point, r is the position vector of a point on the scattering rough interface with the unit vector n of the normal to the surface, and $r_1 = |r_o - r|$. In Eq. (5.51) the scattered field measured at a field point $r_o = \{x_o, y_o, z_o\}$ can be determined when the scattered pressure $p_s(r)$ and its derivative with respect to the surface normal are known on the scattering interface. The Kirchhoff approximation is used to determine these quantities. The Kirchhoff approximation, which is often called the tangent plane method, assumes the acoustic pressure on any point Q of a rough scattering surface is equivalent to the acoustic pressure present on a plane tangent to the rough surface through Q. The easily understandable physical basis for Eq. (5.51) combined with the Kirchhoff approximation and the lack of restriction in validity to small hk values, are some of the main reasons for the extensive use of the Helmholtz–Kirchhoff method to calculate rough surface scattering. However, Helmholtz–Kirchhoff methods are generally limited in use by constraints on the radius of curvature r_c of the rough surface and the grazing angle θ. These limitations are provided in Eq. (5.52) based on Fig. 5.13 with $\theta = \pi/2 - \theta_i$:

$$kr_c \sin^3 \theta \gg 1. \tag{5.52}$$

Eq. (5.52) shows that the severity of the restriction, connected with the radius of curvature of the surface roughness, depends on the angle of incidence. Therefore, the governing factor is the rate of change of the surface gradient and not the absolute magnitude of the surface height or the surface gradient. Also, the Helmholtz–Kirchhoff method does not take into account self-shadowing by the roughness or multiple scattering.

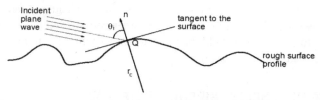

FIGURE 5.13

The two-dimensional geometry for scattering of a plane wave incident onto a rough surface with surface profile as shown in the figure. The local radius of curvature is r_c at the point Q on the rough surface, and the plane wave is incident at the angle θ_i with the normal n to the tangent through Q.

Shadowing occurs in two situations:

1. low incident grazing angles, when sections of the surface are in the shadow of the incident field;

2. low scattered grazing angles, when sections of the surface are in the shadow seen from the receiver.

In the first case certain parts of the rough surface will not scatter acoustic energy and in the second case rays scattered by the rough surface will not reach the receiver since they have to intersect the rough surface to reach the receiver. The second case leads to *multiple scattering* since once scattered sound rays are scattered again at the surface roughness which blocks their path to the receiver. The grazing angles are considered low with respect to shadowing influence, when they are approximately less than or equal to the rms slope angle ξ of the rough surface. For a Gaussian roughness spectrum, ξ may be found from the rms slope $s = 2^{-\frac{1}{2}}h/\ell$, through $\xi = \tan^{-1} s$. ℓ is the surface correlation length.

For a very rough, pressure-release, isotropic surface with a Gaussian roughness spectrum which satisfies the constraints associated with the Helmholtz–Kirchhoff method the backscattered intensity I_{bs}, according to Chapman et al. [72], is:

$$I_{bs} = \frac{\ell^2}{16\pi\,\sigma^2\,r^2\,\sin^4\theta}A_M\,e^{-\frac{\ell^2\cot^2\theta}{4\sigma^2}}, \tag{5.53}$$

where σ^2 is the surface roughness variance, r is the distance between the scattering rough surface and the field point, A_M is the area of the mean plane of the scattering surface, θ is the grazing angle and ℓ is the surface roughness correlation length. A characteristic feature of Eq. (5.53) is that the backscattered intensity does not depend on the frequency of the scattered sound.

A comprehensive study of the limitations of the Kirchhoff approximation has been reported by Thorsos [73] for a rough surface with a one-dimensional roughness height variation and with a Gaussian roughness spectrum satisfying Eqs. (5.45) and (5.46). This study showed that the applicability of the Kirchhoff approximation is determined by the magnitude of $k\ell$. This is somewhat in contrast to the conventional wisdom that the radius of curvature for any part of the rough surface must be larger than the acoustic wavelength, λ. Moreover, Thorsos [73] has shown that validity of the Kirchhoff approximation is strongly influenced by the relationship between the incident and scattered grazing angles and the rms slope s of the rough surface. As a result, the scattering cross section σ_s per unit scattering angle, per unit surface length is:

$$\sigma_s = \frac{k}{2\pi} \left[\frac{1 + \sin\theta_i \sin\theta_s + \cos\theta_i \cos\theta_s}{\sin\theta_i + \sin\theta_s} \right]^2 \Phi$$

where

$$\Phi = \int_{-\infty}^{+\infty} \cos\{kx(\cos\theta_i + \cos\theta_s)\} \left[e^{-\chi^2\left(1 - e^{\frac{\chi^2}{\ell^2}}\right)} - e^{-\chi^2} \right] dx. \qquad (5.54)$$

In Eq. (5.54) θ_i and θ_s are grazing angles with respect to the plane mean surface of the rough surface for incident and scattered waves, respectively, as shown in Fig. 5.12, and $\chi = hk (\sin\theta_i + \sin\theta_s)$. A shadowing correction factor can be developed, based on the magnitude of θ_i and θ_s, that can be multiplied by σ_s from Eq. (5.54) to account for shadowing in the Kirchhoff approximation. It reduces errors; however, it does not totally remove them, particularly at high scattering angles.

Many rough surfaces found in the sea do not satisfy the constraints for using either the perturbation or Helmholtz–Kirchhoff methods for calculating scattering. Rocks, either distributed randomly as boulders over a flat seafloor or constituting the whole seafloor, introduce discontinuities or radii of curvatures smaller than the wavelength. The reflection and scattering from some of these surfaces can be described by theories developed by Biot [74,75] for rigid bosses, i.e., hemispherical round knobs, on a rigid surface or by Twersky [37] using a Green's function formulation. However, these theories have only been developed for pressure-release (Dirichlet) boundaries and rigid (Neumann) boundaries.

Seafloor scattering was modeled in 2014 using first-order perturbation theory due to interface scattering from a rough seafloor, volume scattering from a heterogeneous sediment layer, or a combination of interface and volume scattering by Steininger et al. [76] and growing evidence that seafloor scattering is dominated by heterogeneities—a fluctuating continuum or discrete objects like shells, rocks, and bubbles—within the sediment volume as opposed to the seafloor roughness has been put forward by Holland et al. [77].

5.4.3 SCATTERING FROM SURFACES WITH TWO SCALES OF ROUGHNESS

A rough sea surface and rough seabed normally have a relatively broad spatial roughness spectrum. A widely used model for scattering from these surfaces is the composite roughness model, discussed by Brekhovskikh and Lysanov [1], Kur'yanov [78], McDaniel and Gorman [79], and Jackson et al. [80]. This model avoids many of the perturbation approximation and Helmholtz–Kirchhoff method shortcomings by combining these methods and treating the rough surface as a sum of small-scale and large-scale rough surfaces. In this combination, the large-scale rough surface satisfies the Helmholtz—Kirchhoff method constraint, i.e., the radii of curvatures are comparable to or larger than the acoustic wavelength, while arbitrary roughness heights and slopes are permitted, and the small-scale rough surfaces satisfy the perturbation approximation constraints, i.e., the rms-roughness heights are much smaller than the acoustic wavelength and that the surface slopes are sufficiently small, while the radii of curvatures can be arbitrary. A successful combination of the two methods requires the surface wave number spectrum to decay sufficiently fast with increasing wave number, that a large-scale surface can be defined on which the Kirchhoff approximation holds, and a small-scale surface can be defined on which the perturbation approximation can be used. Not all real surfaces permit such a separation in small- and large-scale roughness contributions. A model for high-frequency backscattering from a seabed using the composite roughness approach to a two-fluid rough boundary is given by Jackson et al. [80] for high and low grazing angles.

The scattering strength S in Eq. (5.2) may be written as:

$$S = 10 \log \left(\frac{r^2 I_{sca}}{I_{inc} A} \right) = 10 \log \sigma, \qquad (5.55)$$

where σ is the scattering cross section per unit area, and where r is the range from the scattering surface to the field point, where I_{sca} is measured, and A is the area of the scattering rough surface. If backscattering is assumed to be produced by small-scale roughness with a local grazing angle for the incident sound waves dependent on the slope of the large-scale roughness, as shown in Fig. 5.14, the local grazing angle θ_ℓ is expressed in terms of the true grazing angle θ through the approximation $\theta_\ell \sim \theta + \xi_x$, where ξ_x is the x-component of the large-scale slope (expressed as an angle), if grazing angles are less than 70 degrees. Also the rms slope s of the large-scale surface roughness is assumed to be small, $s < 0.1$, and the slope is assumed to be a Gaussian-distributed random variable.

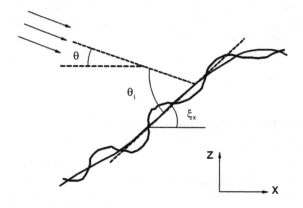

FIGURE 5.14

Decomposition of a rough surface into small- and large-scale components in the x–z plane.

According to Jackson et al. [80] the backscattering cross section σ_{bs} per unit area {dimensionless} for grazing angles 70 degrees and less is obtained by averaging the small-scale backscattering cross section, i.e.:

$$\sigma_{bs} = \Phi(\theta, s)(s\pi^{1/2}) \int_{-\infty}^{+\infty} \sigma_{ss}(\theta + \xi_x) e^{-\frac{\xi_x^2}{s^2}} d\xi_x, \tag{5.56}$$

where $\Phi(\theta,s)$ is a function which accounts for shadowing by the large-scale rough surface.

This function is:

$$\Phi(\theta, s) = \frac{1 - e^{-2Q}}{2Q} \text{ with}$$

$$Q = \frac{\pi^{1/2} e^{-t^2} - t(1 - erf(t))}{4t} \tag{5.57}$$

$$\text{and } t = \frac{\tan \theta}{s},$$

where erf denotes the error function. In Eq. (5.56), $\sigma_{ss}(\theta + \xi_x)$ is the small-scale backscattering cross section as a function of the local grazing angle.

For an incident acoustic field with grazing angles exceeding 70 degrees, it is possible to replace the entire roughness model by the Kirchhoff approximation at steep grazing angles, since the Kirchhoff constraint from Eq. (5.52) is less stringent at steep grazing angles. When Jackson et al. [80] made this replacement the backscattering cross section per unit area became:

$$\sigma_{bs} = \frac{g^2\left(\frac{\pi}{2}\right)}{8\pi \sin^2\theta \cos^2\theta} \int_{-\infty}^{+\infty} e^{-qu^{2\alpha}} J_0(u) u \, du \tag{5.58}$$

with

$$q = \frac{\sin^2\theta}{\cos^{2\alpha}\theta} C_h^2 \, 2^{1-2\alpha} k^{2(1-\alpha)},$$

where u is the magnitude of the radial distance along the integration path, $J_0(u)$ is the zeroth-order Bessel function of first kind and the function $g(\pi/2)$ is the plane wave reflection coefficient for normal incidence expressed by:

$$g\left(\frac{\pi}{2}\right) = \frac{\rho_s c_s - \rho_w c_w}{\rho_s c_s + \rho_w c_w}, \tag{5.59}$$

where ρ and c denote the density and the velocity of sound in seawater, index w, and in the seabed, index s, respectively.

The parameters C_h and α have been evaluated in Jackson et al. [80] and numerical simulations show that $\alpha = 0.5$ to 0.75 gives a reasonable approximation for roughness data from surfaces with a power-law roughness spectrum, where $\alpha = 0.5$ corresponds to the power-law exponent $\gamma = 3.0$, a spatial spectrum rich in high frequencies, and $\alpha = 1.0$ corresponds to $\gamma = 4.0$, a much "smoother" bottom.

5.5 SCATTERING FROM CURVED, ROUGH SURFACES

Plane wave scattering by an elastic sphere with a rough surface is an example, where the Helmholtz–Kirchhoff method, due to the assumptions inherent in the Kirchhoff approximation, is applicable only in a small área of the rough sphere. Most of the rough sphere will have sound incident under very low grazing angles and does not satisfy the conditions given in Eq. (5.52). The scattering includes shadowing, multiple scattering, and surface waves excited and propagating on the surface which are not covered by the Kirchhoff approximation. However, if the rms-roughness height h is small compared to the acoustic wavelength λ and the gradient of the roughness height (the slope) is small compared to 1, the perturbation approximation can be used to calculate the scattering from a sphere with a rough surface. In addition, the analytical solution for scattering from a sphere with a smooth surface, as discussed by Faran [81] has been available for many years. The literature contains a significant amount of data on scattering from smooth spheres of various diameters arising from such applications as transducer calibrations based on backscattering from spheres, as discussed by Foote [18]. Scattering of sound from spheres forms an illustrative example on the use of the scattering theory.

The scattering geometry given Fig. 5.15 is used in Sun and Bjørnø [82] and Bjørnø and Sun [83], where the acoustic plane wave is incident in the z-direction in a Cartesian frame and spherical coordinates are used in the sphere-centered reference frame.

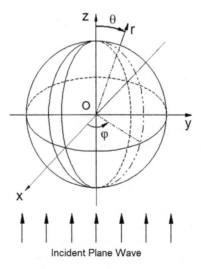

FIGURE 5.15

The geometry for scattering of plane acoustic waves by an elastic sphere with a rough surface.

The pressure amplitude function P_{inc} of the incident plane wave may be written as:

$$P_{inc} = P_0 \sum_{n=0}^{\infty} (2n + 1)i^n j_n(kr)P_n(\cos \theta), \qquad (5.60)$$

where P_0 is the incident wave's pressure amplitude, $k = 2\pi f/c$ is the acoustic wave number, c is the sound velocity in the water around the sphere, f is the incident wave's frequency, j_n is the nth-order spherical Bessel function, and $P_n(\cos \theta)$ is the nth-order Legendre polynomial. Due to the symmetry in the scattering geometry, shown in Fig. 5.15, the incident and scattered waves are independent of the spherical coordinate φ.

The scattered (outgoing) acoustic waves with the amplitude function, p_{sca}, based on scattering from *a smooth sphere* can now be determined by using resonance scattering theory, as discussed by Neubauer [84], and the form function from Hay and Mercer [85]:

$$p_{sca} = P_0 \sum_{n=0}^{\infty} c_n[j_n(kr) + i\, y_n(kr)]P_n(\cos \theta) \qquad (5.61)$$

with

$$c_n = -(2n + 1)i^{n+1} \sin \eta_n e^{-i\eta_n}.$$

In Eq. (5.61), y_n denotes the nth-order spherical Neumann function and η_n is the phase shift of the nth partial wave, as discussed by Faran [81]. If r is large, when the field point for measuring the scattered wave is situated at some distance from the sphere, Eq. (5.61) for the scattered wave pressure reduces to:

$$p_{sca} = -P_0 \frac{e^{ikr}}{kr} \sum_{n=0}^{\infty} (2n + 1) \sin \eta_n e^{-i\eta_n} P_n(\cos \theta). \qquad (5.62)$$

which in terms of the *form function*, f_∞, from Hay and Mercer [85], becomes:

$$p_{sca} = P_0 a f_\infty \frac{e^{irk}}{2r} \qquad (5.63)$$

where $f_\infty = -2/(ka)\sum_{n=0}^{\infty}(2n + 1)\sin \eta_n\, e^{-i\eta_n} P_n(\cos \theta)$.

For backscattering, i.e., $\theta = \pi$, f_∞ becomes $f_\infty = -\frac{2}{ka}\sum_{n=0}^{\infty}(2n + 1)(-1)^n \sin \eta_n\, e^{-i\eta_n}$.

In Eq. (5.63), a is the smooth sphere's radius, and the form function f_∞ can be divided into two parts, one representing the motion of a rigid, mobile sphere, and one representing the natural vibration of the elastic sphere.

When roughness is introduced on the sphere's surface and the conditions for using the perturbation approximation are satisfied, incident and scattered field can be developed into a Taylor series referenced to the smooth sphere's reference surface, $r = a$, i.e.,

$$f(r_0, \theta) = f(a, \theta) + \zeta f'(a, \theta) + \frac{\zeta^2}{2}f''(a, \theta) + ..., \text{ where } f = p_{inc} \text{ or } p_{sca}, \quad (5.64)$$

where (r_0, θ) is a point on the rough surface, (a, θ) is a point on the mean (smooth) surface, and $\zeta = \zeta(\theta)$ is the roughness height above the mean surface. The scattered field is given by contributions from several terms that represent the perturbation approximation orders:

$$p_{sca} = p_{sca,0}(r, \theta) + p_{sca,1}(r, \theta) + p_{sca,2}(r, \theta) + \cdots. \qquad (5.65)$$

Standard boundary conditions that relate the continuity in pressure, displacement, and shear stress on the sphere's surface are used to develop the zeroth through nth orders in the perturbation approximation. The zeroth-order term in Eq. (5.65) represents scattering from a smooth, elastic sphere, as given by Eq. (5.61) or (5.62). Based on Helmholtz integral equation, the first-order perturbation contribution to scattering from the rough sphere is:

$$p_{sca,1}(\mathbf{r}, \theta) = \int_S \zeta(\theta_0) \Big[p''_{inc}(a, \theta_0) + p''_{sca,0}(a, \theta_0) \Big] G(\mathbf{r}, \mathbf{r}_0) \, dS, \qquad (5.66)$$

where double prime is the second derivative with respect to r, and the integration is performed over the smooth sphere's reference surface, S. The subscript 0 refers to a point on the rough sphere, and the position vector $\mathbf{r}$ is the field point where scattered field measurements are performed. The position vector $\mathbf{r}_0$ is a position on the rough sphere's surface. The half-space Green's function, $G(\mathbf{r}, \mathbf{r}_0)$, in the far field (e.g., $r \gg a$ and $r \gg \lambda$) is:

$$G(\mathbf{r}, \mathbf{r}_0) = e^{ikr} \frac{e^{-ika(\cos\theta\cos\theta_0 + \sin\theta\sin\theta_0)}}{2\pi r}. \qquad (5.67)$$

In the first-order perturbation approximation, the field scattered from a sphere with small roughness is given by superimposing the field scattered from a smooth sphere, represented by the zeroth-order term in Eq. (5.65) with the first-order term given by Eq. (5.66). Due to the random nature of surface roughness, perturbation contributions to the scattered field arising from surface roughness can only be evaluated by ensemble averaging Eq. (5.66) over the whole rough surface, which leads to a coherently scattered acoustic field. If the surface roughness satisfies the condition $\langle \zeta \rangle = 0$, then Eq. (5.66) shows that to this order, $\langle p_{sca,1}(\mathbf{r}) \rangle = 0$, i.e., the first-order perturbation approximation does not contribute to changes in the coherently scattered field, and only contributes to the incoherently scattered field, as shown by evaluating the average scattered acoustic intensity. Thus the first-order perturbation approximation does not conserve energy since the coherent field is unchanged while the incoherent field is changed by the rough surface.

Since the first-order perturbation contribution does not conserve energy, the second-order perturbation contribution to the scattered field is required. Using standard boundary conditions on the sphere and the Helmholtz integral equation the coherent part of the second-order perturbation contribution to the total scattered acoustic field is:

$$\langle p_{sca,2}(r, \theta) \rangle = k^3 h^2 P_0 a^2 \frac{e^{ikr}}{4\pi r} \int_0^\pi \sin\theta_0 F_1 F_2 \, d\theta_0, \qquad (5.68)$$

where F_1 and F_2 are given by Eqs. (5.69) and (5.70), i.e.:

$$F_1(\theta_0) = \sum_{n=0}^{\infty} (2n+1) i^n \left\{ j_n''' + \left[y_n''' - i j_n''' \right] \sin \eta_n e^{-i\eta_n} \right\} P_n(\cos \theta_0) \qquad (5.69)$$

and

$$F_2(\theta, \theta_0) = e^{-ika(\cos \theta \cos \theta_0 + \sin \theta \sin \theta_0)}. \qquad (5.70)$$

The coherent part of the second-order perturbation contribution, Eq. (5.68), to the total scattered acoustic field is given in terms of the form function $f_{\infty}^{sca,2}$ by the expression:

$$< p_{sca,2}(r, \theta) > = P_0 a f_{\infty}^{sca,2} \frac{e^{ikr}}{2r} \qquad (5.71)$$

with

$$f_{\infty}^{sca,2} = k^2 h^2 \frac{ka}{2\pi} \int_0^{\pi} \sin \theta_0 F_1 F_2 \, d\theta_0. \qquad (5.72)$$

For backscattering, i.e., $\theta = \pi$, $\sin \theta = 0$, $\cos \theta = -1$, and Eq. (5.72) reduces to:

$$f_{\infty}^{sca,2} = -\frac{2}{ka} \sum_{n=0}^{\infty} (2n+1)(-1)^n \sin \eta_n e^{-i\eta_n} \left\{ -D_n^2 k^2 h^2 \right\} \qquad (5.73)$$

where D_n^2 is:

$$D_n^2 = (-i)^n \left[\frac{ka \, j_n'''(ka)}{\sin \eta_n e^{-i\eta_n}} + ka \, y_n'''(ka) - ika \, j_n'''(ka) \right] \cdot \int_{-1}^{+1} ka \, P_n(x) e^{ikax} \, dx.$$

By use of $kh \ll 1$, the backscattering form function for the elastic, rough sphere is:

$$f_{\infty} = -\frac{2}{ka} \sum_{n=0}^{\infty} (2n+1)(-1)^n \sin \eta_n e^{-i\eta_n} e^{-D_n^2 k^2 h^2}. \qquad (5.74)$$

When Eq. (5.74) is compared with Eq. (5.63), the form function for the backscattering from the smooth sphere, it is obvious that the last term, $\exp(-D_n^2 k^2 h^2)$, in Eq. (5.74) expresses the rough surface scattering contributions to the smooth surface scattering given by Eq. (5.63).

The backscattering cross section σ_{bs} for the rough sphere can now be written by using the backscattering form function as:

$$\sigma_{bs} = \frac{a^2}{4} |f_{\infty}|^2. \qquad (5.75)$$

Fig. 5.16 shows the influence of rms-roughness heights h when it is normalized by the smooth sphere radius a.

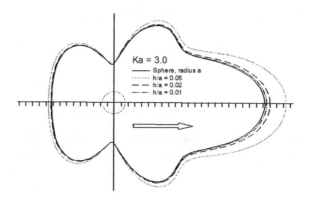

FIGURE 5.16

Directivity patterns for scattering from a rough, solid cast iron sphere with different rms-roughness heights *h* normalized by the smooth sphere radius *a* for $ka = 3$. The *solid line* is the smooth sphere directivity pattern, *dashed lines* are directivity patterns for $h/a = 0.01$ and 0.02, respectively, and the *dotted line* is the directivity pattern for $h/a = 0.05$.

*Adapted from Figure 3, Bjørnø, L., Scattering of Plane Acoustic Waves at Elastic Particles with Rough Edges, Hydroacoustics, **6**, pp. 7–18, 2003, with permission of Journal Hydroacoustics.*

Fig. 5.16 shows that an increase in the scattered sound amplitude in the backward, and, in particular, in the forward direction, relative to the directivity patterns from a smooth sphere, is due to roughness. For increasing *ka* values the forward scatter increases. The increase in the scattered field amplitude in nearly all directions for increasing surface roughness may partly be explained by one weakness in the perturbation approximation that energy conservation is not fully satisfied. In addition, the perturbation approximation may not be suitable for the special geometry of a spherical surface.

Fig. 5.17 presents numerical results for the backscattering form function for a solid cast iron sphere calculated by using Eq. (5.74) for a series of *h/a* values. The solid line is the smooth sphere backscattering form function. The deviations between the smooth sphere backscattering form function and the rough sphere values for increasing *ka* values can be explained as follows. For a given *h/a* value, the rms-surface roughness height *h* is a constant. Since the backscattering form function is a function of the incident wave's acoustic frequency, a change in frequency changes *ka* and *kh*. As the *ka* value increases, the apparent, or effective, roughness of the spherical surface increases, i.e., the sphere appears rougher. From Fig. 5.17 it may be seen that, for $ka < 10$ (i.e., for $kh < 1.0$ with $h/a = 0.1$) the change in the backscattering form function due to the surface roughness is small, which reflects the nature of the perturbation approximation. For $ka > 10$ (i.e., for $kh > 1$ and $h/a = 0.1$), the deviations between the form functions by smooth and rough spheres with increasing frequency increases, with lower form function values for the rough sphere. For *ka* values over 20 (i.e., for $kh > 2$ for $h/a = 0.1$), the deviation between the smooth and the rough sphere form function increases dramatically, with much

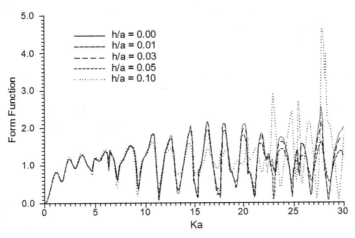

FIGURE 5.17

The backscattering form function for the rough surface of a cast iron sphere as a function of *ka*, for a series of rms-roughness heights *h/a*, where *a* is the smooth sphere radius (which for a rough sphere is the mean sphere surface radius).

*From Figure 2, Bjørnø, L., Scattering of Plane Acoustic Waves at Elastic Particles with Rough Edges, Hydro-acoustics, **6**, pp. 7–18, 2003, with permission of Journal Hydroacoustics.*

higher form function values for the rough sphere, since with increasing *kh* values, the conditions for using the perturbation approximation are no longer satisfied. Fig. 5.18 shows a comparison between measured and calculated backscattering form functions for a rough cast iron sphere reported by Bjørnø [8]. The ultrasonic

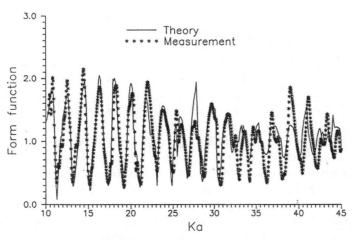

FIGURE 5.18

Measured and calculated backscattering form function for the rough surface of a cast iron sphere.

*From Figure 6, Bjørnø, L., Scattering of Plane Acoustic Waves at Elastic Particles with Rough Edges, Hydro-acoustics, **6**, pp. 7–18, 2003, with permission of Journal Hydroacoustics.*

frequencies used to measure the backscattering from the 25 mm diameter cast iron sphere were in the 200–800 kHz band.

In 2015, a finite element (FE) model for scattering of acoustic signals from fluid and elastic rough bottom surfaces has been developed by Isakson et al. [86]. This model is compared to models based on the perturbation theory and on the Kirchhoff approximation for various grazing angles. For a fluid-like bottom the authors found agreement between the results of the FE and the Kirchhoff approximation except at small grazing angles. For the elastic case, the perturbation theory predicts the FE results well, except at the shear wave intromission angle. At this intromission angle the FE model shows a more realistic solution, probably due to multiple scattering being accounted for in the FE model, but not in the perturbation theory for the modeled surface roughness.

5.6 REVERBERATION

Reverberation is one of the most important factors that may influence and deteriorate the reception of underwater acoustic signals, such as echoes from targets and underwater communication. Since reverberation is created by the same physical processes that cause echoes to form, the characteristics of the two signals are very similar which frequently make it very difficult to remove or suppress reverberation. Reverberation is normally heard as a long irregular and slowly decaying signal with several sharper peaks caused by backscattering from stronger scatterers in the water column, on or near the sea surface, or on or in the seabed. Surface and bottom bounces of transmitted sound signals also cause temporal fluctuations in the reverberation level, which in deep water can be reduced by using sonars with sufficient vertical directivity. In shallow water reflections from the surface and bottom produce level fluctuations very early after the sonar pulse's emission. Also, the sound velocity profile, in particular in deep water, has a substantial influence on the reverberation signal and its duration. If downward refraction occurs, the bottom reverberation dominates. Upward refraction amplifies the sea surface's reverberation influence and makes the reverberation characteristics dependent on the wind speed and sea state. Therefore, reverberation qualities are closely connected with the scattering—in particular the backscattering—qualities of the acoustic transmitting and receiving system's environment.

Measurements of volume and surface, i.e., sea surface and bottom, reverberation have been carried out extensively using explosives as sources of transient omnidirectional signals, as discussed by Chapman and Harris [39]. In shallow water it is necessary to use sonar systems with a high degree of directivity to obtain reliable data by reducing the backscatter from the sea surface when seabed backscatter measurements are being made.

Over the decades prior to 2016 several equations that relate the reverberation level RL {unit: dB} to the backscattering strength from volume and surface reverberation sources have been developed. The reverberation intensity is a function of time after the emission of the acoustic signal, normally a short duration pulse and is the sum of backscattered signal intensities from individual scatterers. Multiple scattering is normally not included in the reverberation calculation, but as mentioned in Section 5.2.2, it may have an influence when multiple scattering objects have a separation small compared with their size and acoustic wavelength. Also, reverberation expressions assume the sound signals travel to the scatterers and back to the (monostatic) receiver via the same path, and that the acoustic signal duration is small compared to the time it takes to reach the scatterers. For uniform scattering of a CW pulse—a tone burst—from scatterers in an infinite, nonrefracting underwater environment as shown in Fig. 5.19A, the equivalent plane-wave *volume reverberation level* RL_v is:

$$RL_v = SL - 20 \log r - 2\alpha r + 10 \log \psi + 10 \log \left(\frac{c\tau}{2}\right) + S_{bv}, \qquad (5.76)$$

where RL_v is the rms-pressure level of the reverberation {unit: dB rel. 1 µPa}, SL is the source level of the transmitted pulse measured at 1 m source distance

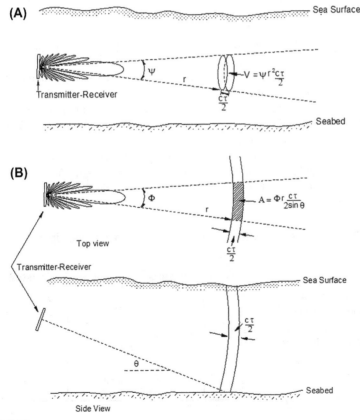

FIGURE 5.19

Reverberation caused by backscattering from (A) volume scatterers in the volume V situated in the water column and from (B) surface scatterers of area A of the seabed.

{unit: dB rel. 1 µPa}, and r is the distance {unit: m} between the source and the scatterers contributing to the reverberation, c is the velocity of sound in seawater {unit: m/s}, and S_{bv} {unit: dB} is the volume scatterer's backscattering strength per unit volume (i.e., the target strength of 1 m^3 of the scattering volume). r is a function of time t through $r = ct/2$, which introduces the time dependence of the reverberation level. Sound propagation is assumed to follow a spherical propagation law. α {unit: dB/m} is the absorption coefficient for sound propagation in seawater and τ {unit: s} is the transmitted tone burst duration, or for a wideband pulse τ is the reciprocal of the bandwidth. ψ {unit: steradians} is an ideal solid angle beam width for the source–receiver combination—the monostatically operating sonar system—including azimuth and elevation, and is the aperture angle for an ideal beam pattern with a flat response over the angle, ψ, and no response outside ψ. Expressions for the ideal solid angle ψ for various transmitter–receiver systems provided by Urick [2] are given in Table 5.7, where λ is the wavelength {unit: m}.

Similar to Eq. (5.76), Eq. (5.77) for *the surface* (i.e., seabed or sea surface) *reverberation level*, RL_s, can be used for monostatic operation of a sound transmitter–receiver system transmitting a CW pulse into a semi-infinite medium, i.e.:

$$RL_s = SL - 30 \log r - 2\alpha r + 10 \log \Phi + 10 \log\left(\frac{c\tau}{2\cos\theta}\right) + S_{bs}, \qquad (5.77)$$

where Φ {unit: radians} is the ideal, plane angle beam width of the source–receiver system and θ is the grazing angle. S_{bs} is the backscattering strength {unit: dB} of 1 m^2 of the scattering surface. Urick [2] provides several expressions for the angle, Φ, which may be found under the same assumptions as for Ψ in Eq. (5.76). These expressions are given in Table 5.8.

To illustrate the use of Eq. (5.77) the expected surface reverberation level contribution can be calculated under the following operational conditions for a sonar and its environment. The reverberation is produced by a seabed consisting of fine sand

Table 5.7 Transducer Array Reverberation Level Dependence Upon Solid Angle, ψ

Description	10 log ψ	Constraints
Circular, plane transducer array of radius a in an infinite baffle	$20 \log\left(\frac{\lambda}{2\pi a}\right) + 7.7$	$a > 2\lambda$
Rectangular array, with horizontal side length a and vertical side length b, in an infinite baffle	$10 \log\left(\frac{\lambda^2}{4\pi ab}\right) + 7.4$	a and $b \gg \lambda$
Vertical line array of length l	$10 \log\left(\frac{\lambda}{2\pi l}\right) + 14.2$	$l > \lambda$

Table 5.8 Surface Reverberation Level Dependence Upon Plane Angle Beam Width, Φ

Description	10 logΦ	Constraints
Circular plane transducer array of radius a in an infinite baffle	$10 \log\left(\frac{\lambda}{2\pi a}\right) + 6.9$	$a > 2\lambda$
Rectangular transducer array, with vertical side length a, in an infinite baffle	$10 \log\left(\frac{\lambda}{2\pi a}\right) + 9.2$	a and $b \gg \lambda$
A horizontal line array of length l	$10 \log\left(\frac{\lambda}{2\pi l}\right) + 9.2$	$l > \lambda$

insonified at a 20 degrees grazing angle by an echo-ranging sonar. The sonar emits a 25 kHz tone burst with a 0.5 ms duration and produces a 226 dB source level relative to 1 µPa at 1 m. The sonar is a 0.3 m diameter circular plane array in a baffle, and the sonar is 50 m above the seabed. The backscattering strength $S_{bs} = -29$ dB and the distance r between the sonar transducer and the scattering surface is $r = 50/\sin\theta = 146$ m. The ideal, plane angle beam width Φ of the source–receiver system is $\Phi = 17.9$ degrees. From Chapter 4 the absorption coefficient $\alpha = 6.9 \cdot 10^{-4}$ Np/m in seawater at 25 kHz. The expected surface reverberation level RL_s from Eq. (5.77) is then $RL_s = 226 - 65 - 0.2 - 5 - 4 - 29 = 122.8$ dB rel. 1 µPa.

Because omnidirectional sources and receivers, such as explosives and hydrophones, are used frequently in the investigation of scattering leading to reverberation over a broad frequency range, a reformulation of the expressions for the reverberation level RL in terms of the acoustic energy per unit area at a known distance from the explosive sound source must be made. Chapman and Marshall [87] provide the following expression for *narrow band reverberation produced by layered scatterers in the ocean volume* as:

$$RL_v = E + 10 \log \int_{z1}^{z2} \Omega(z) \ dz - 30 \log t + \kappa, \tag{5.78}$$

where RL_v is the rms-pressure level of the reverberation in the frequency band being considered {unit: dB rel. 1 µPa}, E is the energy level per unit area in the same frequency band at a 100 m distance from the explosive source {unit: dB rel. 1 µPa$^2 \cdot$ s}, and t is time {unit: s}. The volume scattering coefficient Ω is a function of the depth, z, and may be expressed by: $\Omega = (I_{sca}r^2)/(I_{inc}V)$, where I_{sca} is the scattered intensity at the distance r, while I_{inc} is the incident acoustic wave intensity scattered by the volume V. The constant κ depends on the measurement conditions and for an infinite ocean $\kappa = 42$ dB.

Since echo detection from a target is strongly influenced by the presence of reverberation and the target is frequently surrounded by reverberation producing scatterers, which produce interfering echoes, the so-called *clutter*, of no interest to the target detection, see Stanton et al. [88], it is of interest to calculate the *echo-to-reverberation level ratio*. If it is assumed that the transmission loss for the echo, given by Eq. (5.8), is the same as the transmission loss for reverberation, i.e., the two return signal contributors are at the same distance from the monostatic source–receiver system, then combining Eq. (5.8) and the reverberation level Eqs. (5.76) and (5.77) provide:

$$EL - RL_{v,s} = TS - S_{bv \text{ or } bs} - 10 \log(A \text{ or } V), \tag{5.79}$$

where EL is the rms-pressure level of the echo {unit: dB rel. 1 µPa}, RL_v or RL_s are the reverberation levels from Eqs. (5.76) and (5.77), respectively, and TS is the target strength {unit: dB}. S_{bv} or S_{bs} denote the scattering strength of the volume or the surface, expressed in Eq. (5.76) or (5.77), respectively. The factors A and V are characteristics of the sonar system and are expressed by: $A(r) = (\Phi r)(c\tau/(2 \cos \theta))$ and $V(r) = (\psi r^2)(c\tau/2)$, as shown in Fig. 5.19. Eq. (5.79) shows the interesting feature that the echo-to-reverberation ratio is independent of source level and transmission loss; however, the reverberation level is a function of range r since the insonified volume $V(r)$ and/or surface $A(r)$ are functions of r.

As an example consider a cylindrical target with a 2 m length and a 0.5 m diameter on the sandy seabed insonified under beam aspect by a 25 kHz sonar, it will have a target strength $TS = 9.2$ dB, as discussed in Section 5.2.1.4. The echo level above the reverberation level found from Eq. (5.79) is 25.6 dB.

Since r is a function of time, t, the *volume reverberation level* expressed by Eq. (5.74) decreases with *time* as $-20 \log(t/t_o)$, where t_o is the time duration of the acoustic signal. The surface reverberation provided by Eq. (5.77) shows a more rapid decrease in level with time, i.e., $-30 \log(t/t_o)$, compared to volume reverberation. The range, r, dependence of the echo-to-reverberation ratio for small target detection when influence from volume and surface reverberation is present, varies from a target echo with the r-dependence, $-40 \log r$, to surface reverberation with the r-dependence, $-30 \log r$, to volume reverberation with the r-dependence, $-20 \log r$. The influence of *ambient noise*, which is discussed in Chapter 6, adds to the detection difficulties created by reverberation since the ambient noise at a position in the sea is essentially constant in level, independent of r, and only shows long-term variations, i.e., no level variations over the time it takes for the sound signal to pass between the sonar system and the target.

Eqs. (5.76)–(5.79) only provide a good approximation to the reverberation level under conditions consistent with their derivation. Spherical spreading is generally assumed; however, at long distances in shallow water a cylindrical spreading is more appropriate. Shadow and convergence zone effects, as discussed in Chapter 3, are not included. At longer ranges the results from the reverberation level expressions are less reliable due to the increasing number of scatterers contributing to the reverberation's time history, the increasing number of sea surface and seabed bounces and the produced multipath arrivals, as discussed in Chapter 2 where contributions from more than one scatterer arrive simultaneously at the receiver, make it difficult to separate the individual contributions to the reverberation level.

The sources of reverberation, backscattering from the large number and broad variety of types of scatterers in the water column and at its boundaries, form the basis for a substantial variation in reverberation amplitude with time due to reverberation. Extensive experimental measurements have confirmed that the instantaneous reverberation amplitude variation, like many other underwater acoustic signals, follows a Rayleigh distribution [6], since a large number of signals are mixed and all signals are sinusoidal with the same frequency discussed by Abraham and Lyons [89], but with random phases equally distributed between $-\pi$ and $+\pi$, which is discussed in Chapter 6. However, reverberation amplitude statistics do not always exhibit a Rayleigh probability distribution function, and other statistical distribution functions like the k-distribution have to be used. This is the case when the sonar's range-bearing resolution cell size is small, which, for instance, is found at higher grazing angles for parametric arrays or wide aperture arrays such as a synthetic aperture sonar combined with wide bandwidth transmitted waveforms, or if the backscattering seabed is inhomogeneous, for instance, due to patchiness in space and time. This patchiness can be a result of living shellfish not uniformly distributed on the seabed, or by motion of seagrass (e.g., *Posidonia oceanica*) due to swells or currents, which shows acoustical impedance contrasts due to trapped gas bubbles. The patchiness is most frequently found by seabeds in shallow water, and the nonuniformity of the seabed environment leads to interfering backscattered signals and produces discrete reverberation or target like echoes, the *clutter*. At higher grazing angles narrow sound beams include fewer patches of a seabed, while lower grazing angles increase the number of patches with influence on the reverberation and thus drive the amplitude statistics toward the Rayleigh distribution, as discussed by Lyons et al. [90]. The deviation from a Rayleigh statistical distribution may influence the development detection algorithms for distinguishing between mine and nonmine debris on a seabed and adds to the detection and identification difficulties in reverberation limited environments.

Since *RL* depends on the source level *SL* as shown in Eqs. (5.76) and (5.77), an increase in the source level to improve the echo-to-reverberation level ratio only leads to an increase in the reverberation level without any improvement in the echo-to-reverberation level ratio. An improvement in this ratio can be achieved by reducing the signal duration or, which is the same, increasing the signal bandwidth, if the sonar system permits, which lowers the amplitude level of the undesired backscatter contributions. Substantial improvements in echo-to-reverberation level ratio can be obtained by using a sonar system with a higher directivity to obtain spatial selectivity where contributions to the reverberation level from scatterers outside the direction between the echo-producing target and the sonar system are considerably reduced.

The moving sea surface, moving scattering objects in the water column, such as fish schools and the moving platform where the sonar system is mounted, may cause

Doppler effects that lead to frequency shifts in the mean reverberation frequency and frequency spreading. Also tidal water and water currents, which cause the sea surface and scatterers to move in the water column, will broaden the reverberation's spectral content.

Measured data have produced evidence that volume reverberation in shallow water varies more rapidly in time and space than in deep water. This is particularly caused by the presence of swim bladder—bearing fish of various sizes. In general, volume reverberation is rarely the dominant reverberation type except when backscattering contributions from long ranges in calm, deep waters are the only reverberation sources. In general, in most cases in deep water sea surface reverberation is dominant. The sound refraction introduced by a near surface bubble layer may strongly enhance the air—sea interface reverberation contributions in the low- to moderate-frequency range. Since the bubbles are produced by breaking waves, reverberation shows strong wind-speed dependence as discussed by Kieffer et al. [91]. In shallow water (i.e., for water depths less than 200 m) and low wind speeds, the seabed appears to be the strongest contributor to reverberation.

The *detection threshold (DT)* is a function of the echo-to-reverberation level ratio, when detection is reverberation limited. Before classification and further action can be taken on a target, the first step is its detection. The detection threshold *DT* {unit: dB} for a signal of interest, such as a target echo buried in noise, is defined by Urick [2] as the ratio of the signal power in the receiver bandwidth to the noise power in a 1-Hz bandwidth, i.e., the noise spectrum level measured at the receiver terminals. Therefore, the required signal to noise ratio must be decided on a realistic basis generally involving a probability criterion. The decision process requires the threshold to be set to a level that when it is exceeded, the decision "target is present" is made, and no false alarms occur, according to Picinbono [92]. To set this threshold correctly is difficult. If the threshold is too high, only strong echo targets are detected and targets are missed, and if it is set too low, too many false alarms are produced.

For reverberation limited target detection by active sonar, the relation between the detection threshold and the echo and the reverberation levels is:

$$DT = EL - RL_{v,s},\qquad(5.80)$$

which expresses the *minimum detectable echo level* in a reverberation background.

If an expression for the detection level of a target echo in a background of broadband Gaussian noise is used, according to Urick [2], the reverberation amplitude in general is Rayleigh distributed and the reverberation is not broadband, but may show a frequency spread caused by factors mentioned earlier. The detection threshold under reverberation limitations is:

$$DT = 5 \log \frac{d_i w}{t} - 10 \log w'\qquad(5.81)$$

where d_i is the detection index connected with the detection probability, as given by Fig. 12.7 in Urick [2], w is the bandwidth of the receiving system, t is the echo signal's time duration, and w' is the reverberation bandwidth, where in general $w' > w$.

According to Kroenert [93] the *reverberation limited detection threshold* DT_R may be written as a function of the reverberation bandwidth w_R {unit: Hz} and the detection index d_i:

$$DT_R = 10 \log \left(\frac{d}{2w_R} \right),$$
(5.82)

DT_R is inversely proportional to the measured reverberation bandwidth, which should be made as large as possible to obtain the smallest detection threshold. When a pulsed monochromatic wave (a CW pulse) is used with pulse length t {unit: s}, the approximate signal bandwidth is $1/t$ {unit: Hz}. Since the reverberation bandwidth is closely connected with the signal bandwidth, the reverberation limited detection threshold for a CW pulse will be about $10 \log(d_i t/2)$, i.e., $t \sim 1/w_R$. The CW-pulse duration, t, should be as short as possible to increase the reverberation limited detection performance. The wide reverberation spectrum obtained due to the short pulse duration provides the improved reverberation limited detection performance, not the reduced reverberation power available due to the short pulse length. However, according to Urick [2], the *noise limited detection threshold* DT_N may be expressed by:

$$DT_N = 10 \log \left(\frac{d}{2t} \right),$$
(5.83)

which shows that the noise limited detection threshold decreases with increasing pulse length for a CW signal. The contradiction between the reverberation and the noise limited detection thresholds explains why a wideband FM-slide transmission, i.e., a linear frequency-modulated pulse, is the preferred detection waveform since it provides the long pulse length necessary for noise limited detection and the wide bandwidth required for reverberation limited detection, regardless of the receiver type. The FM mode requires less signal amplitude for detection than a CW mode, and the FM mode has been shown by Kroenert [93] to outperform the CW mode by up to 15 dB for 90% detection probability.

5.7 SYMBOLS AND ABBREVIATIONS

Table 5.9 is a list of symbols and abbreviations used in this chapter along with the page where the symbol or abbreviation is defined and first used.

Table 5.9 List of Symbols and Abbreviations

Symbol or Abbreviation	Description	Page of First Usage
a	Radius of a sphere or radius of curvature {unit: m}	303
a_s	Mean particle radius for particles in suspensions {unit: m}	318
c_g	Velocity of sound in gas {unit: m/s}	316
c_p	Specific heat at constant pressure {unit: J/°C}	307
c_s	Velocity of sound in an elastic sphere {unit: m/s}	305
c_v	Specific heat at constant volume {unit: J/°C}	307
c_w	Velocity of sound in water {unit: m/s}	305
d_i	Detection index {dimensionless}	353
f	Frequency {unit: Hz} or wave field function {dimensionless}	306
f_∞	Form function {dimensionless}	310
$f_p(a)$	Probability function for scatterers with dimension a	313
f_r	Resonance frequency {unit: Hz}	306
g	Gravitational constant {$= 9.81$ m/s^2}	317
h	Rms-roughness height {unit: m}	298
$j_n(x)$	nth-order spherical Bessel function {dimensionless}	341
k	Acoustic wave number {unit: m^{-1}}	298
ℓ	Correlation length {unit: m}	333
$n(a)$	Number of bubbles of radius between a and da	315
p_g	Ambient pressure {unit: Pa}	317
p_0	Ambient pressure acting on a bubble at sea surface {unit: Pa}	307
p_{bs}	Backscattered peak pressure amplitude {unit: Pa}	302
p_{inc}	Peak pressure incident on the target {unit: Pa}	302
r	Distance between scattering surface and field point {unit: m}	303

Table 5.9 List of Symbols and Abbreviations—cont'd

Symbol or Abbreviation	Description	Page of First Usage
r_c	Radius of curvature of a rough surface {unit: m}	335
w	Wind speed (unit: m/s) or bandwidth {unit: Hz}	316
$y_n(x)$	nth-order spherical Neumann function {dimensionless}	341
z	Water depth {unit: m}	307
A	Area of a scattering rough surface {unit: m^2}	338
E	Bulk modulus of elasticity {unit: Pa}	316
H	Water column height {unit: m} or DSL thickness {unit: m}	314, 317
I_{bs}	Backscattered acoustic intensity {unit: W/m^2}	302
I_e	Echo intensity {unit: W/m^2}	300
I_{inc}	Incident plane wave intensity at scattering surface or volume (unit: W/m^2)	300
I_r	Reference intensity, 1 W/m^2	300
I_{sca}	Sound intensity scattered from a unit area {unit: W/m^2}	299
$J_0(x)$	Zeroth-order Bessel function of first kind {dimensionless}	339
K	Spatial wave number {unit: m^{-1}}	333
L	Fish length or cylinder length {unit: cm}	309
N_s	Average number of scatterers per unit volume {dimensionless}	313
$P_n(x)$	nth-order Legendre polynomial {dimensionless}	341
$\boldsymbol{R}$	Sound amplitude reflection coefficient for normal incidence {dimensionless}	329
R_a	Rayleigh parameter {dimensionless}	298
$\boldsymbol{R_t}$	Target's intensity reflection coefficient {dimensionless}	300

Continued

Table 5.9 List of Symbols and Abbreviations—cont'd

Symbol or Abbreviation	Description	Page of First Usage
$R(\theta)$	Smooth surface amplitude reflection coefficient {dimensionless}	319
S	Scattering strength {unit: dB for 1 m^2 or 1 m^3}	299
S_b	Backscattering strength {unit: dB for 1 m^2 or 1 m^3}	302
S_{bc}	Column backscattering strength {unit: dB per m^3}	314
S_{bv}	Backscattering strength of an average unit volume {unit: dB per m^3}	314
S_{vl}	DSL volume scattering strength {unit: dB per m^3}	317
$S(\Omega)$	Frequency spectrum {dimensionless}	322
V_f	Bubble plume void fraction in breaking waves {dimensionless}	324
α	Absorption coefficient {unit: dB/m}	348
β	Void fraction, free parameter or slope versus w and f of the scattering strength curve	316
δ	Attenuation function {dimensionless} or variance of the facet slopes	306
δ_r	Attenuation function at resonance {dimensionless}	316
γ	Ratio of specific heats at constant pressure and constant volume {dimensionless}	307
γ_e	Scattering enhancement factor {dimensionless}	318
ζ	Roughness height above the mean surface of a rough curved surface {unit: m}	332
η_n	Phase shift of the nth partial wave {dimensionless}	341
θ	Angle of oblique incidence or grazing angle {unit: radians or degrees}	320
κ	Spatial roughness wave number {unit: m^{-1}}	320

Table 5.9 List of Symbols and Abbreviations—cont'd

Symbol or Abbreviation	Description	Page of First Usage
λ	Wavelength {unit: m}	298
μ	Constant in Lambert's law {dimensionless}	321
ν	Facet strength {dimensionless}	329
ξ	Facet slope or rms slope angle of a rough surface {dimensionless}	324
ρ_s	Density of an elastic sphere {unit: kg/m^3}	305
ρ_w	Density of water {unit: kg/m^3}	305
σ_{at}	Total acoustic absorption cross section {unit: m^2}	300
σ_{bs}	Backscattering cross section {unit: m^2}	302
σ_{et}	Total acoustic extinction cross section {unit: m^2}	300
σ_s	Acoustical scattering cross section {unit: m^2}	300
σ_{se}	Effective acoustical scattering cross section {unit: m^2}	300
σ_{st}	Total acoustic scattering cross section {unit: m^2}	300
τ	Transmitted tone burst duration {unit: s}	348
φ	Scattering angle {unit: radians or degrees}	321
Ω	Frequency {unit: Hz}	322
AUVs	Autonomous underwater vehicles	301
CW	Continuous wave	312
DT	Detection threshold level {unit: dB}	352
EL	Echo level {unit: dB}	300
RL	Reverberation level {unit: dB}	346
RL_s	Surface reverberation level {unit: dB}	348
RL_v	Volume reverberation level {unit: dB re 1 µPa}	347
SL	Source level {unit: dB}	301
TL	Transmission loss {unit: dB}	301
TS	Target strength {unit: dB}	301
DSL	Deep scattering layer {dimensionless}	313
erf	Error function {dimensionless}	339

REFERENCES

[1] Brekhovskikh, L. and Lysanov, Yu., *Fundamentals of Ocean Acoustics*, Springer Series in Electrophysics, Vol. 8, Springer-Verlag, Berlin, 1982.

[2] Urick, R.J., *Principles of Underwater Sound*, 3rd Edition, McGraw-Hill Book Company, New York, 1983, Peninsula Publishing.

[3] Medwin, H. and Clay, C.S., *Fundamentals of Acoustical Oceanography*, Academic Press, Boston, 1998.

[4] Neubauer, W.G., *Pulsed circumferential waves on aluminium cylinders in water.* J. Acoust. Soc. Am., **45**, (5), pp. 1134–1144, 1969.

[5] Überall, H., Ahyi, A.C., Raju, P.K., Bjørnø, I.K. and Bjørnø, L., *Circumferential − wave phase velocities for empty, fluid-immersed spherical metal shells.* J. Acoust. Soc. Am., **112**, (6), pp. 2113–2720, 2002.

[6] Rayleigh, J.W.S., *The Theory of Sound,* Vol 2. Dover Publications, New York, 1945.

[7] Bjørnø, I. K. and Bjørnø, L., *Modelling of multiple scattering in suspensions.* In: Underwater Acoustics, J.S. Papadakis (Ed.), IACM-FORTH, Heraklion, Crete, pp. 87–92, 1996.

[8] Bjørnø, L., *Scattering of plane acoustic waves at elastic particles with rough surfaces.* Hydroacoustics, **6**, pp. 7–18, 2003.

[9] Devin, C., *Survey of thermal, radiation, and viscous damping of pulsating air bubbles in water.* J. Acoust. Soc. Am., **31**, (12), pp. 1654–1667, 1959.

[10] Love, R.H., *Dorsal-aspect target strength of an individual fish.* J. Acoust. Soc. Am., **49**, (3B), pp. 816–823, 1971.

[11] Love, R.H., *Resonant acoustic scattering by swimbladder-bearing fish.* J. Acoust. Soc. Am., **64**, (2), pp. 571–580, 1978.

[12] Kjærgaard, N., Bjørnø, L., Kirkegaard, E. and Lassen, H., *Broadband analysis of acoustical scattering by individual fish.* Rapp. T-V Reun. Cons. Int. Explor. Mer., **189**, pp. 370–380, 1990.

[13] Reeder, D.B., Jech, J.M. and Stanton, T.K., *Broadband acoustic backscatter and high-resolution morphology of fish: Measurement and modelling.* J. Acoust. Soc. Am., **116**, (2), pp. 747–761, 2004.

[14] Ramirez, M. and Bjørnø, I.K., *Experimental investigations of acoustic backscattering from particles in water.* In: Underwater Acoustics, L. Bjørnø (Ed.), The European Community Press, Luxembourg, pp. 1005–1009, 1994.

[15] Bjørnø, L. and Kjeldgaard, M., *A wide frequency band anechoic water tank.* Acustica, **32**, (2), pp. 103–109, 1975.

[16] Sanders, S., Jane's Fighting Ships 2004–2005, Jane's Information Group Inc., Alexandria, Virginia, 2004.

[17] Bjørnø, I.K. and Bjørnø, L., *Significance of multiple scattering from ensembles of elastic objects in water.* Proc. Undersea Defence Technology, Nexus Publishers, UK, pp. 296–299, 1995.

[18] Foote, K.G., *Target strength of fish.* In: Encyclopedia of Acoustics, M.J. Crocker (Ed.), Vol. 1, Chapter 44, John Wiley & Sons Ltd., pp. 493–500, 1997.

[19] Trevorrow, M.V., *Measurements of near-surface bubble plumes in the open ocean with implications for high-frequency sonar performance.* J. Acoust. Soc. Am., **114**, (5), pp. 2672–2684, 2003.

[20] Thorpe, S.A., *On the clouds of bubbles formed by breaking wind-waves in deep water, and their role in air-sea gas transfer.* Phil. Trans. R. Soc. Lond., **A 304**, pp. 155–210, 1982.

[21] King III, W.F., *Sound propagation in wakes.* J. Acoust. Soc. Am., **53**, (3), pp. 735−745, 1973.

[22] Bradley, D.L., Culver, R.L., Di, X. and Bjørnø, L., *Acoustic qualities of ship wakes.* Acta Acust. United AC, **88**, pp. 687−690, 2002.

[23] Prosperetti, A., *Bubble-related ambient noise in the ocean.* J. Acoust. Soc. Am., **84**, pp. 1042−1054, 1988.

[24] Kozhevnikova, I.N. and Bjørnø, L., *Experimental study of acoustical emission from bubble cloud excitation.* Ultrasonics, **30**, (1), pp. 21−25, 1992.

[25] Deane, G.B., Preisig, J.C. and Lavery, A.C., *The Suspension of Large Bubbles Near the Sea Surface by Turbulence and Their Role in Absorbing Forward-Scattered Sound.* IEEE J. Ocean. Eng., **38**, (4), pp. 632−641, 2013.

[26] Hall, M.V., *A comprehensive model of wind-generated bubbles in the ocean and predictions of the effects on sound propagation at frequencies up to 40 kHz.* J. Acoust. Soc. Am., **86**, (3), pp. 1103−1117, 1989.

[27] Wu, J., *Bubbles in the near-surface ocean: a general description.* J. Geophys. Res., **93**, pp. 587−590, 1988.

[28] Kozhevnikova, I. N. and Bjørnø, L., *Sound propagation in upper layers of the ocean comprising bubble clouds.* In: Acoustic Signal Processing for Ocean Exploration, J.F.M. Moura & I.M.G. Lourtie (Eds.), Kluwer Academic Publishers, pp. 69−76, 1993.

[29] Wood, A.B., *A Textbook of Sound,* Bell & Sons Ltd., London, 1949.

[30] Leighton, T.G., *The Acoustic Bubble*, Academic Press, London, 1996.

[31] EU-MAST Project: *TRIDISMA − 3D Sediment transport Measurements by Acoustics.* Contract No. MAS3-CT95-0017.

[32] Thorne, P.D and Buckingham, M.J., *Measurements of scattering by suspensions of irregularly shaped sand particles and comparison with a single parameter modified sphere model.* J. Acoust. Soc. Am., **116**, (5), pp. 2876−2889, 2004.

[33] Phillips, O.M., *The Dynamics of the Upper Ocean*, Cambridge University Press, Cambridge, 1969.

[34] Pierson, W.J. and Moskowitz, L., *A proposed spectral form for fully developed wind seas based on the similarity theory of S.A. Kitaigorodskii.* J. Geophys. Res., **69**, pp. 5181−5190, 1964.

[35] Eckart, C., *Scattering of Sound from the Sea Surface.* J. Acoust. Soc. Am., **25**, (3), pp. 566−570, 1953.

[36] Urick, R. J. and Hoover, R.M., *Backscattering of Sound from the Sea Surface: Its Measurement, Causes and Application to the Prediction of Reverberation Levels.* J. Acoust. Soc. Am., **28**, (6), pp. 1038−1042, 1956.

[37] Twersky, V., *On Scattering and Reflection of Sound by rough Surfaces.* J. Acoust. Soc. Am., **29**, (2), pp. 209−218, 1957.

[38] Garrison, G.R., Murphy, S.R. and Potter, D.S., *Measurement of the Backscattering of Underwater Sound from the Sea Surface.* J. Acoust. Soc. Am., **32**, (1), pp. 104−111, 1960.

[39] Chapman, R.P. and Harris, J.H., *Surface Backscattering Strengths Measured with Explosive Sound Sources.* J. Acoust. Soc. Am., **34**, (10), pp. 1592−1597, 1962.

[40] Chapman, R.P. and Scott, H.D., *Surface Backscattering Strengths Measured over an Extended Range of Frequencies and Grazing Angles.* J. Acoust. Soc. Am., **36**, pp. 1735−1737, 1964.

[41] Ogden, P.M. and Erskine, F.T., *Surface and volume scattering measurements using broadband explosive charges in the Critical Sea Test 7 experiment.* J. Acoust. Soc. Am., **96**, (5), pp. 2908−2920, 1994.

[42] Ainslie, M.A., Harrison, C.H. and Burns, P.W., *Reverberation modelling with INSIGHT.* In: Underwater Acoustic Scattering, Proc. Inst. Acoust., **16**, (6), pp. 105−112, 1994.

[43] Ellis, D.D. and Crowe, D.V., *Bistatic reverberation calculations using a three-dimensional scattering function.* J. Acoust. Soc. Am., **89**, (5), pp. 2207−2214, 1991.

[44] McDonald, B.E., *Echoes from vertically striated subresonant bubble clouds: A model for ocean surface reverberation.* J. Acoust. Soc. Am., **89**, (2), pp. 617−622, 1991.

[45] Henyey, F.S., *Acoustic scattering from ocean microbubble plumes in the 100 Hz to 2 kHz region.* J. Acoust. Soc. Am., **90**, (1), pp. 399−405, 1991.

[46] Gragg, R.F. and Wurmser, D., *Low-frequency scattering from intermediate bubble plumes: Theory and computational parameter study.* J. Acoust. Soc. Am., **94**, (1), pp. 319−329, 1993.

[47] Sarkar, K. and Prosperetti, A., *Backscattering of underwater noise by bubble clouds.* J. Acoust. Soc. Am., **93**, (6), pp. 3128−3138, 1993.

[48] Neighbors, T.H. and Bjørnø, L., *Anomalous Low Frequency Sea Surface Reverberations.* Hydroacoustics, **4**, pp. 181−192, 2001.

[49] Neighbors, T.H., Mayer, W.G. and Bjørnø, L., *Sea surface reverberation scale modeling.* In: Proc. 4th European Conference on Underwater Acoustics, A. Alippi & G.B. Cannelli (Eds.), CNR-IDAC, Rome, Italy, pp. 789−794, 1998.

[50] McDaniel, S.T., *Sea-surface reverberation fluctuations.* J. Acoust. Soc. Am., **94**, (3), pp. 1551−1559, 1993.

[51] McDaniel, S.T. and Gorman, A.D., *Acoustic and radar sea surface backscatter.* J. Geophys. Res., **87**, pp. 4127−4136, 1982.

[52] Mellen, R.H., *Underwater Acoustic Scattering from Arctic Ice.* J. Acoust. Soc. Am., **40**, (5), pp. 1200−1202, 1966.

[53] Milne, A.R., *Underwater Backscattering Strength of Arctic Pack Ice.* J. Acoust. Soc. Am., **36**, (8), pp.1551−1556, 1964.

[54] Brown, J.R., *Reverberation under Artic Ice,* J. Acoust. Soc. Am., **36**, pp. 601−603, 1964.

[55] Brown, J.R. and Milne, A.R., *Reverberation under Arctic Sea-Ice.* J. Acoust. Soc. Am., **42**, (1), pp. 78−82, 1967.

[56] Hayward, T.J. and Yang, T.C., *Low-frequency Arctic reverberation. I: Measurements of under-ice backscattering strengths from short-range direct-part returns.* J. Acoust. Soc. Am., **93**, (5), pp. 2517−2523, 1993.

[57] Hamilton, E.L., *Compressional waves in marine sediments.* Geophysics, **37**, pp. 620−646, 1972.

[58] Hamilton, E.L., *Acoustic properties of sediments.* In: Acoustics and Ocean Bottom, A. Lara-Sáenz, C. Ranz-Guerra and C. Carbó-Fité (Eds.), Instituto de Acústica, CSIC, Madrid, pp. 3−58, 1987.

[59] Hamilton, E.L. and Bachman, R.T., *Sound velocity and related properties of marine sediments.* J. Acoust. Soc. Am., **72**, (6), pp. 1891−1904, 1982.

[60] McKinney, C.M. and Anderson, C.D., *Measurements of Backscattering of Sound from the Ocean Bottom.* J. Acoust. Soc. Am., **36**, (1), pp. 158−163, 1964.

[61] Wong, H-K. and Chesterman, W.D., *Bottom Backscattering near Grazing Incidence in Shallow Water.* J. Acoust. Soc. Am., **44**, (6), pp. 1713−1718, 1968.

[62] Jackson, D.R. and Briggs, K.B., *High-frequency bottom backscattering: Roughness versus sediment volume scattering.* J. Acoust. Soc. Am., **92**, (2), pp. 962–977, 1992.

[63] Gensane, M., *A Statistical Study of Acoustic Signal Backscattering from the Sea Bottom.* IEEE J. Ocean. Eng., **14**, pp. 84–93, 1989.

[64] Mourad, P.D. and Jackson, D.R., *A model/data comparison for low-frequency bottom backscatter.* J. Acoust. Soc. Am., **94**, (1), pp. 344–358, 1993.

[65] Lyons, A.P., *The Potential Impact of Shell Fragment Distributions on High-Frequency Seafloor Backscatter.* IEEE J. Ocean. Eng., **30**, (4), pp. 843–851, 2005.

[66] Williams, K.L., Jackson, D.R., Thorsos, E.I., Tang, D. and Briggs, K.B., *Acoustic Backscattering Experiments in a Well Characterized Sand Sediment: Data/Model Comparisons Using Sediment Fluid and Biot Models.* IEEE J. Ocean. Eng., **27**, (3), pp. 376–387, 2002.

[67] Rayleigh, J.W.S., *On the dynamical theory of gratings.* Proc. Roy. Soc., **A79**, pp. 399–416, 1907.

[68] Rice, S.O., *Reflection of Electromagnetic Waves from Slightly Rough Surfaces.* Comm. Pure Appl. Math., **4**, pp. 351–378, 1951.

[69] Ogilvy, J.A., *Theory of Wave Scattering from Random Rough Surfaces.* Adam Hilger Publ., IOP Publishing Ltd., 1991.

[70] Thorsos, E.I. and Jackson, D.R., *The validity of the perturbation approximation for rough surface scattering using a Gaussian roughness spectrum.* J. Acoust. Soc. Am., **86**, (1), pp. 261–277, 1989.

[71] Morse, P.M. and Feshback, H., *Methods of Theoretical Physics.* Vol. 1, McGraw-Hill, New York, 1953.

[72] Chapman, R.P., Bluy, O.Z. and Hines, P.C., *Backscattering from rough surfaces and inhomogeneous volumes.* In: Encyclopedia of Acoustics, M.J. Crocker (Ed.), Vol. 1, John Wiley & Sons Inc., 1997.

[73] Thorsos, E.I., *The validity of the Kirchhoff approximation for rough surface scattering using a Gaussian roughness spectrum.* J. Acoust. Soc. Am., **83**, (1), pp. 78–92, 1988.

[74] Biot, M.A., *Reflection on a Rough Surface from an Acoustic Point Source.* J. Acoust. Soc. Am., **29**, (11), pp. 1193–1200. 1957.

[75] Biot, M.A., *On the Reflection of Acoustic Waves on a Rough Surface.* J. Acoust. Soc. Am., **30**, (5), pp. 479–480, 1958.

[76] Steininger, G., Dosso, S.E., Holland, C.W. and Dettmer, J., *Estimating seabed scattering mechanisms via Bayesian model selection.* J. Acoust. Soc. Am., **136**, (4), pp. 1552–1562, 2014.

[77] Holland, C.W., Steininger, G. and Dosso, S.E., *Discrimination between discrete and continuum scattering from the sub-seafloor.* J. Acoust. Soc. Am., **138**, (2), pp. 663–673, 2015.

[78] Kur'yanov, B.F., *The scattering of sound at a rough surface with two types of irregularities.* Sov. Phys. Acoust., **8**, (3), pp. 252–257, 1963.

[79] McDaniel, S.T. and Gorman, A.D., *An examination of the composite-roughness scattering model.* J. Acoust. Soc. Am., **73**, (5), pp. 1476–1486, 1983.

[80] Jackson, D.R., Winebrenner, D.P. and Ishimaru, A., *Application of the composite roughness model to high-frequency bottom backscattering.* J. Acoust. Soc. Am., **79**, (5), pp. 1410–1422, 1986.

[81] Faran, J.J., *Sound scattering by solid cylinders and spheres.* J. Acoust. Soc. Am., **23**, (4), pp. 405–418, 1951.

[82] Sun, S. and Bjørnø, L., *Scattering of plane waves from elastic spheres with surface roughnesses*. In: Underwater Acoustics, L. Bjørnø (Ed.), The European Commission, Luxembourg, pp. 171−176, 1994.

[83] Bjørnø, L. and Sun, S., *Use of the Kirchhoff Approximation in Scattering from Elastic, Rough Surfaces*. Acoust. Phys., **41**, (5), pp. 637−648, 1995.

[84] Neubauer, W.G., *Reflection and vibrational modes of elastic spheres*. In: Acoustic Resonance Scattering, H. Überall (Ed.), Gordon & Breach, Philadelphia, pp. 31−48, 1992.

[85] Hay, A.E. and Mercer, D.G., *On the theory of sound scattering and viscous absorption in aqueous suspensions at medium and short wavelengths*, J. Acoust. Soc. Am., **78**, (5), pp. 1761−1771, 1985.

[86] Isakson, M.J. and Chotiros, N.P., *Finite Element Modeling of Acoustic Scattering From Fluid and Elastic Rough Interfaces*. IEEE J. Ocean. Eng., **40**, (2), pp. 475−484, 2015.

[87] Chapman, R.P. and Marshall, J.R., *Reverberation from Deep Scattering Layers in the Western North Atlantic*. J. Acoust. Soc. Am., **40**, (2), pp. 405−411, 1966.

[88] Stanton, T.K., Dezhang, C., Gelb, J.M., Tipple, G.L. and Baik, K., *Interpreting Echo Statistics of Three Distinct Clutter Classes Measured With a Midfrequency Active Sonar: Accounting for Number of Scatterers, Scattering Statistics, and Beampattern Effects*. IEEE J. Ocean. Eng., **40**, (3), pp. 657−665, 2015.

[89] Abraham, D.A. and Lyons, A.P., *Novell physical interpretations of k-distributed reverberation*, IEEE J. Ocean. Eng., **27**, pp. 800−813, 2002.

[90] Lyons, A.P., Abraham, D.A., Akal, T. and Guerrini, P., *Statistical evaluation of 80 kHz shallow-water seafloor reverberation*. SACLANT Undersea Research Centre Report No. SR-270, September 1997.

[91] Keiffer, R.S., Novarini, J.C. and Zingarelli, R.A., *Finite-difference time-domain modelling of low to moderate frequency sea-surface reverberation in the presence of a near-surface bubble layer*. J. Acoust. Soc. Am., **110**, (2), pp. 782−785, 2001.

[92] Picinbono, B., *General detection and estimation theory in an adaptive context*. In: Underwater Acoustics and Signal Processing, L. Bjørnø (Ed.), D. Reidel Publishing Comp., Holland, pp. 355−378, 1981.

[93] Kroenert, J.T., *Discussion of detection threshold with reverberation limited conditions*. J. Acoust. Soc. Am., **71**, (2), pp. 507−508, 1982.

Ambient Noise

6

L. Bjørnø[†]

UltraTech Holding, Taastrup, Denmark

Definitions

Ambient noise is the residual noise background measured at a hydrophone when individual noise sources cannot be identified, or ambient noise is the natural noise environment at a measurement site. Noise is unwanted sound in the ocean, since it generally interferes with the operation of sonar or other underwater sound registration equipment. The ambient noise level is the intensity, in decibel, measured at a measurement site using a nondirectional hydrophone and referred to the intensity of a plane wave having a root mean square (rms) pressure amplitude of 1 μPa. In spite of being measured in different frequency bands, ambient noise levels are always reduced to a 1 Hz frequency band and named the *ambient noise spectral density levels* {unit: dB rel 1 μPa/Hz, when measured as acoustic pressure referenced to a 1 Hz bandwidth; this is often written as dB rel 1 μPa/Hz$^{1/2}$ or dB rel 1 μPa/Hz in publications}.

6.1 PHYSICS AND PHENOMENA

Underwater ambient noise covers the *frequency range* from below 1 Hz to several hundreds of kilohertz. Over this broad frequency range, data show that ambient noise has different characteristics at different frequencies, with the spectral slope showing different behavior with varying external conditions, such as the wind speed. Pioneering work on measurements and the description of ambient noise was carried out by a group of acousticians headed by V.O. Knudsen, who investigated ambient noise in the frequency range from 200 Hz to 50 kHz. Their results, first published in 1948 by Knudsen et al. [1], are summarized in a series of curves describing ambient noise spectra, known as the *Knudsen spectra.*

From a physical and conceptual viewpoint the ambient noise level in the sea is best understood in terms of the characteristic source mechanisms. Fig. 6.1, derived from Wenz [2], illustrates the diversity of ambient noise sources. Each key source mechanism is discussed in Section 6.2. The difference between deep- and shallow-water ambient noise spectra is examined in Section 6.3. Ambient noise directivity including coastal influences on deep- and shallow-water areas is discussed in Section 6.4. Ambient noise coherence and the sources of self-noise are covered in Sections 6.5

[†]30 March 1937–24 October 2015.

Applied Underwater Acoustics. http://dx.doi.org/10.1016/B978-0-12-811240-3.00006-0

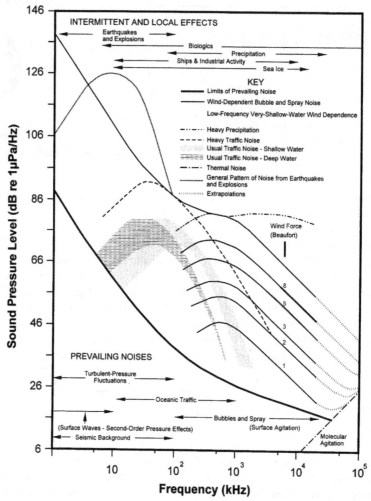

FIGURE 6.1

Ambient noise spectral density levels from 1 Hz to 100 kHz.

Reproduced from Wenz, G.M., Acoustic ambient noise in the ocean: Spectra and sources, J. Acoust. Soc. Amer., 34, (12), 1936, 1962, with the permission of the Acoustical Society of America.

and 6.6, respectively. A fundamental exposition of statistics for ambient noise and signal fluctuations in underwater acoustics is provided in Section 6.7. A list of parameters used in this chapter along with the page of first usage is given in Section 6.8, and a list of major references is provided at the end of the chapter.

6.2 SOURCES OF AMBIENT NOISE
6.2.1 TIDES AND HYDROSTATIC EFFECTS OF WAVES

At very low frequencies tides and surface waves cause large amplitude hydrostatic pressure changes. The magnitude of the tidally produced pressure changes around a hydrophone in water may be demonstrated by the fact that a 1 m increase in the water height will lead to an increase in pressure of 10^4 μPa (i.e., 200 dB rel 1 μPa). Since tidal motion is about 2 cycles/day, the tidal motion spectrum is of minor interest in relation to ambient noise. However, tidal motion may influence ambient noise measurements by changing the temperature of the hydrophone environment, which may cause pyroelectric effects in the piezoelectric hydrophone materials that produce false measurements. In turn, tidal currents can cause flow-induced vibrations of the hydrophone and its support.

Studies of the influence of the flow around a hydrophone on noise generation by Strasberg [3] have shown that the spectral density level is dependent on the flow velocity if the hydrophone is in a fixed position in the water column, is resting on the seabed, or is drifting freely with the water current. The *spectral density level* L_p of the pressure fluctuations {unit: dB rel 1 μPa/Hz} may be calculated by:

$$L_p = 124.79(117) + 27 \log U - 17 \log f \tag{6.1}$$

for a hydrophone held between the surface and the bottom in a turbulent flow,

$$L_p = 116.45(100) + 57 \log U - 17 \log f \tag{6.2}$$

for a hydrophone resting on the bottom with a turbulent boundary layer flow above it, and

$$L_p = 71.91(67) + 17 \log U - 27 \log f \tag{6.3}$$

for a neutrally buoyant hydrophone floating with the mean speed of the current. In Eqs. (6.1)–(6.3), U denotes the mean flow velocity {unit: m/s} and f is the frequency {unit: Hz}. For mean flow velocity flow rates U in knots use the constants in parentheses. The spectral density levels using Eqs. (6.1)–(6.3) are shown in Fig. 6.2 across a band from 0.1 Hz to 10 Hz, when the flowrate is 1.03 m/s (~2 knots). The substantial deviation shows the importance of taking flow-induced noise into consideration, when ambient noise measurements below 5 Hz are performed.

A characteristic frequency f of flow-generated noise sources is determined by the flow velocity U and a typical body dimension L {unit: m}—for instance, the diameter of a circular cylinder or the thickness of the flow boundary layer on a surface, through a dimensionless frequency, the so-called *Strouhal number* S_t, given by $S_t = fL/U$. For dynamically similar flows the Strouhal number is a constant, which can be determined through fluid dynamical measurements.

Surface waves are also sources of hydrostatic pressure changes in the sea; however, the pressure amplitude falls off rapidly with increasing depth and decreasing wave length of the surface waves, and therefore the importance of surface waves as a source of hydrostatic pressure changes in deep water is low. This is not necessarily true in shallow water where a rough surface may have a dominating influence on pressure-sensitive hydrophones at low frequencies.

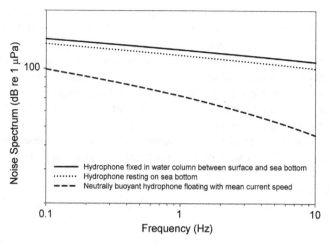

FIGURE 6.2

Hydrophone noise spectrum dependence on turbulent flow conditions.

6.2.2 SEISMIC ACTIVITIES

Because the earth is in a constant state of *seismic activity*, earth unrest causes low-frequency sound in the sea. Seismic activity ranges from large-scale intermittent sources, such as individual earthquakes (seaquakes), distant volcanic eruptions, seismic explorations, and sea mining to *microseisms* (i.e., the near constant background activity measured by seismographs), as discussed by Webb [4]. According to Urick [5], microseisms that have a nearly regular periodicity of 1/7 Hz and vertical amplitude of 10^{-6} m will cause a 120 dB rel 1 μPa pressure amplitude in the sea. Seismic activity may also be found at frequencies above 10 Hz.

Also contributions from man-made activities such as industrial plants, road transport, and construction work in coastal regions as seismic waves may propagate into shallow-water areas close to the shore.

In 1904 it was proposed by Weichert [6] that a close relation should exist between microseisms and ocean wave activities. The most favored mechanism today is nonlinear interactions between surface waves. Noise related to *ocean microseisms* dominates underwater acoustic spectra at frequencies below 4–5 Hz. Ocean surface waves traveling in opposite directions in the vicinity of a storm or as a result of a reflection from a coast can generate a standing wave field. Unlike progressive surface waves, for which the pressure effects decay exponentially with depth, a standing surface wave field produces a mean second-order pressure effect at twice the frequency of the surface waves, which is unattenuated with depth. The pressure amplitude produced by nonlinear surface wave interactions is proportional to the amplitude product of the interacting waves. Theoretical and experimental studies of the relations between ocean surface waves and microseisms have been reported by Longuet-Higgins [7], Hasselmann [8], and Hughes [9]. Kibblewhite and Evans

[10] also found that the variance density spectral level of the vertical component of the seafloor displacement should be proportional to the fourth power of the frequency of interacting ocean waves and to the square of the variance density spectral level of the ocean wave components producing the exciting pressure field.

According to Dünnebier et al. [11], low-frequency noise produced by microseisms may more readily be observed in boreholes than on or close to the seafloor, since the lower noise levels in the boreholes more than compensate for signal losses.

6.2.3 TURBULENCE

The irregular and random motion in turbulent currents of large or small scales is able to produce underwater noise. The scales range from oceanic turbulence phenomena, such as *gyres* and *eddies*, to turbulent flow due to the wind's influence on the sea surface. The pressure changes associated with turbulence may be measured far from the turbulent region and appear as a part of the ambient noise. However, the noise radiated from the turbulent region is not likely to be significant due to its *quadrupole character* and rapid falloff with distance. The mean square−radiated sound pressure from a quadrupole noise source depends on the particle velocity u {unit: m/s} of the turbulent flow in the eighth power. If the particle velocity u {unit: m/s} is low, i.e., the *acoustic Mach number* $M_a = u/c \ll 1$, where c {unit: m/s} is the sound velocity, the quadrupole noise level in water will be very low. A pressure-sensitive hydrophone may pick up the turbulent pressure when measurements are performed in the turbulent flow region. However, the pressure fluctuation is pseudosound, due to its nonpropagating character. The spectrum of the fluctuations will depend on the wave number spectrum of the turbulence, the hydrophone's size and shape, and the hydrophone's motion in the flow. Wenz [2] suggested that pressure levels between 115 and 150 dB rel 1 µPa are related to flow velocities between 0.02 and 0.3 m/s.

Low-frequency underwater noise may also be produced by turbulent pressure fluctuations in the *atmosphere* near the ocean surface, as discussed by Isakovich and Kuryanov [12] and Wilson [13]. The induced noise field in the sea is related, on a one-to-one frequency basis, with the fluctuations in the exciting turbulence field. Wilson [13] concluded that the atmospheric turbulence is the dominating source of wind-generated noise above 5 Hz. However, experimental results covering underwater ambient noise in the frequency range below 10 Hz are sparse due to experimental difficulties related to hydrophone installation, measurement time periods being too short, and the use of local test sites where limited environmental data are available.

6.2.4 SURFACE PHENOMENA

In many instances, low-frequency underwater sound is dependent on the *wind speed* for deep and shallow water. The Knudsen spectra based on many observations provide relations between *sea state* or *wind force*, and the level of underwater ambient noise. The frequency range over which wind speed, via various noise-producing mechanisms, has an influence on ambient noise is very broad, according to Andreev [14]. Above about 500 Hz local wind speed is the dominant factor controlling wind- and wave-generated noise. At frequencies below 500 Hz distant wind-dominated noise sources may contribute.

Wind blowing over a rough sea surface may generate sound which penetrates into the water in addition to fluctuating forces exerted on the sea surface by the wind's turbulence.

Spectrum levels in the frequency range of 8.4 Hz to 3 kHz measured off Nova Scotia in shallow water by Piggott [15] provide the following relation between *wind speed V* {unit: m/s} and the *noise level L(f)* {unit: dB rel 1 µPa} at a frequency *f* {unit: Hz}, expressed by:

$$L(f) = 100 + A(f) + 20n(f)\log(V) + 7n(f) \tag{6.4}$$

where $A(f)$ is a frequency-dependent threshold level equal to about 62 from May through December and about 58.5 from January through April. Eq. (6.4) can be used with the wind speed V {unit: miles/hour} by dropping $7\,n(f)$ from Eq. (6.4). The factor $n(f)$ is given by:

$$n(f) = 2.1 \quad f < 50 \text{ Hz}$$

$$n(f) = \left(2.1 + \frac{0.9}{7}\right) - \frac{0.9}{350} f \quad 50 \text{ Hz} < f < 400 \text{ Hz} \tag{6.5}$$

$$n(f) = 1.2 \quad f \geq 400 \text{ Hz}$$

for frequencies from few hertz to 2000 Hz. These results show that the sound pressure level is approximately proportional to the square of the wind speed below 50 Hz.

Ambient noise studies have frequently used buoyed hydrophones designed to move with the water mass and supplied with a system to minimize the influence of surface motions. Bottom moored systems have been widely used with the hydrophone in a tripod on the bottom or in a buoy anchored on the bottom. Most data analyses have been made in one-third octave bands, corrected for bandwidth to provide average spectrum levels.

Surface phenomena leading to low-frequency ambient noise in the sea are more or less related to the influence of the wind. This also concerns sound sources connected with breaking waves, nonlinear wave–wave interactions, and bubbles.

6.2.4.1 Breaking Waves

Wave breaking is a widespread phenomenon on the wind-driven sea surface that occurs over a wide range of length scales and appears to play a major role in surface layer mixing and in underwater ambient noise generation. However, the mechanisms within the breaking waves that actually generate sounds are not yet fully understood.

On a large scale, the most common type of wave-breaking process is related to the *spilling breaker*, which can be induced by steady flow over an obstacle. These breakers entrain air bubbles at the lower end of the "roller." Among the progressive waves, another wave-breaking type is characterized by the occurrence of *plunging breakers*. Plunging breakers will, before their short plunging state, demonstrate a bend-over, cusp shape with jetting fingers along the leading edge. These fingers develop into streaks which appear to lengthen and ultimately to collapse into a violent moving region of bubbles and water along the leading edge. According to Kerman [16] the first detected sound should occur with the appearance of an air–water cloud (bubble cloud) pushed ahead by the waves. The produced sound level should also increase with the increasing size of the air–water cloud. The final stage of the breaker shows several foam lines produced in succession, each probably associated with a burst of sound. The produced sound level gradually decreases with these foam lines and a turbulent pool is left behind the wave. In a mixed sea, characterized by short surface waves riding on longer waves, the orbital compression by the long waves compels the short waves to steepen and to break near the longwave crests.

There is currently no clear criterion to indicate when a wave will break or that a wave will develop into a spilling or plunging breaker. The general breaker at sea will possess characteristics of both. Visual observations by Banner and Peregrine [17] suggest that vigorously plunging breakers are quite rare. Typical fetch-limited breaking waves inject bubbles to depths typically 20–30 cm and are too shallow to be associated with an ordinary plunging breaker.

Standing waves break in a different manner, according to Longuet-Higgins [18]. The crests can become unstable, throwing droplets into the air, or overturning symmetrically on either side. In extreme cases the wave trough collapses, ejecting a high-velocity water jet.

Prior to wave breaking, low-frequency sound in the 2–200 Hz frequency range may be generated by the interaction between surface waves and turbulence, which occurs in sea states low enough that breaking waves do not occur, as discussed by Carey and Browning [19].

Wind-generated waves, when they break, also produce noise through the action of surf on beaches. Measurements by Cato [20] of surf zone–generated noise performed 5 km from a beach in Australian northern waters showed a power spectral density fluctuation between 67 and 82 dB rel 1 mPa^2/Hz (from 50 Hz to 500 Hz), due to a 1 m surf.

6.2.4.2 Nonlinear Wave–Wave Interaction

Nonlinear interaction between ocean waves has been of interest to seismologists and oceanographers since this interaction mechanism most probably leads to a self-stabilization of the ocean wave spectrum. The second-order effect involved in the surface wave motion by two waves progressing in opposite directions and thereby forming a standing wave has been demonstrated by Cato [20] to be the dominant generation mechanism for noise in the frequency range from 0.1 to 5 Hz. Experimental evidence by Kibblewhite and Ewans [21] has shown that the largest wind-related noise levels occur when a 180 degrees shift in the direction of a long duration wind brings a growing sea into direct opposition with the one already established. While the two wave fields interact, the underwater low-frequency noise level remains very high and then rapidly drops to a level some 20 dB lower as the new wave field becomes dominant. This reflects the nonlinear wave–wave interaction influence on the underwater noise level. Recently, it has been demonstrated conclusively by Wilson et al. [22] that nonlinear surface wave interactions are the dominant mechanism for generating deep ocean ultralow frequency noise in the band 0.2–0.7 Hz.

6.2.4.3 Bubbles

Surface waves breaking in shallow water produce bubble clouds that persist below the sea surface as identifiable acoustic targets for periods of several minutes. The average bubble cloud penetration into the water column is frequently 10–15 m, while maximum penetrations up to 25 m have been observed. The bubble clouds are carried and dispersed by the near-surface turbulence and may serve to identify fluid regions directly affected by the breaking waves which produce them. The distribution of the bubble clouds after their formation is a function of the properties of the bubbles and the turbulence. Bubble clouds in the uppermost layer of the ocean may also be produced by precipitation; breakdown of organic materials; and ship traffic, where bubbles in wakes may persist for hours, as discussed by Bjørnø [23]. Besides the obvious acoustic influence from the bubble clouds, they play an important role in the air–sea interaction processes, such as in the exchange of gas, production of sea salt aerosols, electrical charge exchange, and chemical fractioning in addition to the net upward flux of organic materials and bacteria, as discussed by Kerman [24].

Past observations have shown that the size distribution of bubbles after normalization by depth and wind dependence follows a power-law dependency on radius. Estimates of the slope with increasing bubble radius vary from −3.5 to −5 according to Kolovayev [25] and Johnson and Cook [26]. The number of bubbles also decreases with the depth z, approximately as $e^{-z/k}$, where k {unit: m} is a scaling length. As shown by Wu [27] the total number of bubbles increases very rapidly with wind speed U_{10} (measured at 10 m height above the sea surface) {unit: m/s}. A relation showing as rapid an increase as $(U_{10})^{4.5}$ has been suggested. The bubbles present near the sea surface resonate over a broad frequency range, and their resonance frequencies may be calculated using the *Rayleigh–Plesset* equation. The Rayleigh–Plesset equation, which describes the radial dynamics of a spherical bubble of radius $R(t)$ {unit: m} in an incompressible liquid of density ρ {unit: kg/m^3}, may be expressed by:

$$R\frac{d^2R}{dt^2} + \frac{3}{2}\left(\frac{dR}{dt}\right)^2 = \frac{1}{\rho}\left(p_i - p_o[1 + F(t)] - \frac{2\sigma}{R}\right) \tag{6.6}$$

where σ_s {unit: N·m} is the surface tension and p_i {unit: Pa} and p_o {unit: Pa} are the internal pressure in the bubble and the static pressure, respectively. The excitation function, due to variations in ambient pressure caused by waves or turbulence, is given by the dimensionless function $F(t)$. A rough estimate, based on Eq. (6.6), of the natural frequency of the bubble oscillations is given by:

$$\omega_o = \frac{1}{R_o}\left[\frac{3\gamma p_o}{\rho}\right]^{1/2} \tag{6.7}$$

where R_o {unit: m} is the equilibrium radius of the bubble, γ is the ratio of specific heats, and ω_o {unit: Hz} is the bubble vibration angular frequency. From Eq. (6.7) a bubble in water with a radius of about 0.03 m has a natural frequency of 100 Hz.

The changes in the bulk compressibility of water due to presence of bubbles lead to a considerable variation in the speed of sound in a bubble—water mixture. If the volume fraction of air in water, β, is considered to be small, the *bubble—water mixture speed of sound* may be found from:

$$c_m = c_w\left[1 + \beta\frac{\chi_a}{\chi_w}\right]^{1/2} \tag{6.8}$$

where c {unit: m/s} and χ {unit: m·s/kg} denote the speed of sound and the adiabatic compressibility, respectively. The indices m, a, and w denote mixture, air, and water, respectively. For $\beta = 0.001$ the speed of sound in the mixture will be about 320 m/s. If a bubble cloud in water is assumed to have a linear dimension, L {unit: m} this cloud may be considered as a system of coupled oscillators with the frequency, f_0 {unit: Hz} of the lowest mode given by:

$$f_0 = \frac{c_m}{L} \tag{6.9}$$

which for naturally occurring values of L may lead to rather low frequencies, as discussed by Kozhevnikova and Bjørnø [28,29].

Because most bubbles produced near the sea surface are small, the individual bubble contribution to the low-frequency noise level in the sea is small. However, the collective oscillations of the bubbles in the bubble cloud near the sea surface may generate low-frequency sound. It is well-known that a small amount of bubbles in water significantly changes the bulk compressibility of the water while not substantially changing the density. These changes lead to the considerable variation in the speed of sound in the bubble—water mixture expressed by Eq. (6.8). Kozhevnikova and Bjørnø [28,29] observed that *pulsations of large size bubble clouds* formed near the sea surface may give rise to emission of ambient noise at the comparatively low frequencies given by Eq. (6.9).

Flow of bubbles and cavitation are among the most efficient hydrodynamic noise sources because they are monopole sources. Underwater flow noise sources in the absence of bubbles or cavitation are mainly dipoles and are dominated by the vibration of contiguous surfaces such as ship hull plates, as well as, the flow dipoles of local character. The mean square radiated sound pressure from monopoles is dependent on the fourth power of the flow particle velocity, u {unit: m/s} and the mean square radiated sound pressure from dipoles depends on the sixth power of u.

Therefore, for low flow Mach numbers, M_a, underwater quadrupole noise is irrelevant when compared with dipole and monopole produced flow noise.

While turbulence-generated noise in pure water due to its quadrupole character is a very weak source of sound, the presence of bubbles will amplify the turbulence produced noise by converting the sound source into a monopole source. According to Prosperetti [30] the magnitude of this intensity amplification can be estimated to be on the order of the ratio between the sound velocity in water to the sound velocity in the bubble—water mixture to the fourth power $(c_w/c_m)^4$. In this way, through the presence of bubbles, turbulence can contribute significantly to the ambient noise level at frequencies up to several tens of hertz.

6.2.5 PRECIPITATION

In particular, *rain*, and also *hail* and *snow*, falling on a sea surface is the significant contributor to the ambient noise level in the sea. Frequently, the underwater sound spectrum generated by rain has a shape that can be distinguished from spectra from other sources of sound in the sea, and the relationship between spectral levels and rainfall can be quantified. Rainfall is a climatic factor of great importance and, therefore, measurement of rainfall has a high priority. However, it has been estimated that about 80% of the Earth's precipitation occurs over the ocean where the smallest number of weather stations are located. Nystuen [31] has proposed underwater sound measurements as a way for determining the amount of rain falling on the sea surface. A prospective future procedure for rainfall measurements could include underwater ambient noise measurements at certain geographical locations combined with the use of satellite observations and the use of weather radar. However, there is a long way to go before reaching this goal. Several attempts to describe the underwater noise spectra produced by rainfalls of various magnitudes have been made over the years, as discussed by Heindsmann et al. [32], Franz [33], Bom [34], Scrimger et al. [35], and Nystuen and Farmer [36]; however, a considerable difference remains between rain noise data collected by various scientists.

Possible sources of underwater sound produced by the impact of *single droplets* of rain are the transient introduction of the droplet into the water, the secondary splashes by water droplets thrown up by the entry, oscillations of air bubbles trapped near the surface, and oscillations of cavities open to the atmosphere.

Individual contributions to the underwater noise spectrum from these sources are strongly influenced by factors related to the droplet size, shape, and movement before impact. The most important factors are the equivalent raindrop diameter, the raindrop size distribution, the shape of the raindrops, the wind velocity and its profile, surface tension and raindrop temperature, and frequently the conditions under which the measurements took place. Underwater sound generated by *multiple raindrop impacts* on a water surface are influenced by the interaction between the individual impacts, which include factors, such as water droplets falling on a nonplanar (randomly shaped) water surface, droplets falling in and out of phase leading to phase cancellation, and resonance. The noise generated by multiple raindrop impacts is possibly the field of rain noise generation, where most research is still needed. Extensive experimental studies have to be performed; however, a theoretical (numerical) basis still has to be established. There is experimental evidence for the strong influence of *surface tension* on the noise level produced by *real rain* (multiple impacts), as discussed by Bjørnø [37], Crum et al. [38], and Pumphrey et al. [39]. The results show that a considerable part of the *noise produced by raindrops* falling on a sea surface is caused by the *pulsation of bubbles trapped near the surface*, as shown in Fig. 6.3 from Pumphrey et al. [39]. For instance, the characteristic spectral amplitude of rain noise around 14 kHz is caused by formation and pulsations of small bubbles produced by raindrops having a diameter between 0.8 and 1.1 mm. These bubbles are formed near the water surface by closure of cavities produced by impact of raindrops and under the influence of surface tension. Only raindrops in this size range result in bubble formation for all impacts. By reducing the water surface tension by adding small amounts of detergents to water, Pumphrey et al. [39] have shown that a remarkable reduction in the rain noise spectrum took place since the characteristic spectral amplitude around 14 kHz, shown in Fig. 6.4 from Pumphrey et al. [39], disappeared.

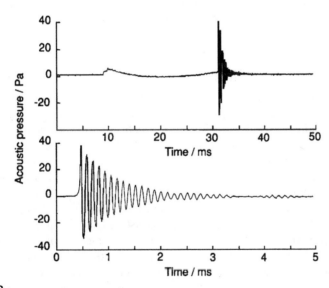

FIGURE 6.3

Sounds produced by regular entrainment of 3 mm rain drops impacting a water surface with a 2 m/s velocity. The upper trace shows the initial impact occurring at 8 ms and the bubble sound starting after 32 ms. The lower trace expands the time axis of the bubble sound.

*Reproduced from Pumphrey, H.C., Crum, L.A. and Bjørnø, L., Underwater sound produced by individual drop impacts and rainfall, J. Acoust. Soc. Amer., **85**, (4), 1518, 1989, with the permission of the Acoustical Society of America.*

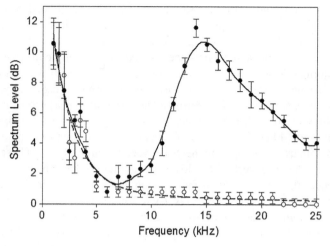

FIGURE 6.4

Acoustic power spectrum level of real rain falling into clean water (*closed circles*) and wither with sulfo detergent added (*open circles*). The dB reference is arbitrary.

Reproduced from Pumphrey, H.C., Crum, L.A. and Bjørnø, L., Underwater sound produced by individual drop impacts and rainfall, *J. Acoust. Soc. Amer.,* **85**, *(4), 1518, 1989, with the permission of the Acoustical Society of America.*

6.2.6 BIOLOGICAL ACTIVITY

Biological noise sources are many and varied. The frequency range of various biological sounds is very wide, i.e., from about 20 Hz to more than 300 kHz. The principal sources of low-frequency biological sound are marine mammals, mainly whales and porpoises, as pointed out by Ewans [40]. Some of the sounds are used for echolocation, for food detection, and for communication between the species; however, the functions of many sounds are not known. Because of their biological origin, such sounds have diurnal and seasonal cycles. *Groups of mammals*, such as whales, can generate loud, continuous sounds which can raise the ambient noise level significantly. Their source levels may exceed 180 dB rel 1 µPa.

One of the best known biological noise sources is the *snapping shrimp* which abounds in shallow waters in latitudes less than 40 degrees. They were first described for frequencies up to 20 kHz by Knudsen et al. [1]. However, measurements by Carey [19] have shown that shrimps are a dominant source of noise from frequencies of few kilohertz to at least 300 kHz. Their source level may be as high as 190 dB rel 1 µPa.

In most *shallow waters* there is a general background of biological noise, but the most pronounced effect is due to the *choruses* that result when large numbers of animals are calling at the same time. Such choruses cover various parts of the spectrum, and they typically increase the spectrum levels by 20 dB or more. The choruses may be considered to belong to one of two general categories. The first category is the diurnally varying type of chorus, which occurs few hours per day at approximately the same time of the day. The second one is the evening chorus that occurs for a few hours after sunset prevails. However, similar choruses are sometimes observed near sunrise and occasionally at other times of the day. The other type of sound chorus results from fish and is often related to spawning. These, less predictable choruses, may occur for longer periods of the day and often have pronounced seasonal relations, which differ considerably between the species. Many fish use their gas-filled swim bladder and their drumming muscles to generate sound and are able to produce a considerable sound pressure level.

6.2.7 ICE NOISE

Underwater noise *under the ice* differs in level and in characteristics from that measured under open-water conditions. Noise under the ice is a superposition of a large number of independent noise events, each presumably caused by *ice fractures* of one kind or another. Ice can be fractured by over-thrusting, and consequent flexural straining in response to horizontal internal ice stresses. As pointed out by Dyer [41], ice can also be fractured by turning moments on pressure ridges in response to opposing wind and current stresses. Acoustical radiation caused by fractures produced by local ice motion is supposed to lie, at least partially, in the frequency range of 1–300 Hz. The noise has been found to be spiky and impulsive when the air temperature decreases and tensile fractures are formed in solid, shore-fast ice. However, under rising temperatures, the spiky character of the noise disappears and a more Gaussian-like amplitude distribution has been found by Milne and Canton [42].

Ice *vibration* may produce low-frequency sound. Wind blowing over the rough ice surface becomes turbulent and transmits varying pressures to the ice and through it into the water. Wind-produced sound is more prominent, under a noncontinuous ice cover than under a continuous one. Vertical ice motions measured with seismometers mounted on the ice have been shown by Greene and Buck [43] to generally correlate with 50 Hz ambient noise levels at 60 m depth. Another type of noise due to the ice motion is an almost sinusoidal signal at 7 Hz and with a variable amplitude, which is attributed by Milne [44] to standing wave patterns between the ice-covered surface and the seabed.

6.2.8 SHIPPING

A major source of underwater ambient noise, primarily in the frequency range of 10–500 Hz, is shipping. *Distant ship traffic* is a principal source of low-frequency noise in the range of 20–150 Hz. Such traffic may take place at distances up to 1000 km or more, which competes with distant storms as a source of low-frequency noise. Even if ship traffic generates sound over a broad frequency band, the propagation over distances of several hundreds of kilometers or more attenuates sound at higher frequencies. Only low-frequency sound is received from distant shipping. This means that the spectral composition of the sound received from ship traffic at short distances is different from sound produced by distant shipping. Noise from ships, which can be identified as individual sources of noise, does not contribute to ambient noise. These ships instead produce distortions in the received ambient noise signals.

The acoustic power produced by a ship is only a small fraction (frequently less than 1/1,000,000) of the mechanical power used for moving the ship. According to Ross [45] a modern submarine proceeding at slow speed produces on the order of 10 mW acoustic power, while surface ships generally radiate from 5 to 100 W of acoustic power. Since each ship operates in a wide variety of configurations ranging from unloaded to fully loaded, at various speeds and in varying mechanical conditions, the sound radiated by a ship can vary over a wide frequency band. Shipping noise is the major contributor to ambient noise in a number of geographical locations. In particular, near a number of ports, the ambient noise depends upon the time of day when the observation takes place. A 6—8 dB higher noise level is measured during periods of high traffic activity compared to periods of low activity. The concentration of traffic in busy shipping lanes is a predominant factor contributing substantially to the permanent noise level in the seas.

The principal *sources of radiated sound from ships* are (1) the propulsion system; (2) the propeller; (3) the auxiliary machinery; (4) the hydrodynamic effects, such as wake turbulence; and (5) the hull movements. Three types of power plants are now commonly used in merchant ships: (1) geared steam and gas turbines; (2) direct-drive, slow-speed diesel engines; and (3) geared medium-speed diesel engines. Direct-drive, slow-speed diesel-powered ships constitute nearly two-thirds of all ships at sea and in particular ships operating in shallow-water areas. Steam and gas turbine—driven ships constitute about 25% of the ships at sea. The *propulsion system* contains, in general, large rotating shafts; gears; bearings; and, depending on the ship type, reciprocating engines, steam and gas turbines, or electric drive motors. A small unbalance in one of these devices results in oscillating forces that are transmitted through the machine structure, the foundations and the hull, into the water. Amplification due to structural resonances may take place along this transmission line. The acoustic signals generated in this way are normally narrow band tones at the systems' rotational frequencies and their harmonics, and the radiated noise power increases as the fourth power of rotational speed. Gear noise is normally dominated by impacts of the gear teeth and may be characterized by tones at multiples of the tooth contact frequency. Broadband signals in the propulsion system are generated by friction forces.

The dominant *surface-ship radiated noise sources* are *propeller cavitation*, *propeller blade tonals*, and *propeller singing*. It is estimated that 80—85% of the noise power radiated into the water by surface ships comes from propeller cavitation. According to Ross [45] and Gray and Greeley [46] there are two types of radiation from cavitating propellers: low-frequency tonals and a broad continuum comprising frequencies up to 100 kHz. Tonals are radiated up to the first 10 harmonics of the blade frequency and are usually dominant at frequencies below about 40—50 Hz. The continuum controls the spectrum above 50 Hz, generally peaking between 50 and 150 Hz. Above 150 Hz, the spectrum decreases with frequency at about 6 dB per octave. The tonals and continuum are modulated by the shaft rotational frequency. The continuum is even more strongly modulated by the blade frequency, according to Ross [45]. *Cavitation*, i.e., stable and transient, may form in the low-pressure regions on the propeller blades as surface cavitation on the propellers leading-edge suction side or as tip-vortex cavitation. Surface cavitation is normally noisier than the tip-vortex cavitation. Because the onset of cavitation is related to the ambient pressure and to the speed of the propeller, cavitation noise decreases with depth and increases with speed. Also, the shape of the propeller blades has a strong influence on the onset of the cavitation, and even small variations in geometry may separate propellers with and without cavitation. A *cavitation index* κ, which sometimes has been used to give a rough estimate for the onset of cavitation, is defined by:

$$\kappa = \frac{P - P_v}{\frac{1}{2}\rho U^2} \tag{6.10}$$

where P and P_v denote the ambient hydrostatic pressure {unit: Pa} and the vapor pressure in the water, respectively, while ρ {unit: kg/m^3} and U {unit: m/s} are the water density and flow velocity, which is the relative velocity between a propeller blade and its surrounding water. The dimensionless number κ that indicates when the hydrostatic pressure falls below the vapor pressure and a local "boiling" of the water is produced forms the most relevant parameter for determining the onset of cavitation and cavitation produced noise. When $\kappa < \kappa_I$, where κ_I is a cavitation inception index, cavitation will occur. κ_I is a function of the hydrodynamic design of the propeller blade, the flow field in which the propeller operates, viscous effects in the water, the amount of free and dissolved gas in the water and the amount of particles in the water to form nuclei for cavitation.

While cavitation noise is a major component of the noise signature of surface ships and of submarines operating close to the sea surface, submarines traveling at sufficient depths may avoid propeller cavitation since the ambient hydrostatic pressure increases with depth and keeps $\kappa \geq \kappa_I$.

In addition to propeller cavitation noise, *propeller blade tonal* components at the fundamental frequencies of the blade rate—most frequently below 20 Hz—may be produced by vibrational excitations of the propeller blades by intermittent cavitation clouds. The occurrence of *singing propellers*, visible as spectral lines in the noise spectra, is assumed to be caused by the coincidence of vortex shedding at the blade's trailing edge and the blade resonance frequencies. Singing propellers cover the frequency range from below 100 to above 1000 Hz. Also amplitude modulations of the cavitation spectra at the propeller blade rate may frequently be used to characterize certain ship types and their propellers.

Auxiliary machinery, such as pumps, compressors, blowers, air conditioning equipment, hydraulic control systems, and electrical generators primarily produce tonal components due to dynamic unbalances in rotating components. Since these devices normally are operating at constant speed, the sound produced is, in general, relatively stable in amplitude and frequency.

The *hydrodynamic noise sources* also include such sources as cavitation produced along the ship's hull, in valves, pipe bends, and hydraulic machinery. This broadband noise is transmitted through machine structures, along pipes, through bulkheads and sea valves into the water. Water flow past struts may induce structural vibrations through processes of unbalanced vortex shedding off the trailing edge of a strut. These vibrations radiate sound into the sea. Turbulent flow along the hull structure may couple pressure fluctuations back to the hull, thus producing vibrations and sound radiation. In general, the interaction between water flow along the ship and the ship's hull surface will lead to unsteady surface forces and to pressure fluctuations in the water, which can induce surface vibrations. These vibrations may produce other surface noise sources at impedance discontinuities along the hull surface—for instance, at positions where surface plates are connected to bulkheads or where the shape of the hull changes.

Low-frequency sound radiation from *hull motions* may involve the whole hull. The hull may experience a rigid-body motion in which it retains its shape and either vibrates in position in response to an external alternating force, or rotates about an axis. Moreover, the hull may vibrate in a beam like flexural mode (whipping) and it may vibrate in a dominantly longitudinal mode, in which the two ends move out of phase in an accordion-like motion. At somewhat higher frequencies, but still below 300 Hz, whole compartments may resonate and emit sound as a cylindrical shell vibrating in a rigid cylindrical baffle. Low-frequency radiation from hull structures is much more important for submerged vehicles, for which image cancellation, as found for surface ships, is much reduced and for which the propeller cavitation may be absent.

Since the noise radiated is a function primarily of ship size and speed, ships with the highest propulsion powers are probably the noisiest. Ross [45] provides the following rough estimate of the total *overall noise power level* L_s {unit: dB rel 1 μPa at 1 m} produced by a surface ship:

$$L_S = 134 + 60 \log \frac{U_S}{10} + 9 \log D \qquad (6.11)$$

where U_s {unit: knots} is the ship speed and D {unit: tons} is the ship's displacement. Eq. (6.11) is applicable for frequencies above 100 Hz and for ships with displacements less than 30,000 tons.

According to Ross [45], for *propeller cavitation*, the total *overall noise pressure level*, L_p {unit: dB rel 1 μPa at 1 m}, which depends on the propeller tip speed, U_p {unit: m/s} and the number of propeller blades, B, the noise measured in the band from 100 Hz to 10 kHz for an individual surface ship over 100 m length, with U_p in the range of 15−20 m/s, is given by:

$$L_P = 175 + 60 \log \frac{U_P}{25} + 10 \log \frac{B}{4} \qquad (6.12)$$

Generally, the total acoustical power radiated below 100 Hz exceeds that radiated above 100 Hz by about 6 dB.

As mentioned in the preceding paragraphs the steady-state noise spectra from ships generally can be classified in two types: (1) *broadband noise* with a continuous spectrum, such as noise from cavitation and (2) *tonal noise* containing discrete line components caused by the propulsion system, gears, modulation of broadband noise, singing propellers, etc. Ship noise is usually a combination of continuous and tonal noise. Transient or intermittent noise can be caused by impacts, doors slamming, items dropped on the floor, unsteady flow, etc.

The noise sources associated with a specific ship are dependent upon the fabrication details, mechanical and electrical systems, and the propulsion subsystem. The contribution of individual noise sources to the overall noise power level may be enhanced or reduced depending upon the age of the ship, how it is maintained, and the use of vibration isolation mounts. Two-stage mounting systems now being used extensively in ship designs produce good vibration reduction results over broader frequency bands. The integration of all noise sources forms the "noise fingerprint" of the ship.

6.2.9 OTHER MAN-MADE (ANTHROPOGENIC) SOURCES

The exploitation of the seas is ever increasing. In addition to shipping, fisheries, offshore oil and gas prospecting, exploitation of mineral resources, marine tourism, defense-related operations, port constructions, explorations, and research activities have shown increasing trends since World War II. As a result, the ambient noise level in the seas due to man-made sources has since World War II increased 12–15 dB and in some locations even more.

In recent years, a new, major, low-frequency source in particular in shallow waters has raised the ambient noise levels below 100 Hz, occasionally by as much as 20 dB. The source is the explosion-like pulses used during *seismic surveying* and produced by sources, such as boomers, air guns, and underwater explosions. These sources cover a broad-frequency range with a considerable power in the low-frequency acoustic regime. One seismic profiler may transmit the same acoustic power into ambient noise as nearly 300 merchant ships. The significance of offshore oil exploration and exploitations as an ambient noise source is enhanced since these activities are most frequently located on the shallow-water parts of the continental shelf which may offer optimum conditions for propagation of sound to distant receivers. Drilling, communication, and transport activities related to offshore work produce sound having some of the same features as shipping. Also, leisure activities in coastal regions and the use of low-frequency sonar systems show an increasing contribution to the ambient noise level.

The use of underwater explosions and other high-intensity sound sources like electric discharges and air guns involve high-intensity bubble pulsations at low frequencies. Depending on the explosion depth from 10% to 50% of the energy available in the explosive charge is radiated as sound, as discussed by Bjørnø [47]. While the detonation process (i.e., a fast combustion process started by a shock wave) in a chemical explosive charge leads to the radiation of a shock wave in water characterized by a steep front pressure rise to the peak pressure P_m {unit: Pa} and an exponential pressure decrease with the time constant θ {unit: ms} after the peak pressure, the shock wave contributes substantially to the high-frequency signal spectrum:

$$P_m = 506 \times 10^5 \left(\frac{Q^{1/3}}{r}\right)^{1.1} \quad \text{and} \quad \theta = 87 Q^{1/3} \left(\frac{Q^{1/3}}{r}\right)^{-0.23} \tag{6.13}$$

where Q {unit: kg} is the weight of the explosive and where r {unit: m} is the distance between the explosion site and the point where the measurement of the pressure time history from the explosion takes place, according to Bjørnø [47].

The low-frequency signal spectrum from an underwater explosion is due to the pulsation of the combustion gas bubble formed by the detonation process. The frequency content of the pulsating gas bubble depends on the size of the explosive charge and the depth at which the detonation process takes place. An increased detonation depth causes a higher fundamental pulsation frequency and an increasing number of bubble pulsations before the gas bubble migrates to the sea surface, where a "blow-out" of the combustion gas terminates the pulsation process. The time interval τ {unit: ms} between the shock front of the pressure wave from an underwater explosion and the peak of the first bubble pulse is given approximately by:

$$\tau = 2.11 \times 10^3 \ Q^{1/3} \ (d + 10)^{-5/6} \tag{6.14}$$

where d {unit: m} is the depth of the underwater explosion below the sea surface. This time interval characterizes the low-frequency part of the spectrum from an underwater explosion.

The same gas bubble pulsation characterizes the sound production process after use of an air gun, but with substantially lower-pressure amplitudes than after a detonation process. The characteristic frequency range for the high-power, low-frequency signals from an air gun used in seismic investigations in water is 10–200 Hz.

6.2.10 SEDIMENT FLOW–GENERATED NOISE

Interparticle collision of mobile seabed materials influenced by flow near the seabed, where the highest concentration of suspended sediments is found, may lead to contributions to a wideband ambient noise spectrum around 10 kHz, as identified by Thorne [48]. Experimental data show an ambient noise level of about 70 dB rel 1 μPa/Hz. The experiments were performed over a seabed consisting of quartz sand overlaid with gravel, and the flow was caused by strong tidal currents.

6.2.11 THERMAL NOISE

At elevated frequencies thermal noise becomes a major part of underwater ambient noise. Thermal noise produced by molecular activity in the sea and by electronic components, such as resistors in the instruments, used for registration of underwater sound places a limit on hydrophone applicability at high frequencies, as discussed by Mellen [49]. The power spectral density of the thermal noise can be described using *Boltzmann's equation*. The thermal noise in the sea is produced through molecular momentum reversals at the surface of an acoustic sensor, where the kinetic energy of the molecules depends on the absolute temperature multiplied by Boltzmann's constant. The forces from the molecular interactions with the sensor surface are temporally and spatially uncorrelated. This interaction between the molecules and the sensor surface is not real noise. It is better described as pseudosound. Mellen's expression [49] for the equivalent noise spectral density level *NL* {unit: dB rel 1 µPa/Hz} at ordinary sea temperatures produced by thermal noise is

$$NL = -15 + 20 \log f \qquad (6.15)$$

where f is the frequency {unit: kHz}. Eq. (6.15) is valid for $f \gg 1$ Hz.

Thermal noise increases with frequency at 6 dB/octave and is the frequency-dependent threshold for the minimum observable sound pressure level at frequencies near and above 100 kHz in the oceans. Thermal noise produced by the sea is normally considered to be a contributor to ambient noise, while thermal noise produced by electronic devices used for registration of underwater sound more correctly belongs to *self-noise*, which is noise produced by the platform carrying the instruments used for acoustic measurements, as discussed in Section 6.6.

6.3 SPECTRA OF AMBIENT NOISE

The diversity of sources contributing to the ambient noise level in the sea results in a broad noise spectrum. While other noise sources may be of major importance, individually or together, at certain locations or at certain times, the most prevalent sources are shipping and wind. Moreover, most sources are related to the sea surface. The description of spectra of ambient noise may appropriately be divided into *deep-water spectra* and *shallow-water spectra*.

Due to the influence of attenuation on the noise (attenuation caused by divergence, absorption, and scattering), the noise spectra may be divided into three distinct frequency regimes when analyzing deep-water ambient noise: (1) below 150 Hz (influence of the whole ocean basin); (2) 150–1000 Hz (influence of numerous surface zones); and (3) above 1 kHz (dominating influence of the local surface zone only). The high-frequency regime is defined as that for which the attenuation is so high that only the closest intersection of the sound rays with the sea surface contributes significantly to the noise. In shallow waters, this regime includes frequencies above 10 kHz; however, in deep water the high-frequency regime applies down to about 1 kHz. Below about 150 Hz, the attenuation maybe so low that it no longer controls the ocean area contributing to the ambient noise level at a measurement site. At these low frequencies even sources located at the edges of the ocean basin may make significant contributions to the ambient noise level.

6.3.1 DEEP-WATER SPECTRA

As shown in Fig. 6.5 adapted from Ref. [5] the deep-water spectrum may be divided into five characteristic bands, depending on the prevailing noise sources. *Band I*, which applies to frequencies below 1 Hz, is still not totally explored. Noise sources are assumed to be of **hydrostatic and seismic** origin. Measurements to be performed at the very low frequencies place strict demands on equipment since many sources of errors may influence the results. An example is the "self-noise" caused by flow around the hydrophone and its supporting structure, as discussed in Section 6.2.1.

In *band II*, from 1 to 20 Hz, measurement difficulties may influence the results. This frequency region is characterized by a spectral slope of −8 to −10 dB/octave. The most probable noise source in this band is **turbulence and shipping**. Wind influence is small.

In *band III*, from 20 to 500 Hz, the ambient noise spectrum flattens out and forms a plateau. This frequency region is mostly dominated by shipping and offshore noise. Noise from distant **shipping and offshore activities** on the continental shelf contribute to the level and the directivity of the ambient noise in deep water in region III. The region III spectrum level is strongly influenced by shipping in the area where measurements take place.

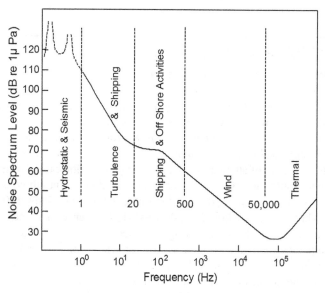

FIGURE 6.5

Deep-water ambient noise spectrum levels divided into five characteristic bands according to the major noise source in each band.

Adapted from Figure 7.4 from Urick, R.J., Principles of Underwater Sound, (3rd. Ed.), McGrawHill Book Comp., 1983, with permission of Peninsula Publishing.

Band IV, ranging from 500 Hz to 50 kHz, is the original Knudsen spectra region. In this region the spectral curves have a slope of −5 to −6 dB/octave. The major sources of noise are **wind**-controlled activities at the sea surface (local sources). A variation in the noise level of more than 30 dB depending on the wind speed is found in region IV.

Band V, comprising frequencies above 50 kHz, is dominated by **thermal** noise, originating in sea molecular motions. The spectrum level increases at 6 dB/octave.

6.3.2 SHALLOW-WATER SPECTRA

While the spectra of ambient noise measured at deep waters are relatively well defined without too great variations from time to time and from place to place, the same is not the case for *shallow-water* ambient noise spectra. Here, location and time have a considerable influence on the spectral composition due to the interaction between noise contributions coming from abroad variety of sources. The ambient noise levels and spectra show considerable differences when measured in various coastal regions, as shown in Fig. 6.6, or when measured in bays and harbors.

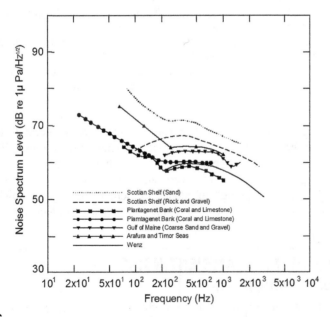

FIGURE 6.6

Shallow-water, wind-driven ambient noise spectra measured at a number of locations, as discussed by Ingenito and Wolf [50] and including data from Piggott [15]. Seabed materials on the locations are given in the figure. The spectra are related to wind speeds between 3.5 and 5.5 m/s (sea state 2).

*Reproduced from Ingenito, F. and Wolf, S.N., Site dependence of wind-dominated ambient noise in shallow water, J. Acoust. Soc. Amer., **85**, (1), 141, 1989, with the permission of the Acoustical Society of America.*

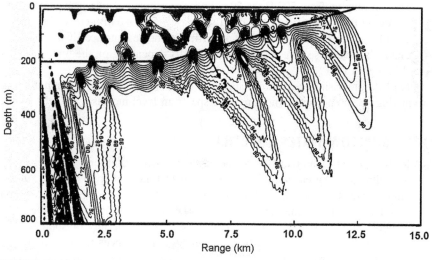

FIGURE 6.7

Numerical demonstration of modes by up-slope propagation of sound. Sound is radiated into the seabed at short ranges, as well as, on the slope. Three modes are initially propagating up the slope; however, due to mode angle steepening, supercritical incidence for the modes guided up the slope is obtained. This leads to successive mode cutoff where the mode energy is radiated into the bottom at cutoff depths.

Reproduced from Jensen, F.B. and Kuperman, W.A., Sound propagation in a wedge-shaped ocean with a penetrable bottom, *J. Acoust. Soc. Amer.,* **67**, *(5), 1564, 1980, with the permission of the Acoustical Society of America.*

The most active *sources in shallow water* are (1) shipping and industrial activities, (2) wind effects, and (3) biological noise. Also intermittent sources like precipitation may have a strong influence. However, in shallow waters the influence of distant shipping on the noise level is very small due to the shielding effect caused for instance by the "stripping" of modes by sound propagating up-slope from deep to shallow waters, as shown in Fig. 6.7 from Jensen and Kuperman [51].

6.4 DIRECTIVITY OF AMBIENT NOISE
6.4.1 NOISE PROPAGATION

The directivity of ambient noise is influenced by the characteristics of the dominant noise sources and by propagation conditions in the sea, which contribute substantially to ambient noise *anisotropy*. In general, low-frequency noise originated at great distances arrives at a hydrophone via primarily horizontal paths, whereas high-frequency noise originates at the sea surface and arrives at the hydrophone via a nearly vertical path. Due to the anisotropic character of ambient noise, its power spectral density $P_N(f)$ {unit: W/Hz} measured omnidirectionally is composed of contributions from all directions, $P_N(f,\theta,\varphi)$, and may be expressed by:

$$P_N(f) = \int_0^{2\pi} \int_0^{2\pi} P_N(f,\theta,\phi)\cos\theta \, d\theta d\varphi \qquad (6.16)$$

where θ {unit: rad} denotes the vertical angle measured from the horizontal plane and φ {unit: rad} is the azimuth angle.

Over broad regions the sources related to activities at the surface of the sea are the most important to the ambient noise spectra. Seismic effects, turbulence, nonlinear wave interactions, and hydrostatic effects are prevalent only at very low frequencies. Therefore, the measurements of ambient noise directivity in general take place at frequencies above 20 Hz, where surface sources, distant shipping, storms, and other sources influence the deep-water spectra.

The sea surface is a pressure-release surface, and, therefore, the basic pressure radiation pattern from a near-surface source is a cosine square function having its maximum straight down and a zero in the horizontal direction like a dipole. For sea-surface radiation a cosine function of the form given by Eq. (6.17) may be used for the intensity radiation patterns:

$$I(\theta) = I_0 \cos^m\theta \qquad (6.17)$$

where I_0 is the acoustic intensity radiated by a small area of the sea surface in the downward direction ($\theta = 0$) and m is an integer. As pointed out by Urick [5], most measurements have confirmed a value of $m = 2$, indicating the dipole character of the radiation due to the image source with a phase shift of π introduced by sound reflection at the sea surface, the so-called Lloyd's mirror effect ($m = 0$, denotes the monopole). If the sea surface is flat, and therefore a perfect pressure-release reflector, and in the absence of refraction effects due to bubbles near the sea surface, the zero will be very deep due to perfect cancellation of the direct and the surface-reflected paths. However, the sea surface is usually rough with layers of microbubbles present near the surface, and therefore the cosine pattern approaches a minimum rather than going to zero, as discussed by Neighbors and Bjørnø [52]. Moreover, incomplete knowledge about the surface decoupling effect at near-grazing angles for a rough surface limits the ability to model sea surface−generated ambient noise.

A study of the *vertical anisotropy* of the ambient noise arrivals at a measurement location may he hased on an examination of the contribution to the total noise power arriving in a number of vertical sectors. The surface sources that contribute noise to a particular vertical sector are those located where the ray bundle from the receiver in the direction of the limiting rays for the sector intersects the sea surface. It is important to realize that the same ray bundle may intersect the surface in several places. Noise received at each angle comes from distributed surface sources located in each of the zones where the ray paths intersect the surface. The number of zones depends on the noise attenuation; however, in general, all zones must be considered out to distances where the frequency-dependent absorption loss becomes dominant. Due to the long distance propagation of distant shipping noise, this noise has been exposed to several surface and bottom bounces during its propagation. This is one of the main reasons for the arrivals of ambient noise below the horizontal axis. The vertical noise contributions are wind speed dependent.

In deep water the vertical directionality of distant shipping noise is given in Fig. 6.8 for a frequency of 75 Hz and for various wind speeds measured at depths between 850 m and 2650 m by Soritin and Hodgkiss [53]. The deep-water directionality shows that most sound power is contained within the deep-water sound channel and therefore the major arrivals are nearly horizontal and are distributed over an angle of about ±20 degrees around horizontal.

The total intensity level produced by sources situated in all sea surface zones may be calculated using Eq. (6.18). By assuming a perfect dipole character of all

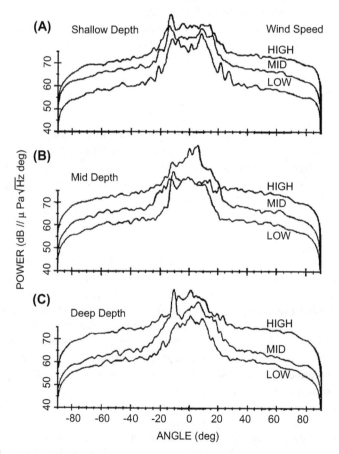

FIGURE 6.8

Deep-water spatial distribution of ambient noise at 75 Hz as a function of depth and wind speed. The depths are 850 m (shallow), 1750 m (mid), and 2650 m (shallow). The wind speeds are 3, 7, and 11 m/s, respectively.

Reproduced from Soritin, B.J. and Hodgkiss, W.S., Fine-scale measurements of the vertical ambient noise field.
*J. Acoust. Soc. Amer., **87**, (5), 2052, 1990, with the permission of the Acoustical Society of America.*

noise sources at the sea surface, the total intensity produced by all noise sources may be expressed through:

$$I = 2\pi I_0 \int_0^{\pi/2} \sin\theta \cos\theta \, e^{-\left(\frac{2\alpha d}{\cos\theta}\right)} d\theta \tag{6.18}$$

where α is the sound pressure absorption coefficient {unit: Np/m} and where d is the measurement point depth below the sea surface {unit: m}. When absorption is insignificant, Eq. (6.18) becomes $I = \pi I_0$.

The vertical directionality of ambient noise is strongly influenced by seabed reflections. Therefore, potential geoacoustic parameters can be inferred by inversion of the ambient noise received and the seabed reflection loss. Compressional and shear wave speeds can be found by comparing the upward- with the downward-going noise [22]. When absorption is insignificant, which is realistic at lower frequencies in shallow water, the total intensity, I, produced by all noise sources, when only return signals from sea surface—generated noise that are normal and near normally incident on the seabed are included, is $I = pI_0(1 + R^2)$, where R^2 is the plane wave intensity reflection coefficient at the water—seabed interface, discussed in Chapter 2.

Busy shipping lanes, harbor activities, and onshore and offshore construction contribute to the *horizontal anisotropy* of the ambient noise and add time dependence to the ambient noise level, in particular in the horizontal plane. The horizontal anisotropy and its time dependence are influenced by the propagation conditions in the sea, which may vary over days or seasons.

The *coastal enhancement effect* may influence the ambient noise level in deep waters. The mechanism behind the coastal enhancement effect is an angle transformation for the propagating noise by reflection from outwardly sloping bottoms, as shown in Fig. 6.9, where sound rays from a source in shallow water (on the

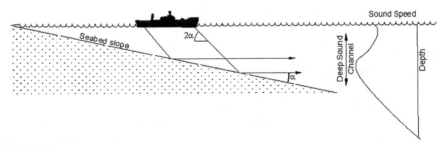

FIGURE 6.9

Costal enhancement effect that shows a down slope conversion taking place when the ray paths are reflected at a sloping bottom. Upon reflection the angle of the reflected ray is less than the angle of the incident ray by twice the slope angle. Refraction of the noise into the deep sound channel may take place when rays enter the channel at a near horizontal angle after reflection on the sloping bottom at a depth nearly equal to the depth of the deep sound channel depth.

continental shelf) propagate out into the deep basin. The importance of the coastal enhancement effect for low-frequency ambient noise stems from the places where shipping lanes converge over shallow sloping bottoms, such as the Strait of Gibraltar and the southwestern approaches to the English Channel. Noise from these areas often dominates the low-frequency spectra for receivers located in the deep sound channel. The coastal enhancement effect may explain the vertical arrival structure measured in the seas and features of the horizontal arrival structure.

6.5 COHERENCE OF AMBIENT NOISE

In spite of some individual deterministic features due to the sources of sound in the sea, in general underwater noise is considered to possess a random character best described by statistical means, as is discussed in Section 6.7. Moreover, as is evident from Section 6.2, the ambient noise process is not stationary in space or in time. However, some self-similarity of the noise exists over shorter distances or times. Since underwater sound-receiving systems normally consists of more than one hydrophone and one channel in the receiving electronic system, the relationship between noise in each channel is important and can be expressed in terms of the *correlation* between channels. The time-averaged cross-correlation coefficient of noise measured at the same time with two hydrophone channels a distance apart in the sea expresses the *spatial coherence* of the noise signal. Coherence may be defined as the relationship between the phases of two noise signals at the same frequency. If the phase difference between the signals is constant, the signals are *coherent*. If the phase difference changes, the signals are *incoherent*. The coherence of underwater signals is strongly influenced by inhomogeneities in the signal propagation path, and a certain loss of coherence always takes place. *Spatial coherence* is the coherence between signals measured at the same time by two hydrophone channels at different locations. *Temporal coherence* is the coherence between two signals measured at different times by the same hydrophone channel. *Spatial—temporal coherence* is the coherence between two signals measured at different times by two hydrophone channels at different locations.

A spatial model for an ambient noise field may be simulated by many random noise sources distributed on the surface of a large sphere and measured with an omnidirectional hydrophone located at the center of the sphere. The noise at the position of the hydrophone is a summation of contributions from a large number of statistically independent plane waves with a random orientation. In spite of the directivity of ambient noise in broader frequency regions, the assumption of an *isotropic* (i.e., the energy density is the same in all directions) model for the noise is less complicated mathematically and is very useful and of practical interest. For instance, the transition region between distant shipping-dominated and surface source-dominated ambient noise often shows a nearly isotropic noise distribution. Also, ambient noise above 50 kHz may be considered isotropic.

For measurements of a monochromatic acoustic signal carried out by two hydrophones spaced a distance d {unit: m} apart and with a time delay τ {unit: s}, in an *isotropic* noise field the *spatial correlation coefficient*, ρ_c, or the normalized time-averaged product of the output of the two hydrophones, according to Cron and Sherman [54] and Burdick [55], is given by:

$$\rho_c(d, \tau) = \frac{\sin\left(\dfrac{2\pi d}{\lambda}\right)}{\dfrac{2\pi d}{\lambda}} \cos(\omega\tau) \tag{6.19}$$

where λ {unit: m} is the wavelength of the acoustic signal and $\omega = 2\pi f$, where f {unit: Hz} is the frequency. Eq. (6.19) shows that total incoherence is obtained when the two hydrophones are spaced a multiple of half wavelengths from one another, i.e., for $\sin(2\pi d/\lambda) = 0$.

According Cron and Sherman [55], for noise generated at the sea surface, the horizontal correlation coefficient for the isotropic noise field may be written as:

$$\rho_c(d, \tau) = \frac{2J_1\left(\dfrac{2\pi d_s}{\lambda}\right)}{\dfrac{2\pi d_s}{\lambda}} \cos(\omega\tau) \tag{6.20}$$

which shows that total incoherence is obtained for a hydrophone separation of $d_s = 0.61\lambda$. J_1 is the Bessel function of first kind and order 1.

For sea surface–generated noise with directivity given by Eq. (6.17), and with $m = 2$, i.e., a dipole radiation pattern, the spatial correlation functions for two hydrophones separated by the distance d and with the line between the hydrophones either parallel to or vertical to the sea surface are shown in Fig. 6.10, while Fig. 6.11

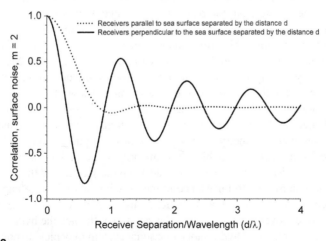

FIGURE 6.10

Effect of receiver orientation on the spatial correlation of surface-generated ambient noise with a dipole-like intensity radiation pattern.

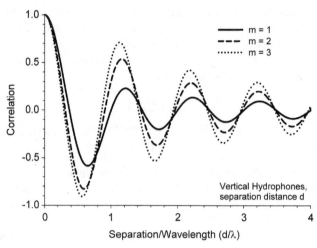

FIGURE 6.11

Spatial correlation of surface-generated ambient noise when hydrophone separation *d* is vertical and noise source directivity is given by $\cos^m$, where $m = 1$, 2, and 3.

shows the effect of the noise source directivity, represented by various values of *m*, on the spatial correlation function for sea surface—generated noise, when the line between the two hydrophones is vertical.

6.6 SELF-NOISE

Self-noise is noise produced by the measurement platform itself during measurements using instruments for underwater sound registration. Self-noise is, for instance, radiated noise from the platform, flow-generated noise in and around the platform, local cavitation and bubble flow, mechanical vibrations of transducers in their mounting positions, electrical interference between electronic instruments operating at the same time, and thermal noise of an electronic origin. Measurements of self-noise by a ship's sonar system will always include contributions from ambient noise, but ambient noise will only dominate at low platform speeds.

The individual *contributions* to measured self-noise by a sensor system, like a single hydrophone, a hydrophone array, or a sonar system, depend on the sensor position and the measurement geometry. A towed array of hydrophones may pick up water-borne noise from platform machinery, from propellers, and in shallow water, in particular, these noise source reflected signals from the seabed and the water surface. Moreover, flow noise due to turbulence in the flow boundary layer along the towed array may contribute self-noise, and accelerations transferred to the hydrophones by the meander-like movements of the towed array may influence the hydrophone measurements. A hull-mounted hydrophone or a sonar system may pick up noise from (1) the platform machinery and ship's propeller with noise from these sources being reflected and scattered at the seabed; (2) transient events in the ship like doors slamming, items dropped on the floor, and in submarines even oral communications; (3) hull plate vibrations on the sensor-mounting position that transfer accelerations to the sensing element; (4) turbulent flow along the ship's hull; and (5) from pulsations in the flow boundary layer of bubbles entrained by bow waves. Simultaneously operating electronic equipment aboard the platform may also contribute to the self-noise level. Turbulent flow boundary layer pressure fluctuations and local cavitation on a sonar dome may contribute to self-noise by exciting compressional and flexural waves in the sonar dome structure, in particular at high platform speeds. The many sources of self-noise have different characteristics, and they interact in a complicated manner, which normally will make self-noise prediction, calculation, and abatement a difficult task.

Radiated noise from a platform is in general measured by use of an *underwater acoustic test range*, where the platform to be tested for self-noise undergoes systematic measurements by passing arrays of hydrophones mounted on the seabed and cabled to registration equipment on land or by passing a vertical array of hydrophones with a radio link to a shore station. By controlling the platform's direction and speed relative to the hydrophone array and by measuring the distance between the platform and the array, it is possible to monitor the radiated noise level, its spectral composition, and frequently its directivity. Through the measurements performed at a test range it is possible to identify and rank individual sources of platform noise. From propagation loss calculations it is possible to estimate the noise source strength referenced to a distance 1 m from the source. This information can then be used to estimate the level of the self-noise caused by the identified noise sources at the position of a hydrophone or a sonar transducer. However, only very systematic measurements of contributions from individual self-noise sources, by successive isolation of the individual source contributions, when possible, may form a basis for self-noise reduction.

Also, interference between acoustic signals from instruments being simultaneously operated aboard the platform may contribute to self-noise. The short distance set by the platform size between two or more transmitting and receiving transducers may frequently cause a signal transmitted by one transducer to be picked up by another transducer at a sound pressure level that jams the receiving transducer or at least strongly exceeds the echo level the receiving transducer is measuring. The difficulties caused by transducer interference can be reduced by time gating received signals, by operating the transducers in different frequency bands and using bandpass filters to reduce or remove potential signal spillover. Alternatively, transducer operation may be time synchronized, which reduces measurement efficiency since as one transducer has to wait for all or most return signal to be received by the other transducer before starting its own transmission/reception operation. Transducer directivity may help reduce the interference, but is most frequently not sufficient to remove it. Transducer directivity may also counteract the interference reduction when an echo-sounding sonar with its directivity steered downward is used and the reflected and scattered signals from the seabed are returned to other receivers on the platform.

6.7 AMPLITUDE DISTRIBUTIONS FOR UNDERWATER NOISE

As pointed out in Section 6.2 the oceans by nature are unstable and inhomogeneous. This leads to *temporal* and *spatial variability* on many scales ranging from time-scales relative to signal duration and spatial scales relative to an acoustic wavelength to days and months and basin-wide variations. This variability influences acoustic signal transmission by adding fluctuations to signal amplitudes and phases. Swells, currents, and internal waves cause Doppler effects and result in band broadening. As discussed in Chapter 5, multipath propagation and scattering cause amplitude and phase fluctuations.

Swells cause temporal variations in an acoustic field due to signal reflections from a moving water surface, which has a time-varying shape. The surface motion influences the signal spectrum by adding sidebands whose magnitude is determined by the swells' propagation velocity. Swells also cause transmitting and/or receiving platforms to move, which causes variations in the signal path length, which affects the signal's phase. *Currents* found in the sea with scales ranging from few mm/s to several m/s cause temporal variability of the sound velocity profiles which (1) influence long-range sound propagation; (2) produce Doppler effects; and (3) cause small, time-dependent fast fluctuations in the signal amplitudes, called scintillations, due to scattering of the acoustic signals by turbulent regions produced by the flow.

Longer periodicity temporal variations in sound propagation are caused by *internal waves* and *tides*. Internal waves are caused by vertical instabilities in seawater density, according to Flatte [56], which may produce several m/s temporal variations in the sound velocity with the periodicity ranging from few minutes up to hours. Fluctuations in sound propagation conditions caused by tides have 12 h periods; however, the flow produced by tides in shallow waters and in straits, bays, and in funnel-like water regions as found in many fjords may show velocities of several m/s. These flow conditions can contribute substantially to time-dependent and spatial variations in the amplitude and phase of sound signals propagated through these regions. Temporal variations in sound propagation conditions can also be caused by temperature variations throughout a day in upper parts of the oceans due to solar heat. Sonar signal propagation conditions may change considerably from morning to afternoon conditions which in shallow water may complicate water region sonar surveys.

In spite of the fact that sound propagation conditions vary substantially in the vertical direction, a nearly horizontal stratification can be assumed for sound propagation in the horizontal direction, which facilitates sound propagation modeling. Several factors influence the ideal assumption of a stratified medium and lead to spatial variability of the acoustic signals. These factors include changes in *seabed topography* and in *seabed materials* with range from the measurement site; changes with range in the *sound velocity profile*; and *scatterers* in the water column, such as fish schools, plankton, and thermal microstructures.

The seabed's topography and material variations are slow at greater depths, such as the abyssal plains and continental shelf, where 1 degree slopes or less are found and the seabed material is the same over tens of kilometers. However, substantial topographic changes with slopes exceeding 10 degrees or more may be found in sea regions, which include canyons, ridges and seamounts, and continental slopes. These topographic variations cause acoustic signal reflections and scattering, which cause spatial variability. In shallow waters and coastal regions, considerable variations in seabed materials over shorter distances, such as tens of meters, may occur and cause spatial variations in sound propagation conditions.

The sea's temporal and spatial variability has a decisive influence on noise, as well as acoustic signal propagation, by changing the amplitude and phase of noise/signals. The character of the noise/signals, for instance their bandwidth, is influenced by the variability. The noise may be characterized as *narrowband* noise when it is monochromatic, i.e., it is represented by only one single frequency, or nearly monochromatic, and it may be characterized as *broadband* noise if it comprises frequencies over a broader interval. If the noise power is constant over a broad frequency range, the noise is called *white noise*; however, in many cases the power spectral density, i.e., the noise power in 1 Hz bands, will show a gradient, often negative, and the noise is called *colored*. If the noise power spectral density reduces at 6 dB/octave, the noise is termed *pink*. Fig. 6.12 shows narrowband noise in a monochromatic noise signal with a substantial amplitude variation. This figure also gives examples of broadband noise as "white noise" and "pink noise." The time domain plot shows the first 0.02 ms time history of a 50 ms noise signal whose amplitude spectra is shown to the right of the time history.

Even though narrowband noise has limited applications in underwater acoustics, since most noise/signals are broadband, narrowband noise has been used extensively in underwater acoustic modeling of the changes in a monochromatic signal caused by factors producing temporal and spatial variability. Broadband noise can be considered to be composed of a sum of narrowband spectral components with the individual changes to the narrowband components due to temporal and spatial variations in the sea constituting the total changes in the broadband signal. The mean acoustic power $\langle P \rangle$ of broadband noise covering the frequency band Δf {unit: Hz}, when the power spectral density of the noise is $P_N(f)$ {unit: W/Hz}, can be written as:

$$\langle P \rangle = \int_{\Delta f} P_N(f) df \qquad (6.21)$$

which for white noise is written as $\langle P \rangle = P_{No} df$, where P_{No} is the constant power spectral density of the white noise. Eq. (6.21) may be expressed in dB by using a 1 W reference power.

Due to the *random* character of underwater noise created by noise sources and propagation conditions, *statistical methods* are frequently used to describe underwater noise characteristics. A random event, also called a *stochastic* event, is an event steered by an ensemble of random variables. Their randomness makes the event unpredictable. If a random event's probability of occurrence is independent of time, i.e., the random variables are all independent of time, the random event is said to be *stationary*. A random event can, for instance, be the variation in underwater noise

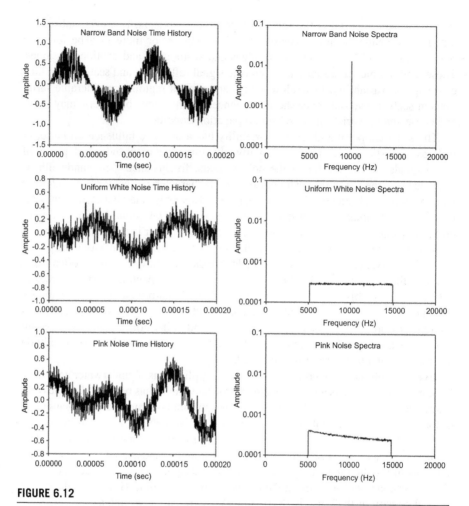

FIGURE 6.12

Time waveform and spectrum with narrow band noise, white noise, and pink noise.

pressure amplitude which is determined by random variables, such as the temporal and spatial variability of the sea. According to Wenz [2], omnidirectionally measured, low-frequency noise data show statistical stationarity over a period less than 1 h; however, for periods of several hours it may be nonstationary. For periods of several days the noise may again appear to be stationary. The main reason for these shifts in stationarity is temporal fluctuations on various scales in the ocean, which influence long-range propagation of noise.

The *probability density function* (*pdf*) of a random variable denotes the probability the random variable obtains a certain value. If the *pdf* for a random variable *x* is

$h(x)$, the probability dH that the random variable falls between x and dx may be written as:

$$dH = h(x)dx \tag{6.22}$$

and the probability, H_{ab}, that a random variable x falls in the interval $a < x < b$ is:

$$H_{ab} = \int_a^b h(x)dx \tag{6.23}$$

The *mean value*, M_x. also called the *expected value*, of a random variable is the *first moment* of the *pdf* and may be written as:

$$M_x = \int_{-\infty}^{\infty} x \, h(x)dx \tag{6.24}$$

Moments centered around the mean value M_x, the so-called *central moments*, are of particular interest in noise statistics since they describe characteristic features of the spectral distribution of noise, such as its variance, skewness, and kurtosis. The second central moment, called the *variance*, σ^2, may be written as:

$$\sigma^2 = \int_{-\infty}^{\infty} (x - M_x)^2 h(x)dx \tag{6.25}$$

with σ being the *standard deviation* of the random variable from the mean value. The higher-order central moments, of which in particular the third, the *skewness*, and the fourth, the *kurtosis*, are of interest, may be expressed for $n = 3$ and 4 by:

$$\Phi^n = \int_{-\infty}^{\infty} (x - M_x)^n h(x)dx \tag{6.26}$$

If the random function x is a function of time, t, the *auto-correlation function*, R_{xx}, of two time signals measured at the same position, but with a time delay τ is given by:

$$R_{xx} = \lim_{T \to \infty} \left(\frac{1}{2T}\right) \int_{-T}^{T} x(t)x^*(t - \tau)dt \tag{6.27}$$

and the *cross-correlation function*, R_{xy}, of two time signals, $x(t)$ and $y(t)$ measured at the same time, but at two different positions may be expressed by:

$$R_{xx} = \lim_{T \to \infty} \left(\frac{1}{2T}\right) \int_{-T}^{T} x(t)y^*(t - \tau)dt \tag{6.28}$$

where the asterisk (*) denotes the complex conjugate of the time function.

One of the most widely used *pdf*s is the *Gaussian distribution*, also called the *normal distribution*. This *pdf* describes the statistical qualities of many underwater acoustic signals. If a random variable x follows a Gaussian distribution, its *pdf*, $h(x)$, may be expressed by:

$$h(x) = \frac{1}{\sigma\sqrt{2\pi}} e^{\left\{-\frac{(x-M_x)^2}{2\sigma^2}\right\}} \tag{6.29}$$

Eq. (6.29) can describe the instantaneous amplitude of a random underwater signal, in particular a narrowband signal. Due to the simplicity of Eq. (6.29) and ease of use, it has been used to describe signal amplitude distributions by underwater signals after the influence of multipath propagation and scattering. However, a better probability distribution function for signals exposed to multipath propagation without a predominant path, or exposed to volume or interface scattering, when the scattering amplitude distribution can be considered as a sum of a great number of nearly the same scattering amplitude contributions, is the *Rayleigh distribution* [57]. The *Rayleigh distribution*, which results from adding a number of signal contributions of nearly the same frequency (i.e., narrowband signals), with comparable amplitudes, but with random phases equally distributed over the interval from $-\pi$ to $+\pi$, is given by:

$$h(x) = \frac{x}{\sigma^2} e^{\left\{ -\frac{x^2}{2\sigma^2} \right\}} \tag{6.30}$$

where x for instance can be the observed pressure amplitude of the underwater signal.

If one of the signals dominates and possesses a nonrandom character, for instance the signal from a certain path in a multipath propagation of signals, a better probability distribution function is found in the *Rice distribution* [58], which describes the sum of a signal and random noise of narrowband character. The Rice distribution is given by:

$$h(x) = \frac{x}{\sigma^2} I_0 \left(\frac{xa}{\sigma^2} \right) e^{\left\{ -\frac{x^2 + a}{2\sigma^2} \right\}} \tag{6.31}$$

where a is the amplitude of the dominating coherent signal, and $I_0(\cdot)$ is the modified Bessel function of first kind and order zero (i.e., $I_0(\nu) = J_0(i\nu)$, where $i = \sqrt{-1}$). The Rice distribution is frequently more useful than the Rayleigh distribution since a coherent signal imbedded in noise that has nearly the same bandwidth as the coherent signal is a daily problem in underwater acoustics. This problem characterizes signals suffering incoherent scattering contributions from volume inhomogeneities or scattering contributions from rough interfaces with small roughness heights, i.e., for the Rayleigh parameter $R_a \ll 1$, discussed in Chapter 5. The Rice distribution can be considered a generalized version of the Rayleigh distribution, and it reduces to this distribution if the power of the coherent signal decreases. If the power of the noise signals decreases, the *pdf* $h(x)$ in Eq. (6.31) becomes the Gaussian distribution, Eq. (6.29). Fig. 6.13 shows the Rayleigh distribution function for $\sigma^2 = 1$ and the Rice distribution function for various values of $\tau = a^2/2\sigma^2$, as a function of the random variable x, which could be the amplitude of an underwater acoustic signal.

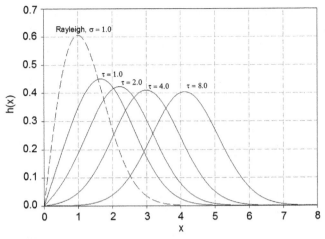

FIGURE 6.13

The Rayleigh distribution and the Rice distribution for the variance $\sigma^2 = 1$ and for $\tau = 1$, 2, 4, and 8, as a function of the random variable x.

6.8 SYMBOLS AND ABBREVIATIONS

Table 6.1 is a list of symbols and abbreviations used in this chapter along with the page where the symbol or abbreviation is defined and first used.

Table 6.1 List of Symbols and Abbreviations

Symbols or Abbreviations	Descriptions	Page of First Usage
L_p	Pressure fluctuation spectral density level {unit: dB rel 1 μPa/Hz}	365
f	Frequency {unit: Hz}	365
U	Mean flow velocity {unit: m/s}	365
L	Typical body dimension {unit: m}	365
S_t	*Strouhal number* (dimensionless)	365
u	Particle velocity	367
c	Sound velocity {unit: m/s}	367
M_a	Acoustic Mach number (dimensionless)	367
V	Wind speed {unit: m/s}	368
$L(f)$	Noise level {unit: dB rel 1 μPa}	368
k	Scale length {unit: m}	370
U_{10}	Wind speed 10 m height above the sea surface {unit: m}	370

Continued

Table 6.1 List of Symbols and Abbreviations—cont'd

Symbols or Abbreviations	Descriptions	Page of First Usage
$R(t)$	Spherical bubble radius {unit: m}	370
ρ	Incompressible liquid density {unit: kg/m^3}	370
σ_s	Surface tension {unit: N·m}	370
p_i	Bubble internal pressure {unit: Pa}	370
p_o	Static pressure {unit: Pa}	370
R_o	Equilibrium bubble radius {unit: m}	371
γ	Ratio of specific heats	371
ω_o	Bubble vibration angular frequency {unit: Hz}	371
β	Volume fraction of air in water	371
c_m	Sound speed in mixture {unit: m/s}	371
c_w	Sound speed in water {unit: m/s}	371
χ	Adiabatic compressibility {unit: m·s/kg}	371
f_0	Coupled oscillator lowest mode frequency {unit: Hz}	371
κ	Cavitation index (dimensionless)	377
P	Ambient hydrostatic pressure {unit: Pa}	377
P_v	Vapor pressure in the water {unit: Pa}	377
U	Flow velocity {unit: m/s}	377
κ_l	Cavitation inception index (dimensionless)	377
L_s	Surface ship noise power level {unit: dB rel 1 μPa at 1 m}	378
U_s	Ship speed {unit: knots}	378
D	Ship displacement {unit: tons}	378
L_p	Noise pressure level {unit: dB rel 1 μPa at 1 m}	378
U_p	Propeller tip speed {unit: m/s}	378
B	Number of propeller blades (dimensionless)	378
P_m	Detonation peak pressure {unit: Pa}	380
θ	Detonation time constant {unit: ms}	380
Q	Explosive weight {unit: kg}	380
r	Range from detonation point {unit: m}	380
τ	Time from explosion to peak of the first bubble pulse {unit: ms}	380
d	Underwater explosion depth {unit: m}	380
NL	Noise spectral density level {unit: dB rel 1 μPa/Hz}	381
$P_N(f)$	Power spectral density {unit: W/Hz}	385
α	Sound pressure absorption coefficient {unit: Np/m}	387
d	Measurement point depth below the sea surface {unit: m}	387

Table 6.1 List of Symbols and Abbreviations—cont'd

Symbols or Abbreviations	Descriptions	Page of First Usage
ρ_c	Spatial correlation coefficient (dimensionless)	389
λ	Acoustic signal wavelength {unit: m}	389
J_1	Bessel function of first kind and order 1 (dimensionless)	389
d_s	Hydrophone separation {unit: m}	389
Δf	Frequency band {unit: Hz}	393

REFERENCES

[1] Knudsen, V.O., Alford, R.S and Emling, J.W., *Underwater ambient noise*, J. Marine Res., **7**, 410, 1948.

[2] Wenz, G.M., *Acoustic ambient noise in the ocean: Spectra and sources*, J. Acoust. Soc. Amer., **34**, (12), 1936, 1962.

[3] Strasberg, M., *Nonacoustic noise interference in measurements of infrasonic ambient noise*, J. Acoust. Soc. Amer., **66**, (5), 1487, 1979.

[4] Webb, Spahr C., *The equilibrium oceanic microseism spectrum*, J. Acoust. Soc. Amer., **92**, (4), 2141, 1992.

[5] Urick, R.J., *Principles of Underwater Sound*, (3rd. Ed.), McGrawHill Book Comp., 1983, Peninsula Publishing.

[6] Weichert, E., *Verhandlungen der Zweiten Internationalen SeismologischenKonferenz*, Geol. Beitr. Geophys. Ergänzungsband, **2**, 41, 1904.

[7] Longuet-Higgins, M.S., *A theory of the origin of microseisms,* Phil. Trans. Royal Soc. London, A-243, 1, 1950.

[8] Hasselmann, K., *A statistical analysis of the generation of microseisms*, Review of Geophysics, **1**, (2), 177, 1963.

[9] Hughes, B., *Estimates of underwater sound and infrasound produced by nonlinearly interacting ocean waves*, J. Acoust. Soc. Amer., **60**, (5), 1032, 1976.

[10] Kibblewhite, A.C. and Evans, K.C., *A study of the ocean and seismic noise at infrasonic frequencies,* In: Ocean Seismoacoustics, T. Akal & J.M. Berkson (Eds.), Plenum Press, pp. 731−741, 1986.

[11] Dünnebier, R.J., Cessaro, R.K. and Anderson, P., *Geo-acoustic noise levels in deep ocean boreholes*, In: Ocean Seismoacoustics, T. Akal & J.M. Berkson (Eds.), pp. 743−51.

[12] Isakovich, M.A. and Kuryanov, B.F., *Theory of low frequency noise in the ocean*, Soviet Phys. Acoust., 16, 49, 1970.

[13] Wilson, J.H., *Very low-frequency (VLF) wind-generated noise produced by turbulent pressure fluctuations in the atmosphere near the ocean surface*, J. Acoust. Soc. Amer., **66**, (5), 1499, 1979.

[14] Andreev, N.N., *On the voice of the sea*, Akademia Nauk, SSSR, 23, 625, 1939.

[15] Piggott, C.L., *Ambient noise at low frequencies in shallow water off the Scotian Shelf*, J. Acoust. Soc. Amer., **36**, (11), 2152, 1964.

[16] Kerman, B.R., *Audio signature of a breaking w*ave, In: Sea Surface Sound, B.R. Kerman (Ed.), Kluwer Academic Publishers, pp. 437−448, 1988.

[17] Banner, M.L. and Peregrine, D.H., Wave breaking in deep water, In: J.L. Lumley, M. van Dyke & H.L. Reed (Eds.), Annual Review of Fluid Mechanics, **25**, pp. 373−397, 1993.

[18] Longuet-Higgins, M.S., *Mechanisms of wave breaking*, In: Sea Surface Sound, B.R. Kerman (Ed.), Kluwer Academic Publishers, pp. 1−30, 1988.

[19] Carey, W.M. and Browning, D., *Low frequency ocean ambient noise, Measurement and theory*, In: Sea Surface Sound, B.R. Kerman (Ed.), Kluwer Academic Publishers, pp. 361−376, 1988.

[20] Cato, D.H., *Features of ambient noise in shallow water*, Proc. International Conference on Shallow-Water Acoustics, Beijing, pp. 385−390, 1997.

[21] Kibblewhite, A.C. and Ewans, K.C., *Wave-wave interaction, microseisms and infrasonic ambient noise in the ocean*, J. Acoust. Soc. Amer., **78**, (3), 981, 1985.

[22] Wilson, D.K., Frisk, G.V., Lindstrom, T.E. and Sellers, C.J., *Measurement and prediction of ultralow frequency ocean ambient noise off the eastern U.S. coast*, J. Acoust. Soc. Amer., **113**, (6), 3117, 2003.

[23] Bjørnø, L., *A constructive, critical evaluation of contributions to acoustics of wakes*, Applied Research Laboratory, The Pennsylvania State University, Technical Report, TR 02-001, Dec. 2001.

[24] Kerman, B.R., *A model for interfacial gas transfer for a well-roughened sea*, J. Geophys. Res. **89**, 1439, 1984.

[25] Kolovayev, A.A., *Investigation of the concentration and statistical size distribution of wind-produced bubbles in the near-surface ocean,* Oceanology, **15**, 659, 1976.

[26] Johnson, B.D. and Cooke, R.C., *Bubble population and spectra in coastal waters. A photographic approach*, J. Geophys. Res., **84**, 3769, 1979.

[27] Wu, J., Bubble population and spectra in near-surface ocean, J. Geophys. Res., **86**, 457, 1981.

[28] Kozhevnikova, I. and Bjørnø, L., *An experimental study of acoustical emission from bubble column excitation*, Ultrasonics, **30**, 21, 1992.

[29] Kozhevnikova, I. and Bjørnø, L., *Near sea surface bubble cloud oscillation as potential sources of ambient noise*, In: Natural Physical Sources of Underwater Sound, B.R. Kerman (Ed.), Kluwer Academic Publishers, pp. 339−347, 1993.

[30] Prosperetti, A., *Bubble dynamics in ocean ambient noise*, In: Sea Surface Sound, B.R. Kerman (Ed.), Kluwer Academic Publishers, pp. 437−448, 1988., pp.151−172.

[31] Nystuen, J.A., *Rainfall measurements using underwater ambient noise*, J. Acoust. Soc. Amer., **79**, (4), 972, 1986.

[32] Heindsmanm, T.E., Smith, R.H. and Arneson, A.D., *Effects of rain upon underwater noise levels*, J. Acoust. Soc. Amer., **27**, (2), 378, 1955.

[33] Franz, G.J., *Splashes as sources of sound in liquids*, J. Acoust. Soc. Amer., **31**, (8), 1080, 1959.

[34] Bom, N., *Effects of rain on underwater noise level*, J. Acoust. Soc. Amer., **45**, (1), 159, 1969.

[35] Scrimger, J.A., Ewans, D.J., McBean, G.A., Farmer, D. and Kerman, B.R., *Underwater noise due to rain, hail and snow*, J. Acoust. Soc. Amer., **81**, (1), 79, 1987.

[36] Nystuen, J.A. and Farmer, D., *The influence of wind on the underwater sound generated by light rain*, J. Acoust. Soc. Amer., **82**, (1), 270, 1987.

[37] Bjørnø, L., *Underwater ambient noise generated by raindrop impacts*, Proc. Nordic Acoustical Meeting 88, Tampere, Finland, 201−204, 1988.

[38] Crum, L.A., Pumphrey, H.C., Prosperetti, A. and Bjørnø, L., *Underwater noise due to precipitation*, J. Acoust. Soc. Amer., **85**, S151, 153, 1989.

[39] Pumphrey, H.C., Crum, L.A. and Bjørnø, L., *Underwater sound produced by individual drop impacts and rainfall*, J. Acoust. Soc. Amer., **85**, (4), 1518, 1989.

[40] Ewans, W.E., *Vocalization among marine mammals*, In: Marine Bio-Acoustics, **2**, W.N. Tavolga (Ed.), Pergamon Press, pp. 160−169, 1967.

[41] Dyer, I., Ice source mechanisms: *Speculations on the origin of low frequency Arctic Ocean noise*, In: Sea Surface Sound, B.R. Kerman (Ed.), Kluwer Academic Publishers, pp. 513−532.

[42] Milne, A.R. and Canton, J.H., *Ambient noise under Arctic sea ice*, J. Acoust. Soc. Amer., **36**, (5), 855, 1964.

[43] Greene, R. and Buck, B.M., *Arctic Ocean ambient noise*, J. Acoust. Soc. Amer., **36**, (6), 1218, 1964.

[44] Milne, A.R., *Shallow water under-ice acoustics in Barrow Strait*, J. Acoust. Soc. Amer., **32**, (8), 1007, 1960.

[45] Ross, D., *Mechanics of Underwater Noise*, Pergamon Press, 1987.

[46] Gray, L.M. and Greeley, D.S., *Source level model for propeller blade rate radiation for the world's merchant fleet*, J. Acoust. Soc. Amer., **67**, (2), 516, 1980.

[47] Bjørnø, L., *A comparison between measured pressure waves in water arising from electrical discharges and detonation of small amounts of chemical explosives*, Trans. ASME, J. Eng. Industry, **92**, Ser. B, (1), 29, 1969.

[48] Thorne, P.E., *Seabed generation of ambient noise*, J. Acoust. Soc. Amer., **87**, (1), 149, 1990.

[49] Mellen, R.H., *Thermal-noise limit in the detection of underwater acoustic signals*, J. Acoust. Soc. Amer., **24**, (5), 478, 1952.

[50] Ingenito, F. and Wolf, S.N., *Site dependence of wind-dominated ambient noise in shallow water*, J. Acoust. Soc. Amer., **85**, (1), 141, 1989.

[51] Jensen, F.B. and Kuperman, W.A., *Sound propagation in a wedge-shaped ocean with a penetrable bottom*, J. Acoust. Soc. Amer., **67**, (5), 1564, 1980.

[52] Neighbors, T.H. and Bjørnø, L., *Bubble plume shape influence on backscattering from the sea surface*, Proc. '97 International Conference on Shallow-Water Acoustics, Acoustics Institute Press, Academia Sinica, Beijing, pp. 433−438, 1997.

[53] Soritin, B.J. and Hodgkiss, W.S., *Fine-scale measurements of the vertical ambient noise field*. J. Acoust. Soc. Amer., **87**, (5), 2052, 1990.

[54] Cron, B.F. and Sherman, C.H., *Spatial-correlation functions for various noise models*, J. Acoust. Soc. Amer., **34**, (11), 1732, 1962.

[55] Burdic, W.S., *Underwater Acoustic System Analysis*, Prentice-Hall Inc., Englewood Cliffs, 1984.

[56] Flatte, S.M., *Sound transmission through a fluctuating ocean*. Cambridge University Press, Cambridge UK, 1979.

[57] Rayleigh, J.W.S., *The Theory of Sound*. Dover Publications, New York, 1945.

[58] Rice, S.O., *Mathematical Analysis of Random Noise*. Bell Sys. Tech. J., **24**, (46), Art. 3.10, 1945.

Shallow-Water Acoustics

7

J.F. Lynch, A.E. Newhall

Woods Hole Oceanographic Institution, Woods Hole, MA, United States

7.1 WHAT IS SHALLOW-WATER ACOUSTICS?

Before discussing any applications, it would be useful to define just what we mean by "shallow-water acoustics," as this seemingly simple term actually does have few definitions.

The first definition, and the most obvious, is the continental shelf region between the beach/shore and the continental shelf break (see Fig. 7.1). The ocean can be divided into three regions: the shelf, slope/rise, and abyssal regions. The shelf region, which we can (operationally) take as the area out to 200 m depth, is usually taken as shallow water. The slope and rise regions between the shelf and deep water are also discussed in this chapter, as they are very important regions, but ones that have been rather ignored (not without reason) until recently.

The second definition is an acoustics definition, i.e., the water depth defined by being 10 acoustic wavelengths or less, at whatever acoustic frequency one considers. This is the regime here one can efficiently use normal mode theory as a descriptor of the acoustic field, and where ray theory begins to break down. While not wrong, this is a rather limiting definition of shallow water, as it removes consideration of medium- and high-frequency acoustics, which definitely have a number of important and interesting applications in shallow water. (For our purposes in this chapter, we consider 10−1000 Hz as low frequency acoustics, 1000−20,000 Hz as medium frequency acoustics, and 20,000 Hz to 5 MHz as high frequency acoustics.)

A third definition of shallow-water acoustics is acoustics in the region where both surface and bottom interactions are of importance. Upward refracting deep-water environments (e.g., the Mediterranean Sea or the Arctic Ocean) can often have significant surface interactions, but shallow-water environments emphasize both.

A fourth, and rather holistic, definition of shallow water is by using *all* its defining environmental characteristics: its bathymetry, geology, biology, physical oceanography, chemistry, and of course acoustics. All these ingredients contribute to the unique "soundscape" of shallow water, and in a sense this is the most gratifying and least limiting of the definitions.

Applied Underwater Acoustics. http://dx.doi.org/10.1016/B978-0-12-811240-3.00007-2

403

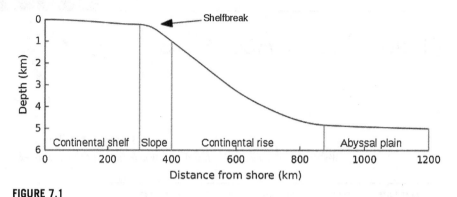

FIGURE 7.1

Typical transition between the shoreline and the deep ocean.

Another way to define shallow-water acoustics is by its rather specific set of applications. The UN Atlas [1] estimates that 40% of the world's population lives within 100 km of the ocean. Moreover, the majority of the world's largest cities (population 2.5 million or more) are located on the coasts. Given the importance of the sea to humanity, it is not surprising that there are many, many applications of coastal (shallow-water) acoustics.

7.1.1 MILITARY APPLICATIONS

Probably the most well-known application of underwater acoustics is for defense purposes, and shallow-water acoustics is no exception.

Anti-submarine warfare (ASW) may be the best publicly known military application, due to the many popular novelizations and movies that have appeared since 1940s. Submarines, like any ship, need to transit to and from their shore bases on a routine basis. And while missile submarines may prefer to hide in deeper water, attack submarines often operate near shipping lanes and hostile ports. Shallow water may offer less hiding opportunities for submarines, but it is a part of their working environment. The operation of naval sonars both by submarines and by those who hunt them is an important part of shallow-water acoustics. These sonars tend to work in the low-to-medium frequency range (10 Hz–20 kHz).

Another well-known shallow-water acoustics application is mine warfare. Access to and egress from ports is economically and militarily vital to most countries, and mines present a low-cost, effective way of either blocking a port or protecting it. Moreover, mines are often indistinguishable from the myriad other pieces of rock and rubble found on the seabed, especially near human habitation. Mine warfare uses a wide variety of acoustic frequencies, from low (e.g., for targeting) to high (e.g., for classification). To date, dolphins are perhaps the best mine hunters to be found, but modern autonomous underwater vehicle (AUV) technology, combined with imaging sonars, is fast becoming a replacement.

In addition to the above two military applications, one could also add the port security applications of swimmer detection and AUV detection, as these "small targets" are worrisome. Mid-to-high frequency acoustics (few kilohertz to few hundred kilohertz) tend to be most relevant here.

7.1.2 DUAL-USE APPLICATIONS

Many applications of shallow-water sonar have dual (civilian and military) usage. Let us look few representative ones.

One of the classic uses of sonar has been for obstacle avoidance, with the 1912 Titanic tragedy being a prime motivator. By 1914, sonars could detect icebergs at a distance of 2 miles [2], and since then sonars have been an essential tool for underwater obstacle detection, whether in the upward, downward, or sideward directions. These tend to be mid-frequency systems.

Another vital use of sonar in shallow water, for almost any sizable marine craft, is bottom echo sounding to avoid running aground. One only need to look as far as the nearest marine supply store to find very modern sonar technology available for this purpose. Here again, mid-frequency systems are favored.

A third dual-use of sonar in shallow water is for acoustic communications. In a military context, this can be for message transmission, or for command and control of small vehicles and instruments. Civilian uses include the transmission of data from scientific experiments, underwater communications between divers, and the same uses as the military. These systems can be anywhere from low to mid-frequency.

7.1.3 OCEAN SCIENCES APPLICATIONS

Mid- to high-frequency acoustics, at this point in time, has become an integral part of many oceanographic tools and instruments. Acoustic Doppler current profilers (ADCPs) [3] and acoustic current meters (ACMs) [4] are standard instrumentation for profiling and precision current measurements. Backscattering sonar images and measures bubbles, turbulence, suspended sediments, and hydrothermal plumes, among other things. Mid-frequency acoustic tomography can measure ocean temperature, currents, and relative vorticity over a coastal region. At lower frequency, acoustic thermometry can measure regional temperature variation, which may become important as climate change alters coastal ocean temperature [5].

In the geology and geophysics area of ocean sciences, mid-frequency multibeam echo sounders have become a highly developed technology that has successfully mapped large areas of the coastal ocean bottom [6]. Numerous acoustic techniques over a variety of frequencies are used to image near surface (defined here as about 1−2 acoustic wavelengths in depth) stratigraphy and chirp sonar used near the bottom can penetrate significantly deeper [7]. Interface waves (e.g., Scholte waves) can be used to determine both compressional wave and shear wave properties of surficial coastal sediments [8]. And, of course, low- to mid-frequency seismic profiling is more and more commonly used to probe deep into the ocean bottom for coastal gas and oil resources.

For marine biology, the coastal and shallow regions are by far the most important homes for marine life. Generally, the acoustic frequency used to study this biology scales directly with the size of the organism. High-frequency backscatter is used to study phytoplankton and zooplankton, mid-frequency sound is effective in looking at fish (both individually and in schools and shoals), and low- to mid-frequency sound is useful in tracking and identifying vocalizing marine mammals.

7.1.4 COMMERCIAL APPLICATIONS

Many of the military and scientific applications noted above easily transition into the commercial sector. Fish finders for commercial and recreational fishing, listening to whales as part of a "whale watch," and oil and gas exploration are just three obvious examples of how acoustic applications can find their way into the commercial sector. Also, sales of sonar equipment for military and scientific work is a big commercial interest itself. There are probably many more key commercial applications of shallow-water acoustics that we have not mentioned, but these representative ones should suffice as examples.

7.2 PHYSICS AND PHENOMENA

One might think, at first glance, that the water depth should not be a major factor in describing ocean acoustics, other than changing the depth parameter in the wave equation. This would be, to be blunt, just plain wrong. To see how shallow-water acoustics differs from "generic" ocean acoustics and to discuss the physics in a methodical fashion, we look at the sonar equation (which subsumes the wave equation and much signal processing) to examine individual points.

7.2.1 SOURCE LEVEL TERM

When Robert Urick's classic applied sonar text *Principles of Underwater Sound* was first published in 1967 [2], the only source considerations that the sonar engineer had to worry about were physical ones, e.g., the size, weight, efficiency, etc. of the source, and how adequate the signal output parameters (e.g., intensity, bandwidth, and signal shape) were for the intended usage. However, today, an "augmented" version of Urick's text would have an additional important consideration for shallow-water environments—marine life concerns, which limit what, when, and where one can transmit in shallow, coastal waters. Over the past 25 years, marine mammal concerns have led to considerable restrictions on the operation of underwater sound sources for research, exploration, and even military purposes. Although the source level (SL) of restriction varies widely internationally, most of the countries that are sophisticated users of underwater sound have some rules in place.

The levels and frequencies that see the most restriction in shallow water are those that match the auditory response curves of the marine life (mostly mammals, though that too is changing) that is found on the continental shelf, and also on the slopes and in canyons. Although these curves are hard to obtain, especially for animals in the wild, some auditory curves have been obtained for captive animals [9]. Moreover, the vocalizations that the animals make in the wild can also provide some clue as to the frequency ranges they hear (see Fig. 7.2). (Note the use of "clue"—the vocalizations cannot be taken as a direct proxy, as animals often are sensitive to other signals besides their own vocalizations.) Generally, one sees signals from about 10 Hz to tens of kHz from marine mammals, so it is this band that has felt the restrictions in SL. The level restriction is a rather difficult issue to deal with, in that its determination depends crucially on what levels can harm the animals, which can include both physical damage as well as disruptions to feeding, mating, and migration.

In addition to the level and frequency restrictions, there are the "time and place" restrictions. Marine sanctuaries are obviously off limits, but outside of these, one has to know where the animals are likely to be found and when. This is another nontrivial issue to consider, and much research has been invested in addressing it. A good example of such research is the Office of Naval Research (ONR)-sponsored "ESME" (Effects of Sound on the Marine Environment) project, which looks at the issues we have discussed in an overview sense and has created a usable demo for public use [10].

Given the new, restricted environment that sound sources must work in, what does that tell us about "practical usage?" Military users tend to have a very specialized relationship with governments and government rules, so that it is hard to say exactly what restrictions apply in this case; such rules vary from country to country, and also will change in wartime or conflict conditions. Seismic exploration sources, which are common on continental shelves worldwide, are now subject to rigorous Environmental Impact Statements (EISs) that examine the overall effect of the sources used on the survey area. Research sources, despite their generally lower levels and smaller periods of usage, have also been heavily restricted, In the United States, for example, explosive sources, which used to be common, are now banned. Lower-frequency air gun sources and arrays are also highly regulated. Sources of all manner tend to be held to 160–180 dB/rel μPa at 1 m maximum level in coastal regions. This poses a problem, in that for a given experiment or application, this lower level might not produce adequate signal-to-noise ratio (SNR) at the receiver.

Luckily, two remedies are available to help with lower available SL. First, one may use a receiving array to boost overall system gain, a common device. And second, one may use the processing gain obtained by cross-correlating a broadband signal that is sent at a low level over a long period of time. Common signals used for shallow-water research are broadband (frequency modulation, FM) sweeps and phase-encoded pseudorandom noise (PRN) sequences. These signals, as they are transmitted with a low SL, often reach the receiver below the noise level. By using a cross-correlator at the receiver end, the signal can be boosted by 20–40 dB in practice. One can also time integrate continuous wave (CW) signals and cross-correlate other signals—the

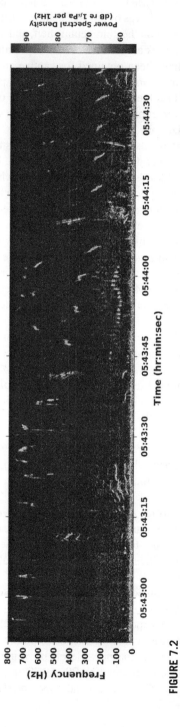

FIGURE 7.2

Spectrogram of 2 min of marine mammal vocalizations in the 10–800 Hz band, seen in Cape Cod Bay, Massachusetts, during summer 2012.

Courtesy of A. Newhall and Y.T. Lin, Woods Hole Oceanographic Institution, private communication.

broadband signals discussed above are just two of the more popular variants. The amount of time one can integrate or cross-correlate over, the "coherence time of the medium," depends on the ocean sound speed fluctuations and the acoustic system frequency and can stretch anywhere from seconds to minutes. To see how this "decoherence" or "decorrelation" time (as it is also called) works physically is simple to show. If we consider the phase φ of an acoustic signal of frequency ω sent on a straight-line path over a distance D, we get $\varphi = \omega t = \omega D/c$, where t is the travel time and D is the distance. If the ocean sound speed now changes by an amount $c \rightarrow c + \Delta c$, then the phase changes to $\varphi = \frac{\omega D}{c}\left(1 - \frac{\Delta c}{c}\right)$. The phase change $\Delta \varphi = \frac{\omega D \Delta c}{c^2}$ will be unknown and unaccounted for at the receiver end if the sonar operator does not have knowledge (measurements) of how the ocean sound speed changed. Thus, any processors that use coherent additions of phase will start to incur substantial error when $\Delta \varphi \rightarrow \frac{\pi}{2}$. We also see that this phase error increases linearly with frequency, range, and sound speed perturbation.

7.2.1.1 Example: Integrating Pseudorandom Noise Sequences and Frequency Modulation Sweeps for Signal Gain

As mentioned previously, two types of signals are quite popular for boosting signals in shallow water (as well as in deep water). They are the PRN sequences (commonly called m-sequences) and FM sweep signals. It is worth looking at each in a little more detail.

PRN sequences are very useful for moored source/receiver configurations, as their correlates give flat side-lobe levels and a very narrow peak, but they are also very susceptible to degradation by relative source/receiver motion effects. Given an integer value for m, the m-sequence is a string of binary digits (0s and 1s) of length $(2^m - 1)$ and contains a nearly equal number of 1s and 0s (with one more of one or the other). M-sequences have many very pretty mathematical properties, but the ones we care about here are their practical ones—how are they implemented and how much gain can they give? Acoustically, the digits can be formed by using phase codes, where a positive sine wave can be a 0 and a negative sine wave (phase going in opposite direction) can be a 1. The number of sine waves of a given frequency used to create each digit relates to the available bandwidth of the source and is easily calculated. For instance [11], three cycles of a 250 Hz carrier signal gives a pulse (digit) length of 12 ms, a reasonably useful length for resolving multipaths at the receiver end. If one uses a sequence length of 1023 digits, this gives a gain of $10 \log(1023) = 30.1$ dB. One can also repeat the sequences N number of times to get an additional $10 \log (N)$ dB gain. This would seem to give unlimited potential for gain; however, the total amount of time one can usefully correlate these sequences is given by the coherence time of the medium, and if one tries to do correlation beyond this, one actually sees a drop in the gain achieved. In our typical shallow-water experiments, we have correlation times of half a minute to several minutes available, depending on which ocean processes are dominant in the area. This provides the limit on the number N of sequences used.

Almost complementary to the m-sequences (in some ways) are FM sweep signals, which do not have as nice peak width or side-lobe properties as the m-sequences, but have the very useful property of "Doppler tolerance" and also are not as difficult to mechanically implement as the phase codes. Thus, sweeps are useful for moving systems.

The mathematics of a linear-frequency sweep follows Duda [12]. For a linear sweep signal, the signal and its phase as a function of time t are, respectively, given by:

$$s(t) = A(t)e^{i\theta(t)} \tag{7.1}$$

and

$$\theta(t) = \frac{bt^2}{2} + \omega_0 t \tag{7.2}$$

where ω_0 is the center (angular) frequency.

The amplitude $A(t)$ gives one the latitude to window the signal in time. The phase derivative leads to a linear instantaneous frequency, i.e.,

$$f(t) = \frac{d\theta}{dt} = bt + \omega_0 \tag{7.3}$$

where b is the time constant of the sweep (units T^{-1}).

The power spectral density $S(f)$ of the signal s is obtained from the inverse formula

$$S(f) = A^2(g(f)) \tag{7.4}$$

where we are mapping frequency to time by using $g(f) = (f - \omega_0)/b$, the inverse function of $f(t)$.

The power spectral density $S(f)$ of the signal is thus

$$S(f) = A^2 \frac{(f - \omega_0)}{b} \tag{7.5}$$

When cross-correlated, one gets a roughly 1/BW (1/bandwidth) time resolution for the main peak, and side lobes that are determined by the taper used, the usual result. However, if a Doppler shift is introduced, it just slightly shifts the sweep in time, and the correlator only misses a small fraction of the total waveform, which means that only a small amount of gain is lost. Moreover, the continuous frequency sweep is easier to accomplish mechanically than the phase jumps of the m-sequence, which require the transducers to instantaneously reverse their mechanical motion.

The gain of the FM sweep comes from the total time—bandwidth product of the transmission being greater than 1, i.e., after picking a bandwidth (which determines the basic pulse width via T*BW = 1), one transmits for as long as one can, given the medium decorrelation time. Thus, for instance one can transmit a 100 Hz bandwidth sweep, which gives a 10 ms pulse width, over 1 s, which gives a T*BW product of 100 for the total transmission, and a gain of 10 log (T*BW) = 20 dB. We can repeat this (and sum it) as long as possible until the medium makes the stacking incoherent, after which one is again losing gain. Thus, if we think we have a 100 s coherence time (which is typical), we get another 10 log (100) = 20 dB gain, which gives a total gain of 40 dB.

We see that these two techniques give us 30—40 dB of correlator gain, which can mean the difference between using a 200—220 dB SL in shallow water, which is harder to get permits for these days, and a 160—180 dB SL, which is more within current limits.

7.2.2 ARRAY GAIN TERM

Although the term in the sonar equation is called "array gain," which deals with the SNR improvement one can obtain by using an extended array as opposed to a "point" source or receiver, we are not concerned here about that property of arrays. Rather, we are looking at some of the other advantages of array processing in shallow water. However, we keep the standard sonar equation name for the array term, just for consistency.

Arrays can be used for both sources and receivers, and to complete the story on sources, let us consider them first. Towed air gun source arrays are used in horizontal planar (x—y plane) configurations in coastal seismic exploration to direct sound downward into the seabed. This is a fairly efficient way to direct the energy to where it is needed and wanted. Vertical source arrays are used in some active sonar military applications to direct sound horizontally below the critical grazing angle, so that losses of energy to the bottom are minimized. (Below the critical grazing angle, sound is losslessly, total-internally reflected by the bottom. Above that angle, one gets transmitted as well as reflected energy due to bottom interaction, with the transmitted energy quickly being lost.) Research groups also use vertical source arrays spanning a significant fraction of the water column to selectively excite particular normal modes or rays [13]. The evolution of the particular mode or ray into other modes and rays ties directly to the range-dependent properties of the shallow-water waveguide.

Receiving arrays too have horizontal and vertical configurations of interest in shallow water. Horizontal line arrays can easily be laid on the seabed in many shallow water locations, allowing one to deploy a sizable, stable instrument. On the other hand, such an instrument is highly vulnerable to fishing activity, particularly dragging, and so must be used carefully and (if possible) in cooperation with local fishermen. Towed horizontal arrays in shallow water need to be shorter than in deep water, especially if used close to the seabed. However, they are still used, and are even incorporated onto small AUVs for survey work. One can also use a combination vertical and horizontal L-array configuration to advantage, as was done in the Shallow Water 2006 (SW06) experiment [14]. This allowed, among other things, very detailed looks at the coherence structure of various modal arrivals versus environmental variability, as seen in Figs. 7.3A and B, and also the long distance, 3-D tracking of sei whales in the experimental region [15].

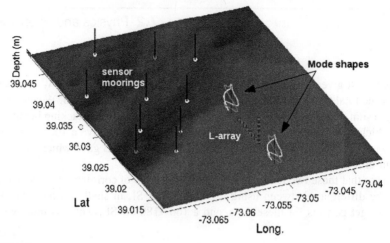

FIGURE 7.3A

The L-array from SW06, an array of oceanographic sensors to the left of it, and the acoustic normal mode shapes expected at the L-array acoustic modes one, two, and three are yellow (gray in print versions), green (dark gray in print versions), and magenta (white in print versions), respectively.

From Duda, T.F., and Collis, J.M., Acoustic field coherence in four-dimensionally variable shallow water environments: Estimation using co-located horizontal and vertical line arrays, *in Proceedings of 2nd International Conference on Underwater Acoustic Measurements: Technologies and Results, J. S. Papadakis and L. Bjorno, eds, 2007.*

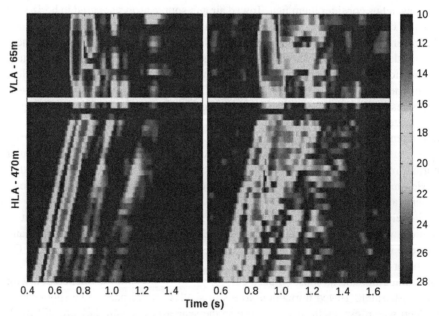

FIGURE 7.3B

The actual vertical line array output in the top panels and the horizontal line array output in the lower panels. The left panels show regular, clean arrivals in calm conditions and the right panels show more scattered arrivals in the presence of a strong oceanic perturbation, due to nonlinear internal waves in this case.

From Duda, T. F., and Collis, J.M., Acoustic field coherence in four-dimensionally variable shallow water environments: Estimation using co-located horizontal and vertical line arrays, *in Proceedings of 2nd International Conference on Underwater Acoustic Measurements: Technologies and Results, J. S. Papadakis and L. Bjorno, eds, 2007.*

Arrays for higher-frequency acoustics work can easily be attached to bottom tripods in shallow water, ensuring a stable configuration. Tripods are commonly used for coastal acoustics, as they can support a large amount of equipment and the batteries needed for power for comparatively long periods. However, as with any bottom-mounted equipment in shallow water, fishing activity has to be considered.

7.2.2.1 Examples: Mode Filtration Techniques in Shallow Water

One very often wants to look at individual multipath arrivals, be they ray or mode, to accomplish various tasks, such as source localization or tracking, or inversions to determine waveguide environmental parameters. We look at acoustic modal arrivals here, which are pertinent to lower-frequency acoustics. (We would assume that most readers are familiar with at least the rudiments of acoustic normal mode theory. However, if you are not, we refer you forward a few sections in this chapter to our section on "Normal Modes and Shallow Water," which shows a basic modal system.)

These techniques are discussed in the context of receiving arrays, but can generally be adapted to transmitting arrays as well. There are five general techniques we discuss here: (1) time separation of modes, (2) vertical array amplitude shading, (3) vertical array steering, (4) horizontal array steering, and (5) horizontal array focusing. Readers looking for more detail on ray and mode filtration are referred to the book by Katznelson et al. [16].

7.2.2.1.1 Time Resolution of Modes

The simplest way to resolve modes, if it is possible, is in the time domain, by letting the differential travel times of the modes sort them out. Mathematically, we have

$$\Delta t_{n,n+1} = t_n - t_{n+1} = \frac{R}{v_n^g} - \frac{R}{v_{n+1}^g} = R\left(\frac{v_{n+1}^g - v_n^g}{v_n^g v_{n+1}^g}\right) \tag{7.6}$$

where v_n^g is the group velocity of the nth mode, and R is the source–receiver separation.

If we have that $\Delta t_{n,n+1} \geq 1/\text{BW}$, then the nth and $n + 1$st modes are resolved. This ignores the slightly different dispersion of the two modes, but is a reasonable first approximation.

It should be noted that, even though it may not be possible to temporally resolve two neighboring modes, a combination of spatial and temporal filtering might be able to resolve them, even if neither technique could do so alone.

7.2.2.1.2 Amplitude-Shaded Vertical Array Mode Resolution

One of the most popular ways of resolving modes experimentally is with an amplitude-shaded vertical array. In this technique, one matches the sensitivity of the hydrophone output at a given depth to a specific mode's amplitude at that depth.

Because shallow-water depths are at most about 200 m at the shelf-break, it is not a hard task to make a hydrophone array that extends vertically across the whole water column. Given that extension, we can then use the orthogonality property of the normal modes to separate them apart.

Mathematically, we can write the pressure at the receiver range r as the sum of normal modes

$$p(r,z) = \sum_n a_n \varphi_n(r,z) \tag{7.7}$$

where we have neglected phase factors that are not needed in this argument. If we now operate on $p(r,z)$ with $\int_0^\infty \varphi_m(r,z)dz$, we obtain

$$\int_0^\infty p(r,z)\varphi_m(r,z)dz = \sum_n a_n \int_0^\infty \varphi_n(r,z)\varphi_m(r,z)dz$$
$$= \sum_m a_n \delta_{mn} = a_m \tag{7.8}$$

This a_m is just the amplitude of the mth mode, which we have now projected by having the receiver sensitivity/response be the same as the shape of the mth mode. In essence, this is a very basic technique, although there are many subtleties that can and should be added to make it more robust. We discuss few of them here, and refer the reader to the literature for more detail.

One of the first things we note is that, in the real world, a vertical array usually does not extend to the upper 10 m of the water column, to avoid sloshing around due to wave action and to avoid ship strike. And under no circumstances does the array extend into the seabed, where there is still some modal energy. Thus our orthogonality integral is truncated and is only approximate, which leads to leakage of energy from the mode we are examining to other modes.

Another point is that one only needs as many hydrophones as one has normal modes to filter the modes, and that these can be arbitrarily spaced [13]. This is true in the theoretical limit of infinite SNR only! Try this in the real world, and you will not be so pleased with the result. Generally, the more evenly spaced the hydrophones are placed in the water column, the better. A good spacing to capture N modes is to use an even vertical Nyquist sampling with $\Delta z_N = \pi/\gamma_N$, where γ_N is the vertical acoustic wave number, which has units L^{-1}.

Two more points need to be considered carefully when filtering modes this way. First, one needs to have the correct vertical sound speed profile (including in the bottom) to be able to generate the correct replica mode for the orthogonality relation. The bottom properties can be hard to obtain, and the water column can change rapidly in time, so this is not trivial. Generally, experiments in shallow water have a string of thermistors attached to the vertical acoustic array (as sound speed is most sensitive to temperature), and a bottom survey is done in the array area. And second, the mooring motion of the array (especially tilt), if not corrected for, can distort what one thinks the pressure field is, and also create a mismatch in the orthogonality relation. This can be remedied by using an active navigation transponder network to monitor mooring motion.

There are other caveats, but these are the main ones. We see that what looked like a trivial method actually needs some care for its implementation!

7.2.2.1.3 Vertical Array Steering

Another way to use a vertical array to aid in mode filtering is to steer it. This is a rather inferior method, in that due to the shortness of the array, the beams are fat. Moreover, one needs to know the vertical water column sound speed, as in the shading technique, and the technique also is prey to mooring motion. Vertical array steering is useful for determining up versus down going energy, but is not so good past that.

7.2.2.1.4 Horizontal Array Steering

A very common and popular way of filtering modes is with a beam-steered horizontal array, which can either be towed, lying on the bottom, or created by a synthetic aperture. In Fig. 7.4, we show both this configuration and the previously discussed vertical steered array configuration, both of which look at the grazing angles of the individual modes in the waveguide. The horizontal array is the better configuration of the two, in that it can be made arbitrarily long, and thus can resolve closely spaced modes. (The wave number resolution available for a line array is roughly π/L, where L is the array length.)

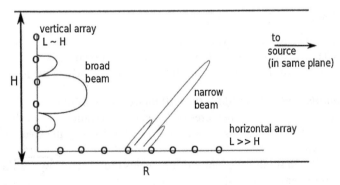

FIGURE 7.4

Resolution achievable by vertical and horizontal linear arrays in shallow water. The horizontal array has the advantage of having greater length available.

The steering phases of the beam former just need to match/cancel the horizontal wave number (times range) of a particular normal mode to give maximum response to that mode, that is, $\varphi_n^{steer} = -k_n R_{element}$, so that

$$P_{array\ output} = \sum_{elements} \frac{a_n e^{ik_n R_{elem}} e^{-ik_n R_{elem}}}{\sqrt{k_n R_{elem}}} = \sum_{elements} \frac{a_n}{\sqrt{k_n R_{elem}}} \qquad (7.9)$$

We note that there is also a dependence of this modal steering on the angle α between the array orientation (e.g., end fire vs. broadside) and the elevation steering angle, which we dub β in this case [15]. the beam former has maximum output at $\beta = \alpha$, which for our shallow-water waveguide case is $k_n cos\alpha = kcos\beta$, which gives discrete modes at $cos\ \beta_n = \frac{k_n}{k} cos\ \alpha$. From this, we see that steering mode filtering works best at end fire, and fails at broadside. It also should be mentioned again that such steered horizontal array mode filtration techniques are closely related to the Hankel transform wave number integration techniques, for which some a typical reference for shallow water acoustics is [17].

7.2.2.1.5 Focused Array Mode Filtration

If a horizontal array is broadside to a source which is in the near field of the array, then one can use phase focusing to resolve normal modes [18,19]. The focusing phase for each mode is $\varphi_n^{focus} = \frac{-k_n \rho^2}{2r_0}$, where ρ is the distance along the array and r_0 is the distance from the source to the center of the array. This phase produces the maximum focused array output at range points $r_{0,n} = \frac{k}{k_n} r_0$. This difference in the "range of focus" for each mode just reflects the additional path length that each mode travels compared to a straight-line path to the receiver. The focused horizontal line array in a sense is the complement to the steered horizontal line array, i.e., it works well at broadside, but fails at end fire.

7.2.3 TRANSMISSION LOSS TERM

By far the biggest difference between deep and shallow water occurs in the transmission loss (TL) term, which involves the acoustic propagation from source to receiver. The detailed nature of the acoustic waveguide enters here, which we look at next.

7.2.3.1 Simple Geometric Spreading Intensity Arguments

Commonly used (and abused) measures of acoustic propagation loss in shallow water are the geometrical spreading laws for sound intensity, i.e., the spherical, intermediate, and cylindrical spreading laws, often called the 20 log R, 15 log R, and 10 log R laws. Based solely on geometric arguments (i.e., ducting by the waveguide surfaces), these intensity spreading laws fit easily into the sonar equation context, and one can see very nice examples worked out in Urick's text [2]. However, these geometric arguments are oversimplifications, and depending on one's intended usage of the sonar equation, can lead to erroneous results. To begin with, the transition from 20 log R to 15 log R to 10 log R is a continuous one, in that the surface of an expanding ray does *not* instantly become a cylinder once the ray encounters a boundary. Rather, the part of the energy that has not yet been reflected maintains a spherical spreading law, which produces the gradual transition to a cylindrical surface at $R \to \infty$. Also, and perhaps more importantly, the intensity that one sees in real waveguides shows distinct convergence (focusing) and shadow (defocusing) zones due to multipath interference, an effect that is not present in the simple spreading laws. This particular oversight has actually caused much angst in the consideration of exposure levels for marine mammals for environmental studies. The levels one sees by just using the spreading law alternately over- and underestimates the levels that the animals actually see. A third cautionary note when using these simple spreading laws is that, due to shallow-water acoustics admitting some strong 3-D propagation effects, the spreading law can actually be 0 log R in some cases, i.e., there is *no* geometrical spreading! This occurs when there is 3-D ducting of sound due to lateral boundaries, a case that might be more familiar in the context of a very loud grammar school hallway in the period between classes. We examine how this effect is produced shortly, in a section that concentrates on 3-D effects.

7.2.3.2 Popular Propagation Theories and Their Application(s) to Shallow Water

As propagation theory has been discussed in some detail in Chapter 2 of this book (and many others), we just look at what the peculiarities of shallow water are as regards the application(s) of such theories.

There are five major propagation theory variants we discuss here: (1) rays, 2) normal modes, (3) 3-D vertical mode and horizontal rays, (4) parabolic equation (PE), and (5) wave number integration. We look a bit at the strengths and peculiarities of each one as regards shallow water, and show some examples. However, let us briefly mention how one may access the computer codes that are the most usual implementations of these theories. As these are discussed fully in Chapter 3 of this book, we stress "briefly."

Probably the most useful reference for many of the standard models used by ocean acousticians is *Computational Ocean Acoustics* [20]. Many of the models that are described there and in this chapter are available online at the Ocean Acoustics Library (http://oalib.hlsresearch.com). These models are available free, for the general use of the acoustics community and are often employed by the authors. Software for calculating acoustic fields in range-dependent and range-independent environments using rays, normal modes, PE, and wave number integration can all be found there.

7.2.3.2.1 Ray Theory

The first, and most obvious, propagation theory one thinks of is ray theory. If we observe our $H \geq 10\lambda$ rule of thumb, we see that frequencies as low as 75 Hz in 200 m of water (the shelf break) and 15 kHz at 1 m depth (the surf zone) are reasonably amenable to a ray theory description. This is a pretty useful range of frequencies for applied systems. But what is new and different about ray theory in shallow water? "Standard ray theory" codes solve what are called "N×2-D" types of problems, where a fully 3-D ocean and seabed is divided into N-range-dependent radial slices, and azimuthal coupling between the slices is ignored. Is this good enough for shallow water? The answer is, as perhaps expected, "mostly." 3-D acoustics is encountered due to a number of effects in shallow water: sloping bathymetry, fronts, surface wave swell, nonlinear internal waves, and bottom ripples are common producers of 3-D effects.

The question of when and where one needs to use 3-D acoustics is a useful one to address, in that fully 3-D ray codes are harder to use and more computationally intensive than conventional N×2-D codes, and should only be used where appropriate. Unhelpfully, if one runs both a 2-D and 3-D ray code in a region with large out-of-plane sound speed gradients or bottom slopes, and finds horizontal deviations of greater than a Fresnel zone $R_F = (\lambda R)^{1/2}$ from the original 2-D ray, then use of the 3-D form is justified, as the deflected path is a new, separate, horizontal ray. This is inelegant, but it works. However, there are simpler calculations to use to make such a decision for nonbottom-interacting and bottom-interacting rays, which we look at next.

A liberal way to look at where 3-D water column sound speed changes may be important is to look at the *maximum* horizontal gradients one gets with various ocean features. For fronts, eddies, and nonlinear internal waves, the gradient is of the order of about 10 m/s over about 100 m, corresponding to about 2°C temperature change over that distance. This would give a horizontal critical angle for rays encountering that "barrier" of about 6.5 degrees, which is rather high, but is in keeping with our generous estimation strategy. Thus, one could produce lateral ray excursions Δd of up to $\Delta d = R\tan\theta_{crit}$, where R here is the straight-line path the ray would take without a horizontal sound speed change. If, at a given range being considered, this is a Fresnel zone width, then 3-D acoustics should be considered. Specifically, this gives

$$R = \frac{\lambda}{\tan^2\theta_{crit}} \tag{7.10}$$

This "deflection criterion" (which can be from either reflection or refraction) is crude, but it is easy to estimate. For example, for 100 Hz, one would see 3-D effects at a range of 1.1 km. This being a liberal estimate, this is about the smallest distance over which one might see such effects.

Historically, it was bathymetric steering that was the first 3-D effect noticed in shallow water, and it is also perhaps the more common one. In looking at 3-D bathymetric effects, we both parallel our previous example and follow the treatment found in Brekhovskikh and Lysanov [21]. Consider Fig. 7.5 where we send a ray on an ostensibly along-shelf, isobathymetric, straight-line path between a source and a receiver. However, if there is a downward slope in the y-direction, that ray can be deflected by bathymetric reflections in the offshore direction, as also shown in Fig. 7.4. Let us, following Brekhovskikh and Lysanov, consider an isovelocity water

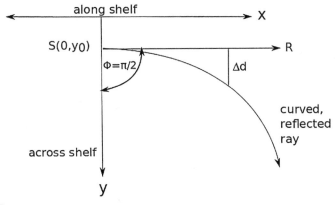

FIGURE 7.5

Bathymetric steering of an initially along-shelf ($\varphi = \pi/2$) ray. When the deflection across shelf is of order Δd, then one should consider the curved path a new, 3-D ray.

"coastal wedge" with bathymetric depth h given by $h = \varepsilon y$, where ε is a small slope, and source depth h_0 given by $h_0 = \varepsilon y_0$, where $(0,y_0)$ is the source position in the (x,y) plane. The deviation of the curved path from the straight-line path, Δd, is defined by $\Delta d \equiv y - y_0$ along the curved path. If we further define, following Brekhovskikh and Lysanov, φ_0 as the initial angle of the ray path to the across shelf direction (90 degrees in Fig. 7.5), and χ_0 to be the (vertical plane) grazing angle of a ray, we may then directly use their ray equation, which is

$$y^2\left(1 - \cos^2\chi^0 \sin^2\varphi_0\right) = y_0^2 \sin^2\chi^0 + \left[x\left(1 - \cos^2\chi^0 \sin^2\varphi_0\right)\left(\sin\varphi_0 \cos\chi^0\right)^{-1}\right.$$
$$\left. - y_0 \cos\chi^0 \cos\varphi_0\right]^2 \qquad (7.11)$$

Substituting in the value $\varphi_0 = \pi/2$ as per Fig. 7.5, and simplifying, we get

$$y^2 = y_0^2 + x^2 \tan^2\chi^0 \qquad (7.12)$$

Thus, we get a y deflection for each x, given some χ^0 and y_0^2 values. We now can look at numbers for this sideways (hyperbolic path) deviation due to bathymetry.

For a source at 100 m depth, and a slope of $\varepsilon = 0.01$, we get $y_0 = 10^4$ m. These numbers are typical of the shelf-break regime, where we go from 100 m water depth to 200 m water depth in about 10 km. Now, let us look in Fig. 7.6 at ray grazing angles of 5, 10, 15, and 20 degrees, typical "totally internally reflected" ray angles for shallow water.

7.2.3.2.1.1 *Example: Ray Wander, Spread, and the Resolution of Ray Multipaths*
In a great many applications, one wants/needs to resolve individual ray and/or mode multipaths. As discussed earlier, both time resolution (from differential dispersion) and angular resolution (from array processing) can be used to help separate multipaths. However, the cases discussed earlier were for a mostly ideal world, where

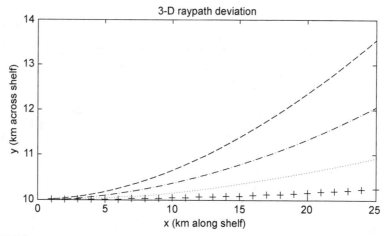

FIGURE 7.6

Deviation of rays in the across shelf direction due to sloping bathymetry. From the least deviation to the greatest, the rays are at 5, 10, 15, and 20 degrees grazing angle for this example. The source is located at (0, 10) km.

the discussion of noise and scattering effects was largely disregarded to first order. Here we look at scattering effects, and how they can degrade multipath resolution. Two effects that we discuss in particular are called "wander" and "spread." Their effects are easy to discuss via simple scattering models, although as always, a detailed description involves more complex models.

We first look at the arrival times for rays in a simple, isovelocity waveguide, where the bottom and surface are flat. This will serve as the "background waveguide" for the cases where surface and bottom boundary scattering are included. We also look at source and receiver on the bottom, just to avoid "partial cycle paths" that add calculational complexity, but not more insight. The scattering will be assumed to be from the surface alone, but it is easy to transcribe what is done here to the bottom.

For the background waveguide, we can show from the geometry that the grazing angle for an eigenray that makes N_C complete ray cycles over a horizontal (S/R) distance R in a water column of depth H is

$$\sin \theta_g = \frac{H}{D} = \frac{H}{\left[H^2 + \left(\frac{1}{2} \frac{R}{N_C} \right)^2 \right]^{1/2}} \tag{7.13}$$

The travel time for each eigenray is

$$T_{N_c} = \frac{2N_C}{c} \left[H^2 + \left(\frac{1}{2} \frac{R}{N_C} \right)^2 \right]^{1/2} \tag{7.14}$$

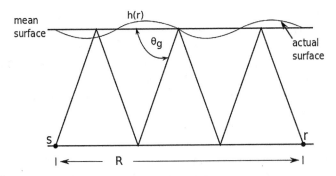

FIGURE 7.7

Geometry for looking at the travel time "wander" of an acoustic pulse due to surface roughness scattering.

From Eq. (7.14), we can get the travel time difference between neighboring eigenrays with $N_C - 1$, N_C, $N_C + 1$.

In Fig. 7.7 we show the geometry for the surface scattering that causes wander, the "jiggle" of the arrival time of a pulse without any change of shape or distortion of the pulse. This is caused by the surface height variation $h(r)$ from the mean at the point of scattering. If the scattering points are far enough from each other to be de-correlated, we just need the probability density function of the surface height to describe the wander. In the plane wave approximation to the local scattering, one sees a local time delay (or advance) at each scattering point [21]:

$$\Delta t = \frac{2h}{c \sin \theta_g} \tag{7.15}$$

where θ_g is the grazing angle of a selected ray with the mean surface. If we consider Gaussian height statistics for the surface roughness (usually a good assumption for ocean surface waves), we know that the mean of h, and thus $\langle \Delta t \rangle$, is zero and that the second moment of the time fluctuation is

$$\sigma_{\Delta t}^2 = \frac{B^2}{\sqrt{2\pi}\sigma_h} \int_{-\infty}^{\infty} x^2 \exp\left\{\frac{-1}{2}\left(\frac{x}{\sigma_h}\right)^2\right\} dx \tag{7.16}$$

where σ_h is the variance of the surface height and

$$B = \left(\frac{2}{c \sin \theta_g}\right) \tag{7.17}$$

This gives the result

$$\sigma_{\Delta t}^2 = B^2 \sigma_h^2 \tag{7.18}$$

which for n interactions (adding their variances) gives

$$(\sigma_{\Delta t})_{rms} = B\sigma_h \sqrt{n} \tag{7.19}$$

This is a good result for a very large number of realizations, where the statistical average makes sense. For a small number of boundary interactions, the more usual case, one really is just seeing the individual realization, which is generally not known. However, for a small consistency check, we can use the "Drunkard's walk" to approximate the statistics, i.e., take that at each interaction, the height is likely to be $\pm\sigma_h$, the root mean square (rms) height deviation. Over n steps, the likely random walk deviation is $d \sim \pm\sqrt{n}\sigma_h$, which gives

$$(\sigma_{\Delta t})_{rms} = B\sigma_h\sqrt{n} \qquad (7.20)$$

This is the same as the previous result, not surprisingly.

Let us look at some typical numbers. For $c = 1500$ m/s, $\theta_g = 10$ degrees, $\sigma_h = 1$ m, and $n = 4$, we get 15 ms wander, which is very much on the order of what one sees in coastal acoustics.

Physically, if $(\sigma_{\Delta t})_{rms}$ becomes of the same size as the undisturbed travel time spacing between neighboring multipaths, then it is impossible to identify a particular multipath, and so schemes based on multipath identification break down.

Another physical effect that can impede multipath identification and resolution is "spread" of an arrival. The spread is different physically from the wander in that it represents the distortion of a pulse shape by scattering to different grazing angles. In this case, one can look at how the local slope at the scattering point changes the cycle distances of the rays (see Fig. 7.8). Let us look at the local angle change at each interaction for a given eigenray, which is the local slope. From Ogilvie [22], the rms slope for a Gaussian distribution is given by

$$\sigma_{slope} = \frac{\sigma_h\sqrt{2}}{\lambda_0} \qquad (7.21)$$

where σ_h is the rms wave height and λ_0 is the horizontal correlation length of the roughness. Thus, instead of having the travel time corresponding to θ_g, the travel

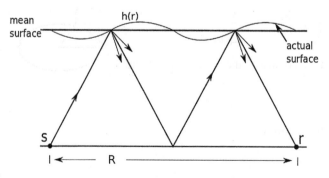

FIGURE 7.8

Geometry for considering the travel time "spread" of a ray due to rough surface scattering.

time can correspond to angles anywhere between $\theta_g \pm \frac{\sigma_h \sqrt{2}}{\lambda_0}$ for each surface interaction. Let us add this angle perturbation independently each time a ray hits the surface. Perturbing the "background" eigenray by $\pm \Delta\theta = \frac{\sigma_h \sqrt{2}}{\lambda_0}$ at each surface interaction gives for the new path length for each leg

$$L_\pm = H/\sin(\theta \pm \Delta\theta) \tag{7.22}$$

Using typical coastal ocean wave numbers of $\sigma_h = 1\,m$ and $\lambda_0 = 100\,m$, we see that $\Delta\theta \sim 0.014(\mathrm{rad})$, or order of 1 degree. For such small $\Delta\theta$, we have

$$\sin(\theta \pm \Delta\theta) \cong \sin\theta\,\cos\Delta\theta \tag{7.23}$$

From this, we can get the length differences

$$\Delta L_\pm = \frac{H}{\sin(\theta \pm \Delta\theta)} - \frac{H}{\sin\theta} \tag{7.24}$$

If we restrict ourselves to rays where $\Delta\theta \ll \theta$, which is a good portion of the sector 0–20 degrees (below the typical bottom critical grazing angle), we can further expand this as

$$\Delta L_\pm = \pm\frac{H\Delta\theta}{\theta^2} \tag{7.25}$$

Finally, the travel time perturbations are obtained, simply by dividing $\Delta L_\pm$ by c.

On the average, plus and minus length and travel time perturbations are equally likely, so that the average of the spread is zero. However, it is the variance (or rms) spread that is more important here. Again, we can invoke "random walk" statistics for the individual surface interactions, which gives an rms travel time spread Δt_{rms} of

$$\Delta t_{rms} \sim \frac{\sqrt{n}H}{\theta^2} \frac{\sigma_h 2\sqrt{2}}{\lambda_0}. \tag{7.26}$$

where n is the number of surface interactions, and the added factor of 2 comes from taking the difference between the $\pm\Delta\theta$ perturbations. For four surface interactions, and a 10 degree ray, we get $\Delta t_{rms} \sim 2.7$ ms, not hugely different in magnitude from the wander number! Again, if the spread of a multipath becomes of the same size as the undisturbed travel time spacing between neighboring multipaths, then it is impossible to identify which particular multipath one is looking at.

There are few caveats, however, to using ray theory for estimating acoustic pulse spread and wander. First, there are the numbers for σ_h, the rms wave height, and λ_0, the correlation length. These depend on having good oceanographic data and/or models, and our answers can be sensitive to having these inputs entered correctly. Second, our ray picture of these processes works best when we either have a large number of surface interactions, or if we average smaller numbers of interactions over many independent wave realizations. Third, we have ignored the finite size of the ray, given by the Fresnel zone width. One must average the surface over this width, which lessens roughness effects. But, those things being said, wander and spread of order of a few milliseconds to a few tens of milliseconds are typically what is seen.

An interesting final note is that the wander of an arrival in ray theory corresponds to adiabatic mode scattering in mode theory, where the energy in a mode stays in that mode, whereas the spread corresponds to coupled mode theory, where the energy can be spread to many neighboring modes. Another interesting point is that one can, and does, have wander and spread acting simultaneously.

7.2.3.2.2 Normal Modes and Shallow Water

Normal mode calculations are best used for water depths of $D \leq 10\lambda$, but can also be used for deeper waveguides if one does not mind more calculational effort. The normal mode solution to the wave equation is a full-wave, nonapproximate solution to the Helmholtz equation, but in practice, it suffers from two calculational difficulties: (1) it leads to a coupled equation system that can be tedious to compute and (2) there is a modal continuum that must often be taken into account (especially when scattering couples the modes strongly) that is also difficult to compute. However, these can be dealt with, and moreover the modal solution often provides much physical insight, as well as experimentally useful quantities (e.g., mode travel times and mode shapes). The modal solution has been used extensively in shallow water since the seminal paper by Chaim Pekeris [23], and so we should look at what its advantages are in shallow water. It should be stated that there is an extensive literature on how modes are used in shallow water, and that our tactic in this chapter is to reference that literature where it exists, and to concentrate here on some slightly more novel ways to use the modal solution.

There are three basic degrees of complexity in the modal solution that should be considered when using modes in shallow water: (1) the simplest range-independent solution, (2) the so-called adiabatic solution for a slowly range varying medium, and (3) the fully coupled mode solution. Let us look at these and where to apply them.

7.2.3.2.2.1 Range-Independent Modes The most basic modal solution is the range-independent solution, good for flat continental shelf areas. The pressure $p(r,z)$ as a function of range and depth is, for a cylindrically symmetric waveguide, and using the asymptotic form of the Hankel function [24]

$$p(r,z) = \frac{\sqrt{2\pi}e^{i\pi/4}}{\rho(z_0)} \sum_{n=1}^{\infty} \frac{u_n(z_0)u_n(z)e^{ik_n r}}{\sqrt{k_n r}} \tag{7.27}$$

where the u_n are the normal mode functions, the k_n are the horizontal wave numbers of the nth mode (and also the modal eigenvalues), and $\rho(z_0)$ is the density of the medium at the source (1, for water.) The infinite sum in the above is a reminder to sum over both the trapped and continuum modes, but in practice, one most often just keeps the totally internally reflected trapped modes, which contain the most energy at long distances. The only real calculational part of the preceding equation is the computation of the normal mode eigenvalues and eigenvectors. There are numerous techniques available for these calculations, and a recommended reference for these is *Computational Ocean Acoustics* by Jensen et al. [20]. These modal solutions generally cover both arbitrary water column and fluid bottom sound speed, density, and attenuation profiles, as is discussed later in this section. Codes that include shear exist, but are relatively rare [25], and for softer sediments, taking shear as a perturbation can be a useful approach [26]. Roughness scattering has also been considered in the modal approach, and some of the early work by Clay and others [13] show good physical insight into how scattering affects modal solutions.

Given that there is much in the literature already discussing the basic properties of modal solutions, we like to pursue a little different topic in this chapter—namely, using some basic "canonical" waveguides (specifically the "hard bottom" and "Pekeris" waveguides) and first-order perturbation theory (FOPT) as springboards for doing some simple estimation of more complicated waveguide effects, including 3-D propagation effects. Being able to get ready insight into what appear to be more complex waveguides is often useful in practical situations.

Let us start with the simplest waveguide, the hard (rigid) bottom waveguide, which consists of an isovelocity water column over a rigid (infinite density) bottom half space. The modal eigenvalue (horizontal wave number) solutions for this waveguide are analytic [13] and are

$$k_n = \sqrt{k^2 - \gamma_n^2} \tag{7.28}$$

where

$$k = \frac{\omega}{c} \quad and \quad \gamma_n = \frac{\left(n - \frac{1}{2}\right)\pi}{H} \tag{7.29}$$

where γ_n is the vertical wave number of the n th mode. The normalized mode function is

$$u_n(z) = \sqrt{\frac{2}{H}} \sin(\gamma_n z) \tag{7.30}$$

This waveguide is useful to consider when we are interested in perturbing the water column, and are not so interested in bottom reflection effects.

The second canonical waveguide of interest is the Pekeris waveguide, which consists of an isovelocity water column over an isovelocity bottom half space. This is a simple extension of the hard bottom model in some ways, and its solutions look very similar. However, one needs to find the eigenvalues for this waveguide numerically, with a simple root finder, rather than having analytic solutions available.

The eigenvalues of the Pekeris waveguide are found by finding the roots of

$$\gamma_n H = \left(n - \frac{1}{2}\right)\pi - \frac{\varphi_n}{2} \quad n = 1, 2, 3... \tag{7.31}$$

where φ_n is the phase of the bottom reflection coefficient for the nth mode, given by

$$\varphi_n = -2tan^{-1}\left[\frac{\sqrt{sin^2\theta_n - \frac{c^2}{c_1^2}}}{m_\rho \cos\theta_n}\right] \tag{7.32}$$

Here, θ_n is the incidence angle of the nth mode, the subscript 1 refers to the bottom medium, and $m_\rho = \rho_1/\rho$.

We see that this is the same as the hard-bottom eigenvalue equation, except that for a nonrigid bottom, the phase of the bottom reflection coefficient is not zero (below critical grazing angle). These roots are easily found.

The mode functions are also similar to the hard bottom, except that they now include a piece that extends into the bottom. They are

$$u_n(z) = A_n \sin(\gamma_n z) \quad 0 \leq z \leq H \tag{7.33}$$

$$u_n(z) = B_n \exp(-\gamma_{1n} z) \quad z \geq H \tag{7.34}$$

where

$$A_n = \sqrt{2\left[\frac{1}{\rho}\left(H - \frac{\sin \gamma_n H}{2\gamma_n}\right) + \frac{1}{\rho_1}\frac{\sin^2 \gamma_n H}{\gamma_{1n}}\right]} \tag{7.35}$$

and

$$\frac{A_n}{B_n} = \frac{\exp(-\gamma_{1n} H)}{\sin \gamma_n H} \tag{7.36}$$

The Pekeris waveguide is useful for including both bottom and water column effects into the perturbation, but is especially useful as a basis for doing linear perturbative inversions for the bottom sound speed, as we show.

We now come to the last piece needed for using these canonical waveguides to solve more complex problems, i.e., the wave number perturbation. The wave number $k = \omega/c$ can be perturbed in various ways, i.e., via the real and imaginary parts of c, or via ω [20]. We are only interested in the real part of c, i.e., perturbations to the sound speed profile.

Following Rajan et al. [26], we have that the wave number perturbation Δk_n is given by

$$\Delta k_n = \frac{-1}{k_n} \int_0^\infty \frac{\Delta c(z)\omega^2}{c^3(z)} \frac{u_n^2(z)dz}{\rho(z)} \tag{7.37}$$

where $\Delta c(z)$ is a perturbation to the background sound speed. We now have the ingredients needed to look at an interesting example.

7.2.3.2.2.1.1 Example: Mixed Layer and Internal Wave Perturbations in a Hard-Bottom Waveguide

Consider the idealized waveguide shown in Fig. 7.9 of a coastal water column in which there are three conditions: (1) an isovelocity condition (state 0); (2) a mixed layer included, with isovelocity water below it (state 1); and (3) a "square wave" approximation to a nonlinear internal wave included, with an isovelocity layer below it (state 2). As simplified as it seems, this picture can actually incorporate much of the acoustics physics that one sees in regions with a mixed layer

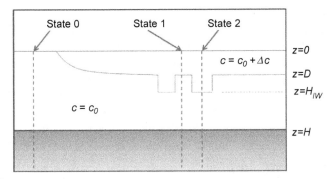

FIGURE 7.9

Idealized waveguide, useful for studying the inclusion of the surface mixed layer and internal waves into the shallow-water scenario.

©2010 IEEE. Reprinted, with permission, from Lynch, J.F., Lin, Y.-T., Duda, T.F. and Newhall, A.E., Acoustic ducting, reflection, refraction, and dispersion by curved nonlinear internal waves in shallow water, IEEE J. Oceanic Eng., **35**, pp. 12–27, 2010.

(ML) and nonlinear internal waves (NLIWs) present. The perturbations to the nth mode wave number of the 0 state by the ML (state 1) and internal waves (IWs; state 2) are given by:

$$k_{1n} = k_{0n} + \Delta k_{1n}, \quad k_{2n} = k_{0n} + \Delta k_{2n} \tag{7.38}$$

We can obtain for Δk_{1n}, using Eqs. (7.30) and (7.37)

$$\Delta k_{1n} = \frac{-2}{k_{0n} H} \frac{\omega^2}{c_0^2} \frac{\Delta c}{c_0} \int_0^D \sin^2(\gamma_{0n} z) dz \tag{7.39}$$

The expression for including IWs is similar, except that the upper limit in the integral becomes H_{IW}. This integral is elementary, and if we also look in the "high mode limit" (i.e., at the modes that interact the most with the near-surface perturbations), we get that the integral is $I = D/2$. Thus, we get for the wave number perturbation the expression

$$\Delta k_{1n} = \frac{-1}{k_{0n}} \frac{\omega^2}{c_0^2} \frac{\Delta c}{c_0} \frac{D}{H} \tag{7.40}$$

Again, one has a similar expression for the internal waves, replacing D with H_{IW} [27].

These expressions show how easy it is to include water column variability to the modal wave numbers, without having to even go to the bother of a complicated mode finding routine. Moreover, they give direct insight into how the environmental parameters affect the modal eigenvalues.

7.2.3.2.2.2 Adiabatic and Coupled Mode Theory Another salient feature of shallow water is that the acoustic environment is extremely variable, both in the water column and in the bottom. Thus, instead of $c(z)$, one most often needs to consider

$c(z,r)$ between a source/receiver pair (and occasionally $c(x,y,z)$ when 3-D effects become important). In this section, we consider the 2-D case, looking at adiabatic and coupled mode propagation in the $r-z$ (S/R) plane.

One begins with the derivation of the coupled mode equations. As this derivation is common, we just present a précis, following *the Computational Ocean Acoustics* presentation (which follows the original derivation by Allan Pierce [29]). We start by writing the Helmholtz equation in cylindrical coordinates (even though the range dependence is not generally cylindrically symmetric—if there is no appreciable out-of-plane 3-D scattering, this makes no difference and is more convenient):

$$\frac{\rho}{r}\frac{\partial}{\partial r}\left(\frac{r}{\rho}\frac{\partial p}{\partial r}\right) + \rho\frac{\partial}{\partial z}\left(\frac{1}{\rho}\frac{\partial p}{\partial z}\right) + \frac{\omega^2}{c^2(z,r)}p = -\frac{\delta(r)\delta(z-z_0)}{2\pi r} \tag{7.41}$$

We next use the fact that the solution can be expressed as a sum of local modes at any range (due to the modes being a complete set) to write the general solution in the form

$$p(r,z) = \sum_m \Phi_m(r)\Psi_m(r,z) \tag{7.42}$$

where the local modes are obtained from the equation

$$\rho(r,z)\frac{\partial}{\partial z}\left[\frac{1}{\rho(r,z)}\frac{\partial \Psi_m(r,z)}{\partial z}\right] + \left[\frac{\omega^2}{c^2(r,z)} - k_m^2(r)\right]\Psi_m(r,z) = 0 \tag{7.43}$$

We can substitute the $p(r,z)$ expansion into the Helmholtz equation, and then uses the projection operator

$$\int (\cdot)\Psi_n(r,z)dz \tag{7.44}$$

on the equation obtained. This projects out many of the terms in the equation (using the orthogonality of the local modes), leaving one with the usual coupled equations:

$$\frac{1}{r}\frac{d}{dr}\left(r\frac{d\Phi_n}{dr}\right) + \sum_m 2B_{mn}\frac{d\Phi_m}{dr} + \sum_m A_{mn}\Phi_m + k_n^2(r)\Phi_n = -\frac{\delta(r)\Psi_n(0,z_S)}{2\pi r} \tag{7.45}$$

The A_{mn} and the B_{mn} are the so-called mode coupling coefficients, which allow energy to be transferred from one mode to another as the sound propagates in range. Usually, the A_{mn} coefficient, which corresponds to very rapid range variation, is omitted, and so we just note that the B_{mn} has the form

$$B_{mn} = \int \frac{\partial \Psi_m}{\partial r}\frac{\Psi_n}{\rho}dz \tag{7.46}$$

Even with this simplification of the mode coupling coefficients, the coupled equations are nontrivial to solve, and it is attractive to try to find conditions where one can ignore the coupling. It is obvious from the form of B_{mn} that this will hold when the range dependence of the modes, and thus the medium, is weak (but still exists!). Mathematically, this corresponds to

$$\left| \frac{B_{mn}}{k_m - k_n} \right| \ll 1 \tag{7.47}$$

In this case, the mode coupling terms can be ignored, and one has an uncoupled equation for weak range dependence

$$\frac{1}{r} \frac{d}{dr} \left(r \frac{d\Phi_n}{dr} \right) + k_n^2(r)\Phi_n = -\frac{\delta(r)\Psi_n(0, z_S)}{2\pi r} \tag{7.48}$$

In the WKB approximation [20], there is a simple solution to this uncoupled equation, namely

$$p(r, z) = \frac{i}{\rho(z_S)\sqrt{8\pi r}} e^{-i\pi/4} \sum_{m=1}^{\infty} \Psi_m(0, z_S)\Psi_m(r, z) \frac{e^{i \int_0^r k_m(r')dr'}}{\sqrt{k_m(r)}} \tag{7.49}$$

This is the so-called "adiabatic mode solution," in which each mode keeps its own energy (ignoring attenuation) during its propagation. In a practical sense, it is an easy solution to use. One just needs to compute mode functions at the source and receiver positions, where one often has good environmental information. The modal eigenvalues versus range are needed for the WKB phase integral, but for simple "canonical" systems such as the hard-bottom and Pekeris waveguides we have considered, these are simple to get and often analytic. Thus, one has another estimational tool, but this time for range-dependent environments.

One of the tacit themes of this chapter is "work smarter, not harder," and so the avoidance of having to do hard, fully coupled mode calculations is to be encouraged. There are two ways to do this: (1) find out where one can use the adiabatic calculation, which is far easier, and which we look at next, and (2) do an alternate calculation for the pressure field (e.g., PE) and then project out the modes to the range of interest. We discuss about that approach later.

In looking at the coupling criterion, we follow Katznelson et al. [16], who give both the mathematical criterion and also a nice example. If we express the range-dependent mode amplitudes $a_l(r)$, which are the solutions to the coupled equations, as a (forward propagating for this purpose) coefficient $C_l(r)$ times the usual WKB solution factors, i.e.,

$$a_l(r) = \frac{C_l(r)}{\sqrt{k_l(r)}} e^{i \int_0^r k_l(r')dr'} \tag{7.50}$$

it can be shown that the coupled mode expression for $C_l(r)$ is

$$C_l(r) = C_l(0) + i \sum_m \sqrt{\frac{k_m(r)}{k_l(r)}} C_m(r) \frac{U_{ml}(r)}{\Delta k_{ml}(r)} [\exp(-i\Delta k_{ml}(r)r) - 1] \tag{7.51}$$

where $\Delta k_{lm}(r) = k_l - k_m$. In this equation, the $C_l(0)$ term is the adiabatic solution, which is augmented by the coupling term, the sum over the m modes. The $U_{ml}(r)$ is essentially the same coupling matrix as the B_{mn} defined earlier, with few minor differences (since different authors tend to absorb different factors into their definitions). Specifically,

$$U_{ml} = \sqrt{\frac{k_l}{k_m}} \int_0^\infty \frac{\rho}{\rho(z)} \Psi_l \frac{\partial \Psi_m}{\partial r} dz \tag{7.52}$$

For the adiabatic solution to work, as we saw in Eq. (7.47),

$$\bar{\kappa} \equiv \frac{|U_{ml}|}{|\Delta k_{ml}|} \ll 1 \tag{7.53}$$

This "nonadiabacity parameter" $\bar{\kappa}$ has a clear physical interpretation. $1/\Delta k_{ml}$ is proportional to the mode cycle distance Δ_{ml} (within a factor of 2π), whereas U_{ml} is inversely proportional to the characteristic scale of environmental variability of the waveguide, as measured by the modal wave number variation, i.e., $k_l(r)$. Thus, $\bar{\kappa}$ measures how quickly the environment changes compared to the mode cycle distance—if it changes slowly over a cycle distance, then the system is adiabatic.

7.2.3.2.2.2.1 Example: Application of the Coupling Criterion to a Canonical Waveguide with Surface Waves

A rather nice example of how to apply the $\bar{\kappa}$ criterion for adiabatic versus coupled behavior is found in Katznelson et al. [16], and we paraphrase it here. Using our hard-bottom waveguide modes, along with a rough ocean surface described by the function $s(r)$, we can write the local modes as

$$\sqrt{\frac{2}{H - s(r)}} \sin \left[\frac{(l + 1/2)\pi}{H - s(r)} [z - s(r)] \right] \tag{7.54}$$

Inserting this in the expression for U_{ml}, we obtain

$$|U_{ml}| = \left| \frac{2s'}{H} \right| \frac{\left(l + \frac{1}{2}\right)\left(m + \frac{1}{2}\right)}{|m - l|(m + l + 1)} \tag{7.55}$$

where s' is the range derivative of $s(r)$. This form nicely exhibits "close coupling," i.e., that U_{ml} is largest for adjacent modes. If we look at modes with $l = m \pm 1$, we get

$$|U_{m,m\pm 1}| \approx 2 \left| \frac{s'}{H} \right| \frac{m^2}{2m} \approx m \left| \frac{s'}{H} \right| \tag{7.56}$$

The coupling is clearly related to the surface slope s', as was mentioned in our previous discussion of multipath spread.

We now look at the cycle distance term in $\bar{\kappa}$. This is directly obtained from the eigenvalue equation, and is

$$|k_m - k_l| = \frac{\pi^2}{2kH^2} |m - l|(m + l + 1) \tag{7.57}$$

This quantity has the largest value (i.e., worst case) for adjacent modes where $(m - l) = \pm 1$, so the adiabacity (or nonadiabacity—take your semantic pick) criterion above reduces to

$$\bar{\kappa} = k \left| \frac{s'H}{2} \right| \ll 1 \tag{7.58}$$

This is a very convenient test for adiabatic versus coupled behavior! All you need is the acoustic frequency (since $k = 2\pi f/c$), the maximum slope you might encounter, and the waveguide depth H. This same form can be used for the bottom, with the criterion being

$$\bar{\kappa} = k \left| \frac{H'H}{2} \right| \ll 1 \tag{7.59}$$

where $H' = dH(r)/dr$.

This coupling criterion relates directly to the ray "time spread," discussed previously, versus the time separation of the unperturbed multipaths. Consider that H' corresponds to some (maximum/characteristic) slope angle χ_b. Also, we note that $M = kH/\pi$ is the number of propagating modes, so the average angle between modes is $\approx \pi/M$. When a "modal ray" is reflected from the rough surface, its angle is changed by order χ_b. Mode coupling will be insignificant if the angle change is less than the angular distance between the modes, a criterion given by $\chi_b \ll \pi/M$. This directly relates (in modal form and in the angular domain) to what we discussed before as spreading due to surface slope (in a ray theory context and in the time domain).

7.2.3.2.3 Vertical Modes and Horizontal Rays

In 1977, Weinberg and Burridge [30] created a hybrid of 2-D ray theory and 1-D vertical (adiabatic) mode theory designed to allow one to easily do 3-D acoustics. While there are better ways available to obtain the 3-D acoustic pressure field, the Weinberg–Burridge theory still has some interesting applications to shallow-water acoustics and can provide considerable insight.

The theory (see *Computational Ocean Acoustics* as a reference containing more detail) starts out by expressing the solution to the 3-D Helmholtz equation as a product of a function in the x–y plane $\Phi_m(x,y)$ and the local vertical normal modes $\Psi_m(x,y;z)$, i.e.,

$$p(x, y, z) = \sum_m \Phi_m(x, y) \Psi_m(x, y; z) \tag{7.60}$$

Substituting this into the 3-D Helmholtz equation and applying the operator

$$\int \frac{(*) \Psi_m(x, y; z)}{\rho} dz \tag{7.61}$$

which exploits the orthogonality of the local modes, we obtain

$$\frac{\partial^2 \Phi_m}{\partial x^2} + \frac{\partial^2 \Phi_m}{\partial y^2} + k_m^2(x,y) + (coupling\ terms) = -\delta(x)\delta(y)\delta(z - z_S) \qquad (7.62)$$

If we can throw away the coupling terms, i.e., make the adiabatic approximation, we obtain a new 2-D Helmholtz equation for the horizontal propagation of the field. This can be solved by ray theory (as Weinberg and Burridge originally did), or by PE, spectral methods, or other ways. The deviation of this horizontal solution from the straight line connecting source and receiver directly shows the 3-D, out-of-plane behavior of each normal mode that constitutes the field. For this reason, it is very interesting.

Calculationally, we must compute all of the local mode eigenvalues, i.e., the $k_m(x,y)$ at each x–y point. This could be onerous for very complicated modes and horizontal variability, but it turns out to be straightforward for the "canonical" and "perturbed canonical" waveguides we discussed previously, as well as useful. If we define a 2-D position vector $\vec{r} = (x, y)$, we have that the index of refraction field in the x–y plane is simply

$$n_m(\vec{r}) = \frac{k_m(\vec{r})}{k_m(\vec{0})} \qquad (7.63)$$

From here, one can do basic 2-D ray tracing to find the trajectories of the "modal rays."

7.2.3.2.3.1 Example: Ducting Between Nonlinear Internal Waves One rather interesting effect that was predicted, and subsequently verified experimentally, is the 3-D ducting of sound between internal wave solitons in the coastal ocean. Computer predictions by Oba and Finette [31], and theoretical predictions by Katznelson and Pereselkov [32] led to the eventual observation (see Fig. 7.10) of strongly ducted energy by Badiey et al. [33].

The ducting of sound by NLIWs can be predicted from the Weinberg–Burridge theory, using the simplified "square internal waves" model depicted earlier, and using the wave numbers for the ML and IW's generated by perturbation theory. (Real solitons have a "sech-squared" shape in the weakly nonlinear limit, described by the Korteweg–deVries equation. Our square wave model in Fig. 7.9 is in the same spirit as the "equivalent square well" one sees in basic quantum mechanics texts.)

If we write that Snell's law (in the horizontal plane!) for an IW duct for vertical mode n is

$$k_{1n} \cos \theta_1 = k_{2n} \cos \theta_2 \qquad (7.64)$$

then the critical angle for trapping mode n within the IW duct is

$$\theta_{1n}^{crit} = \cos^{-1} \frac{k_{2n}}{k_{1n}} \qquad (7.65)$$

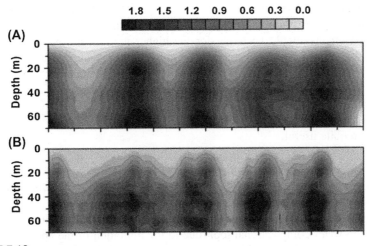

FIGURE 7.10

Measured sound intensity pattern fluctuations (normalized units) due to ducting of sound between nonlinear internal waves on a continental shelf. Panel (A) is from air gun data, and panel (B) is from linear frequency modulated pulses. This ducting can result in fluctuations from the mean level of about 6—8 dB at about 10 km, a very large amount for sonar applications.

*From Badiey, M., Katznelson, B.G., Lynch, J.F., Pereslkov, S.A. and Siegmann, W.L., Measurement and modeling of three-dimensional sound intensity variations due to shallow-water internal waves, J. Acoust. Soc. Am., **117**, 613, 2005.*

The wave numbers are obtained for our model by adding the perturbations to the background wave number for the ML and IWs. Doing this gives a large, but basically very interpretable, form

$$\theta_{1n}^{crit} = \cos^{-1}\left[\frac{\left(k^2 - \gamma_n^2\right)^{1/2} - \dfrac{H_{IW}\Delta c\omega^2}{k_{0n}c_0^3 H}}{\left(k^2 - \gamma_n^2\right)^{1/2} - \dfrac{D\Delta c\omega^2}{k_{0n}c_0^3 H}}\right] \qquad (7.66)$$

where γ_n is the hard-bottom eigenvalue in this example. This form is very physically transparent—it contains all the parameters of the environmental system and the acoustics (frequency) explicitly. Moreover, it works quite well in the real world! Using the parameters from the SW06 shallow-water experiment, this equation predicted a critical angle of about 4—5 degrees for the trapping [28]. Observations of the horizontal angle "wander" due to 3-D effects by Duda and Collis [34] showed just this amount of variability. Moreover, this simple theory predicts that the 3-D IW ducting effects should be larger for the higher modes, that was also clearly observed by Badiey [35]. (The latter effect is intuitively obvious, in that shallow water is very often downward refracting, and the lower acoustic modes are concentrated near the bottom, whereas the IWs are found near the surface.)

Of course, one may use more complicated calculations for these 3-D effects, but in terms of giving a first-order estimate, as well as physical insight, the perturbative forms seem quite useful.

7.2.3.2.4 Parabolic Equation

The workhorse technique for computing the acoustic field for lower-frequency acoustics is the parabolic equation, which has had a meteoric rise to popularity since it was first introduced to ocean acoustics by Tappert in 1977 [36]. Although originally a deep-water computational tool (due to the limitation of the "primitive" PE to ±15° grazing angle), the PE rapidly became useful for shallow-water work when methods for extending the vertical angular regime, most notably the Pade approximation, became available [20]. Quasi-3-D calculations (specifically N×2-D) were also available soon after PE's introduction, and even full 3-D calculations were being pursued by Ding Lee and others [37] during the 1980s. The most appealing part of PE is that it calculates full-wave acoustic fields quickly and efficiently. These fields can then be used for numerous acoustics calculations (e.g., array outputs, noise fields, fluctuation statistics, etc.). And while PE does not provide any insightful decomposition of the acoustic field immediately, it is not hard to use mode filtration techniques on the PE output to create a modal decomposition that is actually quite useful for showing mode coupling and continuum effects. PE is basically narrowband, like mode theory, so that broadband calculations have to be done via Fourier synthesis. This is not difficult, just computer intensive. PE also is excellent for very range-dependent environments. PE also has yet to be fully adapted to the inclusion of shear effects, although much work has been done to include these. One bright note is that, 3-D PE has been extended to very wide azimuthal angles, up to 45 degrees [38], which are needed for shallow-water work. An example of a full 3-D versus an N×2D calculation for the acoustic field in North Mien Hua Canyon, off Taiwan, is shown in Fig. 7.11 as an example of this new PE capability [39]. Our previous work using rays and modes is useful here, in that it can be used as guidance as to whether one needs to use full 3-D PE or not.

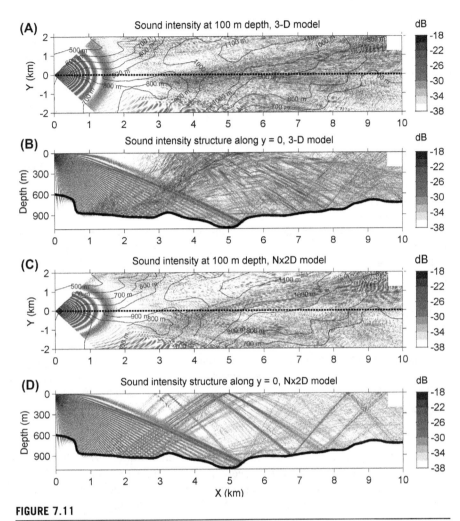

FIGURE 7.11

3-D versus 2-D acoustics calculations using parabolic equation. The top two panels are 3-D calculations, looking at a top and side view. The bottom two panels are the corresponding 2-D calculations. Significant differences are readily apparent. Number contours are isobaths depths.

©2014 IEEE. Reprinted, with permission, from Lin, Y.-T., Duda, T.F, Emerson, C., Gawarkiewicz, G.G., Newhall, A.E., Calder, B., Lynch, J.F., Abbot, P., Yang Y.-J. and Jan, S., Experimental and numerical studies of sound propagation over a submarine canyon northeast of Taiwan, *IEEE J. Ocean. Eng.*, **40**, (1), pp. 237–249, 2014.

7.2.3.2.5 Wave Number Integration

Wave number integration is a technique adapted to cylindrically symmetric (stratified) ocean waveguides, and is based on numerically implementing the Hankel transform pair, i.e.,

$$p(r; z, z_0) = \int_0^\infty g(k_r; z, z_0) J_0(k_r r) k_r dk_r \tag{7.67}$$

and

$$g(k_r; z, z_0) = \int_0^\infty p(r; z, z_0) J_0(k_r r) r dr \tag{7.68}$$

where $g(k_r; z, z_0)$ is the depth-dependent Green's function and $J_0(k_r r)$ is the cylindrical Bessel function.

Very good discussions of the details of the wave number integration technique are found in Frisk [24] and Jensen et al. [20], and the reader is referred to them for a wealth of detail. We just look at some highlights here, as we did for the parabolic equation section.

The biggest task in implementing the wave number integration scheme is determining $g(k_r; z, z_0)$, the depth-dependent Green's function. A number of methods exist for determining this Green's function, and again we refer the reader to Frisk and Jensen et al. for details. Green's function generated has all the details of the propagation through the layered medium contained in it, with one particular piece being very obvious—the normal mode structure of the waveguide. In particular, the denominator $D(k_r)$ of the mathematical structure of the Green's function for a stratified waveguide has the form

$$D(k_r) = 1 + R_B e^{2i\gamma H} \tag{7.69}$$

where R_B is the bottom plane wave reflection coefficient. Eq. (7.69) is the modal eigenvalue equation. Thus, when $D(k_r)$ equals zero, Green's function has poles at the normal modes of the acoustic waveguide [17], including at the "virtual modes" in the continuum [40], as seen in Fig. 7.12. This is not so surprising, in that the Hankel transform is a cylindrically symmetric version of the 2-D Fourier transform of the Helmholtz equation—we are just looking at the system in wave number (momentum) space.

Another direct benefit of the wave number integration technique is that, given Green's function, the plane wave reflection coefficient of the bottom is easily generated. The reflection coefficient has many uses, and being able to calculate it from this type of program is quite handy.

If one just wanted to obtain the pressure field for a stratified fluid medium problem quickly and efficiently, parabolic equation would likely win the race over wave number integration. However, wave number integration handles some cases that PE cannot. Specifically, wave number integration handles stratified media with shear included very well, and also can handle rough surface scattering (with some

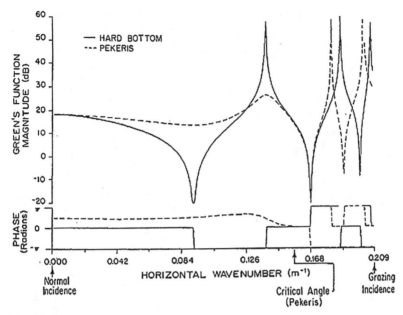

FIGURE 7.12

Depth-dependent Green's functions for a hard-bottom (*solid curve*) and a Pekeris (*dashed curve*) waveguide having the same geometry, water column, and source/receiver positions. The *hard-bottom amplitude curve* shows three trapped modes, whereas the Pekeris waveguide shows two trapped modes and a virtual mode (at the same eigenvalue as the third hard-bottom mode). The phase plot shows jumps of π at the peaks and troughs of the trapped modes, but not in the continuum region of the Pekeris waveguide. These phase jumps are typical of resonance features.

From Frisk, G.V. and Lynch, J.F., Shallow water waveguide characterization ssing the Hankel transform, *J. Acoust. Soc. Am., **75**, 205, 1984.*

restrictions). This includes rough surface scattering with shear included. This makes wave number integration very useful for studies of reverberation in shallow water, an important topic.

7.2.4 AMBIENT NOISE TERM

Shallow water is also distinctive as regards the ambient noise field. As long ago mentioned by Urick [2], the noise in shallow water is a mixture of three dominant components: (1) shipping and industrial noise, (2) wind/wave noise, and (3) biological noise. Not much has changed regarding the categories that Urick specified, but according to many estimates, the shipping and industrial noise has increased over the years, especially in the Northern Hemisphere, where some estimates put the increase as high as 20 dB above the "undisturbed state." This increase in anthropogenic noise has serious implications for both marine mammal acoustics and other sonar uses.

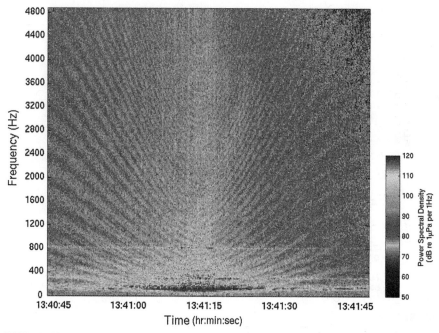

FIGURE 7.13

Broadband acoustic interference pattern due to close passage of a noisy ship. As the ship gets closer, the time difference between the acoustic multipaths becomes smaller, and the broadband interference pattern shifts symmetrically around the point of closest approach.

Data courtesy of Woods Hole Oceanographic Institution.

Regarding shipping noise, there are several newer and interesting topics worth noting, besides just the increase in level over the years. One is the recent interest in using broadband interference noise (often called "bathtub ring" noise) from shipping (see Fig. 7.13) as a source of opportunity for performing inversions for bottom properties [41]. More and more restrictions on research sources, particularly in coastal regions, are pushing the ocean acoustics community to consider using the ambient noise field (including anthropogenic noise) as a useful source, even if it is suboptimal in some ways. Another interesting effect that has been seen in shallow water is the "shallow-water noise notch" (see Fig. 7.14) where the downward refracting sound speed profile insures that there is more noise from near-surface sources at higher grazing angles (particularly near the critical bottom grazing angle, usually 20–30 degrees) than at low angles. This notch, which is useful for sonar purposes, can be filled in by scattering in the medium, and so is an interesting, time-variable phenomenon. A third, rather new, topic that is currently being explored is whether or not there are discernable 3-D effects associated with the coastal noise field, particularly from discrete sources like ships (and any acoustic sources they may use).

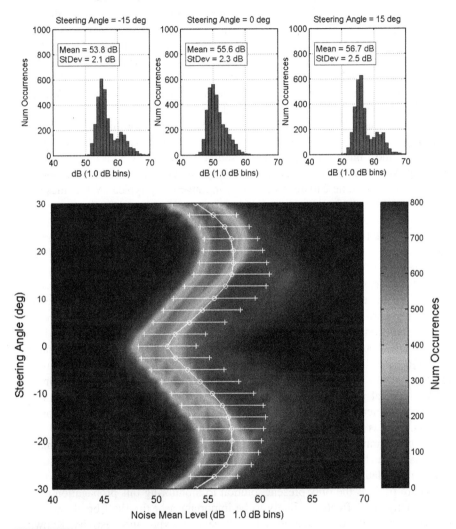

Beamformed Mean Noise Histogram. Source = 224 Hz, 8/1/96
Inactive Oceanogr, 10 hr period, DISK09A, 119 Trans Cycles, 3332 Pings, 10 Hr period

FIGURE 7.14

Shallow-water noise notch (bottom panel) seen in the SW06 experiment of the coast of New Jersey. The top panels show the level distributions seen at various beam-steering angles (−15, 0, and +15 degrees).

Data and analysis courtesy of WHOI and OASIS, Inc.

Given that there are some sharp horizontal sound speed gradients in shallow water, a discrete source can produce 3-D propagation effects; Badiey et al. [35] have, as an example, seen a 3-D horizontal Lloyd's mirror effect in shallow water due to a discrete research source. This is the direct horizontal analogue of the Lloyd's mirror in the vertical plane, where a surface reflected path and a direct path interfere. In the horizontal case, the reflection is off a nonlinear internal wave. Ships also act as discrete sources, and so it is not unreasonable to ask if one can see 3-D ship noise effects when going across a sharp sound speed interface in shallow water. Computer modeling of this is currently being pursued.

Coastal industrial noise is mainly attributable to seismic profiling, oil platforms, and pile driving for wind turbines (the latter being a rather recent entry to the list). By far the most intense industrial noise source is the first one on the list, seismic profiling. Very large, near-surface air gun arrays are currently employed as sources, with overall SLs in the 240 dB rel µPa @ 1 m range being typical. While these systems beam form to keep energy directed into the bottom, there is unavoidable leakage of energy sideways into the coastal ocean, and indeed one can often see air gun signals from sources up to hundreds of kilometers distant. Much work has been done on modeling this leakage, especially as regards Environmental Impact Statements (EIS's) that the exploration companies need to submit to work in given areas. There are still some topics that seem open to work in this area, however. First, there is the amount of bottom attenuation that occurs in a region of interest, which determines how quickly such leaked energy attenuates. By and large, our bottom attenuation estimates have considerable error, which in turn leads to error in the transmission loss (TL) estimates one makes. This error is due to several causes: (1) the need to listen to a source at a large enough distance, so that attenuation is appreciable; (2) the confusion between scattering loss and bottom attenuation, and (3) the general poor knowledge we have of the bottom sediments versus depth and range. Better attenuation estimates would be useful. But are not easy to obtain. Second, there is the oceanographic effect of near-surface ducting, which can allow the energy to travel much further horizontally, and also more directly affect marine mammals, which need to surface to breathe. ML ducts and seasonal near-surface ducts in high-latitude regions are important, but also hard to predict. Daily measurements are perhaps the only present solution to monitoring this propagation condition. Finally, 3-D effects are likely to be important near the shelf break, and near canyons, where there is often considerable seismic exploration activity. Knowing where to apply 3-D, versus N×2D, technology will be a useful bit of knowledge to improve coastal TL estimates for future EISs.

Fixed industrial platforms also contribute to the noise field in the coastal environment. As mentioned, oil rigs and wind turbine installation pile driving are probably the chief contributors at present. Oil rig noise can range from low to high frequency, depending on the machinery in use at a given time (very much like ship noise).

As the rig is at a fixed location, it is easier to measure the local environment and radiated noise well, and indeed this needs to be provided as a usual part of permitting. Pile-driving noise is lower-frequency impulsive noise (most intense from 0 to 1000 Hz), but still has a fair amount of higher-frequency content. This noise can travel well in shallow, coastal waters. As this activity has been increasing of late (as offshore wind turbines become common), this noise has been the focus of much attention. Mitigation of this industrial noise has been an active area in ocean acoustics, and the use of "bubble screens" surrounding the driver, and subsurface pile drivers which lessen the size of the radiating surface have been two among the many ideas tested.

Wind and wave noise are probably reasonably well modeled in shallow water at this time. The empirical Wenz curves [42] give the mean levels well (given the wind field), and theoretical models like the Kuperman–Ingenito model [43] capture the physics of shallow-water wind noise well, including its normal mode structure. Inverses for bottom properties using wind and wave noise are currently being developed, which are discussed in Chapter 2.

Marine biological noise (see Chapter 12) is an important topic, especially given the increasing endangerment of the oceans marine life. The "noise" from marine mammals and fish is of less concern as a "jammer" of sonar systems (as it is intense only very locally), as it is an indicator of animals nearby. In the case of marine mammals, this may necessitate shutting down active sonar systems due to protection laws. In the case of fish, it can indicate that "false targets" (as fish aggregations are often the size of submarines, see Fig. 7.15) might be detected, or that signals might be attenuated by passing through fish schools [44].

FIGURE 7.15

An aggregation ("school") of fish following an REMUS AUV off Cape Hatteras, North Carolina.

Data courtesy of authors and Woods Hole Oceanographic Institution.

Turning to using either passive or active marine life sounds as signal, the prospects are more on the positive side. Vocalizing whales are routinely tracked by passive arrays, either by time-delay cross-correlation or by array beam forming [45]. Smaller marine mammals and dolphins should be possible to track the same way, although the distances that can be examined are smaller due to the higher frequency of the vocalizations. Very small marine biota like snapping shrimp can perhaps be mapped for the "patchiness" of their distribution using their acoustic noise production. Mapping the distribution of marine life is a very important topic, and their "ambient noise" may be a very good tool, among others, with which to do this (see Chapter 12).

7.2.5 REVERBERATION TERM

Noise and reverberation are probably the two most difficult things to try to predict in shallow water. Noise is hard to predict, as we have seen, mainly because the sources of noise are not precisely known spatially, temporally, or in intensity. Reverberation is hard to predict because the position, shape, and composition of the scatterers are generally unknown; moreover, the scattering problem itself can be quite hard to solve. Nonetheless, noise and reverberation are prime considerations in the sonar equation, and we have to find ways to deal with them.

As always, one has the sea surface, sea volume, and sea bottom reverberation to deal with—however, in shallow water, the boundaries are closer and thus more important (on the whole) than in deep water. Moreover, there tends to be more "stuff" in the volume in shallow water than deep water. One encounters bubble, biota, suspended sediments, etc. so that the volume reverberation as well as the boundary reverberation should generally be stronger than in deep water.

Aside from the strength of the interactions, what else (besides basic material that has been treated ages ago, e.g., in Urick) is different about shallow-water reverberation? Let us look, starting from the bottom-up.

At first glance, calculating bottom reverberation is a more or less standard rough surface scattering problem, in which the roughness is described by a frequency-directional spectrum and the bottom material by a simple set of geoacoustic parameters (usually fluid to begin with). A good homogeneous, anisotropic bottom rough surface spectral model that is used in shallow water is the so-called "Goff–Jordan" model [46]. If one combines the input parameters to this spectrum with the water and sediment sound speeds and densities near the water–sediment interface, the basic scattering problem is posed and can in theory be solved. As mentioned, the solution to such a problem is nontrivial, no matter what frequency one looks at or what calculational technique one uses (see COA for examples). But, in shallow water, this is just the beginning. Both surface waves and currents, particularly during storm events, reshape the bottom roughness [47] and nonlinear internal waves have bottom stresses that can change the bottom roughness. Thus the bottom roughness spectrum is space and time dependent in a nontrivial way. Moreover, the material that comprises the bottom is stratified, sorted and mixed in complicated ways due to the (comparatively) rapid geological changes on continental shelves. In addition, it is affected by biology (e.g., bioturbation—the disturbance of the sediment by worms, shellfish, fish, etc.) and contains biological components (shell hash, sea grass, coral reefs, fish and eel burrows, etc.) that can scatter sound. Thus, the "environmental input" one should include in a complete bottom scattering model is far more complicated than one could ever afford to measure, and the scattering by such a complicated medium is in practice next to impossible to deal with fully. Hence, we must carefully consider what elements of the environment need exact inclusion, which can be approximated, and which can be ignored. A simple, although imperfect, rule of thumb is to ignore scatterers and phenomena that are smaller than the acoustic wavelength being considered.

Many shallow-water acoustics experiments have been conducted over the years that have made simple sediment property and surface roughness measurements, and then used these to calculate (or more usually, parameterize) reverberation. An interesting experiment called SAX04 (Sediment Acoustics eXperiment 2004, a follow on to a 1999 experiment similarly named) conducted a kilometer off Fort Walton Beach, Florida, examined very carefully how a coastal bottom environment both looked and changed, and how this affected higher-frequency acoustic scattering, reverberation, and bottom penetration. A detailed (and in some places wryly amusing) overview of the SAX04 experiment is found in Richardson et al. [48]. In this experiment, both manipulated (i.e., designed and controlled) scatterers and natural scatterers were examined. Sand ripples of controlled roughness were created and measured by divers, and the natural subsurface sediments were carefully measured. Moreover, artificial discrete scatterers of controlled size and composition were placed in the experimental field, such as thin aluminum disks that simulated sand dollars. Although there are numerous articles containing very detailed results of this set of experiments, two or three major points strike the author as being particularly interesting in the context of shallow-water reverberation. The first point is that the shallow-water sub bottom material is often very complicated, with inhomogeneities at many scales, and as such needs to be addressed by some sort of parameterized "effective medium" theory. The frequency-dependent attenuation and dispersion of such a medium is a challenge to compute, but is needed for acoustics. Second, the roughness spectrum of the bottom is subject to both slow and fast variability, with bioturbation being an example of a slow process and the tropical storms and hurricanes encountered being examples of the fast processes. At this point in time, sediment transport investigators have developed very good instrumentation for monitoring bottom ripple (and other) roughness. An example of a sector scanning sonar measurement of sediment ripples made by P. Traykovski is shown in Fig. 7.16. Yet another nice result from the SAX experiments is that bottom roughness creates a mechanism for sound to penetrate into the bottom more effectively, thus increasing the possibility of

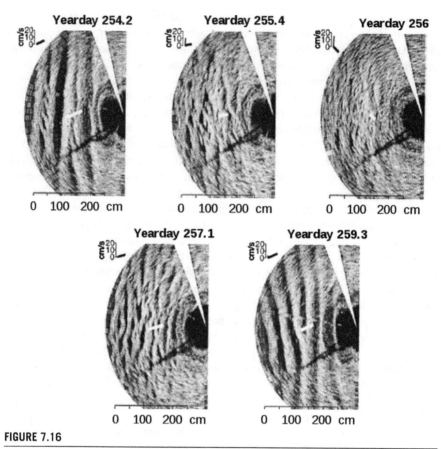

FIGURE 7.16

Bottom roughness/ripple images made with a rotary sector scanning sonar. Stress due to surface waves and near-bottom currents forms, erodes, and reforms the ripple structures over periods of hours to days.

Courtesy P. Traykovski, Woods Hole Oceanographic Institution.

detecting and classifying buried objects using backscattering (which competes with natural reverberation).

The addition of bottom ripples creates a distribution of angles of incidence on the bottom for a given source ray angle, and thus gives the incident sound a chance to be "above critical" at parts of the ripple field. Thus, it can penetrate the bottom. A smooth bottom is totally internally reflecting up to a certain critical grazing angle (see Fig. 7.17), after which sound more easily penetrates the sediment and can scatter and reflect off buried objects.

Finally, another sediment effect well known to transport investigators is the settling of mud and very fine sediment into the troughs of sand ripples. For a given bottom stress, the finer material (e.g., muds and silts) is suspended and moved more

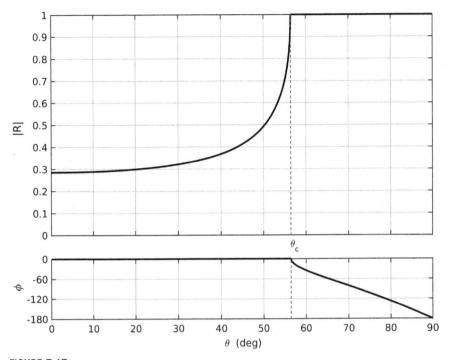

FIGURE 7.17

Typical plane wave reflection coefficient of the bottom versus incident angle. Magnitude is top panel and phase is lower panel. (Grazing angle would be 90 to 0 degrees.)

easily and so tends to be deposited in the nooks, crannies, and troughs of the larger grained material. This can lead to layers several centimeters thick in places, which can be seen readily by high-frequency imaging devices such as side-scan sonars.

7.2.5.1 The Bottom Boundary Layer

Moving slightly up from the water—bottom interface, one encounters the next phenomenon that can perhaps affect acoustics—suspended sediments in the bottom boundary layer. Fortunately for most acoustics work, these sediments are usually order $1-10\,\mu m$ in size, or at most order $100\,\mu m$ sands disturbed by extreme wave events. Such small sediments are usually far into the weak, Rayleigh scattering regime ($ka \ll 1$) for most sonars, and to reach the strong Mie and geometric regimes ($ka \geq 1$), frequencies of $1-10\,MHz$ are needed. Such high frequencies are indeed used for sediment transport studies, but they are not common sonar frequencies, due to attenuation restricting them to very short ranges.

In passing, we note that the use of higher-frequency acoustics to study coastal oceanography and sediment transport is a specialty that most ocean acousticians do not see very often, as its practitioners tend to live in the physical oceanography and marine geology communities. Readers interested in the acoustics and instrumentation peculiar to those areas are recommended to look up the works of Alex Hay (Dalhousie University) and Peter Thorne (National Oceanography Centre and University of Liverpool) via Google Scholar, as they have been two of the leading practitioners (and well published) over the past few decades.

7.2.5.2 Water Column Reverberation

Moving further up into the water column, one reaches the "realm of the pelagic biota," which span a huge range of sizes and scattering strengths. Shallow water, particularly near the shelf break and canyons, is the home for most marine life, and many of these organisms, particular when found in schools or aggregates, can scatter sound significantly. For individual organisms, the first-order rule of thumb for whether or not we need to consider them acoustically is whether or not $ka \geq 1$. If so, then the scatterer deserves further examination. The "physics model" of the scattering by individual organisms is usually rather intricate and complicated, due to their high amount of structure, but such models have been studied [13], and this is a current topic of interest by both the civilian (e.g., fisheries) and military (e.g., anti-submarine warfare, often called ASW) communities. For aggregates of animals [49], the situation becomes more of a statistics study, and one generally needs to understand how many animals' backscatter signals are adding up in a given sonar beam. This also involves knowing some of the large-scale properties of the aggregations, such as their directional orientation, number density, size distributions, species distributions, etc. (see Chapter 12). Moreover, animal distributions are patchy in space and quite variable in time, so that it is very hard to predict a priori what one will see as reverberation from biota.

7.2.5.3 Sea Surface Scattering and Reverberation

As we emerge toward the sea surface, we encounter yet another thorny boundary interaction problem when trying to understand reverberation—the combined sea surface/bubble layer scattering problem. At low wind speeds (<5 m/s), the ocean is not producing breaking/spilling/splashing waves, and so one may describe the sea surface scattering and reverberation rather well just using classic rough surface scattering theory sans bubbles. (Ogilvie [20] is a good full-text reference for this topic, and Brekhovskikh and Lysanov [21] have excellent ocean acoustics oriented chapters.) Even the classical theory of rough surface scattering is complicated enough, but the (nearly) free surface boundary condition and our rather good knowledge of how ocean surface waves work at this point in time give us a fairly solid starting point. Shallow water introduces comparatively minor changes on deep-ocean surface wave theory, due to fetch dependence (distance from the shore) and finite depth effects on the wave dispersion and bottom friction; hence, deep-water theory gives a good starting point. Thus, we can say that this part of the problem is "textbook knowledge" to a good first approximation. And so rather than reproduce well-known results, we just refer the reader to the standard texts.

7.2.5.4 The Sea Surface Plus Bubble Scattering and Reverberation

Things get a bit more exciting when higher wind speeds and wave breaking start to occur. At this point, nature has both changed the "background waveguide" through which acoustic energy travels to the surface and subsequently scatters, via a persistent bubble layer, as well as has introduced very complicated intermittent scatterers—the

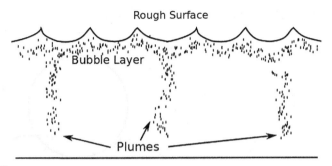

Rough Surface

Bubble Layer

Plumes

FIGURE 7.18

Cartoon representation of the near-surface bubble layer and subsurface plumes, which could be caused by Langmuir circulations or nonlinear internal waves.

bubble clouds and bubble sheets. A very good overview (as per 1994) of the near-surface bubble environment in the ocean is provided in Leighton's *The Acoustic Bubble* [50] and we recommend the reader refer to this text as initial background.

A 2-D schematic view of the bubble layer and rough sea surface combined problem is shown in Fig. 7.18. In this we see some ways that the bubbles can affect backscattering and reverberation. These include (1) changed (enhanced?) Bragg backscatter due to the sound speed dispersion of the near-surface bubble layer and (2) backscattering by bubble plumes and sheets.

Let us, just for insight, consider a physics model of how the persistent, near-surface bubble layer might affect low-frequency backscatter. It is known at this time that an exponentially decaying in depth layer of fine bubbles persists around the surface during storm/wave-breaking events, and that this layer both decreases the local sound speed and also attenuates sound. By decreasing the near-surface sound speed, this bubble layer increases the angle at which an incoming ray/mode interacts with the surface, as a simple consequence of Snell's law. Specifically, if the sound speed just below the layer is c_1 and the sound speed at the sea surface is c_2, where $c_1 > c_2$, then for the incident angle θ_2 at the sea surface, $\sin\theta_2 = (c_2/c_1)\sin\theta_1$. This gives that $\theta_2 < \theta_1$, or equivalently that the ray/mode hits the surface at a higher *grazing* angle. For a given rough surface, higher grazing angle scatter is usually stronger—but the matter does not end there. Backscatter also is typically described by which Bragg scattering resonant component of the surface roughness spectrum it picks out, using the condition:

$$\Delta k_{acoust} = k_{in} - k_{out} = k_{surface} \tag{7.70}$$

where the k stands for horizontal wave number. For backscatter, the outgoing wave is in the negative x-direction, so this equation becomes:

$$\Delta k_{acoust} = k_{in} + k_{out} = k_{surface} \tag{7.71}$$

For a given k_{out}, this increase in the angle k_{in} means that the scattering is picking out a higher $k_{surface}$ from the wave spectrum. Depending on whether or not this surface wave spectral component is stronger or weaker, the acoustic scattering can be stronger or weaker. Finally, there is the matter of the increased attenuation of the sound along the path containing the bubbles. This is given by a simple decay law

$$I = I_0 \exp\left(- n_b \Omega_b^{ext} z\right) \tag{7.72}$$

where I is the intensity, I_0 is the original intensity of a plane wave, n_b is the bubble number density, Ω_b^{ext} is the extinction cross section of a bubble (the sum of the scattering and absorption cross sections, with units of area), and z is the path length. This derivation does not provide a number for the backscattering enhancement (or decrease), but it does point out some of the physics that needs to be considered in doing this problem *if* the sound reaches the surface and scatters. However, it is not a given that the sound will reach the surface, or if it does, that the scattering enhancement seen will describe the 15–20 dB enhancement of backscatter in high sea states where the wind speed is greater than 10 m/s. Indeed, this theory, while looking plausible, probably is only good as a small extension of the lower wind speed regime, where 5 m/s $\leq v_{wind} \leq$ 10 m/s. For higher wind speeds, the bubbles and bubble plumes themselves are likely the main causes of the backscattering.

Leighton describes other theoretical attempts to describe the higher backscatter seen in high sea states where bubbles are present. In the first, the bubble size distribution is invoked to describe both the attenuation and backscatter that is caused by the bubbles alone, independent of the sea surface. Bubbles in the resonant regime are especially important here. The approximately isotropic scatter from the bubbles provides a sufficient backscattered component, see Chapter 5. A second theory described in detail is one by Henyey [51] invoking bubble plumes and columns. These have deep concentrations of small bubbles, which can be provided by Langmuir circulations or, in the case of coastal regions, nonlinear internal waves. Such plumes can be very effective scatterers of sound. Other theories of backscattering by bubbles are also described in Leighton, though space forbids an entire discussion here. However, a common feature of all of them is a fair level of conjecture as to the exact geometry and composition of the bubble scattering layer/plume/column. In this problem, as in many other ocean acoustics problems, our knowledge of the ocean medium ultimately becomes the limiting factor in our acoustics predictions.

7.3 SOME ADDITIONAL TOPICS OF INTEREST IN SHALLOW-WATER ACOUSTICS

In this last part of this chapter, we discuss some assorted topics that do not fit so easily under the sonar equation umbrella, but that give shallow water some of its distinct flavor physically and technically.

7.3.1 ONE- AND TWO-LAYER WATER COLUMN SOUND SPEED PROFILES IN SHALLOW-WATER ACOUSTICS

Given modern CTDs, AUVs, and gliders, it is not so hard for experimenters in ocean acoustics to be able to construct very good range-dependent profiles of the temperature and salinity of the coastal ocean versus depth, and thus the range-dependent sound speed structure. There is still some issue of space–time aliasing of the finer-scale ocean features, but at large scales, the profiles are fairly good. And acoustics codes can handle very detailed range-dependent oceanography profiles these days, so that there is no reason why one should use simplified profiles in calculations. Or is there?

The answer is that there are some very valid reasons for using simplified variants of the ocean (and bottom) sound speed structure, the two foremost being: (1) to understand the basic physics of the shallow-water ocean waveguide and (2) to be able to cleanly isolate the acoustic effects of individual ocean features.

The most basic water column "canonical" profile is the isovelocity profile, famous for its use in the Pekeris model, which represents a well-mixed water column. Such conditions are common in mid- and upper latitudes in the winter. The Pekeris model is perhaps the best "simple physics" model of shallow-water acoustics available, and is both a good pedantic tool and a straightforward estimation device. The two-layer profile is perhaps the most useful of all simplified profiles, in that it can represent a very wide range of coastal ocean features. By making the layer interface a function of range, this model can represent fronts, internal waves (linear and nonlinear), and internal tides rather easily. The bottom sediment half space represents the simplest bottom model.

7.3.2 THE OPTIMUM FREQUENCY

The optimal frequency is a low-frequency ($f < 1000$ Hz) shallow-water acoustic propagation phenomenon that was first reported by Weston [52] based on experimental results. It is basically the tradeoff between the quadratic loss in intensity with frequency due to the water column (which is well known) and the loss due to the bottom, which has a more complicated loss versus frequency behavior.

Looking at propagation in the North Sea between 25 and 6400 Hz, Weston noticed that the overall propagation losses were minimal at about 200 Hz. Further work in the Bristol Channel led Weston to characterize the loss regimes into high loss areas characterized by rocky bottoms (high shear speed) and low loss areas characterized by sand and mud sediments (with much lower shear speeds). At around the same time, Akal [53] refined the optimal frequency dependence for sediments; sand (a harder sediment with a higher sediment shear speed) gave an optimal frequency in the band 400–800 Hz, whereas mud (a softer sediment with a lower sediment shear speed) displayed an optimal frequency around 50–100 Hz. Obviously, the acoustic impedance and the bottom losses, including shear, were determining the optimum frequency.

These results are physically understandable in terms of a simple Pekeris model (plus shear), which gives an analytic approximation for the modal propagation case, which is appropriate for shallow water below 1 kHz. Following Katznelson et al. [16], the modal attenuation coefficient, γ_l, is given by

$$\gamma_l = \frac{sc^2 l^2}{4f^2 H^3} \tag{7.73}$$

where s can include both the attenuation in the sediments and the influence of shear waves, l is the mode number, and H is the water depth. If we look at the water attenuation as well, we get that γ_l has the form

$$\gamma_l = \frac{sc^2l^2}{4f^2H^3} + \frac{0.013(f/1000)^2}{1 + (f/1000)^2} \tag{7.74}$$

For a given mode, one can simply take the frequency derivative to equal zero to get the approximate result

$$f_{opt} = 100 \left(\frac{sc^2l^2}{5.2H^3}\right)^{1/4} \tag{7.75}$$

This is a rather simplified form, and to be more exact, one needs to consider the modal excitations, which are also a function of frequency. Details of such calculations can be found in Katnelson et al. [15]. Also, surface and bottom scattering, which are frequency dependent, can change or even eradicate evidence of this effect.

The existence of an optimal frequency is of interest if one is concerned, for example, with the maximum detection range of a broadband source, or in designing a source for long-range communications.

7.3.3 ARRIVAL STRUCTURES IN SHALLOW-WATER AND RAY/MODE RESOLUTION

When listening to a broadband signal in a shallow-water waveguide, and at a sufficient distance, one hears the time series of multipath arrivals rather distinctly. Moreover, the first arrival tends to be the loudest, with the intensity diminishing afterward. This is in contrast with the deep-water case, where the initial arrivals are weak, and build up to a loud "coda" at the end. Physically, this distinction between the shallow- and deep-water arrival intensity structures is due to the total distance a ray travels versus the sound speed of the water it travels through. (All modes travel the same distance, but their extent in the water column changes versus mode number, so the argument is a little more complicated.) In shallow water, the smallest grazing angle rays travel a smaller distance, but through lower sound speed water. The high grazing angle rays travel a longer distance, but through higher sound speed water. This makes the outcome a seeming toss-up, but in shallow water the race is won by the low angle rays. In deep water, the outcome is the opposite. There can be variations to this "rule of thumb" at intermediate water depth, and for peculiar sound speed profiles, but in general it works well.

Another interesting question to ask when looking at a processed time series plot of acoustic multipath arrivals is: "Am I looking at a ray or a mode?" In one sense, this is an artificial question, in that rays and modes are just constructs we use in solving the wave equation, which is the most basic physics of the system. However, these constructs are also observable physically, when the data are processed properly, and also have a great deal of practical utility. Hence, it is worth looking at what the criteria are for observing rays versus modes.

The basic question to ask is whether the acoustic arrival one sees can be resolved in time as a ray or a mode. As ray resolution is the easier question, we start with that first. Consider a broadband pulse of bandwidth Δf and duration $\tau \sim \frac{1}{\Delta f}$. For this case, we get the criterion

$$\Delta t_n \sim 1/\Delta f \qquad (7.76)$$

where Δt_n is the travel time difference between the nth and $(n+1)$ ray arrivals.

The mode resolution (in time) criterion is a slight bit more involved, as the travel times of the modes vary with frequency due to waveguide and material dispersion effects. We use the time bandwidth product criterion in this case as

$$\Delta t_l \Delta f_l \sim 1 \qquad (7.77)$$

where Δt_l is the travel time difference between adjacent modes and Δf_l is the frequency difference between adjacent modes for a given travel time.

A rather nice example of where one can resolve rays versus modes in a shallow-water Pekeris waveguide is given in Katsnelson et al. [16], and we refer the reader to this text for more detail.

7.3.4 WAVEGUIDE INVARIANT

The "waveguide invariant" is, in many ways, the most variable invariant quantity that one might encounter in shallow water. It is a "modal interference-based intensity invariant," which given the generally fluctuating nature of modal intensities and their interferences, especially in shallow water, is perhaps expecting rather much of nature. But, rather than poke fun at this odd invariant, let us look at it closely.

The intensity pattern versus frequency, range, and azimuth in shallow water is determined by the details of the interference patterns of the normal modes that comprise the field. The basic condition to find a line of constant spectral/intensity level is given by:

$$dI(r, \theta, \omega) = \frac{\partial I}{\partial \omega} \delta \omega + \frac{\partial I}{\partial r} \delta r + \frac{\partial I}{\partial \theta} \delta \theta = 0 \qquad (7.78)$$

This invariance condition produces, for a range and azimuthally independent environment (see Kuperman and D'Spain [54]):

$$\frac{\delta r}{\delta \omega} = -\frac{r}{\omega} \frac{1}{\beta} \qquad (7.79)$$

In Eq. (7.79), β is independent of range and angle, and acts like a true "invariant" of the system. It is theoretically described by

$$\beta = -\frac{\Delta_{lm}^{ph}}{\Delta_{lm}^{gr}} \qquad (7.80)$$

where Δ_{lm}^{ph} is the difference in phase slowness between the lth and mth modes, and Δ_{lm}^{gr} is the difference in group slowness between the lth and mth modes. (Slowness is just the reciprocal of velocity.)

But what is β? What is its numerical value? This can be measured experimentally (and has been, extensively) by looking at the slope of the interference pattern constant intensity "striations" versus range and frequency. This provides experimental verification of the theory, as well as β values. However, a basic range-independent waveguide model can also provide β, as well as some physical insight. We follow the treatment in Brekhovskikh and Lysanov [21] in presenting this.

We begin with the well-known relation between the phase velocity and group velocity of normal modes for an isovelocity ocean with a perfectly reflecting bottom and surface, $v_l u_l = c^2$, where v_l is the phase velocity, u_l is the group velocity, l is the mode index, and c is the speed of sound in water. For this case, for any mode (so we can drop the mode index), we have $du/dv = -v/u$, which has the solutions $v = c/\cos\chi$ and $u = c\cos\chi$. Hence we get for the waveguide invariant

$$\beta = \frac{u}{v} = \cos^2 \chi \qquad (7.81)$$

where χ is the grazing angle of mode l. At small angles, $\beta \approx 1$, which is the common result measured for shallow water.

Without going into more detail (which can be found in Kuperman and D'Spain [54], Brekhovskikh and Lysanov [21], and Katsnelson et al. [16], as good examples), the waveguide invariant actually *does* vary somewhat versus mode number in shallow water, given a non-isovelocity vertical sound speed profile. Also, for a different profile, the waveguide invariant is different—for an n^2 linear profile, where $n(z) = \frac{c_0}{c(z)}$ and $n^2(z) = 1 - 2az$ (where a is the sound speed gradient), $\beta \approx -3$, for example.

There is one more rather interesting use of the waveguide invariant equation that we should mention here. Specifically, it can be used as a way to (approximately) replace a broadband calculation by a simple range average of a narrowband one, which is much easier to do. If we rearrange the preceding frequency/range relation, and take $\beta \approx 1$, we get

$$\frac{\delta r}{r} = \frac{\delta \omega}{\omega} \qquad (7.82)$$

This provides a rather easy "cookbook" way of averaging the intensity over a distance δr at range r get the equivalent of a frequency average over $\delta \omega$ at frequency ω at that range. The interested reader is referred to the article by Harrison and Harrison [55] on this topic.

7.3.5 INTENSITY FLUCTUATION STATISTICS

Intensity fluctuation is a large topic, and also an important one. It is also one that we can claim to have a reasonably good physical understanding of in shallow water. There are many flavors of intensity fluctuation we can look at: broadband pulse peak intensity, broadband pulse integrated intensity, instantaneous narrowband intensity, horizontally and/or vertically integrated (array output) intensity, and so on. Each of these has their uses in practical applications, and their own statistical characterization.

There are two ways that intensity fluctuations are commonly described: via the decibel level and via the scintillation index (SI). The former is useful for attaching numbers to experimental measurements, and the latter is a useful device for understanding the physics underlying the measured fluctuations. We look at both here.

The SI is defined by

$$(SI)^2 = \frac{\langle I^2 \rangle}{\langle I \rangle^2} - 1 \tag{7.83}$$

It generally varies between 0 and 1, but can go above 1 in certain instances. Some plots of the SI for various situations are shown in Fig. 7.19. Consider the simplest situation first. In the "weak scatter" regime, either the scatterers are very weak, or the distance the sound has traversed is too short for much scattering to have occurred. For stronger scatterers or larger ranges, the accumulation of phase error for each individual multipath becomes 2π or greater, i.e., each one is phase random. In the limit of many multipaths of (roughly) unit amplitude, this produces a variance in the intensity of 5.6 dB [56] or equivalently an SI of one. This is the so-called "saturated" regime of scattering, which is quite commonly seen in shallow water. In Fig. 7.17, where saturation is approached from below (i.e., via a slow accumulation), this is called the "weak scattering approach to saturation." This is a common scenario, but it is not the only approach to saturation. In Fig. 7.19, we show a second possibility, i.e., the "focusing regime." In this case, sound is focused (and defocused) by the environment, and is not just phase randomized. A good example of this that we saw previously was the 3-D focusing and defocusing of sound by nonlinear internal waves. This creates fluctuations that can be substantially bigger than 5.6 dB, and thus SIs that are greater than one. Eventually, attenuation and propagation effects past the focusing region reduce this effect, and the SI relaxes to unity (and 5.6 dB). There is yet another scenario, not shown here, where at very large ranges (order 50 km) the SI goes from saturation to exponentially growing with range [57]. This comes about when attenuation is added to the shallow-water waveguide, as it should be. This effect is only predicted for long ranges, and so is not commonly seen.

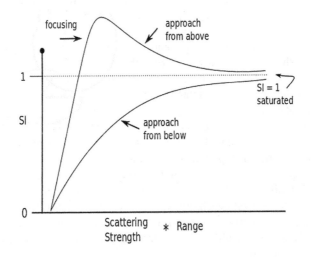

FIGURE 7.19

Approach to saturated scattering, from above and below.

The SI plots give a good qualitative view of the fluctuation physics, but to get a more quantitative answer, one needs the statistics of the fluctuations. In that there are many types of fluctuations, we just restrict ourselves here to simple narrowband and broadband statistics, and not worry about peak statistics, or array statistics, etc.

It is well known at this point in time that for the saturated scattering of many acoustic multipaths, the complex acoustic field can be interpreted as a large number of random, independent contributions. From the central limit theorem, the real and imaginary components of the pressure field will approach zero mean Gaussian random variables. The phases of the components will be uniformly distributed between 0 and 2π. The combined amplitude of the components, a, will display a Rayleigh distribution $w(a)$, i.e.,

$$w(a) = \frac{a}{\sigma^2} \exp\left[\frac{a^2}{2\sigma^2}\right] \tag{7.84}$$

for $a \geq 0$, where σ^2 is the variance.

Turning to the intensity, for saturated scattering it is given [16] by a negative exponential distribution $w(I)$, i.e.,

$$w(I) = \frac{1}{I\sigma_{\log I}\sqrt{2\pi}} \exp\left[-\frac{\left[\log\left(\frac{I}{m}\right)\right]^2}{2\sigma_{\log I}^2}\right] \tag{7.85}$$

where m is the median and $\sigma_{\log I}$ is the standard deviation of the log intensity.

In Fredericks et al. [58], there is a nice example of how different, but simultaneous, ocean processes produce both strong and weak acoustic fluctuations. Looking at data from the 1996 PRIMER experiment off the Mid-Atlantic Bight, an acoustically strong (high ocean frequency) signal was seen from NLIWs in the area, but an acoustically weak (lower ocean frequency) signal was seen from the local front and eddy field. The high-frequency component was seen to show an exponential distribution, whereas the low-frequency oceanography showed a log-normal distribution. Being able to tie well-known ocean processes and their strengths to specific acoustic fluctuation statistics is a huge advantage for operating and improving sonar systems, as one can look at ocean climatology, in situ measurements, and/or ocean numerical models in a given operational area, and from those obtain a reasonable a priori estimate of how a sonar system will perform, or how its performance can be improved by using these statistics in signal processing algorithms.

The next step in looking at fluctuations is to consider broadband signals. In this case, the physical difference from the narrowband result is that all the multipaths do not necessarily interfere with each other, i.e., they may be separated in time. Moreover, the amount of separation may be a function of position in the arrival time structure, so that a "one size fits all" theory of the multipath statistics, as we saw in the narrowband case, does not exist.

One of the first works that generalized the Dyer narrowband result to broadband was that of Makris [59], who considered time—bandwidth products greater than 1. In this case, the standard deviation of the transmission loss goes as $4.34\sqrt{1/\mu}$, where μ

is the time—bandwidth product. For large time—bandwidth products, $\mu > 4$, the averaging out of the fluctuations by the large bandwidth produces a log-normal distribution, consistent with what was seen in the narrowband case.

Colosi and Baggeroer [60] looked at broadband fluctuations in the weak and saturated regimes, and also the approach to saturation. They found that the SI has the limiting behaviors:

$$\text{SI}^2 = 1 + \frac{1}{N}\left(\frac{\langle a^4 \rangle}{\langle a^2 \rangle^2} - 2\right) \quad \text{Narrowband} \tag{7.86}$$

$$\text{SI}^2 = 1 + \frac{1}{N}\left(Q\Phi\frac{\langle a^4 \rangle}{\langle a^2 \rangle^2} - 2\right) \quad \text{Broadband} \tag{7.87}$$

where $Q = \Delta\omega/\omega_0$ is the bandwidth of the system, Φ is the signal phase variability, N is the number of multipaths, and $\langle a^n \rangle$ is the nth moment of the amplitude distribution. The competition between the second moments and the fourth moments of the distribution (second term of each equation) determines whether the SI grows exponentially or not.

Eq. (7.86) is a generalization of the previous results, and consistent with them. Eq. (7.87) clearly introduces the new physics of the system bandwidth and the signal phase variability due to scattering, both of which affect which of the N multipaths are interfering and which are resolvable. This multipath interference is a core reason of why the signal fluctuates.

7.4 SOME NEWER TOPICS

Although shallow-water acoustics is a well-developed area of ocean acoustics at this point in time, there are still some new topics to research. We look, in a broad-brush sense, at three of them, namely (1) canyons, the continental slope, and the shelf break; (2) arctic shallow-water acoustics; and (3) climate change effects on shallow-water acoustics.

7.4.1 THE SHELF BREAK, SLOPE, AND CANYON REGIONS AND THE TRANSITION TO DEEP WATER

The shelf break, slope, and canyon regions are of course not entirely new topics, but it is safe to say that, due to their complicated natures, it is only recently that we can address them rigorously via theory, experiment, and numerical modeling. Geologically, the shelves, slopes, and canyons display complicated bathymetry, material composition, and bottom stratigraphy. Biologically, the shelf-break region and especially the canyon regions are highly active, as evidenced by the large amount of fishing activity. Oceanographically, the shelf-break fronts and their environs are the boundaries between water masses, with very complicated dynamics due to both the water masses and the bathymetric slope. Acoustically, these regions often need fully 3-D acoustics for their characterization—out-of-plane effects are very commonly encountered. For all of these reasons, acoustics in this "transition zone" between deep and shallow water has been harder to attack than the other regions. It should be noted that, as is very often the case in acoustics, the major problem is measuring and characterizing a complicated and hard to access environment. Measurement of very rugged canyon bathymetry is still hard, as is measuring the bottom properties and stratigraphy. The same goes for slopes. The shelf-break front oceanography is both intricate and time varying. Biologics are notoriously hard to locate and quantify on a long-term basis. Thus, even with the best available acoustics codes, we are limited in our predictions of acoustics quantities by the environmental input. One way around this is to try to quantify the uncertainty in our input, and indeed this is where a good amount of current research effort is being devoted.

7.4.2 ARCTIC SHALLOW-WATER ACOUSTICS

Twenty years ago, a fair amount of research work was devoted to understanding acoustic propagation in the Arctic, due of course to the Cold War between East and West. When the Cold War thawed some, around 1990, this effort was largely abandoned. However, a renewal of East—West tensions along with a new factor, climate change, has revived interest in the Arctic, as sad as those reasons may be. Moreover, climate change has had the rather startling effect of having negated the validity of a good portion of the Arctic acoustics knowledge that was gained by previous research. Specifically, it has changed the Arctic Ocean, its ice cover and the connecting seas rather dramatically, so that the sound propagation medium is not the same one as before. Also, due to the melt-back of the ice pack, there will be a northwest passage available to shipping over a good part of the year, so that shipping activity (and its associated noise) will inevitably increase. There is also great interest in the mineral resources in the Arctic, which are now becoming possible to harvest with current technology. Further, the marine ecology of the Arctic Ocean is changing, due to the water warming. Warm water species have already been seen to be migrating northward, as well as cool water species trying to find a cooler habitat.

The two major factors affecting acoustic propagation in shallow water (and deep) in the Arctic are the water column sound speed profile and the ice cover. Due to the ice cover being absent over a large part of the year (it is predicted that the Arctic Ocean will be ice-free in summer rather soon), the albedo of the ocean will decrease, and more heat will be absorbed by the water. This will in turn insure that the ice cover cannot return to its former state. This increased heat will form a warm surface layer, which will reduce the interaction of sound with either the winter ice or summer surface waves. This is a better propagation condition, i.e., a near-surface duct. How deep this duct is compared to the bathymetry will be of great interest to shallow-water acoustics. The ice cover has also changed, with first year ice now predominating. Moreover, the ice is thinner, often has leads and melt pools, and rafts in different ways from before, leading to different types of ridging and fracture behavior. Ice dynamics in the Arctic has opened up again as a research area. Also, with open water now available, the wind field becomes a factor, and so the fetch in shallow-water, as well as the large-scale wind field, need to be considered in the summer. Rather sadly, the Arctic has become a whole new system, and we need to once again measure its physical characteristics to do any work with acoustics.

7.4.3 CLIMATE CHANGE AND SHALLOW-WATER ACOUSTICS

The Arctic is perhaps the most dramatic example of climate change affecting the oceans, but it is not the only one. Indeed, all the world's oceans (including the coastal oceans, which are our interest here) are experiencing climate change. In ice-free waters, the biggest changes that will be seen are: heating, winds, and water mass change. The heating effects versus depth in shallow water will be the biggest effect, most likely. As an example of these, we can cite an experiment in 2012 which actually saw a climate-related shallow-water effect off Cape Hatteras, North Carolina [5]. Thermal profiles for "before and during" conditions are shown in Fig. 7.20. The difference in the water thermal and sound speed profiles is quite large. Moreover, fish species displacement, as discussed foregoing, was clearly noted—we saw warm water fish in an area where we expected cold water species! This change was a temporary one, and the system reverted to its usual state the next year—however, increasing heating of the waters will make such events occur more and more frequently, possibly until a completely new equilibrium state is produced.

Predicting climate change for smaller coastal regions is a challenge to our current climate modeling capabilities, but it is becoming more apparent that we need to start factoring climate change into our predictions of the ocean state, and thus its effect on acoustics.

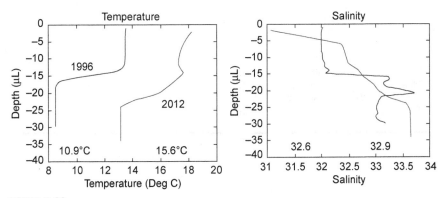

FIGURE 7.20

Large inter-annual changes in temperature and salinity structure in the waters off Cape Hatteras, North Carolina.

Data courtesy of G. Gawarkiewicz, WHOI.

LIST OF ACRONYMS

ACM	Acoustic current meter
ADCP	Acoustic Doppler current profiler
ASW	Anti-submarine warfare
AUV	Autonomous underwater vehicle
BW	Bandwidth
COA	Computational Ocean Acoustics
CTD	Conductivity temperature depth
CW	Continuous wave
dB	Decibel
EIS	Environmental impact Statement
ESME	Effects of sound on the marine environment
FM	Frequency modulation
Hz	Hertz (cycles per second)
IW	Internal wave
KdV	Korteweg de Vries (equation)
KHz	Kilohertz
MHz	Megahertz
ML	Mixed layer
N×2-D	N by two dimensions
NLIW	Nonlinear internal wave
OASIS	Ocean Acoustic Services and Instrumentation Systems
ONR	Office of Naval Research
PE	Parabolic equation
PRIMER	Name of a coastal acoustics experiment, not an acronym per se
PRN	Pseudorandom number
REMUS	Remote environmental measuring units
SAX	Sediment acoustics experiment
SI	Scintillation index
SL	Source level
SNR	Signal to noise ratio
S/R	Source-to-receiver
SW	Shallow water
TL	Transmission loss
UN	United Nations
WHOI	Woods Hole Oceanographic Institution
WKB	Wentzel–Kramers–Brillouin
2-D, 3-D	Two and three dimensional

LIST OF SYMBOLS IN EQUATIONS

$A(t)$	Amplitude as a function of time
A_{SW}	Amplitude of a surface wave
a_n	Amplitude of the nth acoustic normal mode
A_{mn}	Mode coupling coefficient
b	Time constant of an FM sweep pulse
B_{mn}	Mode coupling coefficient
c	Speed of sound in water
D	Distance; depth of mixed layer
f	Acoustic frequency
g	Depth-dependent Green's function
H	Water depth
H_{IW}	Amplitude of a nonlinear internal wave
I	Acoustic intensity
k	Acoustic wavenumber
k_{SW}	Wave number of a surface wave
k_n	Horizontal wave number of the nth acoustic normal mode
$L_{\pm}$	Scattered path lengths
m	Power law exponent for m-sequence
n	Index of refraction
n_m	Index of refraction of the mth mode
N_c	Number of complete ray cycles
p	Acoustic pressure
Q	$\Delta\omega/\omega_0$ (bandwidth of system)
R_F	Fresnel zone radius
R_B	Plane wave reflection coefficient of bottom
$S(f)$	Power spectral density
t	Time
u_n	Normal mode function
U_{ml}	Same as B_{ml} coupling matrix
v^{ph}	Phase velocity
v^{gr}	Group velocity
x,y,z	Usual Cartesian coordinates
β	Waveguide invariant
γ_n	Vertical modal wavenumber; modal attenuation coefficient
δ_{mn}	Kronecker delta
Δ	Small change in a quantity, as in Δc
Δ_{SW}	Depth disturbance due to surface wave
Δk_{ml}	Inverse of mode cycle distance
ε	Small slope parameter
$\theta(t)$	Phase as a function of time
θ_{crit}	Critical reflection angle
θ_g	Grazing angle
$\overline{\kappa}$	Nonadiabacity parameter
λ	Acoustic wavelength
λ_0	Horizontal correlation length
ρ	Distance along an array; density

σ_h	Root mean square surface height
σ_{sl}	Root mean square surface slope
φ	Acoustic phase
φ_{SW}	Phase of a surface wave
φ_n	Nth acoustic normal mode
Φ	Signal phase variability
$\Phi_m(r)$	Radial modal solution
X	Grazing angle of a ray
$\Psi_m(r,z)$	Local modes in range-dependent mode theory
ω	Angular acoustic frequency
Ω_b	Extinction cross section
$\langle \rangle$	Ensemble average

REFERENCES

[1] U.N. Atlas of the Oceans, http://www.oceansatlas.org, 2015.

[2] Urick, R.J., *Principles of Underwater Sound*, Peninsula Publishing, Los Altos, pp. 1–3, 1983.

[3] Wikipedia article, *Acoustic Doppler current profiler*, http://en.wikipedia.org/wiki/Acoustic_Doppler_current_profiler, 2015.

[4] Wikipedia article, *Current meter*, http://en.wikipedia.org/wiki/Current_meter, 2015.

[5] Chen, K., Gawarkiewicz, G.G., Lentz, S.J. and Bane, J.M., *Diagnosing the warming of the Northeastern U.S. Coastal Ocean in 2012: A linkage between the atmospheric jet stream variability and ocean response*, J. Geophys. Res. Oceans, **119**, 2014; http://dx.doi.org/10.1002/2013JC009393.

[6] Wikipedia article, *Multibeam echosounder*, http://en.wikipedia.org/wiki/Multibeam_echosounder, 2015.

[7] Chirp sonar sub-bottom profiling is a commercial technology. A representative sample of the many surveying products available can be found on the Kongsberg web site, at: http://www.km.kongsberg.com/ks/web/nokbg0240.nsf/AllWeb/1AE8CC56C6F31E5 1C1256EA8002D3F2C, 2015.

[8] Holmes, J.D., Carey, W.M. and Lynch, J.F., *Results from an autonomous underwater vehicle towed hydrophone array experiment in Nantucket Sound*, J. Acoust. Soc. Am., **120**, (2), pp. EL15–EL21, 2006.

[9] Richardson, W.J., Greene, C.R., Malme, C.I. and Thomson, D.H., *Marine Mammals and Noise*, Academic Press, San Diego, 1995.

[10] Gisiner, R., Harper, S., Livingston, E. and Simmen, J., *Effects of sound on the marine environment (ESME): an underwater noise risk model*, IEEE J. Ocean. Eng., **31**, (1), pp. 4–7, 2006; http://dx.doi.org/10.1109/JOE.2006.872212.

[11] Munk, W., Worcester, P. and Wunsch, C., *Ocean Acoustic Tomography*, Chapter 5. Cambridge University Press, New York, 1995.

[12] Duda, T.F., *Analysis of finite-duration wide-band frequency sweep signals for ocean tomography*, IEEE J. Ocean. Eng. **18**, (2), 1993; http://dx.doi.org/10.1109/48.219528.

[13] Clay, C.S. and Medwin, H., Acoustical Oceanography: Principles and Applications, Appendix A9, J. Wiley and Sons, New York, 1977.

[14] Duda, T.F., and Collis, J.M., *Acoustic field coherence in four-dimensionally variable shallow water environments: Estimation using co-located horizontal and vertical line arrays*, in Proceedings of 2nd International Conference on Underwater Acoustic Measurements: Technologies and Results, J. S. Papadakis and L. Bjorno, eds, 2007.

[15] Newhall, A.E., Lin, Y.-T. Lynch, J.F., Baumgartner, M.F. and Gawarkiewicz, G.G., *An acoustic normal mode approach for long distance passive localization of vocalizing Sei whales on a continental shelf*, J. Acoust. Soc. Am., **131**, (1), pp. 1814−1825, 2012.

[16] Katznelson, B.J., Petnikov, V., and Lynch, J.F., *Fundamentals of Shallow Water Acoustics*, ONR series book, Springer Verlag, New York, 520 pages, 2012.

[17] Frisk, G.V. and Lynch, J.F., *Shallow water waveguide characterization ssing the Hankel transform*, J. Acoust. Soc. Am., **75**, 205, 1984.

[18] Lynch, J.F., *On the use of focused horizontal arrays as mode separation and source location devices in ocean acoustics. Part I: Theory*, J. Acoust. Soc. Am., **74**, pp. 1406−1417, 1983.

[19] Lynch, J.F., Schwartz, D. and Sivaprasad, K., *On the use of focused horizontal arrays as mode separation and source location devices in ocean acoustics. Part II: Theoretical and numerical modeling results*, J. Acoust. Soc. Am., **78**, (2), pp. 575−586, 1985.

[20] Jensen, F.B., Kuperman, W.A., Porter, M.B. and Schmidt, H., *Computational Ocean Acoustics*, AIP Press, New York, 1994.

[21] Brekhovskikh, L.M. and Lysanov, Y.P., *Fundamentals of Ocean Acoustics,* Springer Verlag, 270 pp., 1991.

[22] Ogilvie, J.A., *Theory of Wave Scattering From Random Rough Surfaces*, Institute of Physics Publishing Ltd., Bristol, 1991.

[23] Pekeris, C.L., *Theory of propagation of explosive sound in shallow water*, Geological Society of America Memoirs, **27**, pp. 1−116, 1948.

[24] Frisk, G.V., *Ocean and Seabed Acoustics: A Theory of Wave Propagation*, PTR Prentice Hall, Upper Saddle River, 1994.

[25] The normal mode code ORCA described at the Applied Research Laboratories, University of Texas website: http://www.arlut.utexas.edu/esl/modeling.html, 2015.

[26] Koch, R.A., Penland, C., Vidmar, P.J. and Hawker, K.E., *On the calculation of normal mode group velocity and attenuation*, J. Acoust. Soc. Am., **73**, 820, 1983.

[27] Rajan, S.D., Lynch, J.F. and Frisk, G.V., *Perturbative inversion methods for obtaining bottom geoacoustic parameters in shallow water*, J. Acoust. Soc. Am., **82**, pp. 998−1017, 1987.

[28] Lynch, J.F., Lin, Y.-T., Duda, T.F. and Newhall, A.E., *Acoustic ducting, reflection, refraction, and dispersion by curved nonlinear internal waves in shallow water*, IEEE J. Ocean. Eng., **35**, pp. 12−27, 2010.

[29] Pierce, A.D., *Extension of the method of normal modes to sound propagation in almost-stratified medium*, J. Acoust. Soc. Am., **37**, pp. 19−27, 1965.

[30] Weinberg, H. and Burridge, R. *Horizontal ray theory for ocean acoustics*, J. Acoust. Soc. Am., **55**, 63, 1974.

[31] Oba, R. and Finette, S., *Acoustic propagation through anisotropic internal wave fields: transmission loss, cross-range coherence, and horizontal refraction*, J. Acoust. Soc. Am., **111**, 769, 2002.

[32] Katznelson, B.G. and Pereselkov, S.A., *Horizontal Refraction of Low Frequency Sound Field due to Soliton Packets in Shallow Water*, Acoust. Phys., 2000.

[33] Badiey, M., Katznelson, B.G., Lynch, J.F., Pereslkov, S.A. and Siegmann, W.L., *Measurement and modeling of three-dimensional sound intensity variations due to shallow-water internal waves*, J. Acoust. Soc. Am., **117**, 613, 2005.

[34] Duda, T.F., Collis, J.M., Lin, Y.-T., Newhall, A.E., Lynch, J.F. and DeFerrari, H.A., *Horizontal coherence of low-frequency fixed-path sound in a continental shelf region with internal-wave activity*, J. Acoust. Soc. Am., **131**, 1782, 2012.

[35] Badiey, M., Katznelson, B.G., Lin, Y.-T. and Lynch, J.F., *Acoustic multipath arrivals in the horizontal plane due to approaching nonlinear internal waves*, J. Acoust. Soc. Am., **129**, EL141, 2011.

[36] Tappert, F.D., *The Parabolic Approximation Method*, in Wave Propagation and Underwater Acoustics, ed. by J.B. Keller and J.S. Papadakis, Lecture Notes in Physics 70, Springer Verlag, New York, 1977.

[37] Lee, D., Botseas, G. and Siegmann, W.L., *Examination of three-dimensional effects using a propagation model with azimuth-coupling capability (FOR3-D)*, J. Acoust. Soc. Am., **91**, 3192, 1992.

[38] Lin, Y.-T., Collis, J.M. and Duda, T.F., *A three-dimensional parabolic equation model of sound propagation using higher-order operator splitting and Padé approximants*, J. Acoust. Soc. Am., **132**, EL364, 2012.

[39] Lin, Y.-T., Duda, T.F, Emerson, C., Gawarkiewicz, G.G., Newhall, A.E., Calder, B., Lynch, J.F., Abbot, P., Yang Y.-J. and Jan, S., *Experimental and numerical studies of sound propagation over a submarine canyon northeast of Taiwan*, IEEE J. Ocean. Eng., **40**, (1), pp. 237–249, 2014.

[40] Tindle, C.T. and Stamp, G., *Virtual modes and the surface boundary condition in underwater acoustics*, J. Sound Vibr., **49**, pp. 231–240, 1976.

[41] Heaney, K.D., *Rapid geoacoustic characterization using a surface ship of opportunity*, IEEE J. Ocean. Eng., **29**, pp. 88–99, 2004.

[42] Wenz, G.M., *Acoustic ambient noise in the ocean: Spectra and sources*, J. Acoust. Soc. Am., **34**, pp. 1936–1956, 1962.

[43] Kuperman, W.A. and Ingenito, F., *Spatial correlation of surface generated noise in a stratified ocean*, J. Acoust. Soc. Am., **67**, pp. 1988–1996, 1980.

[44] Diachok, O., *Bioacoustic resonance absorption spectroscopy*, J. Acoust. Soc. Am., **103**, 3068, 1998.

[45] Steinberg, B.D., *Principles of Aperture and Array System Design: Including Random and Adaptive Arrays*, New York, Wiley-Interscience, 374 pp., 1976.

[46] Goff, J.A. and Jordan. T.H., *Stochastic modeling of seafloor morphology: Inversion of sea beam data for second order statistics*, J. Geophys. Res., **93**, (B11), pp. 13589 – 13608, 1988.

[47] Nielsen, P., *Coastal Bottom Boundary Layers and Sediment Transport*, World Scientific, New Jersey, 324 pp., 1992.

[48] Richardson, M.D., Briggs, K.B., Reed, A.H., Vaughn, W.C., Zimmer, M.A., Bilbee, L.D., and Ray, R.I., *Characterization of the environment during SAX04: Preliminary results*, Proceedings of the International Conference, Underwater Acoustic Measurements: Technologies &Results, Heraklion, Crete, Greece, 2005.

[49] Pitcher, T.J., *Heuristic definitions of fish shoaling behavior*, Anim. Behav., **31**, pp. 611–613, 1983.

[50] Leighton, T.G., *The Acoustic Bubble*, Academic Press, New York, 613 pp., 1994.

[51] Henyey, F.S., *Acoustic scattering from ocean microbubble plumes in the 100 Hz to 2 kHz region*, J. Acoust. Soc. Am., **90**, pp. 399–405, 1991.

[52] Weston, D.E., *Propagation in water with uniform sound velocity but variable-depth lossy bottom*, J. Sound Vibr., **47**, pp. 473–483, 1976.

[53] Akal, T., *The relationship between the physical properties of underwater sediments that affect bottom reflection*, Mar. Geol., **13**, pp. 251–266, 1972.

[54] Kuperman, W.A. and D'Spain, G.L., *Ocean Interference Phenomena and Signal Processing*, AIP Press, 277 pp., 2002.

[55] Harrison, C.H. and Harrison, J.H., *A simple relationship between frequency and range averaging for broadband sonar*, J. Acoust. Soc. Am., **97**, pp. 1314–1317, 1995.

[56] Dyer, I., *Statistics of sound propagation in the ocean*, J. Acoust. Soc. Am., **48**, pp. 337–345, 1970.

[57] Creamer, D., *Scintillating shallow water waveguides*, J. Acoust. Soc. Am., **99**, pp. 2825–2838, 1996.

[58] Fredericks, A., Colosi, J.A., Lynch, J.F., Gawarkiewicz, G.G., Chiu, C.-S., and Abbot, P., *Analysis of multipath scintillations from long range acoustic transmissions on the New England continental slope and shelf*, J. Acoust. Soc. Am., **117**, pp. 1038–1057, 2005.

[59] Makris, N.C., *The effect of saturated transmission scintillation on ocean acoustic intensity measurements*, J. Acoust. Soc. Am., **100**, pp. 769–783, 1996.

[60] Colosi, J.A. and Baggeroer, A.B., *On the kinematics of broadband multipath scintillation and the approach to saturation*, J. Acoust. Soc. Am., **116**, pp. 3515–3522, 2004.

[61] Bowden, K.F., *Physical oceanography of coastal waters*, Ellis Horwood (pub), Chichester, 302 pp., 1983.

APPENDIX 7.A1

A Simple Adiabatic Normal Mode, Range-Dependent Waveguide Model and Its Uses

Let us use the rigid/hard bottom, isovelocity water column modes of Eqs. (7.28)–(7.30) in a range-dependent waveguide, where the depth is given by the background depth H_0 plus a sinusoidal disturbance due to a surface wave, i.e.,

$$H(r) = H_0 + A_{SW}\sin(k_{SW}r + \varphi_{SW}) \equiv H_0 + \Delta_{SW}(r) \qquad (A7.1)$$

We assume (for now) that the range dependence is weak, so that we can use adiabatic, not coupled, mode theory to describe the pressure field. For this case, we write

$$p(r,z) = \sum_n \frac{1}{\sqrt{k_n(r)r}}\varphi_n(z_s)\varphi_n(z)e^{i\int k_n(r')dr'} \equiv \sum_n a_n e^{i\varphi_n} \qquad (A7.2)$$

where the normalized modes are given as

$$\varphi_n(z) = \sqrt{\frac{2}{H_0 + \Delta_{SW}(r)}}\sin(\gamma_n z) \qquad (A7.3)$$

The vertical and horizontal modal wave numbers in Eq. (A7.3) are given by

$$\gamma_n(r) = \frac{\left(n - \frac{1}{2}\right)\pi}{H_0 + \Delta_{SW}(r)} \qquad (A7.4)$$

and

$$k_n(r) = \left(\frac{\omega^2}{c^2} - \gamma_n^2 \right)^{1/2} \tag{A7.5}$$

This waveguide model has a lot of uses as an example. First, it is trivially implemented on a computer, especially if one has access to a mathematics package such as MATLAB or MATHEMATICA. Secondly, after making some very simple approximations (e.g., ignoring small terms in the denominators), many of the quantities of interest are derivable analytically. Thirdly, one can insert antennae (acoustic arrays) into such a simple model waveguide very easily, and explore how they perform. And finally, the numbers that one obtains from this model are actually not so bad when compared to reality.

Hence, let us see what we can do with this simplified waveguide model. In order to be useful to the reader, we provide some of the steps in our examples, and ask the reader to provide the rest as an exercise. These steps are fairly simple.

The first thing one should look at is the "simple field", i.e., the field without the added complications of range dependence. This is done by letting the phase integral in Eq. (A7.2) go to $k_n r$, and also setting any $\Delta_{SW}(r)$ terms equal to zero. The $p(r,z)$ field obtained will consist of real and imaginary components, from which one can obtain amplitudes and phases is the usual way, i.e., $\tan\varphi = Im/Re$ and $A = (Re^2 + Im^2)^{1/2}$.

From just this "background" waveguide, one can look at many useful things. To mention just few here, one can look at how well one can resolve (1) modal arrivals in time versus range; (2) modes with a vertical array versus $\{r, N, z_i, SNR\}$; (3) modes with a horizontal array versus $\{r, N, z_{array}, SNR\}$; (4) modes with a combined horizontal array and travel time; and (5) modes with a combined vertical array and travel time. This is quite a list, but there is much more than can be done, so in some ways it is just representative. We also note that there are some things that this model does not do, such as attenuation, the continuum, and (obviously) mode coupling. One needs at least the Pekeris waveguide modes for the first two, and a fully coupled code for the third.

Let us first look at these five items based on the simple, range-independent version of our model, where we set $\Delta_{SW}(r) = 0$ and so $k_n(r) \rightarrow k_n$. The first task, seeing how well we can resolve mode arrivals in time versus range, can actually be done easily enough analytically (see Frisk's book, Chapter 5), but is also easy to do numerically. For the group velocity of each mode, one just needs to form $v_n^G = (d\omega/dk_n)$ numerically, with a three-point central difference scheme usually being quite sufficient for this. The modal travel time is just $t_n = R/v_n^G$. Assuming some system bandwidth BW, one can see which modes are resolvable in this background waveguide model.

The second task, separating the modes with a vertical array, is actually a chore for which this simple model is well adapted. By making the array element sensitivities match (scale to) the actual mode amplitude, one effectively projects out that mode from the pressure field, which is the sum of all the modes. This is shown in Eqs. (7.7) and (7.8). How to implement this numerically is fairly easy. If one runs the range-independent $p(r,z)$ code without the $e^{ik_n r}$ phase factor (which is just an unnecessary complication in understanding this technique), one obtains

$$p(r, z) = \sum_n \left(\frac{\varphi_n(z_s)}{\sqrt{k_n r}} \right) \varphi_n(r, z) = \sum_n a_n \varphi_n(r, z) \qquad (A7.6)$$

where the bracketed term is identified with a_n. To project out one of the set of modes, say the mth, one just spreads however many hydrophones one has (say I) over the water column (usually uniformly distributed at Nyquist spacing) and then does the numerical integral

$$a_n \sim \int_0^{L \leq H} p(r, z) \varphi_m(r, z) dz \rightarrow \sum_{i=1}^{I} p(r, z_i) \varphi_m(r, z_i) \Delta z_i \qquad (A7.7)$$

Obviously, how one does the numerical integration (e.g., simple quadrature or some other scheme), the length of the vertical array compared to the water column depth H, and the z_i element placing will affect the exact result. This variability in the a_n estimate shows both the numerical and physical errors inherent in the technique.

Another effect that introduces error into the a_n estimate is the finite SNR, which can be examined by introducing noise numerically. This can easily be done in MATLAB by introducing noise into each of the quadrature components of $p(r,z)$. In MATLAB, the statements rnoise = randn(100,1) and imnoise = randn(100,1) will create (100,1) matrices which are 0 mean, normally distributed variables with a variance of 1. These can be used as the first 100 values of the real and imaginary components of the noise field. Note that the noise variance should be scaled to the SL we assume, to get the desired SNR correctly. We leave this to the reader, with the notes that one should be careful of rms noise versus noise variance, and that $|n| = \sqrt{(rnoise)^2 + (imnoise)^2}$, since we are adding quadrature components. Another interesting way to examine the error in the vertical array mode filtering is to look at "modal cross talk," i.e., the amount of amplitude (energy) that leaks into modes other than the mth when one does this projection. To do this, one looks at

$$a_{mn} = \int_0^{L \leq H} \varphi_n(r, z) \varphi_m(r, z) dz \rightarrow \sum_i \varphi_n(r, z_i) \varphi_m(r, z_i) \Delta z_i \qquad (A7.8)$$

where the integral is again approximated numerically. If $L = H$, and Nyquist sampling is used, then $a_{mn} = 0$, but otherwise $a_{mn} \neq 0$. The a_{mn} squared is proportional to the energy in the modal cross talk.

The third topic, horizontal array steering to resolve normal modes, works in the angle/wave number domain. It relies on canceling the phase of the incoming modal plane wave, rather than projecting amplitudes. (And it is even somewhat insensitive to amplitude variations across the array.) In this case, we note that when the modal phases at the array elements, $e^{ik_n r_i}$, are canceled by $e^{-ik_{steer} r_i}$, the array output will be maximized. Specifically, the array output will be

$$\text{OUT} \sim \sum_{j=1}^{J} \sum_{n=1}^{N} a_n e^{ik_n r_i} e^{-ik_{steer} r_i} \qquad (A7.9)$$

which will be a maximum when the phases of the nth mode and the steering phase cancel. One can easily do this sum over the array elements numerically, using k_{steer} as a parameter we vary. We do this using $k_{steer} = k\cos\theta$, and vary the grazing angle theta from 0 to θ_{crit}. When $\theta = \theta_n$, we see a maximum.

The fourth and fifth tasks we can address with our simple range-independent mode model depend on having done the first three, i.e., understanding mode resolution with travel time, a vertical array, and a horizontal array (at end fire). Looking at the fourth task, modal separation with combined travel time and a vertical array, it is most easily (if approximately) done by looking at time and vertical array resolution separately. The set of modes resolved by one technique is *not* the same set as resolved by the other, and the union of the resolved mode sets is thus the answer. In this case, both time and the vertical array do a good job of resolving the low modes (why?), but the combination is not the most effective for resolving high modes. However, modal time resolution is also poorer at close ranges, whereas the vertical array works well there, so there is some advantage versus range.

Turning to the fifth task, i.e., considering time resolution plus a horizontal array, we now have good resolution of the low modes in travel time, and good resolution of the higher modes by the horizontal array, so the system should be more complementary.

Before looking at the "rough surface" variant of these problems, we should note that this toy model is meant, like all toys, for playing with. When looking at our tasks/problems, try varying the center frequency, range, bandwidth, and array configurations to see what happens. In many of these cases, you can also figure out analytically (sometimes via approximations) what is happening numerically. The same will apply to the rough surface example we are looking at next.

The first task to do with the rough surface model is to run it with some suitable set of $\{A_{SW}, k_{SW}, \varphi_{SW}\}$. The amplitude is simple—1 m is a nice round number to start with, and realistic. The phase is also simple—anything between 0 and 2π is fair game, as we look at uniform random phases for our statistics. For the surface wave wave number, we can take some simple results from Bowden's "Physical Oceanography of Coastal Waters" [61], and write

$$k_{SW} = \frac{2\pi}{\lambda_{SW}} = \frac{2\pi}{1.56T^2} \qquad (A7.10)$$

where T is the wave period in seconds. For waves in a coastal region, 5−10 s waves are common, so we can use this range. With this input for the rough surface, we are equipped to code (Eqs. (A7.1)−(A7.5)). Doing this, we can then run the model, and compare it (in magnitude and phase) with the background, "no roughness" model.

We can also look at this via the difference in the models, to see spatially how the roughness induces fluctuations both in the total field and mode by mode. In doing this, again it is worth varying $\{A_{SW}, k_{SW}, \varphi_{SW}\}$ to see what happens.

The first order look at the field with fixed φ_{SW} is very informative, in that it deals with realizations. However, the surface wave field varies quickly in time (over seconds), so if one wants to understand how the $p(r,z)$ field fluctuates, one should next look at the statistics of multiple realizations. We are generally interested in both amplitude and phase/travel time statistics. For amplitude, the easiest way to generate the amplitude (pressure field magnitude) statistics is to run the model multiple times, each time with a random surface wave phase. One then just averages the pressure field magnitude at each point to get the mean and uses the deviations from the mean to get the variance statistics. This can also be done on a mode-by-mode basis, by restricting the sum over modes to an individual mode. Do you find a zero mean fluctuation? Why?

The phase and travel time fluctuations are done similarly, and in this case looking mode by mode is most useful. We define the phase of the nth mode in a given realization to be $\varphi_n = \varphi_n^0 + \Delta\varphi_n$, and also the time wander as $\Delta t_n = \omega\Delta\varphi_n$. In the phase equation, φ_n^0 is the phase one would get in the "zero roughness" case, and $\Delta\varphi_n$ is the phase one gets in a realization that has roughness. The travel time deviation is simply a scalar frequency factor times the phase deviation. Again, one can get the mean and variance of these quantities by taking a number of realizations, and the answers are quite useful for understanding mode resolution by travel time and by array methods. Specifically, the phase variability will affect the horizontal array beam forming, which is done assuming the background model (no roughness) phases, since we do not know the phase realizations. Thus, there is a mismatch between our beam-steering phases and what actually is seen on the array. For the modal amplitude matching by a vertical array, the same type of situation happens. The mode shapes (mode amplitude versus depth) will differ in a rough waveguide from the undisturbed one, the latter of which is used to generate the replicas that are used in the mode projection. As to the travel time, if the rms wander is on the order of or greater than the travel time difference between the unperturbed modes, time resolution becomes unusable. This is because one cannot distinguish which mode one is looking at.

A final topic that should be discussed in dealing with our simple model with roughness is the importance of including mode coupling. Eqs. (7.54)–(7.58) describe the criterion for having "significant" coupling, and given the A_{SW} and k_{SW} one chooses, and the acoustic frequency and waveguide parameters, one can use these equations to test if the inclusion of coupling is needed. Whether it is or not, the adiabatic solution is usually a good first approximation to the field for most cases.

The Seafloor

M.D. Richardson[1], D.R. Jackson[2]

*Marine Geosciences Division, Naval Research Laboratory,
Stennis Space Center, MS, United States[1];
Applied Physics Laboratory, University of Washington,
Seattle, WA, United States[2]*

Definitions

An understanding of seafloor properties and processes is essential for the solution of many problems in underwater acoustics where acoustic—seafloor interactions dominate. This includes an understanding of acoustic propagation, reflection, and scattering models as well as acoustic characterization of seafloor properties. Applied bottom-interacting acoustic problems should employ the simplest seafloor models and sediment characterization that provide results with the accuracy required for the problem. In this chapter, we restrict the discussion to acoustic frequencies between few hundreds of Hz to several hundred kHz, with an emphasis on higher frequencies. Studies of acoustic—seafloor interaction over this range are important for scientific fields associated with acoustical oceanography (e.g., underwater archeology, sediment acoustic classification, bathymetry, marine habitat and fisheries studies, and sediment transport) and underwater acoustics (e.g., antisubmarine warfare, mine warfare, object—such as unexploded ordinance, detection and classification, sediment geoacoustics, seafloor scattering, acoustic propagation, and acoustic communications). These areas can be broadly classified in terms of forward and inverse modeling and include both active and passive sonar systems. The inverse models provide acoustical methods to characterize the environment (acoustical oceanography), whereas forward models utilize inputs from the environment to predict acoustic propagation within the seafloor, scattering from the seafloor, and penetration of acoustic energy into the seafloor (underwater acoustics). Sediment properties of primary interest include sediment mass density, sound speed and attenuation of compressional, shear, and interface waves propagating in sediments, and the statistics of seafloor roughness and volume heterogeneity. Depending on the acoustic frequency the scales of interest in these seafloor properties range from millimeters to tens of meters. Values of these seafloor properties are often lumped into generalized categories based on sediment type.

Applied Underwater Acoustics. http://dx.doi.org/10.1016/B978-0-12-811240-3.00008-4

8.1 BACKGROUND AND HISTORY

The simplest bottom interaction models (e.g., reflection, refraction, and bottom loss) assume the seafloor is a homogenous absorptive fluid with flat interfaces. The seafloor and overlying water column can therefore be characterized by their bulk or mass density and compressional wave speed and attenuation. If sediments support shear waves, elastic propagation models are needed and shear wave speed (or elastic moduli) and attenuation should be characterized. Slightly more complex models require the gradients of bulk density, compressional and shear wave speed and attenuation in multiple sediment layers, as well as geoacoustic properties of the basement. Sandy sediments are porous, and the out-of-phase movement of sand grains and pore fluid can generate a second compressional "slow" wave and frequency dependence in shear and compressional wave speeds, especially at higher acoustic frequencies. Wave attenuations are no longer linear with frequency as assumed in typical viscoelastic models. Frequency-dependent poroelastic behavior requires additional characterization of pore properties such as pore size, permeability, and tortuosity. The seafloor is rarely flat and bottom roughness can lead to frequency-dependent scattering or bottom reverberation. Scattering can occur from large-scale features such as sand ridges, coral reefs, shell deposits, or other hard grounds to fine-scale roughness associated with local hydrodynamic processes, such as sand ripples or biogenic features, including mounds, pits, or other feeding traces. Volume heterogeneity within the sediment can also lead to scattering and reverberation. Heterogeneity includes fluctuations and gradients in sediment bulk density and sound speed, features comparable in size to the acoustic wavelength (such as rocks or shells), and free gas bubbles.

Acoustics has played a major role in seafloor research since the commercial development of echo sounders (fathometers) in the 1920s. Systematic acoustic surveys prior to World War II led to the discovery of numerous seafloor features such as canyons, seamounts, and the mid-oceanic ridge. After World War II, the development of sonar systems was driven largely by anti-submarine warfare (ASW) and more recently mine counter measures (MCM). Acoustic bottom interactions affect passive and active submarine detection and detection of bottom mines on or within the seafloor. Acoustic methods are often used for characterizing underwater environments.

Edwin Hamilton developed his classic approach for development of empirical geoacoustic models in support of ASW during the 1960s and 1980s [1,2]. The parameters in these models include the values and gradients of compressional and shear wave speed and attenuation and bulk density of sediment layers down to and including the basement [3]. More recent interest in high-frequency shallow-water acoustic modeling in support of mine countermeasures has required more detailed seafloor characterization including bottom roughness, volume heterogeneity, and frequency-dependent wave speed and attenuation characterization [4].

The simplest approach is to develop empirical bottom acoustic models based on sediment type (e.g., rock, sand, mud, and muddy sand). Examples include bottom backscattering strength versus grazing angle and bottom loss. Although this approach may in some cases provide adequate predictions for applied applications, a thorough understanding of available physics-based acoustic models and the methods and terminology associated with characterization of sediment properties should be undertaken.

The following sections include descriptions of models for acoustic propagation within sediments and practical models for seafloor refraction, scattering, and bottom loss. The next section discusses the methods and terminology used to characterize sediment physical properties with an emphasis on properties that are most important and can be easily obtained for applied acoustic modeling. The measurement of sediment geoacoustic properties, seafloor roughness, and sediment volume heterogeneity are described in the subsequent sections. This chapter concludes with a discussion of inverse acoustic methods that can be used to characterize the seafloor.

8.2 **THE ORIGIN AND NATURE OF SEAFLOOR SEDIMENTS**

This discussion of the origin and nature of seafloor sediments is restricted to the upper tens of meters of the seafloor where seafloor—acoustic interactions have been studied. The origin, distribution, and nature of seafloor sediments is described in classic texts on marine geology of the seafloor by Kennett [5] and Seibold and Berger [6]. Studies of high-frequency (>10 kHz) seafloor—acoustic interactions in shallow water were summarized in Jackson and Richardson [4]. Underwater acoustic research during the 1950s through 1970s was based on ASW requirements. Edwin Hamilton [3] provided the types of seafloor characterization needed to predict these lower-frequency acoustic seafloor interactions. Seafloor properties required to predict acoustic propagation, reflection, and scattering include depth gradients of sediment density, sound speed and attenuation, and secondarily shear wave speed and attenuation. Sediment types, often characterized by particle grain size and origin were correlated with these sediment properties for acoustic modeling. Seafloor roughness scales of interest were larger-scale, hard-bottom, such as mid-ocean, ridges. Scales of heterogeneity of interest were primarily related to basin-scale changes in depositional processes or a result of turbidly currents. Sediment properties tend to be static over the timescales of interest for these naval applications. In spite of all the published studies, the bathymetry and geomorphology of much of the deeper-ocean seafloor (>100 m) is still poorly known [7—9].

The seafloor properties needed to predict high-frequency acoustic interactions with the seafloor include the same physical and geoacoustic properties; however, more detailed descriptions of seafloor roughness and heterogeneity are needed. In shallower water, seafloor near-surface properties can rapidly change in response to hydrodynamic and biological processes and events. Accurate description of the shallow-water seafloor may require repeated surveys or process-based seafloor modeling.

8.3 **ACOUSTICS OF SEDIMENTS**

Propagation of sound in sediments is an active field of research. The three particular models that have received wide use are described here. In increasing order of complexity these are the fluid, elastic, and poroelastic models.

8.3.1 FLUID MODEL

Derivations of the equations for the fluid model can be found in Refs. [10−12]. This model is appropriate for clayey and silty sediments, although it is often applied to sands as well. Propagation is described by the Helmholtz equation for pressure,

$$\nabla^2 p + \frac{\omega^2}{c_p^2} p = 0. \tag{8.1}$$

The pressure units may be chosen for convenience, for example, pascal or micro pascal. It is assumed that pressure has an $exp(-i\omega t)$ time dependence. The sound speed in the sediment is denoted c_p (unit: m/s), with the subscript p taken from the seismic convention for the P-wave. In the following this wave is referred to either as the *sound wave* or the *compressional* wave. The ratio $k_p = \omega/c_p$ (unit: 1/m) is the wave number, and losses can be included by allowing either the sound speed or wave number to be complex. Defining real and imaginary parts of the wave number as follows:

$$k_p = k_p' + ik_p'', \tag{8.2}$$

it is convenient to define the dimensionless loss parameter:

$$\delta_p = k_p'' \big/ k_p' \tag{8.3}$$

Commonly encountered alternative parameters are loss in decibel per wavelength traveled (dimensionless):

$$\alpha_\lambda = \frac{40\pi\delta_p}{log_e 10}, \tag{8.4}$$

and attenuation, defined as loss in decibel per distance traveled (unit: dB/m):

$$\alpha = \frac{40\pi f \delta_p}{v_p c_w log_e 10}, \tag{8.5}$$

where f is the frequency (unit: Hz).

The complete acoustic description of a fluid sediment is given in terms of sound speed, attenuation, and mass density, ρ (unit: kg/m^3). Measurements of sound speed are typically expressed as the dimensionless ratio, v_p of sediment sound speed to the sound speed, c_w, in the overlying water. Density can be expressed by the analogous ratio, a_ρ. Sound speed, density, and attenuation determine the reflection coefficient for a plane wave at a flat fluid−sediment interface. This is often called the "Rayleigh" reflection coefficient and is given by the expression:

$$R = \frac{z_p - 1}{z_p + 1}, \tag{8.6}$$

where z_p is referred to as the normalized impedance (dimensionless) and is the ratio of angle-dependent acoustic impedances for the two media:

$$z_p = a_\rho k_{zw} / k_{zp}, \tag{8.7}$$

and where k_{zw} and k_{zp} are the z-components of the wave vectors (unit: 1/m) defining the plane waves in the water and sediment, respectively. It is assumed that the plane wave is incident from within the water and is reflected back into the water. In the water

$$k_{zw} = k_w \sin \theta, \tag{8.8}$$

where k_w is the wave number in water, and θ is the grazing angle (the angle between the incoming direction and the horizontal). As a matter of convenience, k_{zw} is taken to be positive, even though the incoming direction is downward. From Snell's law, k_{zp} is

$$k_{zp} = \sqrt{k_p^2 - k_w^2 \cos^2 \theta}. \tag{8.9}$$

Note that k_{zp} is generally complex and, in keeping with the convention adopted for k_{zw}, the sign of the square root is taken so that both real and imaginary parts are positive. It follows that the Rayleigh reflection coefficient is complex. It is often assumed that sound speed is independent of frequency and that attenuation increases linearly with frequency. As is seen in the discussion of the poroelastic model, these are approximations, although often good ones. Fig. 8.1 shows the absolute value of the complex reflection coefficient for parameter choices appropriate to sand and mud seafloors. Sand is an example of a "fast" seafloor, with sediment sound speed greater than that of the overlying water. In the lossless case, this results in perfect reflection ($|R| = 1$) for angles less than the *critical angle*, $\theta_{crit} = \cos^{-1}(1/v_p)$. In the lossy case, the reflection coefficient is still near unity for grazing angles smaller than the critical angle. For "slow" seafloors with

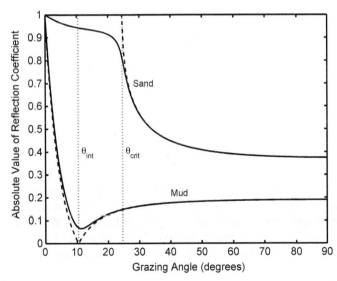

FIGURE 8.1

Reflection coefficients for sand and mud seafloors. The parameters for sand are $v_p = 1.1$, $a_p = 2.0$. The parameters for mud are $v_p = 0.98$, $a_p = 1.5$. For each of these two examples the lossless case ($\delta_p = 0$) is shown as a *dashed line*, and the lossy case ($\delta_p = 0.01$) is shown as a *solid line*. The critical and intromission angles are indicated by *dotted vertical lines*.

From Jackson, D.R. and Richardson, M.D, High-Frequency Seafloor Acoustics. *Springer, 2007, with permission of Springer.*

$v_p < 1$, there is no critical angle. However, the reflection coefficient vanishes in the lossless case at the *intromission angle* $\theta_{\text{int}} = \cos^{-1}\sqrt{\left(a_\rho^2 - 1/v_p^2\right)/\left(a_\rho^2 - 1\right)}$.

Reflection by the seafloor is often quantified by "bottom loss." For a perfectly flat seafloor, the bottom loss (unit: dB) is

$$BL = -20 \log_{10}|R|. \tag{8.10}$$

Measured bottom loss is likely to be larger than predicted by theoretical expressions such as (8.6) and (8.10) due to scattering of sound by seafloor roughness and heterogeneity.

8.3.2 ELASTIC MODEL

Discussions of the theory of wave propagation in elastic media can be found in Refs. [13,14]. Sediments support shear waves that are not included in the fluid model. In such "elastic" media, pressure is not a sufficient descriptor of the forces involved in wave propagation. Instead, the vector displacement of the medium as a function of position can be used as a field variable. The mathematical description of an elastic medium can be simplified if it is isotropic, in which case wave speeds do not depend upon the direction of propagation. In the isotropic case it is convenient to describe the displacement (unit: m) field in terms of scalar and vector potentials (unit: m^2):

$$u(r) = \nabla\phi(r) + \nabla \times \psi(r). \tag{8.11}$$

In a homogeneous medium, these potentials obey Helmholtz equations:

$$\nabla^2 \phi + \frac{\omega^2}{c_p^2}\phi = 0, \tag{8.12}$$

$$\nabla^2\psi + \frac{\omega^2}{c_t^2}\psi = 0. \tag{8.13}$$

Here c_p and c_t are the speeds of compressional ("P") waves and shear ("S") waves, respectively. The particle displacement is parallel to the direction of propagation for compressional waves and orthogonal to the direction of propagation for shear waves (at least in isotropic media). These wave "polarizations" can be deduced from Eq. (8.11). The wave speeds depend upon density and elastic moduli as follows:

$$c_p = \sqrt{\frac{\lambda + 2\mu}{\rho}}, \tag{8.14}$$

$$c_t = \sqrt{\frac{\mu}{\rho}}, \tag{8.15}$$

where λ and μ are called the Lamé parameters (unit: Pa) and ρ is the sediment mass density (unit: kg/m^3). Losses can be introduced by allowing the wave speeds or corresponding wave numbers to be complex, as in the fluid case. The plane-wave reflection coefficient is

$$R = \frac{z_e - 1}{z_e + 1}. \tag{8.16}$$

Bottom loss is again given by Eq. (8.10). In Eq. (8.16)

$$z_e = z_p \cos^2 2\theta_t + z_t \sin^2 2\theta_t \tag{8.17}$$

is dimensionless, as are

$$z_p = \frac{a_\rho k_{zw}}{k_{zp}}, \tag{8.18}$$

and

$$z_t = \frac{a_\rho k_{zw}}{k_{zt}}. \tag{8.19}$$

In these equations, k_{zt} is defined analogously to k_{zp} in the fluid case, but using the shear wave speed. The angle θ_t is the grazing angle of the shear wave in the sediment, obtained using Snell's law, from which one finds

$$\cos 2\theta_t = 2 \left(\frac{c_t}{c_w}\right)^2 \cos^2 \theta - 1, \tag{8.20}$$

where θ and c_w are the grazing angle and sound speed in the overlying water. The shear wave speed, c_t, is complex, as is c_p.

Analogous to the fluid case, elastic seafloors can be characterized by the density ratio, a_ρ, the ratio of compressional wave speed to water sound speed, ν_p, the ratio of shear wave speed to water sound speed, ν_t, and the corresponding loss parameters, δ_p and δ_t. The complex speeds are

$$c_p = c_w \nu_p / (1 + i\delta_p), \tag{8.21}$$

$$c_t = c_w \nu_t / (1 + i\delta_t). \tag{8.22}$$

Fig. 8.2 compares fluid and elastic computations of reflection coefficient for sand. The differences are small due to the small shear wave speed in sand. For rock, with a much larger shear wave speed, the difference between the fluid and elastic models is appreciable as shown in Fig. 8.3.

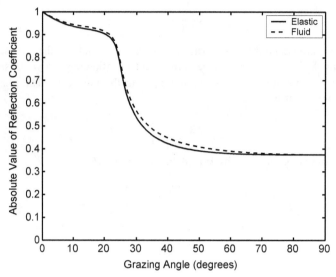

FIGURE 8.2

Magnitude of the complex reflection coefficient for a sand seafloor comparing the elastic and fluid models. The parameters are $\nu_p = 0.1$, $a_p = 2.0$, $\delta_p = 0.01$, $\nu_t = 0.167$, and $\delta_t = 0.01$.

From Jackson, D.R. and Richardson, M.D, High-Frequency Seafloor Acoustics. Springer, 2007, with permission of Springer.

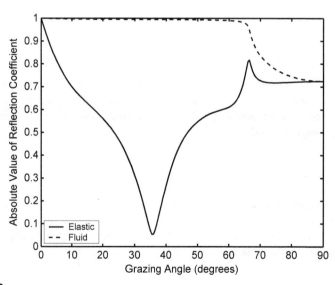

FIGURE 8.3

The magnitude of the complex reflection coefficient for a rock seafloor comparing the elastic and fluid models. The parameters are $\nu_p = 2.5$, $a_p = 2.5$, $\delta_p = 0.02$, $\nu_t = 0.167$, and $\delta_t = 0.1$.

From Jackson, D.R. and Richardson, M.D, High-Frequency Seafloor Acoustics. Springer, 2007, with permission of Springer.

An interesting extension of the elastic model has been developed by Buckingham [15]. In this approach a loss mechanism due to grain slippage is introduced, giving expressions for sound speed and attenuation as functions of frequency in terms of a small number of measured parameters.

8.3.3 POROELASTIC MODEL

Sandy sediments are porous, and generally the sand grains and interstitial fluid do not move together. Biot's poroelastic theory [16−18] has been applied to sands and predicts significant departures from the fluid model [19]. Poroelastic theory is an ambitious approach, since it requires a rather detailed description of the medium structure and predicts the frequency dependence of sound speed and attenuation. It appears that poroelastic theory may provide a reasonable picture of the acoustic behavior of sands. However, differences between the model and data have motivated several recent efforts to either modify poroelastic theory or incorporate porosity effects in other models [20−24]. At this time a consensus has not been reached. In spite of these difficulties, it is likely that future acoustic models for sand will incorporate elements of poroelastic theory. Thus, it is useful to summarize results obtained when this theory is applied to sandy sediments. A full theoretical development is not provided, since the relevant equations are rather complicated, and previously given references provide the necessary background.

In a poroelastic material, there are two longitudinal waves. The "fast" wave is analogous to the fluid model acoustic wave and the elastic model compressional wave. In addition, there is a "slow" wave that has no counterpart in the other models. The speeds and attenuations of these waves can be found from the following expression for the complex wave number (unit: 1/m):

$$k_q = \frac{-b \pm \sqrt{b^2 - 4ac}}{2a}, \quad q = 1, 2. \tag{8.23}$$

The "+" sign corresponds to the fast wave ($q = 1$) and the "−" sign corresponds to the slow wave ($q = 2$). There is also a shear wave due to the slight rigidity of the "frame" comprised of sand grains in contact with each other. This wave has the speed c_t (unit: m/s):

$$c_t = \sqrt{\frac{\rho - \omega \rho_w / \Omega}{\mu}}. \tag{8.24}$$

Some variables in these expressions are secondary parameters, dependent on parameters characterizing the medium through the following series of expressions:

$$a = C_H^2 C_M^2 - C_C^4, \tag{8.25}$$

$$b = -C_H^2 \omega \Omega + 2C_C^2 \omega^2 - \frac{\rho}{\rho_w} C_M^2 \omega^2, \tag{8.26}$$

$$c = \frac{\rho}{\rho_w} \Omega \omega^3 - \omega^4, \tag{8.27}$$

$$C_H^2 = \frac{H}{\rho_w}, \tag{8.28}$$

$$C_M^2 = \frac{M}{\rho_w}, \tag{8.29}$$

$$C_C^2 = \frac{C}{\rho_w}, \tag{8.30}$$

$$\Omega = \omega\frac{\alpha}{\beta} + \frac{iF\eta}{\rho_w\kappa}. \tag{8.31}$$

The complexity of the poroelastic model becomes evident upon the realization that the parameters in Eqs. (8.25)–(8.31) are themselves functions of the material parameters:

$$H = (K_g - K_f)^2/(D - K_f) + K_f + 4\mu/3, \tag{8.32}$$

$$C = K_g(K_g - K_f)/(D - K_f), \tag{8.33}$$

$$M = K_g^2\big/(D - K_f), \tag{8.34}$$

$$D = K_g[1 + \beta(K_g/K_w - 1)], \tag{8.35}$$

$$F = \frac{\varepsilon T/4}{1 - 2iT/\varepsilon}, \tag{8.36}$$

$$T = \frac{-\sqrt{i}J_1\left(\varepsilon\sqrt{i}\right)}{J_0\left(\varepsilon\sqrt{i}\right)}, \tag{8.37}$$

$$\varepsilon = a_0\sqrt{\frac{\omega\rho_w}{\eta}}, \tag{8.38}$$

$$a_0 = \sqrt{\frac{8\alpha\kappa}{\beta}}, \tag{8.39}$$

$$\rho = \beta\rho_w + (1 - \beta)\rho_g. \tag{8.40}$$

Eqs. (8.32)–(8.35) have (unit: Pa). The parameters defined in Eqs. (8.36)–(8.38) are dimensionless. In Eq. (8.37), $J_0\left(\varepsilon\sqrt{i}\right)$ and $J_1\left(\varepsilon\sqrt{i}\right)$ are cylindrical Bessel functions evaluated at complex arguments, and a_0 (unit: m) is a parameter characterizing the size of the pores through which the interstitial fluid flows. In Eq. (8.40), ρ_g is the grain density (unit: kg/m^3). The parameters describing the poroelastic medium are scattered through the equations given earlier and are summarized in Table 8.1. The values in the table are taken from data gathered during SAX99 (Sediment Acoustics eXperiment, 1999) in the Gulf of Mexico in shallow water off the coast of Florida [19].

These parameters can be inserted in Eqs. (8.23)–(8.40) to obtain fast, slow, and shear wave numbers as functions of frequency. These wave numbers are complex with imaginary parts due to viscous losses resulting from fluid flow through the pores of the frame comprised of sand grains. The real and imaginary parts of the

Table 8.1 Biot (Poroelastic) Parameters

Parameter	Symbol	Units	Value
Bulk modulus of grains	K_g	Pa	3.2×10^{10}
Permeability	κ	m^2	2.5×10^{-11}
Tortuosity	α	Dimensionless	1.35
Porosity	β	Dimensionless	0.385
Dynamic viscosity of water	η	kg/m s	0.00105
Mass density of grains	ρ_g	kg/m^3	2690
Bulk modulus of water	K_w	Pa	2.395×10^9
Mass density of water	ρ_w	kg/m^3	1023
Shear modulus of frame	μ	Pa	2.92×10^7
Bulk modulus of frame	K_f	Pa	4.36×10^7

From Williams, K.L., Jackson, D.R., Thorsos, E.I., Tang, D. and Schock, S.G., Comparison of sound speed and attenuation measured in a sandy sediment to predictions based on the Biot theory of porous media. IEEE J. Oceanic Eng., **27**, *pp. 413–428, 2002.*

fast wave number can be used to obtain phase velocity and attenuation that are compared with data from SAX99 in Fig. 8.4. The wave speed data match the model reasonably well except at the lowest frequencies, but the model under-predicts attenuation for frequencies greater than 10 kHz where it tends to follow a linear increase with frequency. The slow wave has such large attenuation and low speed as to make it essentially unobservable in sand.

The plane-wave reflection coefficient in the poroelastic model is a function of frequency and is found using rather complicated expressions [18] that will not be given here. Fig. 8.5 is a comparison between the fluid and poroelastic models for bottom loss, showing that porosity effects are significant in this example.

A simple approach to the application of poroelastic theory to sands has been developed by Williams [25]. This approach is called the effective density fluid model (EDFM) and is based on the observation that the frame moduli are much smaller than the grain and fluid moduli. In this approximation the fast-wave speed is

$$c_1 = \sqrt{\frac{K_{eff}}{\rho_{eff}}}, \tag{8.41}$$

where K_{eff} is the "effective modulus" (unit: Pa)

$$K_{eff} = \left(\frac{1-\beta}{K_g} + \frac{\beta}{K_w}\right)^{-1}, \tag{8.42}$$

and ρ_{eff} is the "effective density" (unit: kg·m^{-3})

$$\rho_{eff} = \frac{\rho\tilde{\rho} - \rho_w^2}{\rho + \tilde{\rho} - 2\rho_w}, \tag{8.43}$$

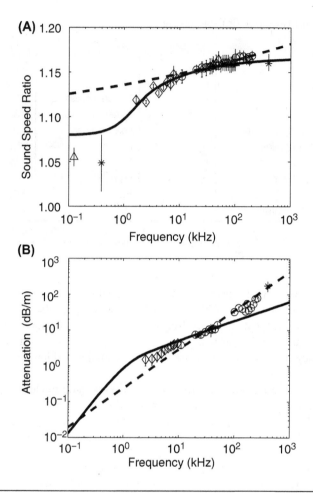

FIGURE 8.4

Comparison of measured sound speed and attenuation with poroelastic models (*solid line* Biot theory; *dashed line* Buckingham model) using parameters of Table 8.1.

From Hefner, B.T. and Williams, K.L., Sound speed and attenuation measurements in unconsolidated glass-bead sediments saturated with viscous pore fluids. *J. Acoust. Soc. Am.*, **120** (5), pp. 2538–2549, 2006 and uses measurements made during SAX99.

with

$$\tilde{\rho} = \frac{\alpha \rho_w}{\beta} + \frac{iF\eta}{\kappa \omega}. \tag{8.44}$$

The effective density approximation provides a practical alternative to the complicated machinery of the poroelastic model and, for sands, gives essentially the same results for fast-wave speed and attenuation as well as reflection coefficient. A comparison between the poroelastic model and EDFM is given in Fig. 8.6. It is evident that the much simpler EDFM formalism adequately approximates the poroelastic model for this sand example.

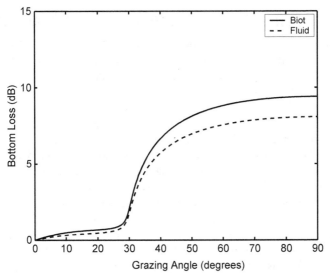

FIGURE 8.5

Comparison of bottom loss in Biot (poroelastic) and fluid models for parameters from Table 8.1. The frequency is 20 kHz.

From Jackson, D.R. and Richardson, M.D, High-Frequency Seafloor Acoustics. *Springer, 2007, with permission of Springer.*

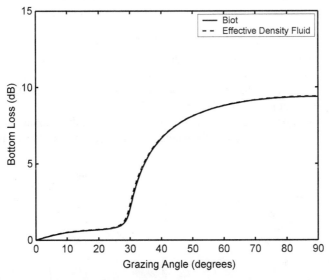

FIGURE 8.6

Comparison of bottom loss in the poroelastic (Biot) model and the effective density fluid model (EDFM) for the parameters of Table 8.1. The frequency is 20 kHz.

From Jackson, D.R. and Richardson, M.D, High-Frequency Seafloor Acoustics. *Springer, 2007, with permission of Springer.*

8.4 MODEL FOR SOUND SCATTERING BY THE SEAFLOOR

Chapter 5 in this book outlines theoretical bases for many practical models for sound scattering by the seafloor, and a more extensive account can be found in Jackson and Richardson [4]. Rather than reviewing the voluminous literature on seafloor acoustic scattering models, this discussion focuses on a model that can be viewed as a successor to the practical model defined in Mourad and Jackson [26] and in APL-UW TR9407 [27]. This earlier model has been described as semiempirical, since sediment volume scattering and scattering by rough seafloors use empirical approaches. The model defined here is taken from Jackson [28] and employs an improved roughness scattering approximation and a physical model for volume scattering. Additional improvements over the older model are the ability to treat seafloors that support shear waves and a generalization from backscattering to bistatic scattering.

It is generally agreed that sound scattering by the seafloor is due to interface roughness and sediment heterogeneity [29–32], with the relative strengths of the two contributions dependent on composition, frequency, and angle. It is convenient to formulate the models in terms of the bottom scattering cross section, σ_b, related to the scattering strength S_b (unit: dB) as follows:

$$S_b = 10 \log(\sigma_b) \tag{8.45}$$

The term "bottom scattering cross section" is a shortened version of the technically correct but cumbersome term "bottom scattering cross section per unit solid angle per unit area." Normally, "cross section" implies a parameter with (unit: m²), so use of the shortened term could lead to the incorrect assumption that σ_b has (unit: m²) when, in fact, it is dimensionless. The bottom scattering cross section is assumed to be the sum of interface roughness and volume heterogeneity contributions:

$$\sigma_b = \sigma_r + \sigma_{vlim}. \tag{8.46}$$

In Eq. (8.46) σ_{vlim} is the equivalent interface scattering cross section due to volume heterogeneity limited to a maximum value in order to avoid violation of energy conservation (see Section 8.4.4). The model to be defined treats bistatic scattering with angular definitions as specified in Fig. 8.7.

The grazing angle of the incident acoustic energy is θ_i, the grazing angle of the scattered acoustic energy is θ_s, and the change in azimuth due to scattering is ϕ_s. This

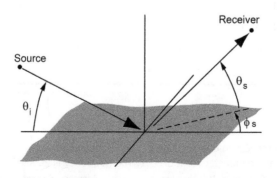

FIGURE 8.7

Angular definitions for bistatic scattering from statistically isotropic seafloor.

angle is referred to as the "bistatic angle." In general, two azimuthal angles are needed, one for the incident and one for the scattered energy. However, the model assumes that roughness statistics are isotropic and volume inhomogeneity is transversely isotropic, so that the bistatic scattering strength depends only on the difference of these two angles. Note that backscattering corresponds to the choice $\theta_s = \theta_i$, $\phi_s = 180$ degrees, and scattering in the specular direction corresponds to $\theta_s = \theta_i$, $\phi_s = 0$ degree.

The seafloor is modeled as an elastic medium, for which the fluid model is a special case. The model allows treatment of poroelastic effects in sands using the previously discussed EDFM. The model developed here [28] combines the small-slope approximation for roughness scattering [33−35] with a perturbation treatment of volume scattering [36,37]. Some typographical errors in the original [28] are corrected here, and this model can be viewed as an improvement on a previous practical model [26,27], using a better roughness scattering approximation, relying less on empirically chosen inputs, and accommodating shear waves. It does not incorporate stratification (layering or vertical gradients) in average seafloor properties. Stratification becomes more important as the frequency is lowered and the acoustic penetration depth increases. Models are available that allow for stratification [38−40], but these do not employ the small-slope approximation. It appears that incorporation of the small-slope approximation in models allowing stratification entails an increased numerical burden [41].

8.4.1 MODEL PARAMETERS

Several input parameters are required, including the speed and density ratios defined earlier (v_p, v_t, a_ρ) and the loss parameters for compressional waves and shear waves (δ_p and δ_t). It is convenient to use the following dimensionless complex speed ratios, obtained using Eqs. (8.21) and (8.22), respectively:

$$a_\rho = \frac{v_p}{1 + i\delta_p},$$
(8.47)

$$a_t = \frac{v_t}{1 + i\delta_t}.$$
(8.48)

Roughness is described by a two-dimensional spectrum of the form:

$$W(K) = \frac{w_2}{K^{\gamma_2}}.$$
(8.49)

The parameter w_2 is the "roughness spectral strength" (unit: $m^{4-\gamma_2}$), and the parameter γ_2 is the dimensionless "roughness spectral exponent," with values ranging from 2 to 4. It is assumed that seafloor roughness statistics are isotropic and Gaussian. Isotropy is implicit in the assumed dependence of the spectrum on the magnitude, K, of the two-dimensional wave vector, $\mathbf{K}$ (unit: 1/m). The spectrum is normalized such that its integral over a given region of K-space is equal to the roughness variance (unit: m^2) for those spectral components. The spectrum is double sided, that is, the wave vector argument spans both positive and negative values.

Volume heterogeneity is described by three-dimensional spectra, following [36]. The density fluctuation spectrum (unit: m^3) is

$$W_{\rho\rho}(\mathbf{k}) = \frac{w_3}{\left[\Lambda^2 \left(k_x^2 + k_y^2 \right) + k_z^2 + L_c^{-2} \right]^{\gamma_3/2}}. \tag{8.50}$$

This is the spectrum for the dimensionless ratio of density fluctuations divided by the mean density. The "volume spectral strength," w_3, has (unit: $m^{3-\gamma_3}$), with the dimensionless γ_3 denoted the "volume spectral exponent." Volume heterogeneity spectra are assumed to be isotropic with respect to directions in the horizontal plane; that is, they are "transversely isotropic," but the spectra have horizontal−vertical anisotropy controlled by the dimensionless parameter Λ. The parameter L_c (unit: m) sets the largest scale of the density fluctuations [36,42]. Gaussian statistics need not be assumed for volume fluctuations, since the first-order perturbation approximation used here is applicable for arbitrary statistics. In addition to fluctuations in sediment density, the model also allows fluctuations in compressional and shear wave speeds. These fluctuations are also normalized by division by the respective means and assumed to have their spectra proportional to the density fluctuation spectrum. These other fluctuations may be correlated with the density fluctuations, with correlations described by appropriate cross spectra. These assumptions lead to the set of model parameters defined by the following equation:

$$W_{\beta\beta'}(\mathbf{k}) = a_{\beta\beta'} W_{\rho\rho}(\mathbf{k}). \tag{8.51}$$

In Eq. (8.51) the dimensionless parameters $a_{\beta\beta'}$ define the proportionality between the density fluctuation spectrum and the other volume fluctuation spectra, $W_{pp}(\mathbf{k})$, $W_{tt}(\mathbf{k})$, $W_{p\rho}(\mathbf{k})$, $W_{t\rho}(\mathbf{k})$, and $W_{pt}(\mathbf{k})$. These are, respectively, the compressional wave fluctuation spectrum, shear wave fluctuation spectrum, (compressional wave)-(density) fluctuation cross spectrum, (shear wave)-(density) fluctuation cross spectrum, and the (compressional wave)-(shear wave) fluctuation cross spectrum. This list exhausts the possible spectra in the present problem, since the cross spectra are symmetric in their subscripts; that is, $W_{\beta'\beta}(\mathbf{k}) = W_{\beta\beta'}(\mathbf{k})$. The price of including this level of generality in the model is the addition of five model parameters a_{pp}, a_{tt}, $a_{p\rho}$, $a_{t\rho}$, and a_{pt}. The parameters a_{pp} and a_{tt} are positive, while the parameters $a_{p\rho}$, $a_{t\rho}$, and a_{pt} are positive or negative depending on whether the corresponding fluctuations are correlated or anticorrelated. The multiplicity of parameters allows flexibility in applying the model. However, it presents a problem in most practical cases, for which many parameters are unknown. In Section 8.4.4 Scattering Model Examples, simple defaults are given for these parameters allowing application when little geoacoustic information is available.

8.4.2 SCATTERING BY SEAFLOOR ROUGHNESS

The small-slope approximation is more accurate than the older small-roughness perturbation and Kirchhoff approximations [43,44] and has been applied to scattering by rough media supporting shear waves [33–35]. The bistatic scattering cross section is

$$\sigma_r(\theta_s, \phi_s, \theta_i) = \frac{k_w^4 |A|^2}{2\pi \Delta K^2 \Delta k_z^2} \int_0^\infty J_0(u) \exp(-qu^{2\alpha}) u \, du, \tag{8.52}$$

where

$$q = \frac{1}{2} C_h^2 \Delta k_z^2 \Delta K^{-2\alpha}, \tag{8.53}$$

$$\alpha = \frac{(\gamma_2 - 2)}{2}, \tag{8.54}$$

and

$$C_h^2 = \frac{2\pi w_2 \Gamma(2 - \alpha)}{2^{2\alpha} \alpha (1 - \alpha) \Gamma(1 + \alpha)}. \tag{8.55}$$

The variables q and α are dimensionless, C_h^2 has (unit: $m^{4-\gamma_2}$), and ΔK is the transverse Bragg wave number, i.e., the magnitude of the difference in the incident and scattered wave vector transverse components:

$$\Delta K = k_w \sqrt{\cos^2\theta_s - 2 \cos\theta_s \cos\theta_i \cos\phi_s + \cos^2\theta_i + b^2}. \tag{8.56}$$

The dimensionless parameter b is assigned the value 0.001 to avoid numerical difficulties that arise if ΔK becomes too small. The variable Δk_z is the corresponding difference in the vertical components of the incident and scattered wave vectors

$$\Delta k_z = k_w(\sin\theta_s + \sin\theta_i). \tag{8.57}$$

The coefficient A (dimensionless) is

$$A = \frac{1}{2}(D_1[1 + R(\theta_s)][1 + R(\theta_i)] + D_2[1 - R(\theta_s)][1 + R(\theta_i)] + D_3[1 + R(\theta_s)]$$
$$[1 - R(\theta_i)] + D_4[1 - R(\theta_s)][1 - R(\theta_i)]) \tag{8.58}$$

where $R(\theta)$ is the elastic reflection coefficient (8.16) evaluated at the incident and scattered grazing angles, and the coefficients D_n are

$$D_1 = -1 + S + \frac{1}{a_p^2 a_\rho \cos2\theta_{ts} \cos2\theta_{ti}}$$
$$- \frac{a_t^2 \left[(a_t^{-2} - 2\cos^2\theta_s - 2\cos^2\theta_i + 2S)S + 2\cos^2\theta_s \cos^2\theta_i \right]}{a_\rho \cos2\theta_{ts} \cos2\theta_{ti}}, \tag{8.59}$$

$$D_2 = -\frac{4a_t^3 \sin\theta_s \sin\theta_{ts}}{\cos2\theta_{ts}\cos2\theta_{ti}} \left[a_p^{-2}\sin^2\theta_{pi}\cos^2\theta_s + (\cos^2\theta_i - S)S \right], \tag{8.60}$$

$$D_3 = -\frac{4a_t^3 \sin\theta_i \sin\theta_{ti}}{\cos2\theta_{ts}\cos2\theta_{ti}} \left[a_p^{-2}\sin^2\theta_{ps}\cos^2\theta_i + (\cos^2\theta_s - S)S \right], \tag{8.61}$$

$$D_4 = \frac{2a_t^4 a_\rho \sin\theta_s \sin\theta_i \sin\theta_{ts} \sin\theta_{ti}}{\cos2\theta_{ts}\cos2\theta_{ti}} \left[2\left(a_t^{-2} - 2S\right)S - 4\cos^2\theta_s\cos^2\theta_i \left(1 - 2a_t^2 a_p^{-2}\right)\right]$$
$$- \left(a_\rho - 1\right)\sin\theta_s \sin\theta_i.$$

$$(8.62)$$

The variable S is

$$S = \cos\theta_s \cos\theta_i \cos\phi_s, \qquad (8.63)$$

and the complex grazing angles θ_{pi}, θ_{ps}, θ_{ti}, and θ_{ts} are found from Snell's law using the complex speed ratios given in Eqs. (8.47) and (8.48).

8.4.3 SCATTERING BY SEAFLOOR HETEROGENEITY

The volume scattering component of the model is taken from Refs. [36,37] and is based on perturbation theory. The contribution of volume scattering to the effective interface scattering cross section is

$$\sigma_v(\theta_s, \phi_s, \theta_i) = -\frac{\pi k_w^4 a_\rho^2}{2} Im\left\{\sum_{\eta,\beta,\eta',\beta'} d_{\eta\beta} d_{\eta'\beta'}^* \frac{W_{\beta\beta'}[(\Delta\mathbf{k}_\eta + \Delta\mathbf{k}_{\eta'}^*)/2]}{\Delta k_{\eta z} - \Delta k_{\eta'z}^*}\right\} \qquad (8.64)$$

where Im denotes the imaginary part, and the $W_{\beta\beta'}$ are the heterogeneity spectra defined earlier. Their arguments are complex, and the algebraic expressions for these spectra yield complex values. The sum enclosed in braces is pure imaginary, at least formally, but there will be a small real part due to numerical error. The β and β' sums run over the three heterogeneity types: density, compressional wave speed, and shear wave speed, (β, $\beta' = \rho, p, t$). The η and η' sums run over the four types of wave conversion caused by volume scattering: compressional to compressional ($\eta = pp$), shear to compressional ($\eta = pt$), compressional to shear ($\eta = tp$), and shear to shear ($\eta = tt$). Symbolically, we can put

$$\eta = qq', \qquad (8.65)$$

where q and q' run over the two types of waves (q, q'=p, t). The Bragg wave vectors for the four types of conversion are

$$\Delta\mathbf{k}_\eta = \mathbf{k}_q^+ - \mathbf{k}_{q'}^-, \qquad (8.66)$$

where

$$\mathbf{k}_q^- = [k_w \cos\theta_i, 0, -k_q \sin\theta_{qi})], \qquad (8.67)$$

$$\mathbf{k}_q^+ = [k_w \cos\theta_s \cos\phi_s, k_w \cos\theta_s \sin\phi_s, k_q \sin\theta_{qs}], \qquad (8.68)$$

are wave vectors for down-going (incident $= -$) and up-going (scattered $= +$) compressional and shear waves.

The dimensionless coefficients $d_{\eta\beta}$ are

$$d_{\eta\beta} = w_\eta D_{\eta\beta} \qquad (8.69)$$

where

$$w_\eta = \Gamma_q(\theta_s)\Gamma_{q'}(\theta_i) \tag{8.70}$$

is dimensionless, with

$$\Gamma_p(\theta) = -\frac{\cos 2\theta_t \sin \theta}{\sin \theta_p}[1 - R(\theta)], \tag{8.71}$$

$$\Gamma_t(\theta) = -\frac{\sin 2\theta_t \sin \theta}{\sin \theta_t}[1 - R(\theta)]. \tag{8.72}$$

The $\Gamma_q(\theta)$ are proportional to transmission coefficients for compressional and shear waves, and $R(\theta)$ is the water-elastic reflection coefficient (8.16). The coefficients $D_{\eta\beta}$ are elements of the three-column matrix

$$D = (D_\rho | D_p | D_t), \tag{8.73}$$

where $D_\rho = D'_\rho + D_t/2$ and

$$D'_\rho = \begin{pmatrix} 1 - b_{pp} \\ b_{pv} \\ -b_{vp} \\ b_{vv} \end{pmatrix}, \tag{8.74}$$

$$D_p = \begin{pmatrix} 2 \\ 0 \\ 0 \\ 0 \end{pmatrix}, \tag{8.75}$$

$$D_t = 2 \begin{pmatrix} 2g^2 b_{pp}^2 - 2g^2 \\ -2gb_{pv}b_{pt} \\ 2gb_{vp}b_{tp} \\ -b_{vv}b_{tt} - b_{vt}b_{tv} \end{pmatrix}, \tag{8.76}$$

with $g = a_t/a_p$ and

$$b_{\delta\delta'} = \mathbf{b}_\delta^+ \cdot \mathbf{b}_{\delta'}^-, \tag{8.77}$$

The $\mathbf{b}_\delta^\pm$ are unit vectors giving the direction of propagation for compressional and shear waves and the direction of particle motion for vertically polarized shear waves. The indices δ and δ' run over the two wave vectors (compressional and shear) and the shear vertical polarization, $\delta,\delta' = p,t,v$:

$$\mathbf{b}_p^\pm = \mathbf{k}_p^\pm/k_p \tag{8.78}$$

$$\mathbf{b}_t^\pm = \mathbf{k}_t^\pm/k_t, \tag{8.79}$$

$$\mathbf{b}_v^\pm = \left[\mp k_{px}^\pm \sin\theta_t^\pm / K^\pm, \mp k_{py}^\pm \sin\theta_t^\pm / K^\pm, K^\pm/k_t\right]. \tag{8.80}$$

In Eqs. (8.79) and (8.80), $\theta_t^{\pm}$ are the complex shear wave grazing angles for the incident wave in the seafloor ($-$) and scattered wave ($+$), the wave vectors $\mathbf{k}_p^{\pm}$ and $\mathbf{k}_t^{\pm}$ are defined in Eqs. (8.67) and (8.68) with $q = p$, t and

$$K^{\pm} = \sqrt{\left(k_{px}^{\pm}\right)^2 + \left(k_{py}^{\pm}\right)^2}. \tag{8.81}$$

8.4.4 SCATTERING MODEL EXAMPLES

Examples are presented using the parameters given in Table 8.2, which were chosen in part to give agreement with an older model [27]. The parameters of Table 8.2 may provide a useful guide for users having little geoacoustic data at hand. For most cases the five parameters a_{pp}, a_{tt}, a_{pp}, a_{tp}, and a_{pt} have been set to zero, as normalized wave speed fluctuations are generally smaller than normalized density fluctuations in unconsolidated sediments. For the seafloor types "rough rock" and "rock," these parameters are assigned values adapted from [37], namely $a_{pp} = 9$, $a_{tt} = 16$, $a_{pp} = 3$, $a_{tp} = 4$, and $a_{pt} = 12$. For soft sediments (medium sand, fine sand, silt), the shear parameters v_t and δ_t are assigned values that render shear effects negligible without introducing singularities into calculations. In all cases the aspect ratio, Λ, has been set to unity. The volume spectrum scale parameter, L_c has been assigned the frequency-dependent value $1000/k_w$ where k_w is the acoustic wave number in water. This default is unphysical, but avoids difficulties that arise in perturbation theory when the volume spectra approach large values as the wave vector argument approaches zero. Even with this choice, the volume scattering cross section sometimes takes on impossibly large values near the specular direction. To suppress this unphysical behavior, the volume scattering cross section will be limited to values less than a preset parameter, σ_0:

$$\sigma_{v\text{lim}}(\theta_s, \phi_s, \theta_i) = \left[\sigma_v(\theta_s, \phi_s, \theta_i)^{-1} + \sigma_0^{-1}\right]^{-1}. \tag{8.82}$$

In the examples to follow, $\sigma_0 = 0.1$, corresponding to a maximum scattering strength of -10 dB. This value was chosen to avoid egregious violations of energy conservation near the specular direction for the rough rock, rock, and cobble cases.

Table 8.2 Parameters Used in Scattering Model Examples

Type	a_ρ	v_p	δ_p	v_t	δ_t	γ_2	w_2	γ_3	w_3
Rough rock	2.5	2.3	0.00174	1.3	0.085	2.75	1.0×10^{-4}	3.0	2.0×10^{-4}
Rock	2.5	2.3	0.00174	1.3	0.085	2.75	2.5×10^{-4}	3.0	6.0×10^{-5}
Cobble	2.5	1.8	0.00137	1.01	0.1	3.0	2.5×10^{-4}	3.0	1.52×10^{-3}
Sandy gravel	2.492	1.337	0.01705	0.156	0.2	3.0	1.8×10^{-4}	3.0	3.77×10^{-4}
Coarse sand	2.231	1.250	0.01638	0.134	0.075	3.25	2.2×10^{-4}	3.0	3.62×10^{-4}
Medium sand	1.845	1.178	0.01624	0.002	1.0	3.25	1.406×10^{-4}	3.0	3.59×10^{-4}
Fine sand	1.451	1.107	0.01602	0.002	1.0	3.25	8.6×10^{-5}	3.0	3.54×10^{-4}
Silt	1.149	0.987	0.00386	0.002	1.0	3.25	1.64×10^{-5}	3.0	4.27×10^{-5}

There is no appreciable effect for the other cases. The present model yields scattering strengths nearly identical to those of an older practical model [27] when applied to softer sediments (coarse sand and finer). The latter model employed an empirical volume scattering parameter, σ_2, which has been related [28] to the physical parameter w_3 used in the present model:

$$w_3 = \frac{80\delta_p\sigma_2}{\pi\log_e 10\left(1 + \delta_p^2\right)^2}.$$ (8.83)

The dimensionless volume scattering parameter σ_2 is the ratio of volume scattering cross section (unit: m^{-1}) to attenuation (unit: dB/m). Expression (8.83) assumes $\gamma_3 = 3$ as in Table 8.1. This choice of power exponent gives σ_v nearly independent of frequency provided δ_p is independent of frequency.

Figs. 8.8–8.17 show model curves for backscattering strength, bottom loss, and bistatic scattering strength for the following cases: rough rock, rock, sandy gravel, medium sand, and silt. All curves are computed for a frequency of 20 kHz. Backscattering strength curves from Ref. [27] are also shown. This comparison is important because the older model provides useful fits to a number of experiments with accompanying geoacoustic data used as model inputs [45]. The rough rock and rock cases differ only in their roughness and heterogeneity parameters. The poorest fit to [27] is obtained in the rough rock case for which roughness is so large that the

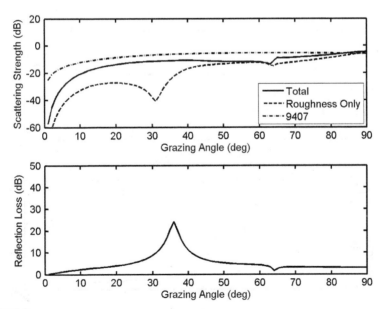

FIGURE 8.8

Model backscattering strength and bottom loss for rough rock using parameters from Table 8.2. Backscattering strength is compared with that from Ref. [27].

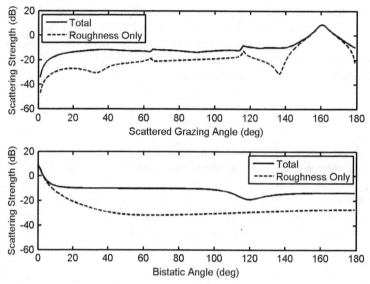

FIGURE 8.9

Model bistatic scattering strength for rough rock using parameters from Table 8.2. The incident grazing angle is 20 degrees.

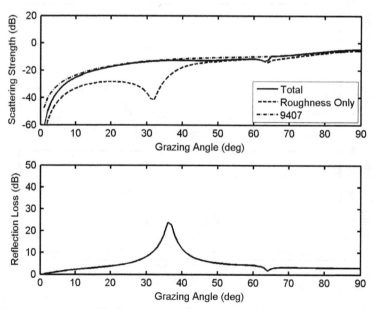

FIGURE 8.10

Model backscattering strength and bottom loss for rock using parameters from Table 8.2. Backscattering strength is compared with that from Ref. [27].

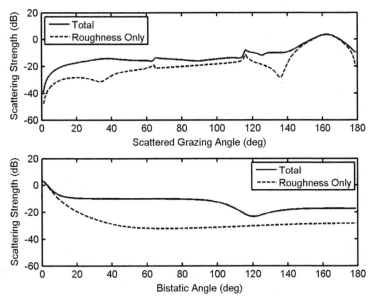

FIGURE 8.11

Model bistatic scattering strength for rock using parameters from Table 8.2. The incident grazing angle is 20 degrees.

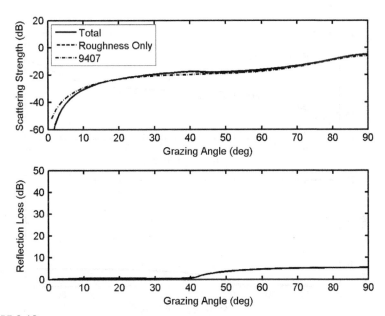

FIGURE 8.12

Model backscattering strength and bottom loss for sandy gravel using parameters from Table 8.2. Backscattering strength is compared with that from Ref. [27].

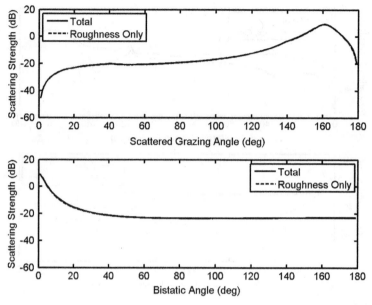

FIGURE 8.13

Model bistatic scattering strength for sandy gravel using parameters from Table 8.2. The incident grazing angle is 20 degrees.

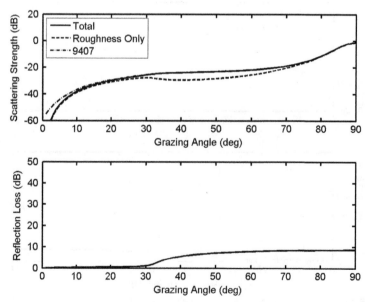

FIGURE 8.14

Model backscattering strength and bottom loss for medium sand using parameters from Table 8.2. Backscattering strength is compared with that from Ref. [27].

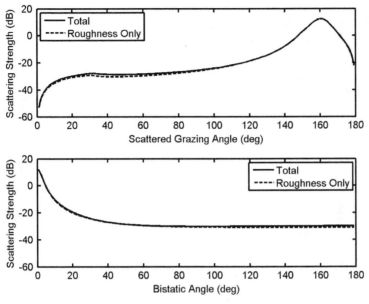

FIGURE 8.15

Model bistatic scattering strength for medium sand using parameters from Table 8.2. The incident grazing angle is 20 degrees.

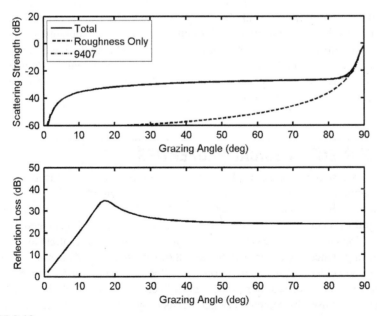

FIGURE 8.16

Model backscattering strength and bottom loss for silt using parameters from Table 8.2. Backscattering strength is compared with that from Ref. [27].

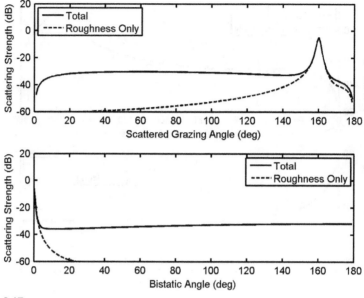

FIGURE 8.17

Model bistatic scattering strength for silt using parameters from Table 8.2. The incident grazing angle is 20 degrees.

small-slope approximation is suspect. The curve from [27] is purely empirical in this case, and the present model should be viewed as essentially empirical in this case as well. The plots show the contribution of roughness to the total. Roughness scattering has a deep minimum for the rough rock and rock cases, due to the interface relative transparency near the 30 degrees grazing angle. This causes a peak in volume scattering which has been adjusted to provide relatively flat curves after addition with the roughness cross section for these two cases.

8.5 SEDIMENT PHYSICAL PROPERTIES

This section discusses sediment physical properties relevant to acoustic—seafloor interaction modeling. These include the statistics of sediment grain size distribution, sediment porosity and bulk density, pore fluid and pore space properties, permeability, grain properties, and descriptors of sediment type. Sediment geoacoustic properties including compressional and shear wave speeds and attenuations and sediment impedance are covered in Section 8.6, followed by a discussion of measurement and statistical characterization of seafloor roughness in Section 8.7, and spatial heterogeneity of seafloor properties in Section 8.8. It should be noted that most techniques used by sedimentary geologists or geotechnical engineers to characterize sediments were not developed with acoustic—seafloor interactions in mind.

8.5.1 GRAIN SIZE DISTRIBUTION

Grain size distribution is the most commonly reported sediment property with more measurements of grain size distribution having been made than for all other sediment physical properties combined. One reason for the wealth of data is that sediment grain size distribution can be determined on disturbed samples, whereas most of the other sediment physical properties described in this section require carefully collected undisturbed samples for laboratory analyses or require in situ measurement techniques. As is discussed later, statistical characterization of grain size distributions may not be the most diagnostic parameter used to model acoustic–sediment interactions; however, they may be the only sediment physical parameter available to the acoustic modeler. Angle-dependent acoustic reflection and scattering from the seafloor are often summarized based on grain-size sediment classification (e.g., Table 8.2 and Figs. 8.8–8.17). Regressions of sound speed versus mean grain size are also commonly reported [46,47]. Values of mean grain size and sorting (standard deviation) are also used to predict other sediment physical properties that are rarely measured such as permeability, tortuosity, and the pore size parameter.

The range of sediment particle sizes in marine sediments covers almost eight orders of magnitude, from boulders hundreds of centimeters in size to clay size particles smaller than one micron (Table 8.3). Grain size distribution has traditionally been reported in the base-2 logarithmic scale devised by Krumbein [48], where $\phi = -\log_2 \cdot d$, where d is the particle diameter in millimeters. This approach allows the natural lognormal distribution of sediment particles to be reported as a normal distribution of phi-transformed particle sizes. This transformation condenses the broad range of grain diameters into a more restrictive range of phi values, expanding the spectrum of very small grain sizes, and facilitating the calculation graphical descriptors of the grain size distribution. This phi transformation is analogous to the well-known (at least to acousticians) sound pressure transformation—the unit decibel.

The classical methods of particle size analysis weigh each size class of sand- and gravel-sized particles retained on standardized wire mesh screens. Silt- and clay-sized particle distributions are determined by pipette methods where particles segregate according to mean diameter based on settling according to Stokes' law. The resultant distributions of particle size weights are plotted as histograms within phi size ranges (e.g., Figs. 8.18 and 8.19), and graphical methods are used to determine the statistics of the sediment distribution (mean, mode, sorting or standard deviation, skewness and kurtosis). Modern automated grain size analysis techniques are based on the classic sieving and pipette methods which are often used to calibrate these techniques [49].

One of the more common methods to characterize sediment particle size distribution was devised by Folk and Ward [50] and includes median (M_d),

$$M_d = \phi_{50,}$$

(8.84)

Table 8.3 Sediment Grain Size Classification Based on Folk [51]

US Standard Wire Mesh #	Millimeters	Micron	Phi (ϕ)	Wentworth Size Class
	1024		−12	Boulder (−8 to −12ϕ)
	256		−8	
	128		−7	Cobble (−6 to −8ϕ)
2 ½	64		−6	
1 1/4	32		−5	
5/8	16		−4	Pebble (−2 to −6ϕ)
0.530	13.2		−3.75	
7/16	11.2		−3.50	
3/8	9.5		−3.25	
5/16	8		−3.00	
1/4	6.3		−2.75	
3.5	5.6		−2.50	
4	4.75		−2.25	
5	4.00		−2.00	
6	3.36		−1.75	
7	2.83		−1.50	
8	2.38		−1.25	Granule (−1 to −2ϕ)
10	2.00		−1.00	
12	1.68		−0.75	
14	1.41		−0.50	
16	1.19		−0.25	Very coarse sand
18	1.00		0.00	
20	0.84		0.25	
25	0.71		0.50	
30	0.59		0.75	Coarse sand
35	0.50	500	1.00	
40	0.42	420	1.25	
45	0.35	350	1.50	
50	0.30	300	1.75	Medium sand
60	0.25	250	2.00	
70	0.21	210	2.25	
80	0.177	177	2.50	
100	0.149	149	2.75	Fine sand
120	0.125	125	3.00	
140	0.105	105	3.25	
170	0.088	88	3.50	
200	0.074	74	3.75	Very fine sand
230	0.0625	62.5	4.00	
270	0.053	53	4.25	
325	0.044	44	4.50	Coarse silt
400	0.037	37	4.75	
450	0.031	31	5.00	
500	0.023	23.4	5.50	Medium silt
635	0.0156	15.6	6.00	
	0.0117	11.7	6.50	Fine silt
	0.0078	7.8	7.00	
	0.0045	4.5	7.50	Very fine silt
	0.0039	3.9	8.00	
	0.0020	2.0	9.00	
	0.00098	0.98	10.00	
	0.00049	0.49	11.00	Clay
	0.00024	0.24	12.00	
	0.00012	0.12	13.00	
	0.000006	0.06	14.00	

The grade scale proposed by Wentworth [52] is logarithmic, where each size class is twice as large as the next smaller size class. Phi (ϕ) is $-\log_2(d)$ where d is the equivalent spherical diameter in millimeters based on Ref. [53]. Equivalent diameter is also expressed in millimeters and microns. The US Standard Wire Mesh (no. of wires/in.^{2}) for sieving granules, sands, and coarse silt-size particles and in inches/opening for pebbles is also given.

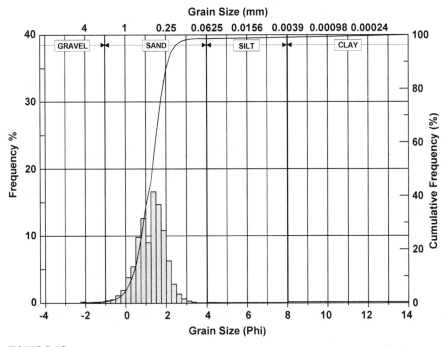

FIGURE 8.18

Histogram of grain size distribution for sandy sediment collected during the SAX99 high-frequency acoustic experiments in the northeastern Gulf of Mexico, October 1999 [54]. This sample contains 0.55% gravel-sized particles, 98.08% sand, 0.39% silt, and 0.99% clay. The values for graphic medium (1.60ϕ), mean (1.27ϕ), standard deviation or sorting (0.7ϕ), skewness (-0.11ϕ), and kurtosis (1.20ϕ) in units of ϕ are calculated from Eqs. (8.84)–(8.88). The silt- and clay-sized fractions accounted for less than 2% of the total particulate weight and do not affect graphic grain size statistics.

mean (M_z),

$$M_z = (\phi_{16} + \phi_{50} + \phi_{84})/3, \tag{8.85}$$

standard deviation or sorting (σ_I),

$$\sigma_I = (\phi_{84} - \phi_{16})/4 + (\phi_{95} - \phi_5)/6.6, \tag{8.86}$$

skewness (Sk_I),

$$Sk_I = (\phi_{16} + \phi_{84} - 2\phi_{50})/[2(\phi_{84} - \phi_{16})] + (\phi_5 + \phi_{95} - 2\phi_{50})/([2(\phi_{95} - \phi_5)], \tag{8.87}$$

and kurtosis (K_G),

$$K_G = (\phi_{95} - \phi_5)/[2.44(\phi_{75} - \phi_{25})], \tag{8.88}$$

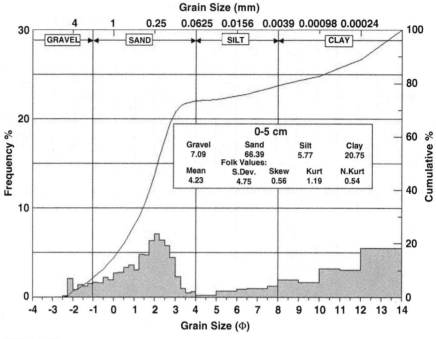

FIGURE 8.19

Sediments (upper 5 cm) from the Singapore Strait exhibit a bimodal grain size distribution. Coarse silt sediments that represent the Folk mean grain size (4.23ϕ) are almost absent from these samples.

where the values of phi (ϕ_x) are graphically determined at cumulative frequencies (unit: %) using plots such as given in Fig. 8.18.

There are several issues that suggest caution when sediment grain size distribution is used to predict acoustic–seafloor interactions. These include issues associated with sediment collection methods, differences in particle size analysis techniques, and the differences among the approaches used for calculating and reporting grain size statistics. Few examples will demonstrate these problems. Sediment samples are often collected remotely using grabs; gravity, piston, or box cores; or in shallow water by divers. These samples are usually much smaller than an acoustic footprint and may not represent the vertical and horizontal variability of the marine environment or adequately characterize larger particles that exist within the acoustic footprint (e.g., shells, gravel, or small rocks). In addition finer grain sizes can winnow out of samples, especially when collected with grabs. Silt- and clay-sized particles (collectively referred to as mud) are subject to a variety of either mechanical or chemical methods of disaggregation prior to pipette or automated size analyses. These methods destroy the structure of naturally occurring aggregates that may be important for acoustic–seafloor interactions and lead to ill-defined lower limits for clay-sized particles. The assumption that larger sand- and gravel-sized particles are spherical is often violated leading to errors in size distribution (see Section 8.5.5). Sediment particle size distribution is often bimodal (e.g., mixtures of sand and mud) and the calculated mean grain size may be nearly absent from the grain distribution (e.g., Fig. 8.19). The statistical description of sediment grain size distribution can vary depending on the graphical or arithmetic (based on moments) methods developed for statistical characterization of grain size distribution. Blott and Pye [55,56] provide excellent reviews of the history and current use of methods and terminology used for sediment particle size characterization. Differences in the boundaries and terminologies used to describe sediment particle classes also exist. Therefore, sediment analysis methodologies should be clearly provided and comparisons of values of sediment properties using different or unknown techniques should be avoided, especially when large data sets comparing acoustic–sediment interactions are developed.

8.5.2 SEDIMENT BULK DENSITY AND POROSITY

Bulk density (ρ) and porosity (β) are measures of the mass and/or volume of sediment particles and pore fluid in fully saturated sediments. Sediment bulk density is the most fundamental sediment property directly affecting acoustic propagation within sediments. Sound speed and bulk density are highly correlated as discussed in Section 8.6, and their product provides a measure of sediment impedance that is used to calculate acoustic reflection from the seafloor and layering within the seafloor. Other similarly related measures reported in the geotechnical literature include water content (w equals the ratio of pore water weight over solids weight) and the void ratio (e equals the ratio pore water volume to solid particle volume). The most common methods to determine values of these properties are to directly measure the total weight of a known volume of sediment (direct measure of sediment bulk density) or to measure the pore fluid weight loss of a known weight of sediment that has been dried for 24 h at 105°C (measurement of sediment water content). These methods assume a relatively undisturbed sample where no previous water loss has occurred and known values of the pore fluid and grain densities. The pore fluid density ρ_w (unit: kg/m^3) can be calculated from the fluid temperature and salinity. The density of the dried grains (ρ_s) (unit: kg/m^3) can be determined using a pycnometer or handbook values. Fractional porosity (β) (unit: dimensionless) can be calculated from water content (w) (unit: dimensionless) and pore fluid and solid densities. For marine sediments a correction for the residue of dried salts should be made [1]:

$$\beta = \beta_0 + 0.036(S_p/35) - 0.0224(S_p/35)\beta_0, \tag{8.89}$$

where S_p is the pore water salinity in parts per thousand and β_0 is the uncorrected fractional porosity. Bulk density can be calculated from fractional porosity and fluid and grain densities using,

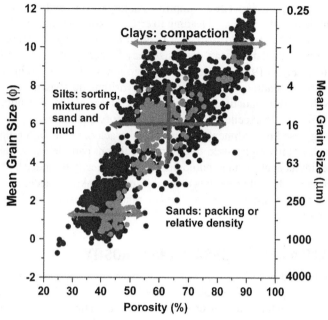

FIGURE 8.20

Scatter diagram of mean grain size versus porosity in surficial sediments (upper meter). The *lighter colored symbols*, which represent carbonate sediments, overlay the *darker colored symbols*, which represent siliciclastic sediment.

From Jackson, D.R. and Richardson, M.D, High-Frequency Seafloor Acoustics. *Springer, 2007, with permission of Springer.*

$$\rho = \rho_w \beta + \rho_g (1 - \beta). \tag{8.90}$$

Nondestructive methods used to measure sediment porosity include electrical resistivity, gamma ray attenuation (core loggers), X-radiography, and high-resolution CT scanning (see Jackson and Richardson [4] for a summary of those techniques). Porosity can be presented as a percent (mostly in the geological literature) or as the fractional value (β) used in most acoustic calculations. The range of typical values of porosity for unconsolidated surficial muddy sediments is 50−90% (bulk density 1837−1185 kg/m^3) depending on consolidation and for sandy sediments 25−50% (2243−1837 kg/m^3) depending on packing (Fig. 8.20). Mean grain size is obviously a poor predictor of near-surface sediment porosity and bulk density. The relationship further deteriorates with depth below the sediment surface as sediments continue to dewater by compaction and increased packing. The most recent and extensive global model of near-surface sediment porosity was developed by Martin et al. [9] and presents porosity on a 5 arc min pixel grid.

8.5.3 **PORE FLUID AND PORE SPACE PROPERTIES**

The density, bulk modulus, and dynamic viscosity of pore fluids in sediments can be important inputs to some sediment propagation models. However, values of these properties are rarely measured as part of the environmental characterization during acoustic experiments. In most cases, values of pore waters are assumed to be the same as the overlaying seawater which is a reasonable assumption. However, the effects of temperature gradients in the seafloor may be required to correctly calculate gradients in sound speed and impedance [57,58]. Numerous very accurate algorithms exist to calculate values of these pore water properties from temperature, salinity, and pressure. In addition, these same algorithms are used to calculate the density, sound speed and attenuation, and overlying water column impedance. In general, these algorithms are overkill for most applications of acoustic seafloor interactions. The presence of organic matter, especially in pore fluids, the spatial and temporal variability of seawater and pore water temperature and salinity, and the lack of sensitivity of acoustic propagation, refection, and scattering models to the natural variability of pore fluid properties suggests that simpler algorithms or tabular values are adequate. For example, the simplified algorithm to calculate sound speed (V_p,) (unit: m/s), denoted c_w in acoustic modeling sections of this chapter, from temperature (T) (unit: °C), salinity (S) (units: ppt), and depth (D) (unit: m), suggested by Medwin [59] is within 1 m/s of the 41 coefficient algorithm published by Chen and Millero [60]:

$$V_p = 1449.2 + 4.6T - 0.055T^2 + 0.00029T^3 + (1.34 - 0.01T)(S - 35)$$
$$+ 0.016D, \tag{8.91}$$

In addition to pore fluid viscosity, two properties associated with pore space geometry (tortuosity and the pore size parameter) are variables that effect permeability and wave propagation [61]. Tortuosity (also referred to as the structure factor) is used to describe the sinuosity of pore space as it affects hydraulic transport. Hydraulic tortuosity, T, used in the Kozeny–Carman equation of permeability is the ratio of the effective hydraulic path length to a straight line distance in the direction of flow. However, the tortuosity factor, α, used in Biot theory is equal to the square of the hydraulic tortuosity. The pore size parameter, a, with units of length, is the size of the parallel channels through which fluid flow passes during the passage of an acoustic wave. In practice tortuosity and pore space parameter are difficult to measure and are either estimated from experimental acoustic data or empirically derived from grain size distribution. Fluid flow–type parameters are only important to acoustic propagation for high permeability granular material. Stoll [61] set the pore size parameter to 1/6 to 1/7 of the mean grain size for sand and silts and the dimensionless structure factor to 1.25 for sand and 3.0 for fine-grained sediments (mud).

8.5.4 PERMEABILITY

Permeability is a property associated with pore fluid flow in sediment relevant to predicting the frequency dependence of wave speed and attenuation using poroelastic theory (e.g., Biot theory described in Section 8.3). Permeability is one of several sediment conductance properties which include electrical, thermal, and soluble material flow through sediments. Sediment permeability or hydraulic conductivity can be measured in the laboratory using falling or constant head methods, measured in situ using flow rate or pressure drop methods, calculated from descriptions of sediment fabric (e.g., Kozeny–Carman equation), or predicted empirically from other sediment properties, such as mean grain size, void ratio, or porosity. Laboratory or in situ methods measure hydraulic flow rates (V) (unit: m/s) which are a product of the hydraulic conductivity (K) (unit: m/s) and the negative gradient of pressure (P) expressed in Eq. (8.92), Darcy' law, where g (unit: m/s^2) is the acceleration of gravity and ρ_f (unit: kg/m^3) is the fluid density. However, poroelastic models use the intrinsic coefficient of permeability κ (unit: m^2) which is independent of pore fluid dynamic (or absolute) viscosity μ given in Eq. (8.93). The hydraulic flow rate can thus be written in terms of the dynamic viscosity and the intrinsic coefficient of permeability, Eq. (8.94):

$$V = -\frac{K}{g\rho_f}\nabla P, \tag{8.92}$$

$$K = \kappa\frac{g\rho_f}{\mu}, \tag{8.93}$$

$$V = -\frac{\kappa}{\mu}\nabla P, \tag{8.94}$$

Values of permeability have a greater range (10 orders of magnitude) and exhibit larger spatial variability than any other seafloor property used in underwater acoustics as shown in Table 8.4.

Differences of an order of magnitude or greater are common within closely sampled sites, at the same site using different measurement techniques, or between measured and modeled estimates of permeability [54]. Direct measurements in muddy sediments often yield permeability values inappropriate for acoustic

Table 8.4 Range of Values for Sediment Intrinsic Permeability and Porosity in Near-Surface Sediments (Upper Meter)

Sediment Type	Grain Size, d (mm)	Permeability, κ (m^2)	Porosity, β
Clay	<0.002	10^{-18}–10^{-15}	0.5–0.9
Silt	0.002–0.05	10^{-16}–10^{-12}	0.35–0.5
Sand	0.05–2.0	10^{-14}–10^{-10}	0.35–0.5
Gravel	>2.0	10^{-10}–10^{-7}	0.25–0.4

Modified from Jackson, D.R. and Richardson, M.D, High-Frequency Seafloor Acoustics. Springer, 2007, with permission of Springer.

applications. Burrows or cracks in sediment might yield locally high values of permeability at scales of centimeters or greater that are not relevant to acoustic propagation where the out-of-phase movement of pore water relative to the frame (granular structure) at near micro scales is important [62].

Structural pore and grain properties have been directly measured using 2-D scanning electron microscopy (SEM) image analyses of resin-impregnated cores [63], and from 3-D images obtained from high-resolution X-ray computed tomography of carefully collected diver cores [64]. Values of porosity and permeability measured using image-based methods were slightly higher than conventional methods (water loss and falling head permeameter), but show considerable promise for laboratory or field experiments that require fine-scale measurements of pore properties.

8.5.5 GRAIN PROPERTIES

The density and bulk modulus of sediment grains and pore fluid are important properties controlling sediment bulk density, impedance, and bulk modulus. A compilation of the density and bulk modulus of the most common mineral constitutions of sediments from handbooks and the peer-reviewed literature can be found in Table 4.7 in Jackson and Richardson [4]. In most cases, values of pure crystalline minerals provide adequate inputs to elastic and poroelastic models of acoustic propagation and scattering. The bulk density (2650 kg/m^3) and bulk modulus ($36-40 \text{ GPa}$) of quartz are the most common values used in geoacoustic modeling [61]. Other common sediments include carbonates (bulk density, 2710 kg/m^3 and bulk modulus, $64-75 \text{ GPa}$), and clay minerals (bulk density, $2600-3000 \text{ kg/m}^3$ and bulk modulus, $6-50 \text{ GPa}$). Care should be exercised when dealing with biogenic sediment types where pore fluids can be present in intraparticulate and interparticulate states. Intraparticulate porosity ranges from 5% in deep sea foraminifera oozes [65,66] up to 10–15% in carbonate sediments from the Florida Keys [67]. Grain morphology, as well as grain size distribution, is an important property controlling packing, fluid flow, and contact mechanics associated with low strain acoustic propagation in sediments [61,68]. However, grain morphology (shape and surface texture) is rarely measured and grains are often assumed to be smooth spherical grains. Naturally occurring grains are rarely smooth and spherical. Blott and Pye [69] provide an excellent review of methods used to measure and describe larger grain shapes.

8.5.6 SEDIMENT TYPE

There are several methods commonly used to classify marine sediments besides the Wentworth grain size classes listed in Table 8.3. Sediments can be characterized by origin, composition, relative grain size contribution, or some combination of these methods. The US Naval Oceanographic Office has over 400 sediment classes in its Master Sediment Database. The following brief description of the classification process is provided to help acousticians understand the mind set of marine geologists and the resultant sediment classification schemes and categories they developed.

The first-order classification is based on origin and includes lithogenous (terrigenous) sediments, derived from weathering igneous and sedimentary rocks; biogenic (carbonate or siliceous) derived from organisms; hydrogenous (authigenic) sediments derived from chemical precipitation from seawater: and cosmogenous from outer space including space dust and meteors [5,6]. Only the terrigenous and biogenic sediments are common enough to be of concern to underwater acoustics.

The second-order classification is based on composition. Terrigenous sediments in the deep sea include sediments transported via turbidity currents or clay-sized particles slowly accumulating on the seafloor as pelagic deposits. Biogenic sediments include the skeletal remains of calcareous (foraminifera or coccoliths) or siliceous (diatoms and radiolarian) plankton, with calcareous oozes being the most common. The calcium compensation depth (CCD) is the depth at which the rate of dissolution of carbonate balances the rate of accumulation. Below the CCD (about 4500 m) carbonates are absent and siliceous oozes and pelagic clays dominate. The most recent worldwide compilation of surficial sediment composition and distribution is provided in an interactive global map [70].

Shallow terrigenous sediments are composed of larger quartz particles (sand) and finer muds (silts and clays) composed of clay minerals. Note that the term "clay" applies to both composition (clay minerals) and a grain size category. Collectively these are classed as siliciclastic sediments. Shallow-water sediments of biogenic origin are predominately carbonate in composition and are derived from the breakdown of shells or from coral reefs. Sediments within each class are further subdivided based on grain size distribution. The most obvious characteristic is mean or median grain size. Common descriptors for the various size classes are provided in Table 8.3. This type of classification only makes sense when the sediment is dominated by a narrow range of grain sizes, such as in Fig. 8.18. However, many types of sediment are mixtures of one or more size classes, often of different origin or composition. To account for these sediment mixtures numerous two- and three-component textural classification schemes have been developed. Two-component systems are based on ratios of two of the following size classes: gravel, sand, or mud (silt and clay). Examples include sandy mud, gravelly sand, muddy sand (see Fig. 8.19), mud, and sand, where the dominant size class is the second or only term in the classification. These classes can be further modified by sediment origin or composition, such as pelagic clays, carbonate sand, or shelly mud. A more complex three-component classification system is based on the ratios of three of the following size classes (gravel, sand, silt, and clay) and is often plotted in the form of a ternary diagram. Two examples are shown in Figs. 8.21 and 8.22. More than 20 schemes have been proposed for the two- or three-component classification of sediments (see Blott and Pye [56], for a historical review). At this point in the discussion it should be obvious that sediment classification schemes overlap and that the same descriptor can be used for different sediment types. This is further complicated by differences in sediment collection, actual grain size analysis, and methods

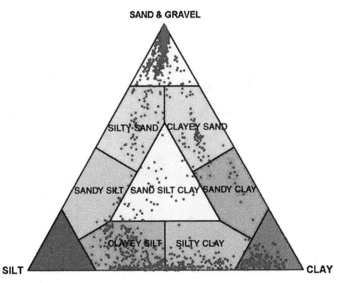

FIGURE 8.21

Ternary diagram of Shepard [71] based on sand–silt–clay ratios.

From Jackson, D.R. and Richardson, M.D, High-Frequency Seafloor Acoustics. Springer, 2007, with permission of Springer.

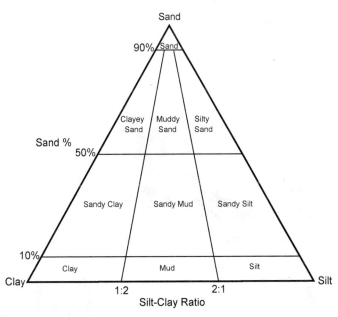

FIGURE 8.22

Ternary diagram from Folk [51] based on sand–silt–clay ratios.

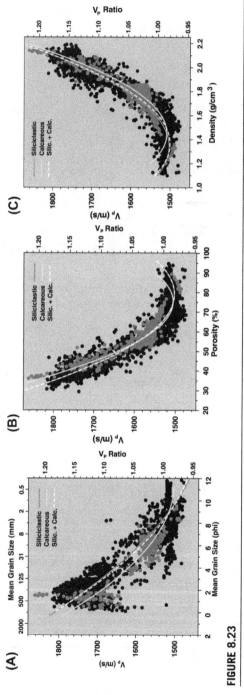

FIGURE 8.23

Sound speed (unit: m/s) and the dimensionless sound speed ratio (V_pR), measured in the laboratory at 400 kHz as functions of (A) mean grain size (unit: phi and mm), (B) porosity (unit: %), and (C) bulk density (unit: g/cm^2). The *lighter symbols* and *solid lines* represent carbonate sediments and overlay the *darker symbols* and *dashed lines* which represent siliciclastic sediments.

for calculating the statistics of grain size distributions. It is therefore important to understand the methodologies used to characterize the sediment type before developing or comparing empirical relationships between sediment type and acoustic properties such as wave speed and attenuation (next section) or scattering strength versus grazing angle relationships for various sediment types. In this chapter the sediment nomenclature follows that used by Hamilton (multiple articles) and Jackson and Richardson [4] based on Shepard's [71] ternary diagram (Fig. 8.21) and the Wentworth grain size scale (Table 8.3).

8.5.7 SUMMARY OF SEDIMENT PROPERTIES

Seafloor properties and descriptions are often used to predict acoustic—sediment interactions such as propagation within sediment, reflection and scattering from the sediment—water interface, shallow-water reverberation, and penetration of acoustic energy into the seafloor. Some properties such as sediment bulk density are fundamental properties of the seafloor that are included in physical models of acoustic—sediment interactions. Porosity, void ratio, or water content can be used to directly calculate sediment bulk density assuming the densities of pore water and grains are known. Other properties such as mean grain size or other grain size distributions statistics are not fundamental properties of acoustic—seafloor models but can be used as empirical predictors of bulk density (Fig. 8.20) and compressional and shear wave speed (Figs. 8.23—8.25) and thus provide trends in acoustic-sediment interactions. Grain size may also be used to predict values of grain and pore parameters used in Biot-type propagation models. Grain size properties are fundamental properties used for modeling sediment transport and other morphodynamic processes at the seafloor. Acoustic—sediment interactions are often summarized by sediment-type categories. These empirical classifications may be useful as first-order summaries of acoustic—sediment interactions such as in scattering models (Table 8.2) or for geoacoustic modeling [1]. However, direct measurement of the relevant sediment properties are needed for site-specific modeling of acoustic—seafloor interactions. This is especially true for determining sediment roughness values, as discussed in Section 8.7 and sediment volume scattering parameters as discussed in Section 8.8, which are spatially and temporally variable and poorly predicted by sediment physical properties such as mean grain size.

8.6 SEDIMENT GEOACOUSTIC PROPERTIES

Edwin Hamilton published his classic approach to geoacoustic modeling during the 1960s to 1980s. These mostly empirical models include values and gradients of compressional and shear wave speed and attenuation and bulk density of sediment layers down to and including the basement. These physical models were mostly developed in support of lower-frequency ASW applications. In his article Hamilton

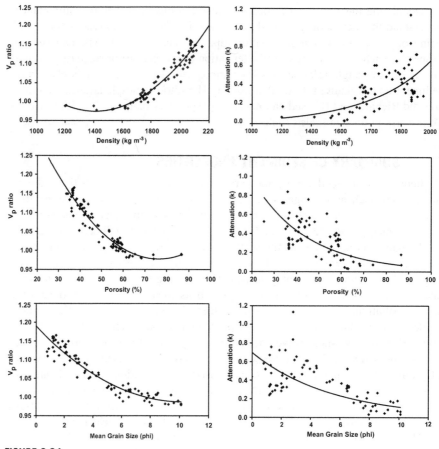

FIGURE 8.24

Sound speed ratio (V_pR (unit: dimensionless)) and attenuation (k (unit: dB/m kHz)) as a function of bulk density, porosity, and mean grain size for average values of in situ sound speed ratio and sediment physical properties at 87 sites where the ISSAMS systems were deployed. The data include both siliciclastic and carbonate sites. The regressions are an update from similar regressions presented in Richardson [81].

From Jackson, D.R. and Richardson, M.D, High-Frequency Seafloor Acoustics. Springer, 2007, with permission of Springer.

[2] provides an excellent summary of his previous work. Near-surface sediment geo-acoustic properties were measured in situ or using laboratory measurements from core samples and included Hamilton's own measurements and an extensive review of the literature. Hamilton provided a series of tables for near-surface sediment properties for different sediment types in common sediment provinces (e.g., continental terrace, hemipelagic, carbonate, and siliceous) worldwide [3,46]. Province-based regressions for sediment geoacoustic properties were also developed with sediment

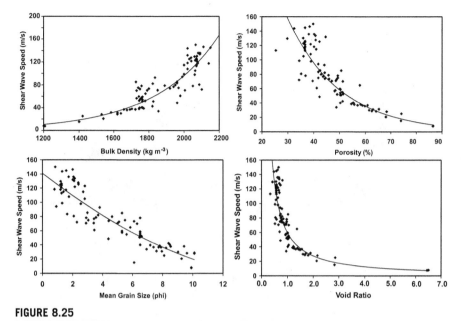

FIGURE 8.25

Empirical relationships between shear wave speed (unit: m/s) and sediment bulk density, porosity, mean grain size, and void ratio for 87 sites.

From Jackson, D.R. and Richardson, M.D, High-Frequency Seafloor Acoustics. *Springer, 2007, with permission of Springer.*

mean grain size the most commonly available sediment parameter (independent variable). Sound velocity gradients were based on wide-angle reflection (sonobouy) methods and are presented as velocity—depth regressions for various provinces. At the time of Hamilton's reviews, very few in situ measurements of shear wave speed or attenuation had been made which is reflected in variability in both surficial values of shear wave properties and gradients. The potential effects of sound speed dispersion and a nonlinear relationship between compressional wave attenuation and frequency were largely ignored by Hamilton. The latest compilations of depth gradients in compressional and shear wave speed and attenuation [72—74] demonstrate the difficulties in predicting gradients of sediment geoacoustic properties based on sediment physical properties. The following sections discuss empirical methods to provide predictions of surficial sediment geoacoustic properties required for modeling acoustic propagation, reflection, and scattering discussed elsewhere in this chapter.

8.6.1 SOUND SPEED AND ATTENUATION

Hamilton and Bachman [46] and later in Bachman [75] provided regressions to predict high-frequency sound speed from easily measured sediment properties such as mean grain size, porosity, and bulk density. These regressions were developed for

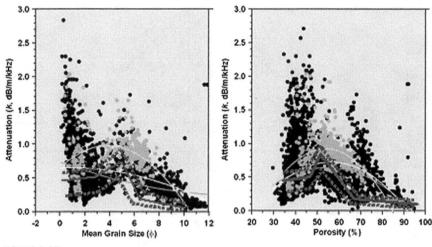

FIGURE 8.26

Attenuation factor (k (unit: dB/m kHz)), measured in the laboratory at 400 kHz, as a function of porosity (unit: %) and mean grain size (unit: phi). The *lighter symbols* represent carbonate sediments and overlay the *darker symbols* which represent siliciclastic sediments. The *red lines* (dark gray in print versions) represent similar fits to data from Hamilton [3,80].

near-surface sediments on the continental shelf primarily from laboratory measurements of remotely collected cores. Sediment sound speed is affected by temperature, salinity, and pressure of sediment pore water. Therefore, Hamilton eliminated these effects by correcting sediment sound speed to a common temperature (23°C), pressure (1 atm) and pore water salinity (35 ppt). Bachman [76] later introduced the dimensionless sound speed ratio (V_pR, denoted v_p in acoustic modeling sections of this chapter), which is the ratio of sediment sound speed to sound speed in water at the same temperature, pressure, and salinity as the independent variable in these regressions. These predictive regressions, including compressional wave speed and attenuation, were updated by Richardson and Briggs [77,78] for a wide variety of shallow-water sediments based of laboratory measurements (400 kHz) of carefully collected diver cores (Figs. 8.23 and 8.26).

The geoacoustic and physical property measurements by Richardson and Briggs [77,79] were coincidently made at 2 cm intervals in the upper 25 cm of sediment from nearly 800 cores collected at 57 siliciclastic and 12 carbonate sites. Sound speed at Hamilton's common temperature (23°C), salinity (35 ppt) and pressure (1 atm) and the dimensionless sound speed ratio (V_pR) are plotted. These data were used to develop a series of regressions for the sound speed ratio and values of sediment porosity, bulk density, and mean grain size (Table 8.5). As suggested from Fig. 8.23, the sound speed ratio is better predicted (higher R-squared value) by bulk density and porosity than by sediment mean grain size for both siliciclastic

Table 8.5 Regressions for Sediment Physical and Geoacoustic Properties for Siliciclastic and Calcareous Sites

Sediment Type	Regression	Number of Points	r^2
Siliciclastic	$V_pR = 1.603 - 0.0156\eta + 0.0001\eta^2$	3905	0.95
Carbonate	$V_pR = 1.760 - 0.0206\eta + 0.0001\eta^2$	609	0.91
All sediments	$V_pR = 1.606 - 0.0158\eta + 0.0001\eta^2$	4514	0.95
Siliciclastic	$V_pR = 1.585 - 0.8991\rho + 0.3352\rho^2$	3905	0.94
Carbonate	$V_pR = 1.878 - 1.2289\rho + 0.4232\rho^2$	609	0.90
All Sediments	$V_pR = 1.649 - 0.9807\rho + 0.3595\rho^2$	4514	0.93
Siliciclastic	$V_pR = 1.184 - 0.0288M_z + 0.0008M_z^2$	2392	0.82
Carbonate	$V_pR = 1.161 - 0.0308M_z + 0.0013M_z^2$	371	0.82
All Sediments	$V_pR = 1.184 - 0.0307M_z + 0.0010M_z^2$	2763	0.82
All Sediments	$k = 0.74 - 0.07M_z - 0.02M_z^2$	2653	0.10
All Sediments	$k = -1.121 + 0.066\eta + 0.0006\eta^2$	4391	0.19

Geoacoustic and physical properties: dimensionless sound speed ratio (V_pR), mean grain size (M_z (unit: ϕ)), porosity (η (unit: %)), bulk density (ρ (unit: g/cm^3)), and attenuation factor (k (unit: dB/m kHz)).

and carbonate sediments. Although values of attenuation (expressed as k (unit: dB/m kHz)) appear slightly higher in coarser sediments, the regressions in Table 8.5 do not provide much confidence in prediction of attenuation from sediment physical properties at these high frequencies (400 kHz). Details on measurement techniques, collection sites, and the data that are used for these regressions can be found in Chapter 5 of Jackson and Richardson [4].

These measurements were later augmented by a series of in situ measurements of compressional and shear wave speed and attenuation collected at 88 sites worldwide using the In Situ Sediment Acoustic Measurement System (ISSAMS) [81] (Figs. 8.24 and 8.25; Tables 8.6 and 8.7). Values of in situ sound speed (38 kHz) were only slightly lower (−7.4 m/s) than values of laboratory sound speed

Table 8.6 Empirical Relationships Among Sediment Physical and Geoacoustic Properties for Measurements Using ISSAMS

Property	Regression	Number of Points	r^2
Density	$V_pR = 1.705 - 1.035 \times 10^{-3}\rho + 3.664 \times 10^{-7}\rho^2$	86	0.92
Porosity	$V_pR = 1.576 - 0.015677\eta + 1.0269 \times 10^{-4}\eta^2$	86	0.91
Mean grain size	$V_pR = 1.19 - 0.03976M_z + 1.9476 \times 10^{-3}M_z^2$	86	0.92
Density	$k = 0.00332e^{0.00241\rho}$	87	0.45
Porosity	$k = 2.153e^{-0.0401\eta}$	87	0.43
Mean grain size	$k = 0.697e^{-0.183M_z}$	87	0.52

Regressions are plotted in Fig. 8.24. Geoacoustic and physical properties are velocity ratio (V_pR (unit: dimensionless)), mean grain size (M_z (unit: ϕ)), porosity (η (unit: %)), bulk density (ρ (unit: kg/m^3)), and attenuation (k (unit: dB/m kHz)).

Modified from Jackson, D.R. and Richardson, M.D, High-Frequency Seafloor Acoustics. Springer, 2007, with permission of Springer.

Table 8.7 Empirical Relationships Among Sediment Physical and Shear Wave Properties for In Situ Measurements Using ISSAMS

Property	Regression	Number of Points	r^2
Density	$V_s = 0.3823e^{0.00284\rho}$	87	0.85
Porosity	$V_s = 690e^{-0.0496\eta}$	87	0.86
Mean grain size	$V_s = 142 - 16.65M_z + 0.455M_z^2$	87	0.82
Void ratio	$V_s = 59.7e^{-1.12}$	87	0.86
Density	$k_s = 322 - 0.136\rho$	18	0.65
Porosity	$k_s = -42.2 + 2.27\eta$	18	0.64
Mean grain size	$k_s = 20.4 + 9.23M_z$	18	0.50

Regressions are plotted in Figs. 8.25 and 8.27. Geoacoustic and physical properties, shear wave speed (V_s (unit: $m \cdot s^{-1}$)), attenuation (k_s (unit: dB/m kHz)), mean grain size (M_z (unit: ϕ)), porosity (η (unit: %)), void ratio (e (unit: dimensionless)) and bulk density (ρ (unit: kg/m^3)). The symbol e in the void ratio regression is the void ratio and not the base of the natural logarithms.

(400 kHz) at the same sites suggesting that sound speed varies little over frequencies covering the range of 38−400 kHz. Sound speed dispersion, especially in sands, suggest that these regressions are not appropriate for acoustic frequencies that are much lower than measured using ISSAMS [19,82]. Values of laboratory measured attenuation, expressed as frequency-independent attenuation (k (unit: dB/m kHz)) were about 30% higher than in situ values suggesting poroelastic effects and scattering losses due to either larger grains or to porosity fluctuations need to be considered [24].

8.6.2 SHEAR WAVE MEASUREMENTS

A variety of laboratory and in situ techniques have been used to measure shear wave speed and attenuation. These include lower-frequency seismic reflection and refraction techniques, seismic tomographic measurements, inversions of interface waves, towed sleds, probes, and laboratory geotechnical consolidation testing as discussed in Chapter 5 in Ref. [4]. The wide variety of techniques, frequencies, and depths make generalizations of shear wave properties needed for acoustic modeling difficult. However, the following generalizations are appropriate. All marine sediments support shear waves; shear wave speed is higher in sand than in mud; depth gradients of shear wave speed follow a power law with an exponent of 0.25−0.33; shear wave attenuation is one to two orders higher than for compressional waves and decreases rapidly within the upper 5 m becoming nearly constant below 10−20 m. In this section recent in situ measurements of surficial sediment shear wave and physical properties are used to develop empirical predictions of shear wave speed and attenuation in surficial sediment, similar to those developed for sound speed and attenuation in the previous section.

Shear wave properties were measured at 87 locations using ISSAMS [83]. Shear wave speed was measured using time-of-flight between bimorph ceramic benders and

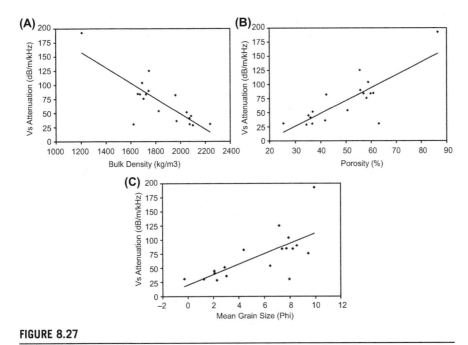

FIGURE 8.27

Empirical relationships between shear wave attenuation {dB m^{-1} kHz^{-1}} and (A) Bulk Density, (B) Porosity, and (C) Mean Grain Size for 19 sites including both carbonate and siliciclastic sediments.

From Jackson and Richardson [4], with permission of Springer.

attenuation (only at 19 sites) was measured by means of transposition methods using two transmitters and two receivers [84]. Multiple measurements were made at depths between 20 and 30 cm below the sediment–water interface at each site with frequencies ranging from 1 kHz in sands to as low as 100 Hz in mud. Shear wave speed increased from less than 10 m/s in high-porosity, low-density mud to greater than 150 m/s in low-porosity, high-density sands, as shown in Fig. 8.25. Shear wave attenuation increased from near 20 dB/m kHz in low porosity, high-density sands to greater than 100 dB/m kHz in high-porosity, low density mud, as shown in Fig. 8.27. These values of shear wave attenuation (k_s) (unit: dB/m kHz) are roughly two orders of magnitude higher than for the sound speed attenuation given in Fig. 8.26. It is also interesting that the empirical trends in attenuation as a function of sediment physical properties behave in an opposite direction for shear and compressional waves. The empirical regressions between sediment shear wave properties and sediment physical properties are presented in Table 8.7.

Gradients of shear wave speed (unit: m/s) and attenuation (unit: dB/m) measured using ISSAMS at 38 kHz for a sandy site in the North Sea [4] are presented in Fig. 8.28. Best fit linear, $V_s = 72.3 + 232d$, and power law, $V_s = 186.58d^{0.27}$, regressions account for 83% and 81% of the variability for shear wave speed

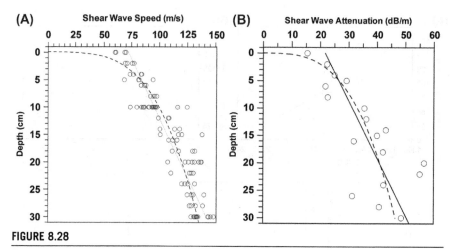

FIGURE 8.28

Gradients of (A) shear wave speed (unit: m/s) and (B) attenuation (dB/m) measured using ISSAMS at 38 kHz for a sandy site in the North Sea.

From Jackson, D.R. and Richardson, M.D, High-Frequency Seafloor Acoustics. *Springer, 2007, with permission of Springer.*

(unit: m/s) where d is depth (unit: m). The power-law relationship provides more realistic values of shear wave speeds at the surface where shear wave speeds should tend to 0 m/s. Best fit linear, $\alpha_s = 21.7 + 96d$, and power law, $\alpha_s = 67.1d^{0.31}$, regressions for shear wave attenuation account for 58% and 62% of the variability, respectively.

Empirical relationships between shear wave speed and sediment physical properties (Table 8.7) combined with a power-law depth dependence of $d^{0.3}$, as shown in Fig. 8.28, are used to provide empirical shear wave speed predictions in the upper meter of sediment based on easily measured sediment physical properties. Predictions of gradients in shear wave attenuation are not presented because of the high variability and uncertainty in empirical relationships presented in Fig. 8.27 and Table 8.7.

The following empirical predictive equations for shear wave speed (V_s) in the upper meter of sediment were generated by combining the most recent empirical relationships between sediment physical properties (Fig. 8.25; Table 8.7) with a depth dependence $d^{0.3}$,

$$V_s = 992e^{-0.0494\eta}d^{0.3}, \tag{8.95}$$

$$V_s = 0.549e^{0.00284\rho}d^{0.3}, \tag{8.96}$$

$$V_s = \left[202.5 - 23.9M_z + 0.6528M_z^2\right]d^{0.3}, \tag{8.97}$$

$$V_s = 85.7e^{-1.12}d^{0.3}, \tag{8.98}$$

where d is depth below the sediment−water interface (unit: m), n is porosity in percent, ρ is bulk density (unit: kg/m³), M_z is mean grain size expressed in phi units, and e in Eq. (8.98) is the dimensionless void ratio not the base e of the natural logarithms.

8.6.3 **INDEX OF IMPEDANCE**

Many acoustic sediment classification systems use the amplitude of the echo returns (reflection) to estimate seafloor acoustic impedance (see Section 8.9). Empirical relationships between seafloor impedance and sediment physical/geoacoustic properties can then be used to map values of seafloor properties such as porosity, bulk density, mean grain size, sound speed and attenuation, and shear wave speed and attenuation. Sediment impedance is the product of sediment sound speed and bulk density. Similar to sound speed, impedance is dependent on pore water temperature, salinity, and pressure (water depth). Therefore, a pore water−independent index of impedance, IOI (unit: g/cm^3) defined as the product of the bulk density (unit: g/cm^3) and the dimensionless sound speed ratio (V_pR), was used to develop empirical relationships between sediment impedance and sediment physical and geoacoustic properties [77,79]. If the acoustic impedance estimated from a sediment classification system is reported in units of kg/m^2s, then values of IOI equal 0.001 times the product of the sediment impedance (unit: kg/m^2s) divided by the sound speed of the in situ pore water (unit: m/s). Previous IOI/sediment property regressions were based on high-frequency (400 kHz) sound speed measurements from small cores; see Figs. 8.23 and 8.26 and Table 8.5. The IOI regressions given here are based on in situ sound speed measurement made at 38 kHz, see Fig. 8.24 and Table 8.6. The lower acoustic frequencies and the fact that geoacoustic and physical property measurement are averaged over larger spatial distances (lower spatial resolution) better match acoustic classification that employs averaging over pings to smooth out time series fluctuations, see Section 8.9. IOI regressions for bulk density, ρ (unit: kg/m^3); porosity, P (unit: %); mean grain size, M_z (unit: ϕ); dimensionless velocity ratio, V_pR; and shear wave speed, V_s (unit: m/s) all have high R^2 values, whereas compressional wave attenuation is poorly predicted from IOI, see Fig. 8.29:

$$\eta = 168.31 - 84.61(\text{IOI}) + 12.29(\text{IOI})^2, \quad R^2 = 0.97 \qquad (8.99)$$

$$M_z = 20.95 - 8.20(\text{IOI}), \quad R^2 = 0.87 \qquad (8.100)$$

$$V_pR = 1.114 - 0.25(\text{IOI}) + 0.11(\text{IOI})^2, \quad R^2 = 0.96 \qquad (8.101)$$

$$V_s = 6.65(\text{IOI})^{3.4}, \quad R^2 = 0.85 \qquad (8.102)$$

$$\rho = 1387 + 3525(\text{IOI}) - 1154.4(\text{IOI})^2 + 154.72(\text{IOI})^3, \quad R^2 = 0.99 \qquad (8.103)$$

The above regressions follow similar trends to the regressions based on higher-frequency (400 kHz) laboratory measurements reported by Richardson and Briggs [78,79]. The in situ based predicted values are slightly lower than the previously reported high-frequency laboratory predictions [77,79] for porosity, mean grain size, the velocity ratio, and slightly higher for bulk density. These difference reflect slightly lower sound speed measured at 38 kHz compared to speeds measured at 400 kHz. The trends for in situ compressional wave attenuation shown in Fig. 8.27 were similar to the attenuation trends at 400 kHz with both exhibiting little predictive value.

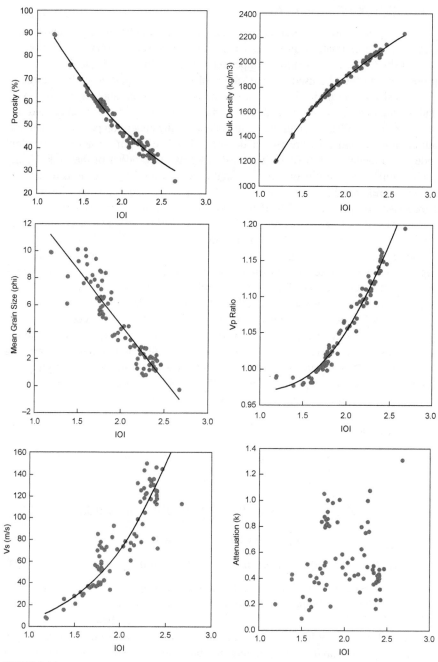

FIGURE 8.29

Empirical relationships used to predict sediment physical and acoustic properties from the index of impedance (IOI (unit: g/cm^3)). Data and regressions are based on averaged in situ sound speed measurements (38 kHz) at 81 sites. Regressions and R^2 values for the above regressions are given in Eqs. (8.99)–(8.103).

8.7 SEAFLOOR ROUGHNESS

Seafloor roughness is a major contributor to scattering, reverberation, and bottom loss over the range of acoustic frequencies considered in this chapter. Penetration of high-frequency energy into the seafloor can also be enhanced by periodic bottom roughness features, such as sand ripples. The morphology of the seafloor encompasses a wide variety of spatial scales from large-scale tectonic features such as the mid-Atlantic Ridge to regional features such as sand ridges, coral reefs, and beach profiles and to fine-scale features associated with sand ripples or biogenic structures. Larger-scale features (morphology) are generally persistent over century-long time intervals and are created by tectonic or geological erosion or depositional processes. These larger bathymetric features are often characterized using hull-mounted single-beam and multibeam sonar [85]. Global bathymetric databases merge these acoustically derived data with satellite altimetry based on gravity anomalies [86]. In some areas, such as the possible crash site of MH-370 in the southeastern Indian Ocean, bathymetric features are very poorly known. At other sites, such as the mid-Atlantic Ridge they have been well characterized [7]. Regional-scale roughness is often characterized by higher-frequency multibeam, side-scan, or interferometric sonar systems [87] and airborne LIDAR [88]. The persistence of these bathymetric features is a function of water depth, sediment type, and larger-scale hydrodynamic events such as storms. Ripple scour depressions or sand ridges may persist for decades in the same location, whereas beach slopes may vary with season. Fine-scale roughness has been characterized at selected sites as part of high-frequency acoustic or seafloor morphodynamic experiments [54]. This scale of micro roughness is often created, modified, or destroyed by competing hydrodynamic and biological processes. The persistence of these features varies with size where centimeter-scale roughness can change over scales of hours and meter-scale features are often persistent for weeks or months [89]. The wide range of spatial scales of roughness and temporal persistence scales greatly complicates methods used to measure or predict seafloor roughness at scales appropriate to the acoustic applications. As was demonstrated earlier in the chapter, roughness features having scales comparable to the acoustic wavelength are responsible for most scattering from the seafloor. Therefore, as a generalization, the higher the acoustic frequency the less the persistent and predictable is the appropriate spatial roughness; and the greater the temporal decorrelation of acoustic returns from the seafloor [89–91].

   The basis (raw data) for statistical characterization of seafloor roughness is the 2-D and 3-D bathymetric maps or, for higher-resolution, higher-frequency applications the raw data includes high-resolution elevation profiles or digital elevation models (DEMs). Bottom scattering models require some sort of statistical characterization of bottom roughness that can include root mean square (RMS) roughness, autocovariance functions, fractal characterization, or some form of power spectral density. In the next two sections we discuss methods to measure and to quantify seafloor roughness appropriate for acoustic bottom interacting models.

8.7.1 MEASUREMENT OF SEAFLOOR ROUGHNESS

A variety of techniques have been used to quantitatively characterize seafloor roughness. Bathymetric surveys during the 1950s through 1980s provided the data needed to statistically quantify large-scale and regional bottom roughness [92]. Many of these systems were hull-mounted single-beam echo sounders (SBESs) which provided bathymetric profiles hundreds of kilometers in length with vertical resolutions on the order of meters to tens of meters. More recently, lower-resolution global bathymetry has been inferred from satellite radar altimetry [7]. A variety of statistical approaches from simple RMS statistics, autocorrelation, spectral analysis, and fractal analyses were used to characterize bottom roughness at these scales. Today's higher-frequency hull-mounted and towed swath interferometric and multibeam echo sounders (MBESs) provide much higher-resolution 3-D bathymetric maps [91,93]. In coastal regions, vertical sampling resolutions better than 10 cm with 0.5−1.0 degree horizontal beam widths are possible using these commercial acoustic systems. Airborne LiDAR systems can achieve horizontal resolutions of 15−30 cm with 2−5 m horizontal resolutions [88]. During the 1980s high-frequency acoustic and stereo photographic techniques were developed to provide fine roughness profiles shown in Fig. 8.30 at millimeter scales over meter-scale path lengths [94]. Digital elevation maps shown in Fig. 8.31 have been measured

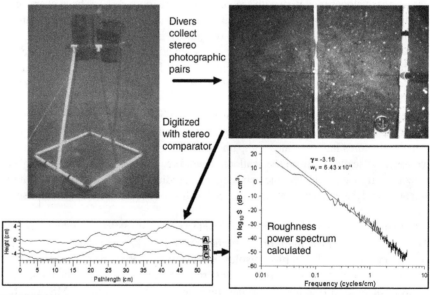

FIGURE 8.30

A summary of photogrammetric techniques used by Briggs to measure seafloor micro topography. Stereo photographs are made with a 35-mm Photosea 2000 camera; paired images are processed with a Benima stereocomparator to obtain relative height profiles; power spectra are estimated from the profiles and are then averaged, see Briggs [94], for details.

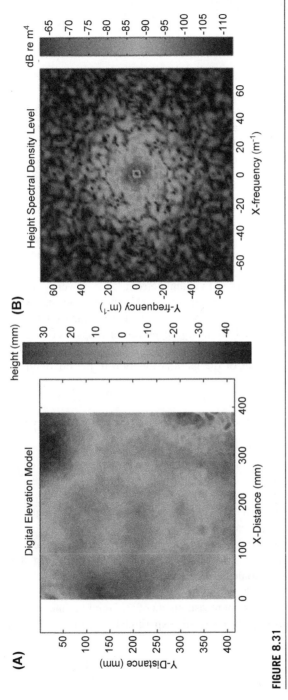

FIGURE 8.31

(A) Digital elevation map calculated from stereo photographs and (B) 2-D roughness spectrum produced from the digital elevation map in Fig. 8.25. The isotropy of the spectrum is evident in this figure [101].

by stereo photography [95], electrical resistivity probing [96], laser line scanning [97,98], mobile laser imaging [99], and high-frequency sector scanning or pencil beam sonar [100].

8.7.2 STATISTICAL CHARACTERIZATION OF SEAFLOOR ROUGHNESS

A variety of interrelated statistical approaches from simple statistics (RMS), auto-correlation, covariance, autocorrelation, lag, fractal, and 1-D and 2-D spectral analysis have been used to characterize bottom roughness. All begin with some sort of measured relief spectrum. For most scattering models roughness is assumed to exhibit spatial stationary and to obey Gaussian statistics.

In the high-frequency scattering model described in Section 8.4, roughness is assumed to follow a simple 2-D power law,

$$W(K) = (w_2/K^{\gamma_2}). \tag{8.104}$$

The parameter w_2 is the "roughness spectral strength," and the parameter γ_2 is the dimensionless "roughness spectral exponent," with values ranging from 2 to 4. This power-law spectral form can only hold over a finite range of scales. Methods and assumptions used to calculate roughness power spectra from digital elevation maps can be found in appendix D.2 in Jackson and Richardson [4]. A method to convert 1-D power spectra derived from transects (see Fig. 8.30) to 2-D power spectra derived from digital elevation maps (see Fig. 8.31) is also provided, but a typographical error in the relevant expression (D.19) must be recognized: The argument of the gamma function in the denominator should be $(\gamma_2 - 1)/2$.

8.7.3 PREDICTION OF SEAFLOOR ROUGHNESS FROM SEDIMENT PHYSICAL PROPERTIES

Sediment roughness is created and altered by a combination of often competing geological, hydrodynamic, biological, and depositional processes. It is therefore not surprising that values of roughness parameters vary spatially and temporally [89–91,102]. Briggs et al. [103] summarized all available roughness data from recent high-frequency acoustic experiments in an attempt to develop empirical predictions of roughness based on mean grain size. A general trend of lower values of RMS roughness at muddy sites was evident in the data shown in Fig. 8.32. However, the values of the dimensionless spectral exponent (γ_2) and the spectral strength (w_2) (unit: $m^{4-\gamma_2}$), overlapped between sands and muds yielding little predictive value as shown in Fig. 8.33. Concurrent measurement of seafloor roughness spectra is recommended during high-frequency acoustic experiments. Perhaps more time-dependent

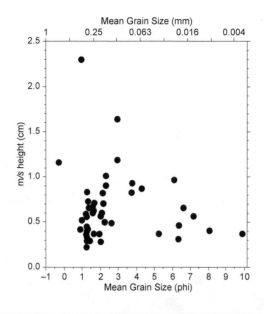

FIGURE 8.32

Plot of RMS roughness (unit: cm) and mean grain size (unit: mm and phi).

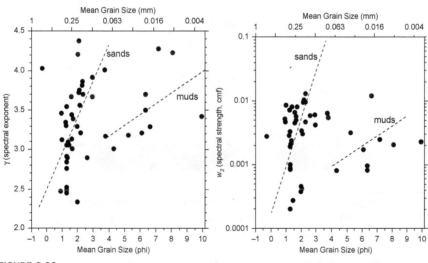

FIGURE 8.33

Plots of 2-D seafloor roughness dimensionless spectral exponent (γ_2) and spectral strength (w_2) (unit: $m^{4-\gamma^2}$) as functions of mean grain size (unit: mm and phi).

models that include the relevant environmental process will allow prediction of sea-floor roughness for different sediment types.

8.8 SEAFLOOR HETEROGENEITY

Spatial variability of values of seafloor physical and geoacoustic properties can be caused by a variety of physical, biological, and biogeochemical processes, as shown in Fig. 8.34. This heterogeneity (or inhomogeneity) leads to acoustic volume scattering as a result of fine-scale fluctuation in sediment properties, discrete scattering from objects such as imbedded shells or rocks, scattering and reflection from layering of sediment strata, or scattering from gas-charged sediments. The 3-D structure

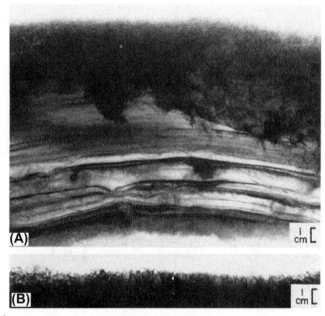

FIGURE 8.34

(A) An X-radiograph of a 2-cm-thick slab of sediment collected from a muddy site in Long Island Sound [104]. The surface top layer (3—5 cm) is well mixed by reworking of the sediment by abundant benthic macrofauna. The deeper layers are alternate layers of coarser more dense (*darker*) sediments and finer less dense (*lighter*) sediments deposited during storms. With time, bioturbation can mix sediments to depths of at least 20 cm. (B) An X-radiograph of the upper 3 cm of the same sediment core showing a dense population of the bivalve *Mulinea lateralis* (exposure time reduced). The gray-scale density in the X-radiograph can be considered a proxy for bulk density.

of sediment can be very difficult to characterize at both fine scales associated with near-surface high-frequency acoustic sediment volume scattering and at geological scales associated with lower-frequency volume scattering, reflection, and propagation. Typical methods of sediment characterization from cores and grab samples do not adequately capture this variability because of the poor spatial resolution of small internal diameter cores at all scales. Ambiguity associated with use of lower-frequency seismic techniques to characterize sediment heterogeneity make model-data comparisons difficult.

For modeling acoustic volume scattering, it is convenient to divide heterogeneity of sediment physical properties into random and nonrandom parts. The nonrandom part is either the average value over a patch of seafloor or a profile or gradient of values of that property often characterized with depth in the seafloor. The random part is the fluctuation about the average or gradient of values of the physical property, often described statistically in terms of variance, covariance, or power spectra, as discussed in Section 8.4. Many high-frequency volume scattering models require spectra describing the spatial fluctuations of sediment bulk density and compressibility (porosity and compressional wave speed). Lower-frequency scattering models generally require only characterization of the nonrandom heterogeneity (gradients, layering, and lateral variability) where acoustic wavelengths are smaller than the patch size.

8.8.1 MEASUREMENTS OF SEDIMENT VOLUME HETEROGENEITY

A variety of direct and indirect methods have been used to characterize sediment heterogeneity for high-frequency volume scattering: traditional laboratory analyses of sediment cores, X-radiography, CT-scanning, optical imaging from sediment profile cameras, electrical resistivity probing and imaging, and high-frequency acoustic imaging or tomography. Jackson and Richardson [4] provides a review of and examples from these techniques. Briggs et al. [105] provides recent comparative examples. None of these techniques alone appears to provide the range of spatial scales and accuracy required for high-frequency acoustic volume scattering models. Many techniques only provide 1-D gradients or 2-D images of sediment properties, whereas scattering models may require a 3-D statistical characterization of heterogeneity, see Eq. (8.50). Multiple techniques are often required to determine the spatial fluctuations of sediment bulk density and compressibility required by many volume scattering models. Statistical distributions of larger-volume discrete scatters, such as shells and rocks, are rarely, if ever, adequately quantified. Attempts to determine power spectral parameters from inversions of measured acoustic scattering have been less than successful. Different characterization techniques often yield very different spectra for the same sediment.

A variety of well-developed seismic exploration techniques have been developed to characterize the distribution and properties of deeper sediment strata. Sediment properties, such as sound speed, attenuation, bulk density, and porosity are routinely measured in the laboratory from gravity, piston, or vibracores. Down-hole or cross-hole acoustic and electrical tomographic techniques are also used to measure sediment physical and geoacoustic properties. These techniques typically only provide characterization of the nonrandom part of the sediment heterogeneity (gradients or average strata properties) which is adequate for many lower-frequency acoustic applications.

Jackson and Richardson [4] provided a compilation of coefficients of variation (CV), which is the ratio of the standard deviation to the mean of any variable multiplied by 100, for sediment physical and geoacoustic properties from shallow-water siliciclastic (62) and carbonate (20) sites, see Appendix C in Ref. [4]. Measurements were made on sediments carefully collected with cores by divers or from box cores. The coefficient of variation was lowest for sound speed (CV = 1.20), followed by bulk density (CV = 2.95), porosity (CV = 5.84), mean grain size (CV = 18.74) and sound speed attenuations (CV = 32.53). Coarser-grained sediments (sands) had lower CV values for grain size, porosity, and bulk density and higher values for sound speed and mean grain size.

8.8.2 GAS IN SEDIMENTS

Pervious descriptions of sediment properties have assumed a two-phase medium: solids and pore fluids. However, the sediment can contain gas dissolved in the pore fluid, in the form of free bubbles, or as a frozen phase (gas hydrates). Gas hydrates are out of the scope of this chapter and the reader is directed to the following reviews [106,107]. Dissolved gas has little impact on acoustic properties, but free gas can greatly alter sound speed and attenuation and result in increased acoustic scattering. Gas, in the form of air, can be entrained in shallow-water sediment due to breaking waves. Photosynthesis in algal mats can trap oxygen bubbles near the sediment–water interface [108] and increased acoustic reverberation has been reported in seagrass meadows when oxygen bubbles form in plant tissues [109]. However, the most striking effects are the result of free methane bubbles in organic-rich muddy sediments of coastal waters [110]. Acoustic terms, such as wipeouts, pull-downs, acoustic masking or blanking, gas horizons, and acoustic turbidity have all been used to describe these gas-charged sediments. Gas concentrations in shallow-water sediments are typically less than 1% by volume but methane volumes as high as 6–9% have been reported for pockmarks [111] and as high as 12% for bays with very high concentrations of organic matter [112]. Methane concentrations of less than 0.01% by volume can have a major impact on compressional wave speed, attenuation, and scattering [113–117]. Fleischer et al. [118] identified over 100 documented cases of free gas bubbles in marine sediments in a 2001 review. The number of documented cases has at least doubled since that review. Most of the acoustic studies are qualitative in nature, and only provide information on the presence, areal distribution, and depth of the gas horizon. However, a number of studies since 1980s have provided data on in situ bubble size distributions with concurrent frequency-dependent measurements of compressional wave speed and attenuation and acoustic scattering [113–117]. The most accurate methods to characterize methane bubble size, shape, and distribution include X-radiography and medical or high-resolution X-ray computed tomography [112,116,119,120]. For the best results, sediments should be maintained in pressure-tight containers until imaged, thus eliminating to effects of pressure release and increased temperature on methane saturation. Bubbles as large as 10 mm equivalent radius were imaged with medical CT scanning in gas-charged sediments from Eckernförde Bay [116].

Most methane in shallow-water sediments is of biogenic origin, the result of metabolic activity of methanogenic bacteria under anaerobic conditions. The diagenetic reaction transport models responsible for methane generation and mechanical fracture models which determine the size, shape, and migration of methane bubbles are well known [121,122]. The concentration of methane in sediments is primarily controlled by the flux rate and quality (reactivity) of organic matter to the seafloor [123]. In muddy sediments, smaller methane bubbles tend to be spherical, whereas larger bubbles tend to be shaped like coins or corn flakes. Methane saturation in sediments is controlled by temperature, salinity, and pressure. The depth and horizontal size of the methane horizon often changes seasonally in response changes in bottom water temperature [124,125]. Rapid local changes in pressure can result in methane ebullition [112,126].

Models of acoustic–bubble interactions in fine-grained sediments developed by Anderson and Hampton [128,129] have been corroborated by laboratory [129] and field [113–117] experiments. Much of the frequency-dependent behavior is related to the methane bubble resonance, with highest predicted attenuation, sound speed, and scattering occurring at frequencies near bubble resonance, as shown in Fig. 8.35. At acoustic frequencies much below bubble resonance "compressibility effects" dominate and sound speed is greatly reduced, increasing the impedance contrast between bubble-free and gas-bearing sediment layers, which results in a reverse polarity reflector in seismic reflection profiles [130,131]. The reduction in compressional wave velocities at these lower frequencies can be used to estimate the distribution of free gas fractional volume. At acoustic frequencies well above resonance, bubbles resonance rarely affects sound speed, and scattering from bubbles dominates attenuation. Minor modification to the Anderson and Hampton [127,128] models that predict resonance bubble frequency, and compressional wave speed and attenuation in sediments with a known distribution of bubble sizes are given in Wilkens and Richardson, [115], Best et al. [117], and Jackson and Richardson [4].

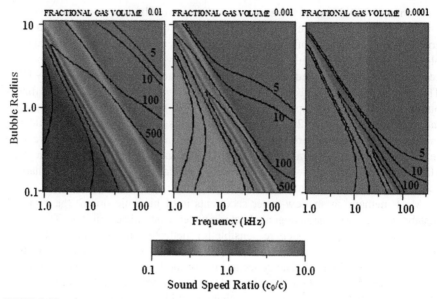

FIGURE 8.35

Sound speed ratio (color scaled where *co/c* is equivalent to the dimensionless V_pR) and attenuation (contours in dB/m) as a function of bubble size and acoustic frequency for three different fractional bubble volume concentrations. Sound speed and attenuation calculations based on the Anderson and Hampton [127,128] model and values of sediment, water, and gas properties are from Wilkens and Richardson [115].

8.9 SEAFLOOR IDENTIFICATION AND CHARACTERIZATION BY USE OF SONAR

Traditionally, the nature of the seafloor has been determined by means of core sampling, photography, and various in situ probing methods. Such measurements are time consuming and expensive when a large area is to be characterized. Acoustic methods have been developed over the past few decades with the aim of providing rapid coverage of large areas. The applications of acoustic methods range from simple identification, such as, for example, hard ground versus soft ground, sediment type (mud, sand, gravel, rock), etc., or more quantitative characterization such as determination of grain size, roughness, acoustic impedance, etc. With high-frequency sonar, one only hopes to determine the properties of surficial sediments, while deeper penetration is possible as frequency is lowered, culminating in seismic measurements capable of revealing deep geological structure. The latter topic is not explored here and attention is given primarily to techniques that operate at frequencies above

10 kHz and that have a monostatic geometry. An extensive review and status report on the use of sonars in determining seabed properties is given in Ref. [132].

Many different types of processing can be used, including (1) texture analysis of sonar images, (2) classification of echo shape, (3) use of second fathometer return, (4) scattering strength, and (5) use of acoustic models to fit data. Of course, combinations of these methods may be used to improve performance. The sonar systems most commonly used for seafloor identification and characterization are single beam, side-scan, and multibeam. Some processing methods are compatible with all three sonar types, some with only one or two. All methods involve extraction of "features" from the echo data, collectively referred to as a "feature vector." The final step in classification is to identify clusters in the multidimensional feature space corresponding to various seabed types, for example, sand, silt, gravel, or whatever division is appropriate to the particular application.

Sonar echoes from the seafloor are quite random, and many approaches to acoustic classification employ averaging over pings to smooth the time series [133]. If the envelope statistics are Rayleigh, the decibel equivalent of the echo has a standard deviation of 5.57 dB [134]. For Rayleigh statistics, the standard deviation of the intensity (proportional to squared envelope) of a single echo is equal to its mean. Averaging the intensities of N pings reduces the ratio of the standard deviation to the mean by a divisor $\sqrt{N}$, so, for $N \gg 1$, the mean intensity will fluctuate with a standard deviation $10 \log_{10} \left(1 + 1/\sqrt{N}\right) \approx 4.34/\sqrt{N}$ dB. Thus, averaging 20 pings should reduce fluctuation to about 1 dB. This assumes that the squared envelope is averaged rather than either the envelope itself or the decibel equivalent of the envelope. Averaging the squared envelope is consistent with most theoretical models that treat average intensity.

Averaging over pings degrades along-track resolution, so there is a trade-off between resolution and random fluctuation of echo features. One must also consider that the sonar platform may have time-varying pitch, yaw, roll, and altitude, and that the seafloor depth and slope may vary. With single-beam systems, the only remedy is to align and average successive pings before averaging. The necessary time shifts may be determined by threshold crossing of the leading edge or by maximizing the cross correlation between members of the group of pings to be averaged. With multibeam systems the same sort of averaging can be performed, but, in addition, platform attitude data can be used to insure that beams are pointing in the desired direction, and the bathymetry provided by the system can be used to determine the true grazing angle of each beam. Fig. 8.36 illustrates alignment and averaging for a multibeam system (Reson 7125) operating at 200 kHz. In this case, five pings have been aligned yielding the RMS envelope shown in Fig. 8.37. No compensation has been made for time-dependent pitch or roll, nor for changes in seafloor topography over the short time (0.7 s) over which these pings were recorded. In actual application [135] four of these five-ping batches are averaged after taking into account differences in roll and topography between batches.

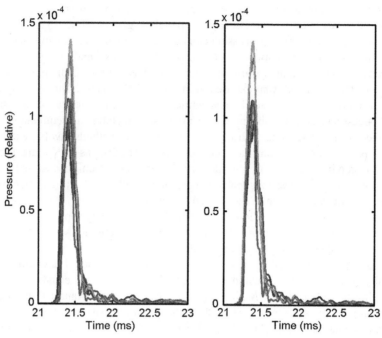

FIGURE 8.36

Echo envelopes for five successive pings for a multibeam system operating at 200 kHz over a sandy seafloor. The left panel shows the envelopes before alignment, and the right panel shows the envelopes after alignment of their leading edges. This beam is about 3 degrees off-nadir with widths of 2 degrees along-track and 1 degree across-track.

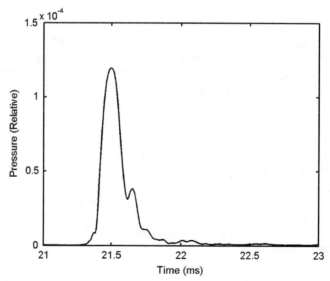

FIGURE 8.37

The RMS envelope obtained from the five aligned envelopes of Fig. 8.36.

The RMS average shown in Fig. 8.37 illustrates some of the aspects of seafloor echoes. The initial peak is predominately due to reflection by the sediment−water interface, while the later portion is at least partly due to scattering from within the volume of the sediment. The pulse length in this example is 150 μs. A longer pulse would cause the echo to smear and merge the interface and volume components. For beams pointed further off-nadir, the time series will be more elongated, and the interface and volume contributions will be intermixed. With a side-scan sonar, the wide beam width in the across-track direction would also prevent separation of interface and volume contributions.

8.9.1 FEATURE CLUSTERING

The features used in classification are extracted from time series exemplified by Fig. 8.37. In some cases, the features are taken directly from the time series [136−139] and involve measures of width and shape, spectral content, energy (integrated, squared envelope), etc. These features will take on differing values as the measurement geometry changes or if different systems are used. This presents a challenge, as the features are not intrinsic to the seafloor. With regard to measurement geometry, the echo width and shape will change as water depth and seafloor slope change. The echo width and shape will also change as the transmitted pulse length changes and as transmitting and receiving directivity change. These problems are circumvented by acquiring ground truth at several locations in the survey area by means of grab sampling, video observation, etc. If sufficient ground truth is available, each cluster in feature space can be associated with a seafloor type. In the case of the QTCView approach [137,138], hundreds of features may be extracted, but these are combined into three essential features by means of principal component analysis before clusters are identified. An example of a survey employing QTCView [140] is shown in Fig. 8.38.

A two-dimensional feature space is employed in the Rox Ann approach, with one feature being the energy (E_1) in the tail of the seafloor echo and the other being the total energy (E_2) in the "second fathometer" echo, which is the later echo resulting from scattering by the seafloor followed by scattering by the ocean surface followed by a second scattering by the seafloor. Fig. 8.39 shows the results of a Rox Ann survey intended to study the impact of trawling on the seafloor [141].

In initial descriptions of this method [142] E_1 was identified as a measure of roughness and E_2 as a measure of hardness. Later analysis and experiment [139] did not support these specific identifications, but indicated that these two features provide useful discrimination of different seafloor types. Discrimination is best when the sea surface is moderately rough [142]. A significant practical advantage of methods such as QTCView and Rox Ann is that they are applicable to a wide range of sonar and seafloor types without appeal to acoustic models. In addition, there is no need for sonar calibration. The primary disadvantage is the need to obtain ground truth, with more ground truth sampling required as the survey area becomes larger and more diverse.

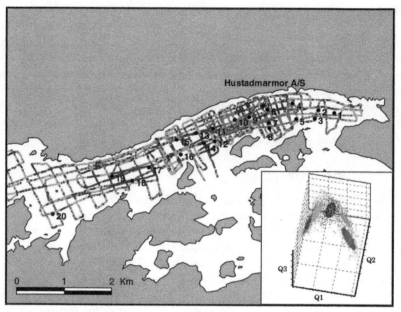

FIGURE 8.38

Habitat survey [140] using the QTCView classification system. The various classes are soft silt—clay (yellow (lighter gray in print versions)), silty sand (turquoise (lighter gray in print versions)), fine sandy silt (blue (gray in print versions)), coarser substrates (pink (lightest gray in print versions)), silt/sand (green (darker gray in print versions)), and coarser substrates (red (dark gray in print versions)). The inset shows clustering in the space of the three first principal components with ellipsoids indicating standard deviations.

Used with permission of Oxford Press.

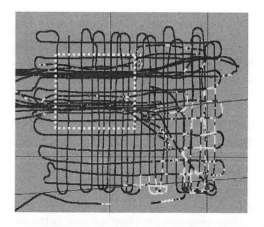

FIGURE 8.39

Rox Ann survey [141] used to identify two different seafloor types: hard-packed sand and mud are shown as *dark lines*, and sand and gravel/stones are shown as *white lines*. The *dashed box* indicates an area chosen for detailed study.

Used with permission of Oxford Press.

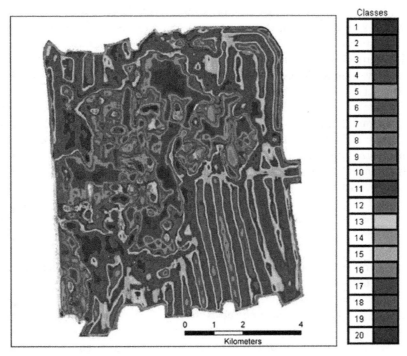

FIGURE 8.40

Classification using texture with multibeam sonar. Discussion of the 20 textural classes is given in the original reference [145].

8.9.2 IMAGE SEGMENTATION

Another approach for seafloor classification employs the texture of images obtained using either side-scan [143,144] or multibeam [145] sonar (Fig. 8.40). This work employs computer-vision segmentation techniques. Before these techniques can be applied, it is necessary to remove trends in the cross-track dependence of image intensity due to propagation and the scattering strength dependence on grazing angle. These methods also have the advantage of not requiring sonar calibration and the disadvantage of requiring substantial ground truth. While side-scan and multibeam sonar provide coverage of wide swaths, use of shallower grazing angles for classification is problematic, since shallower-angle scattering strengths can be quite similar for a wide range of seafloor types, with attendant inversion ambiguity.

8.9.3 REFLECTION

All the methods described earlier employ classification features dependent on measurement geometry and properties of the sonar system employed. That is, the features are not intrinsic properties of the seafloor. Some methods employ an intrinsic feature, the water–seabed reflection coefficient [146–150]. The reflection coefficient provides an estimate of acoustic impedance from which other geoacoustic and geotechnical properties can be determined by means of the regressions presented earlier in this chapter. Inversion for impedance is based on the assumption that the seafloor is flat and without significant stratification on the scale of the pulse resolution. Normal-incidence reflection is usually measured, since measurement at other angles requires two platforms. If the seafloor is layered, it is best to employ sufficient signal bandwidth that the first arrival can be resolved. Inversion in this case is simple: one obtains the normalized impedance using the fluid reflection coefficient from Eq. (8.6). For muds the impedance is the product of sediment sound speed and density, for sands it is the product of sediment sound speed and effective density, and for rock it is the product of compressional wave speed and density. Scattering by the rough interface may bias the measured reflection coefficient downward if all of the scattered energy does not contribute to the receiver output. This will be the case if the scattering strength peak near normal incidence is wider than the combined source–receiver beam pattern. A downward bias will also occur if the pulse length is shorter than the time it takes the scattered echo to die to a small fraction of its peak.

LeBlanc [151,152] developed a method for using reflection by buried interfaces to estimate the slope of attenuation versus frequency. The technique requires wideband data provided by a "chirp sonar" employing matched filtering and careful signal design. Returns from deeper reflectors are deficient in high frequencies due to attenuation and which results in a downward shift of center frequency. A model is employed which ultimately provides an estimate of grain size.

8.9.4 SCATTERING STRENGTH

Another commonly used intrinsic feature is scattering strength, $10 \log \sigma_b$, which is a function of angle. The scattering strength is usually obtained by a simple application of the sonar equation, assuming an ensonified region delineated by the intersection of an annular ring with the beam pattern product. This approach fails near nadir, where the ensonified area has a circular rather than annular boundary, and where the beam patterns and cross section may vary over the ensonified region. To circumvent these difficulties, a generalized version of the sonar equation can be developed. The received intensity (unit: pressure^2) can be written as

$$I(t) = \int_{-\infty}^{\infty} H(t - t')I_0(t')dt', \tag{8.105}$$

where $H(t)$ is the intensity impulse response (unit: $1/m^2$ s) incorporating properties of the environment and sonar, and $I_0(t)$ is the source intensity (unit: pressure$^2 \cdot m^2$) referred to unit range, incorporating both source level and pulse shape. Care must be taken to insure that the source intensity $I_0(t)$ and the received intensity $I(t)$ are normalized in the same fashion. The received intensity must be computed from the system output time series $V(t)$ as $I(t) = |V(t)|^2 / |V_c/P_c|^2$, where P_c is the RMS incident pressure producing output V_c from the system during calibration. The intensity impulse response is

$$H(t) = TRb(\theta)\frac{dR}{dt}\sigma_b(\theta), \qquad (8.106)$$

where T is transmission factor (unit: m^{-4}) including loss due to spreading and attenuation, R is the horizontal range (unit: m), $b(\theta)$ is a dimensionless angle-dependent function incorporating source and receiver directivities, and $\sigma_b(\theta)$ is the dimensionless scattering cross section defined earlier which is assumed to be isotropic (independent of the azimuthal angle). The function $b(\theta)$ is

$$b(\theta) = \int_{-\pi}^{\pi} b_x(\theta, \varphi)b_r(\theta, \varphi)d\varphi, \qquad (8.107)$$

where $b_x(\theta,\varphi)$ is the dimensionless source beam pattern for the square of pressure and $b_r(\theta,\varphi)$ is the equivalent for the receiver. These are expressed in spherical coordinates with z-axis vertical (i.e., earth coordinates). If the sonar is rotated with respect to vertical, it is necessary to transform measured beam patterns from a coordinate system oriented with the sonar to earth coordinates. The functions $b_x(\theta,\varphi)$ and $b_x(\theta,\varphi)$ are normalized to unity in whichever directions the source level and receive sensitivity are defined. It is immaterial whether or not these are the maximum response axes. The function $b(\theta)$ may be regarded as the azimuthal width of the combined source and receiver beam patterns. Many authors use approximations for the azimuthal width, such as assuming "cookie cutter" beam patterns that are constant within the 3 dB limits and zero outside. Expression (8.107) may be used with either theoretical or measured patterns and includes the effects of sidelobes. The beam pattern integral provides an accurate treatment of ensonified area, even near nadir where the ensonified patch is a filled circle rather than a filled annulus. An example of the integration process is shown in Fig. 8.41. The false-color image shows the combined source–receiver pattern (converted to dB for convenience) projected onto a flat plane 10 m below the sonar. The circle shows the integration path for an angle of incidence of 5 degrees with the sonar pitched 2 degrees forward (negative pitch, bow down) and rolled 3 degrees (negative roll, to port). The beam patterns are theoretical patterns for a multibeam system operating at 400 kHz with a 1 degree along-track width and a 0.5 degree across-track width. With single-beam or side-scan sonars, the beam pattern integrand will be much broader, but the principle remains the same.

In Fig. 8.41 far-field beam patterns are used, but this may not be accurate for very narrow beams and low altitudes. Another possible issue is use of the scattering cross section that is a plane wave, hence far-field concept. Although numerous authors have raised concerns that Fresnel corrections must be made if the receiver is in the near field of the ensonified area (viewing it as a large, random acoustic array), these concerns are obviated by including the angular variation of the scattering cross section in the integral over ensonified area as shown in Appendix J in Jackson and Richardson [4]. This is implicit in the formulation presented earlier, so there is no reason to avoid the use of the scattering cross section.

The formalism defined earlier can be used to model intensity due to interface scattering for arbitrary beam patterns and pulse shapes, even near nadir. To make

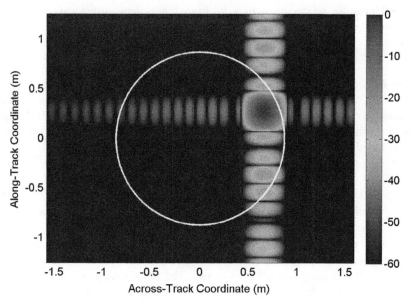

FIGURE 8.41

Illustration of the beam pattern integral. The false-color image is the product $b_x(\theta, \varphi) b_r(\theta, \varphi)$ converted to decibel and projected onto a plane 10 m below the sonar, which has 2 degrees pitch and 3 degrees roll. Theoretical beam patterns are used for a multibeam sonar operating at 400 kHz. The particular beam shown is broadside to the array, and would be at nadir if pitch and roll were zero. The *white circle* shows the beam pattern integration path for an angle of incidence of 5 degrees.

contact with simpler applications of the sonar equation it is useful to define the "energy" (unit: pressure^2·s) of the return signal as

$$E = \int_{-\infty}^{\infty} I(t)dt. \tag{8.108}$$

Then Eq. (8.105) can be used to show that

$$E = E_H E_0, \tag{8.109}$$

where

$$E_H = \int_{-\infty}^{\infty} H(t)dt, \tag{8.110}$$

and

$$E_0 = \int_{-\infty}^{\infty} I_0(t)dt. \tag{8.111}$$

If refraction is negligible,

$$T = (1/r)^4 10^{-2r\alpha/10}. \tag{8.112}$$

The attenuation, α, in this equation, is defined in Eq. (8.5). The slant range r (unit: m) corresponding to time t is

$$r = c_w t/2, \tag{8.113}$$

and the horizontal range is

$$R = r \cos \theta, \tag{8.114}$$

where $\theta = \sin^{-1} h/r$ is the grazing angle corresponding to time t, with h being the altitude of the sonar above the seafloor. With refraction neglected,

$$H(t) = Tb(\theta)rc_w\sigma_b(\theta)/2. \tag{8.115}$$

The usual sonar equation can be obtained by assuming that the pulse length is sufficiently short that

$$I(t) = H(t)E_0, \tag{8.116}$$

where, in Eq. (8.106), the grazing angle and the horizontal range are determined by the time t and are no longer dependent on the integration variable t'. Together Eqs. (8.115) and (8.116) are equivalent to the sonar equation generalized to arbitrary pulse shape but unsuited for application near nadir. For a rectangular pulse of length τ, the energy E_0 is the source intensity multiplied by τ; consequently, Eqs. (8.115) and (8.116) combine to give the commonly encountered form of the sonar equation:

$$I(t) = I_0 rTb(\theta)(c_w\tau/2)\sigma_b(\theta). \tag{8.117}$$

For multibeam systems, a different simplification is useful. If the beams are sufficiently narrow, the scattering transmission factor and the cross section can be considered to be constant over the ensonified area so that Eqs. (8.110) and (8.115) give

$$E_H = TA\sigma_b, \tag{8.118}$$

where σ_b is evaluated at the grazing angle of the beam maximum response axis, and

$$A = \int_0^\infty b(\theta)R dR \tag{8.119}$$

is the effective area for the projection of the beam onto the seafloor, including the effect of sidelobes. Using Eqs. (8.109) and (8.118), one can obtain

$$\sigma_b = E/(TAE_0). \tag{8.120}$$

This can be used to invert for the scattering cross section by using the measured signal energy in place of E and computing T, A, and E_0 from the experimental geometry, seawater properties, and sonar calibration. This equation is applicable near nadir and accommodates arbitrary beam patterns (provided they are narrow enough) and pulse shapes.

Since most sonars used for survey work are not calibrated with regard to source level and receive sensitivity, it is common to resort to relative scattering strength, offset from the true scattering strength by an unknown number of decibel [153].

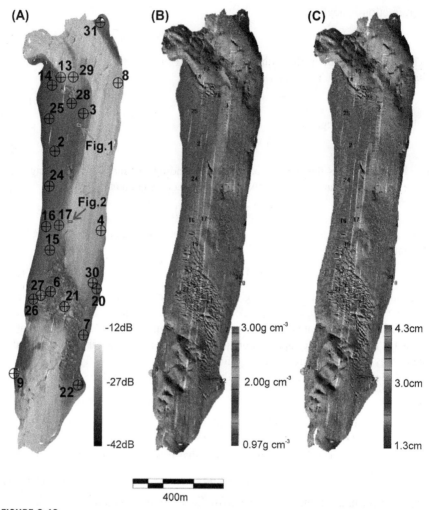

FIGURE 8.42

Inversion using backscattering strength versus angle. Backscattering strength is shown in (A), the index of impedance is shown in (B), and RMS roughness height is shown in (C).

From Fonseca, L. and Mayer, L., Remote estimation of surficial seafloor properties through the application Angular Range Analysis to multibeam sonar data. *Mar. Geophys. Res., 28, pp. 119—126, DOI: 10.1007/ s11001-007-9019-4, 2007.*

This unknown is sometimes determined by a shift to match models or historical data. If a survey includes several distinct seafloor types, adjustment of a single parameter to cover all types is a reasonable strategy. Some authors extract scattering strength features such as slopes and intercepts [154] and then apply clustering algorithms. This approach requires ground truth in order to associate the clusters with seafloor types. Others fit model curves to measured scattering strength [155—159] so that the features extracted are model parameters such as seafloor density and roughness parameters, as shown in Fig. 8.42. If the models provide a reasonable representation of the seafloor types encountered, no ground truth is required. Of course, some environments may have regions that cannot be represented by idealized models.

8.9.5 MODEL FITTING TO ECHO TIME SERIES

An ambitious use of models for inversion of sonar data has been introduced by Pouliquen and Lurton [160] and Sternlicht and de Moustier [161,162]. While this approach has seen only limited practical application [135,163], it offers the hope of requiring minimal ground truth, assuming that the models employed span most seafloor types. This method uses optimization methods to match model predictions to measured receiver intensity time series. This results in determination of several geoacoustic and scattering parameters, although constraints are often introduced to reduce the number of independent parameters [162]. These time series tend to have more complicated behavior than scattering strength curves, hence the expectation that more model parameters can be estimated and with greater accuracy. This expectation is even more plausible with multibeam sonars, which typically provide more than 100 time series, one for each beam.

To illustrate the method, a simple time series model, taken from [135] is presented. Fig. 8.43 shows a computational grid in which each point in the grid contributes a scaled, delayed transmit pulse intensity replica, $I_0(t)$, with the delay corresponding to the round-trip time to the scatterer and back. Each point on the grid represents an annulus having an area $2\pi R\Delta R$, where R is the horizontal coordinate, and ΔR is the horizontal spacing between grid points. These contributions are

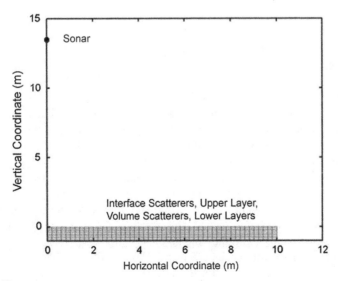

FIGURE 8.43

Computational grid for time series model.

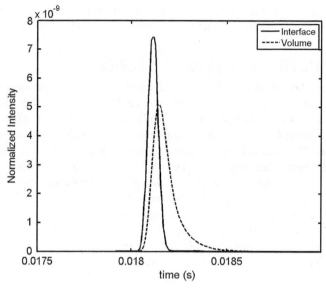

FIGURE 8.44

Model time series for interface and volume scattering for a multibeam system operating at 250 kHz with an along-track beam width 1.6 degrees, and across-track beam width 0.8 degree, for a beam that is steered 8.5 degrees to port. The transmitted pulse is rectangular with smoothed edges and length 50 μs.

weighted by the transmission factor and the appropriate cross section. For the upper (interface) layers, this cross section is $2\pi R\Delta R\sigma$, with σ evaluated at the appropriate grazing angle. For the lower (volume) layers, this cross section is $2\pi R\Delta R\Delta z\sigma_V$, where Δz is the vertical spacing between grid points, and σ_V is the volume scattering cross section mentioned in connection with Eq. (8.83). The transmission factor includes attenuation in the sediment with a possibly unknown attenuation constant and may also include refractive effects. The volume scattering strength may be a function of depth if stratification is known to be important. An example of an intensity time series produced by this model is shown in Fig. 8.44.

In this example, scattering strengths appropriate to sand were used, and refraction at the sand—water interface was ignored as an approximation (valid for the steep angles of interest). The unknown attenuation and volume scattering strength were assumed to be independent of depth. The interface contribution to intensity rises quickly to a maximum, while the volume contribution rises more slowly and has a longer "tail." These differences are in part responsible for the expectation that time series matching will outperform scattering strength matching, in which the delayed arrival of the volume contribution is not taken into account. It should be noted that this separation in the time domain will be lost at angles far from nadir

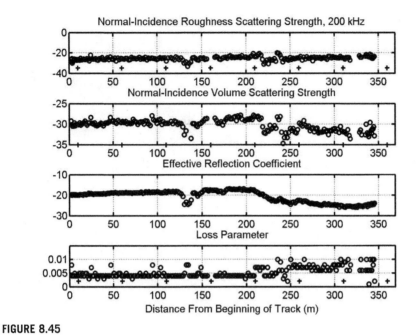

FIGURE 8.45

Inversion results along track shown in Fig. 8.46 [165]. The frequency is 200 kHz, and the + symbols mark the locations shown as markers in Fig. 8.27.

and as beam width and/or pulse length become larger. Still, with appropriate choice of pulse length, the separation may still be observed with wider-beam echo sounders [161]. The final inversion step is to fit model time series (such as the sum of the two curves in Fig. 8.44) to data such as that shown in Fig. 8.37. Fitting may use any of a number of optimization methods. In the approach outlined here, the result is interface scattering cross section versus angle, volume scattering strength, and attenuation (loss parameter δ) as shown in Fig. 8.45. In this figure, an effective reflection coefficient, obtained by an integral of the scattering cross section, is also shown. Fig. 8.46 shows the survey track against a background of backscatter imagery [164] along with markers showing points at which ground-truth data were taken. This includes diver core samples that showed that the seafloor changed (going from West to East) from sand with shell hash cover to shelly mud. The interface scattering strength shows little change along the track, while volume scattering strength, effective reflection coefficient, and loss parameter change significantly as the seafloor type changes. The approach illustrated here employs intrinsic scattering properties of the seafloor, but stops short of inverting for physical properties. By employing physical models for scattering and fitting model time series to data, Pouliquen and Lurton [160] Sternlicht and de Moustier [162], and De and Chakraborty [163] obtain parameters such as sediment acoustic impedance and roughness spectrum parameters. Fitting physical models to time series offers promise of inversion

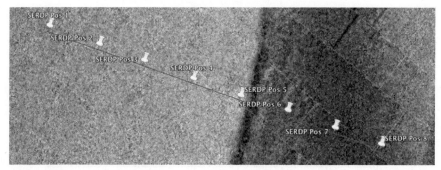

FIGURE 8.46

Ship track along which multibeam data were acquired. The background is a backscatter image at 40 kHz with a 15-dB range from black to white. The push pins mark points at which ground-truth measurements were made. From West (left) to East (right) the seafloor changed from sand with shell hash to shelly mud, with the change occurring in the dark region immediately east of Position 5.

Hefner, B.T., Jackson, D.R. and Ivakin, A.N., Unpublished, 2014, Used with permission of the authors of abstract [165].

requiring minimal ground truth. Advances in this approach will require scattering models that encompass a wider variety of seafloor types as well as tests with extensive physical ground truth.

LIST OF SYMBOLS

A	Ensonified area (unit: m^2)
BL	Bottom loss (unit: dB)
C_h^2	Square of structure constant for interface relief (unit: $m^{4-\gamma_2}$)
D	Water depth (unit: m)
E	Time integral of received intensity (unit: $Pa^2 \cdot m^2 \cdot s$)
E_0	Time integral of source intensity (unit: $Pa^2 \cdot s$)
E_H	Time integral of $H(t)$ (unit: $1/m^2$)
$H(t)$	Intensity impulse response (unit: $1/m^2 s$)
$I(t)$	Echo intensity (unit: Pa^2)
$I_0(t)$	Source intensity (unit: $Pa^2 \cdot m^2$)
K	Magnitude of horizontal component of wave vector (unit: 1/m), hydraulic conductivity (unit: m/s)
K_{eff}	Effective modulus in effective density fluid model (unit: Pa)
K_f	Bulk modulus of sediment frame (unit: Pa)
K_g	Bulk modulus of sediment grains (unit: Pa)
K_G	Kurtosis of grain size distribution (dimensionless)

K_w	Bulk modulus of water (unit: Pa)
L_c	Scale cutoff of heterogeneity spectra (unit: m)
M_d	Median grain size (unit: $\log_2 \cdot$ mm)
M_z	Mean grain size (unit: $\log_2 \cdot$ mm)
R	Reflection coefficient (dimensionless), horizontal range (unit: m)
S	Water salinity (unit: ppt)
S_b	Bottom scattering strength (unit: dB)
Sk_I	Skewness of grain size distribution (dimensionless)
S_p	Pore water salinity (unit: ppt)
T	Temperature (unit: °C), transmission factor (unit: $1/\text{m}^4$)
V	Hydraulic flow rate (unit: m/s)
V_p	Water sound speed (unit: m/s), same as c_w
V_s	Shear wave speed (same as c_t) (unit: m/s)
VpR	Sediment/water sound speed ratio (dimensionless, same as ν_p)
$W(K)$	Roughness spectrum (unit: m^4)
$W_{\rho\rho}(\mathbf{k})$	Spectrum for normalized density fluctuations (unit: m^3)
$W_{\beta\beta'}(\mathbf{k})$	Cross spectra for normalized heterogeneity fluctuations (unit: m^3)
a_0	Pore size parameter (unit: m)
a_p	Ratio of complex compressional wave speed to sound speed of overlying water (dimensionless)
a_t	Ratio of complex shear wave speed to sound speed of overlying water (dimensionless)
$a_{\beta\beta'}$	Coefficients giving strength of heterogeneity cross spectra (dimensionless)
a_ρ	Ratio of sediment density to density of overlying water (dimensionless)
$b(\theta)$	Azimuthal integral of round-trip beam pattern (unit: radians)
$b_r(\theta, \varphi)$	Receiver beam pattern (dimensionless)
$b_x(\theta, \varphi)$	Source beam pattern (dimensionless)
$\mathbf{b}_\delta^\pm$	Unit polarization vectors for wave in elastic media (dimensionless)
c_1	Fast wave speed in poroelastic sediment (unit: m/s)
c_p	Compressional wave speed in sediment (unit: m/s)
c_t	Shear wave speed in sediment (same as V_s) (unit: m/s)
c_w	Sound speed in water (unit: m/s), same as V_p
d	Depth into sediment (unit: m)
e	Void ratio (dimensionless)
f	Frequency (unit: Hz)
k	Hamilton's attenuation factor (unit: dB/m kHz)
k_p	Compressional wave number in sediment (unit: 1/m)
k_q	Wave number for fast and slow waves in poroelastic medium (unit: 1/m)
k_s	Hamilton's attenuation factor for shear waves (unit: dB/m kHz)
k_t	Shear wave number in sediment (unit: 1/m)
k_w	Acoustic wave number in water (unit: 1/m)
k_x	x-component of wave vector (similarly for y) (unit: 1/m)
k_{zp}	z-component of compressional wave vector in sediment (unit: 1/m)
k_{zt}	z-component of shear wave vector in sediment (unit: 1/m)
k_{zw}	z-component of wave vector in water (unit: 1/m)
k_p'	Real part of compressional wave number in sediment (unit: 1/m)
k_p''	Imaginary part of compressional wave number in sediment (unit: 1/m)
p	Pressure (unit: Pa or μPa)

r	Slant range (unit: m)
$\mathbf{u}(\mathbf{r})$	Particle displacement (unit: m)
w_2	Roughness spectral strength (unit: $m^{4-\gamma_2}$)
z_e	Normalized impedance for reflection from elastic medium (dimensionless)
z_p	Normalized impedance for compressional waves (dimensionless)
z_t	Normalized impedance for shear waves (dimensionless)
Γ_p	Transmission coefficient for compressional waves (dimensionless)
Γ_t	Transmission coefficient for shear waves (dimensionless)
ΔK	Transverse component of Bragg wave vector for volume scattering (unit: 1/m)
Δk_z	Vertical component of Bragg wave vector for volume scattering (unit: 1/m)
$\Delta \mathbf{k}_\eta$	Bragg wave vector for scattering between wave types (unit: 1/m)
Λ	Anisotropy parameter for volume heterogeneity spectra (dimensionless)
α	Attenuation (unit: dB/m), tortuosity (dimensionless)
α_λ	Attenuation loss in decibel per wavelength (dimensionless)
β	Fractional porosity (dimensionless)
β_0	Uncorrected fractional porosity (dimensionless)
γ_2	Roughness spectral exponent (dimensionless)
γ_3	Spectral exponent for volume heterogeneity (dimensionless)
δ_p	Dimensionless loss parameter for compressional wave
δ_t	Dimensionless loss parameter for shear wave
η	Dynamic viscosity (unit: kg/m s), porosity (unit: %), index for wave conversion (dimensionless)
θ	Grazing angle in water (unit: radians or degrees)
θ_i	Grazing angle of incident acoustic energy in water (unit: radians or degrees)
θ_s	Grazing angle of scattered acoustic energy in water (unit: radians or degrees)
θ_{crit}	Critical grazing angle (unit: radians or degrees)
θ_{int}	Angle of intromission (unit: radians or degrees)
θ_t	Complex grazing angle of shear wave (unit: radians or degrees)
κ	Permeability (unit: m^2)
λ	Lamé parameter (unit: Pa)
μ	Lamé parameter, shear modulus (unit: Pa), dynamic viscosity (unit: kg/m s)
ν_p	Ratio of compressional wave speed to sound speed of overlying water (dimensionless, same as V_pR)
ν_t	Ratio of shear wave speed to sound speed of overlying water (dimensionless)
ρ	Mass density of sediment (unit: kg/m^3)
ρ_{eff}	Effective density in effective density fluid model (unit: kg/m^3)
ρ_g	Mass density of sediment grains (unit: kg/m^3)
ρ_w	Mass density of water (unit: kg/m^3)
σ_b	Bottom scattering cross section (per unit solid angle per unit area, dimensionless)
σ_r	Contribution to σ_b due to interface roughness (dimensionless)
σ_l	Standard deviation of grain size distribution (unit: $\log_2 \cdot$ mm)
σ_v	Contribution to σ_b due to volume heterogeneity (dimensionless)
σ_{vlim}	σ_v after imposing maximum value (dimensionless)
τ	Sonar pulse length (unit: s)
$\phi(\mathbf{r})$	Scalar potential for particle displacement (unit: m^2)
ϕ_{nn}	nnth percentile of grain size distribution (unit: $\log_2 \cdot$ mm)
ϕ_s	Azimuthal angle of scattered acoustic energy in water (unit: radians or degrees)
$\psi(\mathbf{r})$	Vector potential for particle displacement (unit: m^2)
ω	Angular frequency (unit: rad/s)

REFERENCES

[1] Hamilton, E.L., *Prediction of in-situ acoustic and elastic properties of marine sediments*. Geophysics **36**, pp. 266–284, 1971.

[2] Hamilton E.L., *Acoustic properties of sediments*. In: Acoustics and Ocean Bottom, Lara-Saenz, A., Ranz-Guerra, C. and Carbo-Fite, C. (Eds.), II FASE Specialized Conference. CISC, Madrid, 1987.

[3] Hamilton, E.L., *Geoacoustic modeling of the seafloor*. J. Acoust. Soc. Am., **68** (5), pp. 1313–1340, 1980. Morse, P.M. and Ingard, K.U., *Theoretical Acoustics*, McGraw-Hill, New York, 1968.

[4] Jackson, D.R. and Richardson, M.D, *High-Frequency Seafloor Acoustics*. Springer, 2007.

[5] Kennett, J.P., *Marine Geology*. Prentice-Hall, Englewood Cliffs, N.J., 1975.

[6] Seibold E. and Berger, W.H., *The sea floor: an Introduction to Marine Geology, Third edition*. 356 pp., Springer, Berlin, 1996.

[7] Becker, J.J., Sandwell, D.T., Smith, W.H.F., Braud, J., Binder, B., Depner, J., Fabre, D., Factor, J., Ingalls, S., Kim, S-H., Ladner, R., Marks, K., Nelson, S., Pharaoh, A., Trimmer, R., Von Rosenberg, J., Wallace, G. and Weatherall, P., *Global Bathymetry and Elevation Data at 30 Arc Seconds Resolution: SRTM30_PLUS*. Marine Geodesy **32** (4), pp. 355–371, 2009.

[8] Harris, P.T., Macmillan-Lawler, M., Rupp, J. and Baker, E.K., *Geomorphology of the oceans*. Geology **152**, pp. 4–24, 2014.

[9] Martin, K.M., Wood, W.T. and Becker, J.B. *A global prediction of sediment porosity using machine learning*. Geophys. Res. Lett., 42, 2015.

[10] Morse P.M., and Ingard, K.U., *Theoretical Acoustics*. McGraw-Hill, New York, 1968.

[11] Pierce, A.D., Acoustics: *An Introduction to Its Physical Principles and Applications*. Acoustical Society of America, New York, 1989.

[12] Kinsler, L.E., Frey, A.R., Coppens, A.B., and Sanders, J.V., *Fundamentals of Acoustics, Fourth Edition*, Wiley, New York, 1999.

[13] Landau, L.D. and Lifshitz, E.M., *Theory of Elasticity*. Pergamon, New York, 1970.

[14] Brekhovskikh, L.M. and Godin, O.A., *Acoustics of Layered Media I, Plane and Quasi-Plane Waves*. Springer-Verlag, Berlin, 1990.

[15] Buckingham, M.J., *Wave propagation, stress relaxation, and grain-to-grain shearing in saturated, unconsolidated marine sediments*. J. Acoust. Soc. Am., **108** (6), pp. 2796–2815, 2000.

[16] Biot, M.A., *Mechanics of deformation and acoustic propagation in porous media*. J. Appl. Phys. **33**, pp. 1482–1498, 1962.

[17] Biot, M.A., *Generalized theory of acoustic propagation in porous dissipative media*. J. Acoust. Soc. Am., **34** (9A), pp. 1254–1264, 1962.

[18] Stoll, R.D. and Kan, T.-K., *Reflection of acoustic waves at a water-sediment interface*. J. Acoust. Soc. Am., **70**, pp. 149–156, 1981.

[19] Williams, K.L., Jackson, D.R., Thorsos,E.I., Tang, D. and Schock, S.G., *Comparison of sound speed and attenuation measured in a sandy sediment to predictions based on the Biot theory of porous media*. IEEE J. Oceanic Eng., **27**, pp. 413–428, 2002.

[20] Hefner, B.T. and Williams, K.L., *Sound speed and attenuation measurements in unconsolidated glass-bead sediments saturated with viscous pore fluids*. J. Acoust. Soc. Am., **120** (5), pp. 2538–2549, 2006.

[21] Chotiros, N.P. and Isakson, M.J., *A broadband model of sandy ocean sediments: Biot-Stoll with contact squirt flow and shear drag*. J. Acoust. Soc. Am., **116**, pp. 2011–2022, 2004.

[22] Buckingham, M.J., *On pore-fluid viscosity and the wave properties of saturated granular materials including marine sediments*. J. Acoust. Soc. Am., **123** (3), pp. 1486–1501, 2007.

[23] Chotiros, N.P. and Isakson, M.J., *Shear attenuation and micro-fluidics in water-saturated sand and glass beads*. J. Acoust. Soc. Am., **135** (6), pp. 3264–3279, 2014.

[24] Hefner, B.T. and Jackson, D.R., *Attenuation in sand sediments due to porosity fluctuations*. J. Acoust. Soc. Am., **136** (2), pp. 583–595, 2014.

[25] Williams, K.L., *An effective density fluid model for acoustic propagation in sediments derived from Biot theory*. J. Acoust. Soc. Am., **110**, pp. 2956–2963, 2001.

[26] Mourad, P.D. and Jackson, D.R., *High frequency sonar equation models for bottom backscatter and forward loss*. In: Proc. OCEANS 1989, IEEE, 1989.

[27] *APL-UW High-Frequency Ocean Environmental Acoustic Models Handbook*, Ch. IV, Bottom. APL-UW TR 9407, 1994.

[28] Jackson, D.R., *High-Frequency Bistatic Scattering Model for Elastic Seafloors*. APL-UW TM 2-00, 2000.

[29] Jackson, D.R., Winebrenner, D.P. and Ishimaru, A., *Application of the composite roughness model to high-frequency bottom backscattering*. J. Acoust. Soc. Am., **79**, pp. 1410–1422, 1986.

[30] Jackson, D.R. and Briggs, K.B., *High-frequency bottom backscattering: Roughness vs. sediment volume scattering*. J. Acoust. Soc. Am., **92**, pp. 962–977, 1992.

[31] Lyons, A.P., Anderson, A.L. and Dwan, F.S., *Acoustic scattering from the seafloor: Modeling and data comparison*. J. Acoust. Soc. Am., **95**, pp. 2441–2451, 1994.

[32] Pouliquen, E. and Lyons, A.P., *Backcattering from bioturbated sediments at very high frequency*. IEEE J. Oceanic Eng., **27**, pp. 388–402, 2002.

[33] Yang, T. and Broschat, S.L., *Acoustic scattering from a fluid-elastic-solid interface using the small slope approximation*. J. Acoust. Soc. Am., **96**, pp. 1796–1804, 1994.

[34] Wurmser, D., *A manifestly reciprocal theory of scattering in the presence of elastic media*, J. Math. Phys. **37**, pp. 4434–4479, 1996.

[35] Gragg, R.F., Wurmser, D., and Gauss, R.C., *Small-slope scattering from rough elastic ocean floors: General theory and computational algorithm*. J. Acoust. Soc. Am., **110**, pp. 2878–2901, 2001.

[36] Jackson, D. and Ivakin, A., *Effects of shear elasticity on sea bed scattering: First-order theory*. J. Acoust. Soc. Am., **103** (1), pp. 336–345, 1998.

[37] Ivakin, A. and Jackson, D., *Effects of shear elasticity on sea bed scattering: Numerical examples*. J. Acoust. Soc. Am., **103** (1), pp. 346–354, 1998.

[38] Guillon, L. and Lurton, X., *Backscattering from buried sediment layers: The equivalent input backscattering strength model*. J. Acoust. Soc. Am., **109** (1) pp. 122–132, 2001.

[39] Guillon, L. and Lurton, X., *Backscattering from a layered seafloor with parameters continuously varying with depth*. In: Proc. 6th European Conference on Underwater Acoustics, pp. 37–42, 2002.

[40] Jackson, D.R., Odom, R.I., Boyd, M.L. and Ivakin, A.N., *A geoacoustic bottom interaction model (GABIM)*. IEEE J. Oceanic Eng., **35** (3), pp. 603–617, 2010.

[41] Jackson, D.R., *The small-slope approximation for layered seabeds*. In: Proc. 13th ICA, Montreal, 2013.

[42] Yamamoto, T., *Acoustic scattering in the ocean from velocity and density fluctuations in the sediments*. J. Acoust. Soc. Am., **99**, pp. 866−879, 1996.

[43] Thorsos, E.I. and Broschat, S.L., *An investigation of the small-slope approximation for scattering from rough surfaces: Part I, Theory*. J. Acoust. Soc. Am., **97**, pp., 2082−2093, 1995.

[44] Broschat, S.L. and Thorsos, E.I., *An investigation of the small-slope approximation for scattering from rough surfaces: Part II, Numerical studies*. J. Acoust. Soc. Am., **101** (5), pp. 2615−2625, 1997.

[45] Briggs, K.B., Jackson, D.R. and Moravan, K.Y., *NRL-APL grain size algorithm upgrade*. Naval Research Laboratory, Stennis Space Center, MS. NRL/MR/7430–02-8274, 2002.

[46] Hamilton, E.L. and R.T. Bachman, *Sound velocity and related properties of marine sediments*. J. Acoust. Soc. Am., **72**, pp. 1891−1904, 1982.

[47] Richardson M.D., *In-situ, shallow-water sediments geoacoustic properties*. In: Shallow-water Acoustics, R. Zang and J. Zhou (Eds.), China Ocean Press, Beijing, pp. 163−170, 1997.

[48] Krumbein, W.C., *Size frequency distributions of sediments*. J. Sed. Pet., **4**, pp. 65−77, 1934.

[49] Syvitski, J.P.M., *Principles, Methods, and Applications of Particle Size Analysis, Introduction," in Principles, Methods, and Applications of Particle Size Analysis*. J. P.M Syvitski (Ed.), Cambridge University Press, New York, 1991.

[50] Folk, R.L. and Ward W.C., *Brazos River Bar: A study in the significance of grain size parameters*. J. Sed. Pet., **27**, pp. 3−26, 1957.

[51] Folk, R.L., *Petrology of Sedimentary Rocks*. Hemphill Publishing Company, Austin, Texas, 170 pp., 1974, 1980.

[52] Wentworth, C.K., *A scale of grade and class terms for clastic sediments*. J. Geol. **30**, pp. 377−392, 1922.

[53] Krumbein, W.C. and Pettijohn, F.J., *Manual of Sedimentary Petrography*. Appleton-Century-Crofts, New York, NY, 1938.

[54] Richardson, M.D., Briggs, K.B., Bibee, D.L., Jumars, P.A., Sawyer, W.A., Albert, D.B., Bennett, R.H., Berger, T.K., Buckingham, M.J., Chotiros, N.P., Dahl, P.H., Dewitt, N.T., Fleischer, P., Flood, R., Greenlaw, C.F., Holliday, D.V., Hulbert, M.H., Hutnak, M.P., Jackson, P.D., Jaffe, J.S., Johnson, H.P., Lavoie, D.L., Lyons, A.P., Martens, C.S, McGehee, D.E., Moore, K.D., Orsi, T.H., Piper, J.N., -Ray, R.I., Reed, A.H., Self, R.F.L., Schmidt, J.L., Schock, S.G., Simonet, F., Stoll, R.D., Tang, D.J., Thistle, D.E., Thorsos, E.I., Walter D.J. and Wheatcroft, R.A. *An overview of SAX99: environmental considerations*. IEEE J. Oceanic Eng., **26**, pp. 26−53, 2001.

[55] Blott S.J. and Pye K., *GRADISTAT: A grain size distribution and statistics package for the analysis of unconsolidated sediments*. Earth Surf. Proc. Land., **26**, pp. 1237−1248, 2001.

[56] Blott S.J. and Pyle K., *Particle size scales and classification of sediment types based on particle size distributions: Review and recommended procedures*. Sedimentology, **59** (7), pp. 2071−2096, 2012.

[57] Jackson, D.R. and Richardson M.D., *Seasonal temperature gradients within a sandy seafloor: implications for acoustic propagation and scattering*, pp. 361−368. In: Leighton, T.G., Heald, G.J., Griffiths, H.D. and Griffiths, G. (Eds.), Acoustical Oceanography, Proceedings of the Institute of Acoustics, vol. 23 part 2, 2001.

[58] Wood W.T., Martin K.M., Wooyeol, J. and Sample, J., *Seismic reflectivity effects from seasonal seafloor temperature variation.* Geophys. Res. Lett., **41** (19), pp. 6826–6832, 2014.

[59] Medwin, H., *Speed of sound in seawater: A simple equation for realistic parameters.* J. Acoust. Soc. Am., **58**, pp. 1318–1319, 1975.

[60] Chen, C-T. and Millero, F.J., *Speed of sound in seawater at high pressures.* J. Acoust. Soc. Am., **62**, pp. 1129–1135, 1977.

[61] Stoll R.D., *Sediment Acoustics.* Lecture Notes in earth Sciences. Springer-Verlag, New York, NY, 153 pp, 1989.

[62] Richardson, M.D., Briggs, K.B., Bentley, S.J., Walter, D.J., Orsi, T.H., *The effects of biological and hydrodynamic processes on physical and acoustic properties of sediments off the Eel River, California.* Mar. Geol., **182** (1–2), pp. 121–139, 2002.

[63] Reed, A.H. Briggs, K.B. and Lavoie, D.L., *Porometric properties of siliciclastic marine sand: a comparison of traditional laboratory measurements with image analysis and effective medium modeling.* IEEE J. Oceanic Eng., **27**, pp. 581–592, 2002.

[64] Reed et al., *Physical pore properties and grain interactions of SAX04 sands.* IEEE J. Oceanic Eng., **35** (30), pp. 488–501, 2010.

[65] Bachman R.T. and Hamilton, E.L., *Acoustic and related properties of calcareous deep-sea sediments.* J. Sed. Pet., **52** (3), pp. 733–753, 1982.

[66] Briggs, K.B. Richardson, M.D. and Young, D.K., *Variability in geoacoustic and related properties of surface sediments from the Venezuela Basin, Caribbean Sea.* Mar. Geol., **68**, pp. 73–106, 1985.

[67] Richardson, M.D., Lavoie, D.L. and Briggs, K.B., *Geoacoustic and physical properties of carbonate sediments of the lower Florida Keys.* Geo-Marine Lett., **17** (4), pp. 316–324, 1977.

[68] Buckingham, M.J., *Wave propagation, stress relaxation, and grain-to-grain shearing in saturated, unconsolidated marine sediments.* J. Acoust. Soc. Am., **108**, pp. 2796–2815, 2000.

[69] Blott, S.J and Pye, K., *Particle shape: a review and new methods of characterization and classification.* Sedimentology, **55**, pp. 31–63, 2008.

[70] Dutkiewicz A., Muller, R.D. O'Callagan, S. and Jonasson, H., *Census of seafloor sediments in the world's ocean.* Geology, 2015.

[71] Shepard, F.P., *Nomenclature based on sand-silt-clay ratios.* J. Sed. Pet., **24**, pp. 151–158, 1954.

[72] Kibblewhite A.C., *Attenuation of sound in marine sediments: a review with emphasis on new low-frequency data.* J. Acoust. Soc. Am., **86**, pp. 716–738, 1989.

[73] Stoll, R.D., *Marine sediment acoustics.* J. Acoust. Soc. Am., **77**, pp. 1789–1799. 1985.

[74] Bowles, F.A., *Observations on attenuation and shear wave velocity in fine-grained, marine sediments.* J. Acoust. Soc. Am., **101**, pp. 3385–3397, 1997.

[75] Bachman R.T., *Acoustic and physical property relationships in marine sediment.* J. Acoust. Soc. Am., **78**, pp. 616–621, 1985.

[76] Bachman R.T., *Estimating velocity ratio in marine sediments.* J. Acoust. Soc. Am., **86**, pp. 2029–2032, 1989.

[77] Richardson, M.D. and Briggs, K.B., *On the Use of Acoustic Impedance Values to Determine Sediment Properties*, pp. 15–25. In: Acoustic Classification and Mapping of the Seabed, Pace, N.G.and Langhorne, D.N. (Eds.), Institute of Acoustics, University of Bath, UK, 1993.

[78] Richardson, M.D. and Briggs, K.B., 2004. *Relationships among sediment physical and acoustic properties in siliciclastic and calcareous sediments.* Proceedings of the 7th European Conference on Underwater Acoustics, EUCA 2004, Delft, The Netherlands, pp. 659−664, 2004.

[79] Richardson, M.D. and Briggs, K.B., *Empirical predictions of seafloor properties based on remotely measured sediment impedance.* "High Frequency Ocean Acoustic Conference", March 1−5, 2004, La Jolla CA, 2004.

[80] Hamilton, E.L., *Compressional wave attenuation in marine sediments.* Geophysics, **37**, pp. 620−646, 1972.

[81] Richardson, M.D., *In-situ, shallow-water sediments geoacoustic properties*, In: Shallow-Water Acoustics, Zang, R. and Zhou, J. (Eds.), China Ocean Press, Beijing, pp. 163−170, 1997.

[82] Zimmer, M.A., Bibee, L.D. and Richardson, M.D., *Measurement of the frequency dependence of the sound speed and attenuation of seafloor sands from 1 to 400 kHz.* IEEE J. Oceanic Eng., **35** (3) pp. 538−557, 2010.

[83] Griffin, S.R., Grosz, F.B., and Richardson, M.D., *ISSAMS: A remote in situ sediment acoustic measurement system.* Sea Technol., **37**, pp. 19−22, 1996.

[84] Richardson, M.D., *Attenuation of shear waves in near-surface sediments.* In: High-Frequency Acoustics in Shallow Water, Pace, N.G., Pouliquen, E., Bergem O. and Lyons, A.P. (Eds.), SACLANTCEN Conference Proceedings series CP-45, La Spezia, Italy, pp. 451−457, 1997.

[85] Goff, J.A. and Jordan, T.H., *Stochastic modeling of seafloor morphology - resolution of topographic parameters by Sea Beam data*, IEEE J. Oceanic Eng., **14**, pp. 326−337, 1989.

[86] Marks, K.M. and Smith W.H.F., *An evaluation of publicly available global bathymetry grids.* Mar. Geophys. Res., **27**, pp. 19−34, 2006.

[87] Mayer, L.A., *Frontiers in seafloor mapping and visualization.* Mar. Geophys. Res., **27** (1), pp. 7−17, 2006.

[88] Guenther, G.C., Brooks, M.W. and LaRocque, P.E, *New capabilities of the "SHOALS" airborne lidar bathymeter.* Remote Sens. Environ., **73** (2), pp. 247−255, 2000.

[89] Jackson, D.R., Richardson, M.D., Williams, E.L., Lyons, A.P., Jones, A.D., Briggs, K.B. and Tang, D. *Acoustic Observation of the Time Dependence of the Roughness of Sandy Seafloors.* IEEE J. Oceanic Eng., **34** (4), pp. 407−422, 2009.

[90] Greig A.L., Lyons, A.P. and Pouliquen E., *Comparison of Seafloor Roughness and Scattered Acoustic Temporal Decorrelation.* IEEE J. Oceanic Eng., **34** (4), pp. 423−430, 2009.

[91] Lyons, A.P. and Brown, D.C., *The Impact of the Temporal Variability of Seafloor Roughness on Synthetic Aperture Sonar Repeat-Pass Interferometry.* IEEE J. Oceanic Eng., **38** (1), pp. 91−97, 2013.

[92] Fox C.G. and Hayes, D.E. *Quantitative methods for analyzing the roughness of the seafloor.* Rev. Geophys., **23** (1), pp. 1−48, 1985.

[93] Kraft B.J. and de Moustier, C. *Detailed Bathymetric Surveys Offshore Santa Rosa Island, FL: Before and After Hurricane Ivan (September 16, 2004).* IEEE J. Oceanic Eng., **35** (3), pp. 453−470, 2010.

[94] Briggs, K.B., *Microtopographical roughness of shallow-water continental shelves.* IEEE J. Oceanic Eng., **14**, pp. 360−367, 1989.

[95] Lyons A.L. and Pouliquen, E., *Advances in high resolution seafloor characterization in support of high-frequency underwater acoustics studies: techniques and examples.* Meas. Sci. Technol., **15**, pp. R59−R72, 2004.

[96] Tang, D.J., *Fine-scale measurements of sediment roughness and subbottom variability.* IEEE J. Oceanic Eng., **29**, pp. 929–939, 2004.

[97] Moore, K.D. and Jaffe, J.S., *Time-evolution of high resolution topographic measurements of the sea floor using a 3-D laser line scan mapping system.* IEEE J. Oceanic Eng., **27**, pp. 525–545, 2002.

[98] Wang C.C. and Tang D.J., *Seafloor roughness measured by a laser line scanner and a conductivity probe.* IEEE J. Oceanic Eng., **34** (4) pp. 459–465, 2009.

[99] Isakson, M.J., Chotiros, N.P., Yarbrough, R.A., and Piper, J.N., *Quantifying the effects of roughness scattering on reflection loss measurements.* J. Acoust. Soc. Am., **132** (6), pp. 3687–3697, 2010.

[100] Irish, J.D., Lynch, J.F. Traykovski, P.A., Newhall, A.E., Prada, K. and Hay, A.E., *A self-contained sector-scanning sonar for bottom roughness observations as part of sediment transport studies.* J. Atmos. Oceanic Technol., **16**, pp. 1830–1841, 1999.

[101] Lyons, A.P., Fox, W.L.J., Hasiotis, T. and Pouliquen, E., *Characterization of the two-dimensional roughness of wave-rippled sea floors using digital photogrammetry.* IEEE J. Oceanic Eng., **27**, pp. 515–524, 2002.

[102] Pouliquen, E., Canepa, G., Pautet, L. and Lyons, A.P., *Temporal variability of seafloor roughness and its impact on acoustic scattering, Proceedings of the 7th European Conference on Underwater Acoustics*, EUCA 2004, Delft, Netherlands, pp. 583–588, 2004.

[103] Briggs, K.B., Lyons, A.P., Pouliquen, E., Mayer, L.A. and Richardson, M.D., *Seafloor roughness, sediment grain size, and temporal stability.* In: Underwater Acoustic Measurements: Technologies and Results, Papadakis, J.P. and Bjorno, L. (Eds.), Proceedings of a conference held in Heraklion, Crete, June 28–July 2, 2005.

[104] Richardson, M.D., Young D.K. and Briggs, K.B., *Effects of hydrodynamic and biological processes on sediment geoacoustic properties in Long Island Sound, USA.* Mar. Geol., **52**, pp. 201–26, 1983.

[105] Briggs, K.B., Reed, A.H., Jackson, D.R. and Tang, D.J., *Fine-Scale Volume Heterogeneity in a Mixed Sand/Mud Sediment off Fort Walton Beach, FL.* IEEE J. Oceanic Eng., **35** (3), pp. 471–487, 2010.

[106] Max, M.D., Johnson, A.H. and Dillon, W.P., *Economic Geology of Natural Gas Hydrate.* Springer, The Netherlands, 2003.

[107] Waite, W.F., Santamarina, J.C., Cortes, D.D., Dugan, B., Espinoza, D.N., Germaine, J., Jang, J., Jung, J.W., Kneafsey, T.J., Shin, H., Soga, K., Winters, W.J. and Yun, T.-S., *Physical Properties of hydrate-bearing sediments.* Rev. Geophys., **17**, pp. 38, 2009.

[108] Widmann, R.A. and Huttel, M., *Acoustic detection of gas bubbles in saturated sands at high spatial and temporal resolution.* Limnol. Oceanogr. Methods, **10**, pp. 129–142, 2012.

[109] Wilson, P.S. and Dunton, K.H., *Laboratory investigation of the acoustic response of seagrass tissue in the frequency band 0.5-2.5 kHz.* J. Acoust. Soc. Am., **125**, pp. 1951–1959, 2009.

[110] Judd, A.G. and Hovland, M., *The evidence of shallow gas in marine sediments.* Cont. Shelf Res., **12** (10), pp. 1081–1095, 1992.

[111] Richardson, M.D. and Davis A.M., *Modeling methane-rich sediments from Eckernförde Bay.* Cont. Shelf Res., **18** (14–15), pp. 1671–1688, 1988.

[112] Martens, C.A. and Klump, J.V., *Biochemical cycling in an organic-rich coastal marine basin I. Methane sediment-water exchange processes.* Geochim. Cosmochim. Acta, **44**, pp. 471–490, 1980.

[113] Lyons, A.P., Duncan, M.E., Anderson, A.L. and Hawkins, J.A., *Predictions of the acoustic scattering response of free-methane bubbles in muddy sediments.* J. Acoust. Soc. Am., **99**, pp. 163–172, 1996.

[114] Tang, D., *Modeling high-frequency acoustic backscattering from gas voids buried in sediments.* Geo-Marine Lett., **16**, pp. 261–265, 1996.

[115] Wilkens, R.H. and Richardson, M.D., *The influence of gas bubbles on sediment properties: insitu, laboratory and theoretical results from Eckernförde Bay, Baltic Sea.* Continental Shelf Res., **18** (14–15), pp., 1859–1992, 1988.

[116] Anderson, A.L., Abegg, F., Hawkins, J.A., Duncan, M.E. and Lyons, A.P., *Bubble populations and acoustic interaction with the gassy seafloor of Eckernförde Bay.* Cont. Shelf Res., **18** (14–15), pp. 1807–1838, 1998.

[117] Best, A.I., Tuffin, M.D.J., Dix, J.K. and Bull, J,M, *Tidal height and frequency dependence of acoustic velocity and attenuation in shallow gassy marine sediments.* J. Geophys. Res., **109**, 17 pp., 2004.

[118] Fleischer, P., Orsi, T.H., Richardson, M.D. and Anderson, A.L., *Distribution of free gas in marine sediments: A global overview.* Geo-Mar. Lett., **21**, pp. 103–122, 2001.

[119] Abegg, F. and Anderson, A.L., *The acoustic turbid layer in muddy sediments of Eckernförde Bay, Western Baltic Sea: methane concentration, saturation, and bubble characteristics.* Mar. Geol., **137**, pp. 137–147, 1997.

[120] Boudreau, B.P., Algar, C., Johnson, B.D., Croudace, I., Reed, A.H., Furukawa, Y., Dorgan, K.M., Jumars, P.A., Grader, A.S., Gardiner, B.S., *Bubble growth and rise in soft sediments.* Geology 33 (6), pp. 517–520, 2005.

[121] Katsman, R., Ostrovsky, I. and Makovsky, Y., *Methane bubble growth in fine-grained muddy aquatic sediment: insight from modeling.* Earth Planet. Sci. Lett., **377–378**, pp. 336–346, 2013.

[122] Regnier, R., Dale, A.W., Arndt, S., LaRowe, D.E., Mogollón, J. and Van Cappellen, P., *Quantitative analysis of anaerobic oxidation of methane (AOM) in marine sediments: A modeling perspective.* Earth-Sci. Rev., **106**, pp. 105–130, 2011.

[123] Martens, C.S., Albert, D.B. and Alperin, A.J., *Biochemical processes controlling methane in gassy coastal sediments — Part 1. A model coupling organic matter flux to gas production, oxidation and transport.* Cont. Shelf Res., **18** (14–15), pp. 1741–1770, 1998.

[124] Wever, T.F., Abegg, F., Fiedler, H.M., Fechner, G. and Stender, I.H., *Shallow gas in the muddy sediments of Eckernförde Bay, Germany.* Cont. Shelf Res., **18** (14–15), pp. 1715–1739, 1998.

[125] Mogollón, J.M., Dale, A.W., Heureux I.L., and Regnier, P., *Impact of seasonal temperature and pressure changes on methane gas production, dissolution, and transport in unfractured sediments.* J. Geophys. Res., **116**, 17 pp., 2011.

[126] Jackson, D.R., Williams, K.L., Wever, T.F., Friedrichs, C.T. and Wright, L.D., *Sonar evidence for methane ebullition in Eckernförde Bay.* Cont. Shelf Res., **18**, pp. 1893–1915, 1998.

[127] Anderson, A.L. and Hampton, L.D., *Acoustics of gas-bearing sediments I. Background.* J. Acoust. Soc. Am., **67**, pp. 1865–1903, 1980.

[128] Anderson, A.L. and Hampton, L.D., *Acoustics of gas-bearing sediments, II. Measurements and models.* J. Acoust. Soc. Am., **67**, pp. 1865–1903, 1980.

[129] Gardner, T.N., *An acoustic study of soils that model seabed sediments that contain gas bubbles.* J. Acoust. Soc. Am., **107** (1), pp. 163–176, 2000.

[130] Leighton, T.G. and Robb, G.B.N., *Preliminary mapping of void fractions and sound speeds in gassy sediments from subbottom profiles.* J. Acoust. Soc. Am., **124** (5), pp. EL313–EL320, 2008.

[131] Toth, Z., Spiess, V., Mogollon, J.M. and Jensen, J.B., *Estimating the free gas content in Baltic Sea sediments using compressional wave velocity from marine seismic data.* J. Geophys. Res. Solid Earth, **199** (12), pp. 8577–8593, 2104.

[132] Lurton, X., Lamarche, G., Brown, C., Lucieer, V., Rice, G., Schimel, A. and Weber, T., Backscatter measurements by seafloor-mapping sonars, Guidelines and recommendations, Report by members of the GeoHab Backscatter working Group, May 2015.

[133] Malik, M., Mayer, L., Weber, T., Calder, B. and Huff, L., *Challenges of defining uncertainty in multibeam sonar derived seafloor backscatter.* In: Proc. UA 2013, Corfu, Greece, pp. 1037–1044, 2013.

[134] Dyer, I., *Statistics of sound propagation in the ocean.* J. Acoust. Soc. Am., **48**, pp. 337–345, 1970.

[135] Jackson, D.R., Hefner, B.T., Ivakin, A.N. and Wendelboe, G., *Seafloor characterisation using physics-based inversion of multibeam sonar data.* In: Proc. 11th European Conference on Underwater Acoustics, 2012.

[136] De, C. and Chakraborty, B., *Acoustic characterization of seafloor sediment employing a hybrid method of neural network architecture and fuzzy algorithm.* IEEE Trans. Geosci. Rem. Sens. Lett. **6** (4), pp. 743–747, 2009.

[137] Preston, M., Rosenberger, A. and Collins, W.T., *Bottom classification in very shallow water.* In: Proc. ECUA 2000, Lyon, France, pp. 293–299, 2000.

[138] Galloway, J.L. and Collins, W.T., *Dual-frequency acoustic classification of seabed habitat using the QTC VIEW.* In: Oceans '98, Nice, France, 1998.

[139] Heald, G.J., *High frequency seabed scattering and sediment discrimination.* In: Acoustical Oceanography, Institute of Acoustics 23, Southhampton, UK, pp. 258–267, 2001.

[140] Ellingsen, K.E., Gray, J.S., and Bjørnbom, E., *Acoustic classification of seabed habitats using the QTC VIEWTM system.* ICES J. Mar. Sci., **59**, pp. 825–835, 2002.

[141] Humborstad, O.-D, K.E., Nøttestad, L., Løkkeborg, S. and Rapp, H.T., *RoxAnn bottom classification system, sidescan sonar and video-sledge: spatial resolution and their use in assessing trawling impacts.* ICES J. Mar. Sci., **61**, pp. 53–63, 2004.

[142] Chivers, R., Emerson, N., and Burns, D.R., *Seabed classification using the backscattering of normally incident broadband acoustic pulses.* Hydrograph. J. **26**, pp. 9–16, 1982.

[143] Blondel, Ph., *Automatic mine detection by textural analysis of COTS sidescan sonar imagery.* Int. J. Remote Sens. **21** (16), pp. 3115–3128, 2000.

[144] Huvenne, V.A.I., Blondel, Ph. and Henriet, J.-P., *Textural analyses of sidescan sonar imagery from two mound provinces in the Porcupine Seabight.* Int. Mar. Geol., **189**, pp. 323–341, 2002.

[145] Blondel, Ph. and Gómez Sichi, O., *Textural analyses of multibeam sonar imagery from Stanton Banks, Northern Ireland continental shelf.* Appl. Acoust., **70**, pp. 1288–1297, 2009.

[146] Knott, S.T., Hoskins, H. and LaCasce, E.O., *Estimation of the seabed acoustic impedance structure from normal-incident reflections: Somali Basin.* J. Geophys. Res., **86** (B4), pp. 2935–2952, 1981.

[147] Walter, D.J., Lambert, D.N., Young, D.C. and Stephens, K.P., *Mapping sediment acoustic impedance using remote sensing acoustic techniques in a shallow-water carbonate environment*. Geo-Mar. Lett., **17**, pp. 260–267, 1997.

[148] Lambert, D.N., Kalcic, M.T. and Fass, R.W., *Variability in the acoustic response of shallow-water marine sediments determined by normal-incident 30-kHz and 50-kHz sound*. Mar. Geol., **182**, pp. 179–208, 2002.

[149] Walter, D.J., Lambert, D.N. and Young, D.C., *Sediment facies determination using acoustic techniques in a shallow-water carbonate environment, Dry Tortugas, Florida*. Mar. Geol. **182**, pp. 161–177, 2002.

[150] Schock, S.G., *A method for estimating the physical and acoustic properties of the sea bed using chirp sonar data*. IEEE J. Oceanic Eng., **29** (4), pp. 1200–1217, 2004.

[151] LeBlanc, L.R., Mayer, L., Rufino, M., Schock, S.G. and King, J., *Marine sediment classification using the chirp sonar*. J. Acoust. Soc. Am., **91** (1), pp. 107–115, 1992.

[152] LeBlanc, L.R., Panda, S. and Schock, S.G., *Sonar attenuation modeling for classification of marine sediments*. J. Acoust. Soc. Am., **91** (1), pp. 116–126, 1992.

[153] Canepa, G. and Pace, N.G., *Seafloor segmentation from multibeam bathymetric sonar*. In: Proc. ECUA 2000, Lyon, France, pp. 361–367, 2000.

[154] Hughes-Clarke, J., Danforth, B.W., and Valentine, P., *Areal seabed classification using backscatter angular response*. In: Proc. of conference High-Frequency Acoustics in Shallow Water, Lerici, Italy, pp. 243–250, 1997.

[155] de Moustier, C. and Alexandrou, D., *Angular dependence of 12 kHz seafloor acoustic backscatter*. J. Acoust. Soc. Am., **90** (1), pp. 522–531, 1991.

[156] Matsumoto, H., Dziak, R.P. and Fox, C.G., *Estimation of seafloor microtopographic roughness through modeling of acoustic backscatter data recorded by multibeam sonar systems*. J. Acoust. Soc. Am., **94** (5), pp. 2776–2787, 1993.

[157] Haris, K., Chakraborty, B., De, C., Prabhudesai, R.G. and Fernandes, W., *Model-based seafloor characterization employing multi-beam angular backscatter data – A comparative study with dual-frequency single beam*. J. Acoust. Soc. Am., **130** (6), pp. 3623–3632, 2011.

[158] Fonseca, L., Mayer, L. and Kraft, B., *Seafloor characterization through the application of AVO analysis to multibeam sonar data*. In: Boundary Influences In High Frequency, Shallow Water Acoustics, Pace, N. and Blondel, P. (Eds.), Bath University, UK, pp. 241–250, 2005.

[159] Fonseca, L. and Mayer, L., *Remote estimation of surficial seafloor properties through the application Angular Range Analysis to multibeam sonar data*. Mar. Geophys. Res., **28**, pp. 119–126, DOI 10.1007/s11001-007-9019-4, 2007.

[160] Pouliquen, E. and Lurton, X., *Seabed identification using echo-sounder signals*. In: Proc. European Conference on Underwater Acoustics, Luxembourg, pp. 535–538, 1992.

[161] Sternlicht, D.D. and de Moustier, C.P., *Time-dependent seafloor acoustic backscatter (10–100 kHz)*. J. Acoust. Soc. Am., **114** (5), pp. 2709–2725, 2003.

[162] Sternlicht, D.D. and de Moustier, C.P., *Remote sensing of sediment characteristics by optimized echo-envelope matching*. J. Acoust. Soc. Am., **114** (5), pp. 2727–2743, 2003.

[163] De, C. and Chakraborty, B., *Model-based acoustic remote sensing of seafloor characteristics*. IEEE Trans. Geosci. Rem. Sens., **49** (10), pp. 3868–3877, 2011.

[164] De Moustier, C.P. and Kraft, B., *High-frequency sediment acoustics over transitions from dense shell hash to mud: Repeat surveys at 7 frequencies from 150 kHz to 450 kHz.* J. Acoust. Soc. Am., **134**, pp. 4239, 2013.

[165] Hefner, B.T., Jackson, D.R. and Ivakin, A.N., Unpublished, 2014.

Inverse Methods in Underwater Acoustics

N.R. Chapman
University of Victoria, Victoria, BC, Canada

9.1 INTRODUCTION

The acoustic field in the ocean contains information about physical processes and the structure of the ocean medium through which the acoustic signal has propagated. The process of extracting the information from measurements of physical quantities associated with the acoustic field is known as *inversion*. Simple examples of physical quantities that are measured in underwater acoustics are the amplitude and phase of the sound pressure, or the travel time of an acoustic signal. From this definition of an *inverse problem*, it might be assumed that we can infer properties of the real ocean from our data. However, this is not what is actually possible. Instead, we can at best obtain estimates of parameters of a model designed to approximate the ocean environment. An example, for instance, is a *geoacoustic model* of the ocean bottom consisting of a system of sediment layers, each one described by the sound speed, attenuation, and density of the material in the layer.

In plain words, the inverse problem is model based. In *model-based inversions* in underwater acoustics, the physical model of the ocean is related to the measured quantities through the acoustic (or elastic) wave equation that describes sound propagation. For all but a few simplified ocean waveguide models, analytic solution of the wave equation is not possible and sophisticated numerical techniques such as ray theory approximations, normal mode methods, wave number integral methods, and parabolic equation approximations have been developed for calculating the field in realistic ocean waveguide environments [1]. We implicitly assume that these methods contain the relevant physics of sound propagation in the ocean and are thus able to provide sufficiently accurate predictions. However, the accuracy of the predicted fields in modeling experimental data—and consequently the quality of the model parameter estimates obtained in an inversion—is fundamentally limited by the physical model used to describe the ocean environment. The model may be an inaccurate representation of the real ocean and is very likely an incomplete description. And, in addition to these errors in the theory and model, the experimental data themselves are contaminated by noise.

All these issues place severe constraints on the effectiveness of an inversion in making inferences about the properties and structure of the real ocean. Although the *forward problem* of calculating the acoustic field for a specific ocean environment has a unique solution, the *inverse problem* is inherently nonunique. There

Applied Underwater Acoustics. http://dx.doi.org/10.1016/B978-0-12-811240-3.00009-6

are many models that can provide adequate fits to the experimental data in an inversion, but most of them are unrealistic representations of the real ocean. In practice, prior information about the ocean environment at the experimental site is used to provide realistic constraints in designing the physical form of the model. However, the inversion is fundamentally limited by the model, although there are tests in some approaches that can be applied to eliminate unnecessary model structure. As we will see, the most we can derive from an inversion is the probability of a particular value for a specific model parameter, assuming of course that the form of the model is a reasonable representation of the real ocean.

An important distinction is whether the inverse problem is *linear* or *nonlinear*. The inversion of pressure field data in underwater acoustics is fundamentally a nonlinear inverse problem. However, some of the practical problems are weakly linear or can be linearized locally. Linear methods were developed first and are very appealing because (1) a complete analytical theory exists for linear inverse problems with Gaussian errors [2] and (2) linear methods are computationally much faster. Nonlinear inverse problems require numerical methods that explore multidimensional model parameter spaces in finding solutions.

In practice, the central requirements in enabling model-based inversion are (1) measurements that contain relevant information about the model parameters and (2) methods for estimating values of the model parameters from the data and measures of the uncertainties of the estimates. The first requirement demands careful experimental design to acquire high-quality data. In underwater acoustics, we are primarily interested in using the amplitude and phase information in pressure field data. However, in some instances it may be more practical to use quantities derived from the pressure field, especially if this leads to a linearized inverse problem. Some examples of these observable quantities include intensity or energy density versus range (transmission loss); wave numbers of propagating modes; ocean bottom reflection coefficients or bottom loss; time—frequency dispersion relationships for propagating modes; and travel time of transmitted signals. The second requirement introduces the inversion method that is appropriate for a particular objective, for instance, matched field processing (MFP) for source localization, or Bayesian inference for estimating a geoacoustic profile. It is fundamentally important to stress that the complete solution of the inverse problem involves providing both the estimated values and measures of the errors of the estimates.

Both requirements are critically linked; if there is no relevant information about a particular model parameter in the experimental data, the inversion cannot be successful in providing a meaningful estimate of the parameter. It is unfortunately the case that many inversions published in the literature report only estimated values for model parameters, some of which are not sensitive in the experiments. Without reliable measures of the errors, it is not possible to know if the estimated values of the model parameters are meaningful. Despite the general constraints and limitations discussed in the introductory paragraphs, there has been remarkable success in designing inversion methods that have proved to be very effective in using acoustic field data to characterize the ocean environment in various different applications [3—6].

With this general introduction to inverse problems, we first provide some basic mathematical relationships of the forward and inverse problems in the next section and then focus on describing inversion techniques for three different applications in underwater acoustics. These are (1) localization of a sound source by MFP; (2) estimation of geoacoustic models of the ocean bottom (geoacoustic inversion); and (3) ocean acoustic tomography (OAT). Source localization is described in Section 9.3. Section 9.4 describes the approaches taken for geoacoustic inversion, including linearized methods and nonlinear Bayesian inference by optimization and numerical sampling techniques. The last section, Section 9.5, presents a discussion of tomographic inversion in deep and shallow water.

9.2 SOME BASIC MATHEMATICAL RELATIONSHIPS

The formal definition of an inverse problem can be expressed in terms of the relationship between the model parameters, $\boldsymbol{m} = [m_1, m_2, \ldots m_n]^T$, and the data, $\boldsymbol{d} = [d_1, d_2, \ldots d_k]^T$, where $[.]^T$ denotes the transpose operation. The data are considered to be random variables, so that a measurement represents a particular realization from the random distribution. In underwater acoustics the model parameters may be, for instance, the range and depth of the sound source in a source localization problem, and the data of interest are measurements of the acoustic pressure at an array of hydrophones in the water. The model and data are related through the acoustic or elastic wave equation that describes acoustic propagation. The forward problem that predicts the data that would be observed in an ideal, perfectly accurate experiment in an ocean environment described by a specific set of model parameters can be written as

$$\boldsymbol{d}_0 = F(\boldsymbol{m}), \tag{9.1}$$

where $\boldsymbol{d}_0$ are "perfectly accurate" data. This problem has a stable solution for the acoustic field at receivers in the ocean that is uniquely determined by the physical conditions of temperature and salinity in the water, the depth and geoacoustic properties of the ocean bottom, and the source and receiver geometry. The numerical methods for calculating the field have been tested extensively in benchmarking sessions against simulated waveguide environments of varying complexity and are capable of generating highly accurate solutions [5].

The inverse problem of inferring the set of model parameters that generated the observed data is expressed by the relationship

$$\boldsymbol{m} = F^{-1}(\boldsymbol{d}), \tag{9.2}$$

where F^{-1} is an inverse function. This problem is generally very difficult to solve, assuming that a solution exists. Existence of the solution is addressed by constructing a model that provides an adequate fit to the data, within some measure of uncertainty. However, the solution is nonunique—due to incomplete or inaccurate models and errors in the data, there are many models that will fit the data, and is unstable—small perturbations to the data can lead to large changes in the estimated model parameters.

All measured data contain uncertainty that we can describe as additive noise, $\boldsymbol{n}$:

$$\boldsymbol{d} = \boldsymbol{d}_0 + \boldsymbol{n}. \tag{9.3}$$

The errors can arise from two sources: measurement errors due, for example, to inaccurate readings or ambient noise, and theory or model errors due to inaccurate or incomplete parameterization of the physical system or approximations in the physics of the forward propagation problem. In many cases, theory and model errors can be the dominant source of uncertainty in the inversion. The data errors are not known explicitly, but instead, we can assume that $\boldsymbol{d}$ is a random variable and assign known statistical distributions for the errors. Usually a Gaussian distribution is assumed, but the actual distribution of the theory and model errors is unknown and difficult to estimate.

9.3 SOURCE LOCALIZATION BY MATCHED FIELD PROCESSING

Since the physical system in the ocean includes the sources and receivers, the definition of the inverse problem implicitly includes one of the most common applications of inversion of acoustic field data, source localization by MFP. The concept of MFP has been known for a very long time, from the first simple experiments on time reversal by Parvulescu and others at Hudson Laboratories that were reported in the mid-1960s [7], and the first formal article by Homer Bucker in 1976 [8]. However, the method could not be applied efficiently until numerical propagation models and modern fast computers with large storage capacity became available. This section describes the background of MFP and the evolution of its use in ocean acoustics for source localization.

A harmonic sound source in the ocean creates a unique distribution of the acoustic field in range and depth that can be expressed in terms of the propagating modes in the waveguide [5]:

$$P(r, z) = \frac{e^{i\pi/4}}{\sqrt{8\pi\rho_0(z_s)}} \sum_{m=1}^{M} Z_m(z_s) Z_m(z) \frac{e^{ik_m r}}{\sqrt{k_m r}}. \tag{9.4}$$

It can be seen from Eq. (9.4) that the spatial variation of the complex pressure, P, of the acoustic field contains information about the source/receiver geometry (range, r, and source depth, SD, z_s) and the ocean waveguide model parameters that generate the modes through the modal wave functions, Z_m, and wave numbers, k_m. ρ_0 is the density in the ocean waveguide. The reader can find a discussion of the waveguide model in terms of normal modes in Chapter 3.

MFP was developed first as a method for extracting information about the location of a sound source from the spatial coherence of the acoustic field at an array of hydrophones. In its most basic form, MFP compares measurements of the complex pressure $P(r,z)$ at specific sensor locations with calculated replica fields $Q(r,z)$ for candidate positions in range and depth of the source. If the method for calculating the acoustic field includes the correct physics of sound propagation and if the physical model of the waveguide is a sufficiently accurate representation of the ocean environment, then the calculated field for the correct values of the SD and range (r_s,z_s) will be equal to the measured field (to within a complex constant). This simple description defines MFP in terms of physically intuitive comparisons between measured and calculated acoustic fields. Another intuitive description can be developed in terms of back propagation. In this formulation, MFP is described in terms of back propagation of the field at an array of sensors through the assumed ocean environment; for the correct ocean environment, the back-propagated fields will focus at the source location. It is useful to retain these very physical pictures of MFP in order to understand the more formal developments of MFP as a parameter estimation problem [9,10].

In analogy with the development for a plane wave beam-forming array [11], an expression for the conventional linear matched field (MF) processor can be formulated in terms of a generalized beam former [10]. The power spectrum of a generalized beam former, B, is expressed in terms of the projection of calculated replicas of the acoustic field on the measured fields:

$$B(r,z;\omega) = \frac{\left|Q^{\dagger}(r,z;\omega)P(r,z;\omega)\right|^2}{\left|Q(r,z;\omega)\right|^2\left|P(r,z;\omega)\right|^2}$$

$$= \frac{Q^{\dagger}(r,z;\omega)P(r,z;\omega)P^{\dagger}(r,z;\omega)Q(r,z;\omega)}{\left|Q(r,z;\omega)\right|^2\left|P(r,z;\omega)\right|^2} \tag{9.5}$$

Here, $P = [P_1,P_2,...P_N]^T$ is the vector of pressure field measurements at an array of N elements, $Q = [Q_1,Q_2,...Q_N]^T$ is the vector of calculated replicas at the array, ω is the sound frequency, and $\dagger$ denotes the complex transpose. The output of the beam former in Eq. (9.5) is normalized to a maximum value of unity. Inspection of the expression indicates that the maximum value is attained for the condition that the modeled replica equals the measured field. Fizell has shown that this occurs for the correct location of the source in range and depth [12].

Viewed in this way as the projection of a modeled field on the measured field, the connection with the more physical description of MFP in terms of the correlation between modeled and measured fields is evident. The output of the conventional MF processor is effectively the normalized correlation between the measured and modeled fields at specific locations in depth and range in the waveguide. The quantity $PP^{\dagger}$ in Eq. (9.5) is the data covariance matrix that contains the relative spatial phase information of the signal field across the array of sensors in the off-diagonal terms, as well as the signal power at each sensor in the diagonal terms. The fundamental difference between MFP and conventional beam forming is that the relative phase of the modeled field is determined from the full-field solution to the wave equation instead of from a plane wave approximation.

The linear (Bartlett) processor described here is just one of many MF processors that were developed and used for source localization [10]. Other notable examples include the minimum variance processor that offers higher spatial resolution than the linear processor [9]; the multiple constraint processor, an adaptation of the minimum variance processor that applies constraints to create a wider main lobe so that the processor focuses on a cloud of points in range and depth instead of at a single value [13]; and the Westwood processor that uses only relative phase information [14]. In all applications, the inversion approach involves a systematic grid search to calculate an ambiguity surface of the MF processor output over range and depth, as shown in Fig. 9.1. These calculations can be implemented very efficiently using normal mode methods for range-independent environments because only one calculation of the field is needed to determine all the values over the range and depth grid. The true source location occurs at the ambiguity surface peak, assuming that the ocean wave-guide environment model was correct [12]. For such a condition of perfect localization, the normalized peak value is unity.

The example shown in Fig. 9.1 displays the linear Bartlett MF ambiguity surface based on data from a localization experiment carried out in shallow water (~ 380 m) on the continental shelf off the coast of Vancouver Island, British Columbia. In the experiment, the signal from a 45-Hz continuous wave (CW) source was received at a

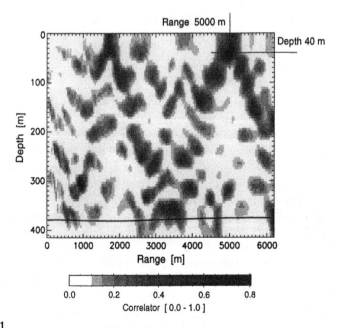

FIGURE 9.1

Matched field ambiguity surface for 45-Hz source from an experiment off the west coast of Vancouver Island, British Columbia. The peak at about 40 m depth and about 5.0 km range indicates the source location.

16-element vertical line array that spanned the water column. The ocean bottom, indicated by the horizontal line at about 380 m in the figure, was weakly range dependent. However, a range-independent assumption was made in calculating the normal modes. The localization search grid covered the range to 6.2 km in steps of 100 m and the depth to 400 m in steps of 1 m. The ambiguity surface peak at a depth of about 40 m and a range of 5.0 km indicates the correct location of the source that was towed in the experiment. The side lobes in the ambiguity surface indicate locations of relatively high correlations.

The peak value at the source location in Fig. 9.1 is not unity, indicating that there is mismatch in the localization problem. Mismatch remains a fundamental issue in MFP due to incomplete knowledge of the ocean environment, and also because correlations exist between some model parameters. The most common example is mismatch in the environmental model; we assume that the ocean environment is known correctly, but this is not generally true. The multiple constraint MF processor mentioned previously was in fact designed to tolerate incomplete knowledge of the ocean environment [13]. The most striking correlation is the relationship between ocean depth and source range. Errors in the ocean depth in calculating the replica fields lead to a "mirage" in localization i.e., a shift in the estimated source range, as explained in a classic article by D'Spain [15].

An insightful approach to address the problem of mismatch in MF source local-ization was developed by Collins and Kuperman who introduced the concept of focalization [16]. Focalization mitigates the impact of an unknown or incompletely known ocean environment by combining objective searches over environmental and geometrical model parameters in the inversion problem to determine an approximate environment that would enable localization. This result, that localization could be obtained with a highly unrealistic model of the environment, follows directly from the fact that the inverse problem does not have a unique solution.

9.4 GEOACOUSTIC INVERSION
9.4.1 GEOACOUSTIC MODELS

Geoacoustic models are used to describe the properties and structure of the ocean bottom in calculations of the acoustic field. The models generally consist of profiles in depth, range and cross-range of the sound speed (c), attenuation (α), and density (ρ) of a layered system of bottom materials. In most cases, the cross-range variation is not significant, but range dependence of the profiles, including changes in depth and the sediment type, along the sound propagation path is common. Knowledge of these physical properties is necessary for constructing geoacoustic models that will enable accurate calculation of the field for various applications such as sonar perfor-mance evaluation and rapid environmental assessment. An example of a geoacoustic model is shown in Fig. 9.2; the form of this model is typical of those used for inver-sions of experimental data.

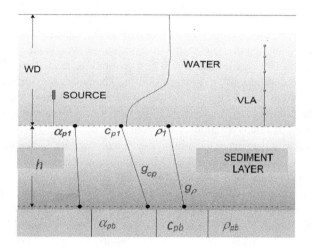

FIGURE 9.2

Geoacoustic model consisting of a simple layered structure of sound speed, c, attenuation, α, and density, ρ. The subscript p refers to the compressional wave, the sediment and basement are denoted by the subscripts 1 and b, respectively.

The geoacoustic model in the figure does not explicitly include shear wave parameters. Although shear wave effects in elastic solid material can be modeled in most numerical propagation codes, the impact of shear wave losses is not significant in most shallow or deep water environments that consist of fine-grained, high-porosity sediment material in which the shear wave speed near the seafloor is very low (<300 m/s). Consequently, in most of the geoacoustic inversions reported in the literature, the bottom is modeled as a fluid. Exceptions to this approach include shallow or deep water environments where elastic solid material is found relatively close (within a few wavelengths) to the seafloor (e.g., thin-sediment basalt regions of the Pacific Ocean). In those environments, the shear wave speed is comparable to or greater than the sound speed in water, and so the coupling with the compressional wave generated in the water is very strong. Inversions of data from such environments must take account of shear wave propagation in the bottom.

The sensitivity of the acoustic field to geoacoustic model parameters was recognized many years ago by researchers who noted that improvements in modeling transmission loss data [17], modal dispersion [18] and bottom loss data [19] could be obtained by adjusting specific model parameters to obtain better agreement with experimental data. The simplicity of the approach is very appealing, and it continues to be applied in some studies [20]. However, the practice of changing selected model parameters in a trial and error fashion is highly subjective, and there is no measure of the uncertainty of the parameter value that provides the best fit to the data. More importantly, it ignores the sensitivities and the impact of errors in other model parameters that are held at fixed values. A more systematic approach of iteration over forward models was suggested by Frisk [3], but the computation time in executing such a grid search over many geoacoustic model parameters was and remains prohibitively long.

Since 1990s, there has been considerable progress in developing objective inversion techniques to estimate geoacoustic model parameters from measurements of the acoustic field—or quantities that can be derived from the acoustic field—in the water. Use of acoustic remote sensing has great appeal because it is an efficient means for characterizing the ocean bottom over large areas, and the estimates are made on material in its natural setting. By comparison, estimates based on point measurements that involve analysis of physical samples of the bottom material are expensive and time consuming and may introduce additional problems in making measurements of the physical quantities in other than in situ conditions. However, as is seen later, the general practice remains that the inferences from inversions of acoustic field data are compared to ground-truth data from physical samples or other in situ measurements. The inversion methods that were developed have been benchmarked in exercises with simulated data [21,22] and have been applied for use with experimental data from many different ocean bottom environments—with varying degrees of success. We describe first the applications of linear inversion methods and then the use of Bayesian statistical inference for nonlinear problems.

9.4.2 LINEAR INVERSIONS FOR GEOACOUSTIC PROFILES

Although the relationship between the pressure and the geoacoustic model parameters is nonlinear, linear relationships can be developed for some observables that are derived from the acoustic field. The problem is linearized in the vicinity of a reference model, m_0, derived from prior knowledge of the local environment, and it is assumed that the unknown model is related to the reference model by a small perturbation. Perturbation inversion has the advantage of fast computational speed in linear methods, but there are some issues that offset this advantage. The most serious concern is that one is never sure that the final model is independent of the reference model. In many cases, the inversion does not converge if the starting model is not close to the solution, or more likely, it converges to a local minimum. Another serious issue is that because the relationship is nonlinear, it can be very misleading to use only the parameter space near the final estimated model to characterize the solution. Nevertheless, if used carefully, the approach can generate remarkably useful models.

An outstanding example of perturbation inversion was reported by Frisk et al. who developed an elegant approach for estimating sound speed profiles in marine sediments by linearizing the relationship between changes in the horizontal wave numbers of propagating modes and changes in the sound speed [23,24]. The method assumes a background model for the sound speed profile $c_0(z)$ that generates a set of horizontal wave numbers, k_{0m}, and corresponding modal functions, $Z_{0m}(z)$, for a sound frequency ω that are solutions of the depth-separated wave equation,

$$\left[\rho_0(z)\frac{d}{dz}\left(\frac{1}{\rho_0(z)}\frac{d}{dz}\right) + k_0^2(z)\right]Z_{0m}(z) = k_{0m}^2 Z_{0m}(z), \qquad (9.6)$$

where $\rho_0(z)$ is the sediment density. The true model is thus

$$c(z) = c_0(z) + \delta c(z), \tag{9.7}$$

and the wave numbers are changed from those for the background model,

$$k(z) = \omega/(c_0(z) + \delta c(z)). \tag{9.8}$$

Applying first-order perturbation theory, an approximation can be obtained for the change in wave number with respect to that for the background model in terms of the change in sound speed [24]:

$$\delta k_m = k_m - k_{0m} = \frac{1}{k_{0m}} \int_0^\infty |Z_{0m}(z)|^2 \frac{k_0^2(z)}{\rho_0(z)} \frac{\delta c(z)}{c_0(z)} dz. \tag{9.9}$$

For a discrete sampling of the sound speed profile in depth, Eq. (9.9) can be cast in terms of a linear relationship between $\delta k(z)$ and the geoacoustic model parameters,

$$\delta k = Gm \tag{9.10}$$

where G is a $N \times M$ matrix consisting of the background sound speed, density, and mode functions; N is the number of discrete samples of the sound speed profile; and M is the number of model parameters [24].

Application of the method requires estimation of the horizontal wave numbers of propagating modes. The basis for this is the Hankel transform relationship between the depth-dependent Green's function and measurements of the variation of pressure with range for a specific sound frequency [23]. Good results have been obtained for experimental data from range-independent waveguides, and an extension of the technique for range-dependent waveguides using a short-time Fourier transform was developed by Becker [25]. Fig. 9.3 shows an example of wave number estimation using this technique applied to data from the Shallow Water '06 (SW06)

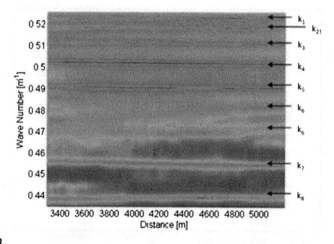

FIGURE 9.3

Modal wave numbers of eight propagating modes that were estimated from SW06 experimental data of sound pressure versus range for a frequency of 125 Hz.

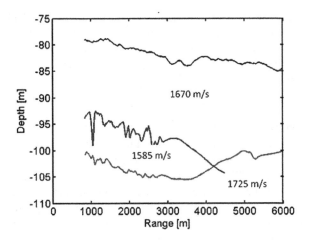

FIGURE 9.4

Chirp sonar depth profile from the SW06 experiment showing the depths of interfaces detected in the survey (*upper curve*: seafloor; *middle curve*: slow-speed "erose" layer boundary; *bottom curve*: "R-reflector" [27] indicating the base of the upper sediment material).

experiment that was carried out on the New Jersey continental shelf [26]. The estimated wave numbers of eight modes that are resolved in the data change slightly with the increasing water depth (WD) along the track. The estimated value of mode 6 is sensitive to a slow speed layer ($\sim$ 15 m below the seafloor) that pinches out and disappears toward the end of the track (Fig. 9.4).

The inverse problem in Eq. (9.10) is ill-posed and requires some form of regularization to obtain a solution. Ballard et al. introduced a simple approach for piece-wise regularization that enabled solution of a discontinuous sound speed profile [28] and used it to invert a range-dependent sound speed profile from the SW06 data. The method requires a priori knowledge of the locations of sound speed discontinuities in the subbottom material. This information was obtained from chirp sonar surveys of the SW06 experimental sites before the experiment, and the resulting section in depth (converted from two-way sonar signal travel time) is shown in Fig. 9.4. The combined inversion of modal wave number data and two-way travel time information was able to estimate the sound speed in the three sediment layers that were defined by the sonar data. However, without this type of additional information, the perturbation inversion can generate only a smoothed approximation to the profile [28].

9.4.3 GEOACOUSTIC INVERSION BY BAYESIAN INFERENCE

The Bayesian formulation of the geoacoustic inverse problem follows from Bayes' rule for measured data and a set of environmental model parameters that is expressed in terms of conditional probabilities [29,30]:

$$P(\boldsymbol{m}|\boldsymbol{d})P(\boldsymbol{d}) = P(\boldsymbol{d}|\boldsymbol{m})P(\boldsymbol{m}). \qquad (9.11)$$

Here, $P(m|d)$ is the conditional probability density function (PDF) of the model, given the experimental data, and $P(d)$ is the PDF of the data for the selected model parameterization. If we assume that the model parameterization is the correct geoacoustic model of the ocean bottom, then for observed data, $P(d) = 1$, and Eq. (9.11) become

$$P(m|d) = P(d|m)P(m). \qquad (9.12)$$

However, in general the correct parameterization of the real ocean bottom is not known, and $P(d)$ can be considered in terms of the likelihood of the parameterization, given the data. In both equations, $P(d|m)$ is the conditional PDF of the data, given a model m, and $P(m)$ is the PDF of the model m. The models are assumed to be random variables and $P(m)$ is interpreted as the distribution of model parameter values that expresses the prior knowledge about the geoacoustic model, independent of the data.

It is evident from Eq. (9.12) that Bayesian inference involves an interaction to combine the information about the model that is contained in the data and the prior knowledge about the model. In an inversion, new information about the model is obtained from the data by performing tests of the ability of candidate models to predict the observed data.

An instructive example of the Bayesian interaction between prior and new information involves the familiar scenario of a visit to a doctor. A patient enters the doctor's office with symptoms that could be related to a number of different causes. The list of possible causes represents the prior knowledge or information, $P(m)$, about the patient's illness. The objective of the visit is to reach a diagnosis of the illness, given by $P(m|d)$ in a Bayesian examination of the patient. The doctor begins the examination by carrying out medical tests. Each test provides new information that the doctor uses to determine which ones of the initial list of causes are more likely. In the Bayesian sense, the result of each test is represented by the conditional probability, $P(d|m)$, and the diagnosis is modified to account for the new information. As more tests are carried out, the initial information continues to be modified so that few of the causes are given higher probability in the final diagnosis.

Returning to geoacoustic inversion, the comparison is straightforward. The inversion is carried out by comparing predictions of the data based on the parameters of specific models against the data themselves. For instance, if the data are pressures at a hydrophone array, the predictions are calculations of the acoustic field using the geoacoustic model parameters. The model parameter values are constrained within specified limits for each model that is tested. If there is little or no information in the experimental data about the proposed model, the resultant probability $P(m|d)$ after the inversion remains close to the original prior probability $P(m)$ which may, for instance, be a uniform distribution for each model parameter. Otherwise, the resultant probability distribution is modified from the prior knowledge by the information about the model contained in the data.

To proceed further it is necessary to examine $P(d|m)$. This conditional PDF expresses the probability that the measured data could have occurred if the ocean bottom was described by a particular geoacoustic model. If the probability is very small, the model is unlikely to be correct. However, if the probability is high, the model is more likely. Thus, $P(d|m)$ can be interpreted as the likelihood of the model given the observed data [31]. From Eq. (9.12),

$$P(m|d) \propto L(m)P(m) \qquad (9.13)$$

where $L(m)$ is the likelihood function for the observed data. The general form of the likelihood function is

$$L(m) \propto exp[-E(m)] \qquad (9.14)$$

where $E(m)$ expresses the mismatch between the data and predictions of the data based on the model. From Eq. (9.12), $P(m|d)$ can be expressed in terms of a generalized mismatch that combines information about the model from both the data and the prior knowledge

$$P(m|d) = \frac{exp(-[E(m) - \log_e P(m)])}{\int exp(-[E(m') - \log_e P(m')])dm'}, \qquad (9.15)$$

where the integration is over the entire model parameter space.

In Bayesian inference, $P(m|d)$ represents the complete solution of the inverse problem and is referred to as the *a posteriori* probability density (PPD). For geoacoustic inversion, it expresses the probability of candidate models in the model parameter space being likely representations of the real ocean bottom. Models with higher probability are expected to be more likely representations of the real environment. It is evident that the PPD is a multidimensional quantity, with dimensions that depend on the number of model parameters in the geoacoustic model that are estimated. The challenge is to interpret the multidimensional PPD in terms of model parameter estimates and uncertainties. This requires numerical computation of properties such as the maximum *a posteriori* (MAP) model estimate, the mean model estimate, the model covariance matrix, and marginal probability distributions. These are defined as:

$$\widehat{m}_{MAP} = Arg_{max}[P(m|d)] \qquad (9.16)$$

$$\widehat{m}_{Mean} = \int m'P(m'|d)dm', \qquad (9.17)$$

$$C = \int (m' - \widehat{m})(m' - \widehat{m})^T P(m'|d)dm', \qquad (9.18)$$

and

$$P(m_i|d) = \int \delta(m_i' - m_i)P(m'|d)dm', \qquad (9.19)$$

respectively. In Eq. (9.19) that expresses the one-dimensional marginal probability distribution, δ is the Dirac delta function. Higher-dimensional marginal probability distributions are defined similarly. Quantified relationships between model parameters can be obtained from the correlation matrix, obtained by normalizing the model covariance matrix (Eq. (9.18)).

Estimating the MAP model (Eq. (9.16)) is an optimization problem to maximize $P(m|d)$ or equivalently, minimize the mismatch $E(m) - log_e P(m)$. Highly efficient numerical algorithms such as simulated annealing (SA) and genetic algorithms (GAs) have been developed for navigating the multidimensional model parameter space and applied successfully in inversions to search for the optimum model.

All the other estimates in Eqs. (9.17)–(9.19) require evaluation of multidimensional integrals to interpret the statistical properties of the PPD and generally involve much more computational effort than the optimization inversions. Numerical methods based on Monte Carlo Markov chain analysis have been developed for efficient evaluation of the various properties [29,30]. Integration of the PPD provides both model parameter estimates and their uncertainties. A useful measure of parameter uncertainty that can be derived from the PPD is the credibility interval, i.e., the $\gamma\%$ highest probability density (%HPD) interval that represents the minimum width interval of model parameter values that contains $\gamma\%$ of the marginal probability distribution.

To implement Bayesian inference, it is necessary to specify the relationship between the observed data and the set of environmental model parameters to define the mismatch function $E(m)$ in Eq. (9.14). The relationship can be interpreted in terms of the mismatch between the measurement (which can be complex valued) and a prediction of the measurement q based on the model:

$$d - q(m) = n. \tag{9.20}$$

As discussed previously, the mismatch n can be interpreted as noise that arises from uncertainty in the experimental data itself, theory errors owing to differences between the environmental model and the real earth, and differences caused by an inaccurate physical theory of sound propagation in the ocean. The statistical distribution of n is generally not known, and the general approach assumes that the errors are Gaussian distributed.

With the assumption of Gaussian errors, the misfit function, $E(m)$, is given by

$$E(m) = \left[(d - q(m))^{\dagger} C_d^{-1}(d - q(m))\right] \tag{9.21}$$

and the likelihood function becomes

$$L(m) = \frac{1}{\pi^N |C_d|} exp\left[-\left[(d - q(m))^{\dagger} C_d^{-1}(d - q(m))\right]\right], \tag{9.22}$$

where $\dagger$ denotes the Hermitian transpose, C_d is the data error covariance matrix, and N is the number of sensors at which data are obtained.

In many applications, the covariance matrix is assumed to be diagonal, $C_d = \sigma^2 I$, where σ is the standard deviation of uncorrelated errors assumed to be the same at each sensor, and I is the identity matrix. For this condition, the likelihood function becomes

$$L(m) = \frac{1}{\pi^N \sigma^{2N}} exp\left[-|d - q(m)|^2 \sigma^{-2}\right]. \tag{9.23}$$

However, the assumption of uncorrelated errors is not usually correct, and can lead to unwarranted overly optimistic estimates of the model parameter uncertainties [31,32].

9.4.4 BAYESIAN MATCHED FIELD INVERSION

Bayesian inference can be applied to any of the types of data that are derived from the acoustic field. We consider here the application with MFP, matched field inversion (MFI) of acoustic pressure data, and discuss examples of optimization inversion to determine the MAP model, and integration of the PPD. The likelihood function derived in Eq. (9.20) is not suitable for use in MFI because the difference $d - q(m)$ requires information about the sound source. For the usual case in MFI that the phase (θ) and amplitude (A) of the source sound pressure are unknown, the modeled data can be expressed as

$$q(m) = Ae^{i\theta}F_\omega(m), \tag{9.24}$$

where F_ω is the forward propagation model used to calculate the replica field at frequency ω for the geoacoustic model m. The dependence on the source can be removed by maximizing the likelihood function (Eq. (9.22)) over θ and A to obtain a covariance-weighted data-model misfit function given by

$$E_\omega(m) = d_\omega^\dagger C_d^{-1} d_\omega - \frac{\left| F_\omega^\dagger(m) C_d^{-1} d_\omega \right|^2}{F_\omega^\dagger(m) C_d^{-1} F_\omega(m)}, \tag{9.25}$$

where d_ω are the observed data at frequency ω. For multi-frequency data the misfits for different frequencies are usually combined incoherently, so that Eq. (9.25) becomes a summation over the number of frequencies.

9.4.4.1 Bayesian Matched Field Inversion by Optimization

In optimization algorithms to determine the MAP model in MFI it is practical to assume spatially uncorrelated data errors at the receiver array. In this case, the covariance matrix is diagonal and the likelihood function then becomes

$$L(m) = \frac{1}{\pi^N \sigma^{2N}} exp\left[-\frac{|d_\omega|^2}{\sigma^2} \left(1 - \frac{\left| F_\omega^\dagger(m) d_\omega \right|^2}{|F_\omega(m)|^2 |d_\omega|^2} \right) \right]. \tag{9.26}$$

It is evident from Eq. (9.26) that the likelihood of a specific model given the data are related to the normalized Bartlett mismatch from MFP,

$$B_\omega(m) = 1 - \frac{\left| F_\omega^\dagger(m) d_\omega \right|^2}{|F_\omega(m)|^2 |d_\omega|^2}. \tag{9.27}$$

To implement Eq. (9.16) it is necessary to specify the prior probability distribution. The form of the prior geoacoustic model determines the structure and properties of the model estimated from the inversion, and so the design of the model requires careful development. In practice, the model is based on knowledge of the local environment obtained from "ground-truth" information such as sediment cores, physical grab samples from the ocean bottom, and high-resolution seismic and chirp sonar surveys. Model structure is usually based on homogeneous or gradient layers of sound speed, attenuation, and density to represent the sediment material in the ocean bottom (Fig. 9.2), and the distribution of model parameter values is assumed to be uniform within the assigned bounds. Thus, $log_e P(m)$ is constant for model parameter values within the bounds and $-\infty$ otherwise. With this assumption, Eq. (9.16) is implemented by minimizing the mismatch function in Eq. (9.27). To complete the prior information, the water sound speed profile is usually taken from measurements at the experimental site and is assumed known in the inversion.

The inversion is carried out by applying a numerical search algorithm to explore the model parameter space to test candidate models that have parameter values within the allowed bounds. However, the dimensions of the model parameter space are generally very large, and extensive regions within the space may contain models with relatively low likelihood of matching the measured data. Grid searches and Monte Carlo search methods that randomly choose models are thus inefficient and impractical, and numerical global search algorithms have been developed that preferentially search more promising regions that contain models that are more likely representations of the real ocean bottom. The most widely used of these methods are SA [34−37] and GAs [38]. The use of the former method is described here.

SA is based on an analogy with the thermodynamic process of annealing a crystal. There are two algorithms that simulate this process: the Metropolis algorithm [39] and the heat bath algorithm [40]. The former method is in more widespread use in underwater acoustics. The Metropolis algorithm is designed to test possible models to determine if the mismatch function is decreased, while a control parameter analogous to temperature in a thermal annealing process is reduced. The control parameter is reduced from an initial high value according to the schedule $T_{j+1} = \gamma^{j+1} T_0$, where T_0 is the initial value and $\gamma \leq 1$. New models are selected at each iteration by perturbing the parameter values according to $m_i^{new} = m_i + \xi \delta_i$, where ξ is a random number from a uniform distribution on $[-1,1]$ and δ_i is the maximum perturbation for the ith model parameter (generally equal to half the span of allowed values). The algorithm assumes that the probability of the change from m_i to m_i^{new} is reversible. If the mismatch function is decreased, the model under test is accepted unconditionally, and the search is likely to continue exploring in the same more promising region of the parameter space. However, if the mismatch function is increased, the model is accepted according to the condition

$$\eta \leq exp[-\Delta B_\omega(m_i)/T] \qquad (9.28)$$

where η is a random number from the interval [0,1], T is the control parameter, and $\Delta B_\omega(m_i)$ is the difference between the values of the mismatch function for the $j+1$th and jth iterations. At high "temperatures," the probability of accepting models that do not decrease the mismatch function is high, and in this way the search process can escape from local minima and explore other regions of the parameter space. The efficiency of the search process depends on the control parameters, and some experimentation is required for each data set. The efficiency is also related to the choice of model parameters, and it is usually more efficient to reparameterize the original model parameters to obtain a new orthogonal basis set consisting of linear combinations of the original parameters [40–42].

Hybrid methods have also been applied for navigating the model parameter space. These are designed to retain some memory of good regions of the parameter space in order to improve the efficiency and speed of global search processes. This approach introduces the benefits of local gradient–based algorithms that are computationally fast but are prone to get caught in local minima, into global search techniques that can explore all regions of the parameter space. One of the most effective hybrid algorithms is Adaptive Simplex Simulated Annealing (ASSA) [43]. This algorithm combines the advantages of the downhill simplex local search technique with SA. The downhill simplex method works locally to evaluate the mismatch function for each model in a set of models that define the simplex and replace the worst one with a new model. The new model is found by a series of prescribed operations that move the simplex to new areas in the model parameter space [43]. The hybrid method provides three significant advantages compared to conventional SA:

- Instead of a single "walker" in the parameter space, ASSA allows a simplex of size $M + 1$, where M is the number of model parameters, to navigate the parameter space.
- The simplex always retains the best M models that have been found.
- The simplex is effectively a shape that adapts to the topology of features in the parameter space such as valleys or ridges that are related to model parameter correlations. Consequently, the search is actually more efficient in model parameterizations for which correlations exist between some parameters.

As in MF source localization, correlations between model parameters are the source of serious concerns, leading to inefficient searches and errors in the estimated parameter values. Re-parameterizing the original set of model parameters during the initial stages of the search process enables more efficient navigation of the parameter space but does not eliminate the basic problem [40–42]. Errors in the geometrical parameters of the experiment are often serious, since the experimental arrangement must be accurately known in calculating the acoustic field at the array. The impact of these errors can be reduced by including parameters such as source range and depth as unknowns in the inversion, at the cost of increasing the computational effort in searching a model parameter space of larger dimensions. This approach supplies a useful consistency check on the quality of the inversions, provided that the inversion generated accurate estimates of the geometric parameters [33]. The impact of errors caused by the correlation between source range and sound frequency could sometimes be mitigated by the use of multi-frequency data (multiple tones or broadband signals) [32,33]. These examples of errors due to mismatch in MFI stress the fundamental issue of nonuniqueness of the solution to the inverse problem.

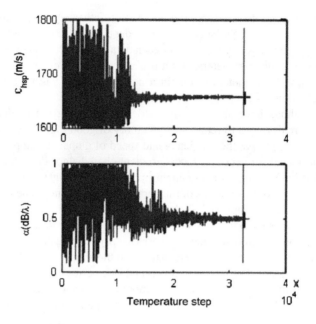

FIGURE 9.5

Annealing history of geoacoustic model parameters (sediment sound speed, c_{hsp}, and attenuation, α) of a half-space geoacoustic model.

Results of optimization inversions using algorithms based on SA are conventionally presented in terms of the annealing history of each model parameter during the search process. An example of the annealing history is shown in Fig. 9.5 that displays plots of the accepted values of two geoacoustic model parameters versus iteration numbers during the search process of the hybrid method ASSA. The inversion assumed a simple half-space geoacoustic model with six model parameters, giving rise to a simplex of seven models. The initial value of the annealing control parameter, T_0, was 0.3, the annealing reduction factor, γ, was 0.99, and the search spanned 36,000 iterations.

As can be seen in Fig. 9.5, the allowed values were well sampled in the initial stages of the search (to about 10,000 iterations). Subsequently, the search fixed on subsets of the allowed values that optimized the cost function and remained in those regions for the remainder of the search. The annealing rates reflect the sensitivity of the model parameter in the experiment: sound speed of the sediment settles in a favored region after about 12,000 iterations, whereas the less sensitive attenuation does not settle until around 20,000 iterations. The spike at the end of the search results from a final "quenching" of the local downhill simplex algorithm to refine the optimal values.

However, the annealing history shows only the rate at which the optimal values were obtained in the search process. Although this gives a rough impression of

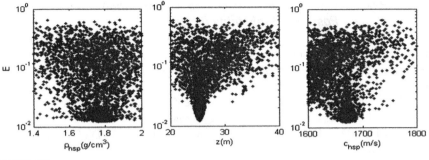

FIGURE 9.6

Scatter plots of mismatch function values for two different geoacoustic model parameters. The center panel shows clustering of accepted models in a favored region of the allowed range of values; the left panel shows a flat scatter indicating that no particular value of this parameter provides a better estimate than any other.

which parameters are more sensitive in the inversion, it does not give a good indication of how well each parameter was estimated. A more informative sense of the hierarchy of sensitivities of the model parameters and a rough, qualitative measure of the uncertainties of the estimated values can be obtained from a scatter plot of the mismatch function values for each model that was tested in the search process. Fig. 9.6 shows scatter plots for three different model parameters in the inversion: the half-space ocean bottom density, ρ_{hsp}; the SD, z; and the ocean bottom sound speed, c_{hsp}. These plots are constructed by plotting the mismatch function values versus model parameter value for models accepted in the search. Plots that appear like "tornadoes" as in the center panel indicate well-estimated parameters with values that cluster in a small region of the allowed range. Those that appear broader at the base, as in the left panel, indicate less sensitive parameters that are not well estimated; the flatness of the display essentially indicates that the experimental data do not contain any useful information about the parameter.

Examination of scatter plots from optimization inversions reveals an inherent weakness of the approach. Optimization inversions always provide an "optimal" estimate for each model parameter, regardless of the sensitivity of model parameters in the experiment. There is always a hierarchy of sensitivities, and the "optimal" values estimated for insensitive parameters do not significantly affect the acoustic field. As a result, inversions can be over-parameterized, with meaningless estimates for some of the model parameters. Optimization inversions do not generate statistically valid measures of the errors in the estimated values and consequently do not provide a complete solution to the inverse problem. However, it usually turns out that the spread of values obtained for a large number of optimization runs (each one with different starting values) is consistent with the error bounds of inversions carried out by Bayesian inference methods to integrate the PPD.

9.4.4.2 Bayesian Matched Field Inversion by Integration of the a Posteriori Probability Density

The complete solution of the geoacoustic inverse problem is provided by numerical integration of the PPD to evaluate integrals that define estimates such as the mean model (Eq. (9.17)) and marginal probability distributions (Eq. (9.19)). Numerical integration involves drawing samples randomly from a distribution that represents the PPD. A straightforward example is Monte Carlo sampling from a uniform distribution. In this case, Q samples are drawn randomly from a distribution given by $g(\boldsymbol{m}) = 1/V$ where V is the volume of the integration. The PPD of the ith model is evaluated as

$$P(\boldsymbol{m}_i|\boldsymbol{d}) = \frac{exp[-E(\boldsymbol{m}_i)]P(\boldsymbol{m}_i)}{\sum_{i=1}^{Q} exp[-E(\boldsymbol{m}_i)]P(\boldsymbol{m}_i)}. \tag{9.29}$$

Then, the mean model, for instance, becomes

$$\langle \boldsymbol{m} \rangle = \frac{V}{Q} \sum_{i=1}^{Q} \boldsymbol{m}_i P(\boldsymbol{m}_i|\boldsymbol{d}). \tag{9.30}$$

Monte Carlo sampling gives an unbiased estimate that converges asymptotically to the PPD. Convergence is monitored numerically by observing that a large increase in the sample size Q does not change the integral, or by comparing two Monte Carlo samplings carried out in parallel. However, the computation time required for convergence in multidimensional models is prohibitively long. Instead, highly efficient techniques based on the Metropolis–Hastings algorithm [44] have been developed that use nonuniform sampling distributions to focus the sampling on promising regions of the model parameter space [29,30]. Although the actual distribution of the PPD is not generally known, a practical and natural choice of an approximate distribution is a Gibbs distribution from classical statistical mechanics. Inspection of Eq. (9.15) indicates that the PPD is proportional to a Gibbs distribution with $T = 1$.

The Metropolis–Hastings algorithm generates a sample set of randomly selected models that in the limit of a large number of samples, closely approximates the PPD. The samples are selected iteratively as the algorithm navigates the model parameter space in a manner similar to that in SA, except that the control parameter equivalent to the temperature is not changed during the iterations. The mean model is simply the average over the sample set, and marginal probability distributions are histograms of individual model parameters.

The presence of correlated data errors caused by unknown theory errors cannot be ignored in MFI by Bayesian integration. Otherwise, model parameter uncertainties will be underestimated, and it will appear from the inversion that the data carry more information about the geoacoustic model than is actually warranted. In this case, the likelihood function should include the covariance-weighted mismatch function,

$$L(\boldsymbol{m}) = \frac{1}{\pi^N \sigma^{2N}} exp\left[-\left[d_\omega^\dagger C_d^{-1} d_\omega - \frac{|F_\omega^\dagger(\boldsymbol{m})C_d^{-1} d_\omega|^2}{F_\omega^\dagger(\boldsymbol{m})C_d^{-1}F_\omega(\boldsymbol{m})} \right] \right], \tag{9.31}$$

and some attempt must be made in the inversion to estimate the full-covariance matrix. This involves making assumptions about the statistics of the data/model mismatch distribution that must be verified by statistical tests. Examples of MFIs that account for correlated data errors can be found in Refs. [32,33].

An example of MFI by Bayesian integration with data from the SW06 experiment is discussed in the following paragraphs to demonstrate the performance of the method [45]. The data from the experiment were seven CW tones over the low-frequency band 53−703 Hz transmitted from a ship that held station at a distance of 1 km from a bottom-moored vertical line array. The array consisted of 16 hydrophones equally spaced at 3.75 m, spanning most of the water column from about 8.2 m above the seafloor. The water depth was about 79 m over the propagation path from the source. Data from the 7 CW tones were combined incoherently in a multifrequency misfit function. The experimental environment is the same environment as the one for the results shown in Section 9.4.2 for the linearized inversion.

Inversion results for the geoacoustic model parameters of a single sediment layer over a half-space bottom model are presented as marginal probability densities in Fig. 9.7. Sensitive parameters that are well estimated have marginal densities that are tightly focused in a favored region within the parameter bounds; marginal densities of parameters for which there is little information in the data are flatter,

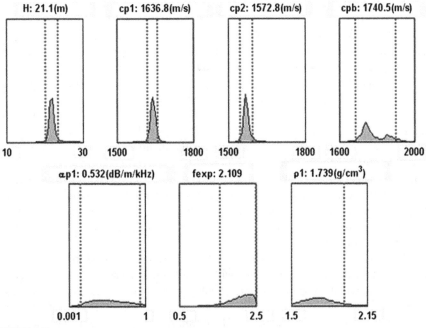

FIGURE 9.7

Marginal probability densities for the model parameters inverted from the SW06 data. The *vertical dotted lines* represent the 95% HPD limits.

indicating that there is no strong preference for any value within the bounds. These shapes are similar to the shapes of the scatter plots from optimization inversions for parameters with similar sensitivities. However, a statistically meaningful measure of the uncertainty can be derived from the marginal densities, such as 95% HPD limits.

Marginal densities for the layer depth (H), the top and bottom sound speed of the sediment layer, c_{p1} and c_{p2}, respectively, and the sound speed in the basement half space, c_{pb}, were tightly focused, indicating that these geoacoustic parameters were well estimated. However, the marginal densities for the other geoacoustic parameters were relatively flat, indicating that the data did not contain significant information about them.

The results shown in the figure are typical of those from other MFIs: the most sensitive parameters are generally the sound speeds in the uppermost layers of sediment (within a few wavelengths of the seafloor). A particularly striking result from this inversion is the accurate estimate of sediment thickness. Ground truth chirp sonar surveys during the experiment revealed a strong subbottom reflector at a depth of about 20 m that was ubiquitous over the experimental area. The inversion was also sensitive to the slow sound speed layer within the sediment above the basement reflector (Fig. 9.4). Although the detailed structure within the sediment could not be resolved with these low-frequency data, the inversion indicated a decrease in sound speed from the top to the bottom of the sediment layer. The presence of the low-speed layer was inferred from the negative gradient of sound speed within the sediment.

Attenuation was interpreted as an intrinsic loss in the sediment, and was modeled in this inversion as frequency dependent, $\alpha_0(f/f_0)^\beta$, where $f_0 = 1$ kHz. The results indicated that the inversion with data from a range of 1 km was not sensitive to attenuation: the marginal densities for the constant, α_{p1}, and the exponent, f_{exp}, were flat. However, the experimental data are affected by other mechanisms that remove energy from the propagation plane, such as scattering. Since the loss accumulates with range, data from greater ranges likely contain more information about attenuation.

Other insights into the estimated model can be obtained from two-dimensional marginal densities. Displays such as shown in Fig. 9.8 reveal model parameter correlations and provide added confidence about the quality of the estimated model. From the figure, there is a clear indication of the correlation between water depth (WD) and range, and also WD and source depth (SD). The negative sound speed gradient in the sediment layer is revealed in the correlation between the top and bottom sound speeds of the layer (c_{p1} and c_{p2}). Other pairs of parameters do not show any strong correlation, as would be expected for pairs such as WD and the thickness of the sediment layer.

MF inversions based on Bayesian integration have been applied to experimental data from many different experiments, with remarkable successes in estimating geoacoustic profiles that compared favorably with ground-truth information for the local environment [46–51]. Most of the experiments were carried out at sites of constant water depth and minimal variability of the ocean sound speed profile and the sediment materials and structure over the track of the experiment. For these conditions, the inversions could be carried out assuming that the sound propagation was independent of range. This is not always the case, particularly if the environment is spatially

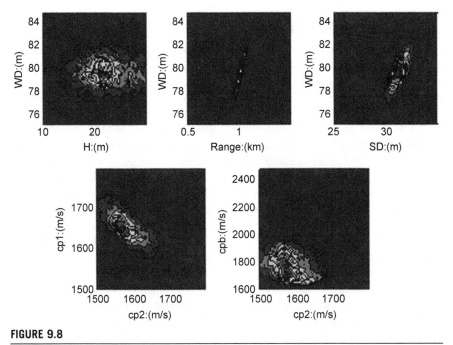

FIGURE 9.8

Two-dimensional marginal densities for the model parameters.

and/or temporally variable. An example of MFI in a strongly range-dependent environment due to the presence of internal waves in a shallow-water continental shelf environment is discussed by Jiang and Chapman [45]. The success of the inversion depended on the assumption that the sound speed variation in the water column could be represented by a single profile based on the observed sound speed variations. However, in conditions when oceanographic data from moored sensors revealed that internal waves passed through the experimental site, inversion using data from signals propagated through the internal wave environment was not successful. Knowledge of the full-range dependence of the sound speed profile is required for inverting such data.

The analysis in Ref. [45] revealed a fundamental weakness of model-based inversions such as MFI. If the environmental variation cannot be modeled sufficiently accurately, the inversion will fail. Although some simple assumptions that involve inverting an effective sound speed profile in the ocean may be effective in conditions of weak variability, the increased computational load of including additional model parameters as unknowns in the inversion is a significant drawback.

As indicated previously, Bayesian inference can be applied for geoacoustic inversion with other types of data. Successful applications have been reported for inversions with bottom loss (reflection coefficient) data [52], time-frequency modal dispersion data [53,54] and also with particle velocity data [55].

9.5 OCEAN ACOUSTIC TOMOGRAPHY

Ocean acoustic tomography (OAT) was developed as an inversion technique for inferring the state of the ocean from precise measurements of the travel time of signals that have propagated through the ocean. Tomography itself is a well-established technique in medicine and seismology that is based on solutions of the inverse problem on slices (τομεζ in Greek) through the medium that are integrated to obtain an image appropriate for the application. In OAT, the slices are propagation paths of sound transmissions at specific angles between sources and receivers deployed in the ocean. Each slice provides a different view of the ocean environment. The tomographic inversion enables reconstruction of the spatial distribution of sound speed inhomogeneities in the volume of ocean within the distributed source–receiver system. A comprehensive treatment of the application of tomography in ocean acoustics for inferring temperature structure and currents is presented in the monograph by Munk et al. [56]. The technique has been applied extensively in deep water, and has been adapted for applications in shallow-water coastal regions. The first two parts of this section focus on deep water applications of OAT in the context of long-term oceanic climate change. The last section deals with acoustic tomography in coastal regions. Although the focus in this section is on travel time inversions, full-field tomography based on MFP was described by Tolstoy et al. [57], and Goncharov and Voronovich who implemented MF tomography in the Norway Sea [58]. This approach uses both phase (travel time) and amplitude information in the inversion.

In applications of OAT for global climate change, the desired knowledge is the evolution of the temperature structure in the ocean. The information provided by OAT from travel time measurements is an integrated measure of the changes in sound speed from assumed background values along acoustic propagation paths. Since there is a near linear relationship between changes in temperature and changes in sound speed, integral measures of the ocean temperature can be extracted from OAT data. This is a unique feature of the measure. It represents a horizontally integrated measure in range and vertically in depth due to the deep cycling of acoustic propagation paths. In addition, the ocean is nearly transparent to low-frequency sound, so measures over basin scale paths can be obtained.

An OAT experiment involves a system of sound sources and receivers arranged to provide a multiplicity of acoustic propagation paths, or vertical slices between the sources and receivers, in the ocean. In many cases, vertical hydrophone arrays are used as receivers.

There are two approaches for inverting the travel time data from an OAT experiment, based on the interpretation of the signal arrival structure in terms of rays or modes. A third method developed by Skarsoulis et al. [59] that uses local peaks in the signal arrival pattern can be used effectively for cases in which the signal arrivals cannot be identified as either rays or modes [60].

9.5.1 INVERSION OF TRAVEL TIMES

In ray acoustics the travel time τ of a specific eigen ray signal arrival is related to the sound speed and current velocity profile in the ocean in terms of an integral along the propagation path:

$$\tau_n = \int_{S_n} \frac{ds}{c(z) \pm v(z)\cos\theta}. \tag{9.32}$$

The variation in sound speed, $c(z)$, is assumed to be only in the vertical slice defining the plane of propagation; $v(z)$ is the current speed, θ is the reception angle, and S_n the path length of the nth eigen ray. When the goal of the inversion is to estimate the current, the problem is solved using reciprocal acoustic transmissions. In the following discussion the focus is on inverting the temperature field, so the current is dropped from Eq. (9.32) to simplify the development.

The inversion of travel time proceeds by assuming that a background sound speed profile $c_0(z)$ is known and the real profile in the ocean differs only very little from the background, as described in Eq. (9.3). The background or reference profile could be, for instance, a historical mean profile for a specific region in the ocean. Linearizing Eq. (9.32) with respect to the background profile, the travel time variation along the ray path S_n is given by

$$\delta\tau_n = \int_{S_n} \frac{\delta c(z)ds}{c_0^2(z)}. \tag{9.33}$$

Assuming that N-specific ray arrivals can be identified in the signal, the integral measure of sound speed along the path and at specific depths in the ocean can be inferred. The problem is generally solved by discretizing the ray path in range and using empirical orthogonal functions (EOFs) to describe the differences in the sound speed profile with depth:

$$\delta c(z) = \sum_m C_m f_m(z), \tag{9.34}$$

where C_m is the amplitude of the mth order EOF, $f_m(z)$. EOF analysis provides a set of orthogonal basis functions that is derived from the data, in this case the covariance matrix of sound speed profile measurements. The EOFs are eigenvectors of the covariance matrix that minimize the residual variance from the mean profile. The use of EOFs reduces the number of unknowns considerably in the inverse problem, since normally only two or three EOFs are sufficient to define the changes in the profile. Similar to the development in Section 9.4.2, this approach leads to a discrete inverse problem that can be expressed in the general form

$$\delta\tau_i = \sum_{j=1}^{J} G_{ij}\delta c_j(z), \tag{9.35}$$

where G_{ij} is an $N \times J$ matrix that relates the changes in travel time and sound speed with respect to the background profile, and N and J are the number of ray arrivals and unknowns, respectively. It is evident from Eqs. (9.33)–(9.35) that the tomographic travel time inversion is based on differential characteristics of the ocean environment.

Ambient noise and physical processes in the ocean such as internal waves introduce uncertainty in the travel time measurements, and, as explained previously, model errors can exist due to inaccuracies derived from the linearization and/or assumptions about the background representation. In the presence of experimental and model errors, the inversion of OAT travel time data becomes an estimation problem. A thorough discussion of the methods for solution in the presence of experimental and model noise and the statistical uncertainty of the estimates is given in Munk et al. [56].

In shallow water, signal arrivals in OAT experiments may be identified as specific modes of propagation. Inversion of modal travel times can also be developed as a linear inverse problem [61,62], and the method has been applied to data from shallow-water experiments by Taroudakis and Markaki [63,64].

9.5.2 ACOUSTIC THERMOMETRY

In deep water applications, OAT provides significant advantages for use in assessment of long-term climate change. The long-range integrated tomographic measure suppresses smaller-scale internal wave and mesoscale noise that cause contamination in point measurement of temperature, and the technique senses information from the full-ocean depth. In comparison, satellite altimetry data provide information only at the surface. High spatial resolution on ocean basin scales can be obtained in the experiments, because the resolution is directly proportional to the number and density of acoustic propagation paths between the sources and receivers. Since around 1990, there have been several successful OAT experiments in ocean basins, including the regional-scale Greenland Sea [65] and Thetis II experiments in the Mediterranean [66], and the ocean basin—scale experiments in the North Pacific (Acoustic Thermometry of Ocean Climate, ATOC) [67] and the Arctic (Acoustic Climate Observation using Underwater Sound, ACOUS) [68]. As an example, results from the ATOC experiment are described here.

The goal of ATOC was to demonstrate the potential of acoustic travel time data for monitoring the long-term variability of the temperature field over the North Pacific Ocean basin. The experimental challenge involved extraction of an expected climate signal of about 10 m°C/year in the presence of mesoscale noise of about 1°C RMS. Data were collected at receivers in the basin over a period from early 1996 to the fall of 1999 from sources off the coast of California (1996–98) and near the Hawaiian Island of Kauai (1997–99). Results showed that signal arrivals for sound transmissions over about 5 Mm range could be resolved and identified with specific acoustic ray propagation paths [67].

The resolved travel time data were used to generate time histories of temperature variability over the nearly 4-year period using a simple model of ocean variability,

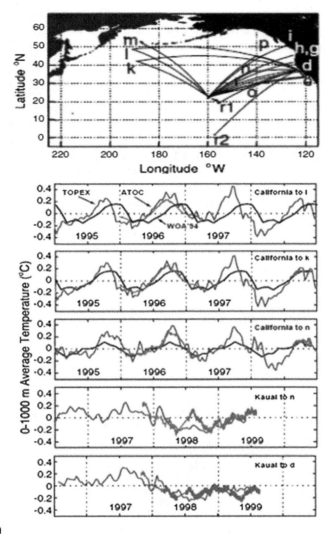

FIGURE 9.9

Upper panel: Map of the ATOC acoustic source and receiver locations in the North Pacific Ocean and the propagation paths of resolved signal transmissions. Sources are at Kauai (r2) and California (d); the other letters are arbitrary names for the receivers. Lower panel: Comparison of the temperature variation derived from altimetry [blue (dark gray in print versions)] and OAT data [red (gray in print versions)]. The altimetry data have been averaged along the acoustic path for these comparisons (B. Dushaw, personal communication).

with uncertainties estimated at about 10 m°C. The arrangement of sources and receivers in the North Pacific Ocean and the comparison with temperature change derived from altimetry data are shown in Fig. 9.9. The temperature change is derived from altimetry data assuming that the variations in sea surface height are due to thermal expansion in the upper 100 m of the ocean. The comparison demonstrates that the integral measure derived from OAT data serves as an effective complementary technique for monitoring temperature variation in the ocean.

9.5.3 ACOUSTIC TOMOGRAPHY IN SHALLOW WATER

Coastal acoustic tomography (CAT) systems have been developed for applications in shallow water to map the horizontal structure of tidal currents [69,70]. CAT systems use multiple vertical line arrays to enclose a volume of water and analyze reciprocal sound transmissions between the receivers. An example of a deployment of eight CAT arrays designed to map strongly nonlinear tidal currents in a section of the Kanmon Strait in Japan [70] is shown in Fig. 9.10. In this application the two-way travel time difference between reciprocal acoustic paths is given by

$$\Delta t_i = \frac{1}{2}\left(t_i^+ - t_i^-\right) = -\int_{\Gamma_i}^0 \frac{\boldsymbol{u}(x,y)\cdot\boldsymbol{n}}{C_0^2}\,ds, \tag{9.36}$$

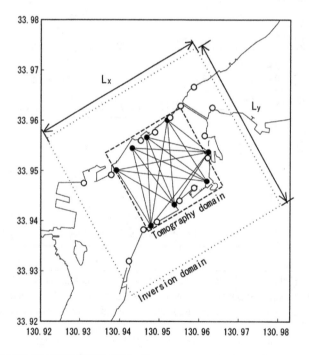

FIGURE 9.10

Sketch of the CAT system and the tomography domain in the Kanmon Strait Positions where boundary conditions are imposed at the coasts with *open circles*. The *solid circles* denote the CAT stations. The tomography and inversion domains are bounded by a *dashed and a dotted rectangle*, respectively (A. Kaneko, personal communication).

(A) **(B)**

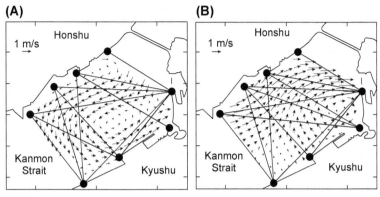

FIGURE 9.11

Horizontal current distributions, obtained at typical tidal phases of 8:00 and 17:00 on March 18, 2003 (A. Kaneko, personal communication). (A) Southwest current (8:00). (B) Northeast current (17:00).

where $t_i^{\pm}$ are the reciprocal travel times for the ith ray path Γ_i, C_0 is the reference sound speed associated with the ray path, u is the current velocity, and n is the unit vector along the ray.

The components of the current are spatial derivatives $u(x,y) = \nabla \times \Psi(x,y)k$, where $\Psi(x,y)$ is a stream function that describes the current field in the water, and k is a unit vector in the vertical direction. Inversion for the unknown currents proceeds by decomposing the stream function in a truncated Fourier series, and solving a linear inversion problem for the Fourier coefficients [70]. In the Kanmon strait deployment, current was monitored on an hourly basis, and an example of the results is shown in Fig. 9.11. The estimated current structure from the CAT data inversion indicates the presence of clockwise and counterclockwise tidal vortices and compares quantitatively very well with conventional measurements by acoustic Doppler current profilers.

REFERENCES

[1] Jensen, F.B., Kuperman, W.A., Porter, M.B. and Schmidt, H., *Computational Ocean Acoustics*, American Institute of Physics, Woodbury, New York, 1994.

[2] Menke, W., *Geophysical Data Analysis: Discrete Inverse Theory*, Academic Press, Orlando, 1987.

[3] Frisk, G.V., *Inverse methods in ocean bottom acoustics*, In: Oceanographic and Geophysical Tomography: Les Houches 1988, Y. Desuabies, A Tarantola and J. Zinn-Justin, (Eds.), North-Holland, Amsterdam, 1990, pp. 439−437.

[4] Caiti, A., Hermand, J-P, Jesus, S.M., and Porter, M.B., (Eds.), *Experimental Acoustic Inversion Methods for Exploration of the Shallow Water Environment*, Kluwer Academic Publishers, Dordrecht, 2000.

[5] Taroudakis, M. and Makrakis, G., (Eds.), *Inverse Problems in Underwater Acoustics*, Springer, New York, 2001.

[6] Caiti, A., Chapman, N.R., Hermand, J-P. and Jesus, S., (Eds.), *Acoustic Sensing Techniques for the Shallow Water Environment*, Springer, Dordrecht, 2006.

[7] Parvulescu, A., *Matched signal ("M.E.S.S.") processing by the ocean*. J. Acoust. Soc. Amer., **98**, pp. 943–960, 1995.

[8] Bucker, H., *Use of calculated sound fields and matched field processing to locate sound sources in shallow water*. J. Acoust. Soc. Amer., **59** (2), pp. 368–373, 1976.

[9] Baggeroer, A.B., Kuperman, W.A. and Schmidt, H., *Matched field processing: Source localization in correlated noise as an optimum parameter estimation problem*. J. Acoust. Soc. Amer., **83**, pp. 571–587, 1988.

[10] Tolstoy, A., *Matched Field Processing for Underwater Acoustics*, World Scientific Publishing, Singapore, 1993.

[11] Cox, H., *Resolving power and sensitivity to mismatch of optimum array processors*. J. Acoust. Soc. Amer., **54**, pp. 771–785, 1973.

[12] Fizell, R.G., *Application of high-resolution processing to range and depth estimation using ambiguity function methods*. J. Acoust. Soc. Amer., **82**, pp. 606–613, 1987.

[13] Schmidt, H., Baggeroer, A.B., Kuperman, W.A. and Scheer, E.K., *Environmentally tolerant beamforming for high resolution matched field processing: deterministic mismatch*. J. Acoust. Soc. Amer., **88**, pp. 1851–1857, 1990.

[14] Westwood, E.K., *Broadband matched-field source localization*. J. Acoust. Soc. Amer., **83**, pp. 2777–2789, 1992.

[15] D'Spain, G.L.D., Murray, J.J., Hodgkiss, W.S., Booth, N.O. and Schey, P., *Mirages in shallow water matched field processing*. J. Acoust. Soc. Amer., **105**, pp. 3245–3265, 1998.

[16] Collins, M.D. and Kuperman, W.A., *Focalization: Environmental focusing and source localization*. J. Acoust. Soc. Amer., **91**, pp. 1410–1422, 1991.

[17] Jensen, F.B. and Kuperman, W.A., *Sound propagation in a wedge shaped ocean with a penetrable bottom*. J. Acoust. Soc. Amer., **67**, pp. 1564–1566, 1981.

[18] Rubano, L.A., *Acoustic propagation in shallow water over a low-velocity bottom*. J. Acoust. Soc. Amer., **67**, pp. 1608–1613, 1980.

[19] Chapman, N.R., *Modeling ocean-bottom reflection loss measurements with the plane-wave reflection coefficient*. J. Acoust. Soc. Amer., **73**, pp. 1601–1607, 1983.

[20] Evans, R.B. and Carey, W.M., *Frequency dependence of sediment attenuation in two low frequency shallow water acoustic experimental data sets*. IEEE J. Oceanic Eng., **23**, pp. 439–447, 1998.

[21] Tolstoy, A., Chapman, N.R. and Brooke, G.E., *Workshop '97: Benchmarking for geoacoustic inversion in shallow water*. J. Comp. Acoust., **6**, pp. 1–28, 1998.

[22] Chapman, N.R., Chin-Bing, S.A., King, D. and Evans, R.B., *Benchmarking geoacoustic inversion methods for range dependent waveguides*. IEEE J. Oceanic Eng., **28**, pp. 320–330, 2003.

[23] Frisk, G.V. and Lynch, J.F., *Shallow water waveguide characterization using the Hankel transform*. J. Acoust. Soc. Amer., **76**, pp. 205–211, 1984.

[24] Rajan, S.D., Lynch, J.F. and Frisk, G.V., *Perturbative inversion methods for obtaining bottom geoacoustic parameters in shallow water*. J. Acoust. Soc. Amer., **82**, pp. 998–1017, 1987.

[25] Becker, K.M. and Frisk, G.V., *Evaluation of an autoregressive spectral estimator for modal wave number estimation in range-dependent shallow water waveguides*. J. Acoust. Soc. Amer., **120**, pp. 1423–1434, 2006.

[26] Tang, D.J., Moum, J.N., Lynch, J.F., Abbot, P., Chapman, N.R., Dahl, P.H., Duda, T.F., Gawarkiewicz, G., Glenn, S., Goff, J.A., Graber, H., Kemp, J., Maffei, A., Nash, J.D. and Newhall, A., *Shallow Water '06: A Joint Acoustic Propagation/Nonlinear Internal Wave Physics Experiment*. Oceanography **20** (4), pp. 156−167, 2007.

[27] Goff, J.A. and Austin, J.A., *Seismic and bathymetric evidence for four different episodes of iceberg scouring on the New Jersey outer shelf: Possible correlation to Heinrich events*. Marine Geology, **266**, pp. 244−254, 2009.

[28] Ballard, M., Becker, K.M. and Goff, J.A., *Geoacoustic inversion for the New Jersey shelf: three dimensional sediment model*. IEEE J. Oceanic Eng., **35**, pp. 28−42, 2009.

[29] Tarantola, A., *Inverse Problem Theory: Methods for Data Fitting and Model Parameter Estimation*, Elsevier, Amsterdam, 1987.

[30] Dosso, S.E., *Quantifying uncertainties in geoacoustic inversion I: A fast Gibbs sampler approach*. J. Acoust. Soc. Amer., **111**, pp. 128−142, 2002.

[31] Gerstoft, P. and Mecklenbrauker, C., *Ocean Acoustic inversion with estimation of the a posteriori probability distribution*. J. Acoust. Soc. Amer., **104**, pp. 808−819, 1998.

[32] Dosso, S.E., Nielsen, P.L. and Wilmut, M.J., *Data error covariance in matched-field geoacoustic inversion*. J. Acoust. Soc. Amer., **119**, pp. 208−219, 2006.

[33] Jiang, Y-M., Chapman, N.R. and Badiey, M., *Quantifying the uncertainty of a geoacoustic model for the New Jersey shelf by inverting air gun data*. J. Acoust. Soc. Amer., **121**, pp. 1879−1894, 2007.

[34] Sen, M.K. and Stoffa, P.L., *Global Optimization Methods in Geophysical Inversion*, Elsevier, Amsterdam, 1995.

[35] Basu, A. and Frazer, L.N., *Rapid Determination of the Critical Temperature in Simulated Annealing Inversion*. Science, **21**, September 1990, pp. 1409−1412, 1990.

[36] Collins, M.D., Kuperman, W.A. and Schmidt, H., *Non-linear inversion for ocean bottom properties*. J. Acoust. Soc. Amer., **92**, pp. 2770−2783, 1992.

[37] Lindsay, C.E. and Chapman, N.R., *Matched Field Inversion for Geoacoustic Model Parameters Using Adaptive Simulated Annealing*. IEEE J. Oceanic Eng., **18**, pp. 224−231, 1993.

[38] Gerstoft, P., *Inversion of seismoacoustic data using genetic algorithms and a posteriori probability distributions*. J. Acoust. Soc. Amer., **95**, pp. 770−782, 1994.

[39] Metropolis, N., Rosenbluth, A.W., Rosenbluth, M.N., Teller, A.H., Teller, E., *Equations of State Calculations by Fast Computing Machines*. J. Chem. Phys., **21** (6), pp. 1087−1092, 1953.

[40] Jaschke, L. and Chapman, N.R., *Matched field inversion of broadband data using the Freeze Bath method*. J. Acoust. Soc. Amer., **106**, pp. 1838−1851, 1999.

[41] Collins, M.D. and Fishman, L., *Efficient navigation of parameter landscapes*. J. Acoust. Soc. Am., **98**, pp. 1637−1644, 1995.

[42] Fialkowski, L.T., Lingevitch, J.F., Perkins, J.S. and Collins, M.D., *Geoacoustic inversion using a rotated coordinate system and simulated annealing*. IEEE J. Oeanic Eng., **28**, pp. 370−379, 2003.

[43] Dosso, S.E., Wilmut, M.J. and Lapinski, A-L., *An adaptive hybrid algorithm for geoacoustic inversion*. IEEE J. Oceanic Eng., **26**, pp. 324−336, 2001.

[44] Hastings, W.K., *Monte Carlo Sampling Methods Using Markov Chains and Their Applications*. Biometrika, **57** (1), pp. 97−109, 1970.

[45] Jiang, Y-M. and Chapman, N.R., *The impact of ocean sound speed variability on the uncertainty of geoacoustic parameter estimates*. J. Acoust. Soc. Am., **125**, pp. 2881−2895, 2009.

[46] Michaloupoulou, Z-H., *Robust multi-tonal matched field inversion: A coherent approach*. J. Acoust. Soc. Am., **104**, pp. 163–170, 1998.

[47] Knobles, D.P., Koch, R.A., Thompson, L.A., Focke K.C. and Eisman, P.E., *Broadband sound propagation in shallow water and geoacoustic inversion*. J. Acoust. Soc. Am., **113**, pp. 205–222, 2003.

[48] Battle, D.J., Gerstoft, P., Hodgkiss, W.S., Kuperman, W.A. and Nielsen, P.L., *Bayesian model selection applied to self-noise geoacoustic inversion*. J. Acoust. Soc. Am., **116**, pp. 2043–2056, 2004.

[49] Koch, R.A. and Knobles, D.P., *Geoacoustic inversion with ships as sources*. J. Acoust. Soc. Am., **117**, pp. 626–637, 2005.

[50] Tollefsen, D., Wilmut, M.J. and Chapman, N.R., *Estimates of geoacoustic model parameters from inversions of horizontal and vertical line array data*. IEEE J. Oceanic Eng., pp. 764–772, 2005.

[51] Jiang, Y-M., Chapman N.R. and DeFerrari, H.A., *Geoacoustic inversion of broadband data by matched beam processing*. J. Acoust. Soc. Am., **119**, pp. 3707–3716, 2006.

[52] Holland, C.W., Dettmer, J. and Dosso, S.E., *Remote sensing of sediment density and velocity gradients in the transition layer*. J. Acoust. Soc. Am., **118**, pp. 163–177, 2005.

[53] Potty G., Miller, J. and Lynch, J.F., *Inversion for sediment geoacoustic properties at the New England Bight*. J. Acoust. Soc. Am., **114**, pp. 1874–1887, 2003.

[54] Bonnel, J. and Chapman, N.R., *Geoacoustic inversion in a dispersive waveguide using warping operators*. J. Acoust. Soc. Am., **130**, pp. EL101–EL107, 2011.

[55] Peng, H. and Li, F., *Geoacoustic inversion based on vector hydrophone array*. Chin. Phys. Lett., **24**, pp. 1977–1980, 2007.

[56] Munk, W., Worcester, P. and Wunsch, W., *Ocean Acoustic Tomography*, Cambridge University Press, 1995.

[57] Tolstoy, A., Diachok, O.I. and Frazer, L.N., *Acoustic tomography via matched field processing*. J. Acoust. Soc. Am., **89**, pp. 1119–1127, 1991.

[58] Goncharov, V.V. and Voronovich, A.G., *An experiment on matched field acoustic tomography with continuous wave signals in the Norway Sea*. J. Acoust. Soc. Am., **93**, pp. 1873–1881, 1993.

[59] Skarsoulis, E.K., Athanassoulis, G.A. and Send, U., *Ocean acoustic tomography based on peak arrivals*. J. Acoust. Soc. Am., **100**, pp. 797–813, 1996.

[60] Kindler, D., Send, U. and Skarsoulis, E.K., *Relative-time inversions in the Labrador Sea acoustic tomography experiment*. Acustica, **87**, pp. 738–747, 2001.

[61] Shang, E.C., *Ocean acoustic tomography based on adiabatic mode theory*. J. Acoust. Soc. Am., **85**, pp. 1531–1537, 1989.

[62] Shang, E.C. and Wang, Y.Y., *On the possibility of monitoring El Nino by using modal ocean acoustic tomography*. J. Acoust. Soc. Am., **91**, pp. 136–140, 1992.

[63] Taroudakis, M.I. and Papadakis, J.S., *Modal inversion schemes for ocean acoustic tomography*. J. Comp. Acoust., **1**, pp. 395–421, 1993.

[64] Taroudakis, M.I. and Markaki, M., *Tomographic inversions in shallow water using modal travel time inversions*. Acustica, **87**, pp. 647–658, 2001.

[65] Worcester, P.E., Lynch, J.F., Morawitz, W.M., Pawlowicz, R., Sutton, P.J., Cornuelle, B.D., Johannessen, O.M., Munk, W.H., Owens, W.B., Schuchman, R. and Spindel, R.C., *Ocean acoustic tomography in the Greenland Sea*. Geophys. Res. Lett., **20**, pp. 2011–2012, 1993.

[66] Send, U., Krahmann, G., Mauury, D., Desaubies, Y., Gaillard, F., Terre, T., Papadakis, J., Taroudakis, M., Skarsoulis, E. and Millot, C., *Acoustic observations of heat content across the Mediterranean Sea.* Nature, **385**, pp. 615−617, 1997.

[67] Worcester, P.E., Cornuelle, B.D., Dzieciuch, M.A., Munk, W.H., Howe, B.M., Mercer, J.A., Spindel, R.C., Colosi, J.A., Metzger, K., Birdsall, T.B. and Baggeroer, A.B., *A test of basin scale acoustic thermometry using a large aperture vertical array at 3250-km range in the eastern North Pacific Ocean.* J. Acoust. Soc. Am., **105**, pp. 3185−3201, 1999.

[68] Mikhalevsky, P.N., Gavrilov, A. and Baggeroer, A.B., *The transarctic acoustic propagation experiment and climate monitoring in the Arctic.* IEEE J. Oceanic Eng., **24**, pp. 183−201, 1999.

[69] Elisseeff, P., Schmidt, H., Johnson, M., Herold, D., Chapman, N.R. and McDonald, M.M., *Acoustic tomography of a coastal front in Haro Strait, British Columbia.* J. Acoust. Soc. Am., **106**, pp. 169−184, 1999.

[70] Yamaguchi, K., Lin, J., Kaneko, A., Yamamoto, T., Gohda, N., Nguyen, H-C, and Zheng, H., *A continuous mapping of tidal current structures in the Kanmon Strait.* J. Oceanogr., **61**, pp. 283−294, 2005.

Sonar Systems

10

L. Bjørnø[†]

UltraTech Holding, Taastrup, Denmark

Definitions

The 20th century has seen a rapid development in electroacoustic transducers. This growth has continued into the 21st century due to an ever-increasing number of places where transducers are applied under water. In general, a transducer is a device that converts energy from one form to another. This includes conversions between electrical, mechanical, magnetic, acoustical, and optical energies, where underwater acoustic signal's transmission and reception utilize these conversions. A transducer acting as an underwater sound source is called a *projector*. An underwater sound signal receiver is called a *hydrophone*. *Sonar*, an abbreviation of the World War II concept, *sound navigation and ranging*, is the process for detecting and locating an object by receiving the sound emitted by the object—the *passive sonar* process— or by receiving the echoes reflected from an object insonified in an echo-ranging process—the *active sonar* process. Underwater sound applications use a transducer for transmission and/or reception. Most underwater acoustic transducers are based on electroacoustics. However, several nonelectroacoustic transducers are used including projectors based on compressed air (air guns), electrical discharge (spark), underwater explosions, hydraulics, hydrodynamics, and hydrophones based on optics. Projectors and hydrophones are connected with a broad variety of electronic devices for amplifying, filtering, processing, and displaying transmitted and received signals. A *sonar system* is a transmission and/or receiving system involving one or more projectors and/or hydrophones and related electronics. Sonar systems often use several hydrophones to form a hydrophone *array* called the *receiver*, and one or more projectors forming the *transmitter*. Projectors may form a *transmitting array* when interconnected in a way, which allows amplitude and phase-controlled emissions from individual projectors. This chapter discusses sonar systems with an emphasis on electroacoustic systems.

Section 10.1 discusses the applications of sonar systems. Section 10.2 provides a detailed examination of sonar system types including transducer materials, projectors and hydrophones, single-element transducer geometries and transducer arrays.

[†]30 March 1937—24 October 2015.

Applied Underwater Acoustics. http://dx.doi.org/10.1016/B978-0-12-811240-3.00010-2
Copyright © 2017 Elsevier Inc. All rights reserved.

Section 10.3 discusses single-beam echo sounders (SBESs) followed in Section 10.4 with a discussion of multibeam echo sounders (MBESs). Sections 10.5 and 10.6 discuss side-scan and synthetic aperture sonars, respectively. Section 10.7 provides an overview of other sonar types followed in Section 10.8 by a brief discussion on transducer calibration. Section 10.9 contains example calculations on the previous sonar types using the sonar equation discussed in Chapter 1. Section 10.10 provides a series of sonar design calculations. A list of symbols and abbreviations used in this chapter with the page of first usage for each and the references for this chapter are provided at the end of the chapter.

10.1 SONAR SYSTEM APPLICATIONS

The spectrum of interest in underwater acoustics ranges from below 1 Hz to far above 1 MHz. This broad frequency range and the demands for acoustic power, bandwidth, size, weight, and depth of operation are a challenge to sonar system manufacturers, and a broad variety of sonar system designs covering various applications are now available.

Underwater *ambient noise* measurements, as discussed in Chapter 6, require hydrophones that cover the spectrum from seismic noise below 1 Hz to thermal noise above several hundred kilohertz. Ambient noise horizontal and vertical directivity is frequently measured by hydrophone arrays with a large number of elements. Noise from shipping and bubbles covers a broader frequency range. Sound sources in offshore activities possess considerable signal amplitudes, thus demanding various bandwidth and dynamic range for the hydrophones used.

Off-shore activities, such as oil and gas prospecting using acoustics and marine seismic; finding and exploiting seabed minerals; route selection and cables and tubes laying in the seabed; dumping rocks to cover laid tubes and cables; device location, control, and operation on the bottom in an oilfield; and positioning production platforms and their auxiliary equipment, are all based on underwater acoustic signals. Acoustic signals can cover frequencies ranging from few tens of hertz to far above 1 MHz. Signal amplitudes can exceed several megapascals, when air guns, boomers, sparkers, and underwater explosions are used.

Underwater navigation of remotely operated vehicles (ROVs), autonomous underwater vehicles (AUVs), swim divers and surface vessels, measurements of vessel speed over the seabed, the use of Doppler log, and the use of active sonar for *collision avoidance* are areas, where sonar systems find a broad application.

Sonar systems are also used in *hydrography* and *bathymetry*. Depth soundings based on echoes, i.e., sound reflected from the seabed, are some of the oldest exploitations of underwater sound. Realistic maps of the seabed in littoral waters and in parts of the World's more than 360 million km^2 deep ocean areas are produced by sonar systems including single-beam, side-scan, and multibeam sonar systems. When accurate and detailed enough, these maps can be used for surface vessel navigation based on depth soundings. Bottom mapping can be used to search for ship and airplane wrecks and historical objects on and in the seabed, as well as studies of sunken cities. Also sub-bottom profiling may be carried out by sonar systems with appropriate directivity and frequencies that penetrate deeper into the seabed.

Modern *fishery* cannot be carried out without using sonar systems to find fish schools, estimate the fish type and the fish school's biomass magnitude, and control the fishing process through acoustic sensors on trawls. Acoustic communication between control systems on the fishing trawler and sensor systems on trawls indicate the fish school's position relative to the trawl opening, the trawl movement, the trawl bottom proximity to the seabed, and the catch size in the trawl. Sonar systems make it possible to carry out an efficient and protective catching operation, as discussed in Chapter 12.

Naval sonar system applications cover a broad range of areas including *passive* and *active sonar systems.* Passive systems can range from a single hydrophone to hundreds of hydrophones in geometric arrays. *Line arrays*, where the hydrophones are positioned along a line, are used to receive signals where the signal's amplitude, phase, and angle of arrival are of interest. Line arrays can be positioned vertically in the water column or mounted horizontally on fixtures along the seabed. They can have several kilometer lengths. Arrays can also be towed behind a surface vessel or a submarine. *Towed arrays* can be used to determine the submarine's position and the bearing. A towed array sonar (TAS) towed up to several hundred meters behind a submarine, may constitute a low-frequency detection component of the submarine's sonar equipment. By using a winch system on the submarine, the TAS can be launched and pulled in again. *Planar arrays* of hydrophones can comprise several hundreds of receiving elements and have a circular, quadratic, or rectangular surface geometry. *Cylindrical* and *spherical* shapes are also used for detection, tracking, and monitoring underwater objects, such as submarines, AUVs, ROVs, swim divers, and moving torpedoes. Passive listening can be performed by using receiving arrays lowered from a helicopter into the sea. These *dipping sonar systems* are very useful for detecting submarines and other underwater activities. *Sonobuoys*, comprised of a hydrophone system and signal processing equipment and positioned at a suitable depth with a connection to a radio transmitter above the water surface are used to passive listen to underwater activity. Sonobuoys are frequently dropped from surface ships or helicopters and may also be deployed from submarines. Passive listening including a *network of hydrophones*, frequently bottom mounted, is used for submarine and surface vessel interception and tracking. *Acoustic mines* and *homing torpedoes* use a hydrophone or hydrophone system to detect targets based on noise signals produced by the target. Therefore, it is of interest to monitor and control a naval vessel's self-noise.

Since the submarines have become more silent and their noise is frequently of a transient character at substantially reduced levels, it is necessary to resort to *active sonar systems* for detection. In an active sonar the projector emits an acoustic signal, frequently a broadband signal, and the return signal, the *echo*, from a target is received by a receiving array either at the same position as the projector (*monostatic mode*) or positioned at a distance from the projector (*bistatic mode*). Active sonar systems measure a target's range (through time measurements) and Doppler components (through frequency measurements) by cross-correlating overlapping received return signal segments with a set of stored references. If several receiving arrays are used to obtain a more accurate determination of target range, bearing, and speed, the operational mode is *multistatic*. Active sonar systems are used for submarine detection and tracking in *anti-submarine warfare* (ASW), mine hunting and sweeping in mine counter measures (MCMs), torpedo detection, and long-range object detection using low-frequency signals in low-frequency active sonar (LFAS). Sonar systems used in active systems comprise single-beam, side-scan, and multibeam sonar systems. Projectors and hydrophones are also used for *underwater communication* between surface vessels and submarines, AUVs and divers. Underwater *navigation* may be carried out by using sonar systems. Active sonar systems with high-frequency directive sonar arrays are used in certain types of homing torpedoes.

Marine geology exploits sonar systems to study sub-bottom profiles by recording echoes from acoustical impedance changes at interfaces between seabed sediment layers. By a process similar to echo sounding the sonar platform's horizontal movement makes it possible to draw a vertical cross section of the seabed structure. This cross section can form the basis for identifying seabed sediment types and determining the amount of sediments available for applications, such as construction materials for roads and buildings. Due to the stronger acoustic signal attenuation in sediments compared to seawater, sound frequencies below 30 kHz, and most frequently few kilohertz, are used for sub-bottom profiling. Also, sonar systems are used to detect manganese–nickel nodules on deep seabeds for mining. The backscattered sound signals from the seabed surface can be used for *seabed identification*, since, as discussed in Chapter 8, they carry information about the seabed material types, such as rock, boulders, gravel, sand, silt, and clay, which from the upper seabed layer. This information can be used for fishery and navigation and as input data to range-dependent acoustic signal propagation model calculations discussed in Chapter 3.

Physical oceanography is a research area that uses sonar systems to produce low-frequency sounds to propagate over great distances, frequently thousands of kilometers, to study the sea on various scales. In the Acoustic Thermometry of Ocean Climate (ATOC) project, sounds at frequencies below 60 Hz were transmitted over thousands of kilometers. In the Heard Island test project for ATOC carried out in January 1991, coded sound signals at a carrier frequency of 57 Hz, produced by a projector with a source level of 206 dB rel 1 μPa at 1 m and with a bandwidth of 14 Hz, were transmitted in the SOFAR (SOund Fixing And Ranging) channel from Heard Island, a small Australian island in the Indian Ocean at 53.4°S and 74.5°E, as discussed by Munk et al. [1]. The sound was received by arrays on the United States East and West Coasts after having traveled south of Africa and through the Atlantic Ocean and south of Australia and New Zealand and through the Pacific Ocean over distances up to 18,000 km. These measurements were performed to determine whether the average sound speed is increasing with time due to global warming. Since the sound velocity in seawater depends on temperature, the accumulated effect over great distances of small changes in the average sea temperature can indicate global warming's influence on the ocean climate. Shorter range sound propagation, over distances from hundreds to thousands kilometers, has been used in *ocean acoustic tomography*, where sound waves are propagated in predetermined directions through the sea, to provide information on fronts, gyres, and other larger-scale temperature variations in the sea by tomographic inversions of travel time difference data. In 2005, *shallow-water tomography* in coastal regions at distances below 100 km and at depths below 30 m using acoustic signals between 1 and 10 kHz was reported by Kaneko et al. [2].

10.2 SONAR SYSTEM TYPES
10.2.1 TRANSDUCER MATERIALS

The heart of projectors and hydrophones in sonar systems today is based on natural crystals (or monocrystals) or ceramic materials. The natural *crystal piezoelectric effect* first observed by the Curie brothers and the *electrostrictive effect* of ceramic are used to emit or receive sound waves in water. Several natural crystals, such as quartz, ammonium dihydrogen phosphate (ADP), lithium sulfate, tourmaline, and Rochelle salt possess piezoelectric qualities, i.e., an electric field applied to the crystal in well-defined directions produces a mechanical deformation. This mechanical deformation is transferred to the water as a sound wave with the same time history as the applied electric field. Vice versa a crystal mechanical deformation caused by a sound wave's pressure variation produces an electric charge with opposite polarity on two sides of the crystal. The electric charge's magnitude follows the sound wave's pressure–time history. This *linear* relationship between mechanical strain and electrical field is a characteristic of natural crystal piezoelectricity. However, natural crystals are not used in a large-scale anymore, since size, shape, cost, and available amounts do not support use in modern sonar systems. Their transduction efficiency in projectors and hydrophones ranges from 40% to around 75%.

Instead, the electrostrictive effect in certain ferroelectric ceramic types based on barium titanate ($BaTiO_3$), lead zirconate titanate or (PZT) ($Pb[Zr,Ti]O_3$), lead magnesium niobate (PMN), lead zirconium niobate (PZN), or lead metaniobate ($PbNb_2O_6$) are used in projectors and hydrophones. The high dielectric and piezoelectric constants, broad variety of shapes available, low material cost and inherent ruggedness are reasons for extended use of these materials. In particular, the high coercive field in $BaTiO_3$ and PZT makes these ceramics applicable for sonar systems, where high power is needed and where they can have nearly any shape and size. Ferroelectric materials with lower coercive fields as PMN, PZN, and their mixtures with lead titanate do not show a residual polarization high enough to allow their use in projectors. However, they can be used in hydrophones. These polycrystalline ceramics are called *piezoceramics*. They are produced by mixing their components, as shown in Fig. 10.1, followed by a calcining, milling, and granulation processes and exposing the product to high pressure and temperature in a mold for ceramic sintering. The ceramic after molding as a rod, disk, cylinder,

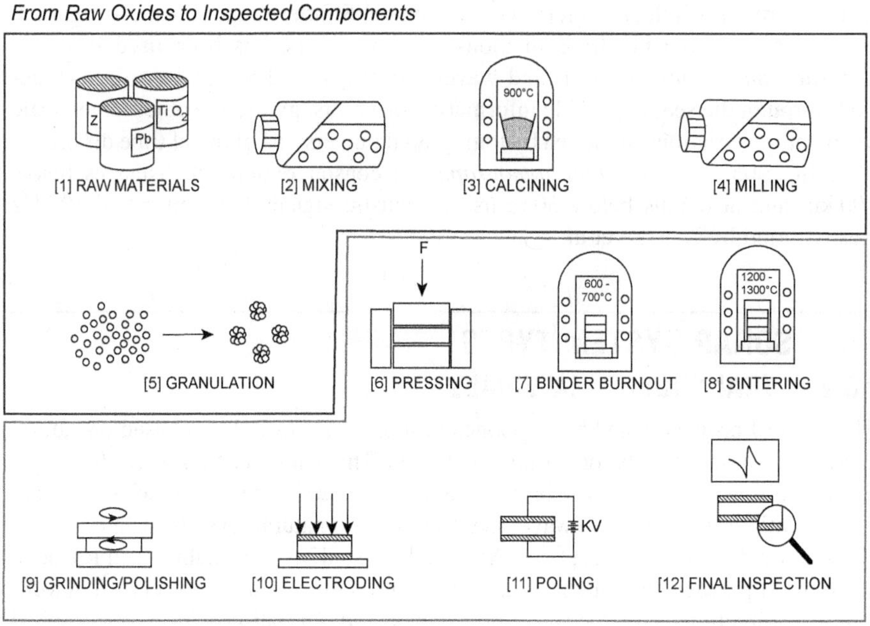

From Raw Oxides to Inspected Components

FIGURE 10.1

Steps in the piezoceramic component manufacturing process from raw materials to finally inspected elements ready for transducer use.

Courtesy of Meggitt A/S, Denmark.

or sphere (normally two half-spheres) is machined to obtain the desired dimensions, since it shrinks during the sintering process at high temperatures. The machining process involves grinding, lapping, polishing, and sometimes cutting with a diamond saw. The machined ceramic is coated with a thin metal layer, most frequently silver, to form electrodes during polarization. To ensure the silver layer has an intimate and strong bond sufficient for soldering electrical connecting wires, the silver is usually fired onto the ceramic surfaces.

An unpolarized ceramic shows no piezoelectric effects since the electric dipole domains are randomly oriented. If an electric field is applied to the unpolarized ceramic, the domains become aligned. Unpolarized ceramics show a *quadratic nonlinear* relation between the electric field and mechanical strain. If a strong and steady electric bias field, of about 4 kV/mm, is applied to the ceramic electrodes at a temperature above the ceramic's *Curie temperature*, the domains become aligned concurrent with substantial changes in the dimensions. When the temperature is lowered from above the Curie temperature to room temperature, with the bias field maintained, the ceramic receives a residual polarization, which produces an approximately *linear*, reciprocal response to a mechanical or electrical drive field. During the *piezoceramic poling* process the high electric field may lead to dielectric breakdown when the ceramics have minor flaws. Poling is more difficult for thick than for thin ceramic pieces. The ability to retain a significant residual polarization, when the bias field is removed, is possessed by materials with a high coercive field, such as $BaTiO_3$ and PZT. Residual polarization is stable and large enough to give these ceramics a strong piezoelectric effect. However, these materials may have depolarization effects if exposed to temperatures near their Curie temperature and high applied electric fields. The Curie temperature is a critical and characteristic temperature for piezoceramic material since the material's dielectric constant and polarization properties decrease progressively above this temperature. The $BaTiO_3$ Curie temperature is about 125°C. Depending on composition, PZT has Curie temperatures between 300 and 500°C. On a smaller-scale depolarization can occur over time as an aging effect.

Fig. 10.2 shows a piezoceramic disk with thickness, x. An applied electric field, $E(t)$, causes the disk thickness to expand Δx in the polarization direction. The $E(t)$ time history is the same as thickness time history of the disk, which is transmitted into the water surrounding the disk as an acoustic signal. If an acoustic signal is incident on the disk, an electric charge is formed with opposite polarity on the surfaces vertical to the direction of polarization. The electric charge magnitude is linearly proportional to the applied acoustic pressure field time history. The transduction efficiency of piezoceramics used in projectors and hydrophones ranges from 50% to over 80%.

Ferromagnetic materials like iron (Fe), cobalt (Co), and nickel (Ni) possess *magnetostrictive* qualities, i.e., they change dimensions if positioned in a

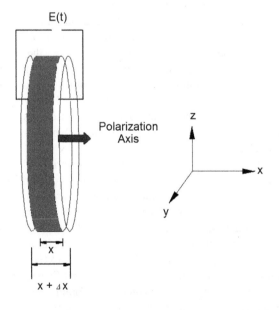

FIGURE 10.2

Piezoceramic disk with thickness *a* driven by an electric field *E* which causes the disk surfaces to displace Δ*a*. Polarization is normal to the disk surfaces.

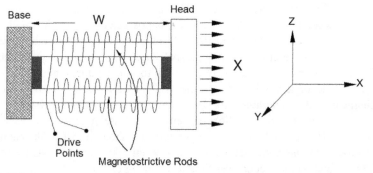

FIGURE 10.3

Magnetostrictive transducer based on two magnetostrictive rods which together with the base and head form a closed magnetic loop.

variable magnetic field. The dimension change's time history follows the magnetic field change. Iron has positive magnetostriction, while nickel has negative magnetostriction. When magnetized an iron bar becomes longer and a nickel bar becomes shorter. Since contractions and expansions are independent of the direction of the electric current that creates the magnetic field, frequency doubling occurs when an AC signal is applied. In addition, to obtain a linear response, the materials require a magnetic bias since the magnetostrictive material's dimension change due to an applied magnetic field is nonlinear. Since magnetic fields only occur in closed loops, it is necessary to produce a transducer shape involving a closed magnetic circuit. When used as underwater sound projectors *magnetostrictive materials* can have various shapes, such as rods, as shown in Fig. 10.3, and rings, i.e., the scroll-wound transducers used in free-flooded rings. Since before World War II they have been used as low-frequency, high-power transducers, which are able to withstand large vibration amplitudes and blows due to their high tensile strength. However, magnetostrictive transducers have a poorer efficiency than PZT-based transducers, their bandwidth is narrower and the driving materials are more expensive. Magnetostrictive transducers are mainly used at frequencies below 50 kHz.

Around 1970 a new type of *magnetostrictive* materials, the *rare earth*–iron *compounds*, were developed and used in transducers in competition with piezoceramics. The strong magneto-mechanical reaction (saturation strain $> 0.3\%$, and about 50 times larger than that of Ni and its alloys) of materials like Terfenol-D (alloy based on terbium, dysprosium, and iron, $Tb_{0.27}Dy_{0.73}Fe_2$) and Galfenol (alloy based on gallium and iron) have made these materials useful as high-power underwater acoustic transducers at frequencies from 10 Hz to 5 kHz, as discussed by Zhu et al. [3], Moffett et al. [4], Kvarnsjö [5], Clark [6], and Sewell and Kuhn [7]. Terfenol-D displays a giant magnetostriction and a low Young's modulus; its electromechanical coupling factor is comparable to that of piezoceramics. According to Moffett et al. [4] a flextensional underwater acoustic projector powered by Terfenol-D radiated 14 kW acoustic power at its resonance frequency of 930 Hz, and it showed a 4.7 mechanical quality factor, Q_m, as discussed in Section 10.2.3. Since Terfenol-D is a brittle material, it is necessary to prestress the transducer's driving material, for example, by using a stress bolt, to prevent the material from working in tension, where it can fracture under high dynamic drive.

As a spin-off of research into magnetostrictive materials metallic glasses were developed. They are amorphous alloys of a metal, such as iron, nickel, or cobalt, and a metalloid, such as boron, silicon, or carbon. They can be produced in thin ribbons, 20–50-μm thick and 25-mm wide, by very fast quenching the molten alloy into a cooled cylinder which preserves the molten alloy's glass-like structure. Coupling coefficients far above 0.7 have been reported by Modzelewski et al. [8]. Due to low-saturation strain and better piezoelectric properties than PZT, metallic glasses are better suited as hydrophones rather than projectors. In spite of high coupling coefficients these materials have not found a broad application in transducers.

Also *electrodynamic* effects, similar to loudspeakers in air, have been used for low-frequency underwater acoustics transducers. These *moving coil transducers*, as shown in Fig. 10.4, have been used for calibration, where their low-frequency, broadband capabilities and low acoustic power in a flat transmitting response above their resonance frequency can be exploited. Since the radiation impedance in water is higher than in air, a rigid piston radiator is normally in contact with the water. For low displacement amplitudes the moving coil transducer has a linear response due to

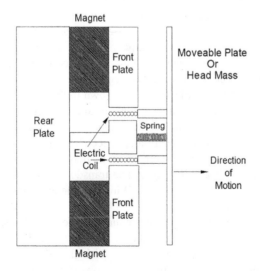

FIGURE 10.4

Moving coil transducer, where the electric coil carries an alternating current and moves in the magnetic field in the groove in the front plate. The coil movement drives the plate, which emits the sound. The spring stabilizes the moveable plate vibration and possesses a very low resonance frequency.

the linear relation between the current in the coil and the force produced. Also *electromagnetic*, or variable reluctance, and *electrostatic* effects have been used in transducers for low-frequency underwater sound. Due to the low forces produced by these effects they can be applied in transducers where the demand for displacement and not for force is important at low frequencies where the radiation impedance is low. The force produced by a transducer depends on its transduction coefficient and on the bias electric field used. For the same transducer size and applied voltage gradient, the force produced by a piezoceramic-based transducer will be 10^6 times higher than the force produced by an electrostatic transducer.

High forces at low frequencies can also be produced by *hydraulic- and mechanical-based transducers*, where hydraulic or mechanical forces are used to drive the transducer pistons in contact with the water. The hydraulic forces are normally governed by a system of valves controlled by electromechanical or piezoelectric devices. The working pressure in the hydraulic systems may be as high as 70 MPa and power outputs of several tens of kilowatts may be achieved. These transducers are generally used as projectors. The acoustic signals propagating over several thousand kilometers in the Heard Island Feasibility Test were generated by driving a circular faceplate of a transducer with a 57 Hz resonance frequency into vibration with hydraulic controlled pistons, which produced a source level of 206 dB rel 1 μPa at 1 m, equal to 3.3 kW acoustic power.

10.2.2 PROJECTORS AND HYDROPHONES

Since most projectors and hydrophones used today are based on piezoceramics due to their transduction qualities and the broad variety of geometrical shapes and sizes made possible by these materials, only *piezoceramic-based projectors and hydrophones* are covered in this chapter. For higher-frequency emission piezoceramic *disks*, *rods*, *plates*, *tubes*, *rings*, and *hemispheres* are driven by an electric field applied directly to their surface electrodes, as shown in Fig. 10.2 for a circular disk. The surface displacement can be directly coupled to the water through an electrical insulation layer. The disks can have circular, quadratic, or rectangular shapes. The rod cross sections can be circular, quadratic, or rectangular. Transducers with these geometries are normally driven at their resonant frequency in a thickness mode, where the thickness is half a wavelength. The transducer's bandwidth is normally narrow and they have a high Q-value. With the electric field applied in the transducer's polarization direction the thickness, i.e., the distance between electrodes, will determine the resonant frequency. Most thickness mode−driven transducers operate from 100 kHz and above. For thicker disks or plates resonant frequencies down to 50 kHz are also produced. Disk-shaped transducers are widely used as hydrophones at frequencies below their lowest resonance frequency.

A typical electromechanical coupling coefficient value can be higher than 0.6, depending on the piezoceramic material composition and the axis direction. The power handling capability of piezoceramics is also high, which usually is very important for an underwater projector. The dielectric loss tangent for piezoceramics is typically less than $5 \cdot 10^{-3}$. Therefore, dielectric losses in piezoceramics are lower than magnetic losses in most magnetostrictive materials. With mechanical quality factors of 1000 and above mechanical losses in piezoceramics are limited. These piezoceramic qualities are advantageous in their use in projectors.

However, *piezoceramics* have several *drawbacks*. They are fragile due to low tensile strength. For projector use they have to be prestressed to a level that allows the material to be under compression through all vibration cycles. The prestressing normally leads to minor changes in the piezoceramic material properties. Another drawback is the difficulty in maintaining the same physical properties for ceramic elements in the same batch or different batches. This may be caused by small changes in the ceramic composition or in the forming or firing procedures used. Also, piezoceramic material properties change with time, i.e., aging effects. For instance, there is a slow decrease in polarization with time and the material tends to return to an unpolarized state. Piezoceramics also show a significant nonlinearity at high power levels. This is due to the linearization process the material is exposed to during polarization, where an originally nonlinear material is linearized for small drive signals. As a projector material the piezoceramic nonlinear behavior leads to an increase in dielectric losses causing ceramic material heating and lower efficiency. For low-frequency projectors designs the relatively high Young's modulus of piezoceramics may be a problem since low-frequency, high-power transmission demands a higher strain.

High voltages may also be generated across the electrodes of piezoceramics through the *pyroelectric effect*. This effect is caused by an increase in ceramic temperature and is due to the same physical mechanisms as the piezoelectric effect in spontaneous ceramic polarization. The PZT material pyroelectric coefficient is of the order $0.2 \cdot 10^{-3}$ coulombs (C)/m^2 °C at 20°C increasing to about $0.8 \cdot 10^{-3}$ C/m^2 °C at 150°C, according to Jaffe et al. [9]. If the charge cannot leak away, high voltages may be created across the ceramics which may cause amplifier and mechanical handling problems with a larger piece of ceramics. For a 0.018 m disk radius and temperature increase from 20 to 21°C, a charge of about $0.2 \cdot 10^{-6}$ C is generated. If the disk is made from Pz28, as shown in Table 10.1, and the disk thickness is $5 \cdot 10^{-3}$ m and its capacitance is about $1.3 \cdot 10^4$ pF then, in the absence of leakage, the voltage across the electrodes is about 15 V.

Larger projectors with improved directivity at frequencies above 50−100 kHz can be produced by mounting a number of *piezoceramic* disks or plates on a backing material, which frequently has lower acoustical impedance than water, such as an epoxy, filled with hollow microglass spheres. This ensures that most of the acoustic effect is emitted into the water. An *array* of disks or plates can be given any shape. The most frequently used is a circular, quadratic, or rectangular planar array. To electrically insulate the piezoceramic transducer elements and ensure a good acoustic radiation from the elements into the water, the array is normally molded into a polymer matrix with the polymer layer thickness between the elements and water much less than a wavelength in the polymer. As shown in Fig. 10.5 a larger outer array of quadratic transducers elements and a smaller inner array of circular

Table 10.1 Piezoceramics Parameters for Various Compositions for Transducer Applications in Projectors and Hydrophones

Symbol	Unit	Pz26	Pz27	Pz28	Pz37	Pz46
$\varepsilon_{3,1}^{\sigma}$		$1.33 \cdot 10^3$	$1.80 \cdot 10^3$	$9.9 \cdot 10^3$	$1.15 \cdot 10^3$	$1.24 \cdot 10^2$
$\varepsilon_{3,1}^{S}$		$7.0 \cdot 10^2$	$9.14 \cdot 10^2$	$5.1 \cdot 10^2$	$8.89 \cdot 10^2$	$1.16 \cdot 10^2$
$\tan\delta$		0.003	0.017	0.004	0.015	0.004
T_c	°C	330	350	330	350	650
k_p		0.568	0.592	0.579	0.375	0.033
k_t		0.471	0.469	0.475	0.549	0.249
k_{31}		0.327	0.327	0.332	0.211	0.021
k_{33}		0.684	0.699	0.687	0.630	0.087
k_{15}		0.553	0.609	0.631		0.045
d_{31}	C/N	$-1.28 \cdot 10^{-10}$	$-1.70 \cdot 10^{-10}$	$-1.14 \cdot 10^{-10}$	$-1.04 \cdot 10^{-10}$	$-2.26 \cdot 10^{-12}$
d_{33}	C/N	$3.28 \cdot 10^{-10}$	$4.25 \cdot 10^{-10}$	$2.75 \cdot 10^{-10}$	$3.81 \cdot 10^{-10}$	$1.91 \cdot 10^{-11}$
d_{15}	C/N	$3.27 \cdot 10^{-10}$	$5.06 \cdot 10^{-10}$	$4.03 \cdot 10^{-10}$		$7.79 \cdot 10^{-12}$
d_h	C/N	$7.24 \cdot 10^{-11}$	$8.50 \cdot 10^{-11}$	$4.68 \cdot 10^{-11}$		$1.46 \cdot 10^{-11}$
g_{31}	Vm/N	-0.0109	-0.0107	-0.0130	-0.0103	
g_{33}	Vm/N	0.0280	0.0267	0.0314	0.0383	0.0174
g_{15}	Vm/N	0.0389	0.0373	0.0373	0.0485	0.0195
e_{31}	C/m²	-2.80	-3.09	-3.60		1.61
e_{33}	C/m²	14.7	16.0	12.4	11.0	2.6
e_{15}	C/m²	9.86	11.64	10.67		0.30
h_{31}	V/m	$-4.52 \cdot 10^8$	$-3.82 \cdot 10^8$	$-7.99 \cdot 10^8$		$1.56 \cdot 10^9$
h_{33}	V/m	$2.37 \cdot 10^9$	$1.98 \cdot 10^9$	$2.76 \cdot 10^9$	$1.8 \cdot 10^9$	$2.58 \cdot 10^9$
h_{15}	V/m	$1.34 \cdot 10^9$	$1.16 \cdot 10^9$	$1.64 \cdot 10^9$		$2.64 \cdot 10^8$
N_p	Hz·m	2209	2011	2187	1813	2468
N_t	Hz·m	2038	1953	1998	1377	2002
N_{31}	Hz·m	1500	1400	1624		1905
N_{33}	Hz·m	1800	1500	1500		
N_{15}	Hz·m	1018	896	964		1207
$Q_{m,p}^{E}$		$2.7 \cdot 10^3$	89	$9.7 \cdot 10^2$	144	$1.7 \cdot 10^3$
$Q_{m,t}^{E}$		$3.3 \cdot 10^3$	74	$1.1 \cdot 10^3$	127	$4.1 \cdot 10^3$
ρ	kg/m³	$7.7 \cdot 10^3$	$7.7 \cdot 10^3$	$7.7 \cdot 10^3$	$6.47 \cdot 10^3$	$6.53 \cdot 10^3$
ν_{12}^{E}		0.334	0.389	0.295	0.385	0.215
s_{33}	m²/N	$1.96 \cdot 10^{-11}$	$2.32 \cdot 10^{-11}$	$1.83 \cdot 10^{-11}$	$3.75 \cdot 10^{-11}$	$4.42 \cdot 10^{-11}$
c_{33}	N/m²	$1.23 \cdot 10^{11}$	$1.13 \cdot 10^{11}$	$1.18 \cdot 10^{11}$	$4.63 \cdot 10^{10}$	$1.03 \cdot 10^{11}$

Data used with permission from the Danish corporation Meggitt A/S materials program.

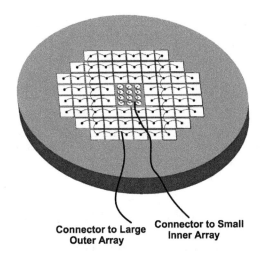

Connector to Large Outer Array

Connector to Small Inner Array

FIGURE 10.5

A planar transducer array consisting of a circular disk inner array of ceramic elements and a quadratic disk outer array. Elements are embedded in an epoxy matrix, frequently an acoustic impedance—reducing material such as flying ash or microscopic air-filled glass spheres, to reduce backward radiation. The array is normally molded into epoxy for water-tightness and surrounded by a stainless steel housing. Quadratic disks are normally separated by an epoxy or cork layer to eliminate vibration transfer between elements.

transducer elements is imbedded into a matrix consisting of a low-acoustic imped-ance material. The thin insulation polymer layer between the transducer elements and the water is not shown. Two sets of transducer elements are frequently used to increase the frequency range for the transducer's operation by varying the element's thickness and to provide flexibility in the transducer directivity by driving the inner or outer array separately or simultaneously. This robust projector construction allows the projector to be used at larger water depths. To obtain improved electrical insulation and preserve good acoustical contact to the water, the whole array in Fig. 10.5 can be embedded in a housing filled with a liquid, which is frequently an oil, such as castor oil.

Higher-frequency acoustic signals in water can also be produced by a *composite material*—based transducer consisting of a *piezoceramic*, such as PZT, embedded in an inert polymer matrix, such as polyurethane, polyethylene, epoxy, or silicone rubber. The piezoceramic material and polymer matrix connectivity form the basis for characterizing these composite materials as a 0-3, 1-3, or 2-2 material. In a 1-3 material the piezoceramic material is connected to the environment in one (1) direction only, while the polymer matrix is connected in three (3) directions. The 1-3 composite piezoceramic material, the most commonly used piezocer-amic-based composite material for projectors, may have the shape of rod with a cylindrical or quadratic cross section, as shown in Fig. 10.6. The rod diameter

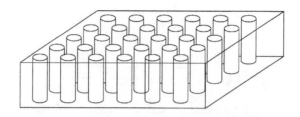

FIGURE 10.6

A 1-3 composite material consisting of piezoelectric cylindrical rods embedded in a polymer matrix. The electrodes, not shown in the figure, are connected with the end surfaces of the cylindrical rods.

may vary from about 25 µm to above 5 mm. The performance depends on the rod volume fraction, rod and polymer materials used, and the distance between rods compared with the composite material wavelength. By varying the material parameters and dimensions it is possible to produce materials with high coupling coefficients, broadband performance through a lower Q-value, and improved impedance matching to water. The better acoustical impedance match of some composite materials makes these materials applicable for hydrophones.

Numerous polymers have been known for many years to be piezoelectric, but not with a large enough effect to make them useful in transducers. Around 1970 strong piezoelectric qualities were discovered in *polyvinylidene fluoride*, also called PVDF or PVF$_2$ with the molecular formula, CH$_2$CF$_2$. As discussed by Vinogradov and Schwartz [10], PDVF is a highly nonreactive, semicrystalline (50–60%), and pure thermoplastic fluoropolymer. This material is available as sheets, tubes, plates, and films. PVDF has four known crystalline phases. For transducers only the nonpolar, non-piezoelectric, α-phase and polar, piezoelectric, β-phase are of interest. The α-phase can be transformed into the β-phase by stretching and annealing the material, followed by a poling process. When a PVDF film at 60–65°C is stretched to 3–5 times its original length, the α-phase is converted into the β-phase which is stabilized by clamping and annealing the film at 120°C. This process is followed by poling the film in a strong electric field, typically higher than 30 MV/m, at a temperature of about 100°C for 1 h, followed by cooling to room temperature with the electric field maintained. Fig. 10.7 shows a poled PVDF film with the hydrogen atoms, H, which have a net positive charge, at the top and fluorine

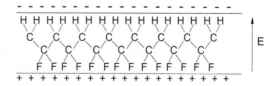

FIGURE 10.7

Hydrogen atoms, H, fluorine atoms, F, and carbon atoms, C, in poled PVDF film.

atoms, F, which have a net negative charge, at the bottom. If an electric field E is put on the PVDF film electrodes in the direction shown in Fig. 10.7, which is the same direction as the poling direction, the film will *contract* in the film plane.

When PDVF and piezoceramics are exposed to the same electric field, PVDF will compress when piezoceramics expand and vice versa. The piezoelectric strain coefficient d (unit: C/N) for PVDF is on the order of 25–35 pC/N, while piezoceramics d-values are several hundred pC/N. Although PVDF is primarily suitable for hydrophone applications, PVDF can find some applications for projectors due to its higher permissible strain compared with piezoceramics, which makes it less susceptible to fracture under high hydrostatic pressures. *Main advantages* of PVDF are its mechanical properties, flexibility, ruggedness, ability to be produced in large sheets, and acoustical impedance close to water's acoustical impedance which reduces the need for impedance matching. Disadvantages of PVDF are a high pyroelectric coefficient, aging effects, and use below 80°C.

PVDF properties can be improved by forming *copolymers* with monomers such as triflouroethylene to produce the most commonly used copolymer, P(VDF-TrFE), which shows improved piezoelectric response due to an improved crystallinity. The copolymers show higher hydrophone figure of merits than pure PVDF and are frequently preferred as transducer material due to their molding qualities.

10.2.3 PARAMETERS OF PIEZOCERAMICS

The piezoceramics parameters of interest are their elastic, dielectric, and piezoelectric constants. Piezoelectric materials overall behavior may be described by their equations of state which connect piezoelectric quantities with mechanical and electrical quantities. The equations of state are normally given on tensor basis, since piezoelectric crystals generally show much different qualities in the directions of their various crystal axes. For piezoceramics, which form the basis for most underwater transducers, the equations of state are considerably simplified since the poling direction is the only axis of symmetry. If the *electric field strength E* (unit: V/m) and the *strain S* (unit: dimensionless) are the two independent variables and isentropic conditions are assumed, the *equations of state* are:

$$T = c^E S - e_t E \tag{10.1}$$

$$D = eS + \varepsilon^S E \tag{10.2}$$

where T is the *stress* (unit: N/m^2), D is the *electric displacement* (unit: C/m^2), c^E is the *elastic modulus* (unit: N/m^2) for constant E, e_t is the *piezoelectric stress coefficient* (unit: C/m^2 or N/Vm), where subscript t denotes a transposition, i.e., the rows and columns in the matrix describing e are interchanged, and ε^S is the absolute *dielectric constant* (unit: F/m) for constant S. If T and E are the independent variables, the equations of state are:

$$S = s^E T + d_t E \tag{10.3}$$

$$D = dT + \varepsilon^T E \tag{10.4}$$

where s is the *elastic compliance coefficient* (unit: m^2/N) and d is the *piezoelectric strain coefficient* (unit: C/N or m/V). Two more piezoelectric coefficients g, the *piezoelectric voltage coefficient* (unit: Vm/N or m^2/C) and h (unit: V/m or N/C) are useful for describing piezoelectric materials.

The four piezoelectric coefficients may be expressed by partial derivatives, where the superscript gives the variable that is held constant:

$$d = \left(\frac{\partial D}{\partial T}\right)^E = \left(\frac{\partial S}{\partial E}\right)^T$$

$$e = \left(\frac{\partial D}{\partial S}\right)^E = -\left(\frac{\partial T}{\partial E}\right)^S$$

$$g = -\left(\frac{\partial E}{\partial T}\right)^D = \left(\frac{\partial S}{\partial D}\right)^T$$

$$h = -\left(\frac{\partial E}{\partial S}\right)^D = -\left(\frac{\partial T}{\partial D}\right)^S$$

The derivatives which form basis for the four coefficients refer to the ratios of infinitesimal changes in the variables, D, T, S, and E. According to Jaffe et al. [9] and the IEEE Standard on Piezoelectricity [11] constant D corresponds to open circuit conditions, constant T corresponds to no force resisting the transducer surface displacement, constant S corresponds to a perfect clamping which prevents any motion, and constant E corresponds to the short circuited case.

The most generally used piezoelectric coefficients are d, which expresses the strain produced by an applied electric field or the electrical charge formed by an applied acoustic pressure, and g, which expresses the electrical field generated by an applied acoustical pressure. d is primarily used for piezoceramics operating in a projector, while g is mostly used for hydrophones. For a description of d and g in various piezoceramic material directions a system of suffixes based on *three orthogonal axes*, 1, 2, and 3 is generally used. Each component of d and g is described by two suffixes, for instance d_{31}, where the first suffix denotes the *poling direction*, 3, and the second suffix denotes the direction for the relevant *strain* or *stress*, 1. The piezoelectric strain coefficient d_{33} expresses the strain in the poling direction produced by an electrical field in the same direction. According to Wilson [12] due to symmetry about the poling direction, $d_{31} = d_{32}$. The same suffixes are used to describe values of g and other constants, such as ε, the dimensionless *relative dielectric constant* (or the permittivity), in various directions. The relation between d_{33} and g_{33} is given by:

$$d_{33} = \varepsilon_{33}^T \, \varepsilon_0 \, g_{33} \tag{10.5}$$

where ε_0 denotes the *free space permittivity*; $\varepsilon_0 = 8.85 \cdot 10^{-12}$ F/m.

For some special applications of *shear waves*, the shear mode of vibration of the piezoceramic material may be excited. This occurs when the electric field is applied in a direction vertical to the poling axis, the 3-axis, to generate a shear stress around the 2-axis. The piezoelectric coefficients are here given the suffix 15, and g_{15} denotes the *shear mode piezoelectric voltage coefficient.*

The parameters mentioned the preceding foregoing can be used to develop several useful piezoceramic parameters, such as the *frequency constant N*, which may be derived from the elastic modulus and density. Since the resonance frequency, f_0, of a particular vibration mode is inversely proportional to the linear dimension L of the ceramic material, $N = f_0 L$, N can be used to calculate other resonance frequencies for the same mode of vibration for other dimensions of the same piezoceramic material.

When a piezoceramic sample is exposed to hydrostatic pressure, i.e., to a uniform stress in all three axes, which is the case, when the acoustic wavelength λ (unit: m) is much longer than any sample characteristic dimension, then the *hydrostatic voltage coefficient g_h* is as follows:

$$g_h = g_{33} + 2g_{31} \tag{10.6}$$

since, due to symmetry, $g_{31} = g_{32}$.

For an evaluation of a piezoceramic material operating in the transmission (projector) mode, as well as in the receiving (hydrophone) mode, a *figure of merit* for the hydrophone application can be expressed by gd (unit: m^2/N).

The *coupling coefficient k* of a piezoceramic material expresses the efficiency with which the material can convert electric energy into acoustic energy and it is closely related to the material's maximum obtainable transducer bandwidth. In physical terms k is the square root of the ratio between the mechanically stored energy and total input electrical energy, and, as discussed by Stansfield [13], is expressed in dimensionless form by:

$$k^2 = \frac{d^2}{\varepsilon^T s^E} = \frac{\varepsilon^T g^2}{s^E} \tag{10.7}$$

By using the *series resonance frequency f_s* (usually called the resonance frequency), i.e., the frequency of maximum motional conductance, and the *parallel resonance frequency f_p* (sometimes called the antiresonance frequency), i.e., the transducer's frequency of maximum motional resistance, as shown in Fig. 10.8 from Yao and Bjørnø [14], the coupling coefficient k of the transducer may be determined by:

$$k^2 = 1 - \left(\frac{f_s}{f_p}\right)^2 \tag{10.8}$$

If sufficiently accurate measurements of the two frequencies can be carried out, the coupling coefficient can be calculated from Eq. (10.8). The coupling coefficient is dependent on which vibration mode of the ceramic in the transducer is excited. The k-values vary with the axis of the piezoceramic material and are indicated by

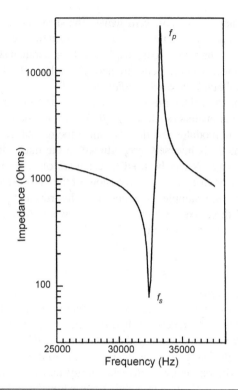

FIGURE 10.8

Impedance as a function of frequency for a Tonpilz transducer in air showing the series resonance f_s and parallel resonance f_p.

©1997 IEEE figure reprinted from Yao, Q. and Bjørnø, L., Broadband Tonpilz Underwater Acoustic Transducers Based on Multimode Optimization, IEEE Trans. Ultrason., Ferroelect. Freq. Control, **44**, (5), pp. 1060–1066, 1997, with permission of the IEEE.

using the same suffixes as other coefficients. k_{33}, for instance, is around 70% for PZT and is the most important parameter for describing the performance of a low-frequency transducer under the assumption the stress distribution through the ceramic is uniform. For many applications this is not the case, and then it is advisable instead to measure the coupling coefficient k_p for the *radial mode* of vibration of a thin disk, where, as discussed by Stansfield [13], k_p is expressed by:

$$k_p^2 = \frac{2\,k_{31}^2}{1 - \nu^E} \tag{10.9}$$

where ν^E denotes *Poisson's ratio* (unit: dimensionless) which has a value between 0.28 and 0.32 for most piezoceramics.

As a dielectric material, the piezoceramic behaves as an imperfect capacitor, where a fraction of the stored energy is lost during each vibration cycle. Piezoceramic material losses in are normally considered to have three components, a dielectric, an elastic, and a piezoelectric or electromechanical component. The dielectric loss factor can be measured from the dielectric displacement phase delay with respect to the applied electric field. The elastic loss factor is determined as the inverse of the mechanical quality factor, the Q_m-value, at the resonance frequency. The piezoelectric loss factor is determined from an analysis of the admittance spectrum, see Zhuang et al. [15]. These losses are characterized by a *loss factor*, tanδ, which describes the losses per cycle of the applied electric or acoustic field. tanδ depends on the ceramic composition. The energy lost in the ceramic contributes to an increase in the ceramic temperature. Excessive losses may reduce the ceramic piezoelectric qualities through depolarization. Since the ceramic composition strongly influences the tanδ magnitude, certain compositions are more applicable for projectors and others for hydrophones. The piezoceramic material loss factor, tan δ, is:

$$\tan \delta = \frac{1}{C_0 \, R_e \, \omega}$$

(10.10)

where C_0 denotes the ceramic's total capacitance (unit: F), R_e is an equivalent resistance (unit: ohm), representing the electric losses, in parallel with the capacitance, and ω is the angular frequency (unit: rad).

According to Stansfield [13] the bandwidth of a transducer is described by its mechanical quality factor, Q_m, which is expressed by:

$$Q_m = \frac{f_s}{f_2 - f_1}$$

(10.11)

where f_2 and f_1 (unit: Hz) denote the frequencies for the 3-dB amplitudes—the half-power frequencies—of the frequency characteristic of the transducer on both sides of the series resonance f_s. The transducer *bandwidth* is normally considered to be $\Delta f = f_2 - f_1$. Q_m measures the ratio between mechanical energy transmitted by the transducer and energy dissipated.

According to Sherman and Butler [16] a relation between the optimum value of Q_m and the coupling coefficient k for *broadband response* of a transducer can be written as:

$$Q_m = \frac{\sqrt{1 - k^2}}{k}$$

(10.12)

For this value of Q_m, the electrical quality factor Q_e is equal to Q_m. As shown by Stansfield [13], in order to obtain the best match between a transducer and its driving amplifier to maintain the maximum power output over the widest possible frequency band, the following approximate relation, $kQ_m \sim 1.2$, should be satisfied. This relationship shows that a wide bandwidth can only be obtained with a transducer possessing a high effective coupling coefficient.

Transducers showing high mechanical Q_m-values in general have narrow bandwidths. The price for increasing the transducer bandwidth is higher losses and a lower Q_m. Low Q_m is generally achieved by matching the transducer impedance to the medium around it. In order to obtain a good performance of a transducer in contact with water over a broad frequency band, the characteristic mechanical impedances of the transducer and water must be similar. High-power transducers normally possess a high Q_m-value and are stress limited at resonance, while broadbanded transducers with lower Q_m-values usually are electric field limited.

The maximum transducer radiated acoustic power at resonance is limited because the storage capacity of its mechanically coupled electric reservoir is limited. According to Woollett [17], the maximum radiated power P_r (unit: W) may be expressed by:

$$P_r = \frac{\eta_{ma}\omega Q_m U_e k^2}{1 - k^2}$$

(10.13)

where η_{ma} denotes the mechanical–acoustical efficiency of the transducer and U_e (unit: joule) denotes the stored electric energy in the transducer which is given by:

$$U_e = \int_V \frac{\epsilon_{33}^S E_p^2}{2} dV$$

(10.14)

where E_p is the peak electric field strength (unit: V/m) and V (unit: m^3) is the volume of the transducer's piezoceramic material. Eq. (10.14) shows that the maximum power radiated is directly proportional to the ceramic volume. The stored electric energy limits are imposed by factors, such as electric insulation breakdown, ceramic depolarization, distortion caused by ferroelectric nonlinearities, and efficiency deterioration due to increased dissipation at high electric fields. Eq. (10.13) shows that increasing the Q_m-value by decreasing the acoustic load results in increased radiated power.

Table 10.1 depicts several parameters for piezoceramics of various compositions to be used for different transducer applications. These compositions include:

- Pz26, for high-power transducers.
- Pz27, for shear transducers and combined transducers.
- Pz28, for very high power and high prestress transducers.
- Pz35, porous PZT for broadband transducers.
- Pz46, bismuth titanate–based ceramic for continuous operation at elevated temperatures above 500°C.

The data are reprinted with permission from the Danish company Meggitt A/S' materials program.

10.2.4 TRANSDUCER GEOMETRIES

Piezoceramic materials in a transducer, either projector or hydrophone, may be given different geometrical shapes depending on the transducer's use, which determines the desired source level, bandwidth, directivity, and the transducer's adaptation to its environment. Some fundamental ceramic shapes in transducers are plates, cylinders, and spheres. Plates and cylinders are used in *piston sources*, where the piezoelectric elements are assembled between a front piston, radiating into the water, and an inertial backing countermass. Piston sources can be given many geometrical shapes.

10.2.4.1 Plates

Plates generally have circular, quadratic, or rectangular shapes. Figs. 10.1 and 10.5 provide examples of circular and quadratic shapes, respectively. Most plates are poled in direction 3, vertical to the plate, with the electrodes positioned on the two surfaces parallel to the 1-2 plane. The electric field driving the ceramic plate is normally in the poling direction to exploit the piezoelectric coefficients d_{33} and g_{33}. However, electrodes may also be on the surfaces parallel to the 1-3 or 2-3 planes, for instance, to generate a shear stress around the 2-axis to exploit the piezo-electric coefficients d_{15} or g_{15}.

10.2.4.2 Cylindrical Elements

Piezoceramic *cylindrical* elements are used in projectors and hydrophones. In many cases the polarization is in the radial direction with the electrodes on the inner and outer cylinder walls. Fig. 10.9 shows a piezoceramic cylindrical single piece ring with a mean radius r_m and a wall thickness t, poled in the radial direction with the outer surface S_o and the inner surface S_i covered by electrodes. The manufacturing, poling processes and tensile strength limit the dimensions of cylin-drical rings produced in pure ceramics to about 0.15 m in diameter. To overcome this size restriction, larger projectors for low-frequency signals build the ring transducer from staves of ceramics alone. In this case, the electrodes are positioned between each ceramic segment, and each segment is poled in the circumferential direction

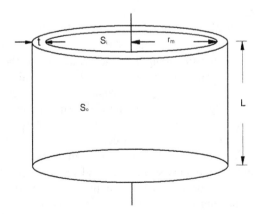

FIGURE 10.9

A piezoceramic, single piece, ring transducer with electrodes on the inner and outer cylindrical surfaces. The poling direction is in the radial direction, normal to the electrodes.

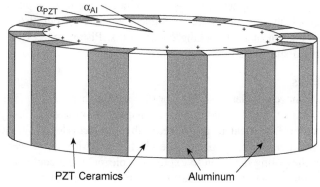

FIGURE 10.10

A ring-shaped transducer consisting of alternate piezoceramic and aluminum staves. The aluminum staves form the electrodes. The poling direction can be in the tangential direction or the radial direction.

exploiting the piezoelectric strain coefficient d_{33}. The poling is arranged in alternating directions around the ring's circumference with an even number of segments. This arrangement is frequently called a *barrel-stave* structure, which permits much larger ring diameters to be produced. When the staves are poled in the circumferential (tangential) direction their thickness vibration mode gives a much higher coupling coefficient than obtained by a single piece ring poled in the radial direction.

A frequently used technique is to replace alternate ceramic staves with metal staves, such as aluminum, as shown in Fig. 10.10, where the aluminum staves form the electrodes. The ceramic elements' poling direction can be in the tangential direction, which is most frequently used, or in the radial direction. This reduces the larger ring transducers' production costs and introduces the possibility to adjust the ring transducer's acoustical characteristics by a suitable selection of materials and dimensions. However, introducing metal staves in the transducer reduces its coupling coefficient. The tensile strength of the ceramics and the joints between pure ceramics and metals require the ring to be prestressed. This is normally done by wrapping glass fiber in an epoxy resin around the ring circumference. Also, cylindrical single piece rings need prestressing if driven at high power. The epoxy resin also forms an electrical insulation between the electrodes and the water.

If a piezoceramic ring is driven in its circumferential (tangential) mode while completely immersed in water, the radial radiation from the outer surface is counteracted by the radiation from the inner surface and end surfaces. In shallow water this unwanted radiation may be prevented by mounting end caps on the ring, which excludes water from the ring's inside. However, it is possible to make the ring's height L large enough to substantially reduce the cancellation effect from the inner surface radiation. The resulting free-flooded ring transducer is virtually independent of hydrostatic pressure, its radiation pattern in the plane perpendicular to the ring axis is omnidirectional and the radiation pattern in the other planes depends on the magnitude of L. The ring's resonance frequency when vibrating in its *radial mode* (or the hoop mode), is reached, when its mean circumference is equal to a wavelength for the ceramic material's longitudinal vibrations. For a ring with a mean radius r_m, poled in the radial direction and driven by electric field in the poling direction, the resonance frequency f_r is given by:

$$f_r = \frac{c_1}{2\pi r_m} \tag{10.15}$$

where c_1, the longitudinal wave velocity in the tangential direction 1, is expressed by:

$$c_1^2 = \left(\rho s_{11}^E\right)^{-1} \tag{10.16}$$

where s_{11}^E is the ceramic elastic compliance coefficient in the 1-direction for constant electric field strength E. The resonance frequency f_t of the ring, when vibrating in its *thickness mode*, i.e., when the wall thickness expands and contracts, is reached when the wall thickness is $\lambda/2$ of the ceramic vibrations. If c_t denotes the thickness wave velocity, f_t may be expressed by:

$$f_t = \frac{c_t}{2t} \tag{10.17}$$

The resonance frequency of the thickness mode is higher than the resonance frequency of the radial mode. The resonance frequencies may also be found by using the frequency constants N discussed in Section 10.2.3.

The cylindrical piezoceramic element in a hydrophone is supplied with electric connectors soldered on the inner and outer electrodes. These connectors are attached to a preamplifier or directly to the coax cable which transports the signals to the signal processing devices. After the electric connections have been established, the ceramic is molded into epoxy or rubber to protect against seawater. Fig. 10.11 shows a hydrophone based on a cylindrical ceramic element. This small,

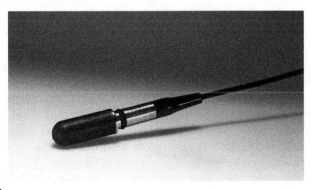

FIGURE 10.11

Broadband, omnidirectional, cylindrical hydrophone. Teledyne-RESON Type 4013.

Courtesy Teledyne-RESON A/S.

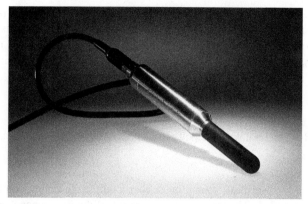

FIGURE 10.12

Broadband, high-sensitivity, low-noise hydrophone with preamplifier. Teledyne-RESON Type 4032.

Courtesy of Teledyne-RESON A/S.

outer diameter 9.7 mm, hydrophone (Teledyne-RESON Type 4013) has a usable frequency range from 1 Hz to 170 kHz and a 700 m maximum operating depth. The horizontal directional response is omnidirectional ± 2 dB at 100 kHz. At 100 kHz its receiving sensitivity is -211 dB rel 1 V/μPa and its transmitting sensitivity is 130 dB rel 1 μPa/V at 1 m. Therefore, at 100 kHz, a 1 μPa sound pressure amplitude on the hydrophone will give a received voltage amplitude of $2.8 \cdot 10^{-11}$ V. For 1 V electrical input amplitude at 100 kHz at 1 m it transmits an acoustical pressure amplitude of 3.2 Pa. This hydrophone has no preamplifier, while the hydrophone (Teledyne-RESON Type 4032) shown in Fig. 10.12 has a pre-amplifier in the housing just after the ceramic elements. The preamplifier has a high input impedance to reduce loss during signal reception. This hydrophone has a 5 Hz to 120 kHz usable frequency range and a 600 m maximum operating depth. Due to the preamplifier, the receiving response is improved to -164 dB rel 1 V/μPa and a 1 μPa sound pressure amplitude will produce a voltage amplitude of $6.3 \cdot 10^{-9}$ V. This hydrophone is omnidirectional in the horizontal plane. Its insulation is nitrile butadiene rubber.

As discussed in Chapter 9, at frequencies below 400 Hz, a low-frequency projector type based on the *Helmholtz resonator* principle with cylindrical shape is frequently used for underwater acoustic tomography. In this case, an open metal tube is excited at one end by a piezoelectric-driven piston source, which vibrates the water column in the tube at the lowest resonance frequency for the open tube. The effective wavelength $\lambda = 4L$ for the lowest resonance frequency of the open tube, where L is the length of the water column in the tube. This projector type is robust, not limited by hydrostatic pressure and cheap. However, it has a very narrow bandwidth and low mechanical–acoustical efficiency and radiated power.

10.2.4.3 Spherical Elements

The piezoceramics with a *spherical* shape is often used in hydrophones. Small projectors may have a spherical shape. Larger spherical projectors are based on an array of circular, triangular, or pentagonal piezoceramic disks. These projectors has a flexible radiation pattern since individual amplitudes and phases may be applied to each disk.

Frequently hydrophones are used to measure the acoustic pressure in a signal at a point in the water space. As a result, the hydrophone has to be small and omnidirectional, which often can be a spherical hydrophone with a small diameter compared with the shortest wavelength of interest. These hydrophones are normally based on two half-spheres of piezoceramics, silvered over the inner and outer surface and poled in the radial direction. The two half-spheres are cemented together along the edge to form a sphere, and an electrically insulated wire is drawn through a small hole in the cemented line and soldered to the inner silver layer. After connecting the electric wires from the inner and outer silver layer to a preamplifier or directly to the coax cable, the sphere is molded into epoxy- or a rubber-type insulation to protect it from the seawater, as shown in Fig. 10.13, where the insulation is nitrile butadiene rubber. This hydrophone (Teledyne-RESON Type 4033) has a usable frequency range from 1 Hz to 140 kHz, an operating depth of 900 m and is omnidirectional ± 2 dB at 100 kHz in the horizontal plane. Due to the influence of the connectors its vertical directional response is $270° \pm 2$ dB at 100 kHz. The receiving and transmitting responses of this hydrophone are -203 dB rel 1 V/µPa at 250 kHz and 144 dB rel 1 µPa/V at 1 m at 100 kHz, respectively. Therefore, at 250 kHz, 1 µPa sound pressure amplitude on the hydrophone leads to a received voltage amplitude of $7 \cdot 10^{-11}$ V, and for 1 V electrical input amplitude at 100 kHz it will transmit at 1 m at 15.8 Pa acoustical pressure amplitude. This hydrophone is not supplied

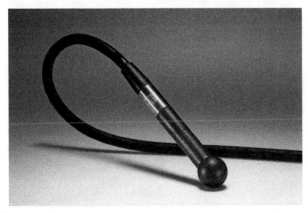

FIGURE 10.13

Broadband, omnidirectional, spherical hydrophone. Teledyne-RESON Type 4033.

Courtesy of Teledyne-RESON A/S.

with a preamplifier, while the hydrophone (Teledyne-RESON Type 4037) shown in Fig. 10.14 has connectors to a differential preamplifier, which increases its receiving response to −193 dB rel 1 V/μPa at 250 kHz. This hydrophone for 1 μPa sound pressure amplitude produces a voltage amplitude of $2.2 \cdot 10^{-10}$ V. The hydrophone has a 1 Hz to 100 kHz usable frequency range and 1500 m maximum operational depth. At 40 kHz it is omnidirectional in the horizontal plane, and its vertical directional response is $270° \pm 3$ dB due to the influence of the mounting support. Spherical hydrophones, due to the inherent strength of a spherical shell, are able to operate at greater depths than cylindrical hydrophones.

According to Wilson [12], the sensitivity M_r (unit: Vm^2/N) of a spherical hydrophone (below resonance) is:

$$M_r = b \frac{g_{33}(\zeta^2 + \zeta - 2) - g_{31}(\zeta^2 + \zeta + 4)}{2(1 + \zeta + \zeta^2)} \tag{10.18}$$

where b is the spherical piezoceramic shell's outer radius (unit: m) and $\zeta = a/b$ with a denoting the inner radius of the shell (unit: m). For a thin shell where $\zeta \rightarrow 1$, Eq. (10.18) reduces to:

$$M_r = -bg_{31} \tag{10.19}$$

In Eq. (10.19) g_{31} denotes the appropriate piezoelectric voltage coefficient when the stress is tangential and therefore normal to the poling direction.

According to Stansfield [13] the capacitance, C (unit F) (below resonance), of the ceramic shell is:

$$C \sim \frac{4\pi \varepsilon b \zeta}{1 - \zeta} = 4\pi \varepsilon b \left\{ \frac{b}{t} - 1 \right\} \tag{10.20}$$

FIGURE 10.14

Omnidirectional, high-receiving voltage response hydrophone with preamplifier and differential balanced signal output to limit DC offset and signal fluctuations due to noise, vibration, and temperature variations. Teledyne-RESON Type 4037.

Courtesy of Teledyne-RESON A/S.

where ε is the dielectric constant (unit: F/m) and $t = b - a$ is the thickness of the spherical shell. Eq. (10.20) shows that the capacitance decreases when the shell thickness increases. The resonance frequency, f_1 (unit: Hz), of the spherical shell is given by:

$$f_1^2 = \frac{2c_{11}}{(2\pi b)^2\{(1 - v)\rho\}} \tag{10.21}$$

where v is Poisson's ratio and ρ is the density of the ceramic material. c_{11} denotes the elastic modulus of the ceramic in the plane normal to the poling direction. Therefore, f_1 represents the resonance frequency connected with the tangential motion of the spherical ceramic. Due to Poisson's ratio this motion represents the transverse contraction, coupled with the radial motion of the sphere. The spherical shell, similar to the cylindrical shell, has another higher thickness resonance frequency for vibrations in the poling direction. The maximum power output from a spherical thin shell hydrophone, $M_r^2 C$ (unit: $(V/Pa)^2F$ or $(V/\mu Pa)^2pF$), can be found from Eqs. (10.19) and (10.20) as:

$$M_r^2 C = 4\pi\varepsilon g_{31}^2 b^3 \left\{\frac{b}{t} - 1\right\} \tag{10.22}$$

For a spherical shell produced from the piezoceramic material, Pz28, given in Table 10.1, the maximum output power is:

$$M_r^2 C = 13.8 \cdot 10^{-11} b^3 \left\{\frac{b}{t} - 1\right\}$$

A thin-walled, spherical hydrophone of Pz28 with an outside radius b of 10^{-2} m and a wall thickness t of 10^{-3} m will have the following properties:

Resonance frequency	$f_1 = 105$ kHz
Capacitance	$C = 7.35 \cdot 10^{-8}$ F
Sensitivity	$M_r = 13 \cdot 10^{-5}$ V/Pa or -197.7 dB re 1 V/μPa
Maximum power output	$M_r^2 C = 12.4 \cdot 10^{-16}$ (V/Pa)^{2}F

10.2.4.4 Tonpilz Transducers

The most general transducer geometry used for projectors is the so-called *Tonpilz* transducer. Its name originates from German, see Hahnemann and Hetch [18], and can be translated as "sound mushroom," since a larger head mass driven by a slender piezoceramic rod gives the transducer an appearance of a mushroom. Fig. 10.15 shows an example of a Tonpilz transducer where a *stack of piezoceramic disks or cylinders*, with stiffness coefficient K (unit: N/m), are assembled between a *head mass* M_1 (unit: kg), which acts like a piston, and a *tail mass* M_2 (unit: kg),

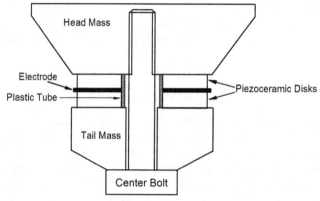

FIGURE 10.15

Tonpilz transducer main components. The ceramics are PZT rings poled in their thickness direction.

© *IEEE 1997, figure adapted from Yao, Q. and Bjørnø, L., Broadband Tonpilz Underwater Acoustic Transducers Based on Multimode Optimization, IEEE Trans. Ultrason., Ferroelect. Freq. Control, 44, (5), pp. 1060–1066, 1997, with permission of the IEEE.*

which acts as a counterweight and provides an inertial backing. The ceramic stack is held under compression by a *center bolt*. A *plastic tube* insulates the bolt electrically from the *electrodes* between the ceramic elements. The ceramic stack could include a much greater number of ceramic disks or cylinders cemented together with electrodes in between, and head and tail masses may vary in magnitude, shape, and material. Normally, the tail mass is heavier than the head mass. The tail mass can be produced from steel, tungsten, or brass, while the head mass can be produced from aluminum, a magnesium alloy, or titanium. The clamping center bolt is used to assure that the ceramics during vibration are not exposed to tensile stress. The head mass' geometrical shape acts as a transformer to adapt the ceramic stack vibration to the movement of the transmitting surface in contact with the water.

Due to the influence of the two masses the Tonpilz transducer is normally used in low-frequency projectors, from a couple of kilohertz to about 50 kHz. The head mass is in contact with the transmission medium and the tail mass operates in air. The center bolt design requirements are low mass and sufficiently bolt strength to produce the required force on the ceramics. However, the bolt should not be so stiff that it restrains the motion of the two masses. These requirements are frequently satisfied by producing the center bolt from high tensile steel and reducing the bolt diameter outside the thread sections. The bolt's stress and the tail mass material, size, and geometry are important factors for tuning the transducer's frequency. For the Tonpilz transducer shown in Fig. 10.15 the angular resonance frequency ω_ℓ (unit: radians/s) vibrating in its longitudinal piston mode, is given by:

$$\omega_\ell = 2\pi \, f_\ell = \left(\frac{K(M_1 + M_2)}{M_1 M_2} \right)^{1/2} \tag{10.23}$$

where K (unit: N/m) is the stiffness coefficient for the stack between the two masses. K is determined by the stack's effective Young's modulus, which includes the influence of the center bolt and joints, the electrodes between the ceramic elements in the stack, and the area and length of the stack. K normally is determined through experimental measurements. When it is assumed that all the transducer's energy is dissipated by the vibration of the head mass and that this dissipation is represented by a mechanical resistor γ, then the mechanical quality factor Q_m of the Tonpilz transducer may be expressed by:

$$Q_m = \frac{\omega_\varrho M_1}{\gamma} \left(1 + \frac{M_1}{M_2} \right) \tag{10.24}$$

For a transducer with low internal losses, γ is approximately equal to the radiation resistance of the head mass acting as a piston source.

The acoustic power radiated from a Tonpilz transducer is limited by mechanical and electrical factors. The electrical factors, such as dielectric loss, ceramic depoling, and electrical breakdown through the ceramic or along its surface, are the most important factors. Therefore, the power transferred to the Tonpilz transducer is limited by the maximum electric root mean square (rms) field strength E (V/m), which safely can be applied. An expression for the acoustic power per unit volume of the ceramic stack radiated at resonance, W_ϱ (unit: W/m^3), is:

$$W_\varrho = \eta_{ma}\,\omega_\varrho\,Q_m k^2 \varepsilon\,E^2 \tag{10.25}$$

where η_{ma} is the mechanical−acoustical efficiency of the Tonpilz transducer, k is the coupling coefficient, and ε (unit: F/m) is the dielectric constant of the ceramic. η_{ma} is the ratio between radiation resistance R_r (unit: Ns/m), arising from the water load on the head mass, and the sum of the internal mechanical resistance loss and the radiation resistance.

Tonpilz transducers are normally covered by an epoxy coating to provide electrical insulation and mounted inside a waterproof housing to provide air loading on the tail mass and avoid backward radiation from the head mass. Since air-filled housings reduce the transducer system's capability, air is sometimes replaced by oil. However, oil may reduce the acoustic transmission efficiency due to energy dissipation in the transducer−oil system.

The use of Tonpilz transducers at frequencies below 1−2 kHz is limited due to difficulties in producing large head mass displacement amplitudes at low frequencies. Since the transducer power output is proportional to $u^2 R_r$, where u (unit: m/s) is the velocity of the surface of head mass in contact with the water, and the radiation resistance is low at low frequencies, the velocity must increase to compensate for the low radiation resistance to preserve the transmitted power level. High vibration amplitudes are needed to achieve a desired output power level at low frequencies. Due to the ceramic stack high stiffness only small strains and there-fore small amplitudes are created despite the large piezoelectric stresses produced in the ceramics. Also, increasing Tonpilz transducer dimensions to produce high power at low frequencies reduces their applicability at frequencies below 1−2 kHz.

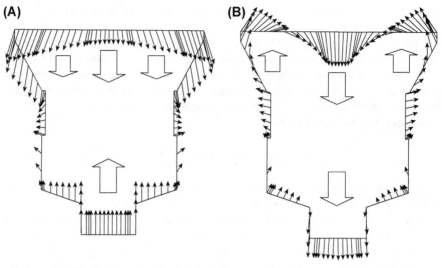

(A) Longitudinal piston mode dominant

(B) Flapping mode dominant

FIGURE 10.16

Surface displacements of the Tonpilz transducer in Fig. 10.15 for two vibration modes in air calculated by means of an FE code: (A) the first piston mode and (B) the first flapping mode related to the vibration of the radiating surface of the head mass. The head mass maximum diameter is 0.056 m.

©1997 IEEE reprinted from Yao, Q. and Bjørnø, L., Broadband Tonpilz Underwater Acoustic Transducers Based on Multimode Optimization, *IEEE Trans. Ultrason., Ferroelect. Freq. Control*, **44**, (5), pp. 1060–1066, 1997, with permission of the IEEE.

Since Tonpilz transducers normally operate at their system resonance frequency, they are in general narrow-band devices with Q_m-values above 3 in spite of the Q_m-lowering effects of the head and tail masses. A long felt desire has been to obtain a broadband transducer with Q_m below 2–3. An attempt to couple other vibration modes to the longitudinal piston mode in Fig. 10.15 to improve the bandwidth of the Tonpilz has been reported by Yao and Bjørnø [14]. Here the first "flapping" mode of the head mass was coupled to the first piston mode, as shown in Fig. 10.16. This figure shows the movements of the surface of the Tonpilz transducer in Fig. 10.15, when the two modes, the first piston (A) and the flapping (B) modes, with resonance frequencies of 35.3 and 42.7 kHz, respectively, were excited. The movements were calculated by a finite-element code, and after building and testing the transducers, Q_m-values below 2 were obtained.

10.2.4.5 Flextensional Transducers

The challenge, to develop projectors for transmitting high acoustic power at low frequencies led to development of impedance transformers based on lever principles to amplify mechanical displacements produced by the piezoceramics. Some of the most widely used impedance transformers are found in flextensional transducers. Five different design categories were classified by Brigham and Royster [19] and two more classes were added by Rynne [20]. Class I, convex flextensional transducer, was first modeled by Royster [21] and is based on a stack of piezoceramic disks or cylinders, which drives a series of "barrel staves" in a flexural or bending mode. The ceramic stack is kept under compression by a stress bolt through the stack's center and the stress bolt is electrically insulated from the ceramics and their electrodes. Class II and III transducers are also based on barrel staves. According to Jones and Christopher [22] they have shapes derived from the convex flextensional class I transducers, are more broadbanded and able to emit higher acoustic power than class I transducers. Class IV transducers are the flextensional transducers that are most intensively studied and most commonly used.

Classes V and VI are planar, circular transducers driven by a stack of piezoceramic disks or a ring transducer. The ring transducer can consist of a number of ceramic disks with metal wedges in between. Steel is frequently used since the steel and ceramic densities are nearly equal. As reported by McMahon [23] the percentage of steel has been varied between 24% and 36%. The circular transmitting shell is either convex as in class V or concave as in class VI transducers and is produced from either steel or aluminum. Two different size convex transducers with steel shells, weighing 30 and 220 kg in air, are reported by McMahon [23]. The 30 kg transducer has a 350 Hz resonance frequency, 35 Hz bandwidth and 204 dB rel 1 μPa at 1 m source level. The 220 kg transducer has a 610 Hz resonance frequency, 160 Hz bandwidth, and 212 dB rel 1 μPa at 1 m source level. McMahon [23] also produced and tested a concave class VI, flextensional transducer that used a tangentially poled PZT ring to drive the concave aluminum shell. This transducer delivered 1200 W acoustic power into water at its 605 Hz resonance frequency and produced a source level of 201.5 dB rel 1 μPa at 1 m. Its 3-dB bandwidth was 107 Hz and the electroacoustic efficiency was 83%.

Class IV, the most used flextensional transducer types, consist of an elliptical cylindrical shell produced from aluminum or steel. Titanium alloys, Kevlar or glass fiber–reinforced epoxy can also be used. Ceramic bars are inserted in one or more stacks along the major axis of the ellipse. The maximum shell bending displacement takes place along the minor axis, the z-direction. The ceramic bars are kept under compression by the elliptical shell into which they have been inserted after increasing the longest main axis by compressing the elliptical cylindrical shell by applying pressure along the z-direction. The x-direction is the bar's normal poling direction. When they are operating in their thickness mode extending the stack, they move the shell surface inward along the z-direction producing a large volume displacement over the whole-shell surface. Due to this manufacturing procedure higher static water pressures uncouple the ceramic stack from the shell, which reduces the transducer's operational depth. However, class IV transducers are able to transmit high acoustic power at low frequencies, as long as operational depth limits are not exceeded. The ceramic stack's acoustic power radiated per unit volume may be calculated by using Eq. (10.25). With reasonable dimensions, frequently smaller than the acoustic wavelength at the transducer's resonance frequency in water, these transducers are able to operate efficiently from below 300 Hz to above 3 kHz. Due to the small transducer size compared with the acoustic wavelength at resonance, the radiated beam patterns are nearly omnidirectional. As pointed out by Brigham [24], several elliptical shells can be put together in the y-direction to form an elliptical cylinder which is then supplied with end plates to make the cylinder watertight.

Class IV transducer design, calculation, and manufacturing are difficult. Several attempts have been made to calculate the electroacoustic and radiation qualities of class IV transducers by Brigham [24], Brigham and Glass [25], Hamonic et al. [26], and Butler et al. [27] using equivalent circuits or finite-element methods. The manufacturing process requires considerable care, which makes flextensional transducers expensive. The acoustical qualities of the class IV transducer depend in particular on the wall thickness and the material used for the shell and the eccentricity a/b of the shell, where a and b denote the half length of the ellipse major and minor axes, respectively. Oswin et al. [28] have shown that maximum eccentricity leads to maximum bandwidth with the lowest acoustic radiation power. Brigham and Glass [25] put forward the rule of thumb, that the eccentricity should not exceed about 3 to avoid severe static and dynamic stresses near the shell sections of highest curvature. The radiated power and bandwidth of the class IV transducer increases with increasing transducer size for a fixed transducer shape. The Q_m-value of class IV transducers vary inversely as the square root of the surface area. For an air-filled transducer the largest bandwidth is obtained for the most eccentric and lightest shell. Operation at increasing water depths requires thicker and larger shells, or the shell may be filled with oil, in particular when the transducer is thermally limited. However, as stated by Brigham and Glass [25], the air-filled design is preferable due to the better power radiation, bandwidth, and electroacoustic efficiency provided by a smaller and lighter transducer. Butler et al. [29] have reported improved acoustic beam patterns, directivity index $DI = 3.4$, and high acoustic output power from a class IV transducer of size less than 1/3 of a wavelength at resonance. This was obtained by simultaneously exciting the shell into an omnidirectional and dipole operation by driving the piezoceramic stack into both extensional and bending modes.

10.2.4.6 Flexural Transducers

Flexural transducers are used as projectors and hydrophones due to their simple construction and low cost. Flexural transducers operate in inextensional bending modes, where the neutral plane does not change its size during the bending operation, as shown in Fig. 10.17. This figure shows a bar, free at both ends and vibrating in the inextensional mode, where the part of the bar above the neutral plane expands, while the part below the neutral plane contracts, and the bar has no net extension. In particular for thin bars, where $t < L/20$ with L (unit: m) denoting the length and t (unit: m) the thickness of the bar, low flexural resonance frequencies can be obtained. For the bar free at both ends, Morse and Ingard [30] give the fundamental bending resonance frequency f_0 (unit: Hz) as:

$$f_0 = 1.028 \frac{tc}{L^2} \tag{10.26}$$

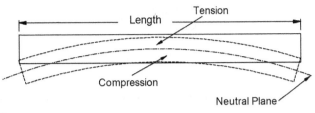

FIGURE 10.17

A dual bilaminar bender transducer design.

where c (unit: m/s), the sound speed in the bar, is given by $c^2 = Y/\rho$, in which Y (unit: N/m^2) is Young's modulus and ρ (unit: kg/m^3) is the bar material density. A bar free at both ends and rigidly clamped at both ends has the same bending resonance frequency. A bar, simply supported at both ends, has a fundamental bending resonance frequency about half the value calculated from Eq. (10.26).

   If two identical bars of a piezoceramic material, poled in the thickness direction, are glued together with a common center electrode in the glue layer, a bilaminar piezoelectric *flexural bar transducer* is produced. If a bilaminar bar transducer is exposed to a force F bending the simple supported bar as shown in Fig. 10.18, the upper bar is exposed to compression, while the lower bar is exposed to tension in the direction vertical to the bar's poling direction. This 31-mode deformation forms a voltage between the center electrode and the electrodes on the free surfaces of the bilaminar bar. If an electric field is put across the electrodes that are parallel wired, and the polarization direction is the same on both bars forming the bilaminar bar, the upper bar will expand and the lower bar will contract, which produces a bending motion of the bilaminar bar. The displacement amplitude of this bending mode is substantially higher than if the bars had been vibrating in the thickness mode, and the natural frequency of the bending mode is much lower than for the thickness mode. The flexural vibrations of piezoceramic bars have been frequently used to produce underwater acoustic projectors and hydrophones. Also the 33-mode of vibration can be exploited in bender bars, but due to poling difficulties shorter bars are required, and the manufacturing is more time consuming.

   One of the major drawbacks of the bender bar is the stress variation across the bar's thickness. The tensile stress forms the limiting factor for applying the bending

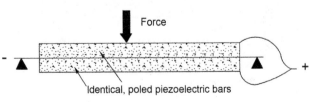

FIGURE 10.18

Bending force on a simple supported, bilaminar piezoelectric flexural bar transducer.

mechanism. The maximum mechanical stress σ_m (unit: N/m^2) in a bar exposed to hydrostatic pressure P_0 (unit: N/m^2) may be expressed by:

$$\sigma_m = \frac{3}{4}\left(\frac{L}{t}\right)^2 P_0 \tag{10.27}$$

For larger values of L/t, the tensile stress limit for the ceramic materials may be exceeded.

The *flexural disk transducer* is excited through the planar radial mode of vibration of the disk characterized by the planar coupling coefficient k_p, given in Table 10.1. The fundamental bending resonance frequency f_0 (unit: Hz) in air of a circular disk of diameter D (unit: m) and thickness t, clamped along the edge is:

$$f_0 = \frac{1.868ct}{D^2\sqrt{1-v^2}} \tag{10.28}$$

where v denotes Poisson's ratio. If the flexural disk is simply supported, f_0 is nearly half the value calculated by Eq. (10.28) using the same disk dimensions. In water, the mass load of the disk will reduce f_0 to f_w, which is given by Woollett [31] as:

$$f_w \sim f_0\sqrt{1 + 0.75\frac{D\rho_0}{2t\rho}} \tag{10.29}$$

where ρ_0 and ρ (unit: kg/m^3) denote the water and the disk density, respectively. If the disk is thin and its density is low, a strong reduction in the bending resonance frequency will occur when the disk is loaded with water.

Bilaminar disks are produced by gluing together two piezoceramic disks poled in their thickness direction and inserting an electrode in the glue layer between the disks. Trilaminar disks can be produced which consist of a metal disk of certain stiffness between two piezoceramic disks. While a disk normally possesses a resonance frequency higher than a bar's resonance frequency, disks are more widely used for flexural transducers since the disk's planar coupling coefficient, k_p, is higher than the bar's coupling coefficient k_{31}. Improved transmitting and receiving sensitivity can be obtained by mounting two bilaminar disks on a support ring as shown in Fig. 10.19. The support ring must be rigid in the axial direction and

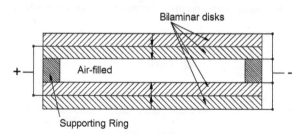

FIGURE 10.19

A dual bilaminar, piezoelectric, bender transducer design. The *arrows* on the disks indicate the poling directions. A support ring separates the two bilaminar transducers.

compliant in the radial direction to permit the motion of the two bilaminar disks. The center of this dual bilaminar bender design is normally air filled to provide the necessary pressure release. This design has been studied extensively by Woollett [31]. For two bilaminar disks of PZT-4 (nearly like Pz27 in Table 10.1) with an overall thickness t and radius a (unit: m), Woollett [31] provided the following theoretical values for the resonance frequency, f_0 (unit: Hz), for a supported edge disk in a plane baffle in air:

$$f_0 = 705 \frac{t}{a^2} \qquad (10.30)$$

With an effective coupling coefficient $k = 0.406$, the ratio of the resonance frequencies in air f_0 and in water f_w, and the mechanical quality factor Q_m are given by:

$$\frac{f_0}{f_w} = \sqrt{1 + 0.103 \frac{a}{t}} \qquad (10.31)$$

$$Q_m = 7.02 \left(\frac{f_0}{f_w}\right)^3 \eta_{ma} \qquad (10.32)$$

The electric field limited acoustic intensity I (unit: W/m^2) is given by:

$$I = 760 \cdot 10^4 \left(1 + 0.103 \frac{a}{t}\right) \left(\frac{t}{a}\right)^2 \eta_{ma}^2 \qquad (10.33)$$

The expressions (10.30)–(10.33) are derived under the assumption that the disks are thin, which limits their depth applicability. Also, flexural disk transducers suffer from the major drawback of low ceramic tensile stress limits. According to Sherman and Butler [16], for a hydrostatic pressure P_0 the induced radial stress σ_r in a simple supported disk of radius a and thickness t maybe found from:

$$\sigma_r \sim 1.25 \left(\frac{a}{t}\right)^2 P_0 \qquad (10.34)$$

Also, waterproofing the individual bilaminar disks can cause practical problems. Normally rubber or epoxy coatings are used to insulate the disks from the seawater. However, the coating's layer thickness has to be thin to prevent the stiffness and mass of the coating from influencing the resonance frequency and disk coupling coefficient. This is a particular problem for thin disks used at low frequencies. Due to simple construction and low costs, flexural bars and disks have in the past been used in many, not too demanding, underwater low-frequency sound generation and reception applications.

10.2.5 ACOUSTIC FIELD QUALITIES OF TRANSDUCERS

Sound is generated in a fluid by any process which produces a nonsteady pressure field. These pressure fields are produced by pulsation or vibration of a boundary surface of the fluid, by action of nonsteady forces on the fluid, and by turbulent motion of the fluid. The acoustic fields produced by various transducer geometries are treated in this section. Several parameters characterize the sound fields produced by *single-element transducers* and by *transducer arrays* such as the emitted source level and the beam patterns. These parameters are closely connected with the transducer's sound radiation qualities, frequency response, bandwidth, impedance, materials, and geometry. The discussion of the acoustic field qualities in this section is divided into fields produced by single-element transducers of various shapes and from transducer arrays.

10.2.5.1 Single-Element Transducers

Single-element transducers may be shaped as disks, plates, cylinders, and spheres, and they are used as projectors and hydrophones. Based on disks and cylinders, piston sources can be manufactured. Piston sources are most frequently mounted in baffles to improve their radiation qualities.

10.2.5.1.1 The Pole Concept

The *pole concept* is a fundamental way to characterize acoustic fields. This concept includes monopoles, dipoles, and higher-order poles, such as quadrupoles. The *monopole* or *point source* radiates sound uniformly in all directions. A pulsating sphere, where the velocity vector everywhere on the sphere's surface has the same magnitude and is vertical to the surface, is an example of a monopole. Such an omni-directionally radiating source could be based on a piezoceramic, air-filled, spherical shell strong enough to sustain the external pressure. Such spheres may have a large radiating area per unit volume and be useful to project low-frequency sound. Also, pulsating bubbles may act as monopole sound sources. In general, volume changes due to *mass fluctuations* give rise to monopoles, which are considered *simple sources*. A simple source is a sphere vibrating harmonically where the sphere radius, a, is much smaller than the emitted sound wavelength, λ (unit: m) i.e., $ka \ll 1$, where $k = 2\pi/\lambda$ is the wave number (unit: m^{-1}). When the wavelength is much greater than any linear characteristic dimension of the radiating body and all surface elements in the radiating body are vibrating in phase, the radiated acoustic field is almost independent of the radiating body shape. Therefore, for $ka \ll 1$ the radiated field from a piston source in a baffle may be considered a monopole acoustic field. Fig. 10.20 shows a spherical monopole of radius a, where mass fluctuations lead to volume fluctuations and to radiation of sound waves. The surface displacement is given by the arrows. The acoustic pressure p (unit: N/m^2) produced by the monopole for $ka \ll 1$ is:

$$p(r,t) = j\rho_0 c U_0 \left(\frac{a}{r}\right) ka e^{j(\omega t - kr)} = \frac{j\rho_0 c Q_s k}{4\pi r} e^{j(\omega t - kr)} \tag{10.35}$$

where $Q_s = 4\pi a^2 U_0$ (unit: m^3/s) is the radiator *source strength*, $j = \sqrt{-1}$ and U_0 (unit: m/s) is the velocity amplitude of the radiator surface's uniform harmonic motion. ρ_0 (unit: kg/m^3) is the fluid density, c (unit: m/s) is the sound velocity, and ω (unit: rad/s) is the angular frequency. r (unit: m) is the distance from the source

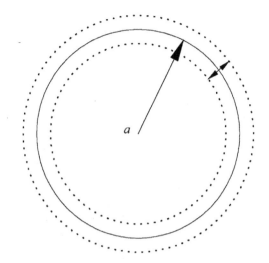

FIGURE 10.20

The monopole source of radius *a*. Volume fluctuations produce sound radiation. The *arrow* indicates the surface displacement.

in the far field, i.e., for $r \gg \lambda$ and t (unit: s) is time. The far-field monopole acoustic intensity I is:

$$I(r) = \frac{1}{8} \rho_0 c \left(\frac{Q_s}{\lambda r} \right)^2 = \frac{\rho_0 Q_s^2 f^2}{8 c r^2} \tag{10.36}$$

where f (unit: Hz) is the frequency of the sound radiated by the monopole. Eq. (10.36) shows that the acoustic intensity drops off as the inverse source distance squared and is directly proportional to the source strength squared. In principle, all acoustic radiation problems related to any source shape may be solved by a superposition of a selected geometry of monopole (or simple) sources.

The *dipole* or *doublet source* shown Fig. 10.21A is equivalent to two simple and equal sound sources pulsating 180 degrees out of phase and situated a small distance d apart. While there will be no resultant mass flow through a spherical surface surrounding the two simple sources, there is a momentum flux resultant. This is equivalent to *a force* on the fluid, which can be provided by the reaction from a solid surface. Acoustic dipole radiation may be found when a simple source is vibrating close to a pressure-release wall, such as the free-water surface, where the simple source produces an out-of-phase *image* on the other side of the wall. The source and image produce dipole radiation in the fluid half-space; see the discussion on the Lloyd's mirror effect in Chapter 2. Another example of a dipole is an oscillating sphere which is free to radiate from both sides. The positive acoustic pressure produced on the advancing side of the sphere is accompanied by a negative pressure of same magnitude on the receding side. If the wavelength is large compared with

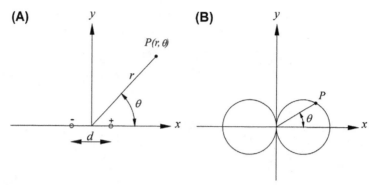

FIGURE 10.21

The geometry of a dipole, or doublet, source with distance d between the two simple sources is shown in (A). The radiation pattern of a dipole source for $\lambda \gg d$ is shown in (B).

the dimension of the dipole source, i.e., the distance d between positive and the negative source, the radiation pattern shown in Fig. 10.21B is a dipole, where the maximum amplitude is found along the dipole axis. For $kd \ll 1$, the acoustic pressure at the field point P in the dipole far field ($kr \gg 1$) is given by:

$$p(r, \theta, t) = \frac{j\rho_0 c Q_s k^2 d \cos \theta}{4\pi r} e^{j(\omega t - kr)} \qquad (10.37)$$

where Q_s (unit: m^3/s) is the source strength of each monopole forming the dipole. Comparing Eqs. (10.35) and (10.37) it can be seen that the dipole pressure amplitude is reduced by the factor $kd \ll 1$, and that a directionality is introduced through $\cos\theta$ which gives maximum acoustic pressure along the x-axis—the dipole axis—and zero acoustic pressure along the y-axis. The eight-shaped radiation pattern of dipoles shown in Fig. 10.21B is produced by the angle θ. The far-field acoustic intensity $I(r,\theta)$ for $kd \ll 1$ is:

$$I(r, \theta) = \frac{\rho_0 c Q_s^2 k^4 d^2 (\cos\theta)^2}{32\pi^2 r^2} = \frac{\pi^2 \rho_0 c Q_s^2 d^2 f^4 (\cos\theta)^2}{2c^4 r^2} \qquad (10.38)$$

A comparison between the intensities expressed by Eqs. (10.36) and (10.38) shows that the monopole intensity is proportional to the square of its frequency, while the dipole intensity is proportional to its frequency in fourth power.

While monopoles and dipoles may occur at fluid boundaries, the *quadrupole* is connected with fluctuating turbulent shear stresses and may occur in a fluid far from boundaries. Moreover, quadrupoles do not involve net volume changes or net forces. A quadrupole is two dipoles combined in one of two ways. If the two dipoles are of the same strength and oriented opposite, while the separation axis of their sources are laying on the same line, such as the x-axis, with a small distance between the two dipoles, then the quadrupole is termed *longitudinal*. The longitudinal quadrupole

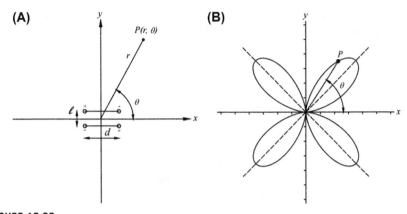

FIGURE 10.22

The lateral quadrupole geometry with the dipoles separated by the distance ℓ is shown in (A). The lateral quadrupole radiation pattern is shown in (B).

will have a far-field intensity proportional to $f^6(\cos\theta)^4/r^2$ with a directional pattern similar to a dipole. The other way of combining the two dipoles is when the dipoles have their axis parallel and separated by a small distance, ℓ, as shown in Fig. 10.22A. This forms a *lateral* quadrupole, which is equivalent to two parallel dipole forces on the fluid of equal magnitude and separated by the small distance ℓ, or as four monopoles with alternating phase at the corners of a square. Such a force system creates a *fluctuating shear stress*, which can be maintained without any boundaries in the fluid. The lateral quadrupole's far-field acoustic intensity is proportional to $f^6(\sin2\theta)^2/r^2$. A lateral quadrupole's directivity pattern looks like a clover leaf, as shown in Fig. 10.22B. While sound is radiated well in front of each monopole source, the sound is canceled at points equidistant from adjacent opposite monopoles. Quadrupoles as sources of sound are of vanishing importance in water.

From the discussions foregoing it can be seen that the total acoustic power produced by the monopole, dipole, and quadrupole is proportional to f^2, f^4, and f^6, respectively. To obtain the same intensity at the same far-field point and frequency by all three pole types requires increased emitted power as the pole order increases. The radiation efficiency ξ_{rad} of the different pole orders depends on the pole order and the ratio between the pole's characteristic size and wavelength. According to Ross [32] it can be written as:

$$\xi_{rad} \sim (ka)^{2n+1} \tag{10.39}$$

where n is the pole order. For $n=0$, the monopole, the radiation efficiency is $\xi_{rad} \sim ka$; for $n=1$, the dipole, the radiation efficiency is $\xi_{rad} \sim (ka)^3$; and for $n=2$, the quadrupole, the radiation efficiency is $\xi_{rad} \sim (ka)^5$. Since most sound sources in water are characterized by $ka \ll 1$, the lowest order pole is the most efficient sound source, and monopoles will, if they exist, dominate the higher-order poles. The fundamental partial differential equation describing the sound propagation from the pole types developed by Lighthill [33] is:

$$\frac{\partial^2 p}{\partial x_i^2} - \frac{1}{c^2}\frac{\partial^2 p}{\partial t^2} = -\frac{\partial m}{\partial t} + \frac{\partial F}{\partial x_i} - \frac{\partial^2 \tau_{ij}}{\partial x_i \partial x_j} \tag{10.40}$$

where the *source terms* on the right side of the equation represent the contributions from the monopole, the dipole, and the quadrupole, respectively. M denotes the rate of mass injection per unit volume, F is the force vector per unit volume and τ_{ij} is the Reynold's stress tensor, which is of importance only for sound produced by turbulence. P denotes the acoustic pressure as a function of the spatial coordinate, x_i, with I taking the values, 1, 2, and 3, and of time, t. Eq. (10.40) reduces to the general wave equation for regions of water free of any acoustic sources.

10.2.5.1.2 Piston Sources

The acoustic power radiated from a *piston source* small compared with the wavelength, i.e., $ka \ll 1$, is identical to the power radiated by a simple source having the same source strength Q_s, since a small piston acts as a volume source with characteristics independent of the exact source shape. A *circular piston source in an infinite, rigid baffle* is covered in several textbooks [34–36] and only the most important characteristics of a circular piston source in an infinite rigid baffle of infinite size are given in this section. The standard treatment derives the pressure in the liquid in front of the piston source produced by an infinitesimal area of the piston. The pressure field from the full piston is calculated by integrating over the piston surface under the assumption, that the velocity of the piston movements is uniform at any time over the whole piston area. This assumption is normally not fully satisfied by real piston sources, where phase variations may be present; however, as a first approximation it leads to realistic and useful results.

The *radiation impedance* Z_r of a vibrating surface, such as a piston source, is the ratio of the force F exerted by the vibrating surface on the medium, for example, water, in front of the surface to the normal velocity u from the surface. Because the force and normal velocity, which is the particle velocity in the water adjacent to the piston source, are not always in phase, the radiation impedance is in general a complex quantity, which may be written as:

$$Z_r = R_r + jX_r = \frac{F}{u} \tag{10.41}$$

where the real and imaginary parts constitute the radiation resistance R_r and radiation reactance X_r, respectively. The radiation impedance magnitude depends on the radiating surface size, shape, and surroundings. According to Stephens and Bate [34], the radiation resistance and reactance of a *circular* piston source of radius a in an infinite rigid baffle, are given by:

$$R_r = \pi a^2 \rho_0 c R_1(2ka) \text{ and } X_r = \pi a^2 \rho_0 c X_1(2ka) \tag{10.42}$$

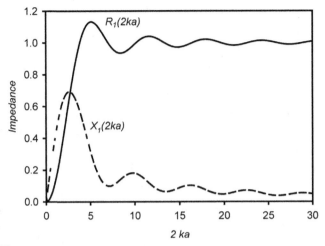

FIGURE 10.23

Impedance functions R_1 and X_1 for a rigid piston source in an infinite and rigid baffle as a function of the argument $2ka$.

where the functions R_1 and X_1 as functions of $2ka$ are given by:

$$R_1 = 1 - 2\,\frac{J_1(2ka)}{2ka} = \frac{(2ka)^2}{2^2 1!2!} - \frac{(2ka)^4}{2^4 2!3!} + \cdots \tag{10.43}$$

$$X_1 = \frac{2H_1(2ka)}{2ka} = \frac{4}{\pi}\left\{\frac{2ka}{1^2 \cdot 3} - \frac{(2ka)^3}{1^2 \cdot 3^2 \cdot 5} + \frac{(2ka)^5}{1^2 \cdot 3^2 \cdot 5^2 \cdot 7} - \cdots\right\} \tag{10.44}$$

where J_1 is the cylindrical Bessel function of order 1 and H_1 is the Struve function of order 1. The impedance functions R_1 and X_1 are shown in Fig. 10.23 as functions of $2ka$ and values of the two functions for the argument $0 \leq 2ka \leq 30$ are given in Table 10.2.

In the *low-frequency limit*, for $ka \ll 1$ or $a/\lambda \ll \frac{1}{2}\pi$, R_r and X_r reduce to

$$R_r = \frac{\pi a^2 \rho_0 c (ka)^2}{2} \quad \text{and } X_r = \frac{8a^2 \rho_0 cka}{3} \tag{10.45}$$

where the low-frequency reactance, which is always positive, behaves like an additional vibrating mass $M_r = X_r/\omega = (8/3)\rho_0 a^3$. This mass is equivalent to the mass of a cylinder of water with the same cross-sectional area as the piston source and length $8a/3\pi$. This "extra mass" reduces the natural frequency and increases the mechanical Q-value of a low-frequency piston source when the piston is small and has a low weight. The expression for R_r in Eq. (10.45) shows that for piston sources small compared to the wavelength, R_r is proportional to the square of the piston area, and for constant piston radius a, the radiation resistance R_r is proportional to the square of the frequency. Since the radiated *acoustic power* Π is given by $\frac{1}{2}u_0^2 R_r$,

Table 10.2 $R_1(2ka)$ and $X_1(2ka)$, $2ka = 0-30.0$

2ka	X₁(2ka)	R₁(2ka)	2ka	X₁(2ka)	R₁(2ka)
0.000	0.000	0.000	15.500	0.097	0.978
0.500	0.209	0.031	16.000	0.102	0.989
1.000	0.397	0.120	16.500	0.101	1.001
1.500	0.547	0.256	17.000	0.095	1.011
2.000	0.647	0.423	17.500	0.084	1.019
2.500	0.691	0.602	18.000	0.072	1.021
3.000	0.680	0.774	18.500	0.060	1.018
3.500	0.624	0.921	19.000	0.051	1.011
4.000	0.535	1.033	19.500	0.047	1.002
4.500	0.429	1.103	20.000	0.047	0.993
5.000	0.323	1.131	20.500	0.051	0.987
5.500	0.230	1.124	21.000	0.058	0.984
6.000	0.159	1.092	21.500	0.064	0.985
6.500	0.116	1.047	22.000	0.069	0.989
7.000	0.099	1.001	22.500	0.071	0.996
7.500	0.104	0.964	23.000	0.070	1.003
8.000	0.122	0.941	23.500	0.065	1.009
8.500	0.146	0.936	24.000	0.058	1.013
9.000	0.166	0.945	24.500	0.050	1.013
9.500	0.178	0.966	25.000	0.043	1.010
10.000	0.178	0.991	25.500	0.039	1.005
10.500	0.167	1.015	26.000	0.037	0.999
11.000	0.146	1.032	26.500	0.038	0.993
11.500	0.122	1.040	27.000	0.042	0.990
12.000	0.097	1.037	27.500	0.047	0.989
12.500	0.078	1.026	28.000	0.051	0.991
13.000	0.066	1.011	28.500	0.054	0.995
13.500	0.063	0.994	29.000	0.054	1.000
14.000	0.068	0.981	29.500	0.052	1.004
14.500	0.077	0.973	30.000	0.048	1.008
15.000	0.088	0.973			

where u_0 is the particle velocity amplitude normal to the surface in a harmonic wave motion of the piston surface, small piston sources are not able to radiate significant acoustic power at low frequencies, without large piston displacement amplitudes. For piston sources small compared with the sound wavelength this is extremely difficult to achieve when using piezoceramics. For a piston source driven at constant normal velocity u, the radiated acoustic power is directly proportional with R_r, which varies substantially with the frequency for small piston sources at low frequencies.

According to Pierce [37], in the *high-frequency limit* for large circular pistons or for high frequencies, $ka \gg 1$, and R_1 and X_1 become:

$$R_1 \sim 1 - \frac{(8/\pi)^{1/2} \cos(2ka - 3\pi/4)}{(2ka)^{3/2}} \tag{10.46}$$

$$X_1 \sim \frac{4/\pi}{2ka} - \frac{(8/\pi)^{1/2} \sin(2ka - 3\pi/4)}{(2ka)^{3/2}} \tag{10.47}$$

In the limit $ka \rightarrow \infty$, Eqs. (10.46) and (10.47) reduce to $R_1 \sim 1$ and $X_1 \sim 0$. The additional vibrating mass is now $M_r = 2\rho a/k^2$, which vanishes for high frequencies. The radiation impedance is real with $Z_r = R_r \sim \pi a^2 \rho_0 c = \rho_0 cS$, and the radiated acoustic power is $\prod \sim \frac{1}{2}\rho_0 c \cdot S \cdot u_0^2$. This acoustic power is the same as a *plane* acoustic wave with the particle velocity amplitude u_0 through a surface area S, when the media *characteristic acoustic impedance* is $\rho_0 c$, where the plane wave is propagating. In plane acoustic waves the acoustic impedance is real and given by the ratio between the acoustic pressure amplitude P and u_0. While the radiation resistance governs the acoustic power propagating from the piston source to distances far from the source, the radiation reactance represents a sort of nonpropagating, but standing wave field close to the piston source.

If the *baffle* around the piston source is *finite* in size and rigid, corrections may be introduced to allow for these effects. If the piston dimensions are large compared with the wavelength λ, i.e., $ka \gg 1$, the emitted acoustic energy is concentrated into a beam in front of the piston and only a small interaction takes place between the acoustic beam and the baffle. The baffle has almost no influence. However, if the piston dimension is the same magnitude or smaller than the wavelength, the pressure field along the baffle has a significant influence on the radiated beam pattern. For very small pistons their emission may be considered as coming from a simple source with an omnidirectional field in front of the piston and no field behind the piston. If the baffle is removed from the piston source, the radiation impedance function R_1 will be reduced to about half of the R_1-value for a baffled piston. As a rule of thumb, baffles larger than 2λ in diameter can be considered infinitely large baffles. If the baffle has finite rigidity, to first order it can be considered to be rigid if the baffle material has a characteristic acoustic impedance ρc high compared with the characteristic acoustic impedance of water.

According to Stansfield [13], the radiation impedance of a *spherical source* of radius a is given by:

$$Z_{rsph} = \frac{4\pi ka^3 \rho_0 c(ka + j)}{1 + (ka)^2} \tag{10.48}$$

For high frequencies, $ka \gg 1$, Eq. (10.48) reduces to a pure radiation resistance, $Z_{rsph} = R_r \sim 4\pi a^2 \rho_0 c$, and for low frequencies, $ka \ll 1$, Z_{rsph} in Eq. (10.48) becomes:

$$Z_{rsph} \sim 4\pi a^2 \rho_0 cka(ka + j)$$

where the radiation resistance is much smaller than the radiation reactance. The additional vibrating mass for low frequencies is $M_r = 3\rho_0 V_{sph}$, where V_{sph} (unit: m^3) is the vibrating sphere volume. Therefore, at low frequencies the additional vibrating mass is 3 times the mass displaced by the sphere.

The *radiation impedance* of a *cylindrical radiator* which is long compared to the wavelength may be expressed by:

$$Z_{rcyl} = j2\pi a h \rho_0 c \frac{H_0^{(2)}(ka)}{H_1^{(2)}(ka)} \tag{10.49}$$

where a and h denote the cylinder radius and length, respectively, while $H_0^{(2)}(ka)$ and $H_1^{(2)}(ka)$ are Hankel functions expressed by:

$$H_0^{(2)}(ka) = J_0(ka) - jN_0(ka) \quad \text{and} \quad H_1^{(2)}(ka) = J_1(ka) - jN_1(ka) \tag{10.50}$$

J_n and N_n are Bessel functions of first and second kind of order n as discussed in Abramowitz and Stegun [38]. N_n is frequently called the Neumann function.

The *acoustic beam patterns* produced in the fluid in front of a flush-mounted rigid circular piston with a flat, infinite, rigid baffle, when the piston is performing harmonic vibrations with the uniform velocity amplitude u_0 (unit: m/s) normal to the piston surface, can be calculated based on the geometry shown in Fig. 10.24. As shown in several textbooks [34–36], the acoustic pressure $p(r,\theta,t)$ (unit: Pa), at the time t (unit: s), is:

$$p(r, \theta, t) = \frac{j\rho_0 c u_0 k}{2\pi} \int_S \frac{e^{j(\omega t - kr_1)}}{r_1} dS \tag{10.51}$$

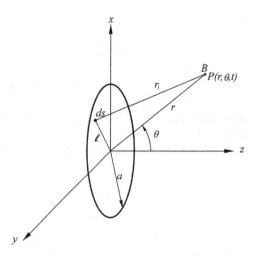

FIGURE 10.24

The geometry used for the derivation of the radiation patterns of a rigid piston mounted on a flat, rigid baffle of infinite extent. *Ds* is an infinitesimal piston area whose vibration produces sound in the field point *B*. The total acoustic pressure *P(r,θ,t)* at *B* is found by integrating the contributions from all infinitesimal piston areas over the piston surface *S*.

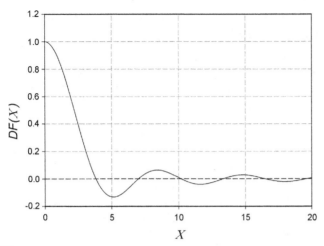

FIGURE 10.25

The acoustic beam pressure amplitude directivity function as a function of the argument $\chi = ka\sin\theta$ for a piston source.

where S (unit: m^2) is the total area of the piston source, and the surface integration is performed over the whole area for $\ell \leq a$, where a is the piston radius (unit: m). ρ_0 is the density (kg/m^3) and c is the sound velocity (unit: m/s) in the fluid.

The integral in Eq. (10.51) can be solved analytically in the *far field* of the piston source, i.e., $r \gg a$, and the acoustic pressure can be written as:

$$p(r, \theta, t) = \frac{j\rho_0 ck}{2\pi r} Q_s e^{j(\omega t - kr)} \left[\frac{2J_1(ka\sin\theta)}{ka\sin\theta}\right] \tag{10.52}$$

where $Q_s = \pi a^2 u_0$ is the piston source strength. Apart from the *pressure directivity function* in the bracket, the acoustic field described by Eq. (10.52) is identical to the acoustic field produced by a hemispherical source in an infinite and rigid baffle. The *pressure directivity function* $DF(\theta) = 2J_1(ka \cdot \sin\theta)/(ka \cdot \sin\theta)$ describes the angular dependence of the acoustic pressure field produced by the piston source. Fig. 10.25 shows the pressure amplitude directivity function as a function of the argument: $\chi = ka \cdot \sin\theta$. The acoustic field intensity produced by the piston source is proportional to p^2. Table 10.3 provides the directivity function values for the acoustic field pressure and intensity for $0 \leq \chi \leq 20$. Fig. 10.25 shows that for some values of $\chi \sim 3.83$, 7.02, 10.15, etc. the directivity factor has zero values, i.e., the acoustic pressure amplitude will be zero at a series of conical nodal surfaces with vertices at $r = 0$. This happens for polar angles:

$$\theta_1 \sim \sin^{-1}\frac{3.83}{ka} = \sin^{-1}\frac{0.61\lambda}{a}; \quad \theta_2 = \sin^{-1}\frac{1.12\lambda}{a} \quad \text{and} \quad \theta_3 = \sin^{-1}\frac{1.62\lambda}{a}$$

Table 10.3 Pressure and Intensity Directivity Function (*DF*) for $\chi = 0-20$

X	DF(X)	DF(X)²	X	DF(X)	DF(X)²
0.00	1.00E+00	1.00E+00	10.25	−3.71E−03	1.38E−05
0.25	9.92E−01	9.84E−01	10.50	−1.50E−02	2.26E−04
0.50	9.69E−01	9.39E−01	10.75	−2.46E−02	6.07E−04
0.75	9.31E−01	8.67E−01	11.00	−3.21E−02	1.03E−03
1.00	8.80E−01	7.75E−01	11.25	−3.72E−02	1.38E−03
1.25	8.17E−01	6.67E−01	11.50	−3.97E−02	1.58E−03
1.50	7.44E−01	5.53E−01	11.75	−3.97E−02	1.57E−03
1.75	6.63E−01	4.40E−01	12.00	−3.72E−02	1.39E−03
2.00	5.77E−01	3.33E−01	12.25	−3.27E−02	1.07E−03
2.25	4.87E−01	2.38E−01	12.50	−2.65E−02	7.01E−04
2.50	3.98E−01	1.58E−01	12.75	−1.90E−02	3.61E−04
2.75	3.10E−01	9.60E−02	13.00	−1.08E−02	1.17E−04
3.00	2.26E−01	5.11E−02	13.25	−2.43E−03	5.92E−06
3.25	1.48E−01	2.20E−02	13.50	5.64E−03	3.18E−05
3.50	7.85E−02	6.16E−03	13.75	1.29E−02	1.67E−04
3.75	1.77E−02	3.14E−04	14.00	1.91E−02	3.63E−04
4.00	−3.30E−02	1.09E−03	14.25	2.37E−02	5.62E−04
4.25	−7.32E−02	5.36E−03	14.50	2.67E−02	7.12E−04
4.50	−1.03E−01	1.05E−02	14.75	2.79E−02	7.78E−04
4.75	−1.22E−01	1.48E−02	15.00	2.73E−02	7.48E−04
5.00	−1.31E−01	1.72E−02	15.25	2.52E−02	6.34E−04
5.25	−1.31E−01	1.73E−02	15.50	2.16E−02	4.66E−04
5.50	−1.24E−01	1.54E−02	15.75	1.68E−02	2.84E−04
5.75	−1.11E−01	1.22E−02	16.00	1.13E−02	1.28E−04
6.00	−9.22E−02	8.51E−03	16.25	5.33E−03	2.84E−05
6.25	−7.06E−02	4.99E−03	16.50	−6.99E−04	4.88E−07
6.50	−4.73E−02	2.24E−03	16.75	−6.41E−03	4.11E−05
6.75	−2.38E−02	5.66E−04	17.00	−1.15E−02	1.32E−04
7.00	−1.34E−03	1.79E−06	17.25	−1.56E−02	2.45E−04
7.25	1.89E−02	3.58E−04	17.50	−1.87E−02	3.49E−04
7.50	3.61E−02	1.30E−03	17.75	−2.04E−02	4.18E−04
7.75	4.94E−02	2.44E−03	18.00	−2.09E−02	4.36E−04
8.00	5.87E−02	3.44E−03	18.25	−2.00E−02	4.02E−04
8.25	6.36E−02	4.04E−03	18.50	−1.80E−02	3.25E−04
8.50	6.43E−02	4.13E−03	18.75	−1.50E−02	2.24E−04
8.75	6.11E−02	3.73E−03	19.00	−1.11E−02	1.24E−04
9.00	5.45E−02	2.97E−03	19.25	−6.76E−03	4.56E−05
9.25	4.52E−02	2.04E−03	19.50	−2.14E−03	4.58E−06
9.50	3.40E−02	1.15E−03	19.75	2.43E−03	5.91E−06
9.75	2.15E−02	4.62E−04	20.00	6.68E−03	4.47E−05
10.00	8.69E−03	7.56E−05			

where λ is the wavelength of the harmonic signal produced by the piston source. The angles θ_1, θ_2, θ_3, ... form the half-angles of the *major lobes* of the pressure field produced by the piston source.

According to Eq. (10.52) the pressure amplitude's individual lobes are inversely proportional to the radial distance from the piston source center. The acoustic wave phase on any spherical wave front within a lobe is the same at all points. However, the waves produced by the piston source are not spherically symmetric since the acoustic pressure also depends on the polar angle θ. Fig. 10.25 shows that the axial pressure for $\theta = 0$ is maximum since the directivity function has the value 1.00 on the piston axis. Based on the directivity function the maximum amplitude of the *first side lobe*, situated between the angles θ_1 and θ_2, is about 17 dB lower than the *main lobe* amplitude found on the piston axis for $\theta = 0$.

The *beam width* is usually specified as the angle enclosed between two directions in which the acoustic field intensity has dropped to an agreed fraction of the axial intensity. The *half-power beam width* of the main lobe, also called the -3 dB *beam width*, can be calculated from the intensity variation given in Table 10.3. When the ratio between the intensity $I(r,\theta)$ at a field point off the piston axis and the intensity $I(r,0)$ is 0.5, the argument $\chi = ka \cdot \sin\theta_{-3dB} = 1.617$. For a piston source with $ka = 20$ and $a/\lambda = 3.18$, the full half-power beam width will be $2\theta_{-3dB} = 9.3°$, since the beam has rotational symmetry around the piston axis. Frequently the -6 dB (intensity ratio $= 0.25$) and -10 dB (intensity ratio $= 0.1$) angles are used to characterize the beam from a piston source. For $ka = 20$ and $a/\lambda = 3.18$, the full -6 dB beam width is $2\theta_{-6dB} = 12.7$ degrees, and the full -10 dB beam width is $2\theta_{-10dB} = 15.7$ degrees, while the intensity nulls are at $\theta_1 = 11$ degrees, at $\theta_2 = 20.6$ degrees and at $\theta_3 = 30.6$ degrees.

It is obvious from the directivity function values given in Table 10.3 that for $ka \gg 1$, the *radiated beam pattern* from a piston source has several side lobes and the half-power beam width of the main lobe is small. The beam pattern main lobe and two side lobes from a piston source for $ka = 12$ are shown in Fig. 10.26, which also shows the half-power beam angle θ_{-3dB}. When $ka < 1$, only the main lobe is present and for very small ka-values the beam pattern is the same as from a simple source in an infinite, rigid baffle. This is due to the fact that the source shape does not influence the radiated beam pattern when $ka \ll 1$, where the source behaves like a simple source.

In the piston source *near field* for $r \sim a$ a strong interference between acoustic signals radiated by different parts of the piston source surface takes place. Near the piston source the radiated signal contributions from various parts of the piston may differ in travel path by $\lambda/2$, which locally causes almost complete cancellation of the acoustic pressure. The acoustic field near the source is calculated by numerical methods since a closed form solution such as Eq. (10.52) cannot be obtained for this part of the acoustic field. A strong pressure amplitude P fluctuation between 0 and $2\rho_0 c u_0$ occurs along the acoustic axis (the z-axis in Fig. 10.26) when $r \geq 0$. This fluctuation after the distance $r_r = \pi a^2/\lambda = A/\lambda$, where A is the piston area, shows a monotonically decreasing behavior, where the pressure amplitude is

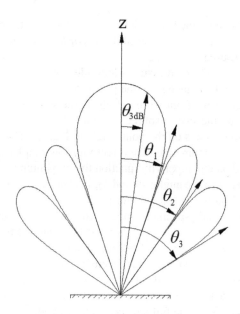

FIGURE 10.26

Beam pattern for a circular, rigid piston source in an infinite rigid baffle for $ka = 12$. The main and side lobe null angles are: $\theta_1 = 18.6$ degrees, $\theta_2 = 35.9$ degrees, and $\theta_3 = 58$ degrees. The half-power beam angle, $\theta_{3dB} = 7.7$ degrees, is shown.

inversely proportional to r similar to a spherically divergent acoustic wave. The distance r_r along the acoustic axis is called the *Rayleigh distance* (named after Lord Rayleigh, 1842–1919), and it is normally used as the transition distance between the *near-field* (also called the *Fresnel region*, named after A.J. Fresnel, 1788–1827) and the *far-field* (also called the *Fraunhofer region*, named after J. von Fraunhofer, 1787–1826).

The acoustic signal beam patterns transmitted by the piston source, Fig. 10.26, show that the acoustic intensity is highest on the acoustic axis. For the intensity the *directivity D* (sometimes called the *directivity factor*) for a source, such as a piston radiator is given by:

$$D = \frac{I_D(r)}{I_O(r)} \tag{10.53}$$

where $I_D I$ is the intensity on the acoustic axis at the source distance r for a directional source, while $I_O I$ is the intensity at the same source distance for an omnidirectional source transmitting the same acoustic power over 4π steradians. The ratio D, therefore, characterizes the increased acoustic power transmitted along the acoustic axis of a directive source over the same power transmitted omnidirectionally from a simple source. The circular piston source's directivity D as a function of ka is:

$$D = \frac{(ka)^2}{\left(1 - \dfrac{J_1(2ka)}{ka}\right)} \tag{10.54}$$

which for $ka \gg 1$ reduces to $D \sim (ka)^2$, which shows the increased directivity with increasing frequency or with increasing source dimension.

Due to the magnitude of D and to facilitate the use of the directivity in sonar equation calculations, it is convenient to define a decibel equivalent to D called the *directivity index DI* (unit: dB) through the expression:

$$DI = 10 \, log \, D \qquad (10.55)$$

The directivity index expresses the *directivity gain* in dB of a transducer, and it is extensively used as a characterizing parameter for transducer transmission and reception. For a large piston source the directivity index is approximately $DI \sim 10 \, log \, (4\pi A/\lambda^2)$, where A is the piston area. As a rule of thumb, the same expression for the directivity index can be used for a quadratic piston source as long as its minimum dimension is greater than $\lambda/2$, and for a rectangular piston source if the length to width ratio is greater than 2, according to Bobber [39]. For a simple source, $D = 1$, and therefore, $DI = 0$.

If all pressure beam patterns angle dependences are gathered in a single function, the *directivity function DF* (θ,φ), where θ is the angle between the direction to the field point B (in Fig. 10.24) and the transducer axis, the z-axis, and φ is the azimuth angle in the transducer plane (not shown in Fig. 10.24), the general transducer directivity D may be written as:

$$D = \frac{4\pi}{\int_{4\pi} DF^2(\theta, \varphi) \sin\theta d\theta d\varphi} \qquad (10.56)$$

Inserting the directivity function for the piston source from Eq. (10.52) into Eq. (10.56), assuming rotational symmetry around the transducer axis, i.e., no φ-dependence, and integrating Eq. (10.56) results in Eq. (10.54) for the piston source directivity.

Like most mechanical systems, the piston source also possesses a *resonance frequency* f_0 determined by its driving mechanism. The mechanism can for instance be based on one or more Tonpilz transducers, on magnetostrictive, hydraulic, or mechanical devices. The piston may form part of the driving system, and this system governs the *frequency response* of the piston source including its spectrum, bandwidth, and Q-value. The resonance sharpness, i.e., if the resonance curve is narrow or broad around the resonance frequency, is described by the piston source Q-value which is given by:

$$Q = \frac{\omega_0}{\omega_2 - \omega_1} \qquad (10.57)$$

where $\omega_0 = 2\pi f_0$ and ω_2 and ω_1 are the two frequencies, respectively, above and below the resonance frequency ω_0, where the average acoustic power has dropped to one-half of its resonance value. These half-power frequencies, also termed the -3 dB *frequencies*, should not be confused with the piston source beam pattern -3 dB angles.

To radiate as much acoustic power as possible it may be necessary to introduce impedance matching layers between the piston source and the medium in front of the piston, for example, water. This *impedance matching* is necessary when the piston material's acoustic impedance $\rho_0 c$ deviates strongly from water's acoustic impedance. The impedance matching is dealt with in more detail in the Transducer Design and Calculations section.

The acoustic power radiated by a piston source depends on the electronic circuits including signal generation and power amplification before the piston source and on the electromechanical qualities of the piston source including its directivity pattern, its Q-value, etc. This radiated power leads to the *source level* produced by the piston source. The source level is the intensity of the radiated acoustic wave in a specified direction, expressed in decibels (dB) relative to the intensity of a plane acoustic wave of *rms* pressure 1 μPa, referred to at a point 1 m from the acoustic center of the piston source. Since the piston source near field, which includes interference between signal contributions from various parts of the piston surface, frequently includes the point 1 m from the piston, the acoustic pressure for the intensity calculation is not measured at the point 1 m from the piston. It is measured in the far field, and then extrapolated back to the point 1 m from the piston source. The source level is an important factor in using the *sonar equation*, which is covered in more detail in Chapter 1, Section 1.6.

Limitations in the source level may originate from *acoustic saturation* effects and *cavitation*. Acoustic saturation effects limit the pressure amplitude which can be transmitted across a distance in front of a piston source and makes the received acoustic pressure at a distance from the piston source nearly independent of its source level. The saturation effects are caused by the nonlinear transfer of the acoustic power at elevated source levels from fundamental frequencies to higher harmonics, where increased absorption takes place. These effects are described in more detail in Chapter 13 on Finite-Amplitude Waves. Cavitation is local bubble formation (boiling), caused by the negative phase part of a higher amplitude acoustic signal which may produce bubble clouds that reflect and scatter acoustic signals and thus reduce the signal amplitude. The threshold for the start of cavitation, near the sea surface is frequently set to 10 kW/m². Since bubble formation is a nucleation process which lasts some time, the cavitation threshold rises with increasing frequency since the periods of negative pressure in an acoustic wave becomes shorter. As shown by Urick [40] an increase in the cavitation threshold based on the acoustic power of more than 30 kW/m² may be found in freshwater when the frequency is increased from 10 to 100 kHz. Below 10 kHz the threshold depends only weakly on frequency. Also, the ambient pressure has a strong influence on the cavitation threshold, which increases with increasing depth below the sea surface. According to Stansfield [13], for a long acoustic pulse at low frequency the cavitation threshold C_t (unit: W/m²) at various depths d (unit: m) is approximately:

$$C_t \sim 3000 \left(1.8 + \frac{d}{10} \right)^2 \tag{10.58}$$

10.2.5.1.3 Hydrophones

As mentioned earlier, a projector which normally operates at its resonance frequency, can also be used as receiver as is frequently done in active sonar systems, where a limited frequency band is exploited. However, for most applications it is desirable to receive underwater sound over a broader frequency band with the receiver sensitivity nearly frequency independent. Untuned piezoelectric hydrophones used well below their resonance frequency provides this possibility, and the potential variation in amplitude and phase of signals received at and near the hydrophone resonance frequency can be avoided. The nearly flat frequency response of hydrophones over several decades below about 60–70% of their resonance frequency forms the basis for signal reception. The hydrophone resonance frequency is directly related to its size. Larger hydrophones can be made more sensitive; however, the cost is a lower resonance frequency. Therefore, a hydrophone should be designed to have its resonance frequency somewhat above the upper limit of the frequency band to be measured. The Q-factor characterizes the hydrophone response at and around resonance and indicates the hydrophones applicability for signal reception. Fig. 10.27 shows a *simple low-frequency hydrophone* with balanced front and tail parts mounted in a housing. The hydrophone is supported at its nodal point and the front piston has an area A_p (unit: m^2) exposed to the acoustic pressure p (unit: Pa). The piezoceramic elements' area is A_e and thickness is t (unit: m). The O-ring provides water-tightness for the hydrophone/housing arrangement. According to Stansfield [13] when feeding into an infinite resistance such a hydrophone has a *pressure sensitivity M* (unit: V/Pa) as a function of angular frequency ω (unit: rad/s) given by:

$$M = \frac{g_{33}t}{2(1-k^2)} \frac{A_p}{A_e\sqrt{\left(W^2 + 1 - (\omega/\omega_0)^2\right)^2 + (\omega/\omega_0 Q_0)^2}} \tag{10.59}$$

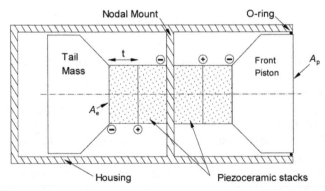

FIGURE 10.27

Schematic exposition of a balanced low-frequency hydrophone supported at its nodal point and mounted in a housing. The $\pm$ symbols indicate the polarity of the voltage signals produced.

In Eq. (10.59) g_{33} (unit: Vm/N) is the piezoelectric voltage coefficient in the poling direction, k is the coupling coefficient, W^2 denotes the bandwidth factor $k^2/(1 - k^2)$, ω_0 is the angular resonance frequency, and Q_0 is the motional Q-factor. The pressure sensitivity in Eq. (10.59) gives the hydrophone output voltage caused by the acoustic pressure at the front piston surface. However, the pressure sensitivity is not always the same as the *free-field sensitivity*. The free-field sensitivity is the output voltage produced by the acoustic pressure which existed at the hydrophone position before the hydrophone was introduced into the acoustic field. The field disturbance caused by the hydrophone is dependent on the hydrophone size and on the measured acoustic wavelength. For small hydrophones, i.e., for $d < \lambda/2$, where d is a characteristic dimension of the hydrophone, such as the diameter of the front piston, the difference between the pressure and the free-field sensitivities is small. However, for $d > \lambda/2$ the acoustic pressure at the piston surface will increase due to the influence of the particle velocity in the acoustic wave, which at larger surfaces of higher acoustical impedances than water will produce a stagnation pressure of nearly 2 times the acoustic pressure in the undisturbed field. The free-field sensitivity may also be influenced by *diffraction effects* in the acoustic field around the hydrophone. The diffraction effects are most significant when $d > \lambda/2$. Therefore the broadband hydrophone dimensions must be kept small compared to the wavelength. The diffraction effects will influence the *free-field frequency response* by inclining downward the response curves around the resonance, see Fig. 10.28. For a broadband hydrophone it is also important to ensure that all resonances are below, or at least outside, the band of interest for the measurements. In practice the input resistance into which the hydrophone feeds its electrical signal will not be infinite since the piezoceramics will have a finite leakage resistance and the input resistance of the first amplifier stage will also have a finite value. This finite resistance influences the low-frequency part of the hydrophone's frequency response curve as shown in Fig. 10.28, where the sensitivity falls off at low frequencies.

According to Stansfield [13] the maximum electric power output from a hydrophone similar to the one shown in Fig. 10.28 is:

$$M^2 C_{lf} = \frac{g_{33}^2}{4} \left(\frac{A_p}{A_e}\right)^2 \varepsilon V_e \tag{10.60}$$

where M and C_{lf} denote the linear hydrophone sensitivity (unit: V/Pa) and the low-frequency capacitance (unit: F), respectively. V_e (unit: m^3) is the piezoceramic elements volume and ε (unit: F/m) is the dielectric constant ($\varepsilon = \varepsilon_r \cdot \varepsilon_0$, where ε_r is the relative dielectric constant and $\varepsilon_0 = 8.85 \cdot 10^{-12}$ F/m). As in Eq. (10.14), the maximum power output is proportional to the hydrophone's piezoceramic material volume. Subject to the limitations in output power expressed by Eq. (10.60), values for hydrophone sensitivity and capacitance can be selected. A higher sensitivity requires a lower capacitance. However, decreasing the hydrophone's capacitance to obtain a higher sensitivity is not always beneficial. Any extra (or "parasitic") load capacitance across the hydrophone terminals such as the cable capacitance and capacitance of the input stage to the first amplifier will reduce the sensitivity.

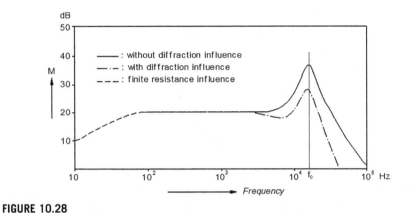

FIGURE 10.28

Frequency response of a hydrophone. The sensitivity M is in dB relative to an arbitrary unit. The influence of finite input impedance into which the hydrophone feeds its signal will influence the low-frequency response. f_0 is the resonance frequency.

Coaxial cables typically have a capacitance of about 100 pF/m. For longer cables the cable capacitance may become comparable to or higher than the hydrophone capacitance. As shown by Rijnja [41] the maximum sensitivity is obtained for C_{lf} equal to the load capacitance. However, when it is not necessary to maximize the hydrophone sensitivity, it is desirable to ensure that the hydrophone capacitance is much higher than the load capacitance to avoid the influence of cable length variations.

Hydrophone sensitivity is not the only factor which influences the hydrophone's ability to measure signal amplitudes down to a desired level and over the desired frequency band. *Self-noise* of the hydrophone, its cabling and amplifier will influence the receiving capabilities. Contributions to self-noise arise from, thermal noise, amplifier noise, and mounting noise. It is a well-known that a resistor at a finite temperature generates voltage fluctuations, known as *thermal noise*. Also, pressure fluctuations due to thermal fluctuations in the sea produce thermal noise. Mellen [42] showed that the thermal noise pressure amplitude p_t (unit: Pa) in a 1 Hz band in the sea is given by:

$$p_t = \sqrt{\frac{4\pi K_b T \rho c}{\lambda^2}} \tag{10.61}$$

where T is the temperature (unit: K) and $K_b = 1.38 \cdot 10^{-23}$ J/K is Boltzmann's constant. From Eq. (10.61) the voltage produced by the hydrophone when subjected to the thermal noise pressure fluctuations can be calculated when the hydrophone sensitivity is known. As shown on Chapter 6, thermal noise in the sea is only of importance near and above 100 kHz.

Another self-noise source is *amplifier noise* caused by the combination of a voltage source across the amplifier input terminals and a current source in series with the input terminals. Since hydrophone produced thermal noise and amplifier noise are uncorrelated, the total receiver system noise is the sum of the two produced noise powers when expressed in the same units.

Mounting noise is due to hydrophone mounting vibrations. Since piezoceramic-based hydrophones also are acceleration sensitive, an *acceleration response* can be expected when the hydrophone-mounting moves. Transducer piezoceramic elements are sensitive to all stresses in the ceramic, regardless of how they are produced. The voltages caused by these stresses are added to acoustically generated voltages. The hydrophone movements can be reduced by balancing as shown in Fig. 10.27 or by decoupling the charge-forming part of the hydrophone from its housing by using compliant mountings. This last procedure can only be used when it is possible to keep the compliant mounting resonance frequency outside the measurement band of interest. This may be a problem for broadband hydrophones. The mounting is particularly important for spherical and cylindrical hydrophones. If they are rigidly attached to the supporting structure an acceleration response may be produced, while a flexible mounting may make it difficult to position the hydrophone in an accurate measurement position. For a flexible mounting the hydrophone acceleration sensitivity may generate a response to the particle velocity of the acoustic field and make the total signal output directional, instead of the desired omnidirectional pressure response.

The water flow past the hydrophone cable produces vortices which couple back to the cable. The vortex production frequency is given by $f_s = Su/d$ (unit: Hz), where u (unit: m/s) is the flow velocity, d (unit: m) is the cable diameter, and S is the dimensionless Strouhal number, which for most cable diameters and flow velocities has a value $S = 0.18$. The vortex shedding produces cable *strumming* which may cause problems when the hydrophone suspension system has a resonance frequency coinciding with the strumming frequency. Cable motion may also cause so-called *triboelectric noise* when friction arises between cable's conductor and screen when the cable is bent. Using low-noise cable will in most cases minimize the triboelectric noise effect. Also, water flow around the hydrophone can lead to *flow noise*, when turbulent pressure fluctuations across the pressure sensitive hydrophone face produce electric signal noise output from the hydrophone, in particular at low frequencies.

One of the most frequently used hydrophone types is the *cylindrical hydrophone*. This hydrophone uses a single piezoceramic tube, such as the hydrophones shown on Figs. 10.11 and 10.12. Although the cylinder does not possess 3-D symmetry, it can have an omnidirectional response if the hydrophone tube element is small compared with the wavelength of the highest frequency sound to be measured. Cylindrical piezoceramic tube elements are normally radially poled and silvered on the outer and inner surfaces where the electrodes are also connected. The cylindrical element's vibration can be in a radial mode, thickness mode, and length mode, all leading to different resonance frequencies. In the *radial mode* the cylinder radius expands and contracts. The resonance frequency is obtained when the cylinder circumference equals one wavelength of the ceramic's longitudinal vibration. This is given by:

$$f_r \sim \frac{c}{2\pi a} \tag{10.62}$$

where f_r (unit: Hz) is the radial mode resonance frequency, c (unit: m/s) is the velocity of longitudinal waves normal to the poling direction in the piezoceramic material, and a (unit: m) is the mean radius of the cylinder.

The *thickness mode* is characterized by an expansion and contraction of the cylinder's wall thickness. The resonance frequency f_t (unit: Hz) of the cylinder wall thickness vibrations is:

$$f_t \sim \frac{c_t}{2w_t} \tag{10.63}$$

where c_t (unit: m/s) is the velocity of thickness waves in the piezoceramic and w_t (unit: m) is the cylinder's wall thickness.

The *length mode* represents an expansion and contraction of the cylinder's length. The length resonance frequency f_L (unit: Hz) occurs when the cylinder's length L (unit: m) is half a wavelength of longitudinal vibrations in the piezoceramic. This is:

$$f_L \sim \frac{c}{2L} \tag{10.64}$$

To use cylindrical hydrophones over a broad frequency band requires that all resonance frequencies are above the measured frequency range. For an omnidirectional cylindrical hydrophone the length mode resonance should be higher than the radial mode resonance, which means that $L < 3a$. Frequently, cylindrical element dimensions may result in *mode couplings* which produce a complicated frequency dependence for the generated voltage. Also, an external pressure applied to the cylindrical hydrophone may cause all dimensions to decrease, and the voltage output from the hydrophone will then consist of the sum of voltages produced by a combination of radial and tangential stresses in the piezoceramic which will involve the piezoelectric voltage coefficients g_{31} and g_{33}.

The two most frequently used cylindrical hydrophone designs are the *end capped* and the *exposed end*. The end-capped cylinder is sealed at both ends and the cylinder is normally air filled. The acoustic pressure on the cylinder is applied in the radial and length directions which lead to longitudinal and circumferential stresses. The low-frequency *pressure sensitivity* (unit: V/Pa) of end-capped hydrophones based on radially poled piezoceramic cylindrical tubes according to Langevin [43] is given by:

$$M = b\left(g_{31}\frac{2+\beta}{1+\beta} + g_{33}\frac{1-\beta}{1+\beta}\right) \tag{10.65}$$

where b (unit: m) is the outer radius of the cylindrical tube and $\beta = a/b$ in which a (unit: m) is the inner radius of the cylindrical tube. For thin-walled tubes, i.e., for $\beta \to 1$, Eq. (10.65) reduces to: $M_t = (3/2)bg_{31}$. For a thin-walled tube the maximum power output from the end-capped hydrophone is given by $M_t^2 C_{lf}$ (unit: $(\text{V/Pa})^2\text{F}$):

$$M_t^2 C_{lf} = \frac{4.5\pi\varepsilon g_{31}^2 Lab^2}{t} \tag{10.66}$$

An exposed-end hydrophone has the inner cylinder wall surfaces shielded; however, the cylinder wall narrow ends are exposed to the acoustic pressure. The cylinder is therefore exposed to a longitudinal stress along the cylinder wall and a circumferential stress in the cylinder wall. An exposed-end hydrophone based on a radially poled piezoceramic tube results in a low-frequency *pressure sensitivity* (unit: V/Pa) given by:

$$M_t = b\left\{ g_{31}(2-\beta) + g_{33}\frac{1-\beta}{1+\beta} \right\} \tag{10.67}$$

which for a thin-walled tube reduces to $M_t = bg_{31}$. The maximum power output from the exposed ends hydrophone $M_t^2 C_{lf}$ (unit: $(V/Pa)^2 F$) is:

$$M_t^2 C_{lf} = 2\pi\varepsilon g_{31}^2 \frac{Lab^2}{t} \tag{10.68}$$

An acoustic pressure measurement at a point in water requires a small omnidirectional hydrophone. This requirement is supported by *spherical hydrophones* that are small compared to the wavelength at the highest frequency of interest. Two types of spherical hydrophones are shown in Figs. 10.13 and 10.14. The pressure sensitivity (unit: V/Pa) below resonance of a spherical hydrophone with its piezoceramic shell silvered over its inner and outer surfaces and poled radially, is given by Eq. (10.18). For a very thin spherical shell the sensitivity is given by Eq. (10.19) and the maximum power output (unit: $(V/Pa)^2 F$) is given by Eq. (10.22). The cylindrical shell major resonance frequency is based on the shell thickness vibration and the *depth limitation* for using a spherical shell hydrophone is based on the permissible stress σ_p in the shell material. Because the stress in the shell roughly can be said to be $2b/t$ higher than the surrounding pressure, the maximum permitted surrounding pressure P_{omax} can be expressed by:

$$P_{omax} = \frac{\sigma_p t}{2b} \tag{10.69}$$

In order to operate the hydrophone at greater depths it is necessary to increase the shell thickness t; however, this reduces the maximum power output from the hydrophone. However, a comparison between the maximum output power expressions for cylindrical and spherical hydrophones at the same resonance frequency show that the spherical hydrophone has a higher potential power output than both cylindrical hydrophone types.

To be able to measure the arrival direction of an underwater acoustic signal, acoustic *vector sensors* have received a considerable interest in recent years. Vector sensors measure the acoustic *particle velocity*, a directional vector, together with the traditional acoustic pressure, a nondirectional scalar. Pressure gradient hydrophones use the dipole or doublet source directivity patterns shown in Fig. 10.21B, which possesses a cosine or "figure of eight" directional response to a plane acoustic wave. In a plane acoustic wave propagating through a fluid the relation among pressure p (unit: Pa), particle velocity u (unit: m/s), and pressure gradient $\partial p/\partial x$ (unit: Pa/m) is given by:

$$p = \rho c u \quad \text{and} \quad \frac{\partial p}{\partial x} = -j\frac{\omega p}{c} \tag{10.70}$$

In Eq. (10.70) ρ (unit: kg/m^3) is the fluid density, ω (unit: rad/s) is the angular frequency, and c (unit: m/s) is the sound speed. Eq. (10.70) show that a constant pressure leads to a constant particle velocity, while the pressure gradient is proportional to frequency and shows the same directional characteristics as the particle velocity.

The pressure gradient may be measured by a pair of hydrophones, and if the hydrophone pair were assumed to be point receivers—a doublet—with a mutual distance d and with equal pressure sensitivity M (unit: V/Pa), the doublet pressure sensitivity M_d by using Eq. (10.37) is given by:

$$M_d = 2\,M\sin\left\{\frac{\pi d}{\lambda}\cos\theta\right\} \tag{10.71}$$

where θ is the angle between the direction to the field point and the doublet axis, shown in Fig. 10.21A. For small values of d compared with λ, Eq. (10.71) may be reduced to $M_d \sim (2M\pi d/\lambda)\cos\theta$. To increase the doublet sensitivity the ratio d/λ should be increased. However, to preserve a response nearly independent of frequency it is recommended that the hydrophone distance is $d < \lambda/6$ at the highest measurement frequency. The $\cos\theta$ term represents the "figure-of-eight" directional response pattern. To obtain reliable and reproducible pressure gradients measurements in the acoustic field, the two hydrophones must be very small and their phase and amplitude characteristics must be nearly equal. Small hydrophones only produce limited output signal amplitudes and the small d/λ values further reduce the output signal amplitudes for decreasing frequencies. A simple pressure gradient hydrophone has been described by Bobber [39]. The difference in pressure on the two sides of a diaphragm drives a piezoceramic disk and the diaphragm forming a bilaminar element in a flexure or bending mode. This produces stretching of the piezoceramic disk in one half of the vibration cycle and compression on the other half. In the frequency range below the flexure resonance, the piezoceramic disk voltage output is proportional to the pressure gradient over the sound propagation distance between the two sides of the bilaminar element.

While the pair of hydrophone pressure gradient sensors may result in errors caused by the finite distance d between the hydrophones and by a phase mismatch between the two hydrophones, procedures and technologies for particle velocities measurements in acoustic fields have attracted much attention since the mid-1950s, when Leslie et al. [44] published their article on particle velocity measurements using a moving coil element mounted inside a hollow brass sphere. *Inertial methods* for particle acceleration and velocity measurements by responding to the water motion when the measuring device is neutrally buoyant have been widely studied. Gabrielson et al. [45] built and tested an *inertial sensor* consisting of a glass microballoon and epoxy composite, to give neutral buoyancy, cast around a small, commercial geophone to give information about the acoustic particle velocity. Their inertial sensor could be used up to several kilohertz. In conjunction with a hydrophone that measures the acoustic pressure, the acoustic intensity, I, also a vector in the water can be determined from the product $I = pu$. The acoustic intensity is valuable for measuring target bearings since the target direction is the negative of the intensity propagation direction. Multiple scattering targets can also be located by using a single *vector sensor* provided that the individual scattered signals are separated in time when received at the vector sensor. The intensity was also measured by use of an underwater pressure-particle acceleration device described by Kim et al. [46], where a neutrally buoyant body consisting of a pair of accelerometers were mounted in a hollow piezoceramic cylinder. The piezoceramic cylinder measured the acoustic pressure, while the rigid body probe's acceleration was measured by the accelerometer pair. The rigid body acceleration is the same as the acoustic particle acceleration when the probe is neutrally buoyant and the acoustic wavelength in water is larger than the sensor dimensions. Due to the high resonance frequency of most accelerometers, the sensor dimension determines the frequency limit for using the sensor. A limiting factor for accelerometers applications is frequently their self-noise. An expansion of this sensor principle to include three accelerometers, one in each Cartesian direction opens the possibility for simultaneous 3-D acceleration measurements. Fiber-optical interferometry can also be used in vector sensors for measuring particle velocity or acceleration.

Today a vector sensor is composed of two or three spatially colocated and orthogonally oriented velocity sensors plus an optional, also colocated, pressure hydrophone, all forming a point-like geometry. Vector sensors provide the possibility for full azimuth/elevation estimation since they do not have the hydrophone array left/right ambiguity. Vector sensors can be used in vertical, as well as towed arrays which have several advantages over traditional pressure hydrophone arrays. Vector sensor arrays have substantially improved directivity over the directivity of pressure hydrophone arrays with the same aperture due to the inherent directional qualities of the individual vector sensor. Vector sensor arrays will possess no left/right or fore/aft ambiguity, no directivity dependence on the inter-hydrophone spatial phase factor as in pressure hydrophone arrays, and they can be used in undersampled (sparse) configurations with uniform geometry, see for instance, Cray and Nuttall [47] and Wong and Chi [48]. Studies in late 2010s on using single vector sensors and *arrays of vector sensors* for direction of arrival estimation, ocean bottom parameter investigations in geoacoustic matched-field inversions, source localization, vector sensor array calibration procedures, and vector sensor detection thresholds can be found in the underwater acoustic measurements conferences [49].

10.2.5.2 Arrays
Up to now the acoustic fields have in general been produced by single-element transducers. However, substantial advantages such as improved directivity, beam steering, beam forming, and shading are found in *arrays* of single-element transducers

(transmitters/receivers). This discussion of array types is divided into *line arrays* including discrete elements and continuous elements, and *planar arrays* including circular, quadratic, and rectangular arrays, each comprising a number of transducer elements organized in various geometries. The section on array qualities includes beam steering and forming, shading, split beam techniques, range, and bearing resolution.

10.2.5.2.1 Array Types

The simplest one-dimensional array is a line of equally spaced, equally weighted (same source strength or sensitivity), and omnidirectional elements, the *discrete element line array*. The element omnidirectionality is based on the assumption that element dimensions are small compared with the sound wavelength and, when transmitting, can be considered as point sources. The receiving array can consist of individual hydrophones on a string or of capped piezoceramic tubes contained in a liquid-filled plastic or rubber hose. Such string array is illustrated in Fig. 10.29, where the array is used as a vertical receiving array with distance d between the centers of each of the N hydrophone elements. N can range from few elements to several hundreds. However, depending on water depth the number of hydrophone elements in a vertical array is normally below 100. While individual elements are omnidirectional, the array will possess substantial directional properties normal to its direction of extension. When the front of a plane acoustic wave of

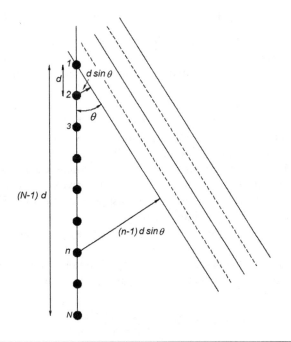

FIGURE 10.29

A plane acoustic wave of wavelength λ incident under the angle θ on a vertical array of N discrete transducer elements.

wavelength λ is incident at an angle θ on the array in Fig. 10.29, the distance normal to the wave front between the first and second array element is $d\sin\theta$. Therefore, the distance normal to the wave front between the first and the nth array element is $(n-1)d\sin\theta$. Since array elements are assumed equally weighted, the array signal output becomes the arithmetic sum of individual array element outputs, and the *pressure directivity function DF(θ)*, also called the normalized directivity pattern, is:

$$DF(\theta) = \left[\frac{\sin\left(\dfrac{Nkd}{2}\sin\theta\right)}{N\sin\left(\dfrac{kd}{2}\sin\theta\right)} \right] \tag{10.72}$$

where $k = 2\pi/\lambda$ is the acoustic wave number. Eq. (10.72) shows the discrete element line array directional sensitivity. For $N = 1$, $DF(\theta) = 1$, which is the omnidirectional character of a single element. $N = 2$ does not produce the classical dipole directivity function unless the two elements are transmitting 180 degrees out of phase and $d \ll \lambda$. The *directivity factor D* of a discrete element line array with N elements is given by:

$$D = \frac{N}{\left\{1 + (2/N)\displaystyle\sum_{n}^{N}\frac{(N-n)\sin(nkd)}{nkd}\right\}} \tag{10.73}$$

which for a large number of elements ($N \gg 1$) and $d = \lambda/2$, reduces to $D = N$. The *directivity index (DI)* (unit: dB) is $DI = 10\ log\ D$ for $N \gg 1$ and $d = \lambda/2$. Since $d = \lambda/2$ only is valid for a single frequency, directivity changes with frequency for the same array.

When the array is oriented in a broadside direction and $d = \lambda/2$, plots of $DF(\theta)$ for $N = 4$, 8, and 16, are shown in Fig. 10.30, which shows the directivity increase as the number of array elements and array length increase. Fig. 10.31 shows plots for an eight-element array with $d = \lambda/4$, λ, and 2λ. This shows that multiple main lobes with the same level, the *grating lobes*, will arise at different angles for increasing element spacing, causing *ambiguity*. Also, the directivity is worse for reduced element spacing. To avoid ambiguity in array design the element spacing d is important since d imposes an upper limit on frequencies where the array can be used without ambiguity. This upper frequency f_0 is given by $f_0 \leq c/2d$, where c is the speed of sound.

A hydrophone array shows an improved signal-to-noise (*s/n*) ratio relative to a single hydrophone. The improvement is given by the *array gain (AG)* (unit: dB), i.e.,

$$AG = 10\ log\left\{\frac{(s/n)_{array}}{s(n)_{single\ hydrophone}}\right\} \tag{10.74}$$

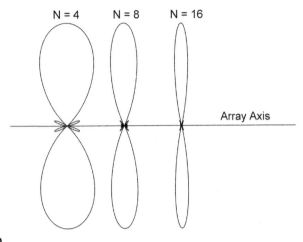

FIGURE 10.30

Impact on $DF(\theta)$ when the number of array elements N is increased and the array element spacing remains constant at $d = \lambda/2$.

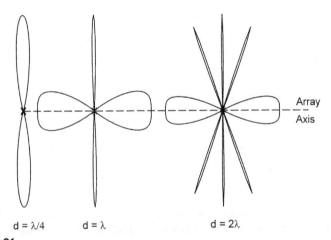

FIGURE 10.31

Effects on an eight-element broadside array with $d = \lambda/4$, $d = \lambda$, and $d = 2\lambda$.

Because s/n values depend on knowledge about the signal and noise coherence along the array, AG is more difficult to determine. By assuming a plane and coherent signal and an isotropic and incoherent noise Eq. (10.74) reduces to the DI for the array.

A graph of the directivity pattern of a discrete element line array calculated as 10 log $(DF(\theta)^2)$ from Eq. (10.72) for $N = 11$ and $d = \lambda/2$ is given in Fig. 10.32. The graph shows that the full -3 dB beam width will be about 10 degrees, and also that the first side lobe level is about 13.3 dB below the main lobe amplitude level,

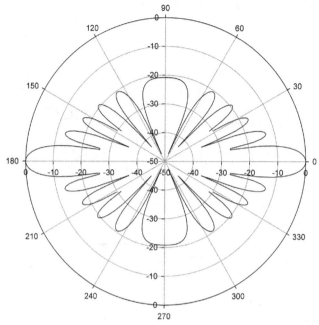

FIGURE 10.32

Directivity pattern of discrete element line array for $N = 11$ and $d = \lambda/2$.

which is less than the first side lobe reduction by a circular piston source as shown in Fig. 10.26.

If the number of discrete line array elements is increased while the distance between elements approaches zero, the factor in the denominator of Eq. (10.72), $\sin\{(kd/2)\sin\theta\} \sim (kd/2)\sin\theta$ and $L = Nd$, where L is the total array length:

$$DF(\theta) = \frac{\sin\{(kL/2)\sin\theta\}}{(kL/2)\sin\theta} \qquad (10.75)$$

which is the pressure directivity function for a *continuous line array.* Eq. (10.75) has the form $\sin(x)/x$, i.e., Sinc(x), Morse and Ingard [50]. Sinc(x) is shown in Fig. 10.33 for values up to $x = 10$ and in Table 10.4 for $0 \le x \le 16$. The beam pattern has nulls for $x = \pm\pi, \pm2\pi, \pm 3\pi, \ldots$. The main and side lobes have peak values at $x = 0, \pm(3/2)\pi, \pm(5/2)\pi, \ldots$.

From the pressure directivity function $DF(\theta)$ in Eq. (10.75) the *directivity index DI* (unit: dB) of the continuous line array may be written as:

$$DI = 20 \, log \left| \frac{\sin\{(kL/2)\sin\theta\}}{(kL/2)\sin\theta} \right| \qquad (10.76)$$

which expresses the array beam pattern intensity in dB, which is given in Fig. 10.34, where $x = (kL/2)\sin\theta$ is used as the normalizing parameter. Insertion of values for

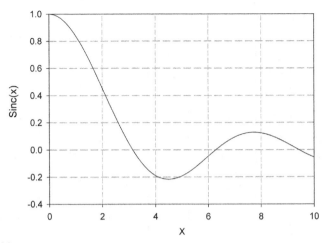

FIGURE 10.33

The Sinc(x)-function for $0 \leq x \leq 10$.

x into Eq. (10.76) shows that the first, second, and third side lobe will have peak values of 13.5, 17.9, and 20.8 dB, respectively, below the main lobe peak. The full -3 dB (the half power), the -6 dB, and the -10 dB *beam widths* for the continuous line array are $2\theta_{-3dB} = 2\sin^{-1}2.78/kL$, $2\theta_{-6dB} = 2\sin^{-1}3.78/kL$, and $2\theta_{-10dB} = 2\sin^{-1}4.64/kL$, respectively.

Based on the directivity index for a discrete element line array, $DI = 10 \, log \, N$, a simple expression for the directivity index for a continuous line for $L \gg \lambda$ and is given by $DI = 10 \, log \, (2L/\lambda)$.

Two-dimensional *planar arrays* may have quadratic, rectangular, or circular shapes. If two discrete element line arrays, one consisting of N elements with element distance d_n and one consisting of M elements with element distance d_m are combined to form a *rectangular planar array* with the dimensions $(N - 1)d_n$ times $(M - 1)d_m$, as shown in Fig. 10.35, according to Urban [51] the *pressure directivity function DF(θ,φ)* is:

$$DF(\theta, \phi) = \frac{\sin\{(Nkd_n/2)\sin\theta\cos\phi\} \cdot \sin\{(Mkd_m/2)\sin\theta\sin \phi\}}{N \sin\{(kd_n/2)\sin\theta\cos\phi\} \cdot M\sin\{(kd_m/2)\sin \theta\sin \phi\}} \qquad (10.77)$$

If all elements are spaced $\lambda/2$ and $N \gg 1$ and $M \gg 1$, the directivity index DI for the planar array with $N \times M$ elements is $DI = 10 \, log \, NM$. If the planar array is baffled, an extra 3 dB is added to the calculated DI value.

For increasing numbers of elements N and M in both array types and for the element distances $d_n \rightarrow 0$ and $d_m \rightarrow 0$, a *rectangular piston source* with the side lengths a and b will have the pressure directivity function given by:

$$DF(\theta, \phi) = \frac{\sin\left\{\frac{1}{2}ka \sin\theta \cos\phi\right\}}{\left\{\frac{1}{2}ka \sin\theta \cos\phi\right\}} \cdot \frac{\sin\left\{\frac{1}{2}kb \sin\theta \sin\phi\right\}}{\left\{\frac{1}{2}kb \sin\theta \sin\phi\right\}} \qquad (10.78)$$

Table 10.4 The Sinc(x)-Function for $0 \le x \le 16$

x	Sinc(x)	X	Sinc(x)	X	Sinc(x)	x	Sinc(x)
0	1	4.1	−0.19958	8.2	0.114723	12.3	−0.0214
0.1	0.998334	4.2	−0.20752	8.3	0.108695	12.4	−0.01336
0.2	0.993347	4.3	−0.21306	8.4	0.101738	12.5	−0.00531
0.3	0.985067	4.4	−0.21627	8.5	0.09394	12.6	0.002668
0.4	0.973546	4.5	−0.21723	8.6	0.085395	12.7	0.010491
0.5	0.958851	4.6	−0.21602	8.7	0.076203	12.8	0.018087
0.6	0.941071	4.7	−0.21275	8.8	0.066468	12.9	0.025386
0.7	0.920311	4.8	−0.20753	8.9	0.056294	13	0.032321
0.8	0.896695	4.9	−0.2005	9	0.045791	13.1	0.038829
0.9	0.870363	5	−0.19178	9.1	0.035066	13.2	0.044854
1	0.841471	5.1	−0.18153	9.2	0.024227	13.3	0.050344
1.1	0.810189	5.2	−0.1699	9.3	0.013382	13.4	0.055252
1.2	0.776699	5.3	−0.15703	9.4	0.002636	13.5	0.05954
1.3	0.741199	5.4	−0.1431	9.5	−0.00791	13.6	0.063174
1.4	0.703893	5.5	−0.12828	9.6	−0.01816	13.7	0.066128
1.5	0.664997	5.6	−0.11273	9.7	−0.02802	13.8	0.068384
1.6	0.624734	5.7	−0.09661	9.8	−0.0374	13.9	0.069929
1.7	0.583332	5.8	−0.0801	9.9	−0.04622	14	0.070758
1.8	0.541026	5.9	−0.06337	10	−0.0544	14.1	0.070873
1.9	0.498053	6	−0.04657	10.1	−0.06189	14.2	0.070284
2	0.454649	6.1	−0.02986	10.2	−0.06862	14.3	0.069005
2.1	0.411052	6.2	−0.0134	10.3	−0.07453	14.4	0.06706
2.2	0.367498	6.3	0.002669	10.4	−0.0796	14.5	0.064476
2.3	0.32422	6.4	0.018211	10.5	−0.08378	14.6	0.061287
2.4	0.281443	6.5	0.033095	10.6	−0.08705	14.7	0.057534
2.5	0.239389	6.6	0.047203	10.7	−0.08941	14.8	0.05326
2.6	0.19827	6.7	0.060425	10.8	−0.09083	14.9	0.048516
2.7	0.158289	6.8	0.072664	10.9	−0.09132	15	0.043353
2.8	0.119639	6.9	0.083832	11	−0.09091	15.1	0.037828
2.9	0.0825	7	0.093855	11.1	−0.0896	15.2	0.032
3	0.04704	7.1	0.102672	11.2	−0.08743	15.3	0.025931
3.1	0.013413	7.2	0.110232	11.3	−0.08443	15.4	0.019683
3.2	−0.01824	7.3	0.116498	11.4	−0.08064	15.5	0.01332
3.3	−0.0478	7.4	0.121447	11.5	−0.07613	15.6	0.006907
3.4	−0.07516	7.5	0.125067	11.6	−0.07093	15.7	0.000507
3.5	−0.10022	7.6	0.127358	11.7	−0.06513	15.8	−0.00582
3.6	−0.12292	7.7	0.128334	11.8	−0.05877	15.9	−0.012
3.7	−0.1432	7.8	0.128018	11.9	−0.05194	16	−0.01799
3.8	−0.16102	7.9	0.126448	12	−0.04471		
3.9	−0.17635	8	0.12367	12.1	−0.03716		
4	−0.1892	8.1	0.119739	12.2	−0.02936		

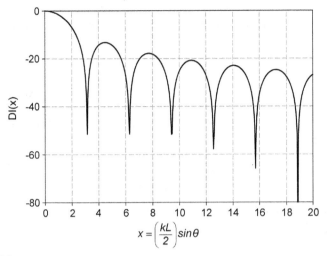

FIGURE 10.34

Directivity index in decibel for a continuous line array as a function of $x = (kL/2)\sin\theta$.

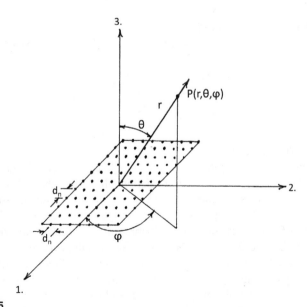

FIGURE 10.35

A planar array consisting of $N \times M$ elements (point sources) with element distances d_n and d_m. The field point $P(r,\theta,\varphi)$ represents the point for acoustic intensity measurement in the field produced by the array.

Based on the rectangular planar array directivity index DI (unit: dB) the directivity index for the rectangular piston source for $a \gg \lambda$ and $b \gg \lambda$ can now be written as:

$$DI \sim 10 \, log \left(\frac{4\pi^2 ab}{\lambda^2} \right) \tag{10.79}$$

While this expression and expressions derived from it are approximate, they are useful for practical sonar design.

For a *quadratic planar piston source* with side length a ($a \gg \lambda$), the pressure directivity function $DF(\theta)$ can be written as:

$$DF(\theta) = \frac{sin^2\{(ka/4)sin\theta\}}{\{(ka/4)sin\theta\}^2} \tag{10.80}$$

and the directivity index (unit: dB) is given by:

$$DI = 20 \, log \left(\frac{2a}{\lambda} \right) \tag{10.81}$$

For a *quadratic planar array* with $N \times N$ elements $(N \gg 1)$, all spaced $\lambda/2$, the directivity index (unit: dB) is given by:

$$DI = 20 \, log \, N \tag{10.82}$$

Three-dimensional *spherical* or *cylindrical arrays* are frequently used in sonar systems mounted in sonar domes for forward looking, anticollision, etc. on submarines and surface vessels. Cylindrical sonar systems frequently form a 120 degrees arc of a circular cylinder. For a cylinder radius R, the arc length s will be $s = 2\pi R/3$. If each row of elements on the 120 degrees arc has N elements and there are M rows, the directivity index (unit: dB) when the baffle formed by the sonar housing or the ship's hull is taken into consideration, according to Waite [52], is:

$$DI = 10 \, log \, NM + 3 \tag{10.83}$$

10.2.5.2.2 Array Qualities

Array directivity has a substantial influence on spatially received and transmitted signals. The array acts as a spatial information filter through its beam patterns. The spatial selectivity is obtained by using *adaptive* or fixed transmit/receive beam patterns. The beam patterns can be changed and "tailored" for specific use by the *beam-forming* process. Beam forming may include *array steering* and/or *array shading*. The general aim of beam forming is to improve the signal-to-noise ratio of echoes arriving from various directions in an omnidirectional noise field.

Beam forming exploits signal interference to change array directivity. In signal transmission, the beam-forming process carried out by the *beam former* controls the phase and relative amplitude of the signal at each array transducer element to establish a pattern of constructive and destructive interference in the acoustic wave front. For signal reception, information from the different array transducer elements is combined in such a way that a preferred radiation pattern is received.

The main and side lobes of an array can be *steered* by introducing *phase or time delays* in series with the array elements. The delays can be introduced mechanically or electronically. The steering of the array beams is performed by the beam former. The steering technology can be used for a transmitting or receiving array. *Preformed beams* formed by electronic beam steering are frequently used in receiving arrays. In passive and active sonar system reception the beam-forming technique involves the combination of delayed signals from each array receiving element at slightly different times, where the element closest to the target will be added to the combined signal after the longest delay. In that way all received signals from all elements reach the output from the array system at exactly the same time producing a much louder signal. The simple beam former only using time or phase delay is frequently called a "delay-and-sum" beam former.

The steering procedure introduces changes in the main and side lobe structure and in the array beam width and produces a beam-steering effect that makes it possible to direct a narrow main lobe at any desired angle relative to the array's line of extension. If all N elements in the vertical line array in Fig. 10.29 have a delay of magnitude $n\tau$, where n is the element number and τ is a constant time delay, the array beam will be turned to θ_t, *the steering angle*, given by:

$$\theta_t = \sin^{-1}\left(\frac{c\tau}{d}\right) \tag{10.84}$$

where c is the sound speed. The line array beam pattern is then be given by the *pressure directivity function DF(θ,θ_t):*

$$DF(\theta, \theta_t) = \frac{\sin\left(\frac{Nkd}{2}(\sin\theta - \sin\theta_t)\right)}{N\sin\left(\frac{kd}{2}(\sin\theta - \sin\theta_t)\right)} \tag{10.85}$$

For $\theta_t = 0$, the pressure directivity function $DF(\theta)$ in Eq. (10.85) reduces to Eq. (10.72). $\theta = \theta_t = 0$ in the direction vertical to the array length is the broadside angle of incidence. The change in the main lobe's -3 dB beam width for changes in θ_t is illustrated in Fig. 10.36 for $N = 10$ elements in a discrete element line array with a $d = \lambda/2$ element separation. The beam pattern given by $10 \log (DF(\theta,\theta_t)^2)$ (unit: dB) broadens for increasing θ_t values. The broadest beam occurs for $\theta_t = \pm 90$ degrees, the end-fire steering angle.

Beam steering using a *time delay* in a vertical, discrete element line array for signal reception, as shown in Fig. 10.29, with individual measurement channels, requires an analogue-to-digital (A/D) converter in series with each channel receiving element followed by a time-delay line. The signals from all delay lines are finally added coherently. The time delay is used to compensate for arrival delays of an obliquely incident acoustic wave at the individual array elements. *Beam forming* by using time delays is independent of frequency, and therefore useful for broadband-receiving systems such as passive sonars.

If the plane acoustic wave front in Fig. 10.29 reaches the array element number 1 at the time $t = 0$, it will reach the nth array element at the time $t_n = \{(n-1)d/c\}\sin\theta$, where c is the sound velocity and θ is the incidence angle. If the time delay produced

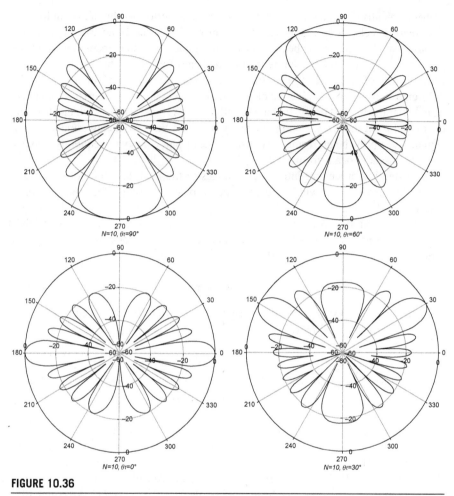

FIGURE 10.36

The beam width pattern (unit: dB) for $N = 10$, $d = \lambda/2$ and steering angles $\theta_t = 0$, 30, 60, and 90 degrees.

by the delay lines in the reception channels of the individual array elements is τ_n and $\tau_n = t_n$, then the signals add in phase for all array channels, provided the signal in each channel is the same except for the individual time delay. The array directional response will now be the same as if the plane acoustic wave is incident from the broadside direction. The directional response of the array from other directions, for instance a direction forming an angle θ_i with the normal to the line array, can be calculated by introducing $\tau_n = \{(n-1)d/c\}\sin\theta_i$. Then the directional response of the array will be a function of θ and θ_i through the time delay $t_n - \tau_n = \{(n-1)d/c\}\{\sin\theta - \sin\theta_i\}$.

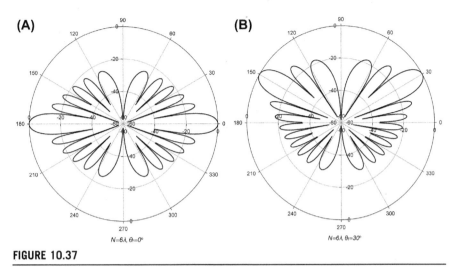

(A)

(B)

FIGURE 10.37

Beam patterns for a 6λ long continuous line array for (A) $\theta_t = 0$ and (B) 30 degrees.

For a *continuous line array* with steering angle θ_t the *pressure directivity function* $DF(\theta,\theta_t)$ is given by:

$$DF(\theta, \theta_t) = \frac{\sin\{(kL/2)(\sin\theta - \sin\theta_t)\}}{(kL/2)(\sin\theta - \sin\theta_t)} \qquad (10.86)$$

L is the line array length, $\theta = 0$ degree is the broadside direction and $\theta = \pm 90$ degrees are the end-fire directions. For the steering angle $\theta_t = 0$ and 30 degrees, Fig. 10.37 shows $10*log(DF(\theta,\theta_t)^2)$ for $L = 6\lambda$. Fig. 10.37 shows that the main beam starts to broaden at increased steering angles, while the first side lobe amplitudes are predominantly unchanged for increasing θ_t.

The time delay for steering an array is applicable for line and planar arrays. Arrays of almost any geometrical shape can be steered to receive signals of any curvature from any direction as long as diffraction effects can be avoided.

Beam steering can also be performed by *phase shifts* in the individual array element's reception channel. Phase shifts are particularly useful when one frequency or a very narrow frequency band is handled. In narrow-band systems the time delay is equivalent to a phase shift and with the small phase shifts from element to element in the array, the array is called a *phased array*. For the steering angle θ_t, the phase difference Φ between each consecutive array element in a discrete element line array with element separation d will be $\Phi = kd\sin(\theta_t)$, where k is the wave number. For N elements in the line array, its pressure directivity function will be the same as in Eq. (10.85). For $N = 15$ and for $d = \lambda/2$, Fig. 10.38 shows the beam pattern's main lobes for the steering angles $\theta_t = 0, 30, 60,$ and 90 degrees. The figure shows

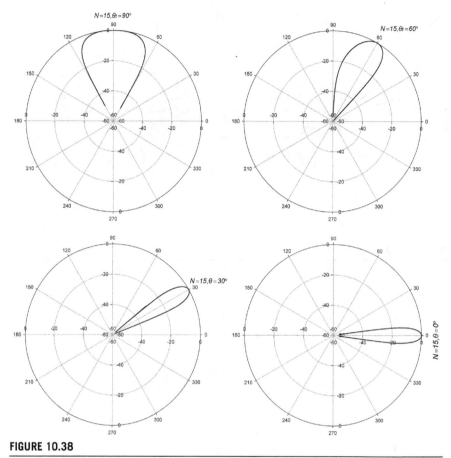

FIGURE 10.38

Main lobes of the beam patterns for a discrete element line array with $N = 15$ and $d = \lambda/2$ for steering angles $\theta_t = 0$, 30, 60, and 90 degrees.

the increase in the -3 dB beam width due to beam broadening at increased steering angles.

In transmit and in particular in receive beam formers the signal from each array element may be amplified by individual "weights" in its measurement channel. By using different *weighting functions*, a desired array sensitivity pattern can be achieved. It is possible to control the array's main lobe, also called the beam, side lobe levels and nulls between the lobes. The use of weighting functions for *array shading* makes it possible to listen for signals in preselected directions. This is very useful when directional noise or attempts to jam the array influence reception. The usual purpose of array shading is to reduce side lobe levels to improve range and bearing resolution.

While *conventional* beam-forming technique exploits a fixed set of time or phase delays and weighting functions to combine the signals received by the individual array elements, the *adaptive* beam-forming technique is more flexible since it combines the information about the element location in space and the wave direction with information about the characteristics of the signals received by the array. Adaptive arrays may reject unwanted signals in the time or frequency domain. They are normally able to automatically adapt the array response to changing receive situations. The fast response change makes the adaptive beam forming computationally demanding. However, today's array systems have data-rate capabilities that permit real-time data processing. The computer capabilities also lead to a shift toward the "wet end" of beam-forming array hardware.

The side lobe level reduction by *array shading* is accompanied by broadening the main lobe and reducing the transmitted power level. The magnitude of the beam broadening depends on the shading (weighting) function used. A number of *shading functions* are discussed under Window Functions in Chapter 11 on Signal Processing. In most applications, array shading is used to obtain a maximum response for the center elements and reduced response for elements toward the ends of the array. Such an array is said to be *tapered* from a high transmitting or receiving sensitivity at the center to low sensitivity at the array ends. A few shading functions are discussed here including rectangular, Gaussian, Hanning, Hamming, and Dolph–Chebyshev shading functions.

In the *rectangular shading function* all array elements have the same "weight" and are multiplied by the same factor. The rectangular shading function does not result in any shading, i.e., the shading function $V(n) = 1$, where $V(n)$ is the weight given the nth array element. The pressure directivity function $DF(\theta)$ is the same as in Eq. (10.72).

The *Gaussian shading function* has a Gauss-shaped form (named after C.F. Gauss: 1777–1855), where the shading function is given by: $V(n) = exp\{-2(\kappa n/N)^2\}$, where the coefficient κ determines the width of the main lobe and N is the number of array elements. For $\kappa \geq 2.5$, the pressure directivity function, $DF(\theta)$, according to Harris [53] is:

$$DF(\theta) = \frac{1}{2}\frac{\sqrt{2\pi}}{\kappa}e^{-1/2(\psi/\kappa)^2}D(\psi) \qquad (10.87)$$

where $\psi = kd\sin\theta$ and $D(\psi) = \dfrac{e^{i\psi/2}\sin\left(\dfrac{N\psi}{2}\right)}{\sin\left(\dfrac{\psi}{2}\right)}$.

Increasing κ values cause reductions in the side lobe level and an increase in the man lobe width compared to the main lobe width without the shading function. Table 10.5 based on Harris [53] shows for various κ-values the reduction in side lobe levels compared with the main lobe, the side lobe level fall off and the beam broadening compared to an unshaded array. The highest theoretical reduction in side lobe levels range from 42 dB for $\kappa = 2.5$ to 69 dB for $\kappa = 3.5$. Due to electroacoustical qualities of the array and array motions, practical side lobe reduction will normally be less than the theoretically determined reduction.

Table 10.5 Impact of Shading Type on Array Main Beam and Side Lobes

Shading Type	Factor	Side Lobe Level Below Main Lobe Level (dB)	Side Lobe Level Fall Off (dB/Octave)	−3 dB Beam Broadening Factor
Gaussian	$\kappa = 2.5$	−42	−6	1.49
	$\kappa = 3.0$	−55	−6	1.74
	$\kappa = 3.5$	−69	−6	
$\cos^\kappa$	$\kappa = 1.0$	−23	−12	1.35
Hanning	$\kappa = 2.0$	−32	−18	1.62
	$\kappa = 3.0$	−39	−24	
Hamming		−43	−6	1.46
Dolph−	$\kappa = 1.5$	−30	0	1.14
Chebyshev	$\kappa = 2.0$	−40	0	1.31
	$\kappa = 2.5$	−50	0	1.49
	$\kappa = 3.0$	−60	0	
	$\kappa = 3.5$	−70	0	

The *Hanning shading function* (named after J.E. von Hann: 1839−1921) is the $\cos^\kappa$-*function* for $\kappa = 2$. The $\cos^\kappa$-*function* has the shading function given by $V(n) = \cos^\kappa(\pi n/N)$, where the κ values are, 1, 2, 3, or 4. For $\kappa = 2$ according to Harris [53] the Hanning shading function is:

$$V(n) = 0.5 + 0.5\cos\left(\frac{\pi n}{N}\right) \tag{10.88}$$

which gives the pressure directivity function $DF(\theta)$:

$$DF(\theta) = 0.5D(\psi) + 0.25\left\{ D\left(\psi - \frac{2\pi}{N}\right) + D\left(\psi + \frac{2\pi}{N}\right)\right\} \tag{10.89}$$

where $D(\psi) = exp\left(i\frac{\psi}{2}\right)\dfrac{\sin\left(\dfrac{N\psi}{2}\right)}{\sin\left(\dfrac{\psi}{2}\right)}$ and $\psi = kd\sin\theta$.

The *Hamming shading function* (proposed by R.W. Hamming: 1915−1998) is a modified Hanning shading function which is given by:

$$V(n) = \alpha + (1 - \alpha)\cos\left(\frac{2\pi n}{N}\right) \tag{10.90}$$

where first side lobe cancellation occurs for $\alpha = 25/46 \sim 0.54$, as discussed by Harris [53]. By using $\alpha = 0.54$, the pressure directivity function can be written as:

$$DF(\theta) = 0.54D(\psi) + 0.23\left\{ D\left(\psi - \frac{2\pi}{N}\right) + D\left(\psi + \frac{2\pi}{N}\right)\right\} \tag{10.91}$$

where $D(\psi)$ and ψ are the same as in Eq. (10.89).

The *Dolph—Chebyshev shading function*, as discussed in Chapter 11, is one of the most frequently used shading function in underwater acoustics. This shading function can provide a narrow main lobe with the lowest possible side lobes. For various factors κ, the side lobe levels below the main lobe level are given in Table 10.5, where a factor $\kappa = 2.5$ reduces the side lobe to -50 dB below the main lobe level, while $\kappa = 4$ brings the side lobe level down to -80 dB below the main lobe level. These reductions in side lobe levels are theoretical values, and in practice the side lobe reductions are smaller. However, Dolph—Chebyshev shading produces a nearly sinusoidal variation in the side lobe structure with the same level for all side lobes.

Because the amplitude of an echo from a target depends on the target location in the transmitting and receiving sonar beams, it is frequently very useful to exploit the *split-beam* principle. Ehrenberg [54] suggested the use of a *dual-beam sonar*, where the transducer was able to produce a wide beam and a narrow beam. This was done by using two circular transducers, one having a small aperture and centered in a circular hole in a circular transducer with a larger aperture. The echo from a target insonified by the wider beam transmitted by the small aperture transducer can be received by the same small aperture transducer, as well as by the narrow beam transducer with the larger aperture. The two echoes were then compared in the signal processing equipment. Also, a split-beam sonar exploiting three or four transducer elements in the same plane is being used.

H. Bodholt [55] describes a three-element split-beam sonar. Fig. 10.39A shows a planar split-beam transducer, where the origin opening angle is 120 degrees for each of the three transducer sections. Each section may consist of an array of single transducer elements, such as Tonpilz transducers, where a taper function is introduced to reduce side lobe levels. For a split-beam sonar it is important to avoid having a large target in a side lobe mistakenly assumed to be a small target in the main lobe. The taper function gives the elements near origin the highest amplitudes, while elements on the transducer edge are given the lowest amplitudes. The section's acoustical centers are marked C_1, C_2, and C_3, with a the distance between the origin and each center. The echo-sounder screen presents a horizontal cross-section in the acoustic field x—y plane produced by the transducer with fish echoes marked as dots at their positions in the beam. Fig. 10.39B shows the coordinate system with the split-beam transducer in the x—y plane and the z-direction as the vertical depth direction. A unit vector s in the target direction has the cosine directions $s = (\cos\nu_x, \cos\nu_y, \cos\nu_z)$. Signals received by the three transducer parts are normally different and the three receiver channels produce the received signals as three digital and complex numbers G_1, G_2, and G_3 from which amplitudes and phase angles can be found. The angles to the target are found from the angular phase difference φ between the three transducer centers:

$$\cos\nu_x = \frac{\varphi_{31} - \varphi_{32}}{\sqrt{3}ka} \quad \text{and} \quad \cos\nu_y = \frac{\varphi_{31} + \varphi_{32}}{\sqrt{3}ka} \tag{10.92}$$

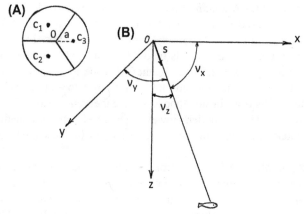

FIGURE 10.39

The three-section split-beam transducer (A) and its coordinate system with the unit vector *s* in the target direction (B).

Drawn by Leif Bjørnø.

where the angular phase differences φ_{31} and φ_{32} can be found from the complex G-values and their complex conjugated G^* through the expressions:

$$\varphi_{31} = arg\left(G_3 \cdot G_1^*\right) + n2\pi \;\; and \;\; \varphi_{32} = arg\left(G_3 \cdot G_2^*\right) + n2\pi \qquad (10.93)$$

where *arg* indicates the argument of the complex quantity and is a phase angle between $-\pi$ and $+\pi$. The phase angle ambiguity $n2\pi$ will normally vanish for targets in the transducer main lobe and within the -10 dB level angle.

10.2.5.2.3 Towed Arrays

During the "cold war," when the Soviets produced quieter submarines the need for passive listening devices operating at low frequencies and with enhanced directivity to permit a more accurate determination of target range, bearing and speed became obvious. At the same time, receiving arrays with higher sensitivity and dynamic range (>120 dB) were needed to support oil company seismic prospecting for new oil resources at greater depths below the seabed when exposed to noisy environments. These requirements made it necessary to look for moveable arrays that satisfied rather strict demands on the received acoustic signal quality. The *towed array* met most of these demands.

Self-noise has been a problem for towed arrays. In the best case the lowest noise limit was set by ambient sea noise, as discussed in Chapter 6. However, in most cases noise from the towing vessel produced by the wave motion against its hull plates, main propulsion plant and auxiliary engine vibrations, and propeller cavitation, limit the towed array's detection range. It is therefore desirable to have the towed array remote from the towing ship and the array should move with minimum drag and maximum stability. Moreover, towed arrays may combat the influence of the thermocline, since the arrays can be towed above or below the thermocline.

A typical towed array may have several hundreds to nearly 2000 hydrophones. The hydrophones may be *piezoceramic* cylinders that permit the *tow wire* to follow their center line through all cylindrical elements or the hydrophones may be spherical- or disk-shaped piezoceramics. Even miniature flextensional transducer types are used as hydrophones. In some special cases the hydrophones are fiber-optic cables wrapped tightly around malleable plastic spools called mandrels. The optical phase difference between signals from a sensor mandrel exposed to the acoustic signals and a reference mandrel represents the acoustic signal energy received. In any case, the hydrophones are small compared to the measured acoustic wavelength and the hydrophones operate below their lowest mechanical resonance.

The *length* of the array may be up to 1000 wavelengths where λ (unit: m) is based on the center frequency for the array. The normal hydrophone separation is $\lambda/2$ which produces an array of about 2000 hydrophones. Each hydrophone is connected to an electronic circuit which may have a preamplifier with sampling and digitizing circuitry. The cables transmit phase and amplitude information from each array hydrophone to the beam-former electronics on the tow vessel. Since one wire for each hydrophone is impractical, towed arrays may include coax cables and use multiplexing. These systems typically run uplink data rates up to and above 12 Mbit/s due to long coax cables' bandwidth limitations. The use of fiber-optic cable connections between the towed arrays and tow vessel has improved the data rate and reduced the need for thick copper wire cable bundles. The hydrophones, communication cables, and tow wires are enclosed in a liquid-filled, flexible plastic or rubber *hose*, with an outside diameter ranging from about 25 mm for *thin-line* towed arrays to above 100 mm for *fat-line* towed arrays. An oil-type liquid is normally used to fill the hose to achieve neutral buoyancy, reduce low-wave number noise and help dissipate internal heat. Solids have been used in towed arrays for seismic studies to avoid oil leakage and pollution. An example is the Australian, "Kariwara," solid-filled towed sonar array, developed by the Defense Science and Technology Organisation (DSTO).

Pressure fluctuations in turbulent boundary layer flow along the towed array surface produce noise, either dipole noise, arising from viscous drag at the surface, or quadrupole noise, arising from fluctuating Reynold's stresses, and, according to Knight [56], this noise may limit the receiving performance of the array. The scale of the turbulence is influenced by the array's diameter and tow speed. Reductions in the array diameter, as found in thin-line arrays, have shown substantial flow noise reductions and improvements in the reception capability. Problems related to storage, deployment, and retrieval of towed arrays are also strongly reduced by reducing the array diameter. The storage of the towed array, when not deployed, can be in a container as shown in Fig. 10.40, where the teardrop-shaped container mounted on the top of a vertical fin on an *Akula Class* Russian attack submarine stores a towed array. The hydrophones may also pick up internal noise produced by waves propagating inside the array from one hydrophone to another, caused by flow in the liquid in the hose or by flexural waves propagating in the hose wall, or by cable strumming

FIGURE 10.40

Teardrop-shaped towed array container mounted on the fin top on a Russian *Akula Class* nuclear-powered attack submarine. The towed array is deployed and retrieved through a hole in the container's tapered end.

Courtesy US Department of Defense.

due to the tow cable being moved through the water. This noise degrades the array reception quality in particular below some hundreds of hertz.

The towed array's *length* is determined by its operational frequency and by the desired directivity index *DI*. The length of the armored *tow cable* connecting the array with the tow vessel is determined by the desired distance between the tow vessel and the towed array, the speed of the tow vessel—more than 10 knots is frequently desired—and the towed array's depth of operation. This tow cable can be several hundred meters long. The front part of the towed array is normally equipped with a *vibration isolation module*, the *VIM*, to reduce the vibration transmitted along the tow cable and to stabilize the tow force that may vary substantially due to sea state, ship motion, etc. To stabilize the movement of the sensing part of the array which is comprised of the hydrophones, and to reduce the influence of noise and turbulence generated by the tow-ship propulsion, a *nonacoustic module (NAM)* is frequently inserted between the *VIM* and the hydrophone section. The hose comprising the hydrophone array is normally terminated by a *tail part* to provide some drag to reduce the meandering of the towed array and to produce a linear motion of the array, as shown in Fig. 10.41.

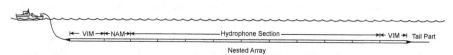

FIGURE 10.41

Schematic towed array showing its characteristic parts.

This linear motion of the array is important for obtaining accurate bearing and range measurements. The position of each hydrophone, which changes continuously over time, must be fairly accurately known. To get information about the accurate positions of the towed array, a number of *heading sensors* and *depth sensors* are used along the hydrophone array, and their signals are utilized to correct for the array's curvature. The sensors used in the thin-line towed arrays, the TB-23 and TB-29 from L-3 Oceans Systems, have resolutions of 0.3 degrees for the heading sensor and 0.5 m for the depth sensor. Temperature sensors are used in towed arrays to determine the sound speed at the operation depth. Also, the time-of-flight of sound signals between transducers clipped on the towed arrays and the individual hydrophones in the array is used to achieve a more accurate knowledge about the spatial position of the individual hydrophones.

Towed arrays are used in both *active* and *passive* systems. In the active system the sound source may be a sonar transducer, for instance of the flextensional type, or it may be an air gun giving a sudden release into the water of compressed air with pressures up to and above 10^7 Pa or based on a propane—air combustion process. The passive receiving arrays, like tactical towed array sonar systems, are normally longer than the active array, due to the desired higher sensitivity, longer detection range, improved low-frequency detection capability, and higher *DI* of the passive arrays. The left/right ambiguity problem in relation to object detection and ranging in the water column is most frequently solved by using two or more arrays towed parallel in the horizontal plane. As discussed by Lemon [57] and Lasky et al. [58], for oil exploration up to 16 horizontal arrays and several sound sources have been towed after a surveillance ship. On a submarine the towed array is able to "see" in a full 360 degrees direction, while the submarines' other sonar systems, the forward looking sonar and the hull-mounted flank array, normally are able to cover horizontal arcs of 260—280 and 150—170 degrees, respectively.

A special towed array developed for the US Navy for ASW is the Surveillance Towed Array Sensor System Low Frequency Active (the SURTASS LFA). This system has active and passive components. The active component, the LFA, is an array of acoustic transmitters normally operating at frequencies between 100 and 500 Hz, each has a source level around 215 dB rel 1 μPa at 1 m and is suspended on a cable beneath a ship. With the array gain the effective source level depending on the number of transducers in the array will become 230—240 dB rel 1 μPa at 1 m. The passive component, the SURTASS, consists of a long array of hydrophones for passive listening. When the target is too quiet to be detected by SURTASS the LFA system can be activated.

Passive towed arrays may operate over broad frequency bands comprising several octaves. While the spacing between the hydrophones normally are kept at $\lambda/2$, groups of hydrophones with different spacing determined by the center frequency of the octave bands, are distributed along the towed array with the highest center frequency spacing at the center of the array and decreasing octave band center frequency spacing symmetrically positioned around the array center. This *nesting* of the higher octave bands in the lower octave bands has turned out to be very useful in producing broadband arrays.

10.3 SINGLE-BEAM ECHO SOUNDERS

Early depth measurements were performed by using a line and a lead weight. However, since about 1920 depth sounding has been performed by using acoustic waves, where a signal from a projector is reflected from the seafloor and received by a hydrophone system, which is frequently the same system as the projector. When the sound velocity in the water is known the acoustic signal's time of flight from the projector to the seafloor and back measures the distance. The standard display known as an *echogram* shows depth on the vertical axis and distance traveled over the seafloor on the horizontal axis. SBESs are available from several manufacturers at various price levels and with a broad variety of capabilities for use by national hydrographic offices, cruise liners, cargo ships, fishing trawlers, and even small pleasure boats and sporting crafts. Teledyne-Odom Inc. (USA), Kongsberg (Norway), Atlas Hydrographic (Germany), L-3 ELAC Nautic (Germany), and Teledyne-Reson A/S (Denmark) are some of the key SBES manufacturers.

An *SBES* consists of a transmitting and a receiving part. The signal a sinusoidal pulse, controlled by a master clock, is amplified through a power amplifier. The pulse power level is determined by water depth and carrier frequency. Frequently, the pulse is fed to the projector terminals through a transformer to step-up the voltage. The projector's (or transducer's) acoustic pulse travels to the seafloor, reflects from the seafloor and returns to the projector (or transducer) where the acoustic signal is converted into an electrical signal, which is processed and displayed. The input to the processing system is *time gated* to keep the transmitted signal from being received directly and to assure the projector transducer has finished ringing down before the received signal arrives. The received signal is passed through a band-pass filter to eliminate out-of-band ambient noise and the ship's self-noise. Then the signal is passed through a *time-variable-gain* (TVG) circuit before the signal envelope is detected and the signal is displayed on a monitor, where different color sequences can be used to represent the amplitudes of the received echo signals. Some SBESs also permit *raw data* to be output for processing and display via a personal computer (PC). Fig. 10.42 shows the footprint on the seafloor formed by the main lobe of the narrow beam of a downward-looking, hull-mounted SBES on a research vessel. The −3 dB beam width of the main lobe is normally taken as the beam limit, which forms the track of footprints on the seafloor in the direction of the movement of the research vessel.

The high amplitude at the front of the echo time history from the seafloor received by an SBES represents the reflection from the seafloor surface by the normal, or near normal, sonar signal. The direct reflection of the normally incident signal produces the strongest echo, while the backscatter from the seafloor surface, the first characteristic echo return, has a lower echo amplitude and represents the seafloor surface roughness features and some backscatter from the seafloor near surface layers. This part of the echo is normally correlated with the seafloor roughness

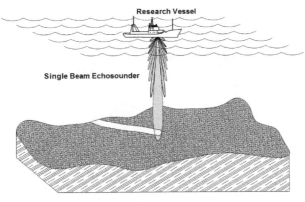

FIGURE 10.42

The downward-looking, hull-mounted single-beam echo sounder and its narrow
symmetrical beam with its footprint on the seafloor along the track of the research vessel.

topography, seafloor materials, and attenuation of the acoustic signals in the near
seafloor surface parts. A rough seafloor surface or the occurrence of larger grain
sizes in the seafloor materials produces a more complex backscattering envelope
with a longer duration, while a flat seafloor produces a sharper, higher amplitude
and shorter duration echo return. The later arrival of the second characteristic
echo part may partly be produced by scattering from substrate layers with different
acoustical impedance in the seafloor and backscattering of seafloor reflected signals
from the sea surface. The second echo is frequently connected with the hardness features of the seafloor, where rock will produce a higher amplitude echo than sand and
mud. Also, the angular response of the seafloor with the highest echo amplitude from
the point, *nadir*, directly below the SBES and lower amplitudes from other parts of
the seafloor surface insonified at higher angles of incidence may contribute to seafloor material classification.

The sinusoidal signal's carrier frequency depends on the echo sounder application. The *frequency* can range from about 10 kHz for deep-water use to above
600 kHz for shallow-water use. While modern SBESs can be tuned by 1 kHz increments to any projector/hydrophone operating band, practical operation is usually
limited to a number of frequency bands. For example, in Teledyne-Reson's Navi-
Sound echo sounders the frequency bands are 28–36 and 190–225 kHz. Some
echo sounders also offer an additional two frequency bands, i.e., 42–60 and
92–110 kHz, for a total of four channels. Echo sounder *resolution* is determined
by pulse length and carrier frequency. The choice of the pulse length, i.e., number
of cycles, depends on the expected bottom type. For a hard bottom, such as rock,
a short pulse length is appropriate. For a soft bottom, such as mud, a longer pulse
length is selected in order to transmit enough acoustic power to control the strength
of the echo returning from the seafloor. Transmitted power control is an important
feature. Echo sounders can normally be tuned in steps from few to several hundred

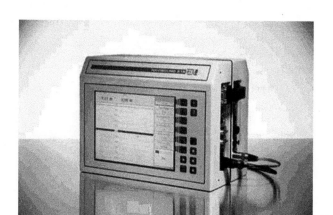

FIGURE 10.43

NaviSound 500 dual channel echo sounder, produced by Teledyne-Reson.

Courtesy of Teledyne-Reson A/S.

watts. Normally, low power levels are used in shallow water and higher power levels are used in deep water.

The NaviSound 500 echo sounder, shown in Fig. 10.43, is used for advanced hydrographic surveys in shallow-water harbors, waterways, and off-shore areas and for dredging operations. This dual channel echo sounder is often used with the TC 2003 transducer shown in Fig. 10.44, which is a high-power, narrow conical beam transducer that operates from 190 to 210 kHz with a 1500 W maximum power output and 2.6 degrees conical beam width. A dual frequency transducer, such as TC 2135 shown in Fig. 10.45, can also to be used with the NaviSound 500. The 13—18 and

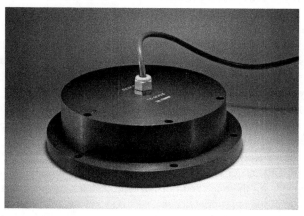

FIGURE 10.44

Teledyne-Reson TC 2003 transducer.

Courtesy of Teledyne-Reson A/S.

FIGURE 10.45

Teledyne-Reson TC 2135 dual frequency transducer.

Courtesy of Teledyne-Reson A/S.

190–200 kHz frequency bands, with 23 degrees conical beam width at 15 kHz and 9 degrees at 200 kHz make the TC 2135 transducer ideal for sediment studies and for hydrography.

Since the echo sounders measure the time it takes for an acoustic signal to travel to the seafloor and back, the depth, H (unit: m), is given by:

$$H = \frac{\tau c}{2} \qquad (10.94)$$

where τ (unit: s) is the signal's round trip travel time from the projector to the seafloor and back, and c (unit: m/s) is the average sound velocity in the water column. For measurement accuracy the sound velocity must be known as accurately as possible. As mentioned in Chapter 1, the sound velocity varies with temperature, salinity, and pressure, with temperature having the greatest effect. In limited areas, such as harbors, temperature, salinity, and pressure do not vary significantly, unless freshwater from a river passes through the harbor. However, in deep water and particularly in estuaries, where a river brings lighter freshwater into the heavier salty seawater, the sound velocity variation is more pronounced. In an estuary, water layers of different density and salinity are formed, and these layers mix over the time due to tidal motion, wind- and wave-generated turbulence, and the magnitude of the river flow volume, which frequently depends on the season. Frequently a build-in "bar check" calibration facility is used to obtain a reliable sound velocity value for the water below an echo sounder. The bar check is performed by lowering a bar to a known depth below the echo sounder projector. Based on the arrival time of the echo from the bar, the echo sounder calculates the sound velocity. The calculated sound velocity is then used to set the sound velocity in the echo sounder. For example, the sound velocity in the NaviSound 500 can vary between 1350 and

1600 m/s in 1 m/s increments. The bar check calibration is only valid in shallow water. In deeper water a *sound velocity profiler* (SVP) is used to obtain a correct average sound velocity.

When the echo sounder's depth range has been set, the echo sounder normally determines the number of soundings per second, with the highest number at short range. The depth range also determines the frequency band to be used.

On the receiver side of the echo sounder it is possible to use an *automatic gain control* (AGC) on the amplifier. The AGC ensures that the received echo is kept at an acceptable level by measuring the returned echo signal strength and adapting the amplification accordingly. If the depth "drops out" due to phase out or air bubbles from the wake of a ship, a "zero filter" can be used online to erase singular values. Also, a first and second mode function is normally valuable in the echo sounder receiver. The first mode will, in the search for a valid seafloor, select the first possible echo, which is useful when the seafloor condition generates multiple traces. The second mode is useful when a soft layer on the seafloor covers a harder sublayer (a second bottom) and the primary signal only shows the interface between the soft layer and the water column.

The ship's speed over the bottom must be taken into consideration when the screen speed and scale is selected, since the ship's speed effectively determines the echo sounder's coverage, especially at depths over 100 m. The ship will have moved a certain distance during the time the sonar signal passes from the projector to the seafloor and back. With a sloping seafloor the effect of the ship's speed is more pronounced since this changes the rate of interaction between the sonar signal and the seafloor. As shown in Fig. 10.46, a record offset is used to set the depth represented by the top line of the recorder screen, and it determines, together with the recording range, the section of the depth profile, which is shown on the recorder screen.

To filter out false depth values caused by turbulence or reflections from the ship's hull and to force the echo sounder to track the true seafloor in areas where the seafloor is covered by soft mud or sea grass the echo sounder may use an *initial lockout function*. This facility will reject any depth less than the initial lockout depth, which is measured from the transducer offset selected by the sonar operator, as shown in Fig. 10.46. The transducer offset is added to the measured depth to give the true depth. The "initial lockout" function should be used with care and only at places where seafloor conditions are known to avoid masking shallow-water readings and reduce the quality of the depth readings. To improve data quality most echo sounders have a *time-gate facility*, which tracks the seafloor with a quality assurance routine. The window width of the "time gate" is determined by how long before the echo sounder expects a return signal from the seafloor based on the last measured depth. Only echo signals falling within the window of the "time gate" are accepted as good quality data. Echo signals outside the window are automatically adjusted and considered to be data of poorer quality. The window width is normally based on the seafloor profile. For gentle undulated seafloor profiles, the width is normally set to a very small value, while a "mountainous" seafloor requires a larger width. The

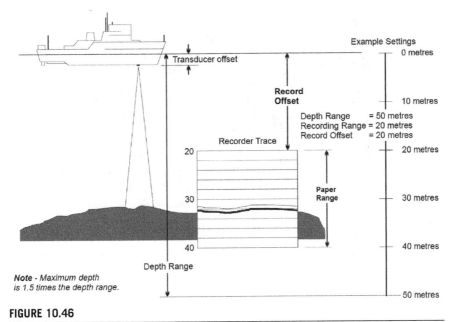

FIGURE 10.46

Examples on screen ranges, including depth range 50 m, recording range 20 m and record offset 20 m.

Courtesy of Teledyne-Reson A/S.

"time gate" is normally set as a percentage of the selected depth range, for example, if the depth is 200 m and the "time-gate" setting is 4%, the "time-gate" window width is 16 m, which is 8 m to either side of the last measured depth value.

The ship's motion and the position of the sonar projector relative to the environment have an influence on the data quality. Tide and heave—vertical movement of the ship—introduce errors in the depth measurement data, which normally must be removed by post processing. If this is not done, the sea surface waves that cause the ship's heave are shown as waveforms on the seafloor. Therefore, most SBESs have an external or internal heave compensator. Also, a ship's rolling and pitching influences depth measurements, since the projector will become angled toward the seafloor, which is a problem for projectors with narrow beam widths. Compensation for rolling and pitching is not generally found on SBESs, while it is mandatory on multibeam sonars due to their narrow individual beams, as discussed in Section 10.4.

Longer range return signals normally have lower amplitude than return signals from nearby objects. The SBES receivers are normally equipped with a TVG function which will dampen the echo return so that nearby signals receive less amplification than signals from greater depths. The damping normally follows a logarithmic scale and is linked to the applied attenuation level in decibels.

FIGURE 10.47

Teledyne-Reson TC 2166 side-looking sonar transducer.

Courtesy of Teledyne-Reson A/S.

Some SBESs are equipped with a *side-looking sonar* (SLS) facility, which is different from the side-scan sonar (SSS) discussed in Section 10.5. All return signals lose amplitude when propagating through the water. For different propagation distances this loss is initially corrected using a TVG function. However, due to the projector beam width, when the acoustic pulse reaches the seafloor it starts to travel sideways and outward and produces return signals from positions to the side of the initial impact position. These return signals are weaker than the initial reflection, and in a way similar to the TVG function. The SLS function will dampen the stronger initial reflection more than the later reflections. An increasing amplification is then applied at the same rate as the signal is lost in the seafloor materials. The SLS facility permits the use of an SLS transducer such as the TC 2166 from Teledyne-Reson A/S, shown in Fig. 10.47. This 200 kHz, dual-sided transducer, has a 90 degrees separation angle between the transducer sides. The transducer has 1.1 and 47 degrees beam angles and a transmitted source level of 171 dB rel 1 µPa/V at 1 m. Therefore, the acoustic pressure amplitude produced by this transducer at 1 m source distance, for a 1 V electrical input amplitude, is 355 Pa. The SLS facility can be used for spot checking or visually scanning a survey area after the depth survey, and it is useful for seafloor control in shallow water, such as harbor basins. Fig. 10.48 is the sonar data from a graphical image of the seafloor structure and objects on the seafloor. Fig. 10.48 shows the wreck of the *SS Lake Illawarra* lying at 35 m below the Tasman Bridge in Hobart, Tasmania. High amplitude reflections are recorded as black, while low reflections are white.

To reduce the influence from rolling and pitching the projector position is normally as close as possible to the ship's center axis and between 1/3 to 1/2 of the ship's length, as measured from the bow. The projector should not be near noise sources on the ship and should avoid the influence of bubble clouds or bubbles trapped on the projector surface.

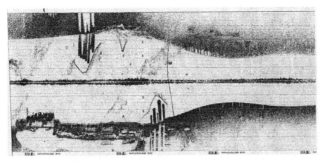

FIGURE 10.48

SLS image of SS *Lake Illawarra*.

Courtesy of Teledyne-Reson A/S.

Because an echo sounder normally accepts the first good return echo from the seafloor, this may lead to errors in the true depth, when the seafloor slopes, as shown in Fig. 10.49. Here the first return signal does not come from directly beneath the projector. Instead the return is from a point offset by the seafloor slope's steepness. Fig. 10.49 shows that the echo sounder's ability to accurately show the seafloor terrain depends on the *projector beam width*, and therefore on its *frequency*. Narrow beams, which normally occur at higher frequencies, will more accurately record the seafloor than broader beams obtained at lower frequencies. The resolution is affected by the pulse length, and if two objects are separated by a distance shorter than half the pulse length, they are shown as one object. When their separation is larger than half the pulse length they are shown as two objects.

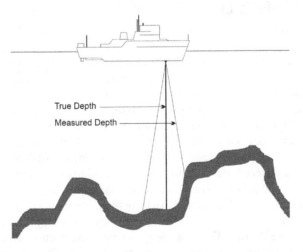

FIGURE 10.49

Influence of a rapidly changing seafloor terrain on the depth measurement.

Courtesy of Teledyne-Reson A/S.

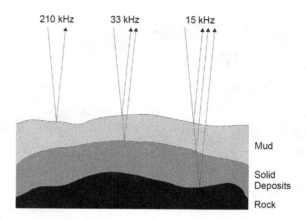

FIGURE 10.50

Influence of frequency on detection of various layers in a multilayered seafloor.

Courtesy of Teledyne-Reson A/S.

When the seafloor consists of several layers of materials, as shown in Fig. 10.50, the echo sounder will identify the various individual layers at low frequencies, while only the topmost layer will be identified at high frequencies. The top layer can sometimes be a very soft material such as mud or sea grass with a smooth and gradual transition from the water's acoustic impedance to the acoustic impedance of the next layer. These top layers are not always detected by an echo sounder.

10.4 MULTIBEAM ECHO SOUNDERS

MBESs can be considered a further development of the SBES. The MBES operates a fan of single beams with small individual widths and with a small angle separating each beam. The best multibeam systems today have 0.5–1 degree beam widths and simultaneously operate more than 800 beams. This makes it possible in hydrography to cover a broad swath along the ship's path with a theoretical maximum swath width of about 7.5 times the water depth with a 150 degree total fan angle. In deep water the swath width decreases from this maximum as a function of depth and seafloor type, for example, rock, sand, and mud. The swath width depends on the noise level and spectrum and the sea state and water column conditions. In deep water the total swath width may exceed tens of kilometers. With sound speeds around 1500 m/s, beams at the fan edge will have arrival times measured in several seconds. This sets a limit on the ship's cruise speed. MBESs have, however, successfully evolved hydrography from 2-D to 3-D. At 10–12 knots speeds surveys can be conducted faster and in more detail, which may lead to considerable cost savings. Although multibeam systems are more complex and expensive than single-beam or SSS systems, they are more cost-effective when their increased swath coverage and survey speed are taken into account. Fig. 10.51 shows an MBES's beam geometry. Part A

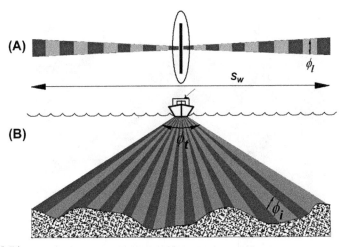

FIGURE 10.51

Schematic beam geometry for a multibeam echo sounder. (A) The beams seen from above with the along-track transmitted beam width φ_ℓ. (B) The fan of narrow received beams with a total across-track fan aperture φ_t and an individual beam width φ_i. The total swath width is S_w.

shows the beams as seen from above with the narrow, along-track transmitted beam width φ_ℓ. Part B shows the fan of narrow received beams with a total across-track fan aperture φ_t and an individual beam width φ_i. S_w denotes the total swath width. The main multibeam system's technical advantage is the ability to accurately detect the incidence angle of different simultaneously arriving echoes. Multiple beams also allow detection of multiple scattering and give improved resolution, for example, the ability to separate the backscattering from two closely spaced targets.

Depending on operating frequency, MBES *applications* can include:

- harbor and breakwater mapping,
- river surveys,
- pipeline and cable route surveys and inspection,
- pre- and post-dredging and condition surveys,
- hydrographic surveys,
- underwater inspection and object location,
- waterside security and MCMs,
- habitat mapping and biomass measurements,
- sediment transport studies and leak detection,
- search and discovery,
- coastal to full ocean depth mapping.

There are three main types of multibeam sonar systems: *high-resolution*, *shallow-water*, and *deep-water* systems. *High-resolution systems* have 300–500 kHz operating frequencies, *shallow-water systems* have 100–300 kHz

FIGURE 10.52

Amalgamation of data from a multibeam system (below water part) and data from a photography (above water part) to show a section of a harbor pier.

Courtesy of Port of London Authority and Teledyne-Reson A/S.

operating frequencies and *deep-water systems* have operating frequencies below 100 kHz, with most systems operating between 10 and 30 kHz. The system size and weight depend on the application. The characteristic dimension of high-resolution systems is less than 0.5 m and the weight is between 10 and 20 kg. The maximum dimension of operational deep-water systems is around 8 m and the weight is between 1000 and 2000 kg. While deep-water systems have to be installed in dry dock, high-resolution systems due to their small size may be mounted on a small vessel either over the side, over the bow, or through a moon pool. The high-resolution systems may also be mounted on a towfish, an ROV, or an AUV. Handheld high-resolution MBESs are available for use by divers, such as combat frogmen or clearance divers, to look for limpet mines on the bottom of a ship or on harbor facilities in low-visibility water. High-resolution systems are also used for "imagery" or forward looking, where they provide information on the return signal strength from volume and surface backscatters. In this mode they are used for obstacle avoidance, navigation, site clearance surveys, underwater inspection, and object location. Amalgamation with data from other sources may also be done as in Fig. 10.52, where a photograph of a part of a harbor pier above water has been "knitted" to an MBES recording of the underwater part of the pier. This process improves the characteristic detail recognition.

A worldwide swath multibeam database is published on the web by Norman Cherkis. By 2004 more than 835 multibeam sonar systems were produced and sold by companies, such as Atlas Elektronik and ELAC (Germany), Seabeam (USA), Kongsberg (Norway), and Teledyne-Reson (Denmark). More than 75% were manufactured and sold by the last two companies. Details about two MBES systems, one high-resolution system (Teledyne-Reson SeaBat 7125, shown on

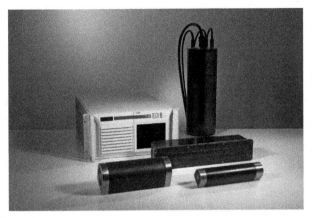

FIGURE 10.53

Teledyne-Reson SeaBat 7125. The projectors in the foreground are 400 kHz (left) and 200 kHz (right). The receiver array is behind the projectors. The link control unit, the black cylinder, and the sonar processor unit with the display are shown in the background.

Courtesy Teledyne-Reson A/S.

Fig. 10.53, [59]) and one deep-water system (Teledyne-Reson SeaBat 7150, where the projector and receiver systems are shown on Fig. 10.54, [60]), are provided in this section to illuminate characteristic features extensively used by MBES types. Both systems offer two operation frequencies, 400 and 200 kHz for SeaBat 7125

FIGURE 10.54

The projector and receiver system in Teledyne-Reson SeaBat 7150 forming a Mills Cross. The eight-element projector array transmits at 12 kHz (large elements) and at 24 kHz (small elements) shown in the upper part of the figure forming the lower part of the transducer "*T*." The eight-element receiver array forms the upper part of the transducer "*T*." Up to 12 elements in each projector array and in the receiver array are available to produce the high resolution.

Courtesy Teledyne-Reson A/S.

Table 10.6 Overview of the Technical Specifications for RESON's SeaBat 7125

Parameters	400 kHz System	200 kHz System
Sonar operating frequency	400 kHz	200 kHz
Across-track beam width	Transmit: > 128 degrees	Transmit: > 128 degrees
	Receive: 0.5 degrees (nadir)	Receive: 1.0 degree (nadir)
Along-track beam width	Transmit: 1 degree	Transmit: 2 degree
	Receive: 27 degrees	Receive: 27 degrees
Number of receive beams:		
Equiangular beams	256	256
Equidistant beams	512	
Swath coverage	128 degrees	128 degrees
Typical range	1–200 m	1–500 m
Ping rate	Up to 50 pings/s	Up to 50 pings/s
Receiver sample rate	34 kHz	34 kHz
Pulse length (continuous wave, CW)	33–300 µs	33–300 µs
Depth rating	400 m with optional 6000 m	
Depth resolution	5 mm	5 mm

Courtesy Teledyne-Reson A/S.

and 24 and 12 kHz for SeaBat 7150. Tables 10.6 and 10.7 give some of the technical specifications for the two systems. The narrow across-track beam width obtained at nadir, which is the point on the seabed vertically below the MBES, increases with increasing incidence angle.

An MBES is normally oriented with the *receiver unit* mounted across-track and the *projector unit* mounted along-track and aft of the receiver unit. The angular-resolution along-track is determined by the projector unit's directivity, and the across-track angular resolution is determined by the receiver unit's directivity. The *receiver–projector configuration* is typically the *Mills Cross* shown in Fig. 10.55 for the SeaBat 7150 in its dual-frequency mode and in Fig. 10.56 for the insonified swath footprint. A rectangular configuration, like the one used in Kongsberg's ME70, is also possible, and this configuration provides the opportunity to use a split-beam technique.

How and where the *multibeam system* is *mounted* is the most important factor in avoiding MBES performance degradation. For high-resolution systems mounted over the side or the bow of the vessel, the depth has to be sufficient to avoid shadows from the keel, hull, or other obstacles, and turbulence from the mounting structure or from surface water waves, which could degrade the system performance. For deep-water systems, which frequently operate in the frequency band where noise

Table 10.7 Overview of the Technical Specifications for RESON's SeaBat 7150

Parameters	12 kHz	24 kHz
Sonar operating frequency	12 kHz	24 kHz
Across-track beam width at nadir[a]	1 degree	0.5 degree
Along-track beam width	1 degree	0.5 degree
Number of across-track beams	256 equiangle beams and 880 equidistant beams	
Swath coverage	Up to 150 degrees, with max. swath width of about 5.5 times the water depth	
Typical depths	200–7000 m	200–4000 m
Ping rate	Up to 15 pings/s	Up to 15 pings/s
Pulse length (continuous wave, CW)	0.5–20 ms	0.3–20 ms
Main sample rate	6 kHz	6 kHz
Bottom detection resolution	200–400 m: 12 cm	200–400 m: 12 cm
	>6400 m: 2 m	>3200 m: 2 m

[a] For 1 degree across-track beam width at nadir, their width will increase to about 2 degree at ±60 degrees from nadir and to about 2.9 degrees at ±70 degrees from nadir.
Courtesy Teledyne-Reson A/S

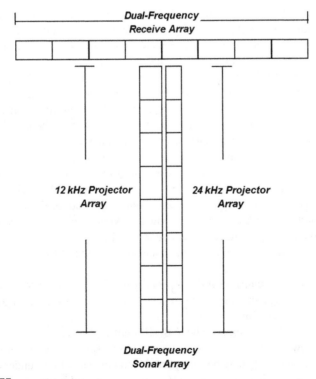

FIGURE 10.55

The Mills Cross formed by the projector arrays and the receiver array in SeaBat 7150, where both arrays consist of eight elements.

Courtesy Teledyne-Reson A/S.

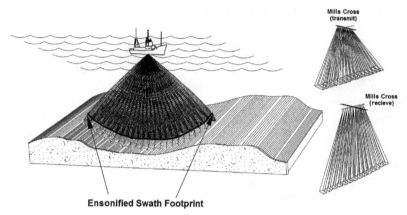

FIGURE 10.56

The narrow, steered beam of the receiving array produces the insonified swath footprint orthogonal to the track of the ship mapping the seafloor. The multibeam sonar transducer forms a Mills Cross with the projector array in the along-track direction and the receiving array in the across-track direction.

is generated by machinery vibrations, propeller and appendage cavitation, hydrodynamic flow, and air bubbles, the quality of the multibeam system performance may be degraded and countermeasures may be required. Mechanical vibrations from the *main propulsion plant* normally do not exceed 2–3 kHz, and since acoustic devices usually are far enough from this vibration source, the main propulsion plant usually does not influence the measurement quality. However, auxiliary machinery may be closer to the multibeam system. Inspection and local acceleration measurements at the installation site should be conducted to determine whether there is the potential of interference between the auxiliary machinery and the multibeam system. *Propeller cavitation* may be one of the main disturbance sources that can influence the quality of measured multibeam signal. Cavitation noise may reach the multibeam system via a direct or a hull-grazing path. In shallow water this noise may arrive via a bottom bounce. Although propellers produce cavitation, the rough edges on the vessel's under-hull may at speeds as low as 8 knots produce *appendage cavitation*. To avoid this noisy, disturbance source, it may be necessary to conduct a careful inspection of drawings or pictures of the under-hull, or a visual inspection while the vessel is in dry dock.

To avoid *hydrodynamic flow noise* externally mounted sonar systems should have well-designed fairings that are as flush mounted to the vessel's hull as possible. Poorly designed sonar fairings can degrade the sonar system's performance by as much as a factor of 4 for speeds above 8–10 knots. Also, cascades of *air bubbles* produced by bow wave breaking even at lower vessel speeds or by breaking surface waves at higher sea states may be transported along the vessel hull underside. When bubble cascades are between the multibeam system and water they produce a transient disruption of direct contact between the receiver–projector system and water column. While only lasting for few seconds these interruptions will reduce MBES recording quality.

10.4.1 MULTIBEAM ECHO SOUNDER STRUCTURE

An *MBES* is normally composed of the following *elements*: a projector unit, a receiver unit, a power link, a high-speed data link between the projector/receiver units and the sonar processor, a sonar processor unit, a user interface, and auxiliary equipment. These components are described in more detail below in terms of the two Teledyne-Reson multibeam systems, SeaBat 7125 and SeaBat 7150. As shown in Tables 10.6 and 10.7 these systems use *equiangular* and *equidistant* beam-forming facilities. As the beam-steering angle increases from nadir, the number of beams per degree of the swath stays constant for the equiangular spacing facility, while the number of beams per degree increases for the equidistant spacing facility. Fig. 10.57 shows the number of beams as a function of the steering angle for a 1 degree nadir beam spacing. If a flat seafloor is assumed, and if all beam footprints, i.e., the effective area of the seafloor insonified by a beam, are assumed to be the same size as the nadir footprint, the center-to-center spacing for each beam intersection with the seafloor is reduced as the steering angle increases. The footprint area is often given as the section cutout of the seafloor by the cone defined by the half-power beam width of a beam in the beam fan. As shown in Fig. 10.58, at ±65 degrees the beam density is about 2.5 times greater than at nadir. To obtain a uniform sounding spacing, the beam center-to-center spacing must decrease which requires a significant increase in the number of beams for the equidistant mode. For example, Table 10.7 shows that at 24 kHz there are 256 beams formed in the equiangular mode, while 880 beams are formed in the equidistant mode. The beam former in SeaBat 7150 is able to generate

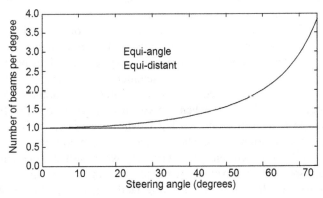

FIGURE 10.57

The number of beams per degree of angle as a function of the steering angle for 1 degree nadir beam spacing.

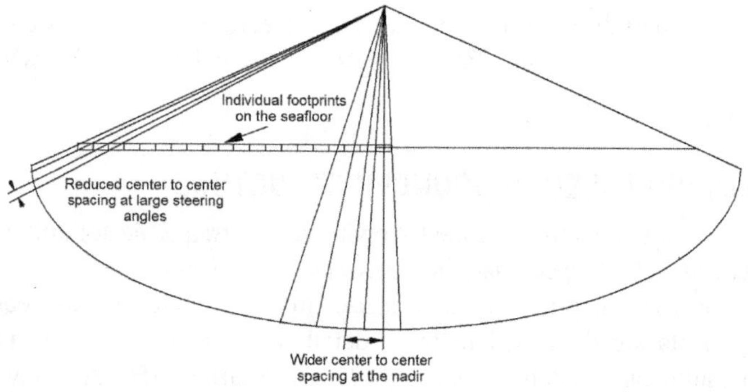

FIGURE 10.58

The reduced center-to-center spacing at large beam angles from nadir, compared to the center-to-center spacing at nadir.

a receive beam centered at any angle within the available swath, and in the equidistant receive mode, a set of coefficients are selected, which form a set of receive beams centered on the midpoints of the 880 footprints.

10.4.1.1 Projector Unit

The *projector unit* includes the transmission electronics and projector array. In general, the electric power transferred to the projector unit ranges from few hundred to several thousand watts. The array characteristic dimensions and transmitted frequency determine the transmitted beam's directivity through the *ka*-value, where k is the wave number and a is a characteristic dimension, such as the array length. The array may consist of several array modules, such as 12 modules to form an 8 m long projector array in the SeaBat 7150. The multibeam projector unit produces a narrow beam in the along-track direction of about 1 degree for SeaBat 7125 at 400 kHz and 0.5 degree for SeaBat 7150 at 24 kHz—and a broad beam illuminating in one transmission the entire across-track sector—128 degrees for SeaBat 7125 and 150 degrees for SeaBat 7150—with a maximum *ping rate* of 50 and 15 Hz, respectively. The *ping rate* is determined by the operator and is based on the selected range for the multibeam system. For the high-resolution system, 50 pings per second set for a 5 m range are sufficient to give 100% seafloor coverage even at a high vessel speed. In some cases, it may be beneficial to use a lower than normal ping rate for a particular range. This may occur when the vessel is moving very slowly and the operator wants to remove redundant data, or when the vessel is operating in an area with high reverberation levels, where echoes from previous pings contaminate signals received from the current ping. For deep-water systems the ship's speed above a certain speed limit, governed by the range of operation, will leave gaps in the seafloor coverage. Therefore, a *multi-ping* facility is operated by SeaBat 7150 to increase the along-track sounding density. The multi-ping comprises four separate frequencies transmitted during each pulse interval and each frequency beam with a different steering angle offset in the along-track direction. The beam-steering angles depend on beam width. For the 24 kHz deep-water systems with a transmit beam width in the along-track direction of 0.5 degree, the four beam steering angles are +0.75, +0.25, −0.25, and −0.75 degrees, respectively. The four pulses are transmitted with an interpulse time interval of about 1 ms. The individual pulse length may be varied between 0.5 and 20 ms. The pulse length, governed by the range of operation, is selected by the operator.

Also, the vessel *pitch* movement influences the seafloor coverage, in particular during intermediate- and deep-water surveys. Pitch is the inclination of the vessel in the longitudinal plane. To maintain the nominal transmit beam vertical, irrespective of the ship motion, a motion sensor with a pitch steering facility is normally introduced. In this case multi-ping steering angles are modified by the vessel's pitch angle to maintain the beam's vertical stabilization. To maintain full seafloor coverage the pitch stabilization in Teledyne-Reson's deep-water system incorporates transmit beam steering up to ±10 degrees from vertical.

10.4.1.2 Receiver Unit

The seafloor return signal is received by the *receiver unit*. This unit includes a receiving transducer array, TVG facility, A/D conversion electronics, filters, demodulation facility, and beam former with steering electronics for swath coverage. The high-resolution system shown in Table 10.6 has an along-track, receiving, 27 degree half-power beam width for 200 kHz and for 400 kHz, while the across-track, receiving, half-power beam width at nadir is 0.5 and 1 degree for the two frequencies, respectively. For 400 kHz the swath is, therefore, covered by footprints that at nadir have 0.5 × 1 degree dimensions. However, several factors may influence the swath coverage quality for deep-water systems. They include noise on the outer beams, i.e., the beams at the highest fan angle, high-amplitude second return signals from structures in the seafloor and high-amplitude nadir return signals due to the multibeam signal's specular reflection at nadir, which may saturate the receiver unit and cause abrupt decreases of up to 40 dB in the backscattered signals from other directions than nadir. In particular, the strong nadir return signal makes it necessary to use the TVG system, which is an analogue amplifier, where the gain is controlled by a digital signal processor. The TVG normally applies a variable gain to the received signal based on range-dependent absorption and spreading losses. However, it may be necessary to introduce other techniques, such as a discontinuity in the TVG operation, to ensure return signals from the seafloor at and around nadir are only coming from a specular reflection ring and not from the real bottom.

Also *roll stabilization* is a part of the receiver unit's signal processing. Roll is the inclination of the vessel in its transverse plane and influences the reception of time, angle, and amplitude in the return signal's information from footprints in the across-track direction. The instantaneous roll angle is read by the beam former and applied to each beam for each sample at its reception time. The roll stabilization keeps the nadir beam of the swath vertical. However, because of the effect on the outer swath beams the stabilization is normally limited to roll angles less than about ±15 degrees.

Dynamically focused beam forming may be used in a number of predetermined focal planes to offset transducer near-field effects. Dynamic focusing increases the vertical resolution while preserving the narrow beams required for horizontal resolution. The dynamic focusing requires short pulse lengths produced by a wide aperture sonar array, i.e., $ka \gg 1$, to form narrow, high-frequency beams with low side lobes.

10.4.1.3 Sonar Processor Unit

High-resolution systems such as Teledyne-Reson SeaBat 7125 use a *link control unit* as a two-way, high-speed data link between the projector/receiver units and the sonar processor unit. Full bandwidth digital data from the receiver is formatted and transmitted to the processor and operator commands from the processor via the link control unit are distributed to the projector/receiver units for implementation. Also the link control unit manages power distribution to the projector/receiver units and monitoring functions. In the most recent high-resolution multibeam systems, such as Teledyne-Reson SeaBat 7125 SV, the link control unit functions are integrated into the sonar processor unit.

The sonar processor is a PC with additional installed hardware and plug-in cards for carrying out the following operations:

- receiving digitized sonar data from the receiver unit;
- receiving operational settings directly through a user interface or from a remote system;
- providing beam-forming and initial processing of data from the acoustical signals before presenting or exporting data to an external system;
- controlling, formatting, and outputting data to external systems, such as making beam-formed data and preprocessed image data available to external systems over a fast Ethernet connection;
- providing an interface for a sound velocity sensor to conduct range measurements and receiver beam forming correctly;
- performing built-in test equipment (BITE) routines and alerting the operator to any conditions requiring a reaction.

Modern sonar processors also allow access to *raw water column and seafloor data* (Snippets) since scientists frequently perform their own signal processing to obtain information not provided by the standard signal processing package. SeaBat systems are able to produce an operator configurable beam data stream to be written to the hard drive in the sonar processor unit or an external hard disk array for individual signal processing. Raw data logging software controls the recording, playback, and data storage functions.

To ensure that the full multibeam system is functioning correctly BITE and a calibration routine may form an integrated part of the sonar processor unit. The BITE normally monitors the internal subsystem status in the multibeam system, such as the projector and receiver units, a possible link control unit and the sonar processor unit. It continuously provides the operator with information about critical temperatures, voltages, and communication status for the whole system. The BITE can also be used to verify an installation, give information about firmware updates, diagnose installation problems, and give information to help localize a potential failure.

An acoustic signal suffers attenuation and distortion from a number of sources during transmission, propagation, and reception. These losses contribute to the inaccuracy of backscattering strength measurements. To improve the target strength estimation and target classification it is necessary to perform *calibration tests* on all array projector and receiver channels to ensure backscattered data accurately represent the absolute reflectivity of the seafloor or objects on the seafloor or in the water column. When this is done, all backscatter data are corrected in the sonar processor unit prior to export. The data are corrected for source level, receiver sensitivity and gain, A/D converter scale coefficient, array weighting, range including TVG application to avoid risk of saturation or loss of too weak signals, absorption coefficient, spreading loss, and footprint area. Since the absorption coefficient varies with depth, an appropriate absorption model, such as the Francois—Garrison model, provided in Chapter 4 of Eq. (4.18), may be used to establish the sonar processor unit's absorption profile for backscattering calculations and gain settings.

The *power* transmitted by the multibeam systems is steered by the sonar processor unit. The power transmitted into the water is controlled by the operator in steps of about 1 dB. For a given power setting, a shorter pulse length provides higher resolution at a shorter range, while a longer pulse length provides maximum range with lower image resolution. To minimize operator intervention, the sonar processor unit normally includes an automatic operation mode, where the range scale, transmitted power, pulse length, and receiver gain are monitored and controlled. The power requirements for deep-water systems are normally few hundred watts for the sonar processor unit while the projector/receiver system may require several kilowatts.

Range and depth filters (*depth gates*) are used to reduce the noise influence and correct for tilted sonar heads to optimize the seafloor detection process. For instance, minimum and maximum nadir depth gates may be used to improve the initial search for the strongest seafloor return signal on the nadir beam. When the initial seafloor detection point is located, the depth gates are propagated to the other beam directions, and seafloor detection from beam to beam is based on the results from the previous beams.

The many individual beams in a multibeam system, the ping rates, and the quantification of received signals lead to a substantial amount of data. The SeaBat 7125 operating at 400 kHz at a range setting of 100 m, with 512 beams per ping and 5 pings per second and with a digitization of 16 bits per sample produces about 155 Gb/h amplitude/phase data. This places stringent requirements on the signal processing hardware/software capacity and speed.

The *user interface* may be a separate instrument in the multibeam system, or an integrated part of the sonar processor unit. A graphical user interface provides the sonar user with a means of configuring and controlling the multibeam sonar system and the monitored data. For example, the sonar image can be displayed on a screen as a sonar wedge or a B-scan mode. The screen is in most cases a separate unit, or it may be an integrated part of the sonar processor unit. The control system includes the projector and receiver unit operation, auxiliary equipment control, BITE, data storage and data export over Ethernet using selected protocols, and selection and operation of data products, such as bathymetry, snippets, and side-scan data.

10.4.1.4 Auxiliary Equipment

The multibeam system must also be able to receive and process data from a variety of auxiliary instruments to produce the most reliable, reproducible, and accurate information about the underwater environment. The most important auxiliary instruments are used to provide the following functions:

- For proper beam steering, the multibeam sonar system requires input from a *sound velocity probe* (*SVP*). Knowledge about the local sound velocity on the measurement site is important for the transmitted and received signal. The SVP continuously reports the local sound velocity to the sonar processor unit. The SVP normally uses a high-frequency pulse signal transmitted over a known distance between the transmitter and a reflector or a receiver. If the reflector is used, the transmitter is the receiver. The pulse propagation time over the known distance gives the average local sound velocity. The measurement accuracy in general is between ±0.05 and ±0.25 m/s, depending on water depth. The local sound velocity is used in the range display and for steering receiver beams.

- Information about the *attenuation profile*, i.e., the contributions from acoustic signal absorption and spreading losses, is necessary for backscattering calculations and for setting the TVG facility. The sound velocity variation with depth, the *sound velocity profile*, can be collected by using an expendable bathythermograph (XBT) profiler or another profiler, or it may be found in databases such as the World Ocean Atlas 2005 [61]. The sound velocity profile can be used to calculate an absorption profile by using an absorption model, such as the Francois–Garrison model, given in Chapter 4 of Eq. (4.18). This calculation can also be performed by the sonar processor unit. Also a *spreading loss coefficient* has to be calculated by the sonar processor unit to take into account the amount of spherical and/or cylindrical spreading losses expected in the ambient water. Accurate knowledge about the attenuation profile is critical to ensure correct calculations of the seafloor backscattering qualities or of object in the water column, and for determining beam profiles.

- *Vessel movement* has a strong influence on the quality of the multibeam system measurements. Since movement cannot be avoided, a correction is necessary. The characteristic movements are *pitch, roll, heave*, and *yaw*. Pitch is the vessel inclination in the longitudinal plane which influences the projected beams. Corrections to steer the beams to remain vertical through a selected range of pitch angles are normally introduced based on continuous pitch angle measurements. In deep-water surveys pitch angle compensation is necessary to maintain full seafloor coverage. Roll is vessel inclination in the transverse plane.

Roll compensation is necessary to keep the swath nadir beam vertical and to keep the real-time seafloor display stable, when the vessel is rolling, as discussed by Pocwiardowski et al. [62]. Heave, which is the vertical movement of the vessel, influences the nadir beam and the swath range. The yaw, which is the erratically deviation of the vessel from its course has to be compensated for, as yaw motions disturb the insonification of the seafloor. Yaw compensation normally takes place in the projector. The multiple ping procedure mentioned earlier can be used for yaw compensation. Sensors for the different vessel movements are available and their signals are processed and exploited via the sonar processor unit. In deep-water operations vessel movements will also produce *Doppler effects*, which influence the multibeam system performance. Doppler effects also have to be compensated for through efficient filtering and vessel motion control.

- *Accurate vessel geographical position* information is necessary when performing bathymetric measurements and for producing hydrographical maps. In general, a differential global positioning system (DGPS) provides the desired accuracy, and the systems data are normally transferred to the multibeam sonar system via the sonar processor unit.

10.4.2 MULTIBEAM ECHO SOUNDER APPLICATIONS

MBESs have applications in areas, such as *bathymetry* soundings; corrected backscatter data time series, *snippets*, as for instance amplitude and phase from the footprints on the seafloor illuminated by a single projector ping; *side scans* forming an image of the seafloor; and high-resolution *forward looking* to avoid collisions and detecting objects in the water column or on the seafloor in front of the vessel. Some of these applications are discussed in more detail in the following subsections.

10.4.2.1 Bathymetry

Bathymetry is the measurement of water depth. Today bathymetry applications include: route surveys for cable and tube laying, production of high-quality hydrography and nautical charts, and investigations of seafloor materials and artifacts on the seafloor. MBESs were primarily developed for bathymetry.

The main parameters utilized in bathymetry are time and angle. The return signal's amplitude and phase as functions of time and angle are used to produce reliable bathymetric information. Since measurements are referenced to the position of the multibeam sonar system, this position has to be known with high accuracy and is influenced by instantaneous movements of the vessel. Depending on the beam angle, seafloor type and topography, and acoustic signal characteristics, the seafloor detection may be based on different techniques. The two most frequently exploited are *center-of-energy* determination and *phase-zero-crossing* detection, or a combination of these two methods.

The center-of-energy determination is used at angles close to nadir (vertical), i.e., up to about ± 10 degrees around nadir, where the echo envelopes have a good signal-to-noise ratio and the return signal time spread is still not excessive. The instantaneous maximum amplitude in the return signals time envelope (the echo) is determined and the arrival time and angle of the maximum amplitude are used for calculations of the seafloor detection, after the necessary corrections mentioned in Section 10.4.1.4. have been introduced. Also, for beams closer to nadir these corrections must take into account the modulation of the echo envelope with the transmitting and receiving array directivity functions. As mentioned in Section 10.4.1.2, a measurement difficulty close to nadir is the high specular reflected signal amplitude which may strongly bias the seafloor detection time. This source of errors is important for narrow beams at low angles of incidence close to nadir.

The phase-zero-crossing procedure is used at higher angles away from nadir, where the echo signal's time histories are long enough to permit a more accurate time and angle determination for the zero phase between two received time signals. This zero phase indicates the arrival time and angle for a return signal from a target or the seafloor detection point on the beam axis. While amplitude information mostly is used near nadir in the center-of-energy procedure, amplitude and phase information are used for range dependence at higher angles, and at the highest angles of the swath, the phase information alone is used. Some multibeam systems are equipped with an autopilot, where the range scale automatically is increased when the seafloor detection goes in and out of range, to secure accurate seafloor detection.

Fundamentally, the sonar image from MBES consists of contributions from time histories related to a great number of narrow beams forming the echo returns from the seafloor. The time histories after beam forming are combined to form a continuous image of the seafloor along the swath and to produce 3-D bathymetry displays. The nadir return has the highest amplitude, and the amplitude may in general decrease with increasing beam tilt angle, unless the beams are backscattered, or maybe directly coherently reflected, from characteristic surfaces formed by the seafloor topography or materials. The time duration of the echo return will also in general increase with increasing beam tilt angle. The influence of the seafloor materials on the swath width is dependent on the type of seafloor materials. Fig. 10.59 shows the swath width for various water depths for Teledyne-Reson SeaBat 7125 operating at 400 kHz on various seafloor materials, for example, rock, sand, and mud. The widest swath occurs on a seafloor consisting of rocks. The backscattering strength of mud is lowest and its attenuation is highest, which results in a narrower swath. Also, for the same seafloor material the swath width depends on the frequency used in the multibeam system. Fig. 10.60 shows the swath width for various water depths, in kilometers, for a SeaBat 7150 operating at 12 kHz and at 24 kHz, respectively, over a sandy seafloor. Due to attenuation, i.e., absorption and spreading losses, the lowest frequency provides the widest swath.

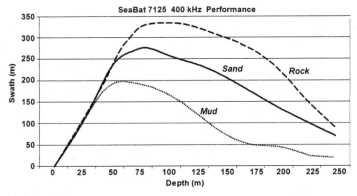

FIGURE 10.59

The swath width (unit: m) as a function of water depth (unit: m) for various seafloor materials, rock, sand, and mud, for SeaBat 7125 operating at 400 kHz.

Courtesy of Teledyne-Reson A/S.

The dynamic range, which is about 150 dB including the contributions from the TVG, of modern multibeam systems, permits simultaneous measurements of bathymetry and backscatter from objects in the water column without sacrificing the sampling rate. This supports the detection and monitoring of targets in the water column near the seafloor, such as various fish types along with high-quality bathymetric measurements.

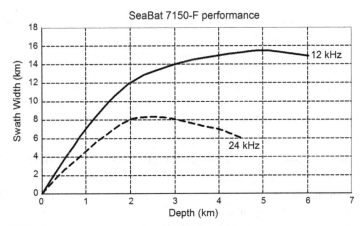

FIGURE 10.60

The swath width (km) performance of SeaBat 7150 over a sandy seafloor as a function of the water depth (km). The *upper curve* (—) is based on the 12 kHz array and the *lower curve* (- - -) is based on the 24 kHz array.

Courtesy of Teledyne-Reson A/S.

10.4.2.2 Snippets

More detailed, single information from each beam "footprint" is comprised in snippets. Bounded by a time window a snippet data sample could include corrected backscatter data, such as amplitude and phase data, from the individual beam footprints on the seafloor insonified by a single sonar ping. The number of snippets in a swath is a function of the number of beams. The length of each snippet depends on the multibeam system's operational mode, beam number, and water depth.

The snippet data correction could include transmitted power, receiver gain, pulse length, beam patterns, absorption and spreading losses, slant range, and grazing angle. The *grazing angle* is the angle of the sound propagation direction relative to the surface of the seafloor. The slant range and grazing angle correction uses the measured bathymetry data from the snippets' area. For identification each snippet package should be given pertinent information, such as time stamp, sequential ping number, sampling rate, sound velocity, transmitted power, receiver gain, attenuation, and range scale.

10.4.2.3 Side-Scan Data

In spite of the fact that an MBES is not a true SSS, several multibeam systems, such as Teledyne-Reson SeaBat 7125 and 7150, are able to produce *side-scan data*. The side-scan data forms an image of the seafloor, which can be used to locate and identify seafloor features and seafloor conditions. Each sonar ping is used to generate a line of data, and each line includes a series of return signal amplitudes as a function of time or range. When these lines are combined and displayed on the multibeam system screen as the vessel moves along its track, a two-dimensional image is formed, which provides a detailed seafloor picture along either side of the vessel.

The side-scan beam-forming process can be done by combining half of the bathymetric beams into two side-scan beams. In this process adjacent pairs of beams are combined by averaging and the brightest points are selected from the averaged beams. This "peak detect" procedure reduces the noise influence on the images. Although side scans cannot be used to accurately measure the true depth, they can provide more detailed images of the seafloor. The side-scan image can then be used with bathymetry to identify seafloor features and to help ensure a survey does not miss small but significant targets.

10.4.3 MULTIBEAM ECHO SOUNDER PERFORMANCE LIMITATIONS

The MBES *performance* is highly dependent on environmental and installation conditions, which may adversely affect the achievable swath width. Figs. 10.59 and 10.60 show swath width dependence on seafloor materials for the 400 kHz mode of the SeaBat 7125 and the frequency dependence for the SeaBat 7150, respectively, for various water depths, i.e., the projector–receiver systems altitude above the seafloor. Also water column salinity and temperature distributions, as well as sound attenuation have the potential to degrade performance. Moreover, the installation conditions may have a considerable influence on the performance, since effects from bubbles, flow and propeller noise, and engine and bulkhead vibrations may reduce the swath width and image resolution.

The effective swath width is determined by several factors. These factors are the:

- water depth;
- total across-track fan aperture;
- maximum propagation range under the influence of local attenuation;
- local signal-to-noise ratio that causes fluctuations in the received signals;
- influence of beam refraction due to the local sound velocity profile;
- seafloor materials;
- stability of the backscattering characteristics of the seafloor;
- vessel movement;
- influence of noise, vibration, and bubbles.

A high ping rate is necessary to obtain a high collected data density across the track. To avoid received signal ambiguity the ping rate is determined by the time of flight for the signals representing the highest beam angle. In this way the arrival of the signal on the highest beam angle triggers the emission of the next projector signal. To avoid gaps in the seafloor coverage in the along-track direction, the vessel speed has to be determined by the width (along-track) of the "footprint" on the seafloor formed by the highest beam angle to prevent the distance traveled by the vessel between two signals exceeding this width. With the notations from Fig. 10.51 the maximum speed should be about $(c/2)\ \varphi_\ell\ \cos(\varphi_t/2)$, where c is the local velocity of sound, which for the SeaBat 7150 operating at 12 kHz, will give a maximum vessel speed of 6−7 knots.

10.5 SIDE-SCAN SONAR

SSS is a tool used to visualize the seabed and objects on, or above, the seabed. The SSS has been produced since the 1950s. The SSS builds a two-dimensional high-resolution picture of insonified structures by using a sonar transducer and the transducer's motion through the water. The transducer can be mounted on a surface vessel bottom, on a pole over the surface vessel side, or on a towed body. Each mounting type has an advantage. The side-scan system can consist of a single transducer radiating the sound signal to one platform side, or it may, in order to cover a broader region of the seabed, consist of two transducers radiating sound to both platform sides during its motion through the water. Although the SSS normally consists of a *single row* of transmitter elements on each platform side, to improve resolution it may include two or more rows of transmitter elements to form a *multirow SSS*.

10.5.1 SINGLE-ROW SSS

A frequently used *single-row*, SSS consists of two single-beam, multielement transducers mounted along the tow direction on each side of a towfish with their beam

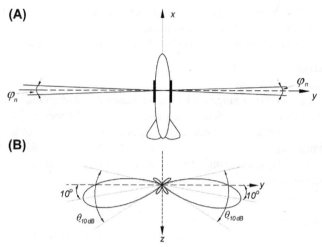

FIGURE 10.61

Two transducers mounted on each side of a towfish. The transducers are tilted 10 degrees below the horizontal plane (the x—y plane in (A)). The horizontal plane half-power aperture angle φ_n is 0.2 degree $\leq \varphi_n \leq 2$ degrees. The vertical plane (the y—z plane in (B)) beam width in perpendicular to the towfish heading is broad and represented by a total -10 dB beam width of 40 degrees $\leq \theta_{-10dB} \leq 60$ degrees.

axis normal to the tow direction. The transducer beams are tilted toward the seabed and are narrow in the horizontal (azimuth) plane and wide in the vertical plane to insonify a swath-like portion of the seabed, as shown in Fig. 10.61, where the tilt angle is 10 degrees. The towfish is normally towed at slow speeds close to the seabed to obtain echo-ranging amplitudes for the insonified seabed surface at *low grazing angles* and high sampling rates. The grazing angle is the angle between the sound path and the seabed surface. One advantage of the SSS operating near the seabed is that shorter distances permit higher frequencies to be used. This provides increased range resolution with improved seabed surface profile vertical resolution. The sound velocity profile has less influence on sound propagation when the SSS operates near the seabed than when the SSS is hull mounted on a surface vessel. The noise influence is also reduced when operating near the seabed in deeper waters. However, moving the towfish near the seabed reduces the swath coverage and the exact towfish position can be difficult to determine. Unless the seabed is horizontal and flat, the image may be geometrically distorted. For instance, when the seabed has a slope the images produced by the SSS on the starboard and on the port sides will have different intensities. For seabed mapping surveys it is necessary to keep the towfish at a constant height above the seabed to maintain a stable grazing angle. However, speed variations by the towing vessel and its turning to add a new route to the mapping operation will cause changes in the towfish height above the seabed.

The SSS is used to detect and identify underwater objects and bathymetric features and for nautical chart production. It is used to distinguish between seabed materials and their textures, detect items dangerous to shipping, give information about cables and pipelines not covered by sediments or rocks vulnerable to ship's anchors, and detect marine archaeology objects of interest. SSS is also used in environmental studies, dredging operations, fishery research, and naval mine detection. In confined areas, such as ports and rivers, it is often impractical to tow an SSS platform. Here the sonar may be hull mounted on a surface vessel or mounted on a pole over the vessel side. Due to the wide SSS vertical beams it can be difficult to use in *confined areas* where multiple reflections from structures, such as piers and other vessels, can disturb image formation. The MBES is frequently found to be more useful and accurate in confined areas.

The fan-shaped acoustic pulses emitted by the SSS and backscattered by the insonified long and narrow seabed strip are received by the sonar transducer as a function of time. This represents the seabed reflectivity along the swath, and through signal processing the desired high-resolution imagery and object detection along the seabed surface is produced. With a correct towfish speed it is possible by signal processing to add the insonified strips side by side to produce continuous seabed maps. Fig. 10.62 shows the echo formation by the SSS produced by backscattering from various types and topographies in a narrow seabed strip. Only the beam produced by the sonar on one side of the towfish is shown. However, a similar beam is produced by the sonar on the opposite side of the towfish. The SSS is moving vertical to and out of the paper plane at an altitude H above the seabed. The point N on the seabed indicates the *nadir* directly below the towfish. The seabed on both sides of *nadir* forms a narrow *blind zone* between the two beams. The point I on the seabed is the first return signal from the seabed. The position of I depends on the vertical

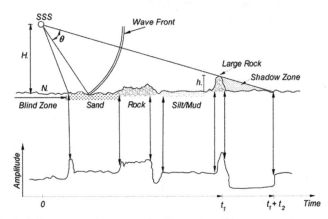

FIGURE 10.62

SSS echo formation. The upper part shows the acoustic field while the lower part shows the echo amplitude time. The echo amplitude is a function of the transducer source level, grazing angle, seabed topography, surface texture, seabed composition, and shadows formed by obstacles on the seabed.

beam width of the sonar beam and may not be well defined. The amplitude value before the first return from the seabed is noise and backscattering from the water column. The backscattered amplitudes from sand, rock, and silt/mud as a function of time are shown on the lower part of the figure. A large piece of rock or an object on the seabed produces a higher amplitude backscattered signal. Since there is no backscattering from the seabed just behind the rock/object, a *shadow zone* is formed with a shape comparable to the rock/object's shape. After the shadow zone a renewed backscattering from the seabed materials occurs. *Shadowgraph effects* are used to determine the height h of the rock/object in Fig. 10.62 based on the time t_1 for the sound to reach the rock/object and the time $t_1 + t_2$ to reach the end of the shadow through:

$$h = \frac{t_2 H}{(t_1 + t_2)} \tag{10.95}$$

The length of the swath footprint, SW, shown in Fig. 10.63, depends on the SSS vertical beam width, beam tilt with respect to the horizontal plane, sonar transducer height above the seabed, sonar transducer source level, signal frequency, backscattering ability of the seabed and objects on it expressed by their *target strength TS*, ambient noise level, and to some extent on the horizontal -3 dB beam angle. The target strength TS is defined as:

$$TS = 10 \, log_{10} \left[\frac{echo \; intensity \; at \; 1 \; m \; from \; the \; target}{the \; incident \; intensity} \right] \tag{10.96}$$

The time delay between each emitted SSS sound pulse is determined by the time for the echo to return from the previous pulse. Since the echo duration depends on environmental conditions and transducer qualities, the next pulse is normally emitted when the received echo has decreased to a preselected level. If the maximum

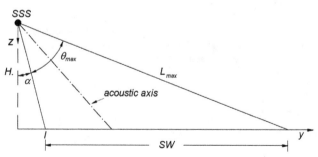

FIGURE 10.63

Beam angles in the vertical plane (the y−z plane) used to calculate the maximum towfish speed V necessary to obtain coherent seabed coverage for seabed mapping surveys. θ_{max}, the maximum beam opening angle, is limited by directions where the acoustic beam level has dropped 10 dB below the acoustic axis level. L_{max} is the maximum reachable range for the sonar beam to produce a usable echo return. SW is the full swath width.

reachable range for the SSS sonar beam, which produces a usable echo, is L_{max} (unit: m), then the time delay t_d (unit: s) between two consecutive SSS sound pulses is: $t_d = 2L_{max}/c$, where c (unit: m/s) is the sound velocity. Since the insonified track in the x-direction normal to the paper plane in Fig. 10.63 increases with the distance from the SSS, see Fig. 10.61, the narrowest track is found around point I in Fig. 10.63, where the pulse first hits the seabed. If the sonar beam opening angle in the x−y plane at point I is φ_{min}, and the maximum beam angle in the vertical plane is θ_{max}, the *maximum speed V* (unit: m/s) of the towfish in the x-direction, to obtain a nearly 100% *coverage* of the seabed without gaps between two consecutive insonified seabed strips, is determined by:

$$V = \frac{c\varphi_{min}\cos(\theta_{max} + \alpha)}{2\cos\alpha} \tag{10.97}$$

where α is the angle in Fig. 10.63 between the vertical direction and the beam direction toward I. If the following values are used for $\alpha = 10$ degrees, $\varphi_{min} = 1$ degrees, and $\theta_{max} = 60$ degrees, while $c = 1500$ m/s, the maximum permitted towfish velocity, V, is $V = 4.55$ m/s or 8.8 knots.

From Fig. 10.62 it can be seen that the single-row side-scan echo return from a rough heterogeneous seabed is complex. The lack of an echo return from the seabed area around nadir and the reduced signal-to-noise ratio by backscattered signals from the outer regions of the side-scan footprint produces an interpretation bias. Seabed materials and their textures in shadowed regions are also impossible to classify, when the SSS is towed in one direction only. Towing in another direction parallel to the first towed path, but offset from this path, may produce the desired information about the seabed behind an obstacle which forms a shadow on the first pass.

The single-row SSS system is normally based on a long and rectangular-shaped sonar transducer. This transducer produces a narrow beam in the horizontal plane to obtain improved spatial resolution. The aperture angle φ_n in the horizontal plane, shown in Fig. 10.61, ranges from 0.2 to 2 degrees. The much broader aperture angle θ in the vertical plane shown in Fig. 10.61 ranges from 40 to 60 degrees. To limit the sea surface influence and reduce backscattering from seabed areas close to nadir, the sonar's acoustic axis is normally tilted about 10 degrees below the horizontal plane. By using the tilt angle and broad vertical beam, it is possible to reach 80 to 85 degrees insonification angles from vertical.

The frequency used by SSS systems normally range from below 100 kHz to above 500 kHz. The higher frequencies restrict the length of the sonar transducer and produce a better resolution at shorter ranges. The lower frequencies give a longer swath width controlled by the transducer qualities, backscattering strength, and signal-to-noise ratio. The swath width determines the pulse repetition rate. Also, very low frequencies have been used in SSS systems, such as the British Geological LOng Range Inclined Asdic (GLORIA) SSS developed for long-range ocean floor texture and topography determination. GLORIA is towed in a "fish" about 200 m behind the research vessel and transmits up to 9 kW acoustic power at 6.4 kHz with about 2 min pulse rate. Under perfect propagation conditions, a full swath width of more than 50 km can be obtained.

The sonar signal emitted by the SSS normally has time duration between 0.05 and 0.2 ms, which permits a spatial resolution better than 0.1 m. An improved range resolution can be obtained by increasing the transducer bandwidth and emitting shorter pulses. The SSS system resolution in the track direction (the x-direction in Fig. 10.61) is reduced with distance from the sonar since the swath width increases with sonar distance. A single-row SSS normally operates a single frequency; however dual-frequency SSS systems have been designed, which may switch between a low frequency for long-range mapping and a high frequency for high-resolution, short-range mapping. The dual-frequency SSS shown in Fig. 10.64 is the EdgeTech 4125 digital, dual-frequency, SSS, which simultaneously operates at 400 and 900 kHz. Also split-beam SSS systems have been produced to take advantage of the qualities of low and high frequencies and their related transducer apertures.

Since sonar operators always want the highest resolution imagery possible, when using dual-frequency operation, they have to sacrifice range, which increases the time to complete a survey. A solution to this problem is to use dynamically focused array technology, which can provide the desired higher-resolution imagery at longer ranges. If the correct electronic delay is applied to a multielement array prior to adding the element signals to form the total array output, a dynamical focusing can be achieved, which allows focusing at each range insonified by the sonar. This technique utilizes modern digital electronics and digital signal processing. While the traditional SSS normally has a length of 0.5−1 m, the multielement array used by dynamical focusing may be up to 4 times longer. The increased sonar length puts constrains on the towfish size and may preclude the use of high-resolution dynamical focusing technology on smaller AUVs.

FIGURE 10.64

The EdgeTech 4125 digital, dual-frequency, side-scan sonar. This sonar operates simultaneously at 400 and 900 kHz. The horizontal plane beam angle is 0.46 degree for the lowest frequency and 0.28 degree for the highest frequency, while the vertical plane beam width is 50 degrees. The tilt angle is adjustable from 25 to 33 degrees. The length of the SSS is 112 cm and its diameter is 9.5 cm.

Image Courtesy of Garry Kozak, GK Consulting, www.2kozak.com.

10.5.2 MULTIROW SSS

The desire to use SSS systems to measure *bathymetric data* has led to the development of more advanced SSS systems based on multiple rows (staves) of transducer elements arranged parallel to each other on the sides of a towfish. While single-row SSS systems can give a rough estimate of seabed relief for a complex seabed topography and texture, the multiple stave arrangement improves the incidence angle and horizontal range estimation accuracy and provides bathymetric data. This harvest of *improved data* is based on exploiting *interferometry* and beam-forming techniques which make data processing and image interpretation more complex. However, the improved data most often justifies the more complex data handling. The interferometry, which leads to enhanced image quality, combines wave data from two sources and accounts for the constructive and destructive interference between the wave data—*constructive*, when the waves are in phase and amplify each other, and *destructive*, when the waves are out of phase and cancel each other.

In the interferometric side scan a single frequency is emitted by a transducer on each side of the towfish and the echo return is received on two—or more—transducers on each towfish side. As far as possible, the transducers should have the same acoustic characteristics. A vertical arrangement of two transducers is shown in Fig. 10.65, to measure the phase and arrival time for the same frequency return signal. The phase difference $\Delta\psi$ between the echoes arriving at transducer (1) and transducer (2) is:

$$\Delta\psi = k(r_1 - r_2) = k\,\Delta x \sin\theta \qquad (10.98)$$

where k (unit: m^{-1}) is the wave number and r_1 and r_2 (unit: m) are the distances between the two transducers and the backscattering object, which is assumed distant enough from the two transducers, for the two signals to be assumed to possess plane wave fronts. The angle θ is the unknown angle of incidence determined by the interferometric system. The phase difference $\Delta\psi$ in Eq. (10.98) is unambiguous as long

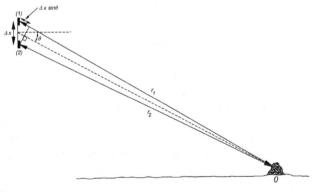

FIGURE 10.65

The range difference $r_1 - r_2 = \Delta x \sin\theta$ observed by a two-transducer interferometric SSS system. The backscattering object on the seabed is marked with O.

as $\Delta x < \lambda/2$. However, with the high frequencies used by multirow SSS, Δx would be so small that precise measurements would be difficult to obtain. Therefore, Δx is normally increased to improve resolution. The towfish movement must also be accurately compensated for in the signal processing to achieve the improved seabed mapping and 3-D object imaging.

The estimation of the angle θ in Eq. (10.98) may be disturbed by multipath effects, where multiple scattering from the seabed and sea surface may distort the measurements. To reduce multipath-induced inaccuracies the computed angle-of-arrival transient imaging (CAATI) signal processing method described by Kraeuther and Bird [63] is used. CAATI increases the number of concurrent angles of arrival, which gives more complex geometries, from which the phase differences can be imaged.

The use of single-row SSS in regions with complex topography is not recommended since this procedure fundamentally assumes a nearly horizontal and flat seabed. However, the multirow SSS with its more sophisticated and expensive techniques and its use of multiple beams is designed for seabed mapping involving complex topography. The multiple beams in *bathymetric side scan*, with beam forming to resolve problems with multipath effects and to obtain bathymetric data, makes multirow side scan a useful technique for surveying seabeds with complex topography, in spite of the additional costs and interpretation difficulties.

10.6 SYNTHETIC APERTURE SONAR

As mentioned in Section 10.5 longer sonar transducers and higher frequencies produce higher resolution at shorter ranges. However, higher frequencies reduce swath width due to sound attenuation. There is a limit on the transducer length a towfish or AUV can carry. The range resolution (RR) normal to the towfish/AUV track can be improved by using broadband pulses, where B (unit: Hz) is the sonar pulse bandwidth and c (unit: m/s) is the sound velocity, i.e.:

$$RR = \frac{c}{2B} \tag{10.99}$$

The horizontal (along-track) resolution (AR) is determined by:

$$AR = R\frac{\lambda}{L_s} \tag{10.100}$$

where λ (unit: m) is wavelength, L_s (unit: m) is sonar length along the track, and R (unit: m) is range normal to the track. As R increases the resolution is reduced, i.e., AR increases. An acceptable along-track resolution is only obtainable near the transducer. Increasing the sonar length is not sufficient to improve the resolution at longer ranges. The *synthetic aperture sonar* (*SAS*) provides a means to obtain improved AR at longer ranges.

The synthetic aperture is formed by a sonar transmitter illuminating a known location while transmitter and receiver move on a known path. This causes the *receiver aperture length*, L_a, to appear significantly longer than the physical length, i.e., $L_a \gg L_s$. By using signal processing that appropriately combines the received acoustic pulses the SAS can form an image with an order of magnitude or more improved resolution. The synthetic aperture technique is widely used in *radar imagery* from airplanes and satellites, where a synthetic aperture radar (SAR) processes the signals received on a short array moved along a known track, as discussed by Skolnik [64]. Cutrona [65−67] suggested using the synthetic aperture technique in underwater acoustics and compared the resolution and signal-to-noise ratio obtained by synthetic aperture and conventional sonars.

The virtual broadside array formed by the SAS illuminates the same target on the seabed with several pulses during the SAS movement along its trajectory, as shown in Fig. 10.66. The synthetic aperture image with improved along-track resolution is produced by coherently processing the data from a substantial number of echoes from the target. To achieve this resolution the target has to be relatively immobile. Fig. 10.66 shows the SAS illuminating a target while the transmitter array moves a distance D, the SAS length, along the straight trajectory. The SAS length is determined by the two positions where the target is just illuminated by the sonar beam within its −3 dB beam width. The SAS *along-track resolution* is:

$$AR_{SAS} = R\frac{\lambda}{2D} \tag{10.101}$$

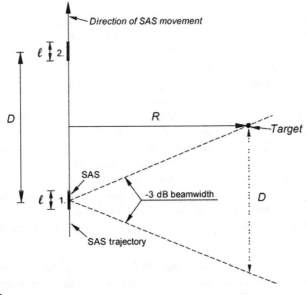

FIGURE 10.66

SAS movement along a straight line trajectory over the distance D. The target is continuously illuminated while the transmitter moves with echoes received from the target at R during movement from position 1 to 2.

According to Cutrona [65], the factor 2 is due to the round trip phase information associated with signal received by each segment of the synthetic aperture as it moves along the illumination track. When the physical length of the transmitter array is ℓ (unit: m), its aperture angle is determined by the ratio, λ/ℓ, and the distance D, based on the geometry in Fig. 10.66, is given by:

$$D = R\frac{\lambda}{\ell} \tag{10.102}$$

Inserting Eq. (10.102) into Eq. (10.101) yields the following simple expression for the SAS along-track resolution:

$$AR_{SAS} = \frac{\ell}{2} \tag{10.103}$$

The along-track resolution is independent of frequency and the across-track range R, and depends only on the transmitter length, ℓ. Thus shorter transmitter arrays provide improved along-track resolution. However, reducing the array length, ℓ, can have a negative influence on the speed the transmitter/receiver array can be moved along the track. To receive the target echo without building grating lobes, the receiving array should not move more than half of the array length, ℓ_r (unit: m) during the time delay between consecutive echoes. If the receiving array speed along the track is v (unit: m/s), the following relation for v may be derived:

$$\frac{2R_{max}}{c} \le \frac{\ell_r}{2v} \quad \text{or} \quad v \le \frac{\ell_r c}{4R_{max}} \tag{10.104}$$

where R_{max} denotes the *SAS maximum range*. Eq. (10.104) shows that R_{max} is proportional to the receiving array length. To increase the along track speed to cover a reasonable seafloor area per hour, ℓ_r has to be increased. A balance is required between the along-track resolution and area covered by the SAS per hour. The theoretical along-track resolution given by Eq. (10.104) is not always achievable. Practical resolutions are frequently 1.5 to 2 times lower.

The frequency-independent SAS along-track resolution given by Eq. (10.104) allows the SAS to use a lower operational frequency than an SSS system to achieve the same resolution. The lower frequency allows the SAS acoustic signal to penetrate slightly into soft bottom sediments, which may enable buried objects detection, which is not possible when using a high-frequency SSS.

For many years the SAS was considered a topic of academic interest compared with the synthetic aperture radar. Some *physical limitations* and *strict requirements* for the SAS slowed the development. These limitations and requirements were as follows:

- Primarily, it was supposed that the platform stability and control carrying the SAS, the towfish for instance, was not high enough to maintain an adequate linear trajectory and, therefore, coherent processing would be impossible.
- The transducer array trajectory accuracy had to be determined better than $\lambda/8$.

- Underwater environment inhomogeneities would produce perturbations of the sound propagation velocity to and from the target, as discussed by R.E. Hansen et al. [68]. Also, since the SAS is near-field imaging, the transition between time and space used during the beam-forming process would require an accurate knowledge about the sound velocity along the signal path to and from the target and sound velocity errors would degrade the image quality due to defocusing.
- The much lower sound velocity compared to the electromagnetic wave velocity would cause a rather low travel speed for the SAS system as seen from Eq. (10.104).

However, in recent years the developments in electronics and signal processing have permitted the development of attitude sensors for more accurate navigation. *Autofocusing* including phase gradient autofocus (PGA) originally developed for SAR and *micronavigation* including inertial navigation allows corrections for sound velocity errors and SAS array movements. Micronavigation aims at correcting errors in the measurement geometry caused by incorrect navigation or an incorrect projection plane where an out-of-plane motion has caused projection errors. Micronavigation can be improved by using a gyrostabilized antenna technique which combines data-driven motion estimates with external attitude sensors, as discussed by Bellettini et al. [69]. R.E. Hansen et al. [70] have shown that by applying near-field beam forming to the SAS antenna, the SAS gain in along-track resolution could be a factor of 10–100 relative to an SSS. Ref. [70] also gives a simple rule-of-thumb for the relation between the deviation, Δz, of the SAS path from a straight vehicle track and the required height accuracy of the seafloor topography, Δh:

$$\Delta z \cdot \Delta h \leq \frac{R}{5f} \tag{10.105}$$

where R (unit: m) is the range normal to the track and f (unit: kHz) is the SAS frequency. For $\Delta z = 0.5$ m and a 100 kHz SAS frequency the required seafloor height accuracy at range $R = 75$ m is $\Delta h \leq 0.3$ m. This height restriction relaxes with increasing range.

A high-resolution synthetic aperture sonar is HISAS 1030 produced by Kongsberg in Norway. The HISAS 1030 is shown in Fig. 10.67 on board the HUGIN 1000 AUV during HISAS 1030 testing in Horten, Norway in 2009. To obtain information about the imaging plane and therefore the bathymetry of the imaged seafloor, the HISAS 1030 is equipped with an interferometry system, where coarse bathymetry is calculated based on the use of a side-scan interferometer. This system's theoretical range-independent resolution (across-track × along-track) is 2 × 2 cm. The practical resolution is 5 × 5 cm at ranges out to about 200 m on both sides of an AUV, which moves at a speed of 2 m/s, and 260 m for a speed of 1.5 m/s. At a 260 m/s speed it is possible, even in cluttered environments, to detect and classify mines and other small objects with an area coverage rate of about 730 m²/s. The SAS operates in a frequency range of 60–120 kHz and its system bandwidth is up to 50 kHz. The sonar dimensions for the transmitter array are 0.32 × 0.18 m and the dimensions for the two receiver arrays are 1.27 × 0.11 m.

FIGURE 10.67

Kongsberg HISAS 1030 where the small transducers are the transmitter. The longer transducers above and below the transmitter are the receivers.

Copyright Kongsberg Maritime AS and Norwegian Defence Research Establishment (FFI).

Data collected using a Kongsberg HISAS 1030 Interferometric SAS on the HUGIN 1000 AUV is shown in Fig. 10.68. This figure shows an interferometric SAS image of the wreck of the 1500 deadweight ton (dwt) oil tanker, HOLMENGRAA, sunk during World War II. The length of the wreck is 68 m, and the width is 9 m. The range to the image center is 95 m and the image size is

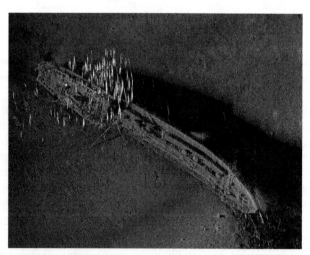

FIGURE 10.68

A HISAS 1030 produced image of the 1500 dwt oil tanker, HOLMENGRAA, sunk during World War II. The image was made at a range of more than 300 m.

Copyright Kongsberg Maritime AS and Norwegian Defence Research Establishment (FFI).

60×78 m. The theoretical resolution in the image is around 3×3 cm, and the resolution in the bathymetry is 18×18 cm. The sonar image is a fusion of interferometric SAS image, bathymetry, and coherence. The red lines on top of the bridge of the ship in the upper left corner of the image represent a school of fish.

10.7 OTHER SONAR TYPES

In addition to the SBESs and MBESs, side-scan and synthetic aperture sonars discussed in previous sections, underwater acoustic transducers are used in a broad variety of applications, such as positioning, navigation, velocity, and depth measurements; military detection and tracking of torpedoes and mines, as well as in their homing facilities; fishery; and communications and data transmission. Transducers used in fishery are discussed in detail in Chapter 12, Bio- and Fishery Acoustics, and Communications are dealt with in Chapter 11, Signal Processing. Although many new sonar types and their transducers have been developed and adapted to specific applications over recent years, this section focuses on few important and extensively used transducers, such as acoustic Doppler current profilers (ADCPs) for velocity measurements, acoustic transponders for underwater intervention, and sonobuoys and variable depth sonars for detection and tracking.

For nearly 30 years commercially available ADCPs have been used to measure water current velocities at a range of depths, sea surface wave motion, and vessel speeds across the seafloor. ADCPs exploit the Doppler effect discussed in Chapter 2, which produces a frequency shift between the transmitted acoustic signal and echo received from objects such as scatterers in the water column, sea surface, or seafloor. For water column scatterers echoes are range gated into bins along three or more beams and then combined to infer the velocity profile. As shown in Fig. 10.69, when the acoustic beam axis forms an angle φ with the vertical direction, the different velocities in the flow velocity profile at various heights above the ADCP produce a series of frequency shifts. The signal frequency, f_0 (unit: Hz), transmitted by the ADCP mounted on the seafloor, is shifted at the vertical distance, z, due to the local flow velocity, $V(z)$ (unit: m/s). The shifted frequency, $f(z)$, is given by:

$$f(z) = f_0 \left(1 - \frac{V(z)\sin \varphi}{c} \right) \tag{10.106}$$

At z_1 the frequency shift f_1 due to flow velocity V_1 is given by $f_0(1 - V_1\sin\varphi/c)$. The frequency shift at bins in various heights z above the ADCP is then given by:

$$\Delta f(z) = f_0 - f(z) = \frac{f_0 V(z)\sin\varphi}{c} \tag{10.107}$$

The frequency shift $\Delta f(z)_r$ in the echo received at the ADCP due to the sound signal traveling to and from the scattering region at z is $\Delta f(z)_r = 2\Delta f(z)$. Therefore, the *flow velocity* $V(z)$ in the acoustic beam at different bin heights z is given by:

$$V(z) = \frac{c\Delta f(z)_r}{2f_0 \sin\varphi} \tag{10.108}$$

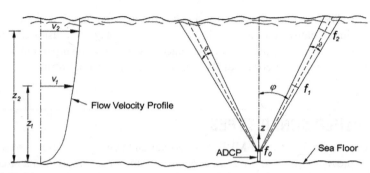

FIGURE 10.69

The flow velocity profile measured by a bottom-mounted ADCP. φ is the tilt angle between the ADCP main axis (the z-direction) and the ADCP transducer head's acoustic axis. θ is the -3 dB beam angle which normally ranges from < 1 to 4 degrees. V_1 and V_2 and f_1 and f_2 are the flow velocities and the Doppler influenced frequencies, respectively, at heights z_1 and z_2.

where c (unit: m/s) is the in situ sound velocity at z. As long as the variability of c as a function of z is not significant relative to $V(z)$ water column variations in c can be ignored.

Backscatters in the water column, which are assumed to be passively following the flow, can be inhomogeneities, such as minor gas bubbles and plankton. To increase the inhomogeneity backscattering level it is necessary to use higher frequencies. ADCPs are available for *frequencies* ranging from about 75 kHz to over 1.2 MHz. The Teledyne RD Instrument's Doppler Volume Sampler (DVS) can operate at 2.4 MHz. High frequencies produce a narrow beam with improved flow velocity resolution in the bins. However, they reduce the range, for example, the maximum DVS range is only 3–5 m. An ADCP operating at 300 kHz has a 300 m maximum range.

The ADCP may have three or more *transducer heads* with the acoustic axis forming an angle φ (normally around 20 degrees) with the ADCP main axis. ADCPs on the seafloor are generally mounted normal to the seafloor. The -3 dB beam width, which depends on the transducer diameter and the frequency, will range from below 1 degree to about 4 degrees. The ADCP has electronics necessary for signal conditioning and processing including an accurate clock for the time-gate facility and a memory. For specific applications the ADCP may include a compass (e.g., a flux-gate type), a pressure sensor for depth information, and pitch and roll sensors when movement is involved. The ADCPs normally operate on long-life batteries unless they are hard wired to shore. The information gathered by the ADCP is stored in memory and after retrieving the ADCP, transferred to a computer for further processing and display of the measured time series of velocity distributions. Information may also be transferred to land via a radio link.

The ADCP qualities which determine range, resolution, and measurement accuracy are the beam absorption and spreading losses, transmitted pulse source level and echo level, pulse bandwidth which controls range resolution, acoustic beam width which is mainly controlled by the transducer size and frequency, sound speed in the water, and quality of the signal processing performed by the ADCP.

The ADCPs are used for shallow- and deep-water *moored applications*. They are used to measure flow velocities with the beams looking vertical or horizontal into the water column. The horizontal looking system is used for measuring flow through channels, rivers, and harbors. The vertical looking systems include bottom- or vessel-mounted ADCPs. The signal frequency depends on deployment depth with decreasing frequency for increasing deployment depth. The bottom-mounted ADCP is used to measure flow in the water column or to monitor surface wave height distributions and direction of surface wave motion. The *vessel-mounted* ADCP may act as a *Doppler velocity log* (DVL), where signals from the DVL are reflected from the seafloor to the DVL. The echo signals are used to monitor the speed of the surface vessel or of a subsea vehicle across the seafloor. If submarines, AUVs, and other subsea vehicles start positions are known together with their compass heading and acceleration determined from the DVL measurements, the vehicles current position may be calculated. As a navigation instrument the DVL is useful for subsea vehicles in situations where a global positioning system (GPS) cannot be used. If the seafloor cannot be reached by the DVL signal due to depth and frequency used, backscattered signals from the water column can be used for navigation, when corrected for the currents in the vehicle's environment. A further DVL development is the *correlation log*, where the echo signals from the seafloor received by elements in an array are correlated to determine the maximum cross-correlation between the echoes. The ratio of the distance between two array elements with maximum cross-correlation and the time delay between echoes gives the vessel speed over the seafloor.

For accurate surface vessel, subsea vehicle and swim diver positioning related to underwater intervention for seafloor oilfield exploitation, seafloor minerals, and archaeological studies in shallow and deep waters, techniques exploiting *acoustic transponders* have been significantly enhanced during recent years. The transponders are a further development of the simple acoustic *pingers*, which transmit acoustic signals, in most cases only a short tone burst at fixed time intervals, to indicate the presence of an underwater object like an AUV, an ROV, or a diver and provide a basis for their location. Pingers are also used to help locate lost valuable cargoes, downed airplanes, and seafloor facilities in offshore oilfields. They typically operate at 8−25 kHz, which is mainly determined by the desired detection range. These transponders are able to transmit and receive coded acoustic signals and are able to "reply" to a recognized coded acoustic interrogation signal by transmitting its own, frequently also coded, acoustic signal. Several acoustic transponders may form an *underwater acoustic positioning system*. These systems may be termed *long baseline* (LBL) systems, *short baseline* (SBL) systems, and *ultra-short baseline* (USBL) systems.

The *LBL system* uses three or more transponders mounted on the seafloor. The relative transponder locations are known or each transponder's global coordinates are known as accurately as possible. A pinger mounted on an AUV can transmit acoustic signals which are received by the baseline transponders which may be connected through cables with a surface vessel or more frequently with a radio link. Based on an accurate clock the pinger signal's arrival time to each baseline transponder is determined and through triangulation the AUV's position is determined. Sometimes, the AUV depth is controlled by a pressure sensor, and its measurement signals are received by a surface vessel. If the AUV has mounted a transponder, i.e., an interrogator, the baseline transponders can reply to the interrogator signals received. By using an accurate clock the replies received by the interrogator are stored for later processing. The stored signals can be used to determine the AUV track. The position determination accuracy can be between 0.1 and 1 m, depending on the clock quality, signal frequency, distances between the baseline transponders and the AUVs, and inhomogeneities influencing the velocity of sound in the water. Procedures are necessary for on-site LBL system calibration.

Recently, fixed positioned transponders on the seafloor have been replaced by self-positioned, free-floating buoys using GPS systems. Such a system of intelligent buoys is flexible since it can be moved from place to place, is easy to calibrate, and is less exposed to multipath effects since only one-way acoustic signals are used.

The *SBL systems* consist for instance of a transmitter (pinger/transponder) mounted on the seafloor or on a subsea vehicle and three or more individual receiver transducers mounted near each other on a platform and which are connected to the same signal processing device. The receiver transducers can be mounted in a row on the surface vessel's bottom to form an antenna, where the individual time of fight of the signal from the transmitter to a receiver determines the relative position of transmitter and receiver. For known receiver positions in the array the transmitter position can be calculated. While an SBL system is easier to use than an LBL system, the SBL system accuracy depends on the array's transducer spacing. For wider spacing the SBL measurement accuracy will be more like the accuracy obtained from an LBL system.

The *USBL system* uses a small transducer array to measure the *distance* through time-of-flight measurements between a transmitter (pinger/transponder) and the array transducers. It also measures the *direction* to the transmitter by measuring the phase shift between the signals received by the individual transducer array elements. The distance and direction information determines the transmitter and its platform position relative to the receiver array. The receiver array may be mounted on the surface vessel bottom or on a rod over the vessel's side. Also, moon pools onboard vessels are used to lower the receiver array into the water. The vessel motion can be accounted for by using GPS, compass and speed measurements and signals from sensors to measure the vessel's pitch, roll, and heaving. Although USBL systems do not require seafloor-mounted transducers, direction measurements are sensitive to influences from inhomogeneities in the underwater environment, which may produce refraction, reflection and multipath effects, in particular when the transmitter is far from the receiver array and the accumulation of the effects amplifies their influence. In many cases, USBL system's accuracy is less than what LBL systems can provide.

Due to the influence on sound propagation in the sea caused by the sound velocity profile and its variations with time and position as discussed in Chapter 1, underwater object detection, identification, and tracking by sonar can become difficult. Special "acoustic shadow zones," where a submarine, AUV, or swim diver may hid and avoid detection, are formed in seas, shallow water, and even harbors, due to temperature and salinity effects. The Baltic Sea is a sea region where strong effects of seasonal temperature variations and fresh- and salt-water mixing from out- and influx through the Danish straits are found. This makes sonar operations at various depths necessary. One way to penetrate into depths with better acoustic reception, i.e., below the surface thermocline, is the use of *sonobuoys* or a variable depth sounder (VDS). The sonobuoy, as its name implies, is a combination of a sonar and buoy. It is a small cylindrical device, frequently about 0.1−0.2 m in diameter and about 1 m long, which can be dropped from an airplane or helicopter or from a surface vessel to listen to underwater activities. Sonobuoys are frequently expendable. After the sonobuoy hits the water surface a float with a UHF/VHF, radio transmitter becomes loose and a hydrophone, or hydrophone array, connected to the radio transmitter sinks together with a position stabilizing load into the water to a predetermined depth. Acoustic signals received by the hydrophone(s) are transmitted to the airplane or surface vessel for further action. Using several hydrophones provides the possibility to find the direction to the underwater object. The sonobuoy can be *passive* and listen to signals from objects, such as propeller noise, machinery vibrations, and transient sounds in the water, or it may be *active* by emitting pings. The ping reflections from the underwater object are received by the hydrophones and transmitted to the airplane or surface vessel. Several sonobuoys may be distributed on the sea surface in patterns to provide better detection, localization, and tracking by simple triangulation techniques. Sonobuoys are also used for other purposes, such as oceanographic studies of temperature and salinity profiles, search, and rescue operations by listening to signals from pingers on crashed airplanes, and to two-way communication between an airplane or a surface vessel and a submarine.

The VDS is also used to overcome the difficulties of hunting for submarines which can hide in acoustic shadow zones formed by the downward refraction of acoustic waves due to the negative sound velocity gradient near the sea surface. Just after World War II it was discovered that a sonar set lowered into the water at appropriate depths gave improved detection ranges for submerged targets, over hull-mounted sonar sets. In a VDS, the sonar transducer arrays for transmission and reception are integrated into a towed body cabled to the surface vessel where the signal processing and display take place. The towed body is normally depth rated to several hundreds of meters to permit the sonar head to reach the best signal propagation depth. The VDS operation depth is determined by cable length and tow vessel speed. VDS are also used for mine hunting, such as the Raytheon developed AN/SQQ-32 variable depth mine-hunting sonar, which in the 1990s became operational for the US Navy. This mine-hunting sonar contributed substantially to the discrimination between real mines and mine-like contacts, and it reduced the false alarm numbers. During recent years helicopter-dipped variable depth sonar has been increasingly used to penetrate into the sea to reach better detection depths. Towed arrays, as discussed in Section 10.2.5.2.3, are also used by Navies to listen to underwater acoustic signals from subsea vehicles operating below the surface thermocline. By towing the array at longer distances from the tow vessel the influence of the vessel self-noise can also be strongly reduced.

10.8 TRANSDUCER CALIBRATION

Some basic transducer characteristics have to be known before use as a hydrophone or projector in order to make reliable and reproducible measurements. The transducer has to be *calibrated*. The calibration is only valid within the ranges of frequency, hydrostatic pressure, and temperature, where it was performed. An extrapolation from these conditions is frequently not possible. Since a calibrated transducer may change its calibrated values, a new calibration before and after performing critical measurements involving the transducer is frequently recommended. The primary requirements of a hydrophone or projector are stability and linearity. The hydrophone sensitivity should in general be independent of frequency, time, and the environmental conditions it is operating under. A projector must produce a controllable, stable sound field when it is used. Projector linearity is desired, but may be sacrificed if high-amplitude signals are important.

10.8.1 DEFINITIONS

For all transducers, hydrophones, and projectors, a receiving and a transmitting response may be defined, see Bobber [39]. The *receiving response M* (unit: V/μPa), the free-field voltage sensitivity, is the ratio of the open-circuit output voltage e_0 (unit: V) across the transducer terminals to the free-field acoustic pressure p_r (unit: μPa) received at the transducer position before the transducer is introduced into the sound field, i.e.:

$$M = \frac{e_0}{p_r} \tag{10.109}$$

The *transmitting current response S_A* (unit: μPa/A) is the ratio of the acoustic pressure p_t produced at a nominal transducer distance of 1 m on the transducer's acoustic axis to the input current i_i (unit: A) at the transducer's electrical terminals. This is given by:

$$S_A = \frac{p_t}{i_i} \tag{10.110}$$

The transmitting response may also be based on the input voltage to the transducer terminals. The *transmitting voltage response S_V* (unit: μPa/V) is the ratio of the transmitted acoustic pressure p_t produced 1 m from the center of the transducer on its acoustic axis to the input voltage e_i (unit: V) applied to the transducers electrical terminals. This is given by:

$$S_V = \frac{p_t}{e_i} \tag{10.111}$$

The transmitting response is referred to the pressure at the distance 1 m from the acoustic center of the projector in a *free field*. A perfect free field is defined as a field in a homogeneous and isotropic medium, without any disturbances arising from boundaries, inhomogeneities such as gas bubbles, marine life and temperature gradients or flow. Such a field is an idealization which is nearly impossible to produce. However, it can under certain circumstances be assumed. Large lakes, pools and larger water-filled tanks can be used as an approximation to a free field. These free field requirements can be met by using a ping or an acoustic impulse short enough to finish the measurements before reflections from boundaries or inhomogeneities are received by the transducer. The reference to a 1 m distance does not mean that the measurement shall be done at a point 1 m from the projectors acoustic center. Frequently this point is in the near field of the projector, or for large cylindrical or spherical projectors the point may be inside the projector. Measurements can instead be done in the far field at a distance, where the acoustic pressure is inversely proportional to distance, and the pressure can then be extrapolated back to the distance of 1 m from the projectors acoustic center.

Two calibration methods, *the primary method* and *the secondary method*, are used. The *primary method* is a fundamental method, which requires measurements of voltage, current, electrical and acoustical impedances, distance, and frequency, while the *secondary method* is a comparison method, where a transducer, normally a hydrophone, calibrated by a primary method, is used as a reference or *standard*. The hydrophones used as standards after being calibrated through a primary method, must maintain stable sensitivity over large temperature and pressure ranges, as well as over long times.

The secondary calibration method is simplest to perform, and it has fewer error sources than the primary method. However, its reliability is strongly dependent on the quality of the standard hydrophone calibration. The secondary calibration method consists of subjecting the hydrophone to be calibrated and the standard hydrophone to the same free-field pressure from a projector, and then comparing the electrical output voltages of the two hydrophones. When the standard hydrophone's receiving response and its open-circuit output voltage are M_s and e_s, respectively, and the same qualities of the hydrophone to be calibrated are M_c and e_c, respectively, M_c may be found from:

$$M_c = M_s \frac{e_s}{e_c} \tag{10.112}$$

Some sources of error may influence secondary calibration methods such as deviations in positioning of the two transducers relative to the projector, standard hydrophone instability, lack of true free-field conditions, lack of a satisfactory signal-to-noise ratio, and failure to measure the voltage under true open-circuit conditions.

Secondary hydrophone calibration can also be done by using a standard projector that produces a known acoustic field. The hydrophone is then placed in this acoustic field beyond the minimum projector distance and the hydrophone's output voltage is measured. Because the presence of the hydrophone disturbs the acoustic field produced by the projector, and therefore the hydrophone's output voltage, i.e., its *free-field sensitivity*, is related to the calculated acoustic pressure in the free field existing before the hydrophone insertion. Depending on the calibration frequency and hydrophone dimension the pressure field across the hydrophone surface may exhibit considerable variations, which makes it difficult to relate the hydrophone output voltage to any particular pressure value. If the hydrophone dimensions are much smaller than the acoustic wavelength, the acoustic pressure across the hydrophone surface may be considered to be constant and the output voltage may be related to the actual acoustic pressure on the hydrophone, which gives the hydrophone's *pressure sensitivity*. The pressure sensitivity can be said to be the hydrophone's sensitivity, when it is exposed to a uniform overall pressure. A small hydrophone, which only causes a vanishing disturbance of the acoustic field, will below resonance have nearly the same free-field and pressure sensitivities. For hydrophones calibrated at higher frequencies a relation between applied pressure and free-field pressure must be known to find the hydrophone's free-field sensitivity, see Stansfield [13].

Two important concepts in relation to transducer calibration are *linearity* and *dynamic range*. A transducer is said to be linear, if its output is proportional to its input, i.e., that the relation between output and input is independent of the absolute values of input and output over a wide range of signal levels. For a projector the output is its free-field pressure and its input is the current or voltage. For a hydrophone the output is its open-circuit voltage and the input is the free-field pressure. The concept definition as for instance transmitting current/voltage response, free-field voltage sensitivity, electrical and acoustical impedances, etc. are only strictly valid for linear transducers. For *nonlinear* transducers the measurement conditions must be completely specified and measured data cannot be extrapolated. Moreover, in a nonlinear transducer formation of harmonics may take place, and the spectral composition of the output and input signals may deviate.

The concept of *dynamic range* is in general only applied to hydrophones. The hydrophone dynamic range describes the signal amplitude range over which the hydrophone can be used to detect and measure acoustic pressures. The dynamic range is normally given in dB for the difference between a maximum and a minimum acoustic pressure level for using the hydrophone. The maximum level may be set by occurrence of nonlinear signal distortion, acoustic saturation or hydrophone damage. The minimum level may be set by the signal-to-noise level being equal to 1 or by the minimum detectable signal level. The dynamic range and its limits are less precisely defined than the concept of transducer linearity, and a transducer may have a maximum level for its dynamic range which is higher than the level where nonlinearity occurs, unless the maximum level is defined as the nonlinearity level.

10.8.2 RECIPROCITY CALIBRATIONS

Reciprocity calibration includes a series of calibration methods. These methods all depend on one transducer to be reciprocal, i.e., the ratio of the transducers receiving response M to its transmitting current response S_A must be equal to a constant, J, called the *reciprocity parameter*:

$$J = \frac{M}{S_A} \tag{10.113}$$

The necessary conditions for a transducer to be *reciprocal* are that it must be linear, passive, and reversible. The *conventional reciprocity calibration method*, also called the three-transducer spherical-wave reciprocity method, is the most widely used. The three transducers, see Fig. 10.70, include T_p, which serves as a projector only; T_r, which is reciprocal and serves as both projector and hydrophone; and T_h, which only serves as a hydrophone. While any one of the three transducers can be the unknown to be calibrated, it is normal to select T_h to be the transducer to be calibrated. All calibration measurements are made under free-field conditions in the transducers far fields to preserve spherical wave conditions for the waves impinging on the transducers. The transducer arrangement with the same separation distance d (unit: m) between all three transducers is shown in Fig. 10.70. If the separation d between the projector T_p and T_r and T_h is large enough to permit spherical divergence of the acoustical energy, the reciprocity parameter J (unit: VA/(μPa)^2 or m^{4}s/kg) may be written as:

$$J = 2\frac{\lambda d_r}{\rho c} = 2\frac{d_r}{\rho f} \tag{10.114}$$

where λ, ρ, c, f, and d_r denote the acoustic wavelength (unit: m), water density (unit: kg/m^3), sound velocity (unit: m/s), acoustic frequency (unit: Hz), and the reference

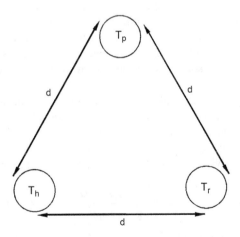

FIGURE 10.70

The conventional reciprocity calibration method utilizing three transducers, the projector T_p, the reciprocal transducer T_r, and the hydrophone to be calibrated T_h, all positioned with the same distance d between their acoustical centers.

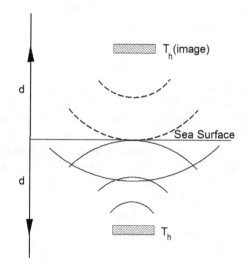

FIGURE 10.71

Self-reciprocity calibration using the sea surface. The transmitting and receiving transducer T_h to be calibrated is at depth d below the sea surface. The image transducer $T_{r(image)}$ is virtually positioned at distance d above the sea surface. The full wave fronts are real acoustic waves and the dashed wave fronts are imaginary.

distance (unit: m) from the transmitter, where the transmitted acoustic pressure is specified in the definition of S_A, respectively. d_r is normally taken as $d_r = 1$ m.

The *first step* in the calibration course is that T_p transmits and produces a free-field sound pressure p_p at the position of T_h and T_r at the distance d from T_p, where $p_p = i_{ip}S_{Ap}d_r/d$. S_{Ap} denotes the transmitting current response of T_p and i_{ip} is the input current at the terminals of T_p. The open-circuit voltages e_{ohp} of T_h and e_{orp} of T_r may now be expressed by $e_{ohp} = M_h p_p$ and $e_{orp} = M_r p_p$, respectively. M_h and M_r denote the receiving responses of T_h and T_r, respectively. Therefore, the ratio $e_{ohp}/e_{orp} = M_h/M_r$.

In the *second step* the reciprocal transducer T_r acts as a projector and produces a free-field sound pressure $p_r = i_{ir}S_{Ar}d_r/d$ at T_h. S_{Ar} denotes the transmitting current response of T_r and i_{ir} is the input current at the terminals of T_r. The open-circuit voltage e_{ohr} of T_h is now $e_{ohr} = M_h p_r$. Since T_r is reciprocal, $M_r = S_{Ar}J$, the unknown free-field voltage sensitivity of the hydrophone M_h (unit: V/μPa) is now be given by:

$$M_h = \sqrt{\frac{e_{ohr}e_{ohp}dJ}{i_{ir}e_{orp}d_r}}$$ (10.115)

If the ratio d/d_r in Eq. (10.115) is combined with J in Eq. (10.114), i.e., J is defined as $J = 2\,d/\rho f$, or for $d_r = d = 1$ m, Eq. (10.115) becomes:

$$M_h = \sqrt{\frac{e_{ohr}e_{ohp}J}{i_{ir}e_{orp}}}$$ (10.116)

Eq. (10.116) is fundamental to reciprocity calibration considered as a true primary method. A simpler, true primary method, is *self-reciprocity calibration*, where only the transducer to be calibrated is required. In this procedure the transmitted signal produced by the transducer is reflected back, for instance by the sea surface, to the transducer which receives its own signal. Like in the Lloyd's mirror effect, see Chapter 2, the image of the transmitting transducer can be thought of as a second transducer at the same distance from the reflecting surface and at the opposite side of the surface as the real transmitting transducer as shown in Fig. 10.71. Theoretically, it must be assumed that the reflection is perfect, which means that the transmitting current response S_A of the image is identical to that of the real transducer. If this assumption is satisfied and the transducer is reversible, the unknown free-field voltage sensitivity M_h (unit: V/μPa) of the transducer can be expressed by:

$$M_h = \sqrt{\frac{e_{oh}J}{i_{ih}}} \tag{10.117}$$

where e_{oh} denotes the open-circuit voltage over the terminals of the transducer during reception, i_{ih} denotes the transmission current, and J is the reciprocity parameter. When the reciprocity parameter is known, M_h can be determined from Eq. (10.117), when measurements of e_{oh} and i_{ih} have been performed.

Also cylindrical waves can be used in a primary calibration method, the *cylindrical-wave reciprocity*. This procedure requires that only cylindrical waves propagate between the projector and hydrophones. This condition exists between two long, thin and parallel, cylindrical transducers, where a two-dimensional signal spreading can be assumed when the cylinder axis distance d (unit: m) satisfies the condition $\lambda/2 < d < L^2/\lambda$. L (unit: m) is the length of the cylindrical transducers and λ (unit: m) is the acoustic wavelength. The calibration setup is similar to Fig. 10.70, where the three spherical transducers are replaced by three cylindrical transducers. Then the free-field voltage sensitivity M_c (unit: V/μPa) for the cylindrical hydrophone can be found from Eq. (10.117), where M_c replaces M_h, and J (unit: VA/(μPa)^2 or m^{4}s/kg) according to Bobber and Sabin [71] is:

$$J = \frac{2L\sqrt{d\lambda}}{\rho c} \tag{10.118}$$

Similar to cylindrical-wave reciprocity, where only cylindrical waves are propagated between the projector and the hydrophones, *plane-wave reciprocity*, where only plane, progressive waves are propagated between the projector and hydrophone can be established. When the distance between two large and plane, circular piston transducers is so small that the plane hydrophone is in the near field of the plane projector, that is when the transducer distance $d < r^2/\lambda$, where r (unit: m) denotes the radius of the plane, circular piston projector, a nondivergent and collimated beam of plane progressive waves exists between the projector and the hydrophone. To avoid standing waves, pulsed signals are used, and therefore d cannot be shorter than some wavelengths. The free-field voltage sensitivity M_p may be found from

Eq. (10.117), where the reciprocity parameter J according to Simmons and Urick [72] is given by:

$$J = \frac{2A}{\rho c} \qquad (10.119)$$

where A is the area of the radiating projector surface, or the area of the collimated beam.

In the low-frequency regime *coupler reciprocity* can be used for absolute calibration of hydrophones. The coupler is a small liquid- or gas-filled chamber which is used to couple projector and hydrophones together acoustically. The chamber dimensions must be small compared with the acoustic wavelength in the liquid or gas, and the chamber walls must have acoustical impedance much higher than the liquid or gas in the chamber. If these conditions are satisfied, the acoustic pressure produced by the projector will essentially be the same everywhere in the chamber. If three transducers are coupled to each other through the liquid or gas in the chamber, and if the transducers are small and, like the chamber walls, possess a low acoustical compliance, the three-transducer reciprocity described foregoing can be used for a reciprocity calibration at low frequencies, where the free-field voltage sensitivity M_c may be found from Eq. (10.117). The reciprocity parameter J will then be:

$$J = 2\pi f \beta V \qquad (10.120)$$

where f is the acoustic frequency (unit: Hz), and V and β denote the liquid or gas volume (unit: m^3) in the chamber and the adiabatic compressibility, or the fractional volume change per unit pressure (unit: Pa^{-1} of the liquid or gas, respectively).

Coupler reciprocity is particularly useful when hydrophone calibration at higher static pressures must be performed. The chamber design and its tightness to liquid or gas are major issues at higher static pressures. The liquid can be water or castor oil, and a coupler, used for reciprocity calibration in the frequency range from 10 Hz to 8 kHz, for temperatures from -2 to 35°C and static pressures from 0 to 105 MPa, is described by Van Buren and Blue [73].

10.8.3 OTHER CALIBRATION METHODS

The frequently used *pistonphone method* for calibration of microphones in air has been adapted to hydrophone pressure calibration in either air or water. Normally, the pistonphone is a small air-filled chamber, where the pressure variation is produced by a small piston driven by a small electric motor or by a moving coil. Since the chamber, like in the coupler reciprocity calibration must be small compared to the wavelength, only calibration at frequencies up to 200–250 Hz can be done in air. While the wavelength for the same frequency is about 4.4 times longer in water than in air, the use of water in the chamber will permit pistonphone calibration at higher frequencies, while still meeting the requirement that the sound pressure is uniform throughout the chamber.

A *water-filled tube* can be used for comparing pressure sensitivities of small hydrophones up to about 3 kHz. The hydrophone is immersed in the water in the tube and the acoustic pressure wave is produced by a projector at the tube bottom. The tube diameter must be small compared with the wavelength such that only plane acoustic waves in the fundamental tube propagation mode are used. This can be done for frequencies far below the cutoff frequency for higher modes. Moreover, the tube wall must be so thick and rigid, that the wall is not acting as an acoustic waveguide.

With increasing transducer size existing test facilities, tanks, ponds, and lakes will not permit calibration to take place through measurements in the transducer's far fields similar to free-field conditions. This limitation drove the development of a *near-field technique*. With an array of small hydrophones, mutually spaced by less than one wavelength, the transmitting array sound field can be sampled in amplitude and phase close to the transmitting array surface. The far-field array beam patterns can be calculated from the sampled data. Through this near-field calibration technique nearly uniform plane acoustic waves can be produced over a broad frequency range, and space, time, and money can be saved. Planar near-field calibration arrays used as projectors and receivers have been developed for the US Navy [73] and a cylindrical near-field calibration array has been discussed by Van Buren [74].

Vector sensors are normally calibrated in terms of pressure rather than particle velocity. Because the relationship between pressure and particle velocity depends on the wave type (plane, cylindrical, or spherical) and on boundary conditions, these conditions must be known to determine the vector sensor's sensitivity. Since plane, progressive waves are specified for determining the free-field voltage sensitivity of transducer, the degree to which plane, progressive waves are established must be known. In a plane, progressive acoustic wave the relationship between the acoustic pressure p (unit: Pa) and particle velocity u (unit: m/s) is $p/u = \rho c$. However, for spherical waves produced by a point source the relation is:

$$\frac{p}{u} = \rho c \sqrt{1 + \left(\frac{\lambda}{2\pi r}\right)^2} \qquad (10.121)$$

where r (unit: m) and λ (unit: m) are the source distance and wavelength, respectively, and where ρ (unit: kg/m^3) and c (unit: m/s) are the water density and sound velocity, respectively.

Because a vector sensor only can be calibrated in a secondary method through direct comparison with a standard pressure hydrophone if plane waves impinge on both transducers, a correction must be introduced when deviations from plane wave conditions occur. This correction can be determined from Eq. (10.121) where the bracket determines the correction factor to be subtracted from the measured vector sensor sensitivity, when it is calibrated by comparison with a standard pressure hydrophone. For $r/\lambda = 0.5$, the correction factor, $10 \cdot log \, [1 + (\lambda/2\pi r)^2]^{1/2} = 0.2$ dB, which shows that for a source-to-transducer distance greater than half a wavelength the correction factor, is negligible. From Eq. (10.121), the particle velocity can be calculated when the vector sensor sensitivity is related to particle velocity instead of the measured free-field acoustic pressure.

One of the most commonly used methods for sonar calibration which was originally developed for SBESs is the *standard target calibration method*. This method involves backscattering from a copper or tungsten carbide ball of known target strength mounted in a water tank or beneath the hull of a ship and insonified by the beam from the sonar. Because the target strength of a ball is well defined as discussed in Chapter 5, knowledge about the product ka, where a (unit: m) is the radius of the ball and k (unit: m^{-1}) is the acoustic wave number, provides a well-defined backscattering pattern, where sonar calibration with a precision of 0.1 dB can be obtained, as discussed by Foote and MacLennan [75]. Calibration at ka-values in the geometrical scattering region where the target strength is frequency independent is a good basis for reliable sonar calibrations. Also MBESs have been calibrated by using a standard target as discussed by Chu et al. [76].

The calibration of a low-frequency, 12 and 24 kHz, *multibeam sonar*, where the sonar dimensions make the standard target calibration method difficult, is reported by Pocwiardowski et al. [62]. The receiver calibration is carried out by using a calibrated internal reference source with the signal transferred to each receiver channel and the complex response of each channel measured. This calibration method requires that the signal magnitude and phase for each individual receiver channel is determined as a function of amplifier gain taking into consideration the effects of TVG. The spread in gains and phases from channel to channel is used to correct raw element data streams prior to the beam-forming operation, as a standard deviation of the phase of 10 degrees may lead to an increase in side lobe levels as high as 7−10 dB.

A method for calibrating of *towed-line arrays* in a sound field in a long water-filled cylindrical tube was reported by Luker et al. [77]. In the tube the sound field may appear as a plane wave field incident at any direction on the line array placed in the tube. The wave field is produced by a number of projectors mounted along an axial line in the tube wall with a center-to-center spacing less than about 1/6 of a wavelength. A line of hydrophones is mounted in another axial line in the tube wall diametrically opposite the projectors. Before inserting the towed-line array into the tube, the projectors are driven, one by one, and the complex received voltage responses of all hydrophones are recorded. This results in a transfer matrix, which after inversion can be used to determine the amplitude and phase with which to drive each projector to obtain any desired plane wave incidence on the towed array. After mounting the towed-line array along the axis of the cylindrical tube, the entire directivity pattern of the array can be obtained by varying the angle of incidence of the plane waves. A further development of the cylindrical calibration chamber for towed-line arrays is reported by Luker and Zalesak [78], where a steel tube of inner diameter 0.289 m and wall thickness of 0.0175 m is used. The tube dimensions permit simulation of water depths up to 410 m and calibration of towed-line arrays over the frequency range from 10 to 600 Hz.

A technique which conveniently can be used for hydrophone calibration and performance evaluation is the *time-delay spectrometry* (*TDS*), reported by Pedersen et al. [79]. This technique uses a frequency signal swept across a preselected frequency band transmitted by a projector and the sweep of a narrow-band filter in the receive signal processing starting when the transmitted signal front arrives at the hydrophone. Optimal performance of the TDS system depends on the filter bandwidth selection, sweep time, sweep rate, frequency range, and signal-to-noise ratio. One advantage of the TDS system is that a free-field condition for the calibration is not necessary since the sweep rate of the filter in general will exclude the influence of reflected signals from the hydrophone environments, since the filter, due to its characteristics will filter out the reflected signals when they arrive at the hydrophone site, producing apparent free-field conditions. The calibration over a frequency range obtained by using the frequency sweep and the TDS technique is for many calibration situations superior to time-gated tone bursts.

Depending on the transducer size and desired frequency band for calibration the calibration can be performed in a water tank, lake, or at sea. *Tank calibrations* have been extensively used since they offer the possibility to control the acoustics of the environment which frequently can be difficult at sea. The tank size sets the limitations for the frequency band over which it provides a reliable and reproducible calibration. Projector signal reflections from the tank sides, tank bottom, and water surface together with the signal frequency limit the maximum usable distance between the projector and the hydrophone to be calibrated. Ideally, the tank dimensions should allow many tens of wavelengths to permit the use of tone burst calibration with time gating on the hydrophone side to isolate the directly received signal from the later arrival of any reverberation due to reflection from the tank walls, bottom, and water surface. The most frequent problem is that resonance transducers often possess a Q-value on the order of 10 or more, which means that the transient rise time to a measurable flat signal pulse and the fall time after the flat signal pulse part last about 10 cycles each. A certain number of cycles of the flat pulse part are also necessary to carry out reliable measurements. This sums up to a pulse of about 40–50 cycles. If the frequency is 10 kHz the length of the pulse will be between 6 and 7.5 m. This dimension and the arrival time for the reverberations, which should be so late that interference with the flat pulse part can be avoided, limit how small a tank is usable for calibration at 10 kHz. For measurements below about 10 kHz, the tank size may become impractically large. Calibrations at lower frequencies will then have to be done in lakes or in other larger bodies of water.

The influence of reverberation can be reduced by lining the tank walls, tank bottom, and water surface with an acoustically absorbing lining, an *anechoic lining*, frequently produced from butyl rubber loaded with cork and aluminum powder to give the lining material an acoustical impedance close to water and to improve absorption due to scattering by the cork particles and due to anelastic effects in the butyl rubber. Such a tank shown in Fig. 10.72 for calibration at frequencies from about 10 to 300 kHz is reported by Bjørnø and Kjeldgaard [80]. The tank lining consists of wedges of the lining material glued on plates of marine plywood.

Lake facilities and their applications are described in an elegant and exhaustive way by Bobber [39]. Even if free-field conditions are met in a water tank or a lake, other factors may influence and limit the calibration measurement accuracy. These factors are: (1) interference from the 50 Hz power line (in United States 60 Hz) and

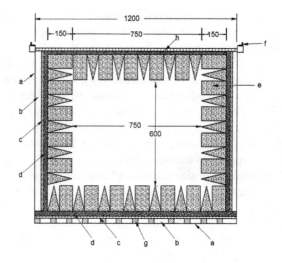

FIGURE 10.72

The *anechoic water tank* described by Bjørnø and Kjeldgaard [80]. Here (A) the steel
tank with wall thickness 0.003 m; (B) a 40 mm water space or quartz sand/water layer;
(C) the marine plywood with plate thickness 0.012 m; (D) a layer of cork and 0.025 m
thick aluminum powder loaded butyl rubber; (E) wedges of loaded butyl rubber with the
base dimensions 0.075 × 0.075 m and height 0.150 m; (F) a track for transducer
positioning equipment; (G) pinewood sticks with cross section 0.03 × 0.03 m and length
0.1 m; and (H) a loaded butyl rubber lined lid of the same sandwich construction found in
the tank bottom and side linings.

Figure reproduced from Bjørnø, L. and Kjeldgaard, M., A Wide Frequency Band Anechoic Water Tank,
Acustica, 32, (2), pp. 103–109, 1975, with permission of Acta Acustica united with Acustica.

from nearby radio stations at frequencies from 0.5 to 2 MHz; (2) interference from
reflections from items not including the tank or lake boundaries such as transducer
adapters and inhomogeneities in the water column, standing waves in the propagation path between the transmitter and receiver and electrical cross talk between the
transmitter and receiver sides; (3) interference with electrical and acoustical noise
from environmental conditions such as nearby acoustical and electrical systems,
traffic, and weather conditions; and (4) gas bubbles in the acoustical signal transmission path, bubbles attached to the transducers or trapped in small holes, cracks, etc.
In particular gas bubbles can be a nuisance for transducer calibration. For tank calibrations a solution is often to wait some days after filling the tank with water before
carrying out measurements. The addition of drops of detergents to the water to
reduce the surface tension on the bubble surface can be useful. Also letting the projector operate for hours before the calibration measurements take place can excite
the gas bubbles and cause them to grow due to rectified diffusion during the pulsation phases and migrate to the water surface. As long as the personnel who carry out
the calibration are aware of these factors, they should be able to find ways to abate
and eliminate them.

10.9 SONAR SYSTEM EXAMPLE CALCULATIONS

This section provides examples of how to use the sonar equations in passive and active operations with the sonar systems described in the previous sections.

Example 10.9.1

Can a target be detected by a vertical hydrophone array using a broadband square law detector under the following conditions?

- The hydrophone array has $N = 11$ elements with an element center-to-center distance equal to $\lambda/2$ at 1 kHz.
- The ambient noise level is 65 dB rel 1 μPa at 1 kHz.
- The detection and false-alarm probabilities are assumed to have the values $P_d = 70\%$ and $P_{fa} = 1\%$.
- The target is emitting broadband noise with $SL = 110$ dB.
- The target–receiver distance, r, is 2000 m.
- The sound propagation is assumed to be spherical.
- The broadband square law detector system has a bandwidth, $B = 2$ kHz and integration time, $T = 1$ s.
- The number of successive time samples necessary to make the detection decision is assumed to be $n = 25$.

The calculation is performed in the following steps:

1. Based on P_d and P_{fa} the ROC curves in Fig. 1.30, Chapter 1 gives $d = 9$ and 5 *log* $d = 4.8$ dB.
2. From Eq. (10.73) for $N \gg 1$, the array directivity index $DI = 10$ *log* $N = 10.4$ dB.
3. From Chapter 4 the absorption coefficient at 1 kHz in seawater can be found to be $\alpha = 6 \cdot 10^{-2}$ dB/km.
4. The transmission loss for spherical propagation is $TL = 20$ *log* $r + \alpha r \cdot 10^{-3}$ dB for α in dB/km, therefore $TL = 66.1$ dB.
5. The bandwidth–time product, $BT = 2000$ and 5 *log* $BT = 16.5$ dB.
6. With the number of successive time samples $n = 25$, 5 *log* $n = 7$ dB.
7. The values in Eq. (1.88), Chapter 1 become $110 - 66.1 - 65 + 10.4 - 4.8 + 16.5 + 7 = 8$ dB ≥ 0. Therefore, the target can be detected by the hydrophone array.

Example 10.9.2

A SBES is mounted on the bottom of a vessel with its acoustic axis pointed vertically downward toward the seafloor. The SBES transducer transmits an acoustic intensity just below the cavitation threshold. The signal reception is influenced by ambient noise and self-noise. What is the maximum vertical distance to the seafloor, where echoes from the seafloor can be received by the SBES when the vessel speed is $V = 22$ *kt*?

The SBES system has the following qualities:

- The echo sounder emits continuous wave (CW) pulses with a carrier frequency $f_0 = 24$ kHz and pulse length $T = 10$ ms.
- The transducer acts as a circular piston source with effective diameter $d_e = 0.25$ m.
- The transducer is situated at a depth $z_t = 4$ m below the sea surface.
- The detection probability $P_d = 90\%$ and the false-alarm probability is $P_{fa} = 0.1\%$ and the number of successive time samples necessary to make the detection decision is $n = 3$.

The environmental conditions are the following:

- The cavitation threshold at $z_t = 4$ m is $I_{ca} = 3.2 \cdot 10^3$ W/m^2.
- The ambient noise level is $NL_a = 36$ dB.
- The sound absorption coefficient in seawater at $f_0 = 24$ kHz is $\alpha = 8 \cdot 10^{-3}$ dB/m.
- The seafloor is plain and consists of silty sand with the density $\rho_s = 1750$ kg/m^3 and the sound speed $c_s = 1620$ m/s.

The calculation is performed in the following steps:

1. The wavelength is now $\lambda = c_w/f_0 = 0.0625$ m and the wave number $k = 2\pi/\lambda = 100.5$ m^{-1}.
2. The half-power beam width is $2\theta_{-3dB} = 2 \cdot \sin^{-1}(1.617/ka) = 14.8$ degrees.
3. The effective transducer area is $A_e = \pi d_e^2/4 = 0.049$ m^2.
4. The acoustic power emitted is $P = A_e I_{ca} = 157$ W.
5. The directivity index is according to Eq. (10.55), $DI = 10 \ log \ D = 10 \ log \ (ka)^2 = 22$ dB as $ka = 12.6 \gg 1$.
6. The emitted source level is given by Eq. (1.77), Chapter 1 since $SL = 10 \ log \ P + 170.8 + DI = 214$ dB relative to the intensity produced by an rms pressure of 1 μPa at 1 m.
7. The self-noise level, NL_s, is given by Eq. (6.20). Thus, when the vessel speed is $V = 22 \ kt = 22 \cdot 1.8532/3600 = 11.3$ m/s, it will be $NL_s \sim (47 + 0.1887V^2) \ 71$ dB.
8. Since the ambient noise and the self-noise are assumed uncorrelated, the total noise level is determined by adding the acoustic energies in the two noise contributions. Since the self-noise level $= 71$ dB $= 10 \ log \ (I_s/I_{ref})$, the intensity $I_s = 8.4 \cdot 10^{-12}$ W/m^2. The ambient noise can be determined by a similar process which yields $I_a = 2.7 \cdot 10^{-15}$ W/m^2. Since $I_s + I_a \sim I_s$, the total noise level is $NL = 71$ dB.
9. The detection threshold DT for a noise limited active sonar is found by using Fig. 1.30, Chapter 1 when $P_d = 90\%$ and $P_{fa} = 0.1\%$, which gives the detection index $d = 21$ and $DT = 5 \ log \ d - 10 \ log \ T - 5 \ log \ n = 24.2$ dB.
10. Based on Chapter 2 the intensity reflection coefficient R_I at the seafloor for vertical incidence is $R_I = [(r_2 - r_1)/(r_2 + r_1)]^2$, where the specific acoustic impedances r_1 and r_2 are $r_1 = \rho_w c_w$ and $r_2 = \rho_s c_s$, where the indices w and s denote the seawater and the seafloor, respectively. The intensity reflection coefficient is then $R_I = 0.091$ and the reflected intensity level is $TS = 10 \ log \ R_I = -10.3$ dB.

11. The two-way transmission loss for spherical propagation of the signals from the SBES is $2\ TL = 40\ log\ r + 2 \cdot \alpha \cdot r \cdot 10^{-3}$ dB.
12. The sonar equation for an active sonar operating under noise limited conditions is given by Eq. (1.88), Chapter 1 as $SL - 2\ TL + TS - NL + DI - DT \geq 0$, which can be rewritten as $2\ TL \leq SL + TS - NL + DI - DT = 130.5$ dB.
13. From this expression the permitted seafloor depth $r \leq 840$ m can be found. Therefore, at the vessel speed of 22 knot a return signal from the seafloor can still be detected if the water depth is less than 840 m.

Example 10.9.3

An SBES used for finding fish schools operates vertically over a seafloor at a depth, $H = 600$ m. What source level SL is it necessary to emit from the SBES to detect (a) the seafloor and (b) the fish school when?

- The echo sounder emits 42 kHz CW pulses and acts like a circular piston source with a diameter, $D = 0.5$ m.
- A fish school at the depth, $H_f = 110$ m has an average distance between the individual fish, $h = 0.1$ m.
- The measured ambient noise level is $NL = 40$ dB rel 1 μPa at 42 kHz.
- The echo sounder system's detection threshold is assumed to have $5\ log\ d = 6$ dB and $5\ log\ n = 3.5$ dB.
- The seafloor target strength is assumed to be $TS_s = 15$ dB and the fish school has a target strength $TS_f = -30$ dB.

Given the above inputs to Eq. (1.92), Chapter 1, the active sonar equation, the remaining inputs can be determined by the following steps:

1. To resolve the individual fish in the fish school the CW pulses must have a duration of $T = 2\ h/c$, where $c = 1500$ m/s. This leads to $T = 0.133 \cdot 10^{-3}$ s.
2. The directivity index DI of the SBES can be found from Eq. (10.55) by using Eq. (10.54) for $ka = 44 \gg 1$. Eq. (10.55) now gives $DI = 33$ dB.
3. When the signal spreads spherically the transmission loss, TL, is given by $TL = 20\ log\ r + \alpha r$, where $\alpha = 10^{-2}$ dB/m at 42 kHz in seawater, and where r (unit: m) is the target distance. This leads to the following values of TL for the signals to (a) the seafloor and (b) the fish school:
   a. $TL_s = 20\ log\ 600 + 10^{-2}600 = 61.6$ dB.
   b. $TL_f = 20\ log\ 100 + 10^{-2}100 = 41$ dB.

Eq. (1.92), Chapter 1, the active sonar equation, gives the following source levels necessary to detect (a) the seafloor and (b) the fish school:

a. $SL_s = 2\ TL - TS + NL - DI + 5\ log\ d - 10\ log\ T - 5\ log\ n = 156.5$ dB.
b. $SL_f = 160.3$ dB.

A 161 dB source level is sufficient for detecting the fish school and the seafloor. Much higher source levels are normally available from SBESs, where maximum SL-values can range from 210 dB to above 230 dB. The practical limit of these SL-values is determined by cavitation, which for the SBES used in this example will be at an emitted power, $P = 400$ W for CW emission. The source level produced by emission of 400 W is: $SL = 10\ log\ P/P_r + 170.8 + DtI = 229.8$ dB *relative to the intensity produced by an rms pressure of* 1 μPa at 1 m. The reference power $P_r = 1$ W.

The pulsed emission and the elevated frequency (42 kHz) will permit even higher emitted power before reaching the cavitation threshold, since the cavitation threshold increases nearly linear with frequency above 10 kHz, as pointed out by Esche [81], and the cavitation is always delayed a fraction of a second due to the time it takes to activate the nuclei for the cavitation. This delay causes an increase in the cavitation threshold for very short sound pulses.

Example 10.9.4

A hull-mounted mine hunting sonar on a vessel operates 160 m above a sandy seafloor. Can a mine be detected in the reverberation limited environment produced by backscattering from the seafloor when?

- A mine is positioned 10 m above the seafloor at a distance, $r_m = 440$ m from the mine hunting sonar measured along its acoustic axis.
- The acoustic axis grazing angle, $\theta = 20$ degrees and the distance along the acoustic axis between the mine hunting sonar and the seafloor is $r_s = 470$ m.
- The mine-hunting sonar emits CW pulses with a carrier frequency, $f = 150$ kHz and duration $T = 10^{-2}$ s. The source level $SL = 205$ dB rel 1 μPa at 1 m.
- The mine-hunting sonar operates as a circular piston source with radius, $a = 0.2$ m and with bandwidth, $B = 10$ kHz.
- The detection threshold is determined by $P_d = 90\%$ and $P_{fa} = 1\%$. The number of successive time samples necessary to make the detection decision is $n = 5$.
- The mine's target strength is $TS_m = -15$ dB and the backscattering strength of 1 m^2 of the sandy seafloor at the grazing angle $\theta = 20$ degrees is $S_{bs} = -30$ dB.
- The absorption coefficient $\alpha = 6 \cdot 10^{-2}$ dB/m at $f = 150$ kHz.

The solution is obtained by performing the following steps:

1. The sonar equation for surface reverberation limited detection, Eq. (1.93), Chapter 1, becomes:

$$SL - 2TL_m + TS_m - RL_s - 5\ log\ d + 10\ log\ BT + 5\ log\ n \geq 0$$

2. The transmission loss between the sonar and the mine is $TL_m = 20\ log\ r_m + \alpha r_m$.
3. The surface reverberation level, RL_s, provided in Eq. (1.80), Chapter 1, is:

$$RL_s = SL - 2TL_s + 10\ log\left[\frac{\Phi r_s c\ T}{2cos\theta}\right] + S_{bs}$$

4. The transmission loss SL_s over the path r_s between the sonar and the seafloor, is

$$TL_s = 20\ log\ r_s + \alpha r_s$$

5. When TL_s is inserted into Eq. (1.79), Chapter 1, we find that

$$RL_s = SL - 30\ log\ r_s - 2\alpha r_s + 10\ log\ \Phi + 10\ log\left(\frac{cT}{2\cos\theta}\right) + S_{bs}.$$

6. From Table 5.8 we find that $10\ log\ \Phi = 10\ log\left\{\frac{\lambda}{2\pi a}\right\} + 6.9$ dB for $a > 2\lambda$.
7. This leads to the following values:

$$30\ log\ r_s + 2\alpha r_s = 30\ log\ 470 + 2 \cdot 6 \cdot 470 \cdot 10^{-2} = 137\ \text{dB}$$

$$TL_m = 20\ log\ 440 + 6 \cdot 10^{-2} 440 = 79\ \text{dB}$$

$$\lambda = \frac{c}{f} = \frac{1500}{15 \cdot 10^4} = 10^{-2}\text{m}$$

$$10\ log\ \Phi = 10\ log\left(\frac{10^{-2}}{2\pi \cdot 0.2}\right) + 6.9 = -14\ \text{dB}$$

$$10\ log\left(\frac{cT}{2\cos\theta}\right) = 10\ log\left(\frac{1500 \cdot 10^{-2}}{2 \cdot 0.94}\right) = 9\ \text{dB}$$

$$RL_s = 205 - 137 - 14 + 9 - 30 = 33\ \text{dB}$$

8. The ROC curves provided in Fig. 1.30, Chapter 1, give $d = 15$ for $P_d = 90\%$ and $P_{fa} = 1\%$, thus

$$DT = 5\ log\ 15 - 10\ log\ 10^{-2} \cdot 10^4 - 5\ log\ 5 = -18\ \text{dB}$$

9. As a result, Eq. (1.93), Chapter 1, becomes:

$$205 - 2 \cdot 79 - 15 - 33 + 18 = 17\ \text{dB} \geq 0$$

Thus, the mine can be detected in the reverberation limited environment for these conditions.

Example 10.9.5

An active sonar is mounted on the hull of a surface vessel and used for detecting torpedoes. Will it be possible to detect a torpedo so early that there will be time enough to take evasive action by the vessel when the following operational conditions are present?

- The hull-mounted active sonar emits CW pulses with carrier frequency $f_0 = 8$ kHz and pulse length $T = 0.1$ s.

- The sonar source level is $SL = 210$ dB and the sonar uses the same transducer to emit and receive.
- The active sonar acts like a piston source with an effective radius, $a = 0.15$ m.
- The torpedo target strength is $TS = -16$ dB.
- The detection probability is $P_d = 99\%$ and the false-alarm probability is $P_{fa} = 1\%$.
- The number of successive time samples necessary to make the detection decision is $n = 4$.
- The ambient noise level is $NL = 50$ dB and the detection is assumed to be noise limited.
- The signal propagation is assumed to be spherical and the acoustic absorption coefficient in seawater at 8 kHz is $\alpha = 3 \cdot 10^{-4}$ dB/m.

The solution is obtained by performing the following steps:

1. The wavelength $\lambda = c/f_0 = 0.188$ m, the wave number $k = 2\pi/\lambda = 33.5$ m^{-1} and $ka = 5$.
2. The directivity index is $DI \sim 10 \ log \ (ka)^2 = 14$ dB.
3. Fig. 1.30, Chapter 1, gives for $P_d = 99\%$ and $P_{fa} = 1\%$, the detection index $d = 23$.
4. The detection threshold $DT = 5 \ log \ d - 10 \ log \ T - 5 \ log \ n = -6.2$ dB.
5. The two-way transmission loss $2 \ TL$ for spherical signal propagation is spherical signal propagation is

$$2TL = 40 \ log \ r + 2 \ \alpha r = 40 \ log \ r + 6 \ r \ 10^{-4}$$

6. Eq. (1.94), Chapter 1, gives: $2TL \leq SL + TS - NL + DI - DT = 164.2$ dB from which the range $r = 9250$ m can be found.

The noise limited detection range $r = 9250$ m should give the vessel time enough to take evasive actions. If the speed of the torpedo relative to the vessel is 50 kt, it will take nearly 6 min from the torpedo detection until it reaches the vessel.

Example 10.9.6

Determine the maximum swath width of an MBES, i.e., the maximum distance to the seafloor at which the MBES can give a detectable return signal under the following conditions:

- Everywhere the sound velocity is 1500 m/s—i.e., isovelocity and no refraction influence.
- The sonar signal contains CW pulses with a duration, $T = 10^{-4}$ s.
- The carrier frequency is $f = 200$ kHz.
- The source level $SL = 200$ dB rel 1 μPa at 1 m.
- The bandwidth is $B = 10$ kHz.
- The detection index is $d = 36$ and the number of successive time samples is $n = 4$.

- The directivity index is $DI = 5$.
- The noise level $NL = 35$ dB.
- The backscattering from the seafloor takes place according Lambert's law with

$$10 \ log \ \mu = -20 \text{ dB.}$$

- The water depth below the MBES is $H = 250$ m.
- The absorption coefficient in seawater at 200 kHz is $\alpha = 8 \cdot 10^{-2}$ dB/m.

The solution is obtained by performing the following steps:

1. The reverberation level due to backscattering from the seafloor is provided by Eq. (1.79), Chapter 1. When Lambert's law, the directivity index, DI, the noise level NL, and the detection threshold DT are inserted into Eq. (1.79), Chapter 1, the sonar equation becomes:

$$SL - 30 \ log \ H + 30 \ log \ sin\theta - \frac{2\alpha H}{sin\theta} + 10 \ log \ \Phi + 10 \ log \left[\frac{cT}{2cos\theta} \right]$$
$$+ \ 10 \ log \ \mu + 20 \ log \ sin\theta - NL + DI - DT \ \geq \ 0$$

2. The minimum grazing angle θ_m for a given water depth H below the MBES, which satisfies the sonar equation, can now be found from the expression:

$$50 \ log \ sin\theta - 10 \ log \ cos\theta - \frac{2\alpha h}{sin\theta} \geq$$
$$30 \ log \ h - SL - 10 \ log \ \Phi - 10 \ log \left(\frac{cT}{2} \right) - 10 \ log \ \mu + NL - DI + DT$$

3. By insertion of available information for the MBES and its environment, the following equation for determination of $\theta = \theta_m$ can be found:

$$50 \ log \ sin\theta_m - 10 \ log \ cos\theta_m - \frac{40}{sin\theta_m} = -70 \text{ dB.}$$

4. Solve for θ_m by taking an initial guess, such as 45 degrees, compare that answer to -70 dB, then use interval halving to make a second guess, and converge on the answer. This approach yields a minimum grazing angle $\theta_m = 40.4$ degrees and the maximum swath angle is 99.2 degrees.

10.10 **SONAR DESIGN CALCULATIONS**

Calculations and design of sonar projectors and hydrophones are in most cases considered a business secret. Only few books on sonar calculations and design have been published, such as the books by Bobber [39], Wilson [12], and Stansfield [13], the youngest of which is more than 20 years old. In 2007 Sherman and Butler [16] published a book dealing with aspects of transducer calculations and design. In general information on sonar calculations and design has to be found in journal articles scattered over many different international journals. These articles frequently are characterized by the lack of some very essential information, which are considered business secrets. Thus, a newcomer in the field has to start on their own and make several mistakes others already have made, but not published. Although transducer calculations and design by many is considered an art for the few, there is a need for a book dedicated to this subject. This section provides a discussion on sonar calculations and design and some hints about what to do and what not to do.

10.10.1 TONPILZ TRANSDUCER AND HYDROPHONE CALCULATIONS

As mentioned in Section 10.2.4.4 on Tonpilz transducers the high-power transfer efficiency of this transducer can only be obtained within a narrow frequency band around the transducer's resonance frequency due to its high mechanical quality factor Q_m. Decreasing the Q_m-value will broaden the bandwidth. In order to maintain a high transfer efficiency, $kQ_m \geq 1$, where k is the electromechanical coupling coefficient, discussed in Section 10.2.3. In a well-designed transducer the effective coupling coefficient can be as high as 0.5, and $Q_m \geq 2$, which gives a bandwidth of 50% around the resonance frequency.

The desire to broaden the bandwidth has led to inclusion of flexural vibrations near the longitudinal piston mode resonance frequency as discussed by Yao et al. [14]. However, it is difficult to keep the same high efficiency over the increased bandwidth and to maintain the impedance within useful limits. It is possible to avoid head mass flexural vibrations by using cylindrical piezoceramic tube elements. The head mass diameter should not exceed half a wavelength in water at the transducer's longitudinal piston mode resonance frequency and the head mass thickness should be more than one-fifth of the head mass diameter to move the lowest flexural mode resonance frequency to values well above the longitudinal piston mode resonance frequency.

For calculation and design of a Tonpilz transducer such as the one in Fig. 10.73, the half-wave resonator formed by the Tonpilz transducer is divided into two quarter-wave resonators, both having the same resonance frequency. In Fig. 10.73, d_h and t_h are the head mass diameter and thickness, respectively, and d_b and t_b are the tail mass diameter and thickness, respectively. w_p and ℓ_{pn} are the diameter and length of the piezoceramic driving elements, respectively. All dimensions have unit m. Indices $n = 1$ and 2 in the length ℓ_{pn} refer to the two parts of the driving elements, i.e., (1) the part connected with the head mass and (2) the part connected with the tail mass. For the quarter-wave resonators the following equations can be used to determine the dimensions of the transducer parts:

$$\tan\left(\frac{2\pi f \ell_{p1}}{c_p}\right) \tan\left(\frac{2\pi f t_h}{c_{m1}}\right) = \frac{\rho_p c_p w^2}{\rho_{m1} c_{m1} d_h^2} \qquad (10.122)$$

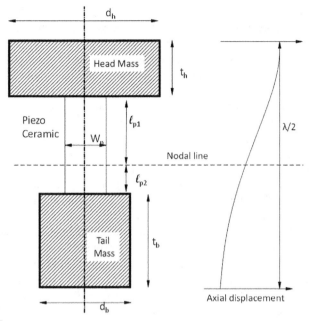

FIGURE 10.73

Tonpilz transducer shown schematically, consisting of a head mass, tail mass, and piezoceramic driving elements. The nodal line, where the displacement is zero, forms the line along which the transducer is fixed to its housing. On both sides of the nodal line the transducer parts form a quarter-wave resonator whose axial displacement is shown in the figure. The transducer's dimensions are defined in the text.

which is used for the head mass and its piezoceramic part and

$$\tan\left(\frac{2\pi f \ell_{p2}}{c_p}\right) \tan\left(\frac{2\pi f t_b}{c_{m2}}\right) = \frac{\rho_p c_p w^2}{\rho_{m2} c_{m2} d_b^2} \tag{10.123}$$

which is used for the tail mass and its piezoceramic part.

In Eqs. (10.122) and (10.123) c_p (unit: m/s) denotes the sound speed in the piezoceramic material and c_{m1} and c_{m2} denote the sound speed in the head mass and tail mass, respectively. Since the head and tail masses most frequently are made from different materials—the head mass from titanium or, when protected from seawater by an epoxy or rubber coating, from aluminum or magnesium, and the tail mass from steel, tungsten, or brass—their sound speed c and density ρ (unit: kg/m³), represented by ρ_{m1} and ρ_{m2} for the head and tail masses, respectively, are different. f (unit: s⁻¹) denotes the common resonance frequency for the two quarter-wave resonators. By inserting appropriate material values in Eqs. (10.122) and (10.123) and selecting the head mass resonance frequency f (unit: Hz) and diameter d_h (unit: m) a little smaller than half a wavelength in water at the frequency, f, and head mass thickness t_h between one-fifth and one-quarter of the head mass diameter, and the piezoceramic diameter w_p to be one-quarter to one-third of the head mass diameter, a calculation of the length of the piezoceramic ℓ_{p1} can be carried out by using Eq. (10.122). The tail mass diameter d_b will normally be smaller than the head mass diameter d_h and the tail mass length t_b will be longer than t_h in order to preserve the diameter w_p for both piezoceramic parts. Selection of d_b and t_b while w_p is preserved, and inserting into Eq. (10.123) will give the length ℓ_{p2} of the lower part of the piezoceramic element driving the transducer.

Example 10.10.1

A Tonpilz transducer has a head mass consisting of titanium with $\rho = 4500$ kg/m^3 and $c = 6070$ m/s. The brass tail mass has $\rho = 8500$ kg/m^3 and $c = 4700$ m/s. The driving piezoceramic elements consist of Pz27 with $\rho = 7700$ kg/m^3 and $c = 4480$ m/s. The material constants for titanium and brass are taken from the Handbook of Chemistry and Physics [82]. The resonance frequency of the transducer is 15 kHz. The diameter of the head mass is $d_h = 0.045$ m, and its thickness is $t_h = 0.01$ m. The diameter of the piezoceramic driving elements is $w_p = 0.015$ m. Eq. (10.141) now gives the length $\ell_{p1} = 0.035$ m of the driving elements. The tail mass diameter is $d_b = 0.040$ m and its thickness is $t_b = 0.05$ m. Since the diameter of the piezoelectric driving elements still is $w_p = 0.015$ m, Eq. (10.142) now gives the driving element length $\ell_{p2} = 0.008$ m, see Fig. 10.73.

To keep the piezoceramic stack of ring-shaped disks or cylinders compressed between the head and tail masses, it is customary to use a *center bolt* as shown in Fig. 10.15. This bolt is normally produced from a material with lower stiffness than the piezoceramic. Since the piezoceramic stack is not able to withstand tensile stress, it is necessary to keep the stack under compressive force during all Tonpilz transducer vibrations. In general, the compressive force applied to the stack is about two times the maximum alternate force in the piezoceramic during transducer operation. The stress in the center bolt during sound transmission from the transducer can be found from the transducer characteristics and the applied electric field E (unit: V/m). According to Stansfield [13] the maximum stress σ_b (unit: Pa) in the bolt can be found from:

$$\sigma_b = \frac{F_b}{A_b} = 2\frac{K}{K_b}E_b k Q_m E \sqrt{\frac{2\eta_{ma}\varepsilon}{E_e}} \tag{10.124}$$

In Eq. (10.124), F_b (unit: N) and A_b (unit: m^2) denote the force axially along the center bolt to keep the piezoceramic under compression and the effective cross-sectional area of the center bolt, respectively. K and K_b (unit: m^2/N) are the stiffness of the piezoceramic stack and the center bolt, respectively. E_b and E_e (unit: Pa) are Young's moduli for the center bolt and the piezoceramic material, respectively. η_{ma} denotes the Tonpilz transducer mechanical−acoustical efficiency. ε is the dimensionless piezoceramic relative dielectric constant.

Example 10.10.2

When the center bolt is made from beryllium copper and the piezoceramic is Pz27 from Table 10.1, the stress σ_b in the center bolt can be calculated for the following typical material qualities:

Beryllium copper: $E_b = 12.8 \cdot 10^{10}$ Pa; $K_b = 2.32 \cdot 10^{-12}$ m²/N.
Pz27: $K = 2.32 \cdot 10^{-11}$ m²/N; $E_e = 1.13 \cdot 10^{11}$ Pa; $\varepsilon = 1130$.
The Tonpilz transducer: $kQ_m = 1.2$; $\eta_{ma} = 0.9$; $E = 2.5 \cdot 10^5$ V/m.

Inserting the values into Eq. (10.143) gives: $\sigma_b = 307$ MPa.

Since the typical yield stress for beryllium copper is above 1000 MPa, it should be safe to apply a stress of 307 MPa. Since the alternating stresses in the center bolt during the vibration of the Tonpilz transducer may lead to formation of fatigue cracks in the bolt material, this should be considered in the transducer's design phase. Since material general fatigue tests are only based on an evaluation of the material's ability to survive 10^8 stress cycles, Stansfield's advice is [13] to not let the bolt stress exceed 35–40% of the tensile strength for materials such as beryllium copper and 45–50% of the tensile strength for steel based bolts.

The production of a *stack of piezoceramic elements*, disks, or cylinders, also causes other problems for the transducer designer. The electrodes—frequently a silver layer—on the elements must be stable and strong enough to withstand soldering and other handling during the production process. The elements must be without cracks, which frequently originate from the burning and poling process, and the element dimensions must be within specified limits. The joints between the elements will give rise to a change in resonance frequency and coupling coefficient. The joints will lead to energy dissipation and therefore to temperature increases depending on the drive level and duty cycle. Serious temperature increases may lead to de-poling and to crack formation in the piezoceramic elements. The contact to electrodes between elements requires that the element's silver layer has been ground to establish an improved contact, avoiding point contacts, between elements and electrodes, and to avoid any substantial deviation in the stack compliance from the elements own compliance. Epoxy resins are most frequently used in joints to improve tensile strength. Before assembling the stack, all components must be carefully cleaned for contaminations arising from earlier handling of the elements and electrodes.

To improve the source level and the bandwidth of the Tonpilz transducers, the piezoceramic, normally a PZT type, can be replaced by new lead magnesium niobate—lead titanate (PMN-28PT) single crystals, where higher piezoelectric coefficients and higher electromechanical coupling are found. A comparison between ceramic PMN-28PT and 81 Vol-% fiber-textured PMN-28PT in Tonpilz transducers by Brosnan et al. [83] shows that the textured type has linearity in source level as a function of the electric drive field up to $E = 230$ kV/m, and the maximum electromechanical coupling coefficient obtained by the 81 Vol-% textured PMN-28PT transducer under the high drive field is $k = 0.69$, which substantially improves the Tonpilz transducer quality.

In order to improve the transfer of sound into the water from a transducer, it sometimes is useful to exploit a *matching layer*. This is in particular used by a plate-shaped transducer, where one or more layers of various materials form a quarter of a wavelength in thickness at the plate's resonance frequency. This can give an impedance transition if the acoustic impedance load due to the water seen from the plate surface is (nearly) the same as the acoustic impedance of the plate material. If the specific acoustical impedance of the plate and the water is Z_p and Z_w, respectively, where $Z_p = \rho_p c_p$ and $Z_w = \rho_w c_w$, then the acoustical impedance Z_m of the matching layer must be:

$$Z_m = \sqrt{Z_p Z_w} \tag{10.125}$$

This matching layer impedance can partly be found by materials like aluminum, titanium, and metal powder loaded epoxy resins. It should be noted that the improved matching and energy transfer between the plate and the water only occurs at a single frequency, the plate's resonance frequency. This type of impedance matching is less efficient in broadband transducers. Goll [84] reports that multiple matching layer transducers with materials like glass and various plastic types have been used for broadband transducers.

10.10.2 EQUIVALENT CIRCUITS

The piezoceramic, underwater acoustic transducer may be considered to consist of three parts: an *acoustical part* with a moving surface in contact with the water; a *mechanical part*, where the body movement is controlled by forces; and an *electrical part*, where the current is controlled by voltage. Therefore, electrical equivalent circuits may represent the acoustical and mechanical parts which make it possible to simulate the whole transducer's function by electrical equivalent circuits. This can be used for transducer design and for evaluating the importance of individual transducers parts on the overall functioning of the transducer.

A simple vibrating mechanical system can be built from physical elements such as masses, springs, and dashpots (resistances), which form a *lumped parameter model*, where it is assumed that each physical element is smaller than one-quarter of a wavelength, masses are perfectly rigid so that no deformation takes place during the transducer operation and springs have stiffness and no mass. For a resonant mechanical system with mass, spring, and resistance, the equation of motion in terms of velocity u (unit: m/s) is:

$$M\frac{du}{dt} + R_m u + K_m \int u \, dt = F_m \sin \omega t \tag{10.126}$$

where M (unit: kg) is the mass moving with the velocity u, R_m (unit: N·s/m) is the mechanical resistance representing the friction in the mechanical system, $K_m = 1/C_m$ (unit: N/m) is the spring stiffness, C_m (unit: m/N) is the spring compliance, and F_m (unit: N) is the driving force. From Eq. (10.126) the resonance frequency f_0 and the mechanical quality factor Q_m can be found as:

$$f_0 = \frac{\omega_0}{2\pi} = \sqrt{\frac{K_m}{M}} \text{ and } Q_m = \frac{\sqrt{MK_m}}{R_m} \tag{10.127}$$

By using the *mechanical–electrical impedance analogies* in Table 10.8, a similar basic equation for a resonant electrical circuit can be written as:

$$L\frac{di}{dt} + Ri + \frac{1}{C}\int idt = U\sin\omega t \tag{10.128}$$

where L (unit: H) is the inductance, R (unit: Ω) is the resistance, C (unit: F) is the capacitance, and U (unit: V) is the voltage. This transition between mechanical and electrical units has been used in the electrical equivalent circuit for a piezoceramic transducer shown in Fig. 10.74. In this figure the capacitance C_m, the inductance L_m, and the resistance R_m form the electrical equivalents to the mechanical values, the elasticity or compliance C_m, the mass M, and the friction R_m.

Table 10.8 Equivalence Between Mechanical and Electrical Components

Mechanical Components	Unit	Electrical Components	Unit
Force (F)	N	Voltage (U)	V
Velocity (u)	m/s	Current (i)	A
Mass (M)	kg	Inductance (L)	H
Friction (mech. resistance) (R_m)	Ns/m	Resistance (R)	Ω
Compliance (elasticity) (C_m)	m/N	Capacitance (C)	F
Stiffness (K_m)	N/m	Inverse capacitance ($1/C$)	F^{-1}
Impedance ($Z_m = F/u$)	Ns/m	Impedance ($Z = U/i$)	$V/A = \Omega$

In general, a transducer can be represented by a four-terminal network with two electrical input terminals and two mechanical output terminals. A conversion between electrical and mechanical quantities takes place with an electromechanical transformation (or transduction) ratio $1:\varphi$, which is the factor that relates the piezoelectrically generated force F to the applied voltage U by $F = \varphi U$. If an alternating voltage $U \cdot \sin\omega t$ is applied to the input terminals, current i (unit: A) is produced in the input stage. At the output stage an alternating force $F \cdot \sin\omega t$ and a velocity u (unit: m/s) of the radiating transducer face are produced, when it is radiating into a radiation impedance $Z_r = R_r + jX_r$, where R_r and X_r are the radiation resistance and the radiation reactance, respectively, and $j = \sqrt{-1}$. For a circular piston source with radius a embedded in an infinite, rigid baffle, Eq. (10.42) gives for Z_r:

$$Z_r = \pi a^2 \rho c R_1(ka) + j\pi a^2 \rho c X_1(ka) \tag{10.129}$$

where $R_1(ka)$ and $X_1(ka)$ can be found from Eqs. (10.43) and (10.44).

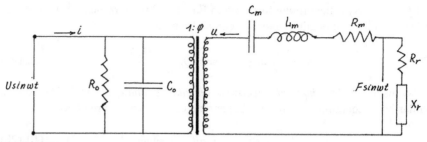

FIGURE 10.74

The lumped elements electrical equivalent circuit which characterizes a piezoceramic driven transducer. The transformation ratio $1{:}\varphi$ relates the mechanical units to equivalent electrical units. The input sinusoidal electrical voltage $U{\cdot}\sin\omega t$ produces the sinusoidal output transducer force $F{\cdot}\sin\omega t$.

Figure drawn by Leif Bjørnø.

The transducer's mechanical impedance Z_m at resonance given by the series circuit impedance shown on the piezoceramic transducer's mechanical side in Fig. 10.74 is:

$$Z_m = R_m + j\left\{\omega L_m - \frac{1}{\omega C_m}1\right\} \tag{10.130}$$

Then the input admittance Y_{in} on the electrical side of the transducer in Fig. 10.74 is:

$$Y_{in} = \frac{i}{U} = \frac{1}{R_0} + j\omega C_0 + \frac{\varphi^2}{Z_m + Z_r} \tag{10.131}$$

where the conversion from mechanical impedances to their equivalent electrical impedances is done by means of the impedance transformation ratio φ^2. φ^2 is generally used to transform all lumped mechanical impedance elements in a mechanical transducer system into their equivalent electrical counterparts. In Eq. (10.131) the first two terms, $1/R_0 + j\omega C_0$, represent the *blocked electrical input admittance* of the transducer, blocked means $u = 0$, and the last term represents the *motional admittance*, which includes the mechanical and acoustical elements. R_0 and C_0 on the electrical input side are the dielectric loss resistance and the capacitance of the dielectric transducer parts, respectively, for blocked transducer conditions.

Equivalent circuits are an easy way to get a first approximation of the significance of individual lumped transducer parts and their contribution to the overall transducer characteristic. However, equivalent circuits more act like a one-dimensional communication line and will not give any information about the physical structure of the lumped elements and their frequently 3-D motion. This limits the equivalent circuit usefulness for design and construction of piezoceramic-based transducers.

Example 10.10.3

For a piezoceramic transducer near its resonance frequency calculate the magnitudes of the lumped elements in its equivalent circuit, such as in Fig. 10.74, and its input admittance Y_{in}. The transducer consists of a solid, cylindrical, piezoceramic rod radiating into water at one end and clamped at the other end. The cylindrical piezoceramic rod, which is poled in its axial direction, has the following qualities:

Diameter $d = 0.03$ m.
Length $\ell = 0.06$ m.
Piezoelectric strain coefficient $d_{33} = 4.25 \cdot 10^{-10}$ m/V.
Mechanical compliance of the rod material $C_{mo} = 2.32 \cdot 10^{-11}$ m²/N.
Relative dielectric constant $\varepsilon = 1130$.
Density $\rho m = 7700$ kg/m³.
$\tan\delta = 0.017$.
Mechanical quality factor $Q_m = 80$.

The velocity of sound in the rod material is: $c = (\rho_m \cdot C_{mo})^{-\frac{1}{2}} = 2365$ m/s.
The specific acoustic impedance of the rod material is: $Z_a = \rho_m \cdot c = 1.82 \cdot 10^7$ kg/m²s.
Since the rod is clamped at one end, its resonance frequency will be $f_0 = c/4\ell = 9854$ Hz.
Since the wavelength in water at this resonance frequency is $\lambda_w = c/f_0 = 0.152$ m and $k = 2\pi/\lambda = 41.3$ m⁻¹, $ka = 0.619$.
The unblocked electrical impedance is $C = \varepsilon \cdot \varepsilon_0 \cdot \pi \cdot d^2/4 = 118 \cdot 10^{-12}$ F.
The dielectric loss resistance is $R_0 = (2\pi \cdot f_0 \cdot C \cdot \tan\delta)^{-1} = 8 \cdot 10^6\ \Omega$.
The blocked transducer capacitance is $C_0 = \varepsilon \cdot \varepsilon_0[1 - d_{33}^2/\varepsilon \cdot \varepsilon_0 \cdot C_{mo}]\ \pi d^2/4\ell = 2.61 \cdot 10^{-11}$ F.
The total mechanical compliance of the rod is $C_m = \ell \cdot C_{mo}/\pi d^2/4 = 1.97 \cdot 10^{-9}$ m/N.
The effective mechanical mass of the rod is $M = \pi d^2\ell \cdot \rho_m/8 = 0.16$ kg.
The mechanical loss resistance in the piezoceramic is $R_{mp} = 2\pi f_0 M/Q_m = 123.8$ kg/s.
The radiation impedance $Z_r = R_r + jX_r$ can be found from Eqs. (10.42)–(10.44) and they give:

$$R_r = \pi a^2 \rho_w c_w[(2ka)^2/2^2 2! - (2ka)^4/2^4 2! \cdot 3! \cdots] = 190 \text{ kg/s}.$$
$$X_r = \pi a^2 \rho_w c_w[(4/\pi)(2ka/3) - (2ka)^3/3^2 5 + (2ka)^5/3^2 5^2 7 \cdots] = 514 \text{ kg/s}.$$

The electromechanical transformation factor is $\varphi = d_{33}\ell/C_{mo} = \underline{1.1}$.
The equivalent electrical components on the mechanical side of the circuit in Fig. 10.74 are now:

$L_{me} = M/\varphi^2 = 0.13$ H.
$C_{me} = C_m\varphi^2 = 2.38 \cdot 10^{-9}$ F.
$R_{me} = R_m/\varphi^2 = 102.3\ \Omega$.

The input admittance will now be $Y_{in} = 1/R_0 + j\omega C_0 + \varphi^2/(Z_m + Z_r) = 10^{-4} - j7 \cdot 10^{-4}\,\Omega^{-1}$ when $Z_m = R_{me} + j\{\omega L_{me} - 1/(\omega C_{me})\} = 102.3 + j1263$ and $Z_r = 190 + j514$.

10.10.3 FINITE-ELEMENT TECHNIQUES

The design of *broadband* piezoceramic-based transducers for underwater sound use has problems not only related to have the exact knowledge about the physical qualities of the transducer materials, but also related to the 2-D or 3-D aspects of the geometrical transducer structure and the excitation of several vibrational modes. The use of simpler one-dimensional models like equivalent circuits will not provide accurate modeling data, and it is necessary to resort to more advanced modeling techniques, such as *finite-element techniques*. Finite-elements techniques have been available for more than four decades for solving engineering problems, see Zienkiewicz [85] and Bathe [86].

The analysis of Tonpilz transducers as a first approximation can be carried out by using equivalent electrical circuits as long as the frequency range of interest has a limited extension in the immediate vicinity of the frequency of the fundamental longitudinal vibration mode. However, when other modes such as the head mass and flexural vibration, see Yao and Bjørnø [14], are of interest in relation to producing a broadband transducer, finite-element techniques are necessary. Also, when mechanical transducer limits are of concern, such as the formation of fractures, influence of nonlinear effects, and fatigue strengths of the various transducer materials, which all are directly related to local structural stress distributions, the use of a finite-element technique is the best solution.

Instead of the equivalent circuit lumped parameter model, the finite-element technique reduces the transducer parts to a three-dimensional array of a large number of discrete elements spatially distributed throughout the transducer. Very complicated transducers can be modeled by using an increasing number of smaller elements, which is computationally intensive. Commercially available model tools include the finite-element codes, ATILA [87], ANSYS [88], and PAFEC [89]. Many research laboratories and industrial companies have developed their own codes for their specific needs. The finite-element codes, which include physical transducer mechanical, electrical potential and piezoelectrical elements, and acoustic radiation elements and elements covering sound propagation in water around the transducer, together with today's fast and powerful computers are a reliable and useful basis for advanced transducer design calculation and predicting transducer qualities which can be verified by calibration.

Each part of a transducer, for instance the piezoceramic driving the head and tail mass in a Tonpilz transducer, is divided into a number of *discrete elements* of selected dimensions and the elements are connected at *nodes* that are moving with individual *displacements*. Each node forms a point of motion and appropriate masses and forces are associated with the nodes. The material stiffness is introduced between the nodes. The mass ascribed to a node receives contributions from the mass of the elements between which the node is established. Since the fundamental theory for finite-element techniques is available in several textbooks and since useful models already are commercially available, the development of computer codes for use of finite-element techniques is not repeated here. However, the fundamental coupled matrix equations used for coupling mechanical and piezoelectric elements in a finite-element transducer model are provided.

The finite-element piezoelectric model based on Eqs. (10.1) and (10.2) in matrix form can be written as:

$$T = c^E S - e_t E \text{ and } D = eS + \varepsilon^S E \tag{10.132}$$

where c^E is the elastic stiffness matrix for constant E, e_t is the transposed piezoelectric constant matrix, where the index t indicates that rows and columns in the matrix e are interchanged, e^S is the clamped permeability matrix, while T, S, and E are vectors representing stress, strain, and electrical field strength, respectively. By introducing the electromechanical transformation ratio matrix φ, the clamped electrical capacitance matrix C^S, the short-circuit stiffness matrix K^E, and the vector quantities F, Q, V, and x, which represent the total force, electrical charge, voltage, and displacement, respectively, Eq. (10.132) according to Sherman and Butler [16] can be rewritten as:

$$K^E x - \varphi V = F \text{ and } \varphi^T x + C^S V = Q \tag{10.133}$$

By introducing the mass matrix M and the mechanical resistance matrix R, the force vector F may be written as $F = F_b - M(d^2x/dt^2) - R(dx/dt)$, where F_b is the blocked acoustic force. By inserting F into Eq. (10.133), the matrix equation for coupling mechanical and piezoelectric elements in a finite-element model can be written as:

$$\begin{bmatrix} M & 0 \\ 0 & 0 \end{bmatrix} \begin{Bmatrix} \dfrac{\partial^2 x}{\partial t^2} \\ \dfrac{\partial^2 V}{\partial t^2} \end{Bmatrix} + \begin{bmatrix} R & 0 \\ 0 & 0 \end{bmatrix} \begin{Bmatrix} \dfrac{\partial x}{\partial t} \\ \dfrac{\partial V}{\partial t} \end{Bmatrix} + \begin{bmatrix} k^E & -\varphi \\ \varphi_t & C^S \end{bmatrix} \begin{Bmatrix} x \\ V \end{Bmatrix} = \begin{Bmatrix} F_b \\ Q \end{Bmatrix} \tag{10.134}$$

Each matrix in Eq. (10.134) consists of four submatrices and each vector consists of two vectors. If the electromechanical transformation matrix φ in Eq. (10.134) is removed, Eq. (10.134) reduces to two separate mechanical and electrical equations.

Complicated sonar transducer 3D forms require an increased number of smaller elements in the finite-element grid structure. Thus, the need for more involved finite-element codes requires more powerful computers to keep computational time within reasonable limits. However, if symmetry is inherent or can be imposed in the transducer model, i.e., if axial symmetry in the transducer is available, a 3-D calculation can nearly be reduced to a 2-D calculation, which substantially reduces the computational effort. Establishing special symmetry elements or symmetry planes have turned out to be useful in transducer calculations and will reduce computational time.

When the transducer is water loaded, the transition at the contact surface between the transducer nodes, where the degree of freedom is the displacement, and the water nodes, where pressure is the degree of freedom, must satisfy the demands for continuity to join the mechanical and the fluid elements. To preserve the pure radiation condition from the transducer and avoid reflection of sound signals back to the transducer, impedance-matched absorbing surfaces at the transition from near- to far-field distance from the transducer or in the far-field can be introduced into the finite-element calculations. For spherical propagation of the sound produced by the transducer in water, special spherical elements in the water, which involve absorption and permit an infinite propagation of the sound field, have been used by finite-element procedures. Positioning the absorbing surface as close to the transducer as possible reduces the number of elements in water and the computation time. However, the fluid field must be large enough that essential parts of the near field can be included, because as shown in Eq. (10.45) an additional vibrating mass should be included for a piston transducer at low frequencies. The coupling between the transducer structure and water at their interface can as shown in [16,86] be expressed through two coupled matrix equations:

$$[M_s]\left\{\frac{\partial^2 x}{\partial t^2}\right\} + [K_s]\{x\} = \{F_s\} + [R]\{P\} \tag{10.135}$$

$$[M_f]\left\{\frac{\partial^2 P}{\partial t^2}\right\} + [K_f]\{P\} = \{F_f\} + \rho[R]^t\left\{\frac{\partial^2 x}{\partial t^2}\right\} \tag{10.136}$$

where [ ] symbolizes a matrix and { } symbolizes a column vector. The indices s and f indicates that the term represents solid (the transducer) and fluid (the water), respectively. M and K denote mass and stiffness, respectively, while F, P, and x denote force, pressure, and displacement, respectively. ρ and R are the fluid density and the coupling at the effective interface area at each node between the solid and fluid. This means that each node at the interface possesses both displacement and pressure degrees of freedom.

It is possible to combine analytical methods in water such as the Helmholtz integral approach with the finite-element method to reduce computation time as shown by the ATILA [87] and PAFEC [89] codes. These codes do not need a large fluid field connected with the transducer since they evaluate acoustic pressure and particle velocity on a closed surface near the transducer, and use these values to calculate the acoustic far-field qualities and transducer directivity by using a Helmholtz integral approach.

A finite-element calculation of the displacement of the nodes of surface elements of a Tonpilz transducer during its longitudinal piston mode vibration at its fundamental frequency and the surface element node displacement during the same transducers "flapping mode" are shown in Fig. 10.16. The finite-element calculation of the displacement and deformation of the elements during the three first vibration modes in air of a Tonpilz transducer is shown in Fig. 10.75; see Yao et al. [14]. This figure shows in (A) the first longitudinal piston mode at $f_1 = 33.5$ kHz; (B) the first head mass flapping mode at $f_2 = 41.6$ kHz; and (C) the first longitudinal piston mode of at $f_3 = 44.9$ kHz. The finite-element center bolt calculation of the impedance variation as a function of frequency for the same Tonpilz transducer in air is shown in Fig. 10.76. By varying the dimensions of the different parts, in particular the Tonpilz transducer head mass, it is possible by using the finite-element code to design a broadband transducer by exploiting its multimode capabilities and the coupling between the modes.

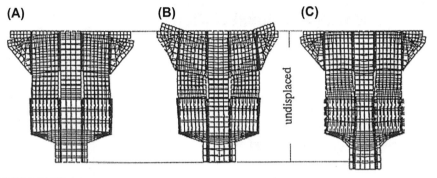

FIGURE 10.75

The finite-element calculation of element's displacement and deformation during (A) the first longitudinal piston mode ($f_1 = 33.5$ kHz), (B) the first head mass flapping mode ($f_2 = 41.6$ kHz), and (C) the first center bolt longitudinal piston mode ($f_3 = 44.9$ kHz).

©1997 IEEE figure reprinted from Yao, Q. and Bjørnø, L., Broadband Tonpilz Underwater Acoustic Transducers Based on Multimode Optimization, *IEEE Trans. Ultrason., Ferroelect. Freq. Control*, **44**, (5), pp. 1060–1066, 1997, with permission of the IEEE.

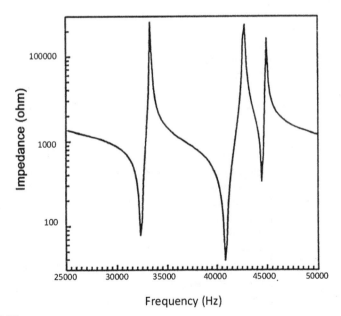

FIGURE 10.76

Finite-element calculations of the impedance variation as a function of frequency for the same Tonpilz transducer as shown in Fig. 10.75.

©1997 IEEE figure reprinted from Yao, Q. and Bjørnø, L., Broadband Tonpilz Underwater Acoustic Transducers Based on Multimode Optimization, *IEEE Trans. Ultrason., Ferroelect. Freq. Control*, **44**, (5), pp. 1060–1066, 1997, with permission of the IEEE.

10.11 SYMBOLS AND ABBREVIATIONS

Table 10.9 is a list of symbols and abbreviations used in this chapter along with the page where the symbol or abbreviation is defined and first used.

Table 10.9 List of Symbols and Abbreviations

Symbol or Abbreviation	Description	Page 1st Usage
SONAR	Sound navigation and ranging	587
SBES	Single-beam echo sounder	588
ROVs	Remotely operated vehicles	588
AUVs	Autonomous underwater vehicles	588
TAS	Towed array sonar	589
ASW	Anti-submarine warfare	590
MCM	Mine counter measures	590
LFAS	Low-frequency active sonar	590
ATOC	Acoustic thermometry of ocean climate	590
SOFAR	Sound fixing and ranging	590
ADP	Ammonium dihydrogen phosphate	591
PZT	Lead zirconate titanate	591
PMN	Lead magnesium niobate	591
PZN	Lead zirconium niobate	591
$BaTiO_3$	Barium titanate	591
PVDF or PVF_2	Polyvinylidene fluoride	600
E	Electric field strength (unit: V/m)	601
S	Strain (unit: dimensionless)	601
T	*Stress* (unit: N/m^2)	601
D	Electric displacement (unit: C/m^2)	601
ε_0	Free space permittivity $8.85 \cdot 10^{-12}$ F/m	602
λ	Acoustic wavelength (unit: m)	603
ρ	Density (unit: kg/m^3)	613
k	Wave number $2\pi/\lambda$ (unit: 1/m)	622
ρ_0	Fluid density (unit: kg/m^3)	622
c	Sound velocity (unit: m/s)	622
ω	Angular frequency (unit: rad/s)	622
r	Distance from the source in the far field (unit: m)	622
p	Acoustic pressure (unit: Pa)	622
t	Time (unit: s)	623
f	Frequency (unit: Hz)	623
J_1	Cylindrical Bessel function of order 1	627

Table 10.9 List of Symbols and Abbreviations—cont'd

Symbol or Abbreviation	Description	Page 1st Usage
H_1	Struve function of order 1	627
K_b	Boltzmann's constant $= 1.38 \cdot 10^{-23}$ J/K	639
A/D	Analog to Digital	653
DSTO	Defence Science and Technology Organisation	661
SURTASS LFA	Surveillance towed array sensor system low-frequency Active	663
TVG	Time variable gain	664
SVP	Sound velocity profiler	668
AGC	Automatic gain control	668
SLS	Side-looking sonar	670
SSS	Side-scan sonar	670
MBES	Multibeam echo sounders	588
BITE	Build-in test equipment	682
Snippets	Raw water column and seafloor data	682
SVP	Sound velocity probe	684
XBT	Expendable bathythermograph	684
DGPS	Differential global positioning system	685
GLORIA	Geological long range inclined asdic	693
SAS	Synthetic aperture sonar	696
CAATI	Computed angle-of-arrival transient imaging	696
RR	Range resolution	696
B	Sonar pulse bandwidth (unit: Hz)	696
AR	Along-track resolution	696
SAR	Synthetic Aperture Rader	697
ℓ	SAS transmitter array physical length (unit: m)	698
PGA	Phase gradient autofocus	699
ADCPs	Acoustic Doppler current profilers	701
DVS	Doppler volume sampler	702
DVL	Doppler velocity log	703
LBL	Long baseline	703
SBL	Short baseline	703
USBL	Ultra-short baseline	703
VDS	Variable depth sounder	705
J	Reciprocity parameter (unit: VA/$(\mu Pa)^2$ Or m^4s/kg)	709
TDS	Time-delay spectrometry	714
PMN-28PT	Magnesium niobate–lead titanate	727

REFERENCES

[1] Munk, W.H., Spindel, A.C., Baggeroer, A. and Birdsall, T., *The Heard Island Feasibility Test*, J. Acoust. Soc. Amer., **96**, (4), pp. 2330–2342, 1994.

[2] Kaneko, A., Yamaguchi, K., Gohda, N. and Zheng, H., *Recent progress in Coastal Acoustic Tomography (CAT)*. In: Proc. 1st International Conference on Underwater Acoustic Measurements: Technologies and Results, J.S. Papadakis & L. Bjørnø (eds), FORTH, Crete, pp. 209–216, 2005.

[3] Zhu, H., Liu, J., Wang, X., Xing, Y. and Zhang, H., *Application of Terfenol-D in China*, J. Alloys Compounds, **258**, pp. 49–52, 1997.

[4] Moffett, M.B., Porzio, R. and Bernier, G.L., *High-Power Terfenol-D Flextensional Transducers*. Report ADA294942, 9th May 1995.

[5] Kvarnsjö, L., *Underwater acoustic transducers based on Terfenol-D*, J. Alloys Compounds, **258**, pp. 123–125, Aug. 1997.

[6] Clark, A.E., *Magnetostrictive Rare-Earth-Fe₂ Compounds*. In: Proc. Power Sonic and Ultrasonic Transducer Design, B. Hamonic & J.N. Decarpigny (eds), Springer Verlag, pp. 43–99, 1988.

[7] Sewell, J.M. and Kuhn, P.M., *Opportunities and Challenges in the Use of Terfenol for Sonic Transducers*. In: Proc. Power Sonic and Ultrasonic Transducer Design, B. Hamonic & J.N. Decarpigny (eds), Springer Verlag, pp. 134–142, 1988.

[8] Modzelewski, C., Savage, H.T., Kabacoff, L.T. and Clark, A.E., *Magnetomechanical coupling and permeability in transversely annealed Metglass* 2605 *alloys*, IEEE Trans. Magnetics, MAG-17, pp. 2837–2839, 1981.

[9] Jaffe, B., Cook, W.R. Jr. and Jaffe, H., *Piezoelectric Ceramics*. Academic Press, 1971.

[10] Vinogradov, A. and Schwartz, M., *Piezoelectricity in Polymers*. In: Encyclopedia of Smart Materials. Vol. 1–2, John Wiley & Sons, pp. 780–792, 2002.

[11] *IEEE Standard on Piezoelectricity*, IEEE Trans. Sonics Ultrason., **SU-31**, (2), 1984.

[12] Wilson, O.B., *Introduction to Theory and Design of Sonar Transducers*. Peninsula Publishing, Los Altos, California, USA, 1988.

[13] Stansfield, D., *Underwater Electroacoustic Transducers*. Bath University Press and Institute of Acoustics, UK, 1991.

[14] Yao, Q. and Bjørnø, L., *Broadband Tonpilz Underwater Acoustic Transducers Based on Multimode Optimization*, IEEE Trans. Ultrason., Ferroelect. Freq. Control, **44**, (5), pp. 1060–1066, 1997.

[15] Zhuang, Y., Ural, S.O., Rajapurkar, S., Amin, A. and Uchino, K., *Derivation of piezoelectric losses from admittance spectra*, Japanese J. Appl. Phys., **48**, pp. 116–121, 2009.

[16] Sherman, C.H. and Butler, J.L., *Transducers and Arrays for Underwater Sound*. Springer Verlag, 2007.

[17] Woollett, R.S., *Power Limitations of Sonic Transducers*, IEEE Trans. Sonics Ultrason., **SU-15**, (4), pp. 218–229, 1968.

[18] Hahnemann, W. and Hecht, H., *Die Grundform des mechanisch-akustischen Schwingungskörper (Der Tonpilz)*, Physik. Zeitschrift, **21**, pp. 187–192, 1920.

[19] Brigham, G.A. and Royster, L.H., *Present Status in the Design of Flextensional Underwater Acoustic Transducers*, J. Acoust. Soc. Amer., **46**, (1), 92, 1969.

[20] Rynne, E.F., *Innovative approaches for generating high power, low frequency sound*. In: Transducers for Sonics and Ultrasonics, M.D. McCollum, B.F. Hamonic and O.B. Wilson (eds), Technomic Publ. Co. Inc., PA, USA, pp. 38–52, 1992.

[21] Royster, L.H., *Flextensional Underwater Acoustic Transducer*, J. Acoust. Soc. Amer., **45**, (3), pp. 671–682, 1969.

[22] Jones, D.F. and Christopher, D.A., *A broadband omnidirectional barrel-stave flextensional transducer*, J. Acoust. Soc. Amer., **106**, (2), pp. L13, 1999.

[23] McMahon, G.W., *The Ring-Shell Flextensional Transducer (Class V)*. In: Power Transducers for Sonics and Ultrasonics, B.F. Hamonic, O.B. Wilson & J.N. Decarpigny (eds), Springer Verlag, pp. 60–74, 1991.

[24] Brigham, G.A., *Analysis of the class-IV flextensional transducer by use of wave mechanics*, J. Acoust. Soc. Amer., **56**, (1), pp. 31–39, 1974.

[25] Brigham, G. and Glass, B., *Present status in flextensional transducer technology*, J. Acoust. Soc. Amer., **68**, (4), pp. 1046–1052, 1980.

[26] Hamonic, B., Debus, J.C., Decarpigny, J.N., Boucher, D. and Tocquet, B., *Analysis of a radiating thin-walled sonar transducer using finite-element methods*, J. Acoust. Soc. Amer., **86**, (4), pp. 1245–1253, 1989.

[27] Butler, S.C., Butler, A.L. and Butler, J.L., *Directional flextensional transducer*, J. Acoust. Soc. Amer., **92**, (5), pp. 2977–2979, 1992.

[28] Oswin, J.R. and Turner, A., *Design limitations of aluminium shell Class IV flextensional transducers*. In: Proc. of the Institute of Acoustics, UK, **6**, Pt 3, pp. 95–100, 1984.

[29] Butler, S.C., Butler, J.L., Butler, A.L. and Cavanagh, G.H., *A low-frequency directional flextensional transducer and line array*, J. Acoust. Soc. Amer., **102**, (1), pp. 308–314, 1997.

[30] Morse, P.M. and Ingard, K.U., *Theoretical Acoustics*. McGraw-Hill Book Comp., 1968.

[31] Woollett, R.S., *Theory of the Piezoelectric Flexural Disk Transducer with Applications to Underwater Sound*. USL Research Report No. **490**, NUWC, Dec. 1960.

[32] Ross, D., *Noise Sources, Radiation and Mitigation*. In: Underwater Acoustics and Signal Processing, L. Bjørnø (Ed.), D. Reidel Publ. Comp., pp. 3–28, 1981.

[33] Lighthill, M.J., *On sound generated aerodynamically. II. Turbulence as a source of sound*. Proc. Royal Soc. (London), A222, pp. 1–32, 1954.

[34] Stephens, R.W.B. and Bate, A.E., *Acoustics and Vibrational Physics*, Edward Arnold (Publ.), London, 1966.

[35] Kinsler, L.E., Frey, A.R., Coppens, A.B. and Sanders, J.V., *Fundamentals of Acoustics*, 3rd Ed. John Wiley & Sons, 1982.

[36] Medwin, H. and Colleagues, *Sounds in the Sea*. Cambridge University Press, 2005.

[37] Pierce, A.D., *Acoustics — An Introduction to its Physical Principles and Applications*, Acoustical Society of America, Woodbury, New York, 1989.

[38] Abramowitz, M. and Stegun, I.A. (eds), *Handbook of Mathematical Functions*, Dover Publications, New York, 1970.

[39] Bobber, R.J., *Underwater Electroacoustic Measurements*. Naval Research Laboratory, Washington D.C., 1970.

[40] Urick, R.J., *Principles of Underwater Sound*, 3rd Ed. McGraw-Hill Book Company, 1983.

[41] Rijnja, H.A.J., *Modern transducers, Theory and Practice*. In: Underwater Acoustics and Signal Processing, L. Bjørnø (Ed.), D. Reidel Publ. Comp., 1981, pp. 225–242.

[42] Mellen, R.H., *The thermal noise limit in the detection of underwater signals*, J. Acoust. Soc. Amer., **24**, (5), pp. 478–480, 1952.

[43] Langevin, R.A., *The electroacoustic sensitivity of cylindrical ceramic tubes*, J. Acoust. Soc. Amer., **26**, (3), pp. 421–427, 1954.

[44] Leslie, C.B., Kendall, J.M. and Jones, J.L., *Hydrophone for measuring particle velocity*, J. Acoust. Soc. Amer., **28**, (4), pp. 711–715, 1956.

[45] Gabrielson, T.B., Gardner, D.L. and Garrett, S.L., *A simple neutrally buoyant sensor for direct measurement of particle velocity and intensity in water*, J. Acoust. Soc. Amer., **97**, (4), pp. 2227–2237, 1995.

[46] Kim, K., Gabrielson, T.B. and Lauchle, G.C., *Development of an accelerometer-based underwater acoustic intensity sensor*, J. Acoust. Soc. Amer., **116**, (6), pp. 3384–3392, 2004.

[47] Cray, B.A. and Nuttall, A.H., *Directivity factors for linear arrays of velocity sensors*, J. Acoust. Soc. Amer., **110**, (1), pp. 324–331, 2001.

[48] Wong, K.T. and Chi, H., *Beam patterns of an underwater acoustic vector hydrophone located away from any reflecting boundary*, IEEE J. Ocean. Eng., **27**, (3), pp. 628–637, 2002.

[49] Proceedings of the 3rd International Conference and Exhibition on *Underwater Acoustic Measurements: Technologies and Results*, Nafplion, Greece, June 2009. J.S. Papadakis and L. Bjørnø (eds), IACM/FORTH, Vol. 1, pp. 19–36, 2009.

[50] Morse, P.M and Ingard, K.U., *Theoretical Acoustics.* McGraw-Hill Book Company, New York, 1968.

[51] Urban, H.G., *Handbook of Underwater Acoustic Engineering.* STN ATLAS Elektronik GmbH, 2002.

[52] Waite, A.D., *Sonar for Practising Engineers.* 3rd Ed. John Wiley & Sons Ltd., 2002.

[53] Harris, F.J., *On the Use of Windows for Harmonic Analysis with the Discrete Fourier Transform.* Proc. IEEE, **66**, No. 1, pp. 51–83, 1978.

[54] Ehrenberg, J.E., *The dual-beam system: a technique for making in situ measurements of the target strength of fish.* In: Proc. IEEE Int. Conf. on Engineering in the Ocean Environment, IEEE, New York, pp. 152–155, 1974.

[55] Bodholt, H., *Split-beam transducers with 3 sections.* In: Proc. Scandinavian Symp. on Phys. Acoust., Ustaoset, Norway, pp. 1–8, 2001.

[56] Knight, A., *Flow noise calculations for extended hydrophones in fluid- and solid-filled towed arrays*, J. Acoust. Soc. Amer., **100**, (1), pp. 245–251, 1996.

[57] Lemon, S.G., *Towed-Array History,* 1917–2003, IEEE J. Ocean. Eng., **29**, (2), pp. 365–373, 2004.

[58] Lasky, M., Doolittle, R.D., Simmons, B.D. and Lemon, S.G., *Recent Progress in Towed Hydrophone Array Research*, IEEE J. Ocean. Eng., **29**, (2), pp. 374–387, 2004.

[59] SeaBat 7125, Product Description, RESON A/S, February 26th, 2008.

[60] SeaBat 7150, Full Ocean Depth Multibeam Echo Sounder. Product Description, RESON A/S, February 2nd, 2005.

[61] World Ocean Atlas 2005, National Oceanographic Data Centre, USA, (http://www.nodc.noa.gov/OC5/WOA05).

[62] Pocwiardowski, P., Yufit, G. and Maillard, E., *Effects of vessel pitch and roll on a seabed backscattering strength estimation.* In Proc. 2nd International Conference and Exhibition on Underwater Acoustic Measurements: Technologies and Results, J.S. Papadakis and L. Bjørnø (eds), Vol. I, IACM − FORTH, Heraklion, Crete, pp. 459–466, 2007.

[63] Kraeutner, P.H. and Bird, J.S., *Seafloor scatter induced angle of arrival errors in swath bathymetry sidescan sonar.* In: Proc. IEEE Oceans'95: MTS/IEEE, New York, **2**, pp. 975–980, 1995.

[64] Skolnik, M.L., *Introduction to radar systems.* McGraw-Hill, New York, 1980.

[65] Cutrona, L.J., Vivian, W.E., Leith, E.N., and Hall, G.O., *A High-Resolution Radar Combat-Surveillance System*, IRE Trans. on Military Electronics, Vol. MIL-5, pp. 127–131, 1961.

[66] Cutrona, L.J., *Comparison of sonar system performance achievable using synthetic-aperture techniques with the performance achievable by more conventional means.* J. Acoust. Soc. Amer., **58**, (2), pp. 336–348, 1975.

[67] Cutrona, L.J., *Additional characteristics of synthetic-aperture sonar systems and a further comparison with nonsynthetic-aperture sonar systems.* J. Acoust. Soc. Amer., **61**, (5), pp. 1213–1217, 1977.

[68] Hansen, R.E., Callow, H.J and Sæbø, T.O., *The effect of sound velocity variations on synthetic aperture sonar.* In: J.S. Papadakis & L. Bjørnø (eds), Proceedings of Underwater Acoustic Measurements: Technologies and Results. 2nd International Conference, F.O.R.T.H., Crete, Vol. I, pp. 323–330, 2007.

[69] Bellettini, A., Pinto, M. and Evans, B., *Experimental results of a 300 kHz shallow water synthetic aperture sonar.* In: J.S. Papadakis & L. Bjørnø (eds), Proceedings of Underwater Acoustic Measurements: Technologies and Results. 2nd International Conference, F.O.R.T.H, Crete, Vol. I, pp. 317–322, 2007.

[70] Hansen, R.E., Callow, H.J., Sæbø, T.O., Synnes, S.A., Hagen, P.E., Fossum, T.G. and Langli, B., *Synthetic aperture sonar in challenging environments: Results from the HISAS 1030.* In: J.S. Papadakis & L. Bjørnø (eds), Proceedings of Underwater Acoustic Measurements: Technologies and Results. 3rd International Conference. F.O.R.T.H., Crete, Vol. I, pp. 409–416, 2009.

[71] Bobber, R.J. and Sabin, G.A., *Cylindrical wave reciprocity parameter,* J. Acoust. Soc. Amer., **33**, (4), pp. 446–451, 1961.

[72] Simmons, B.D. and Urick, R.J., *Plane wave reciprocity parameter and its application to calibration of electroacoustic transducers at close distances,* J. Acoust. Soc. Amer., **21**, (6), pp. 633–635, 1949.

[73] Van Buren, A.L. and Blue, J.E., *Calibration of underwater acoustic transducers at NRL/USRD.* In: B.F. Harmonic, O.B. Wilson and J.-N. Decarpigny (eds), Proceedings of the International Workshop on Power Transducers for Sonics and Ultrasonics, Springer-Verlag, pp. 221–241, 1991.

[74] Van Buren, A.L., *Cylindrical nearfield calibration array,* J. Acoust. Soc. Amer., **56**, pp. 849–855, 1974.

[75] Foote, K.G. and MacLennan, D.N., *Comparison of copper and tungsten carbide calibration spheres,* J. Acoust. Soc. Amer., **75**, (2), pp. 612–616, 1984.

[76] Chu, D., Foote, K.G., Hufnagle, L.C., Hammar, T.R., Liberatore, S.P., Baldwin, K.C., Mayer, L.A. and McLeod, A., *Calibrating a 90-kHz multibeam sonar.* In: Proc. MTS/IEEE Oceans' 2003, San Diego, CA, pp. 1633–1636, 2003.

[77] Luker, L.D., Zalesak, J.F., Brown, C.K. and Scott, R.E., *Automated digital benchtop calibration system for hydrophone arrays,* J. Acoust. Soc. Amer., **73**, (4), pp. 1212–1216, 1983.

[78] Luker, L.D. and Zalesak, J.F., *Free-field acoustic calibration of long underwater acoustic arrays in a closed chamber,* J. Acoust. Soc. Amer., **90**, (5), pp. 2652–2657, 1991.

[79] Pedersen, P.C., Lewin, P.A. and Bjørnø, L., *Application of Time-Delay Spectrometry for calibration of ultrasonic transducers,* IEEE Trans. Ultrason., Ferroelect. Freq. Control, **35**, (2), pp. 185–205, 1988.

[80] Bjørnø, L. and Kjeldgaard, M., *A wide frequency band anechoic water tank,* Acustica, **32**, (2), pp. 103–109, 1975.

[81] Esche, R., *Undersuchung der Schwingungskavitation in Flüssigkeiten*. Acustica, 2, Beiheft AB, pp. 208–216, 1952.

[82] Handbook of Chemistry and Physics, CRC Press, 56th Ed. 1975.

[83] Brosnan, K.H., Messing, G.L., Markley, D.C. and Meyer, R.J., *Comparison of the properties of tonpilz transducers fabricated with <001> fiber-textured lead magnesium niobate-lead titanate ceramic and single crystals*, J. Acoust. Soc. Amer., **126**, (5), 2009, pp. 2257–2265.

[84] Goll, J.H., *The Design of Broad-Band Fluid-Loaded Ultrasonic Transducers*, IEEE Trans. Sonics Ultrason., **SU-26**, (6), 1979, pp. 385–393.

[85] Zienkiewicz, O.C., *The Finite Element Method for Solid and Structural Mechanics*. McGraw-Hill Book Comp., 1967 (Now 6th Ed. 2005).

[86] Bathe, K.J. and Wilson, E.L., *Numerical Methods in Finite Element Analysis*. Prentice-Hall, 1976.

[87] ATILA Finite Element Analysis Simulation Software, Micromechatronics Inc., State College, PA 16803, USA.

[88] ANSYS Inc., Canonsburg, PA 15317, USA.

[89] PAFEC Ltd., Strelley Hall, Nottingham, NG86PE, UK.

Signal Processing

11

D.A. Abraham

Ellicott City, MD, United States

Definitions

In generic terms, *signal processing* is the isolation and extraction of information from observations or measurements of physical quantities. In underwater acoustics applications it is typically the link between data acquisition and an underlying inferential objective.

The goal of this chapter is to present the basic information necessary to apply signal processing concepts and algorithms to underwater acoustics applications. The chapter begins by defining signals and noise in underwater acoustics before describing how they are characterized in time, frequency, and in a probabilistic sense as random processes. The remainder of the chapter covers the core signal processing categories of filtering, detection, and estimation. Within each category the pertinent performance metrics, structured design approaches, implementation, and analysis techniques are discussed. Application of algorithms to many common problems in underwater acoustics applications is illustrated through examples and should be accessible to those with a basic understanding of engineering or data analysis. The tools necessary to develop basic detection or estimation algorithms for new problems are also presented, but generally require a greater familiarity with signal processing and mathematical statistics.

11.1 BACKGROUND AND DEFINITIONS

11.1.1 SIGNALS AND NOISE IN UNDERWATER ACOUSTICS

The *signal* part of signal processing refers to the voltage measurement of a physical quantity obtained or acquired through a transducer for example, the voltage produced by a hydrophone subject to sound pressure. When properly constructed, the voltage signal precisely represents the physical quantity of interest. Signals are often considered measurements over time from a single transducer (i.e., sensor), but are more generally characterized as being over both space and time. The *processing* part of signal processing involves condensing the measured spatiotemporal signals into useful information. For example, a depth sounder estimates the distance from the bottom of a ship to the seafloor to avoid running aground.

Applied Underwater Acoustics. http://dx.doi.org/10.1016/B978-0-12-811240-3.00011-4

As described in other parts of this book, the wavelengths of sound are often the most expedient choice underwater owing to the effectiveness of their propagation. As such, the signals of interest in underwater acoustics applications are typically related to sound pressure as measured by hydrophones. However, both shorter and longer wavelengths and measurements of particle velocity are also used in underwater applications related to remote sensing.

Most underwater acoustics applications can be classified as remote sensing: making inferences about the *state of nature* without direct physical contact. In statistical inference [1], the state of nature represents the underlying truth of the object, phenomenon, or condition being investigated. Remote sensing can be passive, where signals are measurements of indigenous sound, or active, where signals are reflections or scattering created by transmission of a known sound. In both cases, the objective is to learn about some distant state of nature. Examples of the *state of nature* include the number of fish in a fish school, the amplitude and velocity of an internal wave, and the location and type of mine on the seafloor.

The bane of signal processing is generally referred to as noise, loosely defined as anything corrupting or inhibiting the desired inference. *The signal-to-noise power ratio (SNR)* is the most commonly known signal processing performance metric, quantifying when inference is easy (high SNR) or hard (low SNR). Noise comes from many sources including the measurement system (e.g., sensor noise), the acquisition environment (e.g., ambient acoustic noise), or the acquisition conditions (e.g., flow noise). In many applications the distinction between signal and noise is obvious. However, semantics can upend expectations. For example, while clicks from a sperm whale are a signal to a marine biologist, they are interference to an acoustician studying wave-generated ambient noise.

In addition to processing the primary acoustic signals related to remote sensing, other types of signals will often be measured and processed to enable or improve inference. The most common of these are related to the position and orientation of measurement equipment (e.g., the heading of a towed-array receiver).

11.1.2 WHAT IS "SIGNAL PROCESSING" AND WHY DO WE DO IT?

As previously mentioned, signal processing condenses measurements to extract information about some distant state of nature. Signal processing can be described from different perspectives. To an acoustician, it is a tool to turn measured signals into useful information. To a sonar designer, it is one part of a sonar system. To an electrical engineer, it is often restricted to digitization, sampling, filtering, and spectral estimation. A modern underwater acoustic signal processing system can include the following.

- Digitization (sampling in time and quantizing in amplitude)
- Band-pass and base-band filtering
- Beamforming (spatial filtering)
- Matched filtering and/or incoherent integration
- Detection, classification, localization, and tracking

Although measurements are typically acquired jointly in space and time, signal processing is usually partitioned to combine signals across space first and then time. The other components are similarly separable, in part owing to their inherent modularity and also from a paced evolution in computational power. For example, early deployment of beamformers required special hardware so only beam-output data were easily accessible for analysis. Easy access to data enabled advances in the modern signal processing components of *detection, classification, localization, and tracking*.

A broader view of modern signal processing can be found by considering its theoretical foundations, which come from mathematical statistics. Almost all underwater acoustics applications can be categorized as statistical inference, which is comprised of estimation and hypothesis testing [2]. These two branches of statistical inference provide the mathematical framework and tools to derive or design signal processing algorithms extracting the desired information. Examples of point estimation include inferring depth via the depth sounder, internal-wave amplitude, or location of a mine on the seafloor. Hypothesis testing is itself a type of estimation where the state of nature is represented as a finite set (e.g., types of mines or marine mammal vocalizations). The most common is the binary hypothesis test known as a signal detection problem, for which the state of nature is the *presence* or *absence* of an object, phenomenon, or condition. For example, is there a fish school present or not? Is there a communications signal in the measurement or not? Are internal waves present or not? Most underwater acoustics applications combine both detection and estimation: is there a fish school and if so what type are they and how many fish are there in the school.

Each of the signal processing components in the above-mentioned list (perhaps excluding digitization) can be derived through either a detection or estimation paradigm. The importance of casting signal processing as a detection or estimation problem comes from the structured design approaches available within the field of mathematical statistics. The design approaches provide a means for algorithm derivation and dictate when and what optimality criteria are satisfied or if the resulting algorithm is suboptimal. For example, the losses incurred in performing the signal processing operations sequentially as opposed to in a unified manner can be evaluated or unified algorithms can be derived (e.g., see Ref. [3] for combined beamforming and tracking).

Representing signal processing functions as detection or estimation problems has the added benefit of providing unambiguous performance metrics. Early signal processing development and analysis relied on SNR or a similar measure called deflection as the primary performance metric. Because these metrics are derived from second-order statistics of the signal envelope (i.e., average power), they are an unambiguous representation of performance only under certain conditions (e.g., Gaussian-distributed noise). It is important to note the sonar equation [4] is similarly a second-order statistical performance metric and therefore may not accurately represent performance in all scenarios.

11.1.3 STRUCTURE OF THIS CHAPTER

This chapter is structured by the steps necessary to derive or design and apply the signal processing algorithms necessary to achieve a desired inference objective for an underwater acoustics application. The first step is to characterize the signal and noise (Section 11.2), including describing the digitization process and its impact.

Referring to the list of signal processing components in Section 11.1.2, the next function is filtering (Section 11.3). While filtering can be described as part of a detection or estimation process, several practical filtering functions do not need to be [e.g., low-pass filter (LPF) design, digital down-conversion, and windowing].

The remainder of the chapter is split between detection (Section 11.4) and estimation (Section 11.5). Each section describes the relevant performance metrics and the design, implementation, and analysis procedure. Structured design approaches are described and followed by examples commonly encountered in underwater acoustic signal processing.

11.1.4 OTHER RESOURCES

In selecting the material to present in this chapter, a priority was placed on providing tools useful in applying signal processing in underwater acoustics. The brevity inherent in this format precludes detailed presentation of the concepts and their use. When more depth is required, the following texts are recommended.

- Sonar signal processing:
  Nielsen [5], Burdic [6], Knight [7].
- Radar signal processing (not sonar, but good for active systems):
  Peebles [8], Richards [9].
- Digital signal processing and filtering:
  Porat [10], Lyons [11], Oppenheim and Schafer [12],
  Proakis and Manolakis [13], Brigham [14].
- Data analysis and spectral density estimation:
  Bendat and Piersol [15], Kay [16], Marple [17], Cohen [18].
- Beamforming:
  Van Trees [19], Johnson and Dudgeon [20].
- Detection:
  Scharf [21], Kay [22], McDonough and Whalen [23].
- Estimation:
  Scharf [21], Kay [24], Lehmann [25].
- Tracking:
  BarShalom et al. [26], Stone et al. [27].
- Mathematical statistics (hypothesis testing and estimation):
  Lindgren [28], Mukhopadhyay [2], Papoulis [29].

11.1.5 MATHEMATICAL NOTATION

The following notation conventions have been applied to the degree possible. With the confluence of multiple academic disciplines there are inherent contradictions in the notation for which context must be used for resolution.

- Conjugate *, transpose ^{T}, conjugate-transpose ^{H}
- Matrices and vectors are bold: $\mathbf{X}$
- Variables i to n are typically integers, except when $j = \sqrt{-1}$.
- Logarithms: natural is $\log_e(x)$ and base-10 is $\log_{10}(x)$
- Sinc function: $\operatorname{sinc}(x) = \sin(\pi x)/(\pi x)$
- Probability is $\Pr\{\text{event}\}$
- Random variables are capitalized (X) while specific values are lower case (x). Not all capitalized variables are random variables.

11.1.6 LIST OF SYMBOLS AND NOTATION

Symbol	Description	Page of First Usage
B	Number of bits in a quantizer	751
c_w	Speed of sound in water	753
$\delta(t)$	Dirac delta function	754
$\Delta_{\hat{\theta}}$	Bias of $\hat{\theta}$	789
$E[\cdot]$	Expectation operator	759
f_k	Frequency for kth DFT output	757
f_s	Sampling frequency	751
$f_X(x)$	Probability density function	770
$\mathscr{F}\{\cdot\}$	Fourier transform	754
$\mathscr{F}^{-1}\{\cdot\}$	Inverse Fourier transform	754
$I\{\cdot\}$	Indicator function	775
$\cdot\gamma_{XY}(f)$	Coherence function	760
$L(\theta; \mathbf{x})$	Likelihood function of θ	791
P_d	Probability of detection	771
P_f	Probability of false alarm	771
ρ_w	Density of water	753
$R_{XX}(\tau)$	Autocorrelation function	759
$R_{XY}(\tau)$	Cross-correlation function	760
$S_{xx}(f)$ or $S_{XX}(f)$	Spectral density	756
$S_{XY}(f)$	Cross-spectral density function	760
$\sigma_{\hat{\theta}}^2$	Variance of $\hat{\theta}$	789
W	Bandwidth	760
$\hat{\theta}$	Estimator of θ	789

11.1.7 **LIST OF ABBREVIATIONS**

Abbreviation	Description	Page of First Usage
ACF	Autocorrelation function	758
AR	Autoregressive	799
BPF	Band-pass filter	759
BTR	Bearing-time record	804
CA-CFAR	Cell averaging CFAR	787
CCF	Cross-correlation function	801
CFAR	Constant false alarm rate	786
CLT	Central limit theorem	753
CPI	Coherent processing interval	796
CRLB	Cramer-Rao lower bound	791
CW	Continuous wave	781
DCLT	Detection, classification, localization, and tracking	747
DFT	Discrete Fourier transform	757
DT	Detection threshold	772
FFT	Fast Fourier transform	758
FIR	Finite impulse response	758
FM	Frequency Modulation	781
GLR	Generalized likelihood ratio	778
HFM	Hyperbolic frequency modulation	781
HPF	High-pass filter	761
IIR	Infinite impulse response	762
LFM	Linear frequency modulation	781
LLR	Log-likelihood ratio	777
LO	Locally optimal	777
LPF	Low-pass filter	748
LRD	Likelihood ratio detector	777
LTI	Linear time invariant	759
MA	Moving average	799
MAP	Maximum a posteriori	794
MLE	Maximum likelihood estimate	778
MoM	Method of moments	793
MSE	Mean-squared error	789
PDF	Probability density function	770
PSD	Power spectral density	759
RMSE	Root-mean-squared error	790
ROC	Receiver operating characteristic	771
SNR	Signal-to-noise power ratio	746
SQNR	Signal to quantization noise power ratio	751
TVG	Time-varying gain	788
UMP	Uniformly most powerful	777
WSS	Wide-sense stationary	758

11.2 CHARACTERIZING THE SIGNAL AND NOISE

11.2.1 SAMPLING AND QUANTIZING ANALOG SIGNALS

As noted in Section 11.1.1, signals of interest in underwater acoustics are measured using a transducer converting a physical quantity (e.g., sound pressure) into an analog voltage. Although some high-frequency systems perform limited processing on analog measurements, most signal processing is accomplished using digital processors. Digital processors require the analog measurement to be sampled in time and quantized to a finite set of discrete levels.

Sampling in time first requires analog low-pass filtering the measurement to restrict the spectral content to be below the Nyquist [12] frequency $f_s/2$, which is half the sampling frequency f_s {unit: Hz}. Any spectral content above $f_s/2$ aliases to lower frequencies where it is indistinguishable from the naturally occurring low-frequency content. Thus the sampling frequency must be at least twice the maximum frequency of interest.

Quantization represents the voltage of discrete-time measurements in a digital format. For example, a one-bit quantizer uses a "1" to represent positive values and "0" for negative values. For some applications (e.g., detection using minimal computational power), the one-bit quantizer is surprisingly effective. However, most applications require a higher fidelity representation of the signal. Fidelity is usually defined by the dynamic range, which is the ratio of the largest to smallest (nonzero) absolute value representable after quantization. If the maximum amplitude under consideration is A (i.e., the measurement is restricted to be on the interval $[-A,A]$) and B bits are used in the quantizer with one allocated for the sign, the smallest magnitude value is the step size $\Delta = (2A)/2^B = A/2^{B-1}$. With this step size, the most negative value quantized is $-A$ while the most positive one is $A - \Delta$ with zero quantized precisely.[1] The dynamic range is then

$$\mathrm{DR} = 20\log_{10}\left(2^{B-1} - 1\right) \approx 6.021(B - 1) \quad \{\text{unit: dB}\} \qquad (11.1)$$

which exhibits the commonly known 6-dB-per-bit rate of increase (the approximation has error less than 0.5 dB for $B > 5$).

Quantization error is typically treated as a source of random noise. Depending on the quantization method (e.g., see Ref. [12]), the error is uniformly distributed from $-\Delta/2$ to $\Delta/2$ and has variance $\sigma_Q^2 = \Delta^2/12$. The *peak-signal to quantization noise power ratio (SQNR)* is then

$$\mathrm{SQNR} = 10\log_{10}\left\{\frac{\left[\Delta\left(2^{B-1} - 1\right)\right]^2}{\left[\Delta^2/12\right]}\right\} \approx 6.021(B - 1) + 10.792 \text{ dB}. \qquad (11.2)$$

---

[1]An alternative quantization scheme ranges from $-A + \Delta/2$ to $A - \Delta/2$ but does not quantize zero precisely.

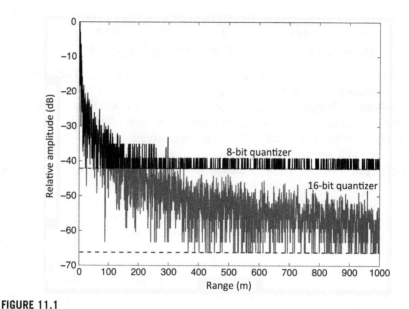

FIGURE 11.1

Simulated active sonar reverberation with 8- and 16-bit quantizers.

While SQNR should clearly be larger than the acoustic noise in a measurement, *B* should also be large enough to accurately represent the smallest signal of interest. For example, active sonar echoes or reverberation can be very loud for short ranges and very weak for long ranges. The impact of not using a large enough dynamic range is illustrated in Fig. 11.1 where 8-bit quantization fails to capture the weaker signals (represented by a toggling of the least significant bits).

When properly sampled in time and quantized, an analog measurement can be reconstructed with minimal error. The two processes are typically done simultaneously using analog-to-digital converters. More detail on discretizing analog measurements may be found in Refs. [11,15,12].

11.2.2 TIME AND FREQUENCY CHARACTERIZATION

In the design and analysis of signal processing algorithms, it is generally more convenient to work with mathematical representations of the physical quantity of interest rather than the discrete-time and quantized levels of digital data. For most of the applications in this chapter, the mathematical signal $x(t)$ will represent an acoustic pressure {unit: μPa} sensed as a function of time {unit: s}. However, the majority of the techniques to be described can be applied to other physical quantities (e.g., particle velocity) and to spatial dimensions in addition to time. The first step in signal processing is characterizing signal and noise into certain categories relevant to processing objectives such as filtering, detection, and estimation.

11.2.2.1 Signal Consistency: Deterministic and Random Signals

The repeatability of a signal determines its consistency. If a signal is identical each time it is observed (not including any concurrently observed noise) or changes in a known manner, it is deterministic. If the signal varies from observation to observation it can be characterized as random if some portion of the change is unknown. While most signals will have some randomness (i.e., they are not perfectly known or completely random), it is often possible to treat the signal deterministically by parameterizing unknown quantities. For example, an active sonar echo could be considered deterministic with an unknown Doppler scale that must be estimated from the data. Random signals abound in underwater acoustics. For example, ambient noise from the ocean surface, sound radiated from a ship moving through the water, or turbulent flow across a transducer. Acoustic propagation through an underwater environment can turn a deterministic signal into a random one as is the case with sound fluctuations [30,31].

Signal processing algorithms are often derived for the extreme cases of deterministic signals and completely random signals owing to the (relative) simplicity of the solutions. It is common to exploit the Gaussian distribution through a *central limit theorem (CLT)* [28] argument for random signals formed from the combination of multiple sources. However, other distributions exist that can account for cases where the CLT does not apply [32].

11.2.2.2 Temporal Characterization

Underwater acoustic signals can be either short in duration relative to the observation time (e.g., an air-gun pulse in a 1-min window) or extend through the analysis window (e.g., a communications signal in a 100-ms window). As described in Section 11.4, detection processing of short or transient signals differs from that for persistent signals.

A signal $x(t)$ can be characterized as either an energy signal or a power signal. An energy signal has finite energy:

$$E = \int_{-\infty}^{\infty} |x(t)|^2 dt < \infty, \tag{11.3}$$

whereas a power signal has finite, but nonzero, average power:

$$P = \lim_{T \to \infty} \frac{1}{2T} \int_{-T}^{T} |x(t)|^2 dt < \infty. \tag{11.4}$$

These energy and average power definitions are mathematical ones; scaling by the appropriate impedance (e.g., the acoustic impedance is the density of water times the speed of sound, $\rho_w c_w$) is required to obtain physical definitions of energy and power. A power signal has infinite energy and an energy signal has zero average power. Transient (finite duration) signals are energy signals while periodic signals are power signals. The following discussion focuses on energy signals; however, similar results exist for power signals through the use of limiting arguments on the temporal extent of the Fourier integral.

11.2.2.3 Spectral Content: The Fourier Transform and Spectral Density

The frequency content of underwater acoustic signals is paramount to many signal processing objectives. For a time-domain signal $x(t)$ {unit of t: s}, the Fourier transform

$$X(f) = \mathscr{F}\{x(t)\} = \int_{-\infty}^{\infty} x(t)e^{-j2\pi ft}dt \qquad (11.5)$$

characterizes spectral content as a function of frequency f {unit: Hz}. For example, consider a time-domain impulse $x(t) = \delta(t)$ where $\delta(\cdot)$ is the Dirac delta function [10] having infinite height at $t = 0$, but is zero at all other times. The Fourier transform of the impulse function is

$$X(f) = \int_{-\infty}^{\infty} \delta(t)e^{-j2\pi ft}dt = 1 \qquad (11.6)$$

exhibiting equal weight at all frequencies.

The inverse Fourier transform is used to obtain a temporal signal from its Fourier transform:

$$x(t) = \mathscr{F}^{-1}\{X(f)\} = \int_{-\infty}^{\infty} X(f)e^{j2\pi ft}df. \qquad (11.7)$$

Suppose a signal places complete emphasis on frequency f_0, resulting in Fourier transform $X(f) = \delta(f - f_0)$. The inverse Fourier transform results in a complex sinusoid with frequency f_0, $x(t) = e^{j2\pi f_0 t}$. Recalling the decomposition of a cosine into positive and negative complex sinusoids,

$$\cos(2\pi f_0 t) = \frac{1}{2}\left[e^{j2\pi f_0 t} + e^{-j2\pi f_0 t}\right], \qquad (11.8)$$

the Fourier transform is seen to place emphasis at both f_0 and $-f_0$,

$$\mathscr{F}\{\cos(2\pi f_0 t)\} = \frac{1}{2}\delta(f - f_0) + \frac{1}{2}\delta(f + f_0). \qquad (11.9)$$

Many of the important properties of the Fourier transform will be described when they are applied in the following sections. A commonly used one in underwater acoustics applications is that of a time delay,

$$\mathscr{F}\{x(t - \tau)\} = \int_{-\infty}^{\infty} x(t - \tau)e^{-j2\pi ft}dt$$

$$= e^{-j2\pi f\tau}X(f) \qquad (11.10)$$

which only alters the Fourier transform by a linear change in phase with frequency.

In most applications, only a small temporal segment of a signal is analyzed. Evaluating only a portion of a signal can be characterized as analyzing the product:

$$x_w(t) = w(t)x(t) \qquad (11.11)$$

where $w(t)$ is a window function extracting the finite duration sample $x_w(t)$. Noting that the Fourier transform of a product is the convolution (*) of the individual Fourier transforms, the Fourier transform of $x_w(t)$ is

$$X_w(f) = W(f) * X(f)$$

$$= \int_{s=-\infty}^{\infty} W(f-s)X(s)ds \qquad (11.12)$$

Suppose the window function extracts duration T centered at the origin, so $w(t) = 1$ when $|t| \leq T/2$ and is zero otherwise. Its Fourier transform is a sinc function, where $\mathrm{sinc}(x) = \sin(\pi x)/(\pi x)$:

$$W(f) = T \, \mathrm{sinc}(fT), \qquad (11.13)$$

with main-lobe width $2/T$ (from zero to zero). For any frequency f, the Fourier transform of the effective signal being analyzed ($x_w(t)$) is

$$X_w(f) = \int_{s=-\infty}^{\infty} T \, \mathrm{sinc}((f-s)T)X(s)ds, \qquad (11.14)$$

which is a weighted average of the Fourier transform of $x(t)$ around f, with most of the contribution coming from the frequency range $[f - 1/T, f + 1/T]$. Thus if T is large enough so $X(f)$ varies slowly on this scale $X_w(f) \approx X(f)$ and the windowing leaves no noticeable impact.

However, consider a sinusoidal signal analyzed in a window of length T. It has Fourier transform

$$\mathcal{F}\{w(t)\cos(2\pi f_0 t)\} = \frac{T}{2}\mathrm{sinc}((f-f_0)T) + \frac{T}{2}\mathrm{sinc}((f+f_0)T), \qquad (11.15)$$

which takes on the shape of the window function's Fourier transform centered at the sinusoid's frequency, illustrating that even large analysis windows can impact signal analysis. The spreading of the sinusoid's impulsive Fourier transform for different analysis window extents is shown in Fig. 11.2 for a 100-Hz sinusoid. A 0.1-s window spreads the sinusoid significantly from 90 to 110 Hz, while increasing the analysis window extent to 1 and 10 s narrows the response by an order of magnitude for each increase. When analyzing sinusoidal signals, clearly the window extent is paramount to the achieved resolution in frequency.

Signals are often characterized as narrowband or broadband; however, the diversity of sounds observed underwater extends well beyond such simple terms. For example, most dolphin whistles are narrowband over short periods of time, but span a broad frequency band over their full duration [33]. Signals with time-dependent frequency content can be evaluated using the short-time Fourier transform (also called a spectrogram) [18,12] or time-frequency analysis with other kernel functions [18].

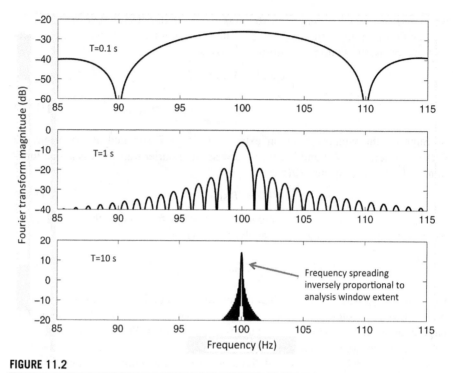

FIGURE 11.2

Fourier transform magnitude for windowed cosine with varying analysis window extents.

Mathematically, the frequency content of a signal is characterized by its spectral density. For deterministic energy signals (random signals are discussed in Section 11.2.4), the spectral density is the squared magnitude of its Fourier transform,

$$S_{xx}(f) = |X(f)|^2. \tag{11.16}$$

If $x(t)$ is a pressure signal, its total energy is

$$E = \int_{-\infty}^{\infty} |x(t)|^2 dt \tag{11.17}$$

$$= \int_{-\infty}^{\infty} S_{xx}(f) df. \tag{11.18}$$

Parseval's theorem (Rayleigh's energy theorem) is used to relate the total energy to the spectral density $S_{xx}(f)$ in Eq. (11.18), which can be described as the energy per unit frequency of the signal [18].

In underwater acoustics, spectral densities are typically [34] conveyed in units of squared pressure per unit of frequency rather than converting to energy or intensity by dividing by the acoustic impedance ($\rho_w c_w$). If the signal $x(t)$ has pressure units (μPa), then $S_{xx}(f)$ has units $(\mu Pa)^2$s/Hz. Dividing by $\rho_w c_w$ with units kg/(m$^2 \cdot$s) results in $S_{xx}(f)/(\rho_w c_w)$ having appropriate units for an energy density spectrum, pJ/(m^{2}Hz).

Evaluation of the power spectral density (discussed in Section 11.2.4) is more common than the energy spectral density. Less common is the usage of amplitude spectral density (i.e., $\sqrt{S_{xx}(f)}$) which results in the confusing "per root Hertz" reference. Spectral densities are usually conveyed in decibels with the appropriate reference unit (e.g., dB//1 μPa^{2}s/Hz). The decibel value of the energy spectral density and amplitude spectral density are identical:

$$10 \log_{10} S_{xx}(f) = 20 \log_{10} \sqrt{S_{xx}(f)}. \tag{11.19}$$

Implicit in the conversion is division inside the logarithm by a unit reference (respectively, 1 μPa^{2}s/Hz or 1μPa$\sqrt{s}/\sqrt{Hz}$). The choice of reference merely indicates a preference for conversion back from decibels to linear units.

11.2.3 DISCRETE FOURIER TRANSFORM

With the exception of some very-high-frequency systems, signal processing in most modern underwater acoustics applications is generally performed after the signal is sampled in time. Sampling in time produces a discrete-time sequence

$$x[n] = x(nT) \tag{11.20}$$

for integer values of n where T is the time {unit: s} between samples for a sampling rate $f_s = 1/T$ {unit: Hz}. Discrete-time sequences are denoted by brackets around the time index to contrast them from continuous-time signals. The primary concern related to sampling in time is aliasing. Aliasing occurs when high-frequency content is inadvertently and irretrievably folded down and added to low-frequency content by sampling at too slow a rate. Because the aliased high-frequency content is indistinguishable from the signal's low-frequency content, the sampled signal is corrupted and cannot be reconstructed to the original continuous-time version. To avoid aliasing, the signal must be low-pass filtered to remove any spectral content above the Nyquist frequency ($=f_s/2$).

Most applications analyze a short period of time, either because the signal naturally has that duration or by segmenting a long-duration signal to produce stationary segments. To characterize the frequency content of a signal containing N samples, the discrete-time equivalent to the Fourier transform is the *discrete Fourier transform (DFT)*,

$$X_k = \sum_{n=0}^{N-1} x[n] e^{-j2\pi kn/N} \tag{11.21}$$

for $k = 0,...,N - 1$. The frequency {unit: Hz} associated with X_k is

$$f_k = \begin{cases} f_s k/N & \text{for} & 0 \le k < N/2 \\ f_s(k - N)/N & \text{for} & N/2 \le k \le N - 1 \end{cases} \tag{11.22}$$

When N is a power of two, the fast *Fourier transform (FFT)* [35,14] efficiently obtains the DFT values $X_0,\dots,X_{N-1}$ with the number of computational operations on the order of $N\log_2 N$ compared to order N^2 operations for a direct implementation. Fast implementations for other factorings exist as well [14].

The DFT is an invertible transform where the time sequence $x[n]$ for $n = 0,\dots,N-1$ can be obtained from the frequency domain values using the inverse DFT:

$$x[n] = \frac{1}{N}\sum_{k=0}^{N-1} X_k e^{j2\pi kn/N}.$$ (11.23)

In some definitions and code implementations of the DFT or FFT, the $1/N$ scale in Eq. (11.23) is incorporated in or with the forward transform of Eq. (11.21), so care needs to be taken when using numerical implementations.

The DFT can be used to approximate the Fourier transform of a signal when it is zero outside of $t \in [0,NT]$ by multiplying by the sampling period T:

$$X(f_k) = \int_{-\infty}^{\infty} x(t)e^{-j2\pi f_k t}\,dt$$

$$\approx T\sum_{n=0}^{N-1} x[n]e^{-j2\pi kn/N}.$$ (11.24)

Because of this it is an integral part of the spectrogram, which can be used to evaluate the time-varying frequency content of a signal. There are a myriad of other applications of the DFT and its fast implementation [14], including fast implementations of Finite Impulse Response (FIR) filters, convolution, and correlation, as well as beamforming and wavenumber analysis for line arrays.

11.2.4 RANDOM PROCESSES: SPECTRA AND CORRELATION FUNCTIONS

Mathematically, underwater acoustic measurements are observations of random or stochastic processes. When represented as a random process, the measurement variable is capitalized: $X(t)$. Complete characterization of a random process is impractical, so several simplifying assumptions are typically made to facilitate design of signal processing algorithms and subsequent data analysis. The two primary assumptions are *wide-sense stationarity (WSS)* and ergodicity. A random process is WSS [36,29,15] if its first and second order statistics are constant with time; that is, it has constant mean and an *autocorrelation function (ACF)*, R_{XX} depending on the delay τ:

$$E[X(t)X^*(t+\tau)] = R_{XX}(\tau)$$ (11.25)

where $E[\cdot]$ is a statistical expectation or ensemble average. Ergodicity implies that ensemble moments (i.e., moments over multiple observations) can equivalently be obtained through averaging a single observation over time.

The *power spectral density (PSD)* represents the frequency content of a WSS random process. As described in Section 11.5.3, it can be estimated by averaging the magnitude-squared Fourier transform or by estimating the ACF and taking its Fourier transform:

$$S_{XX}(f) = \int_{-\infty}^{\infty} R_{XX}(\tau)e^{-j2\pi f\tau}d\tau. \qquad (11.26)$$

That $S_{XX}(f)$ is a *power* spectral density (for a WSS ergodic process) can be shown by starting with the average power, exploiting ergodicity and the inverse Fourier transform relationship between $S_{XX}(f)$ and $R_{XX}(\tau)$ evaluated at $\tau = 0$:

$$P = \frac{1}{T}\int_0^T |X(t)|^2 dt = E[X(t)X^*(t)] = R_{XX}(0) \qquad (11.27)$$

$$= \int_{-\infty}^{\infty} S_{XX}(f)df. \qquad (11.28)$$

Thus if $X(t)$ is a pressure signal with units µPa, then $S_{XX}(f)$ has units $(\mu\text{Pa})^2/\text{Hz}$. Dividing by $\rho_w c_w$ results in $P/(\rho_w c_w)$ having proper units for an equivalent plane-wave intensity of pW/m^2 or $S_{XX}(f)/(\rho_w c_w)$ as an intensity density with units $\text{pW/}(\text{m}^2 \cdot \text{Hz})$.

As an example, consider a zero-mean WSS white noise process with constant PSD, $S_{XX}(f) = \sigma^2$. The corresponding ACF is the inverse Fourier transform of a constant, yielding a Dirac delta function:

$$R_{XX}(\tau) = \sigma^2 \delta(\tau). \qquad (11.29)$$

Thus any two samples of a zero-mean white-noise process are uncorrelated if they are obtained at different times:

$$E[X(t_1)X^*(t_2)] = 0 \text{ for all } t_1 \neq t_2. \qquad (11.30)$$

It is common in underwater acoustics to use filters to limit analysis to a particular frequency band of interest (e.g., third-octave bands). When white noise is passed through a *linear time invariant (LTI)* band-pass filter (BPF; see Section 11.3.2.2) it is considered band-pass noise and the PSD is the product of the noise PSD and the magnitude squared of the filter transfer function. For an ideal BPF isolating the frequency band $[f_1,f_2]$ {unit: Hz}, the PSD is constant within the band and zero elsewhere:

$$S_{YY}(f) = \begin{cases} \sigma^2 & f \in (f_1,f_2) \\ 0 & \text{else} \end{cases}, \qquad (11.31)$$

in this case even rejecting the negative frequency band $[-f_2, -f_1]$. Applying the inverse Fourier transform to $S_{YY}(f)$, the ACF of ideal band-limited noise is

$$R_{YY}(\tau) = \sigma^2 W e^{j\pi\tau(f_1+f_2)} \operatorname{sinc}(W\tau) \tag{11.32}$$

where $\operatorname{sinc}(x) = \sin(\pi x)/(\pi x)$ is the sinc function and $W = f_2 - f_1$ is the bandwidth {unit: Hz}. The average power of the band-limited white noise is $R_{YY}(0) = \sigma^2 W$, which for this scenario is also the product of the PSD and bandwidth. Noting that the sinc function is zero for nonzero integer arguments, sampling the band-limited noise every $1/W$ {unit: s} results in uncorrelated measurements. If it can be further assumed that the random process is Gaussian, data so sampled are independent as well as uncorrelated.

11.2.5 CROSS-SPECTRA AND COHERENCE

The aforementioned autocorrelation function and spectral density represent the properties of a single measurement. In many applications two or more measurements are made. In such situations the cross-correlation function between two measurements $X(t)$ and $Y(t)$ is

$$R_{XY}(\tau) = E[X(t)Y^*(t+\tau)]. \tag{11.33}$$

The corresponding cross-spectral density is

$$S_{XY}(f) = \int_{-\infty}^{\infty} R_{XY}(\tau)e^{-j2\pi f\tau}d\tau. \tag{11.34}$$

Cross-correlation and cross-spectral density functions can be used to describe the autocorrelation and spectral density functions of the output of LTI systems (e.g., filters). The cross-correlation function is also useful in time-delay estimation while the cross-spectral density is used in adaptive beamforming where a matrix form of cross- and auto-spectral densities characterize measurements over multiple hydrophones.

The cross-spectral density is also used to define the coherence between the two measurements. The coherence (squared) function [37,16,15] is

$$\gamma_{XY}(f) = \frac{|S_{XY}(f)|^2}{S_{XX}(f)S_{YY}(f)}. \tag{11.35}$$

The coherence function is always between zero and one and describes how similar two signals are as a function of frequency.

11.2.6 CEPSTRUM

The cepstrum [38,12,39], whose name is a play on the word "spectrum," of a signal $x(t)$ with Fourier transform $X(f)$ is the inverse Fourier transform of the logarithm of the Fourier transform of $x(t)$:

$$\tilde{x}(t) = \mathscr{F}^{-1}\{\tilde{X}(f)\} = \mathscr{F}^{-1}\{\log_e X(f)\}, \tag{11.36}$$

where $\widetilde{X}(f) = \log_e X(f)$. The concept is similarly applied to the spectral density [39] or discrete-time spectrum [12] and has primarily been used to separate two signals combined by a convolution operation in what is called homomorphic deconvolution. For example, a communications signal measured some distance from the transmitter can be described as the transmitted signal convolved with the channel impulse response, $x(t) = h(t)*s(t)$. Because time-domain convolution becomes a product in the frequency domain,

$$\widetilde{X}(f) = \log_e H(f) + \log_e S(f) = \widetilde{H}(f) + \widetilde{S}(f) \tag{11.37}$$

so the cepstrum of the received signal is the sum of the cepstrum of the two components:

$$\widetilde{x}(t) = \widetilde{h}(t) + \widetilde{s}(t). \tag{11.38}$$

When linear filtering applied in the cepstrum domain separates $\widetilde{s}(t)$ from $\widetilde{h}(t)$, the original signal can be reconstructed by reversing the process on the filtered output:

$$s(t) \approx \mathscr{F}^{-1}\{\exp(\mathscr{F}\{\text{filter}[\widetilde{x}(t)]\})\}. \tag{11.39}$$

11.3 FILTERING

Filtering a measurement is often part of a formal detection or estimation process, but can be done for other reasons (e.g., an antialiasing LPF prior to temporal sampling or band-pass filtering prior to base-banding). Prior to temporal sampling, filter design and implementation is inherently analog and often situation specific. As the majority of filtering applications in underwater acoustics occur after sampling, the focus of this section is on discrete-time filters and generic filter types.

11.3.1 FILTER TYPES

The most common filters simply isolate portions of the frequency spectrum. LPFs allow low-frequency signals through their pass band while rejecting spectral content at higher frequencies in the stop band. Fig. 11.3 shows an LPF frequency response and the pertinent design specifications: pass, transition, and stop bands, attenuation, and ripple. Ideal filter characteristics are small ripple, large attenuation, and a narrow transition band.

High-Pass Filters (HPFs) accomplish the opposite of LPFs, rejecting low frequencies while preserving high frequencies. BPFs allow an isolated frequency band through and notch filters reject an isolated frequency band. All-pass filters pass all frequencies with a specified delay.

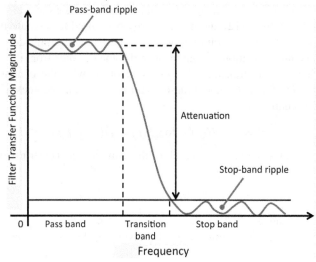

Filter Transfer Function Magnitude

Pass-band ripple

Attenuation

Stop-band ripple

0   Pass band   Transition band   Stop band

Frequency

FIGURE 11.3

Low-pass filter frequency response and design considerations.

11.3.2 PERFORMANCE METRICS, DESIGN, AND IMPLEMENTATION

11.3.2.1 Filtering Performance Metrics

Performance metrics in filter design vary depending on the application. While the primary requirement of a filter is that it pass the signal of interest and reject as much noise and interference as possible, design usually focuses on the frequency response characteristics of the filter. For example, LPF design algorithms specify the maximum ripple in the pass-band, the extent of the transition band, and the maximum acceptable level within the rejection band. Each of these is related to the magnitude of the filter's frequency response or transfer function. The phase response of a filter is next in importance after the magnitude, with the filter's group delay being of primary importance. The group delay describes the time delay a signal incurs when submitted to the filter. While minimizing the delay is important, a constant group delay over the pass band (as occurs for linear-phase filters) is usually the priority.

Filters can also be designed according to other performance metrics. For example, the Eckart filter [40] described in Section 11.3.5 maximizes the SNR at the filter output.

11.3.2.2 Digital Filter Design and Implementation

With rare exception, filters used in underwater acoustics are linear and time invariant (LTI). There are two fundamental types of LTI filters: those with finite impulse response (FIR) and those with infinite impulse response (IIR). The impulse response of an LTI filter is the inverse Fourier transform of its frequency response. IIR filters, which are implemented recursively, have an infinitely long response when subjected to an impulse. When properly designed, the response decays to near zero rapidly. FIR filters have nonzero impulse response only over a finite portion of time. These characteristics can be seen in the example FIR and IIR LPF impulse responses shown in Fig. 11.4. The filters both have 11 coefficients, implying the same computational effort (11 multiplications and additions per output sample) is required for implementation.

Generally, FIR filters are used when processing underwater acoustic signals owing to their linear phase properties within the pass band. When the phase is linear with frequency, the net effect is a time delay (i.e., a constant group delay), whereas nonlinear phase can distort the signal's audible content by subjecting different frequencies to different delays. FIR filters with symmetric coefficients (like the one in the upper panel of Fig. 11.4) have linear phase and therefore constant group delay in the pass band. Digital IIR filters characteristically have a recursive implementation and can produce very sharp transitions from pass band to rejection band. The frequency response of the example LPFs seen in Fig. 11.5 exhibits the sharper transition band of the IIR filter relative to the FIR filter. The nonlinear phase of the IIR filter and the pass-band linear phase of the FIR filter are evident in the phase

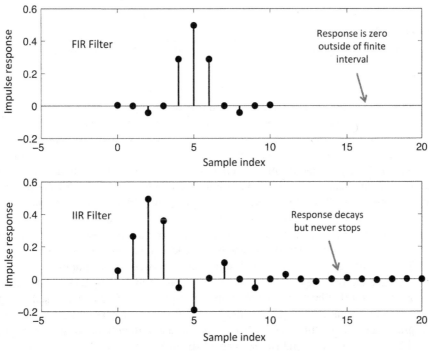

FIGURE 11.4

Finite- and infinite-impulse-response low-pass-filter examples.

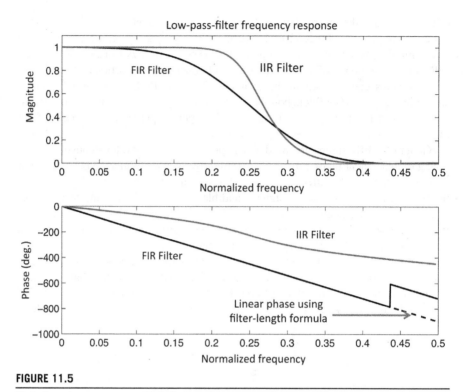

FIGURE 11.5

Finite impulse response and infinite impulse response low-pass-filter frequency response examples.

response. IIR filters are typically only used for smoothing parameter estimates over multiple observations in a context where phase integrity is not paramount.

A common approach to filter design is to use an algorithm to design an LPF and then convert it to one of the other filter types through translation or frequency trans-formation [11,10]. There are many techniques for designing LPF filters, including conversion of analog filter designs, windowing approaches, and truncating the impulse response. The Parks–McClellan algorithm [10] produces an optimum, equiripple, linear phase, FIR filter where the optimality criterion is to minimize the absolute error between the achieved magnitude frequency response and a model filter with unit response in the pass band and the desired attenuation in the stop band.

The primary trade-off in filter design comes between filter impulse-response duration and the sharpness of the transition from pass band to stop band. Sharp tran-sitions require FIR filters with a long impulse response, leading to increased group delay and computational requirements. So-called minimum-phase filters minimize the group delay in the filter design [10]. A useful rule of thumb for the size or length of FIR filters with pass-band ripple near 0.1 dB in Ref. [11] is:

$$\text{Filter length} \approx \frac{f_s A_{\text{dB}}}{22(f_1 - f_0)} \qquad (11.40)$$

where f_s is the sample frequency {unit: Hz}, A_{dB} is the attenuation {unit: dB}, and (f_0, f_1) {unit: Hz} defines the transition band.

The output of an LTI filter is the input convolved with the filter's impulse response. Filter implementation usually depends on the filter type, extent of the impulse response, and duration of the signal to be filtered. LTI IIR filters are typically implemented in the time domain using a recursive implementation where the current output is a linear combination of past output values and current and past input values. LTI FIR filters with short impulse responses are usually implemented in the time domain through convolution. However, if the impulse response is large enough, a computational savings arises when the convolution is accomplished in the frequency domain by a product. The savings comes from the efficient computation of the DFT by the FFT, but is complicated by accomplishing the filter's linear convolution using the DFT's circular convolution. The linear convolution is achieved by zero-padding both the filter impulse response and input signal prior to applying the FFT [10]. The zero-padding is such that the FFT has length greater than the sum of the lengths of the filter impulse response and input signal minus one. The two Fourier transforms are then multiplied prior to an inverse FFT, which produces the filter output. When the input signal is prohibitively long for a single FFT, the process can be accomplished in blocks using an "overlap-and-add" or "overlap-and-save" approach [13].

Many underwater acoustics applications exploit an estimate of time delay (e.g., for estimating range or sound speed). In such applications, the group delay of any filters in the signal processing needs to be taken into account. As previously noted, FIR filters with symmetric (or antisymmetric) coefficients have linear phase. The group delay for these filters is $(N_c - 1)/2$ samples where N_c is the number of filter coefficients [10]. The phase in the pass-band of the FIR LPF filter example of Fig. 11.5 is $-2\pi(f/f_s)(N_c - 1)/2$ {unit: radians} where f/f_s is the normalized frequency and f_s is the sampling rate. By examining the impulse responses in Fig. 11.4, the group delay is sensibly at the center of the symmetric FIR filter impulse response. The shape of the IIR filter impulse response, specifically the majority of the response occurring closer to zero than the FIR filter impulse response, indicates that its group delay is expected to be less than that of the FIR filter. This is also evident in Fig. 11.5 as a smaller average slope in the phase response of the IIR filter than for the FIR filter.

As previously mentioned, analog measurements from transducers are sampled in time and quantized in level. Discrete-time filters are typically designed under an assumption of sampling in time, but with no quantization in level. When such filters are used in finite-precision hardware (particularly those with fixed-point arithmetic), they may not achieve their design specification. Such situations were at one time common in real-time processing systems. However, they are becoming less of a concern with the proliferation of powerful, low-power, floating-point processors.

11.3.3 BAND-PASS SIGNALS: DIGITAL DOWN-CONVERSION

Most underwater acoustic signals are band-pass signals; they occupy a small portion of the frequency spectrum relative to their center frequency. For example, consider an acoustic communications system transmitting a pulsed sinusoid at 18 kHz with duration 1 ms. Accounting for waveforms at other frequencies, motion-induced Doppler, and a transition band in the analog antialiasing LPF, the measured echoes in the sonar would need to be sampled at $f_s = 42$ kHz to avoid aliasing. However, the signal has bandwidth of approximately 1 kHz, much less than its center frequency. If only the band 17−19 kHz is of interest, a bandwidth of $W = 2$ kHz, it may be isolated and resampled to a lower rate through digital down-conversion [11] thereby reducing the computational requirements for processing and storing the signal.

The digital down-conversion process starts by base-banding or demodulating the discrete-time signal ($x[k]$) to shift the center frequency of $f_c = 18$ kHz down to 0 Hz, followed by an LPF passing spectral content below $W/2$ {unit: Hz}:

$$\widetilde{y}[k] = \text{LPF}\left\{e^{-j2\pi f_c k / f_s} x[k]\right\}. \tag{11.41}$$

The final step decimates the signal by extracting every Lth sample:

$$\widetilde{x}[k] = \widetilde{y}[kL], \tag{11.42}$$

where $L = \lfloor f_s / W \rfloor = 21$. The new sampling rate is $f_s' = f_s / L = 2$ kHz, significantly reducing storage and computation requirements. The signal $\widetilde{x}[k]$ is known as the complex envelope and can also be obtained using a Hilbert transform of $x[k]$ followed by decimation.

It is important for the LPF in Eq. (11.41) to reject all spectral content above $W/2$ to avoid aliasing. The filtering rejects the negative-frequency portion of the pass-band signal (which is shifted in the base-banding operation to reside about frequency $-2f_c$) as well as any noise and interference that could alias into the band of interest during decimation. If a real signal is desired as opposed to the complex one in Eq. (11.41), replace f_c by the lower end of the band ($f_c - W/2$) and adjust the LPF to have a cut-off frequency equal to W. Alternatively, if the signal is band-pass filtered it is possible to exploit aliasing to avoid the modulation and filtering steps. By directly sampling or decimating $x[k]$ in a process called band-pass sampling [11], the signal can be folded to a lower frequency band through aliasing. This approach is less common owing to the reduced flexibility in placing the band of interest.

11.3.4 WINDOWING FOR SIDE-LOBE SUPPRESSION

When signal analysis occurs using a short sample in time (e.g., a short-time DFT), the process can be described as evaluating the complete signal after multiplication by a rectangular window.[2] In the frequency domain, the Fourier transform of the

---

[2]The rectangular window is also commonly called a uniform or box-car window.

windowed signal is the convolution between the Fourier transform of the complete signal and the Fourier transform of the window function. For example, an infinite duration complex sinusoid with frequency f_0 {unit: Hz}, when analyzed with a rectangular window of duration T {unit: s}, has a Fourier transform in the shape of a sinc function centered at f_0:

$$X_T(f) = \text{sinc}((f - f_0)T). \tag{11.43}$$

The sinc function has high side-lobes (approximately 13 dB below the peak as seen in Fig. 11.6), arising from the sharp transition at the edges of the rectangular window and the high frequencies necessary to represent it (i.e., the Gibbs phenomena [10]). A plethora of window functions are available [41,11,10,42] to taper the edges, thereby reducing side-lobe levels. Lower side-lobes aid in suppressing strong interfering signals at other frequencies at the expense of a widening of the main lobe and a reduction in SNR. This can be seen in Fig. 11.6 where a rectangular window exhibits a narrow main lobe but relatively high sidelobes compared to when a Hamming window is used on the same signal. The Hamming window of length N has coefficients [10]

$$w[n] = 0.54 - 0.46 \cos\left(\frac{2\pi n}{N-1}\right) \tag{11.44}$$

for $n = 0, \ldots, N - 1$. The Hann window (also often called a Hanning window), which has a higher first side-lobe than the Hamming window, has a more rapid

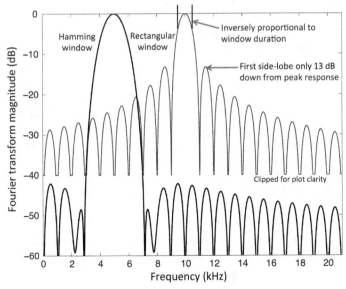

FIGURE 11.6

Fourier transform of sinusoid windowed with a rectangular or Hamming window.

side-lobe level decay and has been shown to be useful [43] when performing signal analysis with overlapping windows. The Hann window function is

$$w[n] = 0.5 - 0.5 \cos\left(\frac{2\pi n}{N-1}\right) \tag{11.45}$$

for $n = 0,\ldots,N-1$, which is similar to the Hamming, but tapers down to precisely zero at $n = 0$ and $n = N - 1$. Note that the window functions presented here produce symmetric windows with the center at $(N-1)/2$; the symmetry is important because it produces linear-phase FIR filters as discussed in Section 11.3.2.2. Some applications of windows (e.g., harmonic analysis with a DFT as noted in Ref. [41]) require forming an asymmetric window function with an even number of coefficients $(\tilde{N})$ by removing the last coefficient formed from an odd window of size $N = \tilde{N} + 1$ in Eq. (11.44) or (11.45).

The Hamming and Hann windows are simple to implement and provide improved interference suppression compared to the rectangular window. However, their main-lobe widths and side-lobe heights cannot be altered for applications requiring more or less suppression or pass-band width. Two common window types allowing such control are the Kaiser and Dolph-Chebyshev windows [11,10]. For a given window size, the Dolph-Chebyshev window minimizes the main-lobe width while constraining the maximum side-lobe to be below a design level. The Kaiser window is similar except for constraining the total energy in the side-lobes rather than their maximum level. Both windows are obtained by first forming the DFT of the window from the appropriate equations (e.g., see Ref. [11]) followed by an inverse DFT operation.

While windows are most commonly applied to filtering and time-series analysis, there are other applications within underwater acoustics. For example, beamforming is a spatial filtering process. When applied to a line-array with equally spaced transducers, it is identical to a short-time DFT applied to one segment of a sampled time-domain signal. The beam pattern of a conventional beamformer for an equally spaced line array is therefore the Fourier transform of the window function (i.e., the spatial equivalent to a finite-window sampling of an infinite duration sinusoid). Nearly all of the knowledge associated with windowing for time-series analysis extends to the beamforming application. One notable difference is the desire to use all of the data from a given array; that is, care should be taken when using window functions that zero-out an element (e.g., the Hann window of Eq. (11.45)). When applied to beamforming, the Hann window should be formed using $N = N_s + 2$ where N_s is the number of sensors and the nonzero window values used to shape the beampattern.

Another window function commonly used in underwater acoustics is the Tukey window. The window is formed by tapering some fraction $\alpha/2$ of the window on each side using the cosine function of the Hann window. The window function has the form [41,42]:

$$
w[n] = \begin{cases} 0.5\left[1 + \cos\left(\pi\left[\dfrac{2n}{\alpha(N-1)} - 1\right]\right)\right] & 0 \le \dfrac{n}{N-1} \le \dfrac{\alpha}{2} \\[2ex] 1 & \dfrac{\alpha}{2} \le \dfrac{n}{N-1} \le 1 - \dfrac{\alpha}{2} \\[2ex] 0.5\left[1 + \cos\left(\pi\left[\dfrac{2n}{\alpha(N-1)} + 1 - \dfrac{2}{\alpha}\right]\right)\right] & 1 - \dfrac{\alpha}{2} \le \dfrac{n}{N-1} \le 1 \end{cases} \quad (11.46)
$$

for $n = 0,\ldots,N-1$, where $\alpha \in [0,1]$. The Tukey window has extremes of the Hann window (Eq. (11.45)) when $\alpha = 1$ and the rectangular window when $\alpha = 0$. It is commonly used in underwater acoustics with a small amount of tapering (e.g., 5–10% or $\alpha = 0.05$ to 0.1) to shape a transmit waveform to limit the transient response of the amplifier.

11.3.5 DATA-DEPENDENT OR ADAPTIVE FILTERING

A final class of filters are those that alter their characteristics based on measured data. Such data-dependent or adaptive filters can be derived to optimize certain criteria such as maximizing SNR or minimizing noise variance. For example, when the signal and noise are zero-mean Gaussian random processes, the Eckart filter [40,6] maximizes the SNR out of the filter by choosing a transfer function magnitude squared:

$$
\left|H(f)\right|^2 = \frac{P_S(f)}{P_N^2(f)} \tag{11.47}
$$

where $P_S(f)$ and $P_N(f)$ are, respectively, the signal and noise power spectra. The difficulty of implementing the Eckart filter lies in knowing the signal and noise power spectra.

In some scenarios it is feasible to sense only the noise source, which allows certain solutions [44] where adaptation is only with respect to the noise. A number of data-adaptive techniques applicable to underwater acoustics arise in adaptive beamforming of sonar arrays [19]. They are particularly beneficial when remote sensing must be done in the presence of a small number of strong interferences coming from unknown directions.

11.4 DETECTION

One of the most basic signal processing functions in underwater acoustics is signal detection: does a measurement contain only noise or both signal and noise? The definitions of signal and noise differ by application. For example, a sonar engineer may consider an active sonar echo as signal and sound arising from a moving sea surface as noise. However, an acoustician studying ambient noise characteristics views the sea-surface-generated sound as signal and considers as noise transient sounds such as active sonar echoes or air-gun pulses. The first step in signal detection is to define what signal is and what noise is for the case at hand.

Formally, signal detection is considered a binary hypothesis test [23,28] where one is to decide if a signal of interest is present in a measurement (the S + N hypothesis) or if it is solely noise (hypothesis N). The signal is usually quantified by an amplitude or strength parameter (e.g., S) such that when $S = 0$ the null or noise-only hypothesis is true. The decision is based on a measurement consisting of some number of data samples represented here as the vector $\mathbf{x}$. The data can be from a single hydrophone or the beamformed output of an array of hydrophones. In the detection examples of this chapter, the data are generally assumed to have been sampled in time and base banded (i.e., down-converted) in frequency as described in Sections 11.2.1 and 11.3.3 so $\mathbf{x}$ is a complex vector with the frequency content of interest centered at zero Hz. Most notably because of the presence of noise (e.g., electronic noise or ambient acoustic noise), the data need to be modeled as random variables. Standard notation from statistical theory [2] entails representing the random variable by a capital letter (e.g., $\mathbf{X}$) while a specific observed value takes on the lower case letter (e.g., $\mathbf{x}$). Random variables are described by their *probability density function (PDF)*, $f_{\mathbf{X}}(\mathbf{x})$. The subscript in the PDF notation represents the random variable to which it applies and is often omitted when unambiguous. For the signal detection problem, the PDF of the data will be different under the N and S + N hypotheses.

Signal detectors comprise a many-to-one function compressing the data $\mathbf{x}$ in a measurement to a scalar decision statistic, $T(\mathbf{x})$. As an illustrative example, consider the energy detector (which will be discussed in more detail in Section 11.4.4):

$$T(\mathbf{x}) = \frac{\mathbf{x}^H \mathbf{x}}{\lambda_0} = \frac{1}{\lambda_0} \sum_{i=1}^{n} |x_i|^2, \tag{11.48}$$

which compresses the n complex data samples in $\mathbf{x}$ to a scalar decision statistic by magnitude-squaring, summing, and scaling by the background noise power λ_0. The decision statistic is then compared to a predetermined decision threshold h and the signal of interest is declared present if $T(\mathbf{x}) > h$. Otherwise the noise-only hypothesis is accepted.

This introductory detection example illustrates two issues arising during the construction and implementation of detection algorithms. First, the background noise power λ_0 as well as other parameters describing the noise or signal are generally unknown. Different approaches for dealing with unknown parameters are described in Section 11.4.2 and an example illustrating how λ_0 can be estimated and incorporated into a detection algorithm is shown in Section 11.4.6. The second issue arises in how to choose the decision threshold h. As described in the following section, the decision threshold normally depends on the probability of false alarm, a detection-performance metric.

11.4.1 PERFORMANCE METRICS, DESIGN, IMPLEMENTATION, AND ANALYSIS PROCEDURE

11.4.1.1 Detection Performance Metrics

Many design requirements and signal processing operations in underwater acoustics utilize SNR as the primary performance metric (e.g., the sonar equation, array design, beamforming, matched-filtering, etc.). The SNR-equivalent metric in detection performance is called detection index [4]:

$$d = \frac{\{E[T(\mathbf{X})|S+N] - E[T(\mathbf{X})|N]\}^2}{\mathrm{Var}\{T(\mathbf{X})|N\}} \quad \{\text{unit: dimensionless}\} \qquad (11.49)$$

or deflection [23] when referring to $\sqrt{d}$. Note that the random vector $\mathbf{X}$ modeling the measured data is used to form a random decision statistic $T(\mathbf{X})$ when describing the performance of a detector. The metric uses the squared difference in the means of the decision statistic under the $S+N$ and N hypotheses relative to the variance under the noise-only hypothesis as a measure of separation. When the decision statistic $T(\mathbf{X})$ follows a Gaussian distribution and signal presence only alters the mean, d adequately reflects detection performance. However, its limitations become apparent when signal presence changes the variance of the decision statistic. In such situations, or when a more precise measure of performance is desired, one must resort to the hypothesis testing roots of detection theory to find more appropriate performance metrics.

The performance of binary hypothesis tests is quantified by the so-called type I and II error probabilities [28]. In signal detection applications these are known as the probability of false alarm P_f and probability of missed detection P_m. The probability of detection $P_d = 1 - P_m$ is more commonly used than the probability of missed detection. The probability of false alarm is the probability one declares a signal is present when it is not; that is, when $T(\mathbf{X})$ exceeds h under hypothesis N,

$$P_f = \mathrm{Pr}\{T(\mathbf{X}) > h|N\}. \qquad (11.50)$$

Similarly, the probability of detection is the probability of correctly declaring a signal is present; that is, when $T(\mathbf{X})$ exceeds h under hypothesis $S+N$,

$$P_d = \mathrm{Pr}\{T(\mathbf{X}) > h|S+N\}. \qquad (11.51)$$

Detection performance is usually displayed on *receiver operating characteristic* (ROC) curves plotting P_d against P_f by varying the decision threshold h. For most detectors, changes in the decision threshold have an inverse effect on P_d and P_f; small values of h lead to high P_d and high P_f while increasing h reduces both probabilities. The theoretical ROC curve for detecting a deterministic signal in Gaussian noise is shown in Fig. 11.7. Ideal performance is achieved when $P_d \to 1$ and $P_f \to 0$, which is the upper left corner of the plot. When SNR becomes very small, the ROC curve tends to $P_d = P_f$ or the "chance line," which represents the worst performance a properly designed detector can achieve. The two points (0,0) and (1,1)

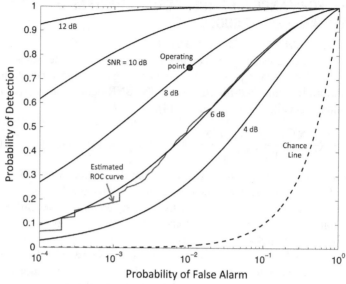

FIGURE 11.7

Theoretical receiver operating characteristic curve for detecting a deterministic signal in Gaussian noise.

are degenerate parts of every ROC curve and represent, respectively, $h \rightarrow \infty$ and $h \rightarrow -\infty$. However, when a detector is implemented the threshold h is usually set to achieve a specific P_f, which when combined with the corresponding value of P_d defines the detector's operating point (shown as $P_f = 10^{-2}$ and $P_d = 0.75$ when the SNR is 8 dB in Fig. 11.7).

11.4.1.2 Required SNR and Detection Threshold

The ROC curves shown in Fig. 11.7 illustrate the improvement in performance expected as SNR (at the detector) increases from 4 to 12 dB. Increasing the SNR moves the ROC curve closer to ideal performance while lowering it reduces performance. An indirect measure of detection performance is the SNR required to achieve a desired operating point; e.g., 8 dB to achieve $P_d = 0.75$ and $P_f = 10^{-2}$ in the example. This measure is useful in system design or analysis and represents the detection threshold (DT) term in the sonar equation, although the naming of DT often leads to confusion and misapplication. It is important to note DT is not the detector or decision threshold h described at the beginning of this section [45]. The former (DT) is a simplified measure of performance depending on a design specification (P_d, P_f) and is used in sonar-equation-based system design, prediction, and analysis. The latter (h) is used to implement or evaluate a detector and generally only depends on P_f.

A useful approximation to the SNR required to achieve an operating point $(\widetilde{S}_{dB})$ was developed by Albersheim [46,9] for a deterministic signal and a decision statistic comprising the incoherent sum of M independent observations:

$$\widetilde{S}_{dB} = -5 \log_{10} M + \left[6.2 + \left(\frac{4.54}{\sqrt{M + 0.44}} \right) \right] \log_{10}(A + 0.12AB + 1.7B) \quad (11.52)$$

{unit: dB} where

$$A = \log_e \left(\frac{0.62}{P_f} \right) \quad \text{and} \quad B = \log_e \left(\frac{P_d}{1 - P_d} \right) \quad (11.53)$$

with $\log_e(\cdot)$ representing the natural logarithm. For the operating point in Fig. 11.7, Albersheim's equation predicts a required SNR of 8.1 dB, within the 0.2-dB error expected over its extensive range of utility. Improved and extended approximations are also available [47–49,45]. The SNR described by Eq. (11.52) is the per-sample SNR after coherent filtering (e.g., matched-filter of signals with known form or band-pass filtering for random signals).

11.4.1.3 Detector Design and Implementation

Detector design starts with a statistical characterization of the signal of interest and the noise expected to be encountered. In particular, design requires the PDFs for the measurement data under the signal-plus-noise (S + N) hypothesis, $f_{\mathbf{X}}(\mathbf{x}|S + N)$, and under the noise-only hypothesis $f_{\mathbf{X}}(\mathbf{x}|N)$. Two commonly encountered statistical characterizations are for deterministic and random signals in additive, zero-mean, white Gaussian noise. In detector design or theoretical analysis, it is common to assume temporal sampling so that the data are statistically independent (e.g., sampling at one over the bandwidth of a spectrally white, band-pass Gaussian, random process). Thus the joint PDF for the complete measurement is the product of the PDFs of the individual samples and only the PDF of an individual data sample needs to be described. If the measured data are represented by the random variables $X_1,\ldots,X_n$, and X_i is independent of X_j for all $i \neq j$, the joint PDF under the S + N hypothesis is

$$f_{\mathbf{X}}(\mathbf{x}|S + N) = \prod_{i=1}^{n} f_X(x_i|S + N). \quad (11.54)$$

Under the noise-only hypothesis, the data PDF is commonly assumed to be complex-Gaussian[3] with zero mean and variance σ_N^2:

$$f_X(x|N) = \frac{1}{\pi \sigma_N^2} e^{-\frac{|x|^2}{\sigma_N^2}}. \quad (11.55)$$

---

[3]A complex Gaussian random variable with variance σ^2 comprises independent Gaussian-distributed real and imaginary parts, each with variance $\sigma^2/2$.

Random signals are typically assumed to be Gaussian distributed with zero mean and variance σ_S^2. When added to complex-Gaussian-distributed noise, the $S + N$ data PDF is also Gaussian but with variance $\sigma_N^2 + \sigma_S^2$:

$$f_X(x|S + N) = \frac{1}{\pi(\sigma_N^2 + \sigma_S^2)}e^{-\frac{|x|^2}{(\sigma_N^2 + \sigma_S^2)}}. \tag{11.56}$$

The deterministic signal usually has a known form (e.g., s_i for the ith sample). When added to complex-Gaussian-distributed noise, the $S + N$ data PDF is similarly complex-Gaussian but with mean equal to s_i:

$$f_X(x|S + N) = \frac{1}{\pi\sigma_N^2}e^{-\frac{|x-s_i|^2}{\sigma_N^2}}. \tag{11.57}$$

Given the statistical characterization of the signal of interest and noise, a many-to-one transformation of the measurement data to a scalar value must be constructed to form what is called the decision statistic, $T(\mathbf{x})$. Ideally, there is no loss of information from the compression (i.e., the decision statistic is sufficient [2] in the statistical sense). Several approaches for designing $T(\mathbf{x})$ will be described in Section 11.4.2. Generally, larger values of $T(\mathbf{x})$ support the $S + N$ hypothesis and smaller values the noise-only hypothesis. An important characteristic of the decision statistic is that it cannot depend on any unknown variables (e.g., SNR or noise power).

In practice, signals and noise can be difficult to characterize statistically with complicated time-varying PDFs depending on a variety of typically unknown parameters. Fortunately, it is often possible to utilize detectors derived for standard theoretical signal models even if the real signals of interest are somewhat different. While most signal detectors are designed assuming Gaussian-distributed noise [22,23], extensions can be found for various non-Gaussian distributions [50] or derived by relying on the more general hypothesis-testing literature [28,2]. Detector design therefore requires compromising between the *mismatch* error encountered when using approximate statistical models for which detector design is straightforward and the *estimation* error incurred by estimating unknown parameters in more accurate models. The more complicated statistical models often result in more complicated detector structures and a more difficult implementation.

Once the decision statistic is formulated, detector implementation is accomplished by evaluating the data compression function $T(\mathbf{x})$ at the observed data $\mathbf{x}$ and comparing it to the threshold h. The unit of measure for h is the same as for $T(\mathbf{x})$ and therefore is not necessarily a physical quantity and is often dimensionless. The threshold h is generally chosen to satisfy a false alarm constraint (i.e., inverting Eq. (11.50) for h given P_f). Often h is set empirically after observing some reasonable quantity of noise-only data or tuned by the user to an appropriate operating point. In signal processing systems with fixed or limited computational resources and downstream false-alarm-reducing algorithms such as trackers and classifiers, detectors may be implemented by ranking all observations within some time period and passing a fixed number related to the capacity of subsequent processing. This essentially implements a time-varying threshold to produce a constant computational load.

11.4.1.4 Analysis of Detection Performance

While theoretical ROC curves exist for simple signal and noise characterizations it is at times necessary to evaluate the performance of detectors on simulated and real data. The process for both cases begins with identifying (or simulating) data under the two hypotheses. For real data this generally requires partitioning measurements based on prior or expert knowledge (e.g., geometric reconstruction of an experiment or expert analysis of the data) to label portions containing signal (x_i^S for $i = 1,...,n_S$) with the remaining data often assumed to only contain noise (x_i^N for $i = 1,...,n_N$). Observations of the decision statistic in the two partitions are then used to form estimates of P_d and P_f as a function of the decision threshold h in the manner of a histogram:

$$\widehat{P}_d(h) = \frac{1}{n_S} \sum_{i=1}^{n_S} I\{T(x_i^S) > h\} \tag{11.58}$$

and

$$\widehat{P}_f(h) = \frac{1}{n_N} \sum_{i=1}^{n_N} I\{T(x_i^N) > h\} \tag{11.59}$$

where $I\{\cdot\}$ is an indicator function returning 1 when the argument is true and zero otherwise. An estimated ROC curve formed from 10^4 independent observations is shown in Fig. 11.7 for the 6-dB SNR case. The Matlab [51] code[4] to obtain $P_d(h)$ or $P_f(h)$ is simply

```
p=1-cumsum(hist(x,h))/n;
```

where the data are in a vector x, the thresholds in a vector h, and n is the number of observations. As described at the Matlab Website [51], the hist function computes a histogram while cumsum performs a cumulative summation.

The probability estimates formed by Eqs. (11.58) and (11.59) are unbiased and, when the number of observations (generically n) is large, approximately Gaussian distributed with variance equal to $p(1 - p)/n$ where p is the true probability. When p is small, which is usually the case for P_f, n_N needs to be correspondingly large (e.g., $n_N > 10/P_f$) to obtain a reliable estimate. The increased variability is evident in Fig. 11.7 at the smaller values of P_f.

---

[4]Matlab [51] is a commercial computational program commonly used for developing, evaluating, and implementing signal processing algorithms. It has a large suite of built-in or add-on subroutines accomplishing or simplifying many of the signal processing functions described in this chapter.

Confidence intervals for the P_f and P_d estimates can be obtained by noting they are single-sample estimates $(\widehat{p})$ of the proportion parameter of a binomial random variable. When the number of observations is large enough, the binomial distribution is approximately Gaussian and yields the $(1 - \alpha)$ 100% confidence interval [2]:

$$\widehat{p} \pm z_{\alpha/2} \sqrt{\frac{\widehat{p}(1 - \widehat{p})}{n}} \qquad (11.60)$$

where $\widehat{p}$ is the estimate and $z_{\alpha/2}$ is the $1 - \alpha/2$ percentile of the standard normal distribution. That is, $z_{\alpha/2}$ satisfies

$$\Pr\left\{Z > z_{\alpha/2}\right\} = \frac{\alpha}{2} \qquad (11.61)$$

when Z is a normal random variable with zero mean and unit variance. Common values of $z_{\alpha/2}$ are found in Table 11.1.

When the data used to estimate the ROC curve are correlated the P_f and P_d estimates are still unbiased. However, the variance of the estimate will be inversely proportional to the effective number of independent samples rather than the actual number of samples (i.e., the variance will be higher than if all the data were independent of each other). Correlated data are most commonly encountered when the complex base-banded signal is sampled at rates greater than the signal bandwidth, in which case the effective number of independent samples can be obtained by the product of the bandwidth and the time extent of the analysis window used to estimate the ROC curve.

11.4.2 STRUCTURED DESIGN APPROACHES

The most common design approach for detectors exploits the Neyman-Pearson lemma [28] and provides a detector whose optimality maximizes P_d while constraining P_f to be at or below some fixed level. The detector decision statistic is formed by taking the ratio of the data PDFs under the S + N and noise-only hypotheses,

$$T_{\mathrm{LR}}(\mathbf{x}) = \frac{f_{\mathbf{X}}(\mathbf{x}|S + N)}{f_{\mathbf{X}}(\mathbf{x}|N)}, \qquad (11.62)$$

Table 11.1 Percentiles of the Standard Normal Distribution for Use in Forming Confidence Intervals

$(1 - \alpha) \cdot 100\%$	α	$z_{\alpha/2}$
75	0.25	1.150
90	0.1	1.645
95	0.05	1.960
99	0.01	2.576

and is often called a *likelihood ratio detector* (LRD). The likelihood ratio in Eq. (11.62) can be manipulated through any monotonic transformation without altering the performance. For example, it is common to take the logarithm and use the *log-likelihood-ratio* (LLR) so the decision statistic for multiple independent and identically distributed measurements can be obtained by summing individual ones:

$$T_{\text{LLR}}(\mathbf{x}) = \log_e \left\{ \frac{f_{\mathbf{X}}(\mathbf{x}|S+N)}{f_{\mathbf{X}}(\mathbf{x}|N)} \right\}$$

$$= \sum_{i=1}^{n} \log_e \left\{ \frac{f_X(x_i|S+N|)}{f_X(x_i|N)} \right\}. \tag{11.63}$$

As previously noted, a key characteristic of a detector decision statistic is that it cannot depend on any unknown parameters. This implies the data PDFs must be completely specified to use the LRD. While it is rare to know data PDFs precisely, they can often be described parametrically with a small number of unknown parameters, for example, signal power S and noise power λ_0. There are multiple approaches to designing detectors for scenarios with unknown parameters. Table 11.2 lists the various cases described here and which design approaches are applicable.

The most desirable approach is to find a decision statistic invariant to the unknown parameter (i.e., $T(\mathbf{x})$ does not depend on the parameter). This is most easily done by an appropriate monotonic transformation of the likelihood ratio; the resulting detector is Neyman-Pearson optimal irrespective of the actual value of the unknown parameter and therefore called a *uniformly most powerful* (UMP) hypothesis test.

When only the signal strength is unknown, a *locally optimal* (LO) detector can be constructed to provide optimal performance for a vanishingly small signal according to

$$T_{\text{LO}}(\mathbf{x}) = \frac{\{\partial f_{\mathbf{X}}(\mathbf{x}; S|S+N)/\partial S\}_{S=0}}{f_{\mathbf{X}}(\mathbf{x}|N)} \tag{11.64}$$

where the notation $f_{\mathbf{X}}(\mathbf{x}; S|S+N)$ represents the joint PDF of $\mathbf{X}$ as depending on parameter S under the $S+N$ hypothesis. An alternative to the LO detector can be obtained by specifying a *design* SNR (e.g., the minimum strength signal of interest)

Table 11.2 Detector Design Options When Various Signal or Noise Parameters Are Unknown

Unknown Parameters	Detector Design Options
None	LRD
Signal strength	UMP, LO, design-SNR, GLR, or Bayesian
Noise or ancillary signal	UMP (unlikely to exist), GLR, or Bayesian

GLR, Generalized likelihood ratio; LO, Locally optimal; LRD, Likelihood ratio detector; SNR, Signal-to-noise power ratio; UMP, Uniformly most powerful.

from which the LRD is formulated. Such a detector is optimal if the observed data exhibit the design SNR and suboptimal otherwise. Performance above the design SNR is often acceptable (e.g., a small loss from the optimal detector), but usually very poor below the design SNR.

When the above-mentioned approaches do not yield a viable detector, the *generalized likelihood ratio* (GLR) detector provides a practical, although usually suboptimal, approach. The GLR is formed by separately maximizing the PDFs under the two hypotheses before forming a ratio:

$$T_{\text{GLR}}(\mathbf{x}) = \frac{\max_{\theta_1} f_{\mathbf{X}}(\mathbf{x}; \theta_1 | S + N)}{\max_{\theta_0} f_{\mathbf{X}}(\mathbf{x}; \theta_0 | N)} \tag{11.65}$$

where θ_1 and θ_0 represent the unknown parameters under, respectively, the S + N and N hypotheses. As described in Section 11.5.2, the parameter values maximizing the PDF are the *maximum likelihood estimates* (MLEs), so the GLR detector has the same form as the LRD but uses the MLE for any unknown parameters.

Unknown parameters not related to the difference between the N and S + N hypotheses are called nuisance parameters. In signal detection problems these can be noise parameters, which will exist under both hypotheses, or ancillary signal parameters (e.g., delay, Doppler or frequency) existing only under the S + N hypothesis. One usually wishes to estimate the signal parameters for each detected signal, so they are often handled by forming a bank of detectors over each parameter's domain and choosing the one with the largest decision statistic, simultaneously providing both the parameter estimate and detector decision statistic. The bank of detectors essentially implements a GLR detector where the maximization in the numerator of Eq. (11.65) over the subject signal parameter is accomplished computationally. When any unknown parameter is estimated and used in a detector, there is a loss in performance relative to when the parameter is known perfectly. This includes template-matching approaches where the "best" template is selected by matching against the data and then used to form a decision statistic. The unknown parameter is the index into the list of templates evaluated.

While not common, when prior information on unknown parameters exists, it can be exploited to form a Bayesian detector [52]. The detector structure is similar to the LRD, where the PDFs are averaged over the unknown parameters prior to forming the ratio:

$$T(\mathbf{x}) = \frac{E_{\theta_1}[f_{\mathbf{X}}(\mathbf{x}; \theta_1 | S + N)]}{E_{\theta_0}[f_{\mathbf{X}}(\mathbf{x}; \theta_0 | N)]}. \tag{11.66}$$

When prior information is not available, it is still possible to utilize the Bayesian approach with what are termed noninformative priors. Such prior distributions properly allocate weight nonpreferentially throughout the domain of the unknown parameters.

11.4.3 DETECTING SIGNALS OF KNOWN FORM: CORRELATION PROCESSING

Many signals in underwater acoustics will have known form with only a few unknown parameters. Examples include active sonar echoes and probe or communications pulses. The general case involves assuming a data segment $\mathbf{x}$ comprises signal with additive noise $\mathbf{v}$:

$$\mathbf{x} = A e^{j\psi} \mathbf{s}(\theta_1) + \mathbf{v} \tag{11.67}$$

where A is the signal amplitude, ψ is the phase, and $\mathbf{s}(\theta_1)$ provides the structure of the signal with parameters in the vector θ_1. The phase of the complex signal (recall it was base-banded) is almost always modeled as uniformly random on $(0, 2\pi)$ owing to imprecise knowledge of arrival time. When signal presence implies $A \neq 0$, the other signal parameters (θ_1) are known, and under certain benign conditions for the noise (zero mean, white, and Gaussian with known variance λ_0), the optimal detector, which in this case is UMP, results in a decision statistic:

$$T(\mathbf{x}; \theta_1) = \frac{\left| \mathbf{s}^H(\theta_1)\mathbf{x} \right|^2}{\lambda_0} \tag{11.68}$$

where the superscript ^{H} is the Hermitian (conjugate transpose) operator. The decision statistic compares the observed data with the known form of the signal in a correlation operation with the result required to be large relative to the noise variance to declare signal presence.

Matched-filtering occurs when correlation processing is applied to a sliding data sample from a time-series and is implemented as an FIR filter whose impulse response is the time-reversed signal. In a detection context, the onset time of the signal is unknown and the matched-filter implements a GLR detector by choosing the time with the largest decision statistic. In the context of matched filtering, correlation processing not only performs the detection process but also serves as an estimation algorithm for the unknown onset time.

As noted in the previous section and demonstrated in the following example, other unknown signal parameters generally require implementation of a bank of detectors spanning the domain of the unknown parameters. The case of unknown noise power is covered under normalization in Section 11.4.6.

11.4.3.1 Example: Doppler Filter Bank

Suppose a pulsed sinusoidal signal with frequency f_0 {unit: Hz} is transmitted and the reflection from a moving target received by a monostatic sonar system. Accounting for the Doppler effect induced by target motion, the frequency of the pulse received by the sonar [6] is

$$f_1 = f_0 \left(\frac{1 - v/c_w}{1 + v/c_w} \right) \quad \{\text{unit: Hz}\} \tag{11.69}$$

where c_w is the speed of sound in water {unit: m/s} and v is the radial velocity {unit: m/s} of the target relative to the sonar. Target motion toward the sonar ("closing") is represented by $v < 0$ and away from the sonar ("opening") by $v > 0$. The detection problem is one of a signal with known form, unknown amplitude, and unknown target velocity (v).

After base-banding the received target echo by a center frequency f_0, the signal $\mathbf{s}(v)$ is simply a pulsed sinusoid with frequency

$$f_1 - f_0 = \frac{-2v/c_w}{1 + v/c_w} f_0 \approx -\frac{2f_0 v}{c_w} \quad \{\text{unit: Hz}\} \tag{11.70}$$

where the approximation assumes the target speed is small relative to the speed of sound in water (i.e., $|v| \ll c_w$). For a specific target radial velocity v, the detector of Eq. (11.68) is a matched-filter isolating frequency $-2f_0 v/c_w$ with temporal duration

$$D(1 + v/c_w)/(1 - v/c_w) \approx D(1 + 2v/c_w) \approx D \quad \{\text{unit: s}\} \tag{11.71}$$

where D is the duration of the transmitted pulse.

When the target velocity is unknown, the GLR detector has the form described by Eq. (11.68) with $\theta_1 = v$ but requires a maximization over v that generally must be carried out computationally. When the largest radial velocity of interest is small relative to c_w, the latter approximation in Eq. (11.71) holds and the echo duration is nearly constant for all v of interest. For this situation, the GLR detector can be implemented through the DFT (see Section 11.2.3) to form

$$T(\mathbf{x}, v_k) = \frac{\left| \mathcal{X}_k \right|^2}{\lambda_0} \tag{11.72}$$

where $\mathcal{X}_k$ is the kth frequency bin of the DFT of $\mathbf{x}$ and v_k is the corresponding radial velocity. The GLR decision statistic is the maximum of $T(\mathbf{x}, v_k)$ over the values of k representing the radial velocities of interest. When the DFT comprising the GLR detector is implemented over consecutive time samples to account for an unknown arrival time, the structure is known as a Doppler filter bank with individual DFT bins called Doppler channels.

To determine the radial velocity associated with each DFT bin, the frequency f_k from Eq. (11.22) is used in Eq. (11.70):

$$v_k = -\frac{c_w f_k}{2f_0} \quad \{\text{unit: m/s}\}. \tag{11.73}$$

The size of the DFT (N_{DFT}) must be large enough to encompass the echo extent, $N_{\text{DFT}} \geq D f_s$. Typically the DFT size will be increased to the nearest power of two to exploit the computational efficiency of the FFT. When the DFT length exceeds the echo extent, the DFT must be zero-padded (i.e., the data submitted to the DFT are set to zero for time-sample indices above $D f_s$). The resulting frequency domain evaluation has finer spacing in frequency, which reduces the mismatch loss from evaluating a finite grid of velocities in forming the maximum. However, it is important to remember the finer spacing in frequency achieved by zero-padding a DFT does not imply a narrower bandwidth when interpreting the DFT as a bank of narrowband filters.

For a concrete example, consider a 1-s duration pulsed sinusoid at $f_0 = 1$ kHz and a radial velocity range of ± 9 m/s (approximately ± 18 kt). The change in frequency of the echo relative to the transmission is

$$f_1 - f_0 = -\frac{2 \cdot 1000 \text{ Hz} \cdot \pm 9 \text{ m/s}}{1500 \text{ m/s}} = \mp 12 \text{ Hz}. \qquad (11.74)$$

After basebanding, suppose the echo is low-pass filtered and the sampling rate decimated to $f_s = 100$ Hz. The DFT size is therefore $N_{DFT} = D f_s = 100$ (or greater) with a corresponding frequency spacing of $f_s/N_{DFT} = 1$ Hz. Thus 25 Doppler channels are required to span the velocity range of interest (equivalently from -12 Hz to 12 Hz). Note that when the number of Doppler channels necessary to span the range of radial velocities of interest is small relative to the DFT size, it may be more efficient to implement the filters individually than to use the FFT.

Finally, because the bandwidth of a 1-s duration sinusoidal pulse is approximately 1 Hz and the Doppler channels are spaced 1 Hz apart, an echo exhibiting a Doppler falling in between Doppler channels will incur a mismatch loss (sometimes called a scalloping loss). As previously noted, increasing the DFT size with zero-padding reduces the mismatch loss. In this example, doubling the DFT size reduces the mismatch loss from 3.8 dB to less than 1 dB, quadrupling it reduces it to less than one-quarter decibel.

11.4.3.2 Pulse Compression, Matched Filtering, and the Ambiguity Function

Waveform design and analysis for active sensing is an extensive research area with complete books devoted to the subject. In underwater acoustics applications, the most common waveforms are the pulsed sinusoid (i.e., a *continuous wave* or CW) and *linear* or *hyperbolic frequency modulated* waveforms (LFM or HFM). The primary distinctions between the CW and frequency modulated (FM) waveforms are in their bandwidth and Doppler sensitivity. CW waveforms are inherently narrowband while FM waveforms can be made wideband. As seen in the previous example, CW waveforms are Doppler sensitive while LFM and HFM waveforms are Doppler insensitive or tolerant [53].

Active sonar performance, with respect to waveform design, is typically limited by the total transmitted energy (in noise-limited scenarios) and bandwidth (in reverberation-limited scenarios) [4]. As such, long waveforms are used to transmit large total energy while FM waveforms are used to combat reverberation.

Correlation processing or matched filtering for active sonar pulses is also known as pulse compression. Pulse compression refers to how matched filtering produces a temporal response narrower than the duration of the original pulse or received echo. Generally, the temporal extent of the compressed pulse is inversely proportional to its bandwidth. For CW waveforms this implies the compressed pulse has about the same length as the transmitted pulse. However, for FM waveforms the compressed pulse can be significantly narrower. The response of the matched filter to variations in time delay and Doppler is quantified by a waveform's ambiguity function [53]. The ambiguity function defines a waveform's Doppler sensitivity, delay-Doppler coupling, and susceptibility to reverberation.

11.4.3.3 Example: LFM Pulse Compression

The pulse compression aspect of matched filtering can be seen in the example of Fig. 11.8. As seen in the upper figure, the measured pulse is a 1-s long, 100-Hz wide LFM waveform (only the real part of the complex base-banded signal is shown) at an SNR of -15 dB. The dashed line is the threshold chosen for $P_f = 10^{-4}$ when testing just the time sample when the echo is received. Prior to matched filtering, no part of the echo is visible nor does the noise shown generate any false alarms.

After matched filtering, the compressed echo response is seen to be much narrower than the original pulse and significantly higher than the decision threshold.

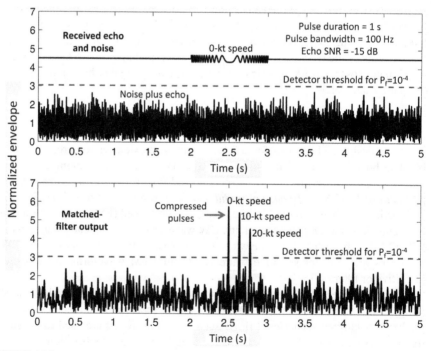

FIGURE 11.8

Example showing how matched-filtering brings out a signal buried in noise. The echo signal in the upper plot is displaced vertically to make it visible. *SNR*, signal-to-noise power ratio.

The probability of detection has gone from being nearly equal to $P_f (P_d = 1.3 \times 10^{-4})$ to almost being one ($P_d = 0.9999$). The matched filter used in the processing assumed a radial target velocity of 0 kt. Using this same matched filter when the target has a radial velocity of 10 or 20 kt results in a small degradation in SNR, owing to the Doppler tolerance of the LFM waveform. Thus an advantage of Doppler tolerance is a reduced need for implementation of a bank of detectors spanning the domain of unknown Doppler of an echo. However, an inadvertent effect lies in the shift in delay when there is mismatch between the radial velocity assumed to form the matched filter and that observed in the measured data. This effect, which is predicted by the waveform ambiguity function, arises from the compression or expansion of the frequency content in an LFM or HFM waveform lining up with a different part of the matched filter designed for a zero-radial-velocity target.

11.4.3.4 Other Applications in Underwater Acoustics

The previously mentioned examples of correlation processing in underwater acoustics entailed cases where a known signal was transmitted. Other examples of correlation processing exist in underwater acoustics. For example, conventional beamforming [6] of an array of hydrophones is a form of correlation processing where the expected form of a plane-wave signal from a particular direction is correlated in spatial dimensions with measurements obtained from the array. Matched-field processing [54] is similarly a type of correlation processing where the signal model is generalized to represent the propagation from a point in space to each element of the array.

Correlation processing is not limited to correlating a known signal form with a measured signal. It can also be applied between two separate measurements [5] to estimate time delays.

11.4.4 DETECTING RANDOM OR UNKNOWN SIGNALS: ENERGY DETECTOR

When the temporal structure of a signal is unknown, the measured data can be described generically as

$$\mathbf{x} = \mathbf{s} + \mathbf{v} \tag{11.75}$$

where $\mathbf{s}$ is the unknown signal and, as before, $\mathbf{v}$ is the noise. With no knowledge about the signal $\mathbf{s}$, it must be modeled as completely random to develop a detector. When the signal arises from the sum of many independent contributions (e.g., multipath propagation with many reflections off random surfaces) it can be assumed to be Gaussian distributed by the central limit theorem. When both signal and noise are zero-mean, white Gaussian random vectors, the decision statistic is an energy detector with the form

$$T(\mathbf{x}) = \frac{\left| \mathbf{x}^H \mathbf{x} \right|^2}{\lambda_0}, \tag{11.76}$$

where λ_0 is the noise variance. Through Parseval's theorem, the detector can equivalently be implemented in the time or frequency domain. For example, if the signal were known to be restricted to a particular frequency band, the decision statistic is the sum of the magnitude-squared DFT bins within the band. The detector estimates the total energy in the band and declares signal present if it is large relative to the noise variance.

The structure of Eq. (11.76) assumes the signal has a particular duration. Often the duration is chosen according to the limits of coherence. When the signal persists beyond the coherence limits a combination of coherent and incoherent averaging is performed,

$$T(\mathbf{x}) = \sum_{i=1}^{m} \frac{\left|\mathbf{x}_i^H \mathbf{x}_i\right|^2}{\lambda_{0,i}} \tag{11.77}$$

where $\mathbf{x}_i$ is the ith of m data segments comprising the signal (i.e., $\mathbf{x}^T = \begin{bmatrix} \mathbf{x}_1^T \cdots \mathbf{x}_m^T \end{bmatrix}$) and $\lambda_{0,i}$ is the noise variance for the ith data segment. The signal is assumed to be coherent within each $\mathbf{x}_i$ with the individual or local decision statistics combined incoherently. The extent of incoherent averaging is often limited by encounter time; for example, a moving source will only be within range or within a single beam for a limited period of time.

Given a large enough time bandwidth product (TW), the SNR required to meet a performance specification (i.e., detection threshold) decreases with coherent averaging according to $10 \log_{10}(TW)$. When m is large enough, however, the reduction for incoherent averaging is only $5 \log_{10}(m)$. As such, detector design should maximize coherent integration prior to any incoherent averaging.

To estimate the noise variance, access to signal-free data is necessary. Depending on the situation, signal-free data generally come from nearby frequencies, time segments (e.g., prior to signal onset), or beams. In each case assumptions are made requiring the data used for noise variance estimation to be essentially the same as the noise within the test sample.

When the noise spectrum is not white (i.e., not constant) within the analysis band and signal-free data exist from which it can be estimated, an Eckart filter [40] can be implemented as described in Section 11.3.5.

11.4.5 DETECTING UNKNOWN SIGNAL ONSET: PAGE'S TEST

The detection problems considered up to now have tested for the presence or absence of signal in a fixed data sample. A slightly different, but very common detection problem occurs when testing for the onset of a signal having unknown duration. Examples from underwater acoustics include the onset of a marine mammal vocalization or a communication signal burst.

A useful detector for this problem, called Page's test [55], is simple to implement and optimal for the *quickest detection* problem [56]. The decision statistic has the form

$$w_k = \max\{0, w_{k-1} + T(x_k)\} \tag{11.78}$$

where k is a time index, x_k is the data measurement at time index k, and the function $T(x)$ is ideally the LLR between the S + N and N hypotheses. The detector is initialized with $w_0 = 0$ and a signal is declared present when w_k first exceeds a decision threshold.

As the LLR is often not available, other detector structures can be used (e.g., the GLR) as long as the average value is negative under hypothesis N and positive under hypothesis S + N:

$$E[T(X)|N] < 0 < E[T(X)|S + N] \tag{11.79}$$

where X is the random variable representing the data. It is common to enforce the condition by biasing $T(X)$ negative under hypothesis N at the expense of poor detection performance for weak signals.

A quickest detection example is shown in Fig. 11.9. The measured data are noise-only until sample 40 where a weak signal (0 dB SNR per sample) begins. The LLR of the measured data are submitted to Eq. (11.78) to produce the Page's test decision statistic shown in the lower panel. The decision statistic is small and often zero under the noise-only condition, but clearly begins to accumulate once the signal commences at sample 40. The decision threshold is chosen large enough so a crossing

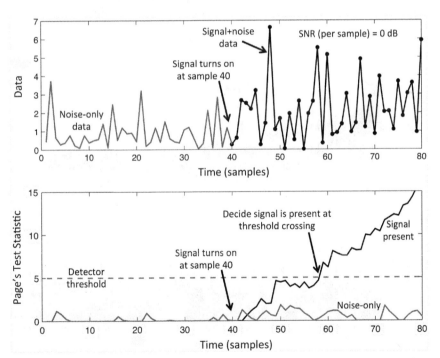

FIGURE 11.9

Example showing detection of the onset of a 0-dB signal-to-noise power ratio (SNR) signal using Page's test.

is rare when there is no signal. An estimate of the starting time of the signal (after it has been detected) is the sample after the most recent time at which the decision statistic was zero.

For signals of finite but unknown duration, a similar process can be applied upon detection to detect the end of the signal. The combined process is called an alternating hypothesis Page's test [57].

11.4.6 NORMALIZING FOR CONSTANT FALSE ALARM RATE

Detection algorithms must be applied in the presence of noise with varying levels. For example, ambient noise from the ocean surface can vary over 20 dB with sea-state and be several tens of decibels higher in the presence of a local interference. The background noise power λ_0 is therefore treated as a nuisance parameter unknown before a measurement is taken. Detection algorithms might estimate λ_0 or form a decision statistic invariant to it. The former are often implemented in a separate process called normalization with a goal of providing a *constant false alarm rate* (CFAR) for a given decision threshold. Consider the active-sonar example shown in the upper panel of Fig. 11.10 where the reverberation power level decays with range. A single, fixed threshold cannot be used to capture the target echoes, which are otherwise clearly visible above the decaying reverberation. Recalling the correlation

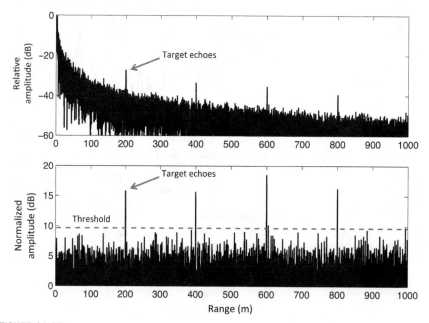

FIGURE 11.10

Example showing constant false alarm rate normalization of active sonar reverberation. Threshold is set for $P_f = 10^{-4}$ when reverberation envelope is Rayleigh distributed.

processor from Eq. (11.68), the matched-filter output must be normalized by the reverberation and noise power before comparison to a threshold. Assuming the reverberation and noise power is stationary over a short time interval, λ_0 can be estimated from leading and lagging windows around each test sample and then used to scale the test sample, approximating the optimal detector of Eq. (11.68). A constant threshold may then be applied to the normalized data to yield a CFAR.

The most common technique for estimating the noise power, called *cell-averaging* (CA) CFAR, is to average the data in the leading and lagging windows around the test sample. Suppose the test sample is x_i (matched filter envelope) around which a buffer of M_b samples is placed to ensure spread signals do not affect the noise power estimation windows, which each contain M_n samples. The noise power estimate for x_i is then

$$\widehat{\lambda}_{0,i} = \frac{1}{2M_n} \sum_{j=1}^{M_n} \left[|x_{i+M_b+j}|^2 + |x_{i-M_b-j}|^2 \right] \tag{11.80}$$

and the normalized matched-filter envelope is

$$z_i = \frac{|x_i|}{\sqrt{\widehat{\lambda}_{0,i}}}. \tag{11.81}$$

The buffer size is set based on the expected temporal spread of the signal. The noise-power estimation window extent is set as large as possible subject to the requirement of (approximate) stationarity within the window.

When a strong signal or interference enters a leading or lagging window, the estimate is biased high and results in a condition known as target masking (i.e., a reduction in P_d for the test sample affected by the corrupted noise power estimate). To mitigate target masking, an order-statistic normalizer can be used where λ_0 is estimated by appropriately scaling one of the ordered samples of data within the leading and lagging windows (e.g., the median). Other approaches for dealing with varying background power include the "greatest-of," "smallest-of," and trimmed mean processors [58]. The trimmed mean normalizer discards some of the largest and smallest samples in the leading and lagging windows prior to forming an average. The greatest-of and smallest-of processors choose between estimates formed solely from the leading or lagging window.

For any normalizer, estimation of λ_0 introduces variability into the detection process and results in an unavoidable, although usually minor, performance loss. The performance loss decreases as the size of the leading and lagging windows increases. For example, using a CA-CFAR normalizer when the leading and lagging windows contain a total of 10 independent samples (see Table 11.3) causes detection threshold to increase over 2 dB when $P_d = 0.5$ and $P_f = 10^{-4}$. The increase, however, is less than 0.5 dB when 50 independent samples are used. Thus there is a trade-off between performance (where large windows are desirable) and robustness to changing power levels (where small windows are desirable).

Table 11.3 Increase in Detection Threshold as a Function of Noise Estimation Window Size for $P_d = 0.5$ and $P_f = 10^{-4}$

Noise-Estimation Window Size (Independent Samples)	Increase in Detection Threshold (dB)
10	2.1
20	1.0
50	0.4
100	0.2

For a Gaussian signal in reverberation or noise with a Rayleigh-distributed envelope, the CFAR loss is [9]

$$L = 10 \log_{10} \left\{ \frac{\left[\left(\frac{P_d}{P_f} \right)^{1/n} - 1 \right] \log_e (P_d)}{\left[1 - P_d^{1/n} \right] \log_e (P_f / P_d)} \right\} \quad \{\text{unit: dB}\} \tag{11.82}$$

where n is the number of independent samples in the noise-power estimation window and (P_f, P_d) describe the operating point. It is common to oversample even the basebanded signal so the sampling rate is greater than the bandwidth of the transmit waveform. When this occurs, the number of independent samples in the noise-power estimation window is approximately

$$n \approx 2 M_n \frac{W}{f_s}. \tag{11.83}$$

Even though Eq. (11.82) is for a Gaussian signal, the results for a deterministic signal are very similar and can generally be used to determine if n is large enough to make the CFAR loss negligible.

Normalization is similar in concept to *time-varying gain* (TVG), which is commonly used to remove a known change such as spreading losses or to reduce the dynamic range of a signal prior to display. TVG differs from normalization in that it is not data dependent and does not produce a CFAR.

11.5 ESTIMATION

Estimation is considered a statistical inference process where a measurement is used to approximate an unknown parameter of interest. For example, one may wish to estimate the time-delay between a signal measured at two sensors, angle of arrival of a signal impinging on an array of sensors, or power spectrum of a received signal. Parameters for which inference can be made include deterministic (i.e., consistent or repeatable) or random parameters, continuous or discrete valued parameters, and scalar or multivariate parameters. The focus of this section is on deterministic, continuous, scalar parameters.

An "estimator" is the mathematical function (or algorithm implementing it) converting the measurement into the parameter estimate. For a scalar parameter of interest θ, the estimator G is a many-to-one mapping converting the measurement $\mathbf{x}$ into the scalar estimate

$$\widehat{\theta} = G(\mathbf{x}) \tag{11.84}$$

where in general the notation $\widehat{\theta}$ represents the estimate of θ. As described at the beginning of Section 11.4, the measured data $\mathbf{x}$ are modeled by the random variable $\mathbf{X}$ and characterized by a PDF $f_{\mathbf{X}}(\mathbf{x})$. A minor deviation of notation occurs when Greek letters are used for parameters and their estimators: they are often not capitalized when formulated as a random variable. To emphasize when the estimator is being described as a function of a random variable, which typically only occurs for theoretical performance analysis, the dependence will be made explicit by using $\widehat{\theta}(\mathbf{X})$ to represent $G(\mathbf{X})$.

11.5.1 PERFORMANCE METRICS, DESIGN, IMPLEMENTATION, AND ANALYSIS PROCEDURE

11.5.1.1 Estimation Performance Metrics

When formulating the estimator as a function the random variable $\mathbf{X}$ modeling the data, $\widehat{\theta}(\mathbf{X})$ is itself a random variable with its own PDF describing the likelihood of all possible values one might observe for $\widehat{\theta}$. Performance metrics for estimation distill various characteristics about the estimator from its PDF. For example, the bias is the average shift or displacement of the estimate from the true value:

$$\text{Bias} = \Delta_{\widehat{\theta}} = E\left[\widehat{\theta}(\mathbf{X}) - \theta\right] = E\left[\widehat{\theta}(\mathbf{X})\right] - \theta \tag{11.85}$$

where the notation $E[\cdot]$ is the expectation or statistical average of the argument. Note the bias is the estimator mean minus the true value. Similarly, the estimator variance represents the spread of the estimator about its mean:

$$\text{Variance} = \sigma_{\widehat{\theta}}^2 = E\left[\left(\widehat{\theta}(\mathbf{X}) - E\left[\widehat{\theta}\right]\right)^2\right]. \tag{11.86}$$

The *mean-squared error* (MSE) combines both the bias and variance to represent the overall disparity between the estimator and the true value

$$\text{MSE} = E\left[\left(\widehat{\theta}(\mathbf{X}) - \theta\right)^2\right] = \Delta_{\widehat{\theta}}^2 + \sigma_{\widehat{\theta}}^2. \tag{11.87}$$

An unbiased estimator where the estimator mean is the true value (i.e., $\Delta_{\widehat{\theta}} = 0$) is desirable, as is a small estimator variance. Because reducing either the bias or variance usually leads to an increase in the other, the MSE is commonly used as a single measure to characterize estimation performance.

It is important to note the units of variance and MSE are the square of the units of θ. As such it is common to use the standard deviation ($\sigma_{\hat{\theta}}$), which is the square root of the variance, or the Root-Mean-Squared Error (RSME). Both of these measures and the bias have the same units as θ.

Another desirable characteristic for an estimator is "consistency." An estimator is consistent if it tends to the true value in the limit when the amount of data grows to infinity. For a consistent estimator, both the bias and variance tend to zero as the data quantity grows to infinity.

11.5.1.2 Estimator Design and Implementation Process

The design process starts with identifying the parameter to be estimated (θ) and the data available for estimation (**x**). The next and perhaps most difficult step in the process is to obtain a statistical characterization of the data with an explicit dependence on θ. This requires characterization of the random variable **X** used to model the measured data **x** through its PDF $f_{\mathbf{X}}(\mathbf{x}|\theta)$, which now depends on θ. While the PDF provides a complete statistical description, it is often difficult to formulate without resorting to simplifying assumptions such as the central limit theorem [2], which can allow use of the Gaussian model when the data are formed from sums of independent random components. In some situations, standard estimation algorithms are employed under implicit assumptions about the data model. For example, power-spectrum estimation is often accomplished without consideration of the inherent implication of wide-sense-stationarity of the random-process model assumed for the data.

Given a statistical description of the data, the parameter estimator is designed using one of the structured design approaches to be described in Section 11.5.2. The result is a many-to-one mapping of the data to each scalar parameter, e.g., $\hat{\theta} = G(\mathbf{x})$. Complications often arise where the estimator depends on nuisance parameters, some of which may only be known approximately (e.g., the speed of sound in water when estimating range from time delay or environmental parameters in matched-field processing).

Conceptually, implementation of the estimator is a straightforward evaluation of the function $G(\mathbf{x})$. However, the estimator may entail multivariate optimizations of complicated functions. The computational load of some estimators can force the use of simpler estimators with potentially suboptimal performance.

11.5.1.3 Estimator Analysis Procedure and the Cramer-Rao Lower Bound

In rare cases the estimator performance metrics described in Eqs. (11.85)–(11.87) can be evaluated analytically. It is more common to resort to a Monte-Carlo [59] analysis using many observations ($\hat{\theta}_1, \ldots, \hat{\theta}_n$) from measurements in a controlled experiment or simulated data (in either case the true value of θ must be known). The Monte-Carlo approach replaces the theoretical expectation operator with an average over the observations, that is,

$$\Delta_{\hat{\theta}} \approx \theta - \frac{1}{n} \sum_{i=1}^{n} \hat{\theta}_i \tag{11.88}$$

and

$$\sigma_{\hat{\theta}}^2 \approx \frac{1}{n} \sum_{i=1}^{n} \left(\hat{\theta}_i - \frac{1}{n} \sum_{j=1}^{n} \hat{\theta}_j \right)^2. \tag{11.89}$$

The number of observations n must be large enough to make the randomness of the Monte-Carlo result negligible relative to the quantity of interest. Common issues encountered when using the Monte-Carlo approach with experimental data include not being able to obtain enough observations, changing conditions over the experiment, and outliers in the data.

The *Cramer-Rao lower bound* (CRLB) provides an alternative to the often difficult task of evaluating the performance of a specific estimator. The CRLB provides the minimum variance attainable for an unbiased estimator given the data represented by the random vector $\mathbf{X}$ and its PDF. The CRLB is [2]

$$\sigma_{\hat{\theta}}^2 \geq \frac{1}{E\left[\left\{ \dfrac{\partial}{\partial \theta} \log_e f_{\mathbf{X}}(\mathbf{X}; \theta) \right\}^2 \right]} = \text{CRLB}. \tag{11.90}$$

Although the CRLB does not provide a means for designing an estimator to meet the bound (such an estimator is call "efficient"), it is useful in sensitivity analyses for evaluating the impact of controllable parameters (e.g., averaging time or array size) on performance limits.

Two limitations of the CRLB lie in its restriction to unbiased estimators and the very loose nature of the bound at low SNR.

11.5.2 STRUCTURED DESIGN APPROACHES

11.5.2.1 *Maximum Likelihood Estimation*

One of the most common estimation design approaches entails maximizing the likelihood of the parameter of interest. The likelihood function of the parameter θ is simply the PDF of the data taken as a function of θ:

$$L(\theta; \mathbf{x}) = f_{\mathbf{X}}(\mathbf{x}|\theta) \tag{11.91}$$

and the MLE is the value of θ maximizing $L(\theta;\mathbf{x})$:

$$\hat{\theta} = \arg\max_{\theta} L(\theta; \mathbf{x}) = \arg\max_{\theta} f_{\mathbf{X}}(\mathbf{x}|\theta) = G(\mathbf{x}) \tag{11.92}$$

where the "arg max" notation represents the argument maximizing the ensuing function.

In the presence of nuisance parameters (e.g., β), the MLE is formed by maximizing over β first:

$$\hat{\theta} = \arg\max_{\theta} \left[\max_{\beta} f_{\mathbf{X}}(\mathbf{x}|\theta, \beta) \right]. \tag{11.93}$$

In some cases this results in a joint optimization and may require numerical evaluation.

A maximum likelihood estimator is consistent (i.e., it converges to the true value as the quantity of data tends to infinity), but not necessarily unbiased (i.e., has mean equal to the true value) or efficient (i.e., meets the CRLB).

Suppose the data under consideration, represented by the random variables $Y_1,...,Y_n$, are known to follow the Rayleigh PDF, for example, as occurs for the envelope of a band-pass Gaussian random process. The Rayleigh PDF has form [60]

$$f_Y(y; \lambda) = \frac{y}{\lambda} e^{-\frac{y^2}{2\lambda}} \tag{11.94}$$

where λ is half the average power of the random variable $E[Y^2] = 2\lambda$. Assuming the data are statistically independent, the MLE of λ is found by choosing the value maximizing the joint PDF

$$f_\mathbf{Y}(\mathbf{y}; \lambda) = \prod_{i=1}^{n} f_Y(y_i; \lambda) = \prod_{i=1}^{n} \frac{y_i}{\lambda} e^{-\frac{y_i^2}{2\lambda}}. \tag{11.95}$$

For many standard probability distributions the maximization can be performed analytically to result in a simple, closed-form estimator. For the Rayleigh-data example, it is easier to maximize (equivalently) the logarithm of the joint PDF,

$$\log_e f_\mathbf{Y}(\mathbf{y}; \lambda) = \sum_{i=1}^{n} \left[\log_e y_i - \log_e \lambda - \frac{y_i^2}{2\lambda} \right]. \tag{11.96}$$

The MLE for λ is now easily found by taking the partial derivative of Eq. (11.96) with respect to λ, equating it to zero, and solving for λ:

$$\widehat{\lambda} = \frac{1}{2n} \sum_{i=1}^{n} y_i^2. \tag{11.97}$$

Noting the relationship between λ and the average power for the ith datum, $E\left[Y_i^2\right] = 2\lambda$, the MLE estimator for λ is seen to unbiased:

$$E\left[\widehat{\lambda}\right] = \lambda. \tag{11.98}$$

However, as previously noted, ML estimators are in general not necessarily unbiased.

Development of the MLE is often done under the assumption of statistically independent data. However, underwater acoustic data are often not statistically independent, with the primary example arising from sampling complex-base-banded band-limited data at a rate higher than the bandwidth. In the above-mentioned example, applying the MLE in Eq. (11.97) to correlated data would still result in a reasonable estimator. However, this is not the case in general so care needs to be taken when an MLE developed under an independent-data model is applied to correlated data.

11.5.2.2 Method of Moments Estimation

A popular alternative to ML estimation is the *method of moments* (MoM). The MoM approach to parameter estimation exploits the (often nonlinear) relationships between unknown parameters and moments of the data. One theoretical moment equation is required for each unknown parameter in the PDF describing the data. The ensuing set of equations is solved for the parameter of interest in terms of the moments. The parameter estimator is then obtained by substituting sample moments (i.e., moments evaluated from the data) for the theoretical moments.

As an example, suppose the measured data $x_1,...,x_n$ are modeled as being Gaussian distributed with unknown mean μ and variance σ^2 and the variance is the parameter of interest. The first two theoretical moments are

$$m_1 = E[X_i] = \mu \tag{11.99}$$

and

$$m_2 = E[X_i^2] = \mu^2 + \sigma^2 \tag{11.100}$$

where X_i is the random variable representing the ith measurement. Solving these two equations for σ^2 results in

$$\sigma^2 = m_2 - m_1^2. \tag{11.101}$$

The MoM estimator for the variance is then formed by replacing the theoretical moments with the sample moments:

$$\widehat{\sigma}^2 = \widehat{m}_2 - \widehat{m}_1^2 \tag{11.102}$$

where

$$\widehat{m}_1 = \frac{1}{n}\sum_{i=1}^{n} x_i \tag{11.103}$$

and

$$\widehat{m}_2 = \frac{1}{n}\sum_{i=1}^{n} x_i^2 \tag{11.104}$$

are, respectively, the first and second sample moments. For this example, the MLE for σ^2 is identical to the MoM estimator. Using the moment equations in Eqs. (11.99) and (11.100), the mean value of the MoM estimator in Eq. (11.102) can be shown to be

$$E[\widehat{\sigma}^2] = \left(1 - \frac{1}{n}\right)\sigma^2. \tag{11.105}$$

Thus the MoM (and for this case also the MLE) estimator for σ^2 is biased. The statistical consistency of $\widehat{\sigma}^2$ is demonstrated in part by the mean value converging to the true value as $n \to \infty$. MoM estimators in general are statistically consistent because the sample moments converge in probability to the true moments of a distribution as the amount of data grows to infinity.

MoM estimators are a practical alternative to MLE, especially when the data PDF is not known precisely as moment equations are often easier to characterize. However, a detriment to MoM estimators can arise when the moment equations are not invertible owing to estimation error introduced by the sample moments. In these situations no parameter estimate can be obtained, leaving one with an undesirable choice of discarding the data or resorting to another estimator.

11.5.2.3 Other Approaches

Other estimator design approaches exist and are appropriate in certain situations. For example, when prior information about a parameter value exists (e.g., the range and bearing of a sonar contact), a Bayesian estimator can probabilistically combine the prior information with that conveyed by the measured data. Bayesian approaches treat the parameters of interest as random variables and require the prior information to be in the form of a PDF, $f_0(\theta)$. Estimation then proceeds from the posterior PDF,

$$f_\theta(\theta|\mathbf{x}) = \frac{f_\mathbf{X}(\mathbf{x}|\theta)f_0(\theta)}{f_\mathbf{X}(\mathbf{x})}, \tag{11.106}$$

which is formed from the prior and data PDFs via Bayes' formula [28]. For example, the Bayesian equivalent to the MLE is the *maximum a posteriori* (MAP) estimator maximizing the posterior PDF,

$$\widehat{\theta}_{\text{MAP}} = \arg\max_\theta f_\theta(\theta|\mathbf{x}). \tag{11.107}$$

Other Bayesian estimators include the posterior mean or median.

Beyond the ML, MoM, and Bayesian estimators, other design approaches generally form a specific cost function from the data PDF and obtain the parameter estimator from the value maximizing (or minimizing) the cost function.

11.5.3 SPECTROGRAM, PERIODOGRAM, AND POWER SPECTRAL DENSITY ESTIMATION

As described in Section 11.2.2, the power spectral density of a random process describes its frequency content. When analyzing a measured signal, it is common to first examine its frequency content. A time-frequency analysis is typically performed using short-time DFTs to construct a spectrogram that can be used to determine the stationarity of the signal. If the frequency content does not change over the analysis window and the signal is assumed to be wide-sense stationary (WSS), the power spectrum can be estimated and provides a second-order statistical characterization of the signal.

The spectrogram segments the analysis window into a set of overlapping subwindows with the underlying assumption of stationarity within each subwindow.

A DFT is computed and displayed for each subwindow to create a plot of the frequency content over time. It is convenient to describe the spectrogram as a bank of narrowband filters applied to the data with each filter identical except for its center frequency. The parameters defining the spectrogram are the subwindow size and shading function, the DFT size, and the overlap between adjacent subwindows. The subwindow size and any shading (e.g., see Section 11.3.4) dictate the width and shape of each of the filters in the aforementioned filter bank. The DFT size dictates the number of filters in the filter bank. When the DFT size is greater than the subwindow size, the balance of the DFT window is "zero-padded"; that is, zero-value time samples are appended at the end of the data in the subwindow to complete the DFT window. The impact of this is to increase the number of center frequencies evaluated; however, it is important to note it does not affect the width (i.e., resolution) of the individual filters and introduces statistical dependence between filter outputs. Having overlap between consecutive subwindows can improve the visual display of time-dependent frequency content, but introduces statistical dependence over time within each filter. The processing applied to form two segments of a spectrogram is shown pictorially in Fig. 11.11, illustrating the window function applied to the short, overlapping time segments followed by a DFT.

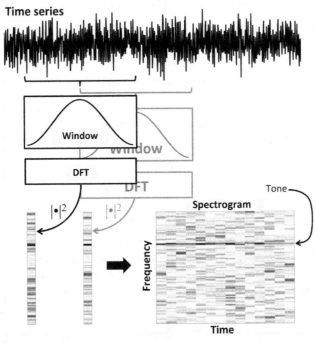

FIGURE 11.11

Example showing time-frequency analysis process.

11.5.3.1 Periodogram for Spectral Density Estimation

Once the stationarity of a measurement has been established, the magnitude-squared DFT outputs can be averaged to form an estimate of the PSD. Given a measurement with a fixed duration (T_m), there is a trade-off between the bias and variance of the PSD estimate. The *coherent processing interval* (CPI) is the temporal extent of the window (T_c) used to isolate the data for one DFT. The number of averages is then related to the total extent of the analysis window and the overlap between successive DFTs. Bias in PSD estimation is inversely proportional to T_c (i.e., the CPI) while variance is approximately inversely proportional to T_m/T_c (i.e., the number of independent CPIs in the analysis window). When using tapering window functions (see Section 11.3.4) overlapping is necessary; however, there is a point of diminishing returns in increasing the overlap owing to an increase in the correlation between successive CPI segments. Generally there is little benefit to overlapping CPIs more than 50–75%, depending on the window function.

Consider the example of a sinusoid with amplitude A, frequency f_0, and random initial phase (ϕ_0) in white Gaussian noise with variance σ_V^2,

$$X(t) = A\,\cos(2\pi f_0 t + \phi_0) + V(t). \tag{11.108}$$

The autocorrelation function of the random process is

$$R_{XX}(\tau) = \frac{A^2}{2}\cos(2\pi f_0 \tau) + \sigma_V^2 \delta(\tau) \tag{11.109}$$

where $\delta(\tau)$ is the Dirac delta function. Taking the Fourier transform results in the PSD

$$S_{XX}(f) = \sigma_V^2 + \frac{A^2}{4}\delta(f - f_0) + \frac{A^2}{4}\delta(f + f_0). \tag{11.110}$$

An example of $X(t)$ is shown in the upper pane of Fig. 11.12 where the signal alone is shown for the first portion, followed by signal plus noise. For this example, the following values were chosen:

$$f_0 = 350 \text{ Hz}$$

$$A = 100 \text{ μPa} = 10^{(40/20)} \text{μPa}$$

$$10\log_{10}\sigma_V^2 = 50 \text{ dB}//1\,\text{μPa}^2/\text{Hz}$$

The peak SNR for the time-domain signal is $10\log_{10}\left(A^2/\sigma_V^2\right) = -10$ dB.

The Welch periodogram [61,17,16] is implemented by averaging the magnitude squared DFT of the windowed CPI segments. If the window function is $w[n]$ and the CPI data for the mth segment is $X_m[n]$ for $n = 0,\ldots,N-1$ and $m = 0,\ldots,M-1$, the PSD estimate is

$$\widehat{S}_{XX}(f_k) = \frac{1}{U_w}\frac{1}{M}\sum_{m=0}^{M-1}\left|\sum_{n=0}^{N-1} w[n]X_m[n]e^{-j2\pi kn/N}\right|^2 \tag{11.111}$$

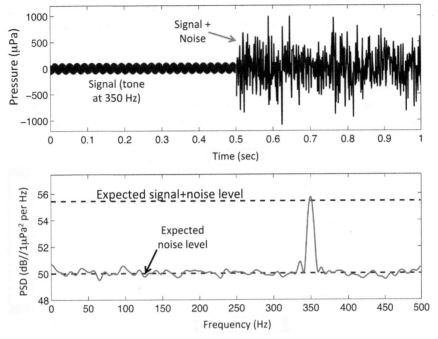

FIGURE 11.12

Example showing spectrum estimation of a sinusoid in noise via Welch periodogram.
FIR, finite impulse response; *IIR*, infinite impulse response.

with units $\mu Pa^2/Hz$ if the data are pressure in units of μPa. If the overlap between adjacent CPI segments is $N - L$ samples, then there is a new CPI after every L sample and

$$X_m[n] = X[n + mL]. \tag{11.112}$$

The inner summation (over n) in Eq. (11.111) is the DFT of one CPI segment for which the kth frequency is

$$f_k = \begin{cases} f_s \dfrac{k}{N} & \text{for} \quad k = 0, ..., \dfrac{N}{2} - 1 \\ f_s \left(\dfrac{k}{N} - 1 \right) & \text{for} \quad k = \dfrac{N}{2}, ..., N - 1 \end{cases} \qquad \{\text{unit: Hz}\} \tag{11.113}$$

where f_s is the sampling frequency. The scaling term U_w is related to the window function according to

$$U_w = \sum_{n=0}^{N-1} w^2[n] \tag{11.114}$$

and acts to remove the effects of the windowing from estimation. The scaling is equivalent to requiring the window function to have unit energy and results in $U_w = N$ for a rectangular, unit-amplitude window.

In applying the Welch periodogram to the example, suppose the narrowest band of interest is approximately $W = 10$ Hz and a Hamming window is used, which has an equivalent noise bandwidth [41] factor of 1.36 (i.e., the bandwidth of the resulting band-pass filter is approximately 1.36 times as large as for a rectangular window function). If the sampling rate is $f_s = 1$ kHz, the DFT size should be

$$N = \left[1.36\frac{f_s}{W}\right] = 136 \tag{11.115}$$

where the function $[\cdot]$ rounds the argument to the nearest integer. Assuming a 50% overlap ($q = 0.5$),

$$L = [N(1 - q)] = 68 \tag{11.116}$$

and if there are $T_a = 30$ s available in the analysis window, the number of DFTs to perform is

$$M = \left\lfloor \frac{\lfloor f_s T_a \rfloor - N}{L} \right\rfloor + 1 = 440 \tag{11.117}$$

where the function $\lfloor \cdot \rfloor$ returns the integer nearest to the argument toward zero (Table 11.4).

The resulting Welch periodogram can be seen in the lower pane of Fig. 11.12 where the white noise is at the appropriate level and the ability of spectral analysis to extract coherent signals with compact frequency support buried in broadband noise is evident from the clearly visible narrowband signal at 350 Hz. When estimating the PSD with the Welch periodogram, the estimate is limited by the window function in that the mean of the estimator is the convolution between the Fourier transform of the window function and the true PSD. As such, what appears to be a narrowband signal centered at 350 Hz is in fact the shape of the Fourier transform of the Hamming window function, sifted out of the convolution by the Dirac delta function in Eq. (11.110) centered at $f = f_0$. The width of the tonal signal 3-dB down from the peak value is approximately 12 Hz, which is close to the desired

Table 11.4 Parameters Used in Example for Periodogram Spectral Density Estimation

Parameter	Value	Note
Analysis bandwidth	$W = 10$ Hz	Dictates window length (~ 100 ms)
Window length	$N = 1.36 f_s/W = 136$	The 1.36 is for the Hamming window
Overlap fraction	$q = 0.5$ and $L = 68$	Related to variance reduction
Number of coherent processing intervals	$M = 440$	Approximately inversely proportional to variance

$W = 10$ Hz and is dictated by the window function extent and shaping (and SNR). Decreasing W would result in a narrower spectral width in the PSD at the expense of a more highly variable estimator when the total analysis window is fixed.

For this example, the expected level of the Welch periodogram at the sinusoid frequency, which includes both signal and noise, can be found by applying the gain attained in the coherent processing (G) to the sinusoid,

$$10 \log_{10} E\left[\widehat{S}_{XX}(f_0)\right] = 10 \log_{10}\left(G\frac{A^2}{4} + \sigma_V^2\right) \approx 55.4 \text{ dB re } 1 \text{ } \mu\text{Pa}^2/\text{Hz} \quad (11.118)$$

The coherent processing gain can be described either in terms of the window function or the analysis bandwidth:

$$G = N\left(\frac{\left(\frac{1}{N}\sum\limits_{n=0}^{N-1} w[n]\right)^2}{\frac{1}{N}\sum\limits_{n=0}^{N-1} w^2[n]}\right) \approx \frac{f_s}{W} \quad (11.119)$$

where the quantity within the parentheses is the equivalent noise bandwidth factor of the window function. For a rectangular window, the gain is the DFT size, $G = N$.

The Welch periodogram is considered a classical or nonparametric spectrum estimator; the spectrum is not assumed to follow any particular functional model. Alternative nonparametric spectrum estimators include variations on the periodogram [17] and a correlogram technique where the ACF is estimated and the Fourier transform relationship between the ACF and the PSD (i.e., Eq. (11.26)) is exploited. The most common parametric spectrum estimation [16,17] approach is to assume the PSD function is comprised of polynomials (as a function of frequency) in the numerator and/or denominator. Such a model can be interpreted as arising from an LTI system and is called autoregressive (AR; polynomial in the denominator only), moving-average (MA; polynomial in the numerator only), or ARMA (both). Spectrum estimation proceeds by estimating the polynomial coefficients (or roots) from the measured data. The primary trade-off arises from desiring high-order polynomials to reduce estimator bias (i.e., mismatch error) at the expense of a greater estimator variability.

11.5.3.2 Zero-Padding the DFT

The examples shown in Figs. 11.11 and 11.12 exploited zero-padding to evaluate the PSD at more frequencies than would a DFT the same size as the window function. The process involves increasing the size of the DFT, often to a power of two to exploit the advantages of the FFT, while enforcing the implicit assumption that the window functions are zero outside of the CPI. If the DFT size is chosen as $\widetilde{N} \geq N$, the Welch periodogram is

$$\widehat{S}_{XX}(f_k) = \frac{1}{U_w}\frac{1}{M}\sum_{m=0}^{M-1}\left|\sum_{n=0}^{N-1} w[n]X_m[n]e^{-j2\pi kn/\widetilde{N}}\right|^2 \quad (11.120)$$

with units $\mu Pa^2/Hz$ if the data are pressure in units of μPa. Note that only the N in the denominator of the exponent of e in Eq. (11.111) has changed to form Eq. (11.120). The corresponding frequency f_k also changes to reflect the finer spacing of the narrowband filters represented by the DFT:

$$f_k = \begin{cases} f_s \dfrac{k}{\widetilde{N}} & \text{for} \quad k = 0, \dots, \dfrac{\widetilde{N}}{2} - 1 \\[2ex] f_s \left(\dfrac{k}{\widetilde{N}} - 1 \right) & \text{for} \quad k = \dfrac{\widetilde{N}}{2}, \dots, \widetilde{N} - 1 \end{cases} \qquad \{\text{unit: Hz}\}. \qquad (11.121)$$

While the zero-padding operation creates more finely spaced narrowband filters, *it does not* make the filters any narrower. That is, the width of the narrowband filters, and therefore the resolution capability of the Welch periodogram, is still related to and restricted by the extent and shape of the window function. The finer spacing of the center frequencies of the narrowband filters implies adjacent filters become more correlated as zero-padding increases.

11.5.4 TIME-DELAY ESTIMATION

A common objective in underwater acoustics is to estimate the location of a sound source. For both active and passive remote sensing this often starts with estimation of the time difference between two signals. For active sensing it is the delay between the source transmission and an echo measured by a receiving sensor. In passive sonar, the signal is generally not known and the time difference is that between measurements from two physically separated sensors as shown in Fig. 11.13. In general, the multipath effects of underwater acoustic propagation are different for the paths from the source to the two sensors as are the ambient noises at each sensor.

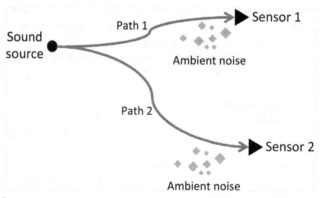

FIGURE 11.13

Time-delay estimation scenario.

The problem can be described as measuring a signal $x_i(t)$ at the ith sensor (for $i = 1,2$) subject to additive ambient noise $v_i(t)$ where the sound source signal $s(t)$ is received with some delay τ_i but affected by a Doppler scale γ_i and a propagation channel characterized by the impulse response $h_i(t)$. Mathematically, the measured signal can be described as

$$x_i(t) = h_i(t) * s(\gamma_i t - \tau_i) + v_i(t) \tag{11.122}$$

where the quantity of interest is the time-delay difference

$$D = \tau_2 - \tau_1. \tag{11.123}$$

Several simplifying assumptions allow derivation of the MLE for D as the time of the peak of the *cross-correlation function* (CCF) between $x_1(t)$ and $x_2(t)$. The simplifying assumptions are: $s(t)$, $v_1(t)$, and $v_2(t)$ are modeled as independent, zero-mean, WSS Gaussian random processes, with constant PSD within the analysis band including the effects of propagation and Doppler, and the analysis time is greater than the delay. The derivation is most easily accomplished by converting the data and their statistical characterization to the frequency domain where it is seen the likelihood function of D depends monotonically on the product

$$\mathscr{F}\{x_1(t)\}\mathscr{F}\{x_2(t)\}^* e^{-j2\pi fD} = X_1(f)X_2^*(f)e^{-j2\pi fD}. \tag{11.124}$$

The product $X_1(f)X_2^*(f)$ in the frequency domain represents a convolution in the time domain. However, the conjugation of $X_2(f)$ represents time-reversal, yielding a correlation between $x_1(t)$ and $x_2(t)$ rather than a convolution. Multiplication by $e^{-j2\pi fD}$ in the frequency domain represents a time-delay D indicating the MLE is the delay in the cross-correlation providing maximum response.

As an example, consider the signal and noise spectra shown in Fig. 11.14. The noise and signal spectra for sensor 1 are shown in the top pane and for sensor 2 in the bottom pane. The full-band signal ranges from zero to 900 Hz while a smaller 100-Hz-wide processing band is also shown centered at 500 Hz. While noise levels are not realistically constant with frequency, the given scenario demonstrates the power of time-bandwidth product. An analysis window of 10-s duration is used to form the sample CCF between the two signals as shown in Fig. 11.15. The peak of the sample CCF for the full-band result is at the correct delay time of 1 s and reveals up to two multipath shortly thereafter. Noting that the sample CCF is the inverse Fourier transform of $X_1(f)X_2^*(f)$, the shape of the CCF will be related to the product of the signal-plus-noise PSDs measured at the two sensors. Exploiting the full bandwidth available in this example results in a narrow CCF (subject to there being enough SNR or time-bandwidth product). When the analysis is restricted to a 100-Hz band centered at 500 Hz, the temporal resolution degrades to approximately 0.01 s (i.e., one over the bandwidth). As seen in Fig. 11.15, while D is estimated with good accuracy (it was ≈ 1 ms off in the example) for the smaller analysis band, the finer multipath are not resolved.

The 100-Hz bandwidth example demonstrates the concept of time-bandwidth product. In many signal processing applications (both detection and estimation),

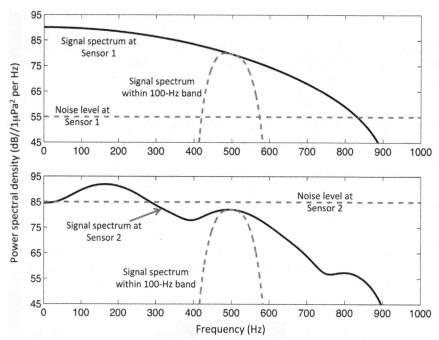

FIGURE 11.14

Time-delay estimation example: power spectral density received at each sensor.

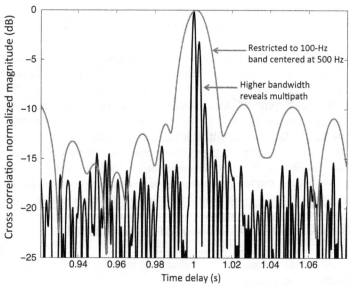

FIGURE 11.15

Time-delay estimation example: sample cross-correlation function for full band and reduced bandwidth.

performance is often dependent on the product between the analysis time and bandwidth being evaluated. In this example, the bandwidth (and approximately level PSD within the band) drives the minimum temporal width of the CCF, and therefore provides a lower bound on estimation performance. The time extent of the analysis window overcomes the high noise level observed at sensor 2 where the noise dominates the signal.

In time-delay estimation, the analysis window is often limited by geometric considerations (e.g., a moving source or receiver). Thus the choice of bandwidth and processing time drive performance. For example, if the signal PSD is known or can be estimated, it can be exploited to improve time-delay estimation. Extensions to more complicated scenarios appropriately dealing with relaxations to the aforementioned simplifying assumptions can be found in Ref. [37].

11.5.5 BEAMFORMING AND ANGLE OF ARRIVAL ESTIMATION

One of the fundamental objectives of array signal processing [19] is estimation of the bearing (more generally angle of arrival) of a sonar contact from a hydrophone array. Angle of arrival estimation is generally performed through a process called beamforming where array data are combined in what is effectively a spatial filter to focus energy coming from specific directions, similar in effect to a parabolic dish. The beamforming process involves applying angle-dependent delays to the measurements from each hydrophone so they align and add constructively for signals emanating from the desired direction while combining destructively for signals from other directions. Under certain simplifying assumptions, beamforming can be derived as a detector for energy arriving from a given direction or as a means to estimate the angle of arrival of a signal. The latter context is considered in this section.

While the physical description of beamforming consisting of delaying each hydrophone's measurement prior to summing is very intuitive, beamforming is often characterized and accomplished in the frequency domain where a delay becomes a phase. For a narrowband signal, the data across an array can be characterized in the frequency domain as a complex vector

$$\mathbf{x}(f) = A\mathbf{d}(f, \theta) + \mathbf{v}(f) \tag{11.125}$$

where A is the signal amplitude, $\mathbf{d}(f,\theta)$ is an array steering vector, and $\mathbf{v}(f)$ is noise and interference. The steering vector encodes the time delays between the sensors for a signal coming from angle θ as frequency-dependent phases. While often not realistic, the noise is typically modeled as Gaussian and spatially "white" (specifically, $\mathbf{v}(f)$ is a multivariate complex Gaussian random vector with zero mean and covariance matrix equal to $\sigma_V^2\mathbf{I}$ where $\mathbf{I}$ is the identity matrix[5]).

Rather than precisely characterizing the signal amplitude, the extremes of a "deterministic" and "random" signal are often evaluated. A deterministic signal

---

[5]The identity matrix is a matrix with ones on the diagonal and zeros elsewhere.

arises when the signal amplitude A is constant over time but unknown. A random signal is represented by modeling A as a zero-mean, complex Gaussian random variable. Under both signal types and the aforementioned assumptions about the noise, the MLE for the bearing has the form

$$\widehat{\theta} = \arg \max_{\theta} \left| \mathbf{d}^{H}(f, \theta) \mathbf{x}(f) \right|^{2} \tag{11.126}$$

where the superscript ^{H} is the conjugate and transpose operator. The function being maximized on the right side of Eq. (11.126) is known as a conventional beamformer with uniform shading. Beamforming is usually performed at a set of fixed angles with the resulting *bearing-time record* (BTR) used to analyze and interpret directionality of the sound field. Under certain conditions, the beamformer output is also an estimate of the spatial power of the signal coming from direction θ. As described in Ref. [19], shading of an array is performed to limit the impact of strong signals or interferences when steering beams near to them at the expense of widening the main-lobe of the beamformer's beam pattern.

Because beamforming can require significant computational effort, it is common to space the beam main-response axes (i.e., the center of each beam) so they overlap at their 3-dB down points. This can result in a coarse angle estimate, so an interpolation or a search near each beam with a local maximum might be performed to increase the precision of the result. A local search or interpolation is the spatial equivalent to zero-padding a DFT (see Section 11.5.3.2); it only improves estimator accuracy by more precisely obtaining the maximum in Eq. (11.126). It does not change the accuracy of the underlying estimation scenario, which is characterized by the proceeding CRLB analysis.

Consider the simulation example for a line array found in Fig. 11.16 where the standard BTR is used to display a contact with an increasing SNR at 80 degrees relative to forward end-fire. Each line in the BTR shows the beam response across all bearings (gray scale) and the peak for that time (circle). The ordinate, which normally represents time in a BTR, shows the per-sensor SNR in decibels [$=10 \log_{10}(A^{2}/\sigma_{V}^{2})$]. The increasing SNR shown in the figure might be indicative of a contact moving closer to the array receiver. The excellent estimation performance at high SNR is seen to degrade, as should be expected, to a completely random estimate when SNR decreases.

As described in Section 11.5.1, estimator accuracy is quantified by the Cramer-Rao lower bound. For a line array comprising equally spaced hydrophones, estimating the bearing of far-field signals (i.e., plane waves) in spatially white noise is mathematically equivalent to estimating frequencies of sinusoids in white noise, for which results can be found in Refs. [16,62]. For a deterministic signal the CRLB for bearing estimation [63] is

$$\text{Var}\left\{\widehat{\theta}_{\text{deg}}\right\} \geq \frac{6 \cdot 180^{2}}{\left(\dfrac{A^{2}}{\sigma_{V}^{2}}\right) n \gamma \left(n^{2}\gamma^{2} - 1\right) \pi^{4} \sin^{2}(\theta)} \quad \left\{\text{unit: deg}^{2}\right\} \tag{11.127}$$

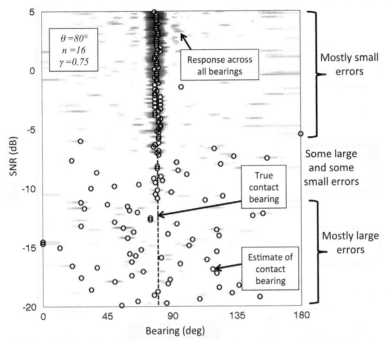

FIGURE 11.16

Bearing estimation simulation for a line array showing large errors at low signal-to-noise power ratio (SNR) and accurate estimation at high SNR.

where θ {unit: deg} is the angle from forward end fire, n is the number of hydrophones in the array, and

$$\gamma = \frac{f}{f_d} \qquad (11.128)$$

is the dimensionless ratio of the operating frequency (f) to the design frequency, $f_d = c_w/(2d)$ {unit: Hz}, d is the intersensor spacing {unit: m}, and c_w is the speed of sound in water {unit: m/s}. From Eq. (11.127), bearing estimation accuracy is seen to improve with array size (n), SNR at the hydrophone level (A^2/σ_V^2), and for angles near broadside to the array $(\theta = 90$ degrees). The example found in Figs. 11.16 and 11.17 $(\theta = 80$ degrees, $n = 16$, $\gamma = 0.75)$ results in a 3-dB-down beam width of 8.5 degrees, yet can achieve an accuracy of 0.3 degree standard deviation at an SNR of 10 dB (at a hydrophone).

As previously noted, the CRLB is usually not a tight bound at low SNR. For example, the CRLB shown in Fig. 11.17 illustrates a constant factor of 10 increase in estimator variance (i.e., factor of $\sqrt{10}$ increase in standard deviation) for every 10-dB decrease in SNR. Simulation results using Monte-Carlo analysis (see Section 11.5.1) show the estimator of Eq. (11.126) is efficient (i.e., it achieves the bound) at

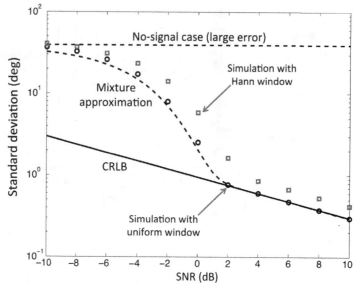

FIGURE 11.17

Example Cramer-Rao lower bound for bearing estimation from a line array. *SNR*, signal-to-noise power ratio.

high SNR, but diverges toward a noise-only result as SNR decreases. At low SNR the estimator encounters what is known as a large-error regime—with increasing probability, the estimate is essentially dominated by noise and equally probable throughout (cosine) angle space. A useful approximation to the variance of a bearing estimate for a line array can be found by mixing the CRLB for the small-error regime with the variance of the noise-dominated result for the large-error regime [62].

Also shown in Fig. 11.17 are the Monte-Carlo results when a Hann window is applied to the array data to reduce the side lobes at the expense of an increase in main-lobe width. The net result is a small increase in estimator variance, which is a small price to pay for resilience to interferences.

REFERENCES

[1] J.O. Berger, *Statistical Decision Theory and Bayesian Analysis*. New York: Springer-Verlag, 1985.
[2] N. Mukhopadhyay, *Probability and Statistical Inference*. New York, NY: Marcel Dekker, Inc., 2000.
[3] R.E. Zarnich, K.L. Bell, and H.L. Van Trees, "A unified method for measurement and tracking of contacts from an array of sensors," *IEEE Trans. Sig. Process.*, vol. 49, no. 12, pp. 2950–2961, 2001.

[4] R.J. Urick, *Principles of Underwater Sound*, 3rd ed. New York: McGraw-Hill, Inc., 1983.

[5] R.O. Nielsen, *Sonar Signal Processing*. Norwood, Massachusetts: Artech House, Inc., 1991.

[6] W.S. Burdic, *Underwater Acoustic System Analysis*, 2nd ed. Prentice Hall, 1991.

[7] W.C. Knight, R.G. Pridham and S.M. Kay, "Digital signal processing for sonar," *Proc. IEEE*, vol. 69, no. 11, pp. 1451–1506, November 1981.

[8] P.Z. Peebles, Jr., *Radar Principles*. New York: John Wiley & Sons, Inc., 1998.

[9] M.A. Richards, *Fundamentals of Radar Signal Processing*. New York: McGraw-Hill, 2005.

[10] B. Porat, *A Course in Digital Signal Processing*. New York: John Wiley & Sons, Inc., 1997.

[11] R.G. Lyons, *Understanding Digital Signal Processing*. Prentice Hall, 2011.

[12] A.V. Oppenheim and R.W. Schafer, *Discrete-Time Signal Processing*, 3rd ed. Upper Saddle River, New Jersey, USA: Pearson Higher Education, Inc., 2010.

[13] J.G. Proakis and D.G. Manolakis, *Digital Signal Processing*, 2nd ed. New York: Macmillan Pub. Co., 1992.

[14] E.O. Brigham, *The Fast Fourier Transform and its Applications*. Englewood Cliffs, New Jersey: Prentice Hall, 1988.

[15] J.S. Bendat and A.G. Piersol, *Random Data Analysis and Measurement Procedures*, 4th ed. Wiley, 2010.

[16] S.M. Kay, *Modern Spectral Estimation Theory and Application*. Englewood Cliffs, NJ: Prentice Hall, 1988.

[17] S.L. Marple, *Digital Spectral Analysis with Applications*. Prentice Hall, Inc., 1987.

[18] L. Cohen, *Time-Frequency Analysis*. Englewood Cliffs, New Jersey: Prentice Hall PTR, 1995.

[19] H.L.V. Trees, *Optimum Array Processing*. New York: John Wiley & Sons, Inc., 2002.

[20] D.H. Johnson and D.E. Dudgeon, *Array Signal Processing: Concepts and Techniques*. Englewood Cliffs, New Jersey: Prentice Hall, 1993.

[21] L.L. Scharf, *Statistical Signal Processing*. Reading, MA: Addison-Wesley Pub. Co., 1991.

[22] S.M. Kay, *Fundamentals of Statistical Signal Processing: Detection Theory*. Prentice Hall PTR, 1998, vol. II Detection Theory.

[23] R.N. McDonough and A.D. Whalen, *Detection of Signals in Noise*, 2nd ed. San Diego: Academic Press, Inc., 1995.

[24] S.M. Kay, *Fundamentals of Statistical Signal Processing: Estimation Theory*. Prentice Hall PTR, 1993.

[25] E.L. Lehmann, *Theory of Point Estimation*. John Wiley & Sons, Inc., 1983.

[26] Y. Bar-Shalom, P.K. Willett and X. Tian, *Tracking and Data Fusion: A Handbook of Algorithms*. YBS Publishing, 2011.

[27] L.D. Stone, R.L. Streit, T.L. Corwin and K.L. Bell, *Bayesian Multiple Target Tracking*, 2nd ed. Artech House, 2014.

[28] B.W. Lindgren, *Statistical Theory*, 3rd ed. New York: Macmillan Pub. Co., 1976.

[29] A. Papoulis, *Probability, Random Variables, and Stochastic Processes*, 3rd ed. Boston: McGraw-Hill, Inc., 1991.

[30] H. Medwin and C.S. Clay, *Fundamentals of Acoustical Oceanography*. Boston: Academic Press, Inc., 1998.

[31] E. Jakeman and K.D. Ridley, *Modeling Fluctuations in Scattered Waves*. Taylor & Francis, 2006.

[32] K.D. Ward, R.J.A. Tough, and S. Watts, *Sea Clutter: Scattering, the K Distribution and Radar Performance*. London: The Institution of Engineering and Technology, 2006.

[33] W.W.L. Au and M.C. Hastings, *Principles of Marine Bioacoustics*. Springer, 2008.

[34] M.A. Ainslie, *Principles of Sonar Performance Modelling*. New York: Springer, 2010.

[35] J.W. Cooley and J.W. Tukey, "An algorithm for the machine calculation of complex Fourier series," *Math. Comp.*, vol. 19, no. 90, pp. 297–301, April 1965.

[36] R.D. Yates and D.J. Goodman, *Probability and Stochastic Processes*. New York: John Wiley & Sons, Inc., 1999.

[37] G.C. Carter, *Coherence and Time Delay Estimation*. IEEE Press, 1993.

[38] B.P. Bogert, M.J.R. Healy and J.W. Tukey, "The quefrency analysis of time series for echoes: cepstrum, pseudoautocovariance, cross-cepstrum and saphe cracking," in: *Proceedings of Symposium on Time Series Analysis*, M. Rosenblatt (Ed.) John Wiley & Sons, 1963, pp. 209–243.

[39] A.V. Oppenheim and R.W. Schafer, "From frequency to quefrency: A history of the cepstrum," *IEEE Sig. Process. Mag.*, vol. 21, no. 5, pp. 95–99, 106, September 2004.

[40] C. Eckart, "Optimal rectifier systems for the detection of steady signals," University of California, Marine Physical Laboratory of the Scripps Institute of Oceanography, La Jolla, California, Tech. Rep. 52-11, March 1952.

[41] F.J. Harris, "On the use of windows for harmonic analysis with the discrete Fourier transform," *Proc. IEEE*, vol. 66, no. 1, pp. 51–83, 1978.

[42] Anon., "Wikipedia page on window functions." [Online]. Available: http://en.wikipedia.org/wiki/Window_function.

[43] P. Baggenstoss, "On the equivalence of Hanning-weighted and overlapped analysis windows using different window sizes," *IEEE Sig. Process. Lett.*, vol. 19, no. 1, pp. 27–30, January 2012.

[44] S. Haykin, *Adaptive Filter Theory*, 3rd ed. Englewood Cliffs, NJ: Prentice Hall, Inc., 1996.

[45] D.A. Abraham, "Detection-threshold approximation for non-Gaussian backgrounds," *IEEE J. Oceanic Eng.*, vol. 35, no. 2, pp. 355–365, April 2010.

[46] W.J. Albersheim, "A closed-form approximation to Robertson's detection characteristics," *Proc. IEEE*, vol. 69, no. 7, p. 839, 1981.

[47] D.A. Shnidman, "Determination of required SNR values," *IEEE Trans. Aero. Elec. Sys.*, vol. 38, no. 3, pp. 1059–1064, 2002.

[48] H. Hmam, "Approximating the SNR value in detection problems," *IEEE Trans. Aero. Elec. Sys.*, vol. 39, no. 4, pp. 1446–1452, 2003.

[49] H. Hmam, "SNR calculation procedure for target types 0, 1, 2, 3," *IEEE Trans. Aero. Elec. Sys.*, vol. 41, no. 3, pp. 1091–1096, 2005.

[50] S.A. Kassam, *Signal Detection in Non-Gaussian Noise*. New York: Springer-Verlag, 1988.

[51] Mathworks, *Documentation Center*. http://www.mathworks.com/help/matlab/index.html, 2014.

[52] H.V. Poor, *An Introduction to Signal Detection and Estimation*. New York: Springer-Verlag, 1988.

[53] D.W. Ricker, *Echo Signal Processing*. Boston, Massachusetts: Kluwer Academic Publishers, 2003.

[54] A.B. Baggeroer, W.A. Kuperman and P.N. Mikhalevsky, "An overview of matched field methods in ocean acoustics," *IEEE J. Oceanic Eng.*, vol. 18, no. 4, pp. 401−424, October 1993.

[55] E.S. Page, "Continuous inspection schemes," *Biometrika*, vol. 41, pp. 100−114, 1954.

[56] M. Basseville and I. Nikiforov, *Detection of Abrupt Changes: Theory and Applications*. Prentice Hall, 1993.

[57] R.L. Streit, "Load modeling in asynchronous data fusion systems using Markov modulated Poisson processes and queues," in: *Proceedings of Signal Processing Workshop*. Washington, D.C.: Maryland/District of Columbia Chapter of the IEEE Signal Processing Society, March 24−25, 1995.

[58] P.P. Gandhi and S.A. Kassam, "Analysis of CFAR processors in nonhomogenous background," *IEEE Trans. Aero. Elec. Sys.*, vol. 24, no. 4, pp. 427−445, July 1988.

[59] S.M. Ross, *Simulation*. Harcourt Academic Press, 1997.

[60] C. Forbes, M. Evans, N. Hastings and B. Peacock, *Statistical Distributions*, 4th ed. Wiley, 2011.

[61] P.D. Welch, "The use of fast Fourier transform for the estimation of power spectra: A method based on time averaging over short, modified periodograms," *IEEE Trans. Audio Electroacoust.*, vol. 15, no. 2, pp. 70−73, June 1967.

[62] D. Rife and R. Boorstyn, "Single tone parameter estimation from discrete-time observations," *IEEE Trans. Inf. Theory*, vol. 20, no. 5, pp. 591−598, September 1974.

[63] L. Brennan, "Angular accuracy of a phased array radar," *IRE Trans. Antennas Propag.*, vol. 9, no. 3, pp. 268−275, May 1961.

Bio- and Fishery Acoustics

12

Ph. Blondel

Department of Physics, University of Bath, Bath, United Kingdom

12.1 INTRODUCTION

Marine life is an important component of biodiversity and an increasingly accessed source of food for the world's over seven billion population. There are 33,200 different fish species, Froese and Pauly [1], more than half of all vertebrate species now identified in the world. In 2010, fish provided more than 2.9 billion people with close to 20% of their protein intake, FAO [2]. Fish capture increases at an average annual rate of 3.2%, and 58.3 million people are actively engaged in fisheries and aquaculture, often in the poorest countries of the world. However, marine ecosystems are increasingly responding to changes in regional climates, in important and sometimes different ways, IPCC [3]. Catastrophic ecosystem collapses, for example, in the Grand Banks (Canada) during the 1990s, and regular overfishing worldwide are leading some to predict "The End of the Line," for example, Clover [4], the advent of "Silent Seas" MCS [5], and global animal loss McCauley et al. [6]. National and international regulations now aim to monitor and manage marine ecosystems, for example, with the European Union Good Environmental Status (with Descriptors 4 "Food Webs" and 11 "Energy, including underwater noise"), European Commission [7]. To a large part all rely, if not predominantly, on the use of acoustic instruments, which are the only ones able to remotely monitor large and deep areas with accuracy.

This chapter aims at presenting two connected domains: fishery acoustics (How does one detect and monitor fish and other marine life?) and bioacoustics (How does marine life use sound? how sensitive are they to other sounds, including human-made sounds?). Section 12.2 sets the scene with a synthesis of the key aspects of marine life which will be relevant to acoustic investigations. Section 12.3 aims to summarize what is known about sound scattering by marine life, based on models and on experiments with dead, captive, or free-ranging animals (using some of the results from Chapter 5). This is used by different active imaging systems, presented succinctly in Section 12.4 (building on material already covered in Chapter 10). Marine animals also produce their own sounds, including for communication and echolocation. The main results are presented in Section 12.5, and how they are used in practice, with passive acoustic monitoring, is detailed

Applied Underwater Acoustics. http://dx.doi.org/10.1016/B978-0-12-811240-3.00012-6

in Section 12.6. Emerging trends and upcoming challenges are finally briefly described in Section 12.7.

There is a large body of supporting scientific and technical literature, and it would be beyond the scope of this chapter to do justice to all of them. Instead, only key references (e.g., for recommendations now used in practice around the world) and good illustrations of important results are presented. The reader desiring more details is invited to consult in particular the textbooks of Medwin and Clay [8], Simmonds and MacLennan [9] for fisheries, Lurton [10] for bioacoustics, and the many references therein.

12.2 MARINE LIFE: FROM WHALES TO PLANKTON

Marine life varies in scale over several orders of magnitude, from small zooplankton micrometers across to 190 metric ton blue whales (Fig. 12.1). Some (e.g., marine mammals) live individually or in small groups, whereas others (e.g., anchovy) form large aggregations of up to millions of animals. Some animals will be large enough to scatter sound as point-like targets, others will only be visible when grouped into layers or schools large enough to be seen as extended targets. For some marine life, most of the scattering will come from the swim bladder (if they have one), and it will vary with depth and behavior. Conversely, other animals might show most scattering from their bodies, or parts of their bodies, varying with orientation respective to the imaging sonar.

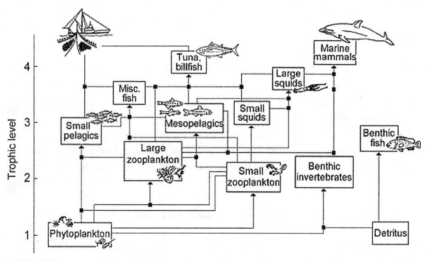

FIGURE 12.1

Marine life varies in scale from phytoplankton (μm and larger) to whales (up to 30 m long), creating complex food webs spanning all trophic levels, EC [11]. The progresses in acoustic sensors, discussed in Chapter 10, mean they can now all be investigated, at all depths.

Table 12.1 Frequency Ranges Most Suitable to Detect Different Animal Sizes

Generic Type	Body Size	Detection Frequency (Body)	Detection Frequency (Internal Organ)
Zooplankton (small)	≪mm	>12 MHz?	N/A
Copepods	≤mm	1.2–12 MHz	N/A
Zooplankton (large), larvae, krill	2–20 mm	120–1200 kHz	1.5–6 to 15–60 kHz
Anchovy, shrimp, etc.	2–20 cm	12–120 kHz	150–600 Hz to 1.5–6 kHz
Cod, tuna, etc.	20–200 cm	1.2–12 kHz	15–60 to 150–600 Hz
Mammals, large fish	>2 m	<1.2 kHz	Variable (up to several kHz)

Adapted from Medwin, H., Clay, C.S., Fundamentals of Acoustic Oceanography, 712 pp., Academic Press, 1998.

The potentially confusing variety of sizes and body compositions can however be approached in a systematic way, looking, for example, at the best frequencies to image their body or their internal organs (e.g., swim bladder or lungs) (Table 12.1). These first answers can then be compared with actual measurements and with models of specific animals, which are presented in Section 12.3. The detection frequency for animal bodies is assuming their body shape can be approximated to a cylinder of radius a, and using $k\,a = 1$ (using the notation of Chapter 5). The detection frequency for internal organs approximates them as spherical bubbles (appropriate for swim bladders of most fish) and is related to their resonant frequency. Both sets of frequencies leave much to be desired, partly because of their underlying assumptions, voluntarily simple, and mostly because of the huge anatomical variations between animals and changes in swim bladder sizes depending on depths and behaviors. Nevertheless, they show the main characteristics: as animals get increasingly larger, the optimal imaging frequencies to detect individuals from their bodies alone get increasingly smaller (i.e., the acoustic wavelengths follow the size variations). If animals have internal organs like swim bladders or lungs with high acoustic contrasts (e.g., if they are filled with gas), they will be easier to detect, at frequencies varying roughly from 100 Hz to several kHz.

12.3 ACOUSTIC SCATTERING BY MARINE LIFE

For the purposes of acoustic scattering, marine animals can be divided into five main groups, based on their sizes and acoustic characteristics:

1. Copepods, zooplankton (small and large), and small crustaceans (e.g., euphausiids like krill), very small and usually visible as large groups or with very high frequencies [12];

2. Fish with a swim bladder (e.g., herring, *Clupea harengus* or cod, *Gadus morhua*), in which the swim bladder can contribute as much as 90% of the backscatter [13];

3. Fish without a swim bladder (e.g., Atlantic mackerel, *Scomber scombrus*), in which the body will be the main contributor to scattering;

4. Larger fish and marine mammals (e.g., seals, dolphins, and whales), for which scattering will be a combination of body, body shape, and the largest internal organs (e.g., lungs).

There has been much work since the 1970s, resulting in hundreds of publications investigating field measurements of specific animals (ideally complemented with direct physical sampling, e.g., with nets), field and laboratory measurements of captive or dead animals (thus restricted to specific orientations and behaviors), as well as analytical and numerical scattering models over a large range of frequencies, for example, Jech et al. [14]. Fig. 12.2 shows a synthesis of expected volume back-scattering variations with frequencies for a range of marine life.

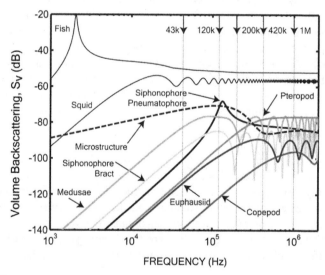

FIGURE 12.2

Volume backscattering varies with frequency, as shown here for the major biological scatterers observed by Lavery et al. [15] in the Gulf of Mexico (the exact dB values assume here a numerical abundance of 1 organism/m^3: actual measurements will scale them depending on actual density, orientation, and behavior of the animals).

Reprinted with permission from Lavery, A.C., Wiebe, P.H., Stanton, T.K., Lawson, G.L., Benfield, M.C., and Copley, N., Determining dominant scatterers of sound in mixed zooplankton populations, J. Acoust. Soc. Am., 122 (6), pp. 3304–3326, 2007. Copyright 2007, Acoustic Society of America.

12.3.1 ZOOPLANKTON SCATTERING

The term "zooplankton" actually covers a large variety of small marine life, including amphipods, krill (classified as euphausiids), copepods, salps, as well as most larval forms of crustaceans or fish (known as "temporary zooplankton" until they grow large enough). These different life forms are key components of marine ecosystems, as they are at the base of most marine food webs.

All species have developed to float in the water column, with structural adaptions like flat bodies, lateral spines, oil droplets or gas-filled floats, and solid or gel-like sheaths [16]. Zooplankton are known to migrate to deeper waters during the day and come up at night, with strong variations in locations and seasons, as well as with the exact types of zooplankton. Because they commonly stay at or close to sharp changes in the water layers (density or temperature), it had been debated how accurately they could be detected with acoustics and distinguished from underlying changes in the water column structure. This controversy has been solved, and a nice historical description of how it was addressed, with satisfactory demonstration that zooplankton had distinct acoustic signatures, can be found in Stanton [17].

The bewildering variety of shapes and sizes of zooplankton means that their acoustic characteristics had to be reduced to simple forms, like fluid-filled spheres and cylinders, for which Helmholtz–Kirchhoff formulations of scattering could be used [8]. Zooplankton can be weak scatterers, if their tissues are materially very close to the surrounding seawater, but they might also have a hard elastic shell (e.g., lateral spines) or gas inclusions (e.g., floats). Laboratory and field measurements have shown variations over four orders of magnitude [17]. Complemented with in situ observations with nets (to identify species and size/type distributions) or cameras (to measure tilt angles, i.e., the orientation of zooplankton to incident acoustic waves), these observations have steadily progressed to offer a general framework in which zooplankton can be modeled as deformed cylinders, with the distorted wave Born approximation [17]. This was validated and found to be valid for all frequencies, sizes, and orientations of weakly scattering[1] zooplankton (e.g., euphausiids).

The backscattering amplitude f_{bs} (in m) (called backscattering form function in other works [14]) is then expressed, using the notation of Stanton [17], as:

$$f_{bs} = \frac{k_1}{4} \int_{r_{pos}} (\gamma_\kappa - \gamma_\rho) \exp\left(\left(2\, i\vec{k_i}\right)_2 \vec{r}_{pos}\right) a \frac{J_1(2\, k_2 a\, cost\, \beta_{tilt})}{\cos(\beta_{tilt})} \left| d\vec{r}_{pos} \right| \quad (12.1)$$

where k_1 and k_2 are the wave numbers of the water and zooplankton body, respectively, γ_κ and γ_ρ are the density and sound speed contrasts of the scatterer, and β_{tilt} is the tilt angle (relative to the incident wave) of the local cross section of the

---

[1]"Weakly scattering" is generally interpreted in the scientific literature [14] as meaning that the material properties of the target are within 5% of the surrounding fluid and there are no shear waves.

cylinder at point r_{pos}, position vector of the axis of the cylinder. J_1 is the Bessel function of the first kind of order 1.

The target strength *TS* (in dB, referenced to 1 m^2) and the differential backscattering cross section σ_{bs} are then derived (using implicit unity normalization factors) as:

$$TS = 10 \log |f_{bs}|^2 = 10 \log \sigma_{bs} \qquad (12.2)$$

Models of other organisms with gas inclusions (pneumatophores) or hard shells are less well constrained, because of the extreme difficulty in acquiring sufficient measurements in realistic conditions to sufficiently validate these models. Zooplankton shells vary in shape, from nearly spherical (e.g., pteropods) to more complex and irregular (e.g., foraminifera and radiolarians) (Fig. 12.3). Lavery et al. [15] investigated different models and compared them with field measurements. A frequently used approach is that designed by Stanton and coworkers [18], with a high-pass dense fluid–sphere model using an empirically derived reflection coefficient. At high *ka* values (where *a* is the external radius of the spherical shell), scattering will incorporate surface waves. At lower values, scattering will mostly depend on the shell volume. For all *ka* values, averaged models of idealized fluid-filled spherical shells were found to match measurements well [15].

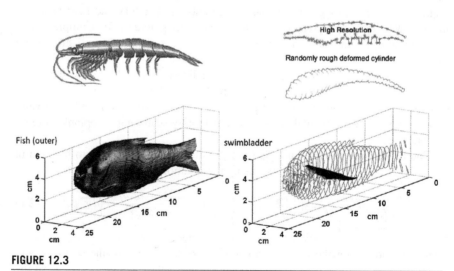

FIGURE 12.3

Numerical and analytical models of marine life are based on high-resolution measurements of the real animals. Top: Euphausiids (e.g., krill) can be modeled as a randomly rough deformed cylinder. Bottom: CT image of the body of a fish (left) and its swim bladder (right).

Adapted from Stanton, T.K., 30 years of advance in active bioacoustics: a personal perspective, Methods in Oceanogr., 1–2, pp. 49–77, 2012. Copyright 2012, with permission from Elsevier.

Some zooplankton types like siphonophores often have gelatinous bodies with gas inclusions, meaning most of their scattering will be affected by the gas (often carbon monoxide). The resonant frequencies of gas bubbles are given by their wave number k_0:

$$k_0 = \frac{\sqrt{3}\,\gamma}{a\,c} \sqrt{\frac{P_0(1 + 0.1\,z)}{\rho}} \tag{12.3}$$

where $\rho = 1027$ kg/m^3, $P_0 = 1.013 \times 10^5$ Pa (surface pressure, corrected for hydrostatic pressure with depth z), and $\gamma = 1.4$ (adiabatic constant). The backscattering cross section for $ka < 0.1$ was calculated by Weston (1967), referenced in Lavery et al. [15] as a function of the bubble radius a and a quality factor Q (typically 5 for fish with a swim bladder, and the same value is used here for zooplankton, in the absence of data to the contrary):

$$\sigma_{bs} = \frac{a^2}{\left(1 - \dfrac{k_0^2}{k^2}\right)^2 + \dfrac{1}{Q^2}} \tag{12.4}$$

At high ka values, this cross section will become independent of frequency and converge toward the geometrical cross section, πa^2.

The exact modeling of acoustic scattering by zooplankton with shells or pneumatophores is still very much an open research question, although the benchmarking of analytical versus numerical scattering models performed by Jech et al. [14] suggest that numerical representations might be often more appropriate.

Values observed for zooplankton scattering in the ocean are usually very small, of the order of -140 dB to -90 dB for single animals (Fig. 12.2). As they form larger and denser aggregations, they will however become visible, forming, for example, the deep scattering layer (DSL), further discussed in a subsequent section. Field measurements combine target strengths as volume scattering strengths (e.g., Fig. 12.2), related to the mode of imaging (beam width, pulse length, etc.) and modulated by target type(s) and density. Volume scattering can be modeled with ensemble-averaging and other techniques, and measurements at several frequencies can often be used advantageously to distinguish zooplankton types. This is presented further in Section 12.3.5.

12.3.2 SWIM BLADDER SCATTERING

More than 80% of fish families have swim bladders, filled with gas (often oxygen, with traces of nitrogen and carbon dioxide). The main role of this organ is to maintain buoyancy and control. Because of their acoustic contrast, swim bladders often contribute up to 90% of the scattering from an individual fish [13]. Scattering between 1 and 25 kHz is often dominated by the resonance of the swim bladder. Although they can have many shapes [17,19] (Fig. 12.3), swim bladders are often best approximated with equivalent spheres, and Eq. (12.3) can be used to identify the peak in scattering as a function of frequency. Fish target strengths are generally small (-30 to -60 dB [10]), and the presence of a (resonant) swim bladder can increase these strengths by $10-15$ dB.

Marine biologists recognize three main types of swim bladdered fish:

1. Physoclistous fish, like cod (*Gadus morhua*) have closed swim bladders, susceptible to pressure variations with depth; these variations can sometimes be slow (e.g., over $1-2$ days for a cod diving 20 m down);
2. Physostomous fish, like herring (*Clupea harengus*) have swim bladders connected to the alimentary canal (and thus the surrounding environment), meaning any excess pressure is vented off; this adaptation is typical of many schooling fish;
3. Very deep fish like orange roughy (*Hoplostethus atlanticus*) have swim bladders filled with oil or fatty tissue (hence lower-density contrasts, decreasing target strengths for a similar volume).

These variations in swim bladder shape, size, and composition should therefore be kept in mind when calculating the expected target strengths of a specific fish, or trying to identify fish types from acoustic measurements. Depending on the frequency used (i.e., if far from the resonance frequency of the gas-filled swim bladder), and on the aspect of the fish relative to the imaging sonar, other parts of its body might also contribute more to scattering than the swim bladder alone (Fig. 12.4). This problem is exacerbated when different fish types are present;

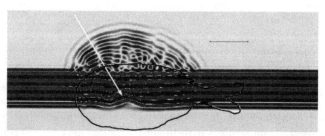

FIGURE 12.4

Numerical simulation of sound scattering for an orange roughy *(From Macaulay, G.J., Hart, A.C., Grimes, P.J., Coombs, R., Barr, R., and Dunford, A.J., Estimation of the target strength of oreo and associated species, Final Research Report for Ministry of Fisheries Research Project OEO2000/01A Objective 1. http://fs.fish.govt.nz/Doc/22654/OEO2000-01A%20oreo%20and %20 Associated %20species%20objective%201%20Final.pdf. ashx, 2002.),* varying from blue (darkest gray in print versions) (low) to red (dark gray in print versions) (high). The swim bladder (located with an *arrow*) is full of wax esters and does not reflect as strongly. The digitized outline of a typical fish sample also shows that some of the scattering comes from other parts of the fish, not only the swim bladder. Copyright: The Ministry for Primary Industries, New Zealand.

Godø et al. [20] show, for example, the variations in scattering strengths for two distinct types of fish, moving between depths, and how this can be better resolved using several frequencies.

12.3.3 FISH BODY SCATTERING

Swim bladders only account for a small portion of a fish body (around 5%), but they can contribute to a large part of its acoustic scattering properties. This will of course vary with the imaging frequency. As the frequency increases, the wavelengths will become closer to the scale of variation of the fish body and any surface structures or strong reflectors (e.g., backbones). Sun et al. [21] measured, for example, the acoustic responses at 220 kHz of yellow perch (*Perca flavescens*) (Fig. 12.5) and northern

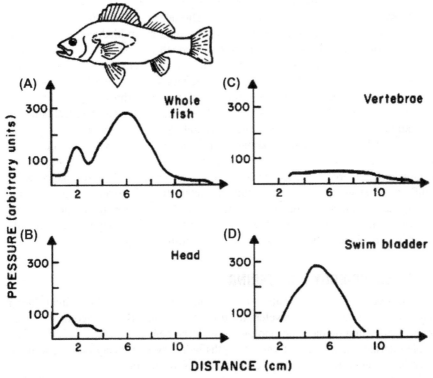

FIGURE 12.5

High-frequency measurements of individual fish (here, a yellow perch imaged at 220 kHz) show that their acoustic response is dominated by the swim bladder (if present) but that other parts of the body also contribute to scattering. For some fish species, the swim bladder might not be the main contributor to scattering.

*Reprinted with permission from Sun Y, Nash R, Clay CS. Acoustic measurements of the anatomy of fish at 220 kHz. J. Acoust. Soc. Am., **78** (5), pp. 1772–1776, 1985. Copyright 1985, Acoustic Society of America.*

hog sucker (*Hypenteeium nigricans*). Relative contributions from the swim bladder were 79% (for the perch) and only 22% (for the hog sucker), whereas the flesh contributed 12% and 45% of the respective returns, the head 6% and 10%, and the vertebrae 3% and 23%. These two sets of results show that the contribution of the swim bladder is not always the main one, and that more information might be needed about fish morphology, aspect toward the imaging sonar (e.g., head-on or broadside), and behavior (e.g., staying at the same depth or not, as it might affect the swim bladder's acoustic properties).

Section 5.2.1.3 presented mathematical expressions of the target strengths of fish as a function of individual length and imaging frequency. Eqs. (5.20) and (5.21) are of the general form: $TS = \alpha \log L + \beta \log f + \gamma$. Tried and tested over many decades and with many types of fish, these equations combine the responses of different parts of a fish body into a simple and useable expression. They can be used with several variations on the values of parameters α, β, and γ, depending on fish type and growth stage (assuming that allometric relations hold).

Many fast-moving fish such as tuna have lost their swim bladders during evolution, as the organ is not able to adjust quickly enough to rapid vertical movements. Their scattering strengths will therefore be smaller and affected by body morphology at scales commensurate with the acoustic wavelength. Observations have shown that the target strengths of different types of fish, or sometimes the same type of fish in different conditions, do not vary linearly with frequency. Gorska et al. [22] studied, for example, the scattering by Atlantic mackerel (*Scomber scombrus*). They were able to model the flesh and backbone with different variations of the deformed cylinder model of Stanton and coworkers [18] (and references therein), explaining the directivity pattern of fish scattering as well as resonance and antiresonance peaks at different frequencies. They also made the interesting observation that, since the swim bladder of physostomous fish compresses with depth, at some point its contribution to scattering will become low enough for their target strengths to be mainly constrained by their body shape and aspect toward the imaging sonar.

12.3.4 LARGE-BODY SCATTERING

The same questions will be pertinent to acoustic scattering by marine animals larger than fish. Depending on the wavelengths used, these animals will become extended targets. Depending on their morphology (e.g., large bony or cartilage structures and sizable internal organs), their target strengths will vary with many factors, some of which have sometimes not been measured in vivo or in situ.

This is, for example, true for sea snakes. Based on their morphology (3—9 m long, usually very thin), their scattering properties are expected to be similar to those of fish, but no measurements have been published so far (2016). Side-scan sonar surveys have shown that other animals like river turtles and crocodiles had strong and distinctive acoustic responses [23], but there are no published target strengths or discussions of which parts of the animals contribute most to the scattering. Marine turtles have attracted more interest, because they sometimes get entangled in fishing nets and also because they are often endangered. Again, there are very few details of their acoustic characteristics. Mahfurdz et al. [24] published tank measurements (at 200 kHz) of sea turtles at different aspects, aiming narrow beams at their head, side, tail, carapace, and plastron (the ventral surface of the shell). Overall, the target strengths measured range between −23 and −17 dB. As expected, the hard shells of the carapace and plastron are more reflective. The tails reflect sound differently, depending on aspect. Turtle heads exhibit the most variability, presumably also because of varying aspects. Interestingly, target strengths increased with turtle age, presumably because of the larger sizes and possibly the thicker shells.

Marine mammals have been much more extensively studied in the last (pre-2016) decades (Fig. 12.6). They cover close to 130 distinct species, extending in size from less than a meter (for juvenile seals) to 30 m (for the blue whale, *Balaenoptera musculus*). These animals are large (relative to the acoustic wavelengths traditionally used to study them) and their relevant characteristics will include the presence of flesh or blubber, through which sound penetrates but is attenuated; large bones, scattering sound and potentially inducing shear waves; internal organs, like lungs or stomachs, filled with mixed gas and/or fluids, introducing strong acoustic discontinuities; and (for some animals and at the highest acoustic resolutions) teeth or tusks. General morphology, from "rounder" to thin and elongated, and the presence of external appendages or flukes, will also affect variations of scattering depending on the aspect of the animal relative to the imaging sound, coming from the side or from head or tail. Because of their larger sizes, variations within animals of one species will be more evident acoustically. These complex contributions to

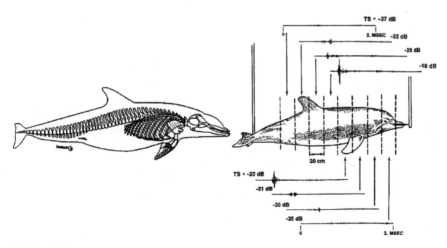

FIGURE 12.6

Acoustic scattering from large animals like dolphins is a complex combination of reflections from their body and internal organs.

*Adapted from Au, W.W.L., Acoustic reflectivity of a dolphin, J. Acoust. Soc. Am., **99** (6), pp. 3844–3848, 1996. Reprinted with permission. Copyright 1996, Acoustic Society of America.*

acoustic scattering are compounded by the large variety of behaviors these animals can exhibit (from breaching whales to rapidly moving dolphins on a group hunt).

Scattering by large animals has been studied in open-water experiments (ocean or close to shore), in tanks/swimming pools, or by measuring dead animals. Some experiments have used wild animals, whereas others have used trained subjects. The growing body of scientific results points to some common features, associated to the aspect of animals relative to the imaging beam, their body morphology, and the effects of greater hydrostatic pressures.

Scattering from animals varies greatly with aspect, being larger when the animals present their broadside to the acoustic beams and lower at tail aspect (by 21 dB for dolphins, as measured by Au [25]). The relative target strength decreases more rapidly toward the tail aspect than toward the head aspect, with echo levels at the head aspect 5 dB lower for dolphins [25]. Similar results were found for fin whales (*Balaenoptera physalus*) [26], gray whales (*Eschrichtius robustus*) [27], humpback whales (*Megaptera novaeangliae*) [28], and killer whales (*Orcinus orca*) [29], inter alia. A compilation of target strengths found in the literature is given in Table 12.2 (Banda, personal communication).

Several studies also investigated variations with frequency. Contrary to what has been reported in some articles or technical reports, the target strengths do not follow Love's curve for fish (Section 5.2.1.3). Target strength seems to decrease (by up to 5−10 dB in some cases) as imaging frequency increases, before leveling off after a certain frequency, at least from measurements on dolphins [25] and humpback whales [30].

Table 12.2 Selection of Typical Target Strengths Reported in the Scientific Literature

Species	TS (Aspect, if Known)
Bottlenose dolphin (*Tursiops truncatus*)	−11 to −24 dB (broadside) 21 dB below broadside values (tail) 5 dB below broadside values (head)
Dusky dolphin (*Lagenorhynchus obscurus*)	−28 dB (broadside)
Fin whale (*Balaenoptera physalus*)	−5 to −10 dB (broadside)
Florida manatee (*Trichechus manatus latirostris*)	−20 to −40 dB
Gray whale (*Eschrichtius robustus*)	−2.9 (tail) to +12.8 (broadside)
Humpback whale (*Megaptera novaeangliae*)	−4 to +7.2 dB (broadside) −3 to +4 dB (head)
Killer whale (*Orcinus orca*)	−50 to −20 dB (head) −50 to −10 dB (broadside) −40 to −10 dB (tail)
Manatee (*Trichechus* spp.)	−39 to −46 dB
Northern right whale (*Eubalaena glacialis*)	−12 to −1 dB
Sperm whale (*Physeter microcephalus*)	−9 to +10 dB

From Banda, personal communication.

Blubber (fat-storage layers used in particular for buoyancy) and tissues contribute significantly to acoustic scattering, as they are distributed around the entire animal body. Blubber thickness varies, from 30 cm for fin whales to a few centimeters for small odontocetes like harbor porpoises (*Phocoena phocoena*) [30]. As these layers become thicker, sound attenuation will increase, and there are some indications that this might increase with frequency (e.g., Au [25]). Bones and similar structures are other sound scatterers, distributed differently according to the species. Detailed measurements of sound velocity and density at different parts of the body were performed for different animals, for example, Cuvier's beaked whale (*Ziphius cavirostris*) [31], bottlenose dolphin (*Tursiops truncatus*) [32], and Yangtze finless porpoise (*Neophocaena asiaeorientalis*) [33]. These values seem to decrease as sound goes further into the body and to change with maturity of the animals, but there are not enough measurements, covering enough species, to draw authoritative conclusions. Because these detailed measurements were often performed on dead animals, McKenna et al. [32] addressed the question of postmortem changes in geometry, density, and sound speed within organs and tissues (melon, bone, blubber, and mandibular fat). They concluded there were no significant changes with time from death, except possibly for blubber.

Lungs are important contributors to acoustic scattering (up to 95%, according to Au [25]), as they are filled with gas. As an example, experimentally measured resonant frequencies of white whale and dolphin lungs were 30 and 36 Hz, respectively. These values were highly damped and far less intense than those predicted using a free spherical air bubble model [34]. Hyperbaric tests on animal bodies revealed that internal organs do not change shape linearly with increasing depths, and also that supersaturation of some gases might occur in the lungs, affecting their acoustic characteristics [35]. Ocean measurements by Bernasconi et al. [26] indicate that, for large whales at least, target strengths decrease with depth, and they decrease faster at the lower frequencies. Lung compression follows Boyle's law down to 170 m but becomes nonlinear or leads to collapse deeper, i.e., at higher surrounding pressures [35]. To complicate matters, for some animals, a greater percentage of the air compressed during a dive will move from the lungs to the nasal and tracheal regions, affecting their acoustic properties. Bernasconi et al. [30] used field observations of diving humpback whales to derive an empirical relation between the target strength *TS* at the surface and at depth *z*:

$$TS(z) = TS(z = 0 \text{ m}) \times \left(1 + \frac{z}{10}\right)^{-0.57} \qquad (12.5)$$

12.3.5 MANY-BODY SCATTERING

Large aggregations of marine life are particularly spectacular when considering zooplankton (Section 12.3.1). Individual animals are weak scatterers, with volume backscattering strengths usually 50 dB lower than fish (e.g., Fig. 12.2). Combining their individual echoes is straightforward if there is no attenuation or multiple scattering, and ensemble-averaging can be used to calculate volume backscattering coefficients. Lavery et al. [15] express them generally as:

$$S_V(f, z_k) = \frac{1}{V_k} \sum_{i=1}^{N_k} \sum_{j=1}^{M_k} \left\langle \sigma_{bs}^{i,j}(f, z_k, T_{i,j}) \right\rangle \tag{12.6}$$

where f is the frequency considered, V_k is the volume of water sampled in depth range z_k, N_k is the number of zooplankton of a particular type in this depth range, M_k is the number of zooplankton taxa, and the averaging term considers the back-scattering cross section of each individual of size i and taxon j in this range, averaged over angular orientations. The parameter $T_{i,j}$ is related to the physical characteristics of the zooplankton type considered.

Their combination results in a sonar-reflective region known as the DSL, discovered during World War II. These strong reflections, sometimes with individual reflections from predators (e.g., fish) show diel vertical movement and are often clearly identifiable (Fig. 12.7). At lower frequencies (<20 kHz), the DSL is caused by the resonance of swim bladders and scattering strengths are highly variable. The resonant frequencies vary during the day, as fish follow the diel migrations of their prey and hydrostatic pressure changes their scattering [10,20]. At higher frequencies (>20 kHz), the DSL visibility depends mostly on population statistics.

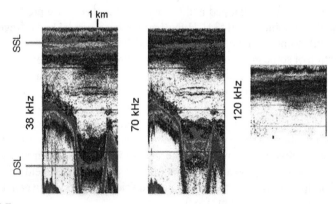

FIGURE 12.7

Dense aggregations of zooplankton can give rise to large regions with distinctive scattering, much higher than that of individual targets. The deep scattering layer (DSL) and a shallow scattering layer (SSL) are presented here at three different frequencies. The measurements were taken in the early morning offshore Hawaii [36], in water 800 m deep. Black horizontal lines are spaced every 200 m down, starting from the sea surface. Volume backscattering strengths are color-coded from -44 dB re 1 m^{-1} [brown (dark gray in print versions)] to -80 dB re 1 m^{-1} (gray).

Redrawn from Domokos, R., Acoustic surveys study abundance and movements of bigeye tuna and their micronektonic forage on the Cross Seamount, NOAA Pacific Islands Fisheries Science Center Quarterly Research Bulletin, https://pifsc-www.irc.noaa.gov/qrb/2009_2/eod2.php, March 2009. Courtesy: NOAA Fisheries.

Fish schools vary greatly in shapes and sizes, typically a few meters vertically and a few tens of meters horizontally, although larger schools can be seen (e.g., Colbo et al. [52]). Section 5.2.2.1 presented the basics of scattering by fish schools, with Eq. (5.24) giving the correspondence between volume density of fish and target strengths. This volume backscatter strength decreases as fish size increases, because the number density has more influence than the individual size. In most cases, though, the fish type or its average target strength (at the frequency or set of frequencies used) is not known. Medwin and Clay [8] recommend addressing this problem by considering the scattered signal as a combination of a fish scattering component σ_f and a random noise σ_n. Because this is analogous to the complex sum of a sinusoidal signal and narrow-band random noise, they recommend addressing it with a Rician probability density function (PDF) of the peak e of the echo envelope:

$$PDF_{Rice}(e) = \frac{2\,e}{\sigma_n} \times \exp\left(-\frac{e^2 + \sigma_f}{\sigma_n}\right) \times I_0\left(\frac{2\,e\sqrt{\sigma_f}}{\sigma_n}\right) \tag{12.7}$$

with $\begin{cases} I_0(x) = J_0(i\,x) \\ I_0(0) = 1 \end{cases}$ (J_0 being the modified Bessel function).

The ratios $\gamma = \sigma_f/\sigma_n$ (equivalent to a signal-to-noise ratio) and the average $\langle\sigma_{bs}\rangle$ can be measured directly. The Rician PDF can then be rewritten with these two parameters as:

$$PDF_{Rice}(e) = \frac{2\,e[1+\gamma]}{\langle\sigma_{bs}\rangle} \times exp\left(-\frac{[1+\gamma]e^2 + \gamma\langle\sigma_{bs}\rangle}{\langle\sigma_{bs}\rangle}\right) \times I_0\left(\frac{2\,e\,\gamma\sqrt{1+\gamma}}{\langle\sigma_{bs}\rangle^{\frac{1}{2}}}\right)$$

$$\tag{12.8}$$

As shown by Medwin and Clay [8], this PDF depends on the ratio of fish length to acoustic wavelength. When it is large, γ is small and the PDF tends to the Rayleigh distribution. When it is small, γ is large and the PDF becomes Gaussian, peaking near $e = \sqrt{\sigma_{bs}(f)}$.

The implications for transducer choices and survey designs are presented further in Medwin and Clay [8] and Colbo et al. [52]. Theoretical models of (spherical) fish schools, incorporating effects of multiple scattering, are presented in Raveau and Feuillade [37] and compared with third-party measurements. The main results are very useful to assess the relative contributions of backscattering and other scattering directions, depending on fish size, fish density, and acoustic wavelength. Very often now, the availability of multifrequency measurements allows making full use of the variations in resonance frequencies between fish types, enabling distinguishing multiple species within the same areas [17] (and references therein). Some manufacturers also offer libraries of multifrequency measurements of different fish types with their own instruments, to which existing data can be matched.

12.4 ACTIVE IMAGING SYSTEMS

12.4.1 SINGLE-BEAM ECHO SOUNDERS

Single-beam echo sounders were presented in Section 10.3. They have been the fundamental tool of fisheries for close to a century now, with the first models (also called fathometers) commercialized in the early 1920s, for example, Balls [38]. The first scientific fishery applications are reported on the German research vessel Meteor in 1926, for example, J.B. [39] and the field has greatly progressed since that time, for example, Medwin and Clay [8] and Lurton [10]. Frequencies depend on the exact applications, and typically range between 12 kHz for deep-water models to up to 200 kHz for shallow-water models with 38 kHz being the most common, Lurton [10].

Fig. 5.2 showed the typical application of echo sounders to detect fish schools. These echo sounders typically transmit short signals (1−10 ms) along a narrow beam (5−15 degrees) and they measure the volume backscattering strength of mid-water targets. Individual echo levels EL are related to the source level SL, the range R to the target, the absorption coefficient α of water at this frequency and its sound speed c, the solid angle of the imaging beam ψ, its signal duration T, and the backscattering strength BS of fish by the equation, Lurton [10]:

$$EL = SL - 40 \log R - 2\,\alpha\,R + 10 \log\left(\frac{\psi R^2 c\,T}{2}\right) + BS \qquad (12.9)$$

The combination of $40 \log R + 2\,\alpha\,R$ corresponds to the time-varying gain. The backscattering component BS can relate to a single fish with the target strength defined, for example, in Eqs. (5.20) and (5.21) or to a fish school given by Eq. (5.24).

Measurements are integrated along the ship's path, creating a two-dimensional view of mid-water targets immediately below the ship (known as echo trace or echogram). Depending on the size of a target relative to the imaging beam, this integration might create the crescent or boomerang shapes well known of early sonars and recreational fishermen. It is an artifact created by the acquisition: as the echo sounder moves from position 1 to 2 to 3 (Fig. 12.8), the target will be imaged at slightly different angles. At position x_1 along the track, it will be at a higher angle from the vertical, as the fish is at the edge of the beam, and its depth will be slightly overestimated. At position x_2, immediately above, the fish will be in the center of the imaging beam, and its depth z_2 will be the correct one. As the ship moves away, the fish, if large enough, will be at the edge of the next beam(s). The crescent shape $z(x)$ follows a simple equation linked to the closest range of the fish:

$$z = z_1 + \frac{(x - x_2)^2}{2\,z_2} \qquad (12.10)$$

This effect will vary with the echo sounder's settings, the vessel's speed and the range: higher ping repetitions, associated with slower speeds and lower frequencies, are more likely to create larger crescent shapes than higher frequencies

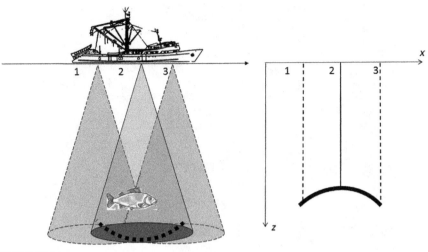

FIGURE 12.8

Targets large enough to be imaged several times, as the sensor moves relative to it, are extremely likely to be at slightly different ranges. This is particularly visible for single-beam echo sounders, resulting in the characteristic crescent shape seen in raw measurements (see text for explanations).

Ship outline taken from Fonteneau, A., Diouf, T., and Mensah, M., Tuna Fisheries in the Eastern Tropical Atlantic, FAO Corporate Document Repository, 2016; http://www.fao.org/docrep/005/t0081e/T0081E04.htm, Copyright FAO.

(e.g., Fig. 12.9). These variations should be corrected during processing. Echo integration is a topic frequently discussed in the processing of fisheries echo sounder measurements. Earlier systems were separate electronic instruments, connected to the output of the echo sounders: they are now assimilated into the software. Echo integrators sum the energy in different parts of the echogram, for example, preselected depth channels or areas specified by an operator, over a specified number of transmissions. The reader interested in the exact details of how this is done in practice, and the physical assumptions behind it, is referred to textbooks, such as Simmonds and MacLennan [9]. The interest of echo integration is well shown in Fig. 12.9, where the water layers closer to the sea surface are strongly affected by turbulence, and those closer to the seabed might be affected by bathymetry, gas-venting, or vegetation (in the shallower areas).

For point-like targets like individual fish, the echo sounder must possess a high enough spatial resolution to detect each of them. The measurement of one such target inside a single beam strongly depends on its angular position (due to beam directivity) and additional techniques are available to better resolve them [10]. Split-beam sounders use different parts of the transducer as interferometers, to determine the angular position relative to the beam axis and compensate for beam directivity at this angle. Dual-beam sounders use two coaxial receiving beams with different

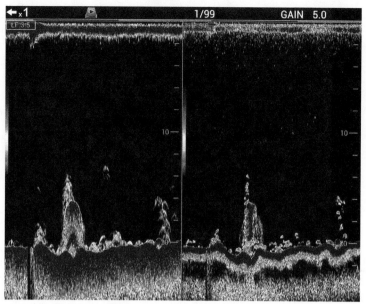

FIGURE 12.9

Single-beam echo sounder measurements of single fish next to a dense school. Left: the lower frequencies show the traditional crescent shapes and overestimate the sizes of the different targets. Right: the higher frequencies show better accuracies (because of the smaller wavelengths), revealing the vertical structure of the fish school and separating distinct groups closer to the seabed. Note also the higher definition of the seabed itself (see Chapter 5 for theoretical justification).

Image from FURUNO, Furuno FISH FINDER Model FCV-1900 technical documentation, *http://www.furuno. com/en/products/fishfinder/FCV-1900#Screenshots, 2016.*

apertures, and echo level differences are used to estimate the angular position of the target.

These two approaches, along with echo integration, rely on accurate calibration of the echo sounders, in transmission and in reception. The equations presented in Section 10.8 are used to measure the reflections from known metal spheres, using the methods of Foote et al. [42], Tomich et al. [43], and Demer et al. [44]. The main interest of calibration, ideally before each survey, is of course the provision of reliable values for the echoes. It is also fundamental to the accuracy of echo integration for fish stock monitoring and biomass quantification. Lurton [10] advises, for example, that transmission and reception sensitivities, or beamwidth patterns, must be accurate to ± 1 dB (a 3-dB bias in level estimations leading to an error by a factor of 2 in biomass estimations).

Typical frequencies used by single-beam echo sounders for fisheries applications are 38, 120, 200, and 420 kHz, as shown by the variety of commercially available sounders using these exact frequencies. Section 12.3 showed the role of frequencies in estimating individual and group backscattering strengths. It is therefore logical to think about using several frequencies to better identify different fish types. Much work has been done in the field, and echograms of the same fish at different frequencies can show striking differences (Fig. 12.5). The different strands of work have been summarized by Korneliussen et al. [45] and implemented in software like EchoView (www.echoview.com). The many applications spawned by this approach are very attractive, but require meticulous preparation of both the sounders to be used and the collection of the measurements. The recommendations of Korneliussen et al. [23], prioritize the following (Fig. 12.10):

- Selection of frequencies with no harmonic interference (e.g., 18, 38, 70, 120, 200, 333, 555, 926, 1543, and 2572 kHz) (Simrad's EK60 uses, for example, seven frequencies simultaneously between 18 and 710 kHz);
- Selection of transducers with similar beam widths, mounted as close as possible with the smaller ones in the center;

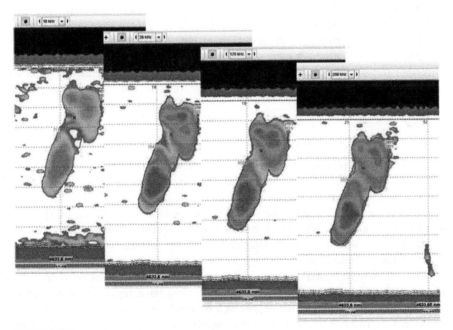

FIGURE 12.10

Marine life echograms at 18, 120, and 200 kHz in the Gulf of Alaska. Note how the definition of fish schools near the seabed and zooplankton and foraging fish at lower depths vary with the imaging frequencies.

*Reprinted with permission from Anderson, C.I.K., Horne, J.K., and Boyle, J., Classifying multi-frequency fisheries acoustic data using a robust probabilistic technique, J. Acoust. Soc. Am. Express Letters, **121** (6), pp. EL230-EL237, 2007. Copyright 2007, Acoustic Society of America.*

- Synchronized transmissions, time-stamped to <10 ms, at power levels without nonlinear effects;
- Use of similar pulse duration and ping rate, digitized sample lengths for all frequencies, and the lowest possible thresholds for data collection;
- "Before and after" calibration for each survey (keeping in mind the earlier *caveat* of Lurton [10] about keeping accuracies below ±1 dB).

These different approaches have been developed to high degrees, providing good estimates of fish densities and biomass and, using multiple frequencies, even potentially identifying the type of fish or the stage of development. But, for physical reasons, the use of a single beam immediately below the surveying vessel will always induce some limitations. The first one is the relatively large beam spread, especially at larger ranges, which limits the resolution of changes within fish schools and also the detection of fish close to the seabed (Fig. 12.8, left, implies that seabed returns will be shaped like an upward crescent, meaning fish sizes and densities might be biased). Second, the monodimensional echo return means measurements must be made at several positions relative to the fish schools, to get accurate ideas of their overall shapes and dimensions. Third, fish are sensitive to the lower-frequency noise of both ships and single-beam echo sounders, and are known to avoid the path of moving vessels, for example, Olsen et al. [47], potentially biasing fish observations close to the vessel, although there are, of course, interspecies differences in avoidance ability/behavior, for example, Colbo et al. [52], and the use of quieter platforms like "silent vessels" or autonomous underwater vehicles, which strongly reduces these biases [48].

12.4.2 SIDE-SCAN SONARS

Single-beam echo sounders provide accurate and very useful measurements of fish and other targets immediately below the surveying vessel, but they do not offer instantaneous coverage athwarthship. Side-scan sonars, conversely, scan the seabed at distances up to 30 km from the ship (for the old GLORIA system, using a frequency of 6.5 kHz, yielding 60-m resolutions useless for fisheries), but generally in the range of a kilometer or less (for frequencies of a few hundreds of kHz, closer to the spatial resolutions of use for fish studies). Side-scan systems are described in detail in Section 10.5, and modern systems can achieve resolutions down to millimeter or centimeter in all environments [49]. They have regularly been used to detect fish and other marine life, either as a by-product of the seabed survey or on purpose. Fig. 12.11 shows two typical examples: a dense school of small targets (fish) and a large, single target (identified as a shark). The images are unprocessed and show the nadir (thick white line) and the water column (black, i.e., with low to nil acoustic returns), with the seabed exhibiting different reflectivities as range from the sonar increases. Fig. 12.11 (left) already shows some returns in the water column: they are small and fuzzy, leading to mottled patterns and bright (reflective) patches as fish density increases. The narrow beam width along track means the

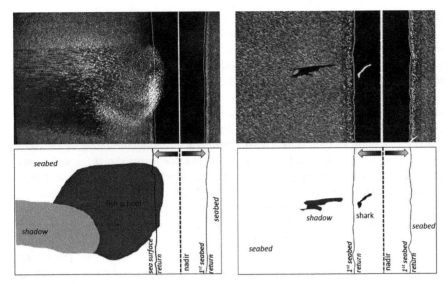

FIGURE 12.11

Side-scan sonar images of a 20-m wide fish school (left) and a hammerhead shark (right), with basic interpretations (bottom sketches). *Arrows* indicate the respective directions of ensonification. Note the role of the shadows in describing the target shapes.

Images courtesy of the Search for the Lost French Fleet of 1565 Expedition, NOAA-OER/St. Augustine Lighthouse and Museum (http://oceanexplorer.noaa.gov/explorations/14lostfleet/logs/july19/media/baitball. html, left; http://oceanexplorer.noaa.gov/explorations/14lostfleet/logs/july19/media/hammerhead.html, right).

fish school creates acoustic shadows at further ranges. From the ranges and lengths of these shadows, it is possible to infer the overall size and height of the fish school, at least in the bottom part of the sonar image. If fish density is not large enough, individual fish or small groupings can be detected, as in the upper part of the image, but it is impossible to identify the exact distribution of fish sizes, the type(s) of fish, or how high they are above the seabed. Fig. 12.11 (right) shows a single target, un-equivocally in the water column and associated with a distinct shadow on the seabed. Based on its size, and the morphology of its shadow, it was identified as a hammer-head shark. It is however doubtful that similarly large targets could be detected if closer to the seabed, or at further ranges, as their acoustic returns would be mixed with those of the geological features below. Side-scan sonars are extremely useful for many tasks (see Ref. [49], for examples), but the limited information they pro-vide about fish means that other tools are often preferable.

12.4.3 MULTIBEAM ECHO SOUNDERS

Multibeam echo sounders are versatile instruments, providing bathymetry and seabed reflectivity (Section 10.4.2.1), "snippets" (Section 10.4.2.2), and pseudo-side-scan imagery of the seabed (Section 10.4.2.3). Snippets record portions of the individual waveforms reflected in each beam sector and increasingly contain water column data, meaning they can be used for imaging of mid-water targets. A wide-ranging review by Colbo et al. [52] shows the many applications of multibeam echo sounders to fisheries, as they directly address the limitations of single-beam echo sounders: higher frequencies and narrower beams, offering higher resolutions, wider coverage across-track, detecting fish some distance away from the surveying vessel, and much improved definition of fish school shapes and distributions.

Fig. 12.12 shows a typical multibeam "ping," acquired by the author in an Arctic fjord using an Imagenex sonar with 260 beams. The raw image shows the seabed as a thicker line 12—14 m deep. Highly backscattering targets in the water column can be seen in close-up, and based on concurrent visual observations in the very clear waters, they are associated to small fish aggregations. A fisheries multibeam survey would cover very large transects, often in much deeper water and therefore covering very large volumes of water. Many studies, summarized in Colbo et al. [52], show that multibeam sounders can determine with confidence school shape parameters (e.g., length, width, height, surface area, volume, and their evolution with time or space as surveys progress), external parameters (depth and distance to thermocline or to seabed), etc.

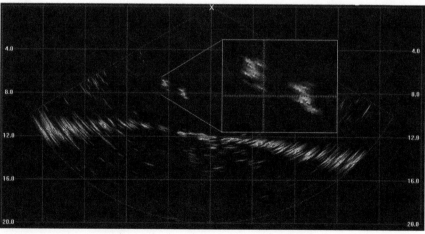

FIGURE 12.12

Raw multibeam image of small groups of fish in a shallow Arctic fjord. The seabed is 12 m deep and picked up as a line of thickness increasing with the outer beams (slightly noisier, owing to the mode of acquisition). The fish show up as clear targets and it is possible to identify distinct reflectors (color coded by increasing reflectivity, from black to red (gray in print versions), scattering as much as the underlying seabed.

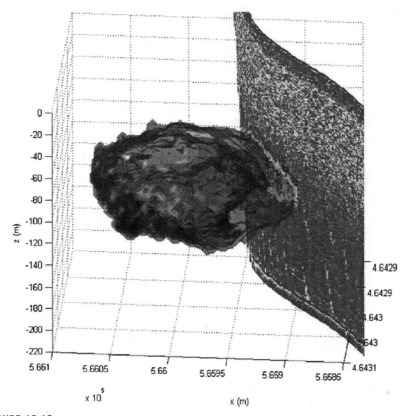

FIGURE 12.13

The wider cover across track of multibeam measurements of a herring school [*red cloud* (gray in print versions)] can be advantageously compared with single-beam measurements along a single transect [*blue segment* (dark gray in print versions)].

*Reprinted from Colbo, K., Ross, T., Brown, C., and Weber, T., A review of oceanographic applications of water column data from multibeam echosounders, Est. Coast. Shelf Sci., **145**, pp. 41–56, 2014, Copyright 2014, with permission from Elsevier.*

Fig. 12.13 shows advantageously the additional information brought by multibeam echo sounders. Repeated single-beam measurements below the survey vessel produced the blue segment, in which the fish school can be recognized as a red "patch" of higher backscatter strengths, but there is no information about the exact shape of this school. Shape is known to evolve with school dimension, but it is also affected by external constraints, from school movement to food availability and predator interaction, and it is of course highly species dependent. Because they cover a large section across track, and have higher resolutions, repeated multibeam measurements show the exact shape of the fish school (red "cloud," extending from the red "patch").

The 3-D measurement of fish schools and densities can ideally be used to estimate the fish biomass through its volume backscattering. The first obstacle is the calibration of the multibeam sonar, often requiring facilities beyond easy and regular reach of survey teams, although some simple approaches using calibration spheres are proposed by Foote et al. [50] and recommended by the ICES for both single-beam and multibeam sounders [[44] (more details about sensor calibration are provided in Section 10.8)]. It should be noted, however, that the absence of such system calibration in the field does not preclude obtaining meaningful results, as shown, for example, in Parsons et al. [51]. The second obstacle is that fish backscatter strongly depends on the orientation of the fish relative to the imaging wave (see Section 12.3). Generally, fish swimming across fan will have reduced scattering volume in the far beams compared to the center beams, but the opposite is true for diving fish, for example, Colbo et al. [52]. Larger targets, like marine mammals, will exhibit stronger returns from their lungs, and smaller targets, like zooplankton, will show backscatter variations associated with internal movements within their aggregations, leading to similar effects. Some authors investigated the difference between multibeam and other measurements, for example, Gurshin et al. [53] with multibeam and split-beam echo sounders for captive cods, Weber et al. [54] with multibeam and single-beam echo sounders, and other references in Colbo et al. [52]. Other effects are linked to the oceanographic setting: Melvin and Cochrane [55] noted, for example, the difficulty of integrating measurements in dynamic and complex environments such as tidal sites.

12.4.4 COMBINING SENSORS

The previous sections showed the relative advantages and limitations of each type of instrument. Modern applications often combine them to address potential data gaps. For example, catch monitoring now often supplements fisheries single-beam echo sounders, to monitor where the fish is in the water column, with additional sensors next to the trawl or seine nets. Along with the useful measurements of water temperature (also used for velocity corrections of acoustic sensors), weight trawl, or flow, high-frequency sounders can be used to monitor the amount of fish closer to the nets (Fig. 12.14). This enables direct imaging closer to the seabed, where fish can be missed during deep trawling because of the higher seabed returns, and in general imaging the amounts and types of fish entering the nets, as they can sometimes be smaller than the resolution achievable with the ship's echo-sounder.

Smaller targets like zooplankton are often at the limit of resolution. Comparisons of acoustic measurements with traditional net sampling and camera observations were, for example, conducted by De Robertis [57], observing euphausiids in Canada and leading to useful observations as to how zooplankton behavior and 3-D organization could influence multibeam measurements. Similar approaches were used by Nichol and Brierley [58] and Cox et al. [59] to look at krill swarms.

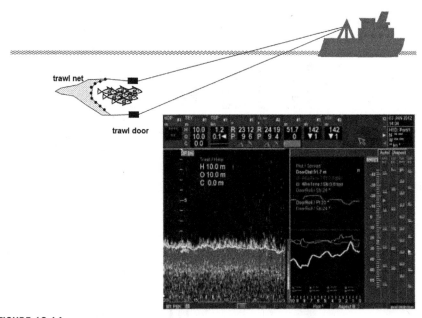

FIGURE 12.14

Additional acoustic sensors can be installed closer to the fish, for example, on the trawl itself.

Scanmar TrawlEye system, Scanmar, Advanced catch systems for increased efficiency and financial gain, *report, http://www.scanmar.no/wp-content/uploads/2016/04/Scanmar-Info Booklet_2011-2012_English.pdf, 2012.*

Multibeam measurements show that swarms cluster, depending on environmental characteristics such as bathymetry and water conditions, and that future surveys may require a stratified design to better measure krill biomass and swarm volumes.

More dynamic environments offer additional challenges, as noted by Melvin and Cochrane [31]. Tidal sites are increasingly used for human activities, including the siting of marine renewable energy structures, and acoustic monitoring of fish and other biological activity is hampered by the rapid changes in the water structure. In the more energetic sites, kolks (vortices within the water column) burst at the surface as "boils," which are short lived (minutes) and of sizes comparable to the water depth. The high currents also affect the deployment of instruments. Williamson et al. [60] showed, however, that it was possible to accurately and reliably monitor these areas by combining multibeam and multifrequency single-beam echo sounders (Fig. 12.15), installed on a platform moored on the seabed. Their studies investigated fish, marine mammals, and diving seabirds, at the same time, comparing acoustic measurements with radar and visual observations from the shore.

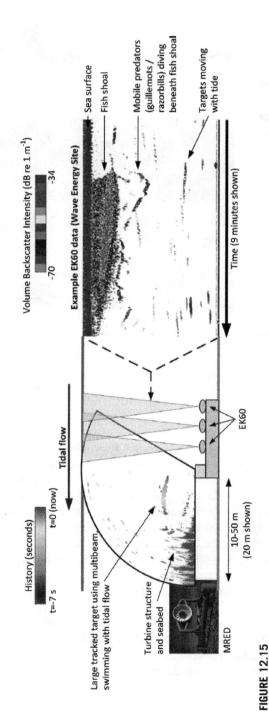

FIGURE 12.15

Different sensor types can be integrated, for example, multibeam with multifrequency single-beam echo sounders. This example shows an innovative approach, with sensors on an autonomous subsea platform imaging the environment around a tidal turbine from the seabed up.

*From Williamson, B.J., Blondel, Ph., Armstrong, E., Bell, P.S., Hall, C., Waggitt, J.J., and Scott, B.E., A self-contained subsea platform for acoustic monitoring of the environment around Marine Renewable Energy Devices —Field deployments at wave and tidal energy sites in Orkney, Scotland, IEEE J. Oceanic. Eng., **41** (1), pp. 67–81, 2015. Creative Commons Attribution 3.0 License.*

12.5 MARINE LIFE AND SOUND
12.5.1 GENERAL POINTS

Sounds are essential to the transmission of information underwater, and as such they are heavily used by all forms of marine life. Even those species not known to produce sound will still use it, or be sensitive to it. The relevant body of literature is vast and constantly updated with new measurements, either on live animals in the field or in the laboratory, and with new observations of behaviors best explained with auditory senses (e.g., flight or avoidance reactions). The masterful synthesis of De Ruiter [61] is recommended to the reader wanting to know the consensus on sound emission by all types of marine animals, and the many different mechanisms by which they receive and can be sensitive to sounds. Backed with considerable references and 25 pages of data about exact frequencies, source levels (SLs), and modes of measurements, this is an essential summary of the key points of sound emission and reception by marine life. The Discovery of Sound in the Sea project (DOSITS) also provides a high-quality educational resource, with multimedia presentations of particular animals (www.dosits.org). The following sections will focus on key points.

The hearing sensitivity of marine animals is measured with audiograms. These are collected either by behavioral analyses, in which trained animals indicate whether they have detected particular signals, or electrophysiological techniques, in which potential differences between electrodes on the subject's body show neural activity for different signals (it can be measured directly as the auditory brainstem response, ABR). Both techniques have inherent biases, insofar as they rely on a few animals that can be trained (behavioral analyses) or animals susceptible of being captured and analyzed. Intraspecific variations are also a concern, therefore, values found in the literature need to be interpreted in view of how many animals of each kind were studied, and how representative they might be. Finally, some animals (like sea otters, turtles, snakes, or polar bears) are amphibious and their auditory systems will be functional both in air and underwater, but with different sensitivities.

Frequency discrimination, directional hearing, and source location abilities will vary with species, and within species. The *hearing threshold* is the lowest level of sound at a particular frequency that an animal can detect. Depending on how loud the sound is, how long it is, and its frequency content, animals might be adversely affected. *Temporary threshold shift* (TTS) is when animals can only detect louder sounds, but this damage is reversed after some time (like a human affected by loud occupational noise or a rock festival). If this threshold does not return to normal levels, this is a *permanent threshold* shift (PTS). To pursue the human analogy, PTS can result from repeated TTS (e.g., listening to loud bands every weekend) or single exposure to a very intense sound (e.g., a close explosion). Noise exposure will also affect animals directly, for example, through injury to the ear and associated structures, or even nonauditory tissues. These injuries can lead to death or equally deleterious long-term effects (e.g., degradation of the ability to echolocate, communicate, or detect predators and preys).

12.5.2 MARINE MAMMALS

Marine mammals can be divided into four main groups: (1) Cetaceans (whales, dolphins, and porpoises), divided into odontocetes (toothed whales) and mysticetes (baleen whales); (2) Pinnipeds (phoceids, like the true seals, otariids, like the eared seals and walruses); (3) Sirenians (e.g., manatees and dugongs); and (4) amphibious mammals like sea otters and polar bears.

Toothed whales produce click sounds (understood as primarily related to echolocation, as their repetition rate increases with proximity to a prey, finishing with a high-rate buzz, similar to that of bats), click-based sounds or pulsed calls, tonal sounds, and other sounds associated with communications, like codas (short rhythmic series of clicks). A wide variety of whistles and other tonal sounds have also been documented (see Ref. [61], for in-depth review). Odontocetes are thought to produce sound using pressurized air, as potentially confirmed by some observations of frequency shifts when some animals go deeper (e.g., reduction of air volume available for sound production). Toothed whales can also produce percussive sounds by beating body parts against the sea surface, for example, when feeding and presumably to affect the prey. Sound levels are generally high, for example, 210−225 dB re 1 μPa at 1 m for broadband echolocation clicks (delphinids) or narrowband, high-frequency clicks at 155−190 dB re 1 μPa at 1 m (harbor porpoises). Baleen whales can also produce sounds by beating body parts against the water surface (or other objects) but other mechanisms of sound production are less well understood [61]. Most vocalizations are thought to be associated with reproduction or communication. They have low frequencies and are repetitive (often stereotypical) and frequency modulated. Local and cultural variations have led to their qualification as "singing," and there have been many studies of how they can be related to specific animal activities. These sounds are loud too, for example, 178−186 dB re 1 μPa at 1 m for 10−100 Hz calls by blue whales (over durations of up to tens of seconds).

Pinniped sounds are equally varied, described as barks, buzzes, grunts, roars, or yelps [62]. Levels range about 135−193 dB re 1 μPa at 1 m, over frequency ranges of tens of hertz to tens of kilohertz. Fig. 12.16 shows how they compare with cetacean calls.

Vocalizations by manatees and dugongs are referred to as whistle-squeaks, chirps, squeaks, chirp squeaks, barks, and trills, again reflecting a high variety. They are made at lower frequencies (<10 kHz), with lower SLs (<150 dB re 1 μPa at 1 m).

Amphibious mammals exhibit a large variety of vocalizations too, but there have been very few measurements of SLs or sensitivity, either in air or underwater (more difficult experimentally). For example, polar bears are known from ABR studies to be sensitive between 11 and 23 kHz (Nachtigall et al., 2007, in Ref. [61]) but there are no measurements of the sound levels they produce (apart from measurements near cubs and mothers in controlled situations). Sea otters are known to vocalize at levels up to 50−113 dB re 20 μPa (up to 7 kHz) in air but there are no reported measurements in water [63]. Measurements of sound emission and reception in amphibious mammals also need to account for the difference between dB levels measured in air (for a reference value of 20 μPa) and those measured in water (for a reference value of 1 μPa), as well as the properties of air versus water and how sound

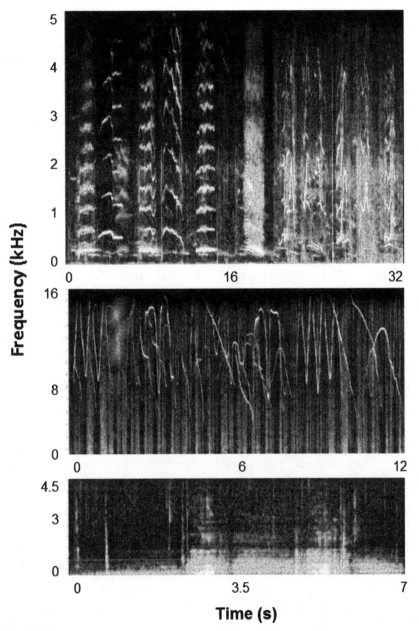

FIGURE 12.16

Marine mammal vocalizations can be quite complex, as shown in this collection of spectrograms collated by NOAA (http://www.nefsc.noaa.gov/psb/acoustics/sounds.html). From top to bottom: humpback whale (*Megaptera novaeangliae*), for 32 s and displayed from 0 to 5 kHz; bottlenose dolphin (*Tursiops truncatus*) (for 12 s and from 0 to 16 kHz); and harbor seal (*Phoca vitulina*) (for 0.6 s only, and from 0 to 4.5 kHz). *Brighter colors* indicate louder sound levels.

Courtesy: NOAA Fisheries.

levels (e.g.) are calculated (see the simple steps outlined in Ref. [64] for in-depth discussion).

The mechanisms of sound reception by marine mammals are still not fully understood [61] and there is sometimes controversy about the exact pathways or combination of mechanisms (e.g., for dolphin hearing), or no information at all (e.g., for polar bears). But the key characteristics of the different species are now known with a higher degree of certainty [61]. For example, toothed whales hear best frequencies between 10 and 100 kHz. Baleen whales can hear frequencies ranging from infrasounds to about 20 kHz. Pinnipeds hear best at 2−40 kHz, manatees at 0.4−20 kHz, and polar bears at 11−23 kHz (in air, but no data is available underwater). Frequency-weighted functions (known as M-level weightings) can be used to account for these distinct hearing sensitivities [65]. Based on a synthesis of field observations (e.g., strandings and postmortems) and PTS and TTS data, the following injury criteria have been proposed by Southall et al. [65] and are now widely regarded as the basis for sound environmental regulations:

- Peak exposure levels should not exceed 230 dB re 1 μPa for cetaceans, 218 dB re 1 μPa (underwater) and 149 dB re 1 μPa (in air) for pinnipeds.
- M-weighted Sound Exposure Levels (SELs, explained in Section 12.6) should not exceed 198 dB re 1 μPa^2 s for cetaceans exposed to pulsed sounds (defined as >3 dB in a 35-ms interval than in a 125-ms interval) and 186 dB re 1 μPa^2 s (in water) for pinnipeds, compared to 215 dB re 1 μPa^2 s for cetaceans exposed to nonpulsed sounds and 203 dB re 1 μPa^2 s for pinnipeds. For pinnipeds in air, these thresholds would be 144 dB re 1 μPa^2 s (pulsed sounds) and 144.5 dB re 1 μPa^2 s (other sounds).

12.5.3 FISH, TURTLES, AND INVERTEBRATES

Other animal groups known to produce and use sounds include fish (with increasing numbers of measurements being reported in the scientific literature), sea turtles and snakes (for which extremely little is known), and invertebrates (like the infamous snapping shrimp, loud enough to affect acoustic communications).

Sounds produced by fish show lower frequencies, mostly below 1 kHz (Fig. 12.17, top). They consist in tonal and pulsed sounds, varyingly called grunts, growls, croaks, chirps, squeals, etc. They are generated by stridulation (friction between hard structures of the fish anatomy), drumming (usually by rapid contraction of muscles near the swim bladder), percussion and involuntary mechanisms (e.g., air movements within the swim bladder of digestion), as well as breathing, feeding, and swimming. Some of these sounds have been demonstrated to be voluntary and associated to fish behaviors like communication. They often vary during the day, with strong variations associated with day or night activities. The New Zealand bigeye (*Pempheris adspersa*) produces short-duration (<8 ms) popping sounds with a peak frequency of 405 Hz and SLs around 115 dB re 1 μPa at 1 m [66]. Fish calls recorded by McCauley and Cato [67] were louder, with "pops" at

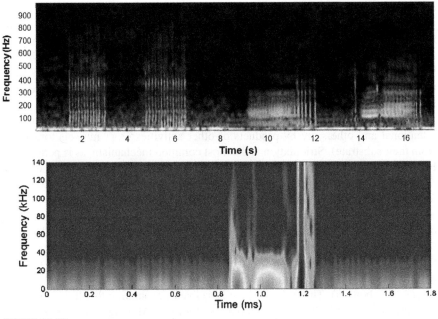

FIGURE 12.17

Top: Vocalization of haddock (*Melanogrammus aeglefinus*), showing a 17-s segment with frequencies between 0 and 800 Hz *(From http://www.nefsc.noaa.gov/psb/acoustics/sounds.html. Courtesy NOAA Fisheries.)*. Bottom: Laboratory measurements of sound emission by snapping shrimps, showing loud, broadband pulses (up to 140 kHz), with typical durations of 0.3 ms [68].

157 dB re 1 μPa at 1 m, "trumpet calls" at 150 dB re 1 μPa at 1 m, and "banging" noises at 144 dB re 1 μPa at 1 m.

Based on hearing mechanisms, fish are usually divided into generalists and specialists [61]. The generalists only receive sound directly, mostly through particle motion, and hear best between tens of hertz and about 1 kHz. Specialists evolved other means of hearing, enabling them to hear better at frequencies of 100 Hz to 2–10 kHz (but up to 100 kHz for some species like the American shad and the gulf menhaden). Some of the specialist species use sound propagation through the swim bladder, meaning their sensitivities would change with depth (depending on how the swim bladder reacts to changes in hydrostatic pressure). Documented changes in hearing sensitivity also occur at specific times, for example, during the breeding season for female midshipman fish (Sisneros and Bass, 2003, in Ref. [61]).

There are very few measurements of sea turtle's hearing, and it seems they hear best at frequencies below 1 kHz [61]. Sea turtles were long thought not to vocalize, but recent research showed the long-necked freshwater turtle (*Chelodina oblonga*) had a variety of underwater vocalizations between 0.1 and 3.5 kHz [69]. For comparison, the giant South American river turtle (*Podocnemis expansa*) can vocalize between 0.03 and 4.5 kHz [70]. No SLs have been documented, but vocalizations from sea turtles are much richer than originally thought, and the range of frequencies might suggest they are also adapted to hearing at frequencies slightly higher than 1 kHz. Based on the data available, sound exposure guidelines for fishes and sea turtles were developed [71], on the line of the injury criteria presented earlier for marine mammals.

Marine invertebrates produce sounds, in particular crustaceans [61]. Some are intentional and others are by-products of other activities (e.g., mussels torn loose from their substrate). Stridulation is the most common mechanism, as is percussion. For American lobster, buzz-like sounds with mean frequency 180 Hz and mean duration of 227 ms have been reported (Henninger and Watson, 2005, in Ref. [61]). Snapping shrimps can be the dominant source of noise in shallow tropical waters, especially in summer and at dusk, and climate change has led them to be noticeable at higher latitudes, e.g., close to UK shores. Sound is produced by cavitation when its large snapper claw is closed extremely rapidly during predation or other encounters. Laboratory measurements by Kim et al. [68] investigated individual snapping shrimps, revealing very loud SLs of 204−219 dB re 1 μPa at 1 m (Fig. 12.17, bottom). These "snaps" are broadband, expanding to 140 kHz in these measurements, and lasting for up to 0.3 ms at a time. Although colonies are small (several tens of animals, up to 300 at most), the combination of individual "snaps," closely spaced, follows Gaussian statistics and is known by submariners as the noise of "frying fat."

Crustaceans are thought to only perceive particle motion, not sound pressure itself, and they are harder of hearing than most generalist fish species. Bottom-dwelling species are also susceptible to vibrations from their substrate [72], below 100 Hz. Cephalopods are sensitive up to 600 Hz. ABR measurements of octopus and cuttlefish showed upper limits of 1 and 1.5 kHz respectively (Hu et al., 2009 in Ref. [61]).

12.6 PASSIVE ACOUSTIC MONITORING

Section 12.5 showed the diversity of sounds produced by marine animals, and passive acoustic monitoring (PAM) can be used to monitor for the presence of certain animals [73,74]. The different approaches are now very well developed [75], and there is a large body of supporting literature, only a few of which can be presented here (the thousands of other references not presented can prove as interesting, and the present selection does not imply any ranking or priority). Further analyses, for example, looking at specific frequencies, can investigate the numbers of animals involved, for each species identified [76], thus quantifying an aspect of biodiversity. PAM can be used to understand specific behaviors, like echolocation clicks of dolphins hunting prey, or use of habitats by particular species [77]. In the most challenging applications, PAM can also monitor acoustic repertoires of particular species [78,79], paving the way to developing automated call detectors (e.g., C-PODs and T-PODs, see later for details). Section 12.5 summarized the broad sensitivity of marine animals to acoustic sounds. PAM can therefore also be used to monitor potential impacts of different types of noise on animal behavior and health [80−82].

The simplest embodiment of a PAM system consists in the deployment of a single hydrophone, deployed from the side of a boat or other platform [83]. Moored hydrophones can be deployed for longer periods, offering the benefit of higher stability and lesser susceptibility to movements induced by weather at the sea surface [84]. The recent and ongoing development of cabled seafloor observatories around the world also offers the possibility of continuous, very long-term, measurements [85]. C-PODs [77] and T-PODs [86] reduce these very large datasets to the desired animal vocalizations, for example, click repetitions. They have often been used in groups, either moored or freely drifting. Directional frequency analysis and recording (DIFAR) sonobuoys have also been used in several studies [87]. Using several hydrophones, at different locations or in arrays, allows the estimation of range and bearing of acoustic sources (Chapter 10). These sources can be animals naturally emitting sounds (e.g., whales or fish), other background sources (Chapter 6), or, in some cases, purpose-built transmitters tagged onto fish or other animals [88].

Metrics and full reporting of all parameters relevant to the analyses and interpretations of the measurements are paramount to PAM (as they are to other types of acoustic measurements, of course). Robinson et al. [89] summarize best practice for underwater noise measurements, and Merchant et al. [90] show how this can be used in PAM reporting. The conversion of raw pressure measurements (often as a voltage or a normalized value in a .wav audio file) to actual pressures has sometimes led to confusion (if some acquisition parameters were not known) and/or reporting of pressures or sound levels with arbitrary units (making intercomparison difficult). The open-source software provided by Merchant et al. [90] (as supplementary information on the publisher's website) addresses this gap, by clearly laying out all acquisition information necessary, and how to translate any measurements into actual pressures and sound levels. Industry and research efforts also routinely include PAMGUARD (www.pamguard.org), an open-source software infrastructure for PAM, with a strong emphasis on cetaceans.

Robinson et al. [89] recommend sound pressure level (SPL) as the best measure of both continuous and pulsed sounds. It is expressed as a ratio of the rms squared sound pressure $<p>$ (over a stated time interval) and the standard reference pressure $p_0 = 1$ µPa used in underwater acoustics:

$$SPL = 20 \log_{10} \left(\frac{\langle p \rangle}{p_0} \right) \qquad (12.11)$$

Its units are therefore dB re 1 µPa.

For pulsed sounds, the other recommended metrics are the peak SPL, the peak-to-peak SPL, the sound exposure level (SEL) for a single pulse, and the cumulative SEL for several pulses, stating clearly how many pulses are considered, how long this is considered for (times t_1 and t_2 in the integral), and the duty cycle of any sampling. Using the same reference pressure of 1 µPa, they can be expressed mathematically as:

$$SPL_{peak} = 20 \log_{10} \left(\frac{p_{peak}}{p_0} \right) \tag{12.12}$$

$$SPL_{pp} = 20 \log_{10} \left(\frac{p_{pp}}{p_0} \right) \tag{12.13}$$

$$SEL = 10 \log_{10} \left(\frac{\int_{t_1}^{t_2} p^2(t) dt}{1 \ \mu Pa^2 s} \right) \tag{12.14}$$

The frequency distribution of animal and background sounds can amount to large ranges (e.g., up to 200 kHz or beyond), and it is often not feasible to consider the measurements at a 1-Hz accuracy. Instead, third-octave and (sometimes and very rarely) twelfth-octave bands are used. Third-octave bands are defined (ANSI, 2009, in Ref. [90]) by their center frequencies f_c (referred to $f_0 = 1$ kHz) and their lower and upper bounds, respectively, f_- and f_+:

$$\begin{cases} f_c = f_0 \times 10^{\frac{i-1}{10}}, & i \in \mathbb{N} \text{ and } \begin{cases} i \geq 1 \text{ for } f_c \geq f_0 \\ i < 1 \text{ for } f_c < f_0 \end{cases} \\ f_+ = f_c \times 10^{\frac{1}{20}} \\ f_- = f_c \times 10^{-\frac{1}{20}} \end{cases} \tag{12.15}$$

Measurements over third-octave bands (or any frequency band) are expressed as spectral density levels: the amplitude values are divided by the frequency bandwidth, resulting in units of dB re 1 µPa^2/Hz (sometimes seen as dB re 1 µPa/Hz$^{1/2}$).

These different metrics can then be used to assess the contributions from the background (using the Wenz curves, seen in Chapter 6, for natural sources) and to compare different sources of noise (e.g., animals vs. shipping or other offshore activities). All sound levels measured in PAM systems are received levels (RL); some applications might require knowledge of the actual SLs, generally at a reference distance of 1 m. Other applications might require understanding of how RL at the range measured would change if measured at another location (e.g., assessing risks to marine mammals at different ranges from a loud source). SL can be obtained by measuring closer to the source, although this is not possible for larger sources (whose contributions cannot be reduced to a single point source at short ranges) or if the survey has already been completed. In most cases, one needs to back-propagate RLs to a generic SL, using the results from Chapters 2 and 3, with those of Chapter 7 as appropriate. Again, clear reporting of the parameters used in modeling acoustic propagation is essential for intercomparison of the final results. This modeling generally requires knowledge of other environmental conditions [91], including bathymetry and seabed types, and information about other loud sources in the vicinity, for example, ships monitored with automatic identification system (AIS) transponders.

12.7 SELECTED PRACTICAL APPLICATIONS
12.7.1 ACTIVE ACOUSTICS: FISH SURVEY

The most common application of fishery acoustics is the surveying of commercial fish stocks [9]. This requires measuring the abundance of the species of interest, locating the largest concentrations of fish and how they vary with seasons and weather, and to determine the age/sex/maturity distribution of these species, to assess commercial viability [92]. These aims have considerable challenges. Measurements of fish numbers is often estimated from surveying along lines, keeping in mind that fish are sensitive to noise and are reputed to swim away from ships. Not all seasons are suitable for surveys, meaning that variations in fish abundance with time and weather need to be accounted. And although the exact type of fish and their maturity/size can be estimated from their acoustic scattering (Section 12.3), these need to be calibrated with trawls or optical imaging.

There are many models of survey design, depending on platform(s) and instrument(s) available and on costs, which can quickly escalate. Optimization of the survey design can now make use of geographical information systems; a workshop convened by ICES in 2005 [93] published a decision tree to adapt a potential survey to measure the abundance of a single fish species[2]. Logistics can sometimes result in one survey being achieved with several vessels, potentially with different instruments. The acoustic measurements are then averaged along specific lengths of the survey track, called the elementary distance sampling unit (EDSU). It should be small enough to capture the key spatial elements of the species of interest, but large enough that intrinsic variability does not obscure statistical analyses. This results in typical EDSU values in the range 1–5 km, with extremes at 0.1 km (for dense schools in tight spaces like a fjord) and 9 km (for sparse deep-sea species) [92]. Water-column echoes from the fish species are isolated through different means (see Ref. [94] for a good illustration), and volume backscattering strengths S_V (Section 12.3) are integrated over depths z to produce a nautical area scattering coefficient (NASC), expressed in m^2/nmi^2:

$$NASC = 4\,\pi(1852)^2 \int_{Z_{min}}^{Z_{max}} 10^{S_V/10} dz \qquad (12.16)$$

[2]www.ices.dk/sites/pub/Publication%20Reports/Expert%20Group%20Report/ftc/2005/wksad05.pdf (Figure 11).

This can then be converted to estimated fish densities ρ by dividing by the expected backscattering cross-section $<\sigma_{bs}>$ (in m^2) of the species of interest:

$$\rho = \frac{NASC}{4\,\pi\langle\sigma_{bs}\rangle} \tag{12.17}$$

Multiplying by the survey area (in square nautical miles) should therefore give an estimate of the abundance.

These estimates need to be tempered with several factors. Although fish tend to avoid larger/noisier ships, recent research shows they do not avoid more silent survey vessels [48]. Gimona and Fernandes [95] discussed how to put error bars on sounder biomass estimates, given the spatial patchiness and other unknowns. Acoustic scattering can be difficult to relate to a single fish species, but multifrequency approaches [45] can resolve ambiguities, especially if associated with in situ sampling. D'Elia et al. [96] illustrated this potential by combining results from eight acoustic surveys in the Central Mediterranean, to estimate abundances of sardine, anchovy, horse mackerel, and other species. These surveys took place over 9 years, using different instruments working at 38, 120, and 200 kHz, some of which were not regularly calibrated. They were ground truthed with daytime pelagic trawls at specific times. Volume backscattering strengths SV and NASC values at 38 and 120 kHz were compared, but the three main species could not be distinguished from their acoustic properties alone (as their swim bladders are very similar in shape and size). Further processing with classification trees [94] and bathymetry parameters (e.g., school depths) improved discrimination between these groups to 85% overall. This thorough examination of multiple, carefully designed surveys shows the steps needed to better estimate fish abundance and compare surveys. Research in mid-2010s [97] investigated how to incorporate multiple sources of uncertainty in a stock assessment, validating their approach with ICES measurements of the Iberian hake stock.

12.7.2 PASSIVE ACOUSTICS: AMBIENT NOISE MONITORING

Accurate knowledge of underwater noise is increasingly required by regulators before approval of offshore projects, from construction of harbor extensions to installation of marine structures, seismic surveying, or new shipping patterns. Understanding baseline noise (before any change in human activity) and how it might evolve later forms a large part of the consenting. Any method that can reliably demonstrate these changes will therefore be welcome. A recent collaboration between the University of Bath and the University of Aberdeen showed how standard PAM could be integrated with automatic identification system (AIS) shipping data, time-lapse video, and meteorological and tide data. This methodology was first presented in Merchant [98], expanded in Pirotta et al. [99] and Merchant et al. [100].

The study site in the Moray Forth (Scotland) is a marine protected area (MPA), with a resident population of bottlenose dolphins (*Tursiops truncates*) and other protected marine mammal species, such as harbor seal (*Phoca vitulina*), harbor porpoise

(*Phocoena phocoena*), and gray seal (*Halichoerus grypus*). The entire area is expected to see developments in shipping density and the expansion of fabrication yards within the MPA, associated to the expansion of the Scottish marine renewable energy industry. Two sites were selected for closer investigation. They were both deep, narrow channels with steep gradients and strong tidal currents. Single PAM devices were deployed for 101 days overall, recording ambient noise at 384 kHz sampling rate, with a duty cycling of 1 min every 10 min. This was synchronized with a time resolution of the AIS data (also recorded every 10 min), provided by third-party network www.shipais.com. Time-lapse footage was recorded from shore, with PAM locations within the field of view of the digital cameras. This allowed identification of vessels with or without AIS, as it is compulsory only for vessels more than 300 GT (gross tons). In two instances, it also identified rigs being moored or towed past (by vessels using dynamic positioning, which produces loud broadband noise). By comparing AIS and acoustic data, it was possible to identify the closest point of approach from a particular vessel to the recording hydrophones, assigning peaks in underwater noise to these vessels (this was possible in about 65% of cases). PAM data was used to calculate 24-h SELs (total SEL, SELs associated to AIS-tracked vessels, and SELs from unidentified sources). Acoustic data (processed with PAMGuard) was also used to detect dolphin clicks and quantify population densities [99]). Additional data used in the analyses included local meteorological data about precipitation and wind speed every 5 min, from the open-access Weather Underground database (www.wunderground.com), and POLPRED tidal computations (at 10 min intervals, to match the acoustic duty cycling). A C-POD was also deployed independently at these two sites, providing additional information about dolphin activity.

The methodology (detailed in Merchant [98]) shows the interest of combining different sources of information, acoustic and nonacoustic. This was aided in this case by the proximity to shore, but recent developments in autonomous surface vehicle technology and station-keeping mean that similar information (from time-lapse footage to environmental conditions) can now be acquired further offshore. The main result of this study was that AIS-operating vessels accounted for the total cumulative sound exposure at 0.1–10 kHz, meaning that noise can be modeled using AIS data alone, and that expected noise levels can be calculated using planned increases in shipping patterns (in time and in space). The European Marine Strategy Framework Directive, European Commission [7], recommends the use of the third-octave bands centered on 63 and 125 Hz to assess shipping impacts on noise pollution. Merchant et al. found that the relationship between shipping and noise was stronger at 125 than at 63 Hz, possibly because of tidal flow noise or low-frequency propagation effects in the shallow waters, presented in Chapter 7. Pirotta et al. [99] also found a direct correlation between the presence of moving vessels and the buzzing activity of bottlenose dolphins, suggesting that dolphins perceive shipping as a clear risk as well as an acoustic masker of their own echolocation signals.

12.7.3 **ACOUSTIC TELEMETRY: FISH BEHAVIOR**

Acoustic tags have been used in many studies of fish behavior [101] and movements of whales [102], and the study selected here looks at the behavior of a relatively large number of fish for a long duration (close to 1 year). MPAs are set up to protect and restore overexploited fish populations. Fish movements occur across different spatial and temporal scales, and they must be understood to assess medium-to long-term MPA efficiency and assist in its management. Aspillaga et al. [103] looked at the white seabream (*Diplodus sargus*) in a northwest Mediterranean MPA. This is a commercially important fish, highly sedentary and known to shape rocky marine ecosystems. Aspillaga's study was one of the longest experiments conducted with this type of fish, and it provided a wealth of new, quantitative information.

MPA is located around the Medes Islands (NE Spain) and is divided into two zones: no-take, established in 1983 and shown in Fig. 12.18, surrounded by a partial reserve buffer, where very limited fishing is allowed. The seabed habitats are varied and white seabream is known to be more abundant in the no-take zone. Water depths vary between 25 and 65 m, and a particularly severe storm (waves up to 14.4 m high) occurred during the study. Forty-one individual fish were caught and tagged; the acoustic tags were very small and surgically inserted inside the animals (using the best ethical procedure available), and they transmitted at 153 dB.

The acoustic monitoring network comprised 27 receivers, most of which are shown in Fig. 12.18. Receivers were moored and placed 8 m below the sea surface: 17 of them within the no-take zone, covering the entire perimeter, and 10 receivers were placed on the mainland shore a few kilometers away. Signal range tests using an acoustic tag as transmitter showed signals could be detected with >90% probability as far as 150 m, after which this dropped to 50%.

During the entire monitoring period, 816,250 valid receptions were recorded by the array. Fish in the no-take zone could be tracked for long periods (329 ± 65 days). The residence index of each fish was calculated as the ratio of the number of days it was detected and the total number of monitoring days. All fish proved to be highly territorial, moving within small home ranges (<1 km^2), with an average residence index of 0.95 ± 0.0. They displayed repetitive diel activity patterns with 95% of all receptions corresponding to depths of 0.4−11 m below the sea surface. Extraordinary movements beyond the ordinary home range were observed for two occasions: (1) during stormy events, fish quickly sheltered to more protected places; (2) during the spawning season, they moved (up to 400 m away) and aggregated in deep areas (>50 m).

These results clearly show the potential of acoustic telemetry with a well-designed receiver array: individual animals could be tracked over long periods (close to a year) with high (submetric) horizontal and vertical accuracy. The data analysis methodology (fully presented in Aspillaga et al. [103] and supplementary material) provides a robust framework for understanding individual behaviors of any type of fish and population dynamics in baseline and extraordinary (e.g., storm) circumstances. These measurements can also later be used in models of how the effects

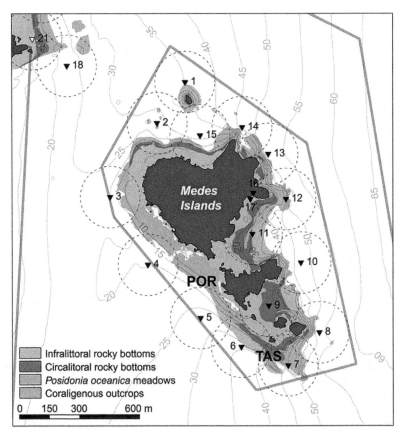

FIGURE 12.18

Marine protected area (MPA) and acoustic receiver array used by Aspillaga et al. [103] to study fish behavior with acoustic telemetry *(Reproduced under Creative Commons Attribution License.)*. The numbers refer to some of the 27 acoustic receivers (numbers not shown were positioned near the mainland, outside the no-take zone and around 1 km away).

of disturbance relate to animal condition/health, environmental variability, bioenergetics, vital rates, and reproductive success *(PCoD* model: *Population Consequences of Disturbance)*.

12.8 CONCLUSIONS: FUTURE DEVELOPMENTS

This chapter aimed to present, through simple concepts and examples, the two connected domains of fishery acoustics and bioacoustics. It started with a simple definition of the different types of marine life (Section 12.2), suitable for acousticians, and highlighting the main parameters of interest. Chronologically, the first question faced was how does one detect and monitor fish and other marine life? Section 12.3 looked at acoustic scattering by different animals, from zooplankton to whales, individually and in combination. The different instruments available to detect these animals were presented briefly in Section 12.4, giving only brief examples. How does marine life use sound? How sensitive are animals to other sounds, including man-made? Differences in physiology led to the separate examination of marine mammals, fish, turtles, and invertebrates in Section 12.5. Guidelines for protection of marine animals were introduced. Monitoring of underwater sound, as well as compliance of human activities with existing (and future) regulations are achieved with passive acoustic monitoring (PAM), presented in Section 12.6 Finally, an arbitrary choice of practical applications was presented in Section 12.7, showing the current trends in the use of fishery acoustics (Section 12.7.1), how traditional PAM can be augmented with suitably chosen additional measurements (Section 12.7.2), and how active acoustics (fish tags) and passive acoustics (widely scattered receivers) can be elegantly used to monitor fish behavior over large temporal and spatial scales while still maintaining a very high resolution (Section 12.7.3).

The extensive bibliography at the end of this chapter is but a short sample of the very high number of peer-reviewed publications, reports, and opinion pieces written about fishery acoustics and bioacoustics. Search engines indicate both subjects have seen a strong linear growth in publications over the last decades, associated with an exponential increase in citation rates, showing how dynamic these two fields are. It is impossible to do justice to all these interesting studies, but it is worth looking at some future developments, and the questions they pose as of 2016:

- Acoustic scattering by marine animals is better constrained, for a wide range of frequencies, but there is still significant work to do in benchmarking analytical and numerical models for different species, and variations within each species, as well as making these models computationally fast and approachable by nonspecialists.
- Target strengths have been measured for many animals in the wild and in the laboratory, but there are still many unknowns, from variations with aspect (e.g., for large animals or even some fish species) to simply any measurement for animals like crocodiles or sea otters.
- Auditory sensitivity of many animals still needs to be measured, in particular for invertebrates and sea turtles.
- Identifying scattering by animals close to the sea surface or to the seabed is still a challenge, and so is the identification of multiple species sharing the same space. Accurate tracking of targets of interest across acoustic beams and quantifying the risks of occlusion of one target by a closer one are also open processing problems (other bioacoustics issues are presented in Ref. [104]).
- New instruments are coming to the fore, from panoramic sonars to multiangle sonars, and coupled with the increased accessibility of autonomous platforms

like autonomous underwater vehicles [105], gliders, and autonomous surface vehicles, new approaches are likely to emerge within the next decade.
- PAM technology has matured but improvements (e.g., for towed arrays, source localization in complex environments, and transformation of RLs into SLs) are still needed.
- National and international policies are developing, with the development of marine protected areas (and similarly protected areas), regulations like the European NATURA-2000 or the Marine Strategy Framework Directive, calibration, operation, and analysis standards (e.g., ISO, MEDIN, and JNCC). How much will they be informed by current measurements, and how much will they shape future capabilities and research efforts?

REFERENCES

[1] R. Froese and D. Pauly (Eds.), FishBase. World Wide Web electronic publication. www.fishbase.org, version (10/2015), 2015.

[2] FAO (Food and Agriculture Organization of the United Nations), "The State of World Fisheries and Aquaculture (SOFIA) - 2014", Rome, 223 pp., 2014.

[3] IPCC; "Climate Change 2014: Impacts, Adaptation, and Vulnerability. Part B: Regional Aspects. Contribution of Working Group II to the Fifth Assessment Report of the Intergovernmental Panel on Climate Change", [V.R. Barros, C.B. Field, D.J. Dokken, M.D. Mastrandrea, K.J. Mach, T.E. Bilir, M. Chatterjee, K.L. Ebi, Y.O. Estrada, R.C. Genova, B. Girma, E.S. Kissel, A.N. Levy, S. MacCracken, P.R. Mastrandrea, and L.L. White (Eds.)]. Cambridge University Press, pp. 688, 2014.

[4] Clover, C., *The End of the Line: How Overfishing is Changing the World and What We Eat*, Ebury Press (UK), 2004.

[5] MCS (Marine Conservation Society), *Silent Seas*, 34 pp., also Available at: http://www.mcsuk.org/information.php/About+MCS/Silent+seas+report, 2008.

[6] McCauley, D.J., M.L. Pinsky, S.R. Palumbi, J.A. Estes, F.H. Joyce, Warner, R.R., *Marine defaunation: Animal loss in the global ocean*, Science **347** (6219):1255641: http://dx.doi.org/10.1126/science.1255641, 2015.

[7] European Commission, *Directive 2008/56/EC of the European Parliament and of the Council of 17 June 2008 establishing a framework for community action in the field of marine environmental policy (Marine Strategy Framework Directive)*, http://eur-lex.europa.eu/legal-content/EN/TXT/PDF/?uri=CELEX:32008L0056&from=EN, 2008.

[8] Medwin, H., Clay, C.S., *Fundamentals of Acoustic Oceanography*, 712 pp., Academic Press, 1998.

[9] Simmonds, E.J., and MacLennan, D.N. *Fisheries Acoustics: Theory and Practice, Second Edition,* Blackwell Publishing, Oxford. 456 pp., 2005.

[10] Lurton, X., *An Introduction to Underwater Acoustics: Principles and Applications*, 680 pp., Springer-Praxis, 2010.

[11] EC, 2016; *Good Environmental Status — Descriptor 4: Food Webs*, http://ec.europa.eu/environment/marine/good-environmental-status/descriptor-4/index_en.htm.

[12] Stanton, T.K., Chu, D., Reeder, D.B., *Non-Rayleigh acoustic scattering characteristics of individual fish and zooplankton*. IEEE J. Oceanic. Eng., **29**, pp. 260–268, 2004.

[13] Foote, K.G., *Target strength of fish*. In: Encyclopedia of Acoustics, M.J. Crocker (Ed.), Vol. 1, Chapter 44, John Wiley & Sons Ltd., pp. 493–500, 1997.

[14] Jech, J.M., Horne, J.K., Chu, D., Demer, D.A., Francis, D.T. I., Gorska, N., Jones, B., Lavery, A.C., Stanton, T.K., Macaulay, G.J., Reeder, D.B., Sawada, K., *Comparisons among ten models of acoustic backscattering used in aquatic ecosystem research*, J. Acoust. Soc. Am., **138** (6): pp. 3742–3764, 2015.

[15] Lavery, A.C., Wiebe, P.H., Stanton, T.K., Lawson, G.L., Benfield, M.C., and Copley, N., *Determining dominant scatterers of sound in mixed zooplankton populations*, J. Acoust. Soc. Am., **122** (6), pp. 3304–3326, 2007.

[16] "Zooplankton - MarineBio.org". MarineBio Conservation Society. Web. http://marinebio.org/oceans/zooplankton/.

[17] Stanton, T.K., *30 years of advance in active bioacoustics: a personal perspective*, Methods Oceanogr., **1–2**, pp. 49–77, 2012.

[18] Stanton, T.K., Chu, D., Wiebe, P.H., Eastwood, R.L., and Warren, J.D., *Acoustic scattering by benthic and planktonic shelled animals*, J. Acoust. Soc. Am., **108** (2), pp. 535–550, 2000.

[19] Macaulay, G.J., Hart, A.C., Grimes, P.J., Coombs, R., Barr, R., and Dunford, A.J., *Estimation of the target strength of oreo and associated species*, Final Research Report for Ministry of Fisheries Research Project OEO2000/01A Objective 1. http://fs.fish.govt.nz/Doc/22654/OEO2000-01A%20Oreo%20and %20 Associated %20species% 20Objective%201%20Final.pdf.ashx, 2002.

[20] Gødø, O.R., Patel, R., and Pedersen, G., *Diel migration and swimbladder resonance of small fish: some implications for analyses of multifrequency echo data.*, ICES J. Marine Sci., **66**, pp. 1143–1148, 2009.

[21] Sun Y, Nash R, Clay CS. *Acoustic measurements of the anatomy of fish at 220 kHz*. J. Acoust. Soc. Am., **78** (5), pp. 1772–1776, 1985.

[22] Gorska, N., Ona, E., and Korneliussen, R., *Acoustic backscattering by Atlantic mackerel as being representative of fish that lack a swimbladder. Backscattering by individual fish.*, ICES J. Marine Sci., **62** (5), pp. 984–995, 2005.

[23] Davy, C.M., Fenton, M.B., *Side-scan sonar enables rapid detection of aquatic reptiles in turbid lotic systems*, Eur. J. Wildl. Res., pp. 123–127, 2013.

[24] Mahfurdz, A., Sunardi, H. Ahmad, S. Abdullah, and Nazuki, S., *Distinguish sea turtle and fish using sound technique in designing acoustic deterrent device*, Telkomnika, **13** (4), pp. 1305–1311, 2015.

[25] Au, W.W.L., *Acoustic reflectivity of a dolphin*, J. Acoust. Soc. Am., **99** (6), pp. 3844–3848, 1996.

[26] Bernasconi, M., Patel, R., Nottestad, L., Pedersen, G., and Brierley, A.S., *Fin whale (Balaenoptera physalus) target strength measurements*, Marine Mamm. Sci., **29** (3), pp. 371–388, 2013.

[27] Lucifredi, I., and Stein, P.J., *Gray whale target strength measurements and the analysis of the backscattered response*, J. Acoust. Soc. Am., **121** (3), pp. 1383–1391, 2007.

[28] Love, R., *Target strengths of humpback whales (Megaptera novaeangliae)*, J. Acoust. Soc. Am., **54** (5), pp. 1312–1315, 1973.

[29] Xu, J., Deng, Z.D., Carlson, T.J., and Moore, B., *Target Strength of Southern Resident Killer Whales (Orcinus orca): Measurement and Modeling*, Marine Technol. Soc. J., **46** (2), pp. 74–84(11), March/April 2012.

[30] Bernasconi, M., Patel, R., Nottestad, L., Pedersen, G., and Brierley, A.S., *The effect of depth on the target strength of a humpback whale (Megaptera novanagliae)*, J. Acoust. Soc. Am., **134** (6), pp. 4316−4322, 2013.

[31] Soldevilla, M.S., McKenna, M.F., Wiggins, S.M., Shadwick, R.E., Cranford, T.W., and Hildebrand, J.A., *Cuvier's beaked whale (Ziphius cavirostris) head tissues: physical properties and CT imaging*, J. Exp. Biol., **208** (12), pp. 2319−2332, 2005.

[32] McKenna, M.F., Goldbogen, J.A., St. Leger, J., Hildebrand, J.A., and Cranford, T.W., *Evaluation of postmortem changes in tissue structure in the bottlenose dolphin (Tursiops truncates)*, Anat. Rec., **290** (8), pp. 1023−1032, 2007.

[33] Wei C, Wang Z, Song Z, et al., *Acoustic Property Reconstruction of a Neonate Yangtze Finless Porpoise's (Neophocaena asiaeorientalis) Head Based on CT Imaging*, B.G. Cooper (Ed.), PLoS One, **10** (4): e0121442: http://dx.doi.org/10.1371/journal.pone.0121442, 2015.

[34] Finneran, J.J., *Whole-lung resonance in a bottlenose dolphin (Tursiops truncates) and white whale (Delphinapterus leucas)*, J. Acoust. Soc. Am., **114** (1), pp. 529−535, 2013.

[35] Moore, M.J., Hammar, T., Arruda, J., Cramer, S., Dennison, S., Montie, E., and Fahlman, A., *Hyperbaric computed tomographic measurement of lung compression in seals and dolphins*, J. Exp. Biol., 214, pp. 2390−2397, 2011.

[36] Domokos, R., *Acoustic surveys study abundance and movements of bigeye tuna and their micronektonic forage on the Cross Seamount*, NOAA Pacific Islands Fisheries Science Center Quarterly Research Bulletin, https://pifsc-www.irc.noaa.gov/qrb/2009_2/eod2.php, March 2009.

[37] Raveau, M. and Feuillade, C., *Resonance scattering by fish schools: A comparison of two models*, J. Acoust. Soc. Am., **139** (1), pp. 163−175, 2016.

[38] Balls, R., *Herring fishing with the echometer*, J. Cons. Int. Explor. Mer, **15** (2), pp. 193−206, 1948.

[39] J.B., *Developments in the use of echo-sounding apparatus*, Nature, no. 2980, Vol. 118, p. 846−848, 1926.

[40] Fonteneau, A., Diouf, T., and Mensah, M., *Tuna Fisheries in the Eastern Tropical Atlantic*, FAO Corporate Document Repository, 2016; http://www.fao.org/docrep/005/t0081e/T0081E04.htm.

[41] FURUNO, *Furuno FISH FINDER Model FCV-1900 technical documentation*, http://www.furuno.com/en/products/fishfinder/FCV-1900#Screenshots, 2016.

[42] Foote, K., Knudsen, H., Vestnes, G., Maclennan, D., and Simmonds, E., *Calibration of acoustic instruments for fish density estimation: a practical guide*, ICES Cooperative Research Report No. CRR-144, 81 pp., ICES, 1987.

[43] Tomich, S.D., Hufnagle Jr., L.C., and Chu, D., *An automated calibration system for fisheries acoustic surveys, Sea Technology*, **53** (3), pp. 37−42, 2011.

[44] Demer, D.A., Berger, L., Bernasconi, M., Bethke, E., Boswell, K., Chu, D., Domokos, R. *et al.*, *Calibration of acoustic instruments*, ICES Cooperative Research Report No. 326. 133 pp., 2015.

[45] Korneliussen, R.J., Diner, N., Ona, E., Berger, L., and Fernandes, P.G., *Proposals for the collection of multifrequency acoustic data*, ICES J. Marine Sci., **65** (6), pp. 982−994, 2008.

[46] Anderson, C.I.K., Horne, J.K., and Boyle, J., Classifying multi-frequency fisheries acoustic data using a robust probabilistic technique, J. Acoust. Soc. Am. Express Letters, **121** (6), pp. EL230-EL237, 2007.

[47] Olsen, K., Angell, J., Pettersen, F., and Lovi, A.K., *Observed Fish Reactions to a Surveying Vessel with Special Reference to Herring, cod, Capelin and Polar co*d, Vol. 300, pp. 131−138, FAO Fisheries Report, 1983.

[48] Fernandes, P.G., Brierley, A.S., Simmonds, E.J., Millard, N.W., McPhail, F. Armstrong, S.D., Stevenson, P., and Squires, M., *Fish do not avoid survey vessels*, Nature, 404, pp. 35−36, 2000.

[49] Blondel, Ph., *Handbook of Sidescan Sonar*, Springer Praxis, 345 pp., ISBN:978-3-540-42641-7, 2009.

[50] Foote, K.G., Chu, D., Hammar, T.R., Baldwin, K.C., Mayer, L.A., Hufnagle Jr., L.C., and Jech, J.M., *Protocols for calibrating multibeam sonar*, J. Acoust. Soc. Am., **117** (1), pp.2013−2027, 2005.

[51] Parsons, M.J.G., Parnum, I.M., and McCauley, R.D., *Visualizing Samsonfish (Seriola hippos) with a Reson 7125 Seabat multibeam sonar*, ICES J. Mar. Sci., **70** (3), pp. 665−674, 2013.

[52] Colbo, K., Ross, T., Brown, C., and Weber, T., *A review of oceanographic applications of water column data from multibeam echosounders*, Est. Coast. Shelf Sci., **145**, pp. 41−56, 2014.

[53] Gurshin, C.W.D., Jech, J.M., Howell, W.H., Weber, T.C., and Mayer, L.A., *Measurements of acoustic backscatter and density of captive Atlantic cod with synchronized 300-kHz multibeam and 120-kHz split-beam echosounders*, ICES J. Marine Sci., **66**, pp. 1303−1309, 2009.

[54] Weber, T.C., Lutcavage, M.E., and Schroth-Miller, M.L., *Near resonance acoustic scattering from organized schools of juvenile Atlantic bluefin tuna (Thunnus thynnus)*, J. Acoust. Soc. Am., **133** (6), pp. 3802−3812, 2013.

[55] Melvin, G.D. and Cochrane, N.A., *Multibeam acoustic detection of fish and water column targets at high-flow sites*, Estuaries and Coasts, **38** (Suppl. 1), pp. 227−240, 2015.

[56] Scanmar, *Advanced catch systems for increased efficiency and financial gain*, report, http://www.scanmar.no/wp-content/uploads/2016/04/Scanmar-Info Booklet_2011-2012_English.pdf, 2012.

[57] De Robertis, A., *Small-scale spatial distribution of the euphausiid Euphausia pacifica and overlap with planktivorous fishes*. J. Plankton. Res., **24** (11), pp. 1207−1220, 2002.

[58] Nichol, S. and Brierley, A.S., *Through a glass less darkly - New approaches for studying the distribution, abundance and biology of Euphausiids*, Deep Sea Res. II, **57** (7−8), 496−507, 2010.

[59] Cox, M.J., Warren, J.D., Demer, D.A., Cutter, G.R., and Brierley, A.S., 2010. Three-dimensional observations of swarms of Antarctic krill (*Euphausia superba*) made using a multi-beam echosounder. Deep-Sea Res. II, **57** (7), pp. 508−518, 2010.

[60] Williamson, B.J., Blondel, Ph., Armstrong, E., Bell, P.S., Hall, C., Waggitt, J.J., and Scott, B.E., *A self-contained subsea platform for acoustic monitoring of the environment around Marine Renewable Energy Devices − Field deployments at wave and tidal energy sites in Orkney, Scotland*, IEEE J. Oceanic. Eng., **41** (1), pp. 67−81, 2015.

[61] De Ruiter, S., *Marine Animal Acoustics*, In: *An introduction to underwater acoustics: principles and applications*, X. Lurton (Ed.), Chapter 10, pp. 425−474, Springer-Praxis, 2010.

[62] Au, W.W.L., and Hastings M.C., *Principles of Marine Bioacoustics*: Springer, NY; 2008.

[63] Ghoul, A. and Reichmuth, C. *Sound production and reception in southern sea otters (Enhydra lutris nereis)*, In: *The Effects of Noise on Aquatic Life*, A.N. Popper and A. Hawkins (Eds.), Springer-Verlag, Berlin, pp. 157−159, 2012.

[64] Finfer, D.C., Leighton, T.G. and White, P.R., *Issues relating to the use of a 61.5 dB conversion factor when comparing airborne and underwater anthropogenic noise levels,* Appl. Acoust., **69** (5), pp. 464−471, 2008.

[65] Southall B. L., Bowles A. E., Ellison W.T., Finneran J. J., Gentry R. L., Greene C. R., Jr., et al., *Marine mammal noise exposure criteria: Initial scientific recommendations*, Aquat. Mamm., **33** (4), pp. 411−509, 2007.

[66] Radford, C.A., Ghazali, S., Jeffs, A.G., and Montgomery, J.C., *Vocalisations of the big-eye Pempheris adspersa: characteristics, source level and active space*, J. Exp. Biol., **218**, pp. 940−948, 2015.

[67] McCauley, R.D. and Cato, D.H., *Patterns of fish calling in a nearshore environment in the Great Barrier Reef*, Philos. Trans. R. Soc. Lond. B Biol. Sci., **355**, pp. 1289−1293, 2000.

[68] Kim, B.-N., Hahn, J., Choi, B.K., and Kin, B.-C., *Acoustic characteristics of pure snapping shrimp noise measured under laboratory conditions*, Proc. Symp. Ultrasonics Electronics, **30**, pp.167−168, 2009.

[69] Giles, J.C., Davis, J.A., McCauley, R.D., and Kuchling, G., *Voice of the turtle: The underwater acoustic repertoire of the long-necked freshwater turtle, Chelodina oblonga*, J. Acoust. Soc. Am., **126** (1), pp. 434−443, 2009.

[70] Ferrara, C.R., Vogt, R.C., and Sousa-Luna, R.S., *Turtle vocalizations as the first evidence of posthatching parental care in Chelonians*, J. Comp. Physcol., **127** (1), pp. 24−32, 2013.

[71] Popper, A.N., Hawkins, A.D., Fay, R., Mann, D., Bartol, S., Carlson, T., Coombs, S., Ellison, W.T., Gentry, R., Halvorsen, M.B., Lokkeborg, S., Rogers, P., Southall, B.L., Zeddies, D.G., and Tavolga, W.N., *ASA S3/SC1.4 TR-2014 Sound Exposure Guidelines for Fishes and Sea Turtles: A Technical Report prepared by ANSI-Accredited Standards Committee S3/SC1 and registered with ANSI*, SpringerBriefs in Oceanography, Springer, 76 pp., 2014.

[72] Roberts, L., Cheesman, S., Elliott, M., and Breithaupt, T., *Sensitivity of macroinvertebrates to substrate borne vibration*, Proc. Inst. Acoust., **37** (1), pp. 234−241, 2015.

[73] Rountree, R.A., Gilmore, R.G., Goudey, C.A., Hawkins, A.D., Luczkovich, J.J., and Mann, D.A., *Listening to Fish*, Fisheries, **31** (9), pp. 433−446, 2006.

[74] Mellinger, D.K., Stafford, K.M., Moore, S.E., Dziak, R.P., and Matsumoto, H., *An overview of fixed passive acoustic observation methods for cetaceans*, Oceanography, **20** (4), pp. 36−45, 2007.

[75] Todd, V.L.G., Todd, I.B., Gardner, J.C., and Morrin, C.N., *Marine mammal observer and Passive Acoustic Monitoring handbook*, Pelagic Publishing: Exeter (UK), 432 pp., 2015.

[76] Harris, S.A., Shears, N.T., and Radford, C.A., *Ecoacoustic indices as proxies for biodiversity on temperate reefs*, Methods Ecol. Evol., **7** (6), pp. 713−724, 2016.

[77] Benjamins, S., Dale, A., van Geel, N., and Wilson, B., *Riding the tide: use of a moving tidal-stream habitat by harbour porpoises*, Mar. Ecol. Prog. Ser., **549**, pp. 275−288, 2016.

[78] Sattar, F., Cullis-Suzuki, S., and Jin, F., *Identification of fish vocalizations from ocean acoustic data*, Appl. Acoust., **110**, pp. 248−255, 2016.

[79] Webster, T.A., Dawson, S.M., Rayment, W.J., Parks, S.E., and Van Parijs, S.M., *Quantitative analysis of the acoustic repertoire of southern right whales in New Zealand*, J. Acoust. Soc. Am., **140** (1), pp. 322−333, 2016.

[80] Hedgeland, D., Pierpoint, C., Wyatt, R., Rypdal, C., and Gubin, D., *An industry perspective of using towed Passive Acoustic Monitoring (PAM) for the detection of marine mammals at sea during seismic surveys*. Society of Petroleum Engineers. International Conference on Health, Safety and Environment in Oil and Gas Exploration and Production, April 12−14, 2010.

[81] Simpson, S., Thompson, P., Blondel, Ph. and others; *Biological Effects of Marine Noise*, a report on a Workshop supported by the Strategic Ocean Funding Initiative (SOFI), the Natural Environment Research Council's Marine Renewable Energy Knowledge Exchange Programme (NERC), and the Underwater Sound Forum (USF), 25 pp., Bristol − Feb. 2012 (http://www.nerc.ac.uk/innovation/activities/infrastructure/offshore/sofi-mreke-workshop/).

[82] A.N. Popper, and A. Hawkins (Eds.), *The Effects of Noise on Aquatic Life II*, Advances in Experimental Medicine and Biology 875, Springer, 1292 pp., 2016.

[83] Wladichuk, J.L, Megill, W.M., and Blondel, Ph., *Passive biosonar: Ambient acoustics in nearshore navigation by grey whales*, Proc. 3rd Underwater Acoustic Measurements Conference, Kos (Greece), p. 1333−1340, 2011.

[84] Merchant, N.D., Witt, M.J., Blondel, Ph., Godley, B.J., and Smith, G.J., *Assessing sound exposure from shipping in coastal waters using a single hydrophone and Automatic Identification System data*, Mar. Pollut. Bull., **64** (7), pp. 1320−1329, 2012.

[85] Merchant, N.D., Blondel, Ph., Dakin, D.T., and Dorocicz, J., *Averaging underwater noise levels for environmental assessment of shipping*, J. Acoust. Soc. Am., **132** (4), pp. EL343-EL349, http://dx.doi.org/10.1121/1.4754429, 2012.

[86] Todd, V.L.G., Pearse, W.D., Tregenza, N.C., Lepper, P.A., and Todd, I.B., *Diel echolocation activity of harbour porpoises (Phocoena phocoena) around North Sea offshore gas installations*, ICES J. Marine Sci., **66**, pp. 734−745, 2009.

[87] Rivers, J., *Blue whale, Balaenoptera musculus, vocalizations from the waters off central California*, Mar. Mamm. Sci. **13**, pp. 186−195, 1997.

[88] Wearmouth, V.J., Southall, E.J., Morritt, D., and Sims, D.W., *Identifying reproductive events using archival tags: egg-laying behaviour of the small spotted catshark Scyliorhinus canicula*, J. Fish Biol., **82**, pp. 96−110, 2013.

[89] Robinson, S.P., Lepper, P.A., and Hazelwood, R.A., *Good Practice Guide for Underwater Noise Measurement*, National Measurement Office, Marine Scotland, The Crown Estate, NPL Good Practice Guide No. 133, ISSN:1368-6550, 97 pp., 2014.

[90] Merchant, N.D., Fristrup, K.M., Johnson, M.P., Tyack, P.L., Witt, M.J., Blondel, Ph., and Parks, S.E., *Measuring acoustic habitats*, Methods Ecol. Evol., **6** (3), pp. 257−265, 2015.

[91] Farcas A, Thompson P.M., and Merchant N.D., *Underwater noise modelling for environmental impact assessment*, Environ. Impact Assess. Rev., **57**, pp. 114−22, 2016.

[92] Fernandes, P.G., *Fisheries Acoustics − Lecture 3: Abundance estimation*, 23 pp., lecture notes, Seiche Course in Underwater Acoustics, 2013.

[93] ICES. 2005. Report of the Workshop on Survey Design and Data Analysis (WKSAD), May 9−13, 2005, Sète, France. ICES CM 2005/B:07, 170 pp., 2005.

[94] Fernandes, P.G., *Classification trees for species identification of fish-school echotraces*, ICES J. Marine Sci., **66**, pp. 1073−1080, 2009.

[95]   Gimona, A., and Fernandes, P.G., *A conditional simulation of acoustic survey data: advantages and potential pitfalls*, Aquat. Living Resour., **16**, pp. 123−129, 2003.

[96]   D'Elia, M., Patti, B., Bonanno, A., Fontana, I., Giacalone, G., Basilone, G., and Fernandes, P.G., *Analysis of backscatter properties and application of classification procedures for the identification of small pelagic fish species in the Central Mediterranean*, Fisheries Research, **149**, pp. 33−42, 2014.

[97]   Scott, F., Jardim, E., Millar, C.P., and Cerviño, S., *An Applied Framework for Incorporating Multiple Sources of Uncertainty in Fisheries Stock Assessments*, PLoS One, **11** (5), 2016.

[98]   Merchant, N.D., *Measuring underwater noise exposure from shipping*, PhD thesis, University of Bath (UK), 147 pp., 2013.

[99]   Pirotta, E., Merchant, N.D., Thompson, P.M., Barton, T.R., and Lusseau, D., *Quantifying the effect of boat disturbance on bottlenose dolphin foraging activity*, Bio. Con., **181**, pp. 82−89, 2015.

[100] Merchant, N.D., Pirotta, E., Barton, T.R., and Thompson, P.M., *Soundscape and noise exposure monitoring in a Marine Protected Area using shipping data and time-lapse footage*, In: *The Effect of Noise on Aquatic Life II*, Adv. Exp. Med. Bio., **875**, pp. 705−712, 2016.

[101] Johnson, M., Aguilar de Soto, N., and Madsen, P.T., *Studying the behaviour and sensory ecology of marine mammals using acoustic recording tags: a review*, Mar. Ecol. Prog., **395**, pp. 55−73, 2009.

[102] Madsen, P.T., Johnson, M., Miller, P.J.O., Aguilar Soto, N., Lynch, J. and Tyack, P., *Quantitative measures of air-gun pulses recorded on sperm whales (Physeter macrocephalus) using acoustic tags during controlled exposure experiments*, J. Acoust. Soc. Am., **120** (4), pp. 2366−2379, 2006.

[103] Aspillaga E., Bartumeus F., Linares C., Starr R.M., and López-Sanz À., et al., *Ordinary and Extraordinary Movement Behaviour of Small Resident Fish within a Mediterranean Marine Protected Area*. PLoS One, **11** (7), 2016.

[104] Dell, A.I., Bender, J.A., Branson, K., Couzin, I.D., de Polavieja, G.G., Noldus, L.P.J.J., et al., *Automated image-based tracking and its application in ecology*, Trends Ecol. Evol., **29**, pp. 417−428, 2014.

[105] Fernandes, P.G., Stevenson, P., Brierley, A.S., Armstrong, F., and Simmonds, E.J., *Autonomous underwater vehicles: future platforms for fisheries acoustics*, ICES J. Mar. Sci., **60** (3), pp. 684−691, 2003.

Finite-Amplitude Waves

13

L. Bjørnø[†]

UltraTech Holding, Taastrup, Denmark

Definitions

The world we live in is *nonlinear*. Relations between characteristic parameters like pressure, density, and temperature in fluids and relations between material constants in solids are nonlinear. Since Robert Hooke (1635–1703) in 1660 put forward his linear elastic relation (named Hooke's law) between force and deformation of solids, attempts have been made to "linearize" the world. The driving force behind the linearization attempts has in particular been lack of fundamental understanding of nonlinearity concepts and lack of tools to handle nonlinear problems. Only the strong development in computer technology, with faster and more powerful computers over the recent 40–50 years, has given access to understand and to exploit nonlinear phenomena. One of these phenomena is *nonlinear acoustics*.

As shown in Chapter 10, the maximum power radiated by a sonar (sound navigation and ranging) system is directly proportional to its piezoceramic volume. Moreover, the stored electric energy in the piezoceramic materials is limited by factors like electric insulation breakdown, ceramic depolarization, and efficiency deterioration caused by increased dissipation at high electric fields. Too large a driving voltage on the piezoceramic materials will make them to respond nonlinearly. This means that the waveform of the pressure signal radiated from the sonar will not be a replica of the driving electrical voltage. The material nonlinearity in the sonar system will most frequently, for high acoustic pressure amplitudes radiated into the water, be accompanied by a series of nonlinear effects along the propagation path for the acoustic signals. Some nonlinear effects may limit the applications of sonar for desired purposes. However, they can also be exploited to enhance the sonar acoustic effects in finite-amplitude waves in water.

Section 13.1 examines nonlinear acoustic phenomena and the associated physics. Section 13.2 focuses on nonlinear underwater acoustics including parametric acoustic arrays. This is followed in Section 13.3 with a discussion of underwater explosions including other sources of high-intensity sound. Section 13.4 provides a list of symbols and abbreviations and the page of first usage for each, and the references for this chapter.

[†]30 March 1937–24 October 2015.

Applied Underwater Acoustics. http://dx.doi.org/10.1016/B978-0-12-811240-3.00013-8

13.1 PHYSICS AND NONLINEAR PHENOMENA

The nonlinear acoustic effects include *generation of harmonics* of an originally sinusoidal signal during its propagation, limitations in the acoustic pressure amplitude to be transmitted over a given distance due to *acoustic saturation*, and other nonlinear effects that may be found in *focused acoustic fields*, which, for instance, can be produced by the sound velocity profile in the water column. Creation of *cavitation* at positions of intense sound, most frequently near the sonar system, also involves nonlinear acoustic effects. The formation of sum and difference sound frequencies when two intense, coexistent sinusoidal signals interact with each other during their propagation in water, forming a so-called *parametric acoustic array* with some advantageous beam patterns, constitutes an exploitation of nonlinear acoustic effects. The jet formation and circulation in the water near an intense sound source, a phenomenon called *acoustic streaming*, and the creation of forces on objects in water by the so-called *acoustic radiation pressure* are also nonlinear acoustic effects.

13.1.1 HARMONIC DISTORTION

The fundamental equations of nonlinear acoustics may be derived from the fundamental equations of fluid mechanics. These equations are an *equation of continuity*, three *equations of motions*, and an *equation of energy*. Moreover, a constitutive equation, the *equation of state*, for the fluid is necessary to describe all types of motion in fluids, see Bjørnø [1]. These six fundamental equations can be reduced to one single, nonlinear fundamental equation, the Burger's equation [2], which describes the propagation of finite-amplitude sound waves in a viscous and heat conducting fluid. The dimensionless Burger's equation is given by:

$$\frac{\partial V}{\partial \sigma} - V \frac{\partial V}{\partial y} = \Gamma^{-1} \frac{\partial^2 V}{\partial y^2} \tag{13.1}$$

where Γ describes the ratio of the influence of nonlinearity to the influence of dissipation. Γ, called *the Gol'dberg number*, was first introduced by Gol'dberg [3] as a criterion such that shock formation should not be likely to take place if $\Gamma < 1$. Γ is expressed through the relation:

$$\Gamma = \left(\frac{B}{A} + 2\right) \frac{u_0 \rho_0 x_0}{b} = \left(\frac{B}{A} + 2\right) Re_a \tag{13.2}$$

where $Re_a = u_o x_c/(b/\rho_o)$ constitutes an *acoustic Reynolds number* analogous to the hydrodynamic Reynolds number. b {unit: kg/(ms)} represents the viscosity and heat conduction effects through the relation: $b = 4\,\mu/3 + \zeta + (1/c_v + 1/c_p)$, where μ and ζ are the shear viscosity and the bulk viscosity {unit: km/(ms)}, respectively, while c_v and c_p are the specific heats at constant volume and at constant pressure, respectively.

In Eq. (13.1) the dimensionless ratio $V = u/u_o$ is the ratio between the local particle velocity u {unit: m/s} in the finite-amplitude sound wave and the particle velocity amplitude u_o {unit: m/s}. $\sigma = (B/2A + 1)M\,x/x_c$ is a dimensionless propagation distance parameter with the one-dimensional distance x {unit: m}, $x = 0$ at the finite-amplitude wave source, and the characteristic distance $x_c = \lambda/2\pi$, where λ {unit: m} represents the wavelength in an originally sinusoidal wave of finite amplitude. $y = c_o(t - x/c_o)/x_c$ is the time relation with t {unit: s} as the time and with c_o {unit: m/s} as the local sound velocity. $M = u_o/c_o$ represents the *acoustic Mach number* on a par with the hydrodynamic Mach number.

The dimensionless ratio B/A is the second-order *nonlinearity ratio*, an important material constant in nonlinear acoustics. This ratio is defined through a Taylor series expansion of the equation of state for the fluid for constant fluid density ρ {unit: kg/m^3} at $\rho = \rho_0$, and by taking into account only terms up to second order [1]. B/A can be expressed by:

$$\frac{B}{A} = 2\rho_0 c_0 \left(\frac{\partial c}{\partial p}\right)_T + \frac{2c_0 T\beta}{c_p}\left(\frac{\partial c}{\partial T}\right)_p \tag{13.3}$$

where p {unit: N/m^2} is the local pressure in a sound wave in the fluid and T {unit: K} is the absolute temperature in the fluid. β denotes the isobaric compressibility {unit: K^{-1}}. Values of B/A for water at different temperatures and pressure are given in Table 13.1, while Table 13.2 gives B/A values for seawater at various salinities and temperatures at $p = 10^5$ N/m^2.

A solution to Eq. (13.1) according to Mendousse [6] can be expressed by:

$$V(\sigma, y) = \frac{2}{\Gamma}\frac{\partial\left(\text{Log}\left[\sum_{n=0}^{\infty} e_n(-1)^n I_n\left(\frac{\Gamma}{2}\right)e^{-n^2\frac{\alpha}{\Gamma}}\cos(ny)\right]\right)}{\partial y} \tag{13.4}$$

where the Neumann factor $e_n = 1$ for $n = 0$, and $e_n = 2$ for $n \geq 1$. I_n is the Bessel function with an imaginary argument.

Table 13.1 Values of B/A for Freshwater Without Gas Bubbles as a Function of Temperature and Pressure (1 bar $= 10^5$ N/m^2)

Pressure (bar)	Temperature					
	0°C	30°C	40°C	50°C	60°C	80°C
1	4.08	5.21	5.49	5.55	5.61	5.74
250	4.90	5.43	5.59	5.62	5.70	5.79
500	5.58	5.63	5.69	5.69	5.75	5.84
1000	6.35	5.83	5.84	5.80	5.83	5.86
2000	6.78	6.08	6.00	5.93	5.88	5.82
5000	6.44	6.14	6.02	5.93	5.77	5.58
10,000	–	–	–	5.36	5.52	5.12

*Reproduced from Hagelberg, M.P., Holton, G., and Kao, S., Calculation of B/A for Water from Measurements of Ultrasonic Velocity versus Temperature and Pressure to 10000 kg/cm², J. Acoust. Soc. Amer., **41**, (3), pp. 564–567, 1967, with the permission of the Acoustical Society of America.*

Table 13.2 Values of B/A for Seawater at Various Salinities and
Temperatures at 1 atm

Salinity (%)	Temperature			
	0°C	10°C	20°C	30°C
33	4.89	5.06	5.21	5.37
35	4.92	5.09	5.25	5.41
37	4.96	5.12	5.27	5.42

*Reproduced from Coppens, A.B., Beyer, R.T., Seiden, M.B., Donohue, J., Guepin, F., Hodson, R.H.,
and Townsend, C., Parameter of Nonlinearity in Fluids. II, J. Acoust. Soc. Amer., **38**, (5), pp. 797–804,
1965, with the permission of the Acoustical Society of America.*

For $\Gamma \gg 1$, the distortion of a finite-amplitude, originally sinusoidal, plane
acoustic wave during propagation through a dissipative fluid can be found from
expression (13.4) and is shown in Fig. 13.1. Fig. 13.1 shows the change of the shape
of a single period of the finite-amplitude plane wave during propagation from the
source at $\sigma = 0$ and through different dimensionless distances σ. In the region
between $\sigma = 0$ and $\sigma = 1$ nonlinear effects are stronger than dissipative effects

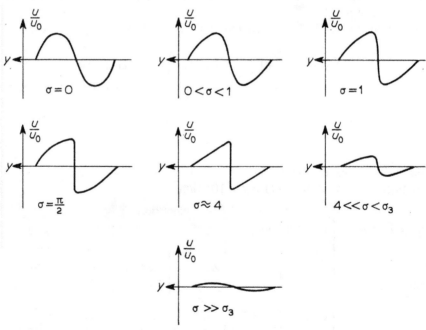

FIGURE 13.1

The distortion of a single period of an originally plane sinusoidal, finite-amplitude wave
during its propagation in a thermoviscous fluid from the source at the dimensionless
distance σ.

Reprinted from Physics Procedia, **3**, (2010), Leif Bjørnø, Introduction to Nonlinear Acoustics, page 9, Copyright
(2010), with permission from Elsevier. http://www.elsevier.com with permission of Elsevier.

and the wave front steepness increases until at $\sigma = 1$ the wave reaches its maximum steepness at the zero crossing. This distance is frequently called the *discontinuity distance*. Only the zero crossing will propagate with the infinitesimal velocity of sound, c_o, while the local phase velocity in the wave will receive contributions from convective (nonlinear) terms of the equation of motion and from effects caused by the nonlinearity of the equation of state of the fluid. The local phase velocity in the compressional part of the wave will be the sum of the local sound velocity, c, and the local particle velocity, u, while in the rarefaction part the local phase velocity will be the difference between these two quantities.

For increasing values of σ the dissipative effects will grow relative to the nonlinear effects and a "sawtooth" wave shape may appear. Due to energy dissipation the wave profile will gradually lose its steepness, and the wave amplitude, and therefore, the nonlinear influence on the wave shape, will be reduced. For values of $\sigma \gg \sigma_3 \sim 0.6\Gamma$, as discussed in Ref. [1], the wave profile returns to its original sinusoidal shape with a strongly reduced amplitude. This forms the so-called "old age region" of the finite-amplitude wave. A Fourier analysis of the wave shape transformation in Fig. 13.1 shows that *higher harmonics* of the original sinusoidal finite-amplitude wave will be formed during the wave shape propagation. Since absorption in most fluids is proportional to the square of the frequency, the transfer of acoustic energy from the fundamental frequency to higher harmonics increases the dissipation of energy in the finite-amplitude wave. The stronger dissipation at the higher amplitudes may also lead to sound beam broadening when the sound amplitude on the acoustic axis is attenuated more than the side lobes.

For sufficiently high values of the original wave pressure amplitude the later amplitude of the distorted, originally sinusoidal wave becomes relatively independent of the initial pressure amplitude. This effect, termed *acoustic saturation*, forms an upper limit for the sound pressure amplitude transmitted over a given distance. Therefore, even for an increase in the finite-amplitude wave amplitude at the source, the same amplitude will be received after wave propagation over a given distance as more and more energy goes into the higher harmonics with their higher rate of dissipation. Acoustic saturation can form an important factor limiting the propagation distance for sonar signals.

13.1.2 FOCUSED SOUND FIELDS

The sound velocity gradients in seawater may lead to refraction of sound beams as discussed in Chapter 2. The refraction of the sound beams may lead to *focused sound fields*, where increases in the acoustic amplitude can take place. The increased amplitude can lead to finite-amplitude effects with formation of higher harmonics and to *excess attenuation*, i.e., attenuation above the linear attenuation of sound waves, as discussed in Ref. [1].

Focused sound fields in seawater may comprise influence of nonlinearity, diffraction, and attenuation. An equation which covers all three effects is the parabolic, *Khokhlov−Zabolotskaya−Kuznetzov (KZK)* equation, which has been used extensively for studies of focused, finite-amplitude, ultrasonic fields in medicine as, for instance, the field in and around the focal point formed by a therapeutic body stone disintegrator, the Lithotripter, as discussed by Neighbors et al. [8,9]. The KZK equation in dimensionless form can be written as:

$$\left[4\frac{\partial^2}{\partial\tau\partial\sigma} - \nabla_\perp^2 - 4\alpha r_0\frac{\partial^3}{\partial\tau^3}\right]P = 2\frac{r_0}{\ell_d}\frac{\partial^2 P^2}{\partial\tau^2} \qquad (13.5)$$

where $P = p/p_o$ is the normalized pressure in the finite-amplitude wave with p_o expressing the pressure at the wave source. $\sigma = x/r_o$, where x is the on-axis distance from the source and r_o ($=\omega a^2/2c_o$) is the Rayleigh distance for a monochromatic source with radius a. $\nabla_\perp^2$ is the two-dimensional transverse Laplacian. $\tau = \omega$ $(t - z/c_o)$ is a retarded time, where ω and t denote the angular frequency {unit: rad/s} and time, respectively. α is the attenuation coefficient, while c_o and ρ_o are the isentropic speed of sound and the fluid density, respectively. ℓ_D ($=\rho_o c_o^3/(B/2A + 1)\omega p_o$) is the discontinuity distance, originally being characterized as the source distance for the first formation of a shock wave in a lossless fluid.

Originally, Eq. (13.5) was derived without thermoviscous loss, e.g., $\alpha = 0$, by Zabolotskaya and Khokhlov [10], while the inclusion of dissipation was developed by Kuznetsov [11]. Eq. (13.5) accounts for nonlinearity, diffraction, and dissipation to an equal order of magnitude, where the 2-D transverse Laplacian term accounts for the diffraction and the term $4\alpha r_0\partial^3/\partial\tau^3 P$ incorporates thermoviscous losses. In the derivation of Eq. (13.5) a well-collimated beam is assumed, i.e., $ka \gg 1$, with $k = \omega/c_o$.

When P is expanded as a Fourier series and inserted into Eq. (13.5), the KZK equation becomes a series of coupled partial differential equations, the numerical solution of which follows a procedure initially developed by Aanonsen et al. [12] and enhanced by Hart and Hamilton [13] for application to focused sound fields. An example of the solution to Eq. (13.5) is given in Fig. 13.2, where the distortion

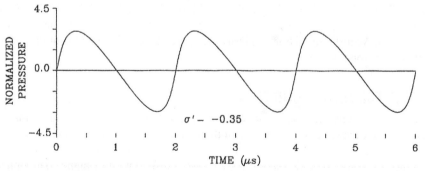

FIGURE 13.2

The time history for the distortion of a focused, originally sinusoidal, finite-amplitude wave during its propagation in a thermoviscous fluid toward the geometrical focal point situated at $\sigma' = 0$. The wave source is situated at $\sigma' = -1$. The figure shows the increase in the wave amplitude due to focusing effects and the wave front steepening due to nonlinear effects during the wave propagation.

*Reproduced from Neighbors, T.H. and Bjørnø, L., Monochromatic Focused Sound Fields in Biological Media.
J. Lithotripsy Stone Dis., **2**, (1), pp. 4–16, 1990 with permission of Futura Publishing Company.*

propagation is calculated for the focusing of an originally sinusoidal, finite-amplitude wave propagating in a thermoviscous fluid. The wave source is situated at the dimensionless distance $\sigma' = -1$, while the geometrical focal point is situated at $\sigma' = 0$. Calculations show that the diffraction effects in the focused wave contribute to an increase in the wave amplitude of the compression part of the wave, while it reduces the amplitude of the rarefaction part. The increasing amplitude, and due to it, the nonlinear effects steepen the waves. Just before the geometrical focal point the wave amplitudes will achieve their maximum value and the increasing dissipative effects will influence the nonlinear process in such a way, that the geometrical focal point will not become the venue for the maximum focused wave amplitudes.

13.1.3 CAVITATION

As mentioned in Chapter 10 on sonar systems, the maximum acoustic output power from a projector is limited by a number of factors introducing electrical, mechanical, and thermal limitations. However, an additional limiting factor is introduced by *cavitation* in the water around and near the projector. At shallow depths this factor may impose more serious limitations on the output power from a projector than other factors.

The alternation acoustic pressure in an acoustic signal produced at the projector face is superimposed on the ambient pressure of the water. In the rarefaction half cycle of the emitted pressure signal, the absolute pressure is reduced below the ambient pressure. With increasing acoustic intensity in the emitted signal, the absolute pressure in the rarefaction half cycle may be reduced to zero or become negative. Dissolved gases in the water, frequently connected with micro particles or existing as free microbubbles, will form nuclei for water rupturing into a large number of frequently visible bubbles, which vibrate and collapse. Cavitation is the overall name for the bubble formation and the vibration and collapse of the bubbles. The collapse phase of the bubbles, where tiny water jets with speeds exceeding some hundreds of meters per second are formed, may lead to erosion effects on the projector face. The same erosion effects from cavitation are frequently found on part of ship propellers which may lead to propulsion system imbalances.

Apart from projector mechanical erosion effects cavitation may also affect acoustic transmission. The formation of a large number of gas bubbles close to the projector face acts as a pressure release surface between the projector and the water. This pressure release surface will strongly reduce the radiation impedance seen from the projector, limit the radiated acoustic power from the projector, and distort the radiated acoustic beam pattern. Increased power transfer to the projector after cavitation formation will only lead to increased internal losses, such as heat formation in the projector. Moreover, due to strongly increased nonlinear acoustic qualities of bubble clouds, see Bjørnø [14], the formation of higher harmonics in the acoustic signal transmitted by the projector may take place and cavitation will also produce a substantial amount of noise which will interfere with the signal transmitted by the projector.

The cavitation threshold, i.e., the acoustic pressure in the signal at cavitation inception, is a function of several factors, such as the amount of gas dissolved in the water, the number of microparticles, the number of microbubbles and their sizes, pollution, temperature, viscosity, and the ambient pressure and thus the depth where the projector is operating. Also, the signal frequency and the duty cycle have a strong influence on the cavitation threshold.

At the sea level, the absolute pressure is reduced to zero if the peak pressure, p_o, in the acoustic signal is 10^5 Pa. If this pressure is used for definition of the acoustic intensity leading to cavitation inception at the sea level, the theoretical cavitation threshold, T, can be defined as:

$$T = \frac{p_0^2}{2\rho c} = \frac{\left(10^5\right)^2}{2 \cdot 1.5 \cdot 10^6} = 0.33 \cdot 10^4 \frac{W}{m^2}$$

which means that the maximum power transmitted by a plane, circular projector face with a diameter of 0.1 m at the sea level to avoid cavitation is below 26 W. Since T is proportional to the ambient pressure squared, the maximum power is transmitted by the projector before cavitation inception increases rapidly with projector depth. For a 30 m depth, this projector will be able to transmit up to 415 W before cavitation inception, unless other factors, such as the number and size of microbubbles influence the cavitation threshold. The theoretical cavitation threshold, T_{dB} {unit dB rel 1 µPa} at the sea level is:

$$T_{dB} = 10 \log\left(\frac{T}{I_0}\right) = 217 \text{ dB rel 1 µPa}$$

when $I_o = 6.5 \cdot 10^{-19}$ W/m² for $p_o = 1$ µPa.

If the totally radiated acoustic power is, P {unit: W}, at the start of the cavitation, which is assumed to be uniformly distributed over a projector's radiating surface area, A {unit: m²}, the relation between P and projector depth, d {unit: m} for the cavitation threshold T is:

$$P = AT\left(1 + \frac{d}{10}\right)^2 \tag{13.6}$$

Due to other factors besides the ambient pressure, such as dissolved gases and the microbubbles number and size, accurate cavitation threshold prediction in seawater is difficult and the scattering of the measured threshold values is substantial. Also, Eq. (13.6) is only valid for low frequencies and long pulses. This is due to the fact that the duration of the rarefaction phase of the signal cycles, controlled by the signal frequency, influences the cavitation threshold since the bubble formation process during cavitation requires a finite time, which also is influenced by the amount of dissolved gas and other factors. Higher frequencies, with a shorter rarefaction phase duration, increase the cavitation threshold. Fig. 13.3 shows the cavitation threshold as a function of the acoustic frequency. As discussed by Urick [15], the curves in Fig. 13.3 are based on measurements published by Esche [16], Strasberg [17], and Flynn [18]. Fig. 13.3 shows that below 10 kHz the cavitation threshold is low, while

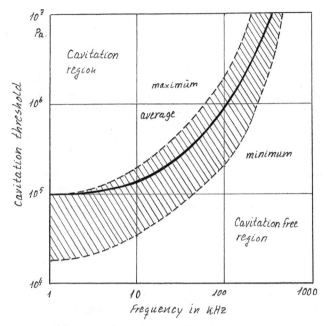

FIGURE 13.3

Cavitation threshold pressure in Pa, as a function of signal frequency (kHz). The *hatched region* around the average curve represents the scattering of the measured results found by various authors. The *hatched region* separates the region, where cavitation always produced, from the cavitation-free region.

Adapted from Urick, R.J., Principles of Underwater Sound, (3rd. Ed.), McGrawHill Book Comp., 1983, Peninsula Publishing, Figure 4.6 with permission of Peninsula Publishing.

above 10 kHz it increases substantially. Also, the pulse length has a strong influence on the cavitation threshold. Longer pulses lead to lower cavitation thresholds, and for pulses of more than 10 ms duration, the cavitation threshold is nearly constant. Pulse lengths less than 10 ms show an increasing cavitation threshold.

13.1.4 ACOUSTIC RADIATION PRESSURE AND ACOUSTIC STREAMING

A propagating sound wave carries acoustic momentum and acoustic energy. The average acoustic momentum carried through a unit area in a fluid per unit of time produces an *acoustic radiation pressure*, p_s {unit: Pa} which, according to Stephens and Bate [19], can be written as:

$$p_s = \rho u_{max}^2 = p_{max}\frac{u_{max}}{c} = \frac{p_{max}^2}{\rho c^2} \tag{13.7}$$

where p_{max} {unit: Pa} and u_{max} {unit: m} denote the maximum pressure and particle velocity, respectively, in a plane acoustic wave, while c {unit: m/s} is the sound velocity. Since the acoustic radiation pressure depends on the acoustic pressure squared, the radiation pressure is more influential at higher than at lower acoustic pressures.

Acoustic streaming is the steady flow which may be associated with sound propagation in a fluid and was first recognized by Faraday. An in-depth exposition of acoustic streaming is given by Nyborg [20]. The streaming velocity is directly dependent on the coefficient of absorption in the fluid, and since the radiation pressure will fall off with distance from a projector due to the absorption, a net force, F {unit: N/m^3} will be exerted on the fluid in front of the projector. This force will accelerate the fluid in a motion away from the projector. However, the motion will be resisted by the fluid viscosity, and in a final steady state a uniform fluid motion will result, provided that a suitable return path for the fluid can be established. The net force F for a plane, circular projector is provided by:

$$F = -\frac{\partial p_s}{\partial x} = 2\alpha E_o(r) \tag{13.8}$$

where x {unit: m} is the direction normal to the projector surface while r {unit: m} is the radial coordinate on the projector surface. α {unit: Np/m} is the absorption coefficient and $E_o(r)$ {unit: J/m^3} is the energy density variation across the projector surface. As shown by Stephens and Bate [19] the streaming velocity $v(r)$ {unit: m/s} as a function of the radial distance r from the center of a plane, circular projector with diameter $2a$ {unit: m} may be written as:

$$v(r) = \frac{\alpha E_o a^2}{\mu} \phi(r) \tag{13.9}$$

where $\phi(r)$ represents the variation of the velocity v across the diameter of the projector and μ {unit: Pa·s} is the coefficient of shear viscosity in the fluid. Acoustical streaming was used by Liebermann [21] to calculate the magnitude of the bulk viscosity coefficient η' {unit: Pa·s}, see Chapter 4, which explained a major part of the deviation found between theoretical and experimental values for absorption of sound in seawater.

13.2 NONLINEAR UNDERWATER ACOUSTICS

The application of nonlinear acoustic phenomena in underwater acoustics, where nonlinear effects mostly had been considered a nuisance, started in the late 1950s when the exploitation of the sum and difference frequencies produced by the interaction between two sound waves of finite amplitudes for production of sound beams was suggested by Westervelt [22,23]. In particular, the difference-frequency beam, proposed by Westervelt was likened to an end-fire array and named the *parametric acoustic array*. Due to its advantageous beam qualities it has found a rather widespread use in underwater acoustics. The *parametric acoustic transmitting array* and the *parametric acoustic receiving array* have been studied extensively over many years, and sonar systems based on the parametric transmitting principles are now commercially available. An early exposition of the parametric acoustic array can be found in Bjørnø [24].

13.2.1 PARAMETRIC ACOUSTIC TRANSMITTING ARRAYS

The generation of sum and difference frequencies by the interaction between two finite-amplitude sound waves has been the subject of discussions for more than 200 years. Helmholtz [25] and Lamb [26] credit the original observation of difference-frequency tones to Sorge in 1745 and Tartini in 1754. A considerable step forward in the studies of difference frequencies and their potential applications was produced by Westervelt's classical article [23] "Parametric Acoustic Array." In this article the nonlinear interaction between two superimposed, high-frequency, primary waves of finite amplitudes was calculated based on the solution to an inhomogeneous wave equation derived from Lighthill's [27] theory for aerodynamic sound generation. Westervelt's inhomogeneous wave equation in a general form may be written as:

$$\nabla^2 p_s - \frac{1}{c^2} \frac{\partial^2 p_s}{\partial t^2} = -\rho_o \frac{\partial q}{\partial t} \tag{13.10}$$

where

$$q = \left(1 + \frac{B}{2A}\right) \frac{1}{\rho_o^2 c_o^4} \frac{\partial p_i^2}{\partial t} \tag{13.11}$$

p_s {unit: Pa} in Eq. (13.10) is the pressure of the scattered—i.e., the difference frequency—wave generated by the collimated primary waves which are designated by the subscript i. t denotes time {unit: s}, while ρ_o {unit: kg/m^3} and c_o {unit: m/s} are the fluid density and velocity of sound, respectively. B/A is the dimensionless second-order nonlinearity ratio, and q is the source strength density of the process responsible for the generation of acoustic energy through the nonlinear interaction between the primary waves.

Inserting Eq. (13.11) into Eq. (13.10) and solving for p_s as a function of the distance R {unit: m} from the observation point to the center of the projector emitting the primary waves and the observation point angular coordinate θ {unit: radian} produces Eq. (13.12) for p_s.

$$p_s(R, \theta) = \frac{\omega_s^2 p_o^2 S}{8\pi \rho_o c_o^4 R} \frac{[1 + B/2A]}{\sqrt{\alpha^2 + k_s^2 \sin^4(\theta/2)}} \tag{13.12}$$

where ω_s is the angular frequency of the difference-frequency wave, p_o denotes the pressure amplitude of each primary wave, S {unit: m^2} denotes the cross-sectional area of the collimated wave interaction zone, k_s is the difference-frequency wave number, and α is the absorption coefficient of the primary waves, and represents the only influence in Eq. (13.12) from the primary frequencies. The α value used is normally the average of the α values for the two primary frequencies. Westervelt's solution to Eq. (13.12) is restricted to the far field of the scattered wave by the condition: $k_s R > (k_s/\alpha)^2$. From Eq. (13.12) the far-field, half-power beam width θ_h {unit: rad} of the difference-frequency wave can be expressed by:

$$\theta_h \sim 2\sqrt{\frac{\alpha}{k_s}} \tag{13.13}$$

This shows that beam narrowing takes place for a decrease in the primary frequency, opposite to the case for a conventional linear projector. Also, a beam width reduction follows a difference-frequency increase. This directivity has the same form as Rutherford scattering in atomic theory.

Eq. (13.12) shows that p_s increases for higher primary wave amplitudes p_o; for increasing difference frequency, i.e., for lower downshift ratio (ω_o/ω_s); for increasing fluid material nonlinearity of B/A; for decreasing fluid density ρ_o; and in particular for decreasing fluid sound velocity. Variations of these parameters have formed the basis for attempts to increase the parametric acoustic conversion efficiency which is expressed by the ratio of sound energy in the difference-frequency wave to the sound energy in the primary waves, see Bjørnø et al. [28].

Because the effective array length can be made very large by selection of primary frequencies, highly directive difference-frequency beams can be produced by a projector small compared to the difference-frequency wavelength. Due to the exponential shading resulting from the primary wave absorption, very low side-lobe levels are usually found in parametric arrays. Fig. 13.4 shows a reduction in the side-lobe levels of 35–40 dB relative to the main-lobe level. The low side-lobe level will reduce the influence of clutter and thus improve the signal-to-noise level, and reverberation will decay quickly, which permits a high pulse repetition rate when the array operates in a pulse mode. Considerable frequency agility characterizes

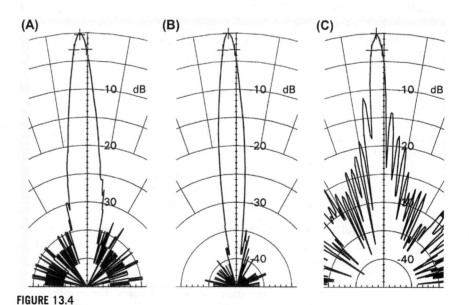

FIGURE 13.4

Parametric beam profiles for Kongsberg Defence Systems' TOPAS PS 40. (A) The primary beam at 40 kHz. (B) The difference-frequency beam with a downshift of 10 to 4 kHz. (C) The difference-frequency beam with a downshift of 4 to 10 kHz. The strongly reduced side-lobe levels at the difference-frequency beams are clearly visible with a reduction of 35 to 40 dB relative to the main-lobe level.

Images provided courtesy of Drs. Arne Løvik and Johnny Dybedal, Kongsberg Defence Systems, Norway.

the parametric array, where large changes in difference frequency can be generated with only small changes in primary frequency, and the projector can continue to operate near its resonance frequency for a large range of difference frequencies. These superior qualities make high-resolution sonar feasible where only small projectors can be deployed, and the lack of side lobes encourages the use of parametric transmitters in reverberation-limited regions of the sea, for instance, in relation to shallow-water communication as discussed by Kopp et al. [29].

Westervelt's [23] model is based on the assumption that the primary fields are collimated plane waves and the extent of the interaction region is limited by viscous absorption, while no nonlinear absorption is present. Also, the observation point is situated far from the interaction region. The influence of the absorption magnitude, i.e., small-signal absorption or nonlinear absorption, and the possibility that most absorption takes place in the near field or in the far field of the primary waves, form four distinct operating regimes for parametric sources. These regimes are:

1. Arrays are limited by small-signal absorption in the transducer's near field. This is the Westervelt case and comprises most experiments based on low-power, high-frequency primary beams [30–32]. For lower primary frequencies the absorption will decrease.
2. Substantial parametric generation may take place in the far field of the primary waves, where the absorption influences the array length. The divergence in the far field of the beams from a piston source, which is most frequently used, leads to reduced primary signal amplitudes. However, far-field absorption is the effect terminating the parametric processes. The far-field absorption-limited array forms the second operating regime for parametric sources. Contributions to the study of this regime are found in Refs. [33–35].
3. The third parametric array operating regime is formed when the array length is limited by nonlinear absorption in the near field of the primary beams. This array type is normally called saturation limited due to the influence of harmonics and shock formation in the primary beams. Since the near field of a piston source may be approximated by a plane-wave field, models for saturation-limited arrays are normally based on plane-wave assumptions. Theoretical and experimental studies of saturation-limited arrays may be found in Refs. [36–38].
4. The fourth regime is of minor practical importance since saturation effects in the far field rarely occur, unless the primary beams are spherically spreading.

Attempts to produce a theoretical treatment dealing simultaneously with all four regimes have been made by several authors [39–42]. The references for the four parametric ray regimes in the years 1960–85 reflect the most productive time in parametric acoustic array research. Ref. [42] is a valuable review of the development in parametric acoustic arrays since it combines research results obtained in the United States and Europe with results obtained in the former USSR.

The physical qualities and the characteristics of the difference-frequency beam in regimes (1) through (3), when the field point is outside or inside the interaction region, show some individual variations. While the field point originally was considered to be outside the interaction region in the parametric array, the limitation of test facility dimensions or measurements near the signal source require a consideration of the field point inside the interaction region. When the field point is inside the interaction region, diffraction effects dominate over viscous effects, and in Ref. [43] a full account of diffraction effects was given through a 3-D scattering integral expressed by:

$$p_s(x,t) = \frac{1 + B/2A}{4\pi\rho_o c_o^4} \int_V R \frac{\partial^2 p_i^2\left(x, t - \dfrac{R}{c_o}\right)}{\partial t^2} dV \qquad (13.14)$$

where R {unit: m} is the distance between the field point for the measurement and the volume element dV of the interaction region. The computational time required for each prediction limits the application of Eq. (13.14). Through a coordinate transformation which reduces Eq. (13.14) to a single integral and by assuming that the primary waves are spherically spreading from their origin, while neglecting viscous absorption and nonlinear attenuation, a close agreement between theory and experimental data was obtained by Rolleigh [44].

The saturation effects by high-amplitude wave interaction in parametric arrays cause an effective shortening of the array length which results in a broadening of the difference-frequency beam, a reduction in the difference-frequency source level, and an increase in side-lobe effects relative to the main beam. A comprehensive study of the parametric array behavior at increasing primary wave amplitudes has been reported by Moffett and Konrad [45]. Fig. 13.4 shows the parametric gain G as a function of the scaled primary source level SL_0^* expressed by:

$$G = SL_- - SL_o \text{ and } SL_o^* = SL_o + 20 \log f_o \text{ {unit: dB}} \qquad (13.15)$$

where SL_o and SL_- denote the rms source level of one primary-frequency component—the two primary-frequency components are assumed to have equal amplitudes—and the rms source level of the difference-frequency signal, respectively. f_o {unit: kHz} is the mean primary frequency. The curves in Fig. 13.5 are for a down-shift ratio, $f_o/f = 10$, and for various degrees of absorption expressed by αR_o {unit: dB}, where R_o {unit: m} is the Rayleigh distance, which expresses collimation length of the primary wave near field. f {unit: kHz} is the difference frequency. The reduction in parametric gain for increasing scaled primary source level at lower absorption due to saturation effects can be seen from Fig. 13.5.

The parametric gain will in most cases be nearly −40 dB or more down from the primary source level, see Fig. 13.6. This figure shows calculated and experimental data for the field point inside and outside the primary wave's interaction region according to Bjørnø et al. [28]. The difference between the sound pressure level of the primary waves and the difference-frequency wave depends on the distance from the projector. Sound pressure level extrapolations show that the two curves for sound pressure levels will intersect at a projector distance of nearly 300 m with a sound pressure level of 110 dB rel 1 μPa. The low conversion efficiency has given rise to several studies focused on improving it. Based on Eq. (13.12) an increase in B/A and a decrease in density and sound velocity of the fluid will result in increased conversion efficiency. Replacing the liquid in the primary waves' near field by ethyl or methyl alcohol—both with higher B/A and lower density and velocity of sound

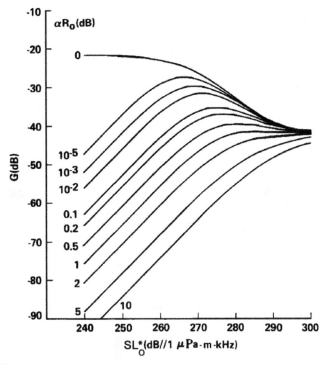

FIGURE 13.5

Parametric gain curves for $f_0/f = 10$ and for various absorption coefficients α {unit: dB/m} and as a function of the scaled primary source level SL_0^* {unit: dB}.

*Reproduced from Moffett, M.B. and Mellen, R.H., Model for Parametric Acoustic Sources J. Acoust. Soc. Amer., **81**, (2), pp. 325–337, 1977, with the permission of the Acoustical Society of America.*

than water—showed increasing conversion efficiency. It was found [28] that it was possible to obtain an average gain relative to the use of water of 10.2 dB in ethyl alcohol and 13.6 dB in methyl alcohol. Moreover, using tone bursts with the burst frequency equal to the difference frequency and with the carrier frequency equal to the average frequency of Westervelt's two primary frequencies increases the conversion efficiency about 2 dB while a 100% modulation of the carrier frequency with the difference frequency will result in more than a 2.5 dB increase in conversion efficiency. Also, Merklinger [37] found conversion efficiency improvements through signal processing procedures. Replacing the fluid in the primary waves' near field with a solid, silicone rubber cylinder, is reported by Ryder et al. [46], who found that the lower sound velocity and the higher B/A of the silicone rubber together with the cylinder's slow waveguide antenna effect in water result in a conversion efficiency increase of 2–5 dB. However, these conversion efficiency improvements were not able to substantially improve the nearly 40 dB in parametric gain.

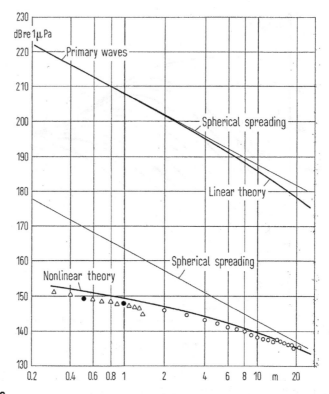

FIGURE 13.6

Measured and calculated primary wave pressure amplitudes and the difference-frequency wave exposed to absorption and spherical spreading as a function of distance to the projector.

Reproduced from Figure 1 Bjørnø, L., Christoffersen, B. and Schreiber, M.P., *Some Experimental Investigations of the Parametric Acoustic Array.* Acustica, **35**, (2), pp. 99–106, 1976, with permission from Acustica – Acta Acustica.

13.2.2 PARAMETRIC ACOUSTIC RECEIVING ARRAYS

The possibility of developing a parametric acoustic receiving array was first suggested by Westervelt [23]. In a parametric receiver, the nonlinear interaction process may take place between a low-frequency signal wave of low intensity and a locally generated high-frequency pump wave of higher intensity. The sum- and difference-frequency signals are then received by a hydrophone on the acoustic axis of the pump wave.

Low- and high-amplitude receiving arrays have been studied, and their difference is in the inclusion of finite-amplitude pump wave effects. Far-field reception was considered by Barnard et al. [47]. They studied the first-order sound field consisting of the interaction between a low-amplitude, spherical, harmonic pump wave and a plane, harmonic signal wave. A comprehensive theoretical and experimental study of the parametric receiving array is reported in Refs. [48,49], including interaction effects between the signal wave and the pump wave in the pump wave near field and far field. Also the misalignment of the pump or the receiver transducer was studied showing that for a misalignment the difference-frequency beam pattern is asymmetrical and is a mirror image of the sum-frequency beam pattern. The influence of transducer vibration either due to platform motion or to transducers not firmly mounted, reported in Ref. [50], showed that transducer motion is a significant factor in parametric receiving array performance. The detrimental effects may be lessened by proper system design. However, one of the main reasons why the parametric receiving array has not received the same attention as the transmitting array may be ascribed to the influence on the reception from transducer motion, water noise at the signal frequency or at the sum and difference frequencies, electronic noise in the equipment, and other factors.

13.2.3 APPLICATIONS OF THE PARAMETRIC ACOUSTIC ARRAY

Around 1985 most theoretical studies which formed the basis for constructing parametric acoustic transmitting arrays had been carried out and the necessary knowledge base was created. Not all aspects of the parametric array had been investigated; unification of theories and relations to other part of the fundamental expressions in nonlinear acoustics were still missing. However, a useful theoretical/numerical basis had been formed and the operation of experimental test models of parametric—transmitting and receiving—arrays in small and large scales had been studied to verify the theoretical/numerical basis. Parametric acoustic array experiments had also been performed before 1985 for detecting biomass in the water column, for seabed studies, subbottom imaging, underwater communication, and transmission of television pictures over shorter distances in a lake. Raytheon, in cooperation with the US Naval Underwater Systems Center (NUSC), had in the late 1960s produced a test model of a parametric subbottom profiler, which in 1970 gave the first qualified pictures of fish schools and seabed structures, thus proving the superior beam qualities of the parametric array. During the 1970s the University of Birmingham carried out studies of harbor bottom profiles in the Netherlands and a large high-power *towed parametric sonar* (TOPS) had been produced and tested by NUSC [51]. This sonar had the following technical data: 0.5×2 m transmitter, 480 mass-loaded elements operating at a mean primary frequency of 24 kHz, and a primary source level of 250 dB rel 1 μPa at 1 m. Its primary 3 dB beam width was 2 degrees $\times$ 8 degrees.

The theoretical/numerical studies of parametric arrays since 1985 have involved substantial contributions from the "Bergen School" in Norway. In a series of papers dealing with sound wave interaction, beam shape influence, near-field contributions, influence of the source field shape, and lack of axis symmetry on the parametrically generated sound field were studied in depth [52,53]. Also, the importance of nonlinear effects on parametric signal generation [54] and the effects of focusing on the nonlinear interaction [55] were studied via solutions to the nonlinear parabolic equation. The interference of the difference-frequency sound after reflection at a surface with the original difference-frequency sound was studied theoretically and experimentally [56]. Also, the influence of misalignment of pump, source, and hydrophone on the performance of the parametric receiving array was quantified in numerical examples which showed that the best performance would be obtained for a good alignment, a high pump frequency, and the hydrophone placed not too far from the source near field [57].

Also the mechanisms of a parametric beam penetration into the seabed, when the nonlinear interaction between the primary waves is terminated at the interface between the water column and the seabed, have been studied theoretically and experimentally for parametric arrays operating in a transient mode [58–60]. Pace et al. [60] used the plane wave spectrum and reflection coefficients at the water column–seabed interface to calculate in 2-D and 3-D the reflected and refracted acoustic fields for conventional (linear) and parametric beams incident on the interface below and above a critical angle of 62 degrees. Figs. 13.7 and 13.8 show 3-D calculations of the contour plot of reflected and refracted fields due to a parametric beam incident on the interface at a 55 degrees and 65 degrees beam axis incidence. The refracted beam strength and direction for incidence beyond the critical angle are caused by the width and position of the plane wave spectrum and their relation with the plane wave reflection coefficients in the vicinity of the critical angle. The plane wave spectrum width around the critical angle leaves a portion of the spectra in the angular range below critical to enable a refracted beam to exist even as the incident beam axis increases beyond critical.

Several successful attempts to exploit the parametric transmitting array for bottom and subbottom profiling occurred in projects under the European Union's former Marine Science and Technology (MAST) program, such as the project "Sediment Identification for Geotechnics by Marine Acoustics" (SIGMA), which provided valuable information about the seabed structure and the relation between seabed material acoustical and geotechnical qualities. The narrow beam qualities

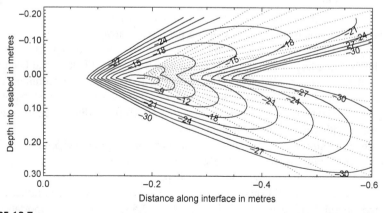

FIGURE 13.7

3-D contour plot of reflected and refracted field due to a parametric beam incident on the water–seabed interface at a 55 degrees beam axis incidence with 3 dB contours. The 5 degrees *dotted lines* are centered in the 5 degrees interval incidence point.

From Pace, N.G. and Bjørnø, L., *The reflection and refraction of acoustic beams at water sediment interfaces.* Acust. Acust., **83**, pp. 855–862, 1997, with permission from Acustica — Acta Acustica.

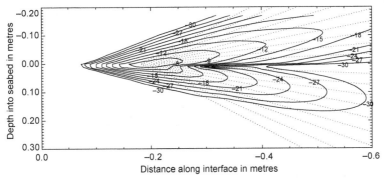

FIGURE 13.8

3-D contour plot of the reflected and refracted field due to a parametric beam incident on the water—seabed interface at a 65 degrees incidence angle with 3 dB contours. The *dotted lines* at 5 degrees intervals are centered in the point of beam axis incidence.

From Pace, N.G. and Bjørnø, L., *The reflection and refraction of acoustic beams at water sediment interfaces.* Acust. Acta Acust., **83**, pp. 855–862, 1997, with permission from Acustica – Acta Acustica.

of the parametric array have also been exploited in relation to underwater communication in the MAST project Acoustic Communication Using Parametric Array (PARACOM) [61], where the interaction between the parametric beam and the water column boundaries was of interest.

An example of the calculation of a parametric array transducer and its sound field is given by Bjørnø [62] in connection with the SIGMA project for seabed studies by a vertically incident parametric beam. The fundamental projector frequency was 60 kHz and the difference frequency was 6 kHz. The projector consisted of four 0.15×0.15 m squared disks with effective side lengths of 0.6 and 0.15 m.

The environmental data were: $\alpha_o = 0.0164$ dB/m (at 60 kHz); water temperature $t = 10°$; a tow-fish ambient static pressure at 10 m depth; $P_o = 2 \cdot 10^5$ Pa; $P_H = 8$ Pa; and salinity $S = 35$ ppt, with a sound velocity $c_o = 1490$ m/s. From Eq. (13.13), the half-power beam width $\theta_h = 1$ degrees and the directivity index DI is about 19 dB. The acoustic power transmitted is 400 W for the primary frequency, which gives a transmitted power per primary frequency of $0.89 \cdot 10^4$ W/m^2 or a sound pressure of $1.63 \cdot 10^5$ Pa. From Section 10.9 in Chapter 10 the primary frequency source level will be 216 dB rel 1 μPa at 1 m, and from Eq. (13.15) the scaled primary frequency source level is $SL_o^* = 252$ dB relative to 1 μPa at 1 m.

Fig. 13.6 gives for the downshift ratio of 10, the gain value of $G = -42$ dB, and SL_- will then be 174 dB rel 1 μPa at 1 m. Fig. 13.6 shows that no acoustical saturation is involved. For an absorption-limited parametric array the length of the interaction region will be $\ell_a = \frac{1}{2}\alpha_o$, which will be $\ell_a = 263$ m. With a depth of operation of 80—120 m, a truncation of the parametric array is caused by the seabed.

The influence of the rectangular shape of the projector represented by its aspect ratio $N = 1{:}4$ can be calculated as follows. The source level coefficient Q, see Berktay [63], may be determined from Fig. 13.9. The scaled Rayleigh distance $r_o = (\lambda_s/\lambda_o)R_o$, where R_o is the Rayleigh distance for the projector, and λ_s and λ_o denote the wavelengths of the difference frequency and the primary frequency, respectively, will

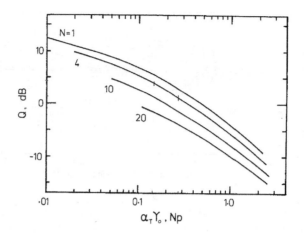

FIGURE 13.9

The source level coefficient Q as a function of the product of the absorption parameter α_T and scaled Rayleigh distance r_o for various projector aspect ratios N.

Adapted from Berktay, H.O., *Parametric sources — Design considerations in the generation of low-frequency signals*. In: Ocean Seismo-Acoustics — Low-frequency underwater acoustics. T. Akal and J.M. Berkson (Eds.), NATO Conference Series, Plenum Press, pp. 785–800, 1986, with permission of Springer.

give $r_o = 36.2$ m, and with the absorption parameter $\alpha_T = 2\alpha_o$, the abscissa value in Fig. 13.9 becomes 0.14, which gives $Q = 4$ dB for the difference-frequency wave. The source level for a fully developed array, i.e., without seabed truncation, but with the projector aspect ratio taken into account, is $SL_- = 165$ dB. The normalized parametric source beam width for a projector aspect ratio of 1:4 is 1.35 degrees and the difference-frequency half-power beam widths in the two directions for a fully developed projector array are $\theta_h = 1.43$ degrees and $\theta_h = 3.39$ degrees.

A few commercially available parametric transmitting systems are produced and sold. The systems are marketed by Atlas Hydrographic GmbH, Germany, by Kongsberg Defence Systems, Norway, and by Innomar Technologie GmbH, Germany. The main application has been bottom and subbottom profiling. *Parasound* produced by Atlas Hydrographic operates with primary waves in the frequency range 18–33 kHz and difference frequencies from 0.5 to 6 kHz. The maximum bottom penetration is above 200 m for the highest power transmitted. Source levels for the 70 kW transmitted power are $SL_o = 245$ dB rel 1 μPa at 1 m and $SL_- = 206$ dB rel 1 μP at 1 m. For a 35 kW transmitted power they are $SL_o = 242$ dB rel 1 μP at 1 m and $SL_- = 200$ dB rel 1 μPa at 1 m.

Kongsberg Defence Systems offers *TOPAS PS Systems:* the PS 18 for water depths from 10 m to full ocean depth; the PS 40 for water depths from 5 m to 1000 m, see Fig. 13.9; and the PS 120, a portable, high-resolution system, for depths between 3 and 400 m. The PS 40 primary frequencies are in the range 35–45 kHz with SL_o ~240 dB rel 1 μPa at 1 m, and parametrically generated frequencies are in the range 1–10 kHz with $SL_o = 185 - 206$ dB rel 1 μPa at 1 m (depending on the difference frequency). Fig. 13.4 shows the TOPAS PS 40 primary and two difference-frequency beam profiles, and Fig. 13.10 shows the subbottom profiles produced by

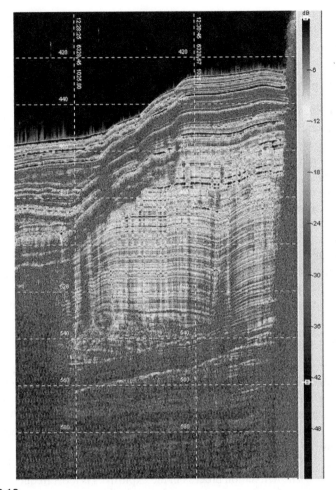

FIGURE 13.10

Subbottom profiles produced by Kongsberg Defence Systems' TOPAS PS 18. Primary frequency: 18 kHz. Primary SL is about 243 dB rel 1 μPa at 1 m. Difference-frequency signal: Chirp 1.5–5 kHz, and 20 ms duration. Difference-frequency SL ~195–207 dB rel 1 μPa at 1 m (*Data from R/V G.O. Sars taken in Trondheim Fjord, Norway.*). Water depth about 350 m. Penetration into the seafloor >120 m. High lateral and range resolution due to the narrow beam and the high bandwidth.

Figure provided courtesy of Drs. Arne Løvik and Johnny Dybedal, Kongsberg Defence Systems, Norway.

TOPAS PS 18. This figure emphasizes the high lateral and range resolution produced by the narrow difference-frequency beam and the high bandwidth used in the Chirp difference-frequency signal covering 1.5–5 kHz with a 20 ms signal duration.

Innomar Technologie's parametric subbottom profiler, *SES*-2000, operates with primary frequencies around 100 kHz and difference frequencies in the range

5–15 kHz, and for deep-water studies the primary frequencies are around 35 kHz and the difference frequencies are between 2 and 7 kHz, which permit sediment penetration up to 150 m.

13.3 UNDERWATER EXPLOSIONS

High-intensity sound in the sea can be generated by using chemical explosives, air guns, or electrical discharges. Chemical explosives have been used as signal sources for single, but reproducible, broadband high-amplitude signals. The characteristic feature of a signal from a chemical explosive is the formation of a *high-frequency shock wave* followed by the *low-frequency pulsation of a gas bubble*. More low-frequency air gun signals have found applications in studies of the seabed, where return signals from layers in the seabed provide information about characteristic structures of interest in prospecting for oil or gas. Electrical discharges between two electrodes have been used for to study shock wave propagation in water, since the signals are very reproducible and the amplitudes and signal time histories can be controlled by the capacity of the condenser bank, the spark gap electrode distance, and other factors. This section deals with the characteristic features of signals based on the use of *chemical explosives*, while other sources of high-intensity sound are also mentioned.

13.3.1 THE SHOCK WAVE

A chemical explosive in most cases consists of the elements C, O, H, and N in various combinations. Some of the explosives used are TNT (*trinitrotoluene*, $C_7H_5O_6N_3$), tetryl ($C_7H_5O_8N_5$), RDX ($C_3H_6N_6O_6$), and PETN ($C_5H_8O_{12}N_4$). Also the explosive pentolite, consisting of 50% TNT and 50% PETN, has been used extensively to produce underwater, high-intensity sound due to its reproducible signals. The energy available in TNT, for instance, is around $4.5 \cdot 10^6$ J/kg. The frequently used SUS (Signal, Underwater Sound) charges are mostly TNT in about 0.82 kg amounts. The detonation process in the explosives is most frequently started by a shock wave produced by an especially sensitive explosive which is used as a primary material to start the detonation process. In the detonation process a detonation front will move with high velocity through the explosive transforming into reaction products under very high pressure, about 10^{10} Pa, and temperatures, above 3000°C. A detonation is characterized by a constant detonation front velocity, which for TNT and tetryl are about 6900 and 7500 m/s, respectively. The calculation of the transition from explosive before the detonation front to reaction products after the front can be done by use of the Rankine–Hugoniot relations and the Chapman–Jouget condition, $D = c + u$, where D is the detonation front propagation velocity {unit: m/s} of propagation and c and u are, respectively, the velocity of sound and the particle velocity in the reaction products after the front; see Cole [64].

When the detonation front reaches the boundary between the explosive and water, a shock wave is transmitted into the water. The time history of the shock wave is characterized by a very fast rise in pressure from the hydrostatic pressure to a peak pressure, P_m {unit: Pa}, with the magnitude depending on the explosive type and the amount in kilograms. The time for the growth in pressure from hydrostatic to the peak pressure is only a small fraction of a second which depends on the type and amount of explosive used. The shock front is followed by an exponential decay in pressure with a time constant θ {unit: s} for the pressure being reduced to P_m/e. Similarities between peak pressures and time histories of shock waves from chemical explosives, and thus their impulses I {unit: Ns/m^2} and energy flux densities E {unit: J/m^2}, were found very early and have given rise to power laws of the form: *constant* $\cdot (W^{1/3}/R)^\alpha$, where W {unit: kg} and R {unit: m} denote the amount of explosive and the distance from the detonation site, respectively, and α is a constant. The influence of the density of the explosive forms an integrated part of the constants. The power laws, which give a compact and reasonably precise method for representing the shock wave data in water, are:

$$P_m = K_1 \left(\frac{W^{1/3}}{R}\right)^\alpha$$

$$\theta = K_2 W^{1/3} \left(\frac{W^{1/3}}{R}\right)^{-\beta}$$

$$I = K_3 W^{1/3} \left(\frac{W^{1/3}}{R}\right)^\gamma \tag{13.16}$$

$$E = K_4 W^{1/3} \left(\frac{W^{1/3}}{R}\right)^\delta$$

where the constants K_1-K_4 and the exponents $\alpha-\delta$ are individual for each type of explosive. For TNT and tetryl the constants and exponents are given in Table 13.3, taken from Bjørnø [65].

Table 13.3 Experimentally Obtained Values of the Constants K_1-K_4 and Exponents $\alpha-\delta$ in Eq. (13.16).

Explosive		TNT	TETRYL
Range $r = R/W^{1/3}$		0.46–11.1	4.31–50
P_m {unit: Pa}	K_1	$521.6 \cdot 10^5$	$506 \cdot 10^5$
	α	1.13	1.10
θ {unit: s}	K_2	$96.5 \cdot 10^{-6}$	$87 \cdot 10^{-6}$
	β	0.18	0.23
I {unit: Ns/m^2}	K_3	5760	5900
	γ	0.89	0.87
E {unit: J/m^2}	K_4	$9.8 \cdot 10^4$	$11 \cdot 10^4$
	δ	2.10	2.12

Table extracted from Bjørnø, L., *A comparison between measured pressure waves in water arising from electrical discharges and detonation of small amounts of chemical explosives.* Trans. ASME J. Eng. Ind., **92**, Ser. B, (1), pp. 29–34, 1970, with permission of the American Society of Mechanical Engineers.

13.3.2 THE GAS BUBBLE

While nearly half the explosive energy goes into the shock wave, for tetryl the energy amount was measured to 46% [65], most of the residual energy is found in the gas bubble. The gas bubble expansion after the shock wave emission is associated with energy emitted in the form of pressure waves which propagates radially from the bubble. Since the pressure in the reaction products of the gas bubble, after the detonation and shock wave emission, is substantially higher than the hydrostatic pressure at the detonation site, the gas bubble expands continuously for a relatively long period. During the bubble expansion its gas pressure decreases gradually to the hydrostatic pressure, but the bubble expansion continues due to the inertia in the outward flowing water. When the pressure in the gas bubble falls below the absolute environmental pressure, the difference between the bubble pressure and the environmental pressure gradually stops the bubble expansion. The bubble boundaries now contract at an increasing rate. This inward motion continues until the gas compressibility and the increasing pressure in the gas bubble stop the inward motion of the bubble boundary and its environment. The bubble now has its first minimum radius and a new expansion and contraction cycle may start. In nearly 80% of the bubble pulsation time, the pressure in the bubble will be below the hydrostatic pressure. Depending on the explosive detonation depth gas bubble oscillations may persist for a number of cycles. In spite of the fact that an oscillating gas bubble, due to Bernoulli effects, will receive a repulsive force from a free surface and will be attracted toward a rigid boundary, the buoyancy effects on the bubble make it migrate toward the water surface. The speed of migration is highest when the bubble has its minimum radius.

The period, T_n {unit: s}, for the gas bubble oscillation depends on the detonation depth, d {unit: m}, and on the amount of explosive W {unit: kg}. The number of oscillations, n, increases with the detonation depth, and the oscillation period T_n is approximately given by Ref. [64]

$$T_n = K_5 \frac{W^{1/3}}{(d + 10.33)^{5/6}} \qquad (13.17)$$

where K_5 is a constant which is about 2.1 for most explosives. Eq. (13.17) assumes that no buoyancy influence is present and that the oscillations take place in the same depth far away from limiting surfaces like the water surface or the seabed.

For each cycle of expansion and contraction, bubble energy will be lost in pressure radiation, flow, and turbulence, and the buoyancy will move the bubble to lower water depths, which all reduce the oscillation period.

An approximate expression for the peak pressure, P_{mp} {unit: Pa}, in the first bubble pulsation, achieved at the first bubble minimum radius, can be found from:

$$P_{mp} = 7 \cdot 10^6 \frac{W^{1/3}}{R} \qquad (13.18)$$

where W {unit: kg} and R {unit: m} are the explosive weight and distance between the detonation site and measurement point, respectively. Eq. (13.18) assumes spherical pressure signal propagation.

While the peak pressure in the first bubble pulse is no more than 10–20% of the peak pressure in the shock wave, the bubble pulse duration is much longer. In fact the impulses of the two pressure waves are nearly equal. This is the reason that the most serious damage on a ship caused by an underwater explosion frequently is due to the gas bubble pulsation, in particular when the explosion takes place under the keel and the gas bubble migrates toward the ship bottom which is already damaged by the shock wave.

The interaction between the sea surface, shock wave, and subsequent gas bubble gives rise to characteristic visible phenomena. When the shock wave is reflected at the water surface, an expansion wave is produced which moves backward into the water column. The water at the surface layer is thrown up with a particle velocity proportional to the arriving shock wave pressure and a rounded dome of whitish water forms on the surface. The white color is caused by the cavitation produced by the expansion wave. At a greater radial distance along the water surface from position of the first encounter between the shock wave and the surface, the shock wave produces a rapidly advancing ring of darkened water due to the shock wave influence on the index of refraction of the water.

When the gas bubble reaches the water surface, characteristic surface phenomena may occur. If the gas bubble is in its high-pressure phase, it will shoot up a narrow plume of spray to considerable heights above the surface. If it is in its expanded phase, only low plume formation will occur since the gas bubble motion is nearly radial, and the plumes are projected outwards in all directions through the spray dome. For detonations close to the water surface the bubble-pulse influence on the high-intensity signal may be avoided due to gas venting out the bubble before its contraction phase, and only the shock wave will be available for the measurements.

13.3.3 OTHER SOURCES OF HIGH-INTENSITY SOUND

Other sources of high-intensity signals are, for instance, *air guns*, in particular for seismic investigations, *electrical discharges* over a spark gap, and *boomers*. These signal sources are given a brief exposition in this section.

An air gun is a mechanical device which releases a high-pressure air bubble under water. Similar to the gas bubble after a chemical explosive detonation, the released air bubble expands and contracts in cycles. During the cycles it emits pressure waves as sources of seismic waves used in reflection seismology in the layers below the seafloor.

The air gun consists of one to several tens of pneumatic chambers, which can form an array that is pressurized with compressed air at pressures normally ranging from 14 to 21 MPa. Air guns are submerged in the seawater and are normally towed behind a vessel at a depth of about 6 m. The air gun is fired by an electrical signal which triggers a solenoid valve, which releases air into a fire chamber causing a piston to move. The piston movement allows air to escape from the main storage chamber into the water, producing a high-pressure air bubble. The release of large bubbles gives low-frequency signals, while small bubbles lead to more high-frequency signals. The pressure variation in the water due to the air bubble pulsation is called the *air gun signature*. The released air bubble is nearly spherical, and its expansion due to the air pressure exceeding the hydrostatic pressure produces a signal with a shocklike front, but not a shock front like the one found by explosive detonation. Similar to the gas bubble from a chemical explosive detonation, the air bubble is exposed to the same mechanism and may go through many expansion/contraction cycles, where it emits pressure waves into the environment. The bubble oscillation amplitude is damped with the increasing number of cycles and the oscillation period is not constant from one cycle to the next. This nonharmonic motion is due to energy transmission into the water at each cycle. The first cycle produces the highest pressure amplitude in the primary wave, frequently exceeding 250 dB rel 1 μPa at 1 m.

Even the largest single air gun will frequently not produce enough energy to permit the seismic signal to penetrate and produce return signals from deeper layers about 5 km below the seafloor. Single air guns produce pressure pulses which oscillate through several cycles, and each cycle produces pressure signals whose reflected signals oscillate through the same cycles as the source signal. This makes it difficult, if not impossible, to separate the primary from the secondary peaks in the return signal from the layers in the seabed. What is needed is a single, well-defined pressure spike from the air gun. The amplitude deficiency and the problems caused by the bubble oscillations can be solved by use of arrays of air guns.

Air gun arrays can contain up to several tens of individual air guns with different size compressed air chambers. The aim is to create the optimum pressure in the initial shocklike pressure wave with minimum bubble reverberation after the initial wave. The multigun array produces increased amplitude due to the increased number of guns. Since different volume air guns produce bubbles with different oscillation periods, an air gun array can be built where the bubble pulses peaks after the primary pulse cancel each other due to being out of phase.

Since the air gun array is towed near the sea surface, the reflection of the bubble pulse pressure signals in the surface, the so-called Lloyd's mirror effect, discussed in Chapter 2, gives rise to a pulse with a negative peak. When the guns and their bubbles have different volumes, the bubble peaks occur at different times. It is possible to tune the array by choosing air gun volumes such that a bubble pulse is canceled by the negative pulse from a smaller air gun. Since the air gun array is towed in the same depth, the negative pulse arrival time is the same in all air gun signatures.

The signature of an air gun array depends on the individual air guns and the stability of the air gun array geometry. Also the synchronization of the repeatability of firing time for the individual air guns will influence the array signature. The time synchronization is normally computer controlled within a time window of 100 μs. The stability of the array geometry involves maintaining of the distance between the individual air guns and keeping their tow depth constant. The tow ship turning and the sea state will influence the stability of the geometry. The sea state, i.e., waves on the surface, will also influence the magnitude of the negative pulse peak produced by the reflection at the sea surface.

When one air gun in an array is fired, its signature may influence the signature of subsequently fired air guns if the distance between the two air guns is small. The first air gun's pressure pulses will change the hydrostatic pressure around the next air gun to be fired. If the air guns are less than 1 m from each other, their bubbles may coalesce into one single bubble. Air guns forming a cluster without the coalescence between the bubbles may produce a much stronger peak pressure during their interaction than tuned air gun arrays.

The high pressure amplitudes and the low-frequency contents of the air gun array signals are so loud that they may disturb, injure, or kill marine life. The impacts of the signals may include temporary or permanent hearing loss, habitat abandonment, disruption of mating and feeding, and even beach stranding and marine mammals deaths. Air gun blasts may also kill eggs and larvae and scare fish from important habitat and thus harm commercial fisheries. The potential damage to life in the sea caused by seismic air gun blasting repeated every 10 s, frequently for 24 h a day, and through weeks at a time, is discussed in Chapter 12 on bio and fishery acoustics.

Electrical discharges over a spark gap establish an electrical spark channel in the water. The time for the spark channel formation is dependent on the voltage gradient over the spark gap and may last from a few to about 100 μs, see Bjørnø [65]. When the conducting path between the electrodes in the spark gap has been established, a considerable rise in the electric current through the spark gap takes place and a fast transfer of a substantial amount of energy to the small volume of water will occur, which causes a rapidly rising temperature followed by a fast expansion of the spark channel. This expansion causes a high transient pressure in the water around the spark channel and a shock wave formation. By changing the capacity C {unit: F} of the condenser bank, voltage V_o {unit: V} over the spark gap, electric circuit induction L {unit: H}, or the spark gap length ℓ {unit: m}, the discharge energy and duration can be controlled. It can be shown empirically [65] that the peak pressure P_m {unit: Pa}, the impulse I {unit: Ns/m^2}, and energy flux density E {unit: J/m^2} are given by:

$$P_m = 6 \cdot 10^5 \frac{\left(\ell V_o^2\right)^{1/3}}{R}$$

$$E = 0.32 \frac{\left(\ell V_o^2\right)^{2/3}}{R^2} \tag{13.19}$$

$$I = 2.1 \frac{\left(\ell V_o^2\right)^{1/3}}{R}$$

where R {unit: m} is the distance from the spark gap center to the field point. Since the pressure in the spark channel can be assumed to be directly proportional to the discharged energy in the water spark gap, it is possible to establish a scaling law similar to the pressure variation produced by a chemical explosive detonation. ℓV_o^2 represents the energy contained in the spark gap on a par with the amount of chemical explosive in kg in Eq. (13.16). The $1/R$ dependence of pressure amplitude given in Eq. (13.19) is due to the longer time to establish the spark channel compared to the interaction time between the explosive detonation wave and the surrounding water. High-intensity signals from electrical discharges show a very high degree of reproducibility.

The *boomer* may produce pressure amplitudes above $7 \cdot 10^7$ Pa. The high-intensity pressure waves can be produced by the discharge of a condenser bank through a flat coil with a flat copper membrane situated, insulated from the coil, above the flat coil and in contact with the water. The discharge will produce eddy currents in the copper membrane which cause the membrane to be pushed away from the coil due to self-induction in the membrane. When the membrane is pushed away from the coil, a pressure wave is established in the water above the membrane.

13.4 LIST OF SYMBOLS AND ABBREVIATIONS

Table 13.4 provides a list of the symbols and abbreviations used in this chapter and the page each symbol or abbreviation first appears.

Table 13.4 List of Symbols and Abbreviations

Symbol or Abbreviation	Description	Page of First Usage
SONAR	SOund NAvigation and Ranging	857
Γ	Gol'dberg number	858
x	One-dimensional distance {unit: m}	859
Re_a	Acoustic Reynolds number	858
b	Represents viscosity and heat conduction effects {unit: kg/(ms)}	858
μ	Shear viscosity {unit: km/(ms)}	858
ζ	Bulk viscosity {unit: km/(ms)}	858
t	Time {unit: s}	859
c_o	Local sound velocity {unit: m/s}	859
σ	Dimensionless propagation distance parameter	859
M	*Acoustic Mach number*	859
B/A	Dimensionless second-order *nonlinearity ratio*	859
p	Local sound wave pressure {unit: N/m^2}	859
T	Absolute temperature {unit: K}	859
e_n	Neumann factor, $e_n = 1$ for $n = 0$, and $e_n = 2$ for $n \geq 1$	859
I_n	Bessel function with imaginary argument	859
c_o	Fluid isentropic speed of sound {unit: m/s}	862

Table 13.4 List of Symbols and Abbreviations—cont'd

Symbol or Abbreviation	Description	Page of First Usage
ρ_o	Fluid density {kg/m^3}	862
ℓ_D	Discontinuity distance, i.e., the source distance for first formation of a shock wave in a lossless fluid	862
T	Theoretical cavitation threshold {unit: W/m^2}	864
T_{dB}	Theoretical cavitation threshold {unit: dB rel 1 μPa}	864
P	Totally radiated acoustic power {unit: W}	864
A	Projector's radiating surface area {unit: m^2}	864
d	Projector depth, d {unit: m}	864
p_s	Acoustic radiation pressure {unit: Pa}	865
p_{max}	Maximum pressure {unit: Pa}	865
u_{max}	Maximum particle velocity {unit: m/s}	865
c	Sound velocity {unit: m/s}	865
p_s	Scattered difference-frequency wave pressure {unit: Pa}	867
θ_h	Half-power beam width {unit: rad}	867
G	Parametric gain {unit: dB}	870
SL_0^*	Scaled primary source level {unit: dB}	870
R	Field point measurement distance {unit: m}	870
f_o	Mean primary frequency {unit: kHz}	870
R_o	Rayleigh distance {unit: m}	870
f	Difference frequency {unit: kHz}	870
NUSC	US Naval Underwater Systems Center	873
TOPS	TOwed Parametric Sonar	873
MAST	Marine Science and Technology	874
SIGMA	Sediment Identification for Geotechnics by Marine Acoustics	874
TNT	Trinitrotoluene ($C_7H_5O_6N_3$)	878
D	Detonation front propagation velocity {unit: m/s}	878
SUS	Signal, underwater sound	878
W	Explosive weight {unit: kg}	880
T_n	The period {unit: s} for gas bubble oscillation	880
P_{mp}	First bubble pulsation peak pressure {unit: Pa}	880
C	Condenser bank capacity {unit: F}	883
V_o	Spark gap voltage {unit: V}	883
L	Electric circuit induction {unit: H}	883
ℓ	Spark gap length {unit: m}	883

REFERENCES

[1] Bjørnø, L., *Introduction to Nonlinear Acoustics*. Phys. Procedia, **3**, pp. 5–16, 2010.

[2] Burgers, J.M., *A mathematical model illustrating the theory of turbulence*. In: Advances in Applied Mechanics, 1, Academic Press, New York, pp. 171–199, 1948.

[3] Gol'dberg, Z.A., *Propagation of plane acoustic waves of finite-amplitude in a viscous, heat-conducting medium*. Soviet Phys. (Acoust.), **4**, pp. 119–120, 1958.

[4] Hagelberg, M.P., Holton, G., and Kao, S., *Calculation of B/A for Water from Measurements of Ultrasonic Velocity versus Temperature and Pressure to 10000 kg/cm^2*, J. Acoust. Soc. Amer., **41**, (3), pp. 564–567, 1967.

[5] Coppens, A.B., Beyer, R.T., Seiden, M.B., Donohue, J., Guepin, F., Hodson, R.H., and Townsend, C., *Parameter of Nonlinearity in Fluids. II*, J. Acoust. Soc. Amer., **38**, (5), pp. 797–804, 1965.

[6] Mendousse, J.S., *Nonlinear dissipative distortion of progressive sound waves at moderate amplitude*. J. Acoust. Soc. Amer., **25**, (1), pp. 51–54, 1953.

[7] Reprinted from *Physics Procedia, 3*, (2010), Leif Bjørnø, Introduction to Nonlinear Acoustics, page 9, Copyright (2010), with permission from Elsevier. http://www.elsevier.com.

[8] Neighbors, T.H. and Bjørnø, L., *Monochromatic Focused Sound Fields in Biological Media*. J. Lithotripsy Stone Dis., **2**, (1), pp. 4–16, 1990.

[9] Neighbors, T.H. and Bjørnø, L., *Focused Finite-Amplitude Ultrasonic Pulses in Liquids*. In: Frontiers of Nonlinear Acoustics, M.F. Hamilton and D.T. Blackstock (Eds.), Elsevier Applied Science, pp. 209–214, 1990.

[10] Zabolotskaya, E.A. and Khokhlov, R.V., *Convergent and divergent sound beams in nonlinear media*. Soviet. Phys. (Acoust.), **16**, pp. 39–42, 1970.

[11] Kuznetsov, V.P., *Equations of nonlinear acoustics*. Soviet Phys. (Acoust.), **16**, pp. 467–470, 1970.

[12] Aanonsen, A.I., Barkve, T., Tjøtte, J.N. and Tjøtta, S., *Distortion and harmonic generation in the nearfield of a finite-amplitude sound beam*. J. Acoust. Soc. Amer., **75**, (3), pp. 749–768, 1984.

[13] Hart, T.S. and Hamilton, M.F., *Nonlinear effects in focused sound beams*. J. Acoust. Soc. Amer., **84**, (4), pp. 1488–1496, 1988.

[14] Bjørnø, L., *Acoustic nonlinearity of bubbly liquids*. Appl. Sci. Res., **38**, pp. 291–296, 1982.

[15] Urick, R.J., *Principles of Underwater Sound*, (3rd Ed.), McGrawHill Book Comp., 1983.

[16] Esche, R., *Untersuchungen der Schwingungskavitation in Flüssigkeiten*. Acustica, **4**, Beih. AB208-AB218, 1952.

[17] Strasberg, M., *Onset of ultrasonic cavitation in tap water*. J. Acoust. Soc. Amer., **31**, (2), pp. 163–176, 1959.

[18] Flynn, H.G., *Physics of Acoustic Cavitation in Liquids*. In: Physical Acoustics, Vol. I, Part B, W.P Mason (Ed.), Academic Press, N.Y., pp. 57–172, 1964.

[19] Stephens, R.W.B. and Bate, A.E., *Acoustics and Vibrational Physics*. Edward Arnold Publ., London, 1966.

[20] Nyborg, W.L.M., *Acoustic Streaming*. In: Physical Acoustics, Vol. II, Part B, W.P. Mason (Ed.), Academic Press, N.Y., pp. 265–331, 1965.

[21] Liebermann, L.N., *The Second Viscosity of Liquids*. Phys. Rev., **75**, pp. 1415–1422, 1949.

[22] Westervelt, P.J., *Scattering of sound by sound*. J. Acoust. Soc. Amer., **29**, (8), pp. 934–935, 1957.

[23] Westervelt, P.J., *Parametric Acoustic Array.* J. Acoust. Soc. Amer., **35**, (4), pp. 535–537, 1963.

[24] Bjørnø, L., *Parametric Acoustic Arrays.* In: Aspects of Signal Processing, Vol. 1, G. Tacconi (Ed.), D. Reidel Publ. Comp., pp. 33–59, 1977.

[25] Helmholtz, H., *Die Lehre von den Tonempfindungen als physiologische Grundlage für die Theorie der Musik.* Brunswick, 1862.

[26] Lamb, H., *The dynamical theory of sound.* Dover, New York, 1960.

[27] Lighthill, M.J., *on the sound generated aero-dynamically. I General theory.* Proc. Roy. Soc. (London), **211A**, pp. 564–587, 1952.

[28] Bjørnø, L., Christoffersen, B. and Schreiber, M.P., *Some Experimental Investigations of the Parametric Acoustic Array.* Acustica, **35**, (2), pp. 99–106, 1976.

[29] Kopp, L., Cano, D., Dubois, E., Wang, L., Smith, B. and Coates, R.F.W., *Potential Performance of Parametric Communications.* IEEE J. Oceanic. Eng., **25**, (3), pp. 282–295, 2000.

[30] Bellin, J.L.S. and Beyer, R.T., *Experimental investigation of the end-fire array.* J. Acoust. Soc. Amer., **34**, (8), pp. 1050–1054, 1962.

[31] Berktay, H.O., *Possible exploitation of non-linear acoustics in underwater transmitting applications.* J. Sound Vib., **2**, pp. 435–461, 1965.

[32] Hobæk, H., *Experimental investigations of an acoustical end-fire array.* J. Sound Vib., **6**, pp. 460–463, 1967.

[33] Muir, T.G. and Blue, J.E., *Experiments on the acoustic modulation of large-amplitude waves.* J. Acoust. Soc. Amer., **46**, (18), pp. 227–232, 1969.

[34] Muir, T.G., *An analysis of the parametric acoustic array for spherical wave fields.* Applied Research Laboratories, Techn. Rep. No. 76-7 (ARL-TR-71-1), Univ. Texas Austin, 1971.

[35] Berktay, H.O. and Leahy, D.L., *Farfield performance of parametric transmitters.* J. Acoust. Soc. Amer., **55**, (3), pp. 539–546, 1974.

[36] Bartram, J.F., *A useful analytical model for the parametric acoustic array.* J. Acoust. Soc. Amer., **52**, (3B), pp. 1042–1044, 1972.

[37] Merklinger, H.M., *Improved efficiency in the parametric transmitting array.* J. Acoust. Soc. Amer., **58**, (4), pp. 784–787, 1975.

[38] Mellen, R.H., Browning, D.G. and Konrad, W.L., *Parametric sonar transmitting array measurements.* J. Acoust. Soc. Amer., **49**, (3B), pp. 932–935, 1971.

[39] Blue, J., *Nonlinear acoustics in undersea communication.* JUA (USN), **22**, pp. 177–187, 1972.

[40] Willette, J.G., *Difference frequency parametric array using an exact description of the primary sound field.* J. Acoust. Soc. Amer., **52**, (1A), 123A, 1972.

[41] Fenlon, F.H., *On the performance of a dual frequency parametric source via matched asymptotic solution to Burger's equation.* J. Acoust. Soc. Amer., **55**, (2), pp. 35–46, 1974.

[42] Novikov, B.K., Rudenko, O.V. and Timoshenko, V.I., *Nonlinear Underwater Acoustics,* Acoustical Society of America Translation Book, 1987.

[43] Muir, T.G. and Willette, J.G., *Parametric acoustic transmitting arrays.* J. Acoust. Soc. Amer., **52**, (5B), pp. 1481–1486, 1972.

[44] Rolleigh, R.I., *Difference frequency pressure within the interaction region of a parametric array.* J. Acoust. Soc. Amer., **58**, (5), pp. 964–971, 1975.

[45] Moffett, M.B. and Konrad, W.L., *Nonlinear sources and receivers.* In: Encyclopedia of Acoustics. Vol. 1. M.J. Crocker (Ed.), John Wiley and Sons, Inc., pp. 607–617, 1997.

[46] Ryder, J.D., Rogers, P.H. and Jarzynski, J., *Radiation of difference-frequency sound generated by nonlinear interaction in a silicone rubber cylinder.* J. Acoust. Soc. Amer., **59**, (5), pp. 1077−1086, 1976.

[47] Barnard, G.R., Willette, J.G., Truchard, J.J. and Shooter, J.A., *Parametric acoustic receiving array.* J. Acoust. Soc. Amer., **52**, (5B), pp. 1437−1441, 1972.

[48] Truchard, J.J., *Parametric acoustic receiving array. I. Theory.* J. Acoust. Soc. Amer., **58**, (6), pp. 1141−1145, 1975.

[49] Truchard, J.J., *Parametric acoustic receiving array. II. Experiments.* J. Acoust. Soc. Amer., **58**, (6), pp. 1146−1150, 1975.

[50] Reeves, C.R., Goldsberry, T.G., Rohde, D.F. and Maki, V.E., *Parametric acoustic receiving array response to transducer vibration.* J. Acoust. Soc. Amer., **67**, (5), pp. 1495−1501, 1980.

[51] Konrad, W.L., *Application of the parametric source to bottom and sub-bottom profiling.* In: Finite-Amplitude Wave Effects in Fluids. L. Bjørnø (Ed.), IPC Science and Technology Press, Guildford, UK, pp. 180−183, 1974.

[52] Tjøtta, J.N. and Tjøtta, S., *Interaction of sound waves. Part I: Basic equations and plane waves.* J. Acoust. Soc. Amer., **82**, (4), pp. 1418−1428, 1987.

[53] Berntsen, J., Tjøtta, J.N. and Tjøtta, S., *Interaction of sound waves. Part IV: Scattering of sound by sound.* J. Acoust. Soc. Amer., **86**, (5), pp. 1968−1983, 1989.

[54] Tjøtta, J.N., Tjøtta, S. and Vefring, E.H., *Propagation and interaction of two collinear finite amplitude sound waves.* J. Acoust. Soc. Amer., **88**, (6), pp. 2859−2870, 1990.

[55] Tjøtta, J.N., Tjøtta, S. and Vefring, E.H., *Effects of focusing on the nonlinear interaction between two collinear finite amplitude sound waves.* J. Acoust. Soc. Amer., **89**, (3), pp. 1017−1027, 1991.

[56] Garrett, G.S., Tjøtta, J.N., Rolleigh, R.L. and Tjøtta, S., *Reflection of parametric radiation from a finite planar target.* J. Acoust. Soc. Amer., **75**, (5), pp. 1462−1472, 1984.

[57] Foote, K.G., Tjøtta, J.N. and Tjøtta, S., *Performance of the parametric receiving array: Effects of misalignments.* J. Acoust. Soc. Amer., **82**, (5), pp. 1753−1757, 1987.

[58] Pace, N.G. and Ceen, R.V., *Time domain study of the terminated transient parametric array.* J. Acoust. Soc. Amer., **73**, (6), pp. 1972−1978, 1983.

[59] Wingham, D.J., *A theoretical study of the penetration of a water sediment interface by a parametric beam.* J. Acoust. Soc. Amer., **76**, (4), pp. 1192−1200, 1984.

[60] Pace, N.G. and Bjørnø, L., *The reflection and refraction of acoustic beams at water sediment interfaces.* Acustica - Acta Acoustica, **83**, pp. 855−862, 1997.

[61] Wang, L.S., Smith, B.V. and Coates, R., *The secondary field of a parametric source following interactions with sea surface.* J. Acoust. Soc. Amer., **105**, (6), pp. 3108−3114, 1999.

[62] Bjørnø, L., *Acoustical field prediction and performance assessment by a vertically down-looking parametric array.* Internal Techn. Rep. No. IA-TR-69, Dept. Ind. Acoust., Techn. Univ. Denmark, February 1998.

[63] Berktay, H.O., *Parametric sources − Design considerations in the generation of low-frequency signals.* In: Ocean Seismo-Acoustics − Low-frequency underwater acoustics. T. Akal and J.M. Berkson (Eds.), NATO Conference Series, Plenum Press, pp. 785−800, 1986.

[64] Cole R.H., *Underwater Explosions.* Princeton University Press, 1948.

[65] Bjørnø, L., *A comparison between measured pressure waves in water arising from electrical discharges and detonation of small amounts of chemical explosives.* Trans. ASME J. Eng. Ind., **92**, Ser. B, (1), pp. 29−34, 1970.

Underwater Acoustic Measurements and Their Applications

14

L. Bjørnø[†]

UltraTech Holding, Taastrup, Denmark

14.1 INTRODUCTION

Chapter 14 is a collection of brief, but informative, articles on topics with a strong component of underwater acoustics in them, or a traditional linkage with the field. The list is not meant to imply completeness nor does it have a "rank ordering." Every reader will want to add at least two to three additional topics and the editors could not agree more with that individual.

The opening paper provides an excellent overview of the complexity of the acoustic signals resulting from installation and operation of renewable energy systems and the consequent efforts required to both measure and interpret those signals. The United Nations' Comprehensive Test Ban Treaty Organization (CTBTO), headquartered in Vienna, Austria, has in its arsenal of tools, a number of hydroacoustic sites in strategic locations in the world's oceans and provides both its own expertise in understanding low-frequency acoustic noise and the opportunity for researchers to take advantage of continuous recording that exceeds in many cases, more than a decade of time. The Characterization of Noise from Ships is both a classic and a modern view of the features of radiated energy from naval vessels. Underwater Soundscapes is a new look at understanding and interpreting noise fields. It is an especially valuable methodology for separating the myriad of oceanic noise sources.

Underwater Acoustic Communications provides an excellent introduction to the issues in that field. Underwater Archaeology is an introduction to a fascinating science that benefits greatly from the ability of high-frequency acoustic energy to image in a competitive way with visual means, but over useful ranges, the artifacts of our forebears. Applications of Underwater Acoustics in Polar Environments is very timely, given the renewed interest in these regions due to accessibility caused by global warming. Tank Experiments, given the cost of at-sea experimentation, have

---

[†]30 March 1937–24 October 2015.

Applied Underwater Acoustics. http://dx.doi.org/10.1016/B978-0-12-811240-3.00014-X

offered a cost-effective alternative and this paper provides a modern review of ongoing facilities and efforts. Acoustic Positioning at Sea provides an overview of an extremely important task when in the field. The dynamics of the ship and the surrounding forcing functions in and on the sea make the exact position determination a daunting task. Ocean Observing Systems and Ocean Observatories, Oceanographers, and Acousticians provides both a complete summary of ongoing efforts as well as a perspective on the long and fruitful relationship between the scientists of both disciplines. Applications of Underwater Acoustics to Military Purposes is a complete, and brief, review of its use in various aspects of Naval Warfare.

14.2 ACOUSTICS AND MARINE RENEWABLE ENERGY DEVELOPMENTS

S.P. Robinson[1], **P.A. Lepper**[2]
[1]*National Physical Laboratory, Teddington, United Kingdom*
[2]*Loughborough University, Loughborough, Leicestershire, United Kingdom*

There is a trend for marine renewable energy to form an increasing part of the energy supply chain over the next decade, especially with the movement toward a low-carbon economy and a sustainable energy supply. Because of this, an increasing variety of marine renewable energy devices are being deployed, with perhaps the highest concentration so far being in northern Europe. This has led to a growth in the interest in the underwater acoustics of these developments during the construction and operational phases, mainly driven by the potential for their interaction with marine life. The possible acoustic interactions include the potential for physical harm, and displacement or avoidance due to radiated noise and vibration [1].

The three main forms of marine renewable energy devices (MREDs) are offshore wind turbines, tidal stream turbines and wave energy converters. Numerous commercially operational offshore wind farms already exist with plans to increase the size and numbers as well as moving further offshore, with considerable associated infrastructural development, including ports, service vessels, and cable laying. Wave and tidal stream energy devices are at a less advanced stage, but there are a number of full-scale prototype systems in advanced testing phases, and many different concepts are currently being developed and tested. The industry as a whole has begun a phase of upscaling for both construction and operation for all three MRED types, a process that seems set to last for the next few decades.

There are many parallels between all three renewable energy sectors. For example, developments involve the deployment of substantial structures into the marine environment, requiring large investment and specialized equipment to install and service them. All of these activities may generate underwater acoustic noise. MREDs invariably use moving components, either above the sea surface or underwater, and these generate underwater noise during operation. There is concern that this noise, if sufficiently high level, will have impact on the marine environment, with particular concern being the impact on sensitive species of marine mammals, fish, and invertebrates that utilize underwater sound as part of their survival strategy. A component of the sound field that may possibly be important is particle motion, to which fish are especially known to be sensitive [2]. Additionally, vibrations that can be transmitted through the sediment might have effects on benthic fauna (fish and invertebrates) [3]. While there are many parallels between the MRED types, there are also fundamental differences. For example, the moving structures that capture energy in tidal stream and some wave energy devices are submerged below the water surface. The designs of wave energy technologies in particular are extremely diverse in size, shape, and foundation.

One way of categorizing the acoustics of marine renewable energy developments is by separately considering the construction phase and the operational phase.

14.2.1 NOISE DURING THE CONSTRUCTION PHASE

Marine percussive pile driving is a significant source of low-frequency, high-energy, impulsive underwater noise. During the pile driving process, a pile is driven into the seabed using a hammer, which is typically powered hydraulically. Such a technique is commonly used to position piles in relatively shallow water for construction of wind farms (although it should be noted that the technique is also used in construction of other structures such as platforms and conductor pipes for the offshore oil and gas industry, bridge supports, and foundations in rivers, estuaries, and harbors, and mooring of offshore installations). Perhaps the most commonly encountered foundation type for offshore wind turbines is a monopile, consisting of a steel cylinder, which is typically 60–70 m long, with a diameter of typically 3–6 m and with a wall thickness of a few centimeters. This is driven into the seabed by a succession of blows of a hydraulic hammer, typically to a seabed penetration of around 25 m. The number of blows required to drive the pile to the required depth varies, but can be several thousand blows in an operation that lasts several hours.

The measurement of marine pile driving noise is made more difficult by the fact that the pile itself is an extended sound source that penetrates the seabed and sea surface, generating sound waves in the water, air, and seabed, and vibrating the seabed surface. In addition, the environment is often shallow coastal water, which gives rise to substantial reverberation, and where bathymetric features and seabed interaction can strongly influence the propagation of the sound. The problem of how to characterize the acoustic radiation from piling in a way, which is independent of environmental conditions, is still the subject of study. In fact, there is not yet agreement over the exact meaning of source level for pile driving [4]. Currently, Working Group 3 of ISO Technical Committee 43 (Sub-Committee 3) is addressing this need for standardization. However, there has been some commonality in the measurements reported so far, and the data illustrate a number of features of the radiated noise. Typically, measurements are made so that both the spatial and temporal variation of the radiated noise is determined. This is often done by making measurements along a radial transect moving away from the pile at a fixed bearing [5]. The left-hand-side plot of Fig. 14.1 shows how the acoustic output level can vary throughout the piling sequence: the upper plot shows the time history of the received signals at approximately 1 km from the pile, and on the lower plot are the Sound Exposure Level (SEL) values for each received pulse. The increase in acoustic pulse energy is often correlated with the hammer energy, which is gradually increased during the "soft start" period [6]. The right-hand-side plot shows the measured

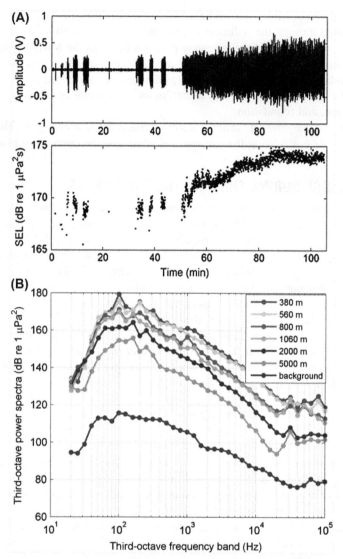

FIGURE 14.1

Example of time history of pile driving sequence and sound exposure level (SEL) for each measured pulse (A) and third-octave band spectra for pulses recorded at ranges from 380 m to 5 km from the driven pile (B).

From Robinson, S.P., Theobald, P., and Lepper, P.A., Underwater noise generated from marine piling. Proc. Meet. Acoust., 17, 070080, pp. 1–10, http://dx.doi.org/10.1121/1.4790330, 2013.

third-octave band spectra for pulses recorded at ranges from 380 m to 5 km from the driven pile. Also shown are the levels for the background noise. It can be seen that the level at 100 Hz is more than 60 dB higher than background at 380 m, reducing to less than 40 dB above background at 5 km. The corresponding values above background at 10 kHz are 45 and 20 dB, respectively.

An alternative way to represent the acoustic output of the source is by stating the received level at a specified range. Using this approach, the SEL values obtained from measurements for a distance of 750 m are typically in the range 172–177 dB re 1 μPa^2s, whereas the peak-to-peak acoustic pressure levels measured at this distance are in the range 200–205 dB re 1 μPa (10–18 kPa) [5]. For accurate prediction of noise radiation during marine pile driving, a good understanding is required of the radiation mechanisms, and progress is being made with modeling the process using finite element techniques [7,8]. With wind turbines being deployed in increasingly deeper water, it is likely that pile driving will give way to other potentially less noisy foundation methods (pin-piled jackets, anchored floating structures, etc.). For harder seabed types, where rock substrates are present (often true for wave and tidal stream energy developments), the foundation is more commonly fixed by drilling into the seabed rather than pile driving, and this process generates less acoustic output [9].

14.2.2 NOISE DURING OPERATION

During operation, it is unlikely that MREDs will emit sufficient noise to cause direct auditory damage to sensitive species, with reported levels being no greater than vessels operating in the vicinity [10,11]. In fact, the increased vessel activity for maintenance activity can be a significant contribution to the noise generated by the development as a whole. However, with regard to the impact of the devices themselves, some level of area avoidance/attraction and masking is possible [12]. In addition, the characteristics of the radiated noise (and therefore the perceivable acoustic signatures) will be an important consideration in how animals interact with the devices. A key example, where auditory cues may play an important role in such an interaction, is how animals are able to detect devices and avoid collision, particularly important for wave and tidal stream systems, where the moving parts are submerged in the water column. To understand these issues, more data on acoustic characteristics for devices during operation are required along with information on existing background noise levels at development sites.

For offshore wind, the operational noise is generally radiated through the foundation and the submerged part of the turbine support, with the gear box providing tonal components within the broadband noise and the radiated noise correlated with wind speed [13,14]. For wave and tidal stream devices, the variety of MRED designs and the distributed nature of the acoustic sources provide a range of source types and signatures. Design concepts for tidal stream energy include horizontal axis turbines, vertical axis turbines, reciprocating hydrofoils, venturi-effect devices, tidal kites, and Archimedes screws; for wave energy, they include attenuators, point absorbers, oscillating wave surge converters, oscillating water column, overtopping, submerged pressure differential, bulge wave, and rotating mass devices (see www.aquaret.com for details on current design concepts). Acoustic sources such as hydraulic hinges, moorings, reciprocating parts, turbines, and gearboxes may be distributed over a device, which can be several hundred meters long. The requirement to measure in the acoustic far-field to determine the source level may necessitate significantly large ranges for measurement, and often in shallow water where the propagation losses are difficult to determine. If the acoustic output is relatively low, there may be difficulty in distinguishing the device noise above the background noise level at ranges greater than a few hundred meters [10].

An issue with characterizing the acoustic output of MREDs is that the output frequently depends on operational state, and this may depend on the wind and sea conditions. This means that an acoustic characterization must involve sampling the acoustic output over time, ideally covering all operating states (and weather conditions). There is currently not a good understanding of the potential influence of the changes in radiated noise relative to background noise on the risk of impact on a range of receptors, in particular how this will influence perception capability, and therefore the collision risk. As with marine pile driving, the concept of acoustic source level is not well established for a distributed source that is attached to the seabed and possibly the sea surface, and where the radiated noise depends on the local environment (seabed, ocean conditions, etc.) [11].

An additional problem when making measurements is the often harsh environment in the vicinity of the developments, particularly for wave and tidal stream energy, where the water column can often be highly energetic with intense lateral, vertical, or oscillatory motion. The fast tidal flow and strong wave action pose severe problems for accurate measurements, motivating the need to explore novel measurement techniques. A major issue in fast tidal flow is flow noise induced at a statically positioned hydrophone. This has led to the use of drifting systems, where the hydrophone is attached to a drogue and drifts with the water, thus minimizing the corruption from flow noise [12]. Often the recorder floats on the surface and has a global positioning satellite (GPS) transponder to enable positional tracking. However, such drifting systems are by their nature mobile, and cannot easily be used for long-term deployments, where the noise levels must be measured in a specified location.

Due to the complexity of measurements required from a wide diversity of MRED types, and due to the complex acoustic environments and the limited understanding of impact on marine life, we are seeing rapid developments in measurement methodologies and equipment used. These are progressing hand in hand with improvements in physical modeling and knowledge gained from biological studies to support the growth in marine renewable industries. This trend is likely to extend over the next decade as the desire to harvest energy from our oceans continues.

14.3 UNDERWATER ACOUSTICS IN NUCLEAR-TEST-BAN TREATY MONITORING

G. Haralabus, M. Zampolli, P. Grenard, M. Prior[1], L. Pautet
CTBTO, Vienna International Centre, Vienna, Austria

Opened for signature in 1996, the Comprehensive Nuclear-Test-Ban Treaty (CTBT) bans nuclear explosions by everyone, everywhere: on the Earth's surface,

---

[1]As of Sep. 2014 with the Netherlands Organization of Applied Scientific Research (TNO).

in the atmosphere, underground and underwater. It is a global deterrent for the development or improvement of nuclear bombs. Since the Treaty is not yet in force, the Preparatory Commission for the Comprehensive Nuclear-Test-Ban Treaty Organization (CTBTO) has been tasked with promoting the Treaty and establishing the CTBT verification regime. As of mid-2016, it is supported by 183 member states (www.ctbto.org).

Although the CTBTO headquarters are in Vienna, its verification regime relies on the International Monitoring System (IMS), which consists of 337 facilities worldwide—some located in the most remote places on the planet—to provide global coverage for signs of nuclear explosions. Over 85% of the facilities are already up and running, providing data in real time to the International Data Centre at the Vienna headquarters. The IMS encompasses four complementary verification technologies: seismic, with 50 primary and 120 auxiliary stations monitoring vibrations in the Earth; infrasound, with 60 stations on the surface, which monitor ultralow-frequency sound waves inaudible to the human ear; radionuclide, with 80 radiation-monitoring stations of which 40 have noble gas detection capability, supported by 16 radionuclide laboratories; and hydroacoustic, with 11 stations monitoring sound waves in the oceans (see Fig. 14.2). The implementation and operation of such a network of stations has posed engineering challenges unprecedented in the history of arms control. The CTBTO has already proved its effectiveness by detecting the 2006, 2009, 2013, and 2016 underground nuclear tests conducted in the Democratic People's Republic of Korea. The IMS is complemented by an on-site inspection capability that can be activated after entry into force of the CTBT.

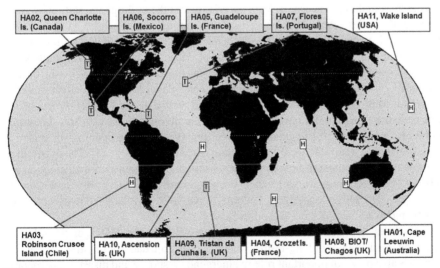

FIGURE 14.2

A world map showing the IMS hydroacoustic component. Cabled hydrophone stations are represented by the letter "H" while T-stations are represented by the letter "T."

14.3.1 **THE HYDROACOUSTIC NETWORK OF THE CTBTO**

As sound propagates very efficiently through water, relatively few hydroacoustic stations are sufficient to cover the world's oceans and ensure that no underwater nuclear explosion goes undetected [15]. 10 of the 11 planned hydroacoustic stations have already been certified as meeting the CTBT's stringent technical requirements.

Of the 11 hydroacoustic stations, 5 are T-stations, which use seismometers to pick up waterborne signals coupled into the Earth's crust. The other 6 are cabled stations, which utilize underwater hydrophones [16]. All the cabled stations have two triplets of underwater hydrophones suspended in the water column in a horizontal triangular configuration with a separation of 2 km, except for HA01 at Cape Leeuwin, Australia, which has only one triplet. The hydrophone depths are as close as possible to the local SOund Fixing And Ranging (SOFAR) channel axis.

The hydrophone sensors record signals in the 1—100 Hz frequency range. The self-noise of the system is required to be 10 dB below the ocean noise for a typically quiet ocean location to maximize the detection range of the hydroacoustic network; the dynamic range is required to be at least 120 dB, with a clipping limit of 185 dB re mPa. The signals acquired from the underwater hydrophones pass through tens of kilometers of underwater fiber-optic cable to a shore station and from there via a real-time satellite link to Vienna.

While the hydroacoustic network's primary purpose is to detect, locate, and identify underwater explosions, secondary benefits come from the civil and scientific applications of the data. Given the remarkable propagation efficiency of low-frequency sound in the ocean, it is not surprising that the hydrophone stations detect many other underwater events, e.g., marine mammal vocalizations [17], iceberg calving [18], underwater volcanoes [19], and seismic exploration. The long-term data acquired by the hydroacoustic system provide an opportunity for studies of the undersea soundscape and long-term ambient noise [20,21], the effect of ocean variability on coherent processing of ambient noise data [22], and the effect of variability on long-distance sound propagation and arrival times [23]. Hydroacoustic data from the IMS network have also been used to investigate the acoustic signature and fault rupture length of tsunamigenic earthquakes, and the coupling of seismic sound generated by submarine events to atmospheric infrasound [24]. The feasibility of using the hydrophone triplets as vector sensors of opportunity for the detection of seismic P- and Rayleigh waves has also been demonstrated [25]. Three-dimensional diffraction and refraction effects become apparent by looking at hydroacoustic signals received from distant natural events.

This brief and nonexhaustive overview shows how the IMS can provide an unprecedented amount of long-term acoustic and seismic data for scientific research. The IMS also plays an important role in emergency response and disaster risk reduction. Data from IMS hydroacoustic and seismic stations are transmitted in near real time to tsunami warning centers around the world established under the auspices of the United Nations Educational, Scientific and Cultural Organization (UNESCO) Intergovernmental Oceanographic Commission and contribute to tsunami emergency response plans.[2]

14.3.2 INSTALLATION AND PERFORMANCE OF THE NEWEST IMS HYDROACOUSTIC STATION: HA03, ROBINSON CRUSOE ISLAND, JUAN FERNÁNDEZ ARCHIPELAGO, CHILE

Following a major engineering and logistical undertaking that spanned 4 years, HA03 was reestablished in 2014. The first HA03 station was destroyed in 2010 by a tsunami that devastated Robinson Crusoe Island and claimed 16 lives. The newly installed HA03 has recorded a multitude of natural signals of scientific interest, which will be available to researchers through the virtual Data Exploitation Centre (vDEC) system[3] at the CTBTO or via their respective National Data Centers, together with data from other IMS stations.

One major natural event recorded by HA03 was the magnitude 8.2 earthquake on Apr. 1, 2014, in Northern Chile. The first of the fast seismic waves traveling through the crust below the ocean and leaking acoustic energy into the water reached the hydrophone at approximately 100 s (see Fig. 14.3A). Most of the acoustic energy received was contained in the sound radiated from the epicenter area into the water near the coast, which traveled toward the hydroacoustic station and arrived at the hydrophone later, as shown in the same figure. The waterborne sound generated by the earthquake was observed to come from the direction predicted on the basis of the event's epicenter, located approximately 1700 km from HA03. HA11 at Wake Island (located 14,000 km away from the event) also received very clear signals from this earthquake.

The calls of some large baleen whales are in the frequency band of HA03 (see Fig. 14.3B). Whales swimming in the vicinity of the triplets can also be tracked by analyzing the time-of-arrival delay at the different hydrophones. Also in April 2014, underwater explosion-like signals generated by bursting underwater gas bubbles emitted from an undersea volcano near the Mariana Islands in the North Pacific Ocean, 15,000 km from Robinson Crusoe Island, were detected by HA03. Naturally the same signals were also picked up by HA11, which was significantly closer to the event than HA03.

While the CTBTO uses its hydroacoustic network to ensure that no underwater nuclear test goes undetected, it is clear that the data offer a veritable wealth of information with the potential to enhance scientific understanding of the oceans.

---

[2]For more information: http://www.ctbto.org/fileadmin/user_upload/pdf/Spectrum/2012/%20Spectrum_18_p33.pdf.

[3]vDEC is a platform that enables researchers to access archived monitoring data and processing software.

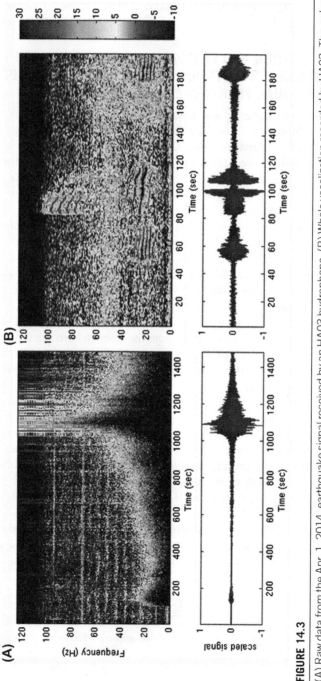

FIGURE 14.3

(A) Raw data from the Apr. 1, 2014, earthquake signal received by an HA03 hydrophone. (B) Whale vocalization recorded by HA03. The *color scale* represents the spectrogram magnitude in decibels (dB) re arbitrary reference.

14.4 CHARACTERIZATION OF NOISE FROM SHIPS

G. Grelowska, E. Kozaczka
Gdansk University of Technology, Gdansk, Poland

14.4.1 GENERAL CHARACTERIZATION OF NOISE PRODUCED BY SHIPS

The passing and underwater movement of objects produce noise of variable intensity, which significantly increase the overall level of noise in the sea [26–29]. This applies to both the sonic and the ultrasonic range. The excessive levels of underwater noise adversely affect the so-called underwater acoustic climate and is the reason why this phenomenon has been extensively investigated for a number of years [28,30]. Disturbances called "shipping noise" are among the important components of the noise in the sea [31].

Among the acoustic waves produced by ships those which should be emphasized are closely connected with maintaining navigational parameters, which means with determination of the stationary or dynamic position of the ship [26,27].

This noise is mostly produced by the main propulsion and auxiliary systems. The noise level depends in particular on the resistance produced by the environment to the moving object, and it is usually connected with forward speed or the rapid change of acceleration in the case of movements related to the ships maneuvers. Apart from that, to increase the safety of shipping, echo-locating systems are used, which allow detection of stationary and moving navigational obstructions and evaluation of the speed of a ship.

Moreover, ships, depending on their type, are equipped with sonars, which increase the underwater noise level by radiation of acoustic pulses of very peak pressure values.

Ships used for investigation of properties of the water environment and of the seafloor have a particular influence on the underwater noise level. In this research, sources of impulse character (known as boomers) are generally applied. This includes also all kinds of underwater explosions and electric discharges.

This generation of pressure waves in water, which can be compared to pressure waves in the air (shock waves), is most harmful, because the low compressibility of water makes these pressure waves very intensive.

There are also destructive disturbances of continuous or impulsive character, which appear in case of research aiming for determination of the properties of sea water and its environment by means of underwater nonlinear sources. Therefore just as in the case of low-frequency sources in air, disturbances produced by parametric sources in water can have adverse impact on the environment.

It remains to stress distinctly that noise propagates in water to considerably greater distances and with considerably lesser losses than in air [32]. In reality, strong local pressure waves are no longer local due to their interaction. Besides, reverberation effects, in the volume as well as on the surface, boost these adverse effects.

Turning our interest to traditional classification of noise produced by ships we have [26,30,33,34]:

- noise generated by dynamical devices placed inside and/or on the surface of the hull, mainly engines, propulsion and auxiliary systems, and systems for transport of mechanical energy—shafting,
- noise produced by ship propellers,
- acoustic effects connected with cavitation by propellers and with flow around the underwater part of the hull.
- The sample characteristic relation of sound pressure level as a function of distance between ship and sensor is shown in Fig. 14.4 [33]. The curves were determined on basis of many repeated measurements performed in the same trial area, and measurements performed, when the ship is crossing the trial area under the same work conditions of its propulsion system. Results of the measurements were subject to relevant statistical processing and approximation procedures. The approximation by means of polynomial of the second or third order is most commonly used to approximate the relation of pressure levels as a function of ship distance, water depth, and forward speed of the ship.

At low speed the ship's service generator is the main source of underwater noise generated by ship. It radiates tonal components that contribute to almost all of the radiated noise power of the ship. This source is independent of the ship speed. Few noise components are loud enough to be contributors to the high-speed signature. The tonal levels of the service diesel generator of the ship are nearly stable in amplitude and frequency [33]. The wide-band energy of noise generated by the service generator of the ship is proportional to the square of generated power [32].

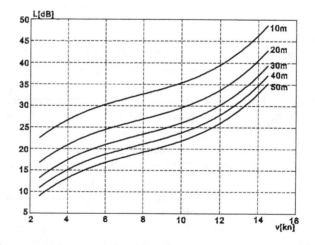

FIGURE 14.4

The characteristic relation of the sound pressure level as a function of the ship speed. Measurements of the ship were performed at the water depth $h = 10, 20, 30, 40,$ and 50 m.

14.4.2 **NOISE GENERATED BY A PROPULSION SYSTEM**

The propulsion engine is the main source of underwater noise for moderate speed of a ship (see Fig. 14.5). Comparison of the spectrum of underwater noise and the spectrum of vibration of the engine allows us to determine the components in underwater noise caused by the engine activity (see Fig. 14.6).

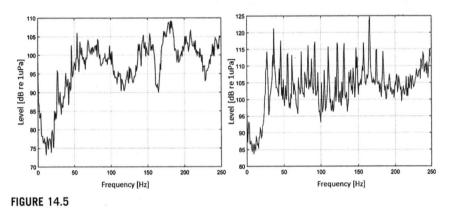

FIGURE 14.5

Examples of acoustic power spectra of vessels, where: left is a ship with turbine propulsion system and right is a ship with conventional propulsion system.

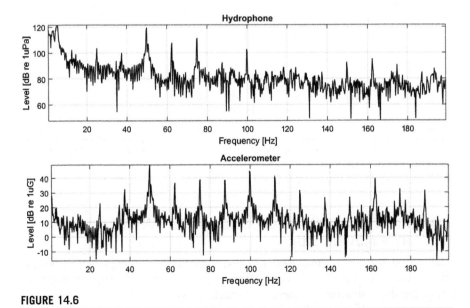

FIGURE 14.6

Spectra of underwater noise produced by small ship motion, measured by hydrophone and spectrum of the vibration of the main propulsion system.

14.4.3 NOISE GENERATED BY A PROPELLER

The most efficient underwater noise source on a ship is the propeller. One part of the noise spectrum is the blade rate, which represents the signal produced by the blade passing frequency and its harmonics. This gives usually the dominant contribution to the low-frequency tonal level at high ship speed, when the propeller is heavily cavitating [35].

In the spectrum of underwater noise produced by ships, components can be distinguished, the origin of which can be directly linked to mechanisms of the ship. In Fig. 14.7, consecutive spectra of underwater noise of a small ship calculated for particular sectors of track length of 1 km and the averaged spectrum for the track are shown.

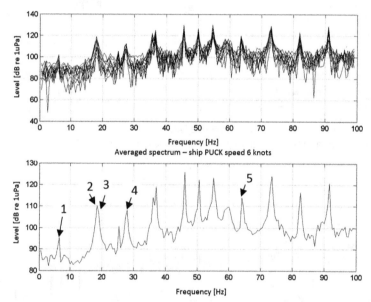

FIGURE 14.7

Consecutive spectra and the average spectrum for 1-km track of a small ship: (1) shaft noise, (2) propeller—fundamental frequency, (3) unbalance of the propeller, (4) combustion of fuel in the cylinders, and (5) propeller—third harmonic.

14.4.4 IDENTIFICATION OF ACOUSTIC WAVES EMITTED BY A MOVING SHIP

Identification of sources of underwater noise generated by a moving vessel, in which various devices are installed, is a complex problem. Vibrational energy propagates through the ship's structural elements and interferes with acoustic waves emitted by various sources, which complicates the source identification.

One of the procedures for identification of ship-generated underwater noise is the examination of its spectrum. On the basis of such investigations characteristic components of the spectrum can be selected, which can be associated with the running ship's mechanisms and devices. Moreover, continuous spectra that reflect work of cavitating ship propellers, turbulent flow in piping systems, fan noise, friction in slide bearings, etc. can give information about the ship operational condition. It is rather difficult in praxis to identify underwater noise. Ship-borne noise is mixed with environmental noise propagating from remote ships, shipbuilding, and port facilities. There are also natural noise sources such as waves, wind, and rain.

To find similarity of the recorded signals of underwater noise and the mechanisms of vibration, coherence functions have been determined. On the diagram of coherence functions, several characteristic components of the function with values close to 1 can be seen (Fig. 14.8).

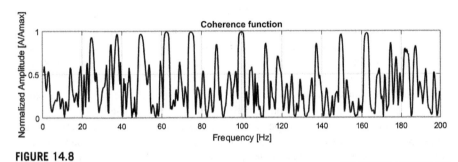

FIGURE 14.8

Coherence function of underwater noise and vibrations (shown in Fig. 14.6).

14.4.5 SUMMARY

In the shallow sea, it is possible to identify underwater acoustic waves associated with propulsion systems and auxiliary mechanisms working in ship. Extensive tests conducted on test ranges have shown that based on measurements of underwater noise, it is possible to identify noise produced by specific components in the ship as, for instance, the operation of the main propulsion systems, the shaft, the propellers, and also operation of electrical installations.

The results of investigations of ship noise conducted under natural conditions are acoustical images of the ship noise, known as spectrograms, and the underwater noise spectra have produced valuable results that allow one to describe the condition of vessels and to rate dynamic parameters of mechanical devices placed both inside and outside the hull.

14.5 UNDERWATER SOUNDSCAPES

J.L. Miksis-Olds

University of New Hampshire, Durham, NH, United States

Interest in underwater soundscapes stems from two overarching applications: (1) those seeking to maximize efforts to detect signals in the marine environment, and (2) those attempting to determine how changes in ocean sound may impact marine ecosystems. The faction of ocean users wanting to maximize detection efforts to gain knowledge about the environment or specific targets of interest include the military, commercial industry, environmental managers, and researchers. Environmental noise level is a major component of the passive sonar equation used in signal detection, so any changes in the soundscape have the potential to interfere with signal detection efforts and are a concern for both soundscape applications.

The term soundscape is currently used in multiple disciplines of study including music, cognitive psychology, acoustic ecology, bioacoustics, and acoustical oceanography. Historical use of the term soundscape has an inherent bias toward humans and terrestrial environments due to the familiarity of people with their terrestrial surroundings. The term soundscape was first used in urban planning to describe the relationship between the acoustic environment and human perception of space in a city setting [36]. Southworth [36] discriminated between sounds informative to humans from background sound. Schafer [37] later used the term soundscape in the context of music composition where natural sounds were used to create music. He used the concept of soundscapes to describe how acoustic characteristics of an area (landscape) reflect natural processes referring to terms such as "soundmarks" and "keynotes" [37]. Soundscape ecology is defined as the combination of all sounds (biologic, abiotic/geophysical, and anthropogenic) emanating from a given landscape to create unique acoustical patterns across a variety of spatial and temporal scales [38]. The conceptual framework and use of soundscapes within the terrestrial environment and disciplines exceeds that of underwater environments. This is largely a function of (1) the ease of recording sound in air compared to underwater, (2) the ability of humans, who are dominantly visual creatures to link sound production to visible sources, and (3) the relative ease of exploring human perception and auditory experience compared to that of aquatic animals in their natural habitat.

From an underwater perspective, a great deal of information related to ocean dynamics and ocean use can be gained simply by listening to the ambient sound. Information contained in soundscapes provide a means for better understanding the influences of environmental parameters on local acoustic processes [39–41], for assessing habitat quality and health [41,42], and for better understanding the impacts and risks of human contributions to the soundscape on marine life. A large number of aquatic species use sound cues contained in local soundscapes to navigate, forage, select habitat, detect predators, and communicate information related to critical life functions (e.g., migration, breeding). To evaluate the impact of anthropogenic contributions to the soundscape on marine animals, it is necessary to better understand soundscape dynamics and how animals are using information contained within soundscapes.

Underwater soundscapes are dynamic in that they vary in space and time within and between habitats. Underwater soundscapes are highly influenced by local and regional conditions, and unlike most terrestrial soundscapes, distant sources can also significantly contribute to local and regional soundscapes because sound propagates such great distances underwater (Fig. 14.9). The underwater soundscape is composed of contributions from human activity (e.g., shipping, seismic airgun surveys) (Fig. 14.9, Anthropogenic Factors), natural abiotic processes (i.e., wind, rain, ice), nonacoustic biotic factors (e.g., animal movement), and acoustic contributions from vocalizing, biological sources (e.g., marine mammals, fish) (Fig. 14.9, Acoustic Behavior). One-way arrows in Fig. 14.9 show that the soundscape is directly influenced in a single direction by anthropogenic and abiotic factors, whereas double-headed arrows indicate that the soundscape is not only influenced by, but also influences, the biological soundscape component [43]. Consequently, the underwater soundscape is not merely a physical parameter of the environment to be measured and quantified. The soundscape is an active component of the feedback loop where changes in soundscape have the potential to impact acoustic behavior and biotic factors, which influences the behavioral ecology of the ecosystem and ultimately further alters the soundscape (Fig. 14.9).

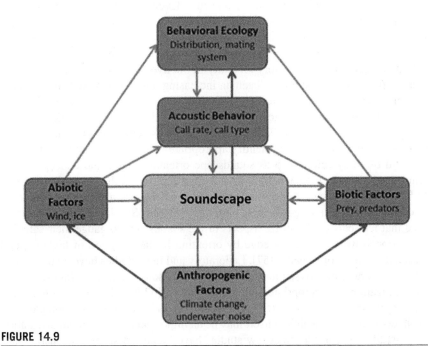

FIGURE 14.9

Soundscape presence within the framework of acoustic ecology.

Adapted from Figure 1 of van Opzeeland, I.C. and Miksis-Olds, J.L., Acoustic ecology of pinnipeds in polar habitats. In: Aquatic Animals: Biology, Habitats, and Threats., D.L. Eder (Ed.), Nova Science Publishers, Inc., New York, pp. 1–52, 2012.

An increase in low-frequency underwater ocean sound was observed over the past 50 years. Although the majority of measurements supporting an increase in ocean sound were made in the North Pacific Ocean [44–47], the ramifications of possible future increases are of global concern. The rise in low-frequency sound (10–200 Hz) has been mainly attributed to an increase in commercial shipping [29,44,46], but shipping increases alone do not fully account for the observed 10–12 dB increase in the 20–40 Hz band from 1965 to 2003 [45,46]. Activities from oil and gas exploration and production, as well as from renewable energy sources, have been increasing [48]. Recovering and increasing whale populations, as well as noise compensation mechanisms employed by marine animals utilizing the low-frequency bands for communication, may also contribute to rising sound levels as the energy contribution of vocalizations increase due to increases in overall population abundance and amplitude as animals vocalize louder to be heard above the noise [49]. Potential contributions to low-frequency sound levels from geophysical sources related to climate change include disintegrating icebergs in the open ocean that generate low-frequency noise with large source levels and are found contributing to the regional noise budget for an extended period [50]. Previous studies have reported a significant increase of ambient noise levels in the North Pacific; however, current studies in the Indian and equatorial Pacific Oceans have not observed a uniform increase in ocean sound levels [20,51]. Very little is known about the global soundscape as a whole, and this needs to be studied. Theory and increasing observations suggest that human-generated noise could be approaching levels at which negative effects on marine life may be occurring [48].

Sound is an important sensory modality in the lives of many marine organisms. Over the past decade, there have been an increasing number of studies that explore how animals use information from the overall environmental soundscape gained via passive listening for communication, orientation, and navigation [38,52–54]. The concept of using ambient or reflected sounds (as opposed to specific communication signals) to direct movement or identify appropriate habitats has been identified as a new field of study referred to as soundscape orientation, and the concept is also included within the broader field of soundscape ecology in the scientific literature [38,53]. It has been speculated that large baleen whales use ambient acoustic cues or acoustic landmarks to guide their migration [55,56]. Similarly, it has been proposed that ice seals could utilize aspects of the soundscape to gauge their safe distance to open water or the ice edge by orienting in the direction of higher sound levels indicative of open water [57]. Laboratory and field studies have demonstrated that both invertebrates (oyster and crab) and fish use soundscape cues for orientation and localization of appropriate settlement habitat [52,54]. Habitats with greater biodiversity are often associated with richer acoustic soundscapes compared to low diversity habitats, which in itself may be an important cue for animal orientation [38,40,41,58,59]. Only a select few studies have identified specific acoustic characteristics of the soundscape that are predictors of behavioral response related to orientation or settlement. Frequencies of 10–40 kHz were identified as strong predictors of ice seal vocal presence in the Bering Sea during the breeding season, yet seals do not vocalize in this frequency range [57,60]. There was a 20—30 dB difference in 10—40 kHz sound levels during solid ice conditions compared to open water or seasonal melting conditions, which may provide a salient acoustic gradient between open water and solid ice conditions by which ice seals could orient so that access to open water for breathing is preserved [57]. Stanley et al. [61] measured the sound intensity level required to elicit settlement and metamorphosis in several species of crab larvae, and Simpson et al. [62] discovered that coral reef fish responded more strongly to the higher frequency components (>570 Hz) of the reef soundscape.

While deconstructing the acoustic environment is providing new information on which soundscape cues are important in soliciting specific animal behaviors, other studies are examining the soundscape in its entirety as an indicator of habitat or overall ecosystem quality and health. Indicators of habitat quality and biodiversity that were developed for terrestrial applications are now being applied to marine habitats and soundscapes [41,42,63]. Rapid acoustic analysis of a habitat's soundscape through the calculated acoustic complexity index, acoustic entropy index, or diversity (acoustic dissimilarity index) is providing a quantitative way to assess biodiversity and compare/contrast soundscapes of different areas [40—42,58] One of the major challenges in applying indices developed in the terrestrial environment to marine systems is distinguishing whether increased levels of entropy or biodiversity were a result of natural biotic signals or increased background noise from human generated or abiotic sources [42,64]. Sound travels further underwater than in air, so noise sources from afar that overlap in frequency of local or regional signals of interest complicate interpretation of the calculated indices and limit the use of filtering techniques. Further development of soundscape-derived ecosystem indicators will provide a useful tool for ecosystem monitoring for a variety of applications.

There is much to be learned from our terrestrial counterparts as the field of underwater acoustics develops its use and framework for defining, visualizing, and comparing acoustic environments. Ocean sound is not linear or often stationary; thus, examining the spectrum as a whole and as the sum of its different parts provides insight to ocean dynamics that would not be identified otherwise [20]. The soundscape can be selectively decomposed and visualized to gain a greater understanding of the sources and environmental dynamics contributing to and shaping the temporal and spatial patterns of the measured sound. To date, application of the underwater soundscape has only taken into account the measured physical component of the soundscape, as making the perceptual link between the soundscape and marine life cognition is not feasible at this point in time. Developing a common vocabulary, measurement parameters, and standard method for displaying acoustic data is critical for a field that strives to understand an environment where sound, as opposed to vision, is the dominant mode of communication and obtaining information, and where the visual link between sound production and source is often limited by distance and the physical barrier of the water surface.

14.6 UNDERWATER ACOUSTIC COMMUNICATIONS

C.C. Tsimenidis

Newcastle University, Newcastle upon Tyne, United Kingdom

In recent years, there has been an immense interest in developing underwater acoustic communication (UAC) systems, most of which are related to remote control and telemetry applications. Other applications include ocean-bottom survey and collection of scientific data acquired by subsea sensors without the need for retrieving the equipment. However, for all these applications the principal function is to achieve reliable communication both in point-to-point links and in network scenarios. In practice, the only feasible method to achieve subsea communications is by means of acoustic signals. Such acoustic links are exposed to adverse physical phenomena governing acoustic wave propagation in the sea. These include ambient noise, frequency-dependent attenuation, temperature and pressure variations, reverberation, and extended multipath sound propagation. Any successful acoustic modem design must consider all these effects to select an appropriate configuration for system-related parameters. The transmitted power level and operating frequency must be considered in conjunction with the ambient noise and transmission range, the utilized modulation scheme, data rate, and the level of diversity must properly match the expected channel conditions related to time and frequency dispersion. Another key task is the choice of multiple-access strategy.

A typical underwater scenario is illustrated in Fig. 14.10, where the communication takes place between a stationary ship and a mobile autonomous underwater

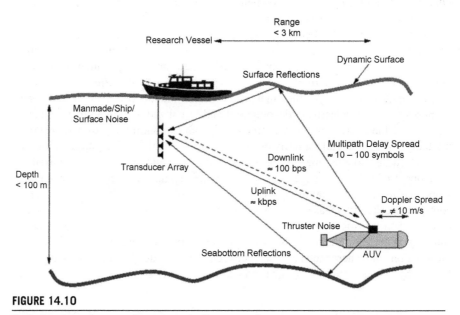

FIGURE 14.10

Typical underwater acoustic communication scenario.

vehicle (AUV). The communication link is operating in asymmetric half-duplex mode due to the limited bandwidth available. Asymmetric here refers to the uplink and downlink data rate requirements. In practice, the downlink requires more reliable connectivity; however, this occurs at lower transmission rates compared to uplink. The demodulation of information-bearing signals at the receiver is implemented using sophisticated multichannel digital signal processing algorithms to remove transmission impairments arising from multipath propagation and Doppler effects [65,66].

The bandwidth available for transmission depends strongly on the transmission range. The latter can be used to classify UAC links according to their range as very short, short, medium, long, and very long. The available bandwidth for each of these links is synoptically illustrated in Table 14.1.

Table 14.1 Underwater Acoustic Communication Classification Using Range and Bandwidth

Link Type	Range (km)	Bandwidth (kHz)
Very short	<0.1	>100
Short	0.1–1	20–50
Medium	1–10	10
Long	10–100	2–5
Very long	>100	<1

The UAC link reception quality is characterized by the signal-to-noise ratio (*SNR*) per received symbol that can be expressed in decibel as:

$$SNR = 10 \log_{10} \frac{E_b}{N_0} \text{ dB},$$

where E_b is the signal energy of a symbol at the output of the channel, $N_0 = 2\sigma_0^2$ is the two-sided noise energy, and σ_0 is the standard deviation of noise samples. Unlike radio frequency–based wireless communications, the noise characteristics exhibit a highly non-Gaussian probability density function. Fig. 14.11 illustrates the time transient and frequency spectrum of noise acquired by a hydrophone near the sea-floor. A closer look reveals the impulsive nature of the ambient noise in the form of signal spikes. The frequency spectrum indicates intensified low-frequency components due to disturbances such as engine noise and man-made noise. In contrast, the spectrum becomes more flat for higher frequencies due to frequency-dependent attenuation [66].

The impulse response of the UAC channel can be characterized as "doubly" spread. That is, it exhibits both delay and Doppler spreading. Delay spread leads to time dispersion and frequency-selective fading. Time dispersion expands a signal in time so that the duration of a received symbol appears to be longer than that of the

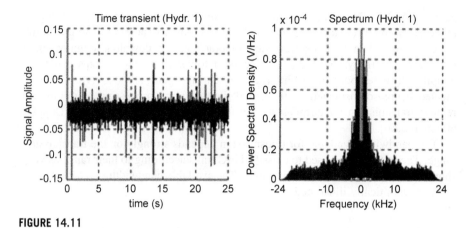

FIGURE 14.11

Noise time transient and frequency spectrum.

transmitted symbol. Frequency-selective fading attenuates the frequency contents of the transmitted signal in such a manner that frequency components that are closely spaced receive the same amount of attenuation; in contrast, if they are far apart they often receive immensely different attenuation. The impulse response characteristics are measured by correlation processing using a wideband chirp signal preceding the actual transmission. Fig. 14.12 illustrates the impulse response of a typical shallow

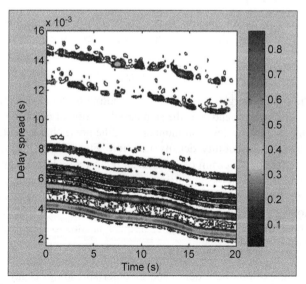

FIGURE 14.12

Time varying impulse response measured over 20 s using a wideband chirp.

water channel with delay-spread times of up to 15 ms. It is worth emphasizing that the observed drift in the impulse response is due to the mobility of the receive vessel. Two multipath dispersion parameters are the root mean square (rms) delay spread and the coherence bandwidth of the channel. Both parameters can be estimated from the impulse response as [66,67]:

$$\tau_{rms} = \sqrt{\frac{\sum_{l=1}^{L}(\tau_l - \tau_0)^2 P_l}{\sum_{l=1}^{L} P_l}} \approx 3.8 \text{ ms}, \quad B_{coh} \approx \frac{1}{\tau_{rms}} = 263 \text{ Hz}$$

where P_l and τ_l represent the power and time of the l-th signal arrival, respectively, and L is the total number of multipath components. Clearly, since the bandwidth of the chirp is $B = 4$ kHz and $B \gg B_{coh}$, we are dealing with a wideband communication channel. The channel exhibits a distinct multipath profile with constant energy arrivals that produce frequency-selective fading that is visibly demonstrated in the frequency response.

Doppler spreading results in two effects, frequency dispersion and time-selective fading. Although all these effects are present in the UAC channel, whether they are apparent to a communications system operating over such channel is dependent on the design of the transmitted signal. Fig. 14.13 depicts the transient of the maximum of the squared cross-correlation coefficient along with its power spectral density. The double-sided Doppler spectrum and the coherence time can be given as [67]:

$$B_d \approx 0.4 \text{ Hz}, \quad T_{coh} = \frac{1}{B_d} \approx 2.5 \text{ s}.$$

The product $\rho = \tau_{rms} B_d$ is known as the spread factor ρ of the channel. Depending on the value of the spread factor, a channel can be characterized as underspread

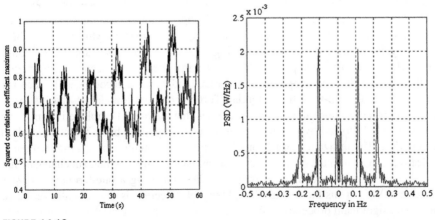

FIGURE 14.13

Maximum squared cross-correlation coefficient (left) and Doppler spectrum (right) of a typical UAC channel.

($\rho < 1$) or overspread ($\rho > 1$). In practice, if the spread factor is $\rho \ll 1$ the channel impulse response can be measured reliably. In contrast, if the channel is overspread the measurement of the channel impulse response is extremely difficult and unreliable. The resulting spread factor of this channel is estimated as [67]:

$$\tau_{rms}B_d \approx 3.8 \text{ ms} \cdot 0.4 \text{ Hz} = 7.6 \cdot 10^{-4} \ll 1.$$

Thus this specific UAC channel can be characterized as underspread, which in turn indicates that phase coherent communications is feasible. The latter is required in linear phase modulation schemes to maximize the transmission efficiency.

In the past two decades, a plethora of receiver structures has been designed to mitigate the multipath and Doppler-induced propagation impairments. The optimal equalizer is a maximum-likelihood sequence estimator utilizing a trellis description of the finite impulse response channel states. At low SNR a maximum a posteriori detector can be utilized; however, in both cases, the computational complexity increases exponentially with channel memory [67]. Efficient, suboptimal approaches have been proposed and successfully demonstrated using the adaptive decision feedback equalizer [68] combined with closed loop carrier phase and timing corrections. However, error propagation of past decisions in the feedback filter can lead to severe performance degradation especially at low SNR levels. To increase SNR, spatial diversity combining and beam-forming receivers have been used; however, these structures are both bulky in nature and difficult to deploy, as well as very costly [69]. Alternatively, the SNR can be increased using temporal diversity techniques such as error control coding (ECC) [70] and spread spectrum [69,71]. Both approaches introduce redundancy in time domain to improve detection; however, this performance boost occurs at the cost of reduced data rate. Furthermore, the introduction of ECC enables the design and utilization of iterative processing algorithms, referred to also as turbo equalization, which dramatically improves performance and makes temporal diversity a feasible alternative to multielement receivers [72].

For short-range and under moderate Doppler conditions, multicarrier-based modulation schemes such as Orthogonal Frequency Division Multiplex (OFDM) using cyclic prefix (CP) or zero padding (ZP) have been the subject of intense research in the last decade [73]. These systems need to be used in conjunction with frequency diversity methods such as ECC, and iterative decoding at the receiver is required to exploit their full potential. Turbo codes and low-density parity check codes represent state-of-the-art channel codes that have been employed to improve performance followed by interleaving to convert impulsive burst errors to random errors [74]. The utilization of CP or ZP in the time domain converts the frequency selective UAC multipath channels into narrow-band, frequency flat, subchannels that can be successfully equalized in frequency domain using a one-tap equalizer per subcarrier [75]. Thus the overall complexity of detection is dramatically reduced; however, significant processing effort is required to remove Doppler-induced effects that manifest themselves as intercarrier interference [76]. Furthermore, signaling overhead in the form of subcarrier interleaved pilots is required to be able to estimate the channel frequency response required for the equalization prior to detection. In cases of extremely frequency selective channels, dense pilot repetition patters are required that can substantially reduce the effective throughput [73]. A further major drawback of OFDM is the high peak to average power ratio (PAPR) of its waveform, which requires either strictly linear power amplifiers to avoid introducing nonlinear distortion and limiting their dynamic range, or utilization of active PAPR reduction methods [75]. Furthermore, the spectral efficiency of the OFDM-based system can be improved if it is used in conjunction with multiple-input-multiple-output (MIMO) systems and employs adaptive modulation and coding along with optimal subcarrier power and bit loading [77]. Future research in this area is heading toward cross-layer approaches that utilize cooperative network coding and OFDM modulated physical layer network coding and asynchronous multiuser MIMO-OFDM techniques [73].

14.7 UNDERWATER ARCHAEOLOGY

A. Caiti, **P. Gambogi**, **D. Scaradozzi**
University of Pisa, Pisa, Italy

Marine archaeological research makes ample use of technological tools from different oceanic engineering fields, as optical imaging, photogrammetry, marine robotics, and, of course, underwater acoustics. Some of the applications of underwater acoustics in the marine archaeological field are well known and obvious, as the use of side-scan sonar in search and localization of seabed relics. Other applications are more subtle and hidden, but by no means less important, as the use of acoustic positioning systems to track divers working over a relic. In the following, the discussion is focused toward the use of acoustic instrumentation in the various stages of the standard procedures applied by marine archaeologists, in particular indicating which instrument is used in which stage and with what requirements. The description of the instrument characteristics and of its operation modality is assumed at this stage known to the reader. Before focusing on acoustic instrumentation and measurements, the requirements from the marine archaeologists will be summarized.

There are many journal papers reporting experimental results and field experience on specific archaeological sites, but not so many comprehensive texts describing in detail the application of acoustic investigation methods to the archaeological domain. One exception is the work of Plets [78], which is a rigorous and comprehensive account of marine geophysical remote sensing techniques applied to the underwater archaeological domain; the classic book of Blondel [79] is also a relevant background for the interpretation of side-scan sonar imagery of seabed relics.

14.7.1 THE WORKING CYCLE OF FIELD MARINE ARCHAEOLOGISTS

The underwater relics of historical significance are a world heritage protected by international conventions, the last in time being the UNESCO Convention on

Underwater Cultural Heritage, 2001. National and regional authorities act in respect of such conventions. Even for the countries that have not (yet) ratified the Convention, there are local regulations most often very close to those of the Convention. One of the implications of the Convention is that marine archaeological sites have not only to be identified, but also continuously monitored and preserved; moreover, such task is not left to the individual's initiative, but it can only be accomplished by competent authorities or under their authorization, monitoring, and supervision. As a consequence, the working cycle of the authorized marine archaeological groups consists of:

* searching for new relicts;
* mapping the found relicts (i.e., producing maps of the relict with precise geometric relations among all the relict components)
* periodically to inspect the known relicts, mapping them again for comparison with previous maps.

In many cases the underwater site is not excavated and the material is not recovered. This is done to preserve the site and to avoid destroying specific information that is carried out by the relict layout. When there is a superior interest to excavate the site, recovering remnants that may be buried within the seafloor, a preliminary detailed map of the site has to be collected and stored. In the following, we will focus on acoustic measurements in large area search and in local area mapping.

14.7.1.1 Large Area Search

Quite often, information about archaeological remnants is vague about the objects to be identified as well as about their localization, so that a broad area must be covered. In other cases, there is not even prior information, but the need to complete a systematic mapping of a given region. In both cases, this is done by acoustical scanning of the seabed morphology. The side-scan sonar is an acoustic instrument par excellence in this experimental activity, although multibeam echo sounders systems can also be employed to provide morphological information. The output of this activity is an area map whose scale is large or very large in comparison to the objects of (possible) interest, which appear as acoustic anomalies. From the analysis of this map, archeologists define a more restricted search area and possibly identify waypoints and targets that deserve special attention. In most cases, the search of relicts or other remnants is done within the continental platform, in areas that geophysicists would consider as shallow water, while for archaeologists a deep water site is any site that cannot be reached by divers. The side-scan sonars to be used in large area search are high-frequency instruments, above (sometimes well above) 100 kHz, and towed relatively close to the seabed, with an elevation from 20 to 60 m above the seabed. Synthetic aperture sonars (SASs), with their increased definition, have a tremendous effect on improving the interpretation of large area maps. The SAS comes, however, with increased costs and operational complexity, and it has been used so far in archaeology as a product showcase for other, high-cost applications more than as a systematic tool. Fig. 14.14 shows side-scan and SAS images of

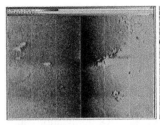

FIGURE 14.14

Images from left to right, side-scan sonar (384 kHz), synthetic aperture sonar (300 kHz), and still video image of a "dolia" wreck site in the Tyrrhenian Sea. In the left image, the wreck site is visible in the lower right of the image. Investigations on the site have been led throughout the years by the Tuscan Superintendence on Cultural Heritage, which has authorized and coordinated several research groups acting on the site. The systems used and the experiments have been described in [81] (left), [82] (center), and [83] (right).

a relict from the Roman imperial time found in the Tuscan Archipelago waters, Tyrrhenian Sea, together with a still video image of the relict. It is interesting to observe the higher definition achieved by the SAS despite its operation at a lower acoustic frequency.

14.7.1.2 Local Surveying and Mapping

If an identified anomaly has confirmed its archaeological relevance, a detailed local mapping of the site is performed. The local mapping involves both acoustic and optical data. Within this activity, geometric relations among the various objects and structures of the site are of paramount importance, because they carry the information on the historical event(s) that have characterized the site. Acoustic local surveys are carried out to map the seabed surface relicts, and to search and image subbottom structures, if any.

The seabed surface local survey is usually carried out with high-resolution multibeam echo-sounding systems that may achieve submetric geometrical accuracy. Fig. 14.15 shows the bathymetry obtained by a 445-kHz multibeam system at an

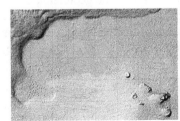

FIGURE 14.15

(Left) High-resolution bathymetry with a 445-kHz system of an amphorae site (lower left of the image); (right) enlarged portion, showing amphorae details. Each amphora length is about 1 m.

amphorae site in the Tyrrhenian Sea [80]. The enlargement shows how the system is capable of imaging even some details of the individual amphora, and to accurately present the spatial relations between the amphorae on the site. The multibeam system was operated from the sea surface over water depths of 30–40 m, spanning a 100 × 50 m area. Differential GPS was employed to have appropriate georeferenced data.

Subbottom imaging is carried out with high-frequency seismic profiling systems. There is a relevant discrepancy between traditional geophysics applications and marine archaeology. To the archaeologist, only the very first strata of sediment are usually of relevance, hence resolution properties of the acoustic instrument are more important than bottom penetration. Therefore chirp systems up to 15 kHz are often used [84], moving at slow speed (1 knot) and along very close transects (1-m distance between parallel transects). With these requirements, operation from a surface platform is difficult; hence, the profiling instrument is also operated from a remotely operated vehicle (ROV), at a shorter distance from the bottom. In turn, this has led to the design of ultrahigh-frequency subbottom systems, as the 140-kHz instrument used by Mindell and Bingham [85]. They report a penetration of 2 m in fine silt; however, the system has so far remained a research prototype. Parametric sonars [86] are also being used in subbottom prospecting for archaeological items [87], since their narrow beam, even at lower frequencies, improves their resolution with respect to standard profilers.

Further to acoustic mapping, an optical (and sometime also magnetic) mapping is performed from close distance. The optical mapping has the purpose to produce data for a photogrammetric reconstruction of the site, fine details imaging, and very small objects and fragments identification. High-resolution camera pictures can be acquired by divers (if the water depth allows) or, better, by an ROV.

Magnetic mapping is performed to further identify and characterize buried objects and structures. Acoustic instrumentation has an important role also in this operational stage, since acoustic positioning systems are used to track the motion of the divers or of the ROV. Both Long Base Line (LBL) and Ultra Short Base Line (USBL) systems can be used for the positioning and tracking task, but in most cases the USBL system is preferred for its easiness of installation and use and reduction in logistic effort. Typically, 30–50 kHz USBL systems are used, with the capabilities of tracking multiple divers simultaneously. The georeferenced information on the diver/ROV position is synchronized with information on the data capture device, so that a 3-D reconstruction is made possible by subsequently fusing all the geometric information available, including those of the multibeam survey [88].

14.7.1.3 Evolving Trends: A Technological Future for the Exploration of Deep Water Archaeological Sites

A clear evolving trend in marine archaeology research is the shift toward deep water relict search and investigation [89]. There are different reasons producing the interest in exploration of deep-water sites. Since most of the sites at diving depth have already been mapped and the historical significance of the deep water findings is likely, the increased availability of instrumentation and technologies make the search and mapping process possible without the need of human divers. In particular, as in other marine operations, underwater archaeology will also employ unmanned vehicles with increasing frequency, either semiautonomous, as the ROV systems, or completely autonomous, as in the case of AUVs.

Underwater acoustics will keep the role as a critical system component in systems for deep-water archaeological investigations, and not as a stand-alone tool. An example is the current SAS systems operated from AUVs, where the sonar data have to be integrated with the vehicle navigation data to properly compose the synthetic image. The sonar—vehicle system is the instrument, and neither sonar nor vehicle can perform without the capabilities of the other. The distinction between acoustic payload and payload carriers is sometimes kept for ease of description, but indeed, it is not strictly accurate. Even more so, in deep water search not only single AUVs will be used for search and mapping, but several AUVs operating as a team will be exploited. This requires exploitation of acoustic communication between the AUVs [90]. In conclusion, acoustics will be a component of complex exploration systems, with the role to gather raw seabed morphological and bathymetric data, providing system localization and georeferences, and enabling communication and cooperation, either in a peer-to-peer or in a networked fashion.

14.8 APPLICATIONS OF UNDERWATER ACOUSTICS IN POLAR ENVIRONMENTS

P. Mikhalevsky[1], **A. Gavrilov**[2]
[1]Leidos Inc., Arlington, VA, United States; [2]Curtin University, Perth, WA, Australia

14.8.1 INTRODUCTION

The unique feature of the Arctic and Antarctic polar oceans that affects underwater acoustics, both the propagation of sound and the ambient noise, is the presence of sea ice that seasonally expands and retreats and ice shelves extending from the land into the ocean. They also have important differences. In the last two decades there has been a significant reduction of both the extent and thickness of the ice in the Arctic while there has been an increase in the extent of the sea ice in the Antarctic. The Arctic Ocean is a Mediterranean basin with limited communication to the world's oceans while the Southern Ocean surrounds the continent of Antarctica and is contiguous with the south Atlantic, south Pacific, and Indian Oceans and acoustically linked to the deep sound channel of the world's oceans.

14.8.2 ARCTIC

Much of the early research dedicated to underwater acoustics in the Arctic over the last half century was driven by military needs to support submarine operations after World War II (WWII) [91]. After the end of the Cold War global climate change emerged as another important focus for underwater acoustics in the Arctic Ocean with the applications of acoustic thermometry and tomography and multipurpose acoustic networks supporting Arctic Ocean observing systems, e.g., Ref. [92].

The Arctic Ocean exchanges waters with the Atlantic Ocean via the Fram Strait and the Barents Sea and with the Pacific via the Bering Strait. This limited oceanic exchange and until recently the near continuous year-round ice cover created a highly stratified and stable sound speed structure in the Arctic. There are three major water masses that affect the sound speed profile in the Arctic Ocean: (1) Polar Water (PW) extends from the sea surface to about 100 m in the Eastern Arctic and 300−400 m in the Western Arctic; (2) Atlantic Intermediate Water (AIW), entering the Arctic from the North Atlantic, extends from below the PW to $\sim$1000 m; and (3) Deep Arctic Water below 1000 m. This stratification creates the classical polar profile of a near bilinear upward refracting sound speed profile, as shown in Fig. 14.16. Also warm Pacific water entering the Arctic through the Bering Strait creates a local sound speed maximum in the PW layer forming a double duct in the Beaufort and north Chukchi Seas.

As shown in Fig. 14.6 the upward refracting profile causes continual reflection and loss of sound energy due to scattering by and coupling into the sea ice. The reflection loss rapidly increases with frequency. Furthermore, the bilinear structure of the sound speed profile forms two sound channels,- near surface and deep water. Low-order modes or lower grazing angle rays are trapped by the near-surface channel and hence experience many more reflections from the sea ice than higher modes, or steeper rays propagating in the deep-water channel. For example, at 20 Hz only mode 1 is trapped by the near-surface channel (Fig. 14.16). The magnitude of the propagation loss is exponentially dependent on the roughness of the sea-ice, which is directly proportional to ice thickness [93]. Fig. 14.16 also illustrates another unique aspect of acoustic propagation and the application of acoustic thermometry and tomography in the Arctic Ocean, namely, the close coupling between acoustic modes and the major Arctic water masses described earlier. The travel speed (group velocity) of a mode is governed primarily by the sound speed in the water layer where most of the modal energy propagates. At 20 Hz mode 1 samples the PW while modes 2 and 3 are most sensitive to changes in the AIW. Applications of acoustic thermometry in the central Arctic in the 1990s detected basin scale warming in the AIW and the influx of warm Atlantic water into the Arctic that were confirmed by submarine and mooring measurements [92,94]. Acoustic thermometry in conjunction with measurements by moorings and gliders has been used in the Fram Strait in a multiyear continuing set of experiments started in 2005 to measure the integral properties of the inflow and outflow of heat through the strait and improve the ice-ocean models of this highly complex exchange between the Arctic and Atlantic waters [95].

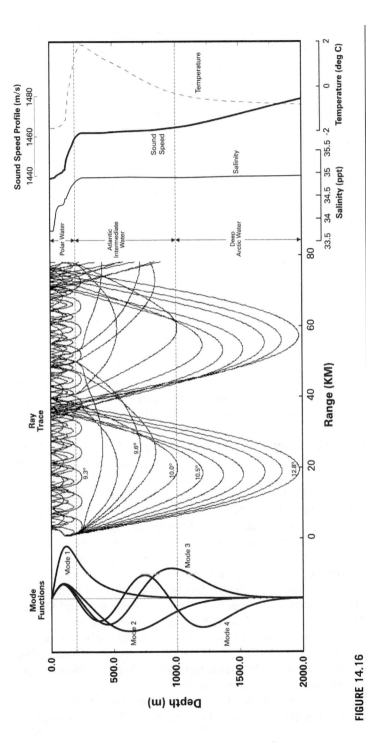

FIGURE 14.16

The Arctic sound speed profile shown on the right was computed from the measured temperature and salinity, also shown, from an ice camp in the eastern Arctic Ocean in April 1994. A ray trace for this sound speed profile is plotted for a source at a depth of 100 m. The mode shapes computed at 20 Hz using this profile are shown on the left. The major Arctic water masses (see text) are indicated.

Reproduced from Figure 2 in Mikhalevsky, P.N., Arctic Acoustics. Encyclopedia of Ocean Sciences, John H. Steele, Karl K. Turekian, Steve A. Thorpe (Eds.), Academic Press, 1, pp. 53–61, 2001, courtesy of Academic Press.

Over the last 30 years there has been a steady decrease in the extent and volume of the sea ice in the Arctic with ~30% reduction of ice extent and reduction in average ice thickness from ~4 to ~1.5 m, e.g., Ref. [96], and forecasts of an ice-free Arctic in the summer months as early as 2020 [97]. The decrease in ice thickness with the reduction of thick multiyear ice causes a corresponding decrease in ice roughness and consequently the sound transmission loss. If the reduction of sea-ice cover continues, modeling has shown that the monotonically increasing propagation loss with frequency, creating the well-known low-pass filter characteristic of Arctic propagation, will persist for some time but with a decreasing magnitude and slope of propagation loss versus. frequency at a given range. However, these modeling results must be verified by new acoustic measurements in the Arctic.

Ambient noise in the Arctic has four major components: ice-generated, biologic, seismic, and anthropogenic, e.g., Ref. [92]. Ice is one of the most important sources of Arctic ambient noise from a few tenths of hertz up to nearly 10 kHz. Arctic ice noise is highly variable with a large range in levels with exceptionally quiet conditions during periods of "ice lockup" when the sea ice is not moving and the water is isolated from the atmosphere and exceptionally noisy conditions when there is active ridging during periods of high winds (see discussion and references in Ref. [91]). With the reduction of ice concentration and thickness throughout the Arctic large ice ridges have already largely disappeared, which could reduce this component of ice noise. At the ice edge in the Marginal Ice Zone (MIZ) regions, ocean waves and swell propagate into the ice-covered waters and cause floe breakup and, coupled with on-ice winds, ridging occurs with resultant high noise levels. With larger areas of open water in the summer months wind and storms will create wind-driven noise characteristic of the temperate oceans.

Marine mammals are prevalent in the Beaufort and Chukchi Seas and in the Fram Strait. Bowhead whales (*Balaena mysticetus*) are the most numerous in the western Arctic and the Fram Strait. Their vocalizations are typically in the 100−400 Hz range. Other whales that have been recorded in the Fram Strait include blue, fin, and sperm whales [92]. Many species of seal, toothed whales such as the beluga and narwhal, as well as walruses frequent the MIZ with vocalizations from a few hundred hertz to about 10 kHz. It is not known what impact the receding ice is having on abundance and distribution patterns of these marine mammals in the Arctic. Long-term passive acoustic monitoring is needed.

Seismic, earthquake, and volcanic noises mainly below 20 Hz occur in the Arctic Ocean. Most of the seismicity is concentrated along the Knipovich and Gakkel Ridges, the latter being one of the Earth's two ultraslow spreading ridges with poorly understood crustal formation. Seismic activity has been intermittently recorded on acoustic arrays, but seafloor broadband seismometers for long-term monitoring are needed to study geodynamics of the Arctic, and possibly help in resolving territorial disputes and assessment of mineral resources.

While shipping noise is one of the dominant ambient noise components in the temperate oceans, it is still largely a small contributor in the Arctic except for episodic events. However, increased trans-Arctic shipping via the Northern route along the Russian coast and in the Northwest Passage through the Canadian archipelago to the western Arctic, fishing, and tourism as the ice recedes will raise the ambient noise levels in the years to come. Oil and gas exploration and production is also expanding. The fastest changing component of the ambient noise in the Arctic is anthropogenic. It is important to monitor the rate of change, frequencies, and levels and the impact on the polar environment (e.g., marine mammals) as this is a new phenomenon in the Arctic Ocean.

The Arctic Ocean is in transition as is underwater acoustics in this polar environment.

14.8.3 ANTARCTIC

Polar waters surrounding Antarctica are much less investigated than the Arctic Ocean with respect to natural sources of ambient ocean noise and underwater sound propagation. This is mainly due to the remoteness of the Antarctic continent from most of the maritime states and shipping routes and its relative insignificance in the naval warfare during the last century. The interest in acoustics of the Southern Ocean and, in particular, polar Antarctic waters, has been driven over the last decades primarily by research related to climate change and studies focused on Antarctic marine fauna, its abundance, and health.

Antarctica is known as one of the major sources of low-frequency underwater noise in the Southern Ocean [98]. Fig. 14.17 shows the fraction of time when coherent low-frequency noise signals were observed at the Cape Leeuwin CTBT hydroacoustic station in Australia in 2002—07, which is drawn from a longer data

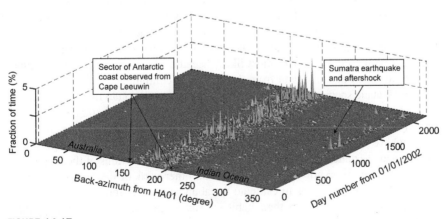

FIGURE 14.17

Fraction of time when coherent low-frequency (<50 Hz) noise signals were observed at the Cape Leeuwin CTBT hydroacoustic station in Australia, versus day of observation and back-azimuth to signal sources.

set than that in Ref. [99]. The main physical sources of noise are icebergs and ice shelves breaking up. Icebergs also produce intense long-lasting low-frequency tonal signals with harmonics, often referred to as "harmonic tremors" that are excited by vibrations of the iceberg body resulting from collisions with the seafloor, other icebergs, or ice shelves [99].

In contrast to the temperate ocean, underwater sound sources at the sea surface in Antarctica are well coupled with the SOFAR channel, as the sound speed minimum is also located near the sea surface in the polar environment. Acoustic energy from a sound source trapped in the near-surface sound channel propagates with some ice scattering losses to the polar front at the Antarctic Convergence, where it dives to a depth of about 1 km and then propagates further to the north without significant absorption and scattering losses in the deep SOFAR channel. For this reason, sounds from Antarctic ice breakup can be heard underwater at the equator.

Another significant source of low-frequency noise in Antarctica and the entire Southern Ocean is the sound of Antarctic blue whale calls. Vocalizations by many remote whales form a whale chorus, which can be seen as band-limited noise from about 17 to 27 Hz in the sea noise spectrum in the Southern Ocean almost year-round [100]. Fin whales most likely also contribute to this noise as the spectrum of their frequency-modulated (FM) calls spans the same frequency band as that of Antarctic blue whales [101].

Favorable conditions of low-frequency sound propagation from shallow sources in Antarctica to the temperate ocean allow observing ice breakup processes over a large sector of the Antarctic coast from a remote hydroacoustic station, which provides additional and cost-effective means to monitor long-term variations in the ice disintegration rate associated in particular with climate change. An analysis of 7-year long recording of sea noise made at the hydroacoustic station off Cape Leeuwin in Western Australia, which is part of the IMS of the Comprehensive Nuclear-Test-Ban Treaty (CTBT), showed that the frequency of occurrence of transient acoustic signals associated with ice breakup events in Antarctica had an obvious seasonal component, but did not reveal any long-term trend that could be associated with climate change, most likely because of too short observation time [102]. Analysis of acoustic data from this station is continuing.

Several whale species, such as blue, fin, humpback, and Antarctic minke whales, inhabiting Antarctic waters during the feeding season in austral summer, became critically endangered due to extensive whaling before the 1970s, especially blue whales whose population had decreased by about 100 [103]. At present, significant effort is made to understand the rate of recovery of the populations of these species. Passive acoustic observations in Antarctica and other parts of the Southern Ocean provide cost-effective means to estimate the abundance of different species and its variation over years, to study their habit in Antarctica and understand migration patterns, e.g., Ref. [104].

14.9 TANK EXPERIMENTS

J.-P. Sessarego, D. Fattaccioli
LMA-CNRS, Marseille and DGA Naval Systems, Toulon, France

14.9.1 INTRODUCTION

Tank experiments can be an interesting alternative to sea trials for validating numerical models because they are not very expensive, they are reproducible, and all parameters can be perfectly controlled. The idea that has led to the setting up of tank experiments is to build a reduced scale model for simulating the real environment. Even if tank experiments are very attractive for the reasons already mentioned, it must be noted that they do not allow getting a perfect mimic of reality. The oceanic environment is a very complex medium, and it cannot be described completely by a simple experiment in the laboratory. To define the scale factor, the two parameters that play a role are distance and wavelength. If d and d', respectively, are a characteristic dimension in the tank and at sea, and λ and λ', respectively, are the wavelengths used in the tank and at sea, then the scale factor F can be defined as:

$$d'/d = \lambda'/\lambda = F$$

The characteristic dimension is, for example, the water depth, H, or the distance, d, between a source and a receiver. Usually the scale factor, F, is between 1 and 10,000.

Indeed, tank experiments can cover very different goals and very different operating modes. Probably the most well-known classical application of tank experiments is the calibration of acoustic systems including their electronics before their use at sea, but tank experiments can also be used to measure various parameters in a controlled, stable, and reproducible situation as, for instance, measurement of sound speed in water or in sediments as a function of temperature and salinity, measurement of attenuation, porosity, electrical conductivity, measurement of a source level, etc.

Perhaps the most interesting application of tank experiments is that they can be used to study separately the factors affecting sound propagation by isolation of the dominant parameters with an effect on propagation as, for instance, a sloping bottom, the coupling between propagation and diffraction for different water depths, influence of bottom materials, and spatial and temporal disturbances. It should be noted that most propagation problems to be studied using scaled models are difficult to solve numerically, even with the use of high-power computers, for instance, due to 3D effects. This is why tank experiments are still attractive, even if over the last 10 years a spectacular progress in reduced computational time and increased storage capacities have been observed.

Finally, tank experiments can be used to demonstrate the validity of new concepts before the design and development of an expensive prototype to be tested in real situation. In the following, after a brief description of the tanks, which have been designed in different laboratories, we will give some examples of applications that have been made.

14.9.2 DESCRIPTION OF DIFFERENT CATEGORIES OF TANKS USED FOR UNDERWATER APPLICATIONS

To cover the wide range of possible applications, scientists have developed different tank concepts, each adapted to one or more specific applications. For example, studies of long-range propagation in 2D or 3D, as well as studies of the coupling between propagation and scattering, must preferably be performed in a very large tank. More specific studies on sediment backscattering or on scattering from small objects can be achieved in tanks of smaller dimensions.

Moreover, for experiments in which the system can be considered as frozen with no surface agitation and no source or receiver motion, one single receiver that can be positioned in the tank by stepping motors or direct current motors can be used. For systems evolving with time, it is necessary to use transmission and/or reception arrays to be able to work in real-time domain. These two concepts lead to the development of very different technologies. In this section the objective is not to present a list of all the different systems existing in the world, but rather to show how tank experiments may be designed and which results can be achieved. We will only give a few examples of existing tank facilities in Europe as an illustration to our discussion.

For real-time application of acoustic tomography and inversion applications, P. Roux in Grenoble [105–107] developed a small water tank (1.5 m × 1 m × 0.6 m) with multichannel, ultrafast, ultrasonic acquisition equipment working at 80 MHz. The scale factor used for these experiments is 10,000, which can be considered as the higher limit for study of underwater acoustic problems. Usually, the frequencies used in these experiments range from 0.5 MHz to several megahertz. A similar tank has been developed at Institut Langevin (Paris) for time reversal applications [108] (Fig. 14.18).

To study evolving phenomena in the tank, P. Roux used two arrays of 64 elements each, one for transmission and the other for reception. With this experimental setup several studies have been performed. Among them shall be mentioned:

- Inversion for surface gravity waves,
- Detection and localization of an anomaly in a waveguide as, for instance, temperature anomaly, target and intruder detection in relation to protection of port areas [103],

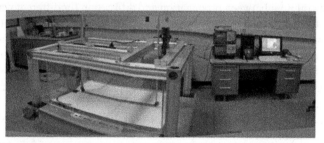

FIGURE 14.18

Picture of the tank developed by P. Roux (ISTerre Grenoble) for tomography and inversion in acoustic waveguides.

- Shallow water acoustic tomography in a fluctuating environment [104],
- Sound focalization in the ocean [105,106].

When the phenomena to be studied can be considered as stationary, measurements can be made with a unique hydrophone, which can be positioned in the water by a carriage moved by stepping motors. Usually, high-quality positioning systems are used and an accuracy of 1/10 mm in translations and 1/100 degrees in rotations are often needed. These constraints can be met, for instance, by a Micro Control equipment, or by an equivalent system. High precision in the hydrophone positioning is necessary when the scale factor is very high ($\sim 10,000$), as at this scale factor 1 mm in the tank corresponds to 10 m at sea. These tanks are in general larger than the tank shown in Fig. 14.19, as, for instance, (3 m $\times$ 1.5 m $\times$ 1 m) for the small tank of LMA, (2.5 m $\times$ 1.5 m $\times$ 1.3 m) for the tank at FORTH/IACM, Greece (Fig. 14.19), and (3 m diameter $\times$ 2 m depth) for the cylindrical tank at Le Havre University, France.

These tanks can be used for different applications:

- Calibration of high-frequency sources (200 kHz–5 MHz) and hydrophones [109],
- Scattering by elastic targets (resonance scattering) [110],
- Sediment studies [111],
- Scattering by objects close to interfaces [112],
- Study of decoherence effects in relation with a perturbation of the wave front, applied to antenna processing [113],
- Scattering by rough surfaces,

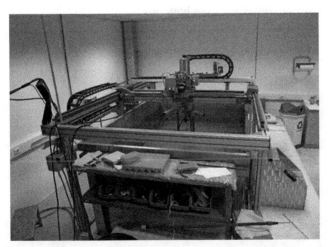

FIGURE 14.19

View of the tank facility at FORTH/IACM, Heraklion, Greece.

- Bottom inversion studies, including estimation of parameters of the bottom,
- Signal propagation through a cloud of bubbles.

While some applications could not be covered by these tank facilities it was decided in France to build a very large tank. This tank was built at LMA in the 1990s with the support from DGA (the French Ministry of Defense). The tank dimensions are 20 m × 3 m × 2.5 m in the deepest part. The bottom of the tank was covered by a 30-cm-thick layer of medium sand, which by the frequencies used (100–500 kHz) corresponds to a semi-infinite bottom. Two independent carriages supporting the system of transmission and reception, respectively, can be moved by brushless motors. Transmitters can be adjusted manually in the X and Y directions and moved automatically in the Z (depth) direction. The receiver can be moved in the X, Y, and Z directions. An acoustic absorbing lining is put on two walls of the tank and behind the transmitter, to avoid any spurious reflection from the walls (Fig. 14.20).

Originally, this tank was built for the study of 2D and 3D long-range propagation problems and to validate numerical codes, but it has also been used for several other studies as, for instance:

- 3D effects by propagation over a sloping bottom [114],
- Coupling between sound propagation and diffraction by a target [115],
- Waveguide invariant demonstration,
- Ultralow-frequency studies,
- Scintillation and source discrimination,
- Detection of buried objects in sediments by either high-resolution techniques or time reversal techniques,
- Synthetic aperture imaging of the bottom,
- New sources for underwater acoustics applications [116].

FIGURE 14.20

View of the oceanic tank (CNRS/LMA, Marseille).

14.9.3 CONCLUSION

The objective of this section has been to show the wide range of applications that can be achieved with laboratory experiments using scale models. For this, we reviewed both the technologies to be adopted and the types of tanks that must be built determined by the applications to be made. These experiments by use of reduced scale models have, due to their repeatability, accuracy, and low costs, made it possible to get quick replies for validation of complex numerical models, and also to give guidance on how to carry out full-scale sea trials.

14.10 ACOUSTIC POSITIONING AT SEA

R.A. Hazelwood

R&V Hazelwood Associates LLP, Guildford, United Kingdom

The well-understood speed of sound in the ocean [117,118] has led to the widespread use of acoustics to position equipment. Two-way transit times between transducers can be accurately timed by the use of transponders, which reply when called. The exploration and exploitation of the resources of the sea depends on this technique. As an example, pipeline connections need to be made underwater. Robotic vehicles are used for many such tasks and cannot directly access the GPS data.

Another good example is the positioning of drilling ships and crane barges. Many new developments occur in water depths over 1 km and arranging anchors in these conditions is difficult and expensive. An alternative is to use a ship equipped with dynamic positioning (DP). Very large ships can be held on station by the use of azimuth thrusters, provided good data can be obtained on their position. The "Pieter Schulte" can lift 48,000 tons, has two 392-m-long hulls, and is 124 m wide. The DP controls 12 thrusters driven by a 95-MW power system [119]. While GPS satellite data are fundamental to this procedure, acoustics also has a major role to play.

The DP control systems need good positional data. Acoustic data streams have some advantages compared with GPS. Often both are used for reliability, but a tighter "fix" can often be achieved by acoustics, even in challenging circumstances. The better the repeatability of the data, the less the energy used by the thrusters, and the better the ship's stability. When working on a large DP crane barge, it hardly feels like being at sea.

The achievement of such repeatable acoustic data has required years of careful development, both for the ship-mounted sensors and the underwater beacons. Computer-based transponders were developed in the 1960s for LBL techniques. A signal from the surface is coded to elicit a response from a specific subsea transponder. This response in turn is coded either for another seabed transponder or the surface unit. The coding was originally devised by using pulses of different frequencies. A "ping" from one unit would initiate a reply "pong" by another. This "ping pong" interchange could be accurately timed, and the two-way "time of flight" converted to a distance using the known speed of sound.

14.10.1 **LBL POSITIONING DEVELOPMENT**

Fig. 14.21 shows a schematic array of acoustic transducers on four seabed transponders and on a side-mounted ship's pole. Ten acoustic pathways are shown, each measurement being made as a two-way time of flight. They provide more than sufficient data to define the five-sided solid figure, and the "redundant" excess data can then be used to assess the consistency and thus the quality of the data.

The intelligence of the computer-based transponders is required to be able to coordinate the measurement of distances across the seabed, and to report these data via a pathway to the surface. However, additional depth data are usually required. Pressure sensors can provide this augmentation, so the system also requires the density of the seawater to provide the depths of each transponder.

The required data telemetry uses a related technique, frequency shift keying (FSK). Typically a stream of "pings" with two frequencies can transmit digital binary data, but additional frequencies are used to ensure the correct timing of the decoder. Such modem (**mo**dulator/**dem**odulator) systems are also used for other systems.

There is also a need to calculate the azimuth direction of the transponder array shown in Fig. 14.21 in relation to north. Each of the transponders may be positioned independently via a "box in" procedure wherein the ship is held on station via GPS sensors at a variety of positions. A sequence of positions is chosen that surround the transponder. Again redundant data are acquired and used to achieve a "least squares best fit" position. This is time-consuming and expensive work, but delivers the required information.

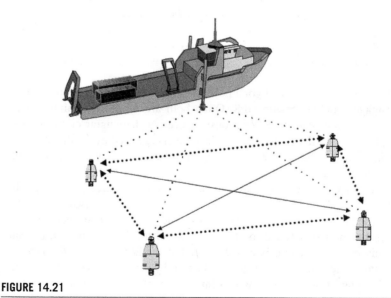

FIGURE 14.21

Schematic array of acoustic transducers on four seabed transponders and on a side-mounted ship's pole.

Once the transponders are located, other vehicles can use this seabed array for their positioning, which do not need GPS sensors and can thus operate underwater. ROVs fitted with a sensor can then be driven to any desired position. They are typically used to maintain underwater structures such as oil and gas platforms. AUVs can also use such an array, although they usually range more widely, and LBL positioning accuracy will degrade as the AUV moves away from the array.

14.10.2 ULTRASHORT BASELINE (USBL) POSITIONING

An alternative array scheme can operate with a single seabed transponder if there is an array mounted on the ship. If the ship's array has three sensors as shown, there will be just enough data, but additional redundant data are usually gathered to measure quality (Fig. 14.22).

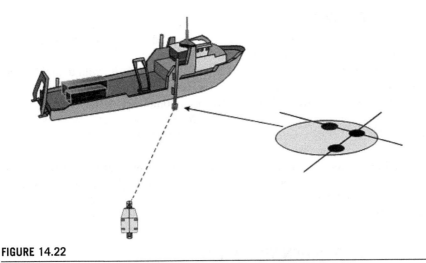

FIGURE 14.22

Single seabed transponder with an array mounted on the ship.

As an example, a fivefold array [116] provides data wherein the redundancy is particularly useful in overcoming potential errors. This direction finding array, combined with the two-way transit time can provide an immediate fix of the transponder in relationship to the ship, but require a precise knowledge of both the ship's position and its attitude. The conversion of a position in "ships frame" to one in the desired inertial frame requires high-quality attitude sensors and careful attention to the matrix rotation mathematics. Use of "direction cosines" (components of the 3×3 attitude matrix) is less fraught with problems than the use of Euler angles [120], but the key issue is to match the attitude sensor data with the mathematics used for conversion.

14.10.3 EFFECTS OF NOISE

The noisy environment under a DP ship provides a major constraint on the achievable accuracy of a USBL position [116]. Since most of the noise is generated by the thrusters it is preferable to mount the direction finding array as far below the hull as possible, with sensor beam shapes tailored to reject the noise. The direction is determined by a very accurate differential measurement of the paths to different sensors in the array, which requires a very stable phase response.

While good design is important, there are fundamental theoretical limits to the conversion of these time differences into directions. Ultimately an LBL array will be more accurate, if more expensive to achieve, and additional seabed transponders are often added to improve performance. This system is less affected by refraction (ray bending) due to the changes in the speed of sound with depth.

14.10.4 IMPROVED CODING

Both positioning and data telemetry can be improved by the use of more complex phase processing. Rather than a single frequency tone burst, the phase can be shifted more rapidly within the pulse. This is phase shift keying (PSK) with digital bits being conveyed by the shifts. FSK receivers can use passive resonant circuits and simple logic, but modern PSK modems require faster computers, with more economical use of transponder battery power. However, the improved telemetry rates and increase in the number of available addresses for transponders have led to widespread use of these more complex systems.

In addition to its use in large systems, this improved telemetry has great benefits for applications such as acoustic releases (reduced risk of accidental triggering) and environmental monitoring (data accessible without the need to recover the instrument).

14.10.5 COORDINATION WITH INERTIAL SENSORS

The correction of a position measured in a tilted ship's frame into an inertial frame may use three accelerometers and three rotation rate attitude sensors. These determine the six degrees of freedom of a rigid body. Marine terminology considers these as pitch, roll, and yaw rotations with surge, sway, and heave translations. Their output can also be integrated to track the translational motion of the USBL array, so that when coordinated in the same instrument, further improvements in positioning can be achieved. This reduces the rigidity required for the mounting on the ship, so that a simple over-the-side pole mount (see Fig. 14.21) may suffice, rather than having to deploy the array through the ship's hull.

Similar systems are useful for AUV navigation, where the inertial guidance can be aided by occasional fixes by acoustic positioning over sparsely covered areas. The drift intrinsic to the integrated output of all inertial systems is thus corrected economically.

14.11 OCEAN OBSERVING SYSTEMS AND OCEAN OBSERVATORIES, OCEANOGRAPHERS, AND ACOUSTICIANS—A PERSONAL PERSPECTIVE

B. Dushaw
University of Washington, Seattle, WA, United States

14.11.1 INTRODUCTION

Twenty years ago the oceanographic community embarked on the grand endeavor of establishing ocean observatories. First, there was an obvious need to transition basic oceanographic research into products and information that would be useful to society. Information on the evolution of the Earth's climate system and warning systems to mitigate natural disasters of atmospheric, oceanographic, or geologic origin are examples. Second, many oceanic processes or systems evolve at decadal to century timescales and require sustained, long-term observations to properly understand them. These two motivations highlight a semantic difference: "Ocean Observing Systems" (OOSs) are operationally focused, while "Ocean Observatories" are research focused, although the difference is often blurred. "Operational" implies the commitment of the significant bureaucracy and management required to deliver promised data, information, and products to society on a sustained basis, e.g., the national weather services, or the CTBTO (www.ctbto.org) hydroacoustic system. Ocean observing systems are global, basin, or regional scales. Examples of OOSs are the Arctic basin system, or the many regional systems along the coasts of the United States. Neptune Canada (www.neptunecanada.ca) is an ocean observatories program that includes research in acoustics. Australia's Integrated Marine Observing System (IMOS, www.imos.org.au/) has been collecting freely available sea noise data at six sites on the continental shelf since 2008. These data have been used by marine biologists for studying marine mammals and fish, e.g., their vocal behaviors, migration patterns, and populations. Natural processes in the ocean have been observed, such as seismic events, volcano activity, and ice disintegration near Antarctica. IMOS observes the general ocean soundscape, from the continental shelf to the Southern Ocean between Australia and Antarctica. In any observing system, data management and archive are formidable issues that must be addressed.

The possible acoustical applications for an ocean observing system are myriad and cross several disciplines, and a review or survey of these applications is beyond the scope of this chapter (see Ref. [121] for bibliographical information). The discussion here therefore addresses mainly tomography and ocean observing systems [122,123], but it applies to acoustics for biological or engineering applications, as well as pure acoustical science.

14.11.2 OCEANOGRAPHERS

Although the value and unique quality of acoustic measurements, tomography in particular, has long been established, general implementation of this measurement type by oceanographers has not been forthcoming. The value of acoustical approaches has been established according to accepted standards of scientific discourse. A rigorous, objective design of an observing system intended to last a century or longer includes essential contributions from both engineering and observational acoustical techniques. Acoustic techniques for ocean observing or engineering offer tremendous opportunities.

For example, in an analysis of a decade of basin-scale acoustic tomography data from the North Pacific, Dushaw et al. [120] made direct comparisons of the tomographic time series to equivalent time series derived from Argo float data and altimetry. The comparison showed significant differences at all timescales, indicating that the existing observing system was badly estimating the large-scale variability of ocean temperature. Similarly, in the 1988 Greenland Sea Project experiment, the rapid sampling of integrated temperature afforded by the acoustic measurements proved to be essential in estimating the net deep water formation during the winter of 1988/1989 [124]. Concurrent, extensive measurements by conductivity, temperature, depth casts proved to be inadequate to the task. The concluding statement of the international OceanObs'09 conference explicitly noted that "the oceans remained seriously undersampled" (www.oceanobs09.net/statement/).

The applications of acoustic tomography (or acoustics in general) for observing systems have not yet had a rigorous, informed, quantitative examination by oceanographers, however. One argument for excluding active acoustic sources from observing systems has been that deploying such technology will attract environmental concerns, law suits, etc. from concern over the impact of sound on marine life. Such an argument is specious, a crass exploitation of this issue to rationalize excluding the acoustical approach. Significant resources have been expended to research this specific issue, with the published conclusion that existing research acoustic sources have no significant biological impact. Acoustic sources have been regularly used to track the positions of instruments. Scientists should be willing to stand up for correct science. Determining secular changes in the ambient sound of the world's oceans, i.e., man-made noise, is a serious motivation for an acoustics component to a global observing system.

One of the shortcomings of acousticians is in making progress on important oceanographic, rather than acoustics, questions. Oceanographers could greatly assist the process of developing roles for acoustics in an ocean observing system. One pressing example is the need for deep ocean measurements, a critical gap in the observing system (www.oceanobs09.net/). Two acoustical applications are evident. First, acoustic rays traverse the deep ocean, hence offer a natural integrating measurement of deep-ocean temperature. Second, implementation of deep Argo floats for the ocean observing system may require a 30-day cycle to conserve their energy. Acoustics offer an obvious means to determine the position of these floats during their month-long drift in the deep ocean. Both of these applications require considerable oceanographic expertise to determine how they can be best employed to address specific oceanographic questions. The need for stronger symbiotic relations between oceanographers and acousticians on these and other questions seems evident. Oceanographers have to do better with respect to acoustics.

14.11.3 ACOUSTICIANS

If oceanographers have been deficient in integrating acoustical techniques into observing systems, acousticians have not done much better. Implementing any technique for observatories requires considerable community organization, planning, and coordination to address such difficult questions as deploying and maintaining long-term observations for community use and the archival and management of those data. Partly as a result of funding limitations, acousticians lack adequate community will and organization to make much headway into ocean observatories. A key aspect of data from an observing system is standardization, which acousticians are only beginning to address.

With respect to tomography, one vice is the perpetual development of new techniques or tools for acoustical observation or data processing, while oftentimes failing to actually use those tools to learn about the ocean. New techniques are fine, but one must publish what one learns about the ocean using those techniques in the mainstream oceanographic journals. This vice is perhaps understandable, since acousticians are not experts in the oceanographic questions that could be tackled using the acoustic techniques. Just as the oceanographers are unfamiliar with the acoustics science, acousticians are unfamiliar with the oceanographic science. Another vice is the oftentimes lengthy time interval between when data are recovered and when the results of analysis get published; this sort of delay is not inherent in the data type and will eventually be remedied. A much closer collaboration between the oceanographers and acousticians is essential.

While ocean observing systems offer great opportunities for putting acoustical techniques to good use, representation of acoustics at conferences associated with setting priorities, planning, and implementing systems has been poor. To some extent this deficiency is a product of funding limitations—effective participation at these workshops is expensive and time consuming. Also, these workshops are often hosted by oceanographic agencies or organizations, hence not on the acoustician's sonar. Without better representation and advocacy at these conferences, however, acoustics will continue to be left out of the process. Acousticians have to do better with respect to oceanography.

14.11.4 FUTURE DIRECTIONS

There are many lines of research that may be readily pursued to better set the stage for implementing acoustical components to the ocean observing systems. Here are a few examples, a by-no-means comprehensive list.

Acousticians do not today have access to reliable low-frequency, deep-ocean source technology; development of such sources has lagged. Meanwhile, passive acoustics as a remote sensing technique has had some positive developments in recent years, and remains a possibly fruitful avenue for investigation. The ability to use the ocean's natural ambient sound sources to quantitatively measure its properties would be a major breakthrough, but more work is required to establish this strategy.

Considerable work remains to be done in establishing the utility of acoustics using quantitative design studies through simulations. One remarkable development available to acousticians is the availability of high-resolution, realistic global ocean models. These models provide a convenient way to compute the environmental effects on acoustic transmissions, and they are a valuable asset for acousticians.

One advantage of research quality acoustic data is that historical data are often as precise as present-day data. With new understanding of acoustic and ocean properties, realistic ocean models, and new computational techniques, new analyses of historical data offer significant reward for little investment. Global-scale acoustic data acquired during the 1991 Heard Island Feasibility Test have yet to be fully exploited (data available from 909ers.apl.washington.edu/~dushaw/heard/index.shtml).

As has been often noted, the tomographic data type, together with other data, is best employed in conjunction with techniques for data assimilation. Data assimilation and tomography have been discussed for many years, yet aside from a few notable instances [125,126], these techniques are not developed to the point of practical or more widespread use. Data assimilation approaches that can routinely handle the acoustic data type are essential.

One region that has garnered support for acoustical applications is the Arctic Ocean. The Arctic is suggested because acoustical applications there are not viewed as in competition with floats and gliders. Nevertheless, the under-ice applications of acoustics are many, from biological measurements, to the positioning of instruments under sea ice, to low-frequency, to Arctic Basin thermometry. An ongoing program for acoustical measurements within the Fram Strait has been conducted by the Nansen Center in Norway over the past 5 years (www.nersc.no). Acoustical technologies as contributions to the Arctic Ocean Observing systems are likely to be fruitful in the near future.

14.12 APPLICATIONS OF UNDERWATER ACOUSTICS TO MILITARY PURPOSES

L. Bjørnø[1]
UltraTech Holding, Taastrup, Denmark

Military applications of developments in underwater acoustics are still increasing. Despite the emergence of many new fields of applications of underwater acoustics, as represented by other sections in this chapter, the military applications are still forming a majority. Even in spite of a decrease in naval budgets and a reduction in the

[1]30 March 1937—24 October 2015.

number of naval vessels, military applications may be said to have pushed forward the development. Among the larger producers of sonar systems and other systems for underwater applications shall be mentioned: Thales Underwater Systems (France), BAe (United Kingdom), STN-Atlas (Germany), Kongsberg (Norway), Teledyne-Reson (Denmark), and in the United States, Lockheed Martin, Raytheon, and Westinghouse. Moreover, a great number of smaller companies produce underwater acoustic equipment for specific use. Two main categories of sonar systems are available, passive and active sonar. While most sonar during WWII were operating at 15–30 kHz or more, long-range detection of submarines under influence of absorption, being proportional with the square of the frequency, has made it necessary to apply low-frequency acoustics. Passive detection, tracking, and identification are being done at frequencies ranging from a few tens of hertz to a few kilohertz. But there are many advantages by the use of higher frequencies in underwater acoustics as, for instance, short-range applications with higher spatial resolution, wider bandwidth, and the use of smaller and cheaper transducers. Today's naval applications includes antisubmarine warfare (ASW), mine hunting and avoidance, depth sounding and bottom mapping, under-ice navigation, torpedo homing, and communication.

14.12.1 PASSIVE SONAR

From frequencies far below 1 kHz passive sonar operates up in the kilohertz range. The radiated noise from submarines has to be reduced considerably to avoid detection. Propagation of low-frequency noise over long ranges improves the application of passive sonar for detection, tracking, and identification of underwater objects. Towed array sonar systems (TASS) are passive sonar systems towed behind a surface ship or a submarine and being extensively use to detect, track, and identify underwater objects. The system improves the passive reception sensitivity due to the higher number of hydrophones in the array, and the possibility of towing the array at a distance from the tow ship reduces the influence of self noise produced by the tow ship. A tactical advantage by use of a TASS in a stratified water column is that a submarine towing a TASS may be positioned in another layer than the TASS, which can make it difficult to detect the submarine. Because the array is not constrained by the size of the tow ship, the array can be made very long, from several tens to several hundreds of meters. Therefore the towed arrays have very narrow beamwidths, or alternatively, can operate at much lower frequencies. The low-frequency capability is particularly advantageous because there are many sources at lower frequencies with higher source levels, and low-frequency signals suffer very little loss from absorption. A weakness of the TASS is the left/right ambiguity in relative bearing, which can be resolved by maneuvering the tow ship. To keep the TASS in a horizontal move, it is necessary to maintain a minimum tow ship speed of over 3 knots. The AN/SQR-19 Tactical Towed Array Sonar provides very long-range passive detection, classification, and tracking of enemy submarines. The TASS is stored on winches onboard surface vessels and submarines (see Fig. 10.40), from where it is deployed during the towing process.

Passive sonar has several advantages. First of all, it is silent. Its stealthiness is very useful. If the radiated noise level from a target is high enough, and all motorized objects make some noise, the target may in principle be detected. The detection will of course depend on the ambient noise level in the area and on the detection technology being used. On a submarine the bow-mounted passive sonar array detects in directions of about 270 degrees centered on the submarine's alignment, the hull-mounted flank array sonar detects in a direction of about 160 degrees on each side of the submarine, while the TASS will detect in a full 360 degrees. The flank array produced by Thales and based on polyvinylidene fluoride has a length of more than 14 m, a bandwidth of more than 3200 Hz, and a bearing accuracy better than 1 degrees. The Target Motion Analysis (TMA) performed by use of a passive sonar will provide the target's range, course, and speed by determination of the direction from which the signal arrives at different times. The resolution in the direction determination is improved by use of cross-correlation of the signals received on different array hydrophones.

As the passive sonar system shall listen to signals from all directions at all times, it requires a wide beamwidth. But at the same time, a narrow beamwidth is required for source location and for ambient noise rejection. These two requirements are met simultaneously by use of a passive beamformer processor. The output of the beamformer processor is frequently displayed as a bearing-time course.

The submarine hunt performed by a surface vessel can be improved by information provided by specialized aircrafts or helicopters operating in the same area as the surface vessel. The passive dipping sonar from a helicopter or buoys dropped from an aircraft, as, for instance, the passive acoustic Directional Frequency Analysis and Recording (DIFAR) sonobuoys, are used by the navies to detect submarines.

Spectral analyses performed by surface ships, submarines, or sonobuoys can give important information on the underwater environment. Listening to underwater sound after its preprocessing through filtration and demodulation can give a trained operator desired information on sources of underwater sound. Low-Frequency Analysis and Ranging (LOFAR) sonobuoys, first used by the US Navy in the 1950s, is a passive acoustic sonobuoy that can detect acoustic environmental signals in the range 5 Hz−40 kHz. For instance, the AN/SSQ-57 LOFAR sonobuoy can work at depths up to 120 m for up to 8 h, and it is an expendable, omnidirectional passive sonar unit. After reception of acoustic signals, the subsurface unit converts the acoustic waves into amplified electronic signals. These signals are sent to a surface unit where they are applied to a FM carrier for very high frequency transmission. This transmission is received by the monitoring vessel, where the signal is recorded, processed, and analyzed. LOFAR systems compare the average background noise to the sound it is receiving, in particular aiming at the low frequencies, where submarine noise differs most from the background noise. This process is different from detecting submarines by listening to the *SNR*.

Measurements of bearing to the place where an underwater sound is originated can be done by use of a DIFAR sonobuoy. These sonobuoys are passively listening and measure the bearing as the angle, with respect to magnetic North, to where a potential target is located, and can be used for search, detection, and classification. The DIFAR sonobuoys detect acoustic energy in the frequency range from 5 Hz to 2.4 kHz, and they can operate at depths up to 300 m for up to 8 h.

Sonar interceptors are passive systems designed to detect and locate the transmission from a hostile active sonar before this sonar can start its detection process. The advantage of the sonar interceptor is that the signal received has only traveled one way from the active sonar to the interceptor, while the active sonar is receiving the weaker two-way signal. Sonar interceptors are particularly useful for antitorpedo defense for submarines. Submarines can use decoys and jammers, but still no "hart kill" submarine launched antitorpedo system is available. The mobile submarine simulator (MOSS) produced by the United States is a heavy-weight, submarine launched decoy, which has the full size as a torpedo and is able to generate strong underwater sounds, very similar to sounds produced by a submarine. Even if it was first brought into service in 1976, it has over the past been progressively improved. The MOSS can be launched either through a standard torpedo tube or through a single dedicated tube.

Modern torpedoes are generally fitted with a passive/active sonar. This sonar can be used to home directly on a target, or it can follow the wake of the target. As torpedoes constitute the most preferred lethal underwater weapons used by naval platforms as surface vessels, submarines, aircrafts, and helicopters, defense against torpedoes has always had a high priority. To meet the torpedo threat it is necessary to achieve early torpedo detection by using the correct noise detection facilities. Also a fast and qualified threat evaluation, where possible ship maneuvers and other options are considered, can be decisive. One option is to use a small noise making towed body acting as a torpedo decoy. However, this only works against passive homing torpedoes, where the decoy acts as a more attractive, but false target. This soft-kill option has been the most widespread in use by the major navies. The hard-kill option, where the antitorpedo system tends to divert and destroy the incoming torpedo, for instance, by explosion of depth charge in the path of the torpedo, has only been used by the Russian Navy. Also the launch of acoustic decoys, which may seduce the incoming torpedo away from the target and break its homing lock, has been used by the Russian Navy. The decoys then draw the torpedo toward a drifting barrage of acoustically fused and programmable depth charges, which will detonate in an attempt to destroy the torpedo.

Mines may be equipped with a sonar system to detect, localize, and recognize a required target. After processing the noise signature of a potential target, the mine may trigger the ignition. The mine may also be equipped with other detectors giving information about the target, as, for instance, magnetic sensors and pressure sensors for detecting the variation in hydrostatic pressure caused by the target. The acoustic mine will, depending on its design, passively listen to activities in its environment or actively send out acoustic signals and listen to and react on the return signal. The acoustic mine may be in free drift, be moored at a certain location and depth, or lie on the seafloor. A deep-water, antisubmarine mine is the United States produced Mark 60 CAPTOR, which can be laid by surface vessels, by submarines, or by aircrafts. When the sonar by this mine detects the sound of a hostile submarine, it launches a Mark 46 torpedo, which tracks the sound until it hits the submarine hull and explodes.

For many years, in particular during the Cold War, the United States operated large sets of passive sonar arrays at various positions in the World's oceans. These arrays called sound surveillance systems (SOSUS) consisted of permanently mounted arrays of hydrophones in the deep oceans, where the very quiet conditions were used for detection of sound propagating over long ranges. The signal processing took place by use of powerful computers ashore. Some SOSUS arrays, after the termination of the Cold War, have been turned over to scientific use.

14.12.2 ACTIVE SONAR

The active sonar of surface vessels is normally used for detection, surveillance, tracking, and identification of submarines. The frequencies used by active sonar range from a few kilohertz and down to and below 1 kHz, and the active sonar can be hull-mounted in a dome below the hull of a surface vessel, be hull-mounted in a bow bulb on a surface vessel or a submarine, or be in a "fish" towed at variable depths behind a surface vessel. On submarines the active sonar is frequently considered to be the secondary sonar as the use of the active sonar will inform the submarine environment about the presence and position of the submarine.

The performance of a hull-mounted sonar on a surface vessel is strongly dependent on the propagation conditions for its acoustic signals near the sea surface. In deeper waters the negative sound velocity gradient causes downward refraction of the signals and the formation of shadow zones, where submarines may hide. As the extent of the shadow zones decreases when the operational depth of the sonar increases, a variable depth sonar (VDS) has turned out to be very useful. The VDS uses large transducers that are towed from a surface vessel on a cable with an adjustable scope. During the tow of a VDS its depth will be determined by its buoyancy, the tow speed, and the cable scope. Two main advantages characterize the VDS. With increasing depth of operation the source level can be increased without cavitation inception, and the VDS can be operated below layers causing limited propagation of sound signals and the surface vessel can take advantage of the deep sound channel while being in the shadow zone of the submarine's sonar.

The low-frequency active sonar increases the detection range and reduces the advantages of stealth technology by submarines. Amplitude levels of up to 250 dB re 1 µPa at 1 m are used in active sonars. High acoustic power levels transmitted at low frequencies demand the use of huge sonar systems, most frequently mounted onboard specially equipped ASW vessels. The reception of the return signal from an underwater object can be received by the transmitter itself, by a towed array or by one or more receivers at greater distance from the transmitter. A bistatic sonar system is a combination of an active system for transmission at one location and a passive system at another location for reception of the return signal from the target. This combination limits the reverberation influence frequently found by a monostatic active sonar system. The transmitting source may also be well outside the weapons range of the target.

Instead of a single receiver in a bistatic sonar system, a multistatic sonar system will consist of a net of receivers distributed over an area of interest, while the whole area is insonified with a powerful transmitter. The receivers may be sonobuoys with radio communication links to processing facilities onboard the surface vessel carrying the transmitter. While the multistatic system is more complex and harder to deploy than a monostatic system (source and receiver colocated), the covert operation of the receivers constitute a considerable advantage. Moreover, multistatic sonar systems provide effective Doppler processing, high precision in target position determination due to triangulation, fewer false alarms, and the antistealth effects by submarine hulls lined with anechoic coatings, is less efficient at low frequencies. The receivers may also demodulate the return signal from the target if frequency modulation is used on transmission. Also pulse compression techniques are most frequently used in the detection process to improve the range resolution by increasing the useful acoustic power. Doppler shifts representing the target speed is measured by use of narrow band transmitted signals.

The dipping sonar is the primary undersea warfare sensor of helicopters. It enables the helicopters to accomplish missions of submarine detection, tracking, localization, and classification by simply "dipping" a sonar into the sea. It also performs operations related to acoustic intercept, environmental data acquisition, and underwater communication. Its advantages are rapid deployment/retrieval and rapid transition between search areas. And advantages over hull-mounted sonars include absence of flow and engine noise and elimination of Doppler shifts due to a moving signal source. Activated sonobuoy systems like the Command Activated Sonobuoy System (CASS), which remains passive until commanded to ping, and the Directional Command Activated Sonobuoy System (DICASS), which as active sonar provides range, bearing, and Doppler information on a submerged contact, is designed to develop and maintain attack criteria. This system can also be command activated to change depth, if desired, and even to scuttle the sonobuoy.

Mines can be laid by purpose-built minelayers, refitted ships, submarines, or aircrafts, and even dropped into a harbor by hand. Simple mines can be very inexpensive, but more sophisticated mines can be very expensive when equipped with several kinds of advanced sensors, and when they are able to deliver a warhead by use of a rocket or a torpedo. Mine Counter Measures (MCM) include the detection, classification, identification, localization, clearance of, and protection against mines. To avoid putting people and valuable equipment in harm's way, the development of AUVs has been promoted with technologies for navigation, communication, advanced sensor systems, and autonomy. Mine hunting is carried out by use of sonar, general hull-mounted sonar, or, for instance, Side-Scan Sonar or SAS. The frequencies used for MCM or mine hunting are high, normally between 100 and 500 kHz, and the sonar provides acoustic images of the seafloor, of mines proud on the seafloor, and mines floating in the water column. When a mine is detected it is inspected and destroyed either by divers or by a, ROV, a remotely controlled unmanned mini submarine. It is a slow, but reliable way to remove mines. Also sea mammals like the Bottlenose Dolphin have been trained to hunt and mark mines.

An updated form for mine sweeping is to use small ROVs that simulate the acoustic and magnetic signatures of larger ships and that are built to survive a mine explosion. Repeated sweeps are frequently required if the mine has its "ship counter" facility enabled.

REFERENCES

[1] Southall, B.L., Bowles, A.E., Ellison, WT., Finneran, J.J., Gentry, R.L., Greene, C.R., Kastak, D., *Marine mammal noise exposure criteria: initial scientific recommendations*. Aquat Mamm., **33**, pp. 411–521, 2007.

[2] Webb, J.F., Fay, R.R and Popper, A.N., *Fish Bioacoustic*. Springer, 2007.

[3] Popper, A.N. and Hastings, M.C., *The effects on fish of human-generated (anthropogenic) sound*. Integrative Zool., **4**, pp. 43–52, 2009.

[4] Ainslie, M.A., de Jong, C.A.F., Robinson, S.P., and Lepper, P.A., *What is the Source Level of Pile Driving Noise in Water?,* In: The Effects of Noise on Aquatic Life, Advances in Experimental Medicine and Biology, **730**, Springer, New York, pp. 445–448, 2012.

[5] Robinson, S.P., Theobald, P., and Lepper, P.A., *Underwater noise generated from marine piling*. Proc. Meet. Acoust., **17**, 070080, pp. 1–10, http://dx.doi.org/10.1121/1.4790330, 2013.

[6] Robinson, S.P., Lepper, P.A, and Ablitt, J., The measurement of the underwater radiated noise from marine piling including characterization of a "soft start" period, Proceedings of IEEE Oceans 2007, IEEE cat. 07EX1527C, ISBN:1-4244-0635-8, 061215-074, 2007.

[7] Reinhall, P.G. and Dahl, P.H., *Underwater Mach wave radiation from impact pile driving: Theory and observation*. J. Acoust. Soc. Am., **130**, (3), pp. 1209–1216, 2011.

[8] Zampolli, M., Nijhof, M.J. J., de Jong, C.A.F., Ainslie, M.A., Jansen, E.H.W., and Quesson, B.A.J., Validation of finite element computations for the quantitative prediction of underwater noise from impact pile driving. J. Acoust. Soc. Am., **133**, (1), pp. 72–81, 2013.

[9] Ward, P.D. and Needham, K., *Modelling the vertical directivity of noise from underwater drilling*, Proceedings of the 11th European Conference on Underwater Acoustics. pp. 1241–1247, 2012.

[10] Robinson, S.P and Lepper, P.A., Scoping study: Review of current knowledge of underwater noise emissions from wave and tidal stream energy devices, The Crown Estate, 2013. Available at: http://www.thecrownestate.co.uk/media/151996/pfow-review-current-knowledge-underwater-noise-emissions-wave-and-tidal-stream-energy-devices.pdf.

[11] Robinson, S.P. and Lepper, P.A., *Review of underwater noise emitted by wave and tidal stream energy devices*. Proceedings of 1st International Conference and Exhibition on Underwater Acoustics, Greece, ISBN:978-618-80725-0-3, pp. 191–198, 2013.

[12] Wilson, B., Lepper, P.A., Carter, C., Robinson, S.P., *Rethinking underwater sound recording methods to work in tidal-stream and wave energy sites*. In: Marine Renewable Energy and Environmental Interactions, Humanity and the Sea, M.A. Shields, A.I.L. Payne (Eds.), http://dx.doi.org/10.1007/978-94-017-8002-5_9, pp. 111–126, Springer Science + Business Media, 2014.

[13] Tougaard, J, Henriksen, O.D., Miller, L.A., Underwater noise from three types of offshore wind turbines: estimation of impact zones for harbor porpoises and harbor seals. J. Acoust. Soc. Am., **125**, pp. 3766–3773, 2009.

[14] Sigray, P. and Andersson, M.H., *Particle motion measured at an operational wind turbine in relation to hearing sensitivity in fish*, J. Acoust. Soc. Am., **130**, (1), pp. 200–207, 2011.

[15] Prior, M.K., Meless, O., Bittner, and P., Sugioka, H., *Long-range detection and location of shallow underwater explosions using deep-sound-channel hydrophones*. IEEE J. Oceanic. Eng., **36**, pp. 703–715, 2011.

[16] Haralabus, G., Pautet, L., Stanley, J.P., and Zampolli, M., *Welcome back HA03*. CTBTO Spectrum, **22**, pp. 18–22, 2014. Also Available at: http://www.ctbto.org/publications/spectrum-publication/.

[17] Miksis-Olds, J.L., Vernon, J.A., and Heaney, K.D., *Applying the dynamic soundscape to estimates of signal detection*, In: Proc. 2nd International Conference and Exhibition on Underwater Acoustics, pp. 863–870, Rhodes, Greece, 2014.

[18] Prior, M.K., Brown, D., and Haralabus, G., *Data features from long-term monitoring of ocean noise*, In: Proc. Int. Conf. Underwater Acoustics and Measurements, pp. 1343–1350, Kos, Greece, 2011.

[19] Heaney, K.D., Campbell, R.L., and Snellen, M., *Long range acoustic measurements of an undersea volcano*, J. Acoust. Soc. Am., **134**, pp. 3299–3306, 2013.

[20] Miksis-Olds, J.L., Bradley, D.L., and Niu, X.M., *Decadal trends in Indian Ocean ambient sound*, J. Acoust. Soc. Am., **134**, pp. 3464–3475, 2013.

[21] Hawkins, R.S., Miksis-Olds, J.L., and Smith, C.M., *Variation in low-frequency estimates of sound levels based on different units of analysis*, J. Acoust. Soc. Am., **135**, pp. 705–711, 2014.

[22] Sabra, K.G., Fried, S., Kuperman, W.A., and Prior, M.K., *On the coherent components of low-frequency ambient noise, in the Indian Ocean*, J. Acoust. Soc. Am. Express Lett., **133**, pp. EL20–EL25, 2013.

[23] deGroot-Hedlin, C., Blackman, D.K., and Jenkins, C.S., *Effects of variability associated with the Antarctic circumpolar current on sound propagation in the ocean*. Geophys. J. Int., **176**, pp. 478–490, 2009.

[24] Evers, L.G., Brown, D., Heaney, K.D., Assink, J.D., Smets, P.S.M., and Snellen, M., *Evanescent wave coupling in a geophysical system: Airborne acoustic signals from the Mw 8.1 Macquarie Ridge earthquake*. Geophys. Res. Lett., **41**, pp. 1644–1650, 2014.

[25] Yildiz, S., Sabra, K., Dorman, L.M., and Kuperman, W.A., *Using hydroacoustic stations as water column seismometers*. Geophys. Res. Lett., **40**, pp. 2573–2578, 2013.

[26] Arveson, P.T. and Vendittis, D.T., *Radiated noise characteristics of modern cargo ship*. J. Acoust. Soc. Am., **107**, (1), pp. 118–129, 2000.

[27] Grelowska, G., Kozaczka, E., Kozaczka, S., and Szymczak, W., *Underwater noise generated by a small ship in the shallow sea*. Arch. Acoust., **38**, (3), 2013.

[28] Hildebrand, J.A., *Anthropogenic and natural sources of ambient noise in the ocean*. Mar. Ecol. Prog. Ser., **395**, pp. 5–20, 2009.

[29] Ross, D., *Ship sources of ambient noise*. IEEE J. Oceanic. Eng., **30**, pp. 257–261, 2005.

[30] Audoly, C., Rousset, C., and Leissing, T., AQUO Project —Modelling of ships as noise source for use in an underwater noise footprint assessment tool. Inter-Noise 2014, Melbourne, Australia, November 16—19, 2014. CD.

[31] Kozaczka, E. and Grelowska, G., Shipping Noise. Arch. Acoust., 29, (2), pp. 169—176, 2004.

[32] Urick, R.J., *Principles of Underwater Sound*. Chapter 10, McGraw-Hill, Inc., New York, 1975.

[33] Kozaczka, E., Domagalski, J., and Gloza, I., *Investigation of the underwater noise produced by ships by means of intensity method*. Pol. Marit. Res., 17, (3), p. 2636, 2010.

[34] Ross, D., *Mechanics of Underwater Noise*. Pergamon Press, New York, 1976.

[35] Kozaczka, E., *Investigations of underwater disturbances generated by the ship propeller*. Arch. Acoust., 13, (2), pp. 133—152, 1978.

[36] Southworth, M., *The sonic environment of cities*. Environ. Behav., 1, pp. 49, 70, 1969.

[37] Schafer, R.M. *Tuning of the World*. Alfred Knopf, New York, 1977.

[38] Pijanowski, B.C., Villanueva-Rivera, L.J., Dumyahn, S.L., Farina, A., Krause, B.L., Napoletano, B.M., Gage, S.H., and Pieretti, N., *Soundscape ecology the science of sound in the landscape*. BioScience, 61, pp. 203—216, 2011.

[39] Miksis-Olds, J.L., Stabeno, P.J., Napp, J.M., Pinchuk, A.I., Nystuen, J.A., Warren, J.D., and Denes, S.L., *Ecosystem response to a temporary sea ice retreat in the Bering Sea*. Prog. Oceanic., 111, pp. 38—51.

[40] McWilliam, J.N. and Hawkins, A.D., *A comparison of inshore marine soundscapes*. J. Exp. Mar. Bio. Ecol., 446, pp. 166—176, 2013.

[41] Staaterman, E., Paris, C.B., DeFerrari, H.A., Mann, D.A., Rice, A.N., and D'Alessandro, E.K., *Celestial patterns in marine soundscapes*. Mar. Ecol. Prog. Ser., 508, pp. 17—32, 2014.

[42] Parks, S.E., Miksis-Olds, J.L., and Denes, S.L., *Assessing marine ecosystem acoustic diversity across ocean basins*. Ecol. Inf., 21, pp. 81—88, 2014.

[43] van Opzeeland, I.C. and Miksis-Olds, J.L., *Acoustic ecology of pinnipeds in polar habitats*. In: Aquatic Animals: Biology, Habitats, and Threats., D.L. Eder (Ed.), Nova Science Publishers, Inc., New York, pp. 1—52, 2012.

[44] Ross, D., *Mechanics of Underwater Noise*. Pergamon Press, New York, pp. 375, 1976.

[45] Andrew, R.K., Howe, B.M., Mercer, J.A., and Dzieciuch, M.A., *Ocean ambient sounds: Comparing the 1960's with the 1990's for a receiver off the California coast*. ARLO, 3, pp. 65—70, 2002.

[46] McDonald, M.A., Hildebrand, J.A., and Wiggins, S.M., *Increases in deep ocean ambient noise in the Northwest Pacific west of San Nicolas Island, California*. J. Acoust. Soc. Am., 120, pp. 711—717, 2006.

[47] Chapman, N.R. and Price, A., *Low frequency deep ocean ambient noise trend in the Northeast Pacific Ocean*. J. Acoust. Soc. Am., 129, pp. EL161-EL165, 2011.

[48] Boyd, I.L., Frisk, G., Urban, E., Tyack, P., Ausubel, J., Seeyave, S., Cato, D., Southall, B., Weise, M., Andrew, R., Akamatsu, T., Dekeling, R., Erbe, C., Farmer, D., Gentry, R., Gross, T., Hawkins, A., Li, F., Metcalf, K., Miller, J.H., Moretti, D., Rodrigo, C., and Shinke, T., *An International Quiet Ocean Experiment*. Oceanography, 24, pp. 174—181, 2011.

[49] Tyack, P.L., *Implications for marine mammals of large-scale changes in the marine acoustic environment*. J. Mammalogy, 89, pp. 549—558, 2008.

[50] Dziak, R., *Fowler, M.J., Matsumoto, H., Bohnenstiehl, D.R., Park, M., Warren, K., and W.S. Lee.*, Life and death sounds of iceberg A53a. *Oceanography,* **26**, pp. 10–12, 2013.

[51] Miksis-Olds, J.L., *Global trends in ocean noise.* In: Effects of Noise on Aquatic Life II, A.N. Popper and A. Hawkins (Eds.), Springer Science + Business Media, LLC, pp. 713–718, 2016.

[52] Simpson, S.D., Meekan, M., Montgomery, J., McCauley, R., and Jeffs, A., *Homeward sound.* Science, **308**, pp. 221, 2005.

[53] Slabbekoorn, H. and Bouton, N., *Soundscape orientation: a new field in need of sound investigation.* Anim. Behav., **76**, pp. e5–e8, 2008.

[54] Stanley, J.A., Radford, C.A., and Jeffs, A.G., *Location, location, location: finding a suitable home among the noise.* Proc. R Soc. B, **270**, pp. 3622–3631, 2012.

[55] Able, K.P., *Mechanisms of orientation, navigation, and homing.* In: Animal Migration, Orientation, and Navigation, S.A. Gauthreaux Jr (Ed.), Academic Press, pp. 283–373, 1980.

[56] Kenney, R.D., Mayo, C.A., and Winn, H.E., Migration and foraging strategies at varying spatial scales in western North Atlantic right whales: a review of hypothesis. J. Cet. Res. Manag. (Special Issue), **2**, pp. 251–260, 2001.

[57] Miksis-Olds, J.L., and Madden, L.E., *Environmental predictors of ice seal presence in the Bering Sea.* PLoS One, **9**, e106998, 2014.

[58] Sueur, J., Pavoine, S., Hamerlynck, O., and Duvail, S., *Rapid acoustic survey for biodiversity appraisal.* PLoS One, **3**, e4065, 2008.

[59] Radford, C.A., Stanley, J.A., Tindle, C.T., Montgomery, J.C., and Jeffs, A.G., *Localised coastal habitats have distinct underwater sound signatures.* Mar. Ecol. Prog. Ser., 401, pp. 21–29, 2010.

[60] Miksis-Olds, J.L. and Parks, S.E., *Seasonal trends in acoustic detection of ribbon seals (Histriophoca fasciata) in the Bering Sea.* Aquat. Mamm., **37**, pp. 464–471, 2011.

[61] Stanley, J.A., Radford, C.A., and Jeffs, A.G., *Behavioural response thresholds in New Zealand crab megalopae to ambient underwater sound.* PLoS One, **6**, e28572, 2011.

[62] Simpson, S.D., Meekan, M.G., Jeffs, A., Montgomery, J.C., and McCauley, R.D., *Settlement-stage coral reef fish prefer the higher-frequency invertebrate-generated audible component of reef noise.* Anim. Behav., **75**, pp. 1861–1868, 2008.

[63] Denes, S.L., Miksis-Olds, J.L., Mellinger, D.K., and Nystuen, J.A., *Assessing the cross platform performance of marine mammal indicators between two collocated acoustic recorders.* Ecol. Inf., **21**, pp. 74–80, 2014.

[64] Depraetere, M., Pavoine, S., Jiguet, F., Gasc, A., Duvail, S., and Sueur, J., *Monitoring animal diversity using acoustic indices: implementation in a temperate woodland.* Ecol. Indic., **13**, pp. 46–54, 2012.

[65] Chitre, M., Shahabodeen S., and Stojanovic, M., *Underwater acoustic communications and networking.* Marine Technol. Soc. J., **42**, (1), pp. 103–115, 2008.

[66] Van-Walree, P., *Propagation and scattering effects in underwater acoustic communication channels.* IEEE J. Oceanic. Eng., **38**, (4), pp. 614–631, September, 2013.

[67] Proakis, J. and Salehi, M. *Digital Communications.* 5th Ed., McGraw-Hill, 2008.

[68] Stojanovic, M., Catipovic, J., and Proakis, J., *Phase-coherent digital communication for underwater acoustic channels.* IEEE J. Oceanic. Eng., **19**, (1), pp. 100–111, January, 1994.

[69] Tsimenidis, C., Hinton, O., Adams, A., and Sharif, B., *Underwater acoustic receiver employing direct-sequence spread spectrum and spatial diversity combining for shallow-water multiaccess networking.* IEEE J. Oceanic. Eng., **26**, (4), pp. 594–603, 2001.

[70] Sozer, E., Proakis, J., and Blackmon, F., *Iterative equalization and decoding techniques for shallow water acoustic channels*. In: Proc. IEEE OCEANS Conf., Honolulu, HI, 2001.

[71] Zvonar, Z., Brady, D., and Catipovic, J., Adaptive equalization techniques for interference suppression in shallow water acoustic telemetry channels. In: The Twenty-Seventh Asilomar Conference on Signals, Systems and Computers, Pacific Grove, CA, 1993.

[72] Blair, B. and Preisig, J., *Multi-channel DFE equalization with waveguide constraints for underwater acoustic communication*. In: 48th Annual Allerton Conference on Communication, Control, and Computing, Allerton, IL, 2010.

[73] Zhou, S. and Wang, Z., *OFDM for Underwater Acoustic Communications*, John Wiley & Sons Ltd., 2014.

[74] Kang, T. and Iltis R., *Iterative carrier frequency offset and channel estimation for underwater acoustic OFDM systems*. IEEE J. Selec. Areas Commun., **26**, (9), pp. 1650–1661, December 2008.

[75] Abdelkareem, A., Sharif, B., Tsimenidis, C., and Neasham, J., *Low-complexity Doppler compensation for OFDM-based underwater acoustic communication systems*. In: IEEE OCEANS Conf., Santander, Spain, 2011.

[76] Tu, K., Fertonani, D., Duman, T.M., Stojanovic, M., Proakis, J.G., and Hursky, P. *Mitigation of intercarrier interference for OFDM over time-varying underwater acoustic channels*. IEEE J. Oceanic. Eng., **36**, (2), pp. 156–171, April, 2011.

[77] Ceballos, P. and Stojanovic, M., *Adaptive channel estimation and data detection for underwater acoustic MIMO OFDM systems*. IEEE J. Oceanic. Eng., **35**, (3), pp. 635–646, July, 2010.

[78] Plets, R., *Underwater Survey and Acoustic Detection and Characterization of Archaeological Materials*. In: The Oxford Handbook of Wetland Archaeology, F. Menotti and A. O'Sullivan (Eds.), pp. 433–449, Oxford University Press, Oxford, 2012.

[79] Blondel, Ph., *The Handbook of Side Scan Sonar*. Springer Praxis, Berlin, 2009.

[80] Drap, P., Seinturier, J., Conte, G., Caiti, A., Scaradozzi, D., Zanoli, S.M., and Gambogi, P., *Underwater cartography for archaeology in the VENUS project*. Geomatica, **62**, (4), pp. 419–427, 2008.

[81] Blondel, Ph. and Pouliquen, E., *Acoustic Texture and Detection of Shipwreck Cargo: Example of a Roman Ship near Elba, Italy*. In: The Application of Recent Advances in Underwater Detection and Survey Techniques to Underwater Archaeology, T. Akal, R.D. Ballard, G.F. Bass (Eds.), The Institute of Nautical Archaeology, Bodrum, 2004.

[82] Bellettini, A., and Pinto, M., *Design and Experimental Results of a 300 kHz Synthetic Aperture Sonar Optimized for Shallow Water Operations*. IEEE J. Oceanic. Eng., **34**, (3), pp. 285–293, 2009.

[83] Caiti, A., Casalino, G., Conte, G., and Zanoli, S.M., *Innovative Technologies in Underwater Archaeology: Field Experience, Open Problems and Research Lines*. Chem. Ecol., **22**, pp. S383–S396, 2006.

[84] Lafferty B., Quinn, R., and Breen, C., *A side-scan sonar and high-resolution Chirp sub-bottom profile study of the natural and anthropogenic sedimentary record of Lower Lough Erne, northwestern Ireland*. J. Archaeological Sci., **33**, pp. 756–766, 2006.

[85] Mindell, D.A. and Bingham, B., *A high frequency, narrow beam, sub bottom profiler for Archaeological Applications*. Proc. IEEE Oceans, 2001, **4**, pp. 2115–2123, Honolulu, 2001.

[86] Caiti, A., Bergem, O., and Dybedal, J., *Parametric sonars for seafloor characterization*. Meas. Sci. Technol., **10**, pp. 1105−1115, 1999.

[87] Wunderlich, J., Wendt, G., and Müller, S., *High-resolution Echo-sounding and Detection of Embedded Archaeological Objects with Nonlinear Sub-bottom Profilers*. Mar. Geophys. Res., **26**, pp. 123−133, 2006.

[88] Scaradozzi, D., Sorbi, L., Zoppini, F., and Gambogi, P., *Tools and techniques for underwater archaeological sites documentation*. Proc. IEEE/MTS Oceans'13, San Diego, 2013.

[89] Foley, B., *Deep Water Archaeology*. In: Encyclopedia of Global Archaeology, C. Smith (Ed.), Springer, pp. 2079−2081, 2014.

[90] Caiti, A., Calabro, V., Di Corato, F., Fabbri, T., Fenucci, D., Munafo, A., Allotta, B., Bartolini, F., Costanzi, R., Gelli, J., Monni, N., Natalini, M., Pugi, L., and Ridolfi, A., Thesaurus: AUV teams for archaeological search. Field results on acoustic communication and localization with the Typhoon. In: Proc. IEEE Mediterranean Control Conf., Palermo, 2014.

[91] Mikhalevsky, P.N., *Arctic Acoustics*. In: Encyclopedia of Ocean Sciences, John H. Steele, Karl K. Turekian, Steve A. Thorpe (Eds.), Academic Press, **1**, pp. 53−61, 2001.

[92] Mikhalevsky, P.N., Sagen, H., A., Worcester, P.F., Baggeroer, A.B., ORcutt, J., Moorre, S.E., Lee, C.M., Vigness-Raposa, K.J., Freitag, L., Arrott, M., Atakan, K., Besczyanska-Möller, A., Duda, T.F., Dushaw, B.D., Gascard, J.C., Gavrilov, A.N., Kerrs, H., Morozov, A.K., Munk, W.H., Rixen, M., Sandven, S., Skarsoulis, E., Stafford, K.M., Vernon, F., and Yuen, M.Y., *Multipurpose acoustic networks in the integrated Arctic Ocean observing system*. Arctic, **68** (50), Supplement 1, pp. 11−27, 2015.

[93] Gavrilov, A.N. and Mikhalevsky, P.N., *Low frequency acoustic propagation loss in the Arctic Ocean: results of the Arctic Climate Observations using Underwater Sound experiment*. J. Acoust. Soc. Am., **119**, pp. 3694−3706, 2006.

[94] Mikhalevsky, P.N., A.N. Gavrilov, M.S. Moustafa and B. Sperry, *Arctic Ocean warning: submarine and acoustic measurements*. Proc. MTS/IEEE Oceans, 2001, **3**, pp. 1523−1528, 2001.

[95] Sandven, S., Sagen, H., Bertino, L., Beszczynska-Möller, A., Fahrbach, E., Worcester, P.F., Dzieciuch, M.A., et al., The Fram Strait integrated ocean observing and modelling system. In: Sustainable Operational Oceanography, *Proc. Sixth Int. Conf. on Euro-GOOS*. 4−6, Sopot, Poland, pp. 50−58, October, 2011. http://eurogoos.eu/documents/conference-proceedings/.

[96] Maslanik, J., Stroeve, J., Fowler, C., and Emery, W., *Distribution and trends in Arctic sea ice age through spring 2011*. Geophys. Res. Lett., **38** (13), L13502, 2011.

[97] Overland, J.E., and Wang, M., *When will the summer Arctic be nearly ice free?* Geophys. Res. Lett., **40**, http://dx.doi.org/10.1002/grl.50316, 2013.

[98] Gavrilov, A. and Li, B., *Antarctica as one of the major sources of noise in the ocean*. Underwater Acoustic Measurements: Technologies & Results, 2nd International Conference and Exhibition, Heraklion, Crete, pp. 1179−1184, June, 2007.

[99] Macayel, D.R., Okal, E.A., Aster, R.C., and Bassis, J.N., *Seismic and intra-annual decrease in the vocalization frequency of Antarctic blue whales*. J. Acoust. Soc, Am., **131**, pp. 4476−4480, 2012.

[100] Gavrilov, A., McCauley, R.D., Gedamke, J., *Steadyinter and intra-annual decrease in the vocalization frequency of Antarctic blue whales*. J. Acoust. Soc. Am., **131**, pp. 4476−4480, 2012.

[101] Širović, A., Hildebrand, J.A., and Wiggins, S.M., *Blue and fin whale call source levels and propagation range in the Southern Ocean.* J. Acoust. Soc. Am., **122**, pp. 1208−1215, 2007.

[102] Gavrilov, A. and Li, B., *Correlation between ocean noise and changes in the environmental conditions in Antarctica.* Underwater Acoustic Measurements: Technologies & Results, 3rd International Conference and Exhibition, Nafplion, Greece, pp. 1199−1200, June, 2009.

[103] Branch, T.A., *Abundance of Antarctic blue whales south of 60°S from three compJlete circumpolar sets of surveys.* J. Cetacean Res. Manage, **9**, pp. 253−62, 2007.

[104] Gedamke, J. and Robinson, S.M., *Acoustic survey for marine mammal occurrence and distribution off East Antarctica (30−80°E) in January−February, 2006.* Deep See Res. Part II: Trop. Stud. Oceanogr., **57**, (9−10), pp. 968−981, 2013.

[105] Marandet, C., Roux, P., Nicolas, B., and Mars, J., *Target detection and localization in shallow water: an experimental demonstration of the acoustic barrier problem at the laboratory scale.* J. Acoust. Soc. Am., **129**, (1), pp. 85−97, 2011.

[106] Iturbe, I., Roux, P., Nicolas, B., Virieux, J., and Mars, J., *Shallow water acoustic tomography performed from a double beamforming algorithm.* IEEE J. Oceanic. Eng., **34**, (2), pp. 140−149, 2009.

[107] Walker, S.C., Roux, P., and Kuperman, W.A., *Synchronized time-reversal focusing with application to remote imaging from a distant virtual source array.* J. Acoust. Soc. Am., **125**, (6), pp. 3828−3834, 2009.

[108] Prada, C. et al., *Experimental detection and focusing in shallow water by decomposition of time reversal operator.* J. Acoust. Soc. Am, **122**, (2), 2007.

[109] Papadakis P., Piperakis, G., Kouzoupis, S., *Calibration of ultrasound transducer heads using short preprocessed ultrasonic pulses.* 2nd International Conference and Exhibition on Underwater Acoustics, Rhodes, pp. 939−946, 2014.

[110] Haumesser, L., Décultot, D., Léon, F., Maze, G., *Experimental identification of finite cylindrical shell vibration modes.* J. Acoust. Soc. Am., **111**, (5), Pt 1, pp. 2034−2039, 2002.

[111] Sessarego, J.-P., Ivakin, A.N., and Ferrand, D., *Frequency Dependence of Phase Speed, Group Speed and Attenuation in Well-Sorted Water-Saturated Sand: Laboratory Experiments.* IEEE J. Oceanic. Eng., **33**, (4), pp. 359−366, 2008.

[112] Sessarego, J-P., Cristini, P., Grigorieva, N., and Fridman, G., *Acoustic scattering by an elastic spherical shell near the seabed.* J. Comp. Acoust., **20**, (1), 2012.

[113] Real, G., Sessarego, J.-P., Cristol, X., Fattaccioli, D., *De-Coherence Effects in Underwater Acoustics: Scaled Experiments.* 2nd International Conference and Exhibition on Underwater Acoustics, Rhodes, pp. 947−954, 2014.

[114] Korakas, A., Sturm, F., Sessarego, J.-P., *Results of matched-field inversion in a three-dimensional wedge-like environment.* Proceedings of the 10th European Conference on Underwater Acoustics, Istanbul, Turkey, 2010, Vol. 1, pp. 357−363, 2010.

[115] Grigorieva, N.S., Fridman, G.M., *Scattering of sound by an elastic spherical shell immersed in a waveguide with a fluid bottom.* Acoust. Phys., **59**, (4), pp. 373−381, 2013.

[116] Brelet, Y., Houard, A., Carbonnel, J., André, Y.B., Jarnac, A., Mysyrowicz, A., Guillermin, R., Sessarego, J-P., Fattaccioli, D., *Femtosecond laser-induced pulsed ultrasound source in water.* Proc. Conf. Lasers and Electro-Optics Europe and International Quantum Electronics Conference, Munich, Germany, May, 2013.

[117] http://resource.npl.co.uk/acoustics/techguides/soundseawater/. This calculates the sound speed for different temperature, salinity and depth.

[118] Hazelwood, R.A., Kelland, N.C., and Smedley, N.J., *One man's signal is another man's noise*. Hydro International, Reed Business bv, Lemmer, The Netherlands, September, 1998.

[119] *Pieter Schulte. Bigger and better.* Offshore Engineer, Supplement "Dutch Offshore", August, 2014.

[120] Wertz, J.R., *Spacecraft attitude determination and control.* Kluwer Academic Publishers, Dordrecht, The Netherlands, 1991.

[121] Dushaw, B., and many others, *A Global Ocean Acoustic Observing Network.* Proceedings of OceanObs'09: Sustained Ocean Observations and Information for Society (Vol. 2), Venice, Italy, September 21–25, 2009, J., Hall, D.E., Harrison, and D., Stammer (Eds.), ESA Publication WPP-306, 2010.

[122] Dushaw, B.D., Worcester, P.F., and others, *A decade of acoustic thermometry in the North Pacific Ocean.* J. Geophys. Res., **114**, C07021. http://dx.doi.org/10.1029/2008JC005124, 2009.

[123] Dushaw, B.D., *Ocean Acoustic Tomography.* Encyclopedia of Remote Sensing, E.G. Njoku (Ed.), Springer, Springer-Verlag Berlin Heidelberg. http://dx.doi.org/10.1007/Springer Reference_331410, 2014.

[124] Morawitz, W.M.L., Sutton, P.J., Cornuelle, B.D., and Worcester, P.F., *Three-dimensional observations of a deep convective chimney in the Greenland Sea during winter 1988/1989.* J. Phys. Oceanogr., **26**, pp. 2316–2343, 1996.

[125] Park, J.-H., and Kaneko, A., *Assimilation of a coastal acoustic tomography data into a barotropic ocean model.* Geophys. Res. Lett., **27**, pp. 3373–3376, 2000.

[126] Lebedev, K.V., Yaremchuk, M., and others, *Monitoring the Kuroshio Extension through dynamically constrained synthesis of the acoustic tomography, satellite altimeter and in situ data.* J. Oceanogr., **59**, pp. 751–763, 2003.

Index